Advances in Experimental Medicine and Biology

Volume 1388

Advances in Experimental Medicine and Biology provides a platform for scientific contributions in the main disciplines of the biomedicine and the life sciences. This series publishes thematic volumes on contemporary research in the areas of microbiology, immunology, neurosciences, biochemistry, biomedical engineering, genetics, physiology, and cancer research. Covering emerging topics and techniques in basic and clinical science, it brings together clinicians and researchers from various fields.

Advances in Experimental Medicine and Biology has been publishing exceptional works in the field for over 40 years, and is indexed in SCOPUS, Medline (PubMed), EMBASE, BIOSIS, Reaxys, EMBiology, the Chemical Abstracts Service (CAS), and Pathway Studio.

2021 Impact Factor: 3.650 (no longer indexed in SCIE as of 2022)

Leonard Shapiro • Paul M. Rea
Editors

Biomedical Visualisation

Volume 12 – The Importance of Context in Image-Making

Springer

Editors
Leonard Shapiro
Department of Human Biology
University of Cape Town
Cape Town, South Africa

Paul M. Rea
School of Medicine, Dentistry and Nursing
University of Glasgow
Glasgow, UK

ISSN 0065-2598 ISSN 2214-8019 (electronic)
Advances in Experimental Medicine and Biology
ISBN 978-3-031-10888-4 ISBN 978-3-031-10889-1 (eBook)
https://doi.org/10.1007/978-3-031-10889-1

This Springer imprint is published by the registered company Springer Nature Switzerland AG
The registered company address is: Gewerbestrasse 11, 6330 Cham, Switzerland

Preface

While highly beneficial advances, improvements, and innovation in digital visualisation technologies in the biosciences continue incrementally, it (currently) still remains the human mind that needs to be applied towards the interpretation and understanding of the images that are generated. To this end, cognitive ability in 3D spatial awareness is critical. It is often analogue, low-technology, hands-on exercises that afford both students and scientists with improved cognitive awareness and the resultant interpretive skills. The chapters in this book therefore explore both high-technology derived visualisation and low-technology mechanisms for educating anatomy students, including clay modelling. Chapters also outline important educational considerations such as visual affordances as well as the use of appropriate imagery for communication to specific audiences. As we embrace blended learning in anatomy education, a timely chapter prompts us to think further about and contribute to the ongoing discourse around important ethical considerations in the use and sharing of digital images of body donors.

My gratitude to Paul Rea for his trust and generosity in giving me the opportunity to bring together the authors for this volume. In doing so, I was able to extend invitations for chapter contributions to many of my colleagues at a number of Universities in my country, South Africa, and to colleagues abroad. Once all of the chapters had come in, I experienced an excitement and surprise at noticing the unpredicted relationships and threads of content that extended through many of the chapters and created linkages between them.

It is my hope that as the authors discover these threads and links, they explore the possibility of collaboration and research with one another. To the readers, I hope that you will appreciate and enjoy reading the chapters that have found their way into this book and that they will stimulate fresh ideas in you! In this, we have the authors of this book to thank for their dedication in putting their thoughts and research into words so that we can share in them.

My thanks and appreciation to Ina Stoeck and Bibhuti Sharma of Springer Nature for their expertise and support throughout the months of work that went into preparing this book for publication. They made it a pleasure!

Thank you to my parents.

Cape Town, South Africa Leonard Shapiro

Contents

Part I Exciting Data and Representation

1 **A Multimodal Social Semiotics Perspective on Teaching and Learning Using Biomedical Visualisations** 3
Natashia Muna

2 **Reasons for Knocking at an Empty House: Visualisation, Representation and Dissemination of Health-Related Public Engagement Media** 23
Ed Young and Anastasia Koch

3 **The Evolution of Scientific Visualisations: A Case Study Approach to Big Data for Varied Audiences** 51
Andrew J. Lunn, Vivien Shaw, and Isabelle C. Winder

Part II Biomedical Education: Theory and Practice

4 **The Affordances of Visual Modes in Pedagogy on the Physics of Motion in Physiotherapy Education** 87
Sumaya Gabriels, Natashia Muna, and Kate le Roux

Part III Making 3D

5 **Mitochondria to Bitter Melon: Understanding the 3D Ultrastructure of the Cell Via 2D Thin Section Reconstruction and the History of Mitochondrial Visualization** 113
David M. Mazierski

6 **Using Molecular Visualisation Techniques to Explain the Molecular Biology of SARS-CoV-2 Spike Protein Mutations to a General Audience** 129
Sarah Iannucci, William Harvey, Joseph Hughes, David L. Robertson, Edward Hutchinson, and Matthieu Poyade

7 **Student-Perceived Value on the Use of Clay Modelling in Undergraduate Clinical Anatomy** 153
Janine C. Correia, Karin J. Baatjes, and Ilse Meyer

Part IV Ethical Considerations

8 Advances in Digital Technology in Teaching Human Anatomy: Ethical Predicaments 173
Kerri Keet and Beverley Kramer

About the Editors

Leonard Shapiro Leonard is affiliated with the Department of Human Biology, at the University of Cape Town, South Africa. He has a keen interest in Anatomy Education and has developed a number of art-based exercises to address and improve students' three-dimensional (3D) spatial awareness and observation ability. His courses and online workshops are in collaboration with lecturers who are actively engaged in improving education methodology in anatomy. These are offered to medical students and lecturers in South Africa and abroad. Leonard has developed a multi-sensory observation method that crucially employs the sense of touch (haptics) coupled with the simultaneous act of drawing. It is called the Haptico-visual observation and drawing (HVOD) method. In anatomy education, the benefits of observing using the HVOD method include the enhanced observation of the 3D form of anatomical parts, the cognitive memorisation of anatomical parts as a 3D 'mental picture', improved spatial orientation within the volume of anatomical parts, and an ability to draw. Leonard contributes to the anatomy education discourse by presenting at anatomy conferences as well as via publications and articles. Leonard graduated in BSocSc and in BA Fine Art (Hons) from the University of Cape Town.

Paul M. Rea Paul is a Professor of Digital and Anatomical Education at the University of Glasgow. He is a Director of Innovation, Engagement and Enterprise within the School of Medicine, Dentistry and Nursing. He is also a Senate Assessor for Student Conduct, a Council Member on Senate, coordinates the day-to-day running of the Body Donor Programme, and is a Licensed Teacher of Anatomy, licensed by the Scottish Parliament. He is qualified with a medical degree (MBChB), a MSc (by research) in craniofacial anatomy/surgery, a Ph.D. in neuroscience, the Diploma in Forensic Medical Science (DipFMS), and an MEd with Merit (Learning and Teaching in Higher Education). He is a Senior Fellow of the Higher Education Academy, a professional member of the Institute of Medical Illustrators (MIMI), and a registered medical illustrator with the Academy for Healthcare Science.
Paul has published widely and presented at many national and international meetings, including invited talks. He has been the lead Editor for Biomedical Visualisation over 11 published volumes and is the founding editor for this book series. This has resulted in almost 70,000 downloads across these

volumes, with contributions from over 300 different authors, across approximately 100 institutions from 16 countries across the globe. He sits on the Executive Editorial Committee for the *Journal of Visual Communication in Medicine*, is an Associate Editor for the *European Journal of Anatomy*, and has reviewed for 25 different journals/publishers.

He is the Public Engagement and Outreach Lead for anatomy, coordinating collaborative projects with the Glasgow Science Centre, NHS, and the Royal College of Physicians and Surgeons of Glasgow. Paul is also a STEM ambassador and has visited numerous schools to undertake outreach work.

His research involves a long-standing strategic partnership with the School of Simulation and Visualisation, The Glasgow School of Art. This has led to multi-million pound investment in creating world leading 3D digital datasets to be used in undergraduate and postgraduate teaching to enhance learning and assessment. This successful collaboration resulted in the creation of the world's first taught MSc Medical Visualisation and Human Anatomy combining anatomy and digital technologies. The Institute of Medical Illustrators also accredits it. It has created college-wide, industry, multi-institutional, and NHS research-linked projects for students. Paul is the Programme Director for this degree.

List of Contributors

Karin J. Baatjes Dean's Division, Faculty of Medicine and Health Sciences, Stellenbosch University, Parow, South Africa

Janine C. Correia Division of Clinical Anatomy, Faculty of Medicine and Health Sciences, Stellenbosch University, Parow, South Africa

Sumaya Gabriels Department of Health and Rehabilitation Sciences, Faculty of Health Sciences, University of Cape Town, Cape Town, South Africa

William Harvey MRC-University of Glasgow Centre for Virus Research, The University of Glasgow, Glasgow, UK

Joseph Hughes MRC-University of Glasgow Centre for Virus Research, The University of Glasgow, Glasgow, UK

Edward Hutchinson MRC-University of Glasgow Centre for Virus Research, The University of Glasgow, Glasgow, UK

Sarah Iannucci The School of Simulation and Visualisation, Glasgow, UK
The Anatomy Facility, College of Medical, Veterinary and Life Sciences, University of Glasgow, Glasgow, UK

Kerri Keet Division of Clinical Anatomy, Department of Biomedical Sciences, Faculty of Medicine and Health Sciences, Stellenbosch University, Stellenbosch, South Africa

Anastasia Koch SAMRC/NHLS/UCT Molecular Mycobacteriology Research Unit & DST/NRF Centre of Excellence for Biomedical TB Research, Department of Pathology, Faculty of Health Sciences, University of Cape Town, Cape Town, South Africa

Beverley Kramer School of Anatomical Sciences, Faculty of Health Sciences, University of the Witwatersrand, Johannesburg, South Africa

Andrew J. Lunn MZool Graduate from School of Natural Sciences, Bangor University, Bangor, UK

David Mazierski Biomedical Communications, University of Toronto Mississauga, Mississauga, Canada

Ilse Meyer Centre for Health Professions Education (CHPE), Faculty of Medicine and Health Sciences, Stellenbosch University, Parow, South Africa

Natashia Muna Language Development Group, Academic Development Programme, Centre for Higher Education Development, University of Cape Town, Cape Town, South Africa
Department of Health Sciences Education, Faculty of Health Sciences, University of Cape Town, Cape Town, South Africa

Matthieu Poyade The School of Simulation and Visualisation, Glasgow, UK

David L. Robertson MRC-University of Glasgow Centre for Virus Research, The University of Glasgow, Glasgow, UK

Kate Le Roux Academic Development Programme, Centre for Higher Education Development, University of Cape Town, Cape Town, South Africa

Vivien Shaw Academic Visitor at Hull York Medical School, Faculty of Health Sciences, University of Hull, Hull, UK

Isabelle C. Winder School of Natural Sciences, Bangor University, Bangor, UK

Ed Young Eh!woza, 1st Floor Section Street Business Complex, Cape Town, South Africa

Part I

Exciting Data and Representation

1 A Multimodal Social Semiotics Perspective on Teaching and Learning Using Biomedical Visualisations

Natashia Muna

Abstract

Over recent years, there has been an explosion of digital technologies and infrastructure, and many biomedical education contexts have taken advantage of these new tools and platforms for teaching and learning visualisation practices. However, there are uncertainties, concerns, and debates in the literature, regarding the impact of digital technology on biomedical visualisation, and education more broadly. In this discussion paper, a multimodal social semiotics perspective is applied to examining visual representations, and the visual language we use to make meaning with these representations, in four different ways. Firstly, in terms of the function, role, and positionality of visual representations within the biomedical sciences. Secondly, within the context of a disciplinary Discourse, wherein visual language functions as just one of many socially specific languages, or modes, such as the written, verbal, and symbolic modes, and through which we animate the discourse and make disciplinary meaning. Thirdly, consideration is given to the meaning-making affordances and limitations of biomedical visualisation practices, affirming that no single mode or media is superior and that multimodal integration is a necessity. Finally, there is a discussion of the approaches to teaching and learning for the acquisition of a disciplinary Discourse, which highlights a need to focus on explicating the hidden curriculum. This discussion presents a perspective that offers some valuable insights into the uncertainties expressed in the literature and empowers educators with some productive pedagogical strategies.

Keywords

Meaning-making · Visual representations · Disciplinary discourse · Multimodality · Social semiotics · Hidden curriculum

The image below (Fig. 1.1) is a multimodal (*written text* as an *image*) visual representation of this chapter, created using WordCloud Generator by MonkeyLearn Inc. (https://monkeylearn.com/wordcloud/). This image has been included (in contrast to the written text of the chapter) to give the viewer/reader a sense of how different modes of representation (written text versus an image, for example) offer different types of meaning-making possibilities, or affordances. Here, a key affordance of the image is that it simultaneously presents an array of concepts

N. Muna (✉)
Academic Development Programme, Centre for Higher Education Development, University of Cape Town, Cape Town, South Africa

Department of Health Sciences Education, Faculty of Health Sciences, University of Cape Town, Cape Town, South Africa
e-mail: Natashia.muna@uct.ac.za

L. Shapiro, P. M. Rea (eds.), *Biomedical Visualisation*, Advances in Experimental Medicine and Biology 1388, https://doi.org/10.1007/978-3-031-10889-1_1

and research that have been written about in this chapter, making them all immediately known, as opposed to having to read the full text of the chapter to encounter these terms. In other words, the image affords meaning about the content of this chapter. However, the image does not offer meaning about the ways we understand and use these terms, individually or in relation to one another, nor how they are relevant to our educational practices. To make those meanings, we must now turn to the affordances of written text.

1.1 Introduction

Visualisations—specifically, visual representations[1]—have long been utilised across the sciences as a means of collecting and recording data, sharing findings, and as a resource for teaching and learning. From early inscriptions to the work of the naturalists, to the development of three-dimensional (3D) virtual models, science has been increasingly utilising the process and production of visualisations as part of what we do. With the explosion of digital technologies and infrastructure over the last 30 years, many have taken advantage of these new tools and platforms for constructing and sharing a widening variety of increasingly sophisticated visual representations.

I was fortunate to experience something of this myself while visiting Karel de Grote University College in Belgium, where I was invited on a tour of their healthcare simulation facility and to experience their virtual reality (VR) lab. As I put on the VR glasses, I found myself standing in a cathedral, only it was not a cathedral, but *a lung*. And as I turned slow awestruck circles, gazing up at the domed 'ceiling' and the translucent alveoli, not dissimilar to stained glass, I cannot deny that there was something almost religious and certainly remarkable about the feeling of 'travelling' into a lung to experience a visual representation, rather than just looking at one.

Despite opportunities like this created by digitally mediated visual representations, the relatively recent and rapid developments in the use of these technologies have resulted in some uncertainty and debate across educational contexts. Biomedical science students tend to report positively that they find the use of digital technologies for teaching and learning visualisations to be engaging and stimulating (Johnson et al. 2012; Huser et al., 2019; Reinke et al. 2021), but other studies report challenges related to students' unfamiliarity with the technology (Moro et al. 2017; Border 2019; Crow and Murray 2020; Stromberga et al. 2021), and the quantitative data remains ambiguous regarding the impact on student success (Meyer and Cui 2020; Stromberga et al. 2021). Some hold the view that digital technologies have led to an overuse and overreliance on visual representations (Matthan and Finn 2020), while questions have also been raised about the impact of digitisation on students' cognitive load (Moro et al. 2017; Border 2019).

As the biomedical and clinical sciences are inherently visual (Bezemer 2017; Reinke et al. 2021), and the affordances[2] of digital technologies have already cemented their position as a dominant influence (Kress and Van Leeuwen 2020), the issues raised above must prompt us to continue developing our understanding. To this end, a multimodal social semiotics perspective has been adopted to examine: the functional roles, power, and positionality that visual representations have within the biomedical sciences; the notion of socially situated languages, and the specialised affordances they have for meaning-making; the affordances of biomedical visualisation practices and the impact of digital technologies on these; and the ways in

[1] The phrase 'visual representations' is used to distinguish between visualisations as a noun (which refers to the visual representations that we see and interact with in the external world around), and visualisation as a verb (the process of constructing an internal or mental representation). The focus of this chapter is on the former.

[2] Affordances refer to the meaning-making possibilities or potentials of something, or put another way, the possibilities and potentials of '...what can be 'said' or 'done'...' with a thing (Kress and Van Leeuwen 2006, p. 123)

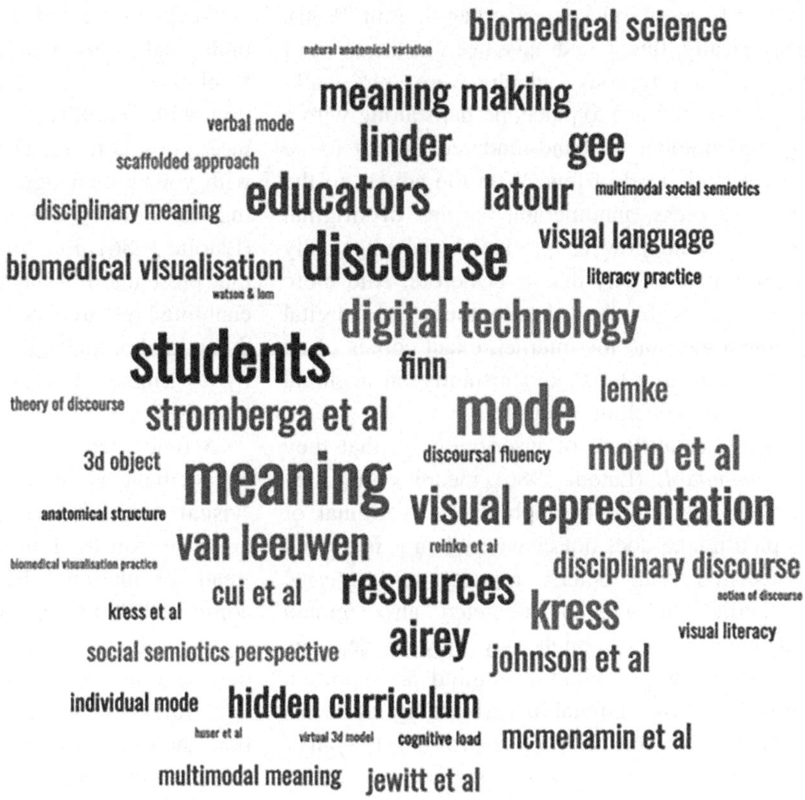

Fig. 1.1 A multimodal visual representation of the written text, 'A multimodal social semiotics perspective on teaching and learning using biomedical visualisations'

which we may teach and learn to become fluent in this multimodal disciplinary language.

1.2 The Functional Roles, Power, and Positionality of Visual Representations

The sociologist and philosopher, Bruno Latour, known for his work in the fields of science and technology, has explored the significance of visual representations to the practice of science, and persuasively argues that only by taking written and visual representations into account (and here we focus on the visual), can we begin to define the modern construct of 'science' (Latour 1986). In fact, Latour suggests that it is the purpose and way that knowledge is recorded within the scientific domain that makes it distinct from indigenous knowledge communities.

The biomedical sciences function predominantly within the positivist paradigm, where the pursuit of universal truth is paramount, and we therefore seek *verifiable* knowledge. It is in this regard that Latour recognises the importance of visual representations, which he refers to as 'inscriptions' (Latour 1986, p. 3). Inscriptions become powerful because they can be shared without alteration, and valuable because they enable us to engage the scientific community in logical reasoning, critique, and debate about our assertions. It is through this process of community engagement that, if our inscriptions are sufficiently convincing (verifiable), our knowledge claims become legitimised and accepted as valid additions to our canon. In other words, inscriptions are not only useful, but *central* to the practice of biomedical science as we know it. In exploring why inscriptions have come to hold this status, Latour identified several relevant and important attributes and affordances of inscriptions.

Firstly, inscriptions are *mobile* and *reproducible* which allows them to be translocated or shared from one place to another (Latour 1986). Historically, this would have been in the form of original inscriptions which were physically moved from place to place, perhaps along with a limited number of hand-rendered copies to be shared with a select few. With the advent of the printing press, innumerable copies of original images could now be produced and physically shared more widely than ever before. And more recently, with the development of digital technologies and the internet, exact copies of an image can now be shared instantly on a global scale at the touch of a button.

Another attribute of inscriptions is that they are *immutable* (Latour 1986), meaning that they remain unchanged by mobility. An original or copied image does not change when it is moved or shared with others. Equally, the internal properties of what is depicted also remain unchanged, even though they may be depicted in different ways. With this in mind, an important affordance of immutability (building on Ivins foundational work in the early 1970s), is *optical consistency*, which is the establishment of a consistent two-way relationship between the real and the represented. When visual representations are constructed, they are done so based on the actual relationship between the observer and the observed (the object of inscription), and the act of inscription records this relationship based on the relative perspectival angle, position, and orientation of the observer and the object (Latour 1986; Kress and Van Leeuwen 2006). As a result, when looking at the inscription, a viewer is afforded the ability to 'travel' from the image to the object, along a linear perspective, and back again (Latour 1986).

Linked to the ability to 'travel', inscriptions also make it possible to inscribe (record) what has been observed, in a *presentable* format (Latour 1986). In other words, inscriptions can present observations and data which become absent once the inscription is moved or shared (mobilised) and in doing so, make these absent things presentable. Underpinning both the ability to make absent things presentable and optical consistency, is the affordance of scalability. For example, by utilising a scalar relationship between the real and the represented, the invisible biological world that exists at the microscopic level can be made presentable. And, in conjunction with mobility, the value of a presentable inscription is that it allows you to engage others with your knowledge claims, without them having to directly observe what is represented (Latour 1986). For this reason, it is also likely that presentability has been a powerful factor in enabling researchers to work in increasingly diversified or niche areas, without losing the ability to engage a wider community with their assertions.

A fundamental factor in enabling community engagement is the establishment of a shared 'visual language' (Latour 1986) that informs how we construct inscriptions and how others 'read' or interpret them. A visual language is composed of the resources—the modes, media, signs, tools, techniques, and conventions[3]—that may be utilised to produce a visual representation, and from which viewers derive meaning about both the nature of the representation and what is being represented, within a community context (Kress and Van Leeuwen 2020). Community members who are fluent in this visual language are able to use the 'grammar' of this language to achieve access to the two-way relationship between the real and the represented, and a disciplinary understanding of the representation (Kress and Van Leeuwen 2006; Airey and Linder 2009).

When this visual language is used with some level of consistency by community members, it allows inscriptions to be *combined* or superimposed upon one another to enable critical comparison (Latour 1986). For example, radiographic images may be overlayed to enable diagnostic comparison. Another function of combinability is the construction of new types of representations. For example, digital technologies have enabled the ability to combine many different representations of a single object, to construct complex 3D models (Cui et al. 2019), which offer multiple perspectives on the object,

[3] These terms will be defined in subsequent sections.

rather than the singular linear perspective of a two-dimensional image. Importantly, inscriptions may also be combined and integrated with different modes of communication, such as written language (Latour 1986), which allows the construction of multimodal representations that incorporate additional information about the object that a visual representation alone could not have communicated.

Collectively, the attributes and affordances of inscriptions, or visual representations, position them as centrally important to the practice of the biomedical sciences. Building on Eisenstein's work on the impact that mobilisation technologies have on inscriptions, Latour (1986) asserts that any technological developments that accelerate mobilisation and increase immutability will be favoured—digital technologies fulfil both criteria. Due to the capability of rapidly mobilising immutable inscriptions, digital technologies have made a significant impact on the pursuit of accuracy. Firstly, because digital technology enables the construction of representations at an increasingly accurate and minute level of detail, thus allowing for greater optical consistency between the real and the represented. Also, increased mobility and widening access enable more robust community engagement. This is important because it is through the comparison of two or more inscriptions that we reach conclusions about features of similarity that verify our knowledge, and relevant differences that motivate us to seek further clarity. As such, visual representations not only play a role in verifying our knowledge, but also in motivating the ongoing pursuit of verifiable knowledge. The key to enabling these affordances across a distributed and diverse community of people lies in utilising a shared visual language.

1.3 A Multimodal Social Semiotics Perspective of Language

To understand the visual language of the biomedical sciences, we must first develop an understanding of language as something which is socially situated within a distinct community, or Discoursal context (Gee 1989). This situated perspective of language allows us to view it as contextualised and central to meaning-making, and therefore learning (Halliday 1993). By situating visual language within the biomedical sciences, we come to recognise that it functions in relation to other forms, or modes, of language, such as verbal and written language (Lemke 2000; Jewitt et al. 2001; Kress and Van Leeuwen 2006; Airey and Linder 2009). This perspective allows us to position digital technologies as an integral resource, and therefore part, of the visual modes (Kress and Van Leeuwen 2006, 2020).

1.3.1 Disciplinary Communities as Discourses

Gee (1989, p6) has convincingly theorised the concept of Discourses,[4] which can be thought of as, 'ways of being...which integrate words, acts, values, beliefs, attitudes and social identities...'. Because there are many 'ways of being', the theory of Discourses provides a framework for recognising socio-culturally organised communities of people who ascribe to a shared set of values, beliefs, and attitudes, while acquiring and mastering a set of contextually situated activities and communication practices with some common intent or purpose. From this perspective, there exist an infinite number of social, cultural, religious, professional, and political Discourses, and it is through participation in Discourse communities that we come to embody particular social identities (Gee 1989). Given that within each academic discipline there are 'ways of being', then a field such as the biomedical sciences can also be thought of as a Discourse (within which more specialised Discourses also exist).

Inherent within any Discourse, is a distinct language (discourse), through which knowledge and ideas are represented and shared—in other

[4] Discourse (distinguished by use of an uppercase 'D') which refers to a socio-cultural community is not to be confused with discourse (lowercase d), which refers to a connected length of language.

words, through which meaning is made (Gee 1989; Kress and Van Leeuwen 2020). At their foundation, all languages are about making meaning, and are composed of a system of signs (such as images, text, symbols, numerals, colours, sounds, etc.) which signify contextualised meanings through a semiotic process (Halliday 1993). The meanings that we choose to make and the way we choose to make those meanings are determined by the context and purpose of the meaning-making—in other words, making meaning is a socially contextualised semiotic process (Halliday 1993; Kress and Van Leeuwen 2020).

The visual language of a Discourse is established through a process of social negotiation among its members. Each Discourse defines both what it is *to see*, and *what* there is to see (Latour 1986). We find this principle at work in the biomedical sciences, for example, in the process of achieving expert consensus during the development of valid 3D anatomical models, to ensure that the most important structures are included (Meyer and Cui 2020). Over time, these consensus practices work to establish and refine the use of visual language modes for meaning-making, which supports the maintenance of optical consistency (Latour 1986). It is this process of signifying meanings through semiotic representations, which transform them into the objects of knowledge that animate the discourse (Goodwin 1994).

1.3.2 Multimodality

Visual representations, defined broadly here, are spatially realised representations, such that all the parts may be simultaneously visually perceived (Kress et al. 2014). Visual representations such as images and 3D objects form an essential part of biomedical discourse, but they cannot do the semantic work of the discourse alone (Kress and Van Leeuwen 2020); an image cannot convey the name of an artery, or represent a blood pressure reading, or convey the meaning of a particular shape. So it becomes evident that the discourse of biomedical science is in fact a 'synergistic integration' of specialised languages and literacy practices (Lemke 2000), which, in addition to visual language, also includes verbal language, reading and writing (the 'academic' literacies; Lea and Street 1998), quantitative language, apparatus use, and gesture or movement (Airey and Linder 2009; Lemke 2000).

As these languages (and their literacy practices) signify meaning in different ways using different signs, they can usefully be conceptualised as 'modes', which are '. . .organized, regular, socially specific means of representation' (Jewitt et al. 2001, p.5). Each mode (speech, writing, images, 3D objects, movement, etc.) has particular affordances for communicating different aspects of meaning, and as such they are not interchangeable. Rather, modes function in an additive and integrative manner, with each mode being selected to complement the others in the construction of multimodal meaning-making complexes (Kress and Van Leeuwen 2020). For example, a biological drawing typically includes an image, written annotation, and quantitative scalar information. Independently, each of these three modes presents some meaning, but it is only through their integration that understanding may be achieved.

As each mode includes all the resources used for making and interpreting signs within that mode (Kress and Van Leeuwen 2020), we must recognise the affordances of digital technology for the visually realised modes of images and 3D objects. In addition to accelerating mobility and increasing immutability of visual representations (Latour 1986), digital technology has also enabled new means of producing representations (Kress and Van Leeuwen 2006), new affordances for representation and meaning-making potential, and thus new forms of knowledge (Halliday 1993).

1.4 A Multimodal Perspective of Biomedical Visualisation Practices

It has been recognised that when teaching and learning biomedical visualisation, there is no single pedagogical strategy, media, or mode which is superior to all others and as such many educators advocate for a multimedia and multimodal approach (Johnson et al. 2012; Moro et al. 2017; Finn and Matthan 2019; Stromberga et al. 2021). This call affirms Halliday's (1993) assertion that teachers have an almost intuitive sense that their students need to learn multimodally, and it aligns with the theory that each mode and media have its own affordances for representing meaning and are thus not interchangeable (Kress and Van Leeuwen 2006). Adopting a multimodal approach requires that we draw from the full range of contextually available tools and languages. In a biomedical context, the tools equate to the resources, apparatus, instruments, hardware, and software that enable us to do our work, while the languages equate to the range of modes that we use to both represent and make meanings.

The biomedical community has long since utilised the affordances of a variety of modes and media for producing visual representations (See Erolin 2019 for a useful historical overview of the use of technology for visualisation). As different media have different affordances for meaning-making (Kress and Van Leeuwen 2006), the manner in which visual representations are produced and used is of consequence. As new tools with novel affordances are developed, if we find that they allow us to do our work better they will be incorporated into the Discourse (Latour 1986), and digital technologies are at the frontier in this regard. Digital technologies have enabled us to create multimedia learning environments which can enhance learning and allow modes to be used to a fuller potential than was previously possible (Stromberga et al. 2021). For example, by utilising the visual mode within a digital space new affordances have become available, such as the ability to render microscopic detail with exceptional accuracy, or instantaneously change the scale at which something is represented.

As each mode has its own affordances for representing certain details of meaning, it is necessary to select the modes and resources used to construct meaning to take advantage of this, and to maximise complementarity in meaning complexes (Kress and Van Leeuwen 2020). For educators to be able to make effective choices, they must understand the multimodal meaning-making affordances of the resources at their disposal. It is thus useful to review the affordances and limitations of some biomedical visualisation practices commonly used in teaching and learning, while considering how digital technologies are impacting on these—a discussion for which anatomy provides a useful example.

1.4.1 Affordances and Limitations of Multimodal Biomedical Visualisation Practices

In medical education contexts, visualisation is primarily intended to exemplify living anatomy, physiological processes, and clinical relevance (Matthan and Finn 2020). As such, realism is of particular importance, and cadaver dissections are considered the 'gold standard' for teaching gross anatomy (Moro et al. 2017).

A body is a multimodal meaning-making complex. Bodies have 3D shape, size, structure, and orientation (Stromberga et al. 2021), as do their component parts—both individually and in relation to one another. Along with being olfactory and auditory (Matthan and Finn 2020), bodies are also haptic, with aspects such as texture and pliability (Johnson et al. 2012). It is also patently obvious that bodies are visual, with dimensional and haptic aspects functioning alongside other signs, such as colour (Johnson et al. 2012), to produce a visual representation. Given the 3D nature of bodies, and the fact that normal human vision occurs in 3D (Cui et al. 2019), dissection affords students a real perspective of the relative location, spatial orientation, and relationship between anatomical structures (Johnson et al. 2012; Moro et al. 2017). Dissection also exposes students to natural anatomical variation (Johnson et al. 2012; Moro et al. 2017; Stromberga et al. 2021), and pathological conditions (Stromberga et al. 2021), while requiring students to utilise the manual mode (Lemke 2004), and thus develop the dexterity required for certain specialisations, such as surgery (Erolin 2019).

Despite these affordances, there are limitations to what can be learnt about anatomy through dissection alone. For example, dissection cannot teach students the specialised written and verbal language and terminology of the discipline. Nor will dissection teach students about the relationship between anatomy, physiology, and clinical relevance (Moro et al. 2017; Stromberga et al. 2021). So, it becomes clear that other multimodal resources and media are required to complement and add to the meanings that can be made from the practice of dissection.

Supplementary resources for teaching and learning typically include textbooks, lecture slides, and notes (Moro et al. 2017), which usually integrate the written and image modes (Kress and Van Leeuwen 2006; Bezemer 2017). Digital technology has also enabled the integration of the verbal mode into these resources through audio recordings and 'voice-overs'. Both written and verbal language are particularly important for naming and explaining. Both have the affordance of being able to effectively present probabilities, generalisations, and abstractions (Kress and Van Leeuwen 2006; Bezemer 2017), making them ideal for explaining the relationship between anatomical structures and physiological functions (Border 2019). While the written text functions as a static record (Kress and Van Leeuwen 2006), the verbal mode is temporally perceived (Kress et al. 2014), and although it is transient, it affords access to other signs such as tone and inflection, which modify the meaning of the words alone.

The mode of image, which is a frequent feature in supplementary resources, is effective at representing meanings which can be iconic (realistic), as with a photograph; conceptual, for example with a figure or diagram; or associative, such as with symbols (Harrison 2003). A key affordance of images, and indeed visual representations more broadly, is that they are spatially realised, and as such allow all the signifiers of meaning to be simultaneously presented as a multimodal meaning-making complex (Kress et al. 2014). This contrasts with the written and verbal modes where meanings are presented sequentially (Kress et al. 2014; Bezemer 2017; Border 2019). For example, physiological functions and processes that occur simultaneously must necessarily be explained in some order. The producers of such resources must thus consider not only which content to include and in what order (Bezemer 2017), but also aspects such as colour, layout, text format, multimodal integration, and accessibility (Huser et al. 2019).

Even with care and attention to the composition and layout of these resources, students may still struggle to interpret the meanings as intended. The sequential nature of such media requires students to relate what they have read or heard to accompanying visual representations, and they may become confused or overwhelmed by extraneous information (Border 2019). The way in which written and visual elements are integrated can also result in meaning-making challenges for students. For example, a line is traditionally used to associate a written label with a structure, but often this does not make clear how much, or which part of the structure is being named (Huser et al. 2019). Another limitation of images in these formats—the printed page or screen—is that they present two-dimensional (2D) representations, of what are typically 3D structures.

When 3D structures are represented monoscopically (in 2D), the process of interpreting the meaning requires the viewer to mentally transform the representation into 3D (Moro et al. 2017; Cui et al. 2019). This is considered a particularly challenging practice for students to master, especially in the absence of depth perception cues that are inherent in stereoscopic (3D) vision due to the relative spatial orientation of structures (Moro et al. 2017; Stromberga et al. 2021), and even students who can visualise in 3D may make erroneous assumptions and conclusions (Moro et al. 2017). As visuospatial ability is considered vital for healthcare professionals (Erolin 2019; Stromberga et al. 2021), and learning in 3D improves spatial awareness (Stromberga et al. 2021), the mode of 3D objects remains essential. In addition to cadavers which are not always accessible (McMenamin et al. 2014; Moro et al.

2017; Stromberga et al. 2021), other types of 3D resources can be utilised.

Objects, like images, can represent realistic or conceptual meanings, and in biomedical contexts there is a strong emphasis on realism in this mode. Two particularly relevant affordances of 3D objects are the opportunity for interaction and depth perception (Kress and Van Leeuwen 2006). Being able to interact with objects, in contrast to static images, means that they can be viewed from varying distances, positions, and most significantly, different angles. For example, wax and plastic models have the affordance of representing the basic or normal shape and orientation of anatomical structures in relation to one another, and so aid with visuospatial learning (McMenamin et al. 2014; Moro et al. 2017; Erolin 2019).

However, although robust, plastic models have low fidelity and usually only represent a limited number of structures in limited detail (Johnson et al. 2012). Plastinated specimens offer an alternative means to produce more detailed, realistic, and accurate 3D models, with the added affordance of representing natural anatomical variation, but the production process requires cadavers, is expensive, and raises some health and safety concerns (McMenamin et al. 2014). Here, the development of 3D printing technologies mitigates some of these limitations and enables new affordances for representation and functionality.

3D printing can produce highly realistic and accurate models of prosected and whole specimens. Large specimens can be printed to scale in parts and manually assembled post-print (McMenamin et al. 2014), and structures often appear clearer than in a real specimen (Erolin 2019), with only the smallest structures being marginally affected by production errors (McMenamin et al. 2014). As this technology develops and becomes more accessible, additional utilities are becoming available, and with them new opportunities for meaning-making. For example, digital colour maps can be developed for an object, and then altered for subsequent prints depending on the desired effect (McMenamin et al. 2014). Typically, 3D prints are rendered in hard plastic, which has haptic qualities similar to bone (Finn and Matthan 2019); however, more advanced multi-material printers can print in different products, for example printing soft tissue in a flexible material (Preece et al. 2013), to produce a wider variety of haptic signifiers such as pliability and texture. It is also becoming increasingly possible to print examples of natural variation (McMenamin et al. 2014). Interaction can also be enhanced, for example by embedding magnets into printed structures that allow them to be assembled and disassembled during teaching and learning (McMenamin et al. 2014). However, the physical materiality of these objects also places limitations on representation and interaction, with physical models typically depicting outer surface features at fixed scalar relationships (Johnson et al. 2012).

Another significant advance that digital technologies have enabled, is the ability to produce virtual representations of 3D objects, in the form of virtual 3D models. This shift in medium—from physical to virtual—has enabled new affordances for both representation and interaction. Virtual 3D models of anatomical structures have the affordance of being able to represent both an outer and inner surface, as well as dimensions of length, width, height, and volume. Within the virtual environment, these models can be rotated in space, and zoomed in or out (Cui et al. 2019), allowing students to interact with many perspectival angles and scales of representation. These dimensional signs afford the ability to quantify the spatial relationships between structures, which is useful for predicting the parameters of structures and in planning surgical procedures (Cui et al. 2019). With virtual models it is also possible to selectively view both the parts and the whole of the model, as well as to virtually assemble and disassemble the model (Cui et al. 2019).

Both the written and verbal modes can also be integrated into virtual models, to add meanings through annotating structures and including 'voice-overs' (Moro et al. 2017; Erolin 2019). With the inclusion of haptic technology into the model, it also affords the ability to actually 'touch' and 'feel' the structure and texture of the object. Even more remarkably, when used in combination with microscopy, haptic technology affords the ability to touch and feel objects that in physical reality are too small to feel (Jones et al. 2003). However, even with the inclusion of interactive features and haptic technology, the manual dexterity developed through the practice of dissection is not easily achieved using models (Erolin 2019). Furthermore, the resource-intensive development of models, along with the ability to share them widely (Erolin 2019; Stromberga et al. 2021), means that there exist limited examples of natural variation (Erolin 2019). But perhaps the most important consideration regarding the utility and affordances of virtual 3D models relates to the medium through which representation and interaction occur.

When 3D models are viewed on a screen, they are seen monoscopically, and in this way function like an image, without the affordance of depth perception cues, thus requiring the student to mentally transform the 2D representation into 3D. However, when represented through the medium of VR, visually the model functions as an object, with depth perception being afforded through cues such as convergence and binocular parallax (Erolin 2019; Meyer and Cui 2020). Thus, with stereopsis, 3D models afford a realistic representation of the orientation and spatial relationships between structures (Moro et al. 2017; Meyer and Cui 2020), and negate the need to mentally transform the representation into 3D, because it is already experienced as such.

Clearly, digital technologies have had a significant impact on the affordances for representation and meaning-making in biomedical visualisation. However, in addition to impacting the modes themselves, the digital medium has also been fundamental to the development of new ways of sharing, accessing, and organising these multimodal teaching and learning resources, in virtual multimedia learning environments (VLEs).

1.4.2 Virtual Learning Environments

VLEs allow greater immutability and mobility of resources that are no longer limited by physical location (Border 2019; Erolin 2019; Crow and Murray 2020). By untethering these resources from their physical locations, opportunities have been made for both collaborative learning among a more distributed and diverse range of participants (Crow and Murray 2020), and more independent student-directed learning (Cui et al. 2019). In this way, more students can access materials where and when it suits them, learn at their own pace, and return to material when needed for practice and revision (Cui et al. 2019; Finn and Matthan 2019).

The benefits of VLEs have motivated the uptake of digitally enabled teaching and learning, as a means to mitigate challenges faced in biomedical education contexts, including: increasing student numbers; decreasing faculty (Moro et al. 2017); financial, ethical, and cultural constraints on cadaver dissection (McMenamin et al. 2014; Moro et al. 2017; Stromberga et al. 2021); and the shift towards problem-based teaching, which has resulted in reduced curriculum time devoted to subjects like anatomy (Johnson et al. 2012; Moro et al. 2017; Cui et al. 2019). While VLEs have allowed for learning opportunities that extend beyond the temporal and physical classroom, which does alleviate some of the practical and logistical challenges faced by educators (Border 2019), VLEs are not without limitations.

Despite their wider availability, access to digital resources and VLEs are still considerably limited by financial and infrastructural constraints (McMenamin et al. 2014; Cui et al. 2019), especially in developing contexts. And, where these resources are being utilised, educators have identified challenges for teaching and learning. For example, when students are unfamiliar with the technology, they tend to focus on learning how to use the tools, rather than on the content being presented (Stromberga et al. 2021). This lack of confidence in the use of digital resources has the potential to inhibit students' engagement (Crow and Murray 2020) and creates a challenge for educators who must manage varying levels of digital literacy among their students (Finn and Matthan 2019). Perhaps most worryingly though, graduates are beginning to report that their anatomical knowledge is inadequate (Johnson et al. 2012; Singh et al. 2022). These concerns indicate the challenges and risks which students and educators may experience while trying to teach, or learn, how to make meaning using the multimodal language of the discipline, or disciplinary discourse (Airey and Linder 2009). It is thus important for us to consider *how* we learn the multimodal languages of a disciplinary discourse.

1.5 Teaching and Learning Discoursal Fluency in the Biomedical Sciences

To become fluent in the discourse, it is necessary to become proficient, or literate, in the practices of each mode (Lemke 2000; Airey and Linder 2009). Although this idea is consonant with our understanding of how we learn to read, write, and speak natural language, it is insufficiently applied to the contextualised use of multimodal language. By returning to Gee's (1989) notion of Discourse we come to see how languages are acquired in teaching and learning contexts, and that this process requires a mobilisation of the hidden curriculum (Hafferty 1998). Against this backdrop, the pedagogic considerations for teaching and learning for multimodal fluency and meaning-making will be discussed, focussing on the development of visual literacy in relation to biomedical visualisation.

1.5.1 Modal Literacy and Discoursal Fluency

Lemke (2004) describes scientific literacy as the ability to make meaning with and integrate the manual and technical, visual, symbolic, mathematical, and verbal modes. Within this context, visual literacy can be understood as the ability to recognise and interpret visual signifiers from a disciplinary perspective, *and* the ability to produce contextually meaningful visual representations (Watson and Lom 2008; Rybarczyk 2011). In other words, to be visually literate you must know how to 'read' the mode from a disciplinary perspective—what to pay attention to and how to interpret what you see or experience to make meaning. Equally, you need to be able to utilise the mode to produce representations—understand how to select media and signifiers to take advantage of their affordances for representing meaning.

As disciplinary experts, educators are fluent in the discourse and so ideally suited and situated to teach students about how to make disciplinary meaning using these modes (Airey and Linder 2009). In this sense, all teachers are language teachers (Halliday 1993). The goal for educators is thus to mobilise their knowledge and make it explicit to students (Airey and Linder 2009); by showing students how to make meaning using the modes individually and together, you are teaching them how to make meaning of the content, but they are also learning the language.

Gee's (1989) theory of Discourse is synonymous with the notion of contextualised language. To become a legitimate member of a Discourse community, you must become literate in each individual mode, and fluent in the integration of modes as they function together to enact the multimodal language of the discipline (Gee 1989; Airey and Linder 2009). Prior to university, students are essentially outsiders to the disciplinary Discourse, but once at university they become apprentices. Gee (1989) argues that the truest learning of a Discourse occurs via acquisition—this is the informal learning, or apprenticeship, that takes place by observing and participating alongside legitimate members of the community.

1.5.2 Acquiring the Discourse

Acquiring a Discourse is always difficult to achieve in a classroom context (Gee 1989, distinct as classrooms are from the legitimate contexts within which Discourses are practiced. Scientists and doctors may teach in classrooms, but they practise in laboratories and clinics (although certainly clinical and laboratory contexts do feature in biomedical education). As such, not all apprentices will become fully fluent, legitimate members of the Discourse. Although fluency is essential, it is not sufficient to enable legitimate membership, as it does not ensure the ability to make meaning with and through the disciplinary modes (Airey and Linder 2009). Whether or not students achieve legitimacy is heavily dependent upon educators' understanding of the process of Discourse acquisition and, accordingly, the teaching practices they employ (Gee 1989; Airey and Linder 2009).

Within teaching contexts, the disciplinary discourse (or at least some of it, depending upon which modes educators are employing) is constantly in use. However, if the discourse is used, but it is never made explicit *how* these modes give rise to meaning, then students may learn to imitate the discourse, without actually understanding how to interpret it to make meaning (Airey and Linder 2009). Thus initially, students may appear to develop fluency through what Airey and Linder (2009, p34) refer to as 'discourse imitation'. However, once they are required to utilise this discourse in unfamiliar ways and contexts, they will struggle, and many will fail. To quote Kress and van Leeuwen (2006, p17), 'To put this point harshly, in terms of... 'visual literacy', institutional education. . .produces illiterates'.

Conversely, when educators do use their teaching to explain how modes give rise to meaning, the classroom holds great value for Discourse acquisition, through its ability to lead to metaknowledge (Gee 1989). Metaknowledge is acquired when we become aware of how the modes of a Discourse work in ways that are distinct from our prior experiences of these modes in other Discourse contexts (Gee 1989). For example, in learning to produce laboratory reports, students may become aware of how the disciplinary ways of using writing and images to develop meanings are different from the writing and image production practices they employed in high school (Kelly-Laubscher et al. 2017). Gee sees this metaknowledge as powerful because it enables students to make conscious choices about how they operate within the Discourse, which includes how they make meaning using the contextualised multimodal language resources.

Along with metaknowledge, students will also utilise what Gee calls 'mushfake' (p. 13). Mushfake, which is a legitimate means to participate in the Discourse, refers to the use of practices that we have mastered in other Discourses, as 'make-do' substitutes for what we have not yet learnt in the Discourse we are acquiring. The role of the educator is thus to guide students away from using these substitutes over time, and towards using the literacy practices of the disciplinary discourse to enable different ways of making meaning (Airey and Linder 2009).

As educators have already mastered the disciplinary discourse, they are ideally positioned to teach students about how to make disciplinary meaning using its multimodal language (Airey and Linder 2009). However, in a late plot twist, it is exactly this mastery of the discourse that makes it so challenging to teach. Once something has been mastered, it requires almost no conscious awareness or explicit thought to perform, essentially it has become 'second nature', or tacit (Gee 1989). And what is tacit to the educator may thus remain hidden from the student.

1.5.3 The Hidden Curriculum

While the theory of Discourse may be new to many biomedical educators, the experience of it is not. In the context of biomedical education, Hafferty (1998, p404) defines the hidden curriculum as '...the commonly held "understandings", customs, rituals, and taken-for-granted aspects of what goes on in the life-space we call medical education'. In other words, he is referring to the '. . .forms of life which integrate words, acts, values, beliefs, attitudes, and social identities. . .' within the context of a Discourse (Gee 1989, p6). In this sense, the notion of a hidden curriculum is synonymous with the notion of acquiring a Discourse—learning through apprenticeship from masters, who themselves are frequently unaware of the things they are doing, and which apprentices must learn from them.

As it is not reasonable, or even desirable, to include every element of the Discourse within the formal curriculum, the hidden curriculum will always remain a part of educational contexts. Matthan and Finn (2020) recognise this in their critique of digitally enabled biomedical visualisation, and develop the argument that given the hidden curriculum, there are multiple risks associated with the uncritical application and adoption of digitally enabled visualisations. While their concerns are valid (there are always affordances *and* limitations), they acknowledge, but do not sufficiently consider an alternative perspective of the hidden curriculum as something which may be desirable (there are always limitations *and* affordances). Despite their narrow framing, their recommendation that educators need to make the hidden curriculum explicit to students, aligns with both Gee (1989) and Airey and Linder's (2009) assertions.

To better understand the hidden curriculum, it is useful to consider the distinction between the informal and hidden curriculums. The informal curriculum refers to everything that is taught, but which is not explicitly articulated or documented in curricula and learning materials (Hafferty 1998; Bezemer 2017), while the hidden curriculum functions at the level of the culture, or Discourse (Hafferty 1998). Bezemer usefully describes the hidden curriculum as '...*the meanings* implied by the embodied, physical actions...' of educators (Bezemer 2017, p. 105, italics added for emphasis). The emphasis on meanings gives us a more specific understanding of the hidden curriculum as those things that are literally invisible—the reasoning and meaning-making processes taking place in the educators' mind. It is this hidden meaning-making that we need to mobilise and make explicit to students, and in doing so, enable their acquisition of metaknowledge (Gee 1989).

Relevant to this discussion is the question of who is most at risk if the hidden curriculum remains hidden? Of particular concern in this regard are the so-called 'non-traditional' students, who are likely to have more diverse experiences, and fewer experiences of western science. The significance of this lies in the power of Discourses to both challenge and influence one another (Gee 1989). For example, the homes of students from more affluent socioeconomic contexts will be more heavily influenced by academic and professional Discourses than the homes of students from indigenous, minority, marginalised, or resource-constrained contexts. As a result, students from privileged contexts will experience fewer conflicts and more similarities between the disciplinary Discourse and the Discourses they have already acquired (Gee 1989). Additionally, educators who themselves hail from these privileged contexts may make assumptions that are only valid for some of their students, about students' prior learning, the availability of meanings in a teaching context, and thus what they do and do not need to make explicit. For example, the current student populace is frequently referred to as being 'digital natives', and yet this will only be true for a subset of students (Finn and Matthan 2019), those who come from privileged contexts that included digital technology as a standard part of their Discourse.

Thus, as educators, we come to recognise that we have a social responsibility to enable widened access and participation through our pedagogy. Indeed, it has been recognised within biomedical contexts that educators have a responsibility to teach students how to make meaning with and about the contextually available language modes, and specifically the use of visual resources (Johnson et al. 2012: Finn and Matthan 2019; Matthan and Finn 2020). If we accept that educators have an important pedagogical and social responsibility to make explicit to students how modes are used individually and together to make meaning, the next question becomes, how can educators do that?

1.5.4 Mobilising the Hidden Curriculum Teaching for Multimodal Visual Fluency

Airey and Linder (2009) recommend that educators adopt a scaffolded approach to teaching students about the disciplinary discourse. Scaffolding requires both channelling of student activity and focussing of attention (Pea 2004). Channelling activity refers to a reduction in task complexity for students, while focussing requires that student attention be explicitly directed to pertinent aspects of the task at hand. In this sense, the task refers to the task of making disciplinary meaning using multimodal resources. As students acquire proficiency and literacy in these practices, the scaffold must be incrementally dismantled, or faded out (Pea 2004), to allow students to use these resources independently and demonstrate their ability to do so.

To understand why a scaffolded approach offers value, we must begin with an important premise of the social semiotics perspective. All communication is multimodal (Jewitt et al. 2001), and while individual modes present contextualised meanings, it is only through the integration of modes that holistic meaning, and thus understanding, can be achieved (Lemke 2000). However, the more modes a discipline has, and the more specialised ways in which these modes are used and integrated, the more complex the discourse, and the learning process facing apprenticed students (Airey and Linder 2009; Moro et al. 2017). Biomedical educators have recognised this multimodal complexity in the literature and the impact of digital technologies on this complexity, with some arguing that integrating too many modes detrimentally increases students' cognitive load (Moro et al. 2017; Reinke et al. 2021), while others posit that the affordances of digital technology for multimodal integration decreases cognitive load (Border 2019; Stromberga et al. 2021). Yet, every day we are all bombarded by a constant stream of multimodal signs, integrated in a wide variety of complex ways, which we use to make meaning about our environments on an almost unconscious level—such is our mastery of the requisite literacy practices. From a social semiotics perspective then, it is not a question of cognitive load, but rather a matter of our ability to 'read' and make meaning from these 'loading' modes within a biomedical education context, as fluently as we do in our daily lives.

When this multimodal complexity is viewed through a social semiotics lens, it is much like passing a shaft of light through a prism and seeing it split into its constituent colours. Through the social semiotics lens, what is initially seen as a single unit of meaning is now understood as something composed of multiple modes, signifiers, resources, and media, each offering particular affordances for meaning-making. Educators need to begin by making students aware of these multiple modes (Lemke 2000) as the fundamental resources through which they will make meanings—what Airey and Linder (2009, p20) refer to as a 'critical constellation of modes'. Airey and Linder (2009) contend that students need to first gain individual modal proficiency, or literacy, before multimodal fluency can be achieved. In other words, educators need to focus on guiding students in their use of individual modes, prior to expecting them to independently, or easily, derive meaning from multimodal complexes, and certainly before expecting them to proficiently utilise multiple modes to produce meaning, for example in assessments (Lemke 2000; Muna et al. 2020). By focussing on the individual modes in meaning complexes, educators channel students' activity and attention and thus reduce the task complexity (Pea 2004). Instead of being required to make meaning from multiple, integrated modes, students can be given an opportunity to make meaning with individual modes in turn. In this way, students come to see how particular resources are used to construct modes, how each mode contributes meaning, and that it is through the integration of these meanings that we achieve understanding.

1.5.5 Seeing for Meaning

Learning to make meaning with a mode involves two steps: firstly, learning what aspects to pay attention to, and secondly, learning what those aspects mean within the disciplinary context. Without explicit guidance, students will direct their attention based on their interest, understanding of the context, and their prior learning and experience (Jewitt et al. 2001; Kress and Van Leeuwen 2006; Bezemer 2017). In this regard, educators have an important role to play in focussing students' attention (Pea 2004) on disciplinarily meaningful features. Equally, educators have an important role to play in guiding the disciplinary meanings that students make using these features (Jewitt et al. 2001; Bezemer 2017).

Becoming a clinician (or any other kind of biomedical professional) is about learning to see through the lens of the profession (Bezemer 2017), which refers to developing 'professional vision' (Goodwin 1994) or 'disciplined perception' (Stevens and Hall 1998). This perception is understood as the contextually situated (disciplinary) ways of recognising, observing, and interpreting features of relevance, a proficiency which must be developed overtime (Goodwin 1994; Stevens and Hall 1998).

By being explicit about what they themselves pay attention to, educators enable students to 'see through the eyes of the discipline'. Through expert guidance and practice, neutral 'aspects' of what is being observed (colour, size, shape, position, orientation, texture, etc.) become meaningful 'features' that are distinctive and immediately apparent to the expert viewer (Stevens and Hall 1998). For example, learning to distinguish natural anatomical variation from pathology requires first knowing what structures and aspects thereof to pay attention to (i.e. recognising the *features*), and then understanding how to meaningfully interpret these observations.

When teaching biomedical visualisation, highlighting the most important features in a functionally relevant way has been found to increase student interest (Johnson et al. 2012). During prosection classes for example, the educator both shows and tells students what to pay attention to (Bezemer 2017). They gesture to structures and manipulate them, while also naming them, and pointing out features of interest. In doing so, they demonstrate to the students what they, as an expert, consider to be noteworthy, teaching them *what* to pay attention to, *why* to pay attention to it, and *how* to make disciplinary meaning from these observations. In this way, the teacher is important, not only for what they say, but also in terms of what they communicate through their actions, body movements, gestures, and expressions (Jewitt et al. 2001). These processes support the development of students' visual literacy.

As students develop literacy in individual modes, they next need to understand how modes function as multimodal meaning complexes (Kress and Van Leeuwen 2006). Here, digital tools such as screencasts, which enable educators to mobilise and integrate modes (Border 2019), offer a valuable resource. Screencasts allow the educator to guide students' attention far more closely through their selection of what content to present, and when and how to introduce and integrate each mode (Border 2019). Resources of this nature are particularly beneficial for students who have few experiences of the practices of scientific meaning-making (Border 2019).

Overtime, this scaffolded approach must fade out (Pea 2004). Dismantling the scaffold allows students to independently make selections about which modes and disciplinary features thereof to pay attention to, to interpret meaning from these, and to transform and develop these meanings by moving between and integrating different modes (Jewitt et al. 2001). Students need to be given opportunities to practise using and making meanings with the discourse if they are to become fluent, which is something only achieved through *participation* (Airey and Linder 2009). For example, students can be given opportunities to create visual representation, which is an important part of visual literacy (Watson and Lom 2008; Rybarczyk 2011), and which has been shown to enhance students' understanding of multimodal resources and how they are used to make meanings (Huser et al. 2019).

Thus, three factors have distinct bearing on how students become visually literate and fluent in the disciplinary discourse more broadly: the existing experiences and literacies students bring with them into the classroom (Gee 1989), the explicit teaching practices used to enable students' acquisition of the Discourse (Gee 1989; Airey and Linder 2009), and opportunities for practice that allow students to develop and negotiate their use of these literacy practices in the process of making disciplinary meaning. If students are to recognise these language modes as important and valuable, they need to see and experience them being explicitly utilised and prioritised across the curriculum, including in assessment practices (Airey and Linder 2009; Matthan and Finn 2020).

1.6 Conclusion

To borrow from Hafferty (1998), the goal of this chapter was to create a structure for examining visual representations, that allows us to reflect on the larger structure of which they are part. In doing so, four broad focus areas have been considered and described to deepen our understanding of biomedical visualisation.

Firstly, we examined the functional roles, power, and positionality that visual representations hold within the biomedical sciences due to their ability to be immutably reproduced, mobilised, and combined, which allows them to make scientific meaning presentable and optically consistent (Latour 1986). The fundamental key that enables these attributes is the establishment of a shared visual language.

Secondly, to unpack the notion of a shared language, a multimodal social semiotics perspective of socially situated language was discussed. Here, Gee's theory of Discourse (Gee 1989) offered a useful framework within which to recognise the situated nature of language and the specialised ways it is used within Discourse contexts, such as the biomedical sciences. By further exploring the nature of situated language we came to recognise that all languages are about meaning-making (Halliday 1993), and are composed of multiple modes, each with their own affordances and limitations (Jewitt et al. 2001; Kress and Van Leeuwen 2006). From this perspective, digital technologies are positioned as part of the resources of visual modes, and as such offer new and valuable affordances for meaning-making in the biomedical sciences (Latour 1986; Kress and Van Leeuwen 2006, 2020).

Thirdly, a social semiotics perspective was applied to a review of the multimodal meaning-making affordances and limitation of common biomedical visualisation practices, to empower educators to make motivated choices about which modes and media can most effectively be utilised to make different meanings (Jewitt et al. 2001). Guiding the notion of choice is a recognition that modes are not interchangeable in terms of the meanings they make, and that all communication is multimodal (Jewitt et al. 2001). As such, the choices educators must make pertain to the development of integrated multimodal meaning complexes (Kress et al. 2014).

Finally, considerations and recommendations for teaching and learning discoursal fluency in the biomedical sciences were discussed. Anchored by an explanation of what it means to be visually literate (Watson and Lom 2008; Rybarczyk 2011), and fluent in the disciplinary discourse (Airey and Linder 2009; Lemke 2000), we returned to Gee's (1989) notion of Discourse and the concepts of acquisition, explicit teaching for metaknowledge, and 'mushfake'. Against this backdrop, the hidden curriculum (Hafferty 1998) was framed as synonymous with the notion of Discourse and learning by acquisition (Gee 1989). Finally, this discussion led to considering how educators can explicitly teach for acquisition (Gee 1989), by adopting a scaffolded approach (Pea 2004; Airey and Linder 2009). The structure of a scaffolded approach is shown to be useful for developing students' individual modal literacies (Airey and Linder 2009), and their disciplined perception (Goodwin 1994; Stevens and Hall 1998), as the precursors to multimodal fluency and their legitimate membership of the disciplinary Discourse (Gee, 1989; Airey and Linder 2009).

Dense though this discussion may have been, it barely scratches the surface of what could be further explored within these focus areas, and it is certainly inadequate to definitively resolve any of the debates and concerns raised in the literature regarding the impact of digital technologies on biomedical visualisation. However, it does offer valuable insights.

In response to concerns that digital technologies have led to an overuse and overreliance on visual representations (Matthan and Finn

2020), the discussion demonstrates that visual representations, digital or otherwise, are fundamental to the biomedical sciences. Indeed, it is not possible to be over-reliant on visual representations, as no other modality can represent the same details of meaning in the same way. As such, visual representations can only add to, and never replace, the other critical modalities that make up the discourse.

In trying to understand why some students find digital technologies highly valuable (Johnson et al. 2012; Huser et al., 2019; Reinke et al. 2021), when others find them unfamiliar and confusing (Moro et al. 2017; Border 2019; Crow and Murray 2020; Stromberga et al. 2021), the discussion highlights that these student experiences depend on factors such as the students' prior experiences, and the pedagogic strategies employed in their classrooms. If students have limited prior experience of digital technologies, they may not have the appropriate literacies that would allow them to 'mushfake', and so participate. Furthermore, for students to acquire digital literacies, educators need to be explicit in their focus on the use of these technologies, and how they are used to make disciplinary meaning.

Finally, there is the question of ambiguous findings across studies measuring student learning based on digitally mediated teaching interventions (Meyer and Cui 2020; Stromberga et al. 2021). In this regard the discussion positions learning as a process of learning to make meaning with and through complex integrations of multiple, contextually situated, language modes and media. Thus, the findings of existing studies need to be interpreted with respect to students' levels of discoursal fluency, and future research in this area should take care to align assessment practices with a language-centred perspective of learning, by including valid assessments of students' modal literacies and discoursal fluency.

Thus, by applying a multimodal social semiotics perspective to teaching and learning biomedical visualisation, much like a new resource which may afford opportunities for new types of meaning-making, it is hoped that this discussion has offered educators new ways of understanding languages as multimodal, socially situated, the vehicle for meaning-making, and central to teaching and learning.

1.7 Postscript

At the beginning of this chapter, a visual representation (Fig. 1.1) was presented which offered meaning regarding the content of the chapter. Having now engaged with that content, it is useful to consider a few more ideas about how viewers/readers may have made meaning with the image, the meanings that are hidden in relation to the image, and further affordances that the image offered.

If we put context, and thus meaning, aside for a moment, the image is simply composed of a selection of written words, none necessarily more significant than the other. However, situated as this chapter is, within an academic context, those fluent in the discourse will have (almost unconsciously) distinguished the names from the concepts, and interpreted the meaning of these names, especially in association with the sign 'et al', as signifiers of the literature this chapter discussed. This unconscious understanding that allows us to make this meaning, is evidence of our literacy and fluency within the academic Discourse, and an example of the kinds of tacit knowledge that we need to mobilise for students.

Another layer of hidden meaning derives from the resource used to create the image. The WordCloud Generator that was chosen, uses an algorithm that takes frequently associated words into account and treats them as a single term, and then ranks terms according to frequency of use (as a total count for the paper). This quantitative measure is then translated into a visual representation using size and scale, such that the most frequently used terms are the largest and those used less frequently are smaller. Without an explicit understanding of how this resource is used to make meaning, the contextual meaning of font size in the image would have remained hidden. That said, readers may have been able to 'mushfake' to achieve a similar meaning, by drawing on knowledge about the common indexical relationship between size and dominance, such that when things are related to one another in terms of size, larger size is frequently associated with dominance.

Finally, and anecdotally, readers may be interested to know about two more affordances that were derived from this image—one intentionally, and one as a surprise. The initial motivation for making the image was to enable the selection of keywords for this chapter. Specifically, a word cloud generator that applies a frequency algorithm was chosen, in order to see which terms were featured most frequently in the text. In addition to this intentional use, the image also afforded the opportunity to check the alignment between the meanings intended and what had been emphasised in the written text. When the image was first constructed, it was noticeable that the term 'social semiotics perspective' was visually represented in very small text, which was a surprise—the intended emphasis and engagement with this term in the text surely warranting a larger font. Upon closer inspection of the image, it was seen that the term was actually duplicated, and with further investigation it became apparent that this was due to slippage between the use of 'social *semiotics* perspective' and 'social *semiotic* perspective', which could then be refined in the written text. Of course, the slippage of an 's' may be entirely trivial, but the point in this instance is that the image afforded meaning about the text which had not been evident through a reading of the text, and which would not have been evident without an accompanying understanding of how the mode and its resources were being used in integration to make meaning.

References

Airey J, Linder C (2009) A disciplinary discourse perspective on university science learning: achieving fluency in a critical constellation of modes. J Res Sci Teach 46(1):27–49

Bezemer J (2017) Visual research in clinical education. Med Educ 51(1):105–113

Border S (2019) Assessing the role of screencasting and video use in anatomy education. In: Biomedical visualisation. Springer, Cham, pp 1–13

Crow J, Murray JA (2020) Online distance learning in biomedical sciences: community, belonging and presence. In: Biomedical visualisation. Springer, Cham, pp 165–178

Cui D, Chen J, Meyer E, Yang G (2019) Anatomy visualizations using stereopsis: current methodologies in developing stereoscopic virtual models in anatomical education. In: Biomedical visualisation. Springer, Cham, pp 49–65

Erolin C (2019) Interactive 3D digital models for anatomy and medical education. In: Biomedical visualisation. Springer, Cham, pp 1–16

Finn GM, Matthan J (2019) Pedagogical perspectives on the use of technology within medical curricula: moving away from norm driven implementation. In: Biomedical visualisation. Springer, Cham, pp 55–65

Gee JP (1989) Literacy, discourse, and linguistics: introduction. J Educ 171(1):5–17

Goodwin C (1994) Professional vision. Am Anthropol 96(3):606–633

Hafferty FW (1998) Beyond curriculum reform: confronting medicine's hidden curriculum. Acad Med 73(4):403–407

Halliday MA (1993) Towards a language-based theory of learning. Linguist Educ 5(2):93–116

Harrison C (2003) Visual social semiotics: understanding how still images make meaning. Tech Commun 50(1): 46–60

Huser C, Marks L, Linn A, Meek S (2019) Student-created online teaching resources for students. In: Biomedical visualisation. Springer, Cham, pp 37–46

Jewitt C, Kress G, Ogborn J, Tsatsarelis C (2001) Exploring learning through visual, actional and linguistic communication: the multimodal environment of a science classroom. Educ Rev 53(1):5–18

Johnson EO, Charchanti AV, Troupis TG (2012) Modernization of an anatomy class: from conceptualization to implementation. A case for integrated multimodal–multidisciplinary teaching. Anat Sci Educ 5(6): 354–366

Jones MG, Andre T, Superfine R, Taylor R (2003) Learning at the nanoscale: the impact of students' use of remote microscopy on concepts of viruses, scale, and microscopy. J Res Sci Teach 40(3):303–322

Kelly-Laubscher RF, Muna N, van der Merwe M (2017) Using the research article as a model for teaching laboratory report writing provides opportunities for development of genre awareness and adoption of new literacy practices. Engl Specif Purp 48:1–16

Kress GR, Van Leeuwen T (2006) Reading images: the grammar of visual design, 2nd edn. Routledge, London, England

Kress G, Charalampos T, Jewitt C, Ogborn J (2014) Multimodal teaching and learning: the rhetorics of the science classroom. Bloomsbury Publishing

Kress GR, Van Leeuwen T (2020) Reading images: the grammar of visual design, 3rd edn. Routledge, London, England

Latour B (1986) Visualization and cognition. Knowledge Society 6(6):1–40

Lea MR, Street BV (1998) Student writing in higher education: an academic literacies approach. Stud High Educ 23(2):157–172

Lemke J (2000) Multimedia literacy demands of the scientific curriculum. Linguist Educ 10(3):247–271

Lemke JL (2004) The literacies of science. In: Crossing borders in literacy and science instruction: Perspectives on theory and practice, pp 33–47

Matthan J, Finn GM (2020) The hidden curriculum of utilisation of imaging and unregulated digital resources within clinical education. In: Biomedical Visualisation. Springer, Cham, pp 145–163

McMenamin PG, Quayle MR, McHenry CR, Adams JW (2014) The production of anatomical teaching resources using three-dimensional (3D) printing technology. Anat Sci Educ 7(6):479–486

Meyer ER, Cui D (2020) Anatomy visualizations using stereopsis: assessment and implication of stereoscopic virtual models in anatomical education. In: Biomedical Visualisation. Springer, Cham, pp 117–130

Moro C, Štromberga Z, Raikos A, Stirling A (2017) The effectiveness of virtual and augmented reality in health sciences and medical anatomy. Anat Sci Educ 10(6): 549–559

Muna N, Paxton M, Frith V, van der Merwe M (2020) Unpacking the multimodal discourse demands of microscopy in a bioscience laboratory. Int J Sci Educ 42(16):2742–2764

Pea RD (2004) The social and technological dimensions of scaffolding and related theoretical concepts for learning, education, and human activity. J Learn Sci 13(3):423–451

Preece D, Williams SB, Lam R, Weller R (2013) "Let's get physical": advantages of a physical model over 3D computer models and textbooks in learning imaging anatomy. Anat Sci Educ 6(4):216–224. https://doi.org/10.1002/ase.1345

Reinke, N.B., Kynn, M. and Parkinson, A.L., 2021. Immersive 3D experience of osmosis improves learning outcomes of first-year cell biology students. CBE—life sciences education, 20(1), p.ar1

Rybarczyk B (2011) Visual literacy in biology: a comparison of visual representations in textbooks and journal articles. J College Sci Teach 41(1):106

Singh R, Yadav N, Pandey M, Jones DG (2022) Is inadequate anatomical knowledge on the part of physicians hazardous for successful clinical practice? *Surg Radiol Anat* 44:1–10

Stevens R, Hall R (1998) Disciplined perception: learning to see in technoscience. In: Talking mathematics in school: studies of teaching and learning, pp 107–149

Stromberga Z, Phelps C, Smith J, Moro C (2021) Teaching with disruptive technology: the use of augmented, virtual, and mixed reality (HoloLens) for disease education. Adv Exp Med Biol 1317:147–162

Watson FL, Lom B (2008) More than a picture: helping undergraduates learn to communicate through scientific images. CBE—Life Sci Educ 7(1):27–35

Reasons for Knocking at an Empty House: Visualisation, Representation and Dissemination of Health-Related Public Engagement Media

2

Ed Young and Anastasia Koch

Abstract

Stigma and mistrust generate significant barriers to the uptake of biomedical, clinical and public health measures to combat infectious diseases. Many pandemics such as HIV, TB and COVID-19 disproportionately affect poorer communities, and the social and public health impact is connected through socio-political histories and contexts. This chapter describes activities and reflections of a South African public engagement programme, Eh!woza, that aims to bring together the biomedicine of disease with its social context and impact. We describe experiences working on tuberculosis-related public engagement programmes in South Africa, and how these approaches were refocused to address the COVID-19 pandemic. We reflect on the lesson learned and considerations around visualising the social impact of disease and making the visualisation of accurate information relatable to younger audiences. Much of the discussion is situated within description and reflection, touching on both the historical and contemporary cultural and political conditions in which infectious diseases have flourished. Finally, the challenges we faced when effectively disseminating media on large-scale digital platforms are highlighted, raising important questions around representation, mass targets, and impactful dissemination of public engagement outputs.

Keywords

Public engagement · Social determinants of health · Stigma · Dissemination · Health-seeking behaviour

2.1 Introduction

As a discipline that is relatively new when considered in the context of more established biomedical and health research fields (Weingart et al. 2021), public engagement could be considered an umbrella term that broadly describes activities that aim to include the perspectives, wishes and opinions of non-specialist groups in health research as a mechanism to achieve greater impact and innovation (Tembo et al. 2021). Participatory and co-production methods feature heavily as tools for effective public engagement (Tembo et al. 2021). The Eh!woza programme was initiated at the end of 2013 through an informal collaboration between a group of scientists, artists, an extracurricular tutoring non-profit

E. Young (✉) · A. Koch
Eh!woza, Cape Town, South Africa

SAMRC/NHLS/UCT Molecular Mycobacteriology Research Unit & DST/NRF Centre of Excellence for Biomedical TB Research, Department of Pathology, Faculty of Health Sciences, University of Cape Town, Cape Town, South Africa
e-mail: ed@ehwoza.com

L. Shapiro, P. M. Rea (eds.), *Biomedical Visualisation*, Advances in Experimental Medicine and Biology 1388, https://doi.org/10.1007/978-3-031-10889-1_2

organisation (NPO), and young people from Khayelitsha, an informal settlement in South Africa with high rates of tuberculosis (TB) and HIV (Moyo et al., 2015). The name of the organisation loosely translates from isiXhosa as, "Hey! Come with us". Over the next 8 years, the project was developed to include multiple public engagement activities (Table 2.1) with co-production as a cornerstone. In 2020, the organisation transitioned from an informal collaborative grouping within the University of Cape Town (UCT) into an independent NPO.

Eh!woza aims to address stigma and misconceptions around TB and HIV (and more recently, COVID-19), by bringing together the biomedical and social attributes of infectious diseases. The first and oldest programme, Eh! woza Doccies (Table 2.1), engages high-school learners from Khayelitsha (an area in Cape Town that is simultaneously heavily impacted by infectious disease and poverty) with biomedical research and provides the opportunity for learners to reflect on the social impact of disease via the production of short documentary films (Young et al. 2018). In doing so, the programme facilitates the co-production of knowledge (Masuku et al. 2018) and simultaneously provides learners with access to an environment where high-impact biomedical research is being conducted, stimulates discussion around disease and health, and improves the distribution of accurate knowledge.

Learner-produced documentary films investigate and depict the impact of disease in neighbourhoods where participants and their families live—in this way, reflecting the lived experience of people impacted by disease. The video teaching model was adapted from a third-year level module at the UCT's Michaelis School of Fine Art and adjusted for high-school learners in disease-affected areas. The result is a series of videos about TB and its social impact that have been produced by 14–17-year-old high-school learners from South Africa's townships (Figs. 2.1, 2.2, and 2.3). The programme represents a multi-disciplinary effort and equally values intellectual and practical input from all contributing stakeholders.

In this chapter, we describe some of the observations made during the public engagement work, with a particular focus on visibility and visualisation within the sphere of public engagement with biomedical research. While considering the current socio-political landscape within South Africa and how it is situated globally, the chapter will draw on reflections of experiences with participants of the programmes, collaborating organisations and describe some of the media produced. Finally, we raise questions around how health-related media may be used to address stigma and misinformation in a way that is genuine and has the most significant impact on target audiences.

2.2 Not a Drama Queen: Setting the Scene for the TB Problem in South Africa

"TB is not a drama queen", declared the then Minister of Health for South Africa, Dr. Aaron Motsoaledi. He was addressing an audience at the Faculty of Health Sciences, UCT 2019 during the third annual Steve Lawn Memorial Lecture (Bernardo 2019). Dr. Motsoaledi was describing, as many had before him, TB as a silent predator that garners noticeably less attention than some of its visceral counterparts such as Ebola, HIV/AIDS (Bernardo 2019) and COVID-19. In what Vidya Krishnan labels as a "phantom plague" (Krishnan 2022) and because of its stealthy character, TB has quietly terrorised South Africans for at least two centuries. It is likely that TB was first introduced into South Africa in the 1700s with the arrival of Europeans and significantly increased in incidence in the 1900s with the expansion of the mining industry (Karim 2009). It primarily impacts poor and overpopulated areas that are at the same time affected by a preventable lack of access to clean water and sanitation (Karim 2009).

The social reality of TB has remained relatively hidden to those dedicating their lives to exploring and understanding the biomedical basis of the disease, potentially due to disinterest

Table 2.1 Core Eh!woza public engagement programmes

Programme name	Start year	Key participants/target audience	Brief description
Eh!woza Doccies	2014	14–17-year-old high-school learners, residents of Khayelitsha and other South African townships; general public.	Eh!woza's flagship project recruits a group of 12–15 high-school learners to an annual intensive workshop programme that combines immersion in state-of-the-art biomedical research with documentary film production. Material produced by learners is used to reach a broader youth audience in schools.
MSF/Muso/ DR-TB Collab	2018	DR-TB survivors, young people in Khayelitsha and other South African townships.	This collaboration with the global humanitarian organisation MSF facilitates storytelling between survivors of drug-resistant TB (DR-TB) and young Khayelitsha-based musicians. The aim is to produce music, music videos and other creative outputs reflecting the experience of having and surviving DR-TB.
Eh!woza Schools	2019	School going youth in Khayelitsha and other South African townships.	Eh!woza schools' programme aims to use Eh!woza-generated media to widely disseminate knowledge through school attending youth and stimulate robust discussion around disease and the social determinants of health.
SciComm	2020	General public (youth-focused).	Compelled by the Covid-19 pandemic, Eh!woza established a SciComm arm to disseminate accurate and visually engaging biomedical information. While social media users have been a key audience, our data guided us to expand into local newspapers, WhatsApp and radio.
Alumni Doccies	2020	Eh!woza alumni trainees, policymakers, general public.	We train previous participants in high-level documentary production. The Khayelitsha-based team has produced over 15 documentaries investigating the impact of COVID-19 in informal settlements. This contextually relevant storytelling highlights the intersection between health and social struggle and uncovers the human reality of disease.

in the lives of those who are exposed to the disease and generally overshadowed by a primary focus on laboratory or clinical research. Lived experiences can become blurred, fictional and less urgent. In recent years, international science-funding bodies aimed to address these challenges and boost public engagement between scientists and affected communities, some strongly encouraging or mandating public engagement in grant applications. In some cases, this endorsement of engagement has been met with disinterest when seeking a piece of the global biomedical research funding share. It is fair to note that there has been an increased interest among younger scientists, although it is difficult to gauge how much appeal is sincere or resumé-driven, aimed at job and funding applications. Apathy towards the social conditions of disease can stifle scientific research if it causes resistance to genuinely and authentically engaging with humans living with treatable diseases. Failure to produce a sustainable roadmap to address stigma and misinformation can render diseases invisible and allows for stigmatised people to be alienated by peers and families due to often treatable infections (Kelly 1999) and could conceivably result in the risk of higher mortality (Drain 2015).

Fig. 2.1 A high-school learner recruited to the *Eh!woza Doccies* programme from extracurricular tutoring programme, IkamvaYouth, collecting interviews on a shoot day | Photo: Eh!woza

A failure to address mistrust among those affected by disease and disengagement from our scientists and political leaders results in scepticism of scientific and medical achievements and difficulty in overcoming the reluctant uptake in life-saving treatment. As observed during COVID-19 vaccine distribution during 2021, there is limited value in developing medical

Fig. 2.2 Eh!woza trainee Alfa Fipaza, guiding *Eh!woza Doccies* participants in camera work | Photo: Eh!woza

Fig. 2.3 *Eh!woza Doccies* participants editing a film about an elderly Khayelitsha resident's experience of TB | Photo: Ed Young, Eh!woza

interventions if large pockets of humanity remain suspicious of its safety and efficacy.

Addressing social hurdles as a principal contributor to epidemics is challenging due to a lack of urgency. A successful example of activism that brought such challenges to the surface is Zackie Achmat's Treatment Action Campaign's (TAC) interventions against the South African government's HIV policies in the early 2000s and lack of access to antiretroviral treatment—a response to an estimated 330,000 premature and preventable deaths as a result of politicised AIDS denialism (Chigwedere et al. 2008).

While sharing parallels with HIV, TB is closely linked to social conditions and more dramatically affects the poor. TB did not and still does not enjoy the vigorous levels of activism that HIV did. TB is an ancient disease, thought to have co-evolved with humans, moving out of Africa with the migration of the earliest modern humans (Comas et al. 2013). TB's more recent historical and current socio-political knots are more complex. Apartheid's forced removals as a method of purging the inner cities under the Group Areas Act (effectively rendering townships "dumping grounds" to serve primarily as a labour resource to the inner cities and surrounding industrial hubs) has been a significant contributor to overpopulation, lack of services, and in turn, the distribution of diseases such as TB (Karim 2009). While displacement predates apartheid to late colonial undertakings such as Cecil John Rhodes' Glen Grey Act adopted in 1984, which introduced a labour tax to force-feed migrant labour to the mining and agricultural industries (Dennis 2015), Richardson and others suggest forced removals as a significant contributor to the Cape Town TB epidemic:

> One underexplored determinant [contributing to TB] has been forced removals, that is, the policies and often violent processes involved in the massive, state-sponsored displacement of people (almost all of them black) from one area to another in South Africa. Starting well before *apartheid* with the Public Health Act of 1897, the Native Reserve Location Act of 1902, and the Native Urban Areas Act of 1923 – and consolidated in the Group Areas Acts of the 1950s and 60s – forced removals uprooted millions of individuals from both developed urban and rural areas to underdeveloped and poorly resourced peri-urban and rural areas

resulting in widespread poverty, disease, and starvation (Richardson et al. 2016)

Also predating the Group Areas Act, and a significant contributor to removals and segregation, was the Natives Land Act of 1913. In his recent book, *1986*, William Dicey says that "Africa looks briefly at history, in particular the Natives Land Act of 1913, whereby 86.3 per cent of the land was given to whites and only 13.7 per cent to blacks". Dicey goes on to quote the *Eminent Persons Group's Mission to South Africa: The Commonwealth Report* (1984), saying that "Thus, a predominantly black country becomes predominantly white, and the black becomes an alien in his own land" (Dicey 2021). This echoes what the ANC (then South African Native National Congress) founding member Sol Plaatje arrived at seven decades earlier that "Awaking on Friday morning, June 20, 1913, the South African Native found himself, not actually a slave, but a pariah in the land of his birth" (Plaatje 2016).

The impact of South Africa's complex history with forced segregation is coupled with the lasting effects of inadequate healthcare provision, large-scale misinformation and lack of access to accurate health messaging. Studies applying population genomics and modelling have been able to demonstrate that the collapse of the Soviet Union resulted in increased rates of drug-resistant TB in the region and that further displacement as a result of armed conflict allowed drug-resistant strains of TB to spread within central Asia (Eldholm et al. 2016). While population genomics has never been applied to the South African TB epidemic, there is likely a causal link to the concentration of disease in low-income areas, largely hidden from public view due to apartheid's spatial design (Karim 2009). With an incidence of 554 per 100,000 people in 2020 (WHO 2021), TB has only recently been surpassed by COVID-19 as the leading cause of death in South Africa. TB rates are heterogenous within South Africa (Osman et al. 2021), with higher rates in areas that are densely populated and exist essentially unchanged from apartheid-style segregation (Fig. 2.4).

The advent of democracy in South Africa did very little to relieve such burdens. As much as one may want to believe that the apartheid state was overthrown solely by a struggle movement, the reality of economic pressures and neoliberal interests played substantial roles in shifting political power, while the perseveration of the economic status quo played a significant role (Wilderson III 2015). The Afrikaner State did not grow a conscience overnight, and neither can the current government—in much the same way that Dr. Eric Williams noted that the British Empire abolished slavery in 1833 when West Indian slavery became an economic nuisance because of taxation, and not because they thought it was the progressive thing to do (Williams 1944). South Africa's high-income sectors have remained largely unaffected. Apart from a minor shift of wealth to a black middle and upper class, today, the country's poor remain destitute. South Africa's existing townships, a carbon copy of apartheid's segregational design, bear the brunt of society's catastrophic social, economic and health crises.

2.3 Eh!woza: The Mechanisms and Growth of a Public Engagement Programme

2.3.1 Core Eh!woza Programmes

Over the past 8 years and through various programmes (Table 2.1), our organisation has focused on developing visual media through its public engagement programme by participants from affected townships heavily burdened by disease and stigma. Access to accurate visual health information is noticeably deprioritised by the state. This is likely because of the indifference, irregular expenditure, and apathy among some members of the state. As formative research when initiating our programme in 2013, we asked participants what they knew and thought about TB and more specifically, what caused it. The response was unanimously almost always that TB was caused by smoking and dust; a

Fig. 2.4 The border between a Strand suburb and an informal settlement near Nomzamo, Cape Town | Photo: Samuel Flans, Eh!woza

misinformed response by learners who understood risk factors to be the cause of disease while unaware of its actual source and mechanisms of transmission.

Little has changed to improve knowledge around infectious diseases within the school curriculum. When asking the same question about TB while recruiting a new cohort of high-school learners in 2021, during a pandemic spreading in much the same way, the answer remained the same. Given the global impact of COVID-19 and coverage by mainstream media, most learners knew what caused COVID-19 and how the disease spread. This highlights the need to circulate accurate information around less "attention-seeking" diseases like TB and make this information relevant and relatable. It is essential that visual media produced is of the highest standard and that storytelling is well developed and as riveting as possible, even when made by 14–17-year-old high-school learners. TB is a characterless topic, and this makes it one of the most dangerous diseases globally. How do you make something uninteresting interesting? How do you make something boring visible in an attempt to decrease loss of life? These questions underpin many of Eh!woza's programmes and speak to some of the core aims of the programme.

As described in the introduction, our core programme, Eh!woza Doccies, combines exposure and engagement with cutting-edge TB biomedical research with the facilitation of documentary film production. In 2018 we implemented a key expansion within our programme. While producing a new set of high-school documentaries, we asked learners to survey Khayelitsha for local young musicians they might know of, who would be interested in providing original music for the learner projects. This was an attempt to secure licensing of soundtracks for the films made during Eh!woza Doccies while simultaneously promoting young musicians on our growing social media platforms. It was a simple idea and it worked. The music was of high production value, and the musicians were keen to be part of the project. As the programme progressed, some artists asked if we were able to assist with shooting music videos as we had access to camera equipment. We agreed in principle, and it appeared to be a valuable trade exchange for music rights. However, the videos were perpetually delayed due to a lack of production capacity, and we conceptualised a way to formalise the project.

We had intended to work with the Khayelitsha mission of Médecins Sans Frontières/Doctors Without Borders (MSF) for some time. MSF had developed support groups for young multidrug-resistant TB (MDR-TB) sufferers and survivors from Khayelitsha. Their project was in its initial phase and existed simply as a WhatsApp-based support group at the time. The group platform aimed to address MDR-related concerns for peers and facilitators while also providing a safe place and emotional support for the group enduring difficulties. We proposed a way to collaborate with MSF on the existing MDR-TB support group project and developed a programme where survivors and patients' personal stories could be shared to influence and inspire the production of music videos for musicians. The MSF/Musos/DR-TB programme (Table 2.1) was developed over an extended period in direct collaboration with the musicians and MSF. The outcome was intended not to be a literal representation of these stories in a music video format but instead more abstract and based on sentiments raised and developed during group sharing sessions. We filmed a pilot video, Innocent Muziq's *Single Parent* (Fig. 2.5), and used this as a foundation to approach the Wellcome Trust to support the project.

The programme started with approximately 13 MDR-TB participants and 6 musicians. Members of the WhatsApp group who were sputum smear-negative, and thus thought to be no longer transmitting TB, attended in-person workshops. Although there was sufficient enthusiasm, the project was challenging as participants carried complex histories, and it was hard to open up to strangers. In conversation with the participants, we trialled various approaches and eventually removed the sterile UCT meeting rooms and replaced them with picnics, trips to the movies, swims at the beach, "braais" (BBQ) at the studio and other more social events.

We discouraged English as primary workshop language, limiting intervention from non-isiXhosa speaking facilitators and creating an intimate environment through which to share experiences. Stories were developed in isiXhosa, the participants' home language and South Africa's second-largest language following isiZulu, generally spoken in Cape Town and the Eastern Cape region. The outcomes were reasonably successful as musicians developed personal stories that were close to home. The participants' experiences did not only focus on the difficulties of the disease itself but on associated social determinants such as living conditions, financial and other hardships faced during and before they were affected by this life-threatening disease.

There were other creative outputs encouraged during these workshops. Poems were developed around TB, mental health and other social burdens, and two additional music videos were produced prior to the COVID-19 pandemic restrictions. In the end, the music used for the videos had little to do with TB, although the visual elements had been influenced by experiences shared during the workshops.

Abospitter's Nquml'umqala (Fig. 2.6) loosely translates to "I will cut your throat". The lyrics are an archetypal message from this rapper to rappers globally that "he is definitely better than them"—petty rap battle-lyrics but in isiXhosa. We developed the visuals during the MSF workshops, and Abospitter was fixed on portraying someone being hunted while nothing was chasing them. Abospitter introduced a character who was not physically trailing the lead but omnipresent, a ghost-like antihero clad in thick plastic with his face covered with traditional beads. When discussing the concept with Abo, he was unsure where the figure emerged from but entertained the idea that the silent killer may have had its origins in the MDR-TB workshops. The idea was abstract and developed at the time of the workshops and potentially an indirect visualisation of the disease and circumstances of their discussions.

Fig. 2.5 *Single Parent*. Innocent Muziq, Eh!woza | Video Still

We developed various tracks with Abospitter and settled on producing a second video he had developed through the MSF project. The concept for *Diminished* was taken more directly from the MDR-TB workshops and looked at addressing issues around mental health and youth suicide. While Abo has completed most of the research around this project, the pandemic has influenced the timeline for bringing this project to fruition, but the intention is to achieve this at a later date.

We developed a second music video for Innocent Muziq, and although *Togetherness* (Fig. 2.7)

Fig. 2.6 *Nquml'umqala*. Abospitter, Eh!woza | Video Still

Fig. 2.7 *Togetherness*. Innocent Muziq, Eh!woza | Video Still

had little to do with the MSF project, apart from Innocent being a participant, we decided to proceed as the track had a strong hook that young people related to well. We produced this video to elevate Eh!woza's credibility among youth in Khayelitsha even further. *Togetherness* was released at the start of COVID-19 lockdown restrictions in South Africa and was one of our most highly watched media products. Ironically, many people who did not understand the language assumed it was a song about the lockdown due to the title, and because it was produced by an infectious disease public engagement organisation.

The experiment of circulating visually powerful videos produced by musicians who were well-known in Khayelitsha with a strong reputation and fanbase was successful and increased the awareness of Eh!woza within the area. By associating the visual impact of music videos and the cultural/social capital of the musicians with a project related to TB, we were able to make TB and infectious disease generally more interesting, albeit in an indirect way.

2.3.2 Situationally Relevant Capacity Development

The importance of the MSF/Musos/MDR-TB project for Eh!woza's growth began to develop ways that differed from our initial aims for the project. The association with young creatives and musicians, no longer in school with more time to produce independent projects, paved the way for developing an essential branch in our organisation. During the 18 months that the MSF project ran, we began implementing an informal training programme at our studio, where workshops are held. We reallocated funds to offer travel costs to and from Khayelitsha daily, and supplied basic daily sustenance while the musicians worked in the studio space. The initial idea was to provide the camera equipment, edit suites, music and other creative software together with a physical space as well as creative guidance to develop music videos and other creative projects. Our thinking was that we would develop concepts over this time, simultaneously training the young creatives, and actually film the projects as the ideas became ready and as production capacity permitted.

We were experimental and non-prescriptive in our approach. The initial training focused on talent and skills development, while the longer-term outlook was to link the work to health. Creative projects such as music videos, comedy skits, and the development of a web series began to emerge. When mentioning this to one of the founding directors of our NPO, Professor Digby Warner, Deputy Director of the Molecular Mycobacteriology Research Unit (MMRU) (Faculty of Health Sciences, University of Cape Town), he seemed enthused and envisioned a school with youth from Khayelitsha working with science and film with drones flying about (one drone had already been purchased to facilitate the production music videos mentioned above). This is precisely what we did and teasingly began referring to the project as "Hogwarts".

We trained the young creatives in camera work, lighting, drone operation and camera stabilisation, filming, photography and sound recording. At the start of the project, there were five musicians, but only Innocent Muziq (Samuel Flans) and Abospitter (Abongile Njamela) remained due to changes in the musicians' commitments and time available to contribute to the project. Later, they were joined by Eh!woza Doccies alumni, Alfa Fipaza, and the three formed the core group of trainees. The principal idea was that the trainee group would ultimately become equipped to film documentaries and music videos independently, while creating music in the studio for personal distribution as well as to be used in documentary films that high-school learners would eventually produce in our Eh!woza Doccies programme (Table 2.1).

An essential forecast of the programme was that the trainees reach a level of independence and eventually take a key role in creative facilitation and tuition during the high-school learner project, *Eh!woza Doccies* (Table 2.1). This vision was stunted by the pandemic's arrival, but again, the initial idea and investment became apparent in a way we had not anticipated—becoming crucial during our efforts to address COVID-19 while our in-person workshop programmes were paused at various levels of COVID-19 restrictions in 2020 and 2021.

2.4 We Pulled Ourselves up by the Bootstraps. How Come You Don't Have Any Boots?

2.4.1 How the COVID-19 Pandemic Exposed and Exacerbated Existing Inequalities

2.4.1.1 Films Uncovering Fears Associated with a New and Unknown Disease

There is a joke: "In the beginning, the Lord said, 'Let there be light'. And there was still nothing. Only now you could see it".

At the start of the pandemic, a necessity emerged for us to distribute critical information in isiXhosa as almost all information about the new disease was distributed in English at the time and not accessible to many communities in local languages. Our aim was to address low levels of trust in government-mandated directives to implement regulations such as physical distancing, mask-wearing and later, the mixed sentiment around the use of vaccines in Cape Town's townships. When visualisation of science information fails to address audiences or is met with significant backlash due to disinformation and distrust in the state, these efforts are defeated and lost on a considerable part of the intended audience.

By the time the pandemic arrived and we were asked to shelter in place, it became apparent that we were losing a year or more of our core high-school programming. Finding our feet in a global crisis, we pivoted and produced Public Service Announcements (PSAs), relaying global messaging from the World Health Organisation (WHO) and other reliable sources. We translated these messages into isiXhosa and had them broadcast over Cape Town TV, a local TV station with a monthly viewership of 2.7 million and viewed primarily by Cape Town's township residents (Cape Town TV n.d.). It was the beginning of the pandemic and an unfamiliar time. The PSAs were delivered by one of our previously mentioned filmmakers in training, Samuel Flans (Innocent Muziq), who lives in Khayelitsha and is a minor celebrity known as a musician, actor and radio DJ, delivering difficult messaging gently but to the point, at a time when very little was known about the novel pandemic.

Being nationally mandated to work from home, we equipped our team with filmmaking gear to explore sentiments about COVID-19 and potential misconceptions that could be fear and reluctance to adhere to non-pharmaceutical interventions in Cape Town's townships, as at the time, there was little investigation around this (Fig. 2.8). The group was issued DJI Osmo Pocket cameras, extremely discreet filming equipment that could shoot at high resolutions and frame rates, with a 3-axis stabilisation system, to create visually impactful films high in production value. We obtained the necessary permits and paperwork to move around the townships responsibly and provided safety training and resources (masks and sanitiser). Many films were made during this time and are described in a recent book chapter, *My Rhodes has no nose: COVID-19 and the two Cities of Cape Town* (Young & Koch 2021). However, a few relevant examples are described below.

COVID FEARS expanded issues raised in a pilot film about COVD-19, *MAKHAZA 2020*. It was produced when restrictions, such as physical distancing, mask mandates and self-isolation, were first implemented. As the lockdown confined the city's residents to their respective homes and apartments, people in townships were required to adhere to similar instructions, often restricting entire families to one-roomed houses, enforced by a visible military presence. Standard COVID-19 regulations seemed unreasonable in congested areas, but the state had offered few alternatives or solutions to residents living in overcrowded environments. Hand sanitiser stocks were depleted at city shopping facilities due to panic and bulk buying. At the same time, these were not readily available or affordable in townships to begin with, and access to soap and running water was not available in several informal settlements. Lack of access made it difficult to comply with handwashing recommendations, and as the virus was believed to spread through physical contact at the time, this increased fear and resentment among residents of townships.

COVID FEARS intended to provide a visual example of such anxieties and address inconsistencies in the state's mostly monolithic approach in combatting the virus. Where mandates and messaging were sometimes unclear, misinformation found fertile ground and became widespread and exacerbated via word of mouth. In response to uncertainty regarding the spread of the disease, a young woman in the film explains that "At least with AIDS you know how you get it" (Fig. 2.9).

2.4.1.2 Films Documenting the Pre-existing Housing Insecurity Exacerbated by COVID-19

Fears intensified and began integrating with reality as the lockdown prolonged. The extended duration was unanticipated, and the reality of job insecurity and loss of income deepened in areas where unemployment rates were already unreasonably high. Lack of income resulted in the inability to afford rent and basic provisions, and numerous backyard dwellers in Khayelitsha unlawfully occupied land belonging to the City of Cape Town. Many residents became reliant on the nationally implemented COVID-19 relief grant of ZAR350 (USD22) per month, while erecting informal structures on the city-owned land as means of survival. Felix Dube and Anél du Plessis estimate that The City of Cape Town, at the time the only major metro run by SA's Democratic Alliance Party (whereas the ANC maintains national government control), demolished over 60,000 informal houses during the lockdown.

Under the Prevention of Illegal Eviction from and Unlawful Occupation of Land Act 19 of 1998 (PIE), South African law defines an "unlawful occupier" as, inter alia, "a person who occupies land without the express or tacit consent of the owner or person in charge, or without any other right in law to occupy such land". PIE imposes stringent measures for the eviction of unlawful

Fig. 2.8 Eh!woza documentary film thumbnail examples

occupiers and stipulates that such evictions can only occur when authorized by a court after a hearing and only at least 14 days after notice has been given to the unlawful occupiers. When eviction of unlawful occupiers is sought by an organ of state, the court must consider the availability of suitable alternative accommodation or other land to which the people will be moved. Thus, a lawful eviction is one that is authorized by a court and carried out in accordance with that court order. However, the COVID-19 pandemic saw a sharp rise in the number of unauthorized evictions in South Africa. The extent of the evictions during the pandemic prompted the government to promulgate regulations to protect unlawful occupiers from eviction during lockdown (Dube & Du Plessis 2021).

Our documentary filmmaking efforts and those of other local press organisations began to expose that during winter, the City carried out evictions without the necessary court orders or alternate

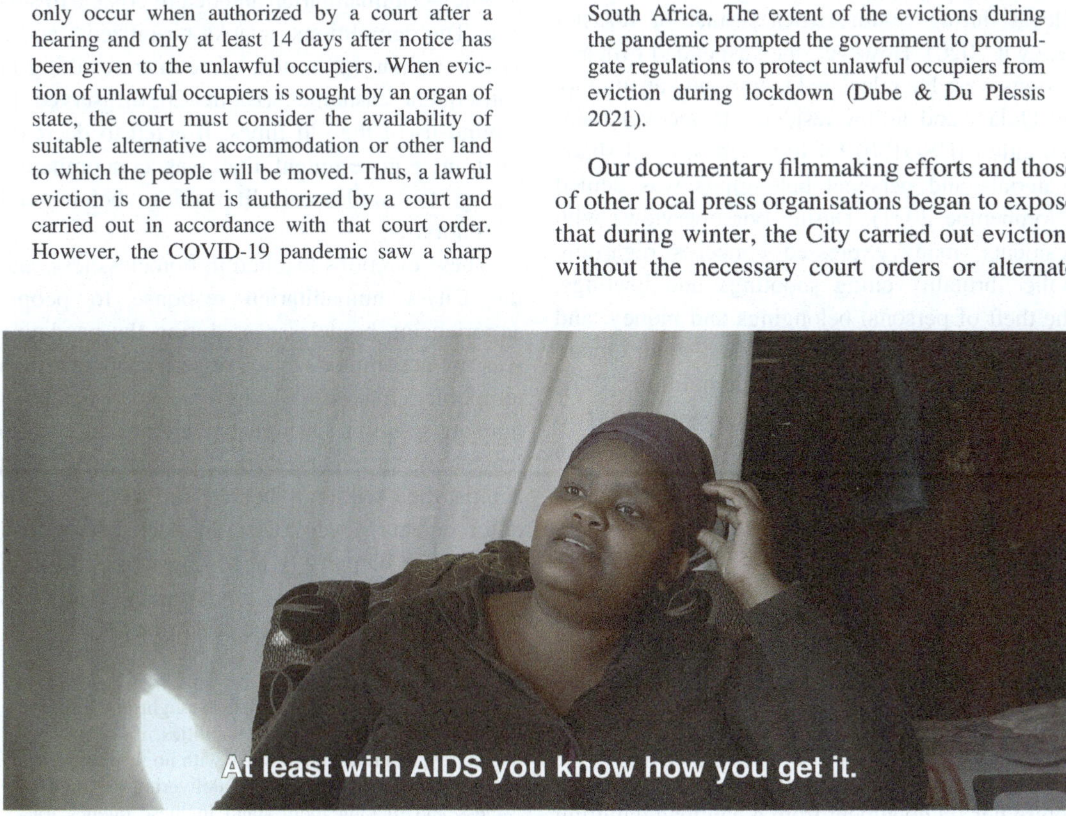

Fig. 2.9 *COVID fears*. Samuel Flans and Alfa Fipaza, Eh!woza | Video Still

solutions, subjecting many citizens to homelessness during the lockdown period. Residents' housing structures were demolished, and often the shacks would be rebuilt at night after law enforcement had vacated the area. Officers were accused by residents of assault with rubber bullets while tearing down shacks and confiscating building materials to avoid a resurfacing of unlawful structures. Evictions went largely unnoticed by the inner-city's middle class and more well-to-do, but garnered some attention after cell phone footage emerged of Bulelani Qolani being dragged out of his shack naked while having a bath (Flans 2019). The footage was widely distributed and viewed online and depicts Qolani hauled from his shack and wrestled by police officers as his home is torn down in seconds (News24 2020a, b).

In a case brought by the South African Human Rights Commission and civic group Housing Assembly, Qolani's argument was heard, and the City of Cape Town's evictions were declared illegal during South Africa's state of disaster (Kiewit 2020). However, the suspended officers' actions were later declared lawful, and an attempt by Qolani and fellow residents to recover compensation (USD130) for lost or destroyed shack materials and personal belongings was denied (Nombembe 2021). During our interviews with residents, many expressed concerns regarding police brutality citing shootings and beatings, the theft of personal belongings and money, and instances of urinating on residents' food. Qolani recalls an incident prior to the demolitions: "The City of Cape Town, they only gave us only two minutes. 1,2,3,4,5... They were standing there in Baden Powell Road. They shot us with rubber bullets, and we ran" (Fig. 2.10).

Produced by our filmmaking crew in training, *THE ANTI-BLACKNESS OF COVID EVICTIONS*, exposes additional instances of police brutality during the City's demolition campaign. The film reveals how a pregnant woman clashed with police officers, begging for her belongings not to be destroyed while her home was being torn down. She pleaded with police to return her ID document from a handbag dangling from the confiscated materials as she needed proof of identity to give birth. Shack theatre founder Mandisi Sindo was arrested and jailed temporarily for attempting to assist her. Destroying structures while occupiers were rebuilding, police also targeted children in occupied areas. Depicted in Fig. 2.11, a 15-year-old teenager recalls that:

> While we were running, they shot me here behind the ear with a rubber bullet. I got into another shack with another man. I sat down and then I heard one of my friends Sikho crying, 'I'm just a child... I'm just as a child.' 'Please, please, no!' They didn't let him go. I came out of the shack. When I got out, they were dragging him by his T-shirt. Others (law enforcement) were beating him

As described by an interviewee in the film, these accounts are reminiscent of security forces terrorising and detaining residents during the state of emergency of apartheid during the 1980s and prior—an indication that we do not need a nefarious apartheid state to affect brutality on exposed people—a humanitarian and health crisis will suffice. The demolitions, attacks on citizens, and a visible military presence, sometimes resulting in consequent casualties (Bailie 2020), served to remind us of the old times. It acted to decrease trust in a government that was concomitantly trying to implement life-saving behavioural regulations.

These evictions resulted in homelessness, and the City's humanitarian response to people experiencing homelessness during the pandemic was to accommodate unhoused residents in a temporary makeshift camp as an emergency housing solution. Occupants were not allowed to leave, and it emerged that people were kept against their will and "herded like cattle" by law enforcement (Meyer 2020). The secretary-general of The GOOD party (a newly formed opposition party contesting local government elections), Brett Heron, described the conditions:

> The whole situation is poorly thought-out. It's an inappropriate site, with no facilities, it's freezing, muddy, and tents are desolate with no furniture and no ground sheets. Buses are delivering the homeless and making them stand in these queues for testing as if they are cattle (Heron in Meyer 2020).

Fig. 2.10 *COVID evictions.* Samuel Flans and Alfa Fipaza, Eh!woza | Video Still

After 2 months, the Strandfontein project was dissolved as a result of the inhumane living conditions, during which detained individuals clashed with law enforcement and reached out to civil society to intervene. During a subsequent probe by the Special Investigating Unit, R42 million ($2.7 million) in irregular expenditure relating to hiring marquee tents for the camp was uncovered (McCain 2022). Coupled with the severe cost of evictions, the unfortunate treatment of humans may have been better substituted with developmental and more permanent housing solutions intending to integrate homeless people back into society. The City has a questionable track record of responding to the poor and unhoused, with similar strategies dating back to the 2010 FIFA World Cup and before (Kelto 2010).

When discussing unwarranted violent behaviour towards the City's exposed citizens, Frank B. Wilderson III's Afropessimism provides a theoretical underpinning to these kinds of

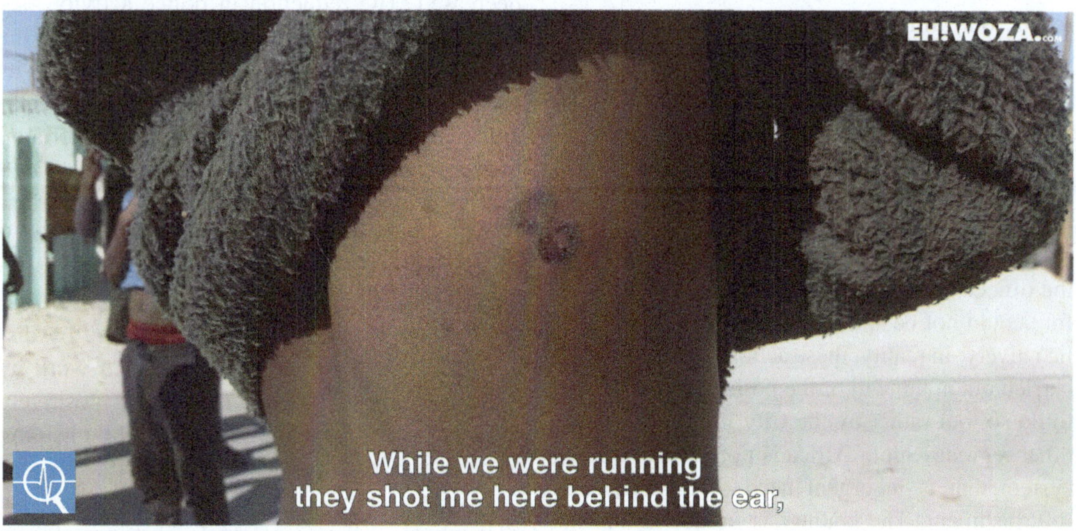

Fig. 2.11 *The anti-blackness of covid evictions.* Samuel Flans and Alfa Fipaza, Eh!woza | Video Still

actions. Wilderson proposes that black society is integral to society as a whole, but merely for the purpose of "human renewal" (white and non-black counterparts), in the form of "gratuitous" and perpetual anti-Black violence, deprived of human capacity.

> Why is anti-Black violence not a form of racist hatred but the genome of Human renewal; a therapeutic balm that the Human race needs to know and heal itself? Why must the world reproduce this violence, this social death, so that social life can regenerate Humans and prevent them from suffering the catastrophe of psychic incoherence—absence? Why must the world find its nourishment in Black flesh? [...] Unlike violence against the working class, which secures an economic order, or violence against non-Black women, which secures a patriarchal order, or violence against Native Americans, which secures a colonial order, the jouissance that constitutes the violence of anti-Blackness secures the order of life itself; sadism in service to the prolongation of life (Wilderson III 2020).

Wilderson cites Orlando Patterson's thoughts on "social death", suggesting that a segment of society cannot be acknowledged as human by the remainder of that society (Patterson 1982).

Through one of our media partners, the Bhekisisa Centre for Health Journalism, media was distributed to News24, South Africa's most prominent news publication, with an average of 1.3 million daily users (News24 2020a, b). As a result, our films garnered more attention, and other news organisations began to pick up on the evictions. Resentment began to emerge against Khayelitsha CAN and the Social Justice Coalition (SJC)—local activist and advocacy organisations—for labelling law enforcement as anti-black. The overwhelming sentiment on social media platforms pointed to the fact that the officers responsible were of colour and therefore could not be anti-black. This exposed a public naivety: in reality, these actions were not being carried out at the will of the police officers tasked to do so, but rather by the City itself. Not unlike other regions, South Africa is home to a policing system that is in urgent need of overhaul and refurbishment. The country is unique in that its engrained policing system had not been adequately addressed after Apartheid and many structural issues linger (Rauch et al. 2001). It is paramount for the system to become a structure that should first and foremost protect rather than vilify vulnerable people, who subsequently became more exposed as a direct result of the pandemic and its economic impact on the country. Cultural theorist and scholar Fred Moten reminds us, "But I also know that what it is that is supposed to be repaired is irreparable. It can't be repaired. The only thing we can do is tear this shit down completely and build something new" (Moten and Harney 2013). Jack Halberstam provides a similar sentiment:

> We cannot be satisfied with the recognition and acknowledgement generated by the very system that denies a) that anything was ever broken and b) that we deserved to be the broken part; so we refuse to ask for recognition and instead we want to take apart, dismantle, tear down the structure that, right now, limits our ability to find each other, to see beyond it and to access the places that we know lie outside its walls (Halberstam in Moten and Harney 2013).

The increased public awareness of the evictions during the pandemic did have some positive outcomes. For a short period, the City deescalated the intimidation of people in eThembeni and surrounding areas. According to residents, confrontations were momentarily decreased. The reduction in police activity may have been partially due to the films revealing the conduct of the officers and the City and its residents raising difficult questions online. Mathematical modelling of the impact of moratoria on evictions in cities in the USA during COVID-19 indicates that allowing people to stay in their homes can mitigate the spread of COVID-19 (Nande et al. 2021). Without formal evidence to suggest so, we hope that the distribution of our films, together with additional coverage by dissemination partners and on-the-ground work by housing activist groups, may have positively impacted the treatment of civilians while helping to prevent further spread of COVID-19.

Weeks later, a sewerage pipe burst at eThembeni, the informal settlement in Khayelitsha mentioned prior and where Qolani had been staying (Gontsana 2020). Residents lived knee-deep in human excrement for three months during winter and during the mandated lockdown (Fig. 2.12). While the City insisted that the water was "clean", children swam in it and got sick. The SJC accused the City of constructive eviction, the act in which a landowner deliberately makes a property uninhabitable so that tenants are forced to leave (Buoncristiani 1969). Some residents felt the pipe could have been intentionally broken to force them to leave. However, the City maintained that the incident was accidental and that the water was clean and that they were not obligated to provide services to land unlawfully occupied. Accidental or not, the burst pipe provided an alternative to policed removals and combined with the lack of urgency to fix it, an effective method of eviction.

Where possible, shacks were relocated to higher ground as residents tried to navigate a difficult situation. Civil society groups provided limited aid where possible, but the winter rains caused the dams of human excrement to fill up as soon as progress was made. These views are documented in the film *eTHEMBENI* (Fig. 2.13). A young woman mentions:

> Whoever says that over there is water, even if they are from the City of Cape Town or any other department, must bring a jug and scoop some. If they drink it, we will know that this is only water and not sewage water, and we will leave the City of Cape Town alone. [...] Because sometimes it feels like there's a Republic of Cape Town and the Republic of South Africa

2.4.1.3 Skills Development While Producing COVID-19 Films Led to Increased Teaching Capacity

The films were developed while training during the lockdown, often relying on WhatsApp as a tuition method during the early morning hours, and time became somewhat abstract during the confinement period. Editing became a nocturnal event. We developed a reliable system, and the films progressed in quality while receiving further guidance from our distribution partner, Bhekisisa. Approximately 14 short documentaries about COVID-19 and the disease's social determinants were produced. Our annual iteration of *Eh!woza Doccies* (Table 2.1) remained delayed by adjustments in alert levels. We decided to run an abridged version of the project in Masiphumelele at the Desmond Tutu Health Foundation's (DTHF) Aerobiology Research Unit and recruited nine grade 9 and 10 learners from the Masiphumelele branch of IkamvaYouth. The original programme has its biomedical component based at UCT's IDM and recruits learners from Makhaza in Khayelitsha. Learners attend six Saturday science workshops comprising of talks and experiments. During these sessions, cutting-edge biomedical science relating to TB is explored in terms of TB basics, vaccinology, and the like. This is followed by a two-week intensive production workshop starting with filming in Makhaza, Khayelitsha. Video editing is then taught at our studio in Paarden Eiland over 2 weeks in the winter school holidays and concluded during public holidays and Saturdays throughout the remainder of the year (Young et al. 2018) (Fig. 2.14).

Masiphumelele is located 42 km from Khayelitsha, and while it is a smaller and newer township, residents face even more congested living conditions. During the Masiphumelele project, and in contrast with our annual learner documentary project previously described (Table 2.1), we completed and finalised the entire programme during the 7 day September school holidays in 2021. We installed our edit suites at the DTHF and operated from the Aerobiology unit's state-of-the-art labs for TB transmission and microscopy workshops. Most learners attend a high school directly across the road, and as Masiphumelele is relatively small, transport and travel of the learners are more flexible. Combining the science and filmmaking workshop locations with a condensed workshop schedule facilitated a quicker and more streamlined workshop programme.

In addition to the logistical improvements, the speed and quality at which we were able to guide learners in film production was primarily due to the trainee filmmaking team, who had developed skills during the COVID-19 filming project, and

Fig. 2.12 *eTHEMBENI.* Samuel Flans and Alfa Fipaza, Eh!woza | Video Still

were able to effectively transfer these newly progressed skills to the high-school learners in a vernacular peer-to-peer teaching environment. This is a departure from our previous English-based tuition, which is sometimes confusing and intimidating for young learners. This saw a radical increase in the quality of the films produced in the 2021 high-school programme.

The nine learners were split into groups, producing three short films (Fig. 2.15). The first film revisited the fire that ravaged Masiphumelele in December 2020, leaving a large section of its residents homeless and destitute during the pandemic. The second film addressed issues of sexual violence and shared personal stories by survivors of this epidemic, violence that escalated during the COVID-19 lockdown period. The third project investigated the stories of two people—a woman who had lost her husband to TB due to treatment hesitancy, and a man that shared his experience of coming to grips and living with

Fig. 2.13 *eTHEMBENI.* Samuel Flans and Alfa Fipaza, Eh!woza | Video Still

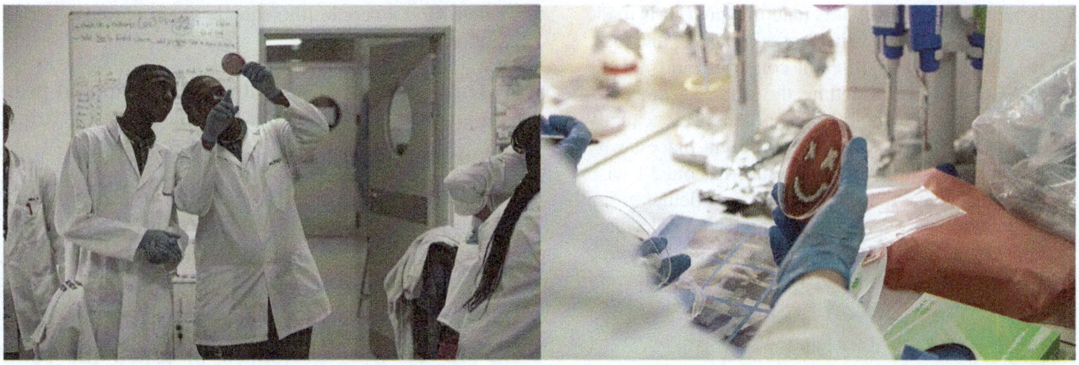

Fig. 2.14 Saturday Science Workshops | Photos: Ed Young, Eh!woza

HIV. The project tries to legitimise and destigmatise living with infectious diseases.

2.5 Algorithms and Echo Chambers: Challenges Around Effective Dissemination of Public Engagement Media

During COVID-19, our online reach and visibility remained limited, and our social media platforms were relatively new and still expanding. In an attempt to increase dissemination, we partnered with previously mentioned CTV, a local community-driven television channel with a broad local viewing audience, together with Bhekisisa and, through them, News24—both of which are prominent national news platforms. More recently, we have also partnered with The Daily Vox for more grassroots and youth-focused distribution.

During the first year of the pandemic, we focused on producing documentaries that uncovered genuine fears surrounding the new and unknown disease. We began documenting parallel concerns, such as the City of Cape Town's unlawful evictions of people occupying land that were a direct social consequence of the pandemic. With anti-vaccine sentiments and distrust rife in the city centres and neighbourhoods, we were eager to know the vaccine-related anxieties the city's township residents faced. We approached the same sector of residents that we had worked with previously and where we had established some degree of trust. During the pandemic's first wave, we had heard instances of stigma where deaths likely due to COVID-19 were attributed to poisonings or similar causes. These kinds of responses by the public are not unlike what has happened during previous epidemics such as Ebola and HIV (Chandler et al. 2015; Drain 2015). To determine what people were currently thinking and feeling about COVID-19 vaccines, we set out to determine the general public's fears regarding vaccine adherence in the township areas. Below are some concerns raised during the interviews:

> I feel like there's a need for a [COVID-19] vaccine, but I don't trust it. The reason for me not trusting it

Fig. 2.15 Masiphumelele high-school documentary project thumbnails

is because of the way it is manufactured. It is made in various places by different companies, and also, it is distributed to different countries, and not the same countries are getting the same vaccine. So, that causes inequality, and then we, actually myself... Since it is made in different places, why are some countries getting a different one? I wish we can get clear reasons for, why a particular group or country, why are they getting a particular vaccine? Because we don't want to see ourselves dying because of a vaccine that is manufactured very cheaply

I don't think I would take it because most people have not been educated about this vaccine. So, probably you wouldn't just jump and go and get vaccinated without getting educated about it and its side effects - what to expect from it after getting it. So, I don't think I'd jump immediately and get it

People are saying that this vaccine is designed to kill us, because the politicians have their own agendas

Some were criticising the vaccine that arrived. They were saying the Chinese have put something in it to depopulate the people in South Africa. So those are the things that people are saying, that when you got the [COVID-19] vaccine, you become weak. Then you only live for a few days, and then you die

It's like they want to colonise us for the second time. That's what people are saying on Facebook, and then I heard that in America, people tried this vaccine, and then they died. And then it is said that that is how we are going to be depopulated by getting vaccinated

Some from the spiritual side, like the pastors, are saying, it's the time of... It is said that there will be 666, and then we will end up using our foreheads when we are shopping. They will put a microchip in us. If you will not allow that, then you are going to die. So, it's like we are being introduced to Satanism

Because these things were said in the bible that they would be there. Now the place that is supposed to help God's people [the churches] are closed. That hurts us

These concerns were already widespread prior to vaccines becoming available in SA. When hosting focus groups with Eh!woza alumni after vaccines became accessible, sentiments remained the same. Participants had moved through our programme, some for two consecutive years, and while having gained additional information and a better understanding of HIV and TB, COVID-19 was scrutinised in much the same way HIV had been during the 1990s when relatively little was known about the disease. When asked whether any of the previous learners would take the vaccine, 100% said they would not. When asked why, social media and word of mouth were cited as primary sources of (mis)-information and distrust that lead to rising fears and anxiety. These concerns, combined with a decrease in trust of public institutions and a general feeling of being "left behind", were key justifications for vaccine hesitancy. When asked if the participants would consider vaccines if offered some form of incentive, such as cell phone data, some were open to the idea.

Vaccine hesitancy remains a significant cause of hospitalisations and deaths, particularly in the context of meagre efforts to disseminate relatable information in poorer and rural areas of South Africa. Vaccine inequity is an equally formidable problem. By May 2021, only "1 percent of the 1.3 billion vaccines injected around the world had been administered in Africa" (Krishnan 2022). By December 2021, only 7% of Africa's citizens had been vaccinated (Adepojo 2021), and in the 6 weeks leading up to Christmas, more vaccines were delivered to the USA, European Union and the United Kingdom than had been delivered to Africa over the entire year of 2021 (Trenchard 2022). While vaccine hesitancy has been suggested as a contributor to low vaccination rates in Africa, large-scale studies have indicated that vaccine acceptance is relatively high in Africa (Solís Arce et al. 2021). Nevertheless, a lack of access to vaccines contributes to hesitancy and mistrust of vaccines on the African continent. The situation is beginning to be remedied by investment in vaccine development and production on the African continent (Trenchard 2022), with South African scientists recently achieving a first by reverse engineering Moderna's mRNA vaccine (Maxmen 2022).

To mitigate against unintentionally amplifying misinformation uncovered in the aforementioned vaccines film, we resolved to interview a health expert to dispel any myths that arose during the initial filming. Some of our distribution partners felt that this was unsafe and that the health expert was at risk of attack on social media platforms. To salvage the project, they recommended scripted questions and proposed filming residents in poorer areas, reading out prescribed questions that would, in turn, be responded to by the expert. As an organisation, we were focused on addressing genuine concerns and fears of people affected by misinformation and felt that scripted messaging disregarded an opportunity for people to raise genuine anxieties that lead to real-life fears and consequences. We reworked the film, retaining the initial interviews with community members but incorporating expert responses, and released a subsequent version under the umbrella of the South African Agency for Science and Technology Advancement (SAASTA)'s National Science Week programme, simultaneously screening the documentary on CTV and distributing online with The Daily Vox.

Work released on mainstream media generates valuable dissemination avenues for young filmmakers producing content from within the City's townships. Having the genuine concerns of the trainees and people from similar backgrounds made visible on platforms usually reserved for high-end journalism is valuable. While it is essential that mainstream media outlets are willing to distribute work from and about townships, they should also share the obligation of actively providing access to accurate information in poorer communities—these activities should run in parallel.

Those with direct access to information in 2022 are predominantly economically secure. Disparities in access to everyday technology significantly contribute to current-day segregation, with access to devices, data, electricity, and technical know-how being primary hurdles in how people access daily information. If we focus on delivering news and accurate information to middle-class and affluent audiences, we should be honest about the reasons for doing so—news platforms are reliant on advertisers, while advertisers rely on news agency readerships to be effective. One example is the mainstream news surveys of South African opinion on current affairs, politics, and health. These results are considerably biased as a notable portion of the South African demographic is excluded from participation due to a lack of digital and financial access. As a result, a large part of the population is omitted from the conversation in much the same way, and for the same reasons as they have been for decades, mirroring the exclusion and segregation predating a democratic SA.

Local news cycles loop online in echo chambers with news feeds populated by algorithms. These loops repeat in familiar circles, with essential information distributed in a way that results in "preaching to the choir". Online platforms develop perimeters around those already in the know, while those in need of accurate information are often excluded and tend to rely on conspiracies and misinformation that is commonly available and popular. While most contemporary media rely on electronic devices and internet data for distribution, a recent survey we conducted in Khayelitsha suggests that a large portion of residents receive information through traditional media such as community-based television, radio, and local newspapers. More work is required to understand this more in-depth. However, reasons could include that older generations prefer traditional media and that lack of access to the internet and data, due to prohibitive costs, exclude people from access to online news. Traditional media, together with in-person engagement through on-the-ground healthcare and social worker and trusted traditional and religious sectors, can act as valuable instruments in disseminating accurate information, increasing confidence and combating infectious diseases.

We addressed some of the COVID-19 related concerns uncovered in films and interviews by producing health-related animations (Fig. 2.16) translated into isiXhosa. The animations have also recently been translated into Sesotho and isiZulu for distribution in rural South Africa in collaboration with Sunshine Cinema—an organisation that disseminates accurate information while stimulating discourse through a network of ambassadors. The ambassadors conduct live screenings and discussions in areas of South Africa that are difficult to access. In addition to these avenues of dissemination, we converted animations into short newspaper adverts that could be placed in township newspapers that are distributed free of charge.

The animations were designed to be visually engaging and appealing solutions to distribute accurate health information, and with the help of artist Mitchell Gilbert Messina, we created simple but visually impactful clips. Our online distribution expanded, and our content started reaching broader audiences. As time progressed, and particularly during the development of animations focusing on vaccines and vaccine hesitancy, our viewership decreased unexpectedly. Implementing the same methods we had always used to increase social media visibility (keyword research and post boosts), our YouTube viewership went from approx. 2000+ views per post, to less than 100 per post (Fig. 2.17). Regardless of efforts in reviving the platforms, views remained consistently low. It seemed that the work, particularly the vaccine work, was negatively affected by platform algorithms.

We have no real explanation for this. One reason might be that social media platforms were concerned about publishing health information that may potentially be inaccurate and therefore potentially dangerous, particularly during vaccine rollout. Without the ability to fact-check vast amounts of health information uploaded daily, it was possibly safer for algorithms to obscure content from less significant sources. However, it is likely more complex than that. Misinformation thrives on social media due to significant public interest in hate-related posts, conspiracies, and complex and coordinated campaigns that ramp up disinformation. In a sophisticated systems network analysis, Johnson and colleagues show that anti-vaccination clusters on Facebook provide much more diverse and captivating reasons for rejecting vaccination than the "monothematic" pro-vaccination groups. While anti-vaccination clusters are smaller, the diversity of themes, together with availability on a wide range of platforms, mean they intertwine with clusters of undecided individuals more easily (Johnson et al. 2020). More worryingly, recent research suggests that Facebook algorithms are designed to facilitate the promotion of misinformation by a small set of highly active users and that Facebook could change this but may not be willing to as it could decrease overall user engagement (Hindman et al. 2022).

Today, online information is mass distributed and sometimes obtained through non-conventional channels. Celebrity status lends impact to comments and responses on social and health-related topics. Recently, artist Nicki Minaj was informed by a cousin that the vaccine had caused a friend to become impotent with swollen testicles as side effects. Minaj tweeted this out to 22 million followers who take her opinion seriously and have the ability to redistribute the information leading to it spreading exponentially (Sun 2021). Low-key "self-described 'news analyst & hip-hop artist' named An0maly who thinks the pandemic is a cover for government control" posted misinformation about the J & J vaccine causing blood clots. Facebook did not remove the statement as it was technically not untrue, although incredibly unlikely. The post was amplified by prominent figures, including Robert Kennedy Jr., resulting in the temporary international halting of the vaccine drive (Parks 2021). Due to the sheer scale and immediacy of possible spread, this type of social engagement contributes directly to the increased spread of COVID, overloaded healthcare facilities and ultimately, loss of life and livelihood.

The internet has intrinsic value, but it is essential to acknowledge that these platforms are not primarily there for us to use, but for them to use us, with the outcome and interest predominantly commercial. In Netflix' *The Social Dilemma*, data analyst Cathy O′ Neil explains that:

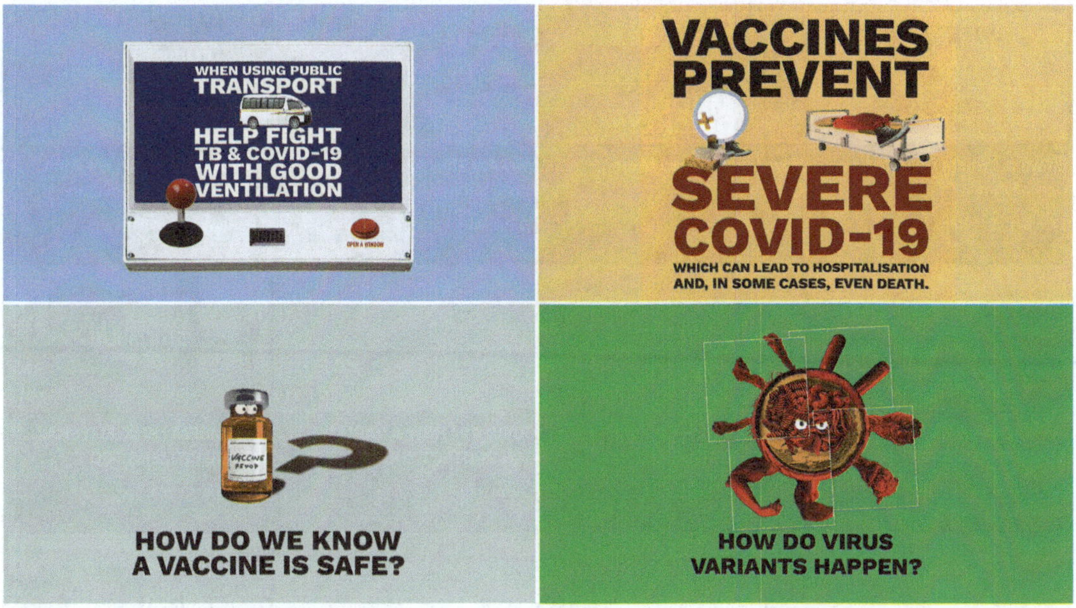

Fig. 2.16 Eh!woza Animation Thumbnails

> I like to say that algorithms are opinions embedded in code. And that algorithms are not objective. Algorithms are optimised to some definition of success. So if you can imagine, if a commercial enterprise builds an algorithm, to their definition of success, it's a commercial interest. It's usually profit (The Social Dilemma 2020)

Regardless of the reasons for content being suppressed, it remains crucial to assess which platforms to distribute high-end health visualisation to avoid ending up in a "tree falling in a forest" scenario. While the sudden drop in viewership was first noted with the release of media related to vaccines specifically, it ended up affecting our social media accounts as a whole. Our organisation does not have the resources and expertise, nor is it strategic focus to interrogate social media and its intended outcomes. It does, however, raise questions about optimal dissemination strategies. An important consideration then becomes how to effectively distribute health-related visual information and prevent resources from being invested in strategies that have low reach and impact. If visual information is created and fails at dissemination level, it serves limited purpose.

2.6 Conclusions

The links between social exclusion, struggle and infectious disease should not be overlooked by clinical and biomedical practice to achieve effective outcomes. The COVID-19 pandemic (as well as HIV before this) has demonstrated that if we are unable to successfully design and implement well-defined strategies to foster trust and decrease stigma, then increases in drug adherence, vaccine acceptance, and the reduction in rates of infections such as HIV, TB and COVID-19 remain unlikely. Some of the work described above highlights that the co-production of public engagement media and programmes is complex yet valuable and can uncover how social factors influence and are influenced by infectious diseases and the knowledge people have about them.

The visualisation of science, stigma and infection is also delicate when targeted to an older, often less educated audience and equally so, when aimed at a younger audience whose primary health information is accessed online, algorithmically determined, and placed across our screens via digital and social platforms. Furthermore, the

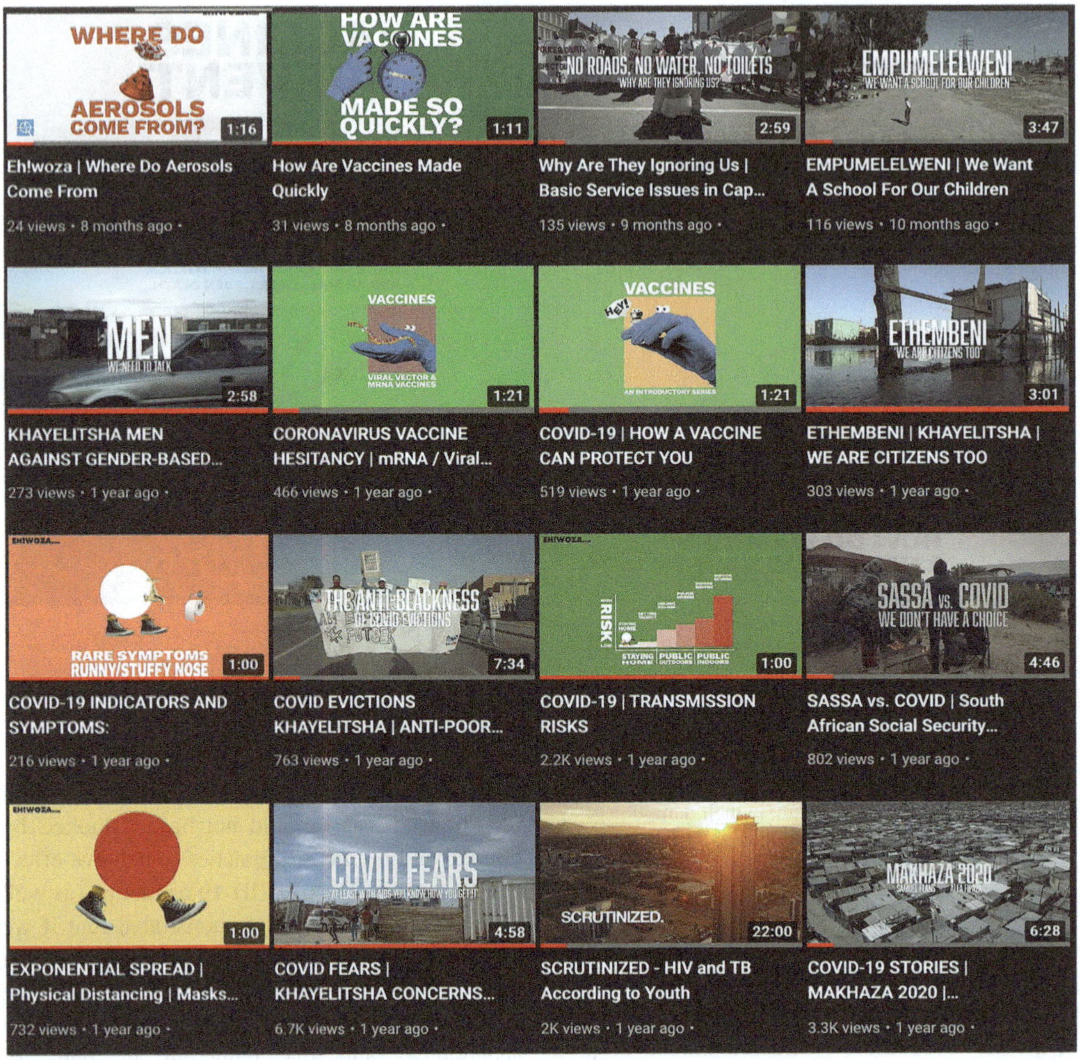

Fig. 2.17 Thumbnails depicting decreased views of Eh!woza's media on YouTube starting approx. one year ago when the organisation implemented COVID-19 vaccine-related animations

complexity of disseminating public engagement media on platforms that can quickly reach mass audiences is often underestimated. This signals a need for new dissemination and theoretical framework approaches to understanding how to intervene in the spreading of misinformation (Johnson et al. 2020). To move the field forward, honest questions should be raised about methods through which we measure the impact of dissemination of visual information within public engagement work—when are digital platforms the most effective platforms to engage young people with health? How do we tailor and target media to reach intended audiences, and how do we combine digital methods with tailored in-person engagements and dialogue to achieve the best possible impact?

Finally, the value of on-the-ground in-person engagement within smaller circles in intimate and safer settings should not be overlooked. These have the potential to ripple through neighbourhoods and communities and have perhaps been undervalued due to the assumed impact of digital platforms that can reach massive audiences speedily.

Acknowledgements Ed Young and Anastasia Koch would like to acknowledge funding from the South African National Research Foundation (UID: 116273) and the Wellcome Trust (219398/Z/19/Z). We would like to thank individuals who have provided consent for images to be included in this manuscript and Bonga Zamisa from the Social Justice Coalition for permission to use their cell phone footage of Bulenai Qolani being dragged out of his shack (Fig. 2.10). We further thank all of the participants of Eh!woza's programmes and collaborators who provided invaluable input to the work described here. The title of this chapter is borrowed from artist Bill Viola's book published to coincide with his exhibition on the 46th Venice Biennale, *Reasons for Knocking at an Empty House (Viola 1995)*, and we thank the artist for his inspiration.

References

Adepojo P (2021) As COVID-19 vaccines arrive in Africa, Omicron is reducing supply and increasing demand Nat Med News https://doi.org/10.1038/d41591-021-00073-x. Accessed 13 Feb 2022

Bernardo C (2019) 'TB is not a drama queen'. UCT News. https://www.news.uct.ac.za/article/-2019-04-05-tb-is-not-a-drama-queen. Accessed 6 Jan 2022

Bailie C (2020) South Africa's military is not suited for the fight against Covid-19. Here's why. Mail & Guardian. https://mg.co.za/coronavirus-essentials/2020-06-02-south-africas-military-is-not-suited-for-the-fight-against-covid-19-heres-why/. Accessed 6 Feb 2022

Cape Town TV (n.d.). https://capetowntv.org/. Accessed 7 Jan 2022

Buoncristiani DM (1969) Partial constructive eviction: the common law answer in the tenant's struggle for habitability. Hastings Law J 21(2):417–432. https://repository.uchastings.edu/hastings_law_journal/vol21/iss2/5

Chandler C et al (2015) Ebola: limitations of correcting misinformation. Lancet 305:1275–1276. https://doi.org/10.1016/S0140-6736(14)62382-5

Chigwedere P, Seage GR, Gruskin S, Lee T-H, Essex M (2008) Estimating the lost benefits of antiretroviral drug use in South Africa. JAIDS J Acquir Immune Defic Syndr 49(4):410–415. https://doi.org/10.1097/QAI.0b013e31818a6cd5

Comas I et al (2013) Out-of-Africa migration and Neolithic coexpansion of *mycobacterium tuberculosis* with modern humans. Nat Genet 45:1176–1182. https://doi.org/10.1038/ng.2744

Drain PK (2015) Ebola: lessons learned from HIV and tuberculosis epidemics. Lancet Infect Dis 15(2): 146–147. https://doi.org/10.1016/S1473-3099(14)71079-5

Dennis P (2015) Abbot Pfanner, the Glen Grey act and the native question. S Afr Hist J 67:271–292. https://doi.org/10.1080/02582473.2015.1094122

Dicey W (2021) 1986. Penguin Random House, South Africa, p 94

Dube F, Du Plessis A (2021) Unlawful occupiers, eviction and the national state of disaster: considering South Africa's emergency legislation and jurisprudence during COVID-19. J Afr Law 65:333–346. https://doi.org/10.1017/S0021855321000401

Eldholm V et al (2016) Armed conflict and population displacement as drivers of the evolution and dispersal of *mycobacterium tuberculosis*. Proc Natl Acad Sci 113:13881–13886. https://doi.org/10.1073/pnas.1611283113

Flans S (2019) COVID EVICTIONS: 'the people who live here are humans, not pigs'. Eh!Woza. https://www.youtube.com/watch?v=Iwzd8DjZupw&t=111s. Accessed 29 Jan 2022

Gontsana M (2020). We are living in liquid shit, say eThembeni families. GroundUp. https://www.groundup.org.za/article/we-are-living-liquid-shit-say-ethembeni-families/. Accessed 27 Jan 2022

Karim SSA (2009) HIV infection and tuberculosis in South Africa: an urgent need to escalate the public health response. Lancet 374:921–933. https://doi.org/10.1016/S0140-6736(09)60916-8

Hindman M, Lubin N, Davis T (2022) Facebook has a superuser-supremacy problem. The Atlantic. https://www.theatlantic.com/technology/archive/2022/02/facebook-hate-speech-misinformation-superusers/621617/. Accessed 12 Feb 2022

Johnson NF et al (2020) The online competition between pro- and anti-vaccination views. Nature 582:230–233. https://doi.org/10.1038/s41586-020-2281-1

Kelly P (1999) Isolation and stigma: the experience of patients with active tuberculosis. J Commun Health Nurs 16:223–241. https://doi.org/10.1207/S15327655JCHN1604_3

Kelto A (2010) South Africa hides its homeless ahead of world cup. NPR. https://www.npr.org/templates/story/story.php?storyId=127593697. Accessed 3 Feb 2022

Kiewit L (2020) High court declares the City of Cape Town's lockdown evictions illegal. Mail & Guardian. https://mg.co.za/news/2020-08-25-high-court-declares-the-city-of-cape-towns-lockdown-evictions-illegal/. Accessed 7 Jan 2022

Krishnan V (2022) Phantom plague: how tuberculosis shaped history. PublicAffairs, New York, p 220

Masuku B et al (2018) Beyond the lab: Eh!Woza and knowing tuberculosis. BMJ Med Humanit 44:285.

https://doi.org/10.1136/medhum-2018-011479. https://www.nature.com/articles/s41467-021-22521-5#citeas292
Maxmen A (2022) South African scientists copy Moderna's COVID vaccine. Nature News https://doi.org/10.1038/d41586-022-00293-2. Accessed 12 Feb 2022
Moyo S et al (2015) Loss from treatment for drug resistant tuberculosis: risk factors and patient outcomes in a community-based program in Khayelitsha, South Africa. PLoS One 10(3):e0118919. https://doi.org/10.1371/journal.pone.0118919
McCain N (2022) City of Cape Town spent R42m irregularly on temporary homeless campsite, SIU finds. News24. https://www.news24.com/news24/SouthAfrica/News/city-of-cape-town-spent-r42m-irregularly-on-temporary-homeless-campsite-siu-finds-20220126. Accessed 29 Jan 2022
Meyer D (2020) 'Cold and cruel': inside Cape Town's controversial homeless camp. The South African. https://www.thesouthafrican.com/news/inside-cape-town-strandfontein-homeless-camp-lockdown/. Accessed 3 Feb 2022
Moten F, Harney S (2013) The undercommons: fugitive planning & black study. Autonomedia/minor compositions, Brooklyn. p 24, 96
Nande A et al (2021) The effect of eviction moratoria on the transmission of SARS-CoV-2. Nat Commun 12: 2274. https://doi.org/10.1038/s41467-021-22521-5
News24 (2020a) News24 hits 20 000 subscribers in first two months. News24. https://www.news24.com/news24/opinions/reader_hub/news24-hits-20-000-subscribers-in-first-two-months-20201123. Accessed 31 Jan 2022
News24 (2020b) WATCH | Khayelitsha man dragged naked from shack, seeks justice. YouTube. https://www.youtube.com/watch?v=EaHDA9oniBM&t=2s. Accessed 30 Jan 2022
Patterson O (1982) Slavery and social death. Harvard University Press, Cambridge
Parks M (2021) The most popular J & J vaccine story on Facebook? A conspiracy theorist posted it. NPR. https://www.npr.org/2021/04/15/987182241/the-most-popular-j-j-vaccine-story-on-facebook-a-conspiracy-theorist-posted-it. Accessed 29 Jan 2022
Plaatje S (2016) Native life in South Africa. P.S King and Son, Ltd., London, p 21
Osman M et al (2021) Health system determinants of tuberculosis mortality in South Africa: a causal loop model. BMC Health Serv Res 21:388 10.1186/s12913-021-06398-0
Rauch J, Shaw M, Louw A (2001) Municipal policing in South Africa: development and challenges. Institute for Security Studies Monograph No 67
Richardson ET et al (2016) Forced removals embodied as tuberculosis. Soc Sci Med 161:13–18. https://doi.org/10.1016/j.socscimed.2016.05.015
Solís Arce JS et al (2021) COVID-19 vaccine acceptance and hesitancy in low- and middle-income countries. Nat Med 27:1385–1394. https://doi.org/10.1038/s41591-021-01454-y
Tembo D et al (2021) Effective engagement and involvement with community stakeholders in the co-production of global health research. BMJ 372. https://doi.org/10.1136/bmj.n178
The Social Dilemma (2020) Netflix. Jeff Orlowski. Exposure Labs
Trenchard T (2022) South Africa hails new COVID jab plant in fight for self-reliance. Al Jazeera. https://www.aljazeera.com/news/2022/1/29/south-africa-hails-new-covid-jab-plant-in-fight-for-self-reliance. Accessed 31 Jan 2022
Sun M (2021) Why is Nicki Minaj tweeting about vaccines and some guy's swollen testicles? The Guardian. https://www.theguardian.com/culture/2021/sep/14/why-is-nicki-minaj-tweeting-about-vaccines-and-some-guys-swollen-testicles. Accessed 29 Jan 2022
Viola B (1995) Reasons for knocking at an empty house: writings 1973–1994. The MIT Press, Cambridge
Weingart P, Joubert M, Connoway K (2021) Public engagement with science—origins, motives and impact in academic literature and science policy. PLoS ONE 16(7):–e0254201. https://doi.org/10.1371/journal.pone.0254201
Wilderson FB III (2015) Incognegro: a memoir of exile and apartheid. Duke University Press, Durham Carolina
Wilderson FB III (2020) Afropessimism. Liveright, New York, p 17,92
Williams E (1944) Capitalism and slavery. UNC Press, Chapill Hill
World Health Organisation (2021) Global Tuberculosis Report. https://www.who.int/publications-detail-redirect/9789240037021. Accessed 12 Feb 2021
Young E et al (2018) Eh!Woza: intersection of art and science to engage youth on tuberculosis. Global Health Innov1 10.15641/ghi.v1i1.520
Young E, Koch A (2021) My Rhodes has no nose: COVID-19 and the two cities of Cape Town. In: Williams O, Tembo D, Ocloo J, Kaur M, Hickey G, Farr M, Beresford P (eds) COVID-19 and co-production in health and social care research, policy, and practice, Co-production methods and working together at a distance, vol 2, 1st edn. Bristol University Press, Bristol, pp 27–34

Ed Young is a PhD candidate at the Molecular Mycobacteriology Research Unit (MMRU, Division of Medical Microbiology, University of Cape Town) and the co-founder and co-director of Eh!woza. Ed has significant expertise in the arts spanning almost twenty years. Ed's cross-disciplinary art practice can broadly be defined as conceptual and ranges from painting, sculpture and installation to video, performance and writing. Over the last ten years, Ed has developed specific expertise in public engagement and science communication, with a focus on uncovering the social determinants of health and how this exacerbates the rates of infectious disease we see in South Africa today.

Anastasia Koch is trained as a biomedical researcher and is one of the co-founders and co-directors of Eh!woza, where she manages the conceptual, strategic and administrative elements of the organisation. Anastasia is particularly interested in developing meaningful and innovative ways of measuring the impact of public engagement work. She maintains a part-time biomedical position as an Honorary Lecturer at the MMRU, with a biomedical research focus on the use of genomics to understand the transmission and pathobiology of *Mycobacterium tuberculosis*.

3 The Evolution of Scientific Visualisations: A Case Study Approach to Big Data for Varied Audiences

Andrew J. Lunn, Vivien Shaw, and Isabelle C. Winder

Abstract

Visual representations of complex data are a cornerstone of how scientific information is shared. By taking large quantities of data and creating accessible visualisations that show relationships, patterns, outliers, and conclusions, important research can be communicated effectively to any audience. The nature of animal cognition is heavily debated with no consensus on what constitutes animal intelligence. Over the last half-century, the methods used to define intelligence have evolved to incorporate larger datasets and more complex theories—moving from relatively simple comparisons of brain mass and body mass to explorations of brain composition and how neuron count changes between specific groups of animals. The primary aim of this chapter is therefore to explore how visualisation choice influences the accessibility of complex scientific information, using animal cognition as a case study. As the datasets concerned with animal intelligence have increased in both size and complexity, have the visualisations that accompany them evolved as well? We first investigate how the basic presentation of visualisations (figure legends, inclusion of statistics, use of colour, etc.) has changed, before discussing alternative approaches that might improve communication with both scientific and general audiences. By building upon the types of visualisation techniques that everyone is taught at school (bar charts, XY scatter plots, pie charts, etc.), we show how small changes can improve our communication with both scientific and general audiences. We suggest that there is no single right way to visualise data, but careful consideration of the audience and the specific message can help, even where communications are constrained by time, technology, or medium.

Keywords

Animal Cognition · Statistical and Graphical Visualisations · Science Communication · Accessible Science · Large Complex Datasets · Visualisation Alternatives

A. J. Lunn (✉)
MZool Graduate from School of Natural Sciences, Bangor University, Bangor, UK

V. Shaw
Academic Visitor at Hull York Medical School, Faculty of Health Sciences, University of Hull, Hull, UK

I. C. Winder
School of Natural Sciences, Bangor University, Bangor, UK

3.1 Introduction

Humans have communicated using pictures since our ancestors first made cave art over 45,000 years ago (Brumm et al. 2021). Today,

L. Shapiro, P. M. Rea (eds.), *Biomedical Visualisation*, Advances in Experimental Medicine and Biology 1388, https://doi.org/10.1007/978-3-031-10889-1_3

outside of the arts, visualisation is gaining importance across domains including journalism, education, science, public information, and workplaces (Kennedy and Engebretsen 2020). Visual tools are particularly important for conveying scientific information. The information being shared often takes the form of abstract ideas and numerical findings that are much easier to understand when represented pictorially. At their best, as Kennedy and Engebretsen (2020, pg. 19) note, "visual representations of statistics and other, often quantitative data can convey complex facts and patterns quickly and effectively". Just as everyone is aware that a picture can paint 1000 words, a complicated statistical finding, if visualised effectively, can be made accessible to a much wider audience than when it is presented only as prose.

To foster scientific literacy, everyone taking high school mathematics or science is taught to make and interpret simple visualisations like bar charts, pie charts, histograms, and boxplots (Fig. 3.1). These visualisations follow some straightforward rules, at least as we first encounter them—pie charts are for showing proportions of a whole, and XY scatter plots for showing the relationship between two variables, for example (Hawkins 2019). Mastering these rules gives the student the tools to read and understand a broad spectrum of classic scientific literature, as well as the necessary background for advanced statistics courses that build on these foundations. In modern life sciences, where testing a hypothesis increasingly requires "big data" (Pal et al. 2020), we rely on developments in computing, statistics, and—ultimately—visualisations, to make sense of our findings.

It is interesting, therefore, that even as life science datasets are increasing in size and complexity, scientists' options when it comes to visualising their outputs seem to have remained more circumscribed. This may be because scientists are not explicitly trained in the creation of innovative imagery (McInerny et al. 2014), with existing courses—like Edward Tufte's famous workshops on Analysing and Presenting Data and Information—primarily taking place outside of academia. Ongoing scientific discussions of the challenges of "big data" in biology, furthermore, often mention challenges with storing, allowing access to, processing, and analysing vast datasets (Alyass et al. 2015; Marx 2013; Pal et al. 2020), but rarely consider the challenges of visualisation. The example at the centre of this chapter comes from the study of brain size and cognition. Humans seem to have absolutely and relatively large brains (Preuss 2017), at least when compared to other animal species. The question of exactly *how* large our brains are, and why they became that way, is one that neuroscientists and zoologists have been exploring for fifty years or more. One might think that we would see almost half a century's worth of development and change in the visualisations used in the scientific literature. In fact (see Sect. 3.2), while the types of data and measurements used have changed, sometimes quite dramatically, the same basic approach to visualisation can be seen in the papers surveyed here, which dated between 1981 (Martin 1981) and 2022 (Lunn et al. 2022). Our primary aim in writing this chapter is therefore to explore this pattern in more detail and show how scientists' *choices* about visualisation—at various stages and scales—influence the accessibility of scientific information for a variety of audiences.

To find out, we first delve deeper into our chosen example (Sect. 3.2). Although scientists studying the brain size–body size relationship and biological indicators of intelligence are using the same broad approach to visualisation, we wondered whether the availability of technology and their choices about symbols, legends, colour, titles, placement, and other factors have an impact on accessibility. How are scientists creating images, and what level of interpretive skill and statistical knowledge do they assume in their audience? Then, in Sects. 3.3 and 3.4, we consider alternative approaches that might improve communication with scientific and general audiences, respectively. The development of new statistics packages, and new functions in

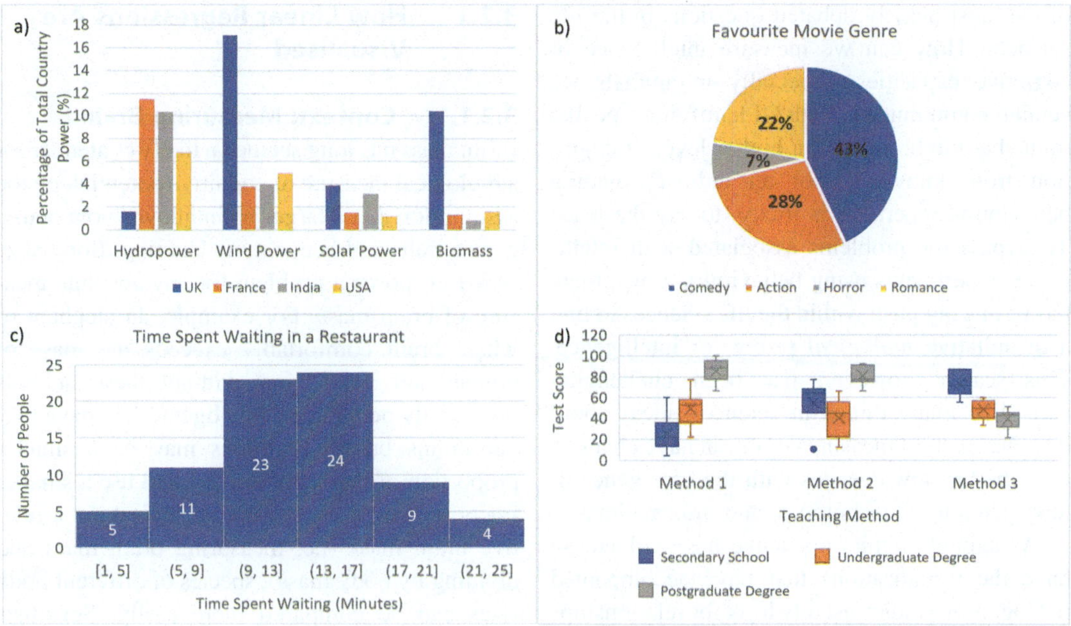

Fig. 3.1 (**a**) Bar chart, used to show quantities in specific categories; (**b**) pie chart, used to show numerical proportions in specific categories; (**c**) histogram, an approximate representation of numerical distribution of a data set; (**d**) box plot, used to show the range and average in different samples, as well as outliers. Example fictitious data sets created by author to demonstrate different visualisations

existing ones, allows writers to handle larger and larger datasets. These have changed the playing field for scientific visualisation. Can new techniques and visualisation approaches improve the way scientists communicate with one another? Do these new techniques also have the potential to help with wider scientific communication (e.g. across disciplinary boundaries or with non-specialists and public audiences)? A discussion of complex relationships relying on statistical values may quite quickly become near meaningless to non-specialist readers, even if they have a keen interest in the subject. The way that data is visualised is thus key to engaging quickly and effectively with a variety of readers. Through this chapter we will demonstrate that there is no universal "right answer" to visualisation, but that more effective visual communication can often be achieved through careful consideration of the intended audience and the message to be communicated.

3.2 Visualisation Case Study: Animal Cognition

The visual representation of complex information is, without a doubt, one of the most important parts of scientific research. By translating numbers and observations into pictures, the data can be understood much more readily and relationships, outliers, and patterns can be shown in a way that is immediately accessible to the reader. Good visualisation is especially important where datasets are large, and relationships can only be described in terms of secondary numbers (statistics). This is because the human brain is not well-equipped to memorise large datasets, let alone to manipulate them, which means mathematics done on paper or on a computer needs translation to become meaningful.

Animal cognition research is a field that has used large datasets for decades, and tackles some of the most heavily debated questions in the life sciences. How can we measure intelligence or cognitive capacities, especially in animals we cannot communicate with? Identifying specific traits that might indicate a higher level of cognition from knowledge of an animal's natural behavioural repertoire is tricky, to say the least. To bypass the problems associated with intelligence experiments using behavioural traits, there has been a big push within the life sciences to find a quantitative *biological* proxy for intelligence. This search, however, has been challenging and—like many things in science—ideas about how we can best measure cognition have changed over the last few decades with the emergence of new technology, techniques, and information.

As animal cognition science has evolved, so have the visualisations that have accompanied it. Cognition studies usually look for relationships between two variables, one typically being some measure of cognition or intelligence, e.g. ability to use different kinds of tools, and the other the proposed biological indicator, e.g. brain volume. Alternatively, studies may look at brain volume and body size, to look for purely biological relationships. Due to the nature of this sort of data (which typically comprises two continuous measures), the broad nature of the visualisations and statistics used have remained very similar for the past ~40+ years. The standard analysis used is a linear regression. Linear regression is a statistical method which tries to predict the value of one thing (e.g. tool use) from a measure of the other (e.g. brain volume). It is usually represented via a graph with two axes and a scattering of points, one per animal. A line is usually added to show the best fitting statistical relationship. Points may fall above and below this line and (very occasionally) on it. In this section, by starting with the literature on this specific relationship and looking at how scientists' visual choices have affected the appearance and the accessibility of their data, we aim to consider how small decisions about colour, style, symbology and figure presentation affect the accessibility of our target area of research. We start with a little contextual information.

3.2.1 How Linear Regressions Are Visualised

3.2.1.1 Context: Measuring Brains

Brain mass has long seemed a likely contender for a biological measure of cognitive potential, on the assumption that a larger brain allows more neural connections and hence more learning. Body size, however, poses a problem for any absolute measure of brain mass. For example, an elephant or whale brain comfortably exceeds the mass of human and ape brains, without these animals necessarily being more intelligent: they have bigger brains but these brains may be a smaller proportion of their total weight and use a smaller fraction of their energy intake. By taking a relative brain mass (i.e. measuring brain mass and dividing by body mass), species of different body sizes can be compared more easily. Scientists generally anticipate that animals with larger brains relative to their body size will be more intelligent. The proportion of body mass made up by the brain decreases as bodies grow larger (Herculano-Houzel 2016; Jerison 1973; Roth and Dicke 2005). Relatively bigger brains do seem to indicate higher levels of intelligence in primates, birds, carnivores, bats, and even some fish like guppies (Benson-Amram et al. 2016; Garamszegi and Eens 2004; Kotrschal et al. 2013; Madden 2001; Ratcliffe et al. 2006; Reader and Laland 2002; Sol et al. 2005). Some authors today prefer to look at absolute brain size (which has not been adjusted for body mass), or at other indicators, but a large quantity of the papers from the past forty years of cognition research have explored or used relative brain size data.

3.2.1.2 Position of Visualisations Within the Article

Given that most scientists working on animal cognition are using the same types of data and the same statistical tools, the fact that they usually choose the same graphs is not surprising. There are, however, some variations in the ways that these graphs are presented in papers from different decades and with different priorities. The first of these relates to where graphs are put into an article, and how they relate to one another.

The following two figures are classic examples of figures from the relative brain mass/animal cognition literature. They each explore relative brain mass in a multitude of species, in the form of a XY scatter plot. Figure 3.2 shows two figures (originally labelled Figs. 1 and 2) from a 1981 paper by Robert D. Martin, which examined the relationship between body weight and brain weight across 548 species of mammals, birds, and reptiles. Martin was interested in how proportional brain mass affects how much energy animals need, and whether knowing more about this could tell scientists something about constraints on the evolution of brains. Figure 3.3 on the other hand shows a figure from Lefebvre et al. (2004). Lefebvre and colleagues showed that innovation rate, which they used as an indicator of higher intelligence, increases in proportion to the size of certain parts of the brain in birds and primates.

The first difference one might see when comparing these two examples is the location of the figures relative to one another within their respective articles. Martin's figures (Fig. 3.2) are located one after the other, with the text explaining the results on the next page. For a reader, this allows easy comparison between figures, but requires that they refer back to them when reading the supporting text. Lefebvre and colleagues (Fig. 3.3) on the other hand display their figure followed immediately by a small section of text discussing the associated results, before presenting their next finding. This allows the reader to see the figure and the supporting text together, without having to refer back several figures and increases the overall readability of the text. However, it assumes that the reader can absorb the content of a figure and remember it for comparison with other visualisations. Each of these approaches has advantages and disadvantages, and different readers may have their own preference. A visually oriented and statistically savvy reader, for instance, may not need all the text detail to get the message of Martin's paper, while someone who prefers the unfolding of a detailed storyline, with visuals as backup, might engage more with Lefebrve et al.'s more incremental approach. Certainly they suggest that the authors of these papers were either operating within different constraints (e.g. numbers of full pages that could be devoted to figures) or else making choices, consciously or otherwise, about which comparisons they wanted to encourage readers to make. Without consulting the authors and journals, it is impossible to say for sure whether each figure was more shaped by choice or constraint. Their differences, however, arguably interact with other small variations in, for example, symbology to shape the reader's experience in quite distinctive ways.

3.2.1.3 Data Point Identification

In Fig. 3.2, Martin uses a solid triangle (▲) for points that represent primates, and a hollow one (△) for other mammals. The same symbols, on another chart, are used for birds (▲) and reptiles (△). This makes it simple for a reader looking at a single chart to tell apart the two groups it compares, but does not facilitate easy comparison between charts—unlike the placement of figures, which seemed at first look to encourage it. This use of symbols prioritises the reader spotting the difference between birds and reptiles (which plot out separately) and primates and mammals (whose symbols overlap). It is much harder to see how birds and mammals would compare. Differences in the two figures' axes and shapes, for instance, make it difficult to imagine them superimposed even if the reader is confident enough in their statistical understanding to be sure direct comparison was possible.

Lefebvre et al. (2004) also use labels to identify specific points, giving information about taxonomic family (Fig. 3.3a) and species (Fig. 3.3b), but instead of shapes they use numbers with corresponding information in the figure legend. Numbers like this allow the interested reader to locate specific animals in the chart, but the bird and primate charts use the same numbering system (1–16), so a given number doesn't always mean the same thing. Number 1, for instance, is either a corvid or a chimpanzee depending on which panel you look at. The primate analysis (Fig. 3.3b), furthermore, only expressly labels

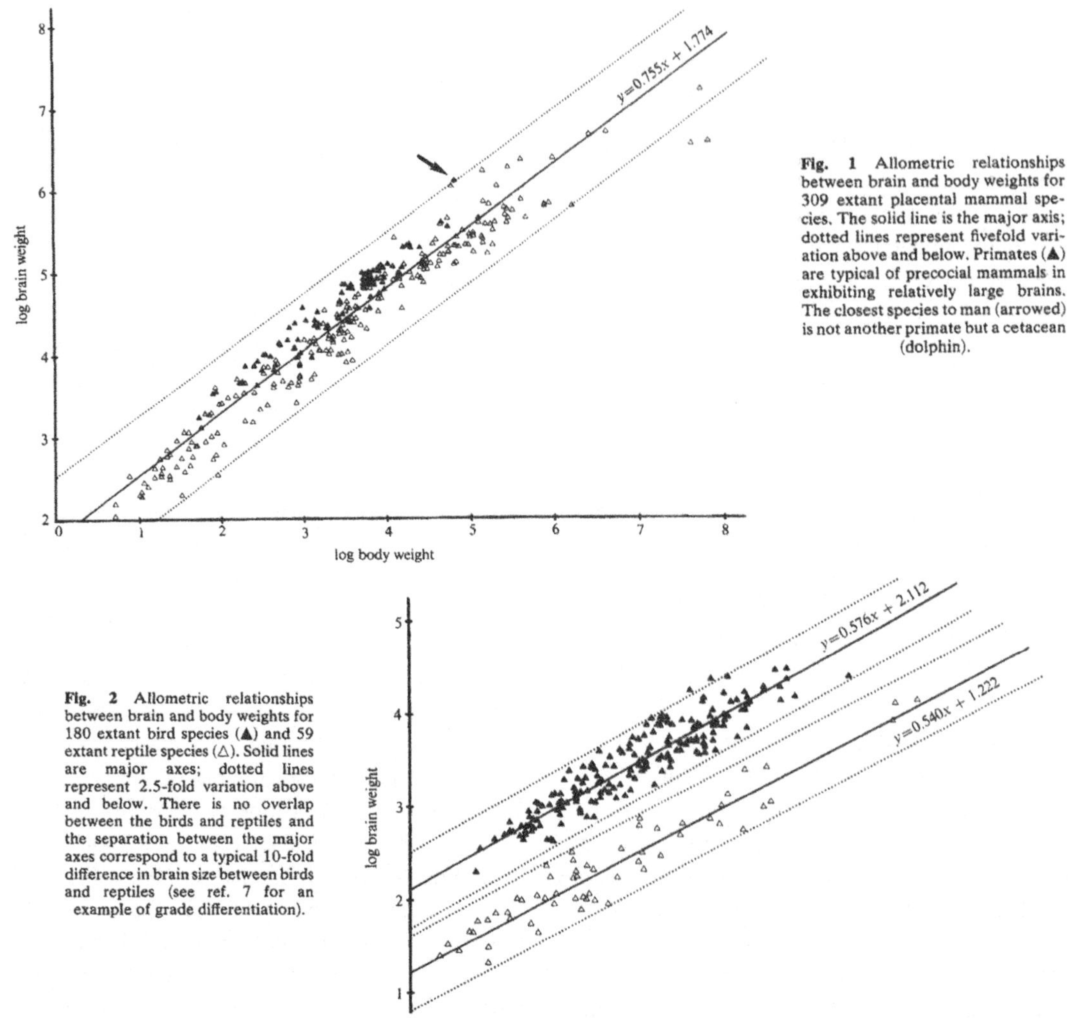

Fig. 3.2 Figures and legends from Martin, R. D., 1981. Two XY scatter plots of body weight against brain weight in primates and other mammals (Fig. 1) and birds and reptiles (Fig. 2), showing the relationships between the body weight and brain weight. Primates and the remainder of mammals are grouped together, while birds and reptiles form two separate distinct groups with no overlap. Reprinted by permission from "Springer Nature": Springer Nature, Nature, *Martin, R. D.,* 1981. *Relative brain size and basal metabolic rate in terrestrial vertebrates. Nature, 293(5827), pp. 57–60.* Copyright ©1981

16 species, saying "some points are not numbered due to space limitations". Labelling all the data points might well have led to visual overload and obscured the pattern the chart is meant to show.

3.2.1.4 Figure Legends

Another important part of the figure which has a huge impact on its accessibility to the reader is the figure legend. The figure legend is a piece of accompanying text that describes the figure and helps the reader understand it without reference to the rest of the manuscript. In 2009, Yu and his colleagues looked at how well people understood information when they were just given a short and highly condensed summary of the paper (the abstract), a particular figure, and the legend that goes with it. They found that people only understood roughly a third of what they were reading

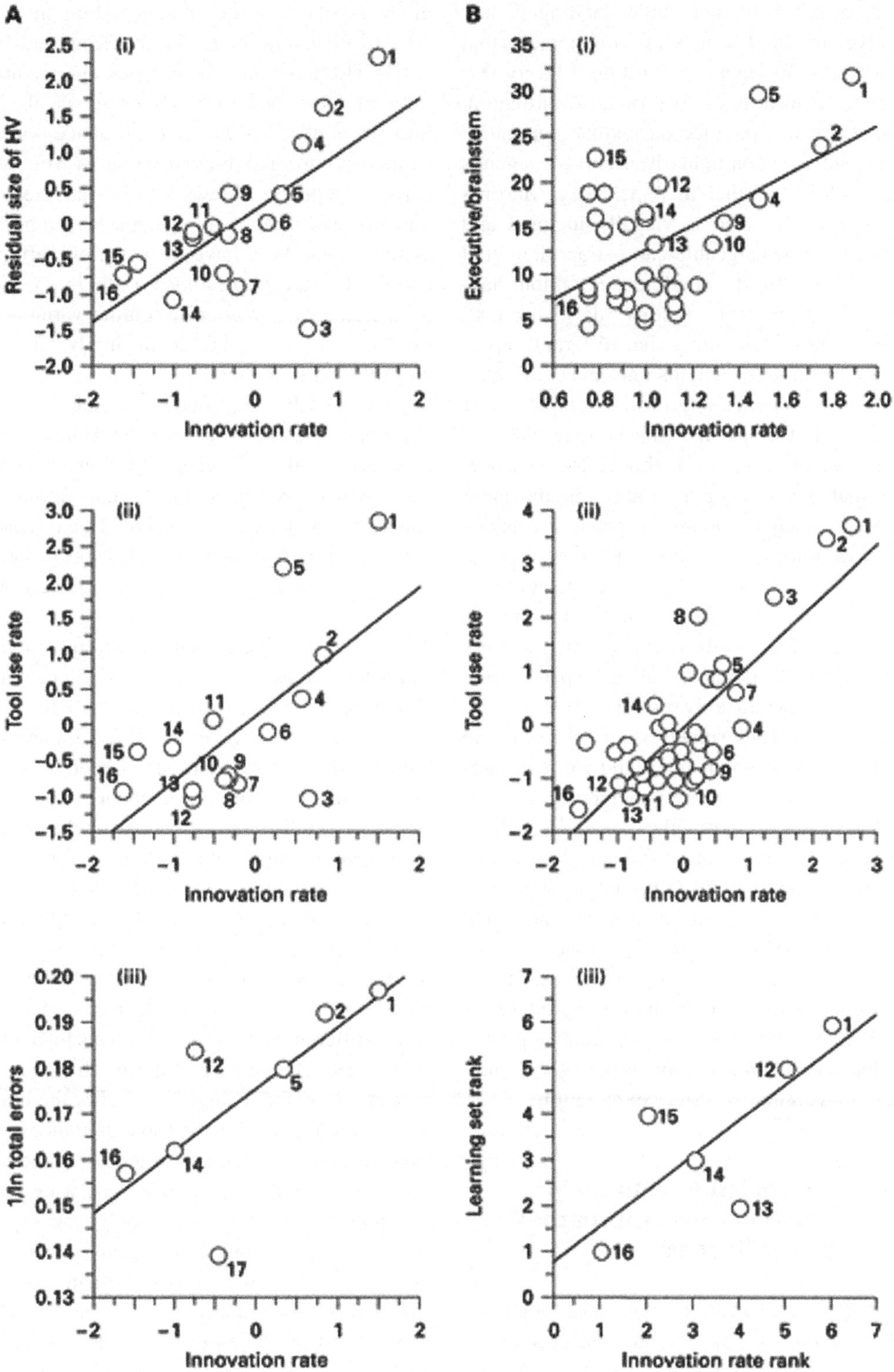

Fig. 3.3 Figure and legend from Lefebvre, L., 2004. A panel of six graphs showing how innovation rate (an indicator of higher intelligence) varies among primates and birds. Here Lefebvre shows that innovation rate is positively correlated with relative brain size. Reprinted by permission from "S. Krager AG": Krager, Brain, Behaviour, and Evolution. *Lefebvre L, Reader S, M, Sol D: Brains, Innovations and Evolution in Birds and Primates. Brain Behav Evol* 2004;*63:233–246. doi: 10.1159/000076784*. Copyright ©2004

by comparison with their understanding if they were given the full text to read. This suggests that improving figure legends would do a lot to help readers to skim articles for crucial information, and might reduce reliance on textual summaries like abstracts. It is particularly useful for legends to stand alone, so that in papers like Martin's (above), a reader who is visually inclined and statistically knowledgeable can extract the detail of key findings from a single page, without having to refer backwards or forwards in the text. This does, however, imply that different papers will require different lengths and levels of detail in their legends depending on the complexity of their figures. Comparing Martin (Fig. 3.2) to Lefevbre et al. (Fig. 3.3) above, for instance, shows that the numbered symbols in the latter necessitate a much longer legend. Decisions about symbology, legends, and other visual elements have knock-on effects for one another.

Despite Yu et al.'s (2009) findings that figures and legends alone, with a text abstract, did not convey the full contents of a paper, visual abstracts are increasingly popular in new academic journals. They sometimes use a key figure from the text, but can also be stand-alone designs a bit like cartoons or tiny conference posters. Visual abstracts are particularly useful for those needing to "preview" an article and decide how much more time to invest in it, and for sharing at meetings or on social media (Ramos and Concepcion 2020). Adopting some of the techniques used in creating visual abstracts might help scientists with ordinary figure legends too—though proponents of the new approach note that maintaining quality is already a challenge, so this requires further exploration.

3.2.2 Known Issues with Brain Size-Body Size Approaches to Intelligence

The simple use of relative brain mass to predict intelligence has its critics. Some smaller animals, such as the elephant shrew (Family: Macroscelididae), appear to have much larger brain sizes relative to their body sizes than we might expect, with their brains making up around 10% of their total body weight (Roth and Dicke 2005). Human brains, by comparison, are around ~2% of their total mass (Bélanger et al. 2011; Snodgrass et al. 2009). If intelligence was only about proportional brain size, then you would expect elephant shrews to be smarter than humans, and there is no evidence to support this assumption. There have been various efforts to correct for these presumed outliers, or adjust measurements to recognise scaling variation and find broader patterns instead of simply calculating proportional brain size.

One such adjusted measure is the encephalisation quotient. In 1973, Harry J Jerison proposed the encephalisation quotient (EQ) which measures the "actual" brain mass compared with the "expected" brain mass, to create a numerical value that attempts to indicate the level of intelligence (Jerison 1973). For our example above, if we used a sample of animals that included both elephant shrews and humans to characterise how brains change as you increase in body size, we can see whether the human value of 2% is *unusually* large or small for a mammal of human size, rather than just larger or smaller overall than that of the elephant shrew. There are thus no universal EQ values as the expected brain size is calculated each time for a study, based on the animals sampled. This means that the results of an analysis of EQs depend on which animals are included in any particular calculation. If, for example, you had a study that looked at the relative EQ of humans and elephants, that would give a different EQ score for humans than a study that compared humans and sheep. The value itself indicates how much larger or smaller the brain of the animal in question is compared to its expected mass, which has been calculated using the larger dataset of all included animals. In a wide sample of mammals, a typical value of 1 (the expected brain mass is the same as the actual brain mass) is usually assigned to cats, with humans obtaining the highest values at around 7 ~ 8 (Jerison 1973). The suggestion therefore is that the human brain is 7 ~ 8 times larger than would be expected, while a cat is a "typical" mammal with a brain exactly as large as we might predict (Roth and

Dicke 2005). However, EQ does not consider other factors such as age, gender, or body fat percentage (Cairó 2011). A study by Minervini et al. (2016), who calculated EQ values for the domestic pig (*Sus scrofa*), demonstrates this by showing how the EQ values varied between piglets (2.42), young adults (0.58), and adult pigs (0.38) without a noticeable change within their behavioural repertoire that would indicate changes in intelligence. Ultimately, in visual terms, the use of EQ means that after fitting the line that describes the overall pattern of a XY scatter plot, instead of comparing which animal falls *absolutely* highest on the y-axis you can look at which falls proportionally highest above the line.

Another alternative to using relative brain measures is using absolute brain size, or the raw data on the mass of a species' brain. The absolute mass of the brain in mammals ranges from <0.1 g in the Etruscan shrew (Naumann 2015) to ~9000 g in some species of whale (Roth and Dicke 2005), and so, on its own, cannot be used as an indicator of intelligence. It might, however, have advantages when comparing animals that are broadly similar in size. In particular, absolute brain mass has been suggested to provide a more accurate indication of higher-order brain function when working within a certain animal group or taxonomic orders (Benson-Amram et al. 2016; Deaner et al. 2007; MacLean et al. 2014; Street et al. 2017). This is because within such a group of animals, all species share the same brain scaling rules and are likely to be similar in size and ecology, making comparisons of actual size differences in the brain more meaningful than when comparing animals of very different groups.

3.2.3 How Linear Regressions Are Visualised (Continued)

3.2.3.1 Figure Axes Titles

The questions scientists ask when working within groups of animals are different to those they might look at on a broader scale, and often focus more on the relationship between brain size and some systematically applied measure of practical intelligence. This is because behavioural or ecological indicators of cognitive capacities that apply to animals with different diets, body plans, and social systems can be hard, but it is much easier to do for groups that are more uniform. As an example, Fig. 3.4 shows a figure from Deaner et al. (2007) (pg. 115), who suggests that "absolute brain size measures were the best predictors of primate cognitive ability" when compared with the encephalisation quotient and relative brain mass measures.

Figure 3.4 contains a lot of complicated information that, unless you are a specialist in the area and have been able to read the accompanying method text in detail, is perhaps harder to understand. For example, a lay reader may not start the paper knowing what unweighted regression, progression index, or global cognition mean. Even if they have read the method, a quick note reminding them of the definition of the progression index or global cognition within the figure legend could aid quick comprehension of the figure. Having only one y-axis label ("Global Cognition") is an effective way of reducing the figure's word count and increasing the clarity. Labelling each of the 6 graphs within the figure with the y-axis would not add anything specifically helpful to the graph but would increase the visual "clutter" of an already complex figure. The x-axes, on the other hand, could have their readability slightly improved. For example, saying "log brain *mass*" instead of just "log brain" does not really impact on the overall impression of the figure, but could reduce the cognitive load (Cook 2006) on the reader who no longer has to refer back to the methods to interpret an abbreviation as well as their definitions. There is an obvious trade-off here between readability and clarity, and perhaps another between house style (e.g. a stated preference for short, succinct legends) and the broader accessibility of the figure. This might be particularly important if a paper is intended to have a wide audience, or to be used for science communication, e.g. as a visual abstract, but it is worth considering at other times too.

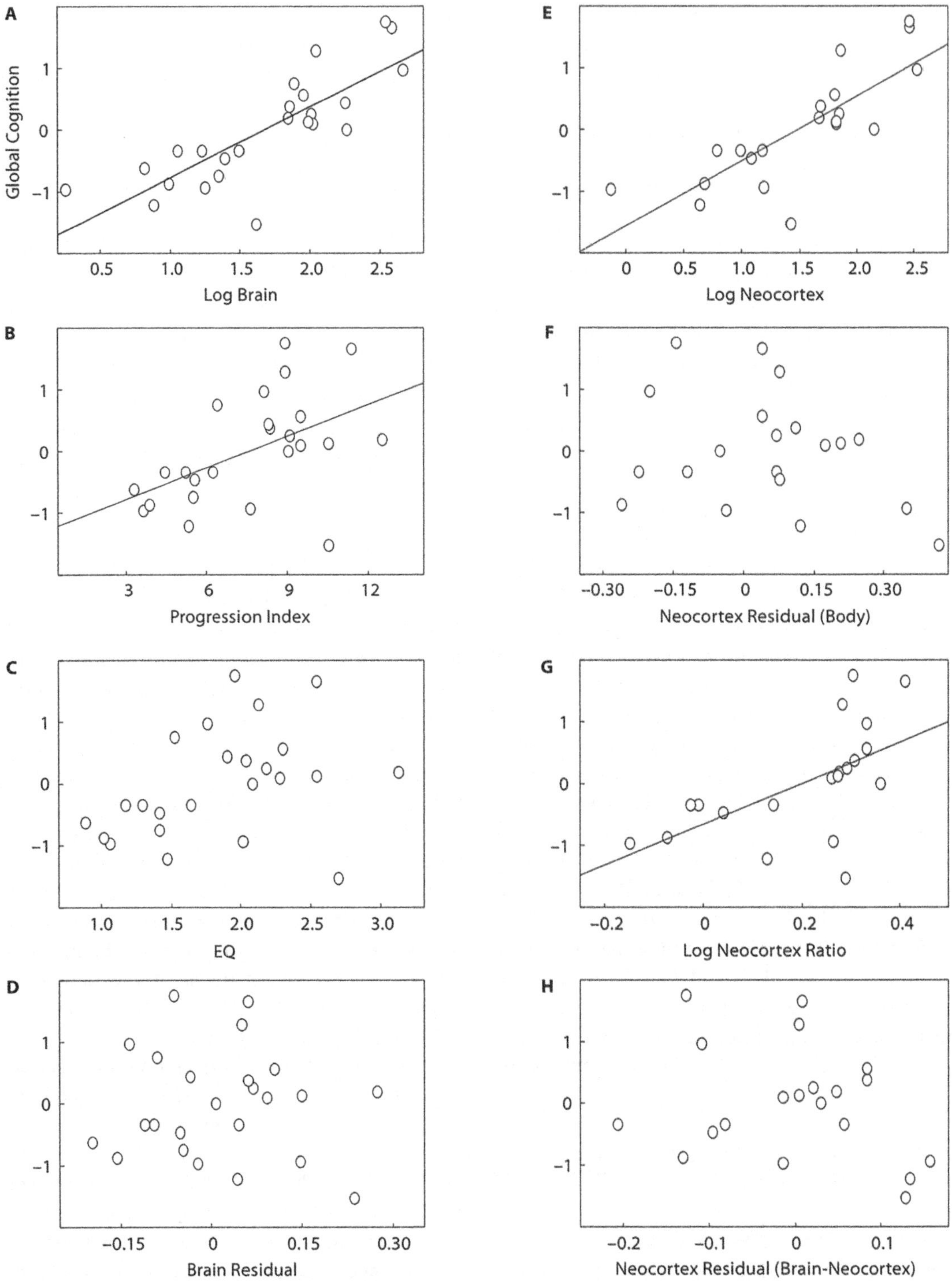

Fig. 3.4 Figure and legend from Deaner, R.O. et al., 2007 who suggest that absolute brain size measures were the best predictors of primate cognitive abilities. The bigger the primate brain, the more intelligent they seem to be. The important graph for this finding is (**a**) and (**b**) as it shows the total size of the brain or neocortex (where intelligence is thought to originate from) having a positive correlation with their cognition score. Reprinted by permission from "Krager Publishers": Krager. Brain, Behavior and Evolution. *Deaner, R.O., Isler, K., Burkart, J., and Van Schaik,*

3.2.3.2 Statistics

The inclusion of statistical information within a figure is quite often used to aid quick understanding of the data, though it arguably only works if the reader already has an understanding of statistics or specific field knowledge. The presentation of statistical visualisations to a general, non-specialist audience is discussed in Sect. 3.4, but even within the scientific literature intended for a more specialist audience, attention to how statistics can be incorporated into visuals is important.

As discussed previously, figures and their legends should ideally be self-contained so all the information the reader requires is present without reference to the remainder of the article. If we look back at Fig. 3.2 (Sect. 3.2.1.2), even though this graph is from 1981, statistics are still included in the form of regression line equations. As figures are, by definition, visualisations, using additional visual features upon the graph is a good way of adding more information. Figure 3.4 does this by adding regression lines to the charts which demonstrate significant differences only, therefore, immediately showing the reader which charts are significant. Within statistics, this could also be the inclusion of the asterisk symbol (*) to indicate the level of significance. Using a note in the legend that states, "*p>0.05 (NS), p<0.05 (*), p<0.01 (**), p<0.005 (***), p<0.001 (****)*", and then labelling the figure with the corresponding symbol would show the reader the level of significance within the figure. This is a good way of conveying a lot of information through a simpler, visual means, and is useful for the reader that already understands what a p-value is. It is also unlikely to be very distracting for those who don't understand the p-value, as the asterisk symbology does not dominate the figures.

3.2.3.3 The Use of Colour

The use of colour in scientific visualisations has become more widespread as online publication has become the norm. Traditionally, all science communication has been through physical journals, usually held in University libraries, and the technology used in both the printing and dissemination of journal articles has undergone a quantum change over the last few decades. Even comparing 1981 (Fig. 3.2) and 2007 (Fig. 3.4) there is a noticeable lack of colour in both our chosen figures, and some journals still require black-and-white versions for print, which might discourage its addition. Colour is, however, not only visually appealing to the reader, but, if used correctly, it can add an additional layer of information to the figure. Similarly to using shapes (as discussed earlier), additional layers of information through the addition of colour allow for more patterns, relationships, and outliers to be visualised on the figure, and can potentially enhance the reader's intuitive ability to grasp patterns.

In 2014, MacLean and his colleagues undertook behaviour experiments (two tests, called the "A not B task" and "Cylinder test") to assess cognitive ability in eight groups of animals. The measurements that they took for both tasks were combined to create a composite graph (Fig. 3.5). The findings suggested that across species, the absolute brain volume (marked as ECV on the figure) was "a robust predictor of performance" across all tests. This method of taking two similar graphs and combining them to create a composite graph can be used if you want to run a series of different tests, but end up with an overall result. It is also a nice example of an animal cognition chart that uses colour effectively to bolster communication of its findings.

Fig. 3.4 (continued) C., 2007. *Overall brain size, and not encephalization quotient, best predicts cognitive ability across non-human primates. Brain, Behaviour and Evolution, 70(2), pp.115–124*. Copyright © 2007, Krager Publishers, Basel, Switzerland

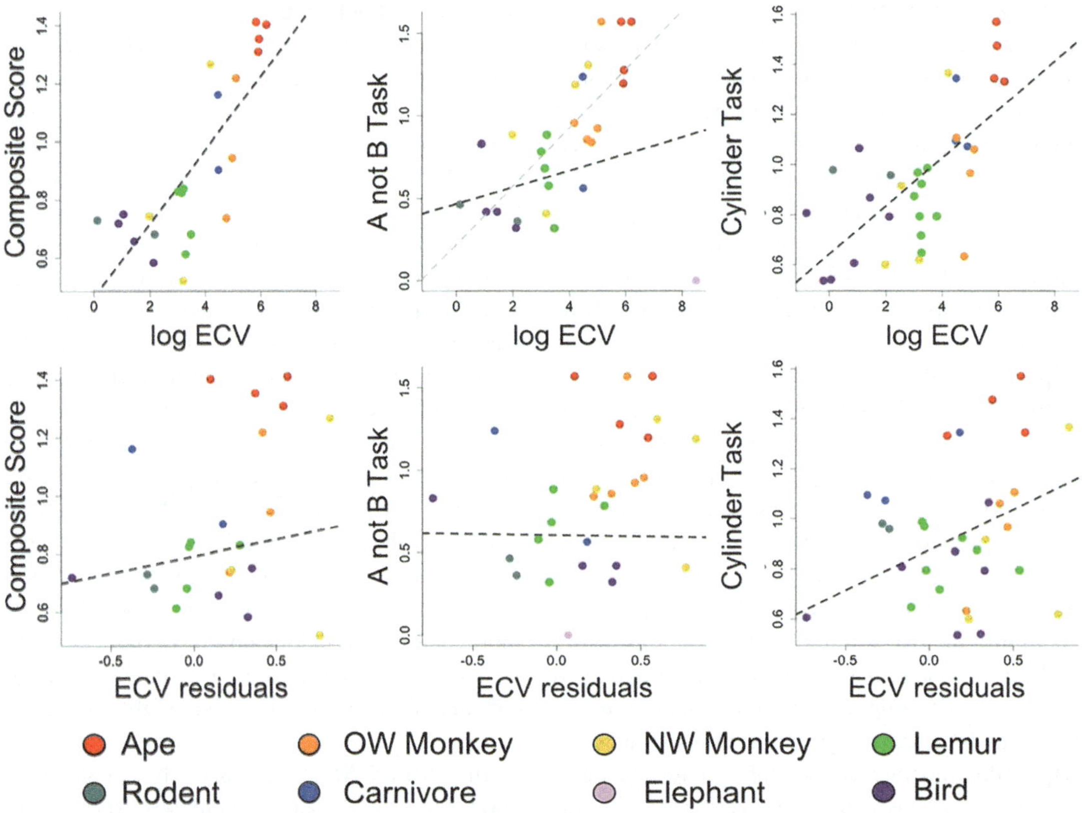

Fig. 3.5 Figure and legend from MacLean, E.L. et al., 2014. Eight groups of animals were tested in two different cognition experiments, the "A not B test" and "Cylinder test". The panel displays results from those two tests and a composite score of both for Endocranial Volume (ECV or Absolute Brain Mass) and Endocranial Volume Residuals. Reprinted by permission from "Proceedings of the National Academy of Sciences": PNAS. MacLean, E.L., Hare, B., Nunn, C.L., Addessi, E., Amici, F., Anderson, R. C., Aureli, F., Baker, J.M., Bania, A.E., Barnard, A.M. and Boogert, N.J., 2014. The evolution of self-control. *Proceedings of the National Academy of Sciences,* 111(20), pp. E2140-E2148

Figure 3.5 clearly demonstrates the benefits of using colour to distinguish between species, especially if compared with the black-and-white symbol approach shown in Fig. 3.2. It is visually appealing, and a reader can immediately see that apes (red) consistently performed highest in both tests while birds (dark purple), lemurs (green), and some New World monkeys (yellow) scored on the lower end. It also clearly identifies elephants (pink) as an outlier point on the top middle graph (log ECV against "A not B task").

So, even just looking at two example figures (Figs. 3.4 and 3.5) that use absolute brain mass as a proxy for intelligence, it has become clear how the presentation of the data has evolved, even over a short time, to become more accessible to readers of both specialist and non-specialist backgrounds. However, there are still problems with using absolute brain mass as an indicator of higher cognition. For those who want to understand cognition more broadly, the hunt is still on: there is still a need to find a single, clear-cut and easily interpretable biological proxy for intelligence. Recent efforts have tended to focus on cortical neuron counts and scaling relationships that describe patterns in brain *composition*, rather than just in size.

3.2.4 Cortical Neuron Count as a Biological Proxy for Intelligence

While absolute brain mass has been shown to correlate with cognitive abilities, it can only be used within those animal groups (taxonomic orders) which share the same brain scaling rules. Scaling, when used in the context of brains, describes how the neuron composition of the brain changes as each brain type changes in volume/mass. There are no universal brain scaling rules. Instead, specific groups like primates or carnivores each have their own scaling patterns. Understanding the differences help us show, for example, that in a rat (or any rodent) as the brain increases in size, the neurons get larger and are added at a slower rate, whereas in humans the neurons remain the same size, so as the brain gets larger, proportionally more neurons are added. Ultimately, this results in more neurons within a primate brain compared to a rodent brain of the same mass, which allows primates to pack in more neurons without the additional costs of a larger (and heavier) brain (Gabi et al. 2010; Herculano-Houzel et al. 2007; Herculano-Housel 2009).

Looking at scaling rules gives us a more nuanced way to assess potential intelligence. The human brain is divided into three different broad sections: the brain stem, which controls basic functions such as heart rate and breathing, the midbrain or cortex that includes feelings and emotions such as fright and flight, and the new or neocortex which is the part on the outside with the distinctive folds in it. This section, the human neocortex, is where abilities and attributes indicative of higher intelligence are thought to originate from. In humans it contains around 16 billion neurons which makes up a massive proportion of the overall brain neuron count (around 19% of a total of 86 billion neurons) (Azevedo et al. 2009). Humans have the largest number of cortical neurons of any species on the planet (Herculano-Houzel 2012), and it is this, the sheer amount of cortical neurons, that has been suggested to account for our enhanced cognitive abilities. Even when compared with brains that are 3 ~ 4 times larger than ours, such as the African elephant's (*Loxodonta africana*) (whose whole brain contains 257 billion total neurons), different mammalian scaling patterns mean that humans still have far greater numbers of cortical neurons (16 billion as opposed to 5.6 billion) (Herculano-Houzel et al. 2014).

If cortical neuron count is an accurate indicator of what currently constitutes "intelligence", it could potentially be used to compare animals across the vertebrates. Non-mammalian vertebrates have an equivalent structure to the neocortex, called the dorsal pallium (Herculano-Houzel 2017). The amount of neurons in either of these two structures thus gives us a new contender for a simple biological proxy for intelligence that has neither of the disadvantages we have already considered. It does not use body mass (such as relative brain mass or the encephalisation quotient do), and so does not have the associated problems discussed earlier in the chapter, and as it bypasses the issues of brain scaling, can be applied much more widely than any absolute measure of brain size. This leads to a new generation of papers looking at cortical neuron counts. This methodology tends to use quite complex datasets, and correspondingly relies on combinations of strategies to make figures more accessible (see below).

3.2.5 Points of Interest and Print/Disability Friendly Figures

3.2.5.1 Flagging Points of Interest

Suzana Herculano-Houzel's 2017 paper (Fig. 3.6) highlights how the number of neurons in the cortex (or pallium) correlates with cognition measures, and how the pallium within birds can pack more neurons than some primate cortices do of similar masses. This specific visualisation is a good way of pulling lots of the previous points made in this chapter together to form a more accessible visualisation despite it including some very complex data.

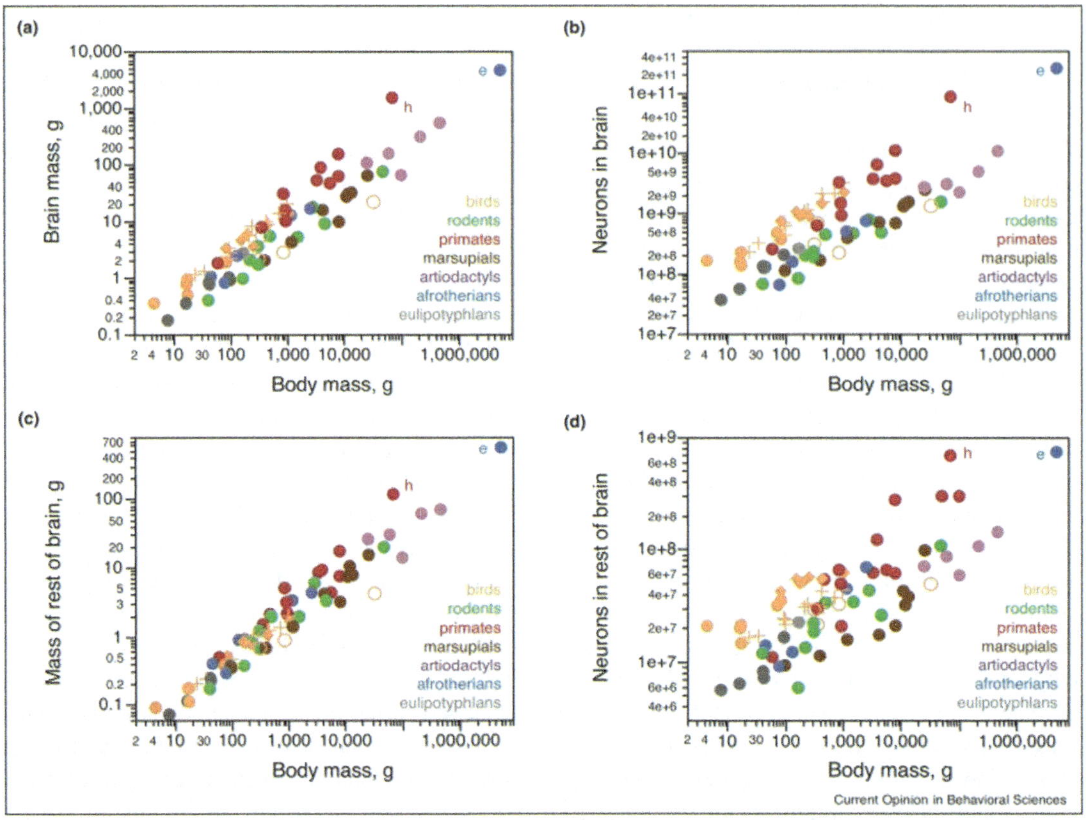

Scaling of numbers of neurons in the cerebral cortex or pallium against numbers of neurons in the rest of brain.
Each point represents one species belonging to the different clades according to the colors in the legend. Among birds, Passeriformes are represented by orange filled circles and lozenges; psittaciformes by orange crosses; and barn owl, emu, pigeon and jungle fowl by unfilled orange circles. **(a)** Scaling of absolute number of neurons in the cerebral cortex with absolute number of neurons in the rest of brain. Exponents, p-values and r^2 values for the different clades: afrotherians, 1.264 ± 0.073, $p < 0.0001$, 0.987; artiodactyls, 1.904 ± 0.172, $p = 0.0016$, 0.976; eulipotyphlans, 0.932 ± 0.142, $p = 0.0072$, 0.935; marsupials, 1.374 ± 0.245, $p = 0.0008$, 0.818; primates, 1.457 ± 0.120, $p < 0.0001$, 0.919; rodents, 1.103 ± 0.130, $p < 0.0001$, 0.900; Passeriformes, 2.089 ± 0.188, $p < 0.0001$, 0.918; Psittaciformes, 2.768 ± 0.212, $p < 0.0001$, 0.950; ensemble of afrotherians (minus the African elephant), eulipotyphlans and rodents, 1.089 ± 0.064, $p < 0.0001$, 0.938 (plotted); all data points together, 1.526 ± 0.093, $p < 0.0001$, 0.776. **(b)** Scaling of the ratio between numbers of neurons in the cerebral cortex of mammals (or pallium of birds) with the number of neurons in the rest of brain. **(c)** Scaling of the residual number of neurons in the cerebral cortex of pallium after regressing onto the number of neurons in the rest of brain, as in (a), with the absolute number of neurons in the rest of brain. **(d)** Scaling of the difference between number of neurons in the cerebral cortex or pallium and rest of brain with the absolute number of neurons in the rest of brain. h, human data point; e, African elephant data point.

Fig. 3.6 Figure and legend from Herculano-Houzel, S., 2017. This figure shows how the number of neurons in the cortex (or pallium) correlate with cognition measures, and how the pallium within birds can pack more neurons than some primate cortices do of similar masses Reprinted by permission from "Elsevier": Current Opinion in Behavioral Sciences. *Herculano-Houzel, S., 2017. Numbers of neurons as biological correlates of cognitive capability. Current Opinion in Behavioral Sciences, 16, pp. 1–7.* Copyright © 2017, Elsevier

Figure 3.6 has a very detailed legend that explains what each part of the plot shows, as well as the significance values for the graph which are in a consistent format—making it easier to read, at least for those readers with the statistical knowledge required to understand this content (see Sect. 3.2.3). It also uses clear, concise axis titles and colour schemes with legends on the figures to allow quick interpretation.

A new visual strategy this figure deploys is the combination of letters and colours as labels to pick out points of interest. The human ("h") and African elephant ("e") are labelled on all graphs

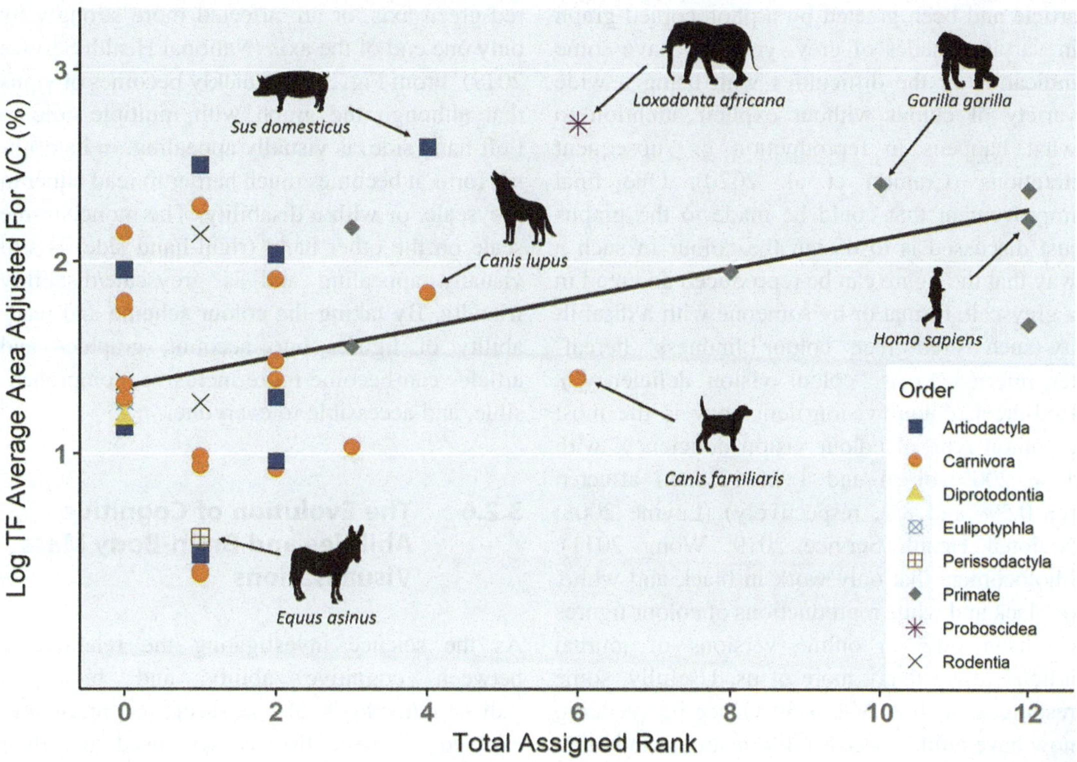

Fig. 3.7 Figure created from Lunn et al. (2022) showing a linear regression analysis of measurements on a feature on cervical vertebra (transverse foramina) which indicates the level of blood flow to the brain, and a ranking system that is indicative of intelligence, where the higher the rank, the higher the proposed intelligence. Transverse foramina measurements are averages between both foramina present, and then displayed as a percentage of the vertebral canal (also present on the cervical vertebrae) to adjust for total organism size across the data set. This figure indicates that the higher the blood supply to the brain (so larger average transverse foramina measure), the more behavioural traits indicative of higher intelligence the species has

while remaining the same colour as their close relatives. Furthermore, the letter labels are discreet enough not to cause visual clutter or distraction, but prominent enough to be easily spotted and looked up in the legend by a curious reader. Identifying points of interest is not necessarily a new feature of figure creation, as (for instance) Martin used an arrow to identify a dolphin in his 1981 figure (Fig. 3.2). But using these identifiers helps place the data in context for the audience, and is a good way of identifying points at either extreme as well as possible outliers. Figure 3.7, by Lunn et al. (2022), shows another example where points of interest are identified through imagery rather than letters. For example, Fig. 3.7 highlights that pigs lie away from the remaining points in their group (Artiodactyls), donkeys are the only member of their group included (Perissodactyla), and wolves and dogs can be rapidly compared by those interested in intelligent carnivores. Both Figs. 3.6 and 3.7 thus manage to highlight *both* the overall patterns their authors found, *and* the placement of specific animals that might be of particular interest to some or all of their readers.

3.2.5.2 Print and Disability Friendly Figures

The use of colours and shapes has become increasingly more common to add additional layers of information to figures, but if you have ever opened an examination paper or journal article and been greeted by a photocopied graph in varying shades of grey, you will have some indication of the difficulties with using a wide variety of colour without explicit attention to what happens in reproduction or subsequent iterations (Crameri et al. 2020). One final improvement that could be made to the graphs just discussed is to design the colour in such a way that the figure can be reproduced and read in a greyscale format or by someone with a disability (such as red-green "colour blindness", hereafter referred to as "colour vision deficiency"). Red-green colour vision deficiency is the most common type of colour vision deficiency, with 1 in 200 women and 1 in 12 men affected (or 0.5% and 8%, respectively) (Levine 2008; National Health Service 2019; Wong 2011). Photocopiers that only work in black and white, or black and white reproductions of colour figures available only in online versions of journal articles affect many more of us. Usefully, some resources, such as Microsoft Office PowerPoint, now have built in accessibility features, and other online sources (such as Coblis (Wickline 2022) which allows a user to see what their visualisations look like to someone with colour vision deficiency) are allowing authors to see their visualisations in different ways, and, hopefully, resulting in more inclusive and accessible visualisations.

Figure 3.8 visualises these differences by using an example, fictitious, data set created by the author to show how colour choice can affect the figure and therefore the reader.

Figure 3.8 shows how a bar chart (right) in a monochrome scale (blue in this example) can be read in its natural form, a greyscale print, and for someone who lacks the full range of colour vision (red-green colour vision deficiency), compared with a graph with a range of colours (left). It is important to note here that the colour vision deficiency graph is simply used as an indication of how someone who is affected by it, *may* see the figure. Colour vision deficiency can affect each person differently, with some people seeing duller versions of the colour while others may be unable to differentiate between the two extremes of the red-green axis, or are affected more strongly by only one end of the axis (National Health Service 2019). From Fig. 3.8, it quickly becomes obvious that although the graph with multiple colours (left-hand side) is visually appealing, in its original form, it becomes much harder to read either in greyscale, or with a disability. The monochrome scale on the other hand (right-hand side) is still visually appealing and is greyscale/disability friendly. By taking the colour scheme and readability of figures into account, graphics and articles can become more inclusive, comprehensible, and accessible to everyone.

3.2.6 The Evolution of Cognitive Abilities and Brain-Body Mass Visualisations

As the science investigating the relationship between cognitive ability and biological indicators/methods of measuring cognition has evolved, so have the datasets used and their visualisations. In particular, there has been a major move away from using relative brain size measures as proxies for intelligence, and towards either restricted but simple measures like absolute brain size or—more recently—measures like cortical neuron count. While the basic display of this data has remained the same, in the form of XY scatter plots representing linear regression analyses, the complexity of charts has changed in line with changes in the datasets used, and as technology has progressed. As science has moved on to online publishing and efficient printing, detailed computer-generated graphs with finite details and additional features (such as colour, etc.) have been added into figures. There also seemed to be a trend in earlier research that assumed much more about the statistical competence of the reader, though some elements even of recent figures (like Fig. 3.6's exponential axis labels) suggest this assumption still holds for academic papers. While many readers will understand the statistics within these articles, making figures and legends more detailed and easier to comprehend improves the accessibility of science

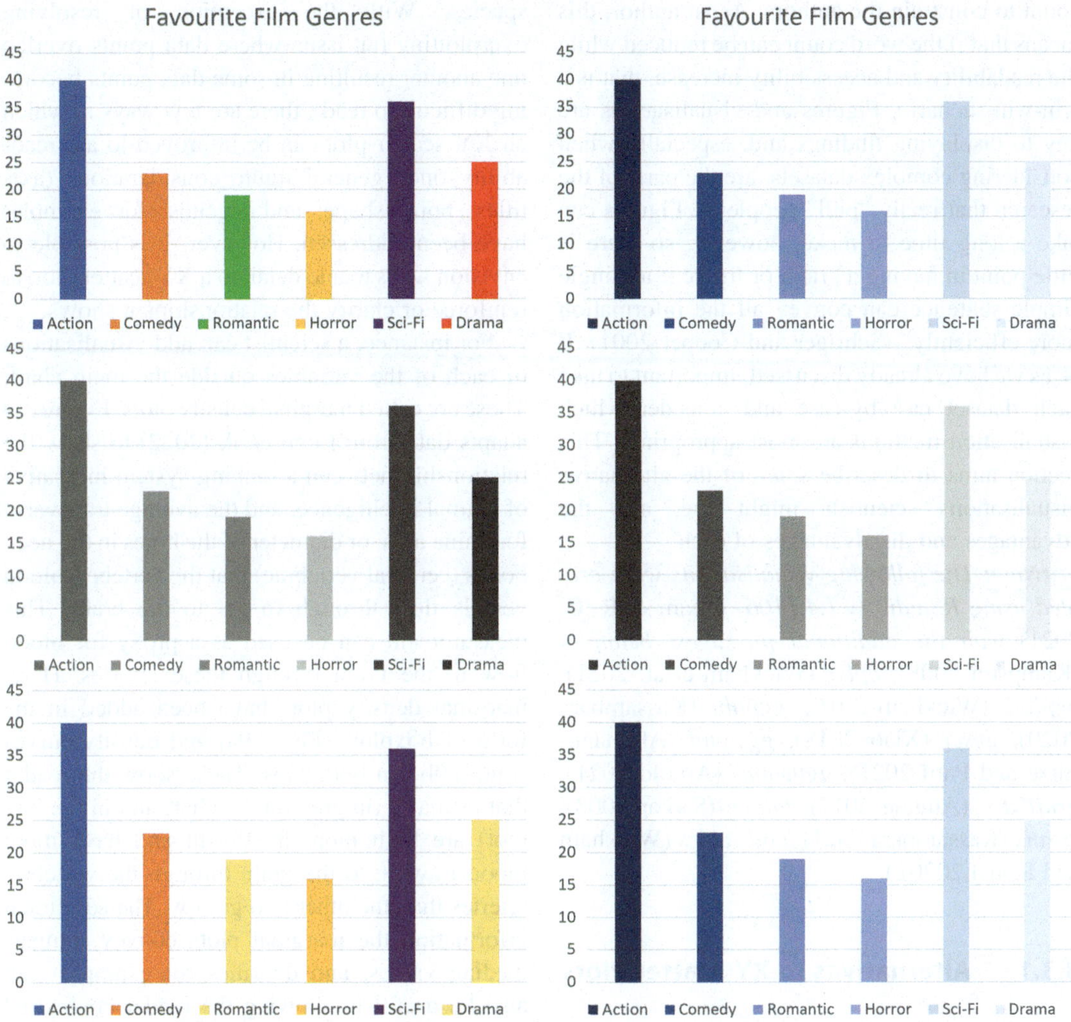

Fig. 3.8 Bar charts with a fictitious data set created by the author to demonstrate how colour choice can affect the figure (and therefore, the reader). The left hand demonstrates a "bad example" with multiple colours, while the right-hand side demonstrates a "better example" with a blue monochrome scale. The first horizontal row represents how the graph would look in its natural colour form. The second horizontal row represents a greyscale version of the same graph. The third horizontal row shows a proposed view for someone with colour vision deficiency—specifically red-green

to anyone and everyone, and fits well with the current trend towards greater inclusivity in general, e.g. via use of colour and greyscale (print) friendly formats. Overall, our explorations of even this small part of the scientific literature has turned up multiple ways in which authors' choices and constraints both affect the accessibility of their visuals.

3.3 Approaching Visualisation Differently

Figures can be invaluable tools for scientists. Scientific writing is typically concise, with articles using a standard introduction, method, results, discussion, conclusion (sometimes called IMRaD) structure, and journals imposing a word count to constrain the authors. As an author, this means that if the word count can be reduced while the readability and accessibility increase, that is a win-win scenario. Figures and visualisations are key to displaying findings and, especially when considering complex datasets, are the parts of the research that really "pull" people in. Figures can take a long time to make, however, so there is little point in having a graph or figure if a "single simple sentence can convey all the information more efficiently" (Schriger and Cooper 2001). It is, as we have already discussed, important to take each dataset case-by-case and consider which visualisation methods are most appropriate. This section aims to describe *some* of the alternative visualisations scientists might use, and the advantages and disadvantages of each.

(*Note: The following visualisations were created using RStudio (v.1.4.1106)* (Team & R. C. 2021) *with the additional packages: beanplot* (Kampstra 2008); *dplyr* (Wickham et al. 2021); *ggplot2* (Wickham 2016); *ggpubr* (Kassambara 2020); *ggsci* (Xiao 2018); *ggsignif* (Ahlmann-Eltze and Patil 2021); *ggthemes* (Arnold 2021); *gridExtra* (Auguie 2017); *lattice* (Sarkar 2008); *rstatix* (Kassambara 2021); *and scales* (Wickham and Seidel 2020).)

3.3.1 Alternatives to XY Scatter Plots

As Sect. 3.2 showed, in the field of animal cognition the use of brain-body mass ratios and specific biological proxies for intelligence usually results in articles that analyse data with linear regressions. In the papers surveyed here, which dated between 1981 (Martin 1981) and 2022 (Lunn et al. 2022), the natural visualisation method has been an XY scatter plot. XY scatter plots, as noted, are good at displaying the relationship between two continuous numerical variables and can be useful for identifying patterns or outliers within the data set. A simple addition to a basic XY scatter plot is the separation of data by a third categorical or numerical value (i.e. through different colours or symbols). In the examples shown in Sect. 3.2, these categorical variables may be groups of animals or species. With the exception of resolving overplotting (an issue where data points overlap one another resulting in some data points becoming difficult to read), there are few ways in which an XY scatter plot can be improved to aid readability once general figure considerations (axis titles, point shape, and legends, for example) have been addressed. However, it is possible to envision ways to add detail to a XY scatter plot, to reinforce or clarify the relationships it shows.

For instance, a scientist can add visualisations of each of the variables outside the main chart. These are called marginal density plots. Figure 3.9 adapts data from Lunn et al. (2022) to show the relationship between a ranking system indicative of animal intelligence, and the average transverse foramina area, or diameter of the holes in the neck bones (cervical vertebrae) that the vertebral blood vessels flow through to get to the brain. This measurement can be used as a proxy for blood flow to the brain through these arteries. Here, marginal density plots have been added in the form of boxplots (Fig. 3.9a) and density curves (Fig. 3.9b) on both axes. These show the reader that primates (in grey on the chart and in the box plot) are both more intelligent and have more blood flowing to the brain through the vertebral arteries than the other two groups. The additional information the marginal plots convey (ranges, median values, and densities for example) can also be useful in showing the reader contextual information about the data that could not easily be incorporated into the symbology of the main plot.

3.3.2 Boxplots and Beanplots

While those working on measuring intelligence and brain size usually use XY scatter plots, it can be informative to consider alternatives. Another typical visualisation approach is through boxplots. Boxplots are designed to show the distribution(s) of values of a continuous variable within one or more groups. As an example, a boxplot could show the range and amount of overlap in ranges of heights of males and females in a single population side by side. Boxplots are good for comparing distributions of large

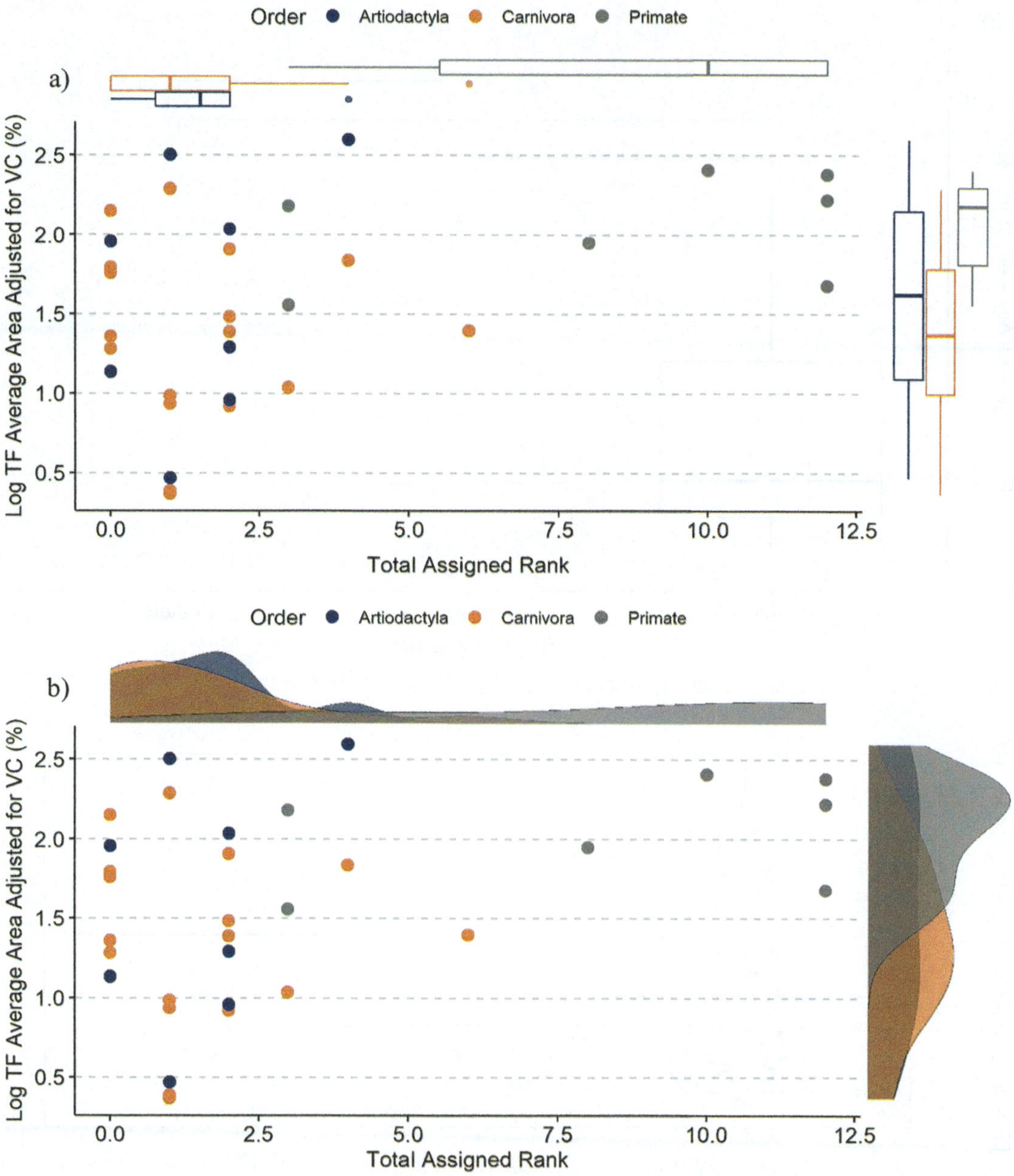

Fig. 3.9 Two XY scatter plots with marginal density plots adapting data from Lunn et al. (2022) to show the relationship between a ranking system indicative of animal intelligence ("Total Assigned Rank") and the logged average transverse foramina area (an accurate indicator of vertebral artery blood flow to the brain), adjusted for overall size through measurements of the vertebral canal on cervical vertebrae. Three taxonomic orders (*Artiodactyla: Blue, Carnivora: Orange,* and *Primate: Grey*) are plotted with two types of marginal density plot (boxplots (**a**) and density curves (**b**)) included to demonstrate the additional information that can be added to standard XY scatter plots

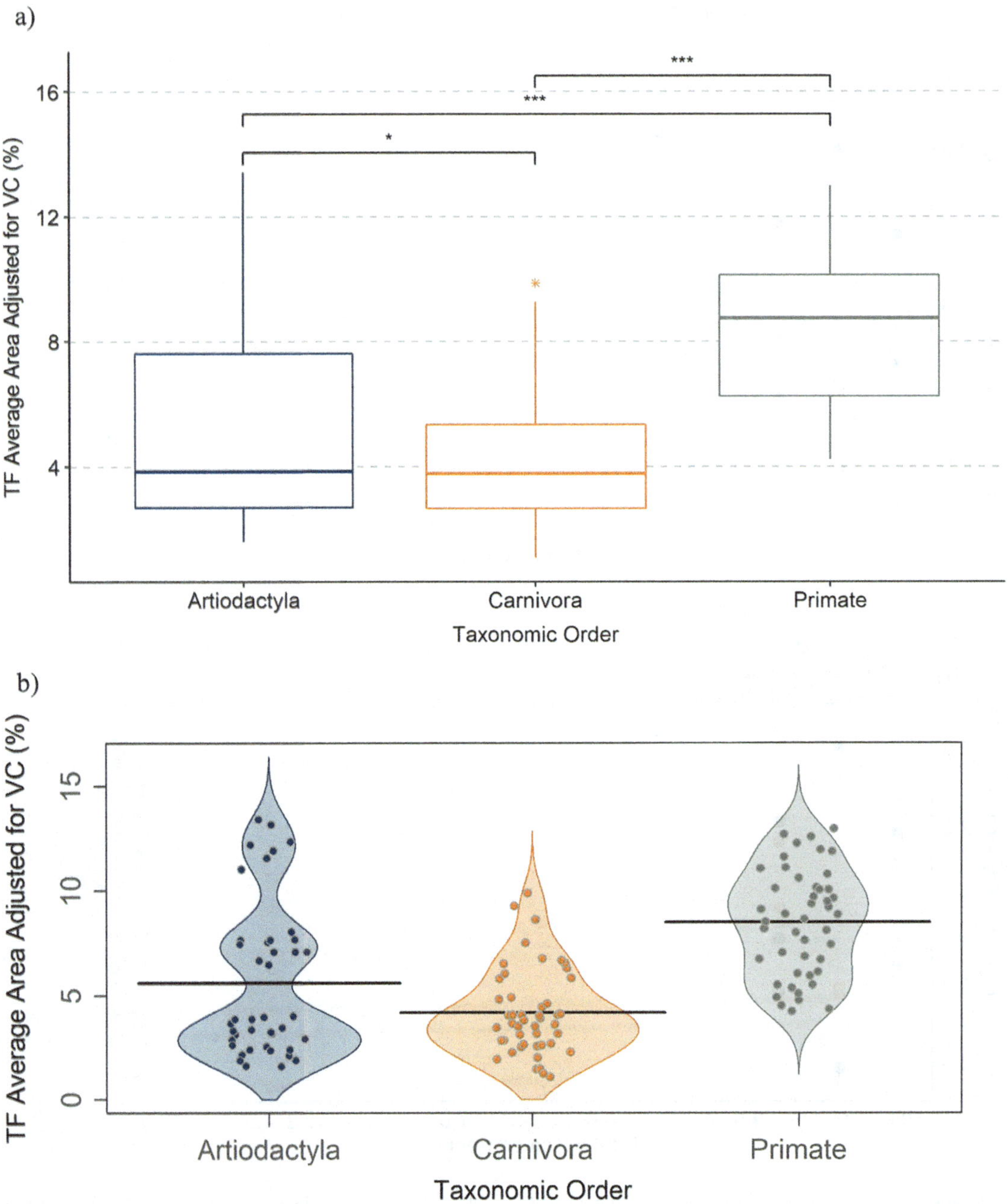

Fig. 3.10 (**a**) Boxplot of taxonomic order (*Artiodactyla: Blue, Carnivora: Orange,* and *Primate: Grey*) with average transverse foramina area (an indicator of blood flow to the brain through the vertebral arteries), adjusted for total organism size through measurements of the vertebral canal. Data adapted from Lunn et al. (2022). A one-way ANOVA test was undertaken and found a statistically significant difference between taxonomic orders and average foramina area ($F(2,132) = 28.588$, $p < 0.005$***). Further Tukey post-hoc analysis showed specifically strong significant differences between Artiodactyl and Primates ($p < 0.005$***) and Carnivores and Primates ($p < 0.005$***), with primates obtaining larger average transverse foramina areas. Between Artiodactyl and Carnivora there was also a statistically significant difference ($p < 0.05$*); however, this was less strong. Outliers are shown through 8-point star symbols. Tukey-HSD post-hoc significance values are shown above each "box" ($p > 0.5 = ns$, $p < 0.5 =$ *, $p < 0.01 =$ **, $p < 0.005 =$ ***). Figure (**b**) Uses the same data (and, therefore, same statistical values) as the boxplot but offers an alternative visualisation

datasets, but do not usually work on smaller datasets and generally do not show the individual data points (Franzblau and Chung 2012; Schriger and Cooper 2001).

Figure 3.10 again adapts data from Lunn et al. (2022) to show how blood flow to the brain varies between three animal groups. Figure 3.10a shows the data visualised as a boxplot, while Fig. 3.10b shows the same information visualised in a beanplot. Beanplots expand upon boxplots to include individual data points as well as the density curve (bean shape). The inclusion of a density curve has already been visualised as a marginal density plot (Fig. 3.9b), but in a beanplot, it is integral to the visualisation. The "shape" of the data can be easily identified through this method as well as the density of data, i.e. where most of the data points fall on the graph. Specifically, a beanplot does not create an artificial box shape around a set of averages and extreme values the way a boxplot does. Instead, each beanplot shape is different, with a wide part of the plot indicating lots of variation, and a narrow part indicating limited variation.

For example, when comparing the two parts of Fig. 3.10, you can instantly see on the beanplot the small cluster of artiodactyl points (even-toed ungulates including pigs, sheep, cattle, giraffes, deer, hippopotami, etc.). These would not be picked out on the boxplot. Furthermore, it is easier to see on the beanplot that more of the primate points fall above the median line than below it (where there is a larger spread). This tells the reader that there are more primates that are above their group average, and that more carnivores and artiodactyls are below their group's average. Ultimately, as for most visualisation choices for figures, the quality and attractiveness of the final output of a boxplot or beanplot is dependent on the specific dataset and author/editor preference—but it is still important to consider all options in light of how much information the author wants or needs to communicate, and how effectively each approach can do this.

3.3.3 Bar Charts and Pie Charts

As well as box- and bean-plots, another common way to visualise measurement data is through the production of bar charts (or graphs) and pie charts. Bar charts are widely used for making comparisons and reading them relies on the ability of the reader to judge the relative height of columns. Being relatively simple to read, however, does not mean they require less careful attention to their construction than other types of visualisations. All the usual figure elements such as labelling, text size, and axes configuration need to be thoughtfully considered, and some features that may visually pull in the reader, such as the use of 3D plots or colour simply for decoration, are best avoided as they may detract from the clarity of information presentation (Franzblau and Chung 2012). Even with simple plots that only compare a few groups, other alterations can be made to increase the clarity of the figure. Figure 3.11 demonstrates just a few of these possible alterations using data from Lunn et al. (2022). For example, in Fig. 3.11b the bars have been reordered from alphabetical order to descending order. This allows values that are close to one another (such as the gorilla and chimpanzee) to be directly compared. Furthermore, the use of colour, here within the same colour palette, increases the visual appeal and separates each bar visually from the next. It is important to note that the green colour palette here does not impede the accessibility as there is no comparison with other primary colours (red for example), and the shading will still be present

Fig. 3.10 (continued) method for the data. Individual data points (dots), the median value (horizontal black line), and density of points (density curve/"bean" shape) are all shown upon the plot

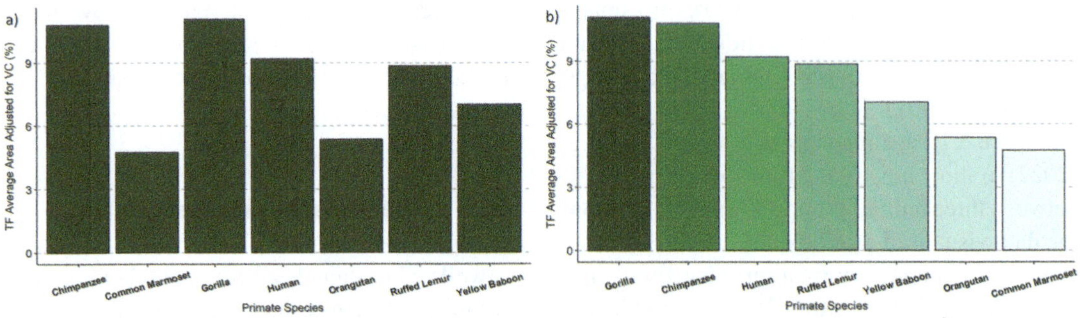

Fig. 3.11 A standard bar chart (ordered alphabetically, as is typical in some statistics package default settings (**a**), and ordered in descending order (**b**)) of seven primate species and the average transverse foramina area (an accurate indicator of blood flow to the brain through the vertebral arteries), adjusted for overall organism size through measurements of the vertebral canal on cervical vertebrae—data from Lunn et al. (2022). The use of colour within the right-hand graph indicates different species

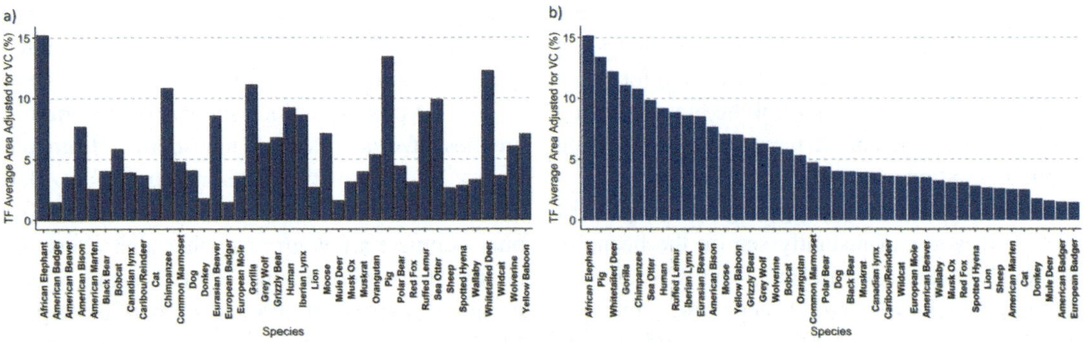

Fig. 3.12 A standard bar chart (left ordered alphabetically (**a**), and right ordered in descending order (**b**)) of 40 mammal species and the average transverse foramina area (an accurate indicator of blood flow to the brain through the vertebral arteries), adjusted for overall organism size through measurements of the vertebral canal on cervical vertebrae—adapted from Lunn et al. (2022)

even if the reader is affected by the green colour (perhaps seeing shades of grey or a dull green instead).

However, while this graph is easy to interpret, some data sets contain many more categories than this one. Figure 3.12 shows bar charts of 40 mammal species and their average transverse foramina area sizes visualised in both an alphabetical order (Fig. 3.12a), and descending order (Fig. 3.12b). While both bar charts clearly label each animal, the sheer number of categories present detract from the overall readability and accessibility of either chart. Unless the intention is to show something about the distribution of scores (e.g. using descending order patterns to highlight an outlier or that some animals have high values and others have low values in a stepped pattern), these charts may not be helping the reader very much. Those interested in specific blood flow values for each species could extract those from a table or appendix, perhaps more easily than from a chart.

.

Is there nothing to be done to visualise this data more clearly? Figure 3.13 suggests otherwise. Using a combination of colour and

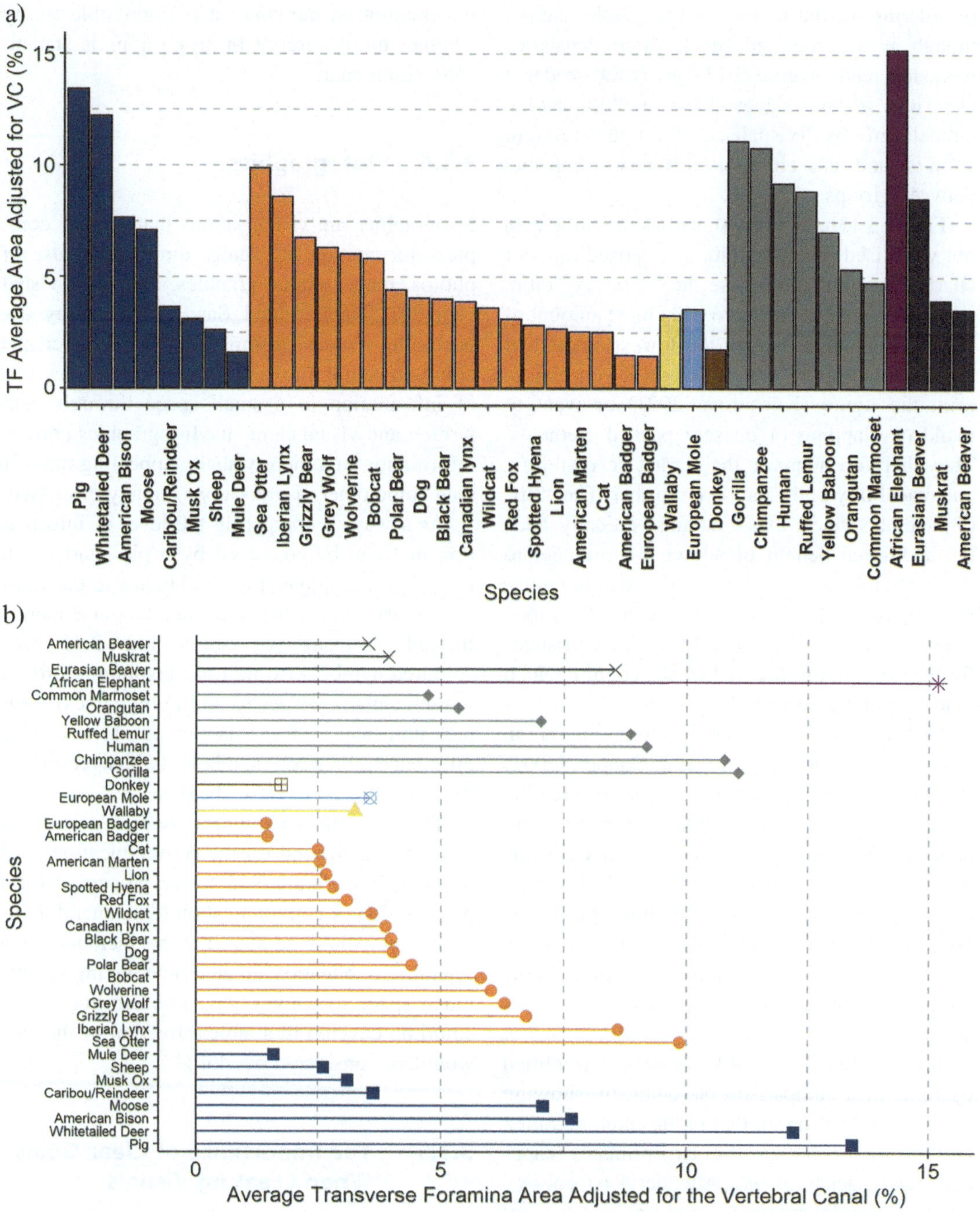

Fig. 3.13 (**a**) A bar chart of 40 mammal species and the average transverse foramina area (an indicator of blood flow to the brain through the vertebral arteries), adjusted for overall organism size through measurements of the vertebral canal on cervical vertebrae—data from Lunn et al. (2022). Data is ordered in decreasing values, as well as by taxonomic order. Figure (**b**) shows the same data but flipped (so the x-axis now runs vertically) and visualised in a stem plot form instead. *Artiodactyl: Dark Blue Square, Carnivora: Orange Circle, Diprotodontia: Yellow Triangle, Eulipotyphla: Light Blue Hollow Circle w/Cross, Perissodactyl: Brown Hollow Square w/Cross, Primate: Grey Diamonds, Proboscidea: Purple 8-Point Star, and Rodentia: Black Crosses*

re-ordering of bars, Fig. 3.13a works better, though it has required much more conscious work to create. Figure 3.13a gives the reader a clear idea of how values change within specific animal groups (visualised through different colours) and also emphasises trends within and between groups.

Figure 3.13b, in contrast, shows the same data but visualised as a stem plot as opposed to a bar chart. Stem plots maximise the "data-ink" ratio, where the intention is to have the most amount of data present while having the lowest amount of "ink" on the graph. This is helpful if you plan to print the figure (Kosourova 2021) or want to avoid overlapping or densely packed elements. They therefore increase the clarity, accessibility, and readability of the data even further. The "flip" of the plot, so the x-axis now runs vertically, adds the additional benefit of allowing the reader to easily read the species present on the figure without having to physically turn their head (as they would for Figs. 3.12 and 3.13a). These changes for the stem plot further reduce the cognitive load placed upon the reader.

Unlike bar charts, solitary pie charts are rare in scientific publications. Schriger & Cooper, (2001) state that "solitary pies have no role in scientific publications since readers should be able to generate the picture from tabular data, making the picture redundant with text and tables". However, an exception to this is when multiple pie charts are used within the same figure (e.g. Figure 3.14). This allows for easy comparisons of patterns within the dataset that would be less easy to spot in a table.

Specifically, Fig. 3.14 shows a simplified example of a comparison pie chart, by showing the neuron distribution (%) in the cerebral cortex and rest of the brain within two primates (capuchin and macaque), one artiodactyl (steenbok), and three carnivores (jackal, red fox, and racoon) all with similar brain masses (between 44 g ~ 70 g). While this figure indicates that the proportion of neurons in the cerebral cortex is larger in primates (a likely result of economic brain scaling rules as discussed in Sect. 3.2.4 (Herculano-Houzel et al. 2006; Herculano-Houzel 2011)), without supporting statistical information on the chart, it is impossible to say whether the difference in proportions is statistically significant.

3.3.4 Infographics

Infographics are visualisations that convey complex information efficiently through the use of photos, maps, charts, graphics, and other visual elements (Naparin and Saad 2017). They are becoming increasingly more common in science communication due to their ability to convey lots of information in a small space through both written and visual elements. Infographics convey information in a more visually appealing manner than some other visualisations already discussed.

As a way of comparing the level of information that can be conveyed by a pie chart or an infographic, compare Fig. 3.15 below to Fig. 3.14 which uses the same data and colour scheme. Instead of using pie charts, Fig. 3.15 uses coloured symbols (to indicate taxonomic order), written elements (species and Latin names), simple statistical values (percentages), and size differences between symbols (which reinforce the percentages) to show the data.

The type of infographics used in any given context will depend on the communication goal of its creator. Figure 3.15 would be perfect for a presentation or communication to a general audience, but would perhaps not be included in a scientific publication. In an article, taking up this much space to convey the same information as could be covered in a table a fraction of the size would be considered overkill.

3.3.5 The Importance of Clear Goals When Creating Visuals

Ultimately, visualisations are created to aid in the comprehension of information and ensure a target audience gets the "message" of a paper or data analysis as the author sees it. Those choosing visualisation strategies need to take into account both their target audience, and their goals—the most important points that they want to

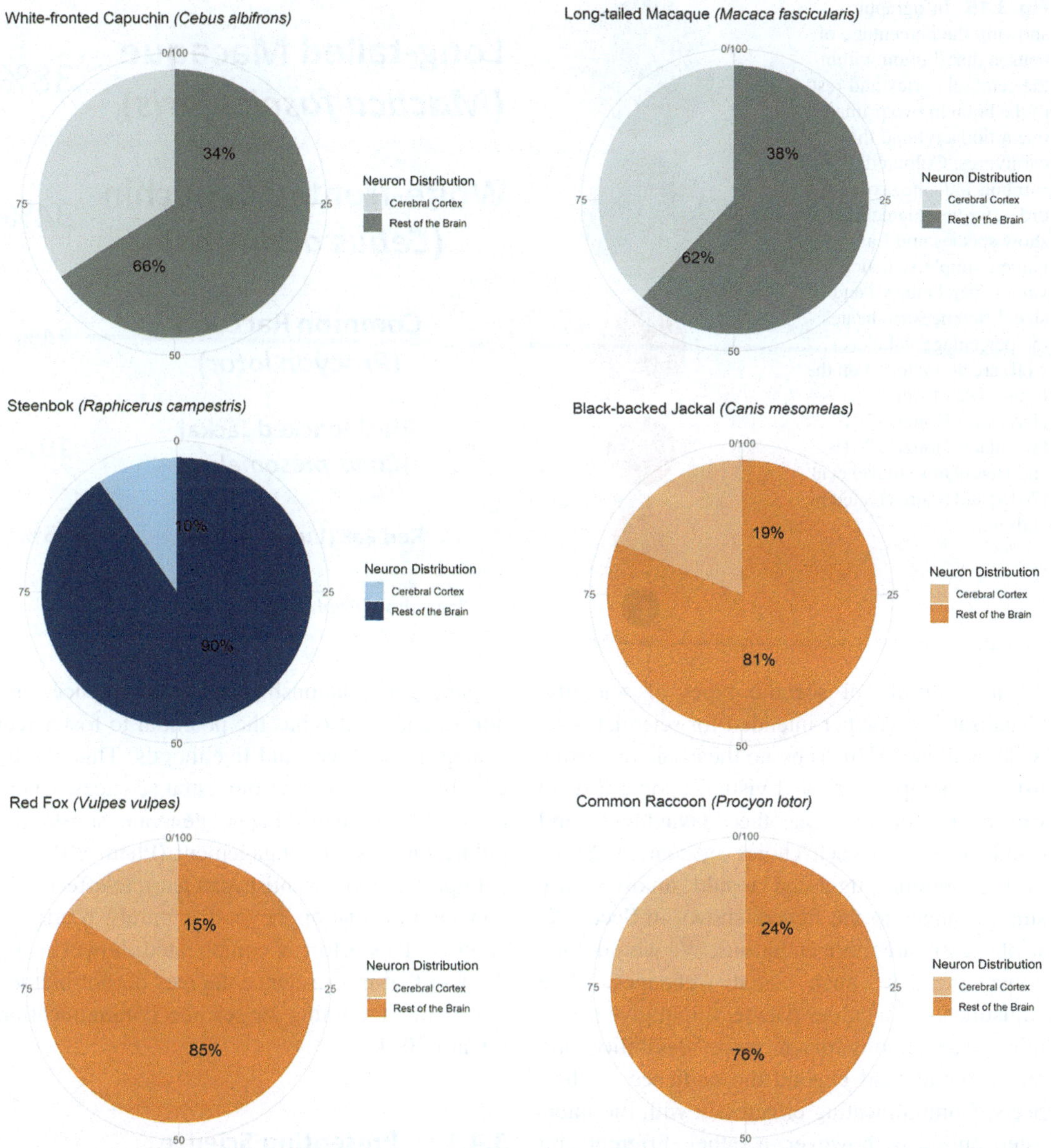

Fig. 3.14 Comparison pie charts showing the percentage neuron distribution in the cerebral cortex and rest of the brain in six species within three taxonomic orders (*Primate: Grey, Artiodactyl: Blue, Carnivora: Orange*). Data from Herculano-Houzel (2017), Herculano-Houzel (2016) and Herculano-Houzel et al. (2015) and figure created by author

communicate to their reader. These two factors together are important because they determine how much detail and complexity a visualisation can reasonably contain without either over- or underestimating the audience's expectations and needs. The example given above of the pie charts and the infographics is a demonstration of choosing both how much information, and which format to present it in to fit different audiences and goals. Both expectations and needs will be substantially different at an academic conference made up of specialists in the same field as the author, and a talk for the general public with a broader interest in science. This section has shown a few ways in which conscious consideration of audience and goal can improve

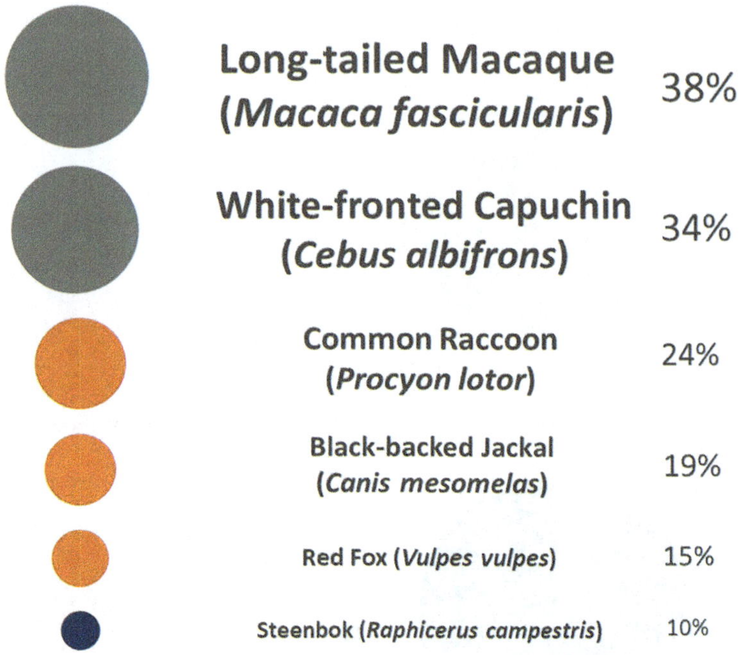

Fig. 3.15 Infographic showing the percentage of neuron distribution within the cerebral cortex and rest of the brain in two primates, one artiodactyl and three carnivores. Coloured symbols indicate taxonomic order, written elements show species and Latin names, simple statistical values (percentages) and size differences (indicates the percentage value as well) are all included on the figure. Data from Herculano-Houzel (2017), Herculano-Houzel (2016) and Herculano-Houzel et al. (2015) and figure created by author

the accessibility of specific types of scientific visualisation, whether intended for scientists or a wider audience. This is by no means an extensive list or a "recipe" for good visualisations; instead we have focused on the advantages and disadvantages of each choice in context. Some of the options suggested would involve only small changes to the figures shown in Sect. 3.2, while others are more ambitious. We would note, however, that such small changes, like adjustments to a colour palette, might give scientific visualisations much more flexibility and reach, within and beyond the audience of their peers. Communicating effectively with the interested public is, however, a rather different aim than communicating with other scientists—and our final section considers how visualisation choices might affect science communication specifically.

3.4 How Scientists Share Information

Public engagement is crucial for building strong connections between science and the public. It can limit the spread of misinformation, encouraging active involvement within science, and help the public see the relevance of scientific research. A stronger relationship between science and non-scientists also has the potential to foster real change in our lives and livelihoods. This can be clearly seen in the environmental sciences, where topics like climate change generate substantial public interest and engagement (Pham 2016). It is important that scientists can learn to effectively convey information beyond a purely academic context. This effort is complicated, however, by the fact that most science degrees do not include any explicit training in science communication (Heard 2021).

3.4.1 Presenting Science

3.4.1.1 Presentations to a General Audience

Only twenty years ago, talks about science were limited to in-person events, and most were aimed primarily, or exclusively, at other scientists. The proliferation of YouTube and specialist organisations like TEDTalks and National Geographic Live have dramatically changed this today. General audiences, we believe, find it easier to engage with talks with a story-like structure and find visualisations designed to tell a story easier to understand than traditional publication-style visualisations which focus on efficiency and accuracy (Dahlstorm 2014; Grainger et al. 2016). This is not to say that efficiency and accuracy are not important in scientific communication (quite the opposite in fact), but rather that if they wish to make information accessible to a wide audience, scientists' default approach to visualisation must change.

Expert talks often use narrative features to great effect in attracting and informing a general audience (Reynolds 2011). Usually, the narrative is communicated orally, with visualisations as backup. Here, we focus on two examples of good practice in our case study field, namely the TEDTalks "What is so special about the human brain?" by Suzana Herculano-Houzel (2013), and "What are animals thinking and feeling?" by Carl Safina (2015).

The visualisations in both talks are kept to a minimum to reduce the possibility of distraction to the audience. Both presenters typically show a single image on any given slide, with images selected to be visually interesting without detracting from the speaker. There is little-to-no text on any slides. For example, Herculano-Houzel uses an illustration to exemplify her point on brain scaling rules, while Safina uses amusing or dramatic wildlife photography as a backdrop for his presentation. The exception to using only one image is when some slides are used to demonstrate specific ideas (as discussed next).

Presentations like these may also combine visual and audio cues to convey specific scientific ideas. In these cases, the images are typically much more than simple decoration. For example, Herculano-Houzel uses a series of visualisations that each build upon one another to explain how primate brains contain a much larger number of neurons than rodent brains of a similar mass. The developing composite visualisation clearly shows how rodent neurons increase in size as the brain grows, while primate neurons always stay the same size, so can be more numerous (Fig. 3.16).

Presentations also tend to use far fewer graphs and statistics than papers. In a scientific article, visualisations are often complex and multi-

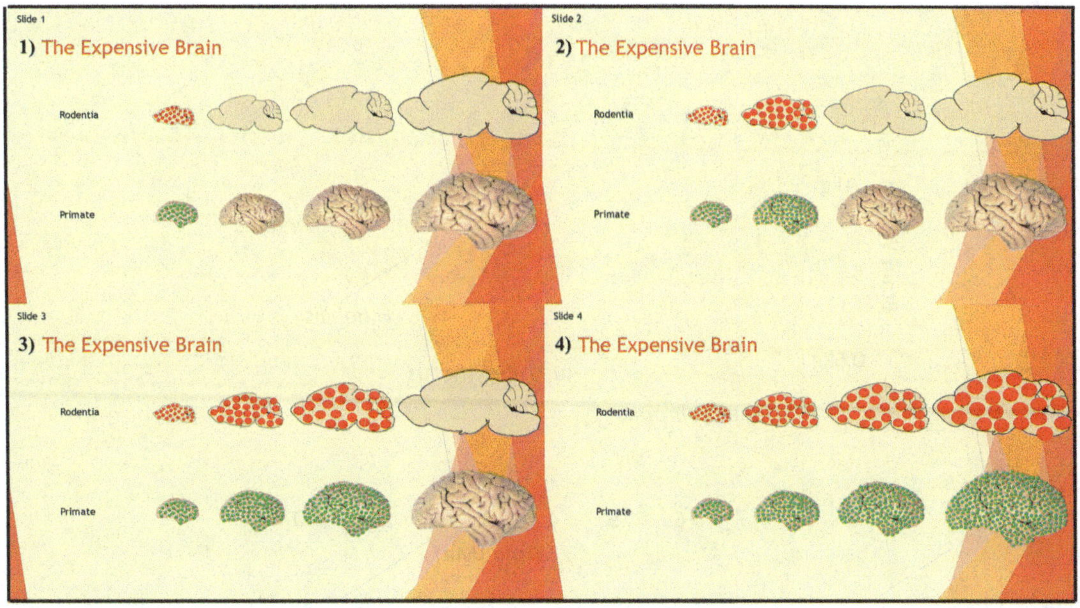

Fig. 3.16 Mock-up visualisations from Suzanna Herculano-Houzel's 2013 TEDTalk "What is so special about the human brain?". The slides show a series of visualisations that build and expand upon one another to explain how primate brains contain a much larger number of neurons when compared to a rodent brain of a similar mass. Coloured circles (*Red = Rodent, Green = Primate*) are used to indicate neurons with the size of the circles indicating neuron size. This visualisation therefore shows that, in rodent brains, as the brain increases in size (between a mouse and capybara for example), the neuron size also increases, while in primates they do not. Therefore, allowing more neurons within a primate brain compared to a rodent brain of a similar mass. Mock-up visualisation created by author

layered, and many require specialist knowledge to interpret. If the primary aim of science communication is to make science accessible to anyone, then it is obvious why they rarely include similar visuals. Graphs are typically only included within a public presentation if they are clear, simple, and are directly relevant to the narrative of the presentation. For example, the only graph Herculano-Houzel shows in her talk is used at the end to demonstrate the effect of cooking on human evolution. Cooking food allowed human ancestors to eat more calories in a set period of time, which in turn gives us more energy to power our brains' higher neuron counts. This graph is a simple line graph that is clearly labelled and looks to have been designed specifically to convey this point—it contains very little extraneous information (Fig. 3.17). It is therefore much less "complete" than a figure from a scientific paper might be but adds significantly to the ability of the talk to communicate a finding. Similarly, in Safina's talk, maps are used to show the geographical distribution of animals. These maps are essential to the conclusions, are very visually appealing, and it would be difficult to convey the talk's core message without them.

Visualisations can also help scientists describe their methods for a wider audience. For example, in a National Geographic Live talk titled "Dolphins: Even Smarter Than You Thought", presenter Brian Skerry (2015) describes a series of behavioural experiments that show increasing evidence of dolphin intelligence. Both publications that he discusses ((Kuczaj et al. 2015) and (Clark and Kuczaj 2016)) contain very detailed, step-by-step methodologies. However, in his public engagement talk, instead of describing his experiments in detail, Skerry explains the broad principles orally and shows pictures of dolphins taking each test. This allows the audience to see how the animals reacted to the experiments without any detail of analysis or

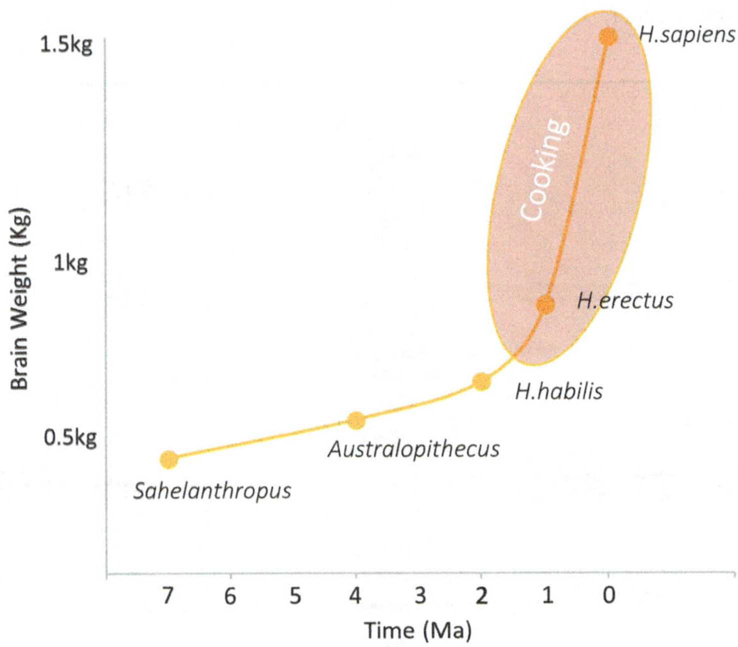

Fig. 3.17 Mock-up visualisation from Suzanna Herculano-Houzel's 2013 TEDTalk "What is so special about the human brain?". A simple line graph to show the effect of cooking on brain size (here in weight, kg) in human evolution. This figure looks to have been designed specifically to convey that the development and evolution of cooking to create calorific dense foods that could be consumed and digested quicker, resulted in rapid brain expansion to a general audience. The axes are easy to understand (described in the talk as "Time in millions of years ago" (x-axis) and "Brain weight (kg)" (y-axis)) and overall aid the speaker in their delivery of the information. Mock-up visualisation created by author

setup that would detract from the core questions being investigated. In this way, science communication presentations are more likely to focus on the main findings and implications of the research described, as these are the parts that contain the narrative and are the most interesting to the audience. If we compare a complex conclusion from Kuczaj et al. (2015):

> However, tugging rate for both dolphins was significantly higher in the presence of the other animal than when interacting with the object alone (Alfonz, $t(22) = 2.86$, $P < .05$; Kimbit, $t(22) = 2.61$, $P < .05$).

And a simpler explanation from the same manuscript:

> Our results suggest that dolphins can cooperate in order to achieve a common goal when given an unfamiliar task.

It quickly becomes obvious how more streamlined conclusions and, by extension, the simpler visualisations that accompany them are more suitable for a science communication presentation to a general audience.

3.4.1.2 Presentations for Younger Audiences

The future of science relies on future generations. It is through education and accessible science in their everyday life, that we hope that young people will be inspired to engage with science and ultimately further human knowledge. Scientific training from a young age also helps develop other skills such as teamwork, communication, and problem-solving. Presentations for younger audiences place an even greater emphasis on narratives and visualisations than those for adult general audiences. They also tend to be shorter. For example, if we look at a TED-Ed talk (TED's education and youth initiative) by Lori Marino et al. (2015) titled "How smart are dolphins?", in theory almost directly comparable to Skerry's dolphin TEDTalk described above, the increased importance of the visualisations in TED-Ed is immediately obvious. The ~5-minute video presents the information and research through a voice-over and continuous animation. The animation keeps the audience engaged and turns complex data into comprehensible information. There are no publication-like visualisations (XY scatter plots, bar chart, etc.) at any point in the presentation, just the animation. Furthermore, within the animated video, technical terms like encephalisation quotient are explained using images and simple spoken sentences rather than equations. The narrative focus and continued visual stimulation help summarise the findings from multiple dolphin intelligence studies in an accessible and memorable way, but one that is clearly different to the approach taken by talks aimed specifically at adult non-specialists.

3.4.2 Science Communication Articles

In addition to presenting specifically for non-specialists, science communicators can also write for each other. The rapid expansion of the internet has led many consumers to become part of a culture that can access information immediately. If you need to find out the name of a film, buy a book, or update your friends and family on your holiday plans, you can do so at the click of a few buttons. Accessing science is no exception. How hot is the sun? Which dinosaurs lived in the Jurassic? Do chimpanzees have a theory of mind? It should come as no surprise to scientists that writing for the internet or wider-circulation magazines can be another way to encourage active involvement within science and limit the spread of misinformation.

Science communication articles can therefore take many forms depending on the author, editor, and type of article. If the information is presented in a news article (The New York Times, The Conversation, and Nature Briefing for example), then the narrative is usually rather short and to the point. As in scientific talks, statistics and graphical visualisations are rare as articles typically focus on the conclusions and any visualisations that are included usually take the form of an easily digestible infographic or picture. News articles often also include quotes from original pieces of work, an eye-catching image to draw in the reader, and bold statements or questions as titles.

A good example of this is the 2012 BBC News Article entitled "Dolphins deserve same rights as humans, say scientists", which only contains 425 words and a single large photograph of dolphins. Outlets that have specific sections for science news may permit longer text but tend to follow the same less visual style. Rapp's (2021) Discover article "Just How Intelligent Are Dolphins?" for instance is over 1200 words long but still only contains one image, a photograph. Askham's 2021 article "Evolving a Bigger Brain Isn't Always About Intelligence", by contrast, is roughly the same length but uses four different images to engage the reader. Three of these are photographs, but with captions that give detail of how the subjects fit with the science discussed, while the fourth is an artist's impression of mammal brain evolution, specifically created to communicate an idea about relationships among animals with different brains. These are very different visual choices to the more complex visualisations of data that Smaers et al. (2021) used in the scientific publication Askham is summarising.

3.4.3 Popular Science Books

The final, and perhaps most involved, format that scientists use to communicate with non-scientists is the popular science book. Popular science books are distinct from talks and articles in that they require sustained engagement from the reader and take a much longer and more detailed look at a subject. Although it might seem intuitively that science book authors will rely primarily on narrative text, many also use visualisations—and in a different way to scientists communicating via shorter media. In a 2016 book "The Human Advantage: How Our Brains Became Remarkable", for instance, Herculano-Houzel guides the reader through her research on brain anatomy to explain how we measure intelligence in animals, and how humans evolved their extraordinary cognitive abilities. As the science behind this is rather complex, both accessible language and visualisations are used throughout to engage the reader. The text focuses more on personal accounts and stories than summaries of projects and develops a strong narrative that helps place the science in context for the reader. In contrast to the scientific presentations and articles discussed earlier, however, this book also contains lots of graphical visualisations and statistical data (in fact, it is on the more complex end of the popular science spectrum). The visualisations are, however, adapted to make them more accessible and visually appealing.

When we compare images from Herculano-Houzel (2016) and the original papers that it is based on, for example, we see several strategies used to make the book visuals more engaging, including the use of wildlife imagery to label points on a graph (the effect is similar to Fig. 3.7's cartoon animals labelling points of interest, see Sect. 3.2.5.1). Other graphical features such as regression lines (lines that best fit the data points to minimise variance) are also used in the book to make the key patterns and relationships within a graph jump out. These are likewise not always used in the corresponding articles, where the reader would instead be anticipated to be able to read the supporting statistical information. Finally, when you compare the figure legend, axis titles, and axis tick labels on Herculano-Houzel's book and article visualisations, they are much simpler in the book. By plainly saying "*Neurons in the cerebral cortex (millions)*" as an axis title (Chap. 4 of Herculoano-Houzel 2016), the reader has a much clearer idea of what the variable is, compared with "*Cx neurons (n)*" (the equivalent title in Herculano-Houzel 2011).

3.4.4 Creating Accessible Science Communication

The importance of accessible science communication cannot be underestimated. Advances in communication technology have allowed important research to reach a much wider audience than it might have previously but necessitates changes in written and visual style to make it engaging and interesting. This section has considered some of the typical forms that science communication work can take, and how each one modifies or develops new visual styles to suit its audience and their needs. Overall, we would suggest that visualisations remain important in all forms of science communication, though their complexity and prominence may vary significantly. The same simple adjustments that can help scientists communicate more easily with their peers can suffice, in some cases, to communicate with a particularly interested general audience, like those willing to read a popular science book. For other more bitesize formats, we see more adventurous and considered visualisations like infographics and animations coming into their own. The same principle of considering both audience and aim seems to hold true for visualisations intended for non-specialists as for specialists and produces some rather different outcomes in each case.

3.5 Conclusion

This chapter has shown that even small changes in how visualisations are presented can have a big effect on the accessibility of the information. Using our example of the science investigating biological indicators of intelligence in animals, it is easy to see how, as the science developed, the visualisations have also changed. Although authors writing for scientific audiences still use the same types of data, analysis, and visualisation, they have also adjusted graphical features like colour, data point identification, titles, legends, etc. to make their visualisations more readable and to foreground different ideas and findings. Even small tweaks, like the addition of extra component graphs or changes to the sequence in which data are presented, can change the core message the reader takes away.

Whenever science is communicated, a meaningful and effective visualisation can only be created by actively considering both the identity and needs of the target audience and the most important points that you wish them to take away (which we have called the authors' goal). Different audiences—members of the same disciplinary community, of other disciplinary communities, from the wider academic world, and from other worlds entirely—have different expectations when they are presented with scientific information. It is not as simple as being able to assume more or less specialist knowledge depending on the reader's background. Different audiences also have different visual repertoires and experiences, and may find specific styles or approaches to visualising data more or less engaging, and more or less informative. Even where background knowledge is present and interest can be assumed to be high (for instance, where an audience has actively chosen to engage with a piece of science communication), visualisations created to communicate one purpose cannot be substituted for those created or adapted to communicate something else. Instead, we have shown that visualisation is an important subject in its own right, and merits scientists' attention even where disciplinary norms would initially suggest choices are limited. When we visualise our datasets well, we can increase their reach and impact substantially.

References

Ahlmann-Eltze C, Patil I (2021) Ggsignif: significance brackets for 'ggplot2'. R package

Alyass A, Turcotte M, Meyre D (2015) From big data analysis to personalized medicine for all: challenges and opportunities. BMC Med Genet 8. https://doi.org/10.1186/s12920-015-0108-y

Arnold JB (2021) Ggthemes: extra themes, scales and Geoms for 'ggplot2'

Askham B, Natural History Museum, London (2021) Evolving a bigger brain isn't always about intelligence. [Online] Available at: https://www.nhm.ac.uk/discover/news/2021/april/evolving-bigger-brain-not-always-about-intelligence.html. Accessed 23 January 2022

Auguie B (2017) gridExtra: miscellaneous functions for "grid" graphics. R package version 2.3

Azevedo FA et al (2009) Equal numbers of neuronal and nonneuronal cells make the human brain an isometrically scaled-up primate brain. J Comp Neurol 513(5):532–541. https://doi.org/10.1002/cne.21974

BBC News (2012) Dolphins deserve same rights as humans, say scientists. [Online] Available at: https://www.bbc.co.uk/news/world-17116882. Accessed 22 January 2022

Bélanger M, Allaman I, Magistretti PJ (2011) Brain energy metabolism: focus on astrocyte-neuron metabolic

cooperation. Cell Metab 14(6):724–738. https://doi.org/10.1016/j.cmet.2011.08.016
Benson-Amram S et al (2016) Brain size predicts problem-solving ability in mammalian carnivores. Proc Natl Acad Sci 113(9):2532–2537. https://doi.org/10.1073/pnas.1505913113
Brumm A et al (2021) Oldest cave art found in Sulawesi. *Science*. Advances 7(3):eabd4648. https://doi.org/10.1126/sciadv.abd4648
Cairó O (2011) External measures of cognition. Front Hum Neurosci 5:108. https://doi.org/10.3389/fnhum.2011.00108
Clark FE, Kuczaj SA (2016) Lateralized behavior of bottlenose dolphins using an underwater maze. Int J Comp Psychol 29(1)., Retrieved from https://escholarship.org/uc/item/4dx6t5z5
Cook MP (2006) Visual representations in science education: the influence of prior knowledge and cognitive load theory on instructional design principles. Sci Educ 90(6):1703–1091. https://doi.org/10.1002/sce.20164
Crameri F, Shepherd GE, Heron PJ (2020) The misuse of colour in science communication. Nat Commun 11: 5444. https://doi.org/10.1038/s41467-020-19160-7
Dahlstorm MF (2014) Using narratives and storytelling to communicate science with nonexpert audiences. Proc Natl Acad Sci 111(Supplement 4):13614–13620. https://doi.org/10.1073/pnas.1320645111
Deaner RO, Isler K, Burkart J, Van Schaik C (2007) Overall brain size, and not Encephalization quotient, best predicts cognitive ability across non-human primates. Brain Behav Evol 70(2):115–124. https://doi.org/10.1159/000102973
Franzblau LE, Chung KC (2012) Graphs, tables, and figures in scientific publications: the good, the bad, and how not to be the latter. J Hand Surg Am 37(3): 591–596. https://doi.org/10.1016/j.jhsa.2011.12.041
Gabi M et al (2010) Cellular scaling rules for the brains of an extended number of primate species. Brain Behav Evol 76(1):32–44. https://doi.org/10.1159/000319872
Garamszegi LZ, Eens M (2004) The evolution of hippocampus volume and brain size in relation to food hoarding in birds. Ecol Lett 7(12):1216–1224. https://doi.org/10.1111/j.1461-0248.2004.00685.x
Grainger S, Mao F, Buytaert W (2016) Environmental data visualisation for non-scientific contexts: literature review and design framework. Environ Model Softw 85:229–318. https://doi.org/10.1016/j.envsoft.2016.09.004
Hawkins D (2019) Biomeasurement: a Student's guide to biological statistics, 4th edn. Oxford University Press, Oxford
Heard S (2021) We're (nearly) all untrained at SciComm. So, some resources. [Online] Available at: https://scientistseessquirrel.wordpress.com/2021/10/26/were-nearly-all-untrained-at-scicomm-so-some-resources/. Accessed 05 February 2022
Herculano-Housel S (2009) The human brain in numbers: a linearly scaled-up primate brain. Front Hum Neurosci 3:31. https://doi.org/10.3389/neuro.09.031.2009
Herculano-Houzel S (2011) Not all brains are made the same: new views on brain scaling in evolution. Brain Behav Evol 78(1):22–36. https://doi.org/10.1159/000327318
Herculano-Houzel S (2012) The remarkable, yet not extraordinary, human brain as a scaled-up primate brain and its associated cost. Proc Natl Acad Sci 109 (Supplement 1):10661–10668. https://doi.org/10.1073/pnas.1201895109
Herculano-Houzel S (2013) What is so special about the human brain? *|TEDGlobal 2013*. [Online]. Available at: https://www.ted.com/talks/suzana_herculano_houzel_what_is_so_special_about_the_human_brain. Accessed 16 January 2022
Herculano-Houzel S (2016) The human advantage: how our brains became remarkable, 1st edn. MIT Press, Camrbridge(Massachusetts)
Herculano-Houzel S (2017) Numbers of neurons as biological correlates of cognitive capability. Curr Opin Behav Sci 16:1–7. https://doi.org/10.1016/j.cobeha.2017.02.004
Herculano-Houzel S et al (2014) The elephant brain in numbers. Front Neuroanat 8:46. https://doi.org/10.3389/fnana.2014.00046
Herculano-Houzel S, Catania K, Manger PR, Kaas JH (2015) Mammalian brains are made of these: a dataset of the numbers and densities of neuronal and nonneuronal cells in the brain of Glires, Primates, Scandentia, Eulipotyphlans, Afrotherians and Artiodactyls, and their relationship with body mass. Brain Behav Evol 86(3–4):145–163. https://doi.org/10.1159/000437413
Herculano-Houzel S, Collins CE, Wong P, Kaas JH (2007) Cellular scaling rules for primate brains. Proceedings of the National Academy of Sciences 104(9): 3562–3567. https://doi.org/10.1073/pnas.0611396104
Herculano-Houzel S, Mota B, Lent R (2006) Cellular scaling rules for rodent brains. Proc Natl Acad Sci 103(32):12138–12143. https://doi.org/10.1073/pnas.0604911103
Jerison HJ (1973) Evolution of the brain and intelligence. Elsevier, New York
Kampstra P (2008) Beanplot: a boxplot alternative for visual comparison of distributions. Journal of Statistical Software, Code Snippets 28(1):1–9
Kassambara A (2020) Ggpubr: 'ggplot2' based publication ready plots
Kassambara A (2021) Rstatix*:* pipe-friendly framework for basic statistical tests
Kennedy H, Engebretsen M (2020) 1. Introduction: the relationships between graphs, charts, maps and meanings, feelings, engagements. In: Kennedy H, Engebretsen M (eds) Data visualization in society. Amsterdam University Press, Amsterdam, pp 19–32. https://doi.org/10.1515/9789048543137-005
Kosourova E (2021) Bar plots: alternatives & specific types. [Online] Available at: https://towardsdatascience.com/bar-plots-alternatives-specific-types-9d10ef250e5. Accessed 16 January 2022

Kotrschal A et al (2013) Artificial selection on relative brain size in the guppy reveals costs and benefits of evolving a larger brain. Curr Biol 23(2):168–171. https://doi.org/10.1016/j.cub.2012.11.058

Kuczaj SA, Winship KA, Eskelinen HC (2015) Can bottlenose dolphins (Tursiops truncatus) cooperate when solving a novel task? Anim Cogn 18(2): 543–550. https://doi.org/10.1007/s10071-014-0822-4

Lefebvre L, Reader SM, Sol D (2004) Brains, innovations and evolution in birds and primates. Brain Behav Evol 63(4):233–246. https://doi.org/10.1159/000076784

Levine T (2008) Using colour in figures: let's agree to differ. Traffic (Copenhagen, Denmark) 10(1):344–347. https://doi.org/10.1111/j.1600-0854.2008.00863.x

Lunn AJ, Winder IC, Shaw V (2022) The vertebral artery blood supply to the brain and its relationship with cognition across the taxonomic classes: Mammalia and Aves in Anatomical Society Summer Meeting Glasgow 2021: Cutting Edge Anatomy. J Anat 240 (4):775–819. https://doi.org/10.1111/joa.13592

MacLean EL et al (2014) The evolution of self-control. Proc Natl Acad Sci 111(20):E2140–E2148. https://doi.org/10.1073/pnas.1323533111

Madden J (2001) Sex, bowers and brains. Proc R Soc Lond Ser B Biol Sci 268(1469):833–838. https://doi.org/10.1098/rspb.2000.1425

Marino L, Kalopaidis M, Bornet J Savva M (2015) How smart are dolphins?–Lori Marino. [Online] Available at: https://ed.ted.com/lessons/how-smart-are-dolphins-lori-marino. Accessed 20 January 2022

Martin RD (1981) Relative brain size and basal metabolic rate in terrestrial vertebrates. Nature 293(5827):57–60. https://doi.org/10.1038/293057a0

Marx V (2013) The big challenges of big data. Nature 498: 225–260. https://doi.org/10.1038/498255a

McInerny GJ et al (2014) Information visualisation for science and policy: engaging users and avoiding bias. Trends Ecol Evol 29(3):148–157. https://doi.org/10.1016/j.tree.2014.01.003

Minervini S et al (2016) Brain mass and encephalization quotients in the domestic industrial pig (Sus scrofa). PLoS One 11(6):e.0157378. https://doi.org/10.1371/journal.pone.0157378

Naparin H, Saad AB (2017) Infographics in education: review on infographics design. The International Journal of Multimedia & Its Applications (IJMAA) 9(4):5. https://doi.org/10.5121/ijma.2017.9602

National Health Service (2019) Colour vision deficiency (colour blindness). [Online] Available at: https://www.nhs.uk/conditions/colour-vision-deficiency/. Accessed 7 January 2022

Naumann RK (2015) Even the smallest mammalian brain has yet to reveal its secrets. Brain Behav Evol 85(1): 1–3. https://doi.org/10.1159/000375438

Pal S et al (2020) Big data in biology: the hope and present-day challenges in it. Gene Reports 21: 100869. https://doi.org/10.1016/j.genrep.2020.100869

Pham D (2016) Public engagement is key for the future of science research. npj Science of Learning 1:16010. https://doi.org/10.1038/npjscilearn.2016.10

Preuss TM (2017) Chapter 8–The human brain: evolution and distinctive features. In: Tibayrenc M, Ayala FJ (eds) On human nature–biology, psychology, ethics, politics, and religion. Elsevier Academic Press, Amsterdam, pp 125–149. https://doi.org/10.1016/B978-0-12-420190-3.00008-9

Ramos E, Concepcion BP (2020) Visual abstracts: redesigning the landscape of research dissemination. Semin Nephrol 40(3):291–297. https://doi.org/10.1016/j.semnephrol.2020.04.008

Rapp J (2021) Just how intelligent are dolphins?. [Online] Available at: https://www.discovermagazine.com/planet-earth/just-how-intelligent-are-dolphins Accessed 22 January 2022

Ratcliffe JM, Fenton MB, Shettleworth SJ (2006) Behavioral flexibility positively correlated with relative brain volume in predatory bats. Brain Behav Evol 67(3): 165–176. https://doi.org/10.1159/000090980

Reader SM, Laland KN (2002) Social intelligence, innovation, and enhanced brain size in primates. Proc Natl Acad Sci 99(7):4436–4441. https://doi.org/10.1073/pnas.062041299

Reynolds G (2011) Presentation Zen: simple ideas on presentation design and delivery. New Riders

Roth G, Dicke U (2005) Evolution of the brain and intelligence. Trends Cogn Sci 9(5):250–257. https://doi.org/10.1016/j.tics.2005.03.005

Safina C (2015) What are animals thinking and feeling? | Mission blue II. [Online] Available at: https://www.ted.com/talks/carl_safina_what_are_animals_thinking_and_feeling. Accessed 16 January 2022

Sarkar D (2008) Lattice: multivariate data visualization with R. Springer, New York

Schriger DL, Cooper RJ (2001) Achieving graphical excellence: suggestions and methods for creating high-quality visual displays of experimental data. Ann Emerg Med 37(1):75–87. https://doi.org/10.1067/mem.2001.111570

Skerry B (2015) Dolphins: even smarter than you thought | Nat Geo Live. [Online] Available at: https://www.youtube.com/watch?v=XZ4hZx6K85Y. Accessed 20 January 2022

Smaers JB et al (2021) The evolution of mammalian brain size. Sci Adv 8(18):eabe2101. https://doi.org/10.1126/sciadv.abe2101

Snodgrass JJ, Leonard WR, Robertson ML (2009) The energetics of encephalization in early hominids. In: Hublin J, Richards MP (eds) The evolution of hominin diets–integrating approaches to the study of Palaeolithic subsistence. Vertebrate Paleobiology and Paleoanthropology. Springer, Dordrecht, pp 15–29. https://doi.org/10.1007/978-1-4020-9699-0_2

Sol D et al (2005) Big brains, enhanced cognition, and response of birds to novel environments. Proc Natl Acad Sci 102(15):5460–5465. https://doi.org/10.1073/pnas.0408145102

Street SE, Navarrete AF, Reader SM, Laland KN (2017) Coevolution of cultural intelligence, extended life history, sociality, and brain size in primates. Proc Natl Acad Sci 114(30):7908–7914. https://doi.org/10.1073/pnas.1620734114

Team, R. C. (2021) R: a language and environment for statistical computing. R Foundation for Statistical Computing, Vienna, Austria

Wickham H (2016) ggplot2: elegant graphics for data analysis. Springer-Verlag, New York

Wickham, H., Francois, R., Henry, L. & Müller, K., 2021. Dplyr: a grammar of data manipulation

Wickham H, Seidel D (2020) Scales: scale functions for visualization. R package version 1(1):1

Wickline M, HCIRN H-CIRN (2022) Coblis–color blindness simulatorcolor blindness simulator. [Online] Available at: https://www.color-blindness.com/coblis-color-blindness-simulator/. Accessed 05 February 2022

Wong B (2011) Point of view: color blindness. Nat Methods 8:441. https://doi.org/10.1038/nmeth.1618

Xiao, N., 2018. Ggsci: scientific journal and Sci-fi themed color palettes for ggplot2

Yu H, Agarwal S, Johnston M, Cohen A (2009) Are figure legends sufficient? Evaluating the contribution of associated text to biomedical figure comprehension. J Biomed Discov Collab 4(1):1–10. https://doi.org/10.1186/1747-5333-4-1

Part II

Biomedical Education: Theory and Practice

The Affordances of Visual Modes in Pedagogy on the Physics of Motion in Physiotherapy Education

4

Sumaya Gabriels, Natashia Muna, and Kate le Roux

Abstract

The concept of angular motion is foundational knowledge in physiotherapy, applied for example in measuring joint range of motion. Yet research points to university students' poor understanding of the concept. Visuals are commonly promoted as valuable for creating opportunities for meaningful learning of the physics of motion. In this chapter we report on a lecturer's use of the affordances of visual modes—images, gestures and objects—in pedagogy on angular motion in the first year of a professional physiotherapy degree programme at a South African university. Using a multimodal social semiotic perspective that positions multimodal language as contextualised, motivated and central to meaning-making, we analysed recorded lectures and an interview with the lecturer. We show that, in organising her pedagogy, the lecturer used affordances of the image mode to create static images representing the spatial arrangement of multiple concepts and their relations simultaneously. This mode was the central site used to promote conceptually meaningful problem-solving, but was strengthened by her use of the affordances of gesture and an object to communicate motion. The lecturer strategically integrated the modes of verbal speech and written text into the visual complex, supporting the development of conceptual meaning and the acquisition of the language modes employed for problem-solving procedures. Her choices were motivated by a deep, contextual understanding of the role of physics in physiotherapy and of the diverse knowledge, language experiences and resources of first-year students. Our findings regarding the affordances of visual modes can inform collaborative education development within the health sciences.

Keywords

Physiotherapy education · Physics pedagogy · Angular motion · Language modes · Visual complex · Multimodal social semiotics

4.1 Introduction

In this introductory vignette, a lecturer in a Physics module for students in a professional

S. Gabriels (✉)
Department of Health and Rehabilitation Sciences, Faculty of Health Sciences, University of Cape Town, Cape Town, South Africa
e-mail: sumaya.gabriels@uct.ac.za

N. Muna
Department of Health Sciences Education, Faculty of Health Sciences, University of Cape Town, Cape Town, South Africa

K. le Roux
Academic Development Programme, Centre for Higher Education Development, University of Cape Town, Cape Town, South Africa

L. Shapiro, P. M. Rea (eds.), *Biomedical Visualisation*, Advances in Experimental Medicine and Biology 1388, https://doi.org/10.1007/978-3-031-10889-1_4

undergraduate Physiotherapy degree programme describes and motivates her pedagogy for teaching kinematics:

> *. . . .up until now when [the students] do physics and they go, 'I walk in a straight line' all they're focused on is, [...] you have a person and they're going from [...] point A to point B, and all they're looking at is this line [from point A to B]. [...] That is all they've been made to focus on up until this point. [...] 'but actually as a physiotherapist, [...] angular [motion] is much more relevant to you because you are going to be looking at the human body', [...] and we're converting all this rotational motion into linear motion.*
>
> *[My pedagogy] is just to give them something visually to look at and go ' Oh yah! When I move, my knee joint is rotating, and my hip joint is rotating and my arms are swinging and my elbow is rotating and my shoulder is, you know everything is rotating.' I think it just gives them actually something concrete to look at.*
>
> *[In the classroom] I swing, I walk and I gesture, I draw and you know, sometimes I throw objects [...] to get them to be able to visualise.* – lecturer in this research study

Human movement is complex, and understanding it is central to physiotherapy practice, a profession that aims to maintain optimal body movement and focuses on rehabilitating movement following an injury or as a result of a medical condition. The nature of motion of human joints is inherently angular as long bones rotate about a fixed axis (Levangie and Norkin 2011). For example, when rehabilitating the shoulder of a cricket bowler who is required to swing their arm as part of the bowling action, a physiotherapist needs to understand that the nature of the movement at the shoulder is angular, the range of the motion of the humerus (a long bone) as it rotates about its axis, and the angular velocity at which the bowler swings their arm. Thus, joint motion, as angular motion, is foundational knowledge for understanding aspects of movement. It is this argument that motivates the lecturer cited in this vignette, to create classroom opportunities for students to understand the concept of angular motion, and its relation to previously learned concepts.

While there is limited literature of students' understanding of angular motion, several studies point to students' poor conceptual understanding of the concept and the possible reasons for this (e.g. Rimoldini and Singh 2005; Mashood and Singh 2012; Mutsvangwa 2020). Evidence of this is seen in the context of the university physics classroom, where basic science concepts such as angular motion are commonly taught. Physics education research suggests that, although students are able to accurately solve the angular motion problems, they do so with limited conceptual understanding (Mashood and Singh 2012). Studies point to how students recruit prior learning of linear motion when working with angular and rotational motion, with poor understanding of the concepts and their relations resulting in the procedural manipulation of—frequently incorrect—symbolic equations (Mashood and Singh 2012; Rimoldini and Singh 2005). Reasons for students' difficulties in learning university physics given by lecturers include students' 'poor mathematical skills' (p. 1512), their 'rote learning' of formulae without understanding the 'big picture' (p. 1512) of physics concepts and their relations (frequently attributed to their schooling), and difficulties working with multiple representations such as diagrams, graphs and algebraic equations (Bray and Williams 2020). Lecturers also suggest that pressures of assessment practices encourage a procedural 'teaching to test approach' (Mashood and Singh 2012, p. 1307). Physics students also report using formulae by rote, without conceptual understanding, not being able to relate physics concepts to the 'real world' (p. 1513), and not understanding their lecturer's language use, which makes physics learning difficult (Bray and Williams 2020).

Beyond the physics classroom, lecturers have observed that in clinical contexts, students struggle to apply concepts of motion as relevant to anatomy (Kelly et al. 2018). Furthermore, it is well recognised that students find it difficult to apply their understanding of physics concepts in health science disciplines such as physiotherapy, pharmacy and occupational therapy (e.g. Kember et al. 2008). A reason for this may be that students do not have a comprehensive understanding of their profession, and therefore the relevance of the science-based courses in their degree programmes, thus focusing their learning on the procedural knowledge valued in assessments. It is therefore argued that it is important to contextualise physics for physiotherapy students (Ingerman et al. 2007; Rayner 2005; Rennie and Parker 1996). These studies point to the importance of pedagogies that contextualise physics concepts in everyday and professional contexts for developing students' conceptual understanding. Indeed, in the introductory vignette, the lecturer identifies her role as a physics lecturer in contextualising the physics concept of angular motion for physiotherapy students.

In the introductory vignette, the cited lecturer also highlights the advantages (or affordances, as defined in Sect. 4.3) of visuals that students can 'look at' for their learning of angular motion for physiotherapy, and describes how she uses images, objects and gestures in her pedagogy to create such opportunities for students. Indeed, since the discipline of physiotherapy, like other biological and health sciences (Bezemer 2017; Reinke et al. 2021), is inherently visual, there is a need for physiotherapy education to create opportunities for students to learn the 'mechanisms of basic science' in relation to 'living' anatomy (Kelly et al. 2018, p. 21). Research points to students' difficulties applying understandings derived from static textbook descriptions of human movement into dynamic representations and highlights the key role of visuals for understanding anatomy and human movement as dynamic (Kelly et al. 2018). This recognition has led to the use of visuals—in various mediums—to support physiotherapy students' understanding of joint movement. These include simulations (Blackford et al. 2015; Gough et al. 2016; da Silva et al. 2021), computer applications (Boucher et al. 1999; Kelly et al. 2018; Restrepo-Lòpez et al. 2021) and physical models (Clement et al. 2017).

This chapter draws on the first-author's case study of pedagogy for angular motion in a Physics module of a Biomechanics course in a professional undergraduate Physiotherapy degree programme at a South African university (Gabriels 2022). Specifically, we focus on pedagogical moments when the lecturer whom we cite in the introductory vignette utilised the particular advantages, or affordances, of visuals such as images, gestures and an object, complemented with written text, verbal speech and symbols, to create opportunities for students to develop conceptual understanding to inform their problem-solving procedures.

Our use of a multimodal social semiotic perspective—primarily following Kress and Van Leeuwen (2006), Kress and Van Leeuwen (2020), Jewitt et al. (2016), Lemke (1998), and Jaipal (2010)—to investigate lecture pedagogy offers a novel contribution to physiotherapy education research on the use of visuals to support student learning. As we explain in our theoretical approach (Sect. 4.3), this perspective positions language as central to meaning-making within disciplinary-specific contexts, such as physics for physiotherapy. We understand language to refer to all the many modes available for use within a given context, such as written text, images, verbal speech, gesture and objects. From a multimodal social semiotic perspective, each mode offers particular affordances for disciplinary meaning-making, while also interacting and working together with other modes toward holistic meaning-making. Finally, as the introductory vignette suggests, in a particular context a lecturer makes motivated pedagogical choices, based on their understanding of the disciplinary context, teaching and learning, and students' needs and experiences.

Thus in this chapter we ask:

1. How does the lecturer use visuals (images, gestures and objects), in particular ways individually and together, to give meaning to the concept of angular motion and related procedures?
2. How does the lecturer use these visuals, complemented by verbal speech, written text and symbols, for holistic meaning of this concept?
3. What motivated the lecturer's choices of multimodal language in her teaching of the physics concept of angular motion in the specific context?

Since we view both language modes and meanings as contextualised, and a lecturer's pedagogical choices as motivated by their understanding of the specific context, we next describe the context of the wider research study (Sect. 4.2). Then, in our theoretical approach (Sect. 4.3), we describe our use of concepts from multimodal social semiotics, followed by the research methodology (Sect. 4.4). The substance of our contribution is provided in our analysis of three key pedagogical moments (Sect. 4.5), and the subsequent discussion thereof (Sect. 4.6).

4.2 Study Context

At the South African university where the research discussed in this chapter was produced, physiotherapy students learn about the concept of joint range of motion (ROM) in the first year of a four-year undergraduate degree programme. This concept is introduced in the Physiotherapy disciplinary course, where the focus is on measuring ROM, while in the Physics module of their Biomechanics course, students are taught about the kinematics of angular motion. In the Physics module, the focus is on angular motion concepts and problem-solving procedures for angular displacement, angular velocity, angular acceleration, as well as the concept of and problem-solving procedures for conversion between linear and angular motion.

The first-year physiotherapy student cohort is diverse in terms of backgrounds, schooling experiences and learning resources. In South Africa, it is recognised that many first-year university students find science or mathematics-based courses challenging (Kizito et al. 2016; Makgato 2007; Linder et al. 2014; Case et al. 2013), including the necessary conceptual understanding of mathematics for problem-solving in physics (Bray and Williams 2020; Ige et al. 2017). Yet, the racial, social, economic, knowledge, language and geographical hierarchies entrenched during the colonial and apartheid eras, endure in contemporary South African schooling and society. Thus, opportunities to learn quality mathematics and science at school are inequitable, in a complex interplay of race, socio-economic status, language, geography and schooling (Spaull 2019; Jansen 2019), with these classifications also being shown to matter in performance in science-based university disciplines (Council on Higher Education 2020; Rooney 2015).

Thus in the transition from school to university, poor overall performance in science programmes, and the socially inequitable nature thereof, should ideally underpin decisions about what is taught and how it is taught in first-year degree programmes.

4.3 Theoretical Approach

4.3.1 Multimodal Social Semiotics

The theory of multimodal social semiotics, founded in and extending the seminal work of Halliday (1993), asserts, firstly, that language is not a neutral 'carrier' or representation of disciplinary meaning, but is central to the process of making meaning in a functional sense. In other words, it is through language that we construct, access and communicate disciplinary knowledge. Secondly, this theory asserts that 'language' should be understood richly, extending beyond verbal and written text to include, for example, images, gestures, symbols and objects (Kress and Van Leeuwen 2006). From this perspective, a mode is understood as an '...organized, regular, socially specific means of representation' (Jewitt et al. 2001, p. 5). A mode includes all of the resources, such as mediums and signifiers, for making meaning within a particular context (Kress and Van Leeuwen 2006). For example, the mode of images includes mediums such as digital or print media and utilises signifiers such as shape and colour. Importantly, meaning is not an inherent property of a mode, but information becomes, or is made, meaningful when it is interpreted within a given social context, such as within the physics classroom and physiotherapy education more broadly (Kress and Van Leeuwen 2006). In addition, each mode has advantages or affordances for effectively representing different types of information in that context, a notion we discuss in detail later in this section. Yet, no mode exists in isolation, but constantly interacts with others in what Kress and Van Leeuwen (2006) refer to as *multimodal meaning-making complexes*. For example, to give meaning to the concept of 'radius' when teaching about angular displacement, a lecturer may annotate an image of a line drawn from the centre of a circle to the circumference with symbols (also naming these symbols verbally and through written text), which they then subsequently use in a symbolic equation.

Thus, conceptualising teaching and learning in physiotherapy as a multimodal communication process, requires lecturers to consider both the affordances of individual modes and the relationships between them as they are integrated in multimodal complexes (Kress and Van Leeuwen 2020) for the construction of disciplinary meaning (Jewitt et al. 2001).

Taking Halliday's seminal work (1993) as a point of departure, researchers have conceptualised multimodal social semiotics in the teaching and learning of a range of science disciplines (e.g. Airey and Linder 2017; Jaipal 2010; Lemke 1998; Linder et al. 2014; O'Halloran 2000). The research reported in this chapter involved putting the theoretical concepts in this scholarship to work in the empirical context of pedagogy for physics in physiotherapy education. In a to-and-fro process that aimed to systematically develop coherence between these theoretical and empirical spaces (Brown and Dowling, 1997), the first-author came to define analytic tools specific for the study context, as summarised in Table 4.1. In the rest of this section we further explicate the multimodal social semiotic perspective introduced here, with reference to this analytic framework. First, we expand on the notion of disciplinary-specific, contextualised meanings (columns A and B), before focusing on the concept of language modes and their affordances (column C). Finally, we briefly consider the notion of motivated choice with regard to how meaning is constructed.

4.3.2 Meaning-Making in the Context of the Physics Classroom

Halliday (1993) argues that language can be used to perform three related types of meanings (called meta-functions): ideational (information about experiences, relationships and things); interpersonal (information about social interactions and relative positionality) and textual (how language is structured in a cohesive and coherent way). Subsequently, these meta-functions have been reconceptualised for the use of multimodal social semiotics in science disciplines. Given the focus of the research of this chapter, Jaipal's (2010) meta-functions, attributed to Lemke (1998), for meaning-making in pedagogy in biology, are productive (Table 4.1). As suggested by the description of each meaning that follows, these meanings are related, yet separating them analytically is necessary for a detailed study of multimodal language use in pedagogy.

Following Jaipal (2010), we use *presentational* (ideational) meaning for the conceptual and procedural content that is being taught. *Conceptual* refers to individual concepts and their relations; '...knowledge that is rich in relationships. It can be thought of as a connected web of knowledge, a network in which the linking relationships are as prominent as the discrete pieces of information' (Hiebert 2013, p. 3). *Procedural* content is about knowing the 'rules' or steps to follow in completing a task (Hiebert 2013, p. 6). From this perspective, accurately solving an angular motion problem using the appropriate symbolic equation requires an understanding of angular motion and its relation to other physics concepts.

Secondly, we use *orientational* (interpersonal) meaning for the stance and position taken towards what is taught, and those involved in the communication process. For example, a lecturer might take a stance towards the physics conventions

Table 4.1 Analytical framework

	A. Meta-functions	B. Meaning explored	C. Language modes
1	Presentational meaning	i • What key concepts (e.g. angular velocity) and related concepts (e.g. radius, circumference, arc length, time) are given meaning? • What mathematical procedures are given meaning (e.g. angular velocity mathematical process)?	• Verbal—How she used; descriptive language (e.g. how fast it spins, the outside of the circle), using formal scientific concept names (e.g. angular velocity, circumference) or commonly used concept names (e.g. 'speed') • Written—The words she wrote on the board • Symbolic language—Symbols and their use in equations • Images—Diagrams, stick figures, arrows, graphs • Gesture—These included how she used her arms or body • Objects
2	Organisational meaning	ii Macro-organisational • How is content organised and sequenced? (e.g. first circle, then, radius, then arc length, lastly angular displacement equation) Micro-organisational • What modes are used? How are the modes used individually or together and sequenced to give meaning to concepts?	
3	Orientational meaning	iii • What stance does lecturer take in relation to knowledge? (e.g. authority on what was important or correct, conventions in the physics course) • How did the lecturer position herself and the students in relation to the knowledge (e.g. students' prior knowledge, positioning students as participants in the discipline, the examples she used, relevance of physics to physiotherapy profession)?	

related to a particular unit of measurement as being important for assessments, and in doing so positions themself as an authority. Importantly, this meaning also orientates students towards the contextual relevance of the knowledge, for example, by choosing examples of human movement when teaching the physics of angular motion to physiotherapy students.

Finally, Jaipal (2010) uses *organisational* (textual) meaning for the pedagogical aspects of meaning, and we consider two levels of organisation. At a macro-level it refers to the order in which the concepts and procedures for angular motion are presented within and across lectures. At a micro-level it is about the lecturer's multimodal language choices for teaching each concept. In other words, choices about which modes to employ and how these modes are intended to work together to create presentational and orientational meanings.

4.3.3 Language Modes for Meaning-Making in the Physics Classroom

With this description of three meanings that are pertinent for this research, we now return to the role of language modes in the meaning-making process. From a multimodal social semiotic perspective, each mode is capable of representing all three types of meaning, but each mode is particularly well suited to representing certain details of meaning in a discipline (Kress and Van Leeuwen 2006; Kress et al. 2014; Jewitt et al. 2016; Lemke 1998). This meaning potential is known as the *affordances* of a mode (Jewitt et al. 2016). In refining the analytic tools for this research, the lecturer's use of the affordances of six modes was identified, and these modes are named in this chapter as verbal talk, written text, symbols, images, gestures and objects (Column C, Table 4.1).

Given the recognised problem of students' poor conceptual understanding of angular motion, a central concept in physiotherapy, and the focus of this volume on visualisation, in this chapter we foreground the modes that are spatially realised and thus offer visual affordances, these being image, gesture and object (Kress et al. 2014). Our particular interest is in how the lecturer worked with the affordances of these three modes, and their interaction (in what we refer to as a *visual* meaning-making complex) in pursuit of creating opportunities for physiotherapy students to develop conceptual understanding of angular motion. However, given the theoretical approach described thus far, we necessarily consider how this visual complex may interact with the written, verbal and symbolic modes for holistic conceptual and procedural meaning. Next in this section, we draw on the literature to identify the affordances of each mode and substantiate our conceptualisation of the visual complex.

In this chapter, we use the *image* mode to include all forms of static, two-dimensional pictorial representations (Kress and Van Leeuwen 2006). A key characteristic of the mode of image, and the visual complex more broadly, is that meanings are spatially realised, such that all the signifiers of meaning, such as colour, shape and size, are presented simultaneously (Kress et al. 2014), with the spatial arrangement of signifiers in relation to one another—the composition of the image—influencing meaning (Kress and Van Leeuwen 2006). As a result, images have the affordance of being able to represent visuospatial meanings, criteria of realism, and conceptual meanings that cannot be easily expressed using other modes (Bezemer 2017; Kress and Van Leeuwen 2020).

Harrison (2003) identifies three ways in which images may function: as iconic, indexical, or symbolic. An image functions *iconically* when it resembles the actual object or person represented, for example, a photograph or a stick-figure drawing of a person. In contrast, an image functions *indexically* when it has no visible resemblance to the meaning it represents, but its meaning is understood because we have come to know the concept being represented by the image in that context. For example in a physics classroom, an arrow image pointing from a word to a symbol is understood as indicating a relationship or association between the word and symbol. Finally, images function *symbolically* when they bear no visual resemblance or conceptual link to an object or person; rather we understand what the image means because of its conventional use within a given context. For example within a physics context, graphs and free body diagrams function as symbolic images.

In the construction of various images, additional resources act as further signifiers of meaning. For example, the signifier of colour performs specific meaning functions in a particular social context. For example, blue may be associated with notions of cold, calm or distance, while red may be associated with heat, salience and foregrounding (Kress and Van Leeuwen 2006).

Another mode offering visual affordances is that of the *object*, which includes 'everyday' items and disciplinary tools (Kress et al. 2014), with the former discussed in this chapter. In contrast to a static, two-dimensional image, an object (being three dimensional) offers the affordances of depth perception, which relates to visuospatial learning (Stromberga et al. 2021), and the opportunity for physical interaction with the mode which can be handled and manipulated in space (Kress and Van Leeuwen 2006). When teaching kinematics in physics, an object in motion can be used to create opportunities for students to access disciplinary meaning, and to position physics concepts in relation to everyday life (Kress et al. 2014).

The third mode with visual affordances that we focus on is that of *gesture*. Gesture refers to the combination of body movements or actions, including facial expressions, that communicate meaning (Kress et al. 2014). In much the same way as images, gestures may function iconically, indexically or symbolically (Allwood and Ahlsén 2015). For example, gesturing a curved motion with one's hand in a way that resembles the arc of a circle, is iconic within a physics context. A pointing gesture that draws attention to something is performing an indexical function. Finally, a gesture that makes meaning through contextualised convention is being used symbolically, such as the right-hand rule commonly used in physics classrooms to determine the direction of vector quantities in angular motion.

We have described the variety of meanings which can be represented by the visual complex, focusing on pedagogy for physics. Yet it is clear that these modes individually and in their interaction, are insufficient to communicate the full richness of meaning of the concepts of angular motion and their related mathematical procedures. As a result, a lecturer must necessarily utilise other complementary modes, such as the verbal, written and symbolic.

The verbal mode refers to the spoken word, or speech, which is sequentially and temporally realised, and aurally perceived (Kress and Van Leeuwen 2006), while the written mode, which is also sequential, is enacted through written text and is visually perceived (Kress and Van Leeuwen 2006, p21). In complement to the visual complex, both verbal speech and written text are well suited to representing meanings related to probabilities and abstractions (Bezemer 2017; Kress and Van Leeuwen 2020) and are important for naming, describing and explaining concepts, symbols and procedures. The affordances of using language (verbal speech and written text) in this way to give meaning to scientific concepts has the potential to facilitate access to knowledge (Tang et al. 2016). In teaching contexts, verbal language is also particularly effective for offering orientational and organisational meanings.

The symbolic mode, for the purposes of this research, refers to symbols such as conceptual symbols, operational signs, and numerals, used in symbolic equations and to annotate images. As suggested by this definition, this is a necessary mode in science disciplines. The symbolic mode has the affordance of representing the relationship between concepts and variables in a contracted form (O'Halloran 2000), which enables the conversions and manipulations required for mathematical procedures and problem-solving.

4.3.4 Motivated Choice in the Development of Multimodal Meaning-Making Complexes

What this discussion of modal affordances makes clear, is that modes are not interchangeable. If a certain mode is not used, the particular details of meaning which it affords may remain inaccessible to students. As such, lecturers must make choices, based on the contextually available resources, about which modes to utilise and how to organise and integrate them, in relation to the meanings they want to make. From a multimodal social semiotic perspective, these choices are always motivated (Jewitt et al. 2001; Kress and Selander 2012) and an understanding of affordances is what enables lecturers to make effective choices in their construction of meaning-making complexes, such that complementary modes are integrated—each adding particular details of meaning for the development of conceptual understanding.

In our attention to how the lecturer used these modes, individually and together in multimodal complexes, we worked between our analysis of the lectures and the lecturer interview data, to understand both how she developed meaning through her pedagogy, and what motivated the choices she made while doing so.

4.4 Methods

Studying a lecturer's pedagogy from a multimodal social semiotic perspective requires detailed attention to the lecturer's action—specifically what modes are used, and how these are used for conceptual and procedural meaning—in relation to the contextual experiences and motivated pedagogical choices. In this section we describe how we used a focused ethnography (Babbie and Mouton 2004; Rashid et al. 2019) to produce the lecture observations and lecturer interview data presented in this chapter.

4.4.1 Participant Selection and Ethical Permissions

The selection of the lecturer was purposive; she was the only lecturer teaching the Physics module on angular motion. Her understanding of the context and pedagogical choices were informed by her background in biomedical engineering, and her experiences when lecturing this course (a role she had done for more than 3 years, at the time of the study), and prior tutoring of both university and school students.

Ethical approval for the study was granted by the research ethics committees in the faculties where the study was conducted and in which the first-author was registered for postgraduate studies. Permission for the study was also granted by the Department of Human Biology, the course convenor and the lecturer. To ensure informed consent, an information sheet was sent to the lecturer, which included the motivation for the study, what participation would involve, potential risks and benefits for the lecturer, how confidentiality and anonymity would be addressed, data management procedures, participant rights and researcher responsibilities. Importantly, guided by a social semiotic perspective, the researcher emphasised that the purpose of the study was to describe and explain the lecturer's contextualised pedagogy, and not evaluate her practice. Based on this negotiation, the lecturer consented in writing to participate in an audio-recorded interview and granted the researcher access to learning materials, including video lecture recordings.

4.4.2 Data Collection

To describe what and how the lecturer used modes to create opportunities for conceptual and procedural meaning-making, observations of the five 40-minute lectures on angular motion were used. Specifically, the researcher observed video lecture recordings, which are regarded as productive in multimodal social semiotic research, allowing the researcher to focus in detail on the range of modes used (Jewitt 2012). The videos used were lecture recordings produced by the university as a learning resource for students, and made available to students on the learning management course site. The audio-visual recording equipment was positioned to focus on the front of the lecture venue, capturing the lecturer who was positioned behind a lectern at the front and centre of the classroom, as well as two wall-mounted white boards with the projection from a document camera.

To understand the lecturer's pedagogical choices in context, an audio-recorded semi-structured interview was conducted with the lecturer. This took place after the researcher had performed a preliminary analysis of the recorded lectures. The interview had three parts. Firstly, the lecturer completed an angular motion problem similar to those used in the course, and then the lecturer took part in a 'talk-around text' process (Lillis 2008) in which she explained her problem-solving steps and answered questions about any conceptual or procedural challenges students may experience when solving such problems. Secondly, the lecturer answered questions about her lecture planning, her experience of having taught the angular motion module and her perceptions of the preparedness of students for university physics and their performance in assessments. Finally, the lecturer viewed 12 video clips from the lecture recordings and answered questions about her pedagogical choices. These clips were selected as rich opportunities to probe the lecturer's use of multimodal language to provide meaning to key concepts and procedures on the topic of angular motion.

4.4.3 Data Production

The process of data production from the lecture recordings was not linear, and involved three related processes. One process involved viewing the recordings to produce an overall mapping of the central presentational meanings, that is, the key concepts and procedures for angular motion (Table 4.2). It was decided to focus the data analysis on the first two lectures as it was clear

Table 4.2 Concepts and procedures

A. Angular displacement		B. Angular velocity		C. Angular acceleration		D. Conversion between linear and angular motion	
Concept	Procedure	Concept	Procedure	Concept	Procedure	Concept	Procedure
1. Circle radius, arc length		1. Angular velocity (omega)		1. Angular displacement Angular velocity recap		1. Conversion of linear to angular motion	
2. Motion about a circle Rotation		2. Quantifying angular velocity	3. Introduces equation for angular velocity	2. Angular acceleration (alpha α)		2. Tangential velocity	
3. Angular displacement (theta Θ) Unit radians	4. Equation for angular displacement	4. Angular velocity as displacement: Time graph			3. Introduces equation for angular acceleration		3. Conversion equations
5. Unit conversion (degrees to radians)	6. Unit conversion degree to radians		5. Mathematical process angular velocity problem-solving example				4 Mathematical process conversion problem-solving example, angular velocity to tangential velocity
	7. Mathematical process angular displacement problem-solving example						

from the overall mapping that key angular motion concepts and procedures were covered in the first two lectures.

A second process, which focused on the macro-organisational meaning, yielded a record (in tabular form) of the sequence in which these meanings were taught. These two productions informed the data preparation for the detailed multimodal social semiotic analysis. Data was prepared in the form of a spreadsheet, with rows for key concepts, related concepts and procedures, and one column for each mode. The lecturer's verbal mode was transcribed verbatim, and all the other language modes were described and screenshots inclulded where relevant (Fig. 4.1). The audio-recording of the lecturer interview was transcribed.

4.4.4 Data Analysis

In the detailed multimodal semiotic analysis of the lecture, the theoretical tools and related questions summarised in the analytic framework (Table 4.1) were applied. This analysis was recorded in a spreadsheet (Fig. 4.2): the rows record the sequence of concepts and procedures on angular motion that was taught by the lecturer, with the columns recording the analysis of the meta-functions (columns A to D) and each mode (columns E to K). This structure was productive for analysing how the lecturer used the affordances of the modes, individually and together in multimodal meaning complexes, to create opportunities for students to access meanings. Looking across the columns for a given concept provides focus on the micro-level detail of how the lecturer used one or more modes to communicate different types of meanings regarding the concept. Looking down the rows provides focus on how, at a macro-level, the lecturer organised the concepts and procedures throughout the two lectures. The presentation of the analysis in the next section, in which we make explicit our theoretical analytic tools, is designed to illustrate our systematic application of these tools.

4.5 Findings

In the Physics module, the angular motion lectures directly followed those on linear motion. Figure 4.3 illustrates the sequence in which concepts were taught (illustrated by the numbering in the figure), with the arrows signifying that the meaning of subsequent concepts were dependent on the understanding of the preceding concepts. As indicated in Table 4.2 and Fig. 4.3, the lecturer started her teaching by focusing on the concept of angular displacement and its use in problem-solving. Over two lectures, she developed the meaning of angular displacement, focusing on underpinning circle concepts, before progressing to explain the concepts of angular velocity, angular acceleration and the process of conversion between angular and linear motion. The final three lectures focused on problem-solving: students were given physics problems that they worked through individually, before the lecturer demonstrated and explained a step-by-step solution. During all lectures, students could ask questions at any point, and the lecturer occasionally posed questions to the class.

In line with the focus of this edited work, we present below excerpts[1] from the lecturer's pedagogy that showcase how she used the particular affordances of the modes of image, gesture and object individually and together in a visual complex, but also put to work and integrated other language modes to give presentational, organisational and orientational meaning to angular motion concepts and procedures. Lecturer interview data is presented where she explains her choices and practice.

4.5.1 Excerpt 1: The Concept of Angular Displacement

Excerpt 1 transcribes the teaching of the concept of angular displacement which is foundational,

[1] The three excerpts are presented as multimodal transcripts, with her verbal speech in plain text and her actions transcribed in italics, with those involving a specific mode given in bold.

Concept/process	Verbal	Written	Image/ Drawing	Gesture/action	Tool/object
Circle	Let's say this is a radius		Draws a circle, draws a line representing a radius, and draws in a second line that is extended to a different point on the circumference of the circle. 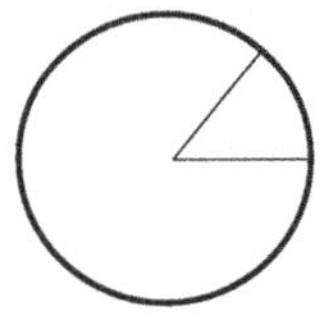		

Fig. 4.1 Content overview spreadsheet

and upon which other concepts of angular motion are developed. Evident in this excerpt is the lecturer's development of a visual complex using the modes of image and gesture. We also see how she complemented these two visual modes with written text, symbols and verbal speech in a broader multimodal complex, to give meaning to the concepts related to the circle, and the relationship between the relevant concepts in an angular displacement equation.

4.5.1.1 Excerpt 1

1a Let's say we have a circle (***draws*** *a circle and places a dot in the centre of the circle,* Fig. 4.4) and we wanted to say this (***draws*** *a straight line from the centre to the circumference representing a radius*) is the radius.

1b So, let's say I (***draws*** *in a second line, that is extended from the centre of the circle to a different point on the circumference of the circle*) rotated about the edge of the circle (***gestures*** *a small arc with her hand*). Let's say I went from here to here (***highlights in red*** *the distance between the two radii, on the circumference*). So, let's say I wanted to know the angle that I moved. It's kinda hard to always be measuring angles. So, they were like, 'surely there is something better?'

1c So, instead of measuring the angle, let's say we have *'r'* radius of our circle (***writes*** *symbol 'r'*) and arc length *'s'* ***writes*** (*symbol 's' alongside the* ***red arc*** *on the circumference*) that we moved.

Row	Meta- Functions				Language modes						
1	A Macro – Organisational Sequence of content	B Presentational	C Orientational	D Micro-Organisational Modes	E Verbal mode	F Specifics related to language use	G Written mode	H Symbolic mode	I Images	J Gesture	K Tools or objects
2	1	Concept: Tangential velocity	*Example* of throwing a ball, that is familiar to most students. Used pronouns *you*, *you've* and *your*, positions learner in the science discourse	Verbal, drawing, symbolic	So, what's gonna happen is you might get an example, where let's say (draws) you are throwing a ball, you've got a ball in your hand. But you can measure the tangential velocity (draws)	Formal Scientific term: 'tangential velocity' Pronouns: *you are, you've got, your, you can*	'V'	V	 Positioned under heading 'How do we convert between angular and linear' Iconic stick figure of arm with ball in hand, upward arrow labelled with 'V'		

Fig. 4.2 Analysis spreadsheet

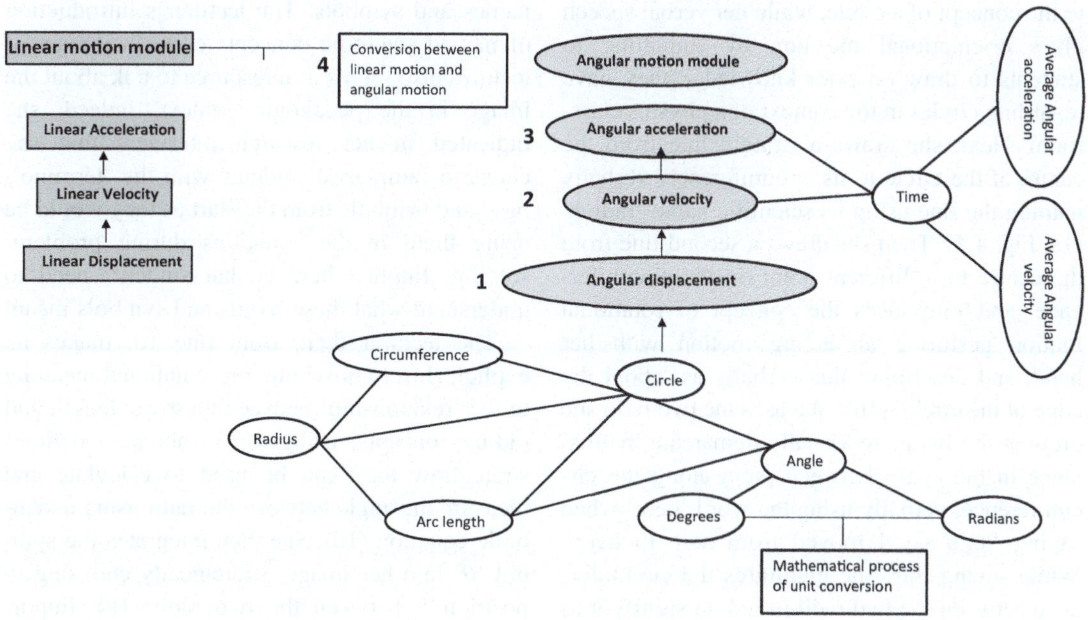

Fig. 4.3 Mapping of linear and angular motion concepts

1d Let's develop a relationship between these two. What if we just said '*s*' over '*r*' gives us an *angle* (***writes** out symbolic equation* $\frac{s}{r} = \theta$*)*.

Ie So, it's more of a ratio, but it represents the angle that we're travelling (***writes** symbol 'θ' in the angle formed by the radii of the circle, **writes** arc length and radius next to symbol 's' and 'r' in the equation and connects the words and symbol with an arrow*).

The lecturer begins by saying 'let's say we have a circle' (1a) and draws a circle on the board. This image gives presentational meaning

Fig. 4.4 Circle and its concepts

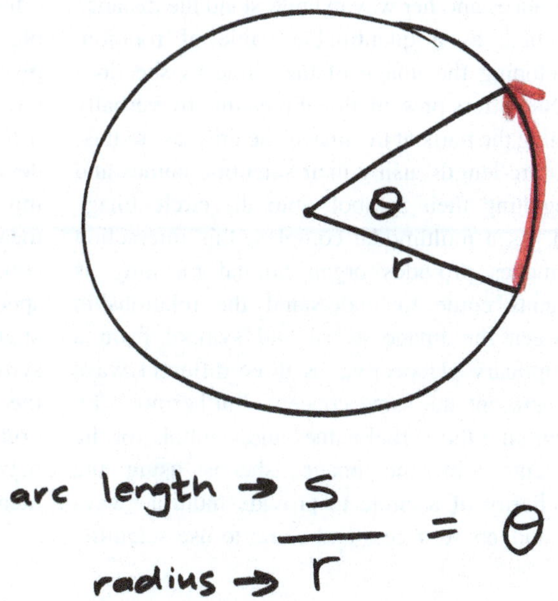

to the concept of a circle, while her verbal speech gives orientational meaning by signalling to students to draw on prior knowledge they have regarding circles in the context of a physics classroom. Next, she draws a straight line from the centre of the circle to its circumference, verbally naming the line using its scientific name 'radius' (1a; Fig. 4.4). Then she draws a second line from the centre to a different point on the circumference, and introduces the concept of rotational motion gesturing an arcing motion with her hand, and describing this verbally as 'about the edge of the circle' (1b). She uses the two radii she drew in the image to visually demarcate the distance in the space between them along the circumference, verbally using the word 'here' when saying 'let's say I moved from here to here'. While saying this, she highlights the circumference between the two radii in red, to signify it as the same motion path she had just gestured with her hand. This interaction of image, gesture and verbal speech in a multimodal complex gives presentational meaning about angular motion. As shown in the introductory vignette produced in the research interview, the lecturer motivated her choice to use image and gesture based on their affordances for providing visualisations of conceptual meaning.

Next, in lines 1c, 1d and 1e, the lecturer introduces another way to understand the distance travelled, as a quantifiable angle of rotation, developing the image of the circle as she does so. She offers presentational meaning by verbally naming the parts of the image she drew as 'radius' and 'arc length' using their scientific names and integrating their symbols into the circle image (1c). As a multimodal complex, this interaction of modes provides organisational meaning as students come to understand the relationship between the image, word and symbol from a disciplinary perspective, as three different ways to represent the same concept. Furthermore, by integrating the verbal names and symbols for the concepts with the image, she is using the familiarity of a circle to provide multiple ways for students to access and come to use scientific names and symbols. The lecturer's introduction of named scientific concepts early in the set of lectures also serves as a resource to talk about the image in the pedagogic context. Indeed, she indicated in her research interview that she chose to familiarise students with the 'terminology' and symbols from the start as they would be using them in the equations during problem-solving. Implied here is that students need to understand what these words and symbols mean.

The lecturer then, from line 1d, makes an explicit shift to providing presentational meaning to the 'relationship' between these arc length and radius concepts, using the symbols to demonstrate how they can be used to calculate and measure the angle between the radii using a symbolic equation (1d). She then integrates the symbol 'θ' into her image, strategically choosing to position it between the two radii (1e). Importantly, this allows her to show that, when the motion path is angular, the distance moved or displaced between two points can be quantified in terms of an angle of rotation. She then writes the full concept names, spatially positioning these next to each symbol 's' and 'r' and linking the written text and symbols indexically with an arrow (Fig. 4.4).

This extract shows how, at the micro-organisational level, the lecturer uses the affordances of image and gesture in a visual complex in which the circle image is central, that gives presentational meaning to the concepts of radius, arc length and angular displacement. Yet, using the affordances of verbal speech for describing and naming these concepts and thinking forward to the need to use these concepts meaningfully in problem-solving, she compliments this visual complex with the verbal speech and symbols early in her pedagogy on angular displacement. Her early inclusion of the symbolic mode (from line 1d) allows her to use the affordances of this mode to develop organisational meaning to the relationship between the concepts, while giving presentational meaning to the equation for angular displacement.

4.5.2 Excerpt 2: Concept of Conversion between Angular and Linear Motion

After explaining angular velocity and its mathematical equation for problem-solving (Fig. 4.3), the lecturer briefly explained angular acceleration. She then proceeded to explain the concept of conversion between angular and linear motion, using an example of a ball being thrown. This excerpt of this teaching moment demonstrates in particular how the lecturer sequentially constructed presentational meaning and then conceptual understanding, again using images as the central site for this process. As she constructed the image, she integrated both iconic and symbolic elements and gestures, to develop a visual complex. As in excerpt 1, she used this visual complex, in interaction with symbols, to give meaning to the relationship between angular and linear motion in a broader multimodal complex.

4.5.2.1 Excerpt 2

2a So, how do we convert? (***writes*** *'how do we convert between angular and linear' on the board*). Ok, so, when we talk about converting you might just be thinking, but I just told you those two are not the same, why would we want to convert at all?

2b So, what's gonna happen is you might get an example, where let's say you are throwing a ball, you've got a ball in your hand (***draws*** *a stick figure image of an arm with an empty circle at one end representing the shoulder and the other end a hand with a ball,* Fig. 4.5a). But you can measure the tangential velocity (***draws*** *in an upward arrow adjacent to the hand and annotates the arrow with the symbol 'v',* Fig. 4.5b).

2c So, you gonna be moving in that direction (***draws*** *in a curved arrow to represent the direction of movement of the arm,* Fig. 4.5c). You gonna be rotating (***gestures*** *swinging arm in rotatory motion*). But then you can maybe measure the speed of the ball in your hand.

2d So, you will be given linear values, but that doesn't mean the ball moves in a linear motion. Does that make sense?

2e So, I'm saying (***writes*** *in symbol ω directly above the curved arrow,* Fig. 4.5d) you rotate like that, but you can measure the tangential velocity or the tangential acceleration.

Before explaining the process of conversion, the lecturer first focuses on the conceptual reason for conversion, which requires her to develop students' understanding of the relationship between linear and angular motion (2a).

From line 2b, she anchors her explanation of this relationship using the example of a ball being thrown during which an arm moves in rotatory motion about the shoulder. Her choice of example allows physics concepts to be related to dynamic human movement, thus providing orientational meaning for students as future physiotherapists. The lecturer explained in the interview that when she first started teaching the Physics module, she reviewed the examples used in the existing course material. She reported often thinking, 'this is so irrelevant' for physiotherapy students, and then purposefully chose more contextually relevant examples. This purposeful selection for relevance aligns with her description of her pedagogy in the introductory vignette, and her goal to help students recognise the fundamental importance

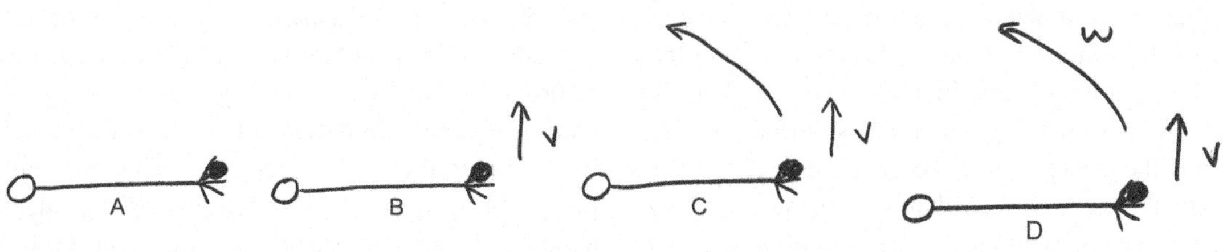

Fig. 4.5 (a, b, c, d). Sequential development of an image in a multimodal complex showing a ball in hand being thrown while arm moves in rotatory motion about the shoulder

of angular motion to the practice of physiotherapy.

She begins (2b), by drawing a stick figure image which functions iconically to represent an arm with a shoulder joint and a hand holding a ball. She explains verbally that the 'tangential velocity' of the ball can be measured when it is thrown. She visually represents the meaning specifically of the scientific concept named 'tangential velocity' by adding a straight upward arrow at the end of the drawn hand holding the ball, to symbolise tangential motion. She then integrates the symbol 'v' into her image, positioning it spatially next to the arrow to represent linear velocity (Fig. 4.5b). This multimodal meaning complex, in which the image is central, but interacts with verbal speech, and a symbol, works to develop an understanding of tangential velocity as a linear quantity.

Next, in line 2c, she gives presentational meaning to angular velocity by drawing a curved arrow above the image to symbolise angular motion, while verbally describing the motion as 'rotating'. Then, she performs an iconic gesture, swinging her arm in a rotatory motion, to dynamically represent the angular motion that the curved arrow symbolises. She then integrates the symbol 'ω' for angular velocity into the image, positioning it next to the curved arrow (Fig. 4.5c, d).

This excerpt demonstrates how macro-organisational meaning is sequentially developed through the integration of modes. As in excerpt 1, the image mode is central. In particular this excerpt foregrounds how, in consideration of human movement as relevant in physiotherapy in her chosen example, she uses the affordances of image to perform both iconic and symbolic presentational meanings. Furthermore, from an organisational perspective, she performs the creation of the image live in the classroom (rather than offering a final, static image). Firstly, this choice allows her to bring the images into interaction with the gesture in a visual complex, but also with the necessary verbal speech and symbols. Thus, she integrates the individual modes and the meanings they perform into wider complexes in her pursuit of conceptual understanding of the relationship between angular and linear motion as a precursor to the conversion process. Secondly, motivated by her concern about students' understanding of this relationship, she strategically sequences the image creation process starting with tangential velocity.

This analysis focuses on a problem-solving example. As in excerpt 1, the lecturer positions the image mode as central for conceptual and presentational meaning: 'I just try to give enough diagrams and draw stuff in, so that they can read and fully understand'. She also explained in the research interview that she organises her teaching in a way that everybody understands: 'something that is maybe really easy to me is really hard for them and I have to teach it in a way that everyone understands...I have to teach it in a kind of structured way'. In line with making the material understandable, the lecturer chose the example based on her understanding of the need to contextualise students' learning and her role as a physics lecturer in physiotherapy education. Yet, in the interview she also indicated that she does not 'shy away' from examples that students 'wouldn't have encountered'. Excerpt 3 which follows offers an example of this.

4.5.3 Excerpt 3: The Concept of a Tangent and Problem-Solving Example Related to Tangential Velocity

Following the teaching moment presented in excerpt 2, the lecturer introduced the conversion equations used to convert between angular and linear velocity (Fig. 4.3). Then she focused on the process of conversion, by giving the students a problem to solve. Excerpt 3 adds to our understanding of how, in pursuit of meaningful use of procedures in problem-solving, she used the affordances of images and gestures in visual complexes, but strengthened her use of the visual by including the mode of object in a visual complex. This excerpt also develops our argument about how she necessarily integrated the verbal and symbolic modes, this time demonstrating in particular how the verbal mode can be used to relate concepts described using a mix of everyday and scientific terms in contextual problems.

The lecturer began by reading the problem example aloud from her lecture notes:

> David uses a slingshot to slay Goliath. If the length of the slingshot strap was ninety centimetres and David spun the slingshot around at six revolutions per second, determine the tangential velocity of the tip of the slingshot i.e., the velocity at which the stone would have been released?

4.5.3.1 Excerpt 3

3a So, what we're saying is (***draws** a stick figure of slingshot strap and stone*), he is swinging it, there's the stone, (***draws** an arrow below the strap of the slingshot, labels it with 90 cm*) that is ninety centimetres. And he is spinning it at (***draws** in a curved arrow above the strap of the slingshot labels it 'ω = 6rev/sec'*) a velocity of six revolutions per second (Fig. 4.6).

3b So, revolutions per second. This is also new to you guys. Have you heard of revolutions per second before? You might have heard of rpm? If you've driven, in your car you have a little rev counter, it measures rpm (***writes** rpm = revolutions per minute,* Fig. 4.7). Rpm is revolutions per minute. Ok, we're now working with revolutions per second.

3c How many degrees is one revolution? (Student response inaudible). How many radians is one revolution? (Student: 2pi) (***writes** 1 rev = 2π rad*). So, what I want to know is, what is the velocity? (***writes** 'v?'*) [she gives students 3 min to solve for angular velocity].

3d Ok. So, the first thing we want to do. We obviously want to convert these things to useful units. So, a lot of you started with omega. So, omega is equal to 6 revs per second (***writes** $\omega = 6\ \frac{rev}{s}$*) and we said one rev is equal to 2 pi radians (*continues to **write** $\omega = 6\ \frac{rev}{s}\ x \frac{2\ \pi\ rad}{1rev}$*).

3e And you gonna see that we gonna cancel out our revolutions (***draws** a red line through 'rev' in the equation,* Fig. 4.7) and we going to have 12 pi rad per second (***writes** 12πrad/sec*). Ok, if you wanna convert that to a value, whatever 12 times pi is, that's fine, you can write it as that as well. Not a problem. But it's quite easy. Your calculator likes to give you the answer as 12pi, and that is much easier to type in than whatever else it is going to give you. I like to leave it with the pi in there.

3f So the next thing we need to figure out is what is our '*r*' value. So, we didn't talk too much about it. So, radius is normally from the centre of the circle to the edge (***gestures** a radius by holding out her arm*). So, if I was swinging a slingshot.

3g Does anybody have something I can swing? Just for visual purposes (*a student hands her a lanyard with a student card on one end*). Oh, that's such a good slingshot. So, if I'm swinging something (***holding** the lanyard at one end of the strap and with the student card hanging freely at the other end, she **swings** the lanyard in a circular motion in front of her*), where's the circle? On the outside right by the student card, right?

3h So, if I say that the strap of the student card is 90 centimetres long, would that be the radius or the diameter? (students: radius). So, the value I

Fig. 4.6 Slingshot drawing

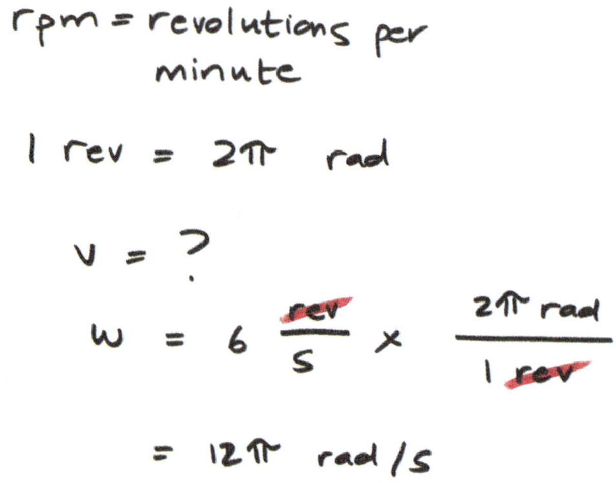

Fig. 4.7 Angular velocity calculation

am giving you here is *this* length (***holds** the lanyard and the student card at each end stretched out in a straight line*). From the centre of the circle to the edge of the circle. This was such a good visual. So, in this case this 90 centimetres equates to our radius.

3i So, in this case (***writes*** $r = 90\ cm = 0{,}9\ m$) radius is equal to 90 centimetres or I'm gonna write it as 0,9 metres. So, now we've got the two values that we were interested in. Now, let's work out '*v*'.

3j We already have the equation we need to use. *V* is equal to omega times *r*, which is equal to 12 pi times 0,9 (***writes*** $v = \omega\ r = (12\ \pi)\ (0{,}9)$, Fig. 4.8). What did you get? You can write 54 over 5 pi. (***writes*** $\frac{54}{5}\pi\ m/s$), or you can write 33,9 (***writes*** *33,9 m/s*), or (***writes*** $= 33{,}9\ m.s^{-1}$).

3k Whatever one you want to write generally, when we start going into linear, we don't leave the pi generally.

The lecturer organises the necessary conceptual and procedural presentational meanings by first using an image to anchor the problem stated verbally (3a). As she creates live in the classroom an iconic stick figure, she integrates symbols and verbal descriptions and names for the information in the problem. For example, labelling what is described verbally as the 'slingshot strap' symbolically as '90 cm', she gives meaning to this strap as the scientific concept radius (3a, Fig. 4.6). Then, by using an image of a curved arrow symbolically, and labelling this symbolically as '$\omega = 6rev/sec$', she provides presentational meaning to the commonly used terms 'swinging' and 'spinning' for angular velocity (3a, Fig. 6). Next, in line 3b, she provides presentational meaning to the unit named verbally in the problem as 'revolutions per minute' by orienting students to the 'rev counter' of a car (3b, Fig. 4.7). Finally, the final step in the creation of the central image involves reminding students that they need to solve for the concept named scientifically as 'velocity', she writes the symbol '*v*?' (3b, Fig. 4.7).

Before she solves for '*v*', the lecturer signals through her verbal questioning that students need to do the procedure of unit conversion (3c) to convert the unit to 'useful units' (3d). She uses the mathematical chain-link method, and illustrates the process to convert from rev/min to radians/sec (3e, Fig. 4.7).

However, motivated by her concern that the procedural steps for solving for tangential velocity are informed by conceptual understanding and her understanding of the affordances of the visual modes for the latter, in line 3f, she uses gestures and an object to give meaning to the concept of radius '*r*' and its numerical value. She organises this meaning by linking it to the verbally named 'slingshot strap' (3a). She begins this by descriptively explaining the scientific concept 'radius' as extending 'from the centre of the circle to the

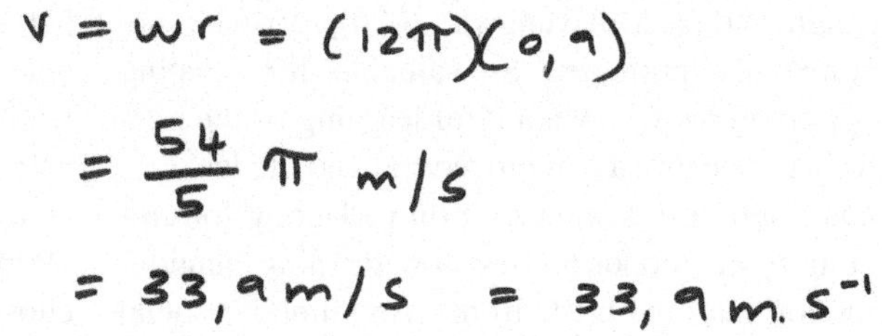

Fig. 4.8 Solving for velocity

edge' (3f). Using a gesture iconically, she visually represents this 'radius' by extending her right arm horizontally, and then combines this with an indexical gesture by running the index finger of her left hand along her extended arm from the shoulder (representing the centre) to the wrist (the edge of the circle) (3f).

Then, in line 3g, she asks students if anybody has something to 'swing', and is offered an object in the form of a lanyard attached to a student's institutional identity card. The lecturer spontaneously uses the visual affordances of this lanyard, and spins it to provide a demonstration mimicking the motion of a slingshot. As she spins the lanyard, she first helps students identify the circle formed (3g), simultaneously describing this verbally as being on the 'outside'. She helps students identify that the visual lanyard strap represents the concept named verbally as 'radius' which, in relation to the problem, would have the value of '90 cm' (3h). This she indicates will be represented as '0,9 m' when substituting the value into the equation (3i). Her use of the object in motion in this way gives presentational meaning to the concept named as 'radius' represented in the static image, but importantly also visually illustrates the motion of the radius mimicking the 'spinning' or 'velocity' necessary for conceptual understanding of angular velocity. Lastly, in lines 3j to 3k, she moves towards using the affordance of symbols and reminds the students that they 'already have an equation' to use. As she writes out the symbolic conversion equation for linear velocity '$v = \omega r$' (Fig. 8), she continues to show them the problem-solving steps, which she writes out (3j-3k, Fig. 4.8).

Thus, in excerpt 3 we see how the lecturer, as in excerpts 1 and 2, uses an image and her live creation thereof in a visual complex with gestures. Also, consistent with her earlier pedagogy, she is organising her pedagogy towards a meaningful understanding of the relationship between concepts in the symbolic equation needed to solve the problem. In particular, this excerpt foregrounds how, given her focus on angular velocity, she strengthens the visual complex by using the affordances of an object to give presentational meaning to the motion of the slingshot. Excerpt 3 also shows how the lecturer utilises the affordances of verbal and symbolic language to relate the everyday terms used in the problem to describe the meaning of and name key scientific concepts.

Reflecting on this problem-solving example in the research interview, the lecturer motivated her choice to use an object, together with her image, by identifying the affordances of the lanyard. She described this as 'easy visual' representing the 'object moving' that allowed her to give meaning to the concepts of radius and velocity:

> Just have the string and you have your object moving and it's an easy visual. And it's very simple in terms of 'this is our point of rotation'. We have our string and then we have our object moving about at a certain velocity. And to determine all our values in this example is very easy to go, 'Oh this is our radius, this is our point of rotation, that's our omega.' You know, it's a very simple example and that is what's nice about it.—lecturer in this research study.

4.6 Discussion and Conclusions

The importance of, and university students' difficulties with, understanding and using the concept of angular motion in physiotherapy and other science disciplines is well recognised (e.g. Rimoldini and Singh 2005; Mashood and

Singh 2012; Mutsvangwa 2020). Visuals are commonly promoted as valuable for creating opportunities for meaningful learning of the necessary conceptual and procedural knowledge. Our contribution to scholarship on pedagogy for creating such opportunities for learning angular motion has involved, from a multimodal social semiotic perspective, positioning multimodal language as contextualised, motivated and central to meaning-making. We put this approach to work in the specific context of first-year university physiotherapy education at a South African university.

In this chapter we have foregrounded the modes of image, gesture and object, that are recognised in the literature as offering visual and spatially realised affordances (Kress et al. 2014). Necessarily, we consider how this visual complex interacts with the written, verbal and symbolic modes in wider multimodal meaning-making complexes. In this section we interpret the findings described in Sect. 4.5 to answer our three research questions—these being describing the lecturer's multimodal pedagogy, and understanding her contextualised and motivated choices in this regard—and bring these into conversation with the research reviewed in the literature. We end with recommendations for education development practice and further research and scholarship.

The lecturer was motivated by her intention to create opportunities for students to use equations meaningfully and thus apply the necessary conceptual understanding in problem-solving procedures. This need has been recognised as important across various contexts (Bray and Williams 2020; Mashood and Singh 2012; Mutsvangwa 2020). The lecturer chose to organise her angular motion lectures to communicate the presentational meaning of key concepts and their relations, before progressing to modelling problem-solving. The former involved organising the relationship of 'the parts and the whole' (Lemke 1998, p. 251), for example, by using the circle and its components to give meaning to the concept of angular displacement, and then building to angular velocity (as summarised in Fig. 4.3). Furthermore, she was motivated by her view of the importance of students being able to relate the physics of motion to their learning and experiences of motion in physiotherapy and other contexts (as argued by Ingerman et al. 2007; Rayner 2005; Rennie and Parker 1996). Thus for orientational meaning, she chose examples involving human movement and references to joint motion. Her understanding of the diverse experiences of her class led her to scaffold the interpretation of these examples in relation to physics concepts.

The findings suggest that the lecturer's micro-level language choices to achieve these two goals were informed by her knowledge of the affordances of visual modes, individually and interaction in visual complexes. As she was looking to enable meaningful procedural problem-solving, she considered how the written, verbal and symbolic modes could be put to work with the visual, in wider multimodal complexes. Thus, as a lecturer, she moved beyond seeing students' difficulties working with multiple representations as a problem (as identified by lecturers in the study by Bray and Williams 2020), and intentionally and strategically used multiple modes as a pedagogic resource for presentational meaning.

In particular, we see that she used the visual mode of image as a central site for the integration of conceptual meaning for problem-solving. She used the practice of creating images live in the lecture to build meaning of individual concepts and their relationships, ultimately realised in a static image representing the spatial arrangement of multiple concepts and their relations simultaneously. The lecture excerpts presented in this chapter also point to similarities in how she used the affordances of image across her drawings, for example the patterned use of the iconic image for the arm and ball in hand, and the strap and stone to develop a 'grammar' for her visual language (Kress and Van Leeuwen 2006; Airey and Linder 2009) with which students could become familiar.

Yet, while placing the image as central in her pedagogy, she consistently organised this mode in a visual complex with gesture. Firstly, the latter afforded her the opportunity to index particular concepts in an image and as discussed below, to integrate her verbal speech about that image. Secondly, and most importantly, when used iconically, her gestures gave presentational meaning to the motion of the static concepts in an image. This is significant, given the importance of physiotherapy students coming to understand physics concepts in relation to dynamic human movement (Kelly et al. 2018).

At times, the lecturer also strengthened her use of the visual for conceptual presentational meaning, by integrating the mode of object, into her consistent use of a visual complex of image and gesture. The object offers the affordance of physical interaction (Kress and Van Leeuwen 2006) to mediate the concept of motion (Kress et al. 2014). Her choice of object, such as the lanyard, also had the affordance of an 'everyday' likeness to the static image used iconically, described verbally or in writing in a physics problem (Kress et al. 2014). Yet, since motion (be it of an object or through gesture) is temporal and thus transient, an image or sequence of images visually represents a more persistent record of the key concepts and their relations simultaneously at a particular moment in this sequence. This supports our argument that the image mode functions as a central site for multimodal meaning-making in the lecture.

Yet the findings point to how, in the pedagogy of physics for physiotherapy, the successive use of these visual complexes alone is insufficient for holistic conceptual and procedural presentational and the necessary orientational meanings, but needs to interact with the verbal, written and symbolic modes. Crucially in pedagogy, visual complexes need to be organised with verbal speech; being sequentially and temporally realised (Kress and Van Leeuwen 2006), the speech gives necessary coherence to the visual. Motivated by her knowledge of the disciplinary specificities of language and also her students' diverse language resources, the lecturer in this chapter used the verbal mode to describe concepts and to name these scientifically while linking to 'everyday' uses of these terms. This practice aimed to create opportunities for students to access and use disciplinary knowledge (Tang et al. 2016). Written text, when integrated consistently into visual complexes (for example representing concept names in writing as seen in excerpt 1) strengthens pedagogy as it has the affordance of allowing what is produced aurally through verbal speech to be visually perceived (Kress and Van Leeuwen 2006). Yet, the lecturer's concern about access returns us once again to the image as central in any multimodal complex, as her use of the visual image itself provides an additional access point to knowledge.

Finally we note that, since the lecturer was teaching toward meaningful problem-solving of angular motion problems in her set of lectures, she needed the affordance of the symbolic mode to represent for example the relationship between key concepts such as radius and arc length in relation to angular displacement, in a contracted form for problem-solving (O'Halloran 2000). Thus, she intentionally put symbols (written and described verbally) in interaction with the modes of image, gesture and verbal speech in a wider multimodal complex.

In concluding, we acknowledge that the research from which this chapter draws is limited to one lecturers' pedagogy in physics lectures, and to one foundational concept in physiotherapy education. Yet we argue that the use of a focused ethnography and a multimodal social semiotic perspective produces deep understandings of this pedagogy. This, we argue, offers a novel contribution to how visuals might be put to work to create opportunities for meaningful learning of kinematics for physiotherapy. While we have emphasised the contextualised nature of language use and focused on a physiotherapy education at one South African university, the literature review indicates that the concern about student learning animating the research in this chapter is certainly not unique to this context. We end by suggesting how our multimodal social semiotic approach might be put to work in collaborative education development between disciplinary and language experts.

Since language use is motivated and conceptual (Jewitt et al. 2001), we argue that certain considerations of pedagogy need to be firmly rooted in a lecturer's disciplinary knowledge relative to physiotherapy. Basic science lecturers, in particular, should have a sense of their role as lecturers in physiotherapy education, their experiences of teaching and learning in that context, as well as their knowledge of the students' needs and the available resources in this context and importantly, the lecturer's ideas about learning and the role of language therein. With this foundation, research data such as that in this chapter can then be used to discuss the affordances of individual modes, but also their interaction in multimodal meaning-making complexes. Given the value of the visual complex, put to work using mediums such as hand-drawn images and objects available in a lecture space (demonstrated in this chapter) there is an opportunity to explore how the affordances of the visual modes can be achieved relatively simply. In addition, further development of the theoretical notion of the visual complex, and its use in further research on lecture pedagogy of other physics concepts and in the study of student problem-solving, has the potential to strengthen this education development.

References

Airey J, Linder C (2009) A disciplinary discourse perspective on university science learning: achieving fluency in a critical constellation of modes. J Res Sci Teach 46(1):27–49

Airey J, Linder C (2017) Social semiotics in university physics education. In: Multiple representations in physics education. Springer, Cham, pp 95–122

Allwood J and Ahlsén E (2015) Meaning potentials in words and gestures. In *Proceedings from the 3rd European Symposium on Multimodal Communication*, Dublin, September 17-18, 2015, https://ep.liu.se/ecp/105/001/ecp16105001.pdf

Babbie E, Mouton J (2004) The practice of social research. Oxford University Press

Bezemer J (2017) Visual research in clinical education. Med Educ 51(1):105–113

Blackford J, McAllister L, Alison J (2015) Simulated learning in the clinical education of novice physiotherapy students. Int J Pract-Based Learn Health Soc Care 3(1):77–93

Boucher B, Hunter D, Henry J (1999) The effectiveness of computer-assisted instruction in teaching biomechanics of the temporomandibular joint. J Phys Ther Educ 13(2):47

Bray A, Williams J (2020) Why is physics hard? Unpacking students' perceptions of physics. J Phys Conf Ser 1512(1):012002. IOP Publishing

Brown A, Dowling P (1997) Doing research/reading research: a mode of interrogation for education. Falmer Press, London

Case J, Marshall D, Grayson D (2013) Mind the gap: science and engineering education at the secondary-tertiary interface. S Afr J Sci 109(7):1–5

Clement SJ, McKee DW, Romano R, Xu J, Lopez JM and Battersby D (2017) June. The internet of simulation: enabling agile model based systems engineering for cyber-physical systems. *In 2017 12th System of Systems Engineering Conference* (SoSE) (pp. 1–6). IEEE

Council on Higher Education (2020) VitalStats: Public education 2018

da Silva MV, Pacagnelli FL, de Almeida LL and Siscoutto RA (2021) PhysioAR: an augmented reality system applied in respiratory physiotherapy for hypertensive patients. *In Proceedings of the Brazilian Symposium on Multimedia and the Web*. pp.37–44

Gabriels S (2022) A multimodal social semiotic analysis of lecturer pedagogy for the physics concept of angular motion in physiotherapy education. [Unpublished Masters dissertation, University of Cape Town]

Gough S, Yohannes AM, Murray J (2016) Using video-reflexive ethnography and simulation-based education to explore patient management and error recognition by pre-registration physiotherapists. Advances in Simulation 1(1):1–16

Halliday MA (1993) Towards a language-based theory of learning. Linguist Educ 5(2):93–116

Harrison C (2003) Visual social semiotics: understanding how still images make meaning. Tech Commun 50(1):46–60

Hiebert J ed. (2013) *Conceptual and procedural knowledge:* The case of mathematics. Routledge. https://doi.org/10.4324/9780203063538

Ige BO, Amosun SL, Hartman N (2017) Widening student access and participation in allied health sciences: a critical reflection. S Afr J High Educ 31(5):65–89

Ingerman Å, Booth S, Linder C (2007) Learning physics as a whole–on supporting students making sense of their studies. Nord Stud Sci Educ 3(2):163–174

Jaipal K (2010) Meaning-making through multiple modalities in a biology classroom: a multimodal semiotics discourse analysis. Sci Educ 94:48–72. https://doi.org/10.1002/sce.203

Jansen JD (2019) Inequality in education: what is to be done? In: Spaull N, Jansen J (eds) South African schooling: the enigma of inequality. policy implications of research in education, vol 10. Springer, Cham. https://doi.org/10.1007/978-3-030-18811-5_19

Jewitt C (2012) *An Introduction to using video for research*. NCRM Working Paper. National Centre for Research Methods. https://eprints.ncrm.ac.uk/id/eprint/2259/4/NCRM_workingpaper_0312.pdf

Jewitt C, Bezemer J, O'Halloran K (2016) Introducing multimodality. Routledge

Jewitt C, Kress G, Ogborn J, Tsatsarelis C (2001) Exploring learning through visual, actional and linguistic communication: the multimodal environment of a science classroom. Educ Rev 53(1):5–18

Kelly D, Hoang TN, Reinoso M, Joukhadar Z, Clements T, Vetere F (2018) Augmented reality learning environment for physiotherapy education. Phys Ther Rev 23(1):21–28

Kember D, Ho A, Hong C (2008) The importance of establishing relevance in motivating student learning. Act Learn High Educ 9(3):249–263

Kizito R, Munyakazi J, Basuayi C (2016) Factors affecting student success in a first-year mathematics course: a South African experience. Int J Math Educ Sci Technol 47(1):100–119

Kress G, Jewitt C, Ogborn J, Tsatsarelis C (2014) Multimodal teaching and learning: the rhetorics of the science classroom. Bloomsbury, London, UK

Kress G, Van Leeuwen T (2020) Reading images: the grammar of visual design. Taylor & Francis Group, Milton

Kress G, Selander S (2012) Multimodal design, learning and cultures of recognition. Internet High Educ 15(4): 265–268

Kress G, Van Leeuwen T (2006) The grammar of visual design, 2nd edn. Routledge, London

Lemke J (1998) Multimedia literacy demands of the scientific curriculum. Linguist Educ 10(3):247–271

Levangie PK, Norkin CC (2011) Joint structure and function: a comprehensive analysis. F. A. Davis Company

Lillis T (2008) Ethnography as method, methodology, and "Deep Theorizing" closing the gap between text and context in academic writing research. Writ Commun 25(3):353–388

Linder A, Airey J, Mayaba N, Webb P (2014) Fostering disciplinary literacy? South African physics lecturers' educational responses to their students' lack of representational competence. Afr J Res Math Sci Technol Educ 18(3):242–252

Makgato M (2007) Factors associated with poor performance of learners in mathematics and physical science in secondary schools in Soshanguve, South Africa. Afr Educ Rev 4(1):89–103

Mashood KK, Singh VA (2012) An inventory on rotational kinematics of a particle: unravelling misconceptions and pitfalls in reasoning. Eur J Phys 33(5):1301–1312

Mutsvangwa A (2020) A study of student teachers' misconceptions on uniform circular motion. J Phys Conf Ser 1512(1):012029. IOP Publishing

O'Halloran KL (2000) Classroom discourse in mathematics: a multisemiotic analysis. Linguist Educ 10(3): 359–388

Rashid M, Hodgson CS, Luig T (2019) Ten tips for conducting focused ethnography in medical education research. Med Educ Online 24(1):1624133

Rayner A (2005) Reflections on context-based science teaching: A case study of physics students for physiotherapy. *Poster presented at the annual UniServe Science Blended Learning Symposium Proceedings*, Sydney, Australia

Reinke NB, Kynn M, Parkinson AL (2021) Immersive 3D experience of osmosis improves learning outcomes of first-year cell biology students. CBE Life Sci Educ 20(1):ar1

Rennie LL, Parker LH (1996) Placing physics problems in real-life context: students' reactions and performance. Aust Sci Teachers J Australian Science Teachers Association 42(1):55–58

Restrepo-Lòpez S, Perez-Restrepo AF and Trujillo-Suarez CA (2021) March. Development and evaluation of computational applications for teaching kinematics in engineering. *In 2021 IEEE World Conference on Engineering Education (EDUNINE).* (pp. 1-5). IEEE

Rimoldini LG, Singh C (2005) Student understanding of rotational and rolling motion concepts. Phys Rev Spec Top-Physics Education Research 1(1):010102

Rooney C (2015) Using survival analysis to identify the determinants of academic exclusion and graduation in three faculties at UCT [Unpublished Masters dissertation, University of Cape Town]. OpenUCT

Spaull N (2019) Equity: a price too high to pay? In: Spaull N, Jansen J (eds) South African schooling: the enigma of inequality. Policy implications of research in education, vol 10. Springer, Cham. https://doi.org/10.1007/978-3-030-18811-5_1

Stromberga Z, Phelps C, Smith J, Moro C (2021) Teaching with disruptive technology: the use of augmented, virtual, and mixed reality (HoloLens) for disease education. Adv Exp Med Biol 1317:147–162

Tang KSK, Ho C, Putra GBS (2016) Developing multimodal communication competencies: a case of disciplinary literacy focus in Singapore. In: Using multimodal representations to support learning in the science classroom. Springer, Cham, pp 135–158

Part III

Making 3D

Mitochondria to Bitter Melon: Understanding the 3D Ultrastructure of the Cell Via 2D Thin Section Reconstruction and the History of Mitochondrial Visualization

5

David M. Mazierski

Abstract

The origin of histology—the study of microscopic anatomy—is intimately connected with the development of the light microscope and improvements in lens design and manufacture.

However, knowledge of the ultrastructure of the cell was hampered by the very nature of light microscopy, which, due to the physical properties of the visible electromagnetic spectrum, could never provide the magnification and resolution for study of the granules seen in cells, which we now know as the organelles. When the electron microscope was developed in the 1930s, a beam of electrons replaced light as the source of illumination, and the inner details of the cell could be observed directly. With thin sections obtained by transmission electron microscopy, cell biologists could embark on the task of reconstructing 3D microstructure via the painstaking stacking of the individual slices.

The three-dimensional visualization of the mitochondrion was particularly challenging, as its convoluted structure could be interpreted in several ways based on differences observed by George Palade at the Rockefeller Institute for Medical Research (NYC), and Fritiof Sjöstrand at the Karolinska Institutet (Stockholm). Palade's interpretation was eventually accepted as correct due to its alignment with the findings of biochemists investigating the cascade of molecular interactions known as the Krebs cycle, responsible for the production of cellular energy in the form of adenosine triphosphate (ATP). However, it can also be argued that Palade's visualization via a physical model of the mitochondrion, which he built with sheets of wax, photographed, and published in 1953, better enabled colleagues to comprehend its unique inner structures known as cristae.

To teach undergraduate science students about this pivotal moment in cell biology and add to their understanding of the reconstruction process, a pedagogical exercise was created in which students are provided with outline drawings of various organic objects cut in random planes of section. Working individually at first, and then in groups, they are tasked with collaborating to devise an accurate description of the shape and texture of the object. After their observations are presented to the class, they are shown a photo of the object prior to its sectioning to determine if their observations were correct.

Keywords

History of histology · History of microscopy · The mitochondrion · Three-dimensional visualization · Undergraduate education

D. M. Mazierski (✉)
Biomedical Communications, University of Toronto Mississauga, Mississauga, ON, Canada
e-mail: d.mazierski@utoronto.ca

L. Shapiro, P. M. Rea (eds.), *Biomedical Visualisation*, Advances in Experimental Medicine and Biology 1388, https://doi.org/10.1007/978-3-031-10889-1_5

5.1 Introduction: The Mitochondrion

In 1948, cell biologist Albert Claude spoke before the Harvey Society of New York, a distinguished gathering of Nobel Prize scholars and scientists, to summarize 10 years of research at the Rockefeller Institute on the components of the eukaryotic cell. He concluded his talk with the statement that "mitochondria may possibly be considered as the power plants of the cell," a striking metaphor that most students of cell biology will recognize as the shorthand descriptor for this organelle (Claude 1948). Likewise, the iconic depiction of the double-membrane organelle with numerous plate-like folds known as cristae is immediately recognizable from hundreds of similar images published in cell biology textbooks. However, it is wrong to assume that it represents the definitive structure of this important constituent of every eukaryotic cell. The persuasiveness of the image and its perpetuation throughout the life sciences can be traced to its visualization via the electron microscope in the 1950s by different researchers in pursuit of the same goal. The uniqueness of those first images and the controversies that arose from their interpretation resulted in a stasis of visual representation that persists in almost all depictions of the mitochondrion to this day (Fig. 5.1).

5.2 The Origins of Light Microscopy and Cytology

While the properties of magnification via droplets of water or glass were described by the Roman philosopher Seneca in 65 AD (Chapman et al. 2020), the first tools explicitly created as compound microscopes appeared around 1590. This discovery is generally attributed to the father and son team of Hans and Zacharias Janssen, spectacle makers in Middelburg, Holland, although other craftsmen working in the developing field of optics have also been named (Bradbury 1967). This was a simple device with a convex lens mounted at each end of two sliding draw tubes. Robert Hooke (1635–1703) refined Janssen's design and designed a microscope (constructed by Christopher White of London) using three lenses capable of approximately 20 to 50 × magnification. His drawings of insects and plants in *Micrographia: or, Some physiological descriptions of minute bodies made by magnifying glasses* (1665) demonstrated the power of these new instruments of observation.

A different type of microscope was created by Antoni van Leeuwenhoek (1632–1723), a draper and microscope enthusiast from Delft. Leeuwenhoek's design utilized a tiny convex bead of glass or polished quartz mounted on a small brass plate with a handle. On the back of the plate, a threaded rod was mounted, terminating in a pinpoint that was at the focal point of the lens, and could be adjusted up and down. Specimens were placed on the tip of the pin and observed through the lens. While simple, these microscopes were capable of up to 270× magnification, more powerful than Hooke's design, and made possible the first written descriptions of blood cells, spermatozoa, and simple one-cell organisms.

5.3 Zeiss and Lens Improvements

The simple microscopes available to the first scientists who used them were faced with challenges that were inherent in the refractive properties of their lenses. Refraction refers to the way in which light is bent as it passes through a lens, which produces the effect of magnification. However, simple lenses suffer from two problems which plagued early microscopists and cytologists for the next hundred years. The first problem is spherical aberration. Light rays entering the periphery of the lens are bent more acutely relative to rays passing through the center of the lens, creating a generally blurry image with only one point in focus. To compensate for this, a diaphragm with a small opening placed in front of the lens will restrict light to only the central part of the lens, increasing sharpness but requiring a much brighter light source to compensate for the smaller amount of light entering the device

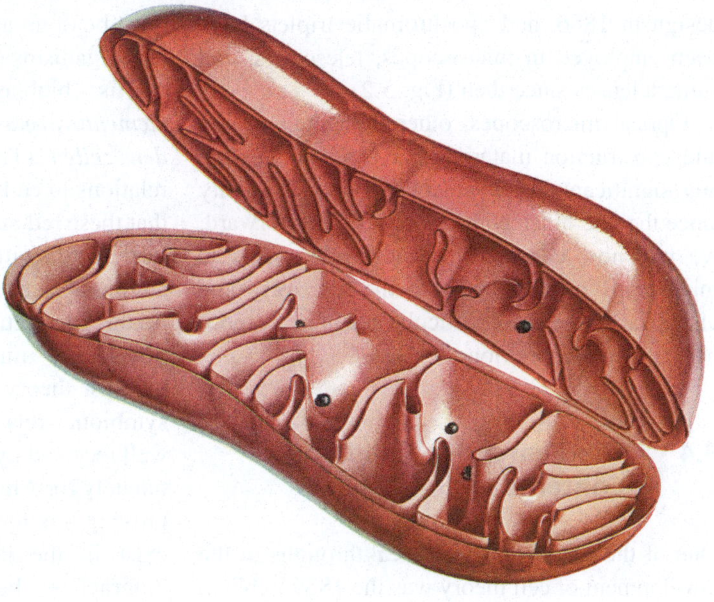

Fig. 5.1 The mitochondrion as illustrated by Jack Kunz in 1964 for *Time Life Books* (Image courtesy Maria Meier)

(photographers will be familiar with this problem, as a "wide open" camera lens will also have a very restricted point of focus, while "stopping down" the internal iris diaphragm to its smallest opening will create images with more detail in front of and behind the point of focus, known as depth of field). As optical design became more sophisticated in the late eighteenth century, combinations of convex and concave lenses that brought peripheral and axial rays of light together at roughly the same point reduced this problem (Bradbury 1967).

The second problem is chromatic aberration. Isaac Newton's experiments in the late seventeenth-century demonstrated that visible white light is made up of different colors, which he could prove by refracting a narrow beam of sunlight through a prism. The fact that different colors of light (which we now know are a function of the specific wavelength of electromagnetic energy) bend, or refract, differently means that the more powerful the lens, the more separation of colors would occur, resulting in blurry images surrounded by fringes of a different color. The problem was reduced at around the same time as spherical aberration was controlled via the use of combinations of convex and concave lenses. In addition, the development of glasses with different refractive indices also contributed to the reduction of chromatic aberration. First attempts were made with lenses composed of glass and water, but John Dolland of England designed a telescope lens that employed a convex lens made from crown glass (low dispersion, low refractive index glass made from alkali-lime silicates) and a convex lens with flint glass (high dispersion, high refractive index glass made with the addition of lead oxide), which he patented in 1759. This combination of lenses is known as an achromatic crown doublet, but did not appear in microscope designs until the early years of the nineteenth century (Bradbury 1967).

Whereas achromat lenses reduce much of the spherical and chromatic aberration, an apochromatic lens (a triplet using three different types of glass) is even more effective in reducing chromatic aberration. This design was pioneered by Ernst Abbe via his research at the Carl Zeiss Optical Works in Jena, Germany. As with achromat lens design, his first efforts used a liquid as the third lens, but after collaboration with a local glass manufacturer, lenses made with the addition of lithium or borate possessing a different refractive property compared to flint and crown glass were used (Bradbury 1967). Abbe patented his design in 1886, and apochromatic triplets have been employed in microscopes, telescopes, and camera lenses since then (Fig. 5.2).

Optical microscopes, other than their design and construction materials, have not undergone any significant improvements in optical quality since the late nineteenth century. Going forward, we will now consider further developments in microscopy as they apply to the discovery and visualization of one particular cellular component, the mitochondrion.

5.4 The Search for Cellular Contents

One of the most important breakthroughs in the development of cell theory was the 1839 publication by Theodor Schwann (1810–1882) entitled *Mikroskopische Untersuchungen über die Uebereinstimmung in der Struktur und dem Wachsthum der Thiere und Pflanzen* (Microscopic investigations on the similarity of structure and growth of animals and plants), including the observation that all animal and plant tissues are composed of cells, and each cell contains a nucleus (Hajdu, 2002). The higher resolving power of the apochromatic lenses of Abbe and Zeiss spurred new investigations into the ultrastructure of the cell. Hypothesis and publications came at an accelerated pace during the last half of the nineteenth century as cytologists sought to understand the form and function of the organelles.

Albert Kölliker (1817–1905) provided the first accurate description of the intracellular bodies we know as mitochondria as early as 1850, which he observed in striated muscle cells and were named sarcosomes by Gustaf Retzius (1842–1919) (Lehninger 1964). Another researcher focussed on the role of "sarcosomes" was the German Robert Altmann (1852–1900). Altmann developed an acid fuchsin picric acid stain that targeted these intracellular bodies, or as he referred to them, bioblasts. They were ubiquitous and numerous in every cell type he examined, not only metabolically active muscle cells, and he deduced that these plentiful intracellular bodies must be of great importance to the life of all cells and organisms (Altmann 1890) (Fig. 5.3).

His bioblast theory, published in *Die elementarorganismen und ihre beziehungen zu den zellen* (The cellular organelles and their relations to cells) in 1890, stated that he believed that these red-staining granules were fundamental units of life with their own genetic and metabolic autonomies, anticipating the role of mitochondria in the production of cellular energy and their distinctive mitochondrial DNA. Altmann's bioblast theory that mitochondria represented a symbiotic relationship with bacteria was not well received by the scientific community. Unfortunately for Altmann, the tools and techniques for proving this forward-thinking hypothesis did not exist in the late nineteenth century, and his biographers believe that the opprobrium he endured from his colleagues was a factor in his early death at the age of 48 (Feja et al. 2016; Cowdry 1953).

By the early twentieth century, mitochondria were observed and described by many authors under diverse names. One of the barriers to understanding the role and importance of mitochondria was due to the fact that they exhibit many forms in different cell types. Their appearance varies so greatly that early cytologists believed that they were observing different particles and as a result, could not even agree as to what they should be called. By the 1880s, there were at least 33 different names for the particles, most of them based on the Greek words for "mitos" (thread) or "chondros" (granule), the two most commonly described shapes. Carl Benda of Germany is credited with the definitive name "mitochondrion" by combining both words in 1898 (Ernster and Schatz 1981). Benda perfected a second staining technique that revealed mitochondria in the somatic cells of many different animals, but he believed that the granules he observed were different from those described by Altmann (Benda 1898). His early work studying spermatozoa led him to believe that mitochondria represented a "motor-organ of the cell" (Cavers 1914) as he found them in great numbers in the tailpiece of sperm. Although later studies

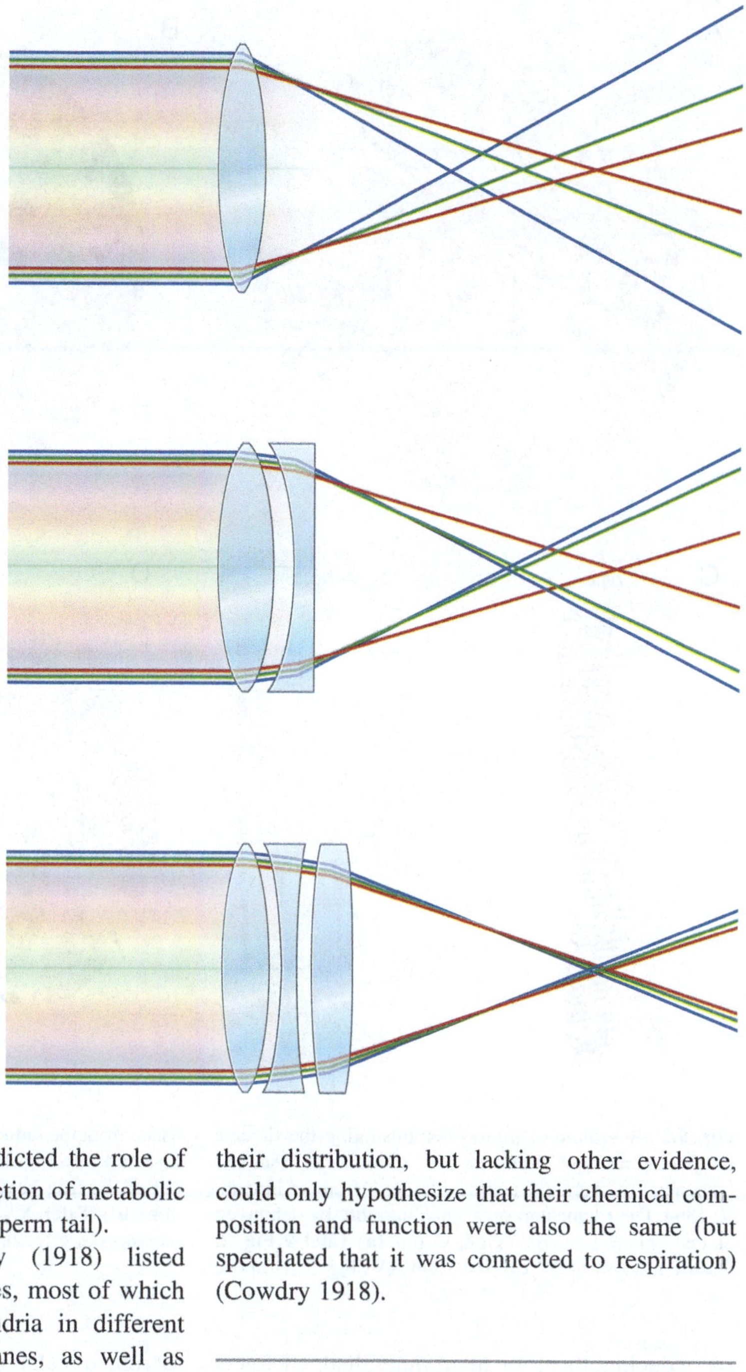

Fig. 5.2 Advances in lens designs, from top to bottom: Single lens displaying chromatic aberration, achromat lens, and apochromat lens

disproved this, it correctly predicted the role of the mitochondrion in the production of metabolic energy (required by the active sperm tail).

In tabular form, Cowdry (1918) listed 91 names for granular structures, most of which we now identify as mitochondria in different forms or cut in different planes, as well as non-mitochondrial organelles such as endoplasmic reticulum, chloroplasts, and Nissl bodies. Cowdry also concluded that mitochondria in plant and animal cells were essentially the same, as they reacted to various fixatives, stains, and dyes in the same way, and appeared universal in their distribution, but lacking other evidence, could only hypothesize that their chemical composition and function were also the same (but speculated that it was connected to respiration) (Cowdry 1918).

5.5 The Fractionation Revolution

While light microscopy had reached the limits of what could be resolved by direct observation, biochemists attempted to answer questions about the function of mitochondria and other organelles

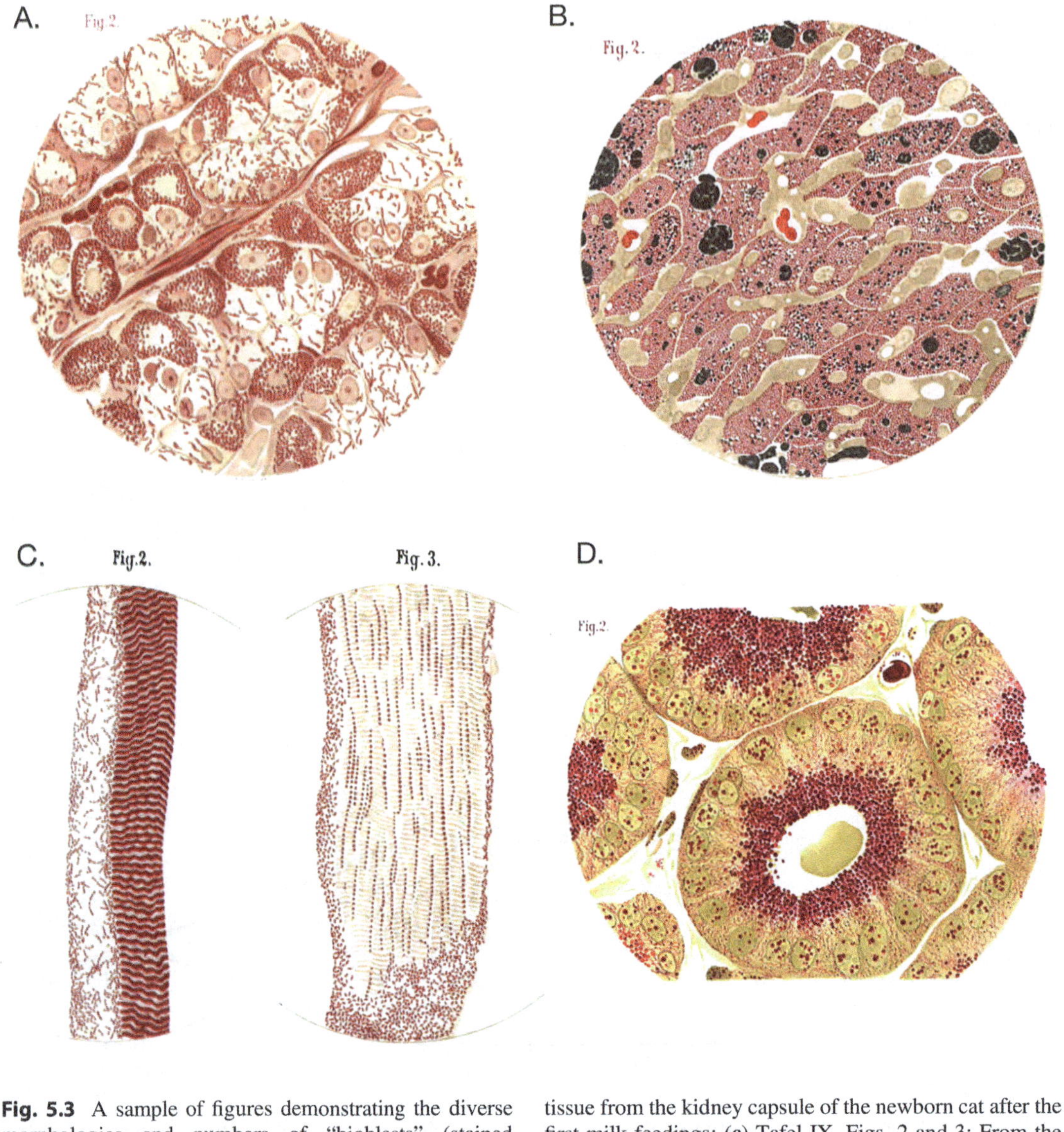

Fig. 5.3 A sample of figures demonstrating the diverse morphologies and numbers of "bioblasts" (stained magenta) in various animal tissues, from Altmann, R. 1894. Die Elementarorganismen und ihre Beziehungen zu den Zellen. Leipzig: Veit & Comp. (**a**) Tafel V Fig. 2: Gastric mucosa of the cat; (**b**) Tafel XVI Fig. 2: Adipose tissue from the kidney capsule of the newborn cat after the first milk feedings; (**c**) Tafel IX, Figs. 2 and 3: From the tail of the frog larvae, two stages of muscle fiber formation; (**d**) Tafel XX Fig. 2: Abdominal gland of *Triton taeniatus* (crested newt). Four different stages of secretion

by reducing them to their individual chemical constituents. In this way, ATP and the Krebs cycle were discovered in the 1920s and 1930s, although how and where these processes took place within the cell was still a mystery.

At the University of Chicago in 1934, Robert Bensley and Normand Hoerr began to put disrupted cells into a centrifuge which would separate out the various organelles by weight in a process referred to as cell fractionation (Rasmussen 1997). The result of the first centrifugation would be a pellet of the heaviest organelles from the disrupted cells forming as sediment at the bottom of the centrifuge tube, typically the nuclei. The remaining fluid, or supernatant, could be removed and centrifuged again, resulting in the removal of mitochondria and lysosomes. Further centrifugation would allow other organelles and cytoplasmic membranes to be separated and isolated. The pellets could then be examined via microscopy, but more importantly, their biochemical makeup and role in cellular function can be ascertained by exposing them to various enzymes and proteins, and subjecting them to biochemical analysis (Ham 1979). This work took on great importance at the Rockefeller Institute for Medical Research (NYC), in the laboratory of Albert Claude with biochemists Rollin Hotchkiss and George Hogeboom (Bechtel and Abrahamsen 2007).

5.6 The Electron Microscope

The highest possible resolution of a light microscope is approximately 0.2 μm, which is a function of the length of visible light waves. Any object smaller than half the wavelength of the visible light source cannot be resolved. In the 1930s, German engineers Ernst Ruska (1906–1988) and Max Knoll (1897–1969) developed a device that used a beam of electrons as the energy source (instead of light), and magnetic coils as "lenses," which could generate an image on a cathode ray screen or photographic plate. As transmission electron microscopes became commercially available after the Second World War, new technologies for preparing even thinner sections, fixatives and embedding media, and stains that would respond to the electron beam had to be developed. Some of the first TEM images were of flat tissue culture cells that could be compared with similar images made with a light microscope to make the case that the new technology was not introducing artifacts and the TEM images were a true representation of the cell (Rasmussen 1997).

At the same time as Claude, Hotchkiss, and Hogeboom were determining the functional role of organelles via biochemistry, George Palade began exploring cellular structure with electron microscopy. By the 1950s, Palade had developed an osmium fixative that became the standard preparation for TEM imaging, while Keith Porter and Joseph Blum built a high-speed ultramicrotome that could prepare extremely thin sections (Bechtel and Abrahamsen 2007).

5.7 Visualization of the Mitochondrion

Palade recognized that a high level of molecular organization must be present to account for the complicated metabolic processes that take place in the mitochondrion, which was beyond the resolution of light microscopy. His examination of serially sectioned mitochondria in many different cell types revealed the presence of an inner and outer membrane, with numerous infolds of the inner membrane that greatly increased the interior surface area, which he named cristae. Metabolically active cells, such as muscle cells, have more mitochondria with more cristae, than mitochondria found in less active cells. Later research would confirm the role that these walls would facilitate the high production rate of ATP by increasing the interior surface area required for oxidative phosphorylation (Bechtel and Abrahamsen 2007).

At the same time, Fritiof Sjöstrand (1912–2011) was embarking on his own electron microscope studies of mitochondria at the Karolinska Institut in Stockholm. While using techniques similar to Palade's, Sjöstrand's goal was to achieve even higher resolution and more detailed images. Based on his measurements and interpretation, Sjöstrand claimed that the wall-like structures, or lamellae, he observed were not folds of the inner membrane but stacks of double-walled plates occupying the center of the organelle. Sjöstrand's microscopy was not conducted alongside the complimentary fractional biochemical research as Palade's was (Fig. 5.4).

Palade and Sjöstrand published important papers on their observations in 1952 and 1953, while also presenting their findings at conferences between 1952 and 1954. The different ways in which Palade and Sjöstand interpreted their

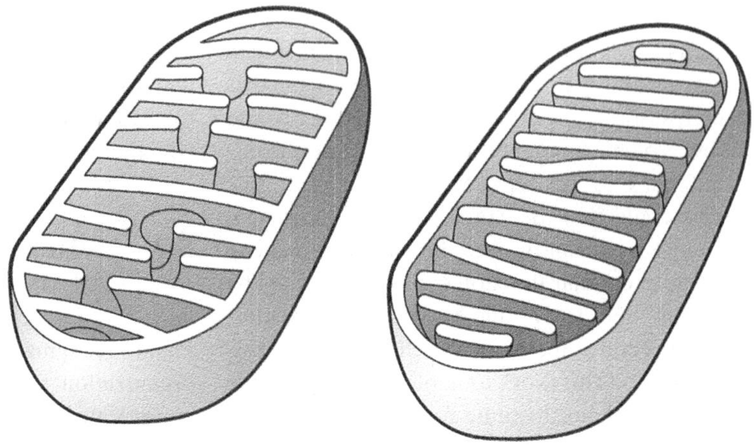

Fig. 5.4 Schematic representation of the mitochondrial cristae, or lamellae, as proposed by Palade (left) and Sjöstrand (right). Author's drawings are based on illustrations after Palade (1953) and Sjöstrand (1953)

micrographs and their subsequent public disagreement about their conclusions had two components. One centered on the different techniques with which their specimens were prepared and the possibility that artifacts were the source of the discrepancies in the resulting images and interpretation. Palade and Porter's technique relied on their extensive research with osmium tetroxide solutions at various pH values to optimize staining while minimizing any alterations to the appearance of the tissue as compared to unfixed material. Sjöstrand, while using the same OsO_4 formula, believed that Palade's technique could result in postmortem artifacts due to the tonicity of the preservative and the time between tissue extraction and preparation, which Sjöstrand avoided by injecting isotonic fixative into living tissue. The other difference was in the philosophy behind the implied functionality of the membranes and their role in energy production. Palade's images and interpretations tied in with the fractional biochemistry work conducted by his colleagues, while Sjöstrand believed he could infer molecular structure and predict biochemical behavior with accurate measurements far more precise than any that could be made from the images Palade had produced (Rasmussen 1997).

Rasmussen (1997) states that Palade's images were important in creating support for his interpretation of mitochondrial morphology and may have played a part in the acceptance of his hypothesis. More importantly, however, was Palade's cooperation with biochemists who interpreted function via the results of fractionation, which proved complementary. Sjöstrand believed that his higher resolution, sharper images would allow for the exact measurement of the molecular components of the membranes and therefore predict the kinds of metabolic activities that could take place, putting him at odds with biochemical evidence. In the end, Palade's model of mitochondrial structure became the accepted standard, and his visualization became as iconic as Claude's "powerhouse of the cell." This successful collaboration between microscopy and biochemistry has been heralded as the beginning of the modern discipline of cell biology (Rasmussen 1997; Bechtel and Abrahamsen 2007) (Fig. 5.5).

5.8 Beyond Palade and Sjöstrand

An extensive survey of recent literature shows that improvements and variations in the techniques used to prepare specimens for electron microscopy, the examination of a wider range of tissues, and considerations of metabolic states and pathologies have revealed that the internal structure is more varied and diverse than originally described by Palade or Sjöstrand. Despite these demonstrations of structural diversity, practically every image of mitochondria found in medical and educational literature to date hews closely to the Palade model. In addition, new imaging

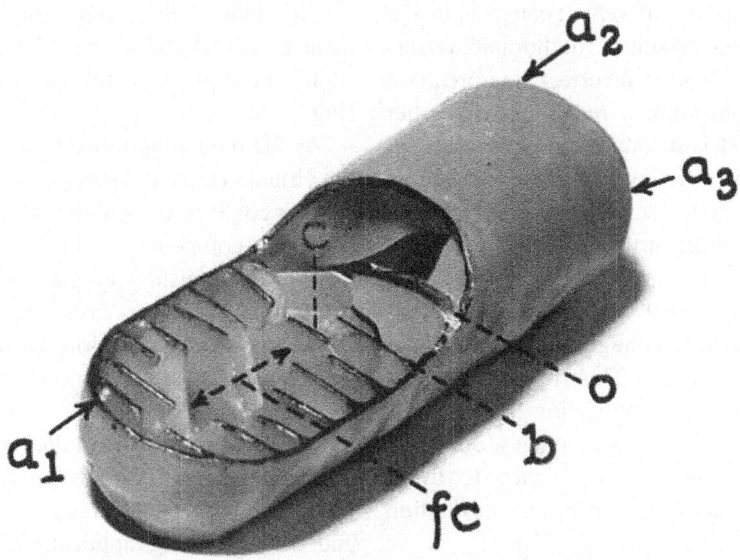

Fig. 5.5 Wax reconstruction of a mitochondrion from Palade (1953) based on his analysis of his transmission electron micrographs. "A photograph of the tri-dimensional model of a mitochondrion. The mitochondrial membrane and the cristae were made out of paraffin plates of equal thickness and the model was cut open in order to show the arrangement of the cristae as deduced from the preceding electron micrographs. Note how the apparent thickness of the cristae varies with the angle of the section: the crista marked **o** appears thicker than that marked **c** because the first was sectioned obliquely and the other normally. A branching crista can be seen at **b**, and the central free channel at **Jc**. The crista marked **c**, which appears 'free' in the middle of this mitochondrial profile, is actually anchored down in the membrane but its attachment is obscured by the other cristae." Labels a1, a2, and a3 refer to planes of section seen in other micrographs in the paper

techniques such as scanning electron tomography (SET) and high-resolution scanning electron microscopy (HRSEM) have revealed the interior of mitochondria at a level of detail unattainable via TEM. Lehninger (1964) provided many examples of the various forms that mitochondria take in single-cell organisms, invertebrates and vertebrates, accompanied by a number of scanning electron micrographs and drawings (including a sketch by Palade). These include tubular, saccular, prismatic, and even crystal-shaped cristae and septae (Lehninger 1964; Krstić 1979).

5.9 Scanning Electron Microscopy

In transmission electron microscopy, a beam of electrons replaces light waves as the source of energy passing through the thinly sectioned and stained material. By contrast, scanning electron microscopy (SEM) employs the beam of electrons to demonstrate surface features and textures by recording the different levels of energy as the beam is reflected off of the specimen. Scanning electron imaging evolved at the same time as transmission electron microscopy, but its implementation was delayed due to the precision required for the creation of the imaging beam (McMullan 1995). SEM specimens must be coated with a layer of reflective metal such as gold or platinum, and placed in a vacuum-sealed chamber for imaging. Select specimens can be quick-frozen with liquid nitrogen and "freeze fractured" before metal coating, to allow for the imaging of internal structures.

Hollenberg et al. (1989) surveyed over 200 TEM and SEM images of mitochondria, including those of Palade and Sjöstrand, as well as prepared dozens of their own high-resolution scanning electron micrographs (HRSEM) at magnifications up to ×45,000. Their conclusion was that the often-illustrated "traditional" cristae model of Palade was an incorrect interpretation due in part to the limited resolution of earlier images combined with the unquestioned acceptance of Palade's hypothesis by subsequent researchers and authors. They observed that in most cases, the inner structure of mitochondria consisted not of walls or shelves, but of tubules that spanned the internal space, with the lumens of the tubules contiguous with the intramembranous space. This hypothesis was supported with their publication of a book containing paired stereoscopic images, complete with a fold-out 3D viewing device to assist readers with visualizing their high-resolution scans.

5.10 Building 3D Models

In the pre-computer era, accurate three-dimensional models of objects, structures, or geography could be created if serial sections or section data, such as contour lines, could be obtained. In cartography, three-dimensional topographic maps can be created from contour maps. A copy of the paper map is printed for every contour line present, and then glued to cardstock, wood, or plastic. The shape defined by the contour is cut out, and the shapes are glued together in order from lowest to highest contour. In this way, a 1:50,000 scale 3D map of Iceland was made in the early 1990s by a team from the Model Workshop in Reykjavík. The 76-m^2 map can be viewed in the Reykjavík City Hall (Fig. 5.6).

Utilizing this reconstruction technique, Hans-Peter Hoffmann and Charlotte J. Avers at Rutgers University used serially sectioned images of yeast cells to demonstrate another unique morphology of mitochondria. They printed out over 300 mitochondrial profiles from 58 transmission electron micrographs (TEM), glued them to foam board, and assembled them together to demonstrate the three-dimensional structure of one giant mitochondrion in *Saccharomyces cerevisiae* which differed greatly from the expected number and size of mitochondria, suggesting that "a reassessment of mitochondrial genetics and biogenesis" should take place (Hoffmann and Avers 1973) (Fig. 5.7).

As 3D modelling moved from the physical to the virtual in the late 1990s–early 2000s, a similar process could be created digitally, albeit still with a manual component. Cameron Slayden, a student in the Augusta (Georgia) University Medical Illustration Graduate Program, took on a 3D reconstruction of a dendrite based on a series of TEMs. For his thesis project, The Visible Dendrite, he manually outlined the region of interest in a series of images that followed a dendrite and its synaptic spines (the blue areas in Fig. 5.8). He brought the slices into Maxon Cinema4D, a high-end 3D modelling application, and joined the sections via a process variously described as skinning, lofting, or sweep blending. This is the same technique used in boat building, where profiles of the boat at any given point are cut out and joined together via the external planks that form the hull. Slayden then reintroduced the TEMs into the animated model, which enabled the viewer to move back and forth through the dendrite to visualize the three-dimensional form of the structure as well as its internal components, such as mitochondria. Serendipitously, the TEMs captured a mitochondrion budding into one of the dendritic spines, a novel incident not previously visualized (Slayden, pers. comm.).

5.11 Segmentation and Beyond

The laborious process of reconstructing 3D structures by isolating and assembling 2D sections has been replaced by newer technologies. One example is 3D image segmentation. Segmentation refers to the process of defining regions of interest (ROI) using masking and threshold tools in software dedicated to this task (Abel et al. 2012). This is now a common process used to generate 3D volumes from CT and MRI data. Sequential images of 2D thin sections are brought together in software where the different densities of physical or cellular elements are identified and outlined as regions

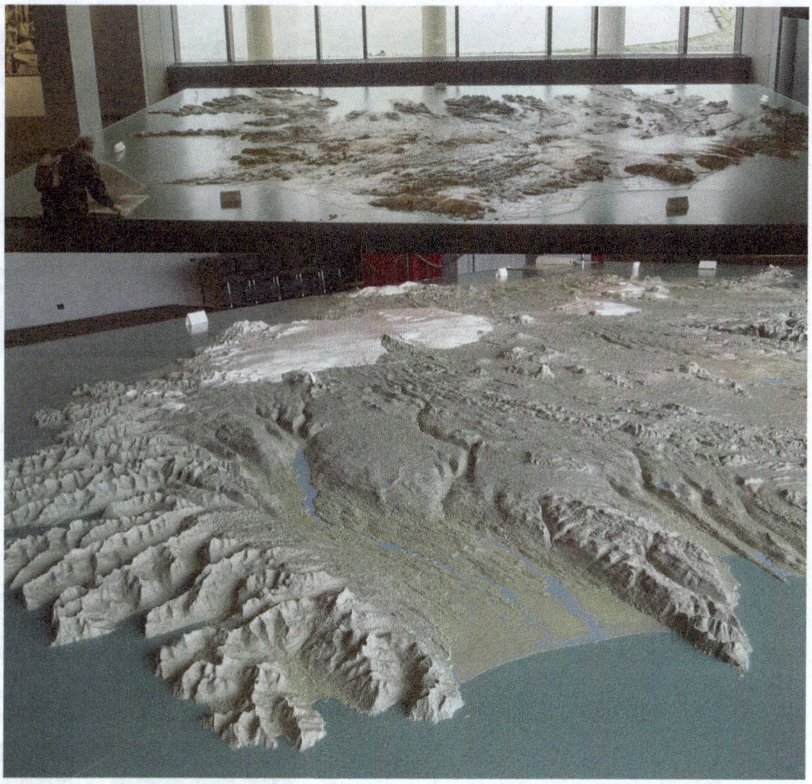

Fig. 5.6 3D map of Iceland created by layering sheets of woodcut along the contours of Iceland's topography. The horizontal scale is 1:50,000 but the vertical scale is 1:25,000, which exaggerates the heights of the terrain

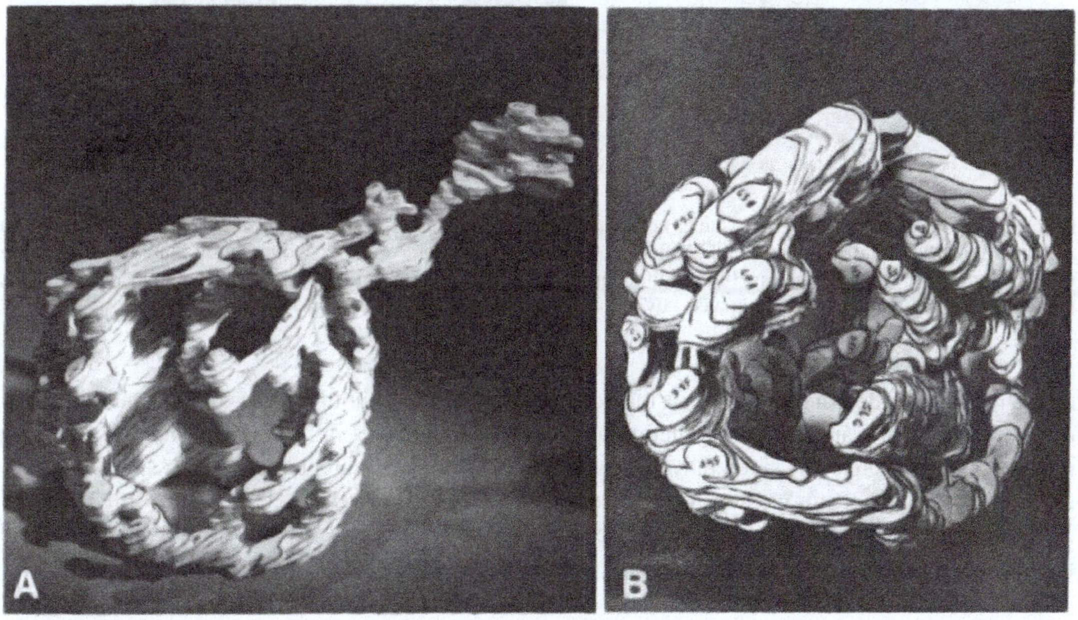

Fig. 5.7 3D model of mitochondrial yeast by Hoffmann and Avers 1973

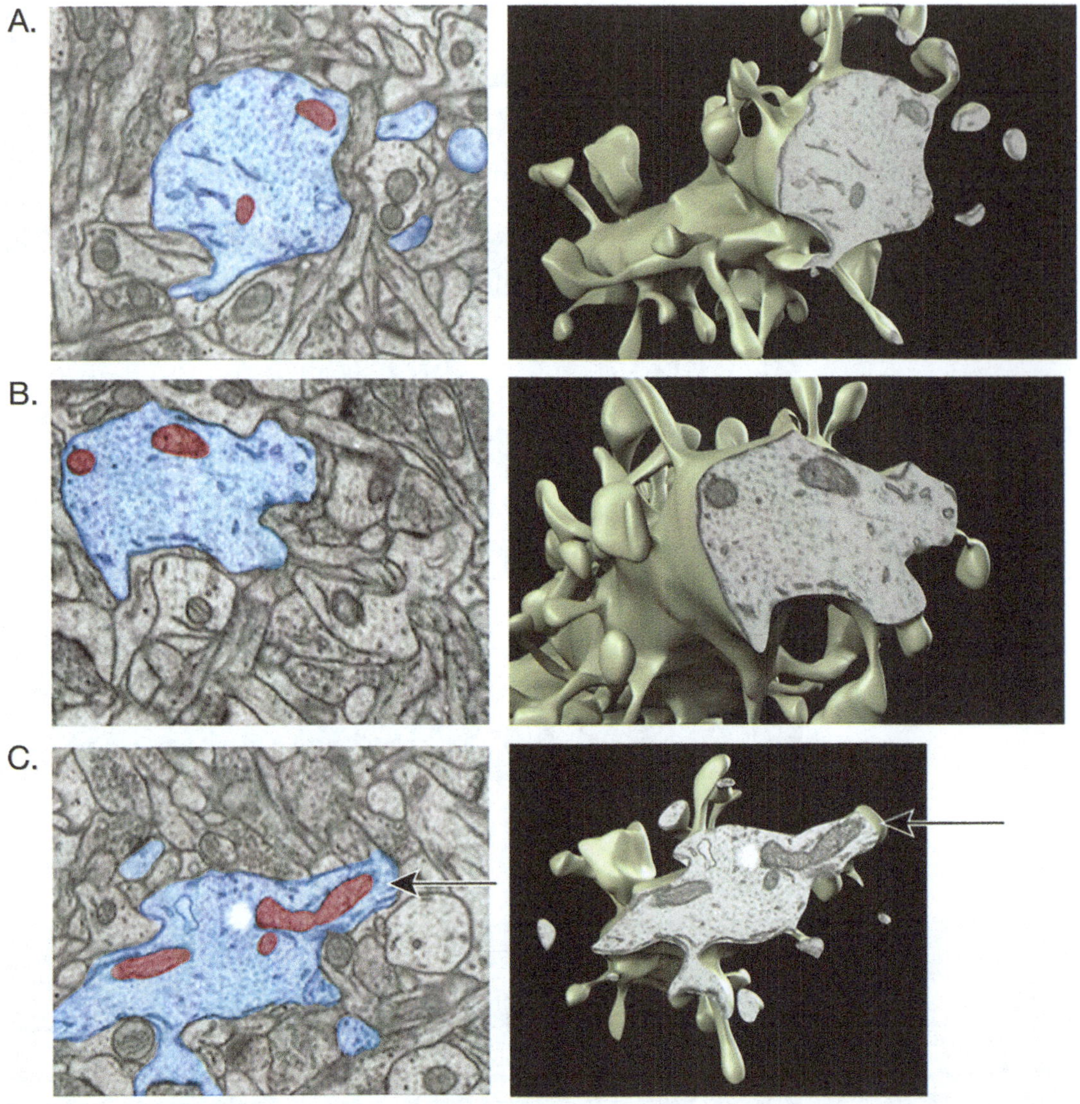

Fig. 5.8 Three sets of images from Cameron Slayden's animated reconstruction of a dendrite. TEM image is on the left (dendrite highlighted in blue, mitochondria highlighted in red); frames from animated "fly through" of the assembled 3D dendrite at the same plane of section on the right. Note mitochondrion budding into a dendritic spine in "C." Images courtesy of Cameron Slayden, MS. © 2004

of interest. Once identified throughout the series of images, the software can then generate a 3D model of the structure being studied by adding the 2D outlines together to create 3D geometry. Another tool that has been used to visualize the three-dimensional nature of mitochondria is fast 100-nm resolution three-dimensional microscopy. Egner et al. (2002) produced stunning animations of mitochondria in live *Saccharomyces cerevisiae* yeast cells that were treated with green fluorescent protein (GFP) which targeted the organelle.

5.12 Microscopy, Medical Visualization, and Undergraduate Education

Despite the technologies currently available, it is important for undergraduate students in the life sciences to understand how these visualization processes evolved, and to obtain an understanding of the challenges faced by the pre-digital imaging researchers who were striving to visualize the three-dimensional structure of organelles which was crucial to solving the mystery of their metabolic function.

In 2015, a curriculum was developed for a new undergraduate course to be offered at the University of Toronto Mississauga campus for the Minor in Health Science Communications in the Department of Biology. HSC200, Introduction to Biocommunication Visualization, serves as a prerequisite for a series of upper-level HSC electives, and prepares students with a grounding in historical and contemporary visualization principles and techniques. One of the presentations covers the history of microscopy and how micrographs have influenced the content and style of contemporary illustration, as well as the role of the microscope in scientific discovery. In particular, the class is designed to convey a sense of the challenges faced by cytologists like Palade and Sjöstrand. An in-class exercise was designed to give the students an opportunity to infer three-dimensional structure from a series of random 2D sections of organic materials.

A selection of fruits and vegetables, including some uncommon or "exotic" types, were selected at specialty grocery stores and farmer's markets. Also included was a chicken wing section (the humerus, or "drumstick" portion). At least three exemplars with similar shapes of each subject were obtained so that they could be sectioned two to three times at different angles (it is difficult to make more than three sections before the structural integrity of the subject is destroyed). A minimum of six outlines of the various planes of section, including interior detail, were drawn, scanned, and placed in a multipage Adobe InDesign document. Each specimen was then coded with a letter, and the random sections were identified by number. Space on each page was provided for students to note their observations.

In the first phase of the in-class exercise, each student is given a page with an individual specimen drawn on it (i.e., A1, A2, and A3). They are told only that the outline drawings are from organic specimens (no scale bar is provided), and the task was not to bias any discussion of form by guessing what the object was. Instead, they are asked to see if they could come up with a description of the object's shape and form, texture, and internal structures based on one piece of evidence. Five minutes are assigned to this phase of the exercise.

For the second phase, the students are asked to organize into small groups based on the letter assigned to each drawing. Each group now has at least six pieces of evidence in the form of outlines of the random sections to analyze, and discussions take place in order to form a consensus about the form of the object. This phase is assigned 10 min.

Finally, a representative from each group is asked to come to the front of the classroom and describe the group's findings while the drawings are projected on the screen behind them. After their brief presentation, a photograph of the object taken before sections were made is projected so all of the students can see how close their prediction is.

Future plans are to complete a University of Toronto Human Ethics Protocol and design a series of before and after challenges to test the efficacy of this exercise in promoting awareness of the difficulties of interpreting three-dimensional structure from two-dimensional data (Fig. 5.9).

5.13 Conclusion

The advent of microscopy in the seventeenth century marked the beginning of hundreds of years of effort to gain an understanding of the basic units of life, the cell, and their mysterious

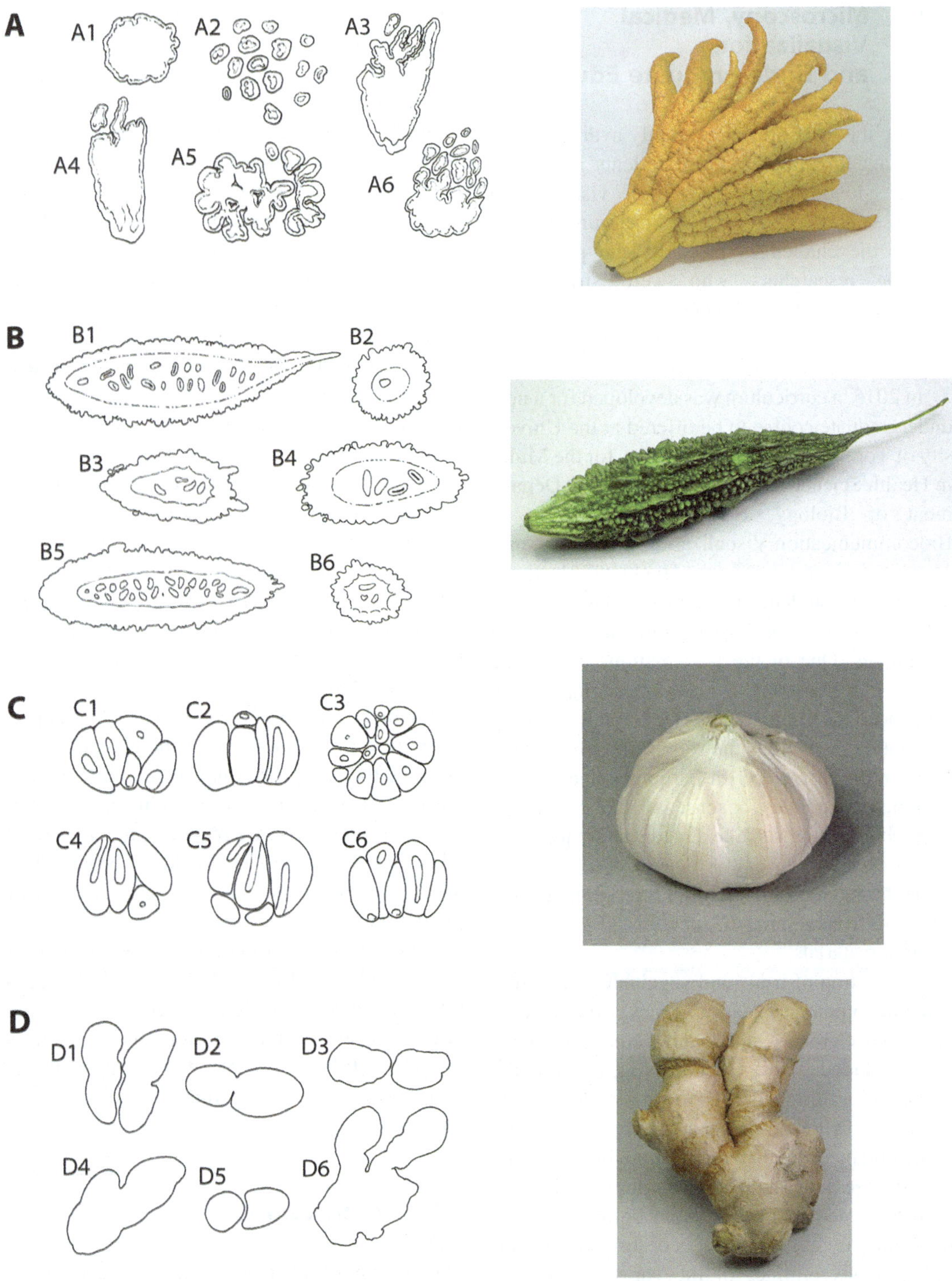

Fig. 5.9 Four examples from the undergraduate laboratory exercise to create an understanding of the difficulties in creating 3D constructions from 2D data. L: line drawings were given to the in-class student groups for them to try and ascertain the 3D form. R: photo of the object cut in random different planes of section to produce the line drawings. A: Buddha's hand (citron); B: Bitter melon; C: Garlic; D: Ginger root. Drawings and photos by the author

contents. It was only through technological advances in the twentieth century and the combined efforts of both microscopy and biochemistry that the challenges faced by early cytologists were overcome. Images played a key role in our understanding of the mitochondrion, the organelle responsible for producing cellular energy that made the evolution of complex multicellular life forms possible. The information captured in the transmission electron micrographs produced by Palade in the 1950s not only supported the evidence of biochemical processes discovered by his colleagues, but provided the basis for his 2D and 3D visualizations that strengthened the acceptance of the Rockefeller Institute's findings and led to the creation of the iconic mitochondrion diagram. Interpretive 3D models led to the construction of physical models based on collecting, compiling, and assembling dozens or even hundreds of layers of data, a task now somewhat simplified by the use of readily available, often inexpensive computer software that automates the task of identifying and segmenting regions of interest.

While technology has advanced to the point where assembling sectional data has become more of an automated task, it is important to understand the basic principles of interpreting 2D section information, as this is still a component of traditional histological study. In addition, laboratory exercises that reinforce this concept also provide a historical perspective on the challenges faced by early microscopists. The experience and perspective gained by laboratory exercises that encourage 2D to 3D mental transformations will yield benefits in future studies when students are faced with greater reconstruction challenges.

References

Abel RL, Laurini CR, Richter M (2012) A paleobiologist's guide to 'virtual' micro-CT preparation. Palaeontol Electronica 15(2):17

Altmann R (1890) Die Elementarorganismen Und Ihre Beziehungen Zu Den Zelle. Metzger & Wittig, Leipzig

Bechtel W, Abrahamsen A (2007) In search of mitochondrial mechanisms: Interfield excursions between cell biology and biochemistry. J Hist Biol 40:1–33

Benda C (1898) Ueber die Spermatogenese der Vertebraten und höherer Evertebraten, II. Theil: Die Histiogenese der Spermien. Archiv für Anatomie und Physiologie 73:393–398

Bradbury S (1967) The evolution of the microscope. Pergamon Press, Oxford

Cavers F (1914) Chondriosomes (mitochondria) and their significance. New Phytol 13(3):96–106

Chapman JA, Lee LMJ, Swailes NT (2020) From scope to screen: the evolution of histology education. In: Rea PM (ed) Biomedical Visualization. Springer Nature, pp 75–108

Claude A (1948) Studies on cells: morphology, chemical constitution, and distribution of biochemical functions. In: The Harvey lectures. Springfield, IL, Charles C. Thomas, pp 121–164

Cowdry EV (1918) The mitochondrial constituents of protoplasm. Carnegie Institution of Washington, Contributions to Embryology, 8(25), 39–160

Cowdry EV (1953) Historical background of research on mitochondria. J Histochem Cytochem 1(4):183–187

Egner A, Jakobs S, Hell SW (2002) Fast 100-nm resolution three-dimensional microscope reveals structural plasticity of mitochondria in live yeast. Proc Natl Acad Sci 99(6):3370–3375

Ernster LA, Schatz G (1981) Mitochondria: a historical review. J Cell Biol 91(3):227–255

Feja C, Löffler S, Korge M (2016) Vergessene Anatomen: Richard Altmann. Ärzteblatt Sachsen 2016(1):40–41

Hajdu SI (2002) A note from history: introduction of the cell theory. Ann Clin Lab Sci 32(1):98–100

Ham AWACDH (1979) Histology, 8th edn. J.B. Lippincott Company, Philadelphia

Hoffmann H-P, Avers CJ (1973) Mitochondrion of yeast: ultrastructural evidence for one giant, branched organelle per cell. Science 181(1401):749–751

Hollenberg MJ, Cormack DH, Lea PJ (1989) Stereo atlas of the cell. B.C. Decker Inc, Toronto

Krstić RV (1979) Ultrastructure of the mammalian cell: an atlas. Springer-Verlag, Berlin

Lehninger AL (1964) The mitochondrion: molecular basis of structure and function. W.A. Benjamin, Inc., New York, Amsterdam

McMullan D (1995) Scanning electron microscopy 1928-1965. Scanning 17(3):175–185

Palade GE (1953) An electron microscope study of the mitochondrial structure. J Histochem Cytochem 1(14): 188–211

Rasmussen N (1997) Picture control: the electron microscope and the transformation of biology in America 1940-1960. Stanford University Press, Stanford

Sjöstrand FS (1953) Electron microscopy of mitochondria and cytoplasmic double membranes. Nature 171:30–31

Using Molecular Visualisation Techniques to Explain the Molecular Biology of SARS-CoV-2 Spike Protein Mutations to a General Audience

6

Sarah Iannucci, William Harvey, Joseph Hughes, David L. Robertson, Edward Hutchinson, and Matthieu Poyade

Abstract

Since the COVID-19 pandemic started in 2019, the virus responsible for the outbreak—SARS-CoV-2—has continued to evolve. Mutations of the virus' spike protein, the main protein driving infectivity and transmissibility, are especially concerning as they may allow the virus to improve its infectivity, transmissibility, and ability to evade the immune system. Understanding how specific molecular changes can alter the behaviour of a virus is challenging for non-experts, but this information helps us to understand the pandemic we are living through and the public health measures and interventions needed to bring it under control. In response to communication challenges arising from the COVID-19 pandemic, we recently developed an online educational application to explain the molecular biology of SARS-CoV-2 spike protein mutations to the general public. We used visualisation techniques such as 3D modelling and animation, which have been shown to be highly effective teaching tools in molecular biology, allowing the viewer to better understand protein structure, function, and dynamics. We also included interactive elements for users to learn actively by engaging with the digital content, and consequently improve information retention.

This chapter presents the methodological and technological framework which we used to create this resource, the 'SARS-CoV-2 Spike Protein Mutation Explorer' (SSPME). It explains how molecular visualisation and 3D modelling software were used to develop accurate models of relevant proteins; how 3D animation software was used to accurately visualise the dynamic molecular processes of SARS-CoV-2 infection, transmission, and antibody evasion; and how game development software was used to compile the 3D models and animations into a comprehensive, informative interactive application on SARS-CoV-2 spike protein mutations. This chapter indicates how cutting-edge visualisation techniques and technologies can be used to improve science communication about complex topics in molecular biology and infection biology to the general public,

S. Iannucci (✉)
The School of Simulation and Visualisation, The Glasgow School of Art, Glasgow, UK

the Anatomy Facility, College of Medical, Veterinary and Life Sciences, University of Glasgow, Glasgow, UK

W. Harvey · J. Hughes · D. L. Robertson · E. Hutchinson
MRC-University of Glasgow Centre for Virus Research, The University of Glasgow, Glasgow, UK
e-mail: William.Harvey@glasgow.ac.uk; joseph.hughes@glasgow.ac.uk; David.L.Robertson@glasgow.ac.uk; Edward.Hutchinson@glasgow.ac.uk

M. Poyade
The School of Simulation and Visualisation, The Glasgow School of Art, Glasgow, UK
e-mail: M.Poyade@gsa.ac.uk

L. Shapiro, P. M. Rea (eds.), *Biomedical Visualisation*, Advances in Experimental Medicine and Biology 1388, https://doi.org/10.1007/978-3-031-10889-1_6

something that is critical to gaining control of the continuing COVID-19 pandemic.

Keywords

SARS-CoV-2 · COVID-19 · 3D modelling · 3D animation · Interactive visualisation · Molecular visualisation

6.1 Introduction

6.1.1 SARS-CoV-2 Spike Protein Mutations

Severe Acute Respiratory Syndrome Coronavirus 2—SARS-CoV-2—is the virus responsible for the COVID-19 pandemic, a world-changing disease outbreak that as of late January 2022 has caused over 5.5 million deaths (World Health Organization 2021). Over the course of the pandemic, SARS-CoV-2 has mutated, resulting in changing viral properties. One of the four viral proteins found inside virus particles is the spike protein which decorates the outer viral particle like a crown or a solar corona (it is these distinctive spikes that give coronaviruses their name) (Fig. 6.1). The spike (S) proteins assemble as trimers, with each monomer divided into S1 and S2 subunits. In its trimeric form, there is a distinct 'head' region of the spike sitting atop a 'stalk' which anchors the spike to the viral membrane (Casalino et al. 2020). The S1 subunit of S contains two important regions of interest (ROIs), which are determinants of infectivity and antigenicity (Casalino et al. 2020).

The first of these ROIs is the receptor-binding domain (RBD), located at the tip of the spike protein (Fig. 6.2a). The RBD facilitates binding to angiotensin-converting enzyme 2 (ACE-2), a protein found on the surface of the host cell (Peacock et al. 2021; Letko et al. 2020). The RBD can change its conformation to make the spike protein 'open' or 'closed', regulating the binding capabilities (Ke et al. 2020; Benton et al. 2020). The second ROI is the N-terminal domain (NTD), located towards the upper sides of the spike protein (Fig. 6.2b). The NTD is one of the targets for the host's immune system (Casalino et al. 2020).

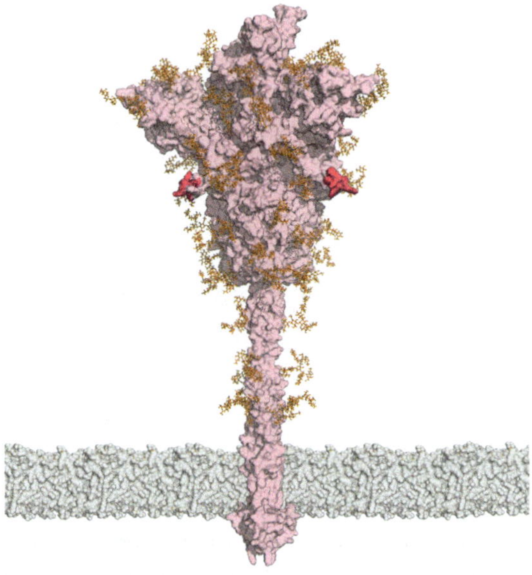

Fig. 6.1 The SARS-CoV-2 spike protein (pink) embedded in the viral membrane (grey). This image was generated in PyMOL using a full-length model of SARS-CoV-2 with one receptor-binding domain in the open conformation (6VSB_1_1_1) (Woo et al. 2020)

The spike protein is chemically attached to sugar molecules called glycans, which are important for protein folding and may assist in 'shielding' the spike protein from searching host antibodies (Fig. 6.2c) (Rossen et al. 1998; Xiong et al. 2018).

Mutations of the spike protein can affect its properties and they are seen as particularly concerning if they could affect the key viral properties of infectivity and antigenicity (the extent to which the virus can be recognised by antibodies or T cells), which could give strains of SARS-CoV-2 the potential to be more harmful to the human population (Letko et al. 2020). As mutations accumulate in SARS-CoV-2, viral genomic surveillance can identify concerning viral variants and monitor their evolution (Harvey et al. 2021). Variants with the potential to increase the impact of SARS-CoV-2—termed Variants of Concern (VOCs)—may require immediate action by public health authorities (Centers for Disease Control and Prevention 2021).

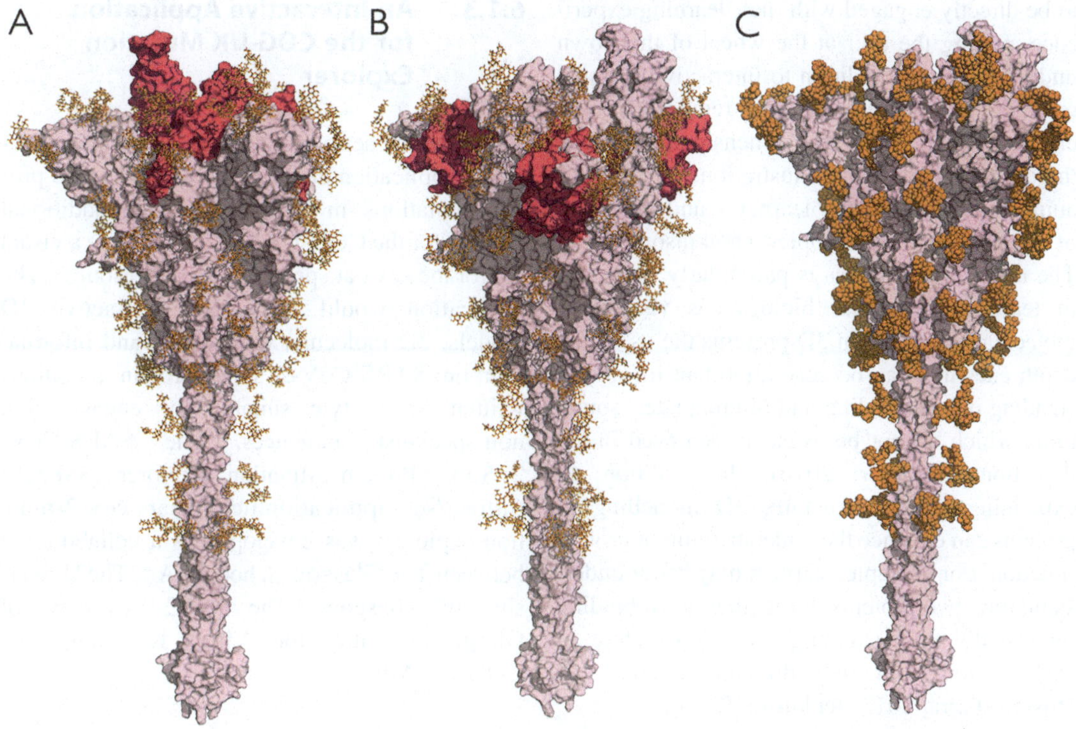

Fig. 6.2 The ROI of the spike protein, including the RBD in magenta (seen in **a**), NTD in magenta (seen in **b**), and glycans in orange (seen in **c**)

Many groups around the globe are leading the charge on sequencing the SARS-CoV-2 genome and tracking these mutations. The COVID-19 Genomics UK Consortium (COG-UK) is one such organisation pioneering this genomic sequencing, helping to facilitate the understanding of SARS-CoV-2 transmission and evolution (COVID-19 Genomics UK Consortium 2021). To share their analyses, COG-UK created the COG-UK Mutation Explorer (COG-UK/ME), an online interface allowing the public to access SARS-CoV-2 mutation and variant data (Wright et al. 2021). Given this resource's potential to impact the public's understanding of SARS-CoV-2 and COVID-19 public health regulations, this tool ideally should be accessible to the general public. However, when we began this project, the website was mainly used to present large amounts of detailed information (such as data tables or protein structures) without explanatory context for non-specialists. Therefore, we wished to use visualisation techniques and technologies to make the data contained within the COG-UK/ME more accessible to a non-expert audience.

6.1.2 Molecular Visualisation in Education

Molecular biology is a highly complex subject to teach students, and even more challenging for a general audience without a scientific background to understand (Jenkinson 2018). However, incorporating emerging technologies and visualisation techniques can be effective in facilitating the understanding of complex scientific topics (Jenkinson 2018). Introducing interactivity into educational tools may lead to more active learning in users (Quintana et al. 2004; Schrand 2008). Active learning allows the user to be directly engaged with their learning experience, putting the user at the wheel of their own understanding. In addition to interactivity, visual resources are being used more regularly in science education. Static two-dimensional (2D) and three-dimensional (3D) illustrations, as well as animations, can aid the learner's understanding of complex scientific topics (Jenkinson 2018). The use of 3D graphics is particularly important in teaching molecular biology, as visualising molecular structures in 3D presents the user with depth cues that can be very important for understanding protein folding and binding sites, something which cannot be as easily achieved in 2D depictions (Höffler 2010). In addition to explaining protein structure, 3D modelling of proteins can enhance the understanding of protein function. For example, learners may better understand how two proteins fit together when binding by visually understanding the 3D structure of each protein's binding domains (Barak and Hussein-Farraj 2013; Jenkinson 2018).

One important feature of understanding molecular biology is the dynamic nature of proteins, as they are perpetually moving and interacting with structures in their environment, and this can be difficult to grasp when using educational materials such as static images and textual descriptions (Robic 2010; Queloz et al. 2016; Jenkinson 2018). Including 3D animations can help improve a learner's understanding of protein dynamics and key processes, better conveying the concept of molecular interactions as non-linear and highly fluid (Barak and Hussein-Farraj 2013; Jenkinson 2018). However, the creator of 3D molecular animations must also maintain a clear focus on the narrative and not depict an entirely chaotic molecular environment (Jenkinson 2018). Considering the potential benefits of 3D molecular visualisation in science education, it is possible that interactive 3D models and animations of SARS-CoV-2 proteins and key viral processes may facilitate the understanding of SARS-CoV-2 mutations and VOCs for non-experts.

6.1.3 An Interactive Application for the COG-UK Mutation Explorer

This research set out to develop an online educational application about SARS-CoV-2 spike protein mutations in order to provide additional context for the COG-UK/ME and provide a visual aid for the key data presented in this resource. The application would incorporate interactive 3D models, 3D molecular animations, and information on SARS-CoV-2 spike protein mutations written in a style suitable for engaged but non-specialist audiences. The SARS-CoV-2 Spike Protein Mutation Explorer (SSPME) (https://sc2-application.itch.io/sars-cov-2-mutation-explorer) was developed as a collaboration between The Glasgow School of Art, The University of Glasgow, The MRC-University of Glasgow Centre for Virus Research, and COG-UK/ME.

6.2 Materials and Methods

6.2.1 Materials

A variety of software was used when designing and developing the SSPME (Table 6.1).

Additionally, existing visualisations and other resources were used for visual inspiration, information on protein structures, and as open-source assets for this development (Table 6.2).

6.2.2 Concept and Workflow

The concept of this project was to act as a visual educational tool, enabling users to better understand SARS-CoV-2 spike protein mutations and to better digest the information presented in the COG-UK/ME. Fig. 6.3 shows a basic overview of the developmental workflow. After solidifying the concept, a Focus Group (FG) study was conducted within the COG-UK/ME in order to

Table 6.1 The software used in the development of this research, including its purpose and developer

Software	Purpose	Developer
Adobe After Effects	Digital visual effects, motion graphics, compositing	Adobe, Inc.
Adobe Illustrator	Vector graphics editor and design	Adobe, Inc.
Adobe Media Encoder	Video media transcoding	Adobe, Inc.
Adobe Photoshop	Raster graphics editor and design	Adobe, Inc.
Arnold Renderer	Advanced Monte Carlo ray tracing renderer	Solid Angle, Autodesk, Sony Pictures Imageworks
Autodesk Maya	3D modelling and animation	Autodesk, Inc.
Color Oracle	Colour blindness simulator	Bernie Jenny, Nathaniel Vaughn Kelso
Molecular Maya	Importing, modelling, rigging, and animating molecular structures	Digizyme, Inc.
PyMOL	Molecular visualisation	Shrödinger, Inc.
UCSF Chimera	Molecular visualisation	Resource for Biocomputing, Visualization, and Informatics (RBVI), UCSF
Unity	Application development engine	Unity Technologies
Visual Studio Code	Source-code editor	Microsoft

determine areas for improvement or refinement to be addressed in the SSPME. Once the FG study was completed, we conducted research into the structural information then available about relevant features of SARS-CoV-2, particularly the spike protein and its interaction partners. Following this, 3D modelling of protein structures and model rigging was performed using the Molecular Maya plug-in for Autodesk Maya. After the molecular models were rigged, the 3D molecular animations were created. Graphic design was used to create additional visual elements for the animations in order to facilitate the visual narrative. Once the 3D models and animations were completed, the interactive application was built in Unity and hosted freely online.

6.2.3 Co-Design Workshop

In order to review the COG-UK/ME and gain valuable insights into areas for improvement that could be addressed in the SSPME, a FG study was conducted. The FG consisted of

Table 6.2 The resources used in the development of this research, including its use and reference

Resource	Use	Reference
Annotation Lab: Coronavirus Animation	Inspiration on animation narrative and storytelling; visual inspiration on demonstrating viral properties	Iwasa et al. (2021)
A Visual Model for SARS-CoV-2 Membrane Fusion	Inspiration on animation narrative and storytelling; visual inspiration on demonstrating viral properties	Khao et al. (2020)
Abstract Texture	Texturing of the virus 3D model	Paulo (2021)
Click Sound	Application audio sounds	Mixkit (n.d.)
Camera Snapshot Sound	Application audio sounds	Soundjay (n.d.)
SARS-CoV-2 Paintings	Visual inspiration on colour scheme and composition; resource for details on relevant proteins to include in the animations	Goodsell (2020a, b, c)
Voiceover Generator	Generation of animation and application of voiceover narratives	Voicebooking (n.d.)

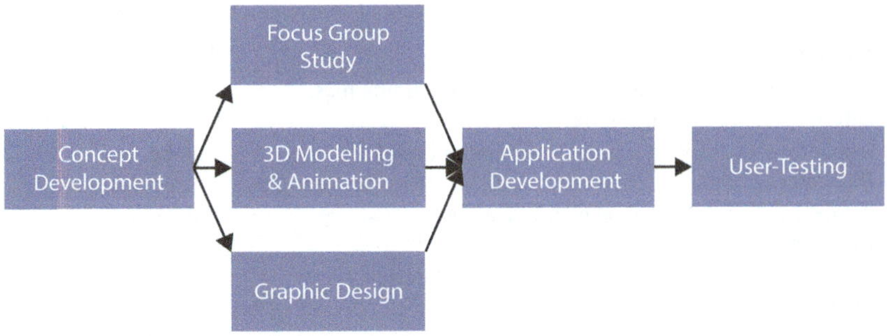

Fig. 6.3 The basic workflow of the development of the SSPME

9 participants, 5 female, and 4 male, who were recruited via social media platforms. Qualitative feedback received from the FG identified several common areas for further development with the current COG-UK/ME resource:

1. Include relevant background information on SARS-CoV-2 viral structure, specifically on the spike protein structure and function.
2. Include information on how mutations of ROI on the spike protein may impact SARS-CoV-2 viral properties.
3. Make the information more geared towards a non-expert audience.
4. Consider accessibility, such as colour blindness, when designing the application.
5. Include visualisations that facilitate the user's understanding of concepts.

This feedback was considered in the design and development process of the SSPME to ensure this resource met the expectations of the general public.

6.2.4 SARS-CoV-2 Structural Details

In order to create scientifically accurate, realistic 3D models of SARS-CoV-2 proteins, structural details were researched and documented for reference. The amino acid (aa) residues of spike protein ROIs were identified for creating individual 3D models of each region for the interactive application model (Table 6.3).

The RCSB Protein Data Bank (PDB) was used to obtain experimentally solved protein structures for use within the 3D modelling software (Berman et al. 2000). The PDB identifications (IDs) were obtained for all proteins to be used in the animations, as these IDs are necessary for importing protein structures directly through Molecular Maya. The Coronavirus Annotation animation resource, an annotated version of a detailed molecular animation of SARS-CoV-2 entry that includes many relevant protein structures, was used as a guide to many of the structures listed in Table 6.4 (Iwasa et al. 2021).

The structural dimensions of SARS-CoV-2 and its structural proteins were researched and summarised in Table 6.5.

These structural details were referenced throughout the development of the SSPME, ensuring that the scaling and organisation of the protein 3D models were scientifically accurate.

6.2.5 3D Modelling

For 3D modelling, two different sets of 3D models were required in the development of the SSPME: a lower-resolution 3D model of the spike protein with individual models of ROI for interactive use in the application and higher-resolution 3D models of the spike proteins and other relevant proteins for the 3D molecular animations. In the creation of all 3D models in this research, the colour blindness simulator Color Oracle was used to test and ensure all 3D model colours were accessible (Jenny and Kelso 2018).

6.2.5.1 Interactive 3D Models

Using both the open and the closed conformations of the spike protein (6VSB_1_1_1, 6VXX_1_1_1; see Table 6.4), selections were

Table 6.3 The amino acid residues of the various regions of interest on the spike protein used in this development

Region of interest	Position in S protein (residue numbers)	Information source
Spike protein 'head'	1–1146	Woo et al. (2020)
Spike protein 'stalk'	1147–1273	Woo et al. (2020)
Receptor-binding domain (RBD)	318–541	Woo et al. (2020)
N-terminal domain (NTD)	13–305	Woo et al. (2020)
Furin cleavage site (FCS)	681–688	Xia et al. (2020)

made in PyMOL of the aa residues correlating to the ROI identified in Table 6.3 (Fig. 6.4). Additionally, the glycans were selected using the PyMOL Generate function (Fig. 6.4). These were all exported as PDB files for later importing into Autodesk Maya using Molecular Maya. In addition, the aa residues of the spike protein excluding the ROI was generated as a PDB file for use as a base spike protein 3D model (aa residues 1–12, 306–317, 542–680, 689–1273). Due to issues importing the glycosylation PDB file into Molecular Maya, it was loaded into UCSF Chimera and, using the Multiscale Model function, an OBJ file was generated, imported into the Autodesk Maya scene, and manually aligned to the spike protein model. The resulting 3D models were exported as FBX files and loaded into Unity for interactive application development. Additionally, a curved 3D plane was created to represent a portion of the viral membrane into which the spike protein model could be inserted into (Fig. 6.5).

6.2.5.2 Animation 3D Models

In order to create the 3D models for use in the animations, the proteins listed in Table 6.4 were imported into Autodesk Maya using Molecular Maya's Import PDB functionality at the lowest resolution preset settings possible. The majority of proteins had a geometry of 25,000 polygon faces or lower, however, the glycans, V-ATPase, and IgA1 had higher polygon face counts due to having more complex structures. Reducing the geometry of these structures might not show the level of detail required to communicate relevant concepts.

When defining the animation contents and narrative, it was decided that a backdrop of

Table 6.4 The RCSB Protein Data Bank identifications and file names for all proteins are needed in this development. * indicates a protein structure that is not full-length and which was modelled by the source before use

Protein	PDB ID/PDB file name	Source
Angiotensin-Converting Enzyme 2 (ACE-2)	6 M17 chains B, E	Yan et al. (2020)
Full-length glycosylated spike (S) protein, 1 RBD open	6VSB_1_1_1*	Woo et al. (2020)
Full-length glycosylated spike (S) protein, all RBDs closed	6VXX_1_1_1*	Woo et al. (2020)
Membrane (M) protein	M_dimer_new_1	Heo and Feig (2020)
Envelope (E) protein	E_protein_1	Heo and Feig (2020)
Lysosome-associated Membrane Protein 1 (LAMP-1)	5GV0	Terasawa et al. (2016)
V-ATPase	5VOX	Zhao et al. (2017)
Transient Receptor Potential Mucolipin 1 (TRPML1)	5WJ5	Schmiege et al. (2017)
Solution structure of human secretory Immunoglobulin A 1 (IgA1)	3CHN	Bonner et al. (2009)
Crystal structure of Pembrolizumab, a full-length Immunoglobulin G 4 (IgG4) antibody	5DK3	Scapin et al. (2015)

Table 6.5 The structural details used in the development of the 3D models for SARS-CoV-2 and its structural proteins

Structure parameter	Value	Source
Diameter of viral membrane	91 ± 11 nanometres (nm)	Ke et al. (2020)
Length of spike protein ectodomain	22.5 nm	Yao et al. (2020)
Number of spike proteins per SARS-CoV-2 virus particle	~26 spike protein trimers	Yao et al. (2020)
Number of spike proteins with RBD open versus closed	'Majority' in closed conformation	Ke et al. (2020)
Spike protein flexibility	Spike protein 'head' freely rotates around 'stalk' with a lean of up to 90 degrees relative to viral envelope normal axis; however, a lean over 50 degrees is less favourable	Ke et al. (2020)
Distribution of spike proteins on SARS-CoV-2 membrane	Random distribution (non-uniform); spike proteins have no apparent relationship to each other	Ke et al. (2020)

mucins would be necessary to depict the respiratory tract (these molecules are commonly found in this environment) where SARS-CoV-2 enacts its viral infection processes. Therefore, a digital painting of mucins was created for use as a backdrop to all animations (Fig. 6.6); this painting was inspired by the portrayal of mucins by David Goodsell in his painting 'Coronavirus' (Goodsell 2020a, b, c). It was also decided that a full SARS-CoV-2 virus 3D model would be necessary in order to provide the viewer with an understanding of the full virus structure, in addition to individual 3D models of proteins for close-up scenes.

6.2.5.2.1 Full SARS-CoV-2 Virion

The structural dimensions detailed in Table 6.5 were used to scale the spike proteins appropriately to the virus particle, which was created as a

Fig. 6.4 Snapshots of the PyMOL scene as region-specific amino acid selections were made. From the left: Receptor-Binding Domain selection, N-Terminal Domain selection, Furin Cleavage Site selection, glycan selection (Woo et al. 2020)

Fig. 6.5 The 3D model set-up in the interactive application, with the 3D spike protein embedded into a portion of the 3D viral membrane

basic sphere object. A 4:1 ratio of spike protein ectodomain length to SARS-CoV-2 outer membrane diameter was used. Since the majority of spike proteins should be in the closed conformation, with an average of 26 spikes per virus particle, 26 spike models were created with 20 in closed and 6 in open conformations. These were randomly dispersed by hand across the viral membrane at random angles generally not exceeding 50 degrees relative to the normal membrane axis, as described by Ke et al. 2020 (Fig. 6.7). Basic 3D model materials were used in colouring the model, and a texture was then applied to the viral membrane so that it resembled a lipid bilayer (Paulo 2021).

6.2.5.2.2 Individual Protein 3D Models

All proteins were imported as described in Sect. 6.2.5.2 and the scene was set up with one spike in focus and two flanking spikes in the background,

Fig. 6.6 The digitally painted mucin background which was used as a backdrop in the 3D molecular animations

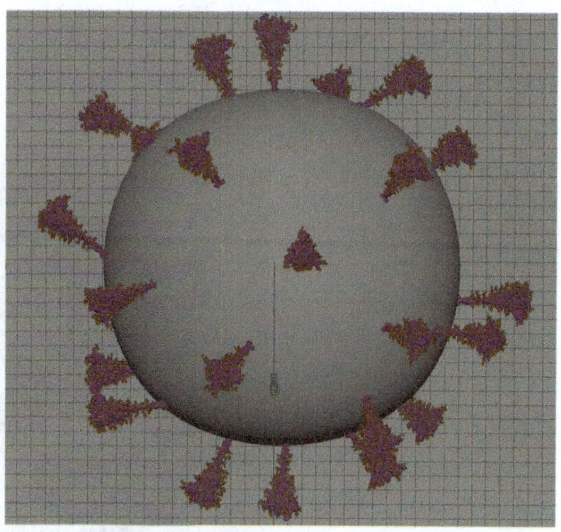

Fig. 6.7 The full SARS-CoV-2 virus 3D model, with 26 spike proteins randomly dispersed and angled by hand across the viral membrane

Fig. 6.8 A screenshot of the main spike protein with two flanking spike proteins, all sitting atop the SARS-CoV-2 viral membrane that is decorated with E and M proteins

all sitting in the viral membrane decorated with E and M proteins (Fig. 6.8). A human cell surface model was also created with associated cell surface receptor proteins (ACE-2, LAMP-1, V-ATPase, TRPML1) randomly distributed by hand around the membrane (Fig. 6.9). Finally, 3D models of relevant host antibodies of isotypes that might be found in respiratory mucus (IgA1, IgG4) were arranged floating intermittently throughout the scene (Fig. 6.10).

6.2.6 Animation

It was decided that two animations could be accomplished within project timelines, one showing the spike protein's RBD and the other the spike protein's NTD. Each animation would begin with a basic introduction to SARS-CoV-2, containing the full virus particle. Then, the animation would focus on one spike, demonstrating key concepts about spike functions that could be affected by spike protein mutations.

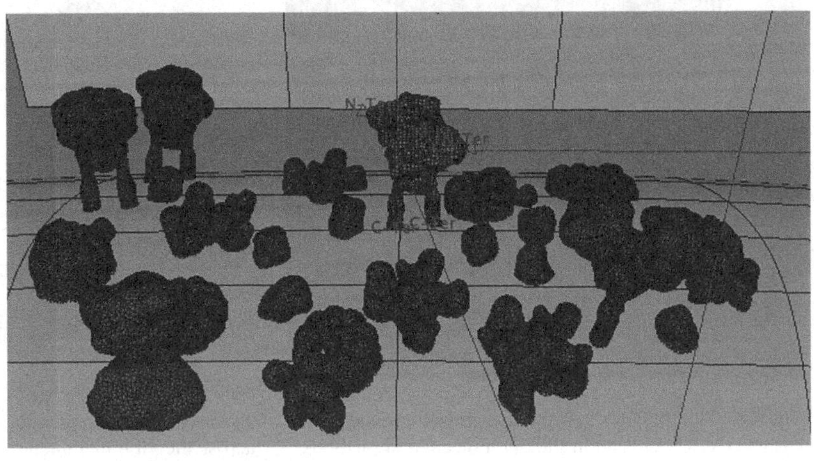

Fig. 6.9 A screenshot of the dispersal of cell surface receptors on the human cell membrane surface

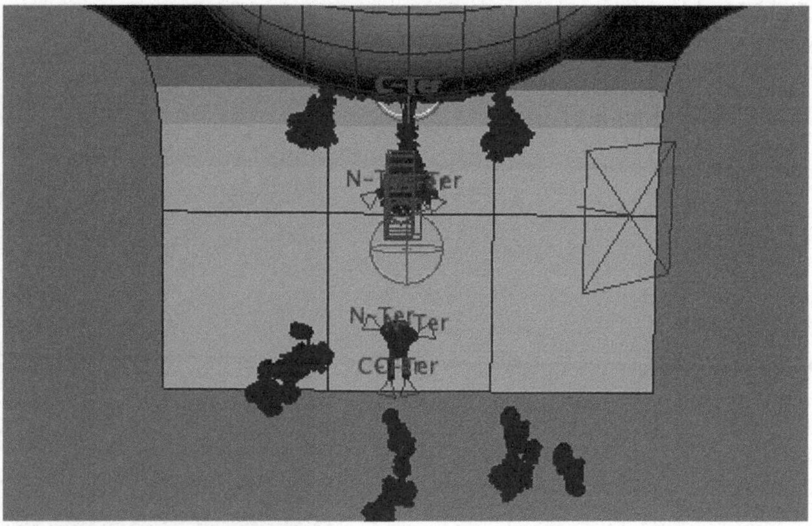

Fig. 6.10 A screenshot of the arrangement of host antibody 3D models (at the top of the image) in relation to the spike protein 3D model (at the bottom of the image) in the 3D modelling scene

6.2.6.1 Molecular Rigging

In order to accurately convey protein dynamics within the animations, molecular 'rigs' were created using Molecular Maya's Rigging Kit. These act as 'skeletons' underneath the molecular mesh, simulating key molecular dynamics (MD) throughout the animations. Rigs were created for each of the three protein chains in the spike, for both open and closed conformations. In order to prevent protein unfolding, Selections and Elastic Networks (ENs) were created for the alpha carbons (CA) of each full chain (aa 1–1273). After testing, EN strength values of 10 were used as they kept the protein's dynamic but bound together. A Field Magnitude (FM) of 100 was used as turbulence in MD, adding more dynamic movement to the entire structure.

A rig specific to the spike protein 'head' was needed to convey the flexibility of this protein region on the 'stalk'. Selections and ENs were created for each protein chain's 'head', and additional ENs were created between 'heads' of each of the three proteins in the trimeric spike to prevent them from disconnecting from one another. Then, a Target was created for each protein chain 'head', with a Handle target and strength value of 10. These Handles could be moved in the 3D scene and animated; the protein chain 'heads' would then follow this motion, creating the desired spike protein flexibility (Fig. 6.11).

Additionally, a rig specific to the spike protein RBD was needed to simulate the movement from open to closed conformations. Since protein chain A in the 6VSB_1_1_1 PDB file was open, a rig was created for this specific protein chain. RBD-specific Selections and ENs were created. The Target was created as a Molecular Maya Chain, which was selected as the protein chain A in the 6VXX_1_1_1 PDB file, as this RBD is in a closed conformation. Another Target was created as its own protein chain A. This would allow for the RBD to be animated as going back and forth between open and closed conformational states.

Finally, a basic rig was created for the ACE-2 PDB file in order to give this structure a realistic MD.

6.2.6.2 Full SARS-CoV-2 Virion Animation

Using the translation and rotation tools only, the full virus models created in Sect. 6.2.5.2.1 were animated to appear as if they are floating in a fluid environment. These tools were animated to different positions and at different rotations in order to convey a sense of randomness to the movement and remove any potential uniformity in the virus movements (Fig. 6.12).

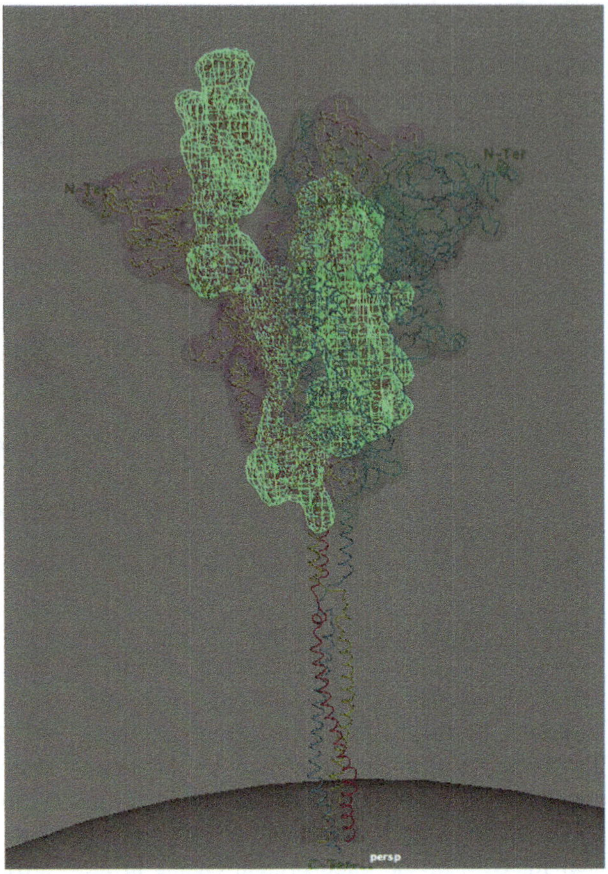

Fig. 6.11 A screenshot example of the Handle created for each protein chain in the spike protein 'head'. The green mesh represents the Handle for one protein chain

6.2.6.3 Individual Spike Protein Animations

Considering the communication aims of the SSPME, there were several important viral processes to animate in the close-up animations of the SARS-CoV-2 spike protein:

1. Spike protein 'head' flexing back-and-forth on the protein 'stalk'.
2. Spike protein glycans exhibiting their characteristic wiggling movement (Grant et al. 2020).
3. Spike protein RBD changing between open and closed conformations.
4. Spike protein RBD binding to ACE-2 on the host cell surface.
5. Antibodies moving around the spike protein but not identifying it.

First, the spike protein 'head' should demonstrate its flexibility by moving back-and-forth on the protein 'stalk'. Using the translation tool, the Target Handle developed in Sect. 6.2.6.1 was animated to depict this movement.

Next, the glycans on the spike protein should exhibit their characteristic wiggling movement (Grant et al. 2020), as they are not stationary structures just sitting on the viral surface. In order to achieve this, the glycan models were duplicated, then manually stretched and moved on to the spike model. Next, a Blend Deformer was used, as this function connects the original mesh to the duplicated, deformed mesh. Fall-Off objects could then be used to alternate between the original and duplicated mesh. These objects were animated back-and-forth, so that the glycan models deformed rapidly between their original,

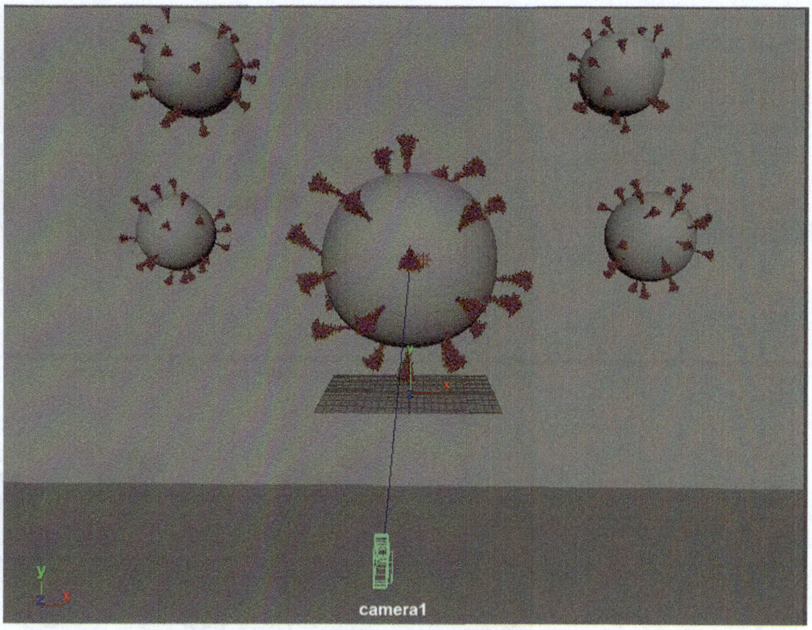

Fig. 6.12 A screenshot of the full virus animation scene composition, with one main virus and four flanking viruses dispersed in the background

normal structure, and position to the warped one (Fig. 6.13). This created the impression of wiggling.

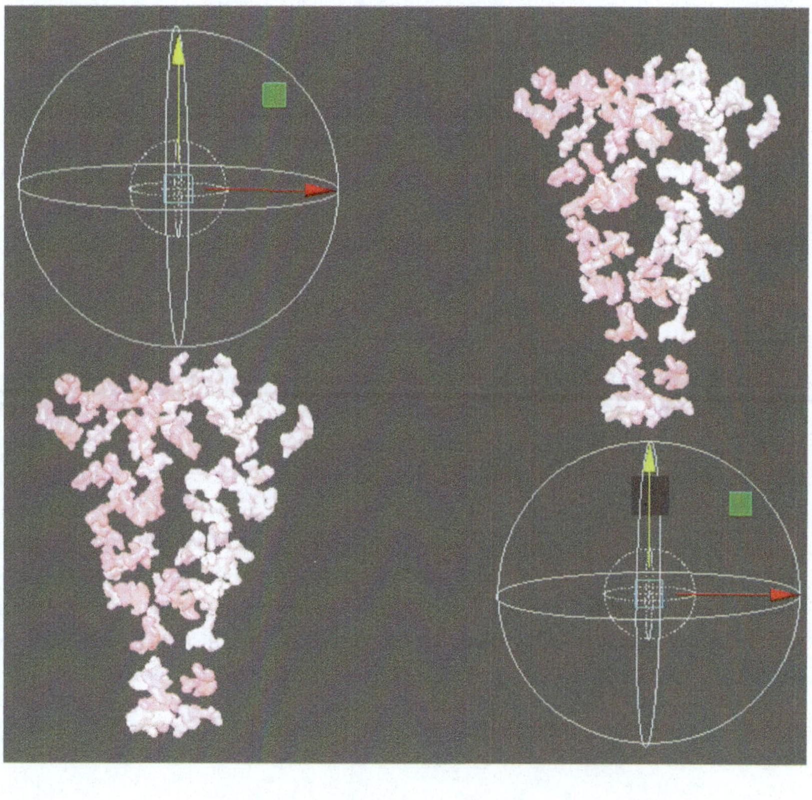

Fig. 6.13 A screenshot of the glycosylation animation process. The spherical objects represent the Fall-Off objects, which pass back-and-forth over the glycan models to rapidly deform and reform its structure, creating the characteristic jiggling movement of the glycans

The next key process to animate was the spike's RBD changing from open to closed conformation. The strength value of the Target

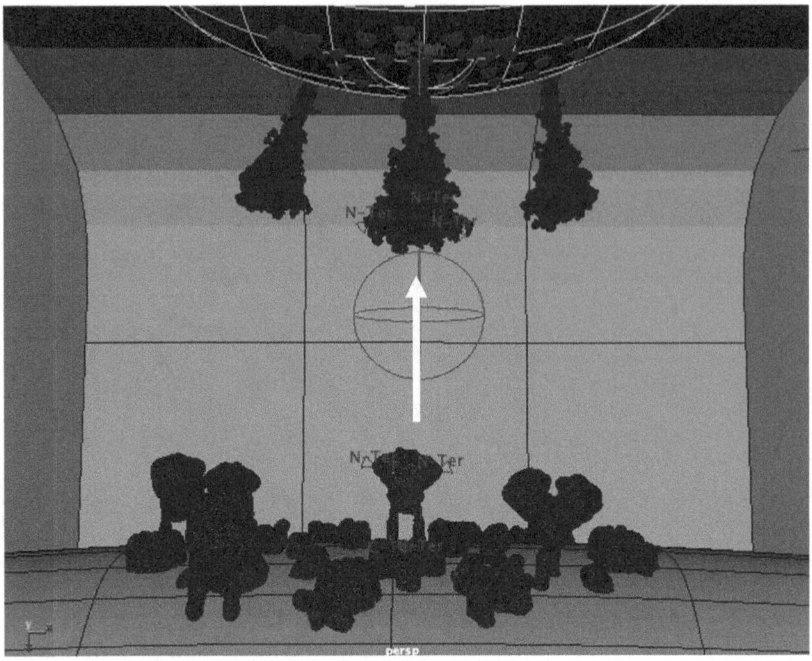

Fig. 6.14 A screenshot of the animation process behind the spike protein RBD binding to the ACE-2 receptor on the human cell membrane. The white arrows indicate the direction of movement in the animation

Selection described in Sect. 6.2.6.1 was animated between 0 and 10 for one protein chain, switching between open and closed conformations of that protein chain.

Next, in order to demonstrate how the spike protein's RBD binds to ACE-2 on the host cell, the ACE-2 model, along with the cell membrane and associated cell surface receptors (Fig. 6.9) were animated to move towards the spike protein. The background plane decorated with mucins was also animated the same distance at the same time, creating the illusion of the spike protein moving towards the ACE-2 receptor and binding (Fig. 6.14).

Finally, for showing the host antibodies floating around the virus and not locating the spike protein, the antibody models were animated using the translation and rotation tools to create random movement.

For all animations, dim background lighting with a foggy environment was added to give visual appeal to the scene, allowing for a more mysterious cellular environment to be created. One additional light was used to highlight the main spike in the centre of the composition. A single camera was set up in front of the main spike to maintain the focus of this specific protein.

6.2.6.4 Animation Post-Processing

The post-processing of the animations was performed in Adobe After Effects, with added 2D graphical elements, voiceover narrative, and subtitles to facilitate and explain the visual narrative.

6.2.7 Interactive Application Development

The application was developed in Unity (2019.4.10f1) with an application structure outlined in Fig. 6.15. The 3D models created in Sect. 6.2.5.1 were imported into Unity and the membrane model was given an image texture to give the appearance of a lipid bilayer (Paulo

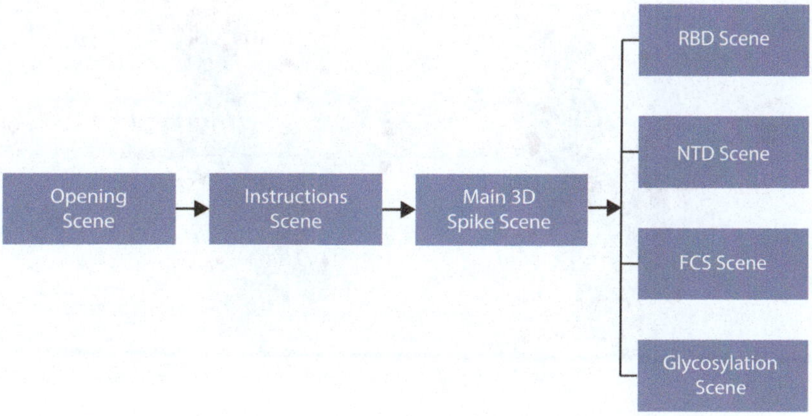

Fig. 6.15 An overview of the structure within the application. Note: From all scenes, the user has the option to return to the Opening Scene

2021). Using PyMOL (as described in Sect. 6.2.5.1), images of the spike with different ROIs highlighted were created and added to the application in the sections detailing ROI-specific information about function and the effects of mutations (Fig. 6.16). Images highlighting specific mutations in the spike 'head' domain that is characteristic of current VOCs—the Alpha, Beta, Gamma, Delta, and Omicron variants—were obtained from the COG-UK/ME and implemented in the VOC information section of the application (Fig. 6.17). The Omicron VOC emerged after the initial release of the application and due to its very rapid spread, details about Omicron were added to the application to ensure that it remained relevant.

Using Unity and Visual Studio Code, a variety of functionality was developed to allow the user to explore the application (Table 6.6). Unity Documentation was referenced for all other basic application functioning (Unity Technologies 2021).

All UI elements were designed in Adobe Illustrator with reference to the colours presented in the COG-UK/ME in order to be visually cohesive with the website. Audio clicks were added to all application buttons in order to provide the user with audio feedback when using the application.

Finally, the text describing relevant scientific concepts was assembled (as discussed in 1 Introduction). Key statements were checked against information published by the CDC and the

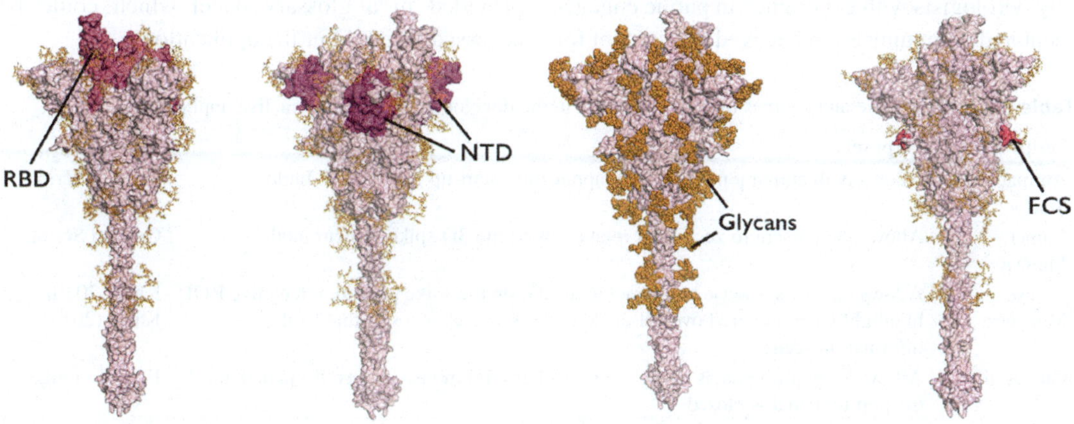

Fig. 6.16 The high-resolution images generated in PyMOL for each ROI within the application. From left to right: Receptor-Binding Domain, N-Terminal Domain, glycosylation, Furin Cleavage Site

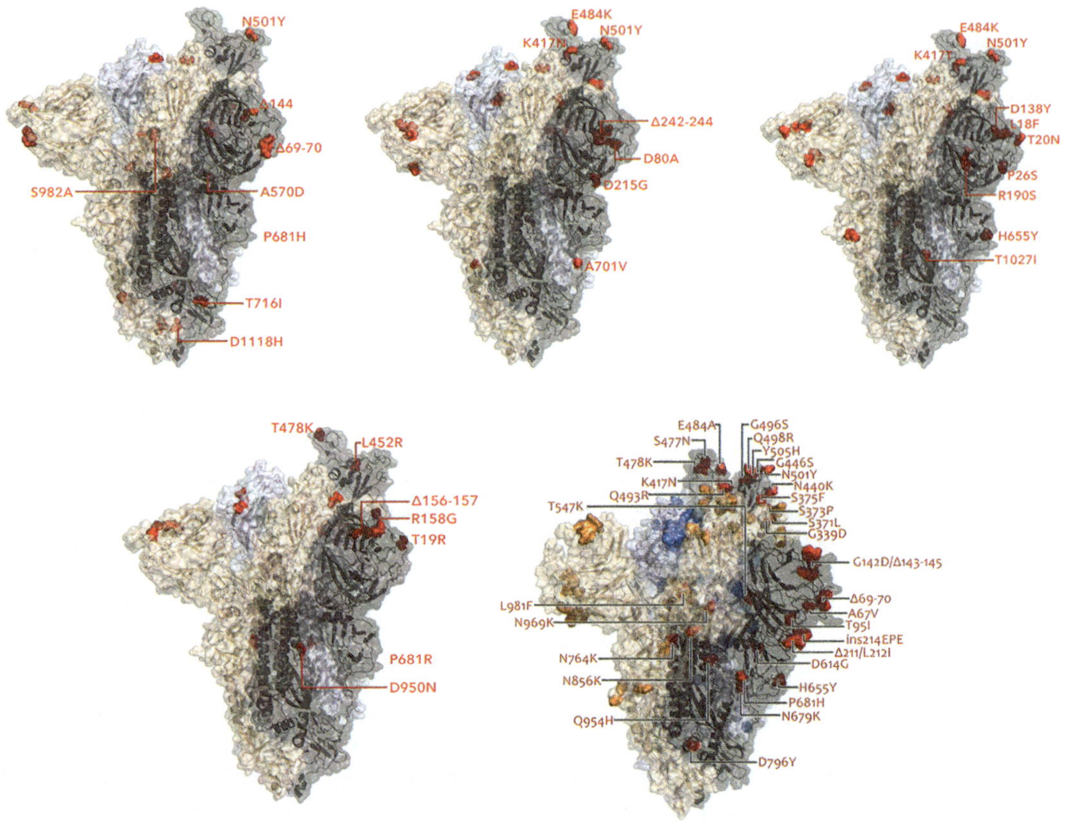

Fig. 6.17 The images of current Variants of Concern (VOCs) used within the VOC menu of the application. Top, from left to right: Alpha, Beta, Gamma. Bottom, from left to right: Delta, Omicron (Wright et al. 2021)

COG-UK/ME (Centers for Disease Control and Prevention 2021; Wright et al. 2021). Additionally, virologists with experience in public engagement and communication checked the content for accuracy and suitability for the intended non-expert audience. Key terminology was provided in a glossary panel which could be accessed throughout the application.

Table 6.6 The purpose and sources of all scripts used in the development of the interactive application

Script	Purpose	Source
Animation Start	Allows both animations to load on application start-up in Web-GL build	Custom Script
Camera Manager	Allows the camera to zoom and rotate around the 3D spike protein model	Custom Script
Mouse Manager	Allows the users' mouse to interact with ROI on the spike protein; interactive ROI highlight when hovered over; if clicked, the scene changes to that ROI's information scene	Khan (2020); Katus (2015)
Quit Active	Allows all pop-up panels to block out background mouse events, if opened, until the pop-up panel is closed	Brian Loranger
Screenshot Handler	Allows the screenshot functionality to work; when pressed, a screenshot of the application scene is automatically saved to the user's downloads folder	Unity Forum (2016)
UI Manager	Controls all UI button functionality within the application	Ouweland (2017)

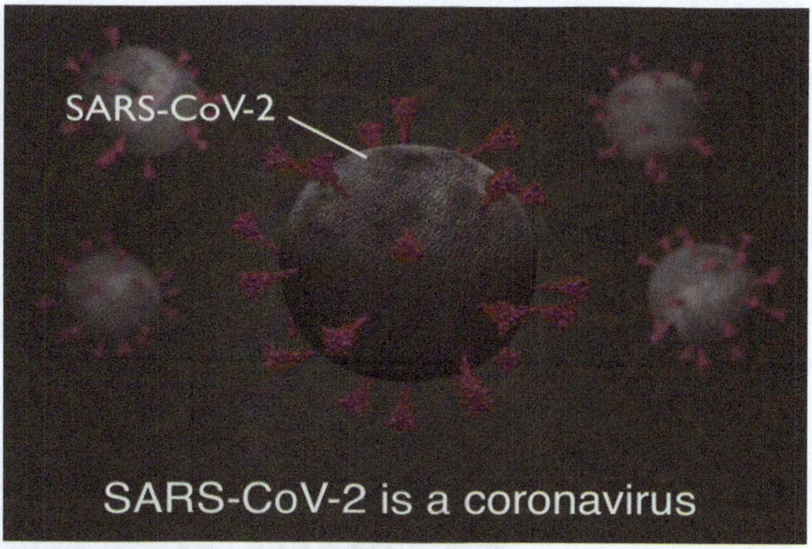

Fig. 6.18 A screenshot of the full SARS-CoV-2 virions floating in the respiratory tract in the 3D animations. Note: the transcript of the narration and 2D labels are shown

6.3 Developmental Results

6.3.1 Animations

Two 3D animations were developed in this research, one focusing on the SARS-CoV-2 spike protein's RBD and the other on the NTD. Both were rendered at the lower-resolution size of 640 × 480 pixels due to time and equipment restraints. Both animations begin with an introductory scene depicting several full SARS-CoV-2 virions floating in the respiratory mucus (Fig. 6.18). The purpose of this first scene is to introduce the viewer to the virus particle and its general structure. For both animations, this is followed by an animation showing two key properties of the spike:

- The spike protein exhibiting its flexible 'head' and stationary 'stalk' (Fig. 6.19a).
- The glycans covering the spike protein (Fig. 6.19b).

Following this, the RBD and NTD animations diverge in narrative but use many of the same rendered stills to go into further detail on viral properties relevant to the ROI. The RBD animation visualises the following three processes:

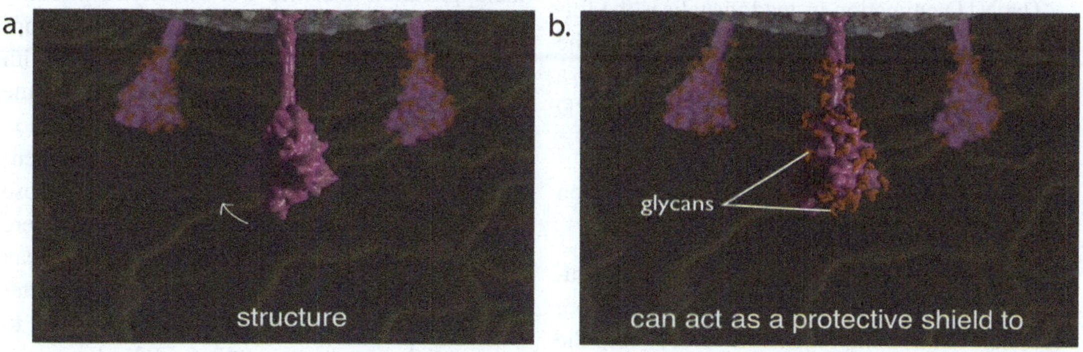

Fig. 6.19 Screenshots of the 3D spike protein animations used in the 3D animations. On the left (**a**), the flexing spike protein. On the right (**b**), the glycosylation movement. Note: the transcript of the narration and 2D labels are shown

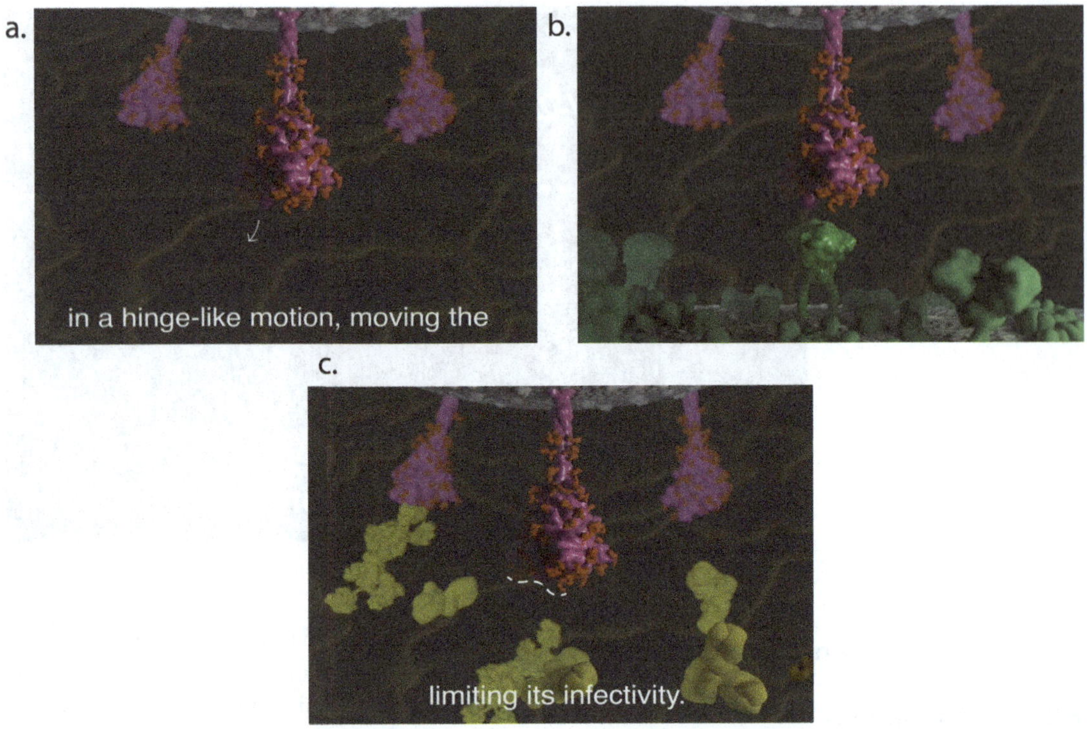

Fig. 6.20 Screenshots of the 3D spike protein animations used only in the RBD animation. On the top left (**a**), the RBD changing from open to closed conformation; on the top right (**b**), the RBD binding to the ACE-2 receptor. On the bottom (**c**), the RBD moving from open to closed conformation to avoid antibody detection. Note: the transcript of the narration and 2D labels are shown

- The RBD moving from open to closed conformation (Fig. 6.20a).
- The RBD binding to the ACE-2 cell surface receptor (Fig. 6.20b).
- The RBD moves to closed conformation to avoid antibody detection (Fig. 6.20c).

The NTD animation instead visualises the following process, using the same frames as the animation described in Fig. 6.20c but with a slightly altered narrative and different 2D graphics:

- The spike protein avoiding antibody detection (Fig. 6.21).

In all animations, overlaid 2D graphics, narration, and subtitles were used to facilitate a more accessible and comprehensible story for the viewer to follow.

6.3.2 The SARS-CoV-2 Spike Protein Mutation Explorer

The SSPME was embedded online in a Web-GL format in order to be accessible to users anywhere in the world with Internet access. The design of the application interface was directly inspired by the COG-UK/ME website at the time of development; the SSPME was therefore created with simplistic, clean scenes in a similar colour scheme to that of the website. The SSPME application flow follows Fig. 6.15, beginning with an Opening Scene (OS) to present the SSPME's purpose to the user and provide appropriate credits where applicable (Fig. 6.22). From the OS, the user can move to the Instructions Scene (IS) to better understand how to use their mouse buttons to navigate the application (Fig. 6.23). From the

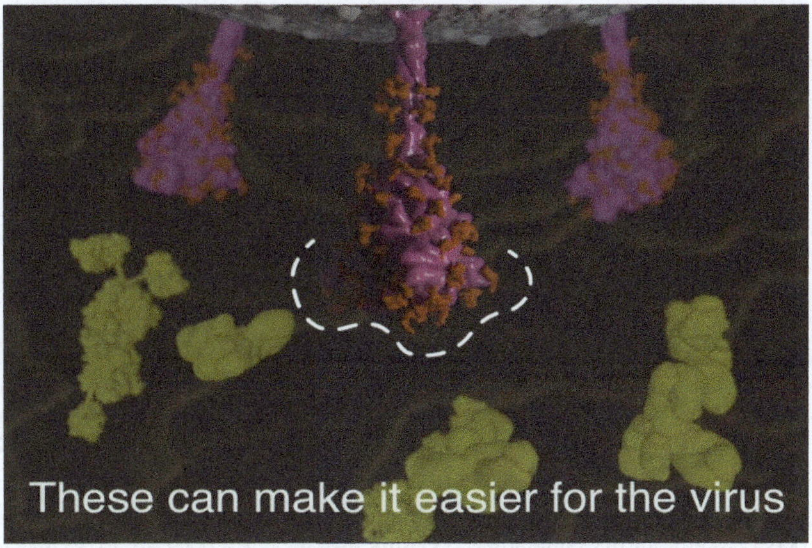

Fig. 6.21 A screenshot of the 3D spike protein animation used only in the NTD animation, showing the spike protein avoiding antibody detection. Note: the transcript of the narration and 2D labels are shown

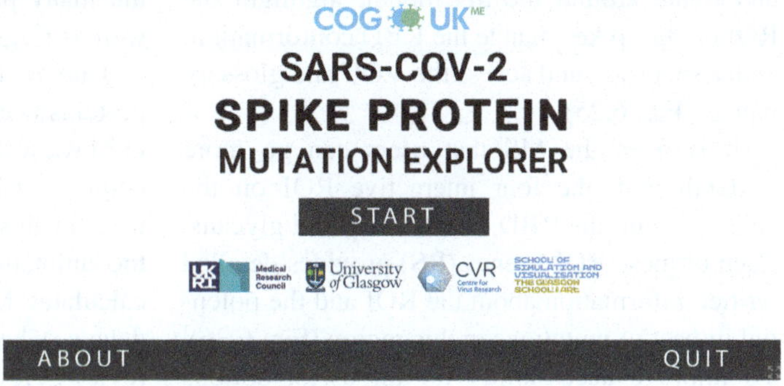

Fig. 6.22 A screenshot of the Opening Scene of the application

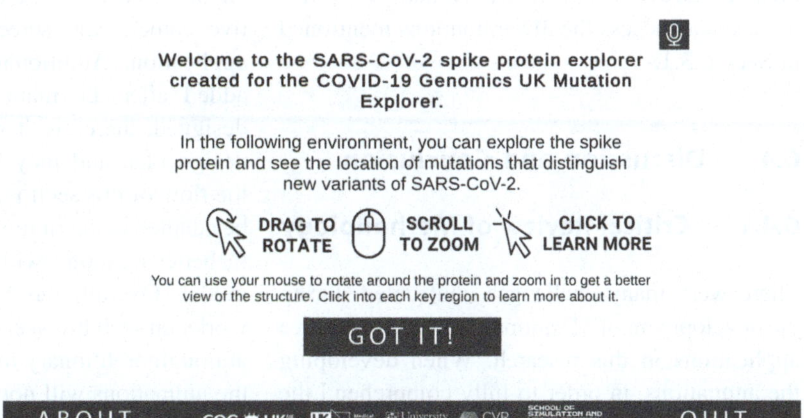

Fig. 6.23 A screenshot of the Instructions Scene of the application

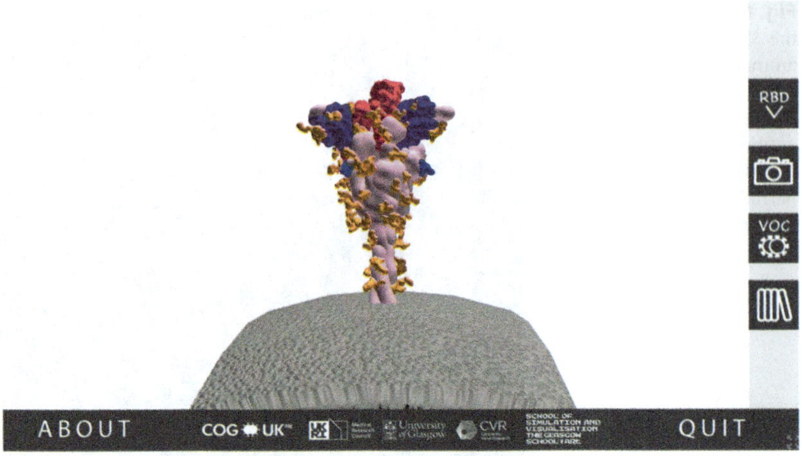

Fig. 6.24 A screenshot of the Main Scene of the application

IS, the user enters the Main Scene (MS) containing the interactive 3D spike protein model (Fig. 6.24). Here, the user can zoom in on and rotate around the 3D model, highlight the ROI on the spike, change the RBD conformation, take a snapshot, and access the VOC and glossary panels (Fig. 6.25).

It is from the MS that users can go more in-depth with the four interactive ROI on the spike protein: the RBD, NTD, FCS, and glycans. Each of these ROI Scenes (RS) provides detailed written information about the ROI and the potential impact of mutations in this region (Fig. 6.26). To improve accessibility for all users, options were given to display the text in 'big' and 'small' (default) options, and an auditory voiceover function was available for the text (Walker 2014). From the RBD and NTD RS, the user can access the 3D animations mentioned in Sect. 6.3.1.

6.4 Discussion and Conclusion

6.4.1 Critical Review of Methodology

There were many challenges confronted during the development of 3D animations and interactive applications in this research. When developing the animations, in order to fully comprehend the functionality and potential of mMaya, the online course *Introduction to Molecular Maya's Rigging Kit* (McGill 2018) was completed. This facilitated the processes involved in creating 3D models of the many proteins used within this research, as well as rigging and animating these 3D models.

Due to the complex nature of the multiple proteins needed in the animations, the 3D scenes in Maya were significant in terms of polygon face count, which was not optimal for the equipment used in this research. This negatively impacted the animation development process as mMaya calculates MD in real time, causing significant delays when any animation process required review. However, this was deemed necessary in order to preserve the realistic appearance of the proteins.

For the SSPME, several scripting issues were encountered early on when creating the interactive camera and screenshot feature within the application. Additionally, the VOC menu was added after the main application was already designed, therefore it was not considered in the design plan and may have negatively impacted the flow of this section. Ideally, this menu would be adapted in the future to be more interactive and in better harmony with the interactive 3D spike model. Overall, the SSPME is functional and works on web browsers such as Google Chrome, although preliminary internal testing showed that the animations will not play correctly in Safari.

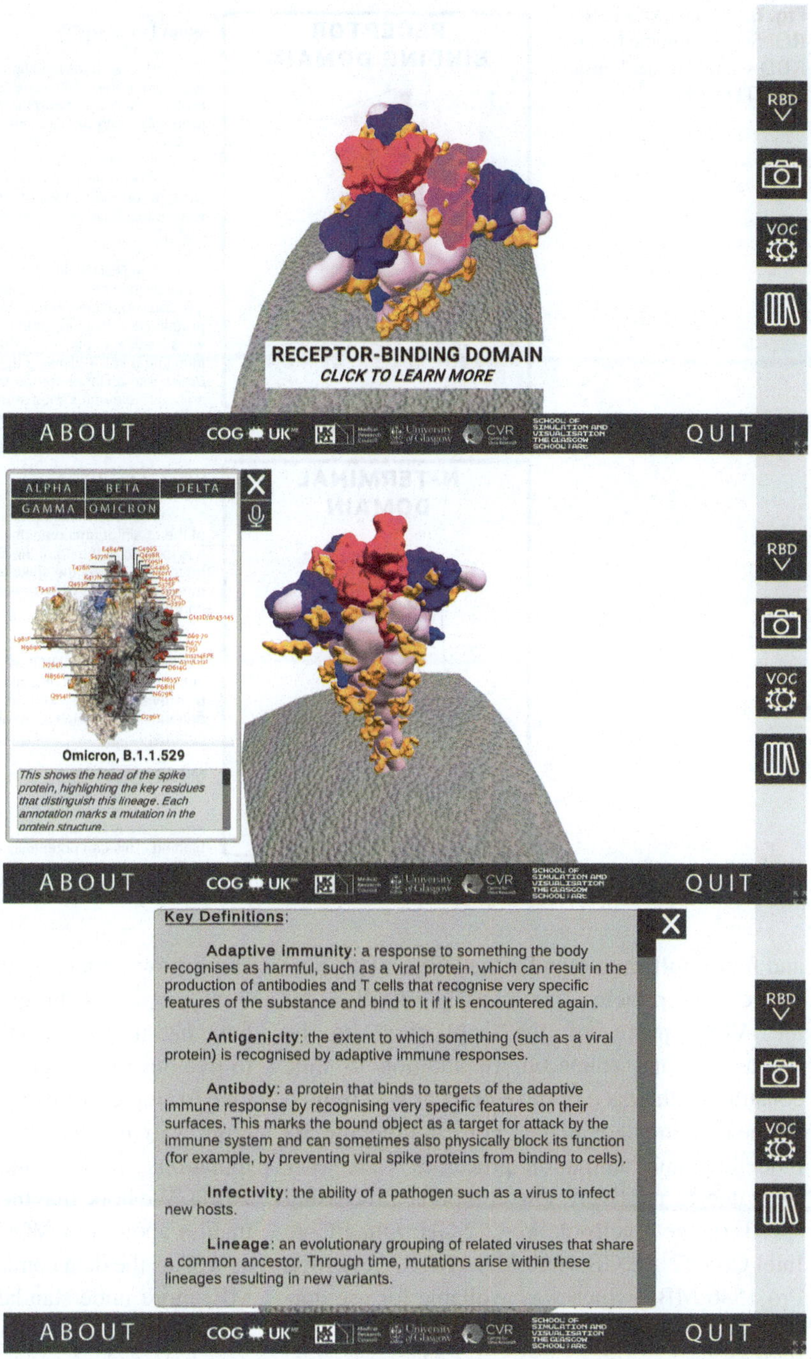

Fig. 6.25 Screenshots of the Main Scene of the application. On the top, the interaction with a region of interest (ROI) on the 3D spike protein. In the middle, the VOC menu opened. On the bottom, the glossary menu opened

6.4.2 Limitations and Future Research

There were several limitations faced during this research. Firstly, this research was conducted during a Master's dissertation and the research was adapting in real-time to communicating about a pandemic. Therefore, the restrictive timeline limited the objectives of the animations and application. This prohibited the creation of a third

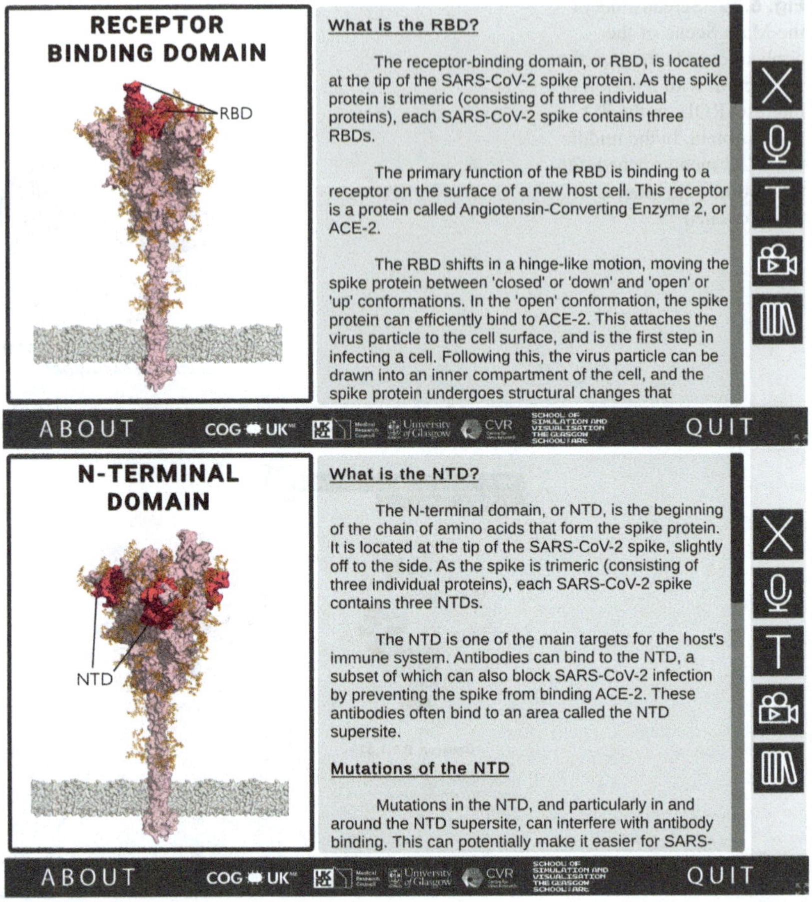

Fig. 6.26 Two of the four ROI Scenes. On the top, the RBD scene. On the bottom, the NTD scene

and fourth full animation for the FCS and glycans of the spike protein. This also did not allow for the VOC menu of the application to be reimagined and enhanced. In addition to time constraints, the pace of development was limited by the equipment available. For example, the 3D modelling and animation processes were very slow due to relatively low computer processing speeds on the MacBook Pro (2.2 GHz Quad-Core Intel Core i7, 16 GB 1600 MHz DDR3, Intel Iris Pro 1536 MB), which was available for use during remote work. In order to counteract this, low-resolution settings were used where possible.

Despite these limitations, this project demonstrates the potential for using three-dimensionality and interactivity in scientific communication and education. The SSPME successfully incorporates a SARS-CoV-2 spike protein 3D model into an interactive application in order to educate users on its various ROIs and about how mutations of these ROIs may impact viral properties, bringing about the VOCs that have had major impacts on the way we have had to live our lives over the course of the pandemic. One example of this is the newly emerging Omicron variant, which has a high number of mutations in the spike protein (Wright et al. 2021). We hope that the SSPME may make information about new VOCs presented in the media, as well as the data contained within the COG-UK/ME, more understandable to a general audience without a molecular virology background. This may help to boost the public's understanding of COVID-19, public health control measures, and the need for the restrictions that were imposed on daily life during the pandemic. Considering how the COVID-19 pandemic has progressed over the last 2 years and how misinformation surrounding infection and disease has spread globally, further research into tools that communicate the science of these topics are vital to combatting further negative development of the pandemic.

6.4.3 Conclusion

Though it has been 2 years since SARS-CoV-2 began its spread across the globe, it is showing no signs of stopping, and may continue to spread without the dedication and collaboration of us all. Promoting scientifically sound information to the general public can only help promote scientific communication and fight the spread of misinformation around SARS-CoV-2 and COVID-19. The SSPME (https://sc2-application.itch.io/sars-cov-2-mutation-explorer) developed in this research is hopefully one of many resources being developed for this very purpose. Between its first release in July 2021 and its current status in January 2022, the SSPME has been viewed over 1700 times by users worldwide and a variety of positive and constructive feedback has been received from user testing, press releases, and social media responses. We intend to continue to update the application as new VOCs emerge so that the SSPME can be as accessible and far-reaching as possible.

Acknowledgements The work described in this chapter originated as a project in the University of Glasgow / the Glasgow School of Art joint taught postgraduate MSc Medical Visualisation and Human Anatomy. We wish to thank Faye Watson, Lois Mason, and Anna Kirk (MRC-University of Glasgow Centre for Virus Research) for their support, advice, and discussion. Work in the Hutchinson group is funded by the UKRI Medical Research Council [MR/N008618/1].

References

Barak M, Hussein-Farraj R (2013) Integrating model-based learning and animations for enhancing students' understanding of proteins structure and function. Res Sci Educ 43(2):619–636

Benton DJ et al (2020) Receptor binding and priming of the spike protein of SARS-CoV-2 for membrane fusion. Nature 588(7837):327–330

Berman HM *et al.* (2000) *RCSB protein data bank.* [Online]. https://www.rcsb.org/ [Accessed 23 12 2021]

Bonner A et al (2009) Location of secretory component on the fc edge of dimeric IgA1 reveals insight into the role of secretory IgA1 in mucosal immunity. Mucosal Immunol 2:74–84

Casalino L et al (2020) Beyond shielding: the roles of glycans in the SARS-CoV-2 spike protein. ACS Cent Sci 6(10):1722–1734

Centers for Disease Control and Prevention (2021). *SARS-CoV-2 variant classifications and definitions.* [Online]. https://www.cdc.gov/coronavirus/2019-ncov/variants/variant-classifications.html. [Accessed 23 12 2021]

COVID-19 Genomics UK Consortium (2021). *COVID-19 Genomics UK Consortium.* [Online]. https://www.cogconsortium.uk/. [Accessed 23 12 2021]

Goodsell D (2020a) *Coronavirus.* [Art] (RCSB Protein Data Bank)

Goodsell D (2020b) *SARS-CoV-2 and neutralizing antibodies.* [Art] (RCSB Protein Data Bank)

Goodsell D (2020c). *SARS-CoV-2 fusion.* [Art] (RCSB Protein Data Bank)

Grant OC, Montgomery D, Ito K, Woods RJ (2020) Analysis of the SARS-CoV-2 spike protein glycan shield reveals implications for immune recognition. Sci Rep 10(1):1–11

Harvey WT et al (2021) SARS-CoV-2 variants, spike mutations and immune escape. Nat Rev Microbiol 19:409–424

Heo L & Feig M (2020) *SARS-CoV-2 protein structure models.* [online]. https://github.com/feiglab/sars-cov-2-proteins/tree/master/Membrane. [accessed 21 05 2021]

Höffler TN (2010) Spatial ability: its influence on learning with visualizations—a meta-analytic review. Educ Psychol Rev 22(3):245–269

Iwasa J *et al.* (2021) *Coronavirus annotation.* [Online]. https://coronavirus-annotation-3.sci.utah.edu/. [Accessed 07 2021]

Jenkinson J (2018) Molecular biology meets the learning sciences: visualizations in education and outreach. J Mol Biol 430(21):4013–4027

Jenny B, Kelso NV (2018) Color Oracle. Monash University, Melbourne, Australia

Katus (2015) *Change Color on mouse over in Unity3D.* s. l.:Youtube

Ke Z et al (2020) Structures and distributions of SARS-CoV-2 spike proteins on intact virions. Nature 588(7838):498–502

Khan MF (2020) *Object Glow Unity—Glow GameObject on Mouse Click C#.* s.l.:Youtube

Khao J, Cheung G, Park J, McGill G (2020) A Visual Model for SARS CoV-2 Membrane Fusion. Digizyme Inc, Brookline, Massachusetts

Letko M, Marzi A, Munster V (2020) Functional assessment of cell entry and receptor usage for SARS-CoV-2 and other lineage B betacoronaviruses. Nat Microbiol 5(4):562–569

McGill G (2018) *Introduction to molecular maya rigging kit.* [Online]. https://clarafi.com/training/introduction-to-molecular-maya-rigging-kit/. [Accessed 21 05 2021]

Mixkit (n.d.) *Free click sound effects.* [Online]. https://mixkit.co/free-sound-effects/click/. [Accessed 2 7 2021]

Ouweland LVD (2017) *Using a Unity button to switch between scenes*. [Online]. https://www.loekvandenouweland.com/content/use-unity-button-to-change-between-scenes.html. [Accessed 2 7 2021]
Paulo J (2021) *Organic Abstract 001*. [Art] (https://3dtextures.me/)
Peacock TP, Penrice-Randal R, Hiscox JA, Barclay WS (2021) SARS- CoV-2 one year on: evidence for ongoing viral adaptation. J Gen Virol 102(4):001584
Queloz AKM, Stern E, Hafen E, Köhler K (2016) Debunking key and lock biology: exploring the prevalence and persistence of students' misconceptions on the nature and flexibility of molecular interactions. Matters Select 2(8):e201606000010
Quintana C et al (2004) A scaffolding design framework for software to support science inquiry. J Learn Sci 13(3):337–386
Robic S (2010) Mathematics, thermodynamics, and Modeling to address ten common misconceptions about protein structure, folding, and stability. CBE-Life Sci Educ 9(3):189–195
Rossen J et al (1998) The viral spike protein is not involved in the polarized sorting of coronaviruses in epithelial cells. J Virol 72(1):497–503
Scapin G et al (2015) Structure of full-length human anti-PD1 therapeutic IgG4 antibody pembrolizumab. Nat Struct Mol Biol 22:953–958
Schmiege P, Fine M, Blobel G, Li X (2017) Human TRPML1 channel structures in open and closed conformations. Nature 550:366–370
Schrand T (2008) Tapping into active learning and multiple intelligences with interactive multimedia: a low-threshold classroom approach. Coll Teach 56(2):78–84
Soundjay (n.d.) *Camera sound effects*. [Online]. https://www.soundjay.com/camera-sound-effect.html. [Accessed 2 7 2021]
Terasawa K et al (2016) Lysosome-associated membrane proteins-1 and -2 (LAMP-1 and LAMP-2) assemble via distinct modes. Biochem Biophys Res Commun 479(3):489–495
Unity Forum (2016) *HOW TO: screenshot in a new tab of the browser (without pop-up blocking)*. [Online]. https://forum.unity.com/threads/how-to-screenshot-in-a-new-tab-of-the-browser-without-pop-up-blocking.440051/. [Accessed 2 7 2021]
Unity Technologies (2021) *Unity documentation*. [Online]. https://docs.unity3d.com/Manual/index.html. [Accessed 2 7 2021]
Voicebooking (n.d.) *Free voice over generator*. [Online]. https://www.voicebooking.com/en/free-voice-over-generator. [Accessed 15 7 2021]
Walker M (2014) *Top tips for accessible apps*. [Online]. https://abilitynet.org.uk/news-blogs/top-tips-accessible-apps. [Accessed 12 07 2021]
Woo H et al (2020) Developing a fully glycosylated full-length SARS-CoV-2 spike protein model in a viral membrane. J Phys Chem B 124(33):7128–7137
World Health Organization (2021). *WHO Coronavirus (COVID-19) Dashboard*. [Online]. https://covid19.who.int/. [Accessed 23 12 2021]
Wright DW *et al.* (2021) *COG-UK mutation explorer*. [Online]. https://sars2.cvr.gla.ac.uk/cog-uk/. [Accessed 09 04 2021]
Xia S et al (2020) The role of furin cleavage site in SARS-CoV-2 spike protein-mediated membrane fusion in the presence or absence of trypsin. Signal Transduct Target Ther 5(1):1–3
Xiong X et al (2018) Glycan shield and fusion activation of a deltacoronavirus spike glycoprotein Fine-tuned for enteric infections. J Virol 92(4):e01628-17
Yan R et al (2020) Structural basis for the recognition of SARS-CoV-2 by full-length human ACE2. Science 367(6485):1444–1448
Yao H et al (2020) Molecular architecture of the SARS-CoV-2 virus. Cell 183:730–738
Zhao J et al (2017) Molecular basis for the binding and modulation of V-ATPase by a bacterial effector protein. PLoS Pathog 13(6):e1006394

Sarah Iannucci is a recent graduate student from the MSc Medical Visualisation and Human Anatomy degree at the Glasgow School of Art and the University of Glasgow, United Kingdom. She is responsible for the conceptualisation, visualisation, methodology, and writing of the research.

William Harvey is a Bioinformatician and Computational Biologist at the MRC-University of Glasgow Centre for Virus Research, the University of Glasgow, United Kingdom. He is responsible for the visualisation and supervision of this research.

Joseph Hughes is a Bioinformatician at the MRC-University of Glasgow Centre for Virus Research, the University of Glasgow, United Kingdom. He assisted with the supervision of this research.

David L Robertson is the Head of Viral Genomics and Bioinformatics at the MRC-University of Glasgow Centre for Virus Research, the University of Glasgow, United Kingdom. He assisted with the supervision of this research.

Edward Hutchinson is a Senior Lecturer at the MRC-University of Glasgow Centre for Virus Research, the University of Glasgow, United Kingdom. He is responsible for the conceptualisation, supervision, funding acquisition, and reviewing of this research.

Matthieu Poyade is a Research Fellow at the School of Simulation and Visualisation, the Glasgow School of Art, United Kingdom. He is responsible for the supervision, providing resources, and reviewing of this research.

Student-Perceived Value on the Use of Clay Modelling in Undergraduate Clinical Anatomy

7

Janine C. Correia, Karin J. Baatjes, and Ilse Meyer

Abstract

Clay modelling is increasingly used as an anatomy teaching method to supplement practical dissection sessions. The use of clay modelling is an active, tactile learning tool utilized to improve student engagement and enhance students' understanding of anatomical relationships in human anatomy. Furthermore, not only does clay modelling engage more senses in the learning process, but it was also found that there are educational advantages to the group interactions that are associated with the construction process to further collaborative learning.

The aim of the study was to explore the effect of building anatomical clay models on students' awareness and understanding of their own thought processes as well as to explore whether clay modelling promoted collaborative learning. This cross-sectional study entailed the use of clay modelling and reflective practice to promote metacognition in third-year BSc students (n = 10), at the University of Stellenbosch's (SU) Division of Clinical Anatomy. The third-year BSc students were asked to build anatomical clay models (in groups), complete a reflective assignment and an evaluation form as part of their assessment of the Applied Anatomy module. The reflective assignment was based on Gibb's reflective cycle. The inclusion of such an assignment would trigger students to reflect upon their learning experiences and thus promote their metacognition.

Ten (10) BSc students volunteered to take part in the study. A mixed method approach was followed; the reflective assignments were qualitatively analyzed, while the evaluation forms were quantitatively (statistically) analyzed. Data obtained from the online evaluation forms indicated agreeable responses confirming that the clay modelling was a valuable learning tool. However, the participants preferred cadaveric dissection instead of clay modelling to learn anatomy. Furthermore, three themes became prominent from the thematic analysis of the reflective assignments, namely, (1) *Advantages of clay modelling*, (2) *Challenges of clay modelling*, and (3) *Suggestions for future practice*.

The research suggests that the inclusion of hands-on clay modelling in the undergraduate clinical anatomy is a valuable learning tool. The participants perceived that it enhanced

J. C. Correia (✉)
Division of Clinical Anatomy, Faculty of Medicine and Health Sciences, Stellenbosch University, Parow, South Africa
e-mail: jcorreia@sun.ac.za

K. J. Baatjes
Dean's Division, Faculty of Medicine and Health Sciences, Stellenbosch University, Parow, South Africa

I. Meyer
Centre for Health Professions Education (CHPE), Faculty of Medicine and Health Sciences, Stellenbosch University, Parow, South Africa

L. Shapiro, P. M. Rea (eds.), *Biomedical Visualisation*, Advances in Experimental Medicine and Biology 1388, https://doi.org/10.1007/978-3-031-10889-1_7

their anatomical knowledge and improved collaborative learning.

Keywords

Anatomical clay models · Anatomy teaching · Reflective practice · Collaborative learning

7.1 Introduction

Anatomy forms the foundation of the basic medical sciences and is needed by all health professionals, in varying degrees, during clinical practice (Sugand et al. 2010). The consequence of a lack of anatomical knowledge is incompetence in healthcare professionals, which can have a detrimental effect in the clinical setting (Sugand et al. 2010). Thus, a solid foundation of anatomical comprehension and insight is critical to the education of health professionals. Cadaver dissection is traditionally considered to be the gold standard for the practical teaching of anatomy; however, dissection is not a uniform learning experience and supplementary novel methods should be incorporated to include the learning preferences of all students. Consequently, novel alternatives are being explored—one of these is the building of anatomical clay models.

Fitzgerald et al. (1979) pioneered the use of plasticine modelling to build their own anatomy models for student learning. Ever since, the use of modelling using an assortment of materials, such as clay, wax, and plasticine, has become progressively more prevalent as an active method of developing students' understanding of anatomy. The use of modelling is an unconventional learning method that supplies a haptic, three-dimensional, and constructive approach to learning anatomy (DeHoff et al. 2011). Active learning using modelling may lead to more understanding gained compared to passive learning practices, such as didactic lectures, and consequently engages students who may have little background knowledge of the topic (Krontiris-Litowitz 2003; Herur et al. 2011; Kooloos et al. 2014).

Clay modelling provides a tactile, and kinaesthetic approach to learning anatomy. According to DeHoff et al. (2011), students need to actively participate in the learning process, to effectively absorb and retain the information and most students learn more effectively when engaging in several senses concurrently. The use of clay models achieves this by incorporating visual, auditory, and tactile senses into the learning process. For this reason, it has been proposed that anatomy lectures should be shortened and should only cover an overview of the relevant structures; the focus should be shifted towards a kinaesthetic and interactive teaching method, such as clay modelling (Krontiris-Litowitz 2003).

Furthermore, not only does clay modelling engage more senses in the learning process than a lecture, but it was also found that there are educational advantages in the group interactions that are associated with the construction process and presentation. Presenting the constructed clay models to fellow students may also further enhance collaborative learning in anatomy (DeHoff et al. 2011). Also, constructivist learning theory, as a theoretical framework, underpins the advantages of clay modelling, according to Curlewis et al. (2021). Clay modelling involves students constructing and manipulating clay models in groups and consequently engaging in an active learning process. This active learning process thus positions the constructivist view of learning as a social activity (Schunk 2008).

7.2 Clay Modelling in Anatomy Education

Several studies have been conducted to test and demonstrate the efficacy of clay models in learning human anatomy (Motoike et al. 2009; Oh et al. 2009; DeHoff et al. 2011; Bareither et al. 2013; Kooloos et al. 2014; Carlson et al. 2019; Zainuddin et al. 2019; Curlewis et al. 2021). Krontiris-Litowitz (2003) used clay modelling to teach functional neuroanatomy and physiology to a group of students and, after analyzing their results in a subsequent exam, found that students who attended the clay modelling workshop scored significantly higher than those who did not. Nicholson et al. (2016) hosted a workshop comprising various new, innovative approaches to learning anatomy—amongst others clay modelling. Students who attended improved their performance significantly in the following tests, compared to non-attendees. When asked for feedback, students indicated that the most useful components of the workshop were body painting and clay modelling.

In a study conducted by DeHoff et al. (2011), it was found that students who used clay modelling as a learning tool did not improve their marks significantly compared to those who used feline dissections. It was found, however, that the group who used clay modelling performed as well as the control group, despite having only had 50% of the study time that the control group had to use plastic models/images. This would suggest that students who use clay models are indeed able to apply what they have learned during the modelling time to other study methods, i.e. plastic models.

Anatomical clay models have also proven useful in the context of learning cross-sectional views for radiological purposes. Students built three-dimensional models of various organs out of soft clay, cut them, and compared the sections to magnetic resonance imaging (MRI) and computed tomography (CT) scans (Oh et al. 2009). There was an overall positive response to this approach. The average score on a subsequent examination of this group of students was significantly higher than that of students who used conventional learning methods. Researchers also found that clay modelling is a useful and valid supplement to conventional anatomical models as retail models are expensive while building your own is cost-effective and can be stored for continued reference (Oh et al. 2009).

On the other hand, even though there are numerous studies outlining the benefit of three-dimensional anatomical model learning, an in-depth literature review conducted by Azer and Azer (2016) concluded that there is no solid evidence to suggest that the use of these models in learning is superior to traditional teaching methods and that further studies are imperative to assess the long- and short-term effects accurately. Having noted this, it could be a more beneficial way of learning for one type of student but not for another. Azer and Azer (2016) found no studies from countries in Africa or South America.

Authors have stressed that anatomy needs to be taught in a multimodal way (Sugand et al. 2010). Cadaveric dissections should not be omitted from a curriculum, as they provide advantages that clay models simply will not be able to replicate, such as allowing students to see the layout of the human body, relationships between different structures/organs, how they interact with one another and how this relates to the anatomy (DeHoff et al. 2011). Dissections allow students to appreciate the physical properties of different tissues, i.e. the textures and colourations (DeHoff et al. 2011). Furthermore, dissections are still necessary to teach healthcare students about pathology as clay models may not necessarily be an adequate learning tool in this regard. Also, as Sugand et al. (2010) pointed out, no human body is perfectly shaped and colour-coded, thus, solely learning anatomy from textbooks and anatomical models is not sufficient, as these do not account for pathological conditions and anatomical variations. A recent systematic review by Curlewis et al., (2021) concluded that the numerous advantages of clay-based modelling surpass the limitations and that it should be used in conjunction with cadaveric practical sessions to improve learning of anatomy.

The aim of the experience of building a clay model is to mimic the effect of manipulating the tissues in a cadaver during dissections and still have the acquisition of a three-dimensional understanding of the human body. An experiment conducted by Kooloos et al. (2014) concluded that students had better anatomical knowledge test scores when they built a clay model compared to those who only watched a video on the same topic. The speculated reasoning for this effect was that the students had more concentration, engagement, and stimulation when building their own model demonstrating an anatomical structure (Kooloos et al. 2014).

The use of modelling is an active, tactile learning practice used to boost student engagement and increase students' understanding of spatial relationships in human anatomy (Bareither et al. 2013). The question thus arises, how the use of clay modelling, whether wax, traditional clay, or plasticine, affects students' metacognition and understanding of their study of anatomy. Metacognition refers to the learner's mindfulness of the development of their learning. It also refers to the capacity of students to control and manage their own learning (Georghiades 2004). Students who exhibit metacognitive approaches in their learning, scrutinizing their understanding, and employing reflective practice, gain a deeper understanding and become autonomous learners (Ross et al. 2006). Reflection is a metacognitive task that establishes a superior understanding of both the self and the status quo so that future actions can be informed by this understanding (Sandars 2009). Gibbs' reflective cycle is one of the most popular theories related to reflective practice (Gibbs 1988). This model allows for the exploration of the dimensions of learning in overlapping prompt areas.

Thus, the study explored the effect of building anatomical clay models on students' awareness and understanding of their own thought processes as well as whether clay modelling promotes collaborative learning.

7.3 Methods

The research was conducted at the Division of Clinical Anatomy, Department of Biomedical Sciences, Faculty of Medicine and Health Sciences, Stellenbosch University. This research involved human participants and was approved by the Faculty of Medicine and Health Sciences, Health Research Ethics Committee (HREC Reference No: N21/04/035), and institutional permission was granted for the study. The study population included all the third-year BSc students enrolled in the applied anatomy module (module code: ANA364).

This study included the use of clay modelling and of reflective practice to promote metacognition in third-year BSc group and occurred during the second semester of 2021. Data was gained from two sources; students' reflection on their learning that was submitted as part of the written aspect of the assignment and an evaluation form. The evaluation form included a five-point Likert scale, and the responses were exported to a Microsoft Excel spreadsheet for regression statistical analysis. This scale consisted of the following options from which to choose, namely, strongly agree (SA), agree (A), neutral (N), disagree (D), and strongly disagree (SD).

The following step was to obtain consent from the respondents who were interested in volunteering for the study. Participation was voluntary and had no impact (positive or negative) on students' status in the course. Ten (10) participants volunteered out of the group of 12 students (83% of class). The volunteering individuals signed the written consent.

As part of the applied anatomy (ANA364) module, the groups were tasked to complete an anatomical clay model of the musculoskeletal system. The students were divided into groups of two or three students to promote collaborative learning. Each group was supplied with a specific part of the skeleton, air drying clay, wood glue, and acrylic paint. The students were encouraged to use recyclable material or upcycle found material (such as wool, string, wooden ice cream sticks, and wood to mount the clay model on) to build the model. The anatomical clay models the students constructed are seen in Figs. 7.1, 7.2, 7.3, 7.4, and 7.5.

Furthermore, as part of their module, the students each submitted a written report with a reflective assignment component. The reflective assignment was based on Gibb's (1988) reflective cycle and was due a week after they had submitted their models, as a reflection on the process. Gibb's (1988) covers six stages: (1) Description of the experience, (2) feelings and thoughts about the experience, (3) evaluation of the experience, (4) analysis to make sense of experience, (5) conclusion about learning and what could have been done differently, and (6) action plan for dealing with similar situations in future. This module ran

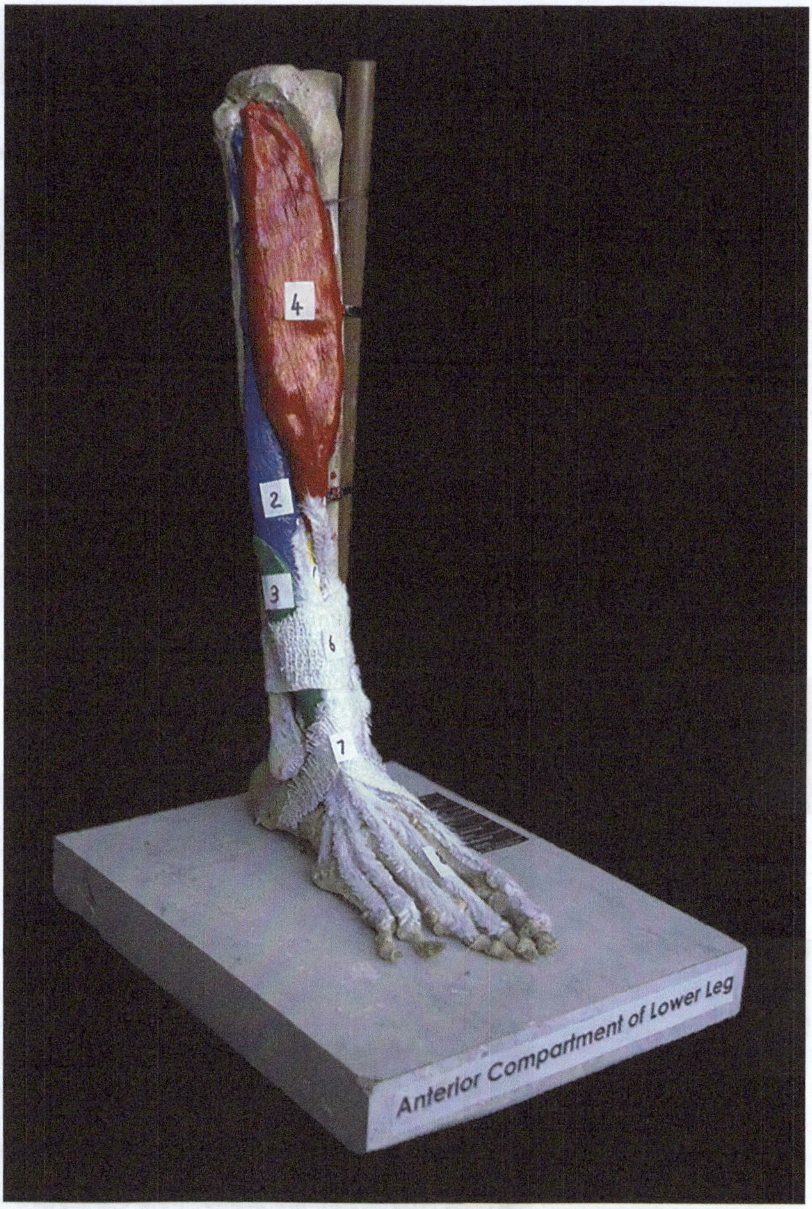

Fig. 7.1 Muscles of anterior compartment of the leg

during a semester giving the students 12 weeks to complete the task between August and November. Grades for this module were calculated as seen in Table 7.1.

A mixed method approach was followed; the reflective assignments were qualitatively analyzed, while the evaluation forms were quantitatively (statistically) analyzed. The reflective tasks were qualitatively analyzed through an inductive approach. The data extracted from the reflective assignments were collected. The analysis of the data obtained from the reflective assignments was conducted by using the six steps of Creswell (2014).

The first step of thematic analysis included sorting and preparing data for analysis by reading through the raw data, and organizing data according to its source (Creswell 2014). Step 2 included developing an overall sense of the data and reflection on the general meaning. This

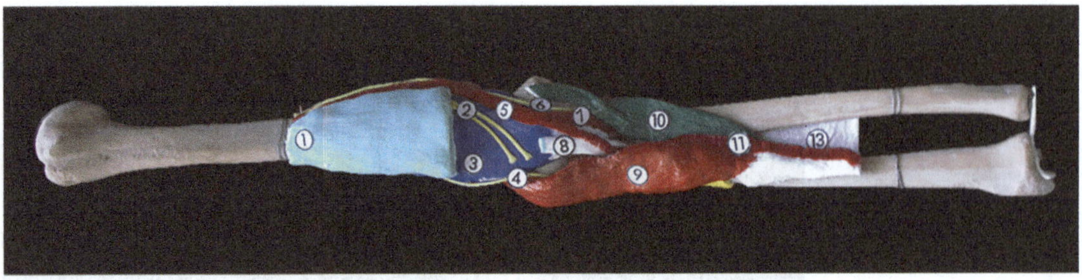

Fig. 7.2 Cubital fossa

phase required all data to be read and an overall sense of information to be gained. The researchers reflected on the general meaning by annotating the transcripts and noting general feelings and thoughts. Step 3 entailed coding the data. Text within the transcripts was highlighted and assigned one or more relevant codes which consist of words or phrases. The codes were developed according to the information brought forward by the participants and were not predetermined. Step 4 included describing and identifying themes (Creswell, 2014). Generated from the coding process are several themes, used to group coded information, forming the main findings of the study. Themes represented the perspectives of multiple participants and were supported by quotations from the transcripts. Step 5 involved presenting the findings. A tabulated

Fig. 7.3 Quadrangular and triangular spaces

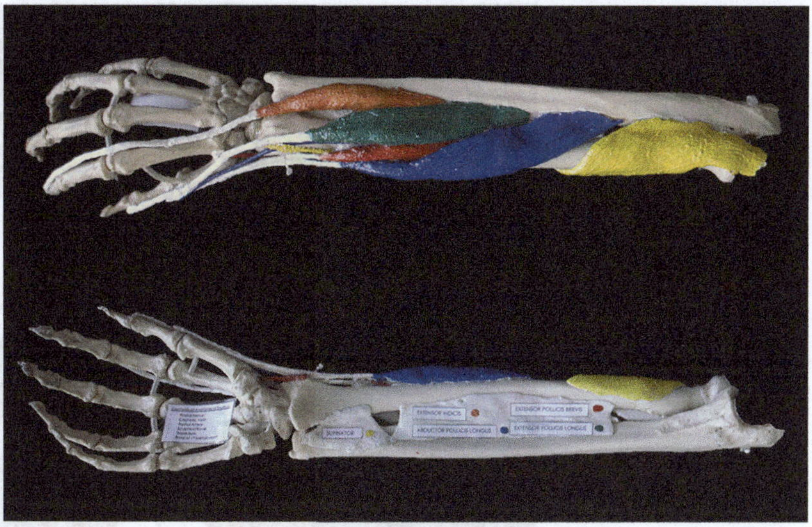

Fig. 7.4 Deep muscular layer of posterior forearm

representation of the themes and sub-themes was provided to summarize the findings. Lastly, step 6 involved interpreting the data. This involved the interpretation of the information by the researchers with influence from the available literature and theories. The researchers documented the interpretation of the information in the discussion section of this study according to the themes.

Reports were randomly assigned to the researchers and thematic content analysis was performed to identify emerging themes, as described above. The process was repeated to identify further emerging themes and categories. Processing entailed general familiarization with the data, the generation of initial codes and validation, the identification of these themes amongst the codes, a review of these themes, the naming of themes, and finally the production of a final report. The frequencies of the emerging themes were documented and reported using descriptive statistics.

7.4 Results

Ten students volunteered to participate in the study. Nine of the ten volunteering students completed the online evaluation form.

7.4.1 Quantitative Results: Evaluation Form

Data displayed in Table 7.2 was obtained from the online evaluation form. Overall, a combined agreeable response indicated that the survey participants found that the process of building the clay models enhanced their ability to visualize the layers of the muscles (SA = 44,44%; A = 33,33%) as well as improved their capability to visualize the muscle origins and insertions (SA = 55,56%; A = 11,11%). When asked if participants found that the clay models were a valuable learning tool in anatomy, the majority (SA = 11,11%; A = 55,56%) agreed that it was a valuable learning instrument. However, mixed responses were received on whether the clay modelling aided them by linking muscles with similar function.

Most participants, however, were neutral (55,56%) or disagreed (D = 22,22%; SD = 22,22%) that the clay modelling project assisted them with their performances in assessments as seen in Table 7.2. This can be because the students mostly focused on the musculoskeletal system during the previous (second) year, as all the clay models were based on this system, on which they had already had written tests and examinations.

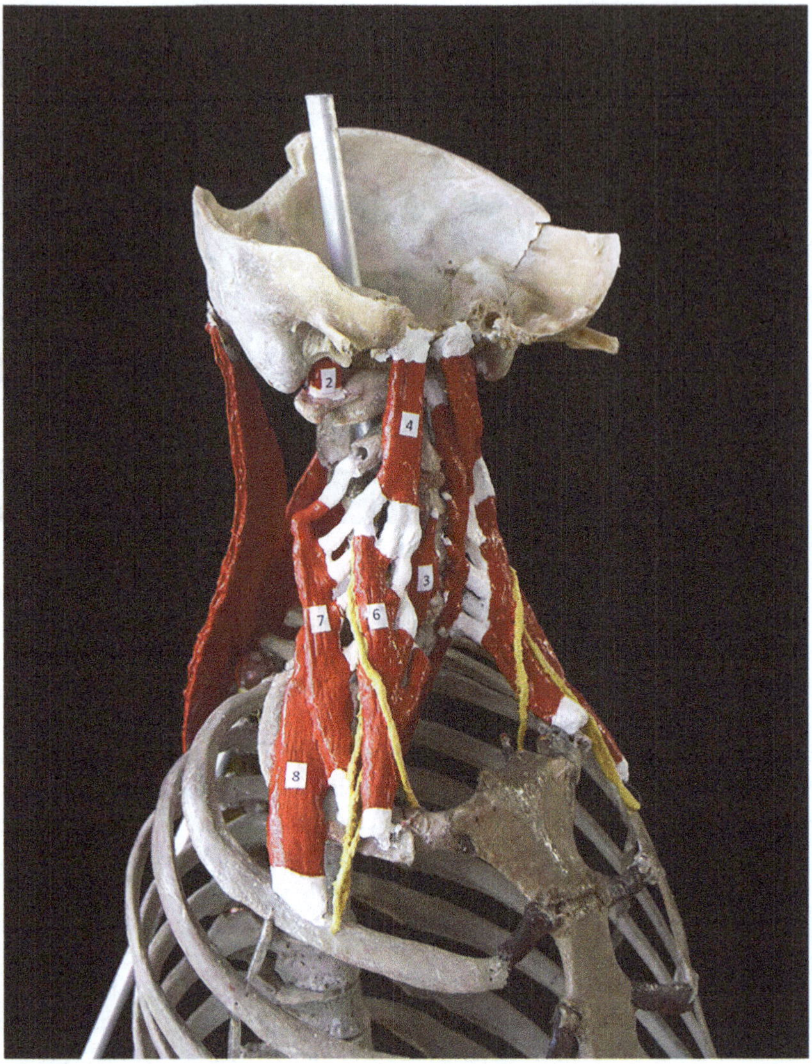

Fig. 7.5 Prevertebral muscles

A combined negative response indicated that seventy-eight percent of survey participants felt that they preferred cadaveric dissection above clay modelling to learn anatomy (SD = 33,33%; D = 44,45%). This is an indication of cadaveric preference for practical learning. Lastly, when asked if they found that the clay modelling promoted collaborative learning, almost all the participants, except one, had a positive response towards enhanced learning through group work (SA = 22,22%; A = 66,67%).

Table 7.1 ANA364 components

1. Group—Submit anatomical clay model with labels (50%).
2. Group project presentation—Each group present their model to the class (20%).
3. Individual—Each student submits a report that includes a "reflection on learning" section of about two pages (30%).

Table 7.2 Responses from the evaluation form

	Strongly agree SA (1)	Agree A (2)	Neutral N (3)	Disagree D (4)	Strongly disagree SD (5)
The clay models improved my ability to visualize the layering of the muscles.	44,45%	33,33%	0%	11,11%	11,11%
The clay models improved my ability to visualize the muscle attachments.	55,56%	11,11%	22,22%	0%	11,11%
The clay models were a valuable learning tool.	11,11%	55,56%	11,11%	11,11%	11,11%
The clay models allowed me to make connections between muscles with similar actions.	11,11%	33,34%	22,22%	22,22%	11,11%
The clay models improved my performance on the anatomy tests/exams.	0%	0%	55,56%	22,22%	22,22%
I prefer using clay modelling as the primary method of laboratory instruction for learning human anatomy rather than dissection.	0%	11,11%	11,11%	44,45%	33,33%
The group work during the clay modelling enhanced my learning through working together.	22,22%	66,67%	0%	0%	11,11%

7.4.2 Qualitative Findings: Reflective Assignment

The data represents the different students' perceptions regarding the effect, awareness, and understanding of their own learning processes on building anatomical clay models. The data was gathered from ten students who volunteered and willingly signed the written consent to participate in the study. Each student submitted a written report with a reflective assignment component. These were thematically analyzed through an inductive approach by using the six steps of Creswell (2014). Processing entailed general familiarization with the data, the generation of initial codes and validation, and the identification of themes amongst the codes. Three themes became prominent, namely, *Advantages of clay modelling*, *Challenges of clay modelling*, and *Suggestions for future practice*. Each theme was analyzed according to the main sub-themes. The themes, sub-themes, and categories that developed from the data are set out in Table 7.3.

7.4.3 Theme 1: Advantages of Clay Modelling

Advantages of clay modelling in this context refer to the benefits identified by students in utilizing clay modelling to enhance their anatomical knowledge, improve collaborative working and develop metacognitive skills. The participants described their experiences as detailed below.

7.4.4 Sub-theme: Educational Value

Educational value is described as the capacity of incorporating clay modelling as a resource, to promote students' learning and development in anatomy.

> I found it very educational and enjoyable. (Participant (P) 3, Female (FM))

> I thoroughly enjoyed the experience as a whole and I value it as a learning experience. (P5, FM)

7.4.4.1 Category: Knowledge Construction

Knowledge construction in anatomy refers to the learning process by which students utilized clay models as a learning tool in applying their prior knowledge to construct meaning and understanding through this experience. They highlighted the learning zone opportunities and the process of reflection to identify their gaps. These opportunities relate to students' learning in a supportive environment, allowing them to make mistakes, to receive feedback and thereby

Table 7.3 Themes, sub-themes, and categories that were developed from the data

Themes	Sub-themes	Categories
1. Advantages of clay modelling.	Educational value	• Knowledge construction. • Improve communities of practice. • Developing metacognitive skills.
	Extracurricular value	• Skills development.
2. Challenges of clay modelling.	Mindset Material Time on task Inexperience	
3. Suggestions for future practice.	Time management Preparation Metacognitive Hybrid approach	

enhance their academic progress. They mentioned their active participation in the learning process and how this experience assisted them in their learning.

> ...it required stepping out of comfort zones, but the opportunity made me reflect more and showed that this skill needs to be worked on (P1, FM)

> I was learning while actively playing... I managed to remember the muscles and their origin, insertions, and functions for the presentation and report without having to refer back to my notes. I would gladly do such a project again, as I gained a significant amount of knowledge from the trial-and-error process, and I believe this experience could help me build an even better model in the future. (P3, FM)

> This task provided me with a challenge on how to learn anatomy using a different style of hands-on learning technique. I felt that I did learn a lot through the process, and I have discovered that I prefer this style of learning to textbook learning. I learnt a lot from the experience, and I am very grateful to have had the opportunity to be challenged by a different type of learning style. (P4, FM)

> I learn better once I visualize it for myself. Building the clay model has enhanced my knowledge and my in-depth understanding. I believe that it will also help me in retaining this knowledge for future purposes. (P7, FM)

> Building the clay models helped me to study and remember the names of the muscles. Building clay models was an effective learning tool. (P8, FM)

> The learning experience gained from physically building an anatomical model is invaluable. You gain a deeper understanding of how intricately all the structures are related to each other and essentially how they all function together. The educational value attached to this learning tool is one that cannot truly be compared to other learning methods. Not only do you tap into your creative side, but you actively engage in a learning process, without feeling exhausted afterwards. (P9, FM)

> The use of clay models to study the structures made it easier to identify the structures. (P10, FM)

7.4.4.2 Category: Improve Communities of Practices

Communities of Practice (CoP) refers to groups of students (2 or 3 in this context) who intentionally collaborate by sharing common goals, ideas, and interact in a process of social learning regarding clay modelling as an alternative technique for teaching anatomy. The students acknowledged that teamwork and thus, collaborative learning had a positive impact on their performances as they relied on each other's strengths. Incorporating collaborative learning enhanced students' skills, such as problem solving, communication, and creative thinking.

> There were collectively decided to best work with everyone's strength, being creative and delegating in a collaborative manner with good communicating skills for the mutual end goal was a good experience. (P1, FM)

> I am not creatively or artistically inclined, however my partner was. This created a great balance where I planned the logistical part and she planned and crafted the muscle shapes exactly as it needed to be. The good experience that I gained from this task was team building and enhancing my creative thinking. (P2, FM)

...it can be hard to talk to your group members about your ideas as everyone else has their own opinion and ideal picture of how they would like the project to work out. My ideas weren't always heard and understood, but it didn't affect me too much as I thought the ideas my group members came up with were good and because building clay models and painting are not my strong points. My group members were accommodating and easy to work with which made the whole experience less stressful and enjoyable. (P4, FM)

There were also some disagreements in the beginning which caused some tension around how we would proceed with the project. Fortunately, we quickly moved past this phase and managed to settle most of our differences with ease. I made a friend in the process and learned that teamwork does not always mean agreeing. I learnt to stand up for the ideas I believed in and agree to ideas in areas I did not feel strongly about. I learnt to play into my strengths and also recognize my weaknesses. (P5, FM)

There was a great sense of teamwork as we all wanted to do as much as possible to make the project a success. Having three people in a group was a positive, because it made it easier to make decisions regarding the model, in a two-person group it is more difficult to come to a consensus on an idea. (P6, M)

...we got the hang of it, and I realised that we work very well in a team... we perfectly complemented each other during this process. This was because she has a completely different way of thinking and that always filled my gaps. She provided the logical manner of approaching this project and I thought of it in a creative way. (P7, FM)

I was excited to be in a group setting as I personally do well in a group setting although I did feel out of my element as I have never had to model anything with clay before. I understood the importance of listening without judgment and open communication. The importance of compromise and planning as not to waste the time of group members who also have other academic demands. I constantly had the fear that our goal driven personalities would clash and interfere with our friendship. Being a part of a team also taught me valuable skills. (P8, FM)

...it brought some relief knowing that I had two other members to work alongside me and that we had more than enough time to take a trial-and-error approach to building this model. I worked very closely with my group members. There was consistent communication and all of us had the same goal in mind. We always took each other's opinions into consideration when discussing anything related to the model... we always managed to find a solution to all the problems we encountered. Throughout the process, you also learn to think of ways to adapt and overcome any problems you experience...working on a project in a group allows you to gain new perspectives and ways of thinking as you ultimately have the same end goal as your group members. The experience I gained during this project will be very valuable towards improving my approach to further projects and group work. The highlight of this project was the continuous support and communication within my group. This is an important aspect of group work, and it determines the success and outcome of the project. I believe that this learning opportunity prepares you for possible future circumstances, especially if you are to be in a career where you constantly work with other individuals. (P9, FM)

7.4.4.3 Category: Developing Metacognitive Skills

Metacognitive skills, in this context, relate to the students' thought processes regarding their learning and problem solving during and after the construction process of clay modelling. The process and reflection assisted them to become aware of their shortcomings and how to improve their learning in future.

...needed more mental capacity that was not initially anticipated...it required stepping out of comfort zones, but the opportunity made me reflect more and showed that this skill needs to be worked on. (P1, FM)

After careful analysis of the situations, we were able to come up with alternatives every time and it was always a better idea than the initial one. (P2, FM)

I would try to be more positive and look at it as another opportunity that can be beneficial to my learning, rather than as a chore I need to complete because I am not good at art or because I don't like group work. (P4, FM)

I started feeling proud of our model. I was proud of all the obstacles we had overcome and all the creative solutions that solved them. Once I let go of those expectations and had the mindset to simply do the best I could do, I felt a weight off my shoulders. (P5, FM)

We had a positive attitude as we went about building the model, so even though we struggled as it was a new experience, we always had a positive outlook. (P6, M)

Problem-solving was gained... (P8, FM)

7.4.5 Sub-theme: Extracurricular Value

Extracurricular value refers to learning that does not fall within the scope of the regular curriculum specifically. Students reported that they developed additional skills, such as teamwork and patience.

7.4.5.1 Category: Skills Development

Students reported that they experienced opportunities during clay modelling to develop additional skills. These are evident in the following quotes:

> . . .patience and perseverance excelled. . . (P1, FM)

> I discovered that I was not as bad at the clay moulding as I thought I would be and can now confidently say that I have learned a new skill. (P4, FM)

> In the process of building a clay model. . .I learned a variety of skills. (P5, FM)

> In an attempt to make sense of the situation and find meaning in it, I believe it has taught me patience as well as how to be kinder to myself. (P5, FM)

> I realised that I am a creative and artistic person and can easily think of great ideas to use in the process. It really challenged me to think in other ways, I felt this exercise to be really enjoyable and taught me some new skills I did not have before. (P7, FM)

> I was also delighted to be able to paint as painting can be used as a form of therapy. The importance of compromise and planning as not to waste the time of group members who also have other academic demands. Being a part of a team also taught me valuable skills. (P8, FM)

> I believe that this learning opportunity prepares you for possible future circumstances, especially if you are to be in a career where you constantly work with other individuals. (P9, FM)

Additional advantages of the clay-based materials were also mentioned by a few students.

> Clay modelling is meant to be a low-cost and easily accessible learning tool. (P3, FM)

Most participants perceived the clay modelling as a valuable tool, both educationally and on an extracurricular level, to improve their understanding of three-dimensional anatomy as well as non-academic skills development.

7.4.6 Theme 2: Challenges of Clay Modelling

Many students highlighted challenges as difficulties, during the construction process of clay modelling. Their mindsets were described as specific feelings of the initial process. Regarding the process, they felt inexperienced, had unclear expectations, were unprepared and perceived it as time consuming. They experienced the process as disordered, complex, and stressful and even a waste of time. Some students, initially, felt disengaged from the process and viewed it as a distraction from other work responsibilities. The construction process often ended in a trial-and-error effort. They felt disappointed as the materials were inadequate to use for modelling, as described below. They also felt that clay modelling was less effective when compared to human dissection. Some expressed their initial feelings as uncertain, overwhelming, apprehensive, and frustrating as they did not think they were artistically inclined.

7.4.7 Sub-theme: Mindset

The mindsets of students refer to how some students perceived their abilities to use clay modelling in this context.

> I was also feeling anxious about having to use clay as I am not an arty person and was worried I might let my group down. In the beginning I felt nervous as I have had many bad experiences with group work prior to this occasion. . . it was a lot of work to try and fit in with my other studies. This made me feel overwhelmed and uneasy at times. (P4, FM)

7.4.8 Sub-theme: Material

Material includes all ingredients that the students needed to be able to mould the clay.

> The first day's work was an absolute disaster as the clay kept breaking. It was also challenging to get the clay to stick. I was feeling discouraged and that I was wasting my time when I could rather be studying or doing other more important work. (P2, FM)

> We struggled to mould the clay into a shape that resembles the muscle, or the clay cracked when it was dry, and we had to rebuild it. (P9, FM)

7.4.9 Sub-theme: Time on Task

Time on task refers to the time that students allowed for and how they managed the task to mould the clay. The students had 12 weeks to complete the task between August and November, and could build the models during the anatomy practical session or schedule time outside of class to work on it. However, some students mentioned that there was not enough time allocated to the task. The students perceived the task as time consuming as they had to repeat this before the completion of the final product as mentioned in the quotes below:

> ...the bad experience was time. I felt although there was ample time, I did not manage it too well. (P2, FM)

> At the beginning of the semester, the task set before us was daunting and I doubted the outcome of our model. The process took numerous sessions to plan and execute the building of the model. It was largely a matter of trial-and-error. We had limited time to work on the model as our schedules were different and we had to work on the other modules that we had. (P6, M)

7.4.10 Sub-theme: Inexperience

A few students commented that they had little knowledge or experience of how to work with clay. This was initially perceived as feelings of incompetence and apprehension.

> I have little to no experience with clay. I felt overwhelmed and uncertain about the task ahead. There was uncertainty...We didn't quite know where to start...some frustration in the beginning. We struggled...we were disappointed...The clay was different to what I was expecting and more difficult to work with than I had anticipated. There were also some disagreements in the beginning which caused some tension around how we would proceed with the project. (P5, FM)

> When we were initially given our topic and instructed to build the muscles of the anterior compartment of the lower leg out of clay, I was apprehensive. I assumed that my artistic lack would result in a poorly built, inaccurate anatomical model and I could not foresee how I would benefit from this project. This resulted in our project taking much longer than we were expecting it to. (P3, FM)

> Before we started, I felt overwhelmed as it was my first-time using clay to build something constructive. (P7, FM)

> The only problem that I encountered with this process was that I lack artistic skills and found it difficult to build the clay model to truly represent the above-named structures. Another possible disadvantage would be that the building of the clay models was carried out incorrectly and thus leads to inaccurate information when learning. (P10, FM)

> The complexity of attaching the clay onto the specimen made it stressful. I am not artistically inclined...the task at hand was time consuming...I found the clay modelling less effective than human dissection as a realistic individual knowing the anatomical structure in real time helps more with memorization and identification of the anatomical structures. (P1, FM)

> The instructions were vague and felt we had to figure out most of what we were supposed to do with regards to our topic, on our own. I was also feeling anxious about having to use clay as I am not an arty person and was worried I might let my group down. (P4, FM)

Although the students found the activity to be valuable, they also mentioned a few challenges in the process. Some participants felt uneasy, apprehensive, and frustrated as they did not think they were artistically inclined. They also perceived the modelling as time consuming and mentioned they preferred cadaveric dissections to learn anatomy, rather than clay modelling.

7.4.11 Theme 3: Suggestions for Future Practice

In their written reflections, students made suggestions to address the challenges experienced during the process of clay modelling. These pertained to clear expectations regarding the planning, time management, materials, and the students' mindsets of the project. The timeline and incorporating the project as part of a hybrid approach to assist students in enhancing their anatomy knowledge, were recommended. These were described as follows:

7.4.12 Sub-theme: Preparation of the Project

Some students commented that they would attend to the preparation of the task by improving their planning of the project and quality of the material in future.

> …more colours in the clay modelling, it made more sense to be realistic about it, and to provide less confusing when compared with the human dissections… would use it as another opportunity to learn from previous errors and put more emphasis on improving my presenting skills and researching every possible aspect of the research paper before concluding it as self-evident. (P1, FM)

> If I had to build another clay model I would definitely plan better and start earlier… paint the tendons to allow it to blend in more… (P2, FM)

> I would make use of clay of better quality. I would explore the option of a different material…I would gladly do such a project again, as I gained a significant amount of knowledge from the trial-and-error process, and I believe this experience could help me build an even better model in the future. (P3, FM)

> I planned incorrectly and attempted to print during load shedding. (P8, FM)

7.4.13 Sub-theme: Metacognition Regarding the Project

Some students became aware of their own mindsets and feelings. They commented on how they would address these in future. These were reflected in the comments below:

> I would change the mindset I go into the project with. I would try to be more positive and look at it as another opportunity that can be beneficial to my learning, rather than as a chore I need to complete because I am not good at art or because I don't like group work. (P4, FM)

> …to put a mesh in the clay to keep it together and prevent it from cracking. Another thing I would do differently is to spend more time working on the model… more time to spend on fine tuning I think the model would be more accurate and neater. (P6, M)

> I also would've liked it if we spent a few minutes on our presentation every time we met instead of doing the presentation in the last few days. (P8, FM)

7.4.14 Sub-theme: Time Management of the Project

The students made suggestions regarding their time management and how they would approach the project differently in future.

> …we should have planned a timeline in order to maximise our time spent on our model without neglecting our other academic commitments… we did not always arrive fully prepared for our group sessions, which also affected our productivity… a better approach would have been to set up a schedule that indicates the number of hours that will be spent on each session as well as materials that will be needed. I would take a more realistic approach to time management and planning the main components of a project in future. (P9, FM)

> …to understand how to correctly mould it for complete accuracy, which further contributed to this being a very time-consuming process…to have given myself some more time to get used to the use of clay and accept that it would not be an easy task. (P7, FM)

> …more helpful earlier in the course of anatomy during the musculoskeletal block. My hope is that it will be an effective learning tool for students in the future. (P5, FM)

7.4.15 Sub-theme: Hybrid Approach

Students suggested a hybrid approach, consisting of a mix of educational activities to learn anatomy. They also suggested the inclusion of an artistic person to enhance their accuracy of the task.

> I found the clay modelling less effective than human dissection as a realistic individual knowing

the anatomical structure in real time helps more with memorization and identification of the anatomical structures. (P1, FM)

I will enlist the help of an artistically individual in order to increase the accuracy of the structures being built. I would suggest a hybrid approach of both cadaveric dissections as well as the building of clay models in order to allow students to maximize their knowledge of the anatomical structures of the body. (P10, FM)

7.5 Discussion

This first-of-its kind study in Africa and South Africa explored the influence of the construction of anatomical clay models on students' awareness and understanding of their own thinking. In addition, the research examined whether clay modelling could promote collaborative learning as well as advance metacognition. Anatomy underpins the basic medical sciences and like others, the authors concur that it should be taught in a multi-faceted manner (Sugand et al. 2010). These measures could, for example, include the use of modelling and multimedia learning tools. Modelling provides a haptic, three-dimensional, and practical method for learning human anatomy (DeHoff et al. 2011). Active learning using clay modelling may lead to more knowledge gained in comparison to passive learning methods, such as didactic lectures as a stand alone. Our participants reflected on this in the study and indicated; "In the process of building a clay model. . .I learned a variety of skills".

The themes that developed from this study related to the educational advantages of clay modelling. These included *advantages of clay modelling, challenges encountered as well as suggestions for future practice*. Our participants experienced the activity as valuable on several levels. It provided a tangible approach to learning anatomy. DeHoff et al. (2011), stated that students need to actively participate in the learning process, to absorb and retain knowledge. This activity engaged the active learning practice by requiring the students to construct the anatomical clay models (Curlewis et al. 2021). In our student group, sentiments, such as, "the use of clay models to study the structures made it easier to identify the structures". were offered. As well as "I was learning while actively playing". Another participant said, "I managed to remember the muscles and their origin, insertions, and functions for the presentation and report without having to refer back to my notes".

Other students in the group commented on the value gained. They stated that the educational value attached to this learning tool is unique and that the learning experience achieved from physically constructing an anatomical model was invaluable. Similarly, a recent systematic review by Curlewis et al. (2021, p. 252) concluded that clay-based modelling is a practical learning technique for human anatomy and should be integrated into the "anatomists" "toolkit". Even though cadaveric dissection provides a hands-on, three-dimensional activity for students, it is a destructive activity rather than a constructive process (DeHoff et al. 2011), and the student perceived the building of the clay model as valuable. The clay modelling activity presented opportunities for students, in groups, to construct new knowledge from an authentic learning event that aligns with constructivist learning theory.

A sub-theme emerged centring on skills development. This was exemplified in the group work. Learning from group tasks as an extracurricular benefit was presented as "The experience taught me to value time management between members and critical thinking came into play when we needed to have a unique presentation". Another comment was "I was excited to be in a group setting as I personally do well in a group setting although I did feel out of my element as I have never had to model anything with clay before". and "There was consistent communication and all of us had the same goal in mind". Thus, learning also occurred in groups using clay-based modelling (Herur et al. 2011; Bareither et al. 2013).

A further theme that was found relates to the clay building process. Here some challenges, obstacles and apprehensions were encountered. The students indicated that they felt inexperienced, unprepared, and regarded the activity as time consuming. "The thought of building an accurate anatomical model out of clay was quite daunting". and "We struggled to mould the clay into a shape that resembles the muscle, or the clay cracked when it was dry, and we had to rebuild it". Despite the initial negative perceptions surrounding the activity, most of the students enjoyed it and found it beneficial to their learning (Curlewis et al. 2021). Some students also indicated that they felt that they were not artistically inclined. Even though anatomical correctness was significant, the students were not required to reproduce exact human anatomy (Curlewis et al. 2021); the students were motivated to rather focus on the actual learning and constructing process of building the models. Therefore, students should not be disheartened if they did not see themselves as an "arty person".

Not all participants experienced the clay modelling as educationally beneficial (Oh et al. 2009; Carlson et al. 2019). One participant commented: "I found the clay modelling less effective than human dissection as a realistic individual knowing the anatomical structure in real time helps more with memorization and identification of the anatomical structures". Similarly, a recent study by Carlson et al. (2019) found that the addition of a hands-on clay modelling activity did not improve student learning, specifically measured by exam performance. This perspective highlights the necessity of exposing the students to multiple learning experiences in coursework.

The overall perspective of the group's experience of educational value was positive, as participants reflected: "I thoroughly enjoyed the experience as a whole and I value it as a learning experience". and "My hope is that it will be an effective learning tool for students in the future". "I would gladly do such a project again, as I gained a significant amount of knowledge from the trial-and-error process, and I believe this experience could help me build an even better model in the future". This was also reflected in the evaluation forms as the majority agreed that it was a valuable learning instrument. This trend was also experienced by Oh et al. (2009).

The group offered several suggestions for application in future research and learning activities. These circled around time management in the course, guidance in terms of planning and clear instructions of tasks, as well as clarification of expectations. They also mentioned that the material was inadequate for clay modelling. These challenges affected the students' mindsets, experienced as feelings of frustration and apprehension which affected their learning significantly. The cognitive load theory (Sweller, 1988) was therefore incorporated to further explore the challenges perceived by the students in this study. This theory implies that learning is most effective when education allows information to be processed effectively (Kirschner, Sweller & Clark 2006) and is therefore relevant to models of instruction, such as clay modelling. Special attention should be paid to avoid cognitive overload to prevent learning being ineffective (Khalil, Paas, Johnson, and Payer, 2005). Cognitive load theorists consider that instructional design will be most effective when it minimizes extraneous load to free up the capacity of working. Special attention should be paid to avoid cognitive overload to prevent learning being ineffective (Khalil, Paas, Johnson, and Payer, 2005). Extraneous cognitive load focuses on how the content is taught and, therefore, does not directly impact on students' learning. This load is thus not necessary for students' learning and can be improved by instructional interventions (Sweller, van Merriënboer & Paas, 1998). The constructional process of clay modelling could be simplified to prevent extraneous load. This includes adequate material, clear instructions, and expectations of the final product of clay modelling. This theory assisted us to make suggestions for the construction process to enhance students' learning of clay modelling as a supplementary method in learning anatomy for future practices.

The limitations of this study relate to the small sample size in one setting being used. Nevertheless, the results of this study resonate with previous research conducted on this topic. Other limitations included the time in the curriculum as well as the timing of the project relating to the assessment of the subject matter were also discordant and may have led to poor motivation for the task. Further studies with larger groups and potential long-term review may prove to enhance a better understanding of this concept and intervention.

7.6 Conclusion

The study confirms that the participants experienced the clay-based modelling in anatomy as a valuable learning tool when compared to traditional passive learning approaches. Students viewed the activity as enjoyable and engaging and felt that they were actively playing while learning anatomy. The students also acknowledged that collaborative learning had a positive impact on their performance as they relied on each other's strengths. Thus, it is clear that active, hands-on learning was not used in isolation but rather in combination with other pedagogies, such as collaborative learning.

Acknowledgements This work was supported by the Fund for Innovation and Research into Learning and Teaching (FIRLT), Stellenbosch University. The team acknowledges Mr. Luke-John Daniels (Division of Clinical Anatomy, Stellenbosch University) who was responsible for capturing the images in the manuscript. Furthermore, the team acknowledges the ANA364 students who participated in the study and credits them for the construction of the impressive clay models.

References

Azer SA, Azer S (2016) 3D anatomy models and impact on learning: a review of the quality of the literature. Health Professions Education 2:80–98. https://doi.org/10.1016/j.hpe.2016.05.002

Bareither ML, Arbel V, Growe M et al (2013) Clay modeling versus written modules as effective interventions in understanding human anatomy. Anat Sci Educ 6:170–176. https://doi.org/10.1002/ase.1321

Carlson D, Chandra S, Hobbs N, Steele J (2019) Clay modeling in a sophomore-level anatomy laboratory: will active learning improve student performance? HAPS Educator 23:358–365. https://doi.org/10.21692/haps.2019.008

Creswell JW (2014) Research design: qualitative, quantitative and mixed methods approaches, 4th edn SAGE

Curlewis K, Leung B, Perera M et al (2021) Clay-based modeling in the Anatomist's toolkit: a systematic review. Anat Sci Educ 14:252–262. https://doi.org/10.1002/ASE.1996

DeHoff ME, Clark KL, Meganathan K (2011) Learning outcomes and student-perceived value of clay modeling and cat dissection in undergraduate human anatomy and physiology. Adv Physiol Educ 35:68–75. https://doi.org/10.1152/advan.00094.2010.-Alternatives

Fitzgerald MJT, Fitzgerald M, Brophy J et al (1979) Purpose-made models in anatomical teaching. J Vis Commun Med 2:71–73. https://doi.org/10.3109/17453057909153892

Georghiades P (2004) From the general to the situated: three decades of metacognition. Int J Sci Educ 26:365–383. https://doi.org/10.1080/0950069032000119401

Gibbs G (1988) Learning by doing: a guide to teaching and learning methods. Oxford Further Education Unit, Oxford

Herur A, Kolagi S, Chinagudi S et al (2011) Active learning by play dough modeling in the medical profession. American Journal of Physiology–Advances in Physiology Education 35:241–243. https://doi.org/10.1152/advan.00087.2010

Khalil MK, Paas F, Johnson TE, Payer AF (2005) Interactive and dynamic visualizations in teaching and learning of anatomy: a cognitive load perspective. The Anatomical Record Part B: The New Anatomist: An Official Publication of the American Association of Anatomists 286(1):8–14

Kirschner PA, Sweller J, Clark RE (2006) Why minimal guidance during instruction does not work: an analysis of the failure of constructivist, discovery, problem-based, experiential, and inquiry-based teaching. Educ Psychol 41(2):75–86. https://doi.org/10.1207/s15326985ep4102_1

Kooloos JGM, Schepens-Franke AN, Bergman EM et al (2014) Anatomical knowledge gain through a clay-modeling exercise compared to live and video observations. Anat Sci Educ 7:420–429. https://doi.org/10.1002/ase.1443

Krontiris-Litowitz J (2003) Using manipulatives to improve learning in the undergraduate neurophysiology curriculum. Adv Physiol Educ 27:109–119. https://doi.org/10.1152/advan.00042.2002

Motoike HK, O'Kane RL, Lenchner E, Haspel C (2009) Clay modeling as a method to learn human muscles: a community college study. Anat Sci Educ 2:19–23. https://doi.org/10.1002/ase.61

Nicholson LL, Reed D, Chan C (2016) An interactive, multi-modal anatomy workshop improves academic performance in the health sciences: a cohort study. BMC Med Educ 16:7. https://doi.org/10.1186/s12909-016-0541-4

Oh C-S, Kim J-Y, Choe YH (2009) Learning of cross-sectional anatomy using clay models. Anat Sci Educ 2:156–159. https://doi.org/10.1002/ase.92

Ross ME, Green SB, Salisbury-Glennon JD, Tollefson N (2006) College students' study strategies as a function of testing: an investigation into metacognitive self-regulation. Innov High Educ 30:361–375. https://doi.org/10.1007/s10755-005-9004-2

Sandars J (2009) The use of reflection in medical education: AMEE guide no. 44. Med Teach 31:685–695. https://doi.org/10.1080/01421590903050374

Schunk DH (2008) Constructivist theory (chapter 6). An Educational Perspective, In Learning Theories, pp 234–277

Sugand K, Abrahams P, Khurana A (2010) The anatomy of anatomy: a review for its modernization. Anat Sci Educ 3:83–93. https://doi.org/10.1002/ase.139

Sweller J (1988) Cognitive load during problem solving: effects on learning. Cogn Sci 12:257–285

Sweller J, van Merriënboer JJG, Paas F (1998) Cognitive architecture and instructional design. Educ Psychol Rev 10:251–296

Zainuddin MZ, Razlan AN, Husin A et al (2019) Sculpting the brainstem: understanding brainstem anatomy using modelling clay. Kelantan International Learning & Innovation Exhibition, In

Part IV

Ethical Considerations

Advances in Digital Technology in Teaching Human Anatomy: Ethical Predicaments

8

Kerri Keet and Beverley Kramer

Abstract

The study of human anatomy has long been inseparable from illustration, as the portrayal of the dissected body alongside a descriptive text supports the understanding of this complex topic. With the advent of new technologies, applications for revealing the details of the human body have increased to include Computed Tomography and Magnetic Resonance Imaging scanning, digital photography and videography, and three-dimensional (3D) printing. In addition, the distribution of the captured image has become widespread in both online teaching and in hard copy textbooks. While these digital images and applications are invaluable for online teaching, particularly during pandemics, anatomists must consider whether the use of the images and data derived from human bodies (either donor or unclaimed) is ethical, as it is not always clear whether permission has been obtained for their creation.

Questions regarding the ethical sourcing and use of images will become more urgent as the utilization of electronic data becomes routine. Particularly concerning is the distribution of images and videos in the public domain on social media, where the context and link with education may be severed. Other considerations include respect for the privacy of the deceased, ownership of images, and commercialization. An initial question that should be posed is "Who is responsible for the oversight of digital and 3D printed models of bodies?" The ethical use of such images needs to be considered by the global anatomy community in order to ensure that anatomists do not retreat into the ethical mire of their forefathers.

Keywords

Digital technology · Illustration · Anatomical education · Dissection · Ethics · Body donation programmes

Ethics is knowing the difference between what you have the right to do and what is right to do. Potter Stewart.

K. Keet (✉)
Division of Clinical Anatomy, Department of Biomedical Sciences, Faculty of Medicine and Health Sciences, Stellenbosch University, Stellenbosch, South Africa
e-mail: kkeet@sun.ac.za

B. Kramer
School of Anatomical Sciences, Faculty of Health Sciences, University of the Witwatersrand, Johannesburg, South Africa

L. Shapiro, P. M. Rea (eds.), *Biomedical Visualisation*, Advances in Experimental Medicine and Biology 1388, https://doi.org/10.1007/978-3-031-10889-1_8

8.1 Introduction

The physical study of the human body and its depiction in graphical form are inseparable (Calkins et al. 1999). As the human body is complex and the understanding of three-dimensional (3D) relations of structures is crucial, drawings and models facilitate its study (Fredieu et al. 2015), especially for students who do not have access to dissection. In the last few decades, digital images, virtual 3D models, and 3D prints have increased in use due to improvements in technology and the extensive scale of access and distribution (Bisht et al. 2019). Technological representations of cadavers available for use range from simple digital photographs created in anatomy departments to commercial virtual dissection software tools, available on virtual dissection "tables," virtual and augmented reality, and Smartphone apps (Bisht et al. 2019; Longhurst et al. 2020; Turhan and Yakut 2020; Uruthiralingam and Rea 2020; Valdez Mayo et al. 2021). Three-dimensional digital models have been created by restacking digital photographs and surface scans obtained from 3D scanners, or scans obtained from Computed Tomography (CT) and Magnetic Resonance Imaging (MRI) (Waldby 2000; Fredieu et al. 2015), overcoming the cognitive skills required to translate a two-dimensional (2D) image into a 3D object (Waldby 2000). These 3D models may be dissected virtually using computer software, or printed on a 3D printer to provide physical models (Fredieu et al. 2015; Bisht et al. 2019). The 3D digital models are described as being superior to commercially available plastic models, as the 3D models are derived from many individual donors and are thus more realistic and accurate in their representation of the range of variations and pathologies present in the population. Commercial models tend to be basic and represent the "normal" anatomy only (Strkalj and Dayal 2019).

The benefits of teaching and learning with digital images and virtual models reside in the ease of access (Attardi and Rogers 2015) and distribution to learners (Zargaran et al. 2020), the ease of reproducibility (Mcmenamin et al. 2014), the use as an adjunct to self-directed learning (Zargaran et al. 2020), the ability to rotate and view structures from many angles, and being able to "undo" a virtual dissection if a mistake is made (Bisht et al. 2019). Virtual medical procedures can even be practised on virtual models (Robb 2000). These benefits are said to be in contrast to dissection which deconstructs the tissues, allowing a dissection to be performed once only, and exposes staff and students to harmful preservation chemicals (Raja and Sultana 2012). Images and models also allow human variations to be recorded for future teaching and learning, and may compensate for a shortage of body donors in some parts of the world (Cornwall 2016, 2017; Strkalj and Dayal 2019). During the coronavirus 2019 (COVID-19), pandemic the distribution of images and virtual models online allowed students to continue their anatomical studies while institutions and dissection laboratories were closed (Jones 2020). Existing online databases of 3D digital models may also be accessible to the public, where digital 3D models created by individuals are uploaded, stored, and shared (Ulguim 2018).

However, the ethical and legal implications of this "electronic immortalisation" (Cornwall 2017) of images from human remains need to be considered, particularly within the context of the past unethical practices and lack of ethical reflection within anatomy (Jones 2019a). Anatomists should apply careful ethical consideration to the way in which human bodies are obtained, treated, and retained for both teaching and research. Body donor programmes, which have become more evident globally in recent years, rely on the trust of their communities to participate and build these programmes.

Within the last few decades, anatomy has transitioned toward becoming a discipline with a stronger ethical foundation, being recognized as the first exposure of health sciences students toward death and the body, ethics, professionalism, and teamwork (Strkalj 2016). Anatomists are starting to question whether it is ethical to make and distribute images obtained from human material without the direct consent of the individual

(Cornwall et al. 2016; Jones 2019b). However, how the use and distribution of these images and 3D prints are ethically perceived by the general population of anatomists has not yet been established. Digital elements may have a higher risk of objectification if they are isolated from their original context. Distancing from physical interaction with the donor and commemoration services (Harries et al. 2018; Jones 2019a; Smith and Hirst 2019), as was the case of online learning during the COVID-19 pandemic may result in dehumanization (Jones 2020).

8.2 Historical Overview of Dissection and Illustration

The portrayal of the human body in images has been influenced by social and technological changes throughout history (Tsafrir and Ohry 2001). Before the development of preservation techniques, drawings, and models were used to allow the long-term study of the body (Fredieu et al. 2015). Images were created on stone, bamboo, silk, and metal before the invention of papyrus and paper (Tsafrir and Ohry 2001). The Greeks are recorded as the first society to dissect the human body in the third century BC, with Galen (130–200 AD) producing the main anatomical text on which anatomists depended until the sixteenth Century (Tsafrir and Ohry 2001; Jones and Whitaker 2009; Ghosh 2015a). However, the anatomy in this text was largely based on dissections of animals and thus differed somewhat from human anatomy (Jones and Whitaker 2009). During the Renaissance period (14th–17th Centuries) in Europe, there was an explosion of anatomical images produced initially for artistic reasons, but it was soon recognized that these were invaluable in facilitating the understanding of anatomical structures (Tsafrir and Ohry 2001). Leonardo da Vinci dissected human bodies and produced approximately 750 drawings of incredible detail, which survive in the Royal Library at Windsor (Tsafrir and Ohry 2001). However, it was Vesalius, considered to be the father of modern anatomy, who produced a massive text, the *De Humani Corpororis Fabrica* in 1542, which contained detailed wood engraved illustrations purportedly by the artist Jan van Calcar (Zampieri et al. 2015). This was a landmark in medical illustration (Calkins et al. 1999) as the plates were accurate and meticulous. The illustrations in the *Fabrica* were based on dissections Vesalius performed on bodies stolen from graves (Jones 2019a) and of a body taken from the gallows (https://web.stanford.edu/class/history13/Readings/vesalius.htm). The *Fabrica* became the foundation of anatomy texts which followed (Tsafrir and Ohry 2001).

Another important and well-known illustrated text that has persisted into modern times is *Gray's Anatomy*, which was first published in 1859 (Tsafrir and Ohry 2001; Strkalj 2008). The original illustrations in *Gray's Anatomy* were made by a demonstrator in anatomy, Henry Vandyke Carter, from bodies which "came out of the hospital mortuary or out of a workhouse where they had died" (Smith 2014). These were the bodies of the unclaimed, of which the dissection was legalized in the 1830s in Britain and the United States of America (USA) in response to public riots against grave robbing (Garment et al. 2007; Halperin 2007; Jones and Whitaker 2009; Ghosh 2015a). Up until this point, the legal supply of bodies was exclusively from executed prisoners, which failed to meet the high demand required for anatomical study (Hildebrandt 2008). Legalizing the use of unclaimed bodies reduced the shortage, thus ending grave robbing. However, the use of unclaimed bodies continued anatomy's association with the exploitation of the poor and marginalized in society (Garment et al. 2007; Halperin 2007; Jones and Whitaker 2009; Ghosh 2015a; Claes 2018). The bodies from which illustrations have been made, despite providing invaluable imagery, are neither acknowledged nor commemorated in some current major health sciences textbooks (Richardson 2008).

Until the advent of photography in the fourth Century (camera obscura) (Taggart 2020) and medical imaging in 1895 (Rubin 2017), illustrations for teaching anatomy were merely portrayals of dissected bodies (Richardson 2008). However, with the advent of photography and the advancement of imaging technology, the illustrations that are used in recent times are no longer "portrayals," which may have been influenced by the artist's interpretation, but are the exact copy of an individual's anatomy. Thus, the question of the provenance of the body and of specific consent to the images and photographs must arise.

8.3 Current Source of Bodies for Anatomical Education and Research

The use of unclaimed bodies persisted as the sole source of anatomical material until the middle of the twentieth Century, when institutions started forming voluntary body donation programmes based upon the consent of the person before their death (Garment et al. 2007). However, in order to meet the demand for dissection in many parts of the world, with the exception of the United Kingdom (UK), Austria, Denmark, France, Sweden, Australia, New Zealand, Canada, Germany, and Switzerland (McHanwell et al. 2008; Riederer et al. 2012; Riederer and Bueno-López 2014; Riederer 2016; Habicht et al. 2018; Jones 2019a; Farsides and Smith 2020), unclaimed bodies are still in use. The use of unclaimed bodies is considered ethically unacceptable as no consent was provided while the person was alive, nor by their relatives after death (Dasgupta 2004; Jones 2012; Jones and Whitaker 2012; Kahn et al. 2017; Caplan and DeCamp 2019; Chia and Oyeniran 2020; Peeler 2020). The International Federation of Associations of Anatomists (IFAA), which is the professional association to which most global anatomical associations subscribe, has recommended that "informed consent must be obtained in writing before any bequest can be accepted" (IFAA 2012). This, therefore "excludes the use of unclaimed bodies" for anatomical training and research purposes (IFAA 2012).

Unclaimed bodies are, however, part of many "legacy collections" including those at universities in South Africa (L'Abbé et al. 2005; Dayal et al. 2009; Alblas et al. 2018; Maass and Friedling 2019; L'Abbe et al. 2021), where skeletal remains have been retained. Digital photographs and micro-CT scans from these collections are used in research and teaching (L'Abbe et al. 2021). While consent may not have been obtained from the individuals of the unclaimed bodies, institutional ethics committees are in place to oversee that their use in research complies with international ethical standards.

In addition, institutions without willed body donation programmes may import bodies from private donation companies in the USA (Habicht et al. 2018). Nigeria is the only known country that still uses bodies of executed persons (Anyanwu et al. 2011; Habicht et al. 2018). During the National Socialist regime (1933–1945), Germany reverted to using the bodies of executed prisoners for anatomical dissection. A high-quality anatomy atlas, Pernkopf's Atlas, resulting from this period is still in use today and has been contested for ethical reasons (Hildebrandt 2009a, b, c, 2019; Seidelman 2012; Ghosh 2015a).

In 1994 the dark history of Pernkopf's Atlas, originally published in 1937, was revealed, as earlier editions of the atlas contained Nazi insignia in some of the artists' signatures (Seidelman 2012). The source of the bodies depicted in the atlas was from the University of Vienna, which was known to have legally received bodies for anatomical teaching and research from Nazi concentration camps (Israel and Seidelman 1996). Eduard Pernkopf, after whom the atlas was named, was dean of the medical faculty and later the rector of the University and an ardent National Socialist (Seidelman 2012). Four artists were used to create the 800 illustrations of dissected bodies for the atlas, which was printed employing the latest technology of four-color offset lithography (Hildebrandt 2009b; Seidelman 2012). The atlas contains arguably the most detailed illustrations of all anatomical publications, including current atlases (Yee et al. 2019a).

The Pernkopf debate regarding the unethical use of prisoners who had not provided consent for their bodies to be used for the atlas was sparked in the 1990s (Israel and Seidelman 1996; Riggs

1998). Printing of the atlas was halted on ethical grounds in 1994 by Elsevier, who subsequently donated the original images to the Medical University of Vienna in 2019 (Czech et al. 2021). The University of Vienna was requested to enquire into the origins of the bodies by the Jewish Holocaust Remembrance Authority, which also requested that the victims be commemorated in future editions (Atlas 2001). It remains unclear who the individual bodies depicted in the atlas were, although it is believed that they may have been executed political prisoners of the Viennese court and not from concentration camps (Atlas 2001).

The question of the morality of the continued use of the atlas also arose from clinicians who considered it superior to existing texts and therefore invaluable for planning surgical procedures (Polak et al. 2017; Yee et al. 2019a, b). The "Vienna Protocol" was published in 2017 (Polak et al. 2017) from a special symposium regarding the treatment of holocaust-era human remains, including the atlas (Polak et al. 2017), and concluded that the atlas may be used if its historical origins are disclosed (Polak et al. 2017). The consequences of the unethical sourcing of the bodies used to produce Pernkof's Atlas include twenty years of ongoing intense debate, the tarnishing of the reputation of the University of Vienna and other German anatomy departments, revisiting of trauma for survivors of the holocaust and their relatives, and the moral dilemma of surgeons in using the atlas (Hildebrandt 2006; Yee et al. 2019b).

The unethical origin of the bodies used to construct teaching resources is not limited to the distant past. In the late 1990s, the body of an executed prisoner was used to produce the digital resource known as the Visible Human Project (VHP) in the USA (Spitzer et al. 1996). The VHP comprises downloadable MRI and CT scans, as well as high-resolution photographs from two whole bodies (Spitzer et al. 1996), with a third body currently being processed (ADInstruments 2019). The bodies were CT and MRI scanned after death, following which they were sectioned into 1 mm (for the male) and 0.3 mm (for the female) slices that were digitally photographed (Ackerman 2017). The datasets (composed of CT and MRI scans and corresponding photographs) are freely available to download from the National Library of Medicine's website (Ackerman 2017). While the datasets were initially intended for researchers who had to obtain a license (Ackerman 2017), they are now publicly available (https://www.nlm.nih.gov/research/visible/visible_human.html). The availability of free datasets over the internet eliminates financial and geographical constraints, allowing educators and researchers in developing countries access to digital anatomical materials that are not available in their home institutions. The datasets are primarily used for anatomical training, surgical planning, and biomechanical modeling (Waldby 2000). However, nonmedical uses include automobile safety research (Ackerman 2017), while nonscientific uses include incorporation into art exhibitions, computer games, advertising, and film special effects, many of which are commercial applications (Waldby 2000; Ackerman 2017). Thus, accompanying advances in technology, the dissected human body has become commodified and used for nonmedical purposes (Waldby 2000).

A concern related to the Visible Human Project is the source of the bodies, particularly the male (Roeggla et al. 1995). Joseph Paul Jernigan was an executed prisoner who donated his body to the Texas State Anatomy Board shortly before execution by lethal injection (Spitzer et al. 1996). Questions around his uncoerced consent have been raised, as prisoners on death row are considered as vulnerable (Hildebrandt 2008). It is also unlikely that he was informed about donating specifically to the VHP, rather giving broad consent for the use of his body in teaching and research (Brown 1999; Atlas 2001). Given the historical disreputable sourcing of cadavers for anatomical training, it is considered ethically unacceptable to use the bodies of executed prisoners, even if they have expressed a desire to donate and the law allows this (Hildebrandt 2008). The IFAA recommendations specifically exclude donations by prisoners who are to be executed (IFAA 2012). In addition, in contrast to the standard practice of body donation programmes, in which the donor is anonymized, the identity of the male donor was revealed by journalists after the National Library of Medicine published the date and time of death, and name of the prison at which he was executed (Hildebrandt 2008). In contrast, while the identity of the first female donor is not known, the issue of voluntary informed consent is also of concern, as first-person consent was not provided by the donor prior to death. Instead, her family donated her body specifically to the VHP after she died from natural causes (Waldby 2000). This raises the issue of proxy or second-person consent for donation, which is no longer permitted by law in some countries, such as the United Kingdom (UK-HTA 2021), Germany, Austria, and Switzerland (Champney et al. 2019), but is permitted in the USA (Dalley et al. 1993), Australia (Aus-AA 1977, 1983; Hutchinson et al. 2020), and in South Africa (SA-NHS 2004; Hutchinson et al. 2020). While proxy donation may be legal, it is not ethical according to IFAA recommendations (IFAA 2012). In contrast, while the identity of the second female donor in the VHP is known, this disclosure was with fully informed consent, as Susan Potter specifically donated her body to the VHP and participated in a National Geographic documentary about her decision to donate (Newman 2018).

8.4 Legislative and Ethical Frameworks for Digital Images and Models of Deceased Bodies

Currently, there is no specific legislative or ethical framework for the sourcing, use, storage, distribution, or deletion of images from deceased human bodies (Cornwall et al. 2016). As legislation generally lags behind technological developments, there is little guidance or regulation for digital images of donors after death, including reconstructed 3D digital and printed models (Thomas 2002; Cornwall 2016, 2017; Cornwall et al. 2016). Much of the existing legislation applies to the protection of living persons only, such as the USA federal code 45 CFR part 46 (Tassé 2011), and the Protection of Personal Information Act 4 of 2013 of South Africa (Republic of South Africa 2013). In a recent description of the use of digital scans of donated and unclaimed individuals, governed by the National Health Act of South Africa to produce 3D images, consent has been interpreted to mean that "a body (and its associated data) can be used for tissue transplantation and research and teaching now and into the future," with "the future" being considered to include "advancement of technology," such as 3D reconstruction, scanning and 3D printing (L'Abbe et al. 2021). Similarly, Australian legislation regarding donated human bodies for the study of teaching anatomy has been interpreted to indicate that the 3D copies of body parts "are produced as a teaching aid and therefore come under the purpose of 'teaching of anatomy'. . . ." (Mcmenamin et al. 2014). While the law may appear to protect the making of the 3D images and printed body parts, this differs from the view of some anatomists, who argue that there are ethical issues around the sourcing and commercial use of images and 3D models (Cornwall 2016, 2017; Jones 2019b). The ethical sourcing, storage, use (including distribution), and disposal of images is relevant to anatomists as it may impact public trust on which body donation programmes depend (Cornwall et al. 2016).

8.5 Ethical Considerations Relating to Digital Technologies

8.5.1 Ethical Status of Digital Images and 3D Printed Models

The ethical status of 3D prints from human bodies is complicated, as while the scans are derived from images of a physical human body, the models are generated from plastic (Smith and Hirst 2019), which may lead to the argument that the model is no longer "human." This is in contrast to plastinated specimens, which are composed of human tissues in which the liquid components have been replaced with silicone polymers (Jones 2019a). The 3D printed models, however, represent the exact likeness of a specific donor (Jones 2019b). Other models representing the exact likeness of a person, such as face masks (Houlton and Billings 2017) and body casts of the human body are considered as "human remains" by the South African Department of Arts and Culture (Western Cape Heritage Resource Management 2018) and would thus require the individual's consent to be created. Jones (2019a, b) provides the following helpful hierarchy to guide the ethical discussion of 3D prints and other technologically derived uses of human bodies, where intact bodies are the highest on the scale of ethical concern and plastic skeletons, the lowest:

> intact body > dissected body > body parts > 2D images of body/body parts > plastinated body (plastinate) > plastinated body parts > computer simulations of the human body and its parts > 3D printed body parts > purely artificial models (including anatomical wax figures; artificial models; plastic skeletons) (Jones 2019a, b).

It is traditional at many institutions to provide health sciences students with a "bone box" which contains a disarticulated skeleton for learning osteology. Some of these skeletons are of dubious provenance and were imported from India (Carney 2011). While it is agreed by most anatomists that the use of real human tissue (cadavers) is the best way to learn and teach anatomy (Kurt et al. 2013; Saltarelli et al. 2014; Pather 2015) due to shortages of material, some universities may provide students with boxes of 3D printed bones instead of real bones (Irish et al. 2017, cited in Smith and Hirst (2019)), or other artificial replicas made from casts of bones (https://boneclones.com/). A 3D osteological collection has recently been produced and is in use at an Australian institution (Strkalj and Dayal 2019). While some researchers and educators have produced 3D prints from donors, although whether specific consent was provided for this use is not clear (Smith et al. 2018), others have used freely available datasets from Internet databases in which the provenance of the material and the corresponding consent has not been declared (Sander et al. 2017; Smith and Jones 2018).

The ethical considerations around the use of images, digital models, and 3D prints of human material are unique and arise in addition to those relating to the physical body (Cornwall 2016). However, until recently, contemporary digital data has received less ethical attention (Cornwall et al. 2016; Harries et al. 2018). Where uncertainty over the status of human-derived images/models/3D prints arises, while these "items" may not be composed of human tissues, they retain their human connection. It is thus important that ethical consideration and proper treatment are afforded to these derived materials as they have the potential to objectify and exploit the body which has been placed with trust in the hands of anatomists (Cornwall et al. 2016; Cornwall 2017; Jones 2019a). Caffell and Jakob (2019) therefore advocate that images and models be treated with the same ethical considerations as human remains.

8.5.2 Informed Consent

Donors generally sign a written and witnessed informed consent form before bequeathing their bodies to a specific institution or anatomical board. Legislative requirements for who may consent to body donation differ between different countries, with some countries requiring first-person consent only, while others allow for consent from relatives or other authorized persons, and yet others permit the use of unclaimed bodies from whom no consent has been obtained (McHanwell et al. 2008; Jones and Whitaker 2012; Riederer 2016).

Currently, there is no consensus on the level of detail that should be provided to donors in the consent process (Zealley et al. 2021). Some argue for full disclosure, detailing all the uses to which the body will be applied, claiming that it is the only means of ensuring truly informed consent, while others state that too much information about the process may deter potential donors (Riederer and Bueno-López 2014; Winkelmann 2016; Farsides and Smith 2020). However, Olejaz and Hoeyer's (2016) interviews with registered body donors revealed that donors knew what would happen to their bodies, and that the extent of this knowledge did not deter their willingness to donate. An argument for "broad" consent, which is a general statement of donation for teaching and research, is that the range of future uses may not be known at the time of donation (Bernanke 2011). Donor's motivations are primarily altruistic although non-altruistic factors, such as the high cost of funerals, may also play a role in their decision to donate (Fennell and Jones 1992; Greene 2003; Papagaroufali 2006; Bolt et al. 2012; Cornwall et al. 2012; Olejaz and Hoeyer 2016; Gürses et al. 2019; Jiang et al. 2020). Several studies have examined the uses for which donors have approved donating their bodies (Greene 2003; Gürses et al. 2019). These uses included teaching and research, long-term retention of organs, parts and bones, as well as their medical history and personal information (Gürses et al. 2019). When asked whether they would prefer to direct their donation toward a specific purpose for which their body is used, many donors left the decision to the institution (Greene 2003; Gürses et al. 2019). In a study of anatomy students' opinions toward body donation and the willingness to consent to the production of images and 3D prints, the majority indicated that they would prefer to donate images of their body rather than donate their physical body (AbouHashem et al. 2017). As yet, however, no studies have investigated registered body donors' attitudes toward images, digital models, and 3D models produced from their donations (AbouHashem et al. 2017). The use of virtual body donors has been recently suggested, where the consented digital images derived from patients are utilized for anatomical teaching (Aso-Escario et al. 2019).

In a review of 110 consent forms of institutional body donation programmes in the USA, Zealley et al. (2021) found that only 10% of programmes obtained consent for images to be taken (Zealley et al. 2021). Although the Human Tissue Authority of the UK states that images and videos fall outside of the Act's scope, obtaining specific consent for images from either the donor or their relatives is strongly recommended (UK-HTA 2021). However, once again there is no guidance regarding the sourcing, use, storage, distribution, disposal, privacy protection, ownership, or commercial uses. Whether consent on donor forms should be extended to include images and 3D prints needs discussion and consensus from anatomists and other stakeholders (Jones 2019b). The possible creation of photographs and digital images is specified in donor forms at some, but not all, institutions in South Africa. Cornwall et al. (2016) and Jones (2019a, b) both suggest erring on the side of caution and obtaining informed consent for the sourcing and use of 3D prints (Cornwall 2016; Jones 2019b).

The extent to which private body donation companies in the USA obtain consent for technological uses is not known, although at least one company mentions images in their consent forms (Innes 2019). Many commercial companies do not declare the provenance of the bodies, nor whether specific informed consent was given, creating an ethical dilemma for anatomists who wish to use these programmes in their teaching (Miller and Lewis 2016; Jones 2019b). Jones (2019b) asserts that informed consent is what allows dissection to be socially accepted instead of being viewed as mutilation of the body, and further argues that it is ethically unacceptable to obtain images and 3D prints from unclaimed bodies as the individuals have not provided any consent for donation (Jones 2019b).

Another consideration for consent is the local as opposed to the global usage of bodies that have been donated to a specific institution. Transport of bodies or parts thereof across borders is generally regulated. Thus, the question arises as to how the electronic transfer of data derived from body donors could or should be regulated, as publication on the Internet is global. Distancing breaks community relationships and increases the risk of commodifying the donor (Champney et al. 2019; Jones 2019b). Jones (2019b) suggests that the distribution of 3D prints to other institutions should be accompanied by a consent statement and acknowledgment of the donor, which will assist with keeping the connection of the 3D model to the donor. This consideration should be extended to include images and digital models.

Anatomists will thus be assured of the provenance of any 3D print or commercial software that they may utilize (Jones 2019b), and donors will be aware of and agree to the possibility of any technologies derived from their bodies being used in a different location from that to which they donated. In relation to the public use, Smith and Hirst (2019) raise the question of whether using 3D printed models of human remains in environments that would not be considered appropriate for bodies, such as in schools or private homes is acceptable.

8.5.3 Privacy

Respect for the donor's privacy and dignity are important aspects of all components of body donation, and in the modern era, should include the sourcing, storage, use, and disposal of images and 3D models (Jones 2019b; Hennessy et al. 2020). The potential for identification of a person extends beyond the face to include tattoos, scars (Jain and Lee 2009), orthopedic (Khartade et al. 2021), and oral (Chugh and Narwal 2017) implants, and any information presented alongside the image, such as age and cause of death. Thus, the secure and protected storage of images and accompanying information in the form of metadata is important (Márquez-Grant and Errickson 2017). Recent digital resources generated from data derived from bodies may disregard the donors' privacy, as the faces of the individuals are visible, with further identifying information accompanying the images. In addition, whether an image of a bone should have different privacy and ethical considerations to an image of a dissected body or part thereof requires discussion, as suggested by the hierarchy proposed by Jones (2019b), as do photographs and realistic line drawings (Harries et al. 2018). A survey of the Portuguese public found that less than half of the respondents considered 3D digital models of skeletal elements to be no different from real bones. However, more than half of the respondents felt that the models should not be in the public domain, with three-quarters agreeing that a description and context should always accompany the model (Campanacho and Alves Cardoso 2018).

8.5.4 Ownership

The important question of ownership of digital images, models and 3D prints from donor bodies requires discussion (Cornwall 2016, 2017). In many countries around the world, the human body is not legally considered as property (Lenk and Beier 2012; Nie and Jones 2019). In South Africa, the head of the department in which the bodies are located is considered to be the legal custodian, but custodianship is not the same as ownership. There are no firm rules regarding ownership (Mahomed et al. 2013) in South Africa, while in the USA, the body cannot be legally owned. However, those in possession of the body have a "quasi-property" right toward the respectful disposition (Rao 2000; Schiefer 2019). In contrast to this view, dissected bodies and parts that were stolen from a UK institution were considered to be the property of that institution, on the basis that they were produced by the labor of the employees of the institution (Herring and Chau 2007).

It is likely that 3D digital and 3D printed models could be categorized as property in the future, subject to copyright legislation, as considerable labor and time go into their production (Cornwall 2016; Harries et al. 2018; Jones 2019b; Smith and Hirst 2019). If so characterized, then should the images and 3D prints, derived and reflecting the characteristics of donors, be considered as purchasable commodities and/or disposable property (Jones 2019b)? To whom would this "property" belong? Do the derived images/3D models belong to the donor, the donor's family, the person who created the image or 3D print, or to the institution to which the donor was physically donated (Smith and Hirst 2019; Hennessy and Smith 2020)? Anatomists should reflect on the provenance of the body from which the images or 3D models were obtained and discuss the issues of ownership and intellectual property rights, with the goal of reaching consensus and developing guidelines that would inform legislation.

8.5.5 Manipulation and Editing of Digital Images and Models

The ethical implications of editing or modifying images from human remains which are subsequently 3D printed also require consideration. According to one online digital database's terms and conditions, individuals uploading digital models give the database the right to "...use, reproduce, publicly display, distribute, adapt, create derivative works of and otherwise exploit User Content for any and all purposes..." (https://sketchfab.com/terms). Modifications could include simple changes in size or involve altering the exact depiction of an individual donor by introducing changes to the structures illustrated in the image (Márquez-Grant and Errickson 2017; Jones 2019b). Alterations could be made for scientific reasons, for example, to illustrate a congenital defect (Jones 2019b), or for artistic and/or decorative purposes, such as 3D printed plant pots in the form of a human skull (https://3dprintingrevolution.com.au/products/skull-planter-pot), vertebrae candle holders, and jewellery derived from human bones, which could then be sold (Errickson and Thompson 2019). This raises further questions regarding whether images and 3D prints derived from human bodies should be sold to the public (Cornwall 2016; Errickson and Thompson 2019). Should the latter digitized items be restricted from trade, as are physical human remains, should they be traded for educational purposes only, or should there be free trade for any purpose, including art or decorative effects? Currently, there are no restrictions placed on who can purchase anatomical textbooks or subscribe to virtual anatomy software. Will 3D prints derived from human bodies also soon be available for sale on popular online shopping platforms?

In addition, if an image or 3D model is altered or modified, would the editor be protected by copyright and/or intellectual property laws, and be entitled to property rights of the modified model (Smith and Hirst 2019), even though the original model was based on a specific person's likeness?

8.5.6 Commercialization

Images from deceased bodies have been used for commercial purposes in textbooks for at least two centuries (Ghosh 2015b). Technological advances have broadened the scope of commercialization to include photographic atlases (both in print and online), online video atlases such as Acland's Video Atlas of Anatomy, digital dissection "tables" (Anatomage), the sale of 3D printed models, and other virtual and augmented reality software programmes, of which there are many (Cornwall et al. 2016; Jones 2019b; Hennessy and Smith 2020). However, as with images distributed on social media, there is often no statement of the source of the bodies and whether informed consent was provided for the use and distribution of the images, including financial gain (Cornwall 2017; Hennessy et al. 2020). While the creators of the datasets and the relatives of the donors may not be receiving any royalties (Chung et al. 2015), companies are profiting from the sale of these virtual cadavers. Unless explicitly informed, donors may not expect that their altruistic donation will yield financial gain for the institution or a third-party company (Jones 2019b). According to Cornwall et al. (2016), commercialization has the greatest potential for exploiting donors and harming the institution's relationship with donors' relatives if consent for commercial uses has not been obtained. Relatives will likely feel exploited and/or entitled to a proportion of the profits, as was the case of the relatives of Henrietta Lacks from whom HeLa cells were derived (Skloot 2010). In addition, there may be local legislative restrictions on commercialization to which adherence is required (Cornwall 2016; Cornwall et al. 2016). Once again, the recommendations of the IFAA (IFAA 2012) state that there should be "no commercialisation in relation to bequests of human remains" for teaching and research. While several authors have addressed the increasing trade of physical human remains and the need for regulation in this legal gray area (Cheney 2006; Champney 2016, 2019; Grow and Shiffman 2017; Champney et al. 2019), less attention has been devoted to the commercialization of virtual models (Cornwall et al. 2016).

8.5.7 Distribution of Images and Digital Models Online

The rapid distribution of images on the global Internet is extensive and swift, with the potential for many copies of the material to be reproduced and published in the public domain, such as on digital databases and social media (Cornwall et al. 2016; Cornwall 2017; Hennessy et al. 2020). Digital data may become immortal, which once again raises questions about who owns the data and who is responsible for the protection of the donor's privacy and curation of the information (Cornwall 2017). The 3D digital model files on public online databases can be downloaded by anyone depending on the permission settings (Ulguim 2018), and used as either digital models or to generate 3D prints (Márquez-Grant and Errickson 2017). Some databases prohibit non-scientific uses and printing of their models, although how this is regulated is not clear (Márquez-Grant and Errickson 2017; Smith et al. 2018). While encouraging data sharing for teaching and research purposes, it is difficult to track or control the use and distribution of digital images (Ulguim 2018; Smith and Hirst 2019). Tracking is far easier with physical parts of human bodies, which are labelled and recorded by an institution in cases of distribution and disposal. As with other public platforms, there may be little contextual information linked to a digitally produced model as metadata, increasing the risk of dehumanization and objectification (Harries et al. 2018; Ulguim 2018). In addition, there is usually little information regarding the provenance of an individual from which the model was generated (Ulguim 2018). Unless there is an accompanying statement that the donor provided informed consent for the distribution of their data online, the ethics of using such a model remains questionable (Smith et al. 2018; Jones 2019b). Anatomists should therefore have a valid educational rationale for obtaining and disseminating images from the deceased (Márquez-Grant and Errickson 2017; Errickson and Thompson 2019; Jones 2019b). Whether images and models should be manipulated and used for nonmedical and commercial purposes are topics that require discussion (Cornwall 2016; Jones 2019b). Even if used respectfully for educational purposes, the foremost ethical issue to consider is whether the donors provided consent for the acquisition of images and 3D prints from their bodies and if so, for what purpose(s).

8.5.8 Public Display of Human Remains: Focus on Social Media

In parallel with technological developments within the discipline of anatomy are the advances in global communication and the development of Web 2.0, which have been increasingly used by anatomists and students as an adjunct to anatomy teaching (Chytas 2019; Douglas et al. 2019; Rai et al. 2019; Curran et al. 2020). Web 2.0 refers to social media websites that allow users to create their own content and "share" content from other users on their own profiles (Hennessy and Smith 2020). Reliance on these sites, which include YouTube (Google LLC, San Bruno, CA), Facebook (Facebook, Inc., Menlo Park, CA), Twitter (Twitter Inc., San Francisco, CA), and Instagram (Facebook, Inc., Menlo Park, CA) among others, for teaching and learning anatomy have increased during the COVID-19 pandemic (Dost et al. 2020; Longhurst et al. 2020; Turhan and Yakut 2020; Caruso 2021; Valdez Mayo et al. 2021). The use of social media for teaching and learning anatomy has been positively rated by the majority of students who utilize these platforms and is said to increase student engagement with content (Jaffar 2012, 2014; Hennessy et al. 2016; Pickering and Bickerdike 2017). Anatomy lecturers may create and upload images from their own content, or use existing content on social media in their teaching (Jones 2016;

Hennessy and Smith 2020). Students also browse social media sites to find images and videos that supplement their learning (Hennessy and Smith 2020). In some cases, students themselves may be creating and disseminating content derived from human remains (Jones 2016). Consequently, concerns have been raised regarding the educational quality of the resources (Chytas 2019; Douglas et al. 2019), and the appropriateness and ethics of displaying human remains on public sites (Jones 2016; Douglas et al. 2019).

Ethical concerns regarding the display of physical human bodies and derived images, and 3D models in the public domain include provenance, informed consent, the reason for the post (educational or voyeuristic), dignity and respect, and privacy of both donors and their relatives (Brooks and Rumsey 2007; IFAA 2012; Jones 2016; Errickson and Thompson 2019; Smith and Hirst 2019). Donors should provide informed consent for the distribution of images in the public domain, as this use falls outside of the scope of teaching and research in a health science setting (Jones 2016; Hennessy et al. 2020). The rationale for posting the image in the public domain, for example, on social media, should be considered carefully as once it is posted, there is no control over how it will be perceived, used, and further distributed (Jones 2016; Márquez-Grant and Errickson 2017; Errickson and Thompson 2019). Images distributed on the Internet may become immortalized as they can never be completely deleted, thus leaving a "digital footprint" (Cornwall et al. 2016; Cornwall 2017). Given these concerns, some authors question whether it is appropriate to post images and videos of deceased bodies on social media, regardless of whether consent was obtained (Hennessy et al. 2020). The importance of this discussion should not be overlooked as the display of physical remains in public, whether in museums or in plastinated exhibitions or for public dissection, is increasingly being contested (Brooks and Rumsey 2007; Jones 2017, IFAA 2021). Unfortunately, many of the images of bodies in the public domain are entirely devoid of information regarding provenance and consent (Barry et al. 2016; Miller and Lewis 2016; Douglas et al. 2019; Hennessy et al. 2020).

Individuals in the health sciences field are anxious regarding the perception of the lay public to images of dissected bodies on social media (Cornwall et al. 2016; Rai et al. 2019; Hennessy et al. 2020), and what impression this will give to potential body donors about the professional and ethical attitudes of body donation programmes (Cornwall et al. 2016). In a survey that investigated the types of physical remains that visitors would prefer not to see in a museum, "medically preserved remains" was rated the highest (Brooks and Rumsey 2007; Harries et al. 2018), which likely refers to dissected body parts preserved in formalin.

Additional disquiet relates to the unethical and unprofessional social media posts of staff and students posting "selfies" of themselves posing with bodies, body parts, or bones (Bond 2013; Mullick 2018; Errickson and Thompson 2019), further removing the link to the altruistic donor or even worse, the unclaimed body. Hennessy and Smith (2020) suggest that given the continued reliance on unclaimed bodies worldwide, it is likely that many images appearing on social media are from unclaimed individuals (Hennessy and Smith 2020), which is unethical as no consent was provided and they have no relatives to speak on their behalf (Lemos et al. 2021).

In response to the increasing prevalence of human remains depicted on social media and rising concerns among anatomists, the IFAA (2012) recommendations state that: "*Limits need to be placed on the extent to which images or other artifacts produced from donations are placed in the public domain, including in social media, both to respect the privacy of the donor (and their surviving relatives) and to prevent arousing morbid curiosity. No individual should be identifiable in images*". As social media usage in teaching health sciences students is likely here to stay, a balance should be struck between the benefits to the students and the harm that may be caused to donors and their families and to donor programmes. Publication on social media, which includes dissemination via WhatsApp (Facebook, Inc., Menlo Park, CA) groups is immediate and thus may lead to a "legally incriminating chain of publication" (Bouter et al. 2020). Shortly before the COVID-19 pandemic drastically increased the use of images and models of cadavers in online teaching and learning, Hennessy et al. (2020) published social media guidelines specifically aimed at anatomists which include a warning regarding "affecting the relationship between the public and body donor programs."

To date, no studies have investigated the public's, donors or relatives of the deceased's opinions toward the global dissemination of images of dissected bodies on social media. It is also not known if donors would condone the wide dissemination of images of their bodies unless it was specifically requested in the informed consent forms (Cornwall et al. 2016; Hennessy and Smith 2020). While some authors justify the dissemination of images on social media by claiming that it maximizes the donor's wishes to be used for educational purposes (Rai et al. 2019), not all images on social media are used in an educational or scientific context (Cornwall et al. 2016; Hennessy et al. 2020).

8.6 The Anatomist's Roles and Responsibilities

Unless carefully considered, the use of digital images, digital models, and 3D printed models of human remains may increase the risk of unethical practices such as objectification and commodification of donors and/or unclaimed bodies (Jones 2019b). Despite existing guidelines from anatomical associations and anatomists, images continue to be published on social media sites without a statement of consent and/or acknowledgment (IFAA 2012; Hennessy et al. 2020). The consequences of the misuse of donated bodies and lack of transparency may affect public trust in the programmes, resulting in a reduction of donations and a shortage of bodies available for teaching and research (Cornwall et al. 2016). Inappropriate uses of donated bodies have resulted in legal action against institutions in the USA (Charatan 2001; Ling 2004). Technological developments thus need to be integrated into the anatomical sciences in a responsible manner, aligned with the ethical and cultural values of society (Cornwall 2016). Informed consent forms for donor programmes should therefore be transparent in their reflection of possible new developments in technology, which may impact the use of donor bodies. Transparency would also safeguard donors against the possibility of exploitation (Cornwall et al. 2016; Jones 2019b). In addition, journal editors and grant funders should play a critical role by requesting transparency in relation to the provenance of the donors used in digital photographs, scans, and 3D models in studies. Oversight committees should be implemented by institutions where they do not exist currently, to assist with the ethical uses of donor bodies (Schmitt et al. 2014) both for teaching and research, with the view to considering and approving the use of donor material in digital imaging, 3D prints and other technologies that may arise in the future.

There is an urgent need for anatomists to discuss and address the ethical questions surrounding digital images and other data obtained from human bodies and form regulatory frameworks that could contribute to legislation (Cornwall 2017). Furthermore, engagement with registered donors and their relatives will provide guidance for body donation programmes regarding their approval and expectations for the use of data derived from their bodies (Cornwall 2016, 2017). Anatomists should embed ethical considerations within their teaching, thereby making students aware of the expected ethical and professional conduct toward digital representations of human remains, in addition to the physical bodies (Caffell and Jakob 2019). As a commitment toward ethical behavior, students could sign a declaration of respectful treatment, which could among other considerations, prohibit the dissemination of images in the public domain (Caffell and Jakob 2019; Champney et al. 2020). In addition, anatomists should be more aware of their own ethical behavior and attitudes toward the deceased including the making of images and 3D prints, as they often serve as role models for students (Caffell and Jakob 2019; Lemos et al. 2021). Any images and models should be used with caution, stating the provenance and whether specific consent was provided, have a strong educational justification for their use, and avoid sensationalism or voyeurism (Caffell and Jakob 2019). Lemos et al. (2021) suggest that access to images should be restricted on a secure digital platform to only the staff and students who are directly involved with the anatomy course. Similarly, students should be instructed that sharing images beyond the educational context in the institution is not ethical. In some cases, institutions prohibit the use of cameras in anatomy laboratories (Dost et al. 2020). Anatomists should challenge those who post images on social media without an accompanying statement of consent (Hennessy et al. 2020).

The ethical obligations of anatomists and body donor programmes have been discussed at length (Champney et al. 2019; Hildebrandt and Champney 2020). These include fully informed consent, respectful treatment of the donor and their family, following the wishes of the donor, and publicly acknowledging the donor's gift (Hildebrandt and Champney 2020). Champney et al. (2019) provided a detailed description of the principles that should be followed in three key overlapping areas; namely toward donors and their families (fully informed consent, transparency, understanding), by the donation organization (respect for the donor's dignity, their family's emotions, and their religion and culture), and toward society (awareness of the cultural and religious context of the society in which they function, the local legislation, and public health concerns).

When considering any ethical issues related to the use of human material, the case of Henrietta Lacks is a pertinent reminder of ethical misconduct for anatomists. Lacks is often referred to as "HeLA" for her cancerous cells that were taken and cultured by scientists without her consent (Skloot 2010). Lacks' cells had an extraordinary capacity to survive and reproduce and they became immortalized. While research undertaken with HeLa cells reinforces much of modern medicine and has produced groundbreaking therapies such as the Salk polio vaccine (Epstein 2016), for example, the failure to ask Lacks' or her family for consent to use her cells is an ethical transgression. Although some may argue that Lacks' cells were taken in a different era of consent, this can never justify what happened. Thus, it is incumbent on anatomists in the present era to learn from the past, to review current practices and regulations for the capture and distribution of electronic data from donated and unclaimed bodies, and to ensure that they will not be found wanting by the next generation of anatomists and donors.

Acknowledgments The authors would like to acknowledge the thousands of human bodies used in the teaching and training of health professional students around the globe. We should do them no harm. We would also like to thank Dr. Brendon Billings, School of Anatomical Sciences, University of the Witwatersrand for his helpful comments on the manuscript.

References

AbouHashem Y, Dayal M, Serafin S, Štrkalj G (2017) Students' attitudes toward body image donation for 3D printing. Clin Anat 30:1005–1006. https://doi.org/10.1002/ca.22978

Ackerman MJ (2017) The visible human project: from body to bits. IEEE Pulse 8:39–41. https://doi.org/10.1109/MPUL.2017.2701221

ADInstruments (2019) To dissect, or not to dissect–how digital cadavers open up new teaching methods for educators. https://www.adinstruments.com/blog/anatomy-month-digital-cadavers-open-new-teaching-methods. Accessed 5 Dec 2021

Alblas A, Greyling LM, Geldenhuys EM (2018) Composition of the Kirsten skeletal collection at Stellenbosch University. S Afr J Sci 114:1–6. https://doi.org/10.17159/sajs.2018/20170198

Anyanwu GE, Udemezue OO, Obikili EN (2011) Dark age of sourcing cadavers in developing countries: a Nigerian survey. Clin Anat 24:831–836. https://doi.org/10.1002/ca.21187

Aso-Escario J, Gil-Begué M, Sebastián-Sebastián C et al (2019) First Spanish program of virtual body donation. Medico-legal issues and academic, clinical care and research interest. Spanish J Leg Med 45:147–154. https://doi.org/10.1016/j.remle.2017.12.002

Atlas MC (2001) Ethics and access to teaching materials in the medical library: the case of the Pernkopf atlas. Bull Med Libr Assoc 89:51–58

Attardi SM, Rogers KA (2015) Design and implementation of an online systemic human anatomy course with laboratory. Anat Sci Educ 8:53–62. https://doi.org/10.1002/ase.1465

AUS-AA (1977) Anatomy act 1977 No 126. New south wales government, NSW Legislation, Sydney, Australia. URL: https://www.legislation.nsw.gov.au/#/view/act/1977/126/full. Accessed 20 January 2019

AUS-AA (1983) Anatomy act 1983 No 164. New South wales government, NSW legislation, Sydney, Australia. URL: https://www.legislation.nsw.gov.au/#/view/act/1983/164. Accessed 19 June 2019

Barry DS, Marzouk F, Chulak-Oglu K et al (2016) Anatomy education for the YouTube generation. Anat Sci Educ 9:90–96. https://doi.org/10.1002/ase.1550

Bernanke DH (2011) Ethical care and use of cadavers: a call for a policy. Anat Sci Educ 4:53–54. https://doi.org/10.1002/ase.201

Bisht B, Hope A, Paul MK (2019) From papyrus leaves to bioprinting and virtual reality: history and innovation in anatomy. Anat Cell Biol 52:226–235. https://doi.org/10.5115/acb.18.213

Bolt S, Eisinga R, Altena M et al (2012) Over my dead body: body donation and the rise in donor registrations in the Netherlands. Omega (United States) 66:57–77. https://doi.org/10.2190/OM.66.1.d

Bond A (2013) Zurich university staff disciplined for posting pictures of body parts on Instagram in Switzerland. In: Dly. Mail online. https://www.dailymail.co.uk/news/article-2383718/Zurich-University-staff-disciplined-posting-pictures-body-parts-Instagram-Switzerland.html. Accessed 7 Oct 2021

Bouter C, Venter B, Etheredge H (2020) Guidelines for the use of WhatsApp groups in clinical settings in South Africa. South African Med J 110:364–368. https://doi.org/10.7196/SAMJ.2020.v110i5.14558

Brooks MM, Rumsey C (2007) The body in the museum. In: Human remains. Guide for museums and academic institutions. AltaMira Press

Brown D (1999) Not just any cadaver would suffice. Washington Post:1–4

Caffell A, Jakob T (2019) 'The dead teach the living': ethical considerations concerning the management of collections of human remains in universities. In: Ethical Approaches to Human Remains. Springer International Publishing, pp 179–209

Calkins CM, Franciosi JP, Kolesari GL (1999) Human anatomical science and illustration: the origin of two inseparable disciplines. Clin Anat 12:120–129. https://doi.org/10.1002/(SICI)1098-2353(1999)12:2<120::AID-CA7>3.0.CO;2-V

Campanacho V, Alves Cardoso F (2018) Portugal residents' opinion regarding the showcasing of threedimensional digital replicas of human bones. In: Digital dilemma. Conference presentation, University College London, 6 October

Caplan I, DeCamp M (2019) Of discomfort and disagreement: unclaimed bodies in anatomy laboratories at United States medical schools. Anat Sci Educ 12:360–369. https://doi.org/10.1002/ase.1853

Carney S (2011) The red market. William Morrow

Caruso MC (2021) Virtual microscopy and other technologies for teaching histology during Covid-19. Anat Sci Educ 14:19–21. https://doi.org/10.1002/ase.2038

Champney TH (2016) The business of bodies: ethical perspectives on for-profit body donation companies. Clin Anat 29:25–29. https://doi.org/10.1002/ca.22643

Champney TH (2019) A bioethos for bodies: respecting a priceless resource. Anat Sci Educ 12:432–434. https://doi.org/10.1002/ase.1855

Champney TH, Broadfield DC, Vargas EH, Hoodiman DR (2020) Establishing and operating a body donation program. Teach Anat:207–214. https://doi.org/10.1007/978-3-030-43283-6_22

Champney TH, Hildebrandt S, Gareth Jones D, Winkelmann A (2019) BODIES R US: ethical views on the commercialization of the dead in medical education and research. Anat Sci Educ 12:317–325. https://doi.org/10.1002/ase.1809

Charatan F (2001) Audit of willed body program implicates former director. West J Med 174:93. https://doi.org/10.1136/ewjm.174.2.93

Cheney A (2006) Body brokers: inside America's underground trade in human remains. Broadway Books, New York

Chia T, Oyeniran O (2020) Ethical considerations in the use of unclaimed bodies for anatomical dissection: a call for action. Ulutas Med J 6:5. https://doi.org/10.5455/umj.20201229101758

Chugh A, Narwal A (2017) Oral mark in the application of an individual identification: from ashes to truth. J Forensic Dent Sci 9:51. https://doi.org/10.4103/jfo.jfds_103_15

Chung BS, Shin DS, Brown P et al (2015) Virtual dissection table including the visible Korean images, complemented by free software of the same data. Int J Morphol 33:440–445. https://doi.org/10.4067/s0717-95022015000200006

Chytas D (2019) Use of social media in anatomy education: a narrative review of the literature. Ann Anat 221:165–172. https://doi.org/10.1016/j.aanat.2018.10.004

Claes T (2018) "By what right does the scalpel enter the pauper's corpse?" dissections and consent in late nineteenth-century Belgium. Soc Hist Med 31:258–277. https://doi.org/10.1093/shm/hkx007

Cornwall J (2017) Body donation and digital technology: the ethical issues. Anatomy 11:42–45. https://doi.org/10.2399/ana.16.061

Cornwall J (2016) The ethics of 3D-printing copies of bodies donated for medical education and research: what is there to worry about? Australas Med J 9:8–11. https://doi.org/10.4066/amj.2015.2567

Cornwall J, Callahan D, Wee R (2016) Ethical issues surrounding the use of images from donated cadavers in the anatomical sciences. Clin Anat 29:30–36. https://doi.org/10.1002/ca.22644

Cornwall J, Perry GF, Louw G, Stringer MD (2012) Who donates their body to science? An international, multicenter, prospective study. Anat Sci Educ 5:208–216. https://doi.org/10.1002/ase.1278

Curran V, Simmons K, Matthews L et al (2020) YouTube as an educational resource in medical education: a scoping review. Med Sci Educ 30:1775–1782. https://doi.org/10.1007/s40670-020-01016-w

Czech H, Druml C, Müller M et al (2021) The Medical University of Vienna and the legacy of Pernkopf's anatomical atlas: elsevier's donation of the original drawings to the Josephinum. Ann Anat 237:151693. https://doi.org/10.1016/j.aanat.2021.151693

Dalley AF, Driscoll RE, Settles HE (1993) The uniform anatomical gift act: what every clinical anatomist should know. Clin Anat 6:247–254. https://doi.org/10.1002/ca.980060409

Dasgupta N (2004) Unclaimed bodies at the anatomy table. J Am Med Assoc 291:122. https://doi.org/10.1001/jama.291.1.122

Dayal MR, Kegley ADT, Štrkalj G et al (2009) The history and composition of the Raymond A. Dart collection of human skeletons at the University of the Witwatersrand, Johannesburg, South Africa. Am J Phys Anthropol 140:324–335. https://doi.org/10.1002/ajpa.21072

Dost S, Hossain A, Shehab M et al (2020) Perceptions of medical students towards online teaching during the COVID-19 pandemic: a national cross-sectional survey of 2721 UK medical students. BMJ Open 10:1–10. https://doi.org/10.1136/bmjopen-2020-042378

Douglas NKM, Scholz M, Myers MA et al (2019) Reviewing the role of instagram in education: can a photo sharing application deliver benefits to medical and dental anatomy education? Med Sci Educ 29:1117–1128. https://doi.org/10.1007/s40670-019-00767-5

Epstein R (2016) The uneasy legacy of Henrietta lacks. In: Hoover Inst. https://www.hoover.org/research/uneasy-legacy-henrietta-lacks. Accessed 24 Nov 2021

Errickson D, Thompson TJU (2019) Sharing is not always caring: social media and the dead. Ethical Approaches to Human Remains. Springer International Publishing, In, pp 299–313

Farsides T, Smith C (2020) Consent in body donation. Eur J Anat 24:239–248. https://doi.org/10.31234/osf.io/48q3x

Fennell S, Jones DG (1992) The bequest of human bodies for dissection: a case study in the Otago Medical School. N Z Med J 105:472–474

Fredieu JR, Kerbo J, Herron M et al (2015) Anatomical models :a digital revolution. Med Sci Edu:183–194. https://doi.org/10.1007/s40670-015-0115-9

Garment A, Lederer S, Rogers N, Boult L (2007) Let the dead teach the living: the rise of body bequeathal in 20th-century America. Acad Med 82:1000–1005. https://doi.org/10.1097/ACM.0b013e318149e986

Ghosh SK (2015a) Human cadaveric dissection: a historical account from ancient Greece to the modern era. Anat Cell Biol 48:153–169. https://doi.org/10.5115/acb.2015.48.3.153

Ghosh SK (2015b) Evolution of illustrations in anatomy: a study from the classical period in Europe to modern times. Anat Sci Educ 8:175–188. https://doi.org/10.1002/ase.1479

Greene JRT (2003) Effects of detailed information about dissection on intentions to bequeath bodies for use in teaching and research. J Anat 202:475–477. https://doi.org/10.1046/j.1469-7580.2003.00176.x

Grow B, Shiffman J (2017) In the U.S. market for human bodies, anyone can sell the donated dead. In: Reuters. https://www.reuters.com/investigates/special-report/usa-bodies-brokers/. Accessed 27 Jan 2022

Gürses İA, Ertaş A, Gürtekin B, et al (2019) Profile and motivations of registered whole-body donors in Turkey: Istanbul University experience

Habicht JL, Kiessling C, Winkelmann A (2018) Bodies for anatomy education in medical schools: an overview of the sources of cadavers worldwide. Acad Med 93:1293–1300. https://doi.org/10.1097/ACM.0000000000002227

Halperin EC (2007) The poor, the black, and the marginalized as the source of cadavers in United States anatomical education. Clin Anat 20:489–495. https://doi.org/10.1002/ca.20445

Harries J, Fibiger L, Smith J et al (2018) Exposure: the ethics of making, sharing and displaying photographs of human remains. Hum Remain Violence An Interdiscip J 4:3–24. https://doi.org/10.7227/hrv.4.1.2

Hennessy CM, Kirkpatrick E, Smith CF, Border S (2016) Social media and anatomy education: using twitter to enhance the student learning experience in anatomy. Anat Sci Educ 9:505–515. https://doi.org/10.1002/ase.1610

Hennessy CM, Royer DF, Meyer AJ, Smith CF (2020) Social media guidelines for anatomists. Anat Sci Educ 13:527–539. https://doi.org/10.1002/ase.1948

Hennessy CM, Smith CF (2020) Digital and social media in anatomy education. Adv Exp Med Biol 1260:109–122. https://doi.org/10.1007/978-3-030-47483-6_6

Herring J, Chau PL (2007) My body, your body, our bodies. Med Law Rev 15:34–61. https://doi.org/10.1093/medlaw/fwl016

Hildebrandt S (2008) Capital punishment and anatomy: history and ethics of an ongoing association. Clin Anat 21:5–14. https://doi.org/10.1002/ca.20571

Hildebrandt S (2009a) Anatomy in the third Reich: an outline, part 1. National Socialist politics, anatomical institutions, and anatomists. Clin Anat 22:883–893. https://doi.org/10.1002/ca.20872

Hildebrandt S (2009b) Anatomy in the third Reich: an outline, part 3. The science and ethics of anatomy in National Socialist Germany and postwar consequences. Clin Anat 22:906–915. https://doi.org/10.1002/ca.20874

Hildebrandt S (2019) The anatomy of murder. Berghahn

Hildebrandt S (2009c) Anatomy in the third Reich: an outline, part 2. Clin Anat 22:894–905. https://doi.org/10.1002/ca.20873

Hildebrandt S (2006) How the Pernkopf controversy facilitated a historical and ethical analysis of the anatomical sciences in Austria and Germany: a recommendation for the continued use of the Pernkopf atlas. Clin Anat 19:91–100. https://doi.org/10.1002/ca.20272

Hildebrandt S, Champney TH (2020) Ethical considerations of body donation. In: Teaching Anatomy. Springer International Publishing, pp 215–222

Houlton TMR, Billings BK (2017) Blood, sweat and plaster casts: reviewing the history, composition, and scientific value of the Raymond A. Dart collection of African life and death masks. HOMO- J Comp Hum Biol 68:362–377. https://doi.org/10.1016/j.jchb.2017.08.004

Hutchinson EF, Kramer B, Billings BK et al (2020) The law, ethics and body donation: a tale of two bequeathal programs. Anat Sci Educ 13:512–519. https://doi.org/10.1002/ase.1922

Innes S (2019) Arizona has a thriving business based on "whole body" donations. In: azcentral. https://www.azcentral.com/in-depth/news/local/arizona-health/2019/06/10/arizona-has-thriving-business-based-whole-body-donations-donate-body-to-science/3579828002/. Accessed 1 May 2021

IFAA. International Federation of Associations of Anatomists (2012) Recommendations of good practice for the donation and study of human bodies and tissues for anatomical examination. https://ifaa.net/recommendations/

IFAA. International Federation of Associations of Anatomists (2021) Public dissection of a donar body. https://ifaa.net/news

Israel HA, Seidelman WE (1996) Nazi origins of an anatomy text: the Pernkopf Atlas. JAMA–J Am Med Assoc 276:1633

Jaffar AA (2012) YouTube: an emerging tool in anatomy education. Anat Sci Educ 5:158–164. https://doi.org/10.1002/ase.1268

Jaffar AA (2014) Exploring the use of a Facebook page in anatomy education. Anat Sci Educ 7:199–208. https://doi.org/10.1002/ase.1404

Jain A, Lee J-E (2009) Scars, marks, and tattoos: a soft biometric for identifying suspects and victims. In: SPIE Int Soc Opt Photonics. https://spie.org/news/1282-scars-marks-and-tattoos-a-soft-biometric-for-identifying-suspects-and-victims?SSO=1. Accessed 12 Jan 2022

Jiang J, Chen Q, Zhang M et al (2020) Effects of commemorations and postdonation services on public willingness to donate bodies in China. Anat Sci Educ 13:218–229. https://doi.org/10.1002/ase.1892

Jones DG (2012) Resurrecting unclaimed bodies. Clin Anat 25:674–675. https://doi.org/10.1002/ca.22077

Jones DG (2016) YouTube anatomy education: sources of ethical perplexity. Anat Sci Educ 9:500–501. https://doi.org/10.1002/ase.1627

Jones DG (2020) Ethical responses to the Covid-19 pandemic: implications for the ethos and practice of anatomy as a health science discipline. Anat Sci Educ 13:549–555. https://doi.org/10.1002/ase.2003

Jones DG (2019a) The ethical awakening of human anatomy: reassessing the past and envisioning a more ethical future. Ethical Approaches to Human Remains. Springer International Publishing, In, pp 73–94

Jones DG (2019b) Three-dimensional printing in anatomy education: assessing potential ethical dimensions. Anat Sci Educ 12:435–443. https://doi.org/10.1002/ase.1851

Jones DG (2017) Human anatomy: a review of the science, ethics and culture of a discipline in transition. Hum Anat–Rev Med Adv. https://doi.org/10.5772/intechopen.68524

Jones DG, Whitaker MI (2009) Speaking for the dead: the human body in biology and medicine. Ashgate

Jones DG, Whitaker MI (2012) Anatomy's use of unclaimed bodies: reasons against continued dependence on an ethically dubious practice. Clin Anat 25:246–254. https://doi.org/10.1002/ca.21223

Kahn PA, Champney TH, Hildebrandt S (2017) The incompatibility of the use of unclaimed bodies with ethical anatomical education in the United States. Anat Sci Educ 10:200–201. https://doi.org/10.1002/ase.1672

Khartade HK, Meshram V, Garg SP, Mishra DK (2021) Identification of skeletal remains by orthopaedic implant: a case report and brief review of the literature. Med Sci Law 61:150–154. https://doi.org/10.1177/0025802420975304

Kurt E, Yurdakul SE, Ataç A (2013) An overview of the technologies used for anatomy education in terms of medical history. Procedia–Soc Behav Sci 103:109–115. https://doi.org/10.1016/j.sbspro.2013.10.314

L'Abbe EN, Kruger GC, Theye CEG et al (2021) The Pretoria bone collection : a 21st century skeletal collection in South Africa. Forensic Sci 1:220–227. https://doi.org/10.3390/forensicsci1030020

L'Abbé EN, Loots M, Meiring JH (2005) The Pretoria bone collection: a modern south African skeletal sample. HOMO–J Comp Hum Biol 56:197–205. https://doi.org/10.1016/j.jchb.2004.10.004

Lemos GA, Araújo DN, de Lima FJC, Bispo RFM (2021) Human anatomy education and management of anatomic specimens during and after COVID-19 pandemic: ethical, legal and biosafety aspects. Ann Anat 233:151608. https://doi.org/10.1016/j.aanat.2020.151608

Lenk C, Beier K (2012) Is the commercialisation of human tissue and body material forbidden in the countries of the European Union. J Med Ethics 38:342–346. https://doi.org/10.1136/jme.2010.038760

Ling A (2004) UCLA Willed Body Program comes under scrutiny as companies sued for the purchase of body parts. J Law Med Ethics 32:532–534

Longhurst GJ, Stone DM, Dulohery K et al (2020) Strength, weakness, opportunity, threat (SWOT) analysis of the adaptations to anatomical education in the United Kingdom and Republic of Ireland in response to the Covid-19 pandemic. Anat Sci Educ 13:301–311. https://doi.org/10.1002/ase.1967

Maass P, Friedling LJ (2019) Documented composition of cadaveric skeletal remains in the University of Cape Town Human Skeletal Collection, South Africa. Forensic Sci Int 294(219):e1–219.e7. https://doi.org/10.1016/j.forsciint.2018.10.007

Mahomed S, Nöthling-Slabbert M, Pepper MS (2013) The legal position on the classification of human tissue in South Africa: can tissues be owned? South African J Bioeth Law 6:14. https://doi.org/10.7196/sajbl.258

Márquez-Grant N, Errickson D (2017) Ethical considerations: an added dimension. In: Human remains: another dimension: the application of imaging to the study of human remains. Elsevier, pp 194–204

McHanwell S, Brenner E, Chirculescu A et al (2008) The legal and ethical framework governing body donation in Europe–a review of current practice and recommendations for good practice. Eur J Anat 12:1–24

Mcmenamin PG, Quayle MR, Mchenry CR, Adams JW (2014) The production of anatomical teaching resources using three-dimensional (3D) printing technology. Anat Sci Educ 7:479–486. https://doi.org/10.1002/ase.1475

Miller GW, Lewis TL (2016) Anatomy education for the YouTube generation: technical, ethical, and educational considerations. Anat Sci Educ 9:496–497. https://doi.org/10.1002/ase.1622

Mullick S (2018) Dental students' selfie with severed heads of corpses kept on a table termed as "disturbing" and "inexcusable." In: Int Bus Times. https://www.ibtimes.co.in/severed-heads-not-donated-yale-study-claim-authorities-after-grotesque-selfie-goes-viral-759568. Accessed 10 Dec 2021

Newman C (2018) Susan potter gave her body to science. Her cadaver became immortal. In: Natl Geogr Mag https://www.nationalgeographic.com/magazine/article/visible-human-susan-potter-cadaver-article. Accessed 5 Dec 2021

Nie JB, Jones DG (2019) Confucian ethics on the commercial use of human bodies and body parts: Yi (righteousness) or/and li (profit)? Anat Sci Educ 12:444–453. https://doi.org/10.1002/ase.1876

Olejaz M, Hoeyer K (2016) Meet the donors: a qualitative analysis of what donation means to Danish whole body donors. Eur J Anat 20:19–29

Papagaroufali E (2006) Playing with one's own death while being alive: the case of registered body-organ donors in Greece. Anthropol Humanism 31:111–123

Pather N (2015) Teaching anatomy: Prosections and dissections. In: Teaching Anatomy. Springer International Publishing, pp 213–221

Peeler M (2020) An unexpected education–unclaimed bodies in the anatomy lab. N Engl J Med 383:2002–2004. https://doi.org/10.1056/NEJMp2028697

Pickering JD, Bickerdike SR (2017) Medical student use of Facebook to support preparation for anatomy assessments. Anat Sci Educ 10:205–214. https://doi.org/10.1002/ase.1663

Polak RJ, Seidelman W, Elbaum L, Hildebrandt S (2017) How to deal with holocaust era human remains: recommendations arising from a special symposium. The International Institute for Holocaust Research

Rai R, Shereen R, Protas M et al (2019) Social media and cadaveric dissection: a survey study. Clin Anat 32: 1033–1041. https://doi.org/10.1002/ca.23421

Raja DS, Sultana B (2012) Potential health hazards for students exposed to formaldehyde in the gross anatomy laboratory. J Environ Health 74:36–40

Rao R (2000) Property, privacy, and the human body. Bost Univ Law Rev 80:359–460

Republic of South Africa (2013) Protection of personal information, act 4 of 2013. Gov Gaz:1–75

Richardson R (2008) The making of Mr. Oxford University Press, Gray's Anatomy

Riederer BM (2016) Body donations today and tomorrow: what is best practice and why? Clin Anat 29:11–18. https://doi.org/10.1002/ca.22641

Riederer BM, Bolt S, Brenner E et al (2012) The legal and ethical framework governing body donation in Europe–1st update on current practice. Eur J Anat 16: 1–21

Riederer BM, Bueno-López JL (2014) Anatomy, respect for the body and body donation–a guide for good practice. Eur J Anat 18:361–368

Riggs G (1998) What should we do about Eduard Pernkopf's atlas? Acad Med 73:380–386. https://doi.org/10.1097/00001888-199804000-00010

Robb RA (2000) Virtual endoscopy: development and evaluation using the visible human datasets. Comput Med Imaging Graph 24:133–151. https://doi.org/10.1016/S0895-6111(00)00014-8

Roeggla G, Landesmann U, Roeggla M (1995) Ethics of executed person on internet. Lancet 345:520. https://doi.org/10.1016/S0140-6736(95)90618-5

Rubin A (2017) History of medical imaging–a brief overview. In: Heal. Beat. https://www.flushinghospital.org/newsletter/history-of-medical-imaging-a-brief-overview/. Accessed 24 Nov 2021

SA-NHA. 2004. National health act No. 61 of 2003. government gazette, Republic of South Africa, Volume 469, No. 35099, 2 March 2012. Pretoria, South Africa: South African Government. 94 p. URL: https://www.gov.za/sites/default/files/gcis_document/201409/a61-03.pdf. Accessed 31 Jan 2022

Saltarelli AJ, Roseth CJ, Saltarelli WA (2014) Human cadavers vs multimedia simulation: a study of student learning in anatomy. Anat Sci Educ 7:331–339. https://doi.org/10.1002/ase.1429

Sander IM, McGoldrick MT, Helms MN et al (2017) Three-dimensional printing of X-ray computed tomography datasets with multiple materials using open-source data processing. Anat Sci Educ 10:383–391. https://doi.org/10.1002/ase.1682

Schiefer A (2019) Robbing the grave: amending the Uniform Anatomical Gift Act to curtail abuses within the whole-body donation industry. Heal Matrix J Law-Medicine 29:371

Schmitt B, Wacker C, Ikemoto L et al (2014) A transparent oversight policy for human anatomical specimen management: the university of California, Davis experience. Acad Med 89:410–414. https://doi.org/10.1097/ACM.0000000000000135

Seidelman WE (2012) Dissecting the history of anatomy in the third Reich-1989-2010: a personal account. Ann Anat 194:228–236. https://doi.org/10.1016/j.aanat.2011.11.013

Skloot R (2010) The immortal life of Henrietta Lacks. Crown Publishers New York

Smith A (2014) Gray's anatomy's forgotten illustrator. https://www.abc.net.au/radionational/programs/archived/bodysphere/gray's-anatomy-a-tale-of-two-doctors/5635018. Accessed 24 Nov 2021

Smith CF, Tollemache N, Covill D, Johnston M (2018) Take away body parts! An investigation into the use of 3D-printed anatomical models in undergraduate anatomy education. Anat Sci Educ 11:44–53. https://doi.org/10.1002/ase.1718

Smith ML, Jones JFX (2018) Dual-extrusion 3D printing of anatomical models for education. Anat Sci Educ 11: 65–72. https://doi.org/10.1002/ase.1730

Smith SE, Hirst CS (2019) 3D data in human remains disciplines: the ethical challenges. Ethical Approaches to Human Remains. Springer International Publishing, In, pp 315–346

Spitzer V, Ackerman MJ, Scherzinger AL, Whitlock D (1996) The visible human male: a technical report. Emerg Infect Dis 3:118–130. https://doi.org/10.1136/jamia.1996.96236280

Strkalj G (2016) Humanistic anatomy: a new program for an old discipline. Nova Biomedical

Strkalj G (2008) A hundred and fifty years of Gray's Anatomy. Adler Mus Bull 34:1–4

Strkalj G, Dayal M (2019) A 3D printed osteological collection and its role in health professions education: a case report from Macquarie University, Sydney. Life Sci J 16:69–72. https://doi.org/10.7537/marslsj160619.11

Taggart E (2020) The history of camera obscura and how it was used to create art. https://mymodernmet.com/camera-obscura/. Accessed 24 Nov 2021

Tassé AM (2011) Biobanking and deceased persons. Hum Genet 130:415–423. https://doi.org/10.1007/s00439-011-1049-y

Thomas CM (2002) Should the law allow sentiment to triumph over science? Massey University, The retention of body parts

Tsafrir J, Ohry A (2001) Medical illustration: from caves to cyberspace. Health Inf Libr J 18:99–109. https://doi.org/10.1046/j.1365-2532.2001.00315.x

Turhan B, Yakut Y (2020) The opinions of physiotherapy students on online anatomy education during Covid-19 pandemic. Anatomy 14:134–138. https://doi.org/10.2399/ana.20.776029

UK-HTA. Human Tissue Authority (2021) Codes of practice and standards C: anatomical examination. https://archive.hta.gov.uk/hta-codes-practice-and-standards-0. Accessed 31 Jan 2022

Ulguim P (2018) Models and metadata: the ethics of sharing bioarchaeological 3D models online. Archaeologies 14:189–228. https://doi.org/10.1007/s11759-018-9346-x

Uruthiralingam U, Rea PM (2020) Augmented and virtual reality in anatomical education–a systematic review. In: Adv Exp Med Biol. Springer, pp. 89–101

Valdez Mayo R, Coquia Adrillana CP, Casil Cruz AL et al (2021) Designing online anatomy education by the "Debug" Guideline: COVID-19 lessons learnt. MedEdPublish 10:1–18. https://doi.org/10.15694/mep.2021.000046.1

Waldby C (2000) The visible human project: Informatic bodies and posthuman medicine. Routledge

Western Cape Heritage Resource Management (2018) Draft national policy on the repatriation and restitution of human remains and heritage objects. https://www.westerncape.gov.za/general-publication/draft-national-policy-repatriation-and-restitution-human-remains-and-heritage-objects. Accessed 27 Jan 2022

Winkelmann A (2016) Consent and consensus–ethical perspectives on obtaining bodies for anatomical dissection. Clin Anat 29:70–77. https://doi.org/10.1002/ca.22651

Yee A, Coombs DM, Hildebrandt S et al (2019a) Nerve surgeons' assessment of the role of Eduard Pernkopf's Atlas of topographic and applied human anatomy in surgical practice. Clin Neurosurg 84:491–498. https://doi.org/10.1093/neuros/nyy159

Yee A, Zubovic E, Yu J et al (2019b) Ethical considerations in the use of Pernkopf's Atlas of anatomy: a surgical case study. Surg (United States) 165: 860–867. https://doi.org/10.1016/j.surg.2018.07.025

Zampieri F, Elmaghawry M, Zanatta A, Thiene G (2015) Andreas Vesalius: celebrating 500 years of dissecting nature. Glob Cardiol Sci Pract 2015:1–20. https://doi.org/10.5339/gcsp.2015.66

Zargaran A, Turki MA, Bhaskar J et al (2020) The role of technology in anatomy teaching: striking the right balance. Adv Med Educ Pract 11:259–266. https://doi.org/10.2147/AMEP.S240150

Zealley JA, Howard D, Thiele C, Balta JY (2021) Human body donation: how informed are the donors? Clin Anat 1–7. doi:https://doi.org/10.1002/ca.23780

The manufacturer's authorised representative in the EU is Springer Nature Customer Service Centre GmbH, Europaplatz 3, 69115 Heidelberg, Germany. If you have any concerns regarding our products, please contact ProductSafety@springernature.com

Printed and bound by CPI Group (UK) Ltd, Croydon, CR0 4YY

28/09/2026

02224091-0019

Applied Cerebral Angiography

Gianni Boris Bradac

Applied Cerebral Angiography

Normal Anatomy and Vascular Pathology

Third Edition

With contributions by Edoardo Boccardi

Springer

Gianni Boris Bradac, MD
Past director of Neuroradiology
Professor Emeritus of Neuroradiology
Department of Neurosciences
University of Turin
Turin
Italy

With contributions by
Edoardo Boccardi, MD
Director of Neuroradiology
Ospedale Niguarda "Ca Granda"
Milan
Italy

ISBN 978-3-319-57227-7 ISBN 978-3-319-57228-4 (eBook)
DOI 10.1007/978-3-319-57228-4

Library of Congress Control Number: 2017948390

Printed on acid-free paper

This Springer imprint is published by Springer Nature
The registered company is Springer International Publishing AG
The registered company address is: Gewerbestrasse 11, 6330 Cham, Switzerland

Preface

This new edition, titled *Applied Cerebral Angiography*, maintains the same structure of the previous work. There is a part with chapters describing the embryology and normal vascular anatomy, including the orbital and the extracranial sector. This is essential for the correct interpretation of the vascular pathology presented in the second part. This work is completely new: the text has been revised and greatly expanded, trying as far as possible to update the literature. Several new figures and drawings have been added. More attention has been given to the endovascular treatment, considering the increasingly important role played by this type of therapy in the various vascular pathologies. A greater space has been dedicated to ischemic stroke in describing its pathogenesis, the anatomo-pathological and angiographic findings influencing the endovascular treatment, which has become today an essential step in the therapy of this pathology. The new possibilities, positive results, and also the limits of this approach have been largely described and discussed.

We hope that also this new edition will be of practical use for all physicians interested in this field.

Turin, Italy

G.B. Bradac
E. Boccardi

Acknowledgements

The book reflects the work done and the experience gained in the neuroradiological units at the Molinette Hospital of the Turin University, at the Niguarda Hospital in Milan, and at the Santa Croce Hospital in Cuneo. It would not have been possible without the involvement in the daily work of many colleagues working for short or long time in these units. To all these doctors, among them we like especially to mention Prof. Mauro Bergui, Dr. Guido Stura (Turin), Dr. Luca Valvassori (Milan), and Dr. Luigi Gozzoli (Cuneo), as well as to the technicians, nurses, and secretaries we express our sincere thanks. We are also grateful for the collaboration among the members of the anesthesiological, neurosurgical, maxillary surgery, otolaryngology, and stroke units.

We are especially grateful to Mario Coriasco, B.Sc. (clinical radiographer and MR technologist), for his help in the technical aspects concerning the manuscript and for the image processing to improve the quality of the figures and to Mr. Günther Hippmann for his effort to correctly represent the schematic drawings.

Finally we would like to express our gratefulness for her support to Dr. Ute Heilmann of Springer Verlag and all the editorial team: Mr. Rakesh Kumar Jotheeswaran, Ms. Lizzy Raj, Mr. Claus-Dieter Bachem, Ms. Anna-Lena Buchholz.

Contents

1 Aortic Arch and Origin of the Brachiocephalic and Cerebral Arteries . . . 1
1.1 Embryogenesis . . . 1
1.2 Variants . . . 5

2 Common (CCA) and Internal (ICA) Carotid Artery . . . 11
2.1 Cervical Segment . . . 11
2.2 Intracranial Segments of ICA . . . 13
2.2.1 Petrous Segment of ICA . . . 13
2.2.2 Cavernous Segment of ICA . . . 15
2.2.3 Supraclinoid Segment of ICA . . . 15
2.3 Congenital Anomalies of the CCA and ICA . . . 25

3 External Carotid Artery . . . 35
3.1 Superior Thyroid Artery (TH) . . . 35
3.2 Lingual Artery (LA) . . . 36
3.3 Facial Artery (FA) . . . 37
3.4 Ascending Pharyngeal Artery (AphA) . . . 38
3.5 Occipital Artery (O) . . . 40
3.6 Posterior Auricular Artery . . . 41
3.7 Internal Maxillary Artery (IMA) . . . 41
3.7.1 Proximal Branches . . . 41
3.7.2 Masticator Space . . . 43
3.7.3 Distal IMA . . . 44
3.7.4 The Terminal Branch . . . 46
3.8 Superficial Temporal Artery . . . 46
3.9 Cervical Arteries . . . 47
3.10 Summary . . . 48
3.10.1 Vascular Malformations . . . 48
3.10.2 Hemangiomas . . . 49
3.10.3 Juvenile Angiofibromas . . . 54
3.10.4 Paragangliomas . . . 54
3.10.5 Meningiomas . . . 56
3.10.6 Other Tumors in the Head and Neck Area . . . 62
3.10.7 General Considerations in Endovascular Treatment in the ECA Area . . . 63

4 Anterior Cerebral Artery 67
4.1 Precommunicating Segment 67
4.2 Distal Segments 70
4.2.1 Infracallosal Segment 70
4.2.2 Precallosal Segment 70
4.2.3 Supracallosal Segment 71
4.2.4 Distal Cortical Branches 71
4.3 Anatomical Variations 72
4.4 Vascular Territories 77
4.5 Angiogram 80

5 Middle Cerebral Artery 81
5.1 M1 Segment 81
5.2 M2, M3, and M4 Segments 82
5.3 Anatomical Variations 89
5.4 Vascular Territories 91
5.5 Angiogram 92

6 Extra- and Intracranial Vertebrobasilar Sector 95
6.1 Extracranial Sector 95
6.1.1 Branches 95
6.2 Intracranial Sector 98
6.2.1 Intracranial Branches of the VA 98
6.2.2 Branches of the Basilar Artery 100
6.2.3 Distal Branches of PICA, AICA, and SCA 108
6.2.4 Variants of Vertebral and Basilar Arteries 108

7 Posterior Cerebral Artery 115
7.1 P1 Segment 115
7.2 P2 Segment 117
7.3 P3 Segment 117
7.4 P4 Segment 118
7.5 Anatomical Variations 118
7.6 Vascular Territories 118
7.7 Angiogram 120

8 Vascular Territories 125

9 Cerebral Veins 129
9.1 Supratentorial Cerebral Veins 130
9.1.1 The Superficial System 131
9.1.2 The Deep System 133
9.2 Infratentorial Cerebral Veins (Veins of the Posterior Fossa) 143
9.2.1 Superior Group 143
9.2.2 Anterior Petrosal Group 148
9.2.3 Posterior Tentorial Group 148
9.3 Dural Sinuses 148
9.3.1 Superior Sagittal Sinus (SSS) 149
9.3.2 Inferior Sagittal Sinus (ISS) 151

9.3.3 Straight Sinus (SS) and Falcine Sinus (FC) 151
9.3.4 Occipital Sinus (OS) and Marginal Sinus (MS) 152
9.3.5 Transverse Sinus (TS) 152
9.3.6 Sigmoid Sinus (SiS), and Jugular Vein 153
9.3.7 Superior Petrosal Sinus (SPS) 153
9.3.8 Inferior Petrosal Sinus (IPS) 153
9.3.9 Sphenoparietal Sinus (SpS) 157
9.3.10 Cavernous Sinus (CS) 157
9.3.11 Superior Ophthalmic Vein (SOV) 158
9.3.12 Inferior Ophthalmic Vein (IOV) 160
9.3.13 Considerations About Vein Variations 160

10 Extracranial Venous Drainage 163
10.1 Orbital Veins 163
10.2 Facial Veins 163
10.3 Retromandibular Vein 164
10.4 Posterior Auricular and Occipital Vein 164
10.5 Deep Cervical Vein 164
10.6 Venous Plexus of the Vertebral Artery 164
10.7 Emissary Veins 164
10.8 Diploic and Meningeal Veins 165
10.9 Internal Jugular Vein 165

11 Aneurysms 167
11.1 Incidence 167
11.2 Type and Location 167
11.3 Anatomopathological Aspects 167
11.4 Pathogenesis 168
11.5 Clinical Presentation 169
11.6 Aneurysm Location 169
11.6.1 Extracranial ICA Aneurysms 169
11.6.2 Intracranial ICA Aneurysms 170
11.6.3 Anterior Cerebral Artery Aneurysms 176
11.6.4 MCA Aneurysms 186
11.6.5 Aneurysms of the Posterior Circulation 186
11.7 Dissecting Aneurysms 196
11.8 Large and Giant Aneurysms 197
11.9 Diagnosis and Treatment 200
11.10 Unruptured Aneurysms 203
11.11 Negative Angiograms in Patients with SAH 203
11.12 Vasospasm 204
11.13 Aneurysms in Children 205

12 Vascular Malformations of the Central Nervous System 207
12.1 Introduction 207
12.2 Classification 207
12.3 AVM 207
12.3.1 Pathogenesis and Pathology 207
12.3.2 Incidence 208

12.3.3 Clinical Relevance 208
12.3.4 Location 209
12.3.5 Diagnosis 209
12.3.6 Treatment 231
12.4 Cavernous Malformations (Cavernomas) 232
12.4.1 Pathology 232
12.4.2 Incidence 232
12.4.3 Location 233
12.4.4 Diagnosis and Clinical Relevance 233
12.5 Capillary Malformations (Telangiectasias) 233
12.6 Developmental Venous Anomaly (DVA) 235
12.6.1 Pathology 235
12.6.2 Incidence 236
12.6.3 Diagnosis and Clinical Relevance 236
12.7 Central Nervous System Vascular Malformation (Part of Well-Defined Congenital or Hereditary Syndrome) . . . 237
12.7.1 Rendu-Osler Syndrome (Hereditary Hemorrhagic Telangiectasias) 237
12.7.2 Sturge-Weber Syndrome (Encephalotrigeminal Angiomatosis) 237
12.7.3 Pathology and Diagnosis 238
12.7.4 Wyburn-Mason Syndrome 238
12.7.5 Klippel-Trenaunay-Weber Syndrome 238
12.8 Arteriovenous Shunts Involving the Vein of Galen 238
12.8.1 Diagnosis and Treatment 239

13 Dural Arteriovenous Fistulas 245
13.1 Incidence 245
13.2 Pathology and Pathogenesis 245
13.3 Clinical Relevance 246
13.4 Location 246
13.5 Diagnosis 246
13.6 Classification 246
13.7 Detailed Considerations in the Most Frequent Fistulas 247
13.8 DAFVs in Pediatric Patients 292

14 Arteriovenous Fistulas 297
14.1 Carotid-Cavernous Fistulas 297
14.1.1 Clinical Presentation 297
14.1.2 Diagnosis and Treatment 297
14.2 Vertebral Arteriovenous Fistulas 300
14.2.1 Clinical Presentation 300
14.2.2 Diagnosis and Treatment 300

15 Ischemic Stroke 303
15.1 Pathology 303
15.2 Location 304
15.3 Pathogenesis of Ischemia and Basic Diagnostic and Therapeutic Approach 304

15.4 Mechanism of Ischemia in the Anterior Circulation 305
15.4.1 Carotid Artery 305
15.4.2 Middle Cerebral Artery 319
15.4.3 Anterior Choroidal Artery 334
15.4.4 Anterior Cerebral Artery 338
15.4.5 Lacunar and Other Deep Infarcts in the Anterior Circulation 340
15.5 Posterior Circulation 341
15.5.1 Subclavian and Innominate Arteries 342
15.5.2 Vertebral Artery 342
15.5.3 Basilar Artery 350
15.5.4 Cerebellar Arteries 358
15.5.5 Posterior Cerebral Artery 362
15.6 Changes in the Venous Sector 364
15.7 Collateral Circulation 364
15.7.1 Collateral Circulation Between Intracranial Arteries 364
15.7.2 Collateral Circulation Between Extracranial and Intracranial Arteries 365
15.7.3 The Vertebrobasilar Sector Deserves a Few More Considerations 366
15.8 General Considerations About Endovascular Treatment 366

16 Spontaneous Dissection of Carotid and Vertebral Arteries ... 371
16.1 Introduction 371
16.2 Pathology and Pathogenesis 371
16.3 Location 372
16.4 Morphological Diagnostic Appearance 373
16.5 Clinical Relevance 374
16.6 Treatment 382
16.7 Dissection and Dissecting Aneurysms in Children 389

17 Other Nonatherosclerotic Vasculopathies 393
17.1 Introduction 393
17.2 Cerebrovascular Fibromuscular Dysplasia 396
17.2.1 Pathology and Etiopathogenesis 396
17.2.2 Diagnosis 396
17.2.3 Clinical Relevance 397
17.3 Moyamoya Disease 401
17.3.1 Pathology and Etiopathogenesis 401
17.3.2 Diagnosis and Clinical Relevance 403
17.4 Takayasu's Arteritis 403
17.5 Sneddon's Syndrome 404
17.6 Reversible Cerebral Vasoconstriction Syndrome (RCVS) 404
17.7 Primary Angiitis of the CNS (PACNS) 406
17.8 Autosomal Dominant Arteriopathy with Subcortical Infarct and Leukoencephalopathy (CADASIL) 406
17.9 Migraine and Stroke 407

18 Cardiac Diseases 409

19 Arterial Occlusive Diseases in Children 415

20 Cerebral Venous Thrombosis 417
20.1 Etiopathogenesis 417
20.2 Location 417
20.3 Diagnosis 418

21 Association of Venous Sinus and IJV Stenosis and Some Clinical Pathological Conditions 425

22 Cerebral Hemorrhage 427

23 Vascular Pathology Involving the Intraorbital Vessels 429

Bibliography 433

Index 477

Short Historical Aspects

In July 1927, Prof. Egas Moniz, director of the neurological clinic in Lisboa, presented at the congress of the Neurological French Society in Paris his experiences concerning a method to study the cerebral vessels that he called "*L'encephalographie arterielle*." The interest for this new method, called later "Cerebral Angiography," was great. Among the several neurological authorities present in the congress, we report the comment of Prof. Babinsky:

Le radiographies qui vient de presenter E.Moniz sont remarquables. Si les observations ulterieures établissent definitivement que les injections auxquelles il a recours sont inoffensives,tous les neurologistes seront reconnaissants a notre eminent collégue de leur avoir procuré un nouveau moyen pouvant permetre *de localizer des tumeurs intracraniennes dont le siege est souvent si difficile a determiner.*

Moniz, in the monograph *Die cerebrale Arteriographie und Phlebographie,* published in 1940, reported his further results confirming the diagnostic value of this technique not only in the tumor pathology but also in other pathological processes especially vascular malformations. In the chapter titled "Hirnarteriosclerose" Moniz reported a few cases of occlusion of the carotid artery and of its intracranial branches making the consideration that this pathology is "*selten ,aber doch haufiger als man annehmen moechte … eine sichere Diagnose dieser Affection war vor der Anwendung der Hirnangiographie nicht moeglich.*"

Since then, great progresses have been made, starting with the introduction of the catheter technique (Seldinger 1953), the subtraction (Ziedses des Plantes 1963), followed by the development of more and more suitable catheters, guide wires, and less toxic contrast media. All these aspects along with the improved technological equipment have characterized the evolution of the cerebral angiography, which has become a very important neuroradiological diagnostic method opening the way for selective and superselective studies and further for the endovascular therapy.

Certainly, the evolution of new methods such as the angio-CT, angio-MR, and ultrasound allows to replace cerebral angiography in many cases today. However, every time the diagnosis is not sufficiently clear or finer details are required to understand the clinical symptoms or to plan the therapy, especially when an endovascular approach is considered, angiography remains today the method of choice.

References

Babinsky J (1927) Revue Neurologique 34:89
Moniz E (1927) Revue Neurologique 34:72
Moniz E (1940) Die Cerebrale Arteriographie and Phlebographie. Springer, Berlin
Seldinger SI (1953) Acta Radiologica (Stock.) 39:368
Ziedses des Plantes BG (1963) Acta Radiol Diagn 1:961

1 Aortic Arch and Origin of the Brachiocephalic and Cerebral Arteries

1.1 Embryogenesis

An important aspect in the embryological development of the cerebrovascular system, as pointed out by Streeter (1918), is that it is not an independent process but it is linked to the progressive development and differentiation of the tissues including the brain parenchyma, its membranes, and the soft tissues of neck and head to which the vascular structure continuously adapts.

The vascular structures develop from primitive vascular arches (Streeter 1918; Congdon 1922; Padget 1948; Barry 1951; Lazorthes 1961; Haughton and Rosenbaum 1974). These are longitudinal vessels developing on each side from a common trunk arising from the heart (truncus arteriosus) which ends in a dilatation (aortic sac) from which arise *the aortic arches*. These have an ascending course forming the primitive ventral (ascending) paired aorta. The vessels then bend dorsally continuing caudally in the paired primitive descending aorta. Both vessels distally fuse together. From these arches arise the brachiocephalic arteries. *In the embryogenesis, six arches in different phases develop and progressively disappear completely or partially*. The first appearing are the arches number one and two which rapidly regress. The third and fourth arches follow. These are the vascular structures which play the most important role in the developing of the brachiocephalic vessels and their future intracranial branches. The fifth arch is not constant. It is frequently incompletely formed, and regresses rapidly after its appearance. From the sixth aortic arch develops the pulmonary artery. In the embryogenesis a vessel (ductus arteriosus) connects each pulmonary artery with the thoracic aorta. This connection closes short after the birth.

The different phases of the development of the extracranial and intracranial cerebral arteries in the embryo have been thoroughly investigated by many authors (Streeter 1918; Congdon 1922; Padget 1944, 1948; Barry 1951; Lazorthes 1961; Arey 1965; Kier 1974; Haughton and Rosenbaum 1974; Lazorthes et al. 1976). A description of this very complex development is summarized, pointing at the more striking aspects considering separately for simplicity, the extra- and intracranial sectors.

At the 4–5 mm stage embryo, approximately 24–29 days of age (Fig. 1.1a), the first and second arches are already completely regressed. The third and fourth arches are developed. From the proximal part of the third arch arises on either side vessels from which develops the future ECA. From its distal part arises the future ICA. The more distal segment of the third arch (ductus caroticus) fuses with the fourth aortic arch continuing on both sides in the dorsal aorta. As described below, connections between ICA and bilateral longitudinal channels (BLA, future vertebro-basilar sector) are present. From the dorsal aorta arise on both sides segmental arteries which contribute later to the formation of the vertebral arteries. In the following stages many changes occur.

G.B. Bradac, *Applied Cerebral Angiography*, DOI 10.1007/978-3-319-57228-4_1

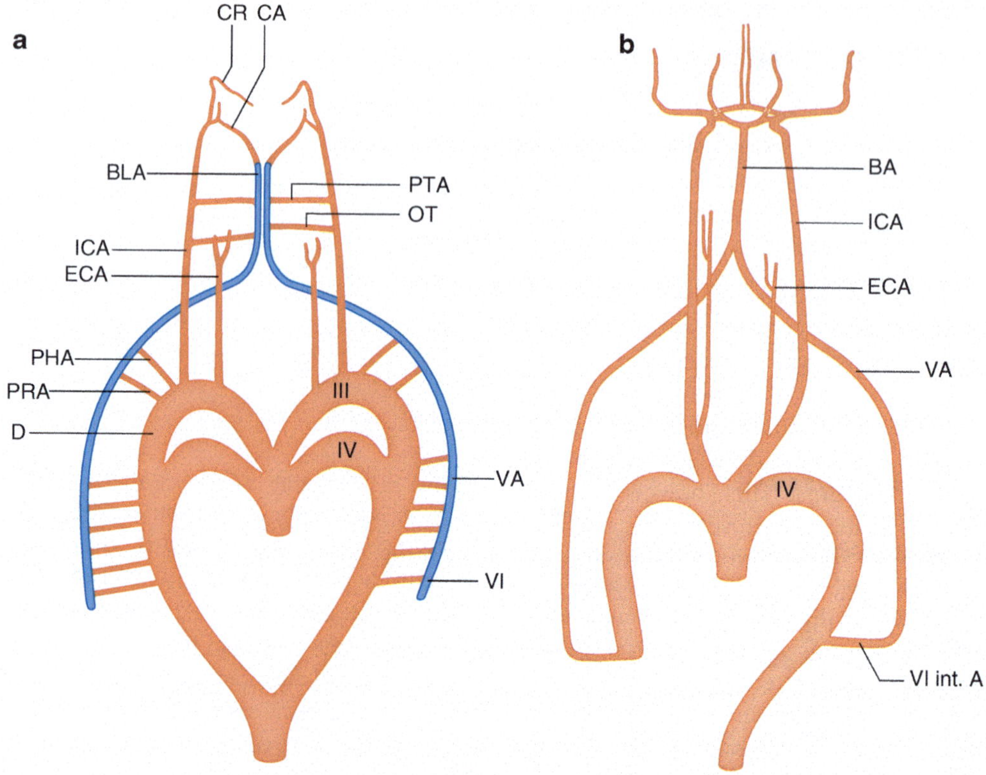

Fig. 1.1 (**a**) Drawing showing the evolution in the early stages (see also text). The first and second arches have already completely disappeared. The ECA arises from the proximal part while the ICA from the distal part of the third aortic arch. Connections between the bilateral longitudinal channels (BLA) and ICA through the trigeminal (PTA) and otic (OT) arteries and between BLA and *ductus caroticus* (D) through the hypoglossal (PHA) and proatlantal (PRA) arteries are present. The *ductus caroticus* connects the third and fourth arches. Intersegmental arteries are connected with a plexiform network from which develop the future vertebral arteries (VA). At this stage, the cranial (CR) and the caudal divisions (CA) of ICA appear. From the caudal division develops the PcomA, which is connected with the distal part of the bilateral longitudinal neural arteries from which develops later the basilar artery. MCA is not developed yet. (**b**) Drawing in advanced evolution (see also text). At this stage, the *ductus caroticus, the hypoglossal, and the proatlantal arteries*, as well *as the trigeminal and otic arteries*, disappear. The right dorsal part of the aorta is regressed. The superior (cervical) intersegmental arteries disappear with exception of the sixth from which arise the vertebral and subclavian arteries. The vertebral arteries (VA) are now developed and connected with the completely formed median basilar artery (BA). The distal branches of the ICA are almost completely formed as well as the circle of Willis. (**c**) Drawing showing the aortic arch and the extra-intracranial cerebral arteries at the end of the embryogenesis. *SA* subclavian artery, *TCT* thyrocervical trunk, *CCA* common carotid artery, *VA* vertebral artery, *ICA* internal carotid artery, *ECA* external carotid artery, *BA* basilar artery. Circle of Willis

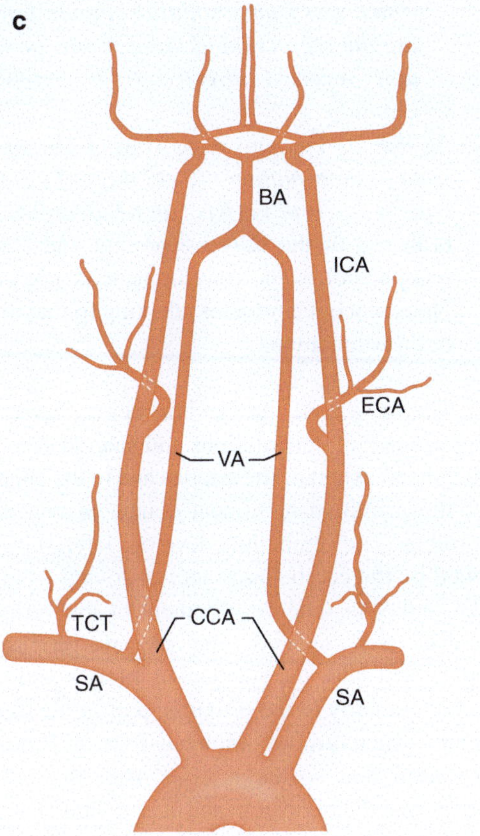

Fig. 1.1 (continued)

At the 20–24 mm stage embryo, approximately 42 days of age (Fig. 1.1b), the anastomoses of the ICA, and ductus caroticus with the BL, disappear (see below). The ECA and ICA arise now as a common trunk, the future common carotid artery (CCA). On the left, the CCA arises from the left fourth aortic arch. From the proximal part of the right fourth aortic arch arises the innominate artery continuing in the right CCA and in the future subclavian artery. The distal part of the right dorsal aorta regresses. On both sides the superior (cervical) intersegmental arteries regress, with exception of the sixth from which develop on either side the subclavian and vertebral arteries (see also Sect. 6.2.4). The more caudally located intersegmental branches become the intercostal arteries.

At this stage the caudal shifting of the heart becomes progressively more prominent leading to further changes of the brachiocephalic vessels. The common carotid artery and the subclavian artery become elongated and progressively located more cranially reaching the adult position. Furthermore the head which is bended ventrally in the first phases of the embryogenesis shifts progressively away from the chest acquiring the typical vertical position.

The final normal aortic arch (Fig. 1.1c) is characterized by the persistence of the left fourth primitive ventral arch from which arise (right to left) the brachiocephalic trunk (innominate artery), the left common carotid artery, and the left subclavian artery. From the brachiocephalic trunk arise the right common carotid artery and the subclavian artery giving off the right vertebral artery. The left vertebral artery arises from the left subclavian artery. Each common carotid artery divides in the external carotid artery (ECA), which supplies the extracranial and meningeal territories and the internal carotid artery (ICA) from which arise the intracranial branches for the cerebral hemisphere and the intraorbital arteries. The vertebral arteries join intracranially the basilar artery. They supply brainstem and cerebellum.

As far as it concerns the development of the intracranial cerebral arteries the most striking aspects are as follows:

1. *At the 4–5 mm stage* (24–29 days of age) (Fig. 1.1a)
   - *Bilateral longitudinal neural arteries (BLA) in the form of plexiform structures develop on the surface of the future brainstem.* The BLA are connected with arteries arising cranially from the ICA (trigeminal and otic arteries) and caudally from branches arising from the ductus caroticus (hypoglossal and proatlantal arteries). The vertebral arteries are in formation. They appear as a plexiform structure arising from intersegmental arteries developing from the dorsal aorta.
   - *There is appearing of the cranial (anterior) and caudal (posterior) division of the ICA.* From the cranial division arises *the anterior choroidal artery* which in this phase of the embryogenesis is well developed

supplying the large choroidal plexus. Opposite to it appear the branches of the primitive ophthalmic artery. More distally develops *the olfactory artery* from which arises a secondary branch, the *anterior cerebral artery*, replacing progressively the regressing olfactory artery. The third branch developing from the cranial division appearing at the 11–12 mm stage (35 days of age) is *the middle* cerebral *artery*. It appears first as a few twigs located between the anterior choroidal artery (AchA) proximally and the developing anterior cerebral artery (ACA) distally. Some of these twigs arise directly from the ACA. From the fusion of these twigs develop the middle cerebral artery.

- From the caudal division arises *the posterior communicating artery (PcomA) from which emerge* at its distal end a diencephalic branch which includes the medial posterior choroidal and a mesencephalic branch from which arises the lateral posterior choroidal artery. In the further evolution, the PcomA continues in *the posterior cerebral artery (PCA)* which progressively extends supplying the posterior part of the cerebral hemispheres. The PcomA (*pars carotica* of PCA) is connected with the cranial part of the bilateral longitudinal neural arteries from which develop the primitive duplicated basilar artery (BA), *which later fuse in the median BA, about the 9 mm stage (32 days of age)*. The cranial part of these longitudinal channels will become the P1 segment (*pars basilaris* of the PCA). This will becomes progressively the predominant flow to the PCA while the PcomA, in the majority of the cases, partially or completely regresses.
- *The formation of the vertebral arteries progresses and is complete at the 12–14 mm stage embryo (35–38 days of age). They join intracranially the formed basilar artery. The connections of the ICA (primitive trigeminal, otic, hypoglossal, and proatlantal arteries) disappear. The flow towards the posterior circulation which was from cranial to caudal, following the regression of these connections, is now inverted directed from caudal to cranial.*

2. *At the 20–24 mm stage* (Fig. 1.1b) *of the embryonic evolution (40–42 days of age) the typical cerebral vessels can be identified.* All branches of the ICA are present and clearly recognizable. The development is not complete, since it continues adapting to that of the brain parenchyma.

The growth of the cerebral hemispheres and appearance of convolutions and sulci lead to further distal extent of the arteries and to the changes of the primitive rectilineal course in one more tortuous. The cerebellum develops later. It is supplied by the cerebellar branches arising from the BA and VAs. The development of the vertebral arteries is completed. The VAs are proximally connected with the subclavian arteries and converge cranially to the proximal part of the basilar artery which is now completely formed. From the vertebral and basilar arteries arise the vessels supplying the brainstem and cerebellum. The cerebellar arteries are the latest to develop. *One remarkable development is the formation at the base of the cerebrum of the circle of Willis about the age of 44–52 days of age* (De Vriese 1905; Padget 1944, 1948). This is an anastomotic circle described by Willis in the 1684, and since then called "*circle of Willis*," in which both anterior cerebral arteries are linked by the anterior communicating artery, and each carotid artery is connected through the posterior communicating artery with the respective PCA arising from the basilar artery (Fig. 1.2).

This is a natural well-constructed security system. Its functional value, however, is somewhat unpredictable owing to the many variants present. According to several authors (De Vriese 1905; Padget 1944–1948; Lazorthes 1961; Lazorthes et al. 1976) *the variants of the circle of Willis occur in the postnatal period and through the life due to various hemodynamic changes, among them the compression of the carotid and vertebral arteries by movements of the head and neck.*

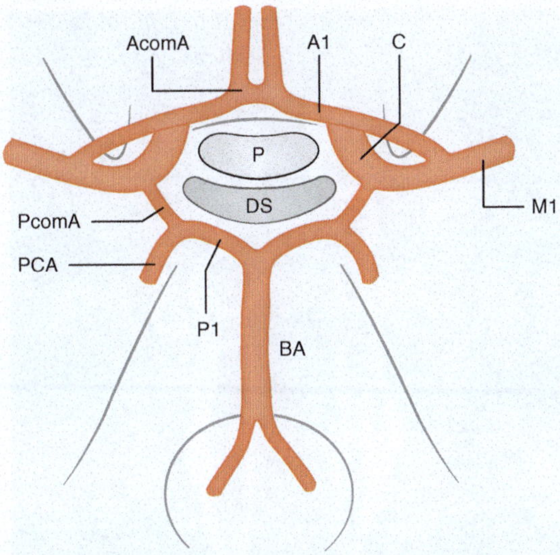

Fig. 1.2 Drawing of the circle of Willis. Internal carotid artery (C), first segment of the anterior cerebral artery (A1), first segment of the middle cerebral artery (M1). *BA* basilar artery, *P1* first segment of the posterior cerebral artery, *PCA* posterior cerebral artery, *AcomA* anterior communicating artery, *PcomA* posterior communicating artery, *P* pituitary gland, *DS* dorsum sellae

1.2 Variants

Some variations of the aortic arch and its branches have already been described in the eighteenth and nineteenth centuries by a few anatomists (Bayford 1789; Tiedermann 1822; Quain 1844; Gray 1859). Angiography, CT angiography, and MR angiography have made possible to demonstrate "in vivo" the great variety of these anomalies, their frequency, and their possible clinical relevance. The progressively better knowledge of the embryogenesis has, furthermore, allowed, at least in some cases, to connect these anomalies with a failure of the normal evolution occurring in specific time of it and involving specific parts of the developing arteries.

Owing to the complexity of the embryonic process, minor variants are the rule. However, these are not recorded in the literature as variants or anomalies. This definition is reserved to more or less complex changes (Adachi 1928; Edwards 1948; Apley 1949; Barry 1951; Lie 1968; Klinkhamer 1969; Haughton and Rosenbaum 1974; Beigelman et al. 1995; Morris 1997; Osborn 1999). In a recent study performed on 2033 patients examined with CT (Mueller et al. 2011), these have been described with a frequency of 13.3% of the cases.

In this chapter are described the variants concerning the aortic arch and the brachiocephalic arteries. Detailed anomalies of the intracranial arteries will be discussed in the specific chapters.

Among the most frequent and more simple anomalies, there are those characterized by the common origin of the left common carotid artery (LC) and the brachiocephalic trunk, and the origin of the LC from the brachiocephalic trunk, instead of arising from the aortic arch. Other variants are the anomalous origin of the left vertebral artery, occurring in about 6% of the cases (Adachi 1928; Uchino et al. 2013b; Mueller et al. 2011) arising from the aortic arch between the left common carotid artery and the left subclavian artery, or more rarely distal to the left subclavian artery. Less frequently, this anomaly involves the right VA, arising from the more proximal segments of the aortic arch or also distal to the left subclavian artery (*vertebral arteria lusoria*). More about the variants of the VAs are described in Sect. 6.2.4.

Among other more complex conditions, there is the right aberrant subclavian artery, arising distal to the left subclavian artery or close to it (*subclavian arteria lusoria*). It is considered to be due to partial persistence of the right dorsal aorta, while its proximal part between the right CCA and the right subclavian artery regresses. It can also be associated at its origin with a small aneurysmal dilatation (Kommerell's diverticulum), perhaps a rudimentary segment of the distal right dorsal aorta. On the aortic arch angiogram the artery appears as the last branch crosses the mediastinum from left to right. Since the first report by Kommerell in 1936 other authors have described this anomaly (Apley

1949; Bosniak 1964; Lie 1968; Klinkhamer 1969; Akers et al. 1991; Freed and Low 1997; Wong et al. 2007; Karcaaltincaba et al. 2009; Uchino et al. 2013b). Another anomaly of the right subclavian artery is that in which the artery arises from the aortic arch separately, proximal to the right CCA. The anomalous subclavian artery can be isolated or associated with other anomalies of the brachiocephalic branches. A frequent association is the common origin of both common carotid arteries and the separate origin of the right CCA proximal to the right subclavian artery. As far as other anomalies concerning the CCA and ICA and VAs see Sects. 2.3 and 6.2.4, respectively.

Extremely rare are other more complex anomalies such as the right aortic arch and the double-aortic arch. In the first case, due to interruption of the left aortic arch at the level of the descending aorta, the brachiocephalic arteries arise from the right aortic arch with a pattern described as a *mirror imaging*. In this condition, from the aortic arch arises (left to right) the left brachiocephalic trunk, which divides distally in the left common carotid and left subclavian arteries, the right common carotid and right subclavian arteries. The left subclavian artery can also arise isolated, distally from the right subclavian artery, appearing on the angiogram as the last branch crossing the mediastinum from right to left. In the double-aortic arch, due to persistence of both arches, commonly each arch gives off a common carotid and a subclavian artery.

Angiographic studies of normal and anomalous aortic arch are presented in Figs. 1.3, 1.4, 1.5, 1.6, 1.7, 1.8, 1.9, and 1.10.

In the majority of the cases these anomalies are asymptomatic, being discovered during a diagnostic study (angio CT-MR or angiography) performed for a cerebral pathology. However, the possibility of such anomalies should be taken into account by the angiographer. Infrequently, respiratory distress and dysphagia can be present, especially in cases of aberrant right subclavian artery and right VA, due to the course of the vessels, crossing the midline in the retro-esophageal space. Congenital heart malformations can be associated especially with a right aortic arch and double-aortic arch. Furthermore, the knowledge of these variants is important in patients in whom aortic arch, esophageal, or anterior neck surgery is planned.

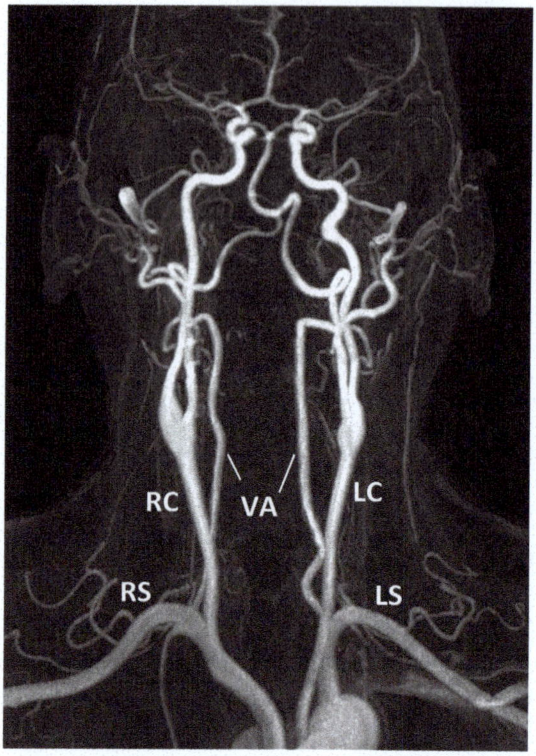

Fig. 1.3 Normal aortic arch, MRI angiography. Brachiocephalic trunk (BR), from which arise the right common carotid (RC) and the right subclavian (RS) arteries. Common left carotid artery (LC), left subclavian artery (LS). Normal origin of both vertebral arteries (VA). That of the right is smaller. The VAs join intracranially the BA. The bifurcation of the two common carotid arteries is well demonstrated as well as the intracranial ICA branches

Some more aspects concerning the embryological development and its abnormalities involving the specific arteries are described later (see Sects. 2.2.3.1, 2.2.3.2, 2.2.3.3, 2.3, 4.3, 5.3, 6.2.4, and 7.5 and Chap. 3).

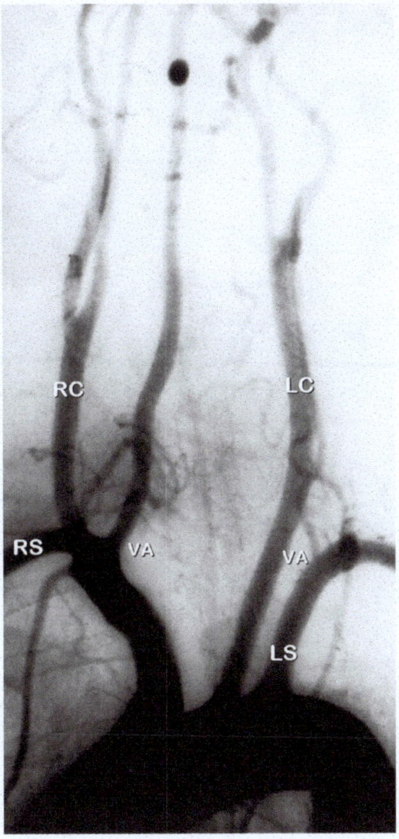

Fig. 1.4 Normal aortic arch angiogram with typical origin of the left and right common carotid arteries (LC, RC). Subclavian arteries (LS, RS). Clear asymmetry of the vertebral arteries (VA). That on the left is hypoplastic

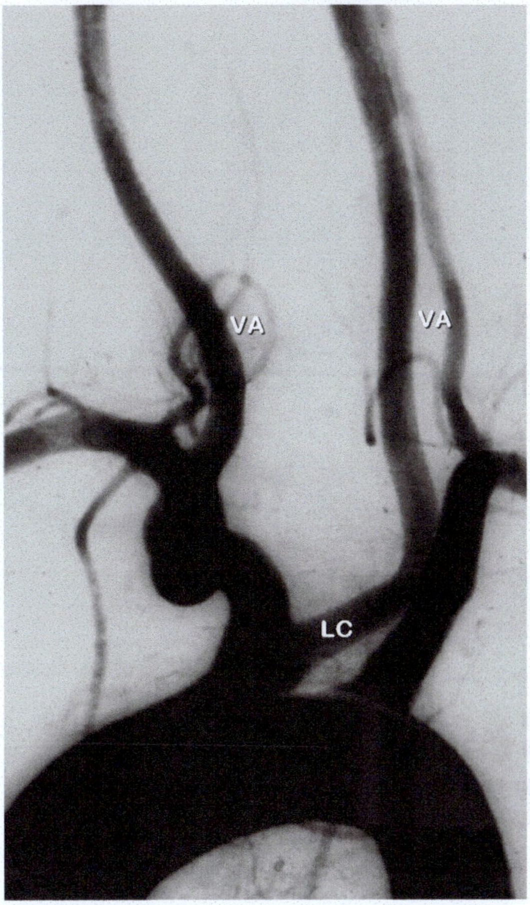

Fig. 1.5 Aortic arch angiogram showing the origin of the left common carotid artery (LC) from the brachiocephalic trunk. The left vertebral artery (VA) is well developed, while that of the right is hypoplastic

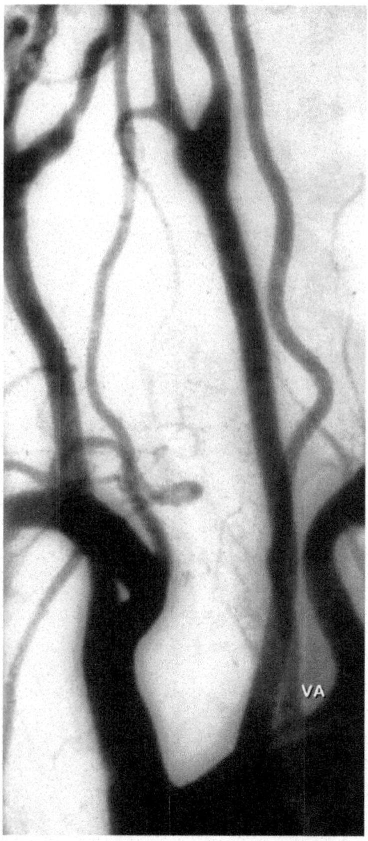

Fig. 1.6 Aortic arch angiogram. Owing to arteriosclerotic elongation of the aortic arch, there is a shifting of the origin of the left common carotid and brachiocephalic trunk towards the heart. Anomalous origin of the left VA (VA) from the aortic arch

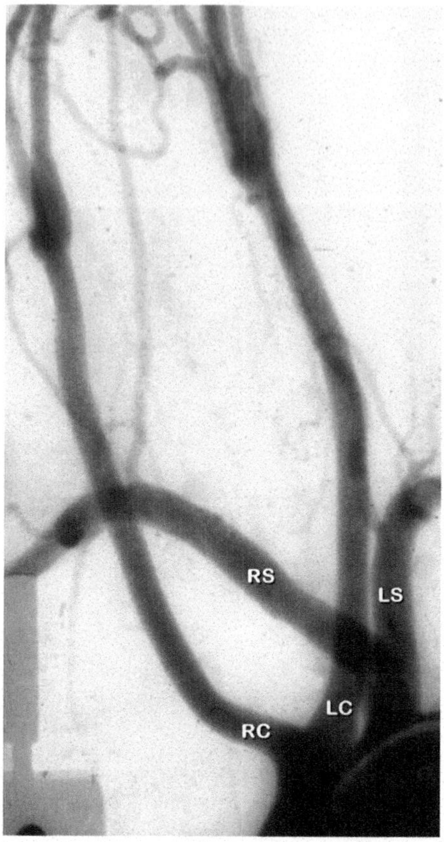

Fig. 1.7 Aortic arch angiogram anomaly. The left and right common carotid arteries (LC, RC) arise as a common trunk. The right subclavian artery (RS) arises distally with a separated or common origin with the left subclavian artery (LS)

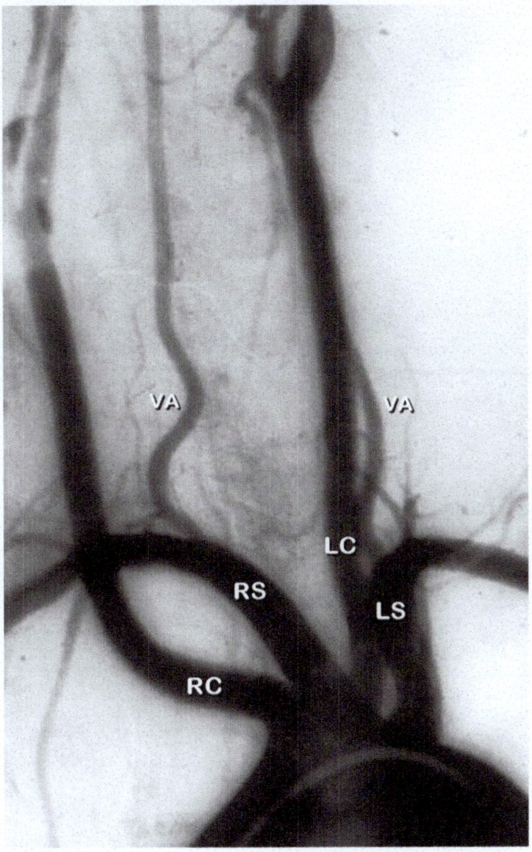

Fig. 1.8 Aortic arch angiogram. The typical brachiocephalic trunk is not formed. The right common carotid artery (RC) is displaced more proximally. It arises from the aortic arch separately from the subclavian artery (RS). This latter arises distally separately or together with the left subclavian artery. Left common carotid artery (LC)

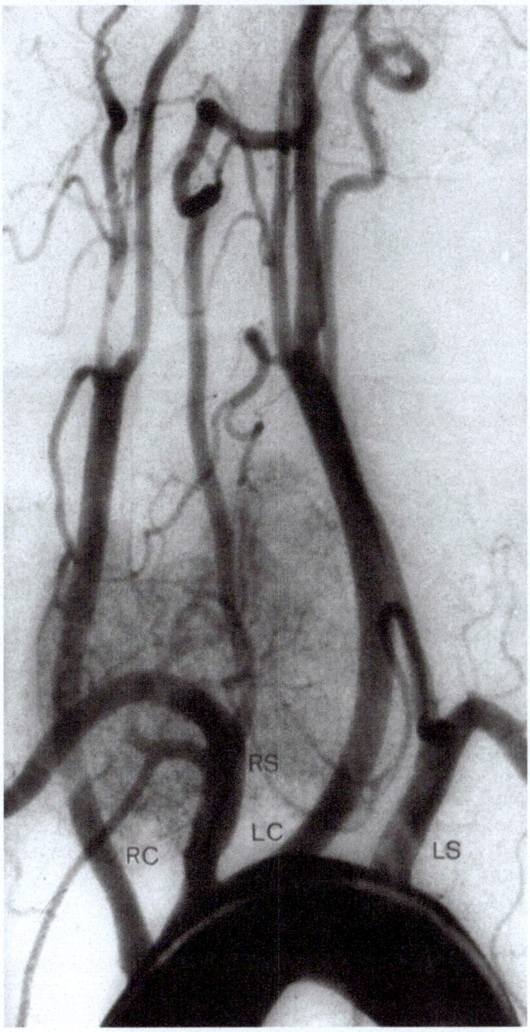

Fig. 1.9 Aortic arch angiogram. The right common carotid artery (RC) arises more proximally. Its origin is separated from that of the right subclavian artery (RS). The typical brachiocephalic trunk is not formed. Left common carotid artery (LC), left subclavian artery (LS)

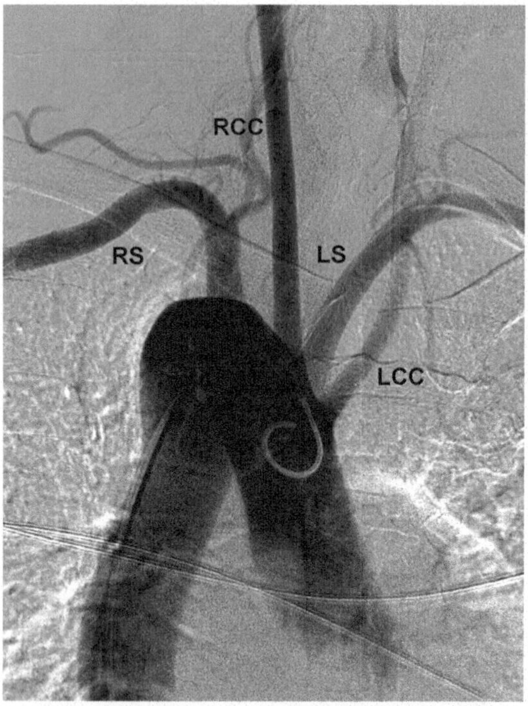

Fig. 1.10 Aortic arch angiogram showing the presence of a right aortic arch. From the ascending aorta arise separately or with a common trunk the left subclavian (LS) and the left common carotid artery (LC). Right common carotid (RC) and right subclavian artery (RS). Pattern which is a "mirror imaging" of the left aortic arch (see also text)

Common (CCA) and Internal (ICA) Carotid Artery

2

2.1 Cervical Segment

Early in the embryogenesis, both primitive proximal external carotid artery (ECA) and internal carotid artery (ICA) arise separately from the third aortic arch: the ECA from its proximal ventral part, and the ICA from the more distal part. The involution of the third aortic arch involving its segment distal to the origin of the ICA (*ductus caroticus*) on both left and right sides results in the formation of a common trunk from which develops on each side the common carotid artery (CCA) continuing cranially in the ICA and ECA. In the further evolution, the left CCA is annexed by the developed left fourth aortic arch, and the right CCA becomes a branch of the brachiocephalic trunk (innominate artery) proximal remnant of the distally regressed right fourth aortic arch (Haughton and Rosenbaum 1974).

The definitive common carotid arteries run cranially in the carotid space, surrounded by the three layers of the deep cervical fascia, called the carotid sheet. Approximately at the level of the hyoid bone, usually between the C4 and the C6 vertebral bodies, each CCA divides into the ICA and ECA.

Cases of a higher bifurcation, up to the first cervical vertebra (Lie 1968), or lower, in the thoracic area (Vitek and Reaves 1973), have been reported. The carotid sheet is a well-defined structure below the carotid bifurcation, though it is incomplete or absent at the level of the oral-nasal pharynx (Harnsberger 1995). The infrahyoid segment of the carotid space contains the common carotid artery and depending on the level of the bifurcation the proximal part of the ICA; the proximal part of the ECA; furthermore the internal jugular vein (IJV); portions of the cranial nerves IX, X, XI, and XII; the sympathetic plexus; and lymph nodes. In the infrahyoid segment, the vessels run in the so-called carotid triangle (Som et al. 2003a) (Fig. 2.1) defined by

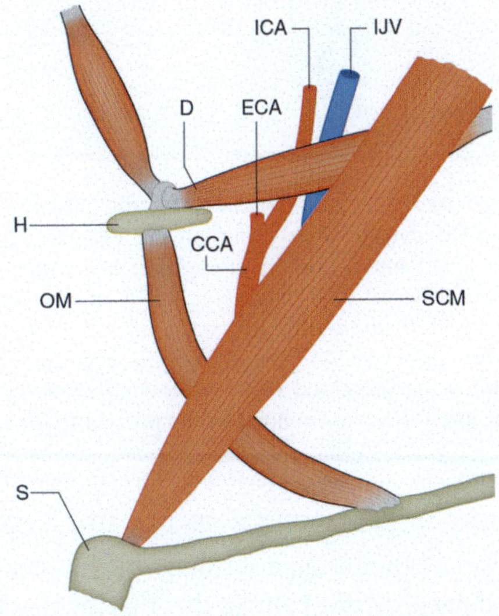

Fig. 2.1 Drawing showing the course of the carotid artery in the carotid triangle. Lateral—oblique view. *SCM* sternocleidomastoid muscle, *OM* superior belly of the omohyoid muscle, *D* posterior belly of the digastric muscle, *H* hyoid bone, *S* sternum, *CCA* common carotid artery, *ECA* proximal external carotid artery, *ICA* infra-suprahyoid segments of internal carotid artery, *IJV* internal jugular vein

G.B. Bradac, *Applied Cerebral Angiography*, DOI 10.1007/978-3-319-57228-4_2

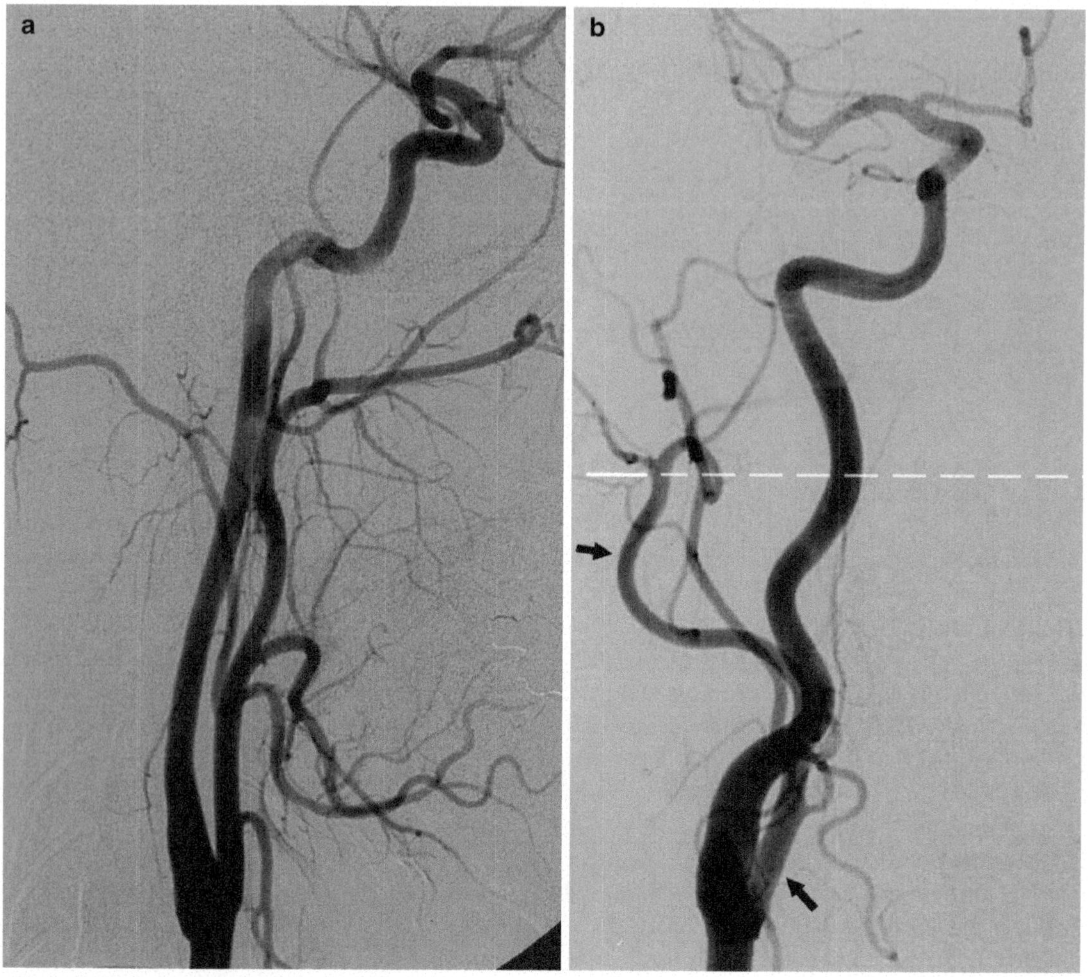

Fig. 2.2 (**a**) Common carotid angiogram, lateral view, showing the course of the external and internal carotid arteries. (**b**) Common carotid angiogram, AP view, showing the course of the external carotid artery (ECA) (*arrows*) first medial and more distally lateral to the internal carotid artery (ICA). The *dotted line* corresponds to the axial plane presented in the following drawing. (**c**) Carotid space (CS), surrounded by the parotid space (PS), the parapharyngeal space (PPS), the retropharyngeal space (RPS), and the perivertebral space (PVS). Masticator space (MS). In the carotid space are indicated the ICA (anteriorly), the internal jugular vein (posteriorly), and the cranial nerves IX, X, XI, and XII. In the anterior part of the parotid space are indicated the ECA and anteriorly to it the retromandibular vein

the sternocleidomastoid muscle laterally and posteriorly, and by the superior belly of the omohyoid and the posterior belly of the digastric muscle inferiorly and superiorly, respectively. In the suprahyoid segment, the ICA is accompanied by the IJV located posterolaterally; the cranial nerves IX, X, XI, and XII; the sympathetic plexus; and the lymph nodes. The ECA runs more laterally.

Near the skull base, the borders of the carotid space (Harnsberger 1995) also called by others (Som and Curtin 2003; Mukherji 2003) the retrostyloid para-pharyngeal space are so outlined: laterally, the parotid space; anteriorly and medially, the para-pharyngeal and retropharyngeal spaces, respectively; and posteriorly, the perivertebral space (Fig. 2.2c).

The first segment of the ICA (carotid bulb) is slightly enlarged, becoming smaller and narrower 1–2 cm distally. The bulb can be enlarged, particularly in older, atherosclerotic patients in whom also tortuosity can develop in the form of

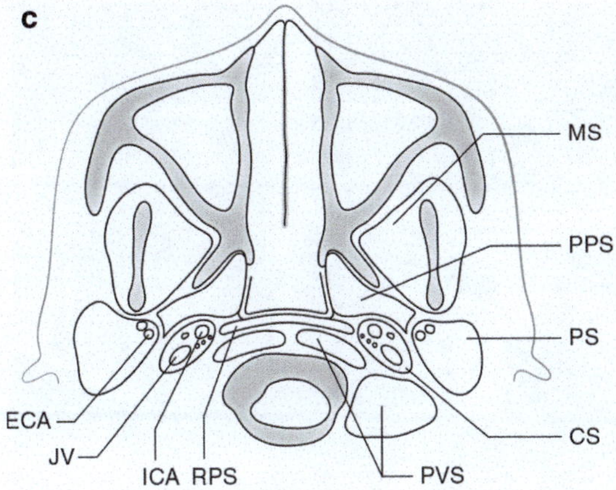

Fig. 2.2 (continued)

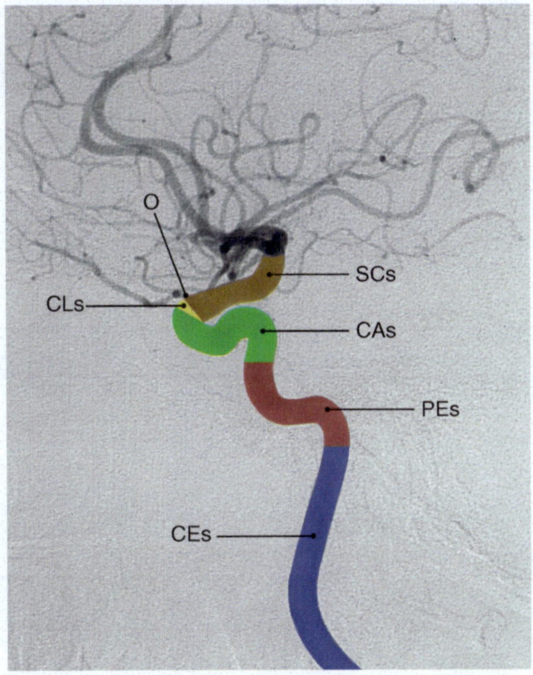

Fig. 2.3 Lateral angiogram on which are approximately indicated the segments of the internal carotid artery. *CEs* cervical segment, *PEs* petrous segment, *CAs* cavernous segment, *CLs* clinoid segment, *SCs* supraclinoid segment, *O* ophthalmic artery

coiling or kinking. Tortuosity is also frequently present as expression of dysplasia. In children tortuosity of ICA is frequent; this commonly disappears later.

At its origin, the ICA lies commonly posteriorly and laterally to the ECA. More distally, it runs medially to the ECA (Fig. 2.2a, b). See also Chap. 3.

2.2 Intracranial Segments of ICA

The intracranial segments of the ICA begin where the artery enters the cranial cavity at the level of the carotid foramen and end at the base of the brain at the anterior perforated substance. It can be subdivided into the petrous, cavernous, and supraclinoid segments (Fig. 2.3).

2.2.1 Petrous Segment of ICA

The ICA enters the base of the skull at the carotid foramen, anteriorly to the jugular fossa and jugular vein. It runs first entirely in the petrous bone, first with a vertical course for about 1 cm, and then horizontally medially directed for about 2 cm. It emerges at the petrous apex, running above the foramen lacerum, continuing in a short vertical segment toward the cavernous sinus. The ICA is surrounded by a fibrous-cartilaginous tissue, inferiorly covering the foramen lacerum and superiorly in continuation with the periosteum of the petrous canal. This segment, which has been called *lacerum segment*, terminates at the level of the petrolingual ligament (Bouthillier et al. 1996; Zyial et al. 1998) which runs between the lingula of the sphenoid bone anteriorly and the petrous apex posteriorly. It cannot be precisely defined neither on angiography nor on CT and MR studies. On the AP view the petrous ridge can be considered the reference point where approximately the ICA enters the cavernous sinus (Shapiro et al. 2014); see Figs. 2.3 and 2.4. Along its petrous course the ICA is surrounded by the sympathetic fibers and by a venous plexus. In its horizontal part it lies anteriorly, medially, and below the tympanic cavity and cochlea and runs below the Gasserian ganglion (Paulus et al. 1977; Bouthillier et al. 1996). It gives off two branches: the *caroticotympanic and the mandibular arteries.*

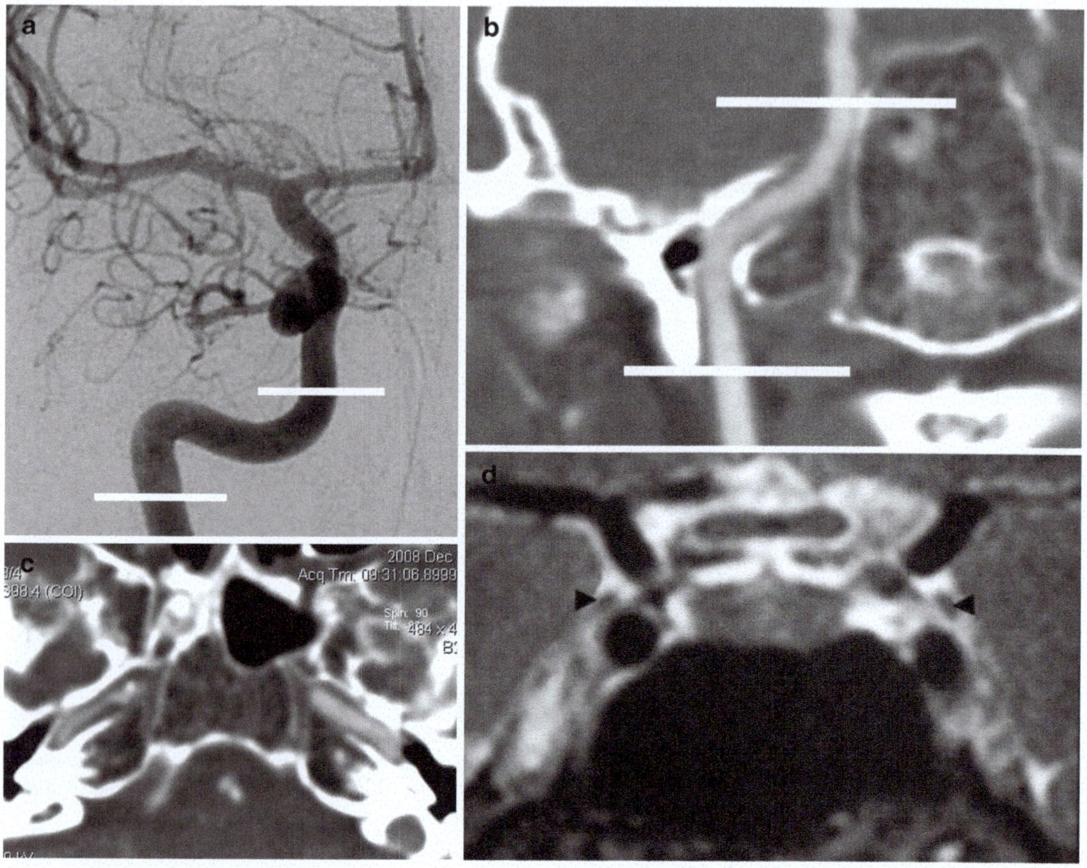

Fig. 2.4 (**a**) Carotid angiogram, AP view. The *lines* define approximately the course of the petrous segment of the ICA continuing into the cavernous segment. (**b**) CT—angiography, coronal reconstruction, showing the course of the petrous segment. (**c**) CT—angiography, showing the horizontal part of the petrous segment of the ICA running above the foramen lacerum. (**d**) MRI, coronal view, sellar, and parasellar area demonstrating the course of the ICA in the cavernous sinus. Cranial nerve III (*arrowheads*)

The caroticotympanic artery is an embryonic remnant of the hyoid stapedial artery (Sect. 2.2.3.1) that supplies the middle ear cavity. When present it arises at the genu between the vertical and horizontal segments running posteriorly. There is a possible anastomosis with the tympanic branch of the ascending pharyngeal artery (see also Sect. 3.4). The caroticotympanic artery can be involved in tumors of the skull base, particularly in tympanojugular paragangliomas. In cases of agenesis of the petrous segment of ICA it is considered to be involved together with the tympanic branch of the AphA in the formation of a kind of collateral circulation toward the distal ICA (see also Sect. 2.3 and Fig. 2.15).

The mandibular artery is a remnant of an embryonic branch arising slightly more cranial than the caroticotympanic artery. It divides into two branches directed anteriorly anastomosing with branches of the IMA: one runs in the pterygoid canal, anastomosing with the vidian artery; the other is more caudally located, anastomosing with the pterigovaginal artery (see also Sect. 3.7.3). The artery is sometimes visible on normal ICA angiogram (Fig. 11.9b) and can be especially involved in the vascularization of angiofibromas (Fig. 3.23).

Apart from the above-described pathological situations the caroticotympanic and the mandibular arteries are commonly not visible on the angiogram.

2.2.2 Cavernous Segment of ICA

In the cavernous segment the ICA runs in the space formed by the separation of a fold of the dura into two layers: the lateral one is the medial wall of the middle cranial fossa; the other is medial, in close contact in its inferior part with the periosteum of the sphenoid bone (periosteal layer). This space, in which run the ICA, venous channels, and nerves, has been called by Taptas (1982) "*the space of the cavernous sinus*." This definition, which distinguishes the space from its contents, is more appropriate than the commonly used "*cavernous sinus*" (see also Sect. 9.3.10). In this space, the ICA is directed first forward and upward, and then curves anteriorly with a horizontal course, making a second curve directed posteriorly and slightly medially to the anterior clinoid process. In its course, laterally to the sella turcica and pituitary gland which is separated by the medial layer of the dura, the artery is surrounded by a venous plexus, and the sympathetic fibers, and has a close relationship with cranial nerves III, IV, and VI and the first and second branch of the trigeminal nerves. The nerves run close to the lateral wall, attached to it by dural sheaths. These latter can fuse together, forming a thin, irregular inner layer adjacent to the external layer of the lateral wall (Umansky and Nathan 1982). Unlike the other nerves, cranial nerve VI runs inside the cavernous sinus.

Due to its "S-shaped" course, the cavernous segment is also called the siphon. It continues in the supraclinoid segment (Figs. 2.3 and 2.4). Before becoming supraclinoid, however, the ICA passes through a kind of dural collar formed by a proximal and a distal dural ring. This is a very short wedge-shaped segment called the *clinoid segment* (Bouthillier et al. 1996).

There are two branches of the cavernous segment: one is *the meningohypophyseal trunk (MHT)*, and the other is *the inferolateral trunk (ILT)*.

The MHT arises from the medial surface of the first part of the cavernous segment. It gives off a branch supplying the neurohypophysis (inferior hypophyseal artery), which is recognizable on the angiogram as a slight blush. It also gives off dural branches for the clivus and tentorium (clival and tentorial branches). The tentorial branch has been called the artery of Bernasconi and Cassinari, who first reported its angiographic visualization in the 1982. These dural branches anastomose with meningeal branches of the contralateral ICA and inferiorly with clival branches of the AphA. There are also possible anastomoses with branches of the middle meningeal artery.

The ILT arises from the lateral surface of the segment running laterally to the sella turcica. It supplies cranial nerves III, IV, and VI and partially the *ganglion Gasseri*. It gives off dural branches for the dura of the cavernous sinus and adjacent area. In the supply of this area, there is a balance between the ICA system, represented by the ILT, and branches of ECA, represented by the middle meningeal artery, accessory meningeal artery, artery of the foramen rotundum, and recurrent meningeal artery branch of the ophthalmic artery. One system can be dominant over the other. Anastomoses are frequently present.

The MHT and ILT are very fine arteries (Fig. 2.5), not always recognizable on the lateral angiogram. They can be dilated and well visible when involved in the supply of pathological processes, especially meningioma and dural arteriovenous fistulas (Figs. 3.28b, 13.10, 13.13, and 13.14).

2.2.3 Supraclinoid Segment of ICA

The supraclinod segment (average length 2 cm) begins where the artery having passed the clinoid segment goes through the dura and enters the subarachnoid space, running posteriorly, superiorly between the anterior clinoid process laterally and the optic nerve located medially and superiorly to the ICA, reaching the lateral side of the optic chiasm. The dural ring surrounding the ICA, where the artery enters the subarachnoid space, is closely adherent to the artery laterally, but it is frequently less adherent medially, forming a thin cavity (carotid cave). Aneurysms arising below the dural ring (intracavernous

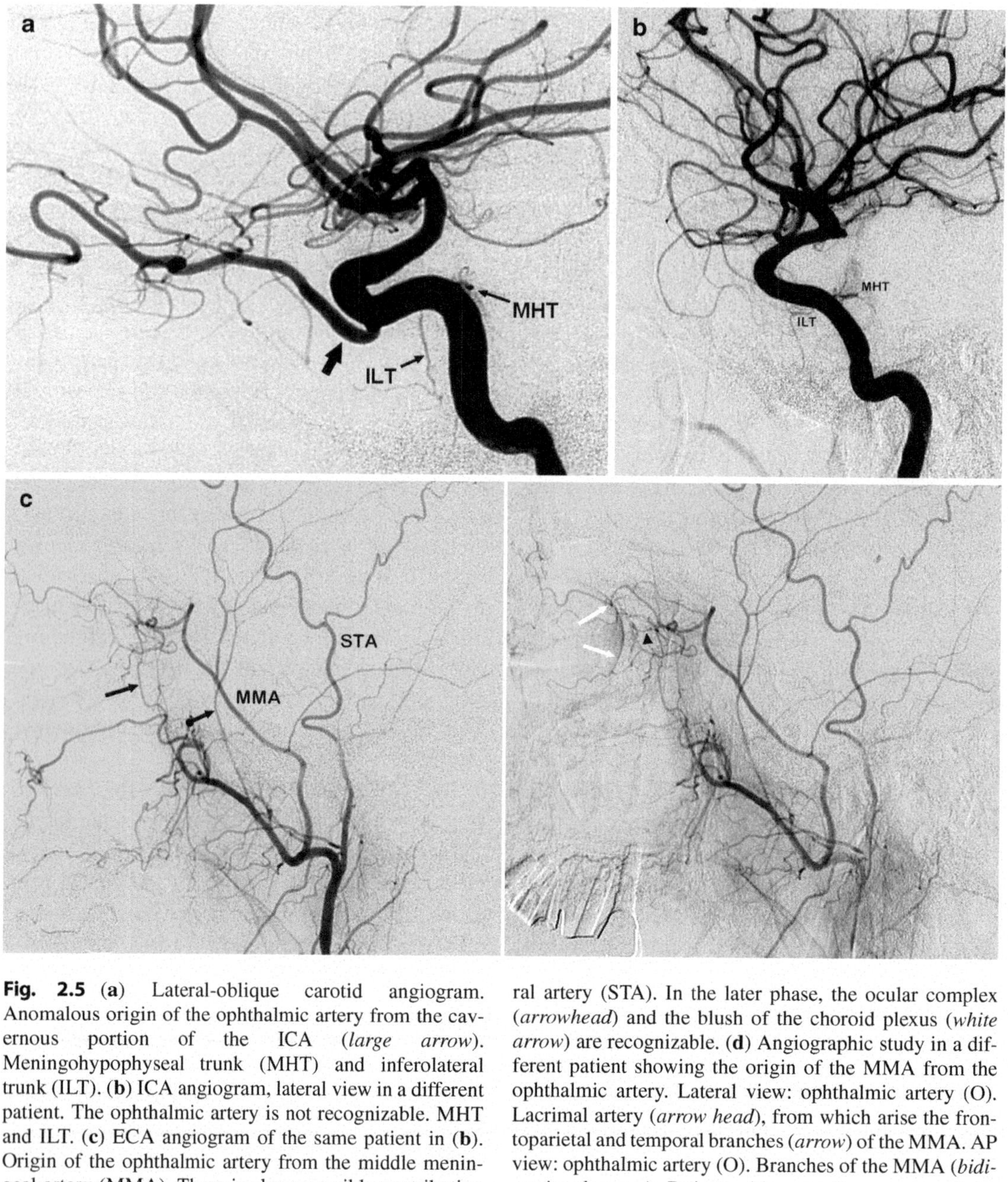

Fig. 2.5 (**a**) Lateral-oblique carotid angiogram. Anomalous origin of the ophthalmic artery from the cavernous portion of the ICA (*large arrow*). Meningohypophyseal trunk (MHT) and inferolateral trunk (ILT). (**b**) ICA angiogram, lateral view in a different patient. The ophthalmic artery is not recognizable. MHT and ILT. (**c**) ECA angiogram of the same patient in (**b**). Origin of the ophthalmic artery from the middle meningeal artery (MMA). There is also a possible contribution from the anterior deep temporal artery (*arrow*). Middle deep temporal artery (*arrow with dot*). Superficial temporal artery (STA). In the later phase, the ocular complex (*arrowhead*) and the blush of the choroid plexus (*white arrow*) are recognizable. (**d**) Angiographic study in a different patient showing the origin of the MMA from the ophthalmic artery. Lateral view: ophthalmic artery (O). Lacrimal artery (*arrow head*), from which arise the frontoparietal and temporal branches (*arrow*) of the MMA. AP view: ophthalmic artery (O). Branches of the MMA (*bidirectional arrow*). Patient with a small aneurysm of the PcomA

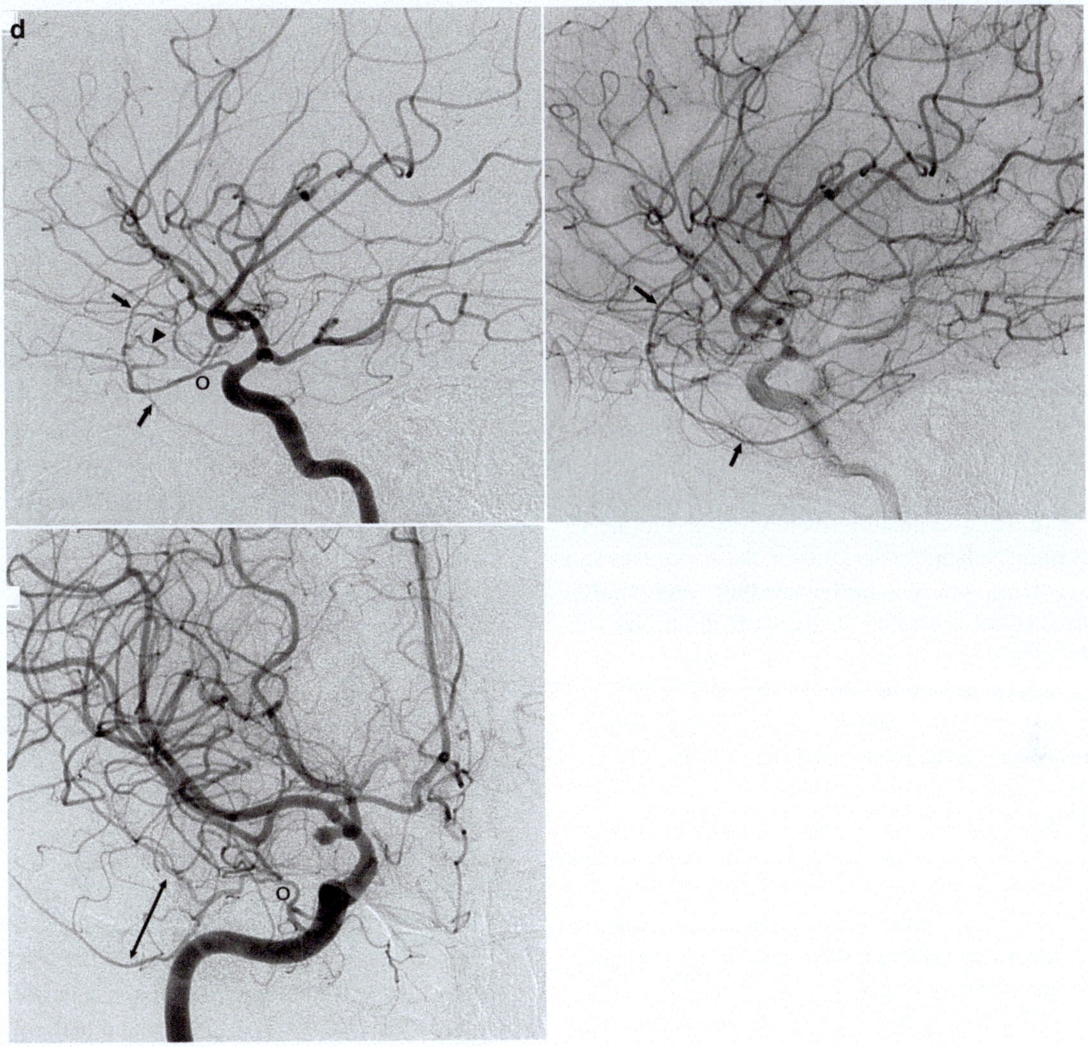

Fig. 2.5 (continued)

aneurysms) can, however, extend superiorly into the subarachnoid space. They are called "*cave aneurysms*" (Kobayashi et al. 1995; Rhoton 2002).

It is not possible for angiographic studies to establish with certainty where the supraclinoid segment begins. This, however, can be approximately located close to the origin of the ophthalmic artery, proximally to it, considering that the artery has commonly an intradural origin (Fig. 2.3). Gibo et al. (1981a) have subdivided the supraclinoid segment in three parts:

- *The ophthalmic segment*, from the origin of the ophthalmic artery to the origin of the posterior communicating artery (PcomA)
- *The communicating segment*, from the origin of the PcomA to the origin of the choroidal artery
- *The choroidal segment*, from the origin of the anterior choroidal artery to the terminal bifurcation of the ICA

At the level of the anterior perforated space (APS), the artery divides into the anterior and middle cerebral arteries.

2.2.3.1 The Ophthalmic Segment

In the ophthalmic segment arise the ophthalmic artery, superior hypophyseal arteries, and other small branches directed toward chiasma and floor of the third ventricle.

The Ophthalmic Artery

The ophthalmic artery (OA) arises from the superior-medial surface of the ICA, commonly very close to the point where the ICA had perforated the dura. It runs below the optic nerve (Hayreh and Dass 1962a, b; Hayreh 1962), pierces again the dura, and enters the orbita, together with the nerve, through the optic canal. Initially the artery runs inferolaterally to the optic nerve (first segment), then crosses the nerve forming a bend under or over the nerve (second segment), and runs further medially and parallel to it (third segment). It gives off three types of branches: *ocular, orbital, and extraorbital.* There is a great variability concerning the origin and course of these branches. Some typical pattern, however, can be recognized (Hayreh 1962).

- The ocular branches include the central retinal artery, the ciliar arteries, and the branches going to the optic nerve. Considering the origin of the optic nerve arteries, they can be classified (Hayreh 1962) into intracanalicular branches arising from the ophthalmic artery close to its origin or in its intracanalicular course, and in intraorbital branches. These latter arise more distally, commonly at the bend where the artery crosses under or over the optic nerve. The central retinal artery and the medial and lateral posterior ciliar arteries, both associated in the supply of the eyeball, arise commonly in the area where the ophthalmic artery crosses under or over the optic nerve.
- The orbital branches include the lacrimal artery, which supplies the lacrimal gland and conjunctiva. An important branch of the lacrimal artery, frequently present, is the recurrent meningeal artery, which runs backwards and passes through the superior orbital fissure, anastomosing with branches of the middle meningeal artery (MMA). There is frequently a balance between these arteries; one can predominate. It can be involved in many pathological processes typically basal meningiomas (Fig. 3.28a), and dural arteriovenous fistulas (Fig. 13.13), and in the supply of angiofibromas and chemodectomas extending toward the orbita and parasellar region (Fig. 3.23). Anastomosis of the lacrimal artery with the anterior deep temporal artery can be an important collateral circulation via the OA in occlusion of the ICA (Fig. 3.12). Other orbital branches are the muscular arteries, which supply the muscle and orbital periosteum.
- The extra orbital branches are numerous. They include the posterior and anterior ethmoidal arteries. The posterior arises from the first segment, and the anterior from the third. These branches have an ascending course and pass through the lamina cribrosa, supplying the dura of the basal anterior cranial fossa. From the anterior ethmoidal artery arises the anterior falx artery, which supplies the falx and anastomoses with the falx branches of the MMA. From the ethmoidal arteries arise branches with a descending course anastomosing with the ascending branches of the sphenopalatine arteries, branches of the internal maxillary artery (IMA). These arteries (Fig. 3.28a) are typically involved in the vascularization of meningiomas of the anterior cranial fossa (Bradac et al. 1990), and in dural arteriovenous fistulas of the region (Fig. 13.19). Involvement in the supply of angiofibroma extending toward the orbita can also occur (Fig. 3.23).

Other arteries of this group are the supraorbital (frequently prominent), the dorsonasal, the medial palpebral, and the supratroclear. These branches anastomose with branches of the ECA, in particular with the facial artery, infraorbital branch of the IMA, and frontal branches of the superficial temporal artery. Such anastomoses may be a collateral way via the OA toward the ICA when the latter is occluded (Figs. 3.12 and 3.13). Furthermore, these branches can be involved in vascular malformations of the craniofacial area (Fig. 3.19).

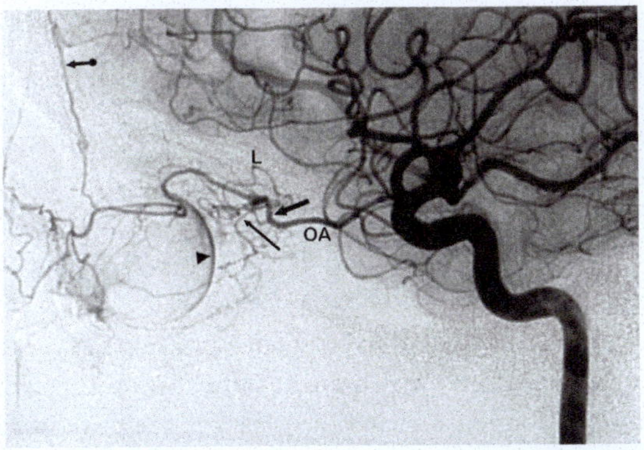

Fig. 2.6 Lateral ICA angiogram. Ophthalmic artery (OA). Bend of the artery around the optic nerve (*large arrow*). From this segment arises the ocular complex comprising the retina and ciliar arteries (*small arrow*). Choroid plexus (*arrow head*), lacrimal artery (L), anterior falx artery (*arrow with dot*)

On the angiogram of the ICA (Vignaud et al. 1972; Huber 1979; Morris 1997; Osborn 1999), the OA is always visible with the exception of the cases in which the artery arises from the MMA. The artery is better defined on the lateral view. From its origin, it runs superiorly for 1–2 mm, and then anteriorly, forming a slight curve with an inferior convexity. About 2 cm from its origin, the OA curves abruptly crossing the optic nerve. The central retinal artery, the ciliar arteries, and the distal branches for the optic nerve arise at the level of the above-described curve (Fig. 2.6). Thus, in embolization procedures involving the OA, the microcatheter should be advanced distally to the above-described curve. The blush corresponding to the plexus of the ocular choroid is always visible as a crescent-shaped structure. The ethmoidal arteries are occasionally evident especially in the lateral view. The anterior falx artery is also easily identifiable, when present, on the lateral angiogram. The other branches are difficult to recognize under normal conditions.

To explain some variations of the OA, it is useful to recall the most important aspects of its embryogenesis (Padget 1948; Hayreh and Dass 1962a, b; HaYreh 1962; Lasjaunias et al. 2001). The definitive OA develops from three sources: *the primitive dorsal OA*, arising from the cavernous portion of the ICA and entering the orbita through the superior orbital fissure; *the primitive ventral OA,* arising from the anterior cerebral artery and entering the orbita through the optic canal; and *the stapedial artery (StA)*, which gives off an orbital branch entering the orbita through the superior orbital fissure.

Inside the orbita and around the optic nerve, an arterial anastomotic circle is formed involving these three arteries. In the further evolution the proximal segment of the primitive ventral OA disappears, arising now from the supracavernous portion of the ICA. This artery will become the definitive OA. The primitive dorsal OA regresses, and the intraorbital branches of the StA are annexed by the definitive OA. In this process, the important changes involving the StA deserve an extra short description.

The StA is the main branch of the hyoid artery, embryonic vessel arising from a segment of ICA which in this stage of evolution is very small and incompletely developed. Later, this segment will become the petrous ICA (see also Sect. 1.1 and Chap. 3). *The StA enters the middle cranial fossa, passing through the tympanic cavity, and divides into intracranial and extracranial branches* (Moret et al. 1977; Lasjaunias et al. 2001). *The intracranial branch (supraorbital artery) is anteriorly directed and supplies the dura of the middle cranial fossa, and extends into the orbita, with a medial and lateral (lacrimal) branch. Both branches enter the orbita through the superior orbital fissure. In some cases, the lacrimal artery penetrates as an isolated branch through the foramen of Hyrtl, located in the greater wing of the sphenoid bone. The second branch of the StA (maxillomandibular artery) is directed extracranially, passing through the foramen spinosum. In the further evolution this branch anastomoses with the primitive ventral pharyngeal artery embryonic vessel representing the proximal external carotid artery. From this connection develops the final internal maxillary artery (IMA) and the middle meningeal artery (MMA). The blood flow is now reversed, being intracranially directed. The intracranial branch of the StA is annexed by the developed MMA, and its intraorbital segment is annexed by the definitive OA.*

As many authors have described (McLennan et al. 1974; Moret et al. 1977; Rodesch et al. 1991b; Morris 1997; Lasjaunias et al. 2001; Perrini et al. 2007), abnormal changes can occur in the embryological evolution of StA and the primitive ventral and dorsal ophthalmic arteries, leading to a series of conditions characterized by different angiographic patterns:

- The proximal part of the primitive ventral OA does not regress and so the OA arises from the anterior cerebral artery (Hassler et al. 1989). This evolution could also explain the origin of the OA from the distal ICA bifurcation as reported by some authors (Parlato et al. 2011).
- The primitive ventral OA disappears instead of the primitive dorsal OA, leading to the cavernous origin of the OA (Figs. 2.5a, 2.7c and 4.11c).
- The proximal segment of the primitive ventral OA does not develop, while the intraorbital part of the StA remains and is connected at the level of the superior orbital fissure with the MMA. In such a condition, the OA is only visible on the ECA, not on the ICA angiogram (Fig. 2.5b, c).
- The lacrimal branch can persist as an isolated branch of the MMA (meningolacrimal artery) entering the orbita through the foramen of Hyrtl and supplying partially the intraorbital structures, while the ocular and neuronal branches arise from the OA. In such cases, the orbital vascularization is partially visible on the ECA and partially on the ICA angiograms. There are commonly no anastomoses between these two systems. In other cases, the MMA gives off a branch, which enters the orbita through the superior orbital fissure and anastomoses with the lacrimal branch normally arising from the OA.
- Another condition is characterized by the origin of the MMA from the OA and so it is only recognizable on the ICA angiogram. This occurs when the intracranial part of the MMA does not develop; the proximal part of the intraorbital-transsphenoidal segment of the StA does not regress and anastomoses with the lacrimal branch of the OA (Fig. 2.5d).
- The MMA originates in the petrous segment of the ICA. This occurs when the first and intracranial segments of the StA do not regress and the extracranial part of the MMA (part of the maxillomandibular) does not develop. On the skull CT, the foramen spinosum is not present and the MMA is only visible on the ICA angiogram. In other cases the persistence of the first segment of the StA can appear as a small branch arising from the petrous segment of the ICA (caroticotympanic artery). See also Sect. 2.2.1.
- Finally, cases of origin of the ophthalmic artery from the basilar artery have been described (Schumacher and Wakhloo 1994; Sade et al. 2004). It is difficult to explain this very rare anomaly considering the classical description of the embryogenesis of the OA. Similarly difficult to explain is the embryological mechanism responsible for the origin of the MMA from the basilar artery as reported by some authors (Seeger and Hemmer 1976; Shah and Hurst 2007; Kumar and Mishra 2012).

Superior Hypophyseal Arteries

The superior hypophyseal arteries (SHA) are formed by a group of small branches arising commonly from the posteromedial surface of the ophthalmic segment of the ICA. The SHA supplies the infundibulum and the anterior lobe of the pituitary gland. Other small branches are involved in the supply of the optic nerve, chiasma, optic tract, and anterior part of the floor of the third ventricle. All these arteries are very small and not recognizable on a normal angiogram.

Supply of the Pituitary Gland

The adenohypophysis is supplied by the superior hypophyseal arteries. These run toward the pituitary stalk, where they connect forming a network of capillaries continuing in venules forming the so-called venous portal system, through which flow the releasing and released-inhibiting hormones from the hypothalamus to the adenohypophysis. The neurohypophysis is supplied by the inferior hypophyseal artery which is a branch of the MHT. There are anastomoses between the branches of the superior hypophyseal and inferior hypophyseal arteries and those of the contralateral arteries.

Each half of the pituitary gland drains into the corresponding cavernous sinus. The further drainage occurs through the inferior petrosal sinuses.

2.2.3.2 The Communicating Segment

From this segment arises the posterior communicating artery (PcomA). The PcomA arises from the posterior surface of the ICA. It runs posteriorly and medially to join the posterior cerebral artery (PCA), in a close relationship with cranial nerve III, which is commonly laterally but some time medially located (Gibo et al. 1981a).

Commonly, the PcomA is slightly smaller than the PCA. It may be, however, very large, continuing directly into the PCA. This variant present in about 30% of the cases is termed the "fetal" origin of the PCA. Indeed, in the embryonic phase, the PCA develops as a posterior extension of the PcomA, caudal division of the ICA. The connection of the PCA with the basilar artery through the P1 segment develops later. In the further evolution, the PcomA (*pars carotica* of the PCA) becomes hypoplastic or regresses entirely, while the P1 segment (*pars basilaris* of the PCA) becomes well developed. This latter evolution occurs in about 70% of the cases (Zeal and Rhoton 1978; Huber 1979; Pedroza et al. 1987). See also Chaps. 1 and 7.

A slight widening of the PcomA at its origin (infundibulum) is not rare. It has been described in 6.5% of normal angiograms (Hassler and Salzmann 1967), and it has been interpreted as an early stage of aneurysm formation. Other studies (Epstein et al. 1970a) made on autopsy specimen have demonstrated neither an aneurysmal nor a preaneurysmal aspect.

The infundibulum appears as a homogeneous conical dilatation of 2–3 mm at the origin of the PcomA. Commonly, it is not considered as a pathological finding. Sometimes, however, the differential diagnosis with a real aneurysm can be difficult and be suspected when the infundibulum has not the typical conical form especially in patients with SAH in whom no other aneurysm can be detected. In these cases, it is also important to exclude, with a short-time angiographic control, the presence of an aneurysm responsible of the SAH not visualized in the first study. This will be discussed in Sect. 11.11. Some authors have reported the very rare evolution of the infundibulum into a saccular aneurysm (Marshman et al. 1998; Cowan et al. 2004; Radulovic et al. 2006; Fischer et al. 2011). In patients with infundibulum and unclear SAH or familiarity of aneurysms, a follow-up in yearly intervals has been proposed (Fischer et al. 2011).

From the PcomA arise many perforating branches. Since the first description by Duret (1874), many anatomical studies have been performed (Foix and Hillemand 1925a, b; Lazorthes and Salamon 1971; Percheron 1976b; Saeki and Rhoton 1977; Zeal and Rhoton 1978; Gibo et al. 1981a; Ono et al. 1984; Pedroza et al. 1987, Tatu et al. 2001). These arteries have been variously termed: tuberothalamic, premammillary, and anterior thalamoperforating arteries. The latter definition seems to us to be the most appropriate and it is the one we will adopt. Many branches are present, also in cases of a smaller PcomA. Among them, there is sometimes a large branch arising in front of or besides the mammillary body (Gibo et al. 1981a; Pedroza et al. 1987). These perforators supply the posterior part of the chiasma, the optic tract, and the mammillary body. They enter the posterior perforated substance, supplying the hypothalamus, subthalamus, and anterior thalamus. Some authors (Gibo et al. 1981a) have found that they can supply also the posterior limb of the internal capsule.

Variations. Anomalous origin of the PcomA is very rare. Its origin from the OA has been reported (Bisaria 1984). Cases of PcomA arising distally to the AchA have also been described (Moyer and Flamm 1992). As far as it concerns its relationship with the anterior choroidal artery, see below in this chapter. The presence of a well-developed or hypo-aplastic PcomA has already been discussed. As it will be better described in the chapters treating the stroke this variation can play an important role as a way of a possible collateral circulation. Indeed the fetal PCA can be involved in occlusions involving the anterior circulation, but can represent a way of collateral circulation when the artery arises from the basilar artery. A different situation occurs in case of occlusion of the basilar artery, in which the fetal PCA can act as an important collateral way.

Depending on its development, the PcomA can be injected only on the ICA angiogram, on the vertebrobasilar angiogram or on both (Figs. 2.7 and 7.5). In the angio-MR, the PcomA, P1, and PCA complex can be well identified (Figs. 7.2 and 7.5d). The perforating branches can be only well examined on the lateral and AP vertebral and sometimes on the lateral carotid angiograms (Figs. 2.7c, 7.8, and 7.9). They

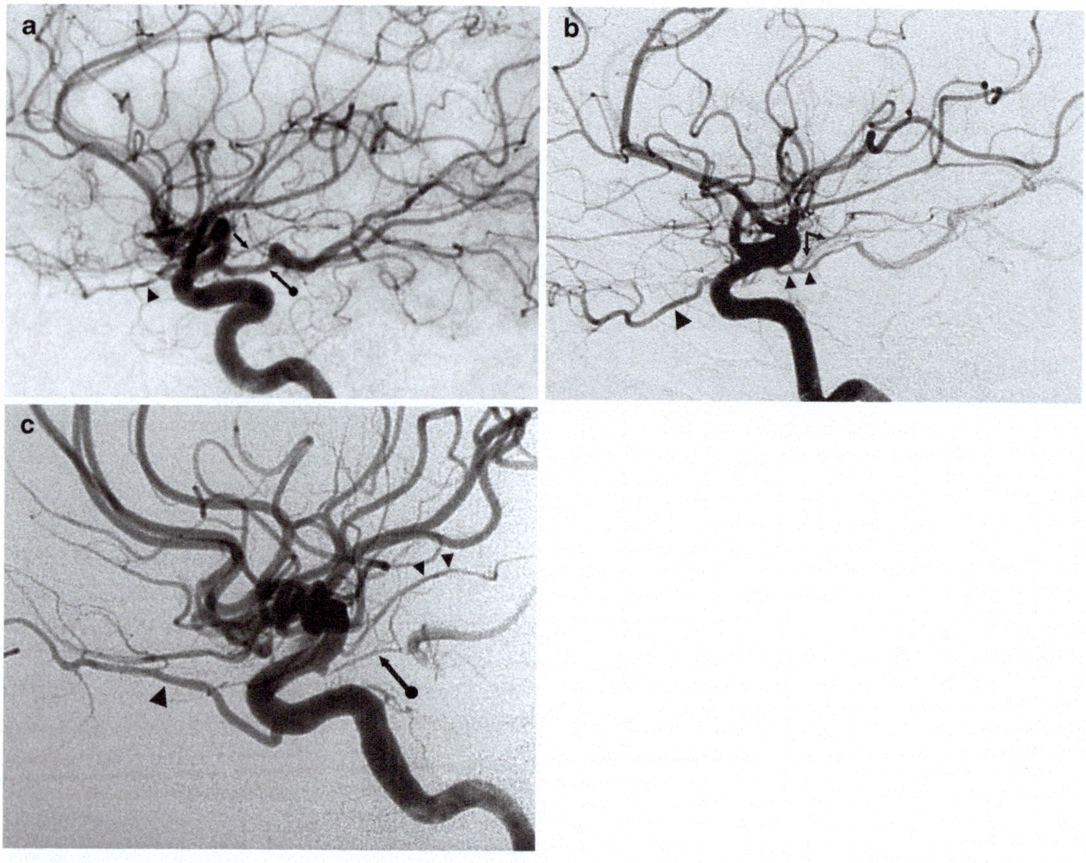

Fig. 2.7 Carotid angiogram. Lateral view in three different cases. (**a**) Large PcomA (*arrow with dot*) continuing in the PCA. Remark the very proximal origin of the temporal and parieto-occipital branches of the PCA. Anterior choroidal artery (*arrow*). Ophthalmic artery (*arrowhead*). (**b**) Small PcomA (*arrowheads*) continuing in the PCA. Anterior choroidal artery (*arrow with angle*). Ophthalmic artery (*arrowhead*). Owing to overlap, the anterior choroidal artery erroneously seems to arise proximally to the PcomA. This extremely rare condition can occur and should be identified by complementary projections. (**c**) Small PcomA (*arrow with dot*) continuing in the PCA. Note the ascending course of a perforating branch. There is a small infundibulum at the origin of the PcomA. Anterior choroidal artery (*arrowheads*). Anomalous origin of the ophthalmic artery from the cavernous portion of ICA (*arrowhead*)

appear as small branches running upwards and slightly backwards.

2.2.3.3 The Choroidal Segment

In the choroidal segment arise the anterior choroidal artery and perforators directly from the ICA.

The Anterior Choroidal Artery

In all cases studied by Rhoton et al. (1979) and Fujii et al. (1980), the anterior choroidal artery (AchA) arose from the posterior surface of the ICA, more laterally to the site of origin of the PcomA, and 2–4 mm distal to it. The AchA can be divided into a cisternal segment, from its origin to the choroid fissure, and a distal segment (Goldberg 1974; Rhoton et al. 1979). The cisternal segment, from which arise the main arteries supplying the parenchyma, has an average length of 25 mm (Otomo 1965; Rhoton et al. 1979). From its origin the AchA runs first medially, below the optic tract crossing it from lateral to medial. The artery runs then laterally around the midbrain in the circumpeduncular cistern. The basal vein lies superiorly and the PCA inferiorly. At the level of the lateral geniculate body, the artery curves sharply crossing the optic tract from medial to lateral, entering the temporal horn through the choroid fissure joining the choroid plexus. The artery extends posteriorly reaching the atrium, where it can anastomose with branches of the posterolateral choroidal artery. Rarely it extends anteriorly toward the foramen of Monro, supplying the plexus and anastomosing with the posteromedial choroidal artery.

From the cisternal segment arise many branches which can be divided into superior, lateral, and medial (Abbie 1933; Carpenter et al. 1954; Rhoton et al. 1979; Duvernoy 1999; Tatu et al. 2001). It can occur that the branches of the superior group do not arise from the main trunk of the AchA, but can have a separate origin directly from ICA. They supply the optic tract and enter the APS posteriorly to the perforators of the distal ICA and A1 segment of the ACA, and medially to those of the MCA. They supply the medial part of the globus pallidus, the tail of the nucleus caudatus, and sometimes the genu of the internal capsule (Goldberg 1974; Rhoton et al. 1979). The most posterior branches arise at the level of the lateral geniculate body, and penetrate the brain to supply the inferior part of the posterior limb of the internal capsule, its retrolenticular segment, and the optic radiations (Rhoton et al. 1979). According to some authors (Hupperts et al. 1994), they can be involved also in the vascularization of the parietal periventricular area. The lateral group supplies the uncus, amygdala, and hippocampus; the medial group the anterolateral midbrain; and the lateral geniculate body (Rhoton et al. 1979).

There is a marked interchangeability in the vascular territory described among the AchA, ICA, PcomA, PCA, and MCA (Rhoton et al. 1979). *Moreover, there are rich anastomoses between the Acha, PcomA, and PCA through branches on the surface of the temporal lobe near the uncus and of the lateral geniculate body. Other anastomoses occur with the choroidal arteries. All these aspects make it difficult to predict the effect of occlusion of the AchA* (Theron and Newton 1976; Rhoton et al. 1979; Friedman et al. 2001).

The artery is commonly well recognizable on the anteroposterior (AP) and lateral angiograms. On the lateral angiogram, the artery runs backward, forming an upward convex curve (cisternal segment). It runs further posteriorly entering the choroid fissure (plexal segment). The plexal segment in the temporal horn and atrium shows a typical blush in the later arterial-capillary phase. On the AP angiogram, the AchA runs first medially and then laterally, surrounding the cerebral peduncle, mixing with perforators of the middle cerebral artery (Figs. 2.7, 2.8, and 2.9).

Variations. Many anomalies concerning the origin and development of the AchA have been described. A few cases of origin from PcomA or middle cerebral artery (Carpenter et al. 1954; Morello and Cooper 1955; Sjoegren 1956; Otomo 1965; Herman et al. 1966; Theron and Newton 1976; Lasjaunias and Berenstein 1990) or from the ICA proximal to the PcomA (Moyer and Flamm 1992; Nishio et al. 2009), as well as a

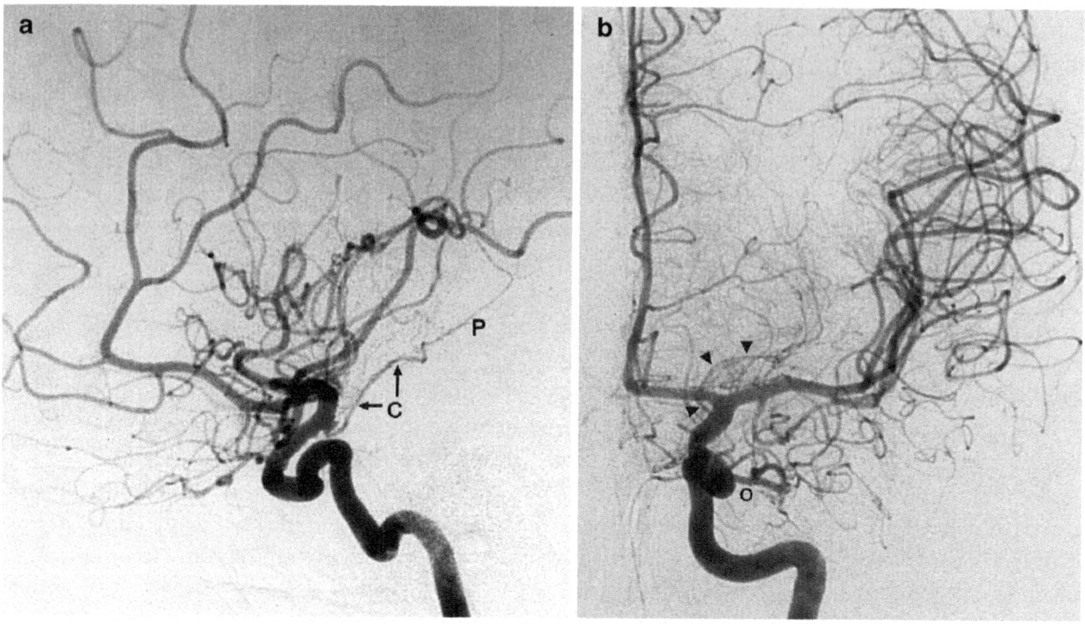

Fig. 2.8 (**a**) Lateral carotid angiogram. Anterior choroidal artery with its cisternal (C) and plexal segments (P). (**b**) AP view. Course of the anterior choroidal artery (*arrowhead*). Ophthalmic artery (O)

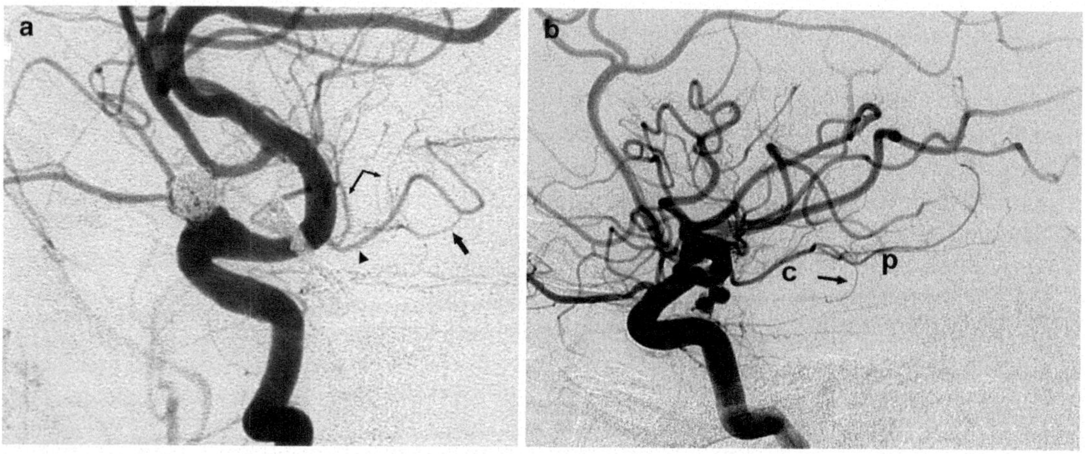

Fig. 2.9 (**a**) Carotid angiogram (oblique view) in a patient with carotid-ophthalmic aneurysm treated with coils. Well-developed AchA (*arrowhead*) from which arise large perforators (*arrow with angle*). Uncal branch (*arrow*). The clip projecting on the ICA was used to treat a contralateral aneurysm. (**b**) Lateral carotid angiogram in patient examined for a ruptured aneurysm of ICA. There is a well-developed AchA. Cisternal (C) and plexal (P) segments. Uncal branch (*arrow*). (**c**) Lateral carotid angiogram. Large PcomA continuing in the PCA (fetal type) (*arrow*). Above there is a well-developed artery, corresponding to a hyperplastic AchA (*arrow with dot*), which takes over partially the vascular territory of the PCA. Above the AchA there is a small branch (*arrowhead*) corresponding probably to a perforating branch arising from the AchA or directly from the ICA. (**d**) Carotid angiogram (oblique view). Large PcomA (*arrow*) continuing in the PCA. Above it there is a well-developed, hyperplastic AchA (*arrowhead*), which takes over the parietal vascular territory of the PCA

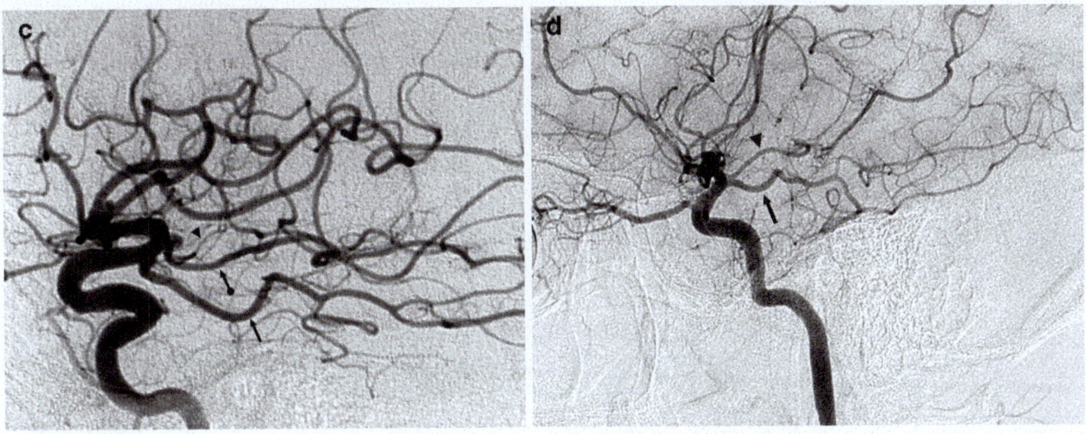

Fig. 2.9 (continued)

case of aplasia (Carpenter et al. 1954), have been reported. Other authors (Saeki and Rhoton 1977) found a few cases of duplicated AchA with two branches arising separately from ICA.

Another anomaly concerns development of the AchA, which can take, in accordance with the studies of some authors (Blackburn 1907; Bergquist 1975; Takahashi et al. 1980, 1990), a hyperplastic or a hypoplastic form. In the hyperplastic form the AchA is well developed taking over partially or almost completely the vascular territory of the PCA. On the angiogram the AchA appears as a large vessel running superiorly to the inferiorly located PcomA continuing in the PCA. The vascular territory normally supplied by the PCA is partially or almost completely taken over by branches of the hyperplastic AchA. This pattern has been interpreted in the past as expression of a duplicated PcomA or PCA. This anomaly probably occurs. However, in the majority of these cases, this pattern is due to a hyperplastic AchA due to an anomalous development and persistence of the anastomoses between AchA and PCA and PcomA normally present in the embryogenesis. The angiographic pattern can change depending on the site of the persistent anastomoses. If these are proximal the PcomA and PCA are hypoplastic and their vascular territory is largely replaced by a well-developed AchA. If the anastomoses are more distal the vascular territories are taken over partially by the AchA and partially by the PCA. In the hypoplastic form of the AchA, only its cisternal segment is developed. Examples are presented in Figs. 2.9a–d and 2.12.

The Perforators of ICA

The perforators of ICA arise from the choroidal segment of the ICA, typically from its posterior wall, distal to the origin of the AchA (Rosner et al. 1984; Mercier et al. 1993). They enter the APS in its posteromedial part overlapping with perforators arising from AchA and supply the genu of the capsula interna, its posterior limb, and the adjacent part of the pallidum. They can replace perforators of the AchA and pars of perforators of the MCA and vice versa. The perforators are rarely evident on the angiogram.

2.3 Congenital Anomalies of the CCA and ICA

- Extremely rare cases of *absence of the CCA* on the right or left have been reported. It has been suggested (Haughton and Rosenbaum 1974) that in these cases, the dorsal part of the third aortic arch (*ductus caroticus*) does not regress. Consequently, there is no formation of the CCA and the ICA and ECA arise separately from the third aortic arch. An anomalous origin of the CCA can also occur (see also Sect. 1.2).

- *Aplasia and hypoplasia of ICA* are other rare anomalies. Tode is reported to be the first to have described an aplasia of the ICA in 1787 on an autoptic examination. Verbiest (1954) described this anomaly on an angiographic study. Later many authors have described this anomaly which in the majority of the cases is discovered incidentally in patients studied for other reasons. It can be uni- or bilateral (Lie 1968; Lee et al. 2003; Mahadevan et al. 2004). Association with a hypoplastic common carotid artery as well as with cerebral aneurysms and with other severe pathologies such as PHACES, and neurofibromatosis can occur (Lee et al. 2003; Mahadevan et al. 2004). Today congenital forms of hypo-aplasia can already be suspected in CT or MR showing, respectively, a small or absent carotid canal and reduction or absence of blood flow. In angiographic studies various types of collateral circulation may be demonstrated, which correspond probably to the phase of the embryological evolution in which the anomaly develops (Cali et al. 1993; Lee et al. 2003). Among the most frequent reported in the literature, there are *that occurring through the circle of Willis involving AcomA and PcomA, and that through the presence of a persistent primitive trigeminal artery* (Lie 1968; Given et al. 2001; Lee et al. 2003).

  A particular form of collateral circulation is characterized by *the persistence of the primitive maxillary artery*, an embryonic vessel which arises at the cavernous portion of the ICA, and runs on the floor of the sella, forming an intrasellar anastomosis connecting both ICAs. Cases in which the anastomosis runs posterior to the dorsum sellae have also been reported. Many authors have been described this rare anomaly (Kishore et al. 1979; Staples 1979; Elefante et al. 1983; Alexander et al. 1984; Tracy 1987; Cali et al. 1993; Midkiff et al. 1995; Chen et al. 1998; Gozzoli et al. 1998; Lajaunias et al. 2001; Given et al. 2001), which should be taken into consideration in particular in patients in whom a transsphenoidal surgery for intrasellar adenoma is planned. Cases associated with hypopituitarism have been reported (Mellado et al. 2001; Moon et al. 2002).

  According to De La Torre and Netzky (1960), the primitive maxillary artery is an embryonic vessel arising from the cavernous portion of ICA, giving off branches extending to the anteromedial and lateral adjacent areas anastomosing with branches of the middle meningeal artery. The artery normally regresses and becomes partially incorporated on both sides in the inferior hypophyseal artery branch of the MHT (Padget 1948). Parkinson (1964), in a large autopsy study, has shown the presence of anastomoses connecting both cavernous carotid arteries through the inferior hypophyseal artery running on the floor of the sella and through the clival arteries, both branches of the MHT. In cases of aplasia of ICA, either the inferior hypophyseal or the clival arteries can enlarge, becoming an important collateral circulation toward the absent ICA. An example is shown in Fig. 2.10.

  Another very rare collateral is represented by the so-called *rete mirabile*. This is a well-known arterial network occurring in other vertebrates to compensate a rudimentary or absent ICA, but is a rarity in the humans. It is formed mainly by branches of the ECA, especially by those arising from the internal maxillary and ascending pharyngeal arteries. The appearance of the network also on the angiogram of the ICA suggests the involvement of persistent embryological vessels such as the stapedial artery, or neovascularization through the *vasa vasorum*. The *rete mirabile* occurs typically in cases of segmental aplasia of the ICA involving its petrous segment allowing the reconstitution of the distal ICA (Minagi and Newton 1966; Danziger et al. 1972; Mahadevan et al. 2004; Lin et al. 2013). Cases of associated aplasia of the transdural vertebral artery with a distal filling of the basilar artery through a rete mirabile involving PICA and spinal arteries have also been reported (Hyogo et al. 1996;

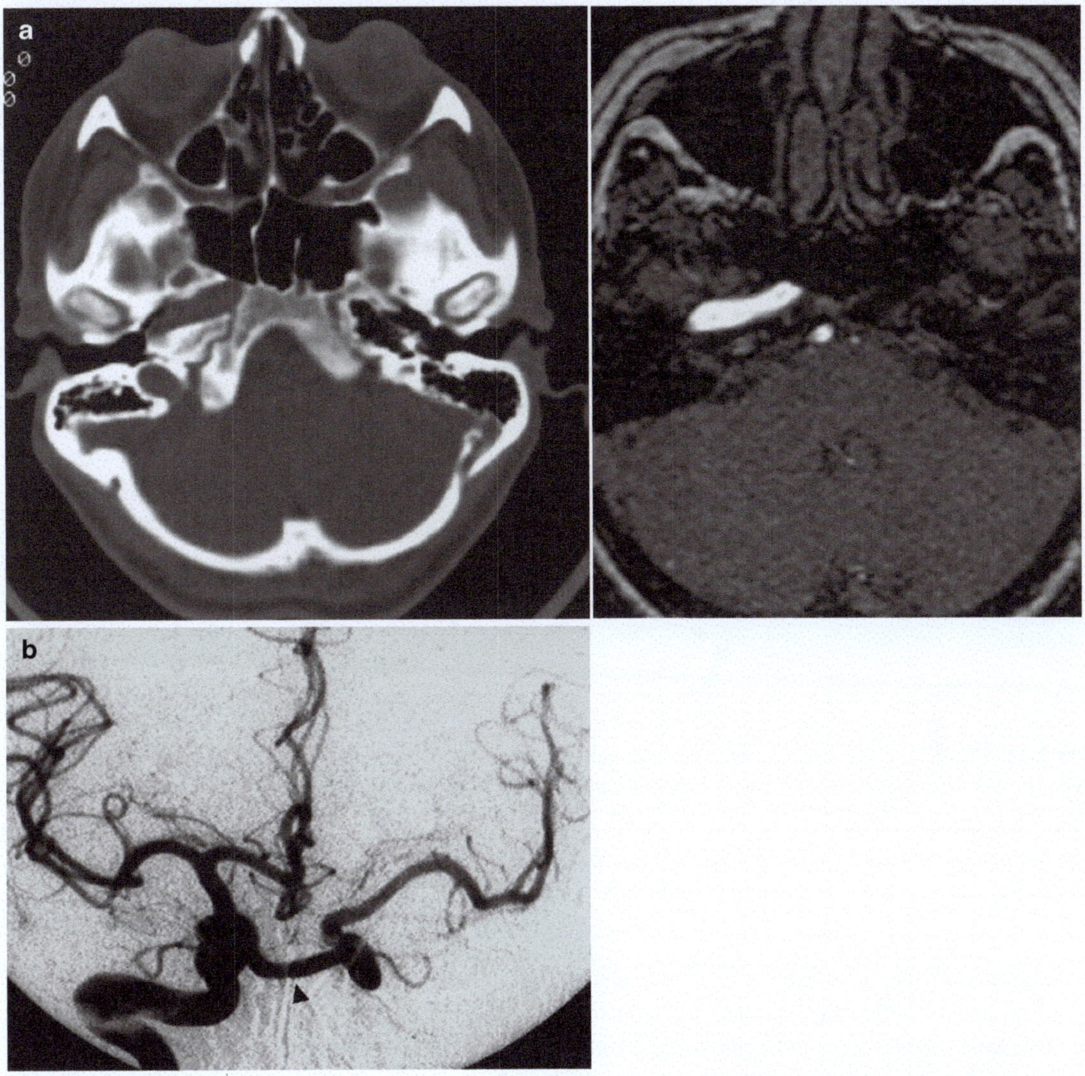

Fig. 2.10 Aplasia of ICA. (**a**) CT and MRI showing, respectively, that the canal of the horizontal portion of the petrous segment of ICA on the left is absent and there is no typical flow signal on the left on MRI. (**b**) Right carotid angiogram. AP view. There is filling of the cavernous portion of the left ICA through intrasellar anastomosis (*arrowhead*) corresponding to the primary maxillary artery (see also Sect. 2.5). Further filling of the middle cerebral artery. Aplastic A1 segment

Mahadevan et al. 2004; Hong et al. 2010a). Association with cerebral aneurysm as well as with other anomalies especially involving the origin of the carotid and vertebral arteries from the aortic arch can occur (Lee et al. 2003). Association with pseudoxhantoma elasticum has been reported (Del Zotto et al. 2012). An example is presented in Fig. 2.11.

- The pathogenesis of the hypo-aplasia of the ICA is not completely clear. It is thinkable that this anomaly occurs in the phase of the embryogenesis when the ICA first appears from the distal segment of the third aortic arch (see also Sect. 2.1).
- *Another very rare form of aplasia is that involving the distal intracranial internal carotid artery.* We observed this anomaly in a child examined for epileptic seizures due to severe cortical dysplasia. The vascular territory of the absent ICA was revascularized mainly through

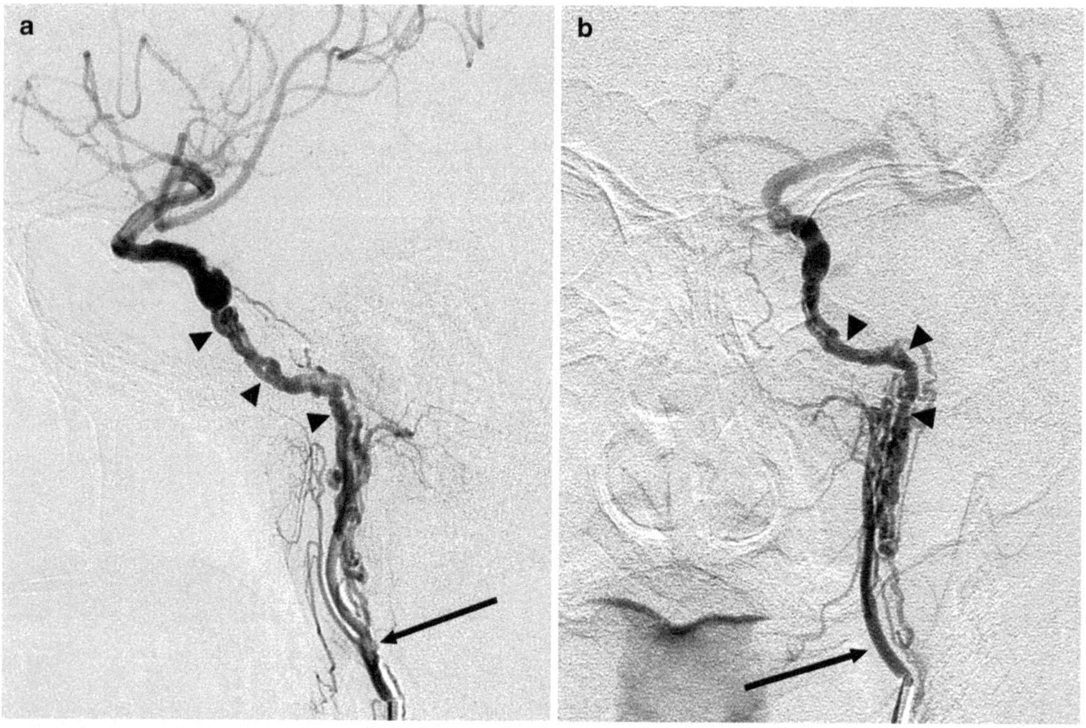

Fig. 2.11 Example of a collateral circulation through *a rete mirabile* incidentally discovered in an old patient. The angiogram of the left common carotid artery showed absence, probably congenital aplasia, of the ICA. A rich network of arteries arising from IMA and especially from the ascending pharyngeal artery (AphA) lead to a reconstruction of the intracranial ICA at its petrous segment. Selective angiogram of the AphA, lateral (**a**) and AP (**b**) views. AphA (*arrow*). Network (*arrowheads*) connecting the AphA with the petrous segment of ICA. A primitive trigeminal artery was present on the right ICA angiogram

anastomoses between the middle meningeal artery and the middle cerebral artery (Fig. 2.12). This kind of aplasia could be due to an anomalous embryogenetic evolution at the level of the distal division of the ICA in cranial and caudal segments (see also Sect. 1.1.).

- *Other anomalies are the embryogenic persistence of the connection between the carotid and* the *vertebrobasilar circulation* (Drawing 2.13c). Considering these in the cranio-caudal direction, the first is represented by the *PcomA* whose different evolution has already been described in Chaps. 1, 2, and 7. The other connections further caudal are *the primitive trigeminal artery (PTA), the otic artery (OT), the hypoglossal (PHA), and the proatlantal (PRA) arteries*. These arteries which regularly develop in the early phase of the embryogenesis (Chap. 1) are transitory vessels which last normally only few days (Padget 1948; Lie 1968) and disappear as their function is replaced by the PcomA and the formed vertebrobasilar system. In some rare cases, however, they do not regress. The most frequent persistent artery, with an incidence of 0.1–0.2% (Lie 1968; Huber 1979; Uchino et al. 2000; Meckel et al. 2013a), is the PTA. It takes its origin from the cavernous portion of the ICA, near the origin of the MTH, sometimes giving off branches for vascular territories normally supplied by the MTH (Parkinson and Shields 1974; Ohshiro et al. 1993; Salas et al. 1998; Suttner et al. 2000). It runs posteriorly passing through or over the dorsum sellae, sometimes having a more medial intrasellar course. This latter condition should be correctly diagnosed especially in the patients in whom a transsphenoidal surgery for pituitary

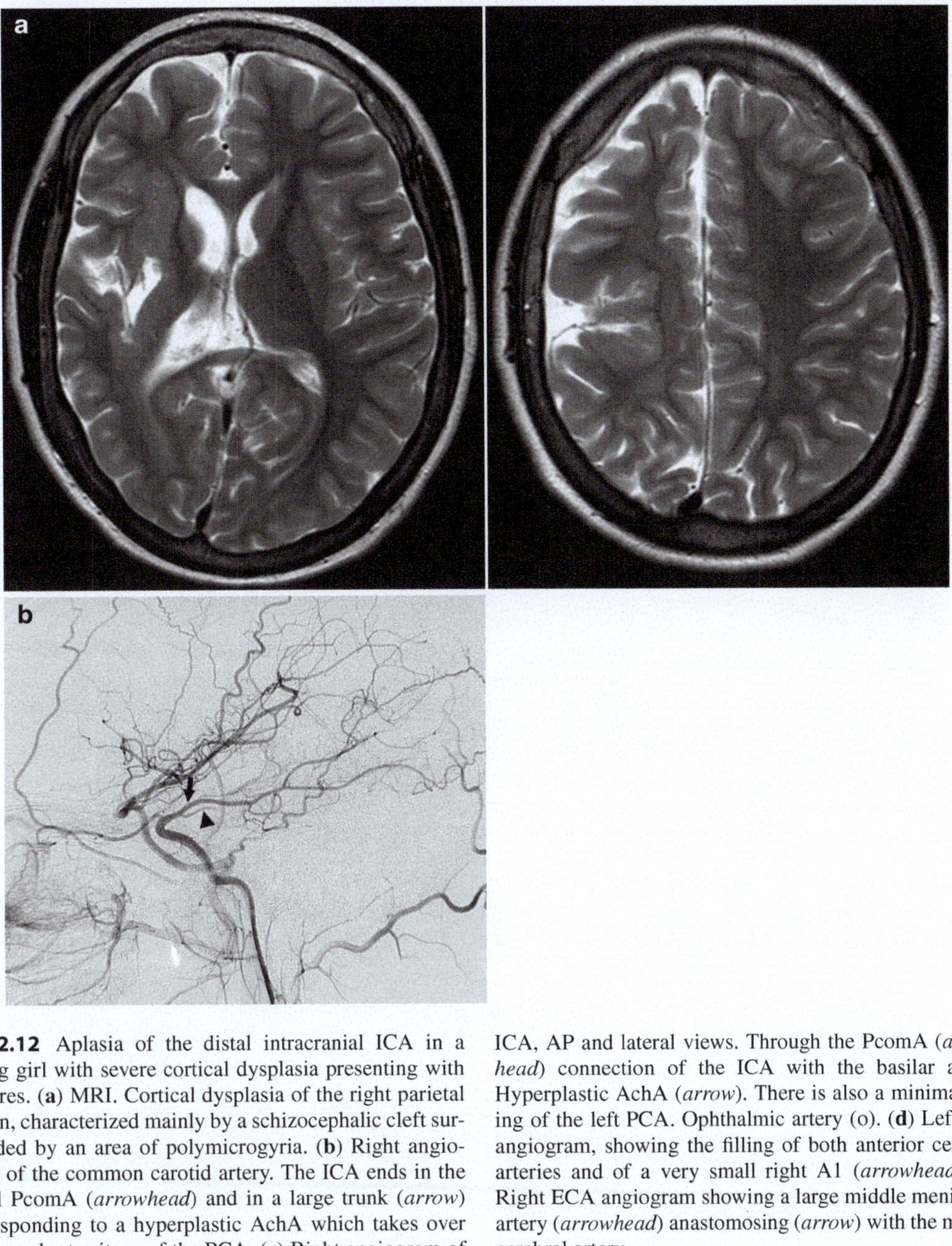

Fig. 2.12 Aplasia of the distal intracranial ICA in a young girl with severe cortical dysplasia presenting with seizures. (**a**) MRI. Cortical dysplasia of the right parietal region, characterized mainly by a schizocephalic cleft surrounded by an area of polymicrogyria. (**b**) Right angiogram of the common carotid artery. The ICA ends in the small PcomA (*arrowhead*) and in a large trunk (*arrow*) corresponding to a hyperplastic AchA which takes over the vascular territory of the PCA. (**c**) Right angiogram of ICA, AP and lateral views. Through the PcomA (*arrowhead*) connection of the ICA with the basilar artery. Hyperplastic AchA (*arrow*). There is also a minimal filling of the left PCA. Ophthalmic artery (o). (**d**) Left ICA angiogram, showing the filling of both anterior cerebral arteries and of a very small right A1 (*arrowhead*). (**e**) Right ECA angiogram showing a large middle meningeal artery (*arrowhead*) anastomosing (*arrow*) with the middle cerebral artery

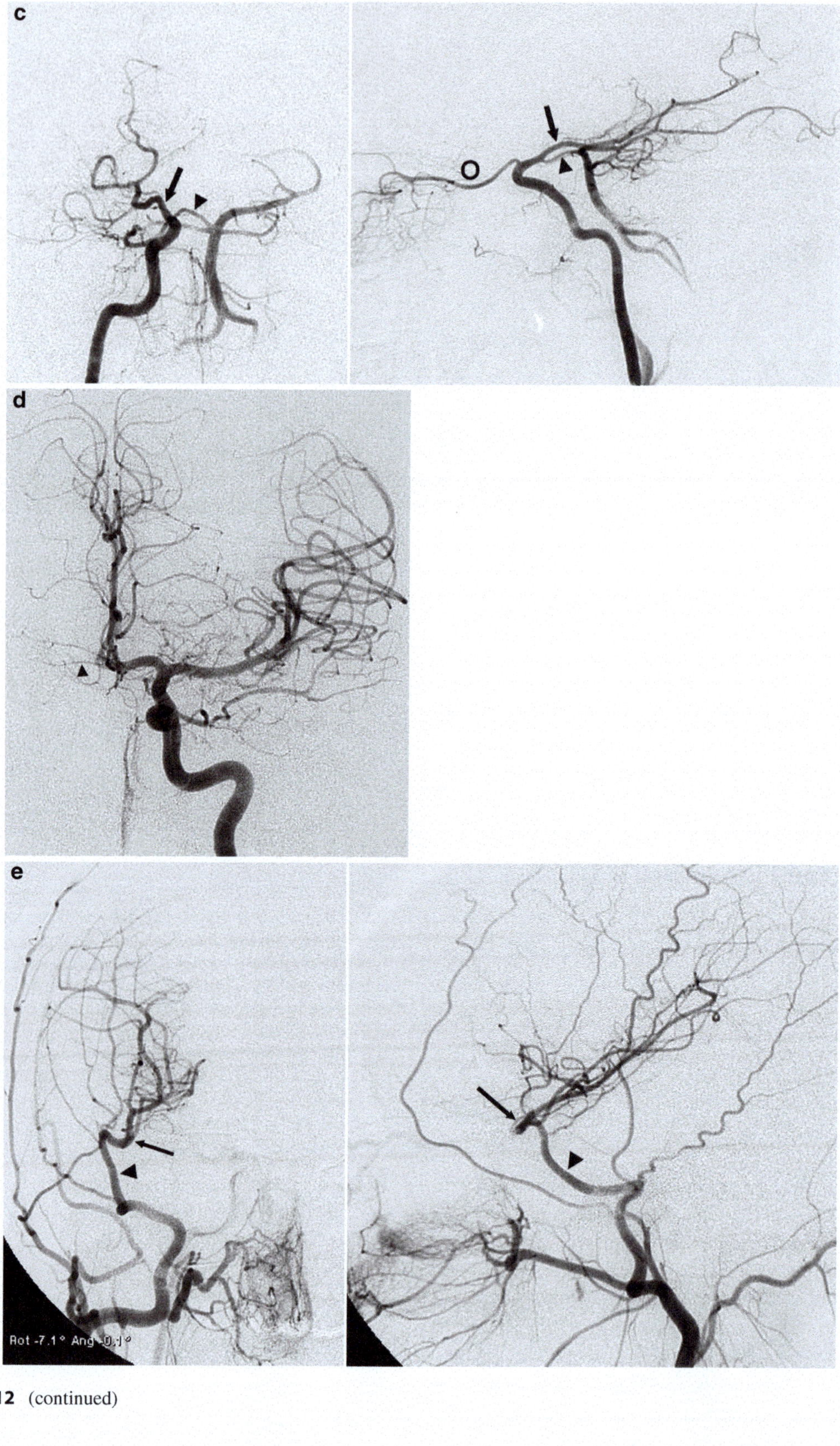

Fig. 2.12 (continued)

adenoma is planned (Lee and Kelly 1989; Richardson et al. 1989; Piotin et al. 1996; Dimmick and Faulderf 2009). Furthermore, as reported by some authors (Ohshiro et al. 1993; Salas et al. 1998; Suttner et al. 2000) PTA can also have a more lateral origin and course giving off, in these cases, branches supplying pons and trigeminal ganglion.

Close to the trigeminal nerve where this leaves the pons, the PTA is connected with the distal basilar artery (BA). Sutton (1950) was the first to report a persistent PTA on an angiographic study. Later two types of PTA have been described in anatomic and angiographic studies by Saltzman (1959) and Wollschlaeger and Wollschlaeger (1964). In one variant the PcomA is absent or hypoplastic and the PTA fills the BA from which arise the posterior cerebral and the superior cerebellar arteries (Fig. 2.13a). In another variant, the PcomA is well developed and continues in the posterior cerebral artery. The PTA supplies only the superior cerebellar artery. There is no connection with the PCA. Commonly the VAs which join the proximal BA and supply the PICA are normal developed. Sometimes they can be hypoplastic.

Association with hypo-aplasia of the BA involving its segment below the entry of the PTA can occur (see for this variant also Sect. 6.2.4).

The PTA can be the site of aneurysm (Ahmad et al. 1994) and cavernous fistula. This latter can be due to the rupture of the aneurysm or have a traumatic cause (Enomoto et al. 1977; Flandroy et al. 1987; Oka et al. 2000; Tokunaga et al. 2004; Geibprasert et al. 2008; Asai et al. 2010; Kobayashi et al. 2011; Meckel et al. 2013). Furthermore, it should be considered that the PTA can be an important collateral circulation from the BA toward the ICA in cases of agenesis or occlusion of the artery. It can, however, be responsible for symptoms due to a vascular steal phenomenon from the basilar artery to the ICA vascular territory, and it can be a way of emboli arising in the vertebra-basilar sector toward the carotid territory or vice versa.

- Other less frequent carotid-vertebrobasilar anastomoses are *the persistent otic, hypoglossal, and proatlantal arteries*. The otic artery connects the petrous ICA with the BA. There are only a few angiographic reports about this anomaly (Reynolds et al. 1980).
- The persistent hypoglossal artery (PHA) arises from the cervical ICA, at the level of C1–C2, and runs dorsally, entering the hypoglossal canal, which is enlarged (recognizable on the CT skull base), joining the vertebral artery, which is commonly hypoplastic or absent in its proximal segment (Kanai et al. 1992; Uchino et al. 2012b; 2013a) (Fig. 2.13b). Cases of PHA arising from the external carotid have also been described (Uchino et al. 2013b). The association with aneurysm has been reported (Brismar 1976; Kanai et al. 1992; De Blasi et al. 2009; Uchino et al. 2013a).
- Another very rare connection is represented by the *persistent proatlantal artery (PRA)*. This is a vessel which arises from the cervical ICA or more rarely from ECA, runs dorsally, reaches the atlas, and runs further above it, connecting with the extradural suboccipital part of the vertebral artery. Padget (1954) and Bloch and Danziger (1974) in their study about the embryonic intersegmental arteries called the vessel running between the occipital bone and atlas, which normally disappears and partially contributes to the development of the extradural transverse segment of the vertebral artery (VA) *suboccipital or proatlantal intersegmental artery* (see also Sect. 6.2.4). In case of persistence of the PRA, some authors (Moffat 1959; Haughton and Rosenbaum 1974) have suggested that this is due to failure of the superior, dorsal part of the third aortic arch, from which the PRA arises, to regress. This embryonic vascular structure known as "*ductus caroticus*" remained connected with the proximal part of the third aortic arch from which develop the ICA and ECA (see also Sect. 1.1). In the further evolution a connection between ICA or ECA and VA through the persistent PRA and in some cases also of the PHA as described by some authors (Suzuki et al.

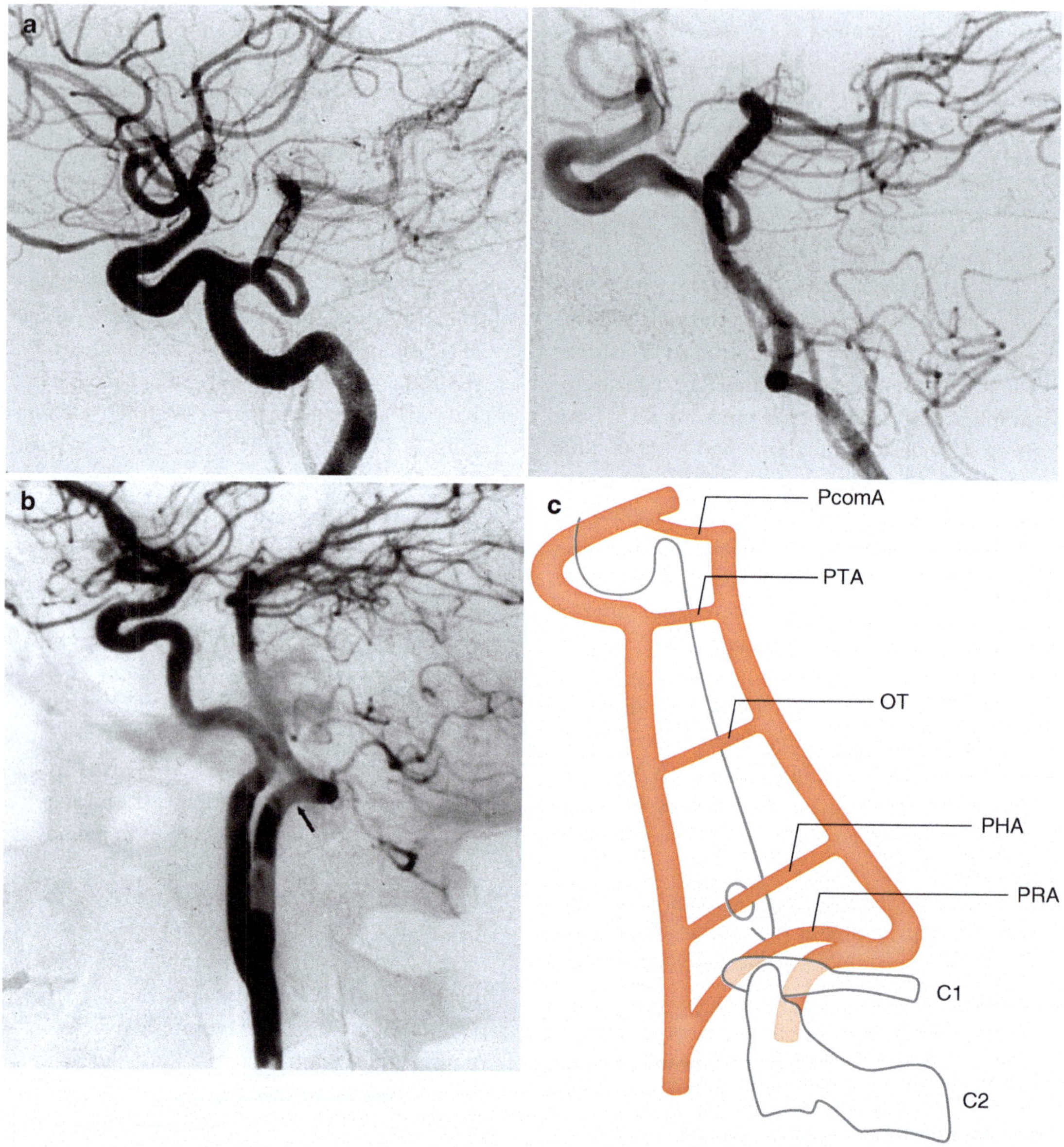

Fig. 2.13 Connections between the ICA and the vertebrobasilar sector. (**a**) Persistent primitive trigeminal artery connecting the cavernous portion of the ICA with the basilar artery. The connection is visible on the carotid and vertebral angiograms. (**b**) Persistent hypoglossal artery arising from ICA, entering the hypoglossal canal (*arrow*), and anastomosing with the vertebral artery. (**c**) Drawing showing the connections between ICA and BA and VA. *PcomA* posterior communicating artery, *PTA* primitive trigeminal artery, *OT* otic artery, *PHA* hypoglossal artery, *PRA* proatlantal artery

1979) can develop. Later other authors (Lasjaunias et al. 1978b; Suzuki et al. 1979) considering that the PRA after its connections with the VA frequently continues in a branch having the aspect and course of that of the occipital artery, and the occipital artery can have an anomalous origin from ICA, have suggested that in both conditions we are dealing with the same vessel, i.e., the proatlantal artery, which could correspond to the proatlantal artery as described by Padget. This can lose its connection with the VA, appearing as an anomalous or a normal occipital artery. Vice versa, the anomalous occipital artery arising from the ICA can connect with the VA appearing as the proatlantal artery. In all cases, in which a PRA is present, the proximal VA is commonly hypoplastic or absent.

- A few other anomalies can occur. *The one is the origin of the superior cerebellar, anterior inferior cerebellar, and posterior inferior cerebellar arteries from the cavernous portion of the ICA* (Scotti 1975; Haughton et al. 1978; Siqueira et al. 1993; Morris 1997; Uchino et al. 2000; Shin et al. 2005; Meckel et al. 2013). This has been interpreted as a partial primitive trigeminal artery persistence (Lasjaunias and Berenstein 1990; Meckel et al. 2013).
- *Another very rare variant involves the PICA*, which arises from the extracranial ICA, enters the posterior fossa through the hypoglossal canal, and supplies the corresponding cerebellar vascular territory without having connections with the vertebral artery. This variant has been interpreted as a partial persistence of the embryonic hypoglossal artery (Andoh et al. 2001; Uchino et al. 2013a). Also cases of origin of PICA from the external carotid artery have been described, and interpreted as anastomosis between the hypoglossal branch of the ascending pharyngeal artery and the PICA and put in the group of variants of the PHA (Lasjaunias et al. 1981; Kim et al. 2009; Uchino et al. 2013a).
- *The internal carotid artery can have an aberrant course*. Among these anomalies there is the *retropharyngeal position* in which the artery runs very close to the pharynx or sometimes appears as a submucosal mass. It is frequently asymptomatic, but in some cases can be responsible of an oropharyngeal pulsatile or foreign body sensation, dysphagia, hoarseness, and respiratory distress. As reported by some authors (Marcucci et al. 2009; Munoz et al. 2010; Gupta et al. 2013) it can be uni- or bilateral. It is important to be aware of the anomaly for the anesthesiologist performing intubation and in cases of oropharyngeal surgery (Fig. 2.14).

  Another type of aberrant course is that in which the artery enters the temporal bone through an enlarged inferior tympanic canaliculus, and thus is located more posteriorly than in normal cases laterally to the jugular bulb and adjacent to the stylomastoid foramen. The distal vertical segment of the petrous ICA protrudes in the middle ear cavity to continue further in the horizontal petrous segment. This anomaly has been interpreted as an agenesis of the terminal part of the cervical ICA, with the formation of a collateral circulation between the enlarged tympanic branch of the ascending pharyngeal artery and the caroticotympanic branch of ICA remnant of the stapedial artery (Lo et al. 1985; Osborn 1999; Sauvaget et al. 2006; Saini et al. 2008). CT of the skull discloses the presence of a soft-tissue mass protruding in the tympanic cavity. The angiographic study shows the anomalous course of the artery, which appears smaller, often narrowed and irregular, clarifying the diagnosis especially differentiating the

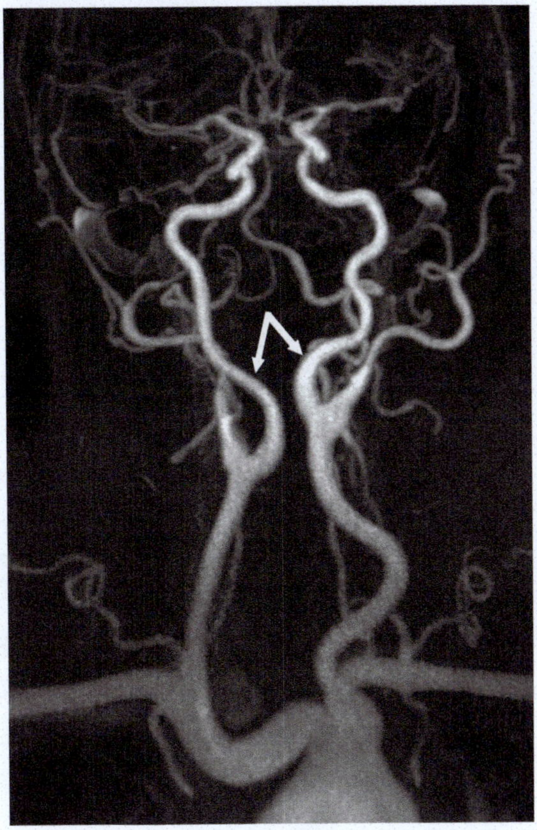

Fig. 2.14 MR angiography with contrast medium. Incidental discovery of retropharyngeal course of both ICA (*arrow with angle*)

anomalous ICA from a suspected small tympanic paraganglioma. An example is presented in Fig. 2.15.

- *A fenestration or duplication of the supraclinoid ICA* has been reported (Yock 1984; Banach and Flam 1993; Rennert et al. 2013). This is a very rare anomaly occurring probably in the embryogenic phase when the ICA divides in its anterior and posterior division (Chap. 1). Failure in this process may explain the presence of this anomaly.

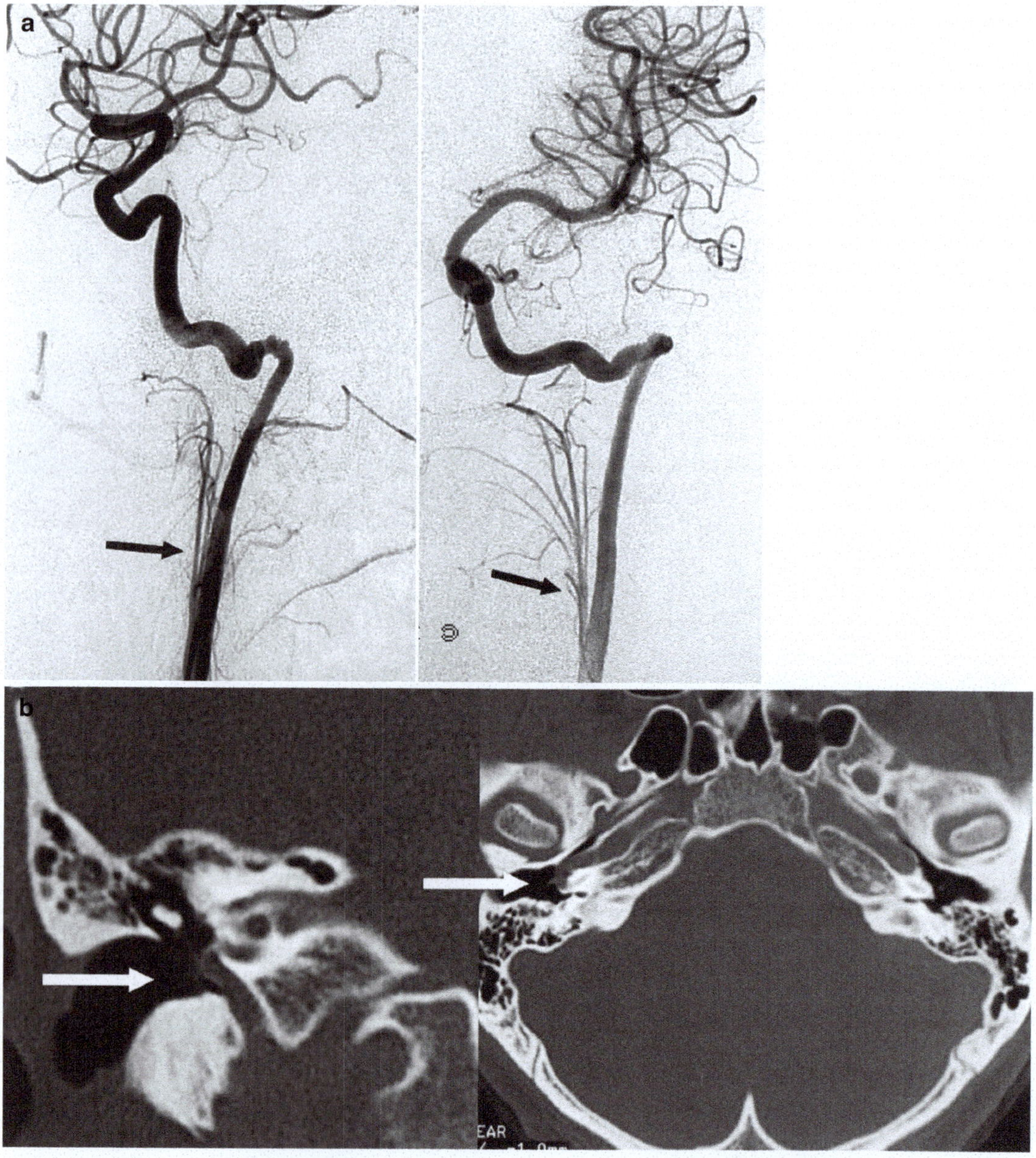

Fig. 2.15 Aberrant course of the ICA. (**a**) Lateral and AP ICA carotid angiogram. Origin of the AphA (*arrow*) from the ICA. The ICA runs more posteriorly and more laterally, respectively, on the lateral and AP angiograms. (**b**) CT, coronal and axial views. The ICA enters the middle ear cavity and is recognizable as a small, rounded, soft-tissue structure (*arrow*)

3 External Carotid Artery

The embryogenesis of the ECA is characterized first by the formation of small network of vessels arising from the first and second arches. These vascular structures rapidly disappear and the definitive ECA develops from the fusion of two vascular structures (Padget 1948; Salamon et al. 1974). One is the ventral pharyngeal artery embryonal vessel developing from the proximal part of the third aortic arch, and the other is the hyoidostapedial trunk which arises from the primitive internal carotid artery developing from the more distal part of the third aortic arch (see also Sects. 2.1 and 2.2.3.1).

The definitive ECA arises from the common carotid bifurcation at the C4 vertebral level. A more proximal or distal origin can occur. The ECA lies in the carotid triangle, initially medial and anterior to the ICA, seldom lateral to it. More cranially, it runs anterolateral to the ICA. The internal jugular vein is located posteriorly to the proximal ECA. More distally, near the skull base, the ECA is located laterally to the jugular vein. (Figs. 2.1 and 2.2). During its course, the ECA gives off several branches and divides near the mandibular condyle within the parotid gland into its terminal branches represented by the internal maxillary artery (IMA) and the superficial temporal artery (STA) (Fig. 3.8).

Among the relatively rare variants of the ECA, there is an isolated origin from the aortic arch (see also Sect. 2.3). Another particular anomaly is that occurring in the so-called nonbifurcating cervical carotid artery (Morimoto et al. 1990; Uchino et al. 2011; Nakai et al. 2012). In this case, the typical carotid bifurcation is not recognizable with its typical trunk of ECA and ICA. The common carotid artery seems to continue directly in the trunk of ECA from which arise its branches. The ICA appears to be a continuation of the ECA. The interpretation of this anomaly is not completely clear. According to some authors (Morimoto et al. 1990; Uchino et al. 2011; Nakai et al. 2012), we are dealing with a segmental agenesia of proximal ICA. The common carotid artery continues into the ECA.

3.1 Superior Thyroid Artery (TH)

The superior thyroid artery arises from the anterior wall of the ECA. It runs inferiorly and somewhat medially toward the thyroid gland. The artery gives off branches for the superior part of the thyroid gland and larynx. It anastomoses with the inferior thyroid artery, a branch of the thyrocervical trunk, arising from the subclavian artery (Figs. 2.2 and 3.1).

G.B. Bradac, *Applied Cerebral Angiography*, DOI 10.1007/978-3-319-57228-4_3

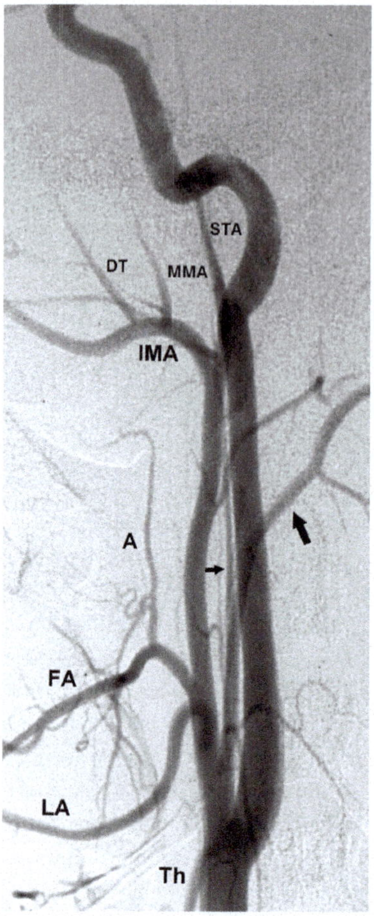

Fig. 3.1 Common carotid angiogram. Lateral view, showing the course of the ECA running anteriorly to the ICA. Some of the main branches are recognizable. Superior thyroid artery (Th), lingual artery (LA), facial artery (FA) from which arises the ascending palatine artery (A), occipital artery (*large arrow*), ascending pharyngeal artery (*small arrow*), internal maxillary artery (IMA) from which arise the middle meningeal artery (MMA) and the middle deep temporal artery (DT), superficial temporal artery (STA)

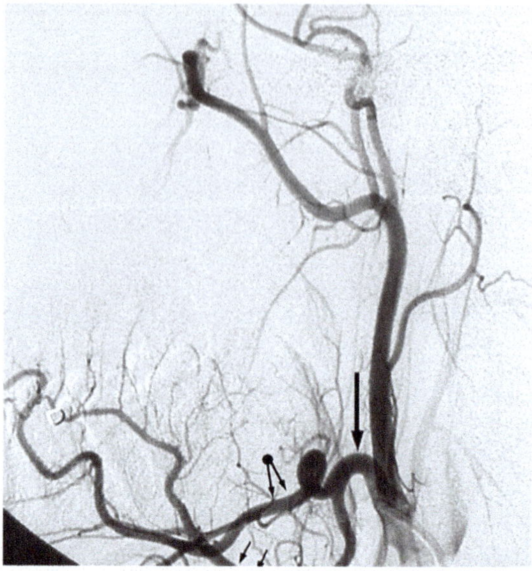

Fig. 3.2 ECA angiogram. Lateral view showing the common trunk of origin (*large arrow*) of the lingual (*small arrow*) and facial (*arrows with dot*) arteries

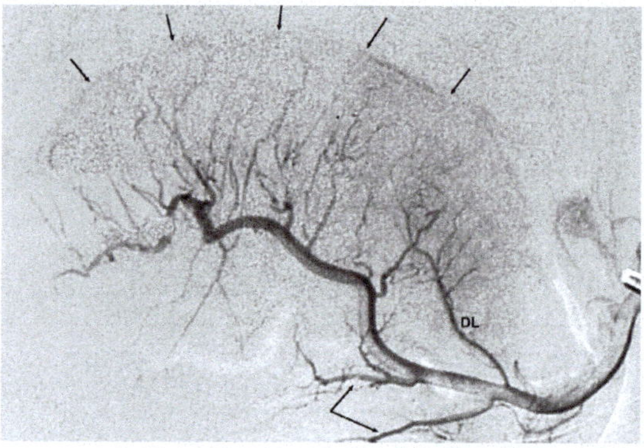

Fig. 3.3 Selective angiographic study of the lingual artery. Branches for the tongue (*arrows*) arising from the main trunk and partially from the dorsal lingual artery (DL). Sublingual branches (*arrow with angle*)

3.2 Lingual Artery (LA)

The LA is the second branch of the ECA arising from its anterior wall. It is not exceptional for the LA to have a common trunk with the facial artery. It runs anteromedially to the pharyngeal muscles toward the tongue. In its course, it gives branches for the sublingual and submandibular glands, and for the mucosa of the pharynx. Its terminal branch is the deep lingual artery which supplies the muscles and lingual mucosa. On the lateral angiogram, the lingual artery has a typical course, first upward, then downward, and finally upward again, forming a gentle superiorly concave curve. The deep lingual artery with its ascending fine branches, the dorsal lingual artery, and the sublingual artery are easy to recognize. The sublingual artery anastomoses through its submental branch with the corresponding branch of the facial artery. Examples of FA are presented in Figs. 3.1, 3.2, and 3.3.

3.3 Facial Artery (FA)

The FA is the third branch arising from the anterior wall of the ECA, frequently as a unique trunk with the lingual artery. It runs forward with an undulating course above the submandibular gland. It curves then around the lower edge of the mandible, continuing anteriorly and superiorly crossing the cheek, ending in the medial angle of the orbit as the angular artery. The latter anastomoses with branches of the ophthalmic artery, establishing a collateral circulation when the ICA is occluded. The FA gives off branches for the submandibular gland, the masseter artery for the masseter muscle anastomosing with masseter branches of the IMA and transverse facial artery, and an ascending palatine artery anastomosing with the descending palatine artery of the IMA and with pharyngeal branches of the AphA. It gives further branches for the mandible, skin, and muscle of the submental area, and cheek, nose, and lip.

The facial artery can be hypoplastic and represented only by the submental artery. In such cases, parts of its vascular territories become replaced by other branches of ECA, especially the lingual, transverse facial, and infraorbital arteries (Djindjian and Merland 1978).

The initially descending and then the obliquely ascending course of the facial artery is easily discerned on the lateral angiogram. Among its branches, the ascending palatine, the masseter, the submandibular, and the submental arteries are the most frequent identifiable. Examples are presented in Figs. 3.1, 3.2, 3.4, and 3.15.

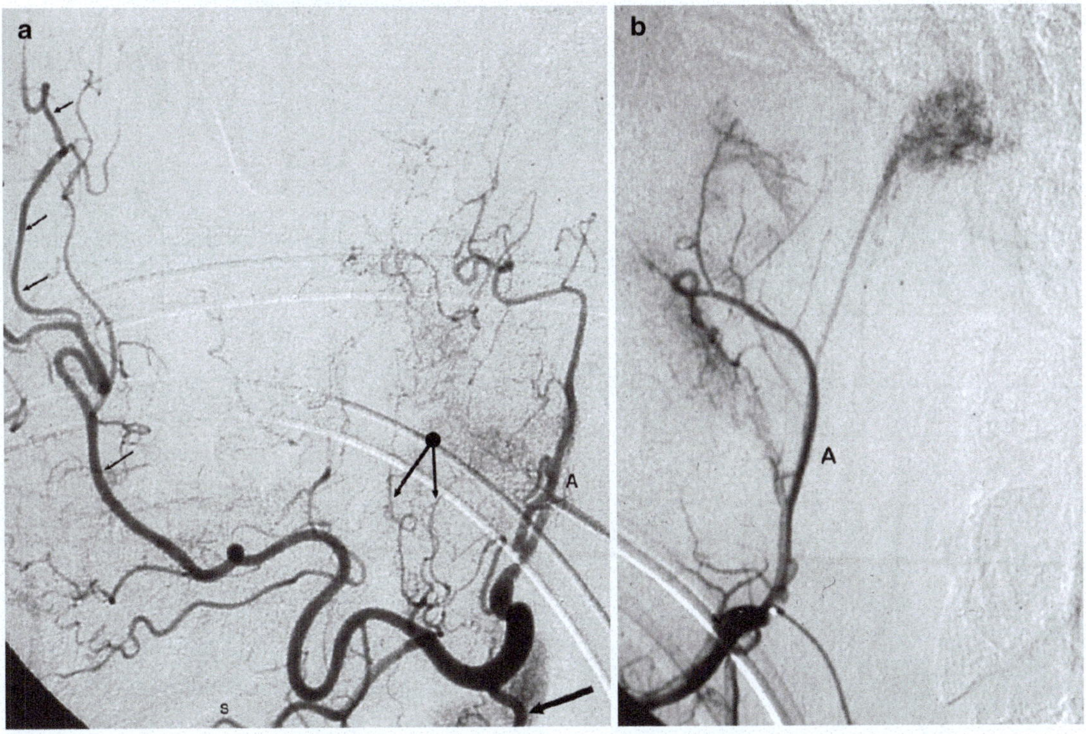

Fig. 3.4 (**a**) Selective angiographic study of the facial artery. Ascending palatine artery (A), branches for the submandibular gland (*arrow*) and masseter muscle (*arrow with dot*). Submental artery (S). Distal branches (*small arrows*) ending in the angular artery. (**b**) Selective study in a different patient. Ascending palatine artery (A), from which arises a branch supplying a large vagal paraganglioma

3.4 Ascending Pharyngeal Artery (AphA)

The AphA is a small vessel that arises from the posterior wall of the ECA, sometimes from the carotid bifurcation or from the proximal segment of the ICA. The AphA can also arise sharing a common trunk with the occipital artery. It runs upward parallel and adjacent to the ICA, posteriorly and medially to it (Figs. 3.5 and 3.6).

The AphA gives off pharyngeal branches for the paramedian mucosa of the naso-oropharynx, which are divided into superior, middle, and inferior branches. The superior branch can anastomose with pharyngeal branches coming from the accessory meningeal and pterygovaginal arteries, both branches of the IMA. The middle branches

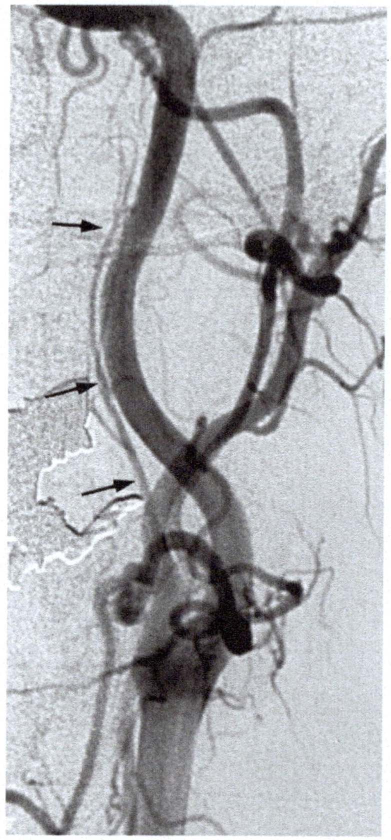

Fig. 3.6 Common carotid angiogram, showing the close relationship of the ascending pharyngeal artery (*arrows*) to the internal carotid artery

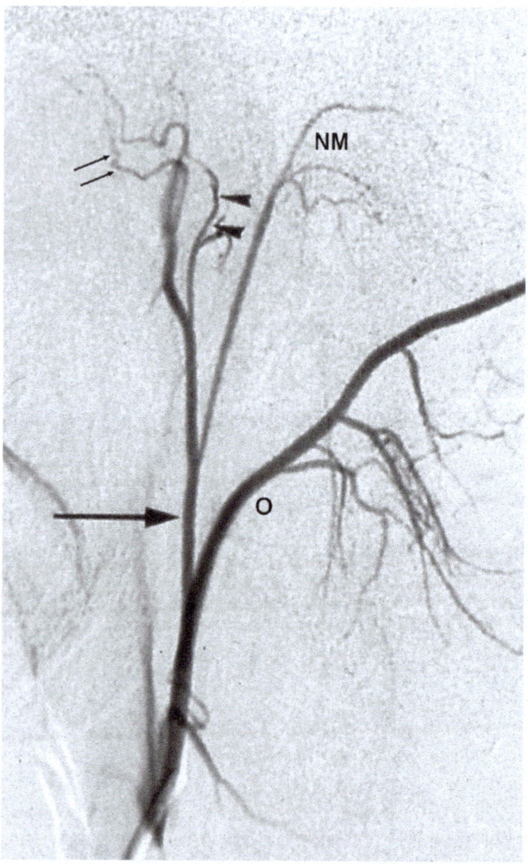

Fig. 3.5 Selective angiographic study, lateral view, of the common trunk of the ascending pharyngeal and occipital arteries. Occipital artery (O), ascending pharyngeal artery (*arrow*) with its tympanic (*arrowheads*), superior pharyngeal (*small arrows*) branches and neuromeningeal trunk (NM)

anastomose with the ascending palatine artery arising from the facial artery and, when present, with the mandibular artery, embryological remnant arising from the ICA.

Other branches are the cutaneomusculospinal arteries which can have anastomoses with the ascending cervical artery and vertebral arteries commonly at the C2–C4 levels.

Near the base of the skull, the AphA divides into its terminal branches, described below:

- Superior pharyngeal branches
- The inferior tympanic artery (Fig. 3.5) which vascularizes the tympanic cavity and can anastomose with the other tympanic branches arising from the stylomastoid artery branch of the occipital artery and tympanic branches arising from IMA
- The neuromeningeal trunk (Fig. 3.5) which can be subdivided into the hypoglossal and

jugular branches (Djindjian and Merland 1978; Lasjaunias and Doyon 1978)

The *hypoglossal branch* enters the hypoglossal canal, running together with the cranial nerve XII supplying it, and with the anterior condylar vein. It gives off branches for the meninges of the adjacent posterior fossa and a descending branch that anastomoses with the ascending radiculomeningeal branch of the vertebral artery, contributing to form the so-called odontoid arch (see also Sect. 6.1.1) (Figs. 3.7c, 3.33, 3.34, 6.2b, c, 13.3a, b, 15.30, and 15.33). The artery also gives off meningeal branches, which extend upward along the clivus (clival branches),

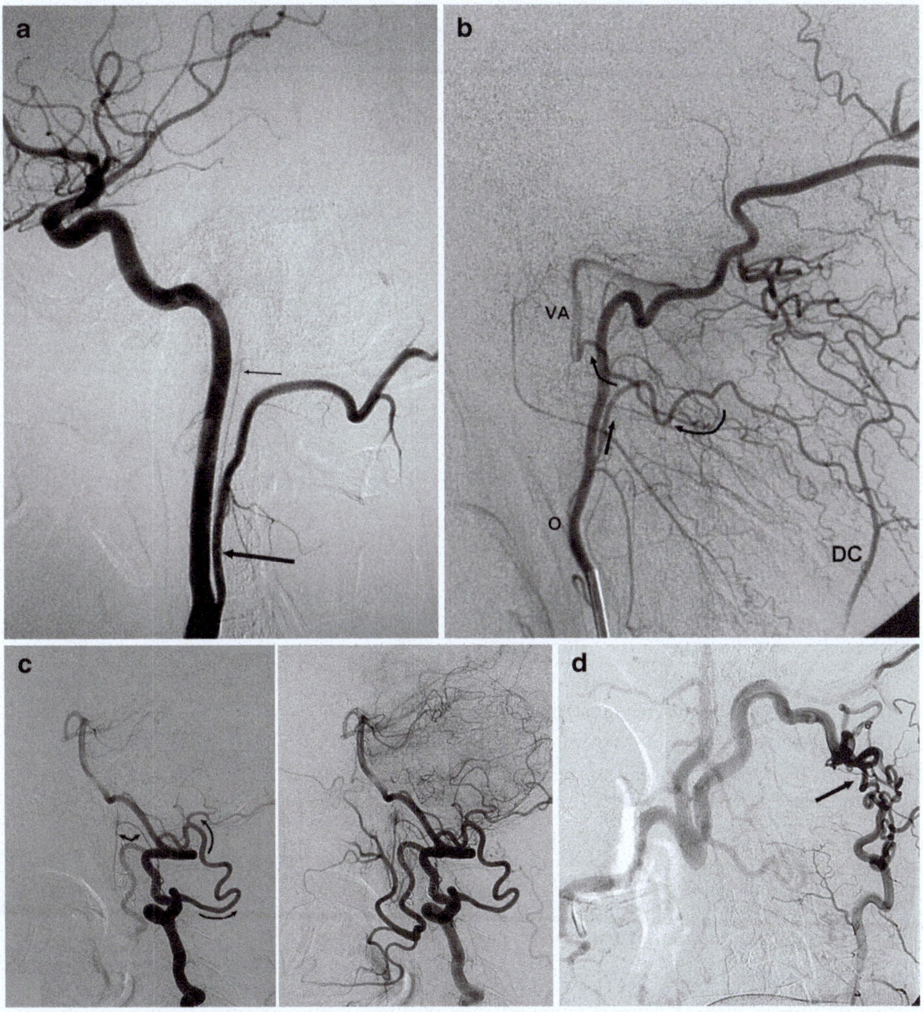

Fig. 3.7 Angiograms in four different patients. (**a**) ICA angiogram, lateral view. The occipital artery (*arrow*) and the AphA (*small arrow*) arise as a common trunk from the ICA. (**b**) Selective angiogram of the occipital artery (O) in another patient. There is a clear anastomosis (*arrows*) between the cutaneous-muscular branch of the ascending segment of the occipital artery with the VA. There are other anastomoses between the horizontal segment of the occipital artery and the deep cervical artery (DC). This latter is retrograde injected and partially also connected to the VA. (**c**) Lateral vertebral angiogram in a different patient (earlier and later phases). The patient had had a surgical ligature of the ECA. Through anastomoses involving cutaneous-muscular branches of the VA and the occipital artery at the C2 level (*arrows*) there is a progressively complete recanalization of the ECA. Odontoid arch (*arrow with angle*) leading to a retrograde filling of the AphA. (**d**) Another example of revascularization of ECA after proximal surgical ligature of the artery. The retrograde injection of the ECA occurs through anastomoses between the deep cervical artery and the occipital artery

anastomosing with the corresponding medial clival branches arising from the MHT (Fig. 3.32). Because of this anastomosis, the neurohypophysis can sometimes be visible on the lateral angiogram of the AphA (Djindjian and Merland 1978).

The *jugular branch* enters the posterior fossa through the jugular foramen, where it supplies cranial nerves IX, X, and XI. It supplies the adjacent dura, anastomosing with the more lateral, descending clival branches of the MHT. It then turns posteriorly to vascularize the dura of the cerebellar fossa and cerebellopontine angle.

The supply of the cranial nerves IX, X, XI, and XII is a very important aspect that should be taken into consideration in the endovascular treatment of lesions involving the AphA. Furthermore, the close relationship of the AphA with the ICA could explain the palsy of the last fourth nerves in many cases of dissection of the ICA (*Bradac et al.* 1990, 2000).

The branches of the AphA may be involved in many pathological processes. Examples are presented in Figs. 3.22, 3.24, 3.25, 3.26, 3.27, 13.3a, b, 13.6, 13.14, 13.15, 13.21, 13.22, and 13.23.

3.5 Occipital Artery (O)

The occipital artery takes its origin from the posterior wall of the ECA, frequently as a common trunk with the APhA (Figs. 3.5, 3.7a, and 3.20). Sometimes it can arise from the ICA (Newton and Young 1968; Teal et al. 1973) or from the vertebral artery through anastomosis at the C1 and C2 level (Lasjaunias et al. 1978b) (Figs. 3.7a, c, 3.20, 6.2b, 6.8a, 6.9a, and 15.33). The artery runs posteriorly and slightly superiorly with an undulating course, in which three segments may be identified. The first ascending, the second horizontal, and the third ascending again. It gives off the stylomastoid artery, several musculocutaneous branches for the neck and posterior part of the head, and meningeal branches.

- The stylomastoid artery arises from the first segment and runs up to the stylomastoid foramen, accompanying the facial nerve and supplying it. It anastomoses superiorly with the superior tympanic artery arising from the petrosquamosal artery, branch of the middle meningeal artery. The stylomastoid artery gives off branches also for the tympanic cavity and adjacent area (posterior tympanic arteries), where they anastomose with the tympanic branches arising from the AphA and IMA. The stylomastoid artery can arise from the posterior auricular artery or directly from the trunk of the ECA. It is a fine branch, not always evident on the lateral ECA angiogram. It can be dilated and is well recognizable when involved in the supply of pathological processes (Figs 3.25, 3.27c, and 13.1).
- Among the musculocutaneous branches, two frequently arise in the ascending segment. A third branch (the splenial artery) arises from the horizontal segment, has a descending course, and supplies the splenius muscle. The muscular branches can anastomose with the corresponding branches of the vertebral artery at the C1 and C2 level (Figs. 3.7b, c, 6.2b, 6.8a, 6.9a, and 15.33), and with the deep cervical artery.
- The mastoid branch (Fig. 3.8) arises from the second segment, approximately near the origin of the splenial artery. It has an ascending

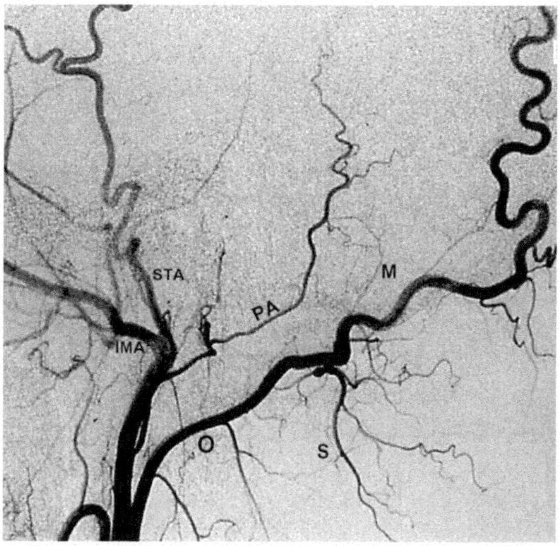

Fig. 3.8 Angiogram of the ECA. Lateral view. Occipital artery (O), with its muscular branches arising from its ascending and horizontal segments. *S* splenial artery, *M* mastoid branch, *PA* posterior auricular artery, *STA* superficial temporal artery, *IMA* internal maxillary artery

course, enters the cranial cavity through the foramen of the mastoid emissary vein, and supplies the meninges of the cerebellopontine angle and cerebellar fossa. These branches are in hemodynamic balance with those arising from the neuro-meningeal trunk of the AphA, with the petrosquamous branch of the middle meningeal artery (MMA), and with the meningeal branches of the vertebral artery (Lasjaunias et al. 1978b). One of the arteries can dominate and completely supply these territories.

- The occipital artery can be involved in several pathological processes (Figs. 3.25, 3.27, 13.4, 13.5, 13.13, 13.14, 13.16, 13.18, and 13.22).

3.6 Posterior Auricular Artery

This is a small branch arising near the terminal ECA (Figs. 3.8 and 3.15), sometimes as a common trunk with the occipital artery. It supplies the pinna of the ear, which on the angiogram appears as an evident blush in the capillary phase of the ECA angiogram.

3.7 Internal Maxillary Artery (IMA)

The internal maxillary artery is the larger of the terminal branches of the ECA. It originates behind the neck of the mandible, within the parotid gland. It runs obliquely forwards and medially in the masticator space, along the border of the lateral pterygoid muscle, and ends in the pterygopalatine fossa, where it divides into its terminal branches. It is always well recognizable on the lateral and AP angiograms (Figs. 3.9, 3.10 and 3.11).

3.7.1 Proximal Branches

Near its origin, the IMA gives off a few thin proximal branches for the mandibular joint, external auditory canal, and tympanic cavity (anterior tympanic artery). The latter can be involved in the vascularization of tympanic chemodectomas

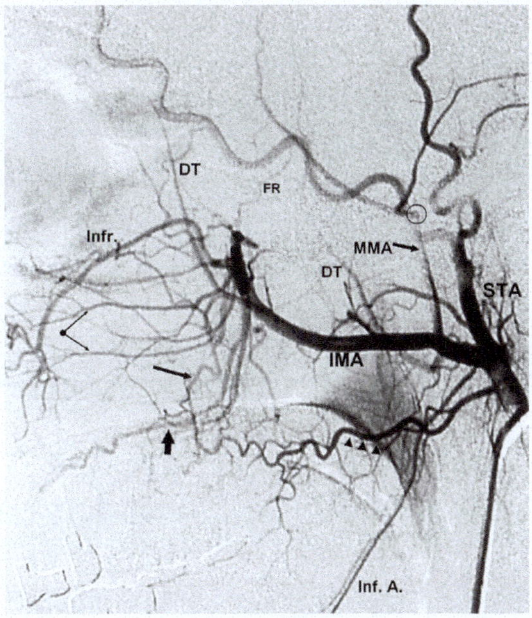

Fig. 3.9 Angiogram of the distal ECA, lateral view. Transverse facial artery (*arrowheads*). Superficial temporal artery (STA), internal maxillary artery (IMA), middle meningeal artery (MMA). The *circle* shows where the MMA passes through the *foramen spinosum*. Middle and anterior deep temporal arteries (DT), inferior alveolar artery (Inf.A.), infraorbital artery (Infr.). Posterior superior alveolar artery (*arrow*), descending palatine artery (*large arrow*), nasal branches of the sphenopalatine artery (*arrow with dot*), small foramen rotundum artery (FR)

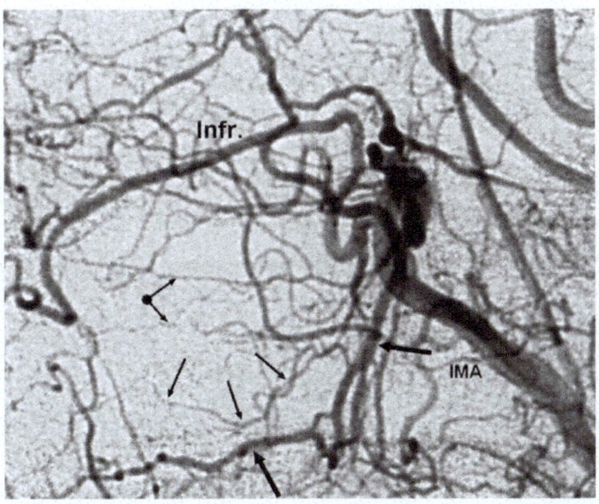

Fig. 3.10 Angiogram of the distal ECA, lateral view, detail. Internal maxillary artery (IMA), infraorbital artery (Infr.), posterior superior alveolar artery (*small arrows*), descending palatine artery (*large arrow*), nasal branches of the sphenopalatine artery (*arrow with dot*)

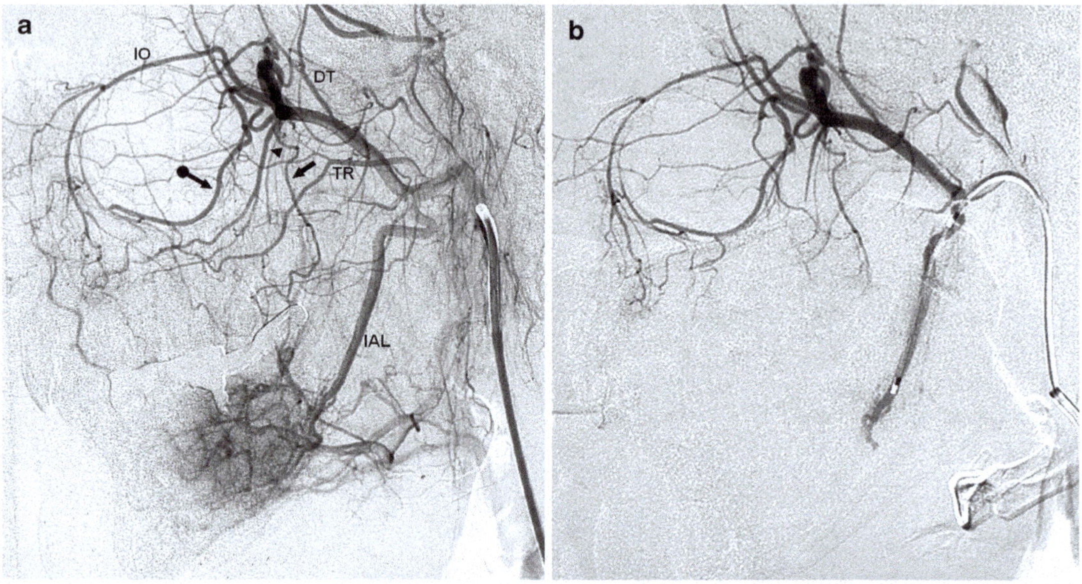

Fig. 3.11 (**a**) Angiogram of the distal ECA, performed to study a capillary-venous malformation of the mandible. Dilated inferior alveolar artery (IAL) supplying the malformation. Transverse facial artery (TR), buccal artery (*arrow*), deep temporal branches (DT), descending palatine artery (*arrowhead*), posterior superior alveolar artery (*arrow with dot*). Infraorbital artery (IO). The distal branches of the sphenopalatine artery are visible running between the infraorbital and posterior superior alveolar arteries. (**b**) Selective catheterization of the IAL, preceding embolization with PVA

(Fig. 3.25). Other more larger branches arising from the proximal segment are the middle meningeal artery (MMA), the accessory meningeal artery, and the inferior alveolar artery.

3.7.1.1 Middle Meningeal Artery

The MMA runs superiorly towards the base of the skull and passes through the foramen spinosum, becoming intracranial, running along the greater sphenoid wing, forming a large convex curve anteriorly and laterally well recognizable on the lateral and AP angiograms, respectively (Figs. 3.9 and 3.15). It is the important artery responsible for the vascularization of the dura, giving off the following branches:

- Branches for the dura of the frontoparietal and temporo-occipital convexity.
- Branches for the frontobasal area, anastomosing with the ethmoidal arteries arising from the ophthalmic artery.
- A meningeal branch can enter the orbita through the superior orbital fissure or through the foramen of Hyrtl and anastomose with the lacrimal artery, branch of the ophthalmic artery.
- Branches for the middle cranial fossa, where they anastomose near the cavernous sinus with the meningeal branches arising from the inferolateral trunk (ILT) meningohypophyseal trunk (MHT), foramen rotundum, and accessory meningeal arteries.
- Close to the foramen spinosum arises the petrosquamous branch, which supplies the dura covering the petrous bone, the tentorium, and extends largely in the posterior fossa where it anastomoses with the meningeal branches arising from the AphA, occipital, and vertebral arteries. The petrosquamous branch gives off a tympanic branch (superior tympanic artery), which runs in the petrous facial canal where it anastomoses with the stylomastoid artery (branch of the occipital or posterior auricular arteries). These arteries can both supply the facial nerve, but there is commonly a predominance of the petrosquamous branch (Lasjaunias

et al. 2001). Furthermore, the petrosquamous branch gives off branches for the tympanic cavity, where they anastomose with the anterior, posterior, and inferior tympanic arteries.

- In the paramedian area of the convexity, the branches anastomose with those of the contralateral side. These branches form an arcade from anterior to posterior, situated at the insertion of the lateral wall of the superior sagittal sinus with the meninges of the vault. This arcade anastomoses anteriorly with the artery of the anterior falx, a branch of the anterior ethmoidal artery, and posteriorly with the falx cerebelli artery. The arteries of the falx itself arise from the arcade as descending branches.

Branches of the MMA are involved in the supply of extra-intracranial tumors (Figs. 3.27 and 3.29), intracranial AVMs, and especially in DAVFs (Figs. 13.1, 13.5, 13.6, 13.9, 13.10, 13.11, 13.12, 13.13, 13.14, 13.15, 13.24, and 13.25).

3.7.1.2 Accessory Meningeal Artery

The artery is frequently small. It can have a common origin with the MMA. The accessory meningeal artery (AMA) gives off branches for the nasopharynx and palate and then it enters the cranial cavity through the foramen ovale, supplying the adjacent dura including part of the cavernous sinus. For the relationship to the ILT see Chap. 2. It is also partially involved in the supply of the trigeminal nerve. The small branch can be recognized on the lateral ECA angiogram (Fig. 3.15). It is involved in many DAVFs (Figs. 13.10, 13.13, 13.14, and 13.15).

3.7.1.3 Inferior Alveolar Artery

This branch of IMA is always visible on the lateral ECA angiogram despite its small caliber. It runs down toward the mandibular foramen supplying then the teeth and adjacent bone. An example of a normal inferior alveolar artery is presented in Figs. 3.9 and 3.15. In Fig. 3.11 there is a dilated artery supplying a vascular malformation.

3.7.2 Masticator Space

In the masticator space the IMA gives off the deep temporal branches (middle and anterior branches) which have an ascending course, forming a mild convex curve. They contribute in the supply of the temporal muscle. The anterior of these arteries gives off branches that run toward the *orbita*, perforate the malar bone, or pass through the inferior orbital fissure. These branches anastomose with the lacrimal artery, branch of the ophthalmic artery. This connection can become an important collateral in occlusion of the ICA (Djindjian and Merland 1978) (Figs. 3.12, 3.13, 3.14, and 11.5).

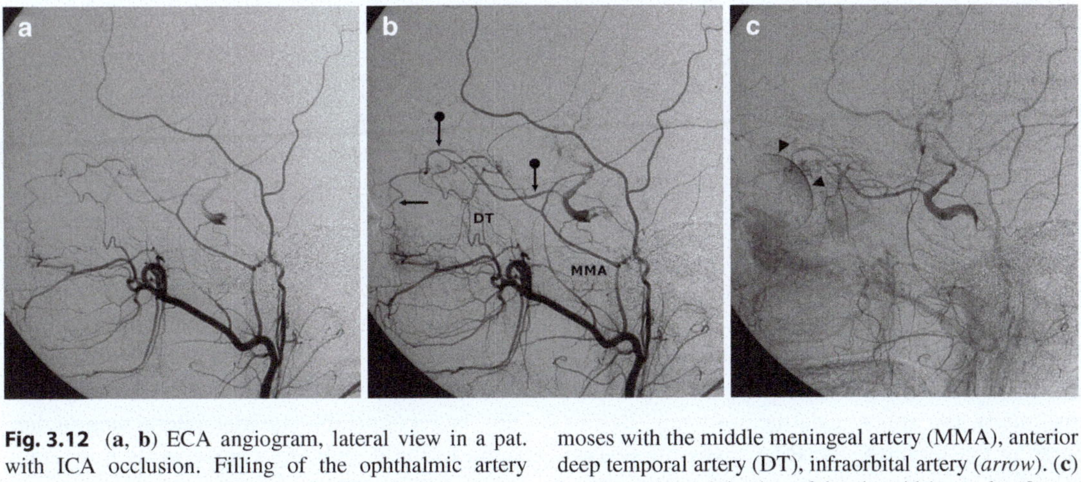

Fig. 3.12 (**a**, **b**) ECA angiogram, lateral view in a pat. with ICA occlusion. Filling of the ophthalmic artery (*arrow with dot*), and further of the ICA through anastomoses with the middle meningeal artery (MMA), anterior deep temporal artery (DT), infraorbital artery (*arrow*). (**c**) In a later phase, injection of the choroid (*arrowhead*)

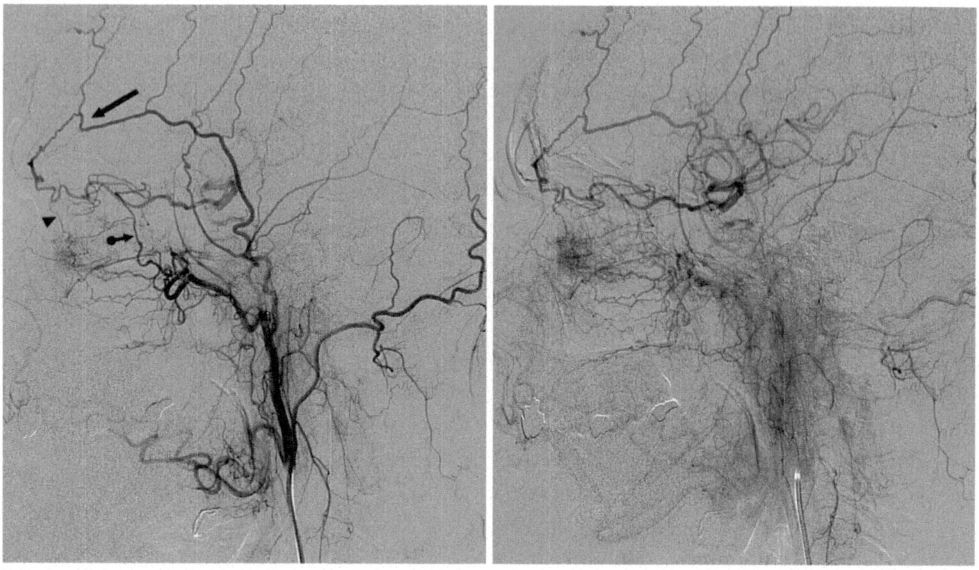

Fig. 3.13 ECA angiogram, lateral view (early and late phases) in a patient with almost complete occlusion of ICA (same case presented in Fig. 17.5). Through the STA (*arrow*), anterior deep temporal artery (*arrow with dot*), as well as fine branches arising from the infraorbital artery (*arrowhead*) retrograde injection of the ophthalmic and intracranial arteries

Other branches in the masticator space are the pterygoid, masseter, and buccal arteries. These arteries are not always recognizable on the angiogram. The masseter artery can be one or made by more than one branches appearing on the lateral angiogram as a fine branch or branches running downwards with and oblique course directed to the corresponding muscle. The buccal artery which supplies the buccinator muscle is more frequently recognizable running downwards with a vertical course anteriorly to the masseter artery (Figs. 3.11 and 3.15). The masseter and buccal arteries anastomose with the facial and transverse facial arteries. The pterygoid arteries are fine branches which cannot be identified with certainty on the angiogram.

3.7.3 Distal IMA

The distal IMA reaches the pterygopalatine fossa, a space bordered anteriorly by the posterior wall of the maxillary sinus and posteriorly by the pterygoid plate, where the artery forms a loop, entering finally the sphenopalatine foramen. Before entering the foramen, the IMA gives off several branches: the posterior superior alveolar artery, the descending palatine artery (great palatine artery), and the infraorbital artery.

The posterior superior alveolar (PSA) artery runs first on the posterior and then on the anterolateral wall of the maxillary sinus. It perforates the maxillary tuberosity running further forwards toward the incisive foramen. It gives off branches for the sinus (antral artery) and teeth. It anastomoses distally with the descending palatine and infraorbital arteries. The posterior superior alveolar artery is a small vessel, sometimes recognizable on the lateral angiogram, with its typical course, first downwards, then forming a curve anteriorly and superiorly (Figs. 3.9, 3.10, and 3.11).

The descending palatine artery (DPA) runs initially downwards along the wall of the maxillary sinus, enters the oral cavity at the greater palatine foramen, and then it forms a typical angle, coursing horizontally and anteriorly through the hard and soft palate supplying it. Anteriorly it ends at the incisive canal, anastomosing with PSA, the nasal branches of the sphenopalatine artery, and the infraorbital artery. The DPA is recognizable on the lateral angiogram owing to the undulating course in its horizontal segment (Figs. 3.9, 3.10, and 3.11).

The infraorbital artery (IFO) enters the *orbita* at the inferior orbital fissure, runs anteriorly, and penetrates the infraorbital canal, where it runs together with the infraorbital nerve. It is involved in the vascularization of the bone structure of the floor of the *orbita* and adjacent muscles and fatty tissue. More distally, it leaves the canal and gives off branches for the cheek, lower lid, upper lip, and nose. The IFO anastomoses with branches of the ophthalmic, facial, transverse facial buccal arteries and PSA and DPA. It can be an important collateral through the ophthalmic artery in cases of occlusion of the ICA (Figs. 3.12 and 3.13). On the lateral angiogram it is always well recognizable with its typical mild superiorly convex curve (Figs. 3.9, 3.10, and 3.11).

Other branches of the distal IMA are the *foramen rotundum artery* (FRA), the *pterygoid (vidian) artery* (VA), and the *pterygovaginal artery* (PVA) (Osborn 1980; Morris 1997; Lasjaunias et al. 2001) (Fig. 3.14a–d). The FRA runs with an oblique supero-posterior course, passes through the foramen rotundum together with the cranial nerve V2, and anastomoses with the ILT. The VA runs posteriorly with a more horizontal course. It passes through the vidian canal, together with the vidian nerve formed by the greater superficial petrosal and the deep petrosal nerves. It anastomoses with the horizontal branch of the mandibular artery, embryonic remnant arising from the petrous segment of the ICA (see also Sect. 2.2). On the lateral ECA angiogram, it is located below the FRA.

The PVA forms a curve directed posteriorly and inferiorly. It gives off branches for then superior nasopharyngeal mucosa anastomosing with

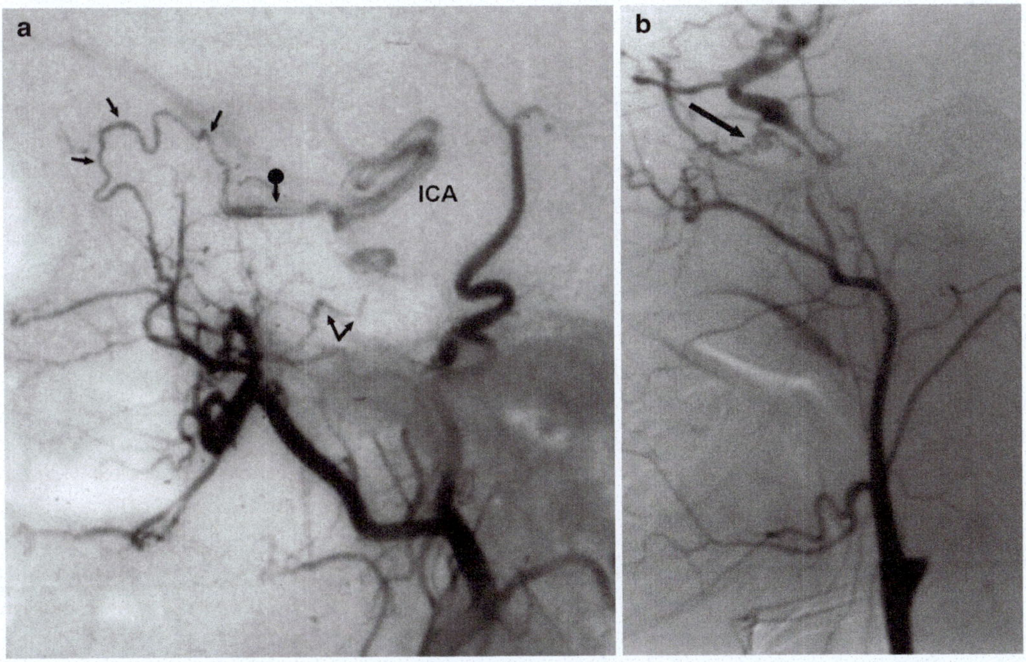

Fig. 3.14 Angiograms in four different patients. (**a**) ECA angiogram, lateral view, in a patient with occlusion of the ICA. Through anastomosis (*arrows*) between the deep anterior temporal artery and the lacrimal branch of the ophthalmic artery, there is injection of the ophthalmic artery (*arrow with dot*) and further of the ICA. Partial injection of the cavernous portion of the ICA, through the foramen rotundum artery (*arrow with angle*). (**b**) Angiogram in a pat. with occlusion of the ICA. Filling of the distal ICA through collateral toward the ophthalmic artery. In addition injection of the distal ICA through the foramen rotundum artery (*arrow*). (**c**) Angiogram of ICA in patient with surgical ligature of ECA. Through the mandibular artery there is filling of the vidian artery (*arrow*), and of the pterigovaginal artery with its typical superior convex curve (*arrow with dot*) and further of the IMA (see Sects. 2.2 and 3.7.3). (**d**) Angiogram of ICA in another patient with surgical ligature of ECA. Through the ILT (*arrow*) filling of the foramen rotundum artery and further of the IMA and indirectly of the pterigovaginal artery (*arrowhead*)

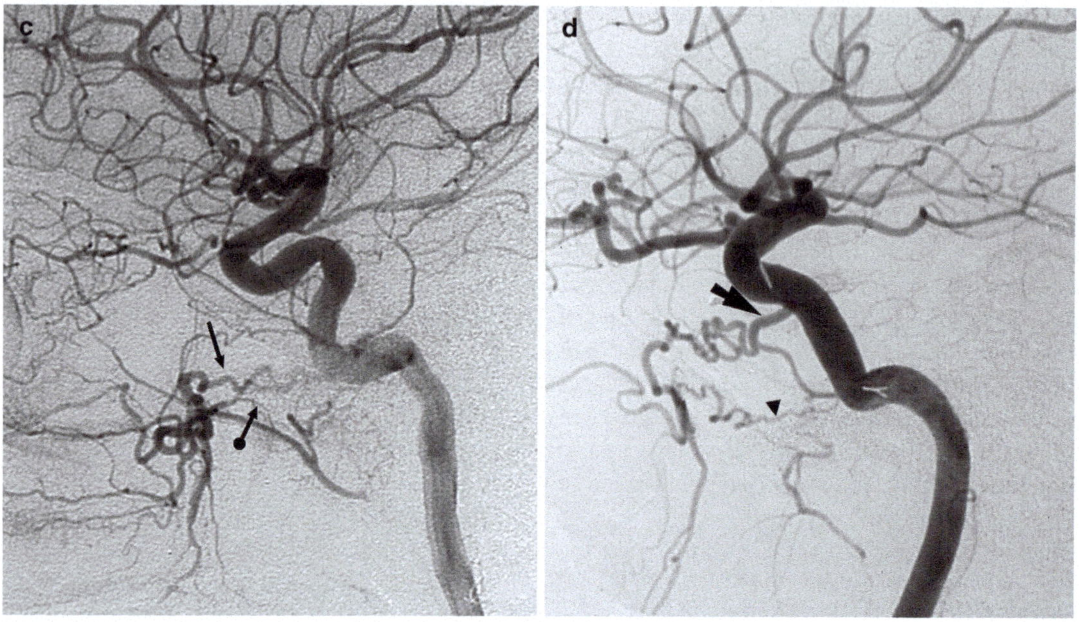

Fig. 3.14 (continued)

branches of the accessory meningeal and AphA. The artery anastomoses posteriorly with the inferior branch of the mandibular artery passing through the cartilage covering the foramen lacerum. It is rarely visible on the ECA lateral angiogram. See also Sect. 2.2.

The FRA, the VA, and PVA can be variously involved in the supply of extra- and intracranial tumors and DAVFs (Figs. 13.6, 13.10, and 13.15). They can play a role in the collateral circulation between ICA and ECA, in cases of occlusion of one of arteries (Fig. 3.14a–d).

3.7.4 The Terminal Branch

The terminal branch is the sphenopalatine artery. This enters the nasal cavity through the sphenopalatine foramen and gives off branches for the medial and lateral nasal mucosa. Anteriorly these branches are connected with branches of the facial, infraorbital, and descending palatine arteries. They anastomose further with the ethmoidal branches of the ophthalmic artery. These branches appear on the lateral angiogram as thin vessels running anteriorly below the infraorbital artery (Figs. 3.9, 3.10, and 3.11). Through their anastomoses with the ethmoidal arteries, they can be involved in the supply of frontobasal meningiomas (Fig. 3.28) and frontobasal DAVFs (Fig. 13.19).

3.8 Superficial Temporal Artery

The superficial temporal artery (STA) is the smaller of the terminal branches of the ECA. It runs upwards and often forms a sharp curve shortly after its origin when it turns over the zygomatic arch. The STA gives off the posterior deep temporal branch supplying the temporal muscle and several branches that run with an undulating course on the scalp, which they supply. This feature is characteristic, and it allows the differentiation of these cutaneous branches from the more rectilinear course of the MMA (Figs. 3.9, 3.15, and 3.29). Other branches are the transosseous which can be well developed when involved in the supply of vascular malformations. Through its anastomosis with the supraorbital branch of the ophthalmic artery, the STA can be involved in the collateral circulation toward the ICA, when this is occluded (Fig. 3.13).

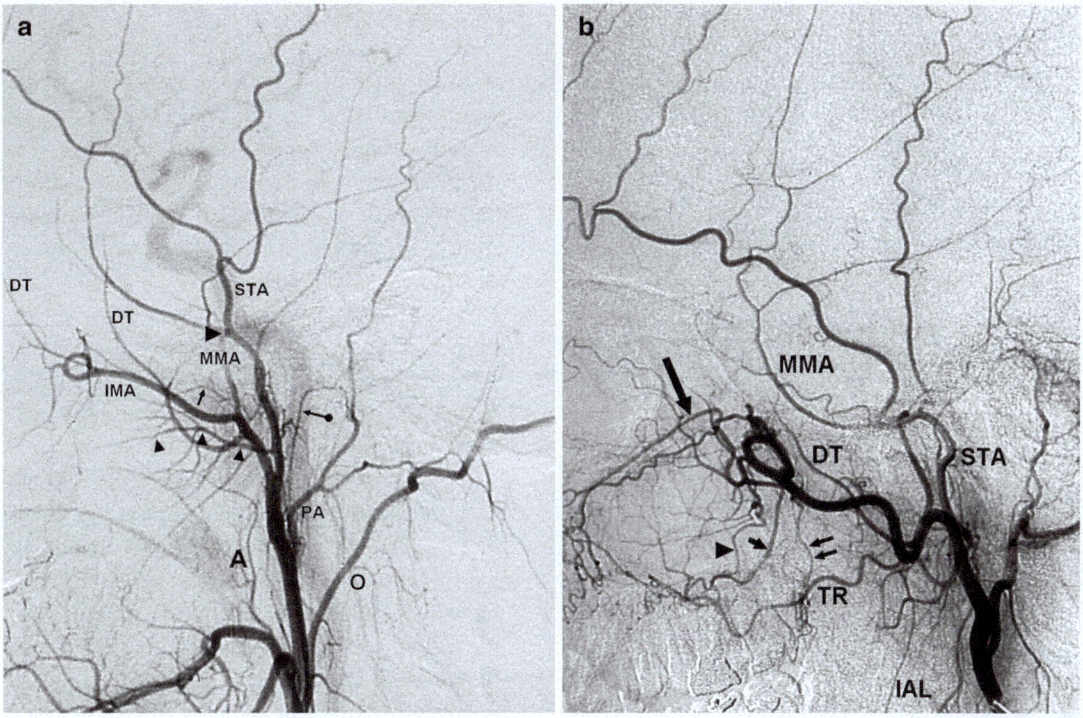

Fig. 3.15 (**a**) External carotid angiogram, lateral view, showing the undulating course of the cutaneous branches of the occipital (O), posterior auricular (PA), and superficial temporal (STA) arteries. The course of the middle meningeal artery (MMA) and middle and anterior deep temporal arteries (DT) appears straighter. The *arrowhead* shows where the MMA enters the cranial cavity through the foramen spinosum. Internal maxillary artery (IMA), accessory meningeal artery (*small arrow*), ascending pharyngeal artery (*arrow with dot*). Ascending palatine artery (A). Horizontal and inferior branches of a small transverse facial artery (*arrowhead*). (**b**) Angiogram of the distal ECA, lateral view, in another patient, showing distal branches of IMA. Infraorbital artery (*large arrow*), posterior superior alveolar artery (*arrowhead*), descending palatine artery (*small arrow*), buccal artery (*two small arrows*), inferior alveolar artery (IAL). Well-developed transverse facial artery (TR)

Another important branch of the STA is the transverse facial artery, which arises from the first segment of the STA (Figs. 3.9, 3.11, 3.15, and 3.30). The artery courses horizontally along the zygoma and supplies partially the parotid gland and the skin of the cheek. It can give off a branch directly inferior to the deep surface of the masseter which it supplies, anastomosing with the masseter branches of the IMA (Salmon 1936). It anastomoses with the facial artery and can occasionally be well developed, especially when the facial artery is hypoplastic.

3.9 Cervical Arteries

In the examination of the upper neck and skull basis, the cervical arteries, represented by the ascending cervical (AC) and deep cervical (DC) arteries, should also be considered. The AC arises commonly from the thyrocervical trunk and the DC from the costocervical trunk both arising from the subclavian artery. The cervical arteries run upwards toward the skull basis, giving off branches supplying the soft tissues and the bone structures of the cervical spinal column. The AC runs more anteriorly, while the DC is more posteriorly located. This is well recognizable on the lateral angiogram. It can occur that one of the cervical arteries, either the AC or the DC, is less developed or absent especially in its distal part, and so its vascular territory is taken over by the other artery. There are many anastomoses, differently developed, between the cervical arteries, external carotid, and vertebral arteries (Lasjaunias 1981; Morris 1997). The most frequently visible are those at the C1–C2 level involving the deep

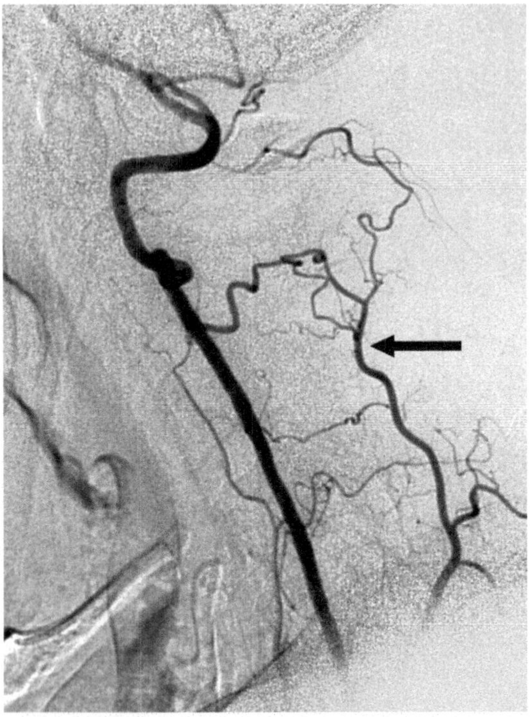

Fig. 3.16 Angiogram of the left subclavian artery in a patient with small vertebral artery. Well-developed deep cervical artery (*arrow*) supplying also the territory of the hypoplastic ascending cervical artery. There are anastomoses between both cervical arteries and with the vertebral artery

cervical, occipital, and vertebral arteries and that on the C2–C4 level between the ascending cervical, ascending pharyngeal, and vertebral arteries (Figs. 3.7b, d, and 3.16).

These anastomoses can play an important role as collateral circulation toward the ECA, ICA, and vertebrobasilar system in case of occlusive diseases (Figs. 3.7c and 15.29) and can be involved in DAVs (Fig. 13.5) and vascular malformations of the head and neck. In the cases in which they are used for an endovascular treatment the presence of their connections especially with the VA should be taken into consideration. Finally it should be remembered that the radiculomedullary arteries continuing in the anterior spinal artery can arise from the ascending cervical artery, instead of taking their origin from the vertebral artery.

3.10 Summary

- ECA supplies the soft tissues and bone of the craniofacial area.
- It represents the main meningeal supply for the intracranial dura mater. Its branches are in balance with the meningeal branches arising from the ophthalmic artery, ICA, and VA.
- It supplies a *few cranial nerves*.
- It can be an *important collateral way* to the intracranial circulation, in cases of an occluded ICA and VA.
- ECA is involved in the vascularization of vascular malformations of the extracranial area and in many richly vascularized tumors in the extra-intracranial area. Endovascular treatment of these lesions is an important part of the therapy. Indication and examples of treatment of extracranial lesions are presented in the following sections. The role of the ECA in the diagnosis and treatment of DAVFs and its contribution in brain AVMs will be discussed in the corresponding chapters.

3.10.1 Vascular Malformations

Vascular malformations represent a congenital pathology that may be already present in pediatric patients or clinically appear later owing to the slow growth. They can be classified into high- and slow-flow malformations. The first are arteriovenous (AV) angiomas in which commonly numerous dilated arteries converge to a vascular network (nidus) where there is a direct shunt with the dilated and early filled veins. The shunts can be sometimes very large, forming huge fistulas. The slow-flow malformations are capillary-venous angiomas, in which the supplying arteries are commonly normal, without an early venous drainage. In these later cases, the angiogram is frequently normal. Pathological blush or sometimes a pathological network can be demonstrated on superselective angiographic studies. Both high- and slow-flow components can be associated in the same malformation.

Lymphangiomas are another more rare vascular malformation involving the lymphatic vessels which can be very dilated appearing as a large cystic space. They are more frequent in childhood and infancy. Lymphangiomas are thought to be due to a failure of the lymphatic sacs to drain into the veins or due to an abnormal sequestration of lymphatic tissue no longer connected with the normal lymphatic channels (Som et al. 2003b). Commonly, they cannot be demonstrated by angiography, unless we are dealing with hemolymphangiomas. These are mixed lesions where lymphatic tissues are associated with another type of angioma.

There is a possible association of all these extracranial vascular malformations with intracranial malformations (aneurysms or angiomas) and with some neurocutaneous pathologies (see Chap. 12).

Location. They can involve the skin of the head and face as well as the mucosa of the oral and nasal cavities. The masticatory muscles more frequently the masseter muscle and bone structures (mandibular and maxillary bone) are more seldom affected; they may be secondarily involved as extensions of the malformation or be the primary location.

Clinical relevance. This varies from esthetic problems to more important clinical symptoms, involving respiration and deglutition up to severe hemorrhage and cardiac failure. Mandibular and maxillary AVM can present as incidental discovery of an osteolytic lesion or with acute life-threatening hemorrhage following dental loosening, dental extraction, or cheek trauma (Benndorf et al. 2001b; Noreau et al. 2001; Fan et al. 2002; Seehra et al. 2006; Siniscalchi et al. 2009; Liu et al. 2009; Churojana et al. 2012; Bergeron et al. 2013; Amat et al. 2015).

Diagnosis and treatment. The diagnosis can sometimes be obtained only by clinical observation. Extensive lesions require CT and MRI. Depending on the site and extension of the malformation, angiography may be necessary to identify the type of lesion and the vessels involved. With regard to the location, the supplying arteries can frequently be suspected. However, other vessels may be indirectly involved, and so a complete study of both ECAs and in some cases also of the cervical arteries is frequently necessary. To exclude intracranial malformations examination of the ICA and VA should also be performed. All these information are necessary to plan the therapy for which in all cases a multidisciplinary approach is mandatory.

Small asymptomatic lesions do not need necessarily a treatment which should be considered in symptomatic or large malformations. The endovascular treatment is in the majority of the cases the first choice of therapy used depending on the site of the malformation PVA, glue, or onyx which can lead to a complete cure in cases of small lesions. In other cases the endovascular treatment precedes the surgical excision or can be a palliative therapy aimed to reduce the volume of the AVM for cosmetic reasons or to reduce pain or the risk of hemorrhage (Thiex et al. 2011; Dmytriw et al. 2014). The endovascular treatment can be associated with injection of a sclerosing agent through direct puncture of the devascularized AVM or injection of glue especially in malformations involving bone structures. Positive results have been obtained in some cases through direct injection of bleomycin, an antiangiogenic drug (Dmytriw et al. 2014). The combination of these techniques depends on the site, extension of the lesion, and experience and attitude of the team involved. Examples are presented in Figs. 3.17; 3.18; 3.19, 3.20, and 3.21. Endovascular therapy commonly performed with injection of PVA can also be performed in patients presenting with dramatic epistaxis not linked to vascular malformations but due to disturbances of the coagulations or linked to a fragility of the microvasculature more frequent in old patients. This therapeutic approach allows in these cases a rapid and complete control of the bleeding.

3.10.2 Hemangiomas

Hemangiomas are real tumors (Mulliken and Glowacki 1982) consisting prevalently of masses of endothelial cells with or without vascular lumens. They can be present at birth, with a strong prevalence among females, or they can appear in the first months of life. Hemangiomas are characterized by a phase of rapid growth, followed by a period of progressive involution which can be

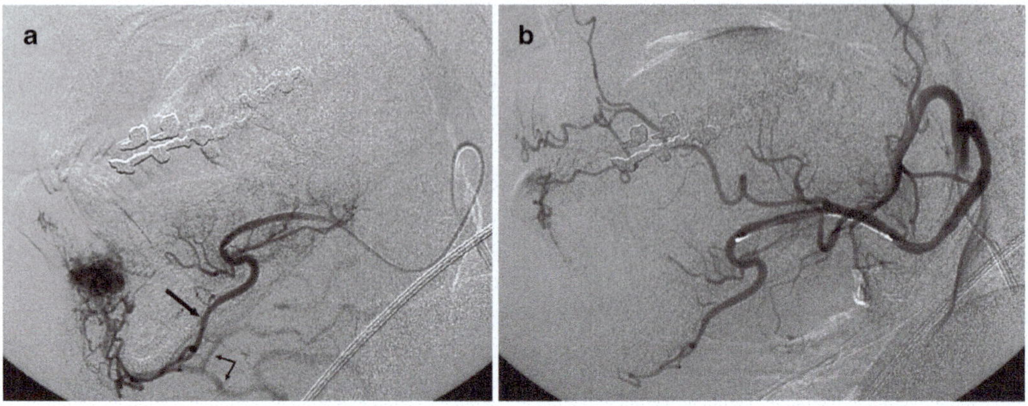

Fig. 3.17 Small arteriovenous angioma of the mental area supplied by the facial artery. (**a**) Lateral angiogram of the selective submental artery (*arrow*) showing the angioma and the early drainage in the submental veins (*arrow with angle*). (**b**) Control angiogram after occlusion of the malformation with polyvinyl alcohol (PVA) particles

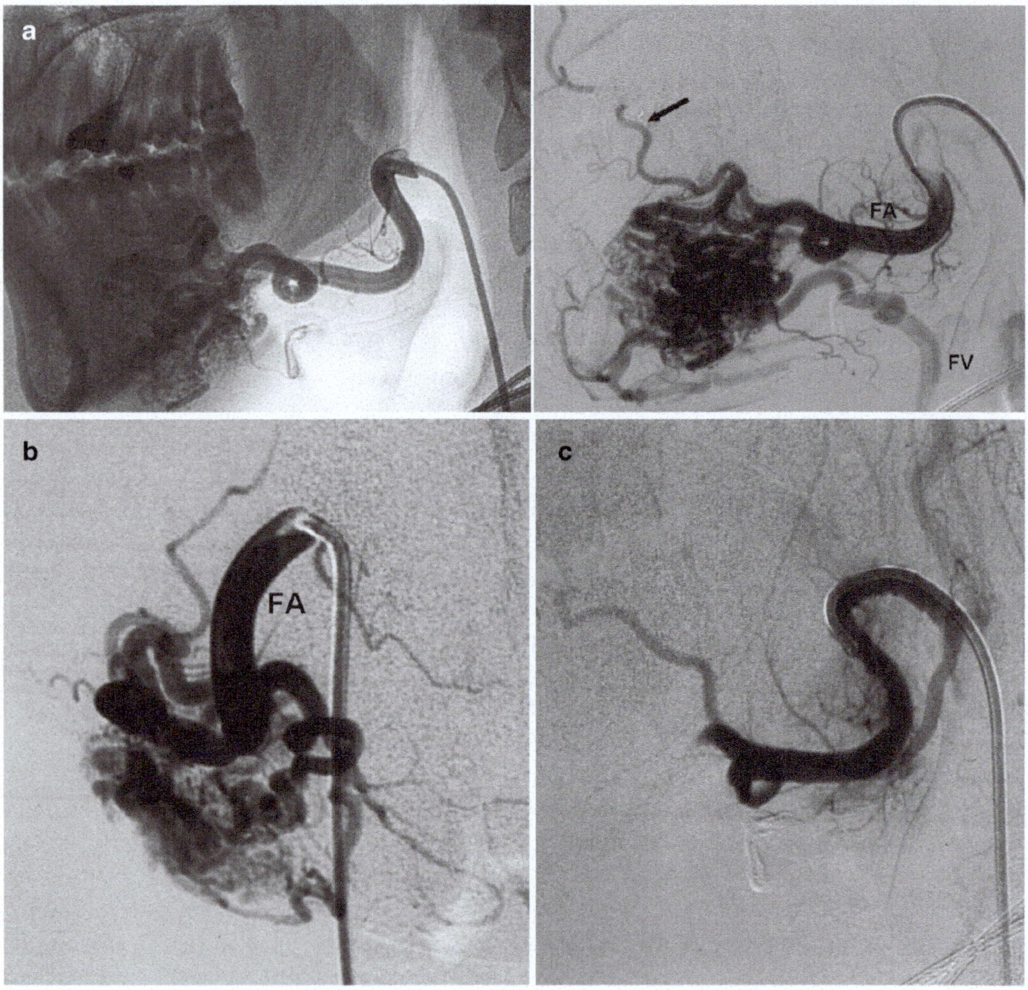

Fig. 3.18 Large arteriovenous angioma of the inferior check supplied by the facial artery. (**a**) Lateral angiogram of the facial artery (subtracted and unsubtracted image). Dilated facial artery (FA) supplying the malformation. Note the normalization of the lumen distally to the malformation (*arrow*). Early drainage in the facial vein (FV). (**b**) AP view of the FA. (**c**) Lateral angiogram after preoperative devascularization of the lesion

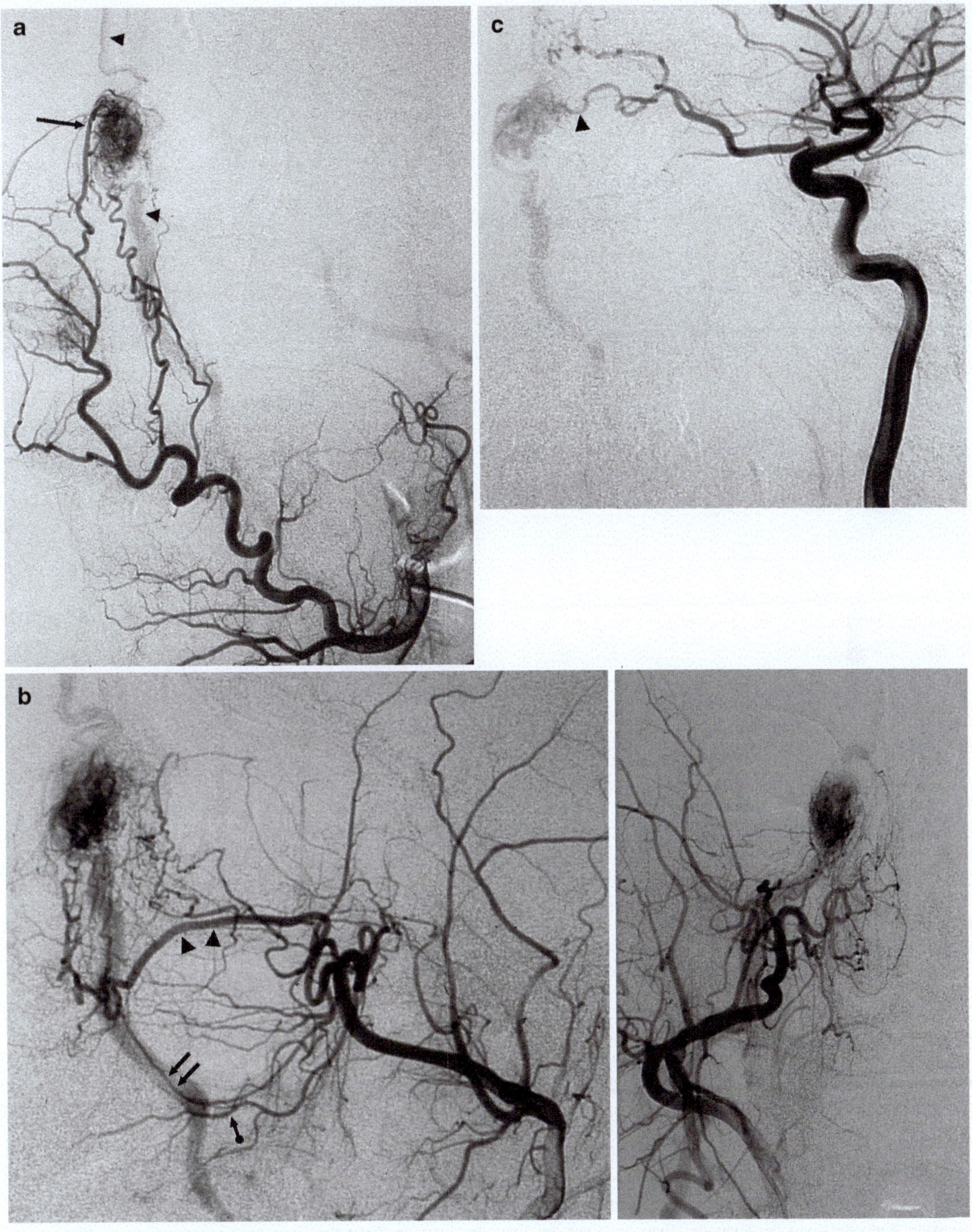

Fig. 3.19 Arteriovenous angioma of the right ala of the nose supplied by branches of the facial artery, internal maxillary artery, and ophthalmic artery. (**a**) Lateral angiogram of the facial artery. The arteriovenous angioma is supplied by the angular artery (*arrow*) and other small branches of the distal facial artery. Early drainage in the facial and frontal veins (*arrowhead*). (**b**) Lateral and AP angiograms of the internal maxillary artery. Clear involvement of the infraorbital artery (*arrowheads*) in the supply of the angioma. There is also a minimal involvement of the posterior superior alveolar artery (*arrows*). Normal descending palatine artery (*arrow with dot*). (**c**) Lateral internal carotid angiogram, showing the partial supplying through the nasal branch of the ophthalmic artery (*arrowhead*). The AVM was devascularized with polyvinyl alcohol (PVA), injected in the ECA branches, followed by surgical excision

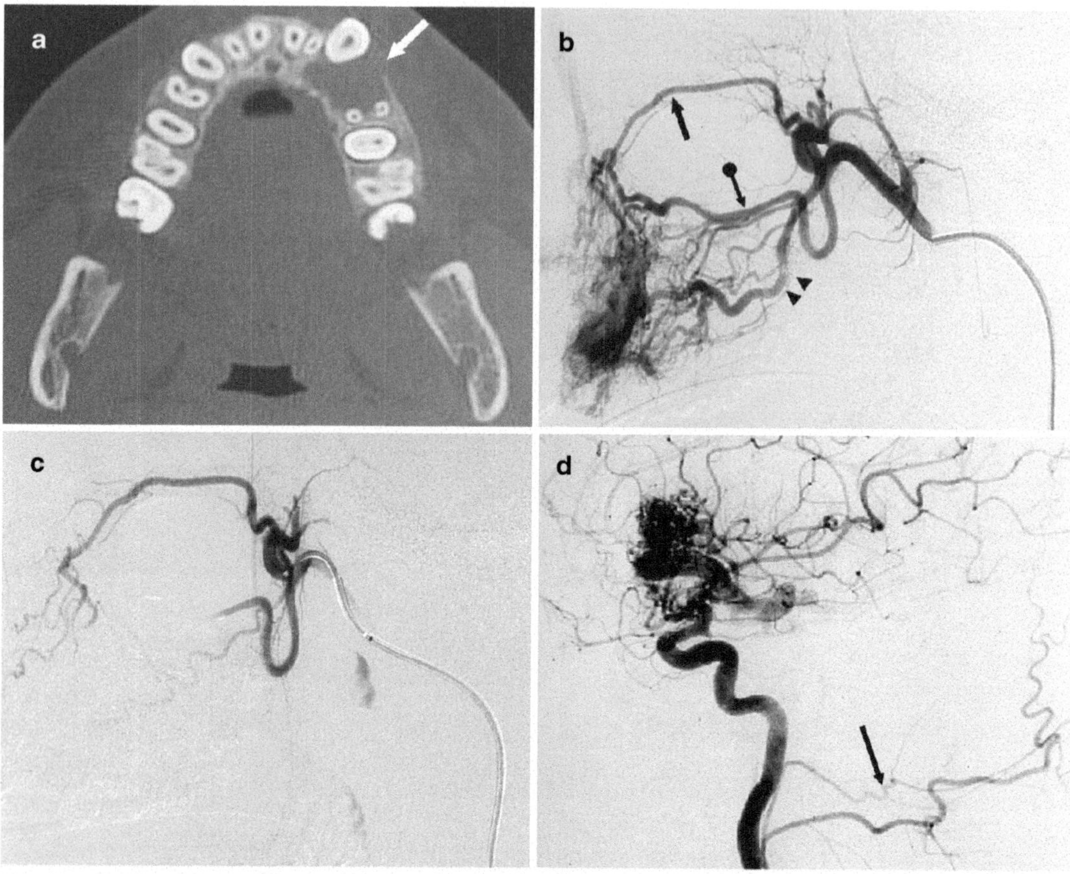

Fig. 3.20 Alveolar maxillary arteriovenous angioma presenting with hemorrhage. (**a**) CT shows the bone changes with displacement of the tooth (*arrow*). (**b**) Lateral angiogram of the internal maxillary artery. The angioma is supplied by branches of the infraorbital artery (*arrow*). Branches of the superior posterior alveolar artery (*arrow with dot*) and descending palatine artery (*arrowhead*) are also involved. (**c**) Acute endovascular treatment with devascularization using polyvinyl alcohol (PVA), followed by surgical curettage. Good outcome also in the long-term controls. (**d**) Internal carotid artery in the same patient showing an asymptomatic arteriovenous malformation. There is a common trunk of the occipital and ascending pharyngeal arteries arising from the internal carotid artery. Well-developed neuromeningeal trunk (*arrow*)

complete at the age of 6–7 years (Pitanguy et al. 1984) in about 95% of the cases. They are located in the head and neck involving prevalently the soft tissues. Involvement of the bone is rare. This occurs typically in the mandible. The clinical relevance varies from an asymptomatic small lesion on the skin to large disfiguring hemangiomas of the face and of the oral cavity leading to disturbances of the vision, deglutition, and respiration, and severe hemorrhages and cardiac failure.

Depending on the clinical condition MRI study and angiography can be indicated. The latter will show the frequently rich vascularized tumor which volume can be reduced by embolization using PVA particles.

Hemangiomas can be isolated or associated with other pathological conditions. Thrombocytopenia (Kasabach-Merritt syndrome) and association with several other abnormalities in the so-called PHACE syndrome can occur. Pascual-Castroviejo (1978) was the first to describe the association of extracranial capillary hemangiomas with vascular and nonvascular intracranial malformation. The definition of "PHACE" (posterior fossa malformation including intracranial hemangiomas, cervicofacial hemangiomas, eye

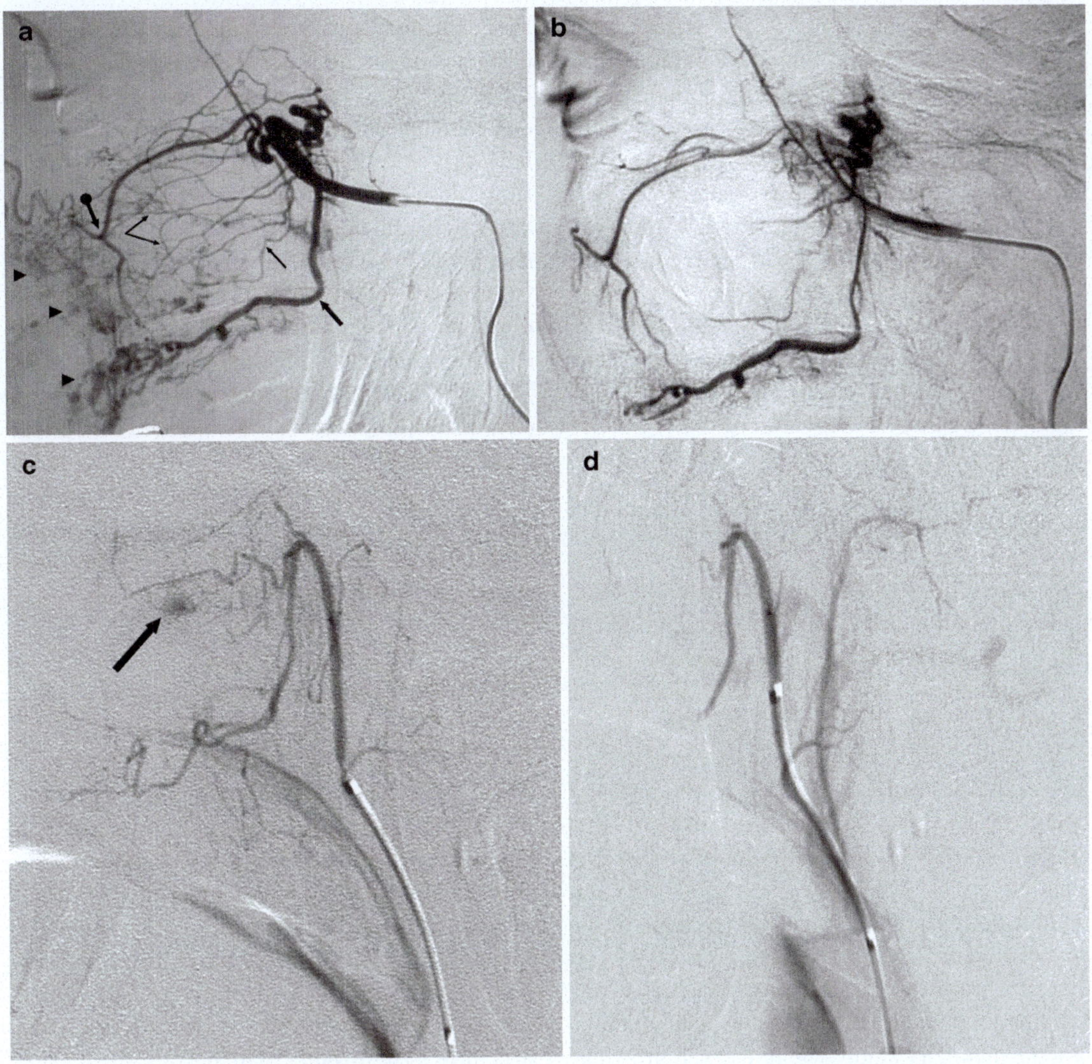

Fig. 3.21 Two examples of Rendu-Osler in middle-aged patients presenting with severe epistaxis. (**a**) Lateral angiogram of the distal IMA. Multiple capillary angiomas presenting with a rich vascularization involving the skin (*arrowhead*) and the nasal mucosa. There is also a possible microaneurysmal dilatation along the descending palatine artery. Several branches are involved. Infraorbital artery (*arrow with dot*), nasal branches of the sphenopalatine artery (*arrow with angle*), superior posterior alveolar artery (*small arrow*). Descending palatine artery (*arrow*). (**b**) Lateral angiogram posttreatment with PVA. (**c, d**) Another patient in whom the small angioma was supplied only by pharyngeal branches of the AphA. (**c**) Angiogram of the AphA showing the small angioma (*arrow*). (**d**) Angiogram posttreatment

abnormalities, arterial anomalies, cardiac defects, and aortic coarctation) was coined by Frieden et al (1996). Later, other authors described similar complex syndromes in various combinations. In addition to the cutaneous hemangioma, similar lesions can be present also in the buccal, parapharyngeal, and subglottis areas. In the posterior fossa, Dandy-Walker malformation, hypoplasia of cerebellum, and hemangiomas in the cerebellopontine angle have been frequently reported as well as intraorbital hemangiomas. Cases of sternal defects have been described (Tortori-Donati et al. 1999; Metry et al. 2001; Poetke et al. 2002; Bhattacharaya et al. 2004; Church and Lowe

2006; Judd et al. 2007; Alsuwaidan 2012). Considering especially the arterial anomalies, these include hypoplasia of the internal carotid artery, its progressive occlusion associated with pattern mimicking the Moyamoya disease, and the persistence of fetal arteries. The etiopathogenesis of PHACE is unknown. Female predominance is present. Steroid therapy leads commonly to reduction or complete involution of the hemangiomas. In the cases in which the vascular tumor is responsible of a severe clinical involvement, angiography can be indicated, followed by devascularization of the tumor with PV.

3.10.3 Juvenile Angiofibromas

These are tumors that develop typically in young males before puberty, indicating an influence of male hormones. Rare cases of angiofibromas in older patients and also females have been described (Osborn 1959).

Juvenile angiofibromas are benign tumors formed by a fibrovascular stroma. The vascular component is very rich and consists of true arterial-capillary vessels and simple endothelial cavities.

Location. The tumor arises at the level of the sphenopalatine foramen, from which it can extend toward the nasal cavity and involve further the ethmoidal cells, the sinus sphenoidalis, and the maxillary sinus. The tumor can grow laterally into the pterygopalatine and further into the infratemporal fossa or superiorly, invading the *orbita* through the inferior orbital fissure. Through the foramina of the base of the skull, it can extend into the middle cranial fossa, with possible involvement of the cavernous sinus. Commonly, juvenile angiofibromas do not primarily destroy the bone structures, which are initially displaced and thinned and only later eroded.

Clinical relevance. The symptoms depend on the growth direction and extension. The most frequent presenting clinical signs are epistaxis and nasal obstruction.

Diagnosis and treatment. Initial examinations are CT and MR, followed by angiography. Pharyngeal branches of the AphA and distal branches of the IMA are the vessels commonly involved. Supply from the petrous and cavernous segments of the ICA through the mandibular artery and the ILT, respectively, and supply from branches of the ophthalmic arteries depending on the extension of the tumor can occur (Figs. 3.22 and 3.23). Preoperative endovascular treatment with devascularization of the tumor is an important part of the therapy (Valavanis and Chistoforidis 2000; Roche et al. 2007). Direct percutaneous embolization with *onyx* has been proposed (Gemmette et al. 2012; Lv et al. 2013; Gao et al. 2013). When the latter technique is used, it is important to perform, during the procedure, angiographic controls especially to identify the filling of dangerous anastomosis (see also Sect. 3.10.7).

3.10.4 Paragangliomas

Paragangliomas belong to a group of tumors in which three types can be distinguished (Lantos et al. 1997): the pheochromocytomas originating in the adrenal medulla, the sympathetic paragangliomas arising from neuroendocrine cells associated with the sympathetic chain, and the parasympathetic paragangliomas (chemodectomas) arising from cells which have chemoreceptor activity associated with paraganglionic structures. These latter will be described here. Women in the fourth and fifth decades of life are predominantly affected. In up to 10% of patients paragangliomas are multiple. Cases in the same family have been reported. A malignant potential evolution with metastasis especially in the lung and vertebral column is very rare, but possible, and is more frequent in younger patients (Zak and Lawson 1982; Ma et al. 2010; Obholzer et al. 2011). Differently from pheochromocytomas, paragangliomas are rarely active.

The tumor can present as a very small nodule or as a large mass reaching the diameter of 10–15 cm. Histologically is appears as an encapsulated mass consisting of epithelial cells with a high vascular stroma.

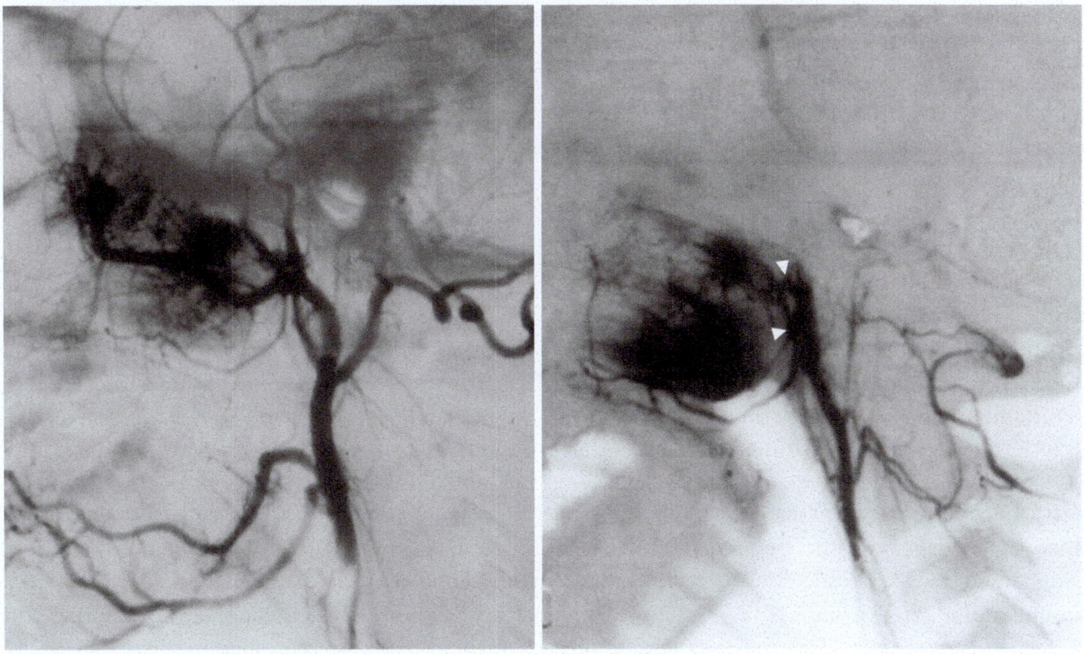

Fig. 3.22 Large juvenile nasopharyngeal angiofibroma, lateral angiogram of the external carotid artery (*left*) and of the ascending pharyngeal artery (*right*), showing the richly vascularized tumors, supplied by distal branches of the IMA and superior and middle pharyngeal branches of the AphA (*white arrows*). On both angiograms there is a partial filling of the vertebral artery through anastomoses with the occipital (*right*) and AphA (*right*). Note also the retrograde filling of the deep cervical artery (*right*). The lesion was preoperatively devascularized with PVA after advancing the catheter distally to the anastomoses

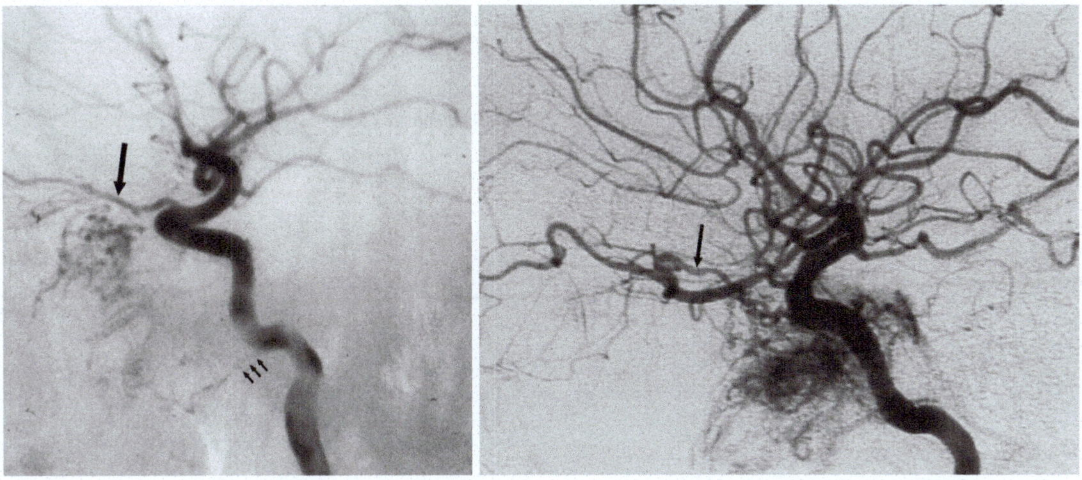

Fig. 3.23 Left ICA angiogram in a large juvenile angiofibroma with an intraorbital extension. On this angiogram, the partial involvement of the mandibular artery (*arrows*) and of branches of the ophthalmic artery (*arrow*) are recognizable. On the *right* another example of a juvenile angiofibroma extending intracranially in the parasellar region. This portion of the tumor is richly vascularized by branches of the cavernous portion of ICA, by the recurrent meningeal artery of the ophthalmic artery (*arrow*) and probably also from the mandibular artery

Location. Most frequently, paragangliomas are located in the temporo-basal area arising from paraganglionic structures in association with the tympanic branch of the glossopharyngeal nerve (tympanic paraganglioma) or from cells located in the adventitia of the bulb of internal jugular vein (jugular paraganglioma). The second location in frequency is the carotid body and vagal paragangliomas arising, respectively, from paraganglionic cells located at the carotid bifurcation or from cells associated with the vagal nerve or ganglion nodosum. Less common are lesions in the larynx, pharynx, lung, orbit, and nose. A curious location is that in the cauda equina where no paraganglionic structures are present. Interesting to note is that this site occurs more frequently in men (Sonneland et al. 1986; Singh et al. 2005; Taschner et al. 2012).

Clinical relevance. The mass effect is the dominant feature in carotid body paragangliomas. Various degrees of involvement of the last four cranial nerves are present in vagal and tympano-jugular lesions. In the latter case, otological symptoms, characterized by hearing loss, tinnitus, and palsy of cranial nerve VII, can occur.

Diagnosis and treatment. Initial examination is by CT and MRI which show the location and extension of the tumor. Angiography including study of the bilateral ICA, ECA, and vertebral artery is an essential diagnostic step. The supplying arteries can vary, depending on the location and extension of the tumor. Some basic features can be summarized as follows.

In paragangliomas of the ICA bifurcation and vagal paragangliomas, the pharyngeal branches of the AphA and branches of the occipital artery are always involved. The ascending palatine branch of the facial artery can also occasionally supply the tumor. In tumors of the ICA bifurcation, infiltration of the adventitia of the ICA can occur. In tympano-jugular paragangliomas the tympanic branches (anterior from IMA, posterior from the stylomastoid, superior from the petrosquamous branch of MMA, and inferior from the AphA) can be involved in various combinations, depending on the extension of the tumor. The vertebral artery may be involved through its extracranial and also intracranial branches in the rare cases of extension of the tumor in the posterior fossa. The ICA can supply the tumor through its caroticotympanic branch; furthermore through extension of the tumor into the carotid canal, there is a possible encasement of this segment of ICA.

The venous drainage almost always occurs in the internal jugular vein, which may also be directly infiltrated and occluded by the tumor. In these cases a retrograde injection of the sigmoid sinus and IPS can occur. These aspects of the venous circulation should always be carefully examined as well as in case of occlusion of the internal jugular vein the presence of a drainage through the contralateral jugular vein and through the venous plexus surrounding the vertebral artery should be demonstrated. Examples of paragangliomas are presented in Figs. 3.24, 3.25, 3.26, and 3.27.

Preoperative endovascular treatment aimed to lead to a devascularization of the tumor is mandatory (Valavanis and Christoforidis 2000). It should be noted that paragangliomas can consist of a single compartment to which converge all supplying arteries, or of multiple compartments each supplied by a specific artery. Compartment can be completely separated or be connected to each other (Moret et al. 1980). This should be taken into consideration in the treatment. The devascularization can be associated, in selective cases, with endovascular occlusion with balloon or coils of the ICA or VA. In some cases, especially in old patients, the endovascular devascularization, reducing the volume of the tumor and improving the clinical symptoms, can be considered as a useful palliative therapy.

3.10.5 Meningiomas

Meningiomas are a relatively common pathology, having a frequency of about 18% of all intracranial tumors. They typically predominate in women of middle to old age and are not rarely multiple. Meningiomas originate from

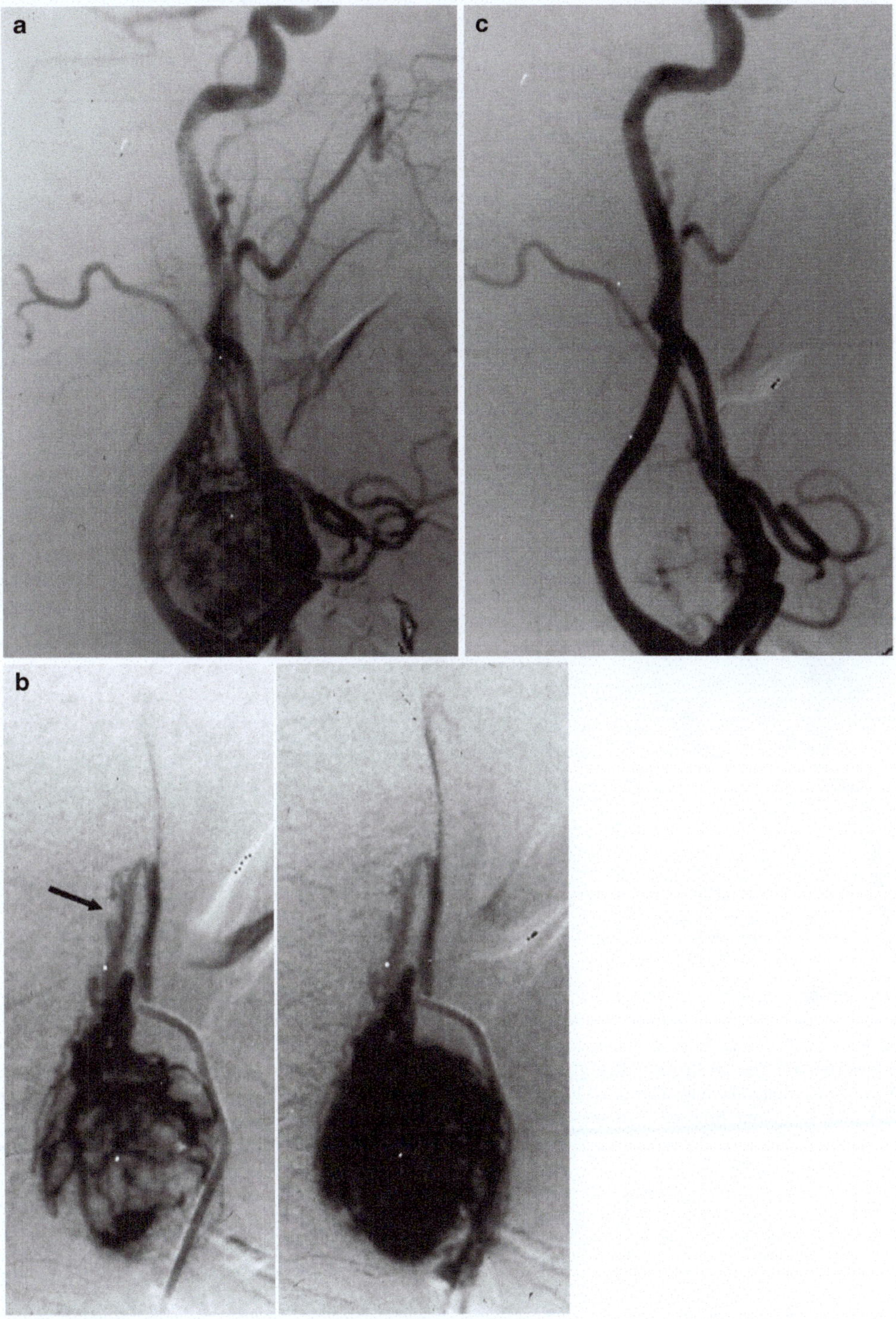

Fig. 3.24 Paraganglioma of the carotid bifurcation. (**a**) Common carotid angiogram, lateral view, showing the typical location of the richly vascularized tumor. (**b**) Selective study of the ascending pharyngeal artery which is the main supplying artery through its inferior pharyngeal branches (*arrow*). (**c**) Control angiogram after devascularization of the tumor with PVA

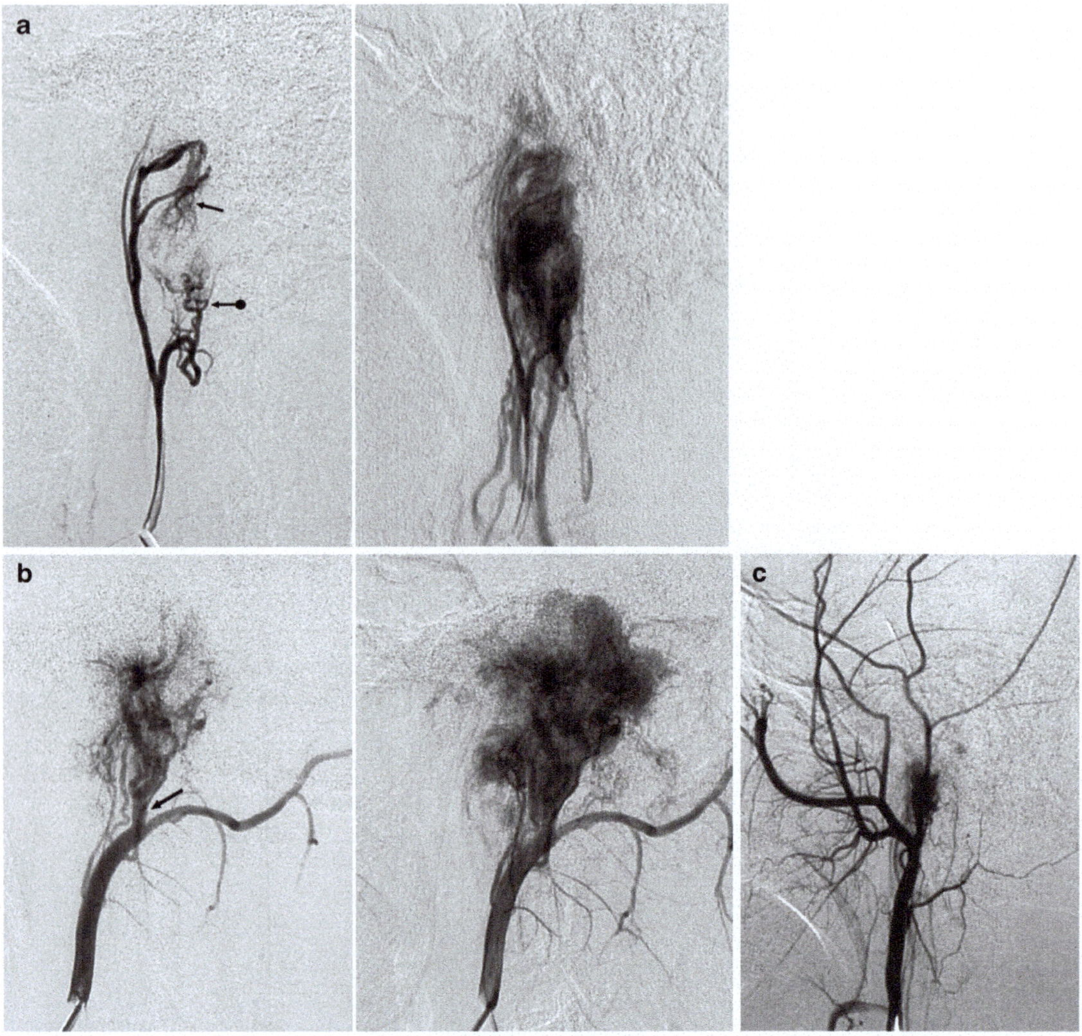

Fig. 3.25 Tympano-jugular paraganglioma supplied by branches of the occipital and ascending pharyngeal artery. A small component came from the anterior tympanic artery branch of the IMA. (**a**) Angiogram of the ascending pharyngeal artery. Lateral view. The pharyngeal (*arrow with dot*) and tympanic branches (*arrow*) supply the tumor. In the later phase, the anterior compartment of the tumor is well defined. (**b**) Angiogram of the occipital artery. Lateral view. A large stylomastoid artery (*arrow*) is involved in the supply of the posterior compartment of the tumor, well defined in the later phase. (**c**) Small supply from the anterior tympanic branch

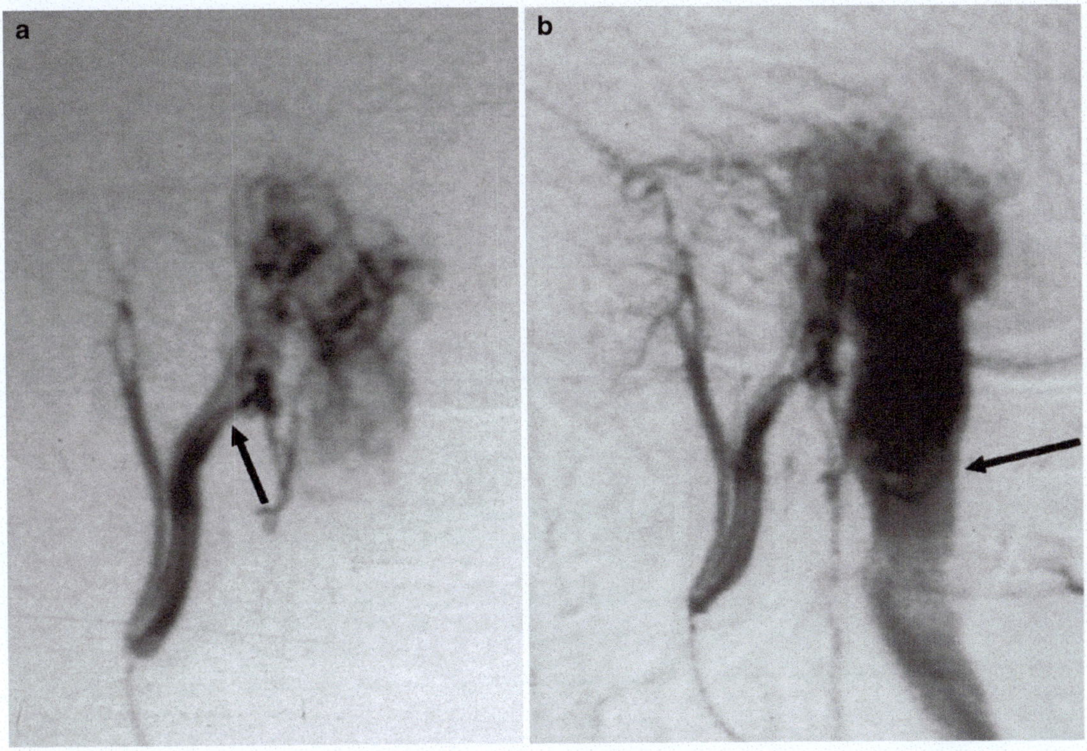

Fig. 3.26 Large tympano-jugular paraganglioma supplied mainly by tympanic branches of the ascending pharyngeal artery. (**a**) Angiogram (oblique view) of the selective catheterization of an enormous dilated ascending pharyngeal artery. Tympanic branches (*arrow*) supplying the tumor. (**b**) Drainage in the jugular vein (*arrow*) infiltrated by the paraganglioma

meningothelial cells (arachnoidal) of the meninges. These cells are thought to be a mixed cell type having mesenchymal and neural crest elements, which explains probably the variety of the histological aspects of the meningiomas (meningothelial, fibrous, transitional, psammomatous, and angiomatous) (Bradac et al. 1990; Schiffer 1993; Lantos et al. 1997). They are commonly benign tumors. Malignant degeneration, however, even if rare, can occur, and this is more frequent in males.

The vascularization can vary. It is particularly rich in the angiomatous type. Essentially there are two groups of supplying arteries (Bradac et al. 1990). The first consists of meningeal branches entering the tumor at the site of the dural attachment supplying the central part of it; the second consists of pial branches of the

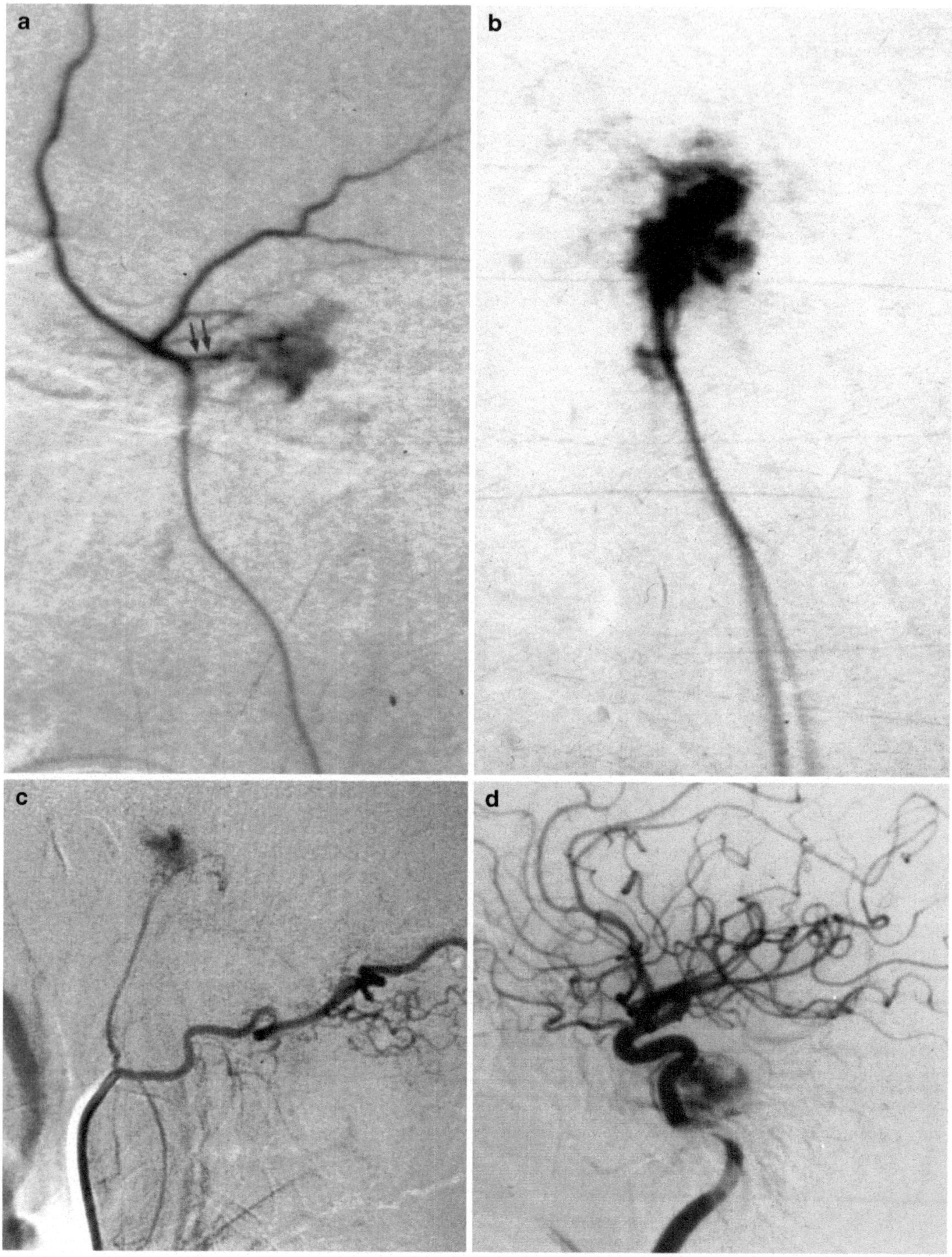

Fig. 3.27 Selective studies of a tympanic paraganglioma supplied by superior tympanic branches (**a**) of the petrosquamous branch (*arrows*) of the middle meningeal artery (*arrows*). There is a second supply arising from the inferior tympanic branch of the ascending pharyngeal artery. (**b**) Small paraganglioma (**c**) in another patient supplied only by the tympanic branch of the stylomastoid artery. Another example (**d**) of a large tympano-jugular paraganglioma with involvement of cavernous and petrous branches of the internal carotid artery

ICA and/or vertebral artery, which vascularize the more peripheral part of the tumor. One vascular sector can exert evident predominance over the other. The origin of the meningeal vascularization depends largely on the site of the primary dural attachment of the meningioma. In addition, the tumor can extend and secondarily involve further dural areas, leading to a recruitment of other meningeal feeders. Preoperative devascularization may be useful in selective cases. Examples are presented in Figs. 3.28 and 3.29.

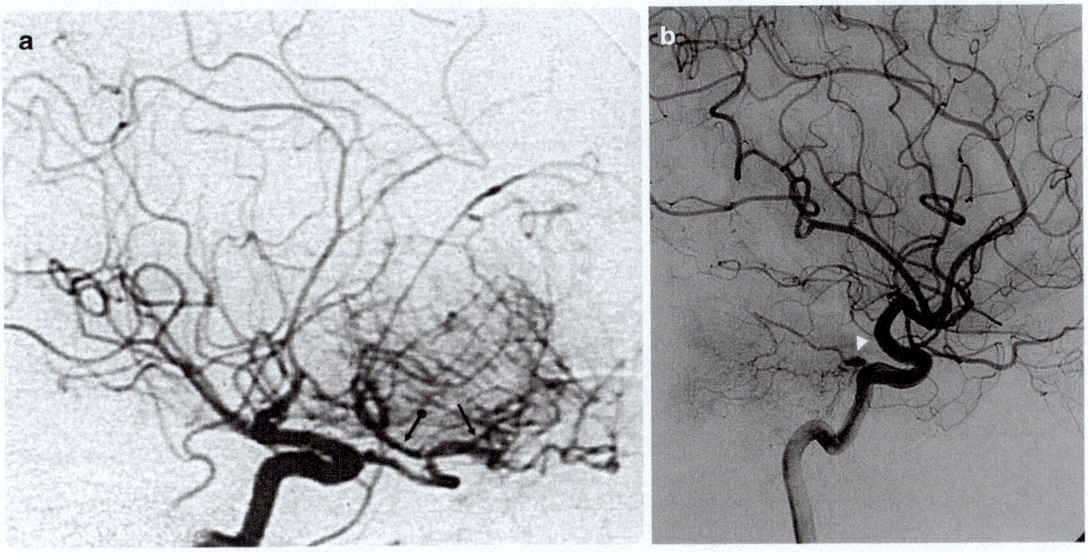

Fig. 3.28 (**a**) Frontobasal meningioma, carotid angiogram, lateral view. The tumor is richly vascularized by branches of the ophthalmic artery represented by the recurrent meningeal artery (*arrow with dot*) and posterior ethmoidal arteries (*arrow*). (**b**) Petroclival meningioma, carotid angiogram, lateral view. The tumor is vascularized by the enlarged branches of the MHT (*arrowhead*)

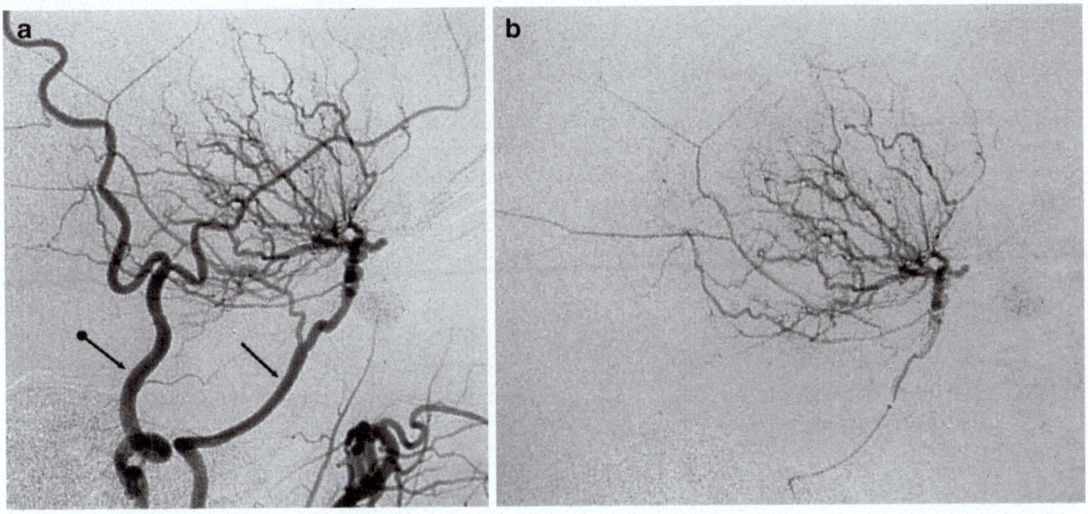

Fig. 3.29 Temporo-basal meningioma, richly vascularized by distal branches of the middle meningeal artery. (**a**) External carotid artery angiogram, lateral view. Middle meningeal artery (*arrow*) supplying the tumor. Superficial temporal artery (*arrow with dot*). (**b**) Selective study of the middle meningeal artery followed by devascularization of the lesion with polyvinyl alcohol. (**c**) Control angiogram posttreatment

Fig. 3.29 (continued)

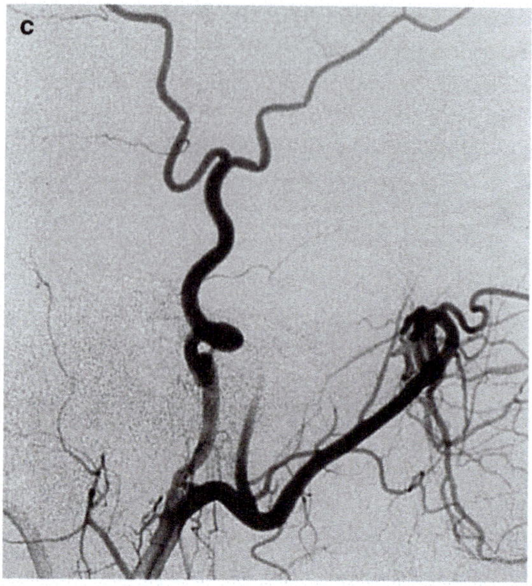

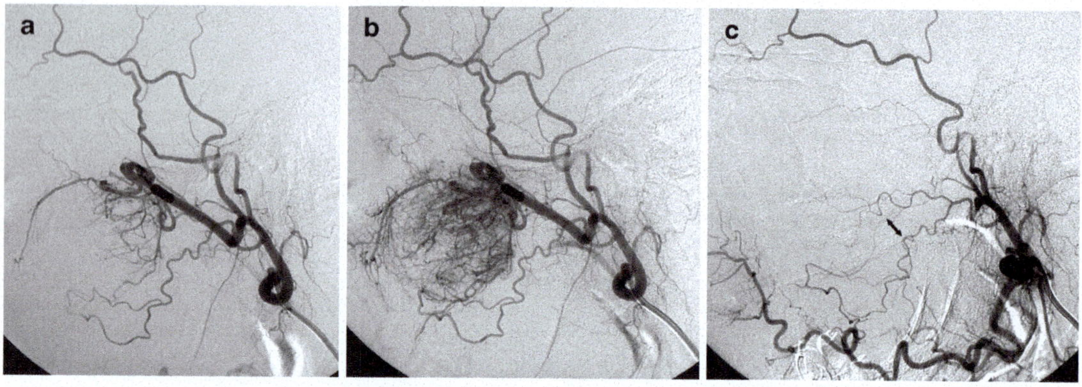

Fig. 3.30 Metastasis of melanoma involving the maxillary sinus with already minimal infiltration of the adjacent areas (soft tissue of the cheek and nasal cavity). The angiogram (**a**, **b**) shows an extremely richly vascularized tumor involving the distal branches of the IMA. There is possible involvement also of the inferior branches of the transverse facial artery. Preoperative devascularization with PVA with occlusion of distal branches and trunk of IMA was performed. (**c**) Control angiogram. Horizontal and inferior branches of the transverse facial artery (*bidirectional arrow*) are now better visible

3.10.6 Other Tumors in the Head and Neck Area

Angiographic study of these lesions is rare. Indications are commonly very large tumors leading to displacement or encasement of ICA or VA or tumors in which a very rich vascularization is suspected. In these latter cases angiography followed by devascularization facilitating further treatment can be very useful (Fig. 3.30).

3.10.7 General Considerations in Endovascular Treatment in the ECA Area

- Abundant anastomoses are present among all the branches of the ECA. In the endovascular treatment of the different pathologies of this sector, especially in cases of vascular malformations, proximal occlusion of the supplying arteries should be avoided, since this would rapidly lead to the formation of a collateral circulation, reinjecting the lesion. *The embolic material should be directly injected into the nidus or close to it. The possible involvement of the ascending and deep cervical arteries should be taken into account.*
- Many anastomoses (dangerous connections) can link the ECA with the ICA and VA (Table 3.1 and drawings 3.31, 3.32, and 3.33 and Fig. 3.34). Anastomoses are also present between the cervical arteries and ECA and VA. These connections are listed in Table 3.1 and should be always taken into consideration when endovascular treatment is planned. The anastomoses can be minimal or large (Morris 1997; Lasjaunias et al. 2001; Geibprasert et al. 2009). They can already be recognizable at the beginning of the treatment or may enlarge and appear during it, following superselective injection or opening of collateral paths as a consequence of occlusion of other arteries. *Frequent angiographic control is mandatory.*
- Another important aspect is the supply of a few cranial nerves from branches of the ECA (Table 3.2). This should be considered whenever a vascular treatment involves these branches. *It is important to take particular caution in the occlusion of these vessels and in the choice of embolic material in these cases.*

Table 3.1 List of most remarkable connections involving ECA

Anastomosis between ECA and ICA	
Foramen rotundum A. (IMA)	ILT (ICA)
Vidian A. (IMA)	Mandibular A. (ICA)
Pterigovaginal A. (IMA)	Mandibular A. (ICA)
Middle meningeal A. (IMA)	MHT and ILT (ICA)
Accessory meningeal A. (IMA)	ILT (ICA)
Clival branches (A.Ph.A.)	MHT (ICA)
Tympanic branches (A.Ph.A.)	Carotido-tympanic A. (ICA)
Anastomosis between ECA and VA	
Hypoglossal A. (A.Ph.A.)	Radiculo meningeal branch (VA) (Odontoid arch)
C2–C4 cut.—musc. branches (A.Ph.A.)	C2-C4 cut.—musc. branches (VA)
C1–C2 cut.—musc. branches (Occ.A.)	C1-C2 cut.—musc. branches (VA)
C2–C4 cut.—musc. branches (Asc.cerv.A.)	C2-C4 cut.—musc. Branches (A.Ph.A.—VA)
C1–C2 cut.—musc. branches (Deep cerv. A.)	C1-C2 cut.—musc. branches (Occ.A.—VA)
Anastomosis between ECA and ophthalmic artery	
Middle meningeal A.	Lachrymal branch (Oph.A.)
Deep ant. Temp. A.	Lachrymal branch (Oph.A.)
Spheno-palatine A.	Etmoidal branches (Oph.A.)
Superficial temporal A.	Supra-orbital branch (Oph.A.)
Facial-infraorbital A.	Nasal branches (Oph.A.)
Ophthalmic artery arising from middle meningeal artery	

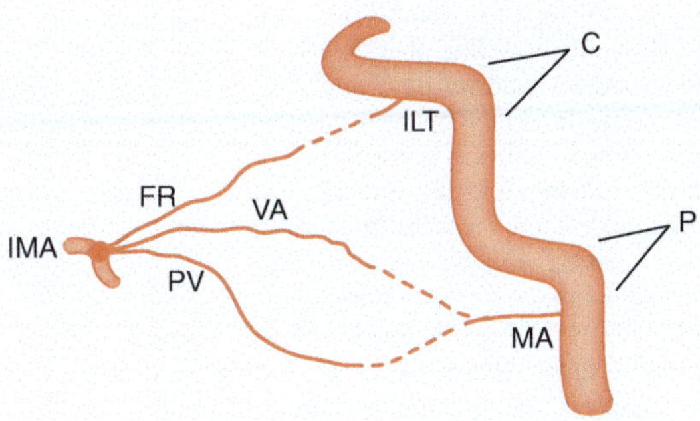

Fig. 3.31 Drawing. Anastomosis between branches of the distal internal maxillary artery (IMA), foramen rotundum artery (FR), vidian artery (VA), pterygovaginal artery (PV) with the inferior lateral trunk (ILT), cavernous portion of ICA (C) and with the mandibular artery (MA), petrous portion of the internal carotid artery (P)

Fig. 3.32 Drawing. Anastomosis between tympanic branch (T) of the ascending pharyngeal artery (APhA) and caroticotympanic branch (CT) of the petrous segment of the internal carotid artery (ICA). Anastomosis between clival ascending branches (CL) of the neuromeningeal trunk (NMT) of the APhA and descending branches of the meningeal hypophyseal trunk (MHT) of the cavernous segment of the internal carotid artery

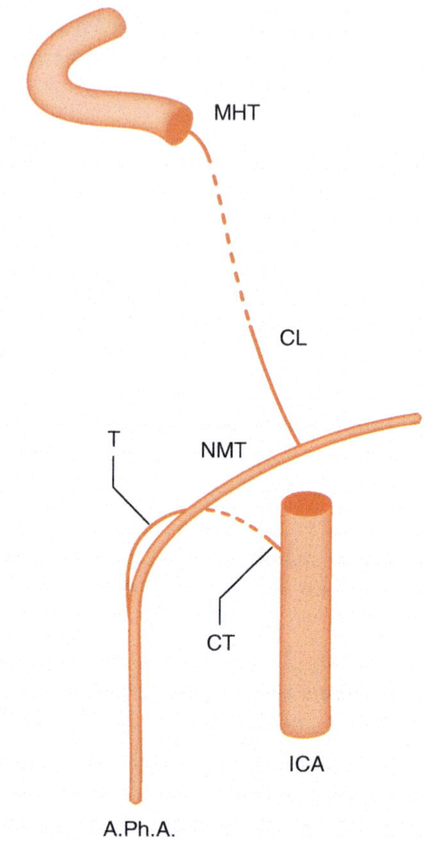

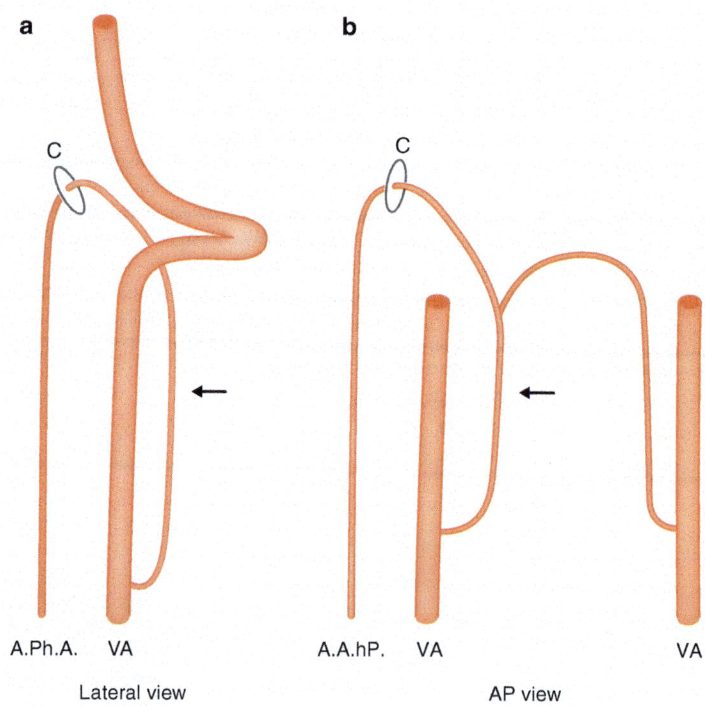

Fig. 3.33 Drawing. (**a**, **b**) Lateral and AP views of the connection between the ascending pharyngeal artery (APhA) and vertebral artery (VA). This is formed by the hypoglossal branch, which passes through the hypoglossal canal (C), and gives off a descending branch anastomosing with the ascending C3 radiculomeningeal branch (*arrow*) of the VA. The AP view demonstrates also the connection on the dorsal surface of the odontoid process of the radiculomeningeal branch of the VA with that of the contralateral vertebral artery. These anastomoses form the so-called arch of the odontoid process

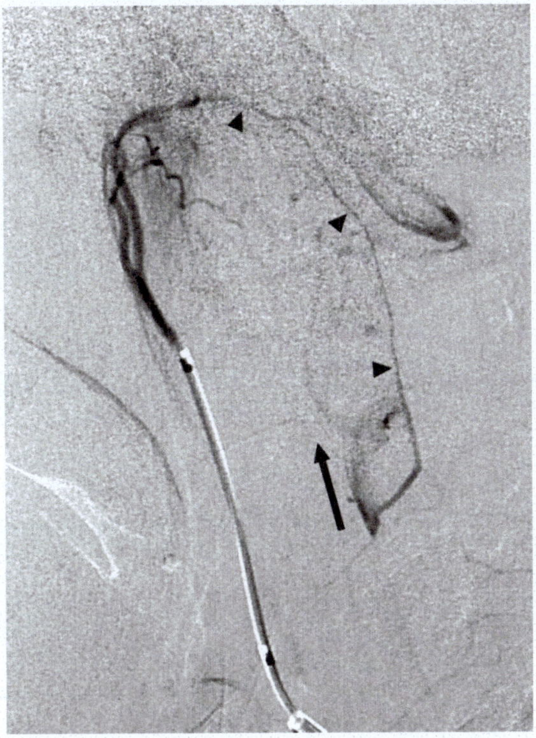

Fig. 3.34 Selective study of the hypoglossal branch of the AphA in a patient with tympano-jugular paraganglioma. Dangerous anastomosis with the radiculomeningeal branch (*arrowhead*) of the vertebral artery

Table 3.2 Branches of ECA supplying cranial nerves

VII cranial nerve	Tympanic branch of the petrosquamous artery (middle meningeal artery) and stylo-mastoid artery (occipital-postero-auricular artery)
IX, X, XI cranial nerves	Jugular branch of the neuro-meningeal trunk of the ascending pharyngeal artery
XII cranial nerve	Hypoglossal branch of the neuromeningeal trunk of the ascending pharyngeal artery

4 Anterior Cerebral Artery

The anterior cerebral artery (ACA) develops as a branch of the primitive olfactory artery which originates from the anterior cranial division of the ICA (see also Chap. 1). The primitive olfactory artery regresses and is replaced by the ACA (De Vriese 1905; Abbie 1934; Padget 1948; Liu and Kricheff 1974).

The final ACA can be divided into several segments (Huber 1979) (Fig. 4.1):

- A1 precommunicating segment: from its origin from the ICA to the anterior communicating artery (AcomA)
- Distally to the AcomA, the ACA continues as the pericallosal artery
- A2 (infracallosal segment)
- A3 (precallosal segment)
- A4 (supracallosal segment)

4.1 Precommunicating Segment

The first part of the ACA is called the A1 or the *precommunicating* segment. It arises at the terminal ICA and it runs medially above the chiasma and optic nerve with a horizontal, sometimes slightly descending, ascending or tortuous course joining the contralateral A1 by the way of the anterior communicating artery (AcomA). The length of the A1 is on average 12.7 mm (Perlmutter and Rhoton 1976) and that of the AcomA is on average 2–3 mm (Marinković et al. 1990; Tao et al. 2006) (Figs. 4.1 and 4.2). Many small arteries arise from the A1 segment and from the AcomA.

Perforating branches are found along the whole A1, but they are more numerous in its proximal part (Perlmutter and Rhoton 1976; Dunker and Harris 1976). Among the more distal branches a few arise also from the AcomA (Stephen and Stilwell 1969; Krayenbühl et al. 1972; Dunker and Harris 1976; Perlmutter and Rhoton 1976; Rosner et al. 1984; Lang et al. 1987; Marinković et al. 1990; Tao et al. 2006).

As far as it concerns the branches of A1, some of these arise from its inferior surface supplying the optic nerve and chiasma. From the superior surface of A1 arise branches supplying the suprachiasmatic anterior portion of the hypothalamus and anterior commissure while other branches enter the medial part of the anterior perforated substance (APS).

The branches of the AcomA are represented by small short perforating arteries in number of 4–5 supplying the lamina terminalis and the preoptic hypothalamic region. In addition to these branches there is a single longer branch, which runs in front of the lamina terminalis towards the rostrum and genu of the corpus callosum where commonly ends divide in two short terminal branches. Rarely it can extend more distally. It is involved in the supply of the lamina terminalis, anterior commissure, preoptic part of hypothalamus, anterior pillars of fornix, septum pellucidum, and part of the corpus callosum. This artery which has been reported to be present in about 90% of the cases (Marinković et al. 1990;

G.B. Bradac, *Applied Cerebral Angiography*, DOI 10.1007/978-3-319-57228-4_4

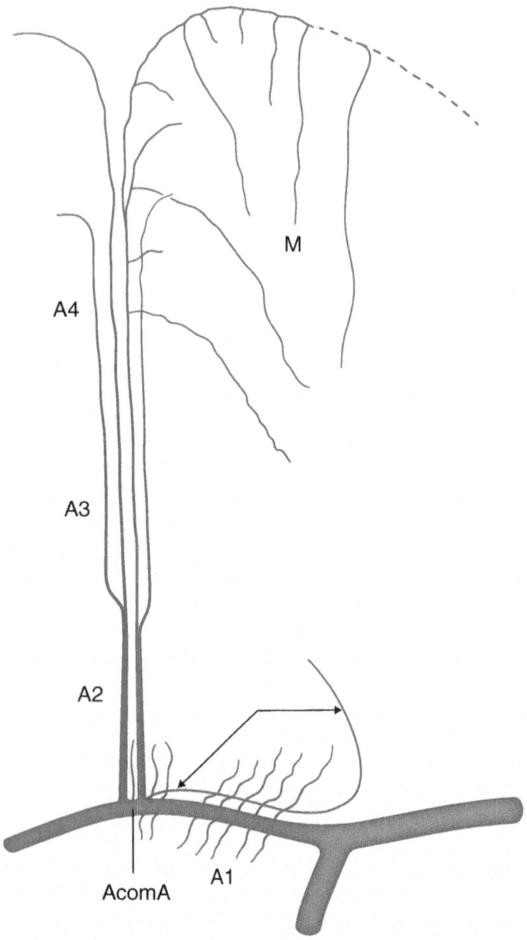

Fig. 4.1 Drawing showing the segments of the anterior cerebral artery. A1, A2, A3, and A4. Artery of Heubner (*arrow with angle*) arising in the angle A1–A2. Other perforators arise from the superior and inferior surface of A1. Anterior communicating artery (AcomA) from which arises the subcallosal artery. Medullary arteries (M, short and long branches) arising from cortical branches. Dashed line indicating leptomeningeal anastomoses with the MCA

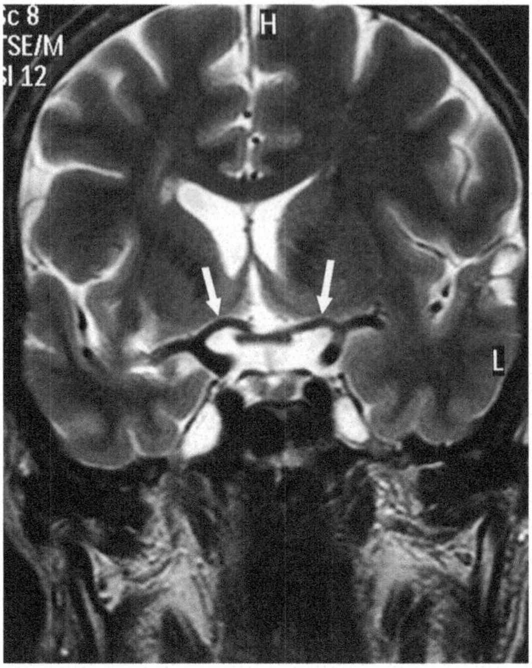

Fig. 4.2 Coronal MRI T2-weighted image showing the relationship of the A1 (*arrow*) to the chiasma

Tao et al. 2006) has been called *subcallosal artery*, which corresponds to the "artère cerebrale antérieure mediane courte" of Lazorthes (1961). Sometimes the artery can be well developed. It can extend between both pericallosal arteries along the corpus callosum supplying it and also the adjacent parts of the cerebral hemispheres. The artery has been called (*median artery of the corpus callosum, median callosal artery, arteria pericallosa triplex*) (De Almeida 1931; Lazorthes 1961; Baptista 1963; Stephen and Stilwell 1969; Perlmutter and Rhoton 1978; Huber 1979, Kakou et al. 2000; Marinković et al. 1990; Tao et al. 2006; Kornieieva et al 2017).

Another important branch is the *recurrent artery of Heubner*. It has been described by this author in 1872, and considered by Padget (1948) as an embryological remnant of the primitive olfactory artery. It is the largest and longest perforating branch. It commonly takes its origin from the distal A1 or proximal A2 segments, rarely from the AcomA. In a few cases, it can have a common origin with the frontopolar artery (Critchley 1930; Westberg 1963; Lazorthes 1961; Lazorthes et al. 1976; Perlmutter and Rhoton 1976; Rosner et al. 1984; Gomes et al. 1984; Tao et al. 2006). In its course, the artery of Heubner runs back parallel to the A1 and M1 segments to enter the APS mainly in its anterior part. It can be a single branch or sometimes be made by multiple branches, supplying the inferior part of the head of the nucleus caudatus, and the adjacent parts of the anterior limb of the internal capsule, and of putamen and pallidum (Perlmutter and Rhoton 1976; Rosner et al. 1984) (Drawing 4.1 and Figs. 4.3, 4.4, 4.5, 4.10, 4.11a–d, 4.13a–d, 5.15 and 11.12a).

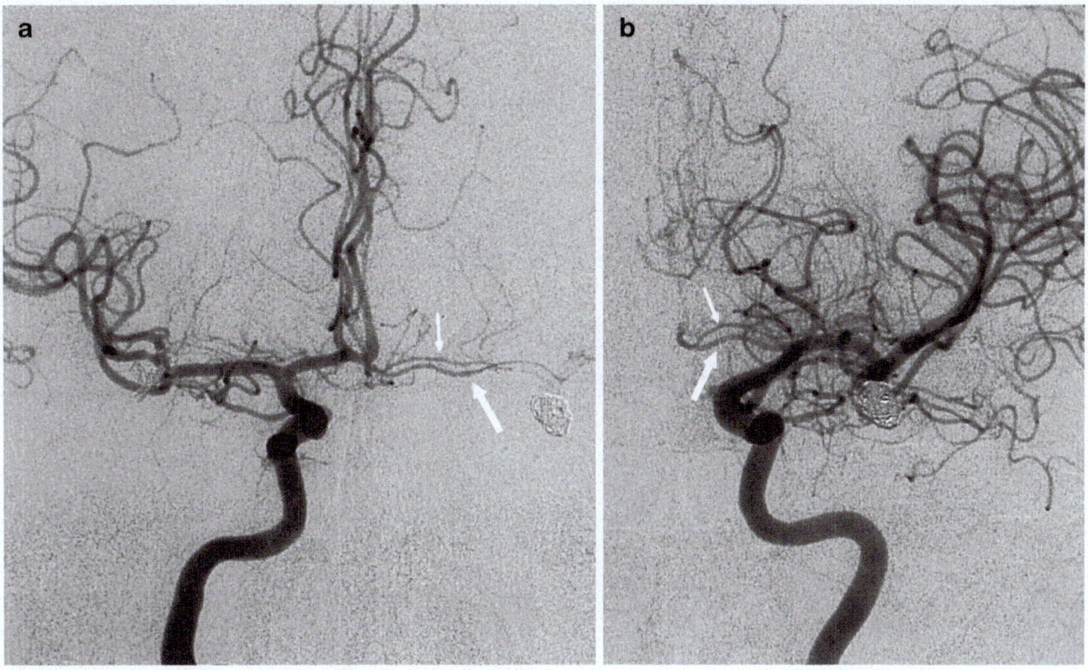

Fig. 4.3 Right (**a**) and left (**b**) AP carotid angiogram in a patient treated with coils for a ruptured left MCA aneurysm. Well-developed A1 on the right. Hypoplastic left A1 (*large white arrow*). Artery of Heubner (*small white arrow*) running parallel to the A1 segment

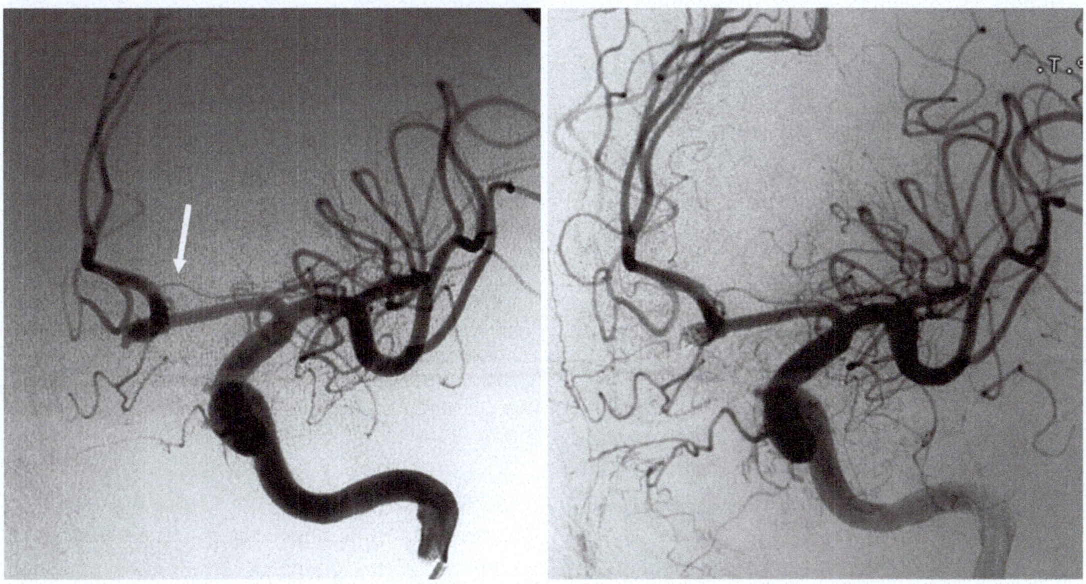

Fig. 4.4 Carotid angiogram oblique view pre- and posttreatment with coils of ruptured aneurysm of the AcomA. Artery of Heubner (*white arrow*) arising close to the aneurysm

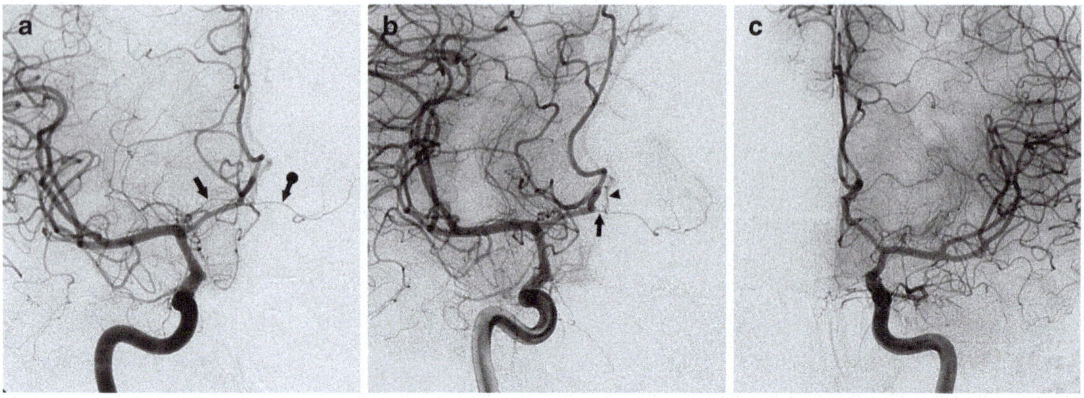

Fig. 4.5 Patient with meningioma of the *planum sphenoidale*. Bilateral carotid angiogram. AP view. Typical shifting of both A1. (**a**) On the right carotid angiogram, the right (*arrow*) and the left (*arrow with dot*) arteries of Heubner are clearly recognizable. (**b**) On the oblique view the origin of the left Heubner artery from the proximal left A2 can be demonstrated. Faint injection of the left A2 (*white arrowhead*). AcomA (*arrow*). (**c**) On the left ICA angiogram the Heubner artery is more difficult to identify

4.2 Distal Segments

Distal to the junction with the AcomA begins the *postcommunicating segment* called the *pericallosal artery*, which can be subdivided, according to its relationship to the corpus callosum, into three further segments (Liu and Kricheff 1974; Huber 1979) (Drawing 4.7 and Figs. 4.6 and 4.8).

4.2.1 Infracallosal Segment

This is also called the A2. It runs into the interhemispheric fissure upwards in front of the lamina terminalis towards the genu of the corpus callosum. It gives off *the orbitofrontal* (infraorbital) and the *frontopolar* arteries which can arise as a single or sometimes as a common trunk. The orbitofrontal artery runs forward and towards the floor of the anterior cranial fossa. The frontopolar artery runs anteriorly on the medial surface of the hemisphere towards the frontal pole.

4.2.2 Precallosal Segment

This, called the A3, is a short segment, which curves around the genu of the corpus callosum (Drawing 4.6 and Fig. 4.7). It gives off the *callosomarginal* artery, which runs from the frontal to the parietal region above the gyrus cinguli in the cingulate sulcus, giving off frontal branches and ending in the paracentral artery. In other cases it can be well developed only in the frontal area or be not recognizable as a trunk but only represented by small isolated branches. In some cases it can be completely absent. In the cases of hypo-aplasia of

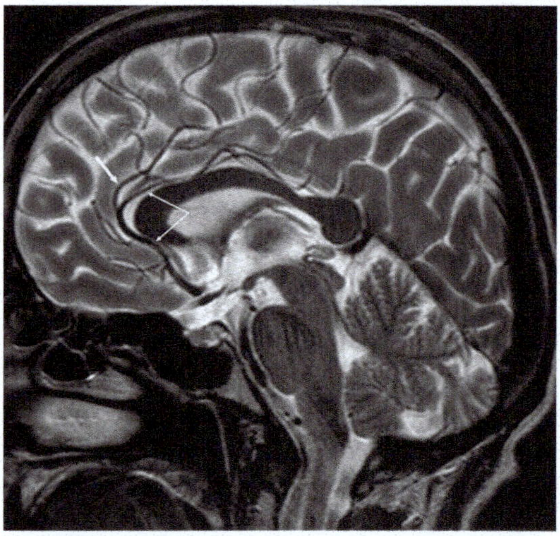

Fig. 4.6 MRI sagittal T2-weighted image. Infra-, pre-, supracallosal segments of the pericallosal artery (*white arrow with angle*). The supracallosal segment runs with an undulating course above the corpus callosum, partially in the pericallosal cistern and partially superior to it. From the precallosal segment arises the callosomarginal artery (*white arrow*)

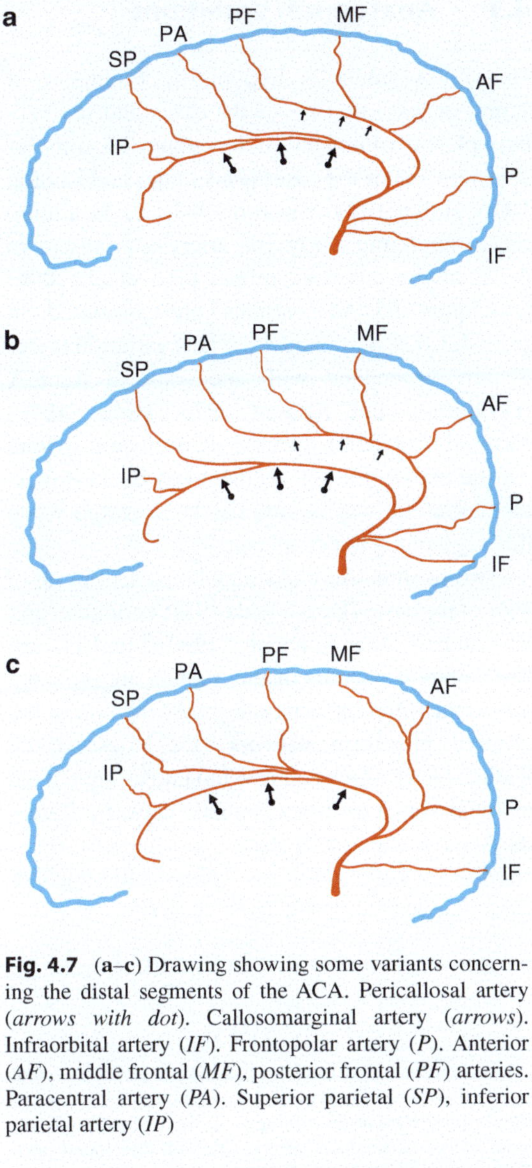

Fig. 4.7 (**a**–**c**) Drawing showing some variants concerning the distal segments of the ACA. Pericallosal artery (*arrows with dot*). Callosomarginal artery (*arrows*). Infraorbital artery (*IF*). Frontopolar artery (*P*). Anterior (*AF*), middle frontal (*MF*), posterior frontal (*PF*) arteries. Paracentral artery (*PA*). Superior parietal (*SP*), inferior parietal artery (*IP*)

the artery its cortical branches are replaced by arteries arising from the presupracallosal segments of the pericallosal artery. Vice versa it can occur such that all the cortical branches arise from the callosomarginal artery in the very rare case of aplasia of the pericallosal artery. The well-developed callosomarginal artery appears on the lateral carotid angiogram running superiorly to the pericallosal artery and on the AP view slightly displaced from the pericallosal artery which runs on the midline.

4.2.3 Supracallosal Segment

This segment is called the A4 (Figure 4.6 and drawing 4.7). It is the more distal segment of the pericallosal artery. It runs posteriorly in the pericallosal cistern, above the surface of the corpus callosum towards the splenium. This segment has been called posterior pericallosal artery and its extent depends on the size of the artery of the splenium, branch of the posterior cerebral artery. Occasionally, it can extend below the corpus callosum towards the foramen of Monro (Stephen and Stilwell 1969; Kier 1974; Perlmutter and Rhoton 1978, Kakou et al. 2000), as in the embryological life and anastomoses with the posterior medial choroidal artery (see also Sect. 12.8). The supracallosal artery can have an undulating course and sometimes shows an upward distension in its midportion. It is commonly well developed, but it can be sometimes hypoplastic especially in its distal segment, or rarely uni- or *bilaterally absent.*

A meningeal branch can arise from the precallosal segment supplying the inferior part of the falx (De Almeida 1931; Moniz 1940; Lasjaunias and Berenstein 1990). It anastomoses with branches of the middle meningeal artery descending along the falx. It can also be connected with the meningeal branch arising from the posterior cerebral artery (see Sect. 7.1).

4.2.4 Distal Cortical Branches

Several arteries arise from the supracallosal segment of the pericallosal artery, and/or callosomarginal artery, and run on the medial surface of the hemisphere (Ring and Waddington 1968; Moscow et al. 1974; Perlmutter and Rhoton 1978). These arteries can have a variable origin; their distal course, however, and their vascular territories are relatively constant. From anterior to posterior, the cortical branches are represented by the group of the *frontal arteries, paracentral branch, and parietal arteries*. The frontal arteries, which can be subdivided into the anterior, middle, and posterior branches, run in the frontal area. The posterior frontal artery terminates in the superior limit of the precentral sulcus. The paracentral artery which frequently arises from the pericallosal artery, more rarely from the callosomarginal artery, or as a common trunk with the posterior frontal and superior parietal artery, runs towards the paracentral lobule and extends to the central sulcus with branches for the precentral and postcentral gyri. The more distal branches are the inferior and the superior parietal arteries. The inferior parietal runs on the inferior surface of the precuneus supplying it. It can occur that the pericallosal artery ends in the inferior parietal artery and it does not continue around the splenium. The other branch is the superior parietal artery, which is commonly larger. It takes its origin commonly from the pericallosal artery, sometimes as a common trunk with the posterior frontal and paracentral arteries. It runs towards the marginal segment of the cingulate sulcus and marks the boundary between the paracentral lobule and precuneus with branches supplying both the paracentral lobule and the superior part of the precuneus.

Examples of the pericallosal and callosomarginal arteries and origin and course of their branches are presented in Drawing 4.7 and in Fig. 4.8.

4.3 Anatomical Variations

One A1 segment is hypoplastic (diameter of 1 mm or less) in 10% of the cases and severely hypoplastic (diameter of less than 0.5 mm) or absent in 1% of the cases (Perlmutter and Rhoton 1976; Huber 1979; Yasargil 1984a, b). In a more recent anatomic study, the artery was absent in 2.2% of the cases examined (Tao et al. 2006) (examples of A1 variants are presented in Figs. 4.3, 4.9, 4.10, and 7.2). This variation is frequently associated with aneurysm of AcomA (Perlmutter and Rhoton 1976; Huber 1979; Yasargil 1984a, b). In case of occlusion of one carotid sector, the A1 segment can be an important collateral way, which can be impaired when this segment is hypo-aplastic.

Two or three AcomAs may be present in up to 40% of the cases (De Almeida 1931; Perlmutter and Rhoton 1976; Marinković et al. 1990) (Fig. 4.11a–d). The different patterns of the AcomA are probably due to the different embryological evolution of the primary plexiform network (De Vriese 1905; Padget 1944) which can develop into a single large or small vessel or into a complex network. Rarely the AcomA is absent. In these cases, the pericallosal arteries of both sides fuse proximally together (Marinković et al. 1990; Tao et al. 2005).

Fig. 4.8 Different patients in whom the various courses of the distal segments of the ACA are shown. (**a**) Well-developed pericallosal artery (P). Callosomarginal artery (CM), giving off frontal branches (*arrows*). The paracentral artery (*arrow with dot*) arises from the pericallosal artery. (**b**) Lateral carotid angiogram. Well-developed pericallosal artery (*arrowhead*), dividing distally in the superior (*arrow with angle*) and inferior (*arrows*) parietal arteries. Callosomarginal artery (*arrowheads*) giving off frontal branches and the paracentral artery. There is a common trunk arising from callosomarginal artery for the posterior frontal (PF) and the paracentral artery (PA). (**c**) Lateral carotid angiogram. Well-developed pericallosal artery. However, its distal segment is hypoplastic (*arrowheads*). From the pericallosal artery arise frontal branches (*double arrow*) and as a common trunk the paracentral artery (PA) and the parietal branches (*arrow with angle*). There is a common trunk for the infraorbital (IF) and the frontopolar (P) arteries arising from the infracallosal segment of the pericallosal artery. (**d**) Well-developed pericallosal artery which with an undulating course terminates in the parietal area (*arrowhead*). The callosomarginal artery as a trunk is not recognizable. From the infracallosal segment of the pericallosal artery arises a well-developed branch (*arrow*) supplying the anterior and middle frontal area. From the supracallosal segment arises another long branch (*arrow with dot*) supplying the posterior frontal and paracentral area. (**e**) Left ICA lateral angiogram. Well-developed callosomarginal artery from which arise all distal branches, while the pericallosal artery is on this side hypoplastic. (**f**) Pericallosal artery (*arrow*). Well-developed subcallosal artery (*arrow with dot*). (**g**) Well-developed callosomarginal artery (*large arrow*) supplying small convexity angioma. Another branch (*double arrow*), probably a dilated paracentral artery, arises from the dilated pericallosal artery

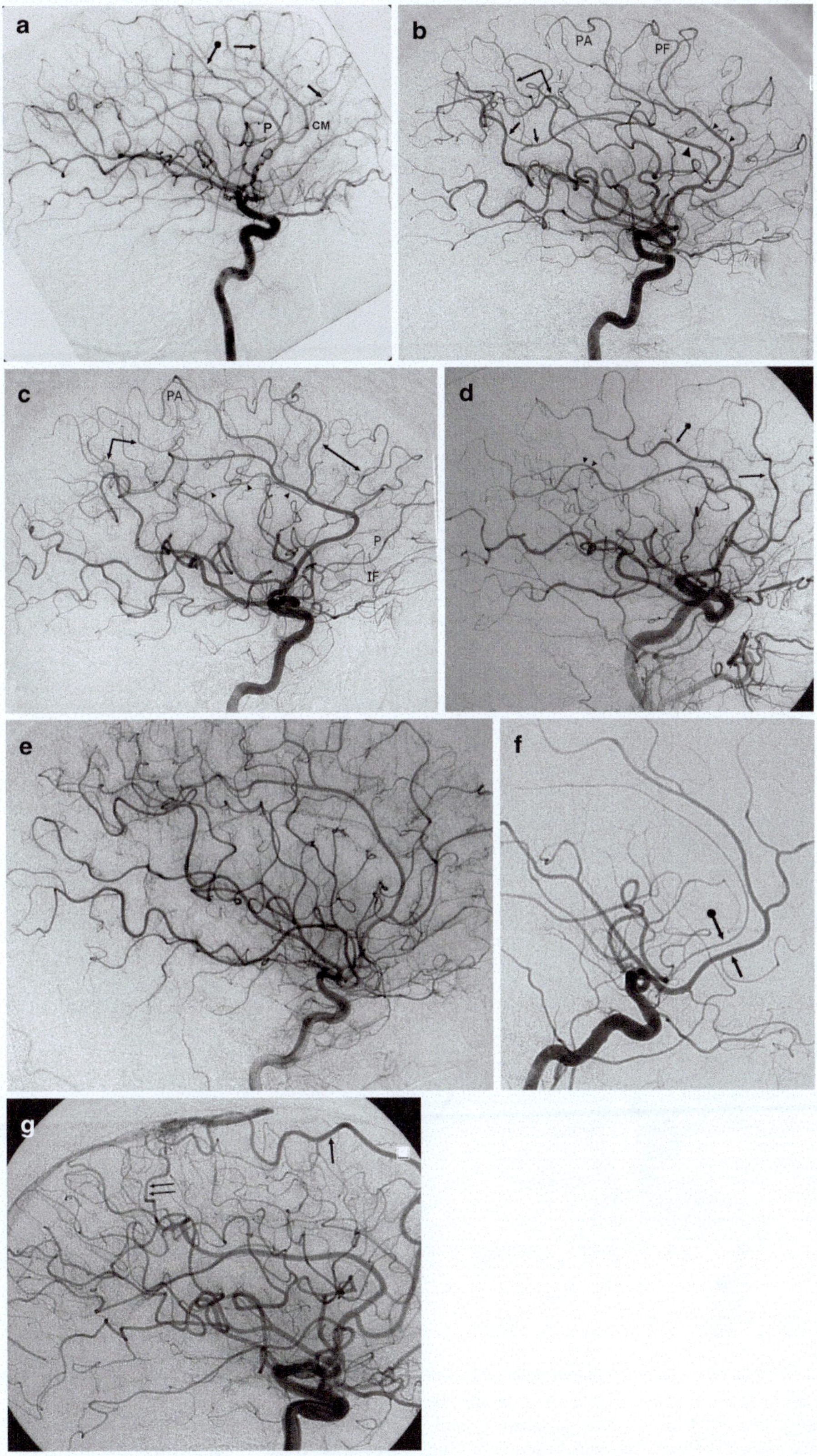
a
P
CM
b
PA
PF
c
PA
P
IF
d
e
f
g

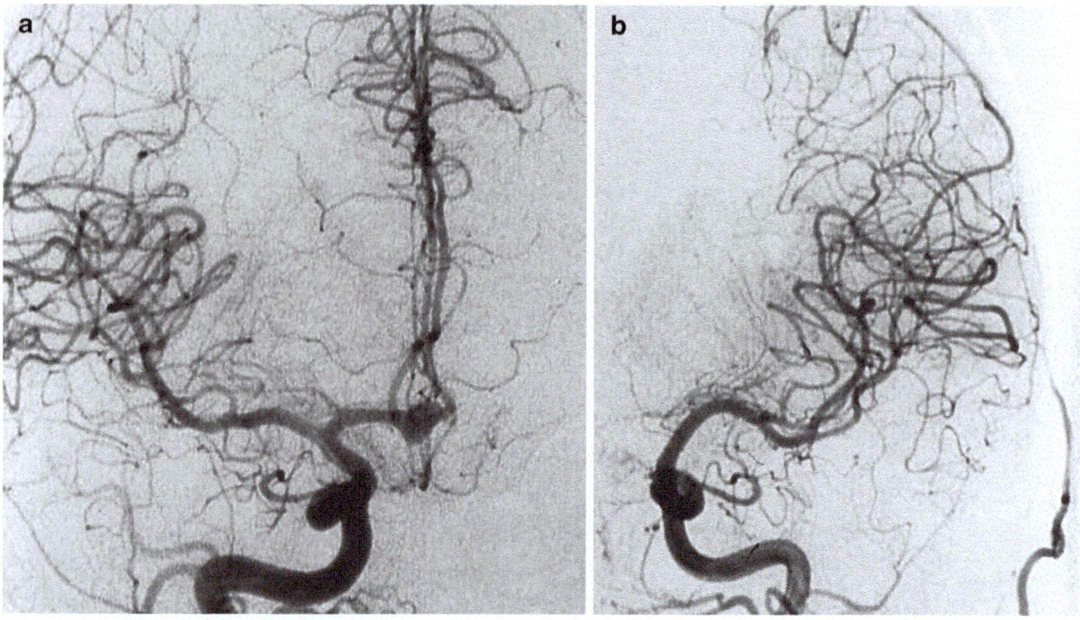

Fig. 4.9 Right and left carotid angiograms. Hypoplastic left A1 segment and a well-developed right A1 in a patient with AcomA aneurysm

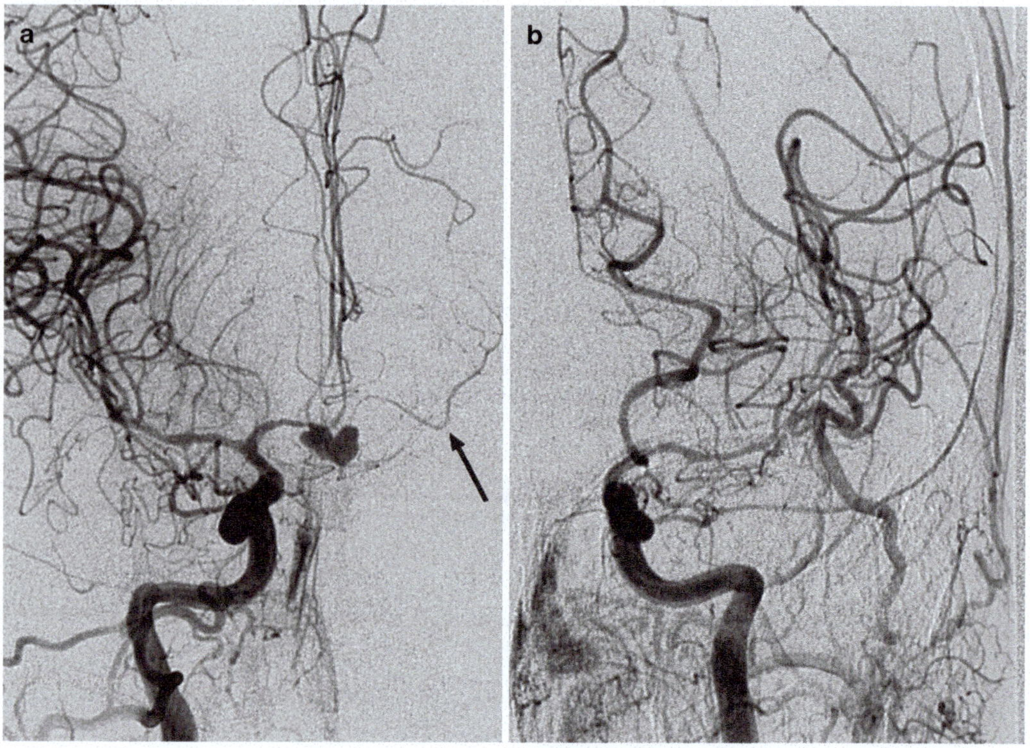

Fig. 4.10 Middle-aged woman presenting with SAH due to rupture of AcomA aneurysm. Right (**a**) and left (**b**) carotid angiograms. There is an irregular-shaped aneurysm with the neck in the angle between A1 and A2 of the right ACA. Owing to the aplasia of the left A1 there is no overlapping and so the left Heubner artery arising from the left proximal A2 is clearly recognizable (*arrow*). The aneurysm was occluded with coils and the already present spasm was treated with medical therapy with a final good result

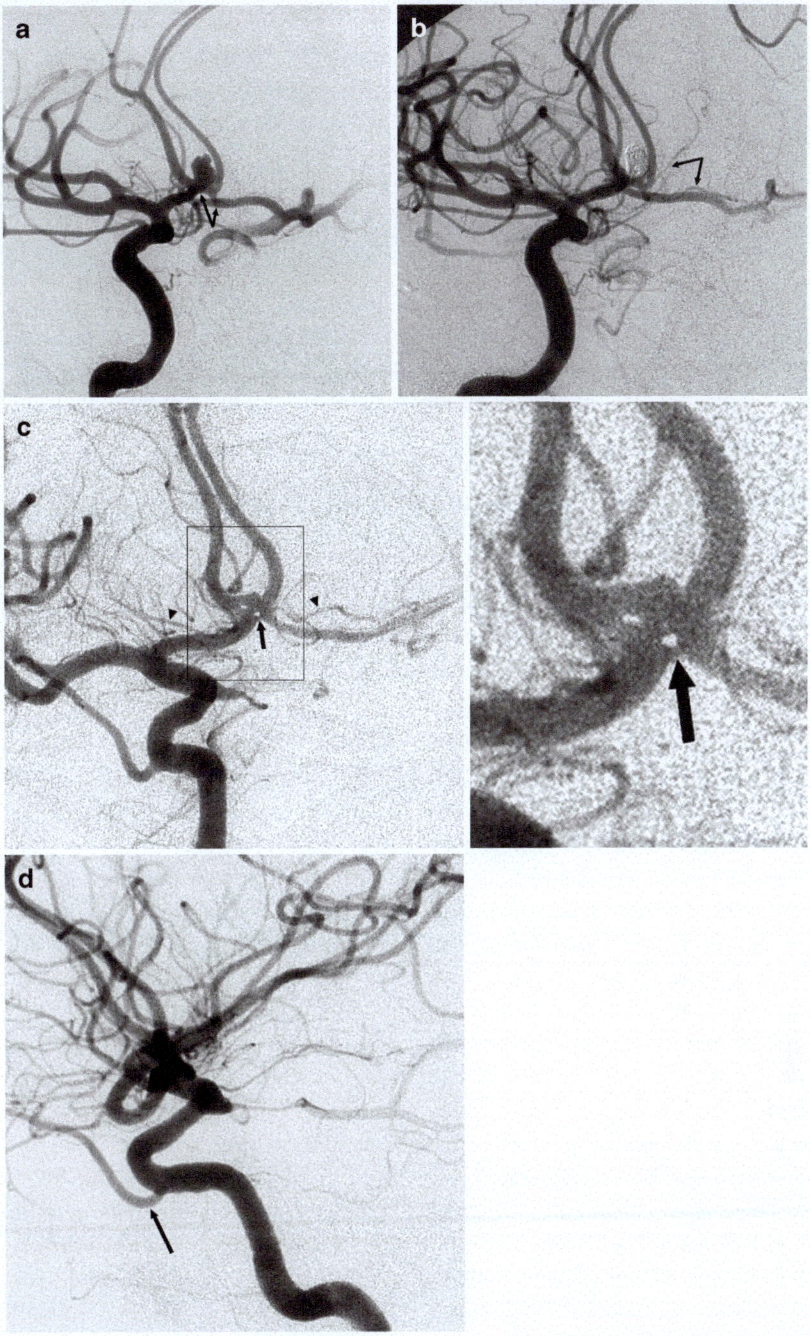

Fig. 4.11 Studies in two different patients. (**a**) Right internal carotid angiogram (oblique view) in a patient with a ruptured AcomA aneurysm. There is a duplication of the anterior communicating artery (*small arrow with angle*). In such cases, many projections are frequently necessary to identify the aneurysm neck. (**b**) Angiogram postcoiling. The left artery of Heubner arising as a common trunk with the frontopolar artery (*arrow with angle*) is well recognizable running parallel to the left A1. (**c**) Carotid angiogram (oblique view) in another patient with plexiform aspect (*arrow*) of AcomA. Detail of the area. The artery of Heubner is well visualized on both sides (*arrowheads*). Origin of the ophthalmic artery from the cavernous portion of the ICA. (**d**) Lateral carotid angiogram in the same patient showing better the anomalous origin of the ophthalmic artery (*arrow*)

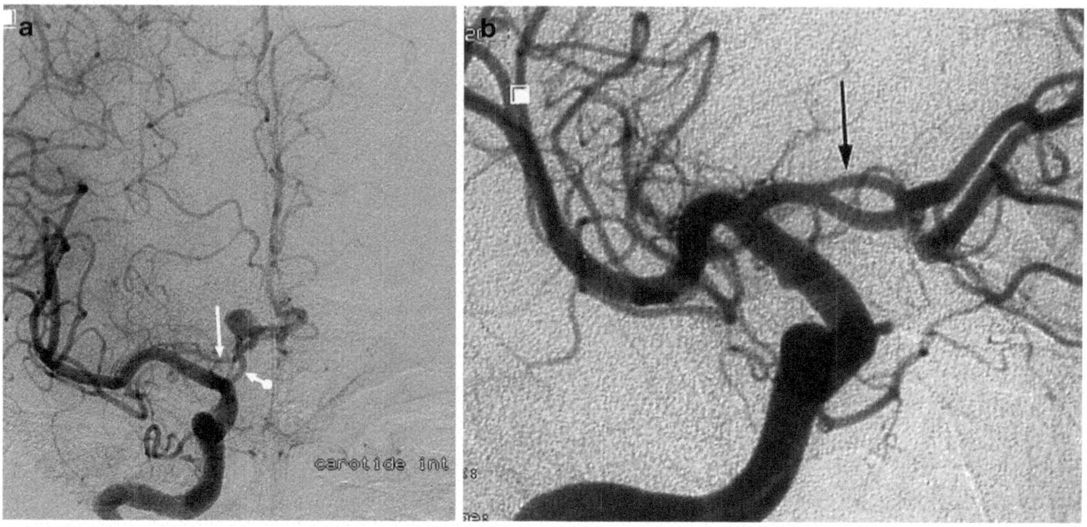

Fig. 4.12 (**a**) Right carotid angiogram, AP view. Incidental discovery of aneurysm in a patient with duplication of A1. One A1 segment arises normally from the distal ICA bifurcation (*arrow*), and the other one arises from the ICA close to the origin of the ophthalmic artery. Where the two branches distally join together, an aneurysm is present. (**b**) Carotid angiogram in another patient with incidental discovery of a duplication of A1. In this case the A1 arises as a normal trunk, but along its course duplicated in two vessels

The artery of Heubner can exceptionally be absent on one side (Westberg 1963; Perlmutter and Rhoton 1976; Tao et al. 2006). In cases of hypoaplasia of the A1, the artery of Heubner can be very large and should not be confused with the A1.

Another very rare variant is the so-called *carotid-anterior cerebral artery anastomosis* in which an anomalous vessel arises from the ICA, close to the origin of the ophthalmic artery. It has an ascending course and passes inferiorly to the optic nerve to reach the A1–A2 junction. It can replace the normal A1, but both the anomalous branch and the normal A1 can be present so we could in this case also speak of a duplicated origin of the ACA. There are no clear embryological explanations for this anomaly. Since the first description of Robinson (1959), many authors have described this variant (Turnbull 1962; Nutik and Dilenge 1976; Bernini et al. 1982; Rosenorn et al. 1985; Milenkovic 1985; Friedlander and Ogilvy 1996; Morris 1997; Kilic et al. 2005; Chakraborty et al. 2006; Wong et al. 2008; Uchino et al. 2012a). The association of aneurysm of the AcomA is frequent. An example is presented in Fig. 4.12a.

The A1 segment may have a normal origin, but along its course it features duplication (two different vessels) or fenestration (one vessel with two lumens) (Wollschlaeger and Wollschlaeger 1974; Perlmutter and Rhoton 1976) (Figs. 4.12b, 5.3b and 5.11a).

The several variations involving the pericallosal, the callosomarginal, and their branches have already been described in Sects. 4.2.2 and 4.2.3. We present here other more rare variations involving these branches. One pericallosal artery can give off a few or sometimes the majority of the branches also for the contralateral hemisphere. As described in Sect. 4.1, together with both pericallosal arteries, there is frequently a third branch arising from the AcomA. It can be small and short ending in the proximal part of the corpus callosum (*subcallosal artery*) (Fig. 4.8f), or be well developed and take over partially the vascular territories of the pericallosal arteries (*median artery of the corpus callosum, arteria pericallosa triplex*—Fig. 4.15). In other cases, the pericallosal artery can be unique (*azygos pericallosal artery*—Figs. 4.13 and 4.14).

Another variant is the presence of an anastomotic connections between the pericallosal arteries more frequently at the level of the genu of the corpus callosum (Yasargyl and Carter 1974; Perlmutter and Rhoton 1976). In all these cases, there is an increased tendency to develop aneurysms at the AcomA or distally.

An extremely rare anomaly is the persistence of the primitive olfactory artery from which the ACA develops. In the case reported by Tsuji et al. (1995), the anomalous artery ran anteriorly in the frontobasal area, forming then a curve posteriorly to reach the corpus callosum. An associate aneurysm was present.

4.4 Vascular Territories

The vascular territories supplied by the perforating branches arising from A1 and AcomA have already been described. These branches are *end arteries*, without anastomoses to each other. Occlusion of these branches, which can occur in case of stroke or in treatment of aneurysms of this area, can lead to various clinical symptoms such dysarthria, contralateral limb clumsiness, abulia alternating to agitation due to involvement of the Heubner artery, and consequent infarct of the head of the nucleus caudatus and adjacent areas (Mendez et al. 1989; Caplan et al. 1990). Memory loss, agitation, and confabulatory state can develop in involvement of the perforators arising from AcomA, especially the subcallosal artery due to ischemia involving fornix and genu of corpus callosum (Gade 1982; Parkin et al. 1988; Mosimann et al. 2012, Rizek et al. 2013; Meila et al. 2015).

As far as it concerns the cortical branches, the orbitofrontal artery supplies the gyrus rectus, olfactory bulb and tract, and medial part of the orbital gyri. The frontopolar artery supplies the medial and partially lateral surface of the frontal lobe. The frontal, paracentral, and parietal arteries supply the medial surface of the hemisphere in the frontoparietal area. These arteries extend also to the cortex of the convexity for about 1.3 cm connecting with the distal branches of the MCA and sometimes with the PCA. Among these branches, the paracentral and partially the superior parietal and posterior frontal arteries are involved in the supply of the medial and superior part of the *primary motor and sensory cortex and of the supplementary area*. The parietal branches supply the area corresponding to the precuneus where they connect with the parieto-occipital branches of the PCA. Along its course the pericallosal artery gives off perforators for the fornix, septum pellucidum, and corpus callosum and branches also for the gyrus cinguli. The branches supplying the corpus callosum can directly penetrate this structure or after their origin from the pericallosal artery can have a course parallel to it before entering the corpus callosum (Perlmutter and Rhoton 1978; Kakou et al. 2000).

From the distal branches, running on the surface of the cortex, arise small arteries that supply the brain parenchyma. These arteries enter the parenchyma with a perpendicular course and can be subdivided into cortical, medullary, and corticomedullary arteries (De Reuck 1972). The first of these supply the cortex and end with horizontal branches in the various cellular layers. The medullary arteries supply the superficial (short medullary branches) and deep white matter (long medullary branches). The latter run towards the ventricular wall. The cortico-medullary arteries have aspects of both. These arteries are *end arteries* that means that there are no connections between these branches nor are there anastomoses with the perforating branches arising from A1 or AcomA (Lazorthes 1961; De Reuck 1972).

In their course, the medullary arteries are surrounded by a thin space (perimedullary spaces, Virchow-Robin spaces). It had been supposed that the perivascular space was continuous with the subarachnoid space. This, however, was not confirmed by electron microscopy performed by Hutchings and Weller (1986). These authors showed that the pia matter represents a clear barrier, separating the subarachnoid space from the subpial space and perivascular space. Neither the subarachnoid space nor the pia matter extends into the brain, accompanying the blood vessels as they run into the perivascular space.

Differently of that occurring in the vascular territory of the perforators, potential leptomeningeal (pial) anastomoses are present at the surface of the hemisphere between distal branches of the ACA, MCA, and PCA. Connections with the MCA can develop in the frontoparietal convexity and in the frontobasal region and with the PCA

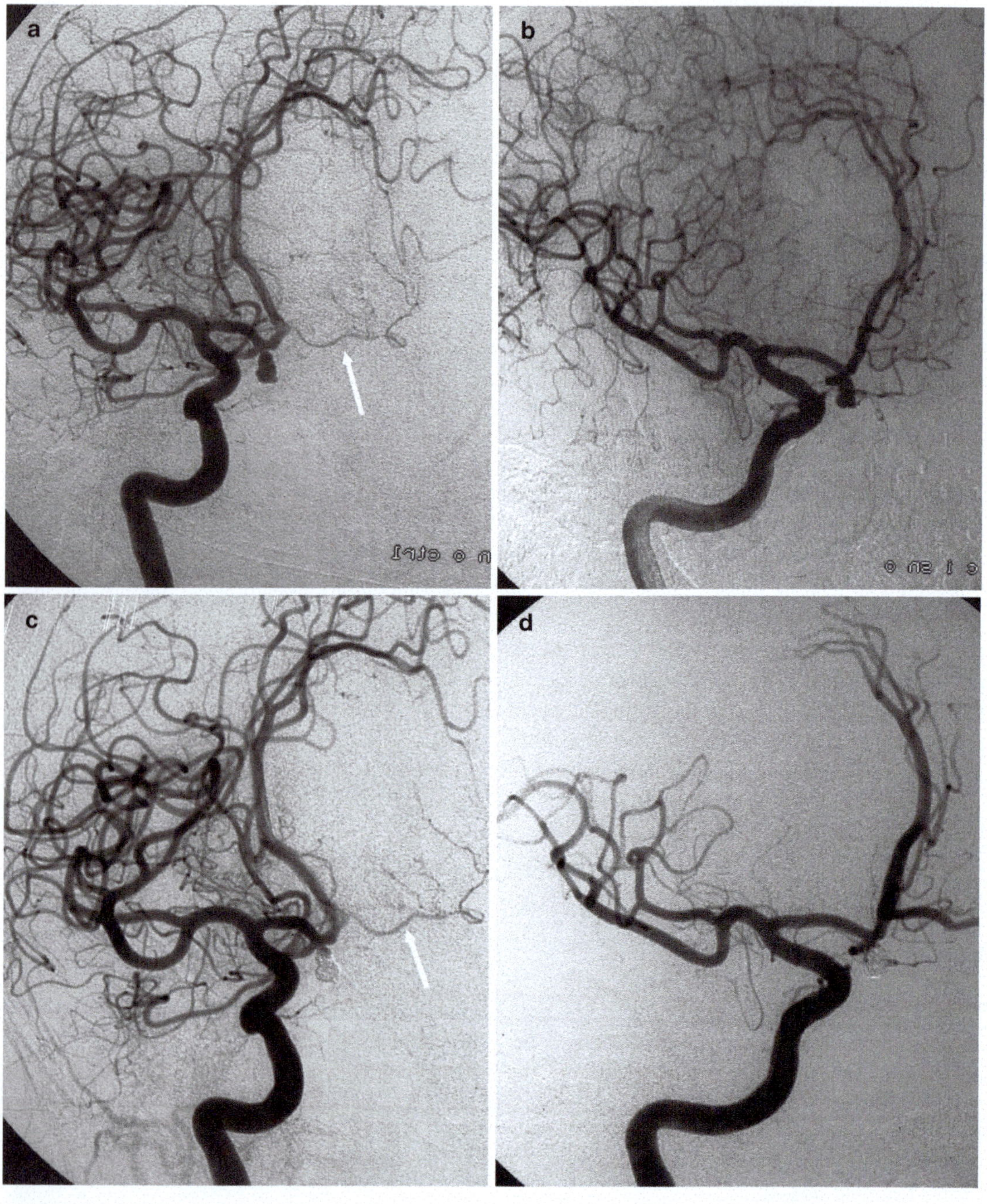

Fig. 4.13 Carotid angiogram with two oblique views (**a**, **b**) in a patient with ruptured aneurysm in the angle A1–A2 of the right anterior cerebral artery. The pericallosal artery is unique (azygos pericallosal artery). Left artery of Heubner (*white arrow*). (**c**, **d**) Control angiogram after occlusion of the aneurysm with coils. Note the Heubner artery (*white arrow*). Both A1 and Heubner artery are recognizable on the angiogram in (**d**)

on the medial surface and convexity of the occipital region. These anastomoses develop in the border regions of the vascular territory of each of the three major arteries. The location of these areas can change in relation to the variable extent of the vascular territory of the arteries (Van der Eecken 1954; Lazorthes 1961).

Anastomoses are also present between the posterior pericallosal artery of the ACA and splenial artery of the PCA.

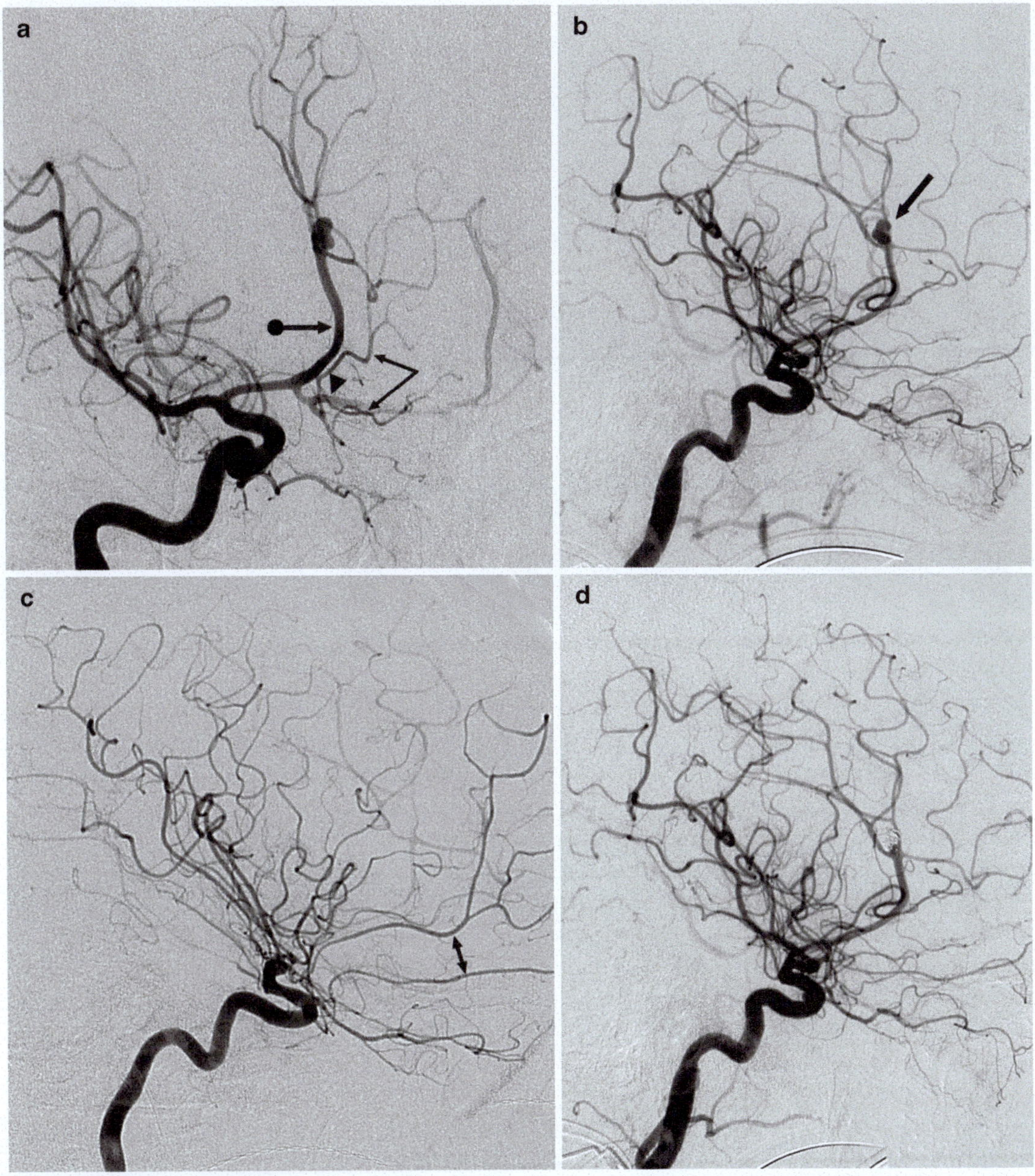

Fig. 4.14 Right carotid angiogram in patient with SAH. (**a**) Oblique view showing the aneurysm at the junction of the pericallosal–calloso-marginal arteries. The pericallosal artery (*arrow with dot*) is unique (azygos artery). It divides distally in branches for both hemispheres. Large right A1. Small left A1 (*arrowhead*) from which arises an infraorbital and a frontopolar branch (*arrow with angle*). (**b**) Lateral right carotid angiogram better demonstrating the aneurysm. (**c, d**) Posttreatment left (**c**) and (*right*) lateral angiograms The unique pericallosal artery is better filled on the right angiogram (**d**) due to the well-developed A1. Occlusion of the aneurysm. On the left angiogram (**c**) only the infraorbital and frontopolar branches (*bidirectional arrow*) are well injected

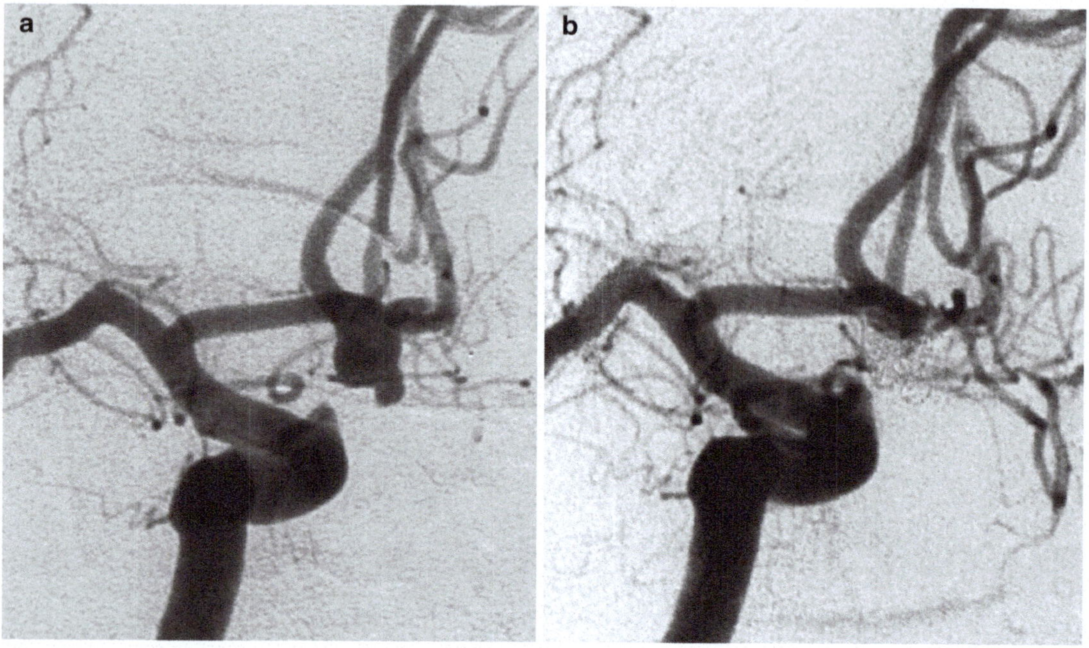

Fig. 4.15 Carotid angiogram (**a**) showing three pericallosal arteries in a patient with ruptured AcomA aneurysm treated with coils (**b**)

4.5 Angiogram

On the angiogram the A1 segment is well recognizable in the AP and oblique views (Figs. 4.3, 4.4, 4.5, 4.9, 4.10, 4.11, 4.13, and 4.15). The AcomA is not always precisely recognizable, even if its presence can be assumed when on the performed angiogram both ACAs are injected. It can be well identified in some cases (Figs. 4.3 and 4.5). In other cases, it can be more difficult to make a precise identification of the AcomA when anomalies are present (Fig. 4.11a–d). Among the perforators only the artery of Heubner can be clearly and frequently identified as a small branch running back to the A1 and proximal M1 with a straight or undulating course visible on the AP or oblique view (Figs. 4.3, 4.4, 4.5, 4.10, 4.11, 4.14, and 11.12). The demonstration of the artery on the angiogram, preceding the endovascular or surgical treatment of aneurysm of the AcomA as well as in surgery for other frontobasal pathologies, can be useful. The precise demonstration of other perforators is commonly more difficult. Sometimes a well-developed subcallosal artery can be recognizable on the lateral angiogram (Fig. 4.8f).

The pericallosal and the callosomarginal arteries can be discerned in the AP view. The course of these arteries as well as of their cortical branches, especially in the frontal area, may be more precisely demonstrated on the lateral angiogram. The variations in origin and course of these branches are great (Drawing 4.7 and Fig. 4.8). What is constant is the close relation of each branch with the vascular territory it supplies. Taking this into consideration we can try to identify the single branches starting from their distal end, which corresponds to its vascular territory, and follow their course retrogradely to their origin.

The angio-MR and angio-CT are other diagnostic possibilities, though the angiogram offers better specific information. Some relationship to the cerebral parenchyma (chiasma, diencephalon, and corpus callosum) can be readily identified with MR (Figs. 4.2 and 4.6).

5 Middle Cerebral Artery

The middle cerebral artery (MCA) arises in the embryogenesis as the latest branch from the cranial division of the ICA and develops to the largest artery supplying a large part of the cerebral hemispheres and basal ganglia. The development of the MCA correlates with the great extension of the cortical mantle in the mammal (Kier 1974). As visible in Drawings 5.1 and 5.2 the artery can be divided into four segments:

- M1 segment, running horizontally in the Sylvian fissure
- M2 segment, running vertically on the surface of the insula
- M3 segment, running laterally and exiting the insular cistern
- M4 segment, which comprises the distal cortical branches

5.1 M1 Segment

The first segment arises at the ICA distal bifurcation and runs horizontally and laterally in the Sylvian fissure below the anterior perforated substance (APS) towards the insula. Also called sphenoidal segment, because it runs parallel 1 cm posterior to the sphenoid ridge, the M1 ends at the anterior border of the *insula*, where it sharply turns superiorly and posteriorly to form the M2 segment. The M1 divides distally into two or three branches (bi-trifurcation). Its length varies, being on average of 16 mm in length (Umansky et al. 1985). It can be very short, and so the bi-trifurcation may be located near the bifurcation of the ICA. M1 gives off two types of important branches (Drawing in Fig. 5.2).

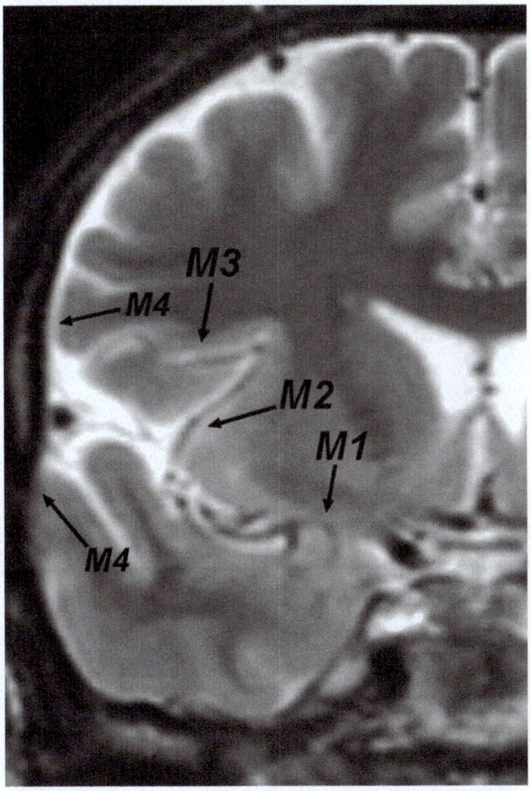

Fig. 5.1 Coronal MRI T2 weighted. The MCA runs first horizontally in the Sylvian cistern (M1), then on the surface of the insula (M2), further laterally (M3) exiting the insular cistern, and reaching the convexity (M4)

G.B. Bradac, *Applied Cerebral Angiography*, DOI 10.1007/978-3-319-57228-4_5

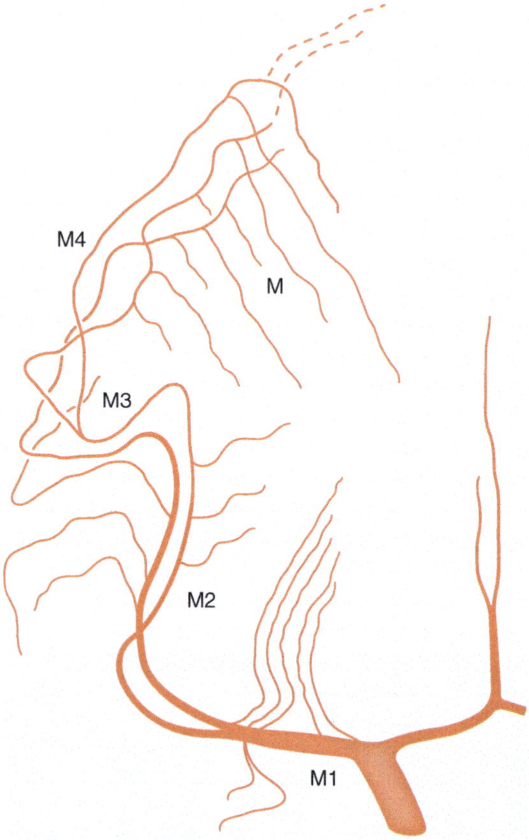

Fig. 5.2 Drawing showing the segments of the MCA on the AP view. M1, M2, M3, and M4. Deep perforators arising from M1. Medullary arteries (M, short and long branches) arising from the cortical branches. *Dashed lines* indicating anastomoses with ACA

Perforators. Since the first anatomical description of these perforating branches by Duret (1874), many detailed studies about these arteries have been performed. Also called the lenticulostriate arteries, these branches arise along the length of the M1 commonly from its superior surface. When the M1 is short, dividing early in its distal branches, the perforators commonly arise from the superior branch and sometimes from a branch of the M2 segment (Kaplan 1965; Westberg 1966; Lazorthes et al. 1976; Rosner et al. 1984; Umansky et al. 1985). The branches can be subdivided into medial, intermediate, and lateral groups. The medial is the group least constant and can be replaced by perforators arising from A1 (Rosner et al. 1984). In the microanatomical study of Umansky et al. (1985), the medial perforators of the MCA were replaced by perforators arising from the A1 segment or by the artery of Heubner in 26% of the cases. In the study of the same authors, perforators originate as a common trunk in about 50% of the cases.

The medial branches have a straight course, the intermediate a slightly tortuous course, and the lateral display the typical S-shaped aspect (Drawing 5.2 and Fig. 5.3). Perforators commonly enter the lateral part of the APS (Rosner et al. 1984; Rhoton 2002) and supply the superior part of the head of the nucleus caudatus and its body. They entirely supply the superior part of the internal capsule as well as the globus pallidus and putamen.

The different aspects concerning the M1 and the perforators should be carefully considered in the surgical or endovascular treatment of aneurysm located in the M1 segment, as well as in the endovascular treatment of occlusion-stenosis of this segment. Examples are presented in Figs. 5.3, 5.4, and 5.5.

Cortical branches: From the M1 segment arises frequently a branch for the pole of the temporal lobe and sometimes also an orbitofrontal branch for the pars orbitalis of the inferior frontal gyrus. In exceptional cases a temporal branch can arise from the ICA.

5.2 M2, M3, and M4 Segments

The M2 segment, also called the insular segment, is formed by the branches arising from the bifurcation of M1 which run on the surface of the insula. They curve then laterally forming the M3 segment directed towards the sylvian-insular cistern, extending finally to the surface of the cortex of the hemisphere forming the M4 segment. Considering their course and the area they supply, the cortical branches can be divided into *the*

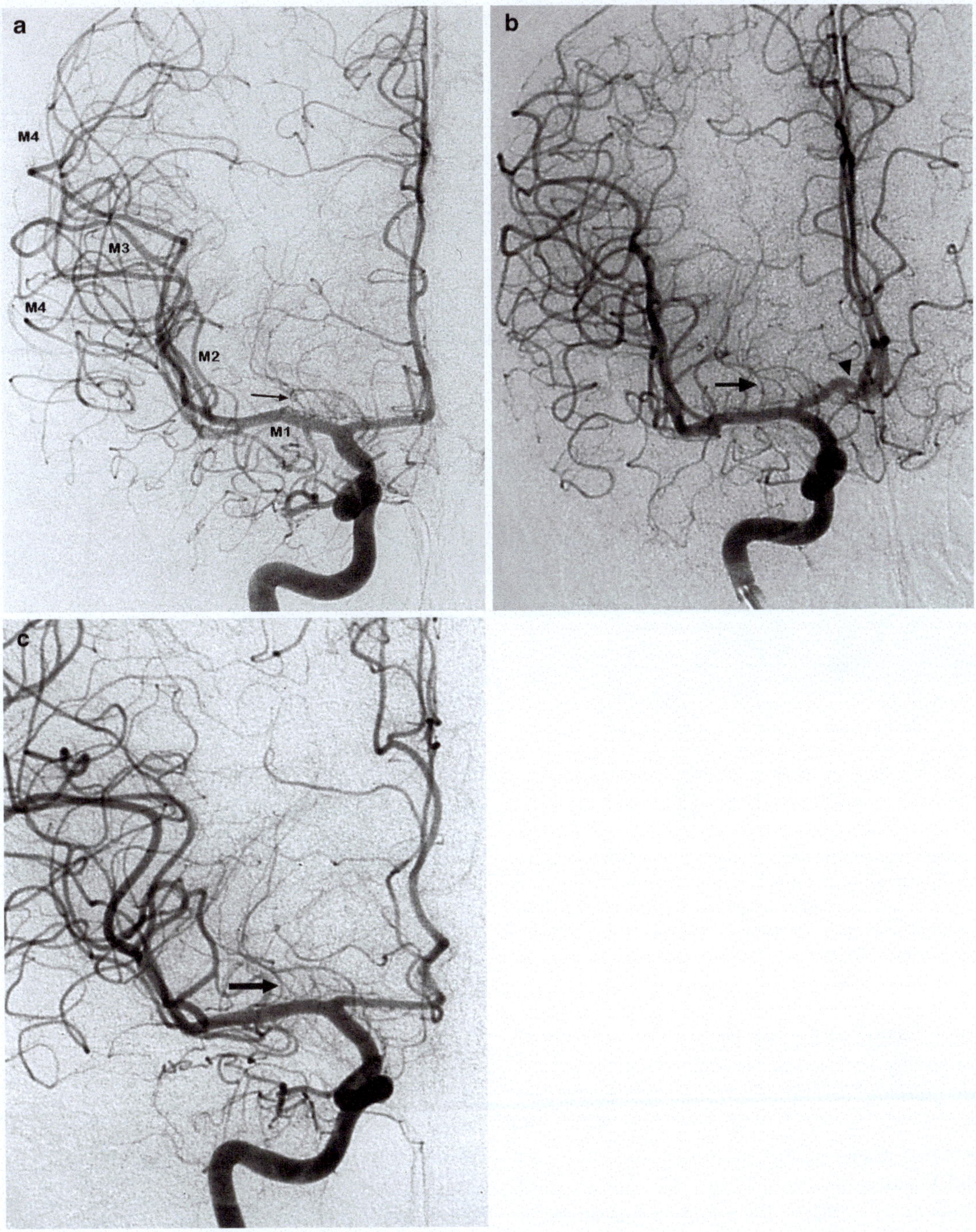

Fig. 5.3 AP carotid angiograms, in three different cases. In (**a**) the different segments (M1, M2, M3, M4) are indicated. In all three cases (**a–c**) the M1 segment is long. The deep perforators are well recognizable. The more distal perforators show the typical S shape (*arrow*). In (**a**) a temporal branch arising from the M1 is also recognizable. In (**b**) there is a fenestration of distal A1 (*arrowhead*)

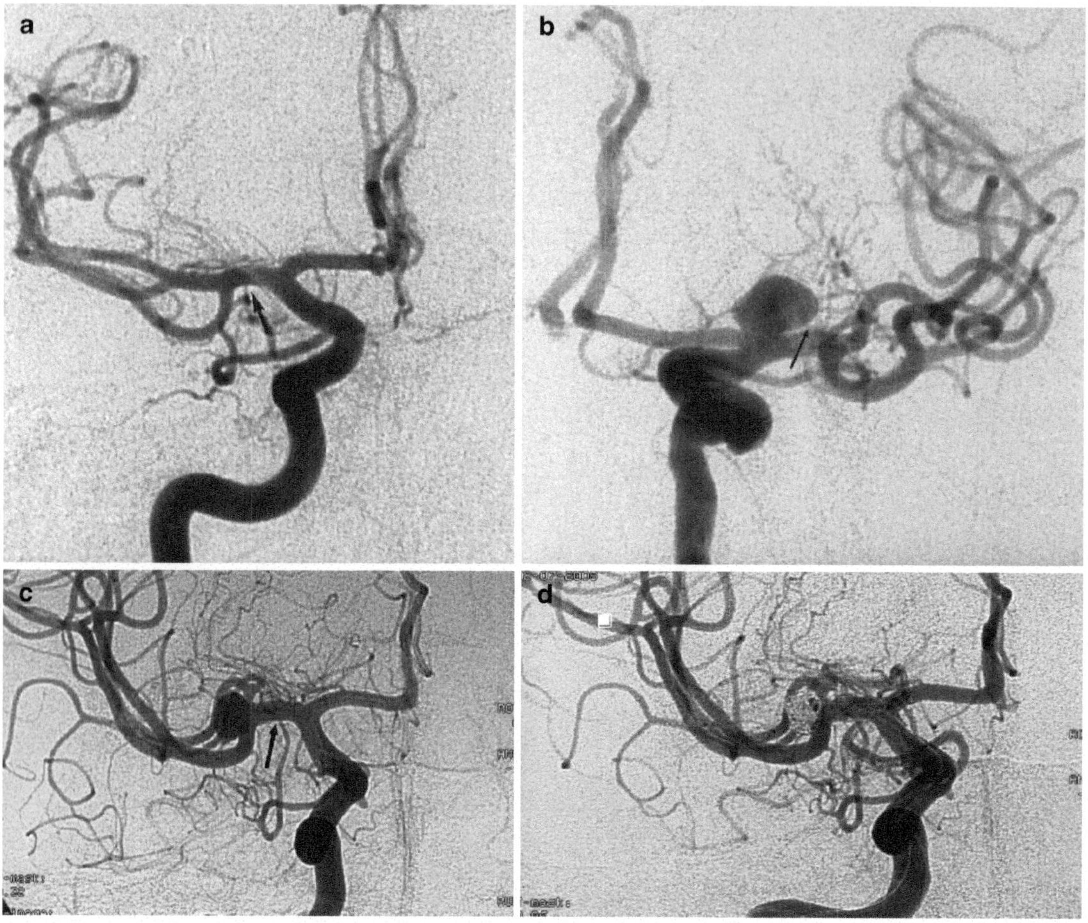

Fig. 5.4 AP carotid angiograms in three different cases. (**a**) Apparently very short M1 (*arrow*) owing to the early origin of a large temporal branch. The main trunk from which arise the perforators runs superiorly and bifurcates more distally. (**b**) Aneurysm with large neck at the carotid bifurcation. Owing to the short M1 the perforators (*arrow*) are close to the aneurysm. (**c**, **d**) The short M1 (*arrow*) gives off perforators that are close to the trifurcation from which arises an aneurysm presenting with SAH. Control angiogram after treatment with coils

orbitofrontal artery, ascending frontal arteries (operculo-frontal and central arteries), ascending parietal arteries (anterior and posterior), descending arteries (temporal branches), and terminal group (gyrus angularis artery) (Ring and Waddington 1967; Ring 1974; Huber 1979; Gibo et al. 1981b). Overlapping variations in course and origin make it difficult in a given case to identify precisely the arteries on the angiogram. On the basis of anatomical and angiographic studies, Michotey et al. (1974) and Salamon and Huang (1976) have *described two basic patterns that can be of some help in identifying the single artery* (Drawing 5.6a, b).

In cases of *bifurcation* of the M1 which is the most frequent pattern, two trunks are present: one anterosuperior and the other posteroinferior. The orbitofrontal, operculofrontal, and central arteries arise from the anterosuperior trunk. The remaining branches take their origin from the posteroinferior trunk. In cases of *trifurcation*, the orbitofrontal and operculofrontal arise from the anterosuperior trunk; the central, parietal, and gyrus angular arteries arise from a middle trunk, and the temporal branches arise from the posteroinferior trunk. Other variants characterized by a division of M1 in multiple trunks can also occur, but they are more rare

(Salamon and Huang 1976; Gibo et al. 1981b, Umansky et al. 1985).

The orbitofrontal artery is the first branch to leave the Sylvian-insular cistern. It is not constant and sometimes can originate directly from the M1 segment. It is directed anteriorly, with a slightly horizontal course, running frequently first in the sulcus separating the pars triangularis from the pars orbitalis of the inferior frontal gyrus and further extending towards its pars orbitalis. It is in balance with the infraorbital branch of the ACA.

The operculofrontal branch leaves the insular cistern and is directed first to the pars triangularis of the inferior frontal gyrus; in its ascending course, it divides further into three or more branches, which resemble a candelabrum. They supply the *Broca area* and the other frontal regions (prefrontal and premotor).

The central arteries. These, which are the most posterior ascending frontal branches, are formed commonly by two branches arising separately or as a common trunk, which run towards

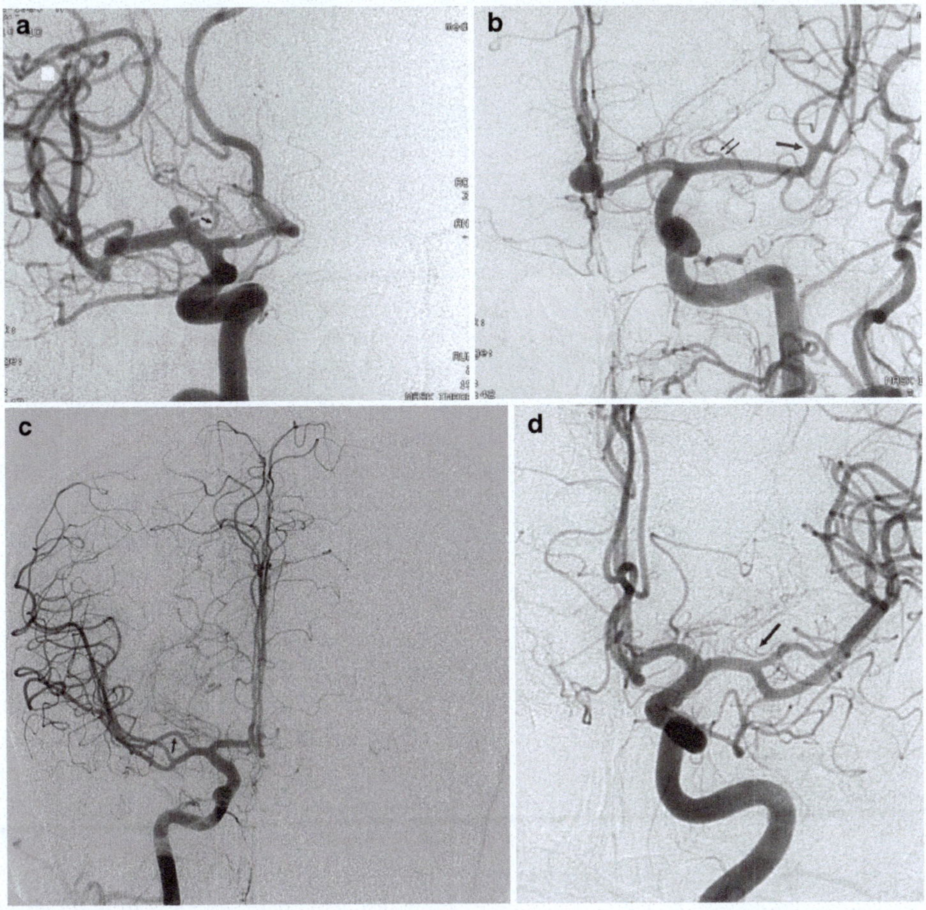

Fig. 5.5 Carotid angiograms in different cases. (**a**) Oblique view in a patient examined for an unruptured aneurysm with a neck on the M1 segment at the junction with a large perforator (*arrow*). This seems to be a large trunk from which all perforators arise. (**b**) AP view, in a patient examined for a ruptured aneurysm of the AcomA. There is a long M1 which trifurcates distally in the insular cistern (*arrow*). Some perforators arise from the proximal part of the long M1 and from the proximal part of A1. Other perforators seem to originate from the temporo-insular branch (*arrows*). (**c**) Example of a very short M1, which divides into a superior branch (*arrow*) from which arise the perforators, and into an inferior branch dividing further distally. (**d**) In this case the short M1 divides into superior and inferior branches. The perforators arise partially from M1 and partially from its superior branch (*arrow*). (**e**) Another patient with an incidental discovery of a middle cerebral aneurysm. Angiogram and detail magnification. The aneurysm is located at the bifurcation of a very short M1. The perforators arise from the superior branch of M1 (*white arrowhead*). At least one perforator seems to arise from the A1. Heubner artery (*black arrowheads*). AchA (*small black point*), hypoplastic PcomA (*two small black points*)

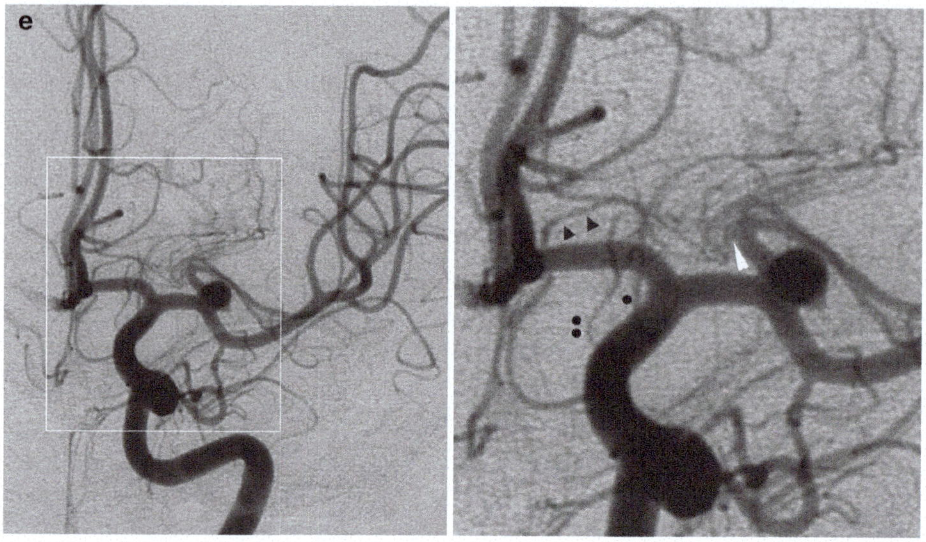

Fig. 5.5 (continued)

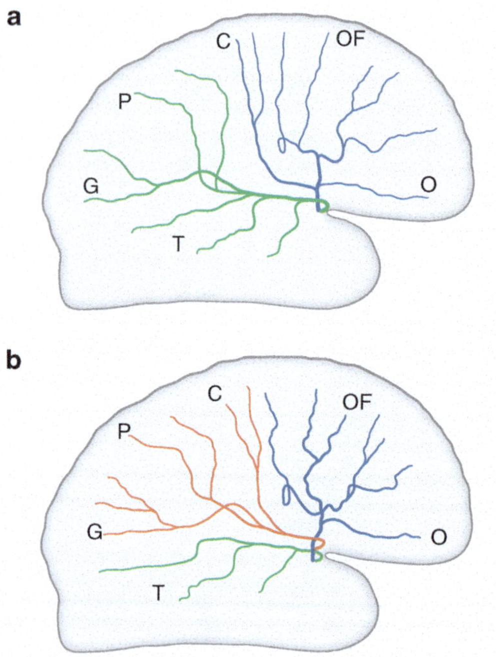

Fig. 5.6 Drawing showing the most frequent pattern of the distal branches of the MCA on a lateral projection (modified from Michotey; Salamon, and Huang). (**a**) In the bifurcation pattern, the orbitofrontal (O), operculofrontal (OF), and central arteries (C) arise from the anterosuperior trunk. The anterior and posterior parietal (P), gyrus angularis (G), and temporal (T) arteries arise from the inferoposterior trunk. (b) In the trifurcation pattern, the orbitofrontal (O) and operculofrontal (OF) arteries arise from the anterosuperior trunk. The central (C), anterior and posterior parietal (P), and gyrus angularis (G) arteries arise from the middle trunk and the temporal (T) branches from the inferoposterior trunk

the central fissure. They supply the precentral gyrus (*primary motor cortex*) and the postcentral gyrus (*primary sensory cortex*). These branches are frequently very fine and their identification on the angiogram can be difficult.

The parietal group consists of two arteries: the anterior parietal and posterior parietal arteries which frequently arise in the posterior part of the Sylvian fissure. The anterior branch is commonly very small, and difficult to be identified on the angiogram. It runs towards the postcentral sulcus and extends further to the parietal lobe. The posterior branch is commonly larger. Differently from the other ascending branches that run in the sulci, it has a superficial course towards the parietal lobe (Salamon and Houang 1976). It marks the borders between the parietal lobe and the posterior located supramarginal lobe. The posterior parietal branch has a less upward and more posteriorly directed course, in comparison to the other ascending branches. This pattern can help in identifying the artery on the lateral angiogram. This can also be useful, considering that the *central arteries* run in the adjacent area located anteriorly. In this context we report also the study of Christoforidis et al. (2002). These authors have suggested that the *central arteries* can be identified on the lateral carotid angiogram in an area projecting between the superior parietal and posterior frontal arteries, branches of the anterior cerebral artery (see Sect. 4.2.4).

The temporal group consists of several branches that emerge from the Sylvian fissure and take a descending course directed inferoposteriorly, supplying the temporal lobe. From their origin, they are termed *anterior*, *middle*, and *posterior*. The latter extends also to the lateral surface of the occipital lobe (temporo-occipital artery).

The angular artery, which is considered the terminal branch of the M4, is commonly that with the larger diameter according to the microsurgical anatomy study of some authors (Gibo et al. 1981). It emerges from the distal part of the Sylvian fissure and runs posteriorly towards the supramarginal and angular gyri. It plays an important role in the supply *of the Wernicke's area.* It is frequently identified on the lateral angiogram, since it forms an upward convex curve, as it turns around the Heschl's gyrus (Ring 1974; Huber 1979). It is not uncommon for the artery to extend with a few branches to the parietal and temporo-occipital areas, thereby replacing the corresponding arteries.

Examples of course of the cortical branches are presented in Figs. 5.6, 5.7, 5.8, 5.9, and 5.10.

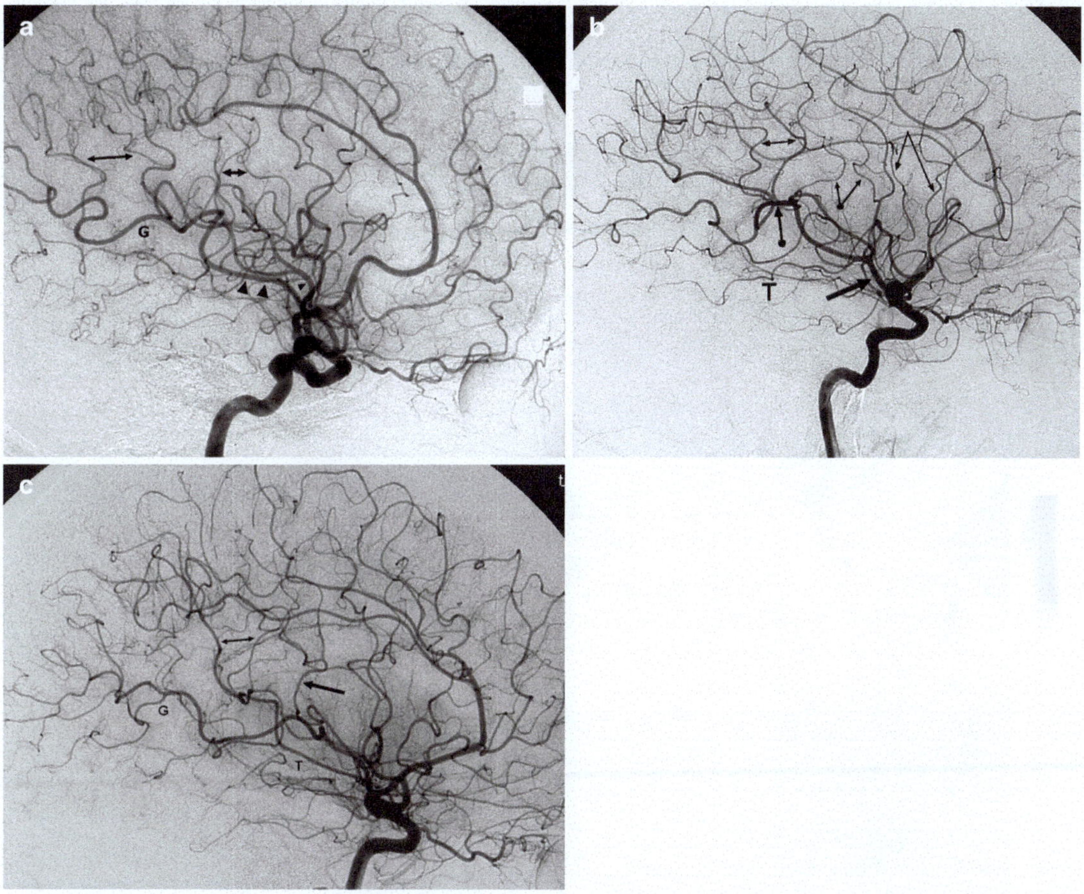

Fig. 5.7 (**a**) Lateral carotid angiogram. From a superior located trunk (*arrowhead*) arise the operculofrontal arteries and two small branches (*big bidirectional arrow*) which run towards the central area. From a large posterior trunk (*arrowheads*) arise a well-developed gyrus angularis artery (G) and two small parietal branches (*smaller bidirectional arrow*). (**b**) Lateral carotid angiogram. Operculofrontal arteries (*arrows with angle*). Common posterior trunk (*large black arrow*) for gyrus angularis, parietal, and temporal arteries. Small parietal branches (*bidirectional arrow*). Anteriorly to these, there are very fine arteries (*arrow with angle and dot*) running towards the central area. Typical curve of gyrus angularis artery (*arrow with dot*). The well-developed gyrus angularis artery gives off a large branch for the parietal area. Small temporal branches (T). (**c**) The parietal (*bidirectional arrow*), the gyrus angular (G), and the temporal (T) arteries seem to arise from a common trunk. It seems that arising also from the same common trunk, running anteriorly to the parietal arteries, there are small branches (*arrow*) directed towards the central area

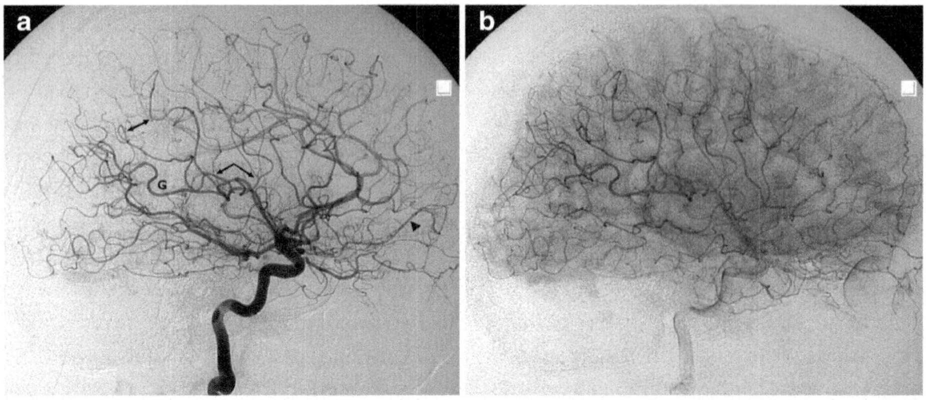

Fig. 5.8 Early and late phases of a lateral carotid angiogram. From a posterior trunk arise the gyrus angularis (G), the parietal branches (*bidirectional arrow*), and the small temporal branches. The anterior parietal branch is particularly well developed. Anteriorly to the parietal arteries there are small branches (*arrow with angle*), running towards the central area. Well-developed orbitofrontal artery (*arrowhead*). There is a fetal posterior cerebral artery

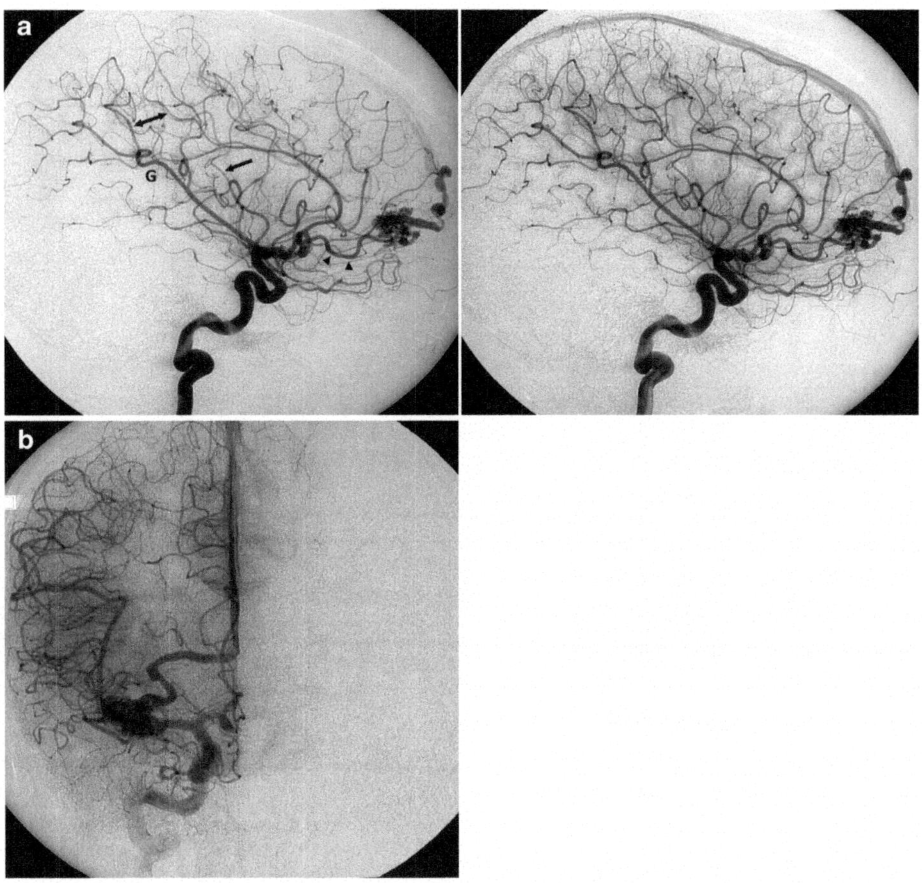

Fig. 5.9 (**a**) Early and later phases of lateral carotid angiogram in a patient with frontobasal AVM supplied by a dilated orbitofrontal artery (*two arrowheads*), draining in the SSS through a single fronto-orbital vein. There is a large posterior trunk continuing in the gyrus angularis artery (G), which gives off also parietal branches (*bidirectional arrow*). Anteriorly from the large trunk arises an artery (*arrow*) which divides distally into two branches directed to the central area. (**b**) Same case, AP view, showing better the supply of the AVM from the branch of the MCA and the unique venous drainage towards the SSS

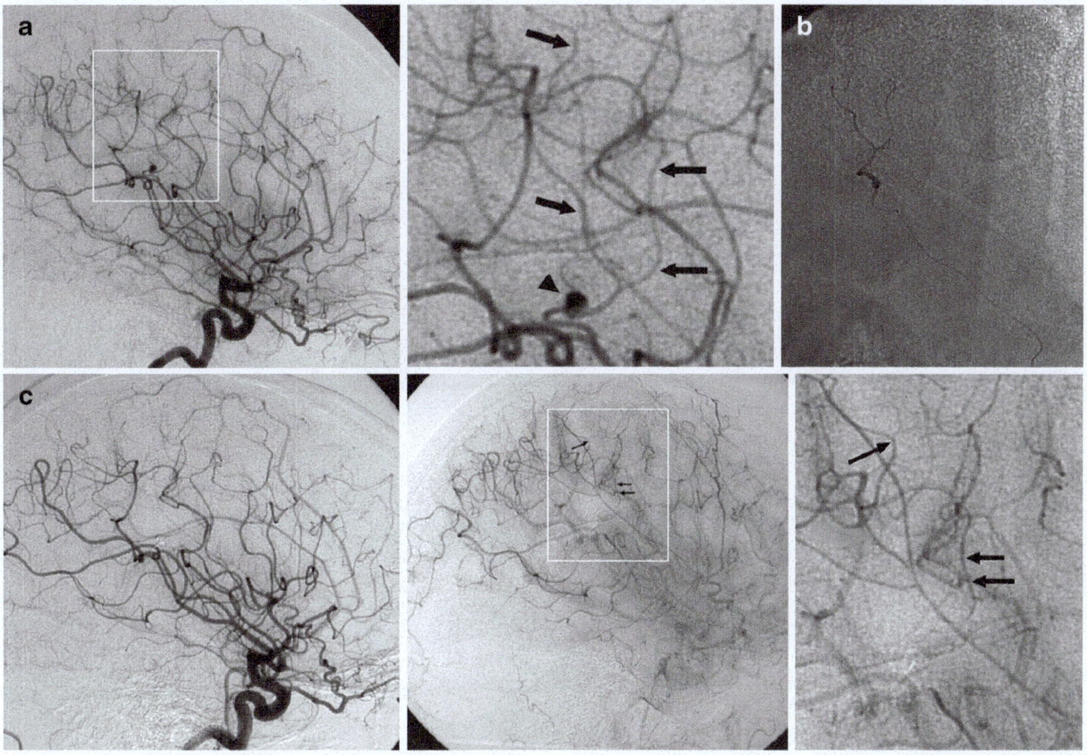

Fig. 5.10 Carotid angiogram in a patient presenting with SAH. (**a**) Lateral ICA angiogram (normal and magnification images), showing a very small probably dissecting aneurysm (*arrowhead*). The aneurysm involves a branch of the MCA which divides distally into two branches (*arrows*), corresponding probably to the central arteries. (**b**) The decision was taken to occlude the aneurysm and the parent artery. Selective study preceding the injection of glue. (**c**) Angiogram posttreatment with magnification showing the retrograde injection of the occluded branches (*arrows*) through opening of leptomeningeal anastomoses with ACA. There is the overlapping of the middle meningeal artery arising from the ophthalmic artery. The patient recovered well without any neurological deficit

5.3 Anatomical Variations

Some variants concerning the M1 segment and perforators have already been described.

Other rare variants are the isolated origin of the temporal branch from the ICA and *fenestration* and *duplication* of the M1 segment. In case of fenestration the vessel is unique, but it is divided into two lumens (Fig. 5.11). The duplication is characterized by the presence of two arteries arising from the distal ICA (Fig. 5.12). This should be differentiated from the condition called *accessory MCA (AMCA)*, term coined by Crompton (1962). Also in these cases two arteries are present but one arises from the ICA and the other from the ACA. Considering that the MCA arises from several twigs originating from the distal cranial division of ICA and partially from the primitive ACA, the persistence of one twig arising from the ACA could explain this anomaly. Some authors (Handa et al. 1970a, b) proposed that the AMCA represents a large recurrent artery of Heubner. But this is no longer accepted by the majority of the authors (Huber 1979; Gibo et al. 1981b; Umansky et al. 1988; Takahashi et al. 1989; Müller et al. 1991; Tacconi

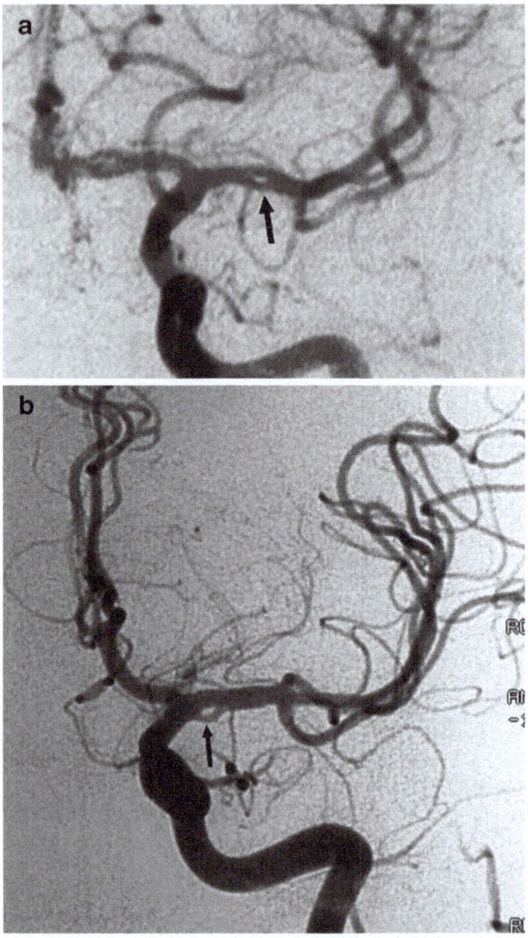

Fig. 5.11 Two examples of fenestration (**a**, **b**) of the M1 (*arrow*). On the angiogram in (**a**) there is also a fenestration of the A1

et al. 1995) since in all the cases studied the AMCA gave off perforators and cortical branches, while the recurrent artery of Heubner ends as a perforator branch. It should be mentioned that in one of the cases reported by Umansky et al. (1988) the Heubner artery originated from the AMCA. The AMCA diagnosed in our angiographic material could arise from the distal or more proximal A1. Examples are demonstrated in Figs. 5.13 and 5.14.

Cases of aplasia of the middle cerebral artery have been reported. This is a very rare condition frequently associated with aneurysm of the ACA (Nakazawa et al. 1985; Han et al. 1993; Amagasaki et al. 1998).

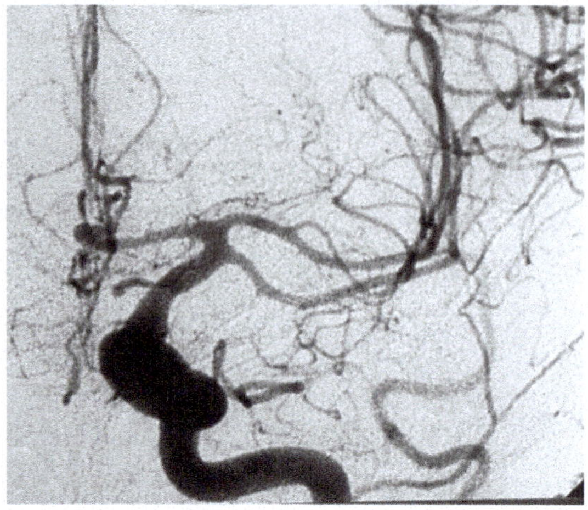

Fig. 5.12 There are two equally well-developed branches of the MCA arising from the internal carotid artery (duplication of the MCA). The perforators arise from the superior branch. Small aneurysm of the AcomA

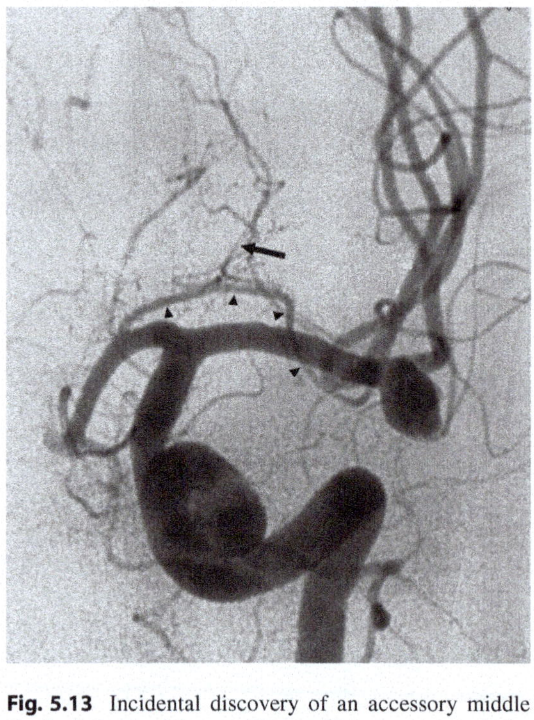

Fig. 5.13 Incidental discovery of an accessory middle cerebral artery in a patient with aneurysm of the MCA. Parallel to the large M1, there is another smaller branch (*arrowheads*) arising from the proximal anterior cerebral artery, running above the main trunk of M1. The majority of the perforators arise from the main trunk close to the trifurcation. At least one perforator seems to originate from the accessory middle cerebral artery (*arrow*)

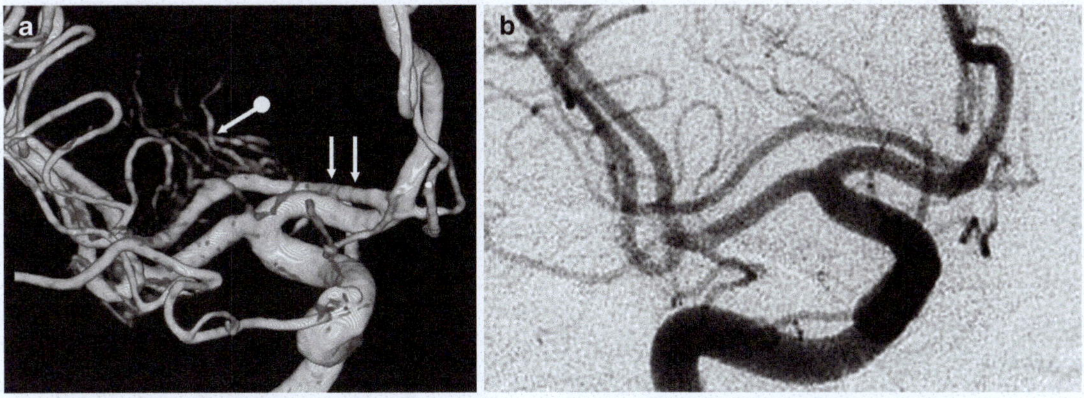

Fig. 5.14 Another example of an accessory middle cerebral artery arising in the distal segment of the A1. (**a**) On the 3D CT angiography the accessory branch (*arrow*) is well recognizable. It gives off the perforators (*arrow with dot*). (**b**) More detailed study on the ICA angiogram

5.4 Vascular Territories

The MCA supplies largely the lateral portion of the frontoparietal and occipital lobes. The distal cortical branches of MCA, however, do not reach the superior margin of the hemispheres. Therefore, there is an area in the paramedian convexity to a variable extent which is not supplied by the MCA but by branches of the ACA and PCA. Further, the MCA supplies the temporal pole, the latero-basal and lateral part of the temporal lobe, and the latero-basal part of the frontal lobe.

The vascular territories of the deep perforators have already been described. These arteries are *end-arteries*; that is, there is no anastomosis between the single branches and no connections with the medullary arteries. This is probably true in the majority of the cases. In a few cases, however, anastomotic connections between perforators of the MCA and the artery of Heubner have been demonstrated (Umansky et al. 1985). We could identify on the angiogram a possible anastomosis occurring between the artery of Heubner and a perforator of the M1 (Fig. 5.15). In another angiographic study a clear connection between the perforators of M1 was recognizable (Fig. 17.7). Other authors (Kodama and Suzuki 1974) have shown in microangiographic studies the existence of a few anastomoses between deep perforators and the medullary arteries.

From the branches running on the surface of the cortex arise branches entering the brain parenchyma having the same angio-architecture already described in Sect. 4.4. Short branches (cortical) supply the cortex, and others supply the white matter (medullary arteries). These latter run in the perimedullary space, some arresting in the superficial part of the white matter (short branches), and others extending to the deep part (long branches) converging towards the periventricular area and basal ganglia. All these branches are *end arteries.*

From the insular branches also arise medullary arteries which supply the subinsular area without, however, anastomosing with the deep perforators supplying the basal ganglia.

Conversely there is a potential collateral leptomeningeal circulation between the superficial distal branches of the MCA and ACA at the convexity in the frontoparietal and the PCA at the occipital convexity. Connections can develop on the temporo-occipital convexity between the MCA and PCA and between MCA and ACA in the frontobasal area. As already described in Sect. 4.4, considering that the vascular territory of the arteries can have a variable extent, the area of the connection (border area) can have a different location.

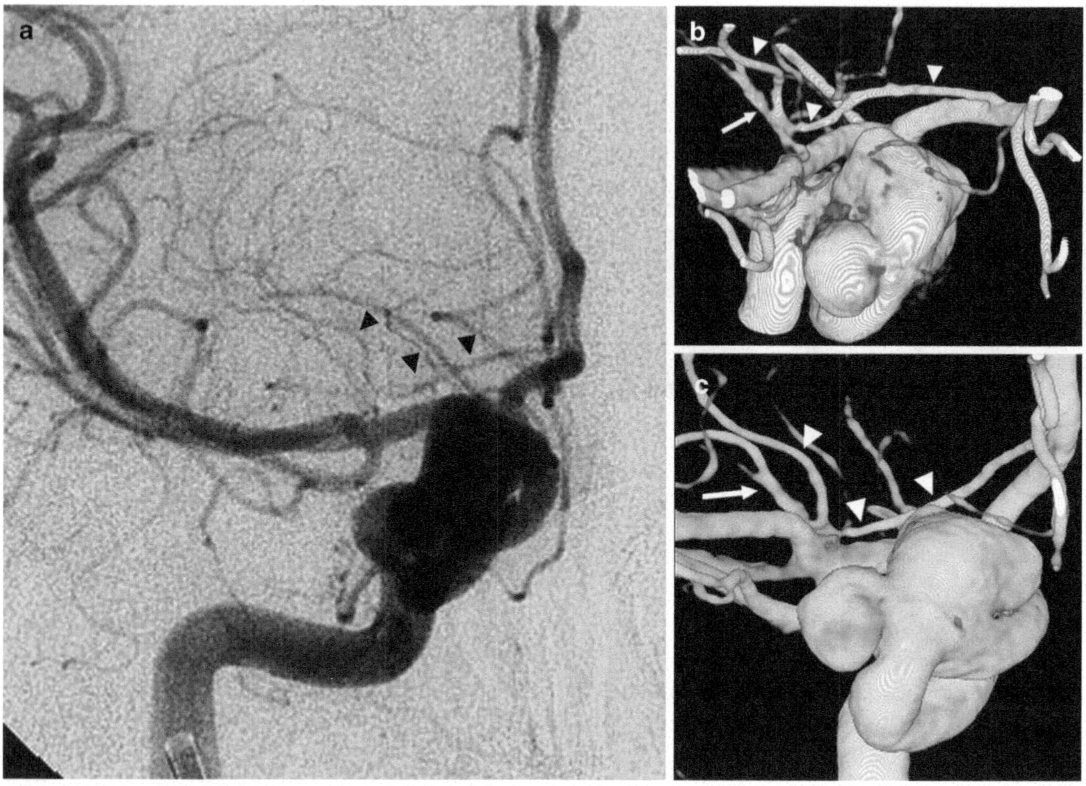

Fig. 5.15 (**a**) AP angiogram in a patient studied for a large aneurysm of the distal ICA. The Heubner artery (*arrowheads*) is well defined. (**b**) 3D angiographic study. The Heubner artery (*arrowhead*) seems to have a connection with a perforator of the M1 (*arrow*). (**c**) In another projection, the connection seems to be confirmed. Heubner artery (*arrowhead*). Perforator of M1 (*arrow*)

5.5 Angiogram

The M1 segment and the perforators are always well recognizable on the AP view. The length of the segment can vary, as already described, and this influences the origin of the perforators (Figs. 5.3, 5.4, and 5.5). On the AP view the M1 segment commonly has a horizontal course; in children it is more frequently directed upward (Harwood-Nash 1974); in older patients it can be very tortuous.

Some typical features of the M2 and M3 segments are recognizable on the AP and lateral angiograms. On the AP view the M2 branches have an ascending course directed towards the top of the insula where they form a loop. The highest and most medial loop corresponds commonly to the most distal branches (parietal and angular arteries). This is also called the *Sylvian point* (Chase and Taveras 1963; Huber 1979; Osborn 1999). The arteries run further laterally (M3 segments), and leave the insular cistern, reaching the convexity of the hemisphere. On the lateral angiogram the loops can sometimes be well recognizable. They are located on an ideal horizontal line marking the superior border of the insula. The most posteriorly located loop corresponds to the Sylvian point.

The arteries of the M4 segment are better demonstrated on the lateral angiogram (Drawing 5.6 and Figs. 5.7, 5.8, 5.9, and 5.10). Due to their great variations of origin and course and overlapping, a certain identification of the single artery can often be very difficult. Some morphologic features have already been described in Sect. 5.2: these can be helpful in identifying the arteries. Furthermore, as has been described for the distal branches of the ACA (Sect. 4.5), there is a great constancy for each artery or group of arteries, independently of the origin, to take, in their distal segment, a course towards the territory they supply. Taking this into consideration, from the practical point of view it is helpful for identifying the artery and their branches to define first, on the lateral angiogram, even if approximately, the vascular territory inside of which run the supplying vessels. The course of the identified artery can be retrogradely followed to the origin examining the early and late phases of the angiogram.

Extra- and Intracranial Vertebrobasilar Sector

6

6.1 Extracranial Sector

The vertebral artery (VA) originates bilaterally from the subclavian artery. It runs posterosuperiorly behind the anterior scalene muscle commonly reaching the foramen of the transverse process of the sixth cervical vertebra. This first segment of the VA has been called V1. The VA runs, further, vertically (V2 segment) through the foramina of the transverse processes from C6 to C2, surrounded by the venous plexus. The spinal nerves lie behind. Between the foramina of C2 and C1, the artery curves laterally and somewhat anteriorly. Exiting from C1 begins the V3 segment, which curves backward, running in the sulcus of the posterior arch of C1. The artery forms then a second upward and forward curve and reaches the foramen magnum, where it penetrates the dura forming the last segment (V4). It is conceivable that these curves protect the VA, allowing it to accommodate more easily movements in the atlanto-occipital region (Fig. 6.1).

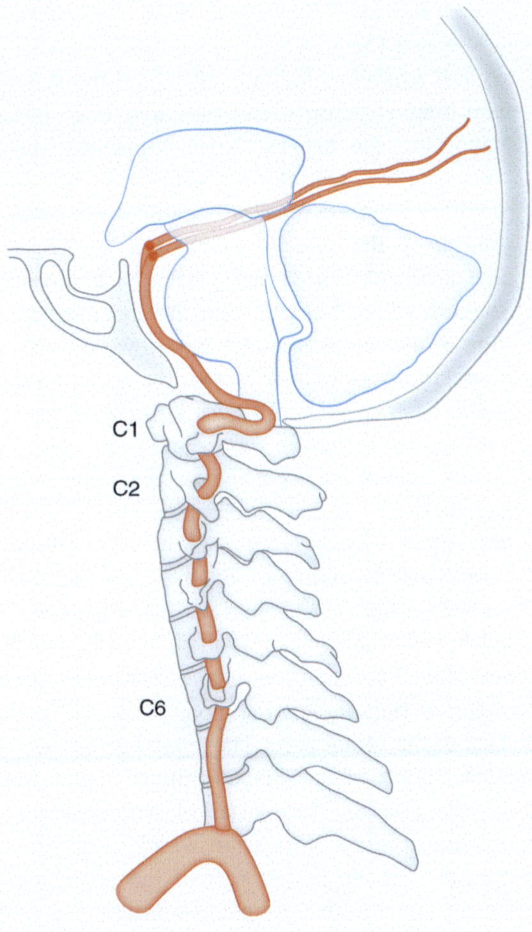

Fig. 6.1 Drawing showing the course of the vertebral artery arising from the subclavian artery and entering the intracranial cavity at the foramen magnum

6.1.1 Branches

In its extracranial course the VA gives off several branches.

- *Cutaneous-muscular branches*. These branches have potential anastomoses with branches of ECA and cervical arteries. The most frequent

G.B. Bradac, *Applied Cerebral Angiography*, DOI 10.1007/978-3-319-57228-4_6

well developed and recognizable on the angiogram are the anastomosis at the C1 and C2 levels with branches arising from the ascending and horizontal segment of the occipital artery and with the deep cervical artery. Other possible anastomoses are that with the AphA and the ascending cervical arteries at the C3–C4 levels. They can be recognized in normal angiograms (see also Sects. 3.4, 3.5 and 3.9) or in cases of occlusions of the VA or ECA, where they act as a collateral circulation towards the occluded sector or will be involved in several pathologies, in particular in DAVF and vascular malformations of the head and neck. Examples are shown in Figs. 3.7b–d, 3.16, 6.2b, 6.8a, 6.9a, 15.29, and 15.33.

- *Spinal branches* include those for the dura and bone structures, and radicular branches supplying the nerves. After supplying the nerves, a few radicular branches (radiculomedullary arteries) reach the spinal cord, continuing in the anterior spinal artery (ASA) and posterior spinal artery (PSA). It is not the aim of this book to treat the vascularization of the spinal cord, for which we refer to a specific work (Thron 1988). We recall only some basic aspects. Three longitudinal axes are involved: the one *is the anterior spinal artery* which runs on the median anterior sulcus of the spinal cord, from which arise the sulcal arteries giving off the perforators supplying a great part of the grey matter (anterior horns and part of the posterior horns) and the anterior part of the white matter. *The other two are the posterior spinal arteries* running on the posterolateral surface of the spinal cord. These are more irregular, sometimes discontinuous, mainly involved in the formation of a superficial reticular pial network from which arise perforators for the dorsal part of the posterior horns and for the dorsal and lateral white matter. The dorsal network is interconnected with its anterior part arising from the anterior axis. The ASA is well recognizable on the angiogram of the VAs and sometimes of the ascending cervical arteries arising from a radiculomedullary artery commonly at the C4–C6 levels. Its ascending and descending segments, anastomosing, respectively, cranially with the descending ASA arising from the intracranial VA and caudally with the ascending segment arising from the radiculomedullary artery of the intercostal artery are well shown (Figs. 6.2a and 16.9). The PSA is commonly not recognizable on angiogram in normal conditions.
- *Meningeal branches* of the distal extracranial VA: Two types of arteries can be present and be recognizable on the angiogram. The one is the meningeal artery running in the anterior part of the spinal canal, and the other runs posteriorly in the posterior fossa. The artery present in the spinal canal has been first described by Greitz and Lauren (1968) and called *anterior meningeal artery*. Further study (Lasjaunias et al. 1978a) has demonstrated that this artery is a radiculomeningeal branch arising from the VA at the level of the third vertebral body that runs cranially in the anterolateral part of the spinal canal and anastomoses with the contralateral branch at the level of the dorsal surface of the odontoid process. The anterior meningeal artery is connected further with the hypoglossal artery, which is a branch of the ascending pharyngeal artery, which enters the spinal canal through the hypoglossal foramen. These connections form the so-called odontoid arch. On the vertebral angiogram the artery is visible as a fine branch located anterior and lateral to the ASA (Drawing 3.33, Figs. 3.7c, 6.2b, c, 15.33, and 15.37; see also Sect. 3.4). The posterior group is represented by the falx cerebelli and posterior meningeal arteries, which originate from the extracranial VA, sometimes from its intracranial segment. They may also take their origin from the posterior inferior cerebellar, from the occipital or AphA arteries. There is a balance among these arteries, and so one can predominate in supplying the dura of the posterior fossa. *The falx cerebelli artery* (*FCA*) runs close to the falx and continues in the meningeal branch along the

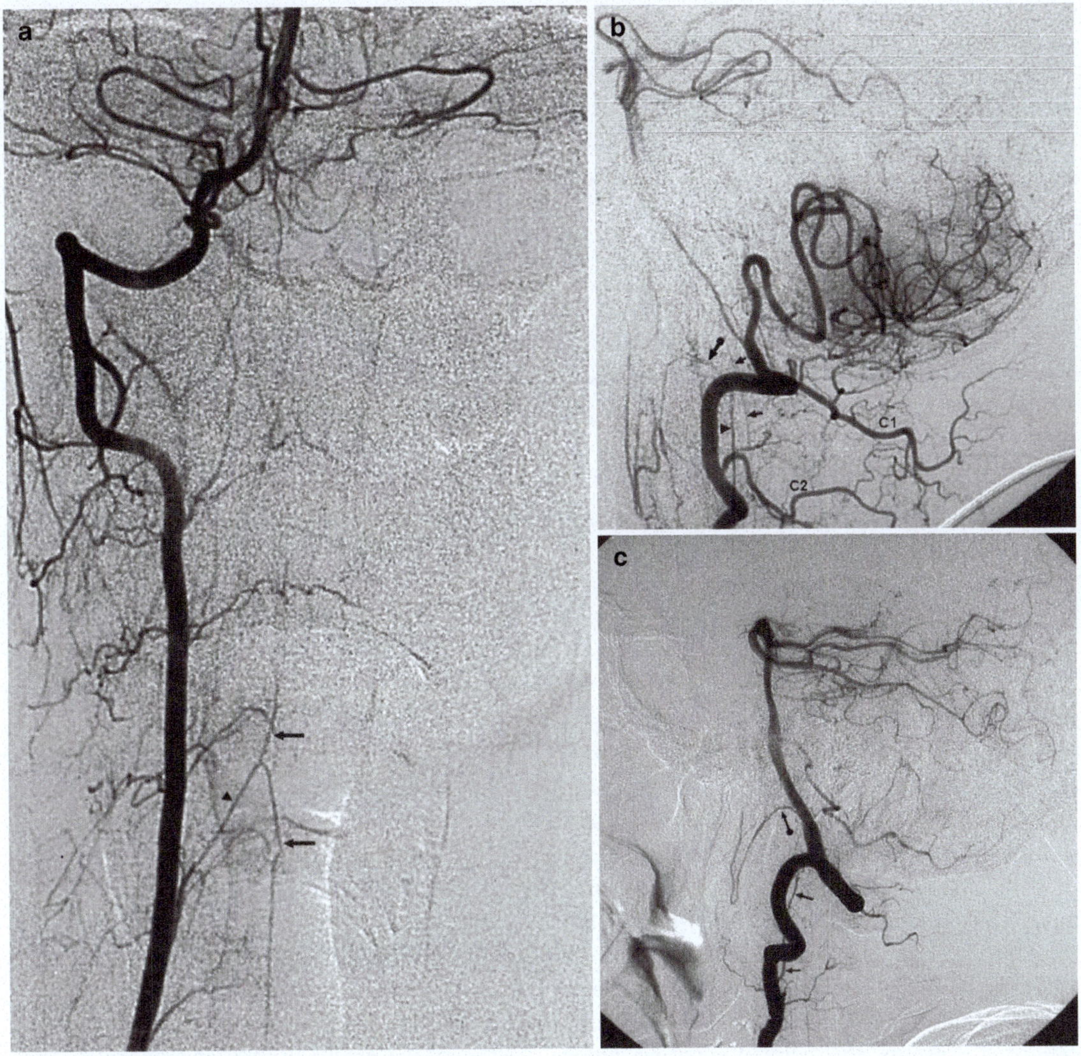

Fig. 6.2 (**a**) AP right vertebral angiogram. Radiculomedullary artery (*arrowhead*) continuing in the ascending and descending segments (*arrows*) of the anterior spinal artery. Presence of a large musculocutaneous branch at the C2 level. (**b**) VA angiogram lateral view. Hypoplastic VA ending in the posterior inferior cerebellar artery. Radiculomeningeal branch of the VA (*arrowhead*) anastomosing with the hypoglossal branch of the AphA (*arrow with dot*). Posteriorly runs the anterior spinal artery (*double small arrows*). Musculocutaneous branches at the C1–C2 levels, presenting possible anastomoses with the occipital and cervical arteries. (**c**) VA angiogram. Another example of a connection between the hypoglossal branch of the AphA (*arrow with dot*) and radiculomeningeal branch of the VA (*arrows*)

straight sinus. It anastomoses superiorly with the paramedian branches of the middle meningeal artery (see Sect. 3.7.1.1). On the lateral angiogram, the FCA runs slightly away from the inner table of the skull, and it is near the midline on the AP view (Figs. 6.3, 13.12, 13.16, and 15.37). *The posterior meningeal artery* runs on the dura covering the cerebellar hemispheres, so it has a more lateral course on the AP; it runs close to the inner table of the skull on the lateral angiogram (Figs. 13.1 and 15.35).

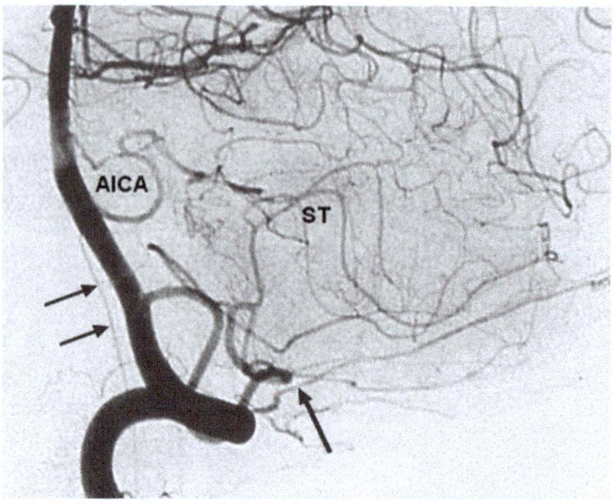

Fig. 6.3 Vertebral angiogram. Lateral view. Falx cerebellar artery (*arrow*). ASA (*double arrow*). Very tortuous segment of the first segments of PICA. Typical pattern of its supratonsillar segment (ST). AICA

6.2 Intracranial Sector

At the level of the foramen magnum, the VA penetrates the dura, where it is sometimes identifiable on the angiogram owing to a small change in its caliber. The artery runs in the subarachnoid space ventrolateral to the medulla oblongata and joins the contralateral artery at the bulbopontine junction to form the basilar artery (BA). The BA runs on the ventral surface of the pons in the median-paramedian position; it sometimes adopts a tortuous course displaying a lateral extension towards the cerebellopontine angle (Fig. 6.4). In the interpeduncular fossa under the floor of the third ventricle, the BA divides into the two posterior cerebral arteries (PCAs). The point of the division can be more caudally located, causing a cranial course of the proximal segments of the PCAs (Fig. 7.8c). The basilar artery may be elongated, especially in hypertensive and atherosclerotic patients, bulging in the floor of the third ventricle. In such cases, the PCAs have first a caudally directed course before they encircle the midbrain. Many arteries arise from the intracranial sector of the vertebrobasilar system (Drawing 6.5a–c). In this chapter the branches of the medulla and pons are described . The branches supplying midbrain, thalamus, and the temporo-occipital area are treated in Chap. 7.

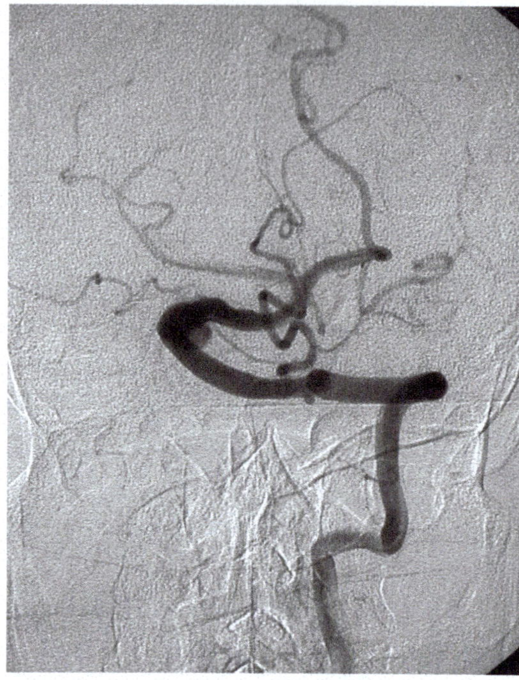

Fig. 6.4 Tortuous basilar artery, probably due to atherosclerosis extending into the cerebellopontine angle, causing facial spasm

6.2.1 Intracranial Branches of the VA

6.2.1.1 Perforators of the VA

Small arteries arise from the posterior surface of the VA (Duvernoy 1999), running on the anterior surface of the medulla (anterior branches), especially in its superior part. Other branches of the VA extend to the lateral surface of the medulla (lateral branches) especially to its inferior and middle portion. Anastomoses, irregular in number, are present between these superficial vessels and branches of the PICA and BA typically on the surface of the olive (see also Sects. 6.2.1.2 and 6.2.2.2). From the superficial arteries arise perforators, which penetrate the medulla supplying it. These branches are not recognizable on the angiogram.

The perforating branches are basically "end arteries" without possibility of a collateral circulation. This is a common feature for all perforators supplying the brainstem which are progressively described in the next chapters.

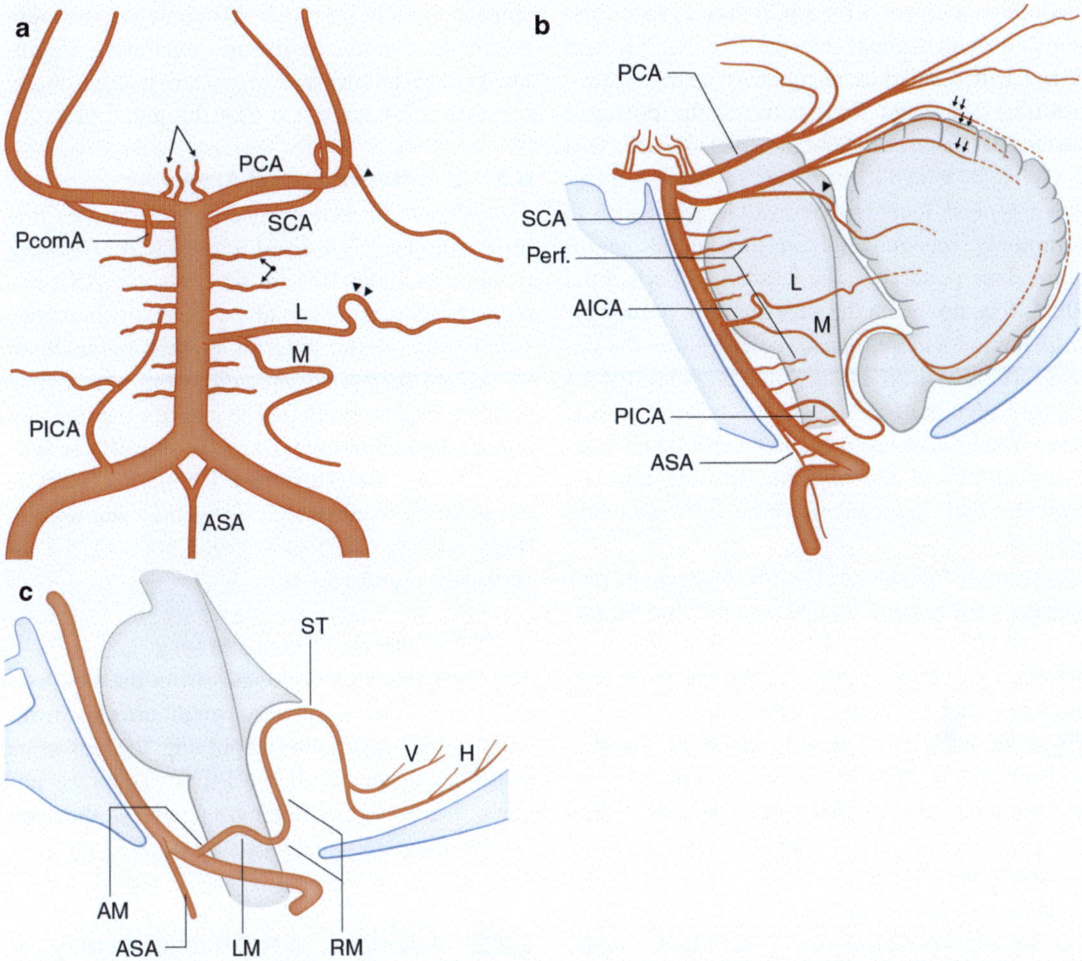

Fig. 6.5 (**a**) Drawing showing the intracranial vertebrobasilar sector in the AP projection. The two vertebral arteries unite in the basilar artery. Anterior spinal artery (ASA). Posterior inferior cerebellar artery (PICA) replacing on the right the absent AICA. Lateral branch (L) and medial branch (M) of the AICA replacing on the left the absent PICA. Loop of the lateral branch in front of the meatus acusticus (*arrowshead*). Superior cerebellar artery (SCA). Marginal branch (*arrowhead*). Lateropontine arteries (*arrows with angle and dot*). Posterior cerebral artery (PCA), posterior communicating artery (PcomA). Posterior thalamoperforating branches (*arrows with angle*). (**b**) Lateral view. Perforators for brainstem (*arrow dividing into two branches*). SCA with its vermian and hemispheric branches (*arrows*). Marginal branch (*arrowhead*). Pial anastomoses between PICA, AICA, and SCA (*dashed line*). Other indications as in (**a**). (**c**) Course of the PICA as seen on the lateral angiogram. Anterior medullary segment (AM), lateral medullary segment (LM), retromedullary segment (RM), supratonsillar segment (ST), vermian (V), hemispheric branches (H), and anterior spinal artery (ASA)

6.2.1.2 Posterior Inferior Cerebellar Artery

The posterior inferior cerebellar artery (PICA) arises bilaterally normally from the first intracranial, intradural segment of the VA. It can have, however, an extracranial origin below the foramen magnum at the C1 level (Margolis and Newton 1974; Fine et al. 1999). It can also rarely arise more proximally at the C1–C2 levels either from a normal VA or from a VA with an anomalous course (see Sect. 6.2.4). It runs anterolaterally to the *medulla*, reaching its posterior border after a variable course, which can be very tortuous, sometimes with a loop which can extend inferiorly to the foramen magnum. In this first part of its course (anterior AM and lateral LM medullary segments) the PICA has a close relationship with cranial nerves IX, X, XI, and XII. Further, it takes an upward course (retromedullary segment RM) between the posterior surface of the *medulla* and the anterior surface of the *tonsilla*, which it surrounds superiorly, forming a typical loop (supratonsillar segment ST) commonly recognizable on the lateral angiogram. This pattern can be less evident when the PICA does not reach the superior pole of the *tonsilla*, having a more inferior course along the lateral surface of the *tonsilla*. Small branches for the choroid plexus of the IV ventricle arise at this level. The PICA runs then downwards on the posterior surface of the *tonsilla*, dividing into the vermian and hemispheric branches (Drawing 6.5a–c).

Vascular Territories. The PICA gives off perforators arising from its LM segment for the *lateral medulla*. There is a balance between these branches and those of the VA which commonly predominate (Duvernoy 1999). Furthermore the PICA supplies *the posterior medulla* together with the PSA, through perforators arising from its RM segment. The distal branches supply the *tonsilla*, inferior vermis, and inferior surface of the cerebellar hemisphere.

Variants. The anomalous origin of the PICA has already been described. The PICA can be well developed and also partially supply the contralateral hemisphere. It can be hypoplastic or absent and replaced partially or completely by a large anterior inferior cerebellar artery. Among other very rare anomalies there are cases of fenestration of PICA as described by some authors (Theodopoulos and Lawton 2000; Lesley 2008; Kumar et al. 2012). As far as it concerns its anomalous origin from ICA/ECA see Sect. 2.3.

On the angiogram, the PICA and its segments can be better studied on the lateral view. The first segments (anterior medullary and lateral medullary) can be characterized by a tortuous course. Commonly easier to identify are the retromedullary and supratonsillar segments, which surround the *tonsilla* and the distal vermian and hemispheric branches. On the AP view, the different segments cannot easily be separately distinguished. Examples are presented in Figs. 6.2b, 6.3, 6.6, 6.7, 6.8a, 6.9, 6.10, 6.11, and 6.12.

6.2.1.3 Anterior Spinal Artery

The ASA is formed by two small branches that arise from each VA distal to the PICA (Drawing 6.5a–c). In about 10% of the cases, the ASA has a unilateral origin. The artery gives off branches which run on the anterior median-paramedian surface of the *inferior medulla*, from which arise perforators supplying it. The anterior median and paramedian *superior part of the medulla* is supplied by the VA (Duvernoy 1999). The ASA is frequently recognizable on the angiogram (Figs. 6.2a, b, 6.3, 6.8a, b, 6.9, 6.10, 6.11, 15.33, 16.9, 16.11, and 16.14).

6.2.1.4 Posterior Spinal Artery

The PSA arises on both sides from the PICA or distal VA. They are very small arteries from which arise perforators supplying the posterior medulla together with the PICA. They are not recognizable on the angiogram, unless they are involved in pathological processes (Fig. 6.8b).

6.2.2 Branches of the Basilar Artery

6.2.2.1 Median-Paramedian Perforators

The median-paramedian perforators are small branches which arise from the posterior surface of the basilar artery (Drawing 6.5b). The majority enter the pons at the level of the basilar sulcus. Others adopt first a transverse path before entering the pons. Some of these arteries' perforating branches are long reaching the floor of the fourth ventricle; others are more lateral and shorter. They supply the median, paramedian, and partially the lateral part of the pons (Kaplan and Ford 1966; Tatu et al. 2001; Duvernoy 1999). These very fine branches are not recognizable on the angiogram.

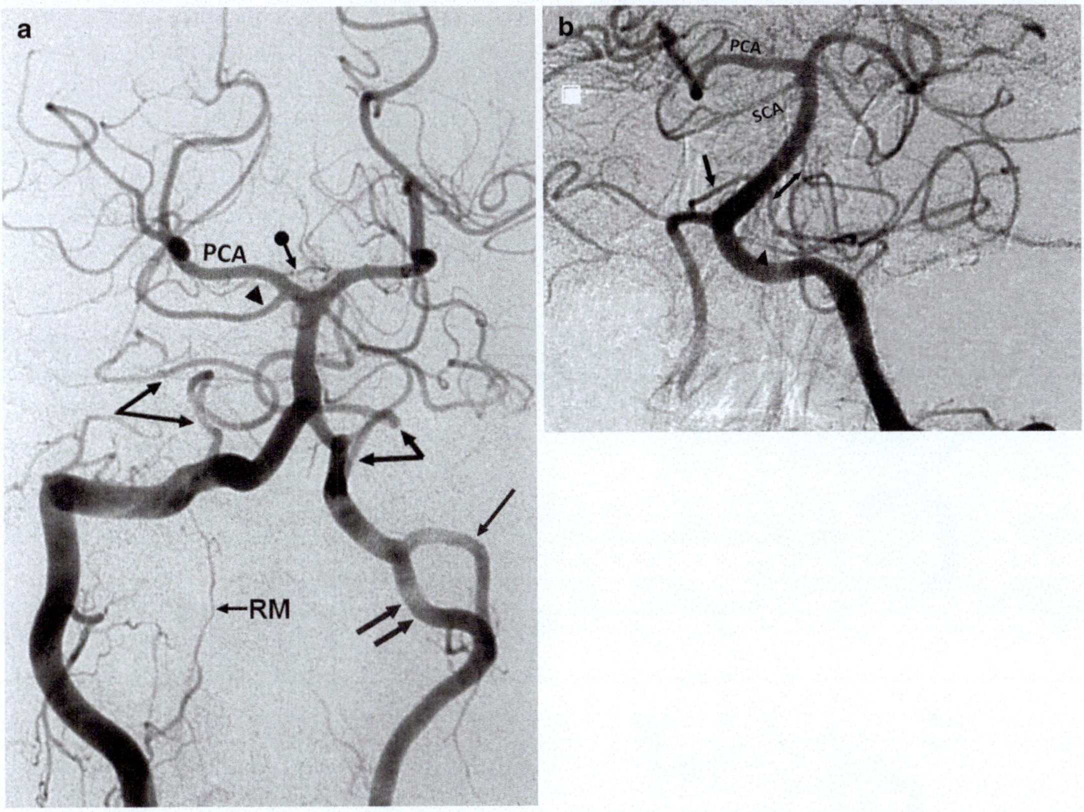

Fig. 6.6 (**a**) Right vertebral angiogram with retrograde filling of the left vertebral artery. The latter is duplicated. The true artery is smaller (*arrow*) than the anomalous branch (*double arrow*). The two branches fuse together continuing in the basilar artery. On the right and on the left there is a well-developed PICA (*arrow with angle*) supplying also the territory of the hypoplastic AICA. From the P1 of the right PCA arises a large posterior thalamo-perforating branch (*arrow with dot*) dividing distally into small branches. The right SCA arises from the right P1 (*arrowhead*). Presence on the right of a radiculomeningeal branch (*white arrow*). (**b**) Left VA angiogram with retrograde filling of the hypoplastic right VA. On the left, the PICA (*arrowhead*) and two AICAs (*bidirectional arrow*) are present. On the right there is a well-developed AICA (*arrow*) supplying the territory of the absent PICA. Superior cerebellar (SCA) and posterior cerebral (PCA) arteries

6.2.2.2 Lateral Pontine Branches

The lateral pontine branches arise from the basilar artery and run laterally towards the middle cerebellar peduncle (MCP). They give off perforators that supply the lateral pons and MCP (Drawing 6.5a, b). A few lateral pontine branches arising from the more proximal segment of the BA extend also to the lateral superior part of the medulla anastomosing with the corresponding branches of VA and PICA especially on the surface of the olive (Akar et al. 1995; Duvernoy 1999). The lateral pontine branches are frequently well visible on the AP angiogram. Example are presented in Figs. 6.9b, 6.11a, 12.13, and 12.14.

6.2.2.3 Anterior Inferior Cerebellar Artery

The anterior inferior cerebellar artery (AICA) arises from the basilar artery, commonly from its first or second segment (Scialfa et al. 1976; Naidich et al. 1976). It runs laterally with a straight course towards the lateral inferior part of

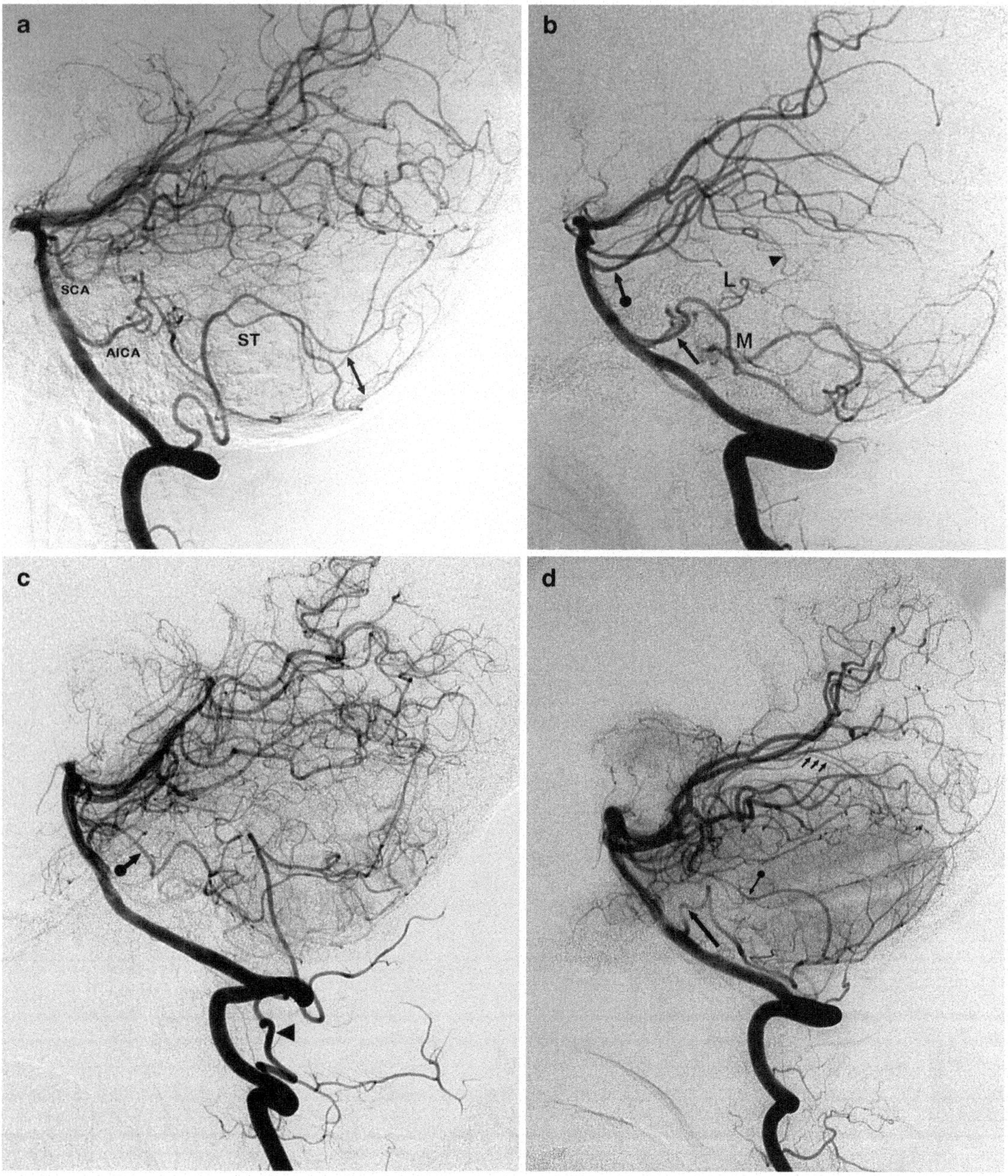

Fig. 6.7 (**a**) Lateral vertebral angiogram. Tortuous first segment of the PICA. Typical supratonsillar loop (ST). Vermian and hemispheric branches (*bidirectional arrow*). AICA. SCA. (**b**) The PICA is absent bilaterally. It is replaced by two well-developed AICAs (*arrow*). The lateral branch (L) is small, while the medial (M) is well developed. SCA (*arrow with dot*). Marginal branch (*arrowhead*) of the SCA. (**c**) Origin of the PICA from C2 (*arrowhead*). The vascular territory of the PICA is partially supplied by the medial branch of the AICA (*arrow with dot*). Well-developed musculocutaneous branches at the C1 and C2 levels, which seem in this case to arise from the anomalous PICA. (**d**) The PICA is absent. Its vascular territory is replaced by a well-developed AICA (*arrow*). Cerebellar branches of the SCA (*arrows*), forming the typical convex curve. Well-developed marginal branch (*arrow with dot*)

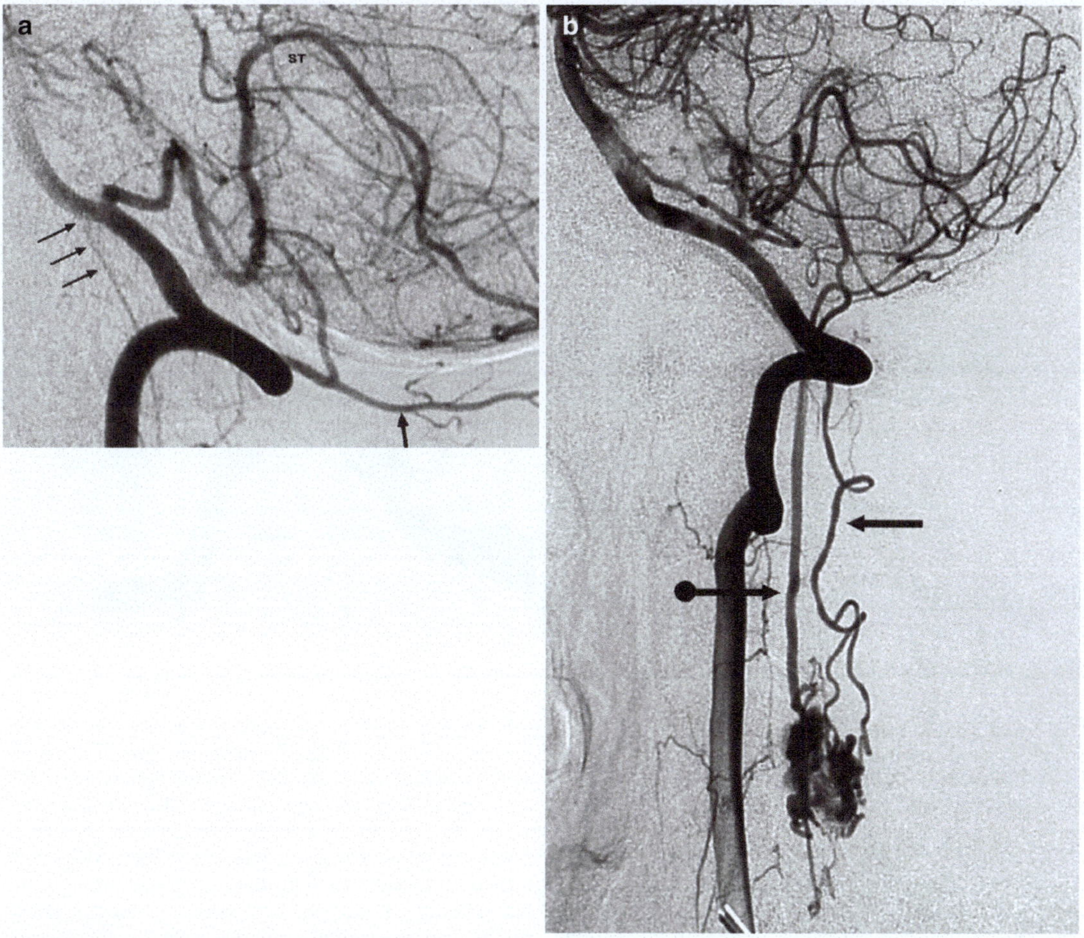

Fig. 6.8 (**a**) Lateral vertebral angiogram (detail). PICA with its typical supratonsillar loop (ST). ASA (*arrows*). Muscular branch at the level of C1, which can potentially anastomose with the corresponding branch of the occipital artery (*arrow*). (**b**) Patient with arteriovenous malformation involving the cervical spinal cord. Lateral vertebral angiogram showing the supply by a very dilated ASA (*arrow with dot*) and PSA (*arrow*)

the pons and the MCP. In its course, the artery crosses the VI cranial nerve; shortly after, it divides into two major branches: the rostrolateral (RL) and the caudomedial (CM) branches (Naidich et al. 1976) (Drawing 6.5a, b). Anterior to the flocculus, the lateral branch reaches the VII and VIII cranial nerves. Here, it describes a rather complex loop, which can partially extend into the meatus acusticus. In this part of its course, the lateral branch gives off the labyrinthine artery. The latter can rarely arise directly from the basilar artery (Naidich et al. 1976; Brunsteins and Ferreri 1990). The lateral branch runs further towards the horizontal fissure of the cerebellum, where it anastomoses with the marginal branch of the superior cerebellar artery. One of these branches can predominate and replace completely the other. The CM branch is frequently smaller. It is directed to the anterior medial surface of the cerebellum. Sometimes, it can arise as a separate trunk from the BA and it can be well developed, supplying partially or completely the vascular territory of the PICA when this is hypo-aplastic.

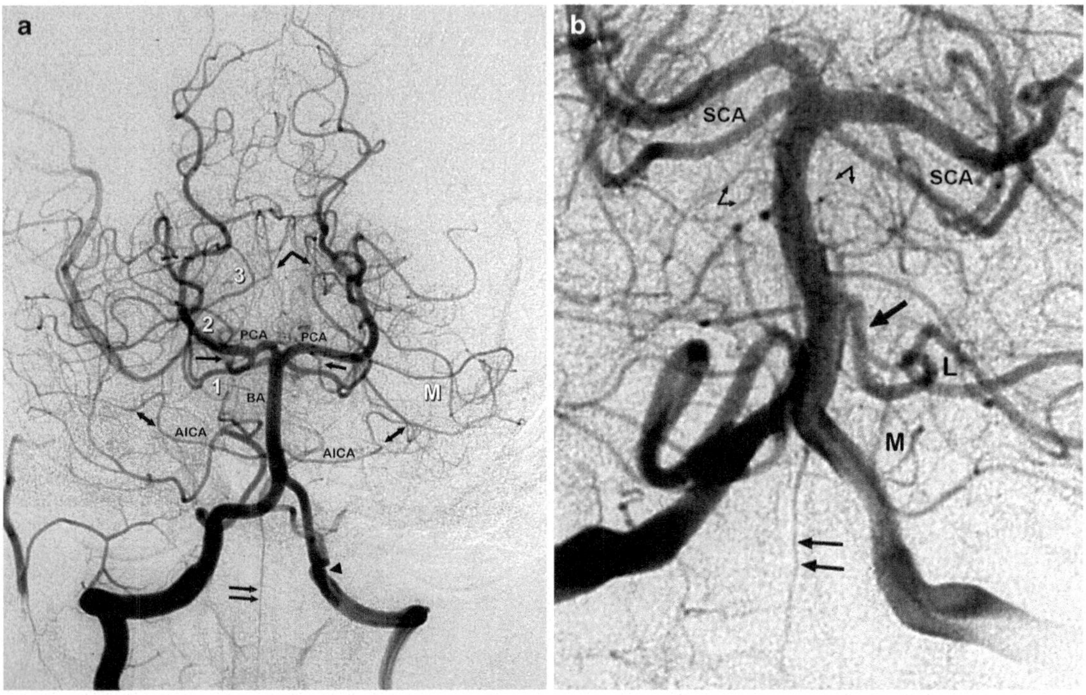

Fig. 6.9 Vertebral angiograms in two different patients. AP view. (**a**) Asymmetric vertebral artery. Basilar artery (BA). Both PICAs are well developed. That on the left probably has an extracranial origin (*arrowhead*). Unilateral origin of ASA (*double arrow*). Bilateral AICA. Note the loop in front of the internal meatus acusticus (*bidirectional arrows*). The numbers *1*, *2*, and *3* indicate, respectively, the pontine, midbrain, and quadrigeminal segments of the SCA. The latter converge to each other (*arrow with angle*) continuing in the terminal vermian branches. Large marginal branch of the SCA (M). From both PCA retrograde injection of the PcomA (*arrow*) which is well developed on the right. This allows a precise identification of the P1. On the right there is a C1 anastomosis with the occipital artery. (**b**) Bilateral voluminous VA. A well-developed AICA bilaterally replaces the absent PICA. The *arrow* indicates the left AICA with its lateral branch (L) and medial branch (M) extending to the territory of the PICA. Superior cerebellar artery (SCA). The lateral pontine arteries of the BA (*arrow with angle*). Unilateral ASA (*double arrow*) arising from the left VA

Vascular Territories. The AICA gives off perforators that supply the lateral part of the pons, the MCP, and sometimes can also contribute to supply the lateral superior part of the medulla in balance with the lateral pontine branches of the BA (Akar et al. 1995; Duvernoy 1999) (see also Sect. 6.2.2.2). The distal branches supply the flocculus and the adjacent anterior parts of the cerebellum. The labyrinthine artery supplies the VII and VIII cranial nerves.

Variants. The sizes of the AICA and PICA are closely correlated. When one is well developed the other can be small or absent. In an angiographic study (Takahashi et al. 1968), in 48% of the cases both the AICA and PICA were well developed. In 40% of cases, one or both AICAs were dominant, replacing partially or completely the PICA. In 10% of cases, the AICA was hypoplastic or absent, and its vascular territory was replaced by a well-developed PICA. In addition to the main trunk of the AICA, an accessory AICA with a separate origin from the BA may be present (Naidich et al. 1976).

Angiogram. The artery is easily recognizable on the lateral and particularly on the AP vertebral angiogram. In this latter view, the lateral branch runs towards the internal meatus acusticus, where it forms a loop which can frequently be clearly identified. At this level arises the labyrinthine

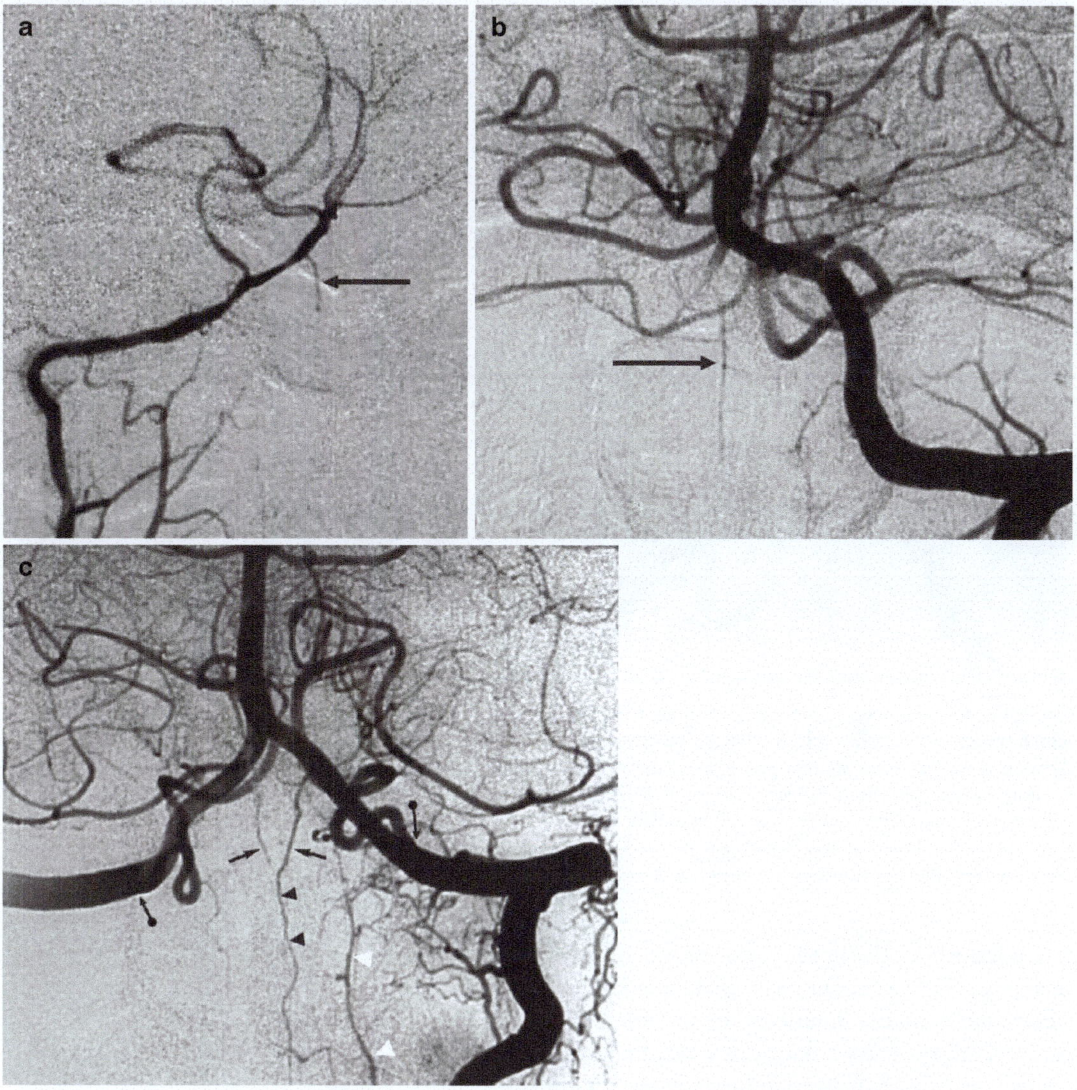

Fig. 6.10 Left and right (**a**, **b**) vertebral angiograms in the same patient. (**a**) Right hypoplastic VA, from which arises the ASA (*arrow*). (**b**) On the angiogram of the voluminous left VA, there is a partial retrograde injection of the right VA with filling of the ASA (*arrow*). (**c**) Left VA angiogram with retrograde injection of the right VA. ASA (*arrowheads*) with its bilateral contributions (*arrows*) from both VAs. Radiculomeningeal artery (*white arrowheads*)

artery, which however cannot commonly be identified. More distal branches extend to the anterior surface of the cerebellar hemisphere. Angiographic examples are presented in Figs. 6.3, 6.6, 6.7, 6.9, 6.11, 6.12, 6.13, 6.14a, b, 7.8a, 12.14, 12.15, and 12.16. The medial branch is particularly well visible when it replaces the hypo-aplastic PICA (Figs. 6.7, 6.9b, 6.12, 6.14, 15.42, 16.11, and 16.13).

6.2.2.4 Superior Cerebellar Artery

The SCA is the most constant among the cerebellar arteries and is almost always present. It arises as a unique trunk, which divides into superior and inferior branches. It can be double or rarely triple (Mani et al. 1968; Hardy and Rhoton 1978). An origin from the P1 may also occur. From its origin, the SCA runs laterally on the surface of the superior part of the pons (*pontine segment*) and

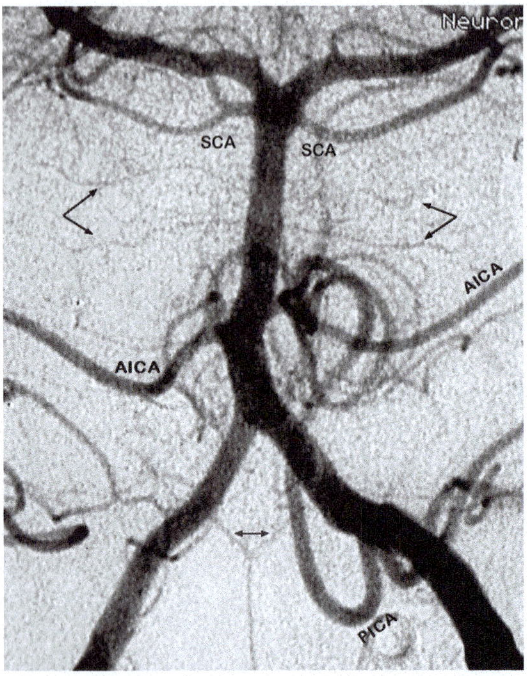

Fig. 6.11 VA angiogram, AP view. Dominant left VA, which gives off a large PICA. There are two well-developed AICAs. That on the right replaces the vascular territory of the right PICA. Bilateral origin of ASA (*bidirectional arrow*). The lateral pontine branches arising from the basilar artery are well visible (*arrow with angle*). Note on the right the early division of the SCA

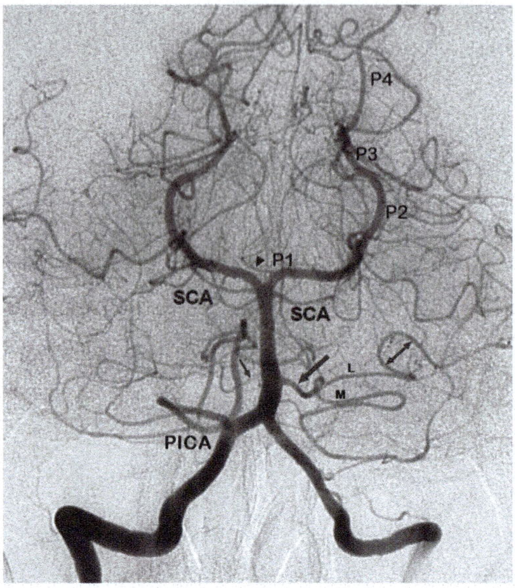

Fig. 6.12 Vertebral angiogram. AP view. Right VA angiogram with reflux in the smaller left VA. On the right, a well-developed PICA replaces mainly the vascular territory of the small AICA (*small arrow*). On the left, there is a well-developed AICA (*arrow*) supplying the vascular territory of the absent PICA. Lateral (L) and medial (M) branches of the left AICA. Note the typical loop of the lateral branch in front of the meatus acusticus (*bidirectional arrow*). The SCA is double on the right. The four segments of the PCA (P1–P4) show their typical course. A large posterior thalamo-perforating branch arising from the left P1 is recognizable (*arrowhead*)

lies inferiorly to the third cranial nerve. It turns then posteriorly and courses around the midbrain (*midbrain segment*) inferiorly to the posterior cerebral artery. The trunk or one of its branches can make a loop caudally coming in contact with the trigeminal nerve (Hardy and Rhoton 1978). It reaches further the quadrigeminal plate (*quadrigeminal segment*), where both the right and left SCA approach to each other and continue each in the vermis branch which is considered the terminal branch (Drawing 6.5a, b).

Branches. In its course around the midbrain, the SCA gives off small branches which supply the superior-posterior part of the pons and midbrain and branches for the cerebellar hemisphere. The first among these latter is the marginal artery, which runs laterally towards the horizontal fissure and supplies the corresponding cerebellar area, in balance with the lateral branch of the AICA. One of these arteries can predominate. In its course around the

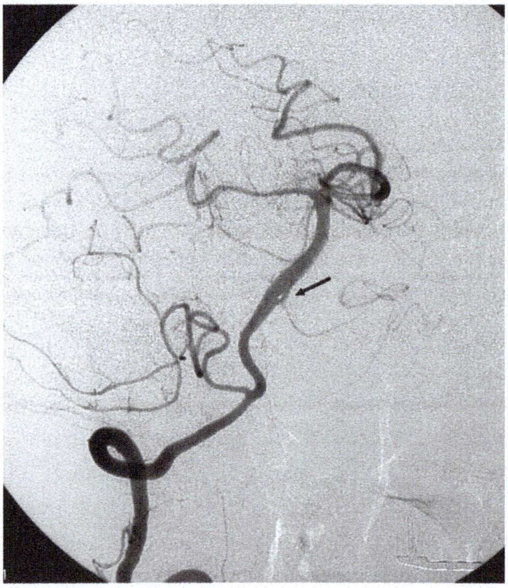

Fig. 6.13 Fenestration (*arrow*) of the basilar artery near the origin of the AICA

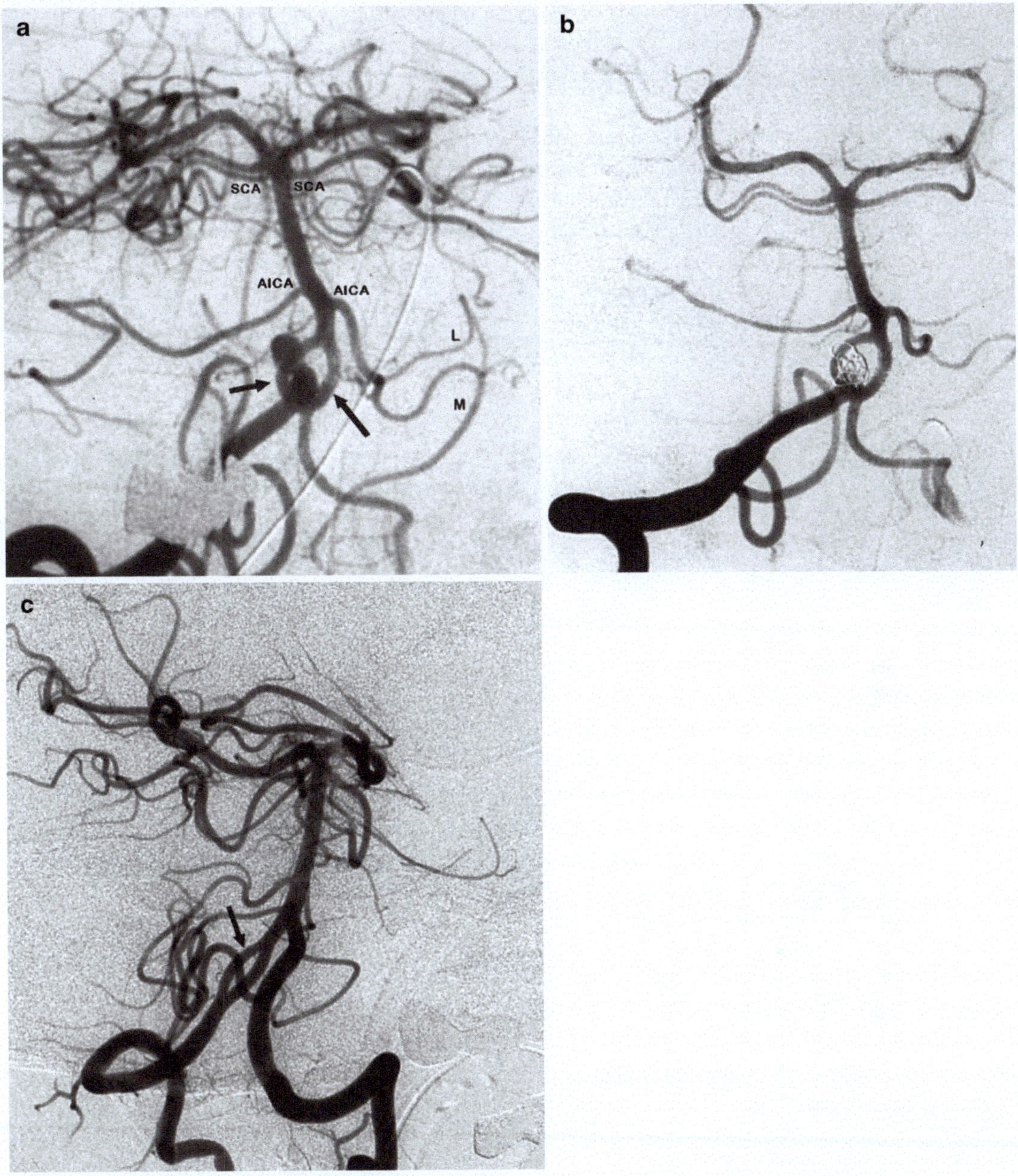

Fig. 6.14 (**a**) Angiogram in a duplicated proximal basilar artery (*arrows*) associated with ruptured aneurysm treated acutely with coils (**b**). There is asymmetry of the vertebral arteries: that of the left is hypoplastic. Bilateral well-developed AICA. That on the left replaces the absent PICA. Lateral branch (L), medial branch (M), and superior cerebellar arteries (SCA) duplicated on the right. One arises from the basilar artery and the other has a common origin with the P1. Duplication (**c**) of the vertebral artery (*arrow*) in another patient. Left vertebral angiogram with reflux in the right duplicated vertebral artery

midbrain, the SCA lies inferiorly to the superior surface of the cerebellum; so to reach it, the branches run first upward, towards the convexity, and then downward on the surface of the cerebellum. The terminal branch is the artery for the superior vermis. The vermis branches of both side approach to each other, running almost parallel.

From the midbrain segment arises a *meningeal artery*, sending small branches to the posterior part of the tentorium (Wollschlaeger and Wollschlaeger 1974).

Vascular Territories. Each artery supplies the posterosuperior part of pons and midbrain. It can also contribute to the supply of the posterolateral midbrain in balance with perforators arising directly from the posterior cerebral artery (PCA), the collicular artery, and the posteromedial choroidal artery (Duvernoy 1999). The distal branches supply the superior vermis and superior cerebellar hemisphere.

Variants. The SCA can arise uni- or bilaterally from the P1 segment of the posterior cerebral artery (Figs. 6.6a and 15.44b). When the SCA is duplicated, one branch can arise from the BA and the other from the P1 (Fig. 6.14b).

On the angiogram. The pontine, midbrain, and quadrigeminal segments are easily recognizable on the AP view (Figs. 6.9, 6.11, 6.12, 6.14, and 12.13). The midbrain segment is identified on the lateral view running inferiorly to the posterior cerebral artery (Figs. 6.7 and 7.9). Because of the overlapping, there can be some difficulties to distinguish the cerebellar branches from those of the PCA. The cerebellar arteries form a superior convex curve with a stepladder appearance. The vermis branches are the superior ones (Figs. 6.7 and 7.9. The marginal branch, when well developed, is easy to identify due to its course downwards to the cerebellar horizontal fissure (Fig. 6.7). On the AP view the cerebellar branches, due to their location below the tentorium, run inferiorly to the temporal branches of the PCA (Figs. 6.9 and 7.5).

6.2.3 Distal Branches of PICA, AICA, and SCA

The angioarchitecture of the distal branches of the cerebellar arteries is similar to that already described at the level of the arteries of the cerebral hemispheres (De Reuck 1972). These arteries run on the surface of the cerebellar cortex connected to each other distally in the so-called border zone through leptomeningeal anastomoses. From the superficial arteries arise with a perpendicular course perforating branches supplying the cortex and branches (medullary arteries) extending towards the white matter and deep nuclei. The penetrating branches are *end arteries* without the possibility of a collateral circulation.

6.2.4 Variants of Vertebral and Basilar Arteries

There is a frequent asymmetry of the size of the VAs. The left VA is dominant in the majority of cases; less frequently the right VA is larger. One VA may be hypoplastic, ending intracranially in the PICA. In cases of unilateral ASA, the presence of a hypoplastic VA does not exclude that the ASA takes its origin from this artery (Fig. 6.10a, b).

Other rare conditions are *duplication and fenestration*. These can be found also in the anterior circulation (Sects. 2.3, 4.3, 5.3, and 7.5) but are reported to be more frequent in the vertebrobasilar sector (Uchino et al. 2012b), involving the extra- and intracranial segments of the vertebral and the basilar artery. The definition of fenestration or duplication as reported in the literature is not always homogeneous. This is due partially to the different terminology used by the authors for the same finding and to the difficulty in some cases to make a differential diagnosis between these two conditions. To simplify matter, from a practical point of view, we define *duplication* as occurring when two clearly separated vessels are present. In the case of *fenestration* the vessel is unique but on one part of its course there is a double lumen.

In this context it is useful to recall some aspects of the embryological development of the vertebrobasilar system (Schmeidel 1932; Padget 1948, 1954) (see also Chap. 1). Each VA is formed by metameric cervical intersegmental arteries arising from the upper part of the primitive dorsal aorta connected with a longitudinal plexiform structure. Later the metameric arteries regress with the exception of that from which arises the vertebral artery, commonly the sixth (Padget 1954). The longitudinal plexiform structure evolves bilaterally forming the final vertebral artery connected proximally through the sixth metameric artery with the subclavian artery and distally with the basilar artery.

As far as it concerns the numbering of the intersegmental arteries, we follow that proposed by Padget (1954). According to this author, the intersegmental artery accompanying the first cervical nerve should not be numbered in the cervical series, since it does not run between two cervical segments. Instead it runs between the occipital bone and the first cervical segment, which is the atlas. It should be called suboccipital intersegmental or proatlantal intersegmental artery, which normally regresses and partially becomes the transverse suboccipital part of the VA (see also Chap. 2, Sect. 2.3). The first cervical intersegmental artery is that running between the first and second cervical segments. *Therefore the intersegmental artery from which develops the stem of the vertebral artery and the subclavian artery is the sixth running between the sixth and seventh vertebral bodies.*

The persistence together with the normal VA of another intersegmental metameric artery explains probably the duplication of the extracranial vertebral artery. The most frequent duplication is at the C1–C2 levels due to persistence of the first intersegmental artery (Fig. 6.6a). The anomalous artery enters the spinal canal joining intracranially the normal VA. It can also occur that only the anomalous vessel persists, while the normal VA does not develop. Another frequent pattern, commonly on the left, is the presence of the normal VA, arising from the subclavian artery and entering the transverse foramen at the C6 level with, in addition, a second VA with an abnormal origin and course (Fig. 6.15). The latter

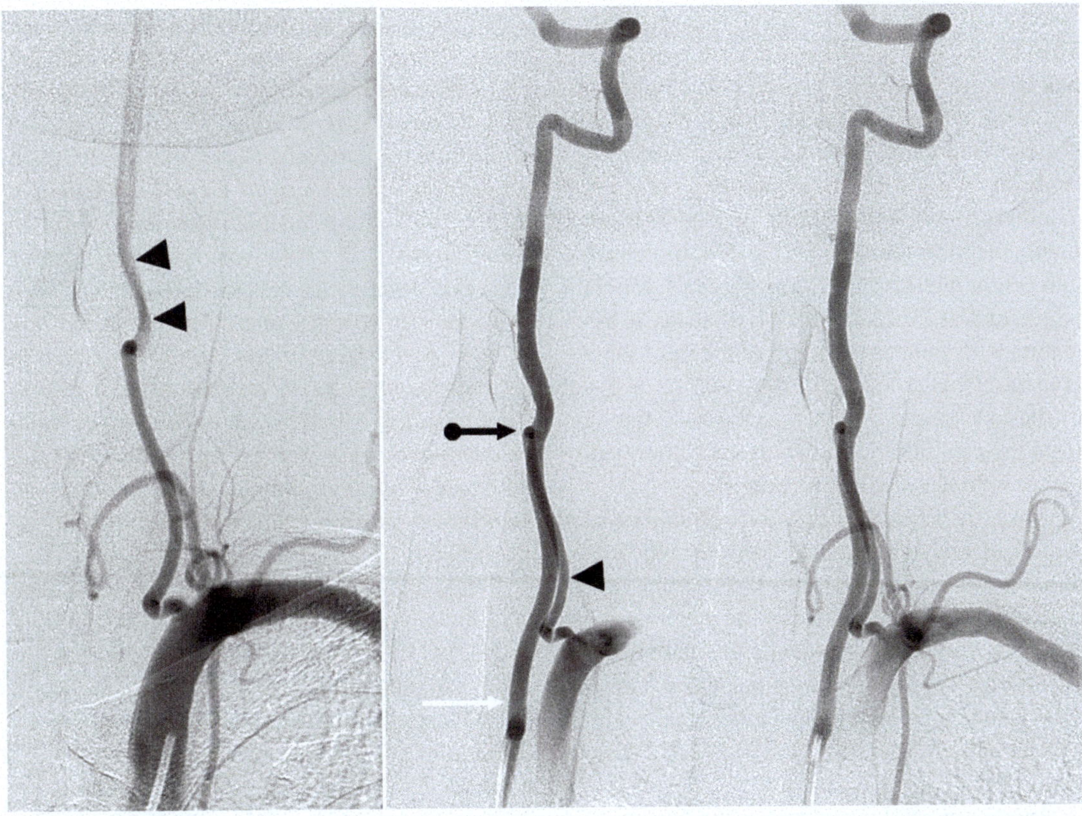

Fig. 6.15 Incidental discovery of a duplicated left VA. (**a**) Left subclavian angiogram showing the normal origin of the left VA. More distally, the contrast medium is more diluted (*arrowheads*). This is due to the presence of a second more voluminous VA having an abnormal origin from the aortic arch. (**b**) Selective angiogram of this second VA (*arrow*), which joins distally (*arrow with dot*) the VA with a normal course and origin (*arrowhead*) which is retrograde filled as well as the subclavian artery. Courtesy of Dr. Gozzoli and Dr. Boghi, Neuroradiology of Cuneo Hospital

is thought to be due to the persistence of the fifth intersegmental artery (Goddard et al. 2001). The abnormal VA arises from the aortic arch proximal to the subclavian artery (SA) and enters the fourth or fifth transverse foramen (TF). Both the normal and abnormal VAs fuse distally together. The abnormal VA can also develop from the seventh intersegmental artery, arising in this case from the aortic arch, distal to the SA and entering the seventh TF. Duplication can occur less frequently also on the right. A bilateral duplication has also been reported (Ionete and Omojola 2006).

Other anomalies of the VA are the single anomalous origin from the aortic arch, commonly on the left occurring in about 6% of the cases. The origin can be proximal or distal to the left subclavian artery, as described above, with entering the fifth–fourth or seventh transverse foramen, respectively (Figs. 1.6 and 15.28). Anomalous origin of the right VA is more rare. This can occur from the aortic arch more proximal between the right subclavian and right common carotid arteries, or more distally between the left common carotid and left subclavian arteries or distal to the left subclavian artery (*vertebral arteria lusoria*). The entering of the TF is commonly the fifth–fourth in the cases of more proximal origin and seventh in the cases of vertebral arteria lusoria. The latter occurs typically in association with anomalous origin of the right subclavian (*subclavian arteria lusoria*) close or distal to the left subclavian artery. Origin of the VAs from the common carotid artery more frequently on the right has also been reported.

A very rare anomaly is the thoracic origin of the VA, arising as a common trunk with the upper thoracic intercostal arteries. It has been interpreted as due to persistence in the upper thoracic region of a connection between the intersegmental arteries supplying the spinal canal and the embryological longitudinal plexus from which develops the VA (Chiras et al. 1982; Stoesslein et al. 1982).

Another anomalous pattern is the fenestration of the extracranial VA, probably due to the partial persistence of the longitudinal plexiform structure (Fig. 6.16).

There is a rich literature dedicated to duplication, fenestration, or anomalous origin and course of the VAs, either in the distal segment of the VA, commonly at the C1–C2 levels (Rieger and Huber 1983; Tokuda et al. 1985; Hashimoto et al. 1987; Takahashi et al. 2003; Uchino et al. 2013b), or at the origin of the VA (Kemmetmueller 1911; Adachi 1928; Suzuki et al. 1978; Eisenberg et al. 1986; Takasato et al. 1992; Lemke et al. 1999; Lasjaunias et al. 2001; Goddard et al. 2001; Ionete and Omojola 2006; Satti et al. 2007; Karcaaltincaba et al. 2009; Dabus and Walker 2010; Meila et al. 2012; Uchino et al. 2013b).

Anomalies of the VAs should always be taken into account especially in patients in whom surgery involving the aortic arch, anterior neck, or craniovertebral junction is planned (*Tokuda* et al. 1985*; Matula* et al. 1997*; Lemke* et al. 1999*; Albayram* et al. 2002*; Takahashi* et al. 2003*; Uchino* et al. 2013b). *In particular, the demonstration of an anomalous right subclavian and anomalous right VA arising distally to the left subclavian artery is important in mediastinal and esophageal surgery* (*Lacout* et al. 2012). *A duplication of the VA at the C1–C2 levels should also be taken into consideration in patients, in whom a puncture of the subarachnoid space is planned. Anomalous cervical course has been considered to be a risk factor favoring dissection of the artery* (*Jackson* et al. 2000). *Also in patients having suffered a stroke in the vertebrobasilar sector, in which one or both VAs cannot be demonstrated in their typical site of origin on the routine angiographic examination, the possibility of an anomalous origin should be considered.*

Fenestration and duplication can be found also in the intracranial VA and BA (Takahashi et al. 1973; Wollschlaeger and Wollschlaeger 1974; Lasjaunias et al. 2001; Tanaka et al. 2006). In MR angiographic study of 3327 patients (Uchino et al. 2012c) this anomaly was diagnosed in 2.77% of the cases involving the intradural segment of the VA, the vertebrobasilar junction, and more frequently the basilar artery in its proximal and middle segments. The anomaly is probably due to failure of the bilateral

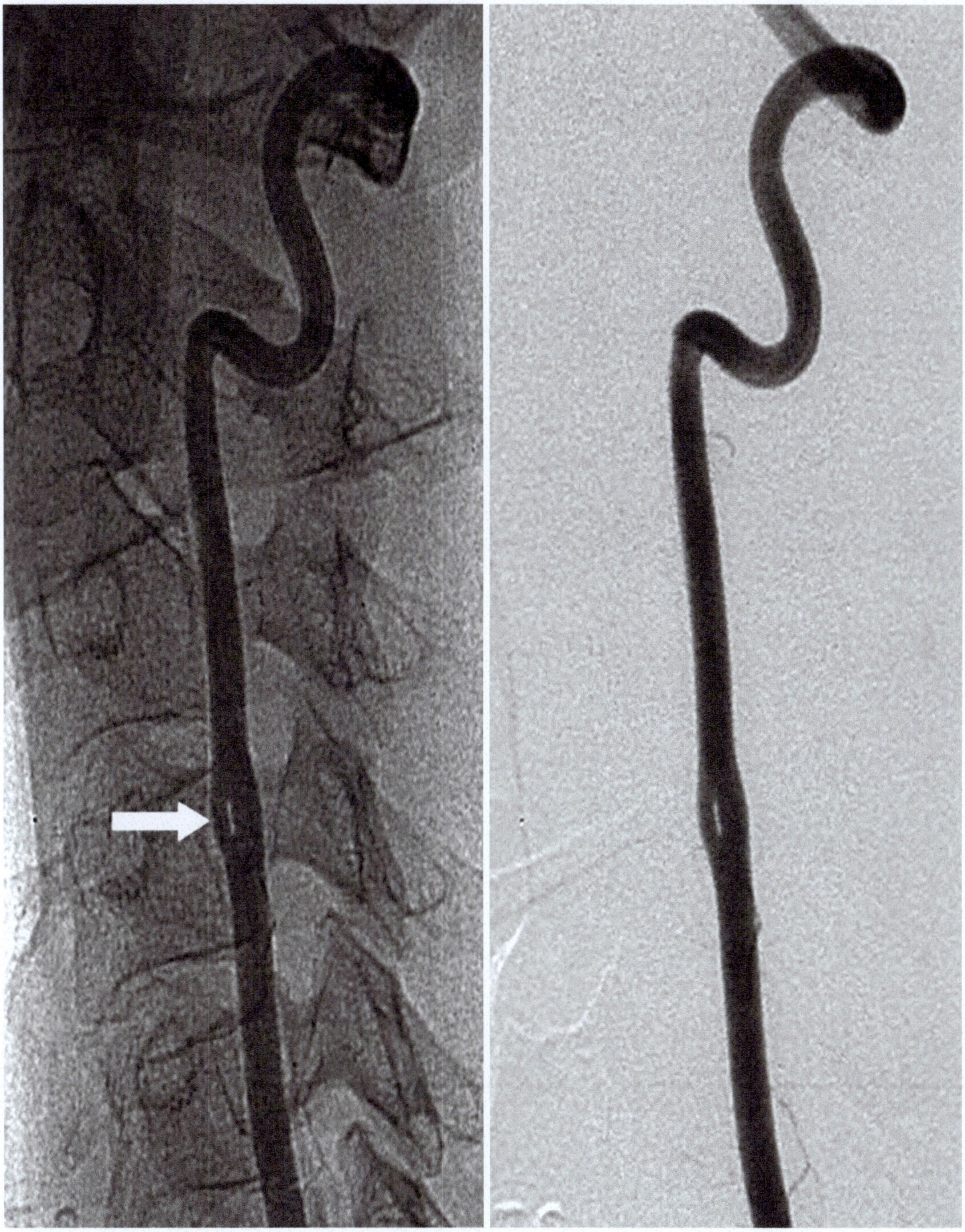

Fig. 6.16 Incidental discovery of fenestration of the vertebral artery (*arrow*) in its extracranial course

plexiform longitudinal channel to develop into a single vessel (Figs. 6.13 and 6.14) (see also embryogenesis, Chap. 1). Aneurysms at the site of the anomaly but also elsewhere are frequent (Campos et al. 1987; Picard et al. 1993; Tasker and Byrne 1997; Uchino et al. 2012b; Krings et al. 2007b; Stark et al. 2013; Trivelato et al. 2016) (Fig. 6.14a, b).

Other anomalies are that characterized by failure of fusion of the two primitive longitudinal channels presenting in the form of two BA due to the persistence of the two primitive longitudinal channels (Goldstein et al. 1999; Ho et al. 2004). In the cases of partial fusion the pattern is that of an apparently short BA. In the caudal failure of fusion the angiographic pattern is that of a higher vertebrobasilar junction. In the cranial failure the pattern is of an apparently lower origin of the PCA. From the distal unfused segments of the BA the SCAs are recognizable.

A segmental aplasia of the BA (Lasjaunias et al. 1979; Ricolfi et al. 1996; Hoh et al. 2004; Burger et al. 2007; Caranci et al. 2012; Bradac 2014) has also been described. In this last condition the VAs join together continuing in the proximal segment of the BA, which is not connected with its distal segment. The latter can be fed like in the first phases of the embryogenesis by the well-developed PcomA. In other cases the BA has no connections neither with the PcomA nor with the VAs. It appears as an isolated trunk supplied by the primitive trigeminal artery. The persistence of the trigeminal artery associated probably to impairment in the development of the VAs at the level of the intersegmental arteries leads to a reduction of the flow from the VAs towards the BA leading to a progressive interruption between these arteries.

A case of segmental aplasia is presented in Fig. 6.17.

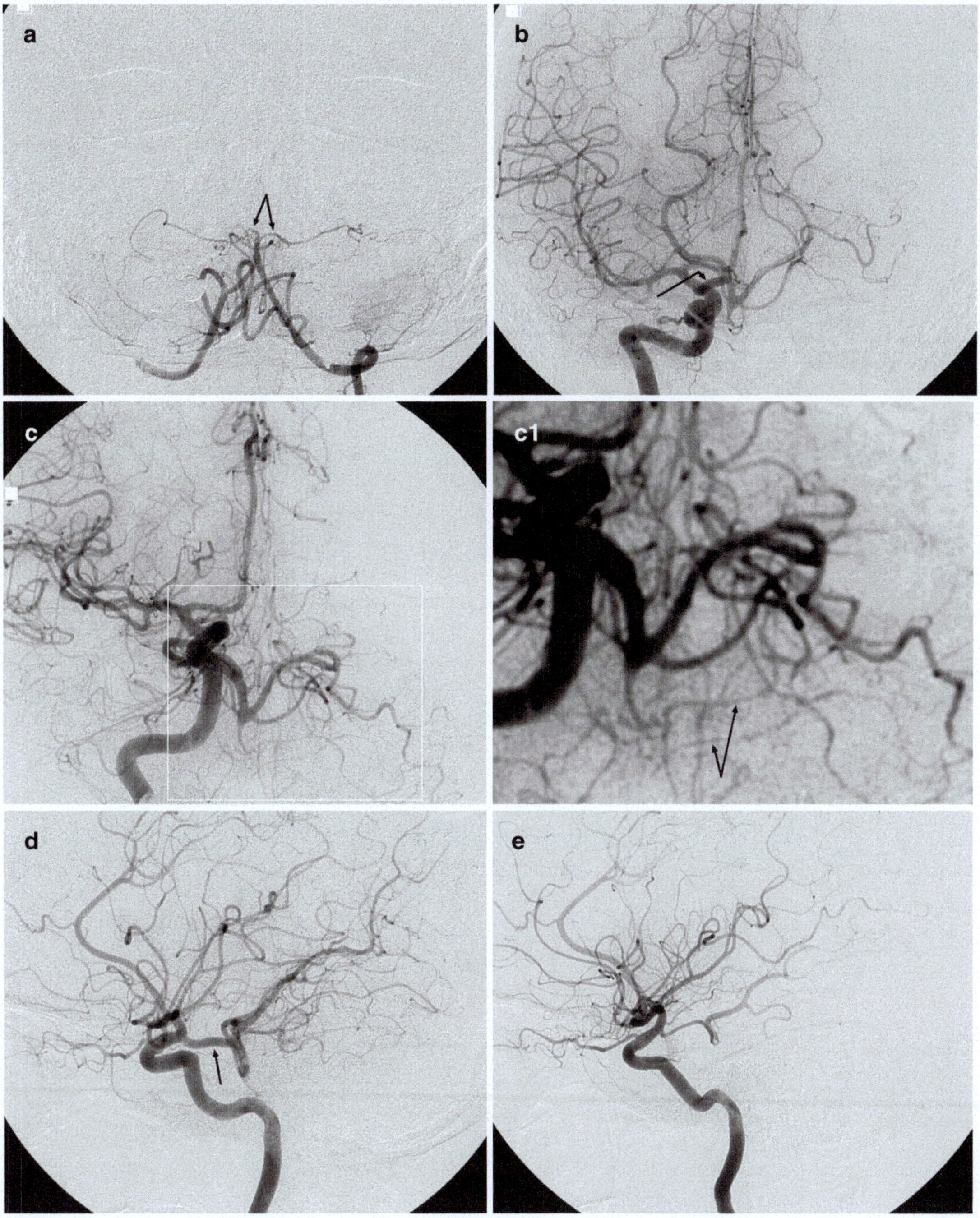

Fig. 6.17 Incidental discovery of a segmental aplasia of the basilar artery in a middle-aged hypertensive patient. The routine MR examination was normal apart from a few micro-ischemic lesions in the white matter of the cerebral hemispheres. An unexpected interruption of the flow in the middle basilar artery was visible on MR angiography that led to conventional angiography. (**a**) Left vertebral angiogram with retrograde filling of the right VA. Both VAs were small. They gave off two well-developed PICAs. The BA was small, ending in the AICAs (*arrow with angle*). (**b**) Right ICA angiogram. AP view. A well-developed PcomA (*arrow*) is connected with the P1. Through this connection, there is a filling of the distal, short BA from which arise with an ascending course both PCAs and SCAs. (**c**) Right ICA angiogram, modified AP view, better showing the filling of the distal BA and of its distal branches. (C1), magnified detail showing the filling of the lateral pontine branches (*arrow with angle*). (**d, e**) Right and left ICA angiograms. Lateral view, showing the PcomA connected with the BA and PCAs

7 Posterior Cerebral Artery

From the caudal division of ICA arises the PcomA from which posteriorly develops progressively the PCA, and anteriorly and caudally a connection with BA. Later in evolution in the majority of the cases the connection of the PcomA with the PCA (also called pars carotica) regresses or becomes hypoplastic, while in the connection of the BA with the PCA (also called pars basilaris) the future P1 becomes predominant (see also Chap. 1 and Sect. 2.2.3.2). The posterior shift of the vascular territories of the PCA (posterior temporal and occipital areas) leads to increased distance between the origin of the PCA and its distal vascular territories. The decrease of this distance by shifting the origin of the PCA from ICA to the BA could be interpreted as a natural evolution to compensate this process (Kier 1974).

- The course of the PCA arising from the BA can be divided into four segments (Huber 1979) (Drawings 7.1 and 7.4)
- P1 (precommunicating segment): from its origin from the BA to the junction with the PcomA
- P2 (ambient segment): around the midbrain towards the quadrigeminal plate
- P3 (quadrigeminal segment): on the surface of the quadrigeminal plate
- P4 (distal segments)

7.1 P1 Segment

This is the first segment, extending from its origin at the basilar artery to its junction with the PcomA (Drawings 7.1 and 7.4 and Figs. 7.2 and 7.6). The P1 segment (pars basilaris of the PCA) is a short segment, with an average length of 7.1 mm (Zeal and Rhoton 1978). It runs in the interpeduncular fossa, where it has a close relationship with the commonly inferiorly located cranial nerve III. The P1 can have a horizontal or a slightly ascending or descending course.

The P1 segment gives off perforating branches called *thalamoperforating pedicles* (Foix and Hillemand 1925a, b) or *paramedian thalamic arteries* (Percheron 1976a, b). It seems to be more simple to call these arteries *posterior thalamoperforating branches*, while the *anterior thalamoperforating branches* arise from PcomA. They enter the posterior perforated substance (PPS), supplying the medial part of the mesencephalon, the medial part of the thalamus, and the posterior part of the hypothalamus (Lazorthes and Salamon 1971; Saeki and Rhoton 1977; Zeal and Rhoton 1978; Duvernoy 1999; Tatu et al. 2001).

They can also take over the supply of the anterior part of the thalamus, normally supplied by the PcomA, when this is absent (Percheron 1976a, b). On the contrary when P1 is absent, its

G.B. Bradac, *Applied Cerebral Angiography*, DOI 10.1007/978-3-319-57228-4_7

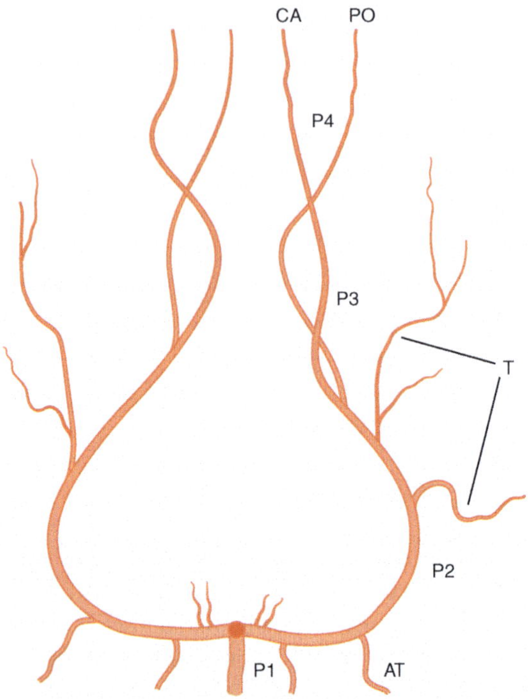

Fig. 7.1 Drawing. Course of the P1, P2, P3, and P4 segments of the posterior cerebral artery (PCA) in the AP view. Anterior temporal artery (AT), middle temporal and temporo-occipital branches (T), calcarine artery (CA), and parieto-occipital artery (PA)

vascular territory can be taken over by the PcomA or by branches arising from the contralateral P1.

The posterior thalamoperforating arteries commonly consist of several small branches, although not rarely a larger trunk, further dividing into small branches, may predominate. They can arise entirely only from one P1, as previously described (Westberg 1966; Percheron 1976a, b; Saeki and Rhoton 1977; Zeal and Rhoton 1978; Castaigne et al. 1981) and also documented more recently (Brassier et al. 1988; Lazzaro et al. 2010). Furthermore, even in cases of hypoplasia of the P1, perforators may arise from it (Zeal and Rhoton 1978). All these variants probably explain the different patterns of midbrain and thalamic infarction in cases of proximal occlusion of the P1. Examples are presented in Figs. 7.8a–c, 6.6, and 6.12.

Another branch which arises from the P1 and sometimes from the P2 is the *collicular* (*quadrigeminal*) *artery*, which runs around the midbrain contributing to the supply of its lateral and pos-

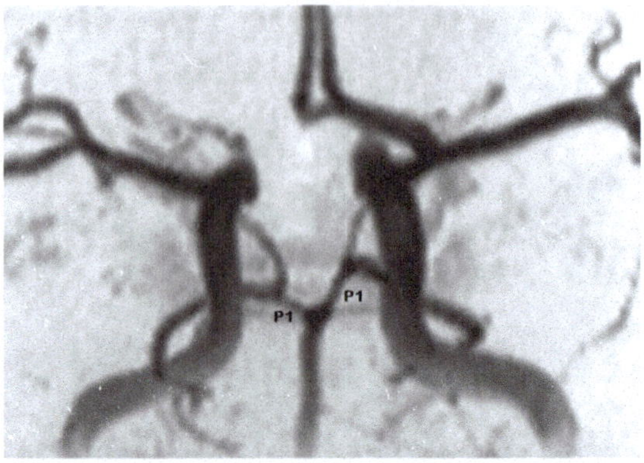

Fig. 7.2 Angio-MRI. The P1 segments are well defined owing to the presence of the posterior communicating artery (PcomA) bilaterally

terior part. There is frequently a second smaller artery, running close and parallel to the collicular artery, called the accessory collicular artery (Zeal and Rhoton 1978; Duvernoy 1999; Tatu et al. 2001).

From the P1 and sometimes from the P2 or the parieto-occipital branch arises the posterior medial choroidal artery, which, after reaching the quadrigeminal plate and giving off branches for the posterior thalamus, courses forward in the roof of the third ventricle running together with the internal cerebral vein towards the foramen of Monro. It supplies the choroid plexus of the third ventricle (Galloway and Greitz 1960; Wackenheim and Braun 1970; Margolis et al. 1974; Fujii et al. 1980) (Figs. 7.3 and 7.4).

Near the origin of the P1 arises *a meningeal branch*, which has been described by Wollschlaeger and Wollschlaeger (1965), who termed it the *artery of Davidoff and Schechter*, in honor of their former teachers. It runs first parallel and inferiorly to the PCA, and then dips under the notch of the tentorium running further medially and posteriorly towards the region where the tentorium and falx join together, vascularizing these dural structures. This meningeal branch is not rare but it is very fine and so not recognizable on the normal angiogram (Wollschlaeger and Wollschlaeger 1965). It can be involved in pathological processes of the region, especially in meningiomas and dural arteriovenous fistula (Weinstein et al. 1974) (Figs. 13.16, 13.17, and 13.18).

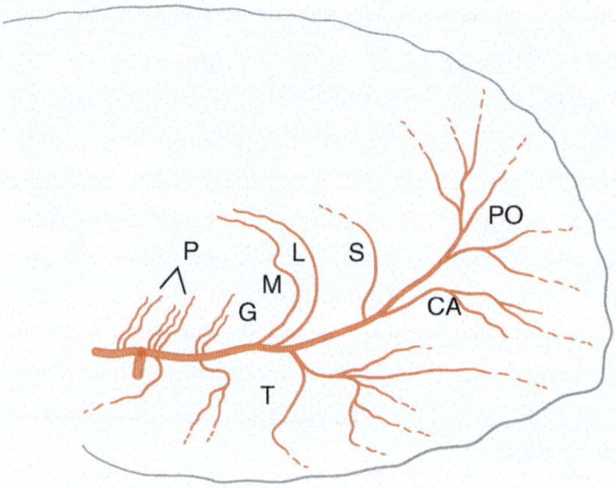

Fig. 7.3 Drawing. Course of the PCA and its branches in the lateral view. Anterior from PcomA and posterior from P1 thalamoperforating arteries (p), thalamogeniculate artery (G), posterior choroidal arteries, medial (M), lateral (L), splenial artery (S), parieto-occipital arteries (PO), calcarine artery (CA), temporal arteries (T), anastomoses indicated by *dashed lines* with the anterior cerebral artery (ACA) and middle cerebral artery (MCA)

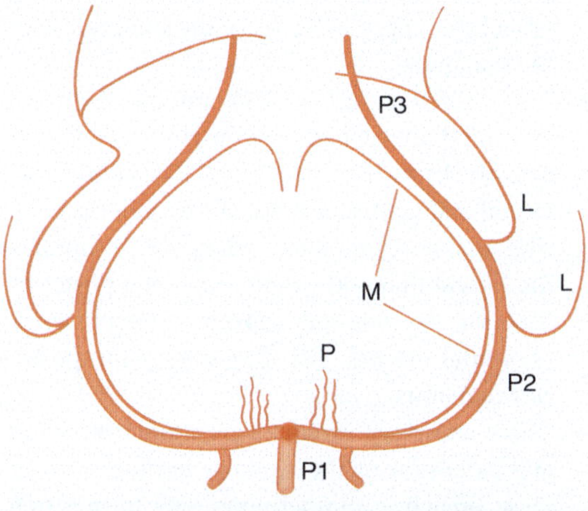

Fig. 7.4 Course of the P1, P2, and P3 segments of the PCA in the AP view. Posterior thalamoperforating branches (P), medial (M), and lateral (L) posterior choroidal arteries

7.2 P2 Segment

Also called the circumpeduncular segment, the P2 segment runs around the midbrain (Drawings 7.1 and 7.4), commonly separated from it by several millimeters, and terminates on the surface of the quadrigeminal plate. The P2 segment lies inferior to the optic tract and basal vein, and superior to the superior cerebellar artery. The parahippocampal gyrus lies laterally, with the free margin of the tentorium above. The P2 segment gives off the thalamogeniculate artery, which can be a single branch or present with two or three vessels supplying the lateral thalamus (Duvernoy 1999; Tatu et al. 2001). The artery can arise in the proximal or more distal P2 segment (Zeal and Rhoton 1978).

Other branches are the peduncular perforators for the lateral midbrain and the posterior lateral choroidal artery. The latter, which can sometimes arise from the parieto-occipital artery, consists of a single or several branches (Zeal and Rhoton 1978). Basically, there is an anterior branch which runs in the circumpeduncular cistern, enters the choroid fissure, and extends forwards to supply the plexus of the temporal horn anastomosing with the anterior choroidal artery (AchA). A posterior branch reaches the pulvinar, gives branches to it, and terminates in the choroid plexus of the lateral ventricle at the level of the atrium, anastomosing with branches of the AchA (Galloway and Greitz 1960; Wackenheim and Braun 1970; Margolis et al. 1974; Fujii et al. 1980; Duvernoy 1999) (Figs. 7.3 and 7.4).

From the P2 arise cortical branches for the temporal lobe. They are also called the inferior temporal arteries and supply entirely the inferior medial surface of the temporal lobe, including the hippocampus and the inferior medial surface of the occipital lobe. These cortical branches are named in relation to the site of origin and vascular territories involved: anterior, middle, and posterior (Zeal and Rhoton 1978). From the posterior temporal artery can arise a few branches supplying the primary visual cortex (Margolis et al. 1974) (Figs. 7.6 and 7.7).

7.3 P3 Segment

The short P3 segment runs in the quadrigeminal cistern on the surface of the quadrigeminal plate, approaching that of the other side at a variable distance, averaging 16 mm (Margolis et al. 1974) (Drawings 7.1 and 7.4).

7.4 P4 Segment

The P4 segment consists of the terminal branches represented by the parieto-occipital, calcarine, and splenial arteries (Drawings 7.1 and 7.3). The parieto-occipital arises either together with the calcarine artery, at the level of the calcarine fissure, or sometimes as a single trunk, more proximally from the P2 segment. The parieto-occipital artery runs posterosuperiorly on the medial surface of the hemisphere and gives off branches for the precuneus, cuneus, and, as reported by some authors (Margolis et al. 1974), also for the primary visual cortex in about 35% of the cases. The calcarine artery arises at the calcarine fissure, usually located laterally to the parieto-occipital artery. Further, it runs medially and inferiorly along the calcarine fissure, supplying the primary visual cortex.

The splenial artery, also called the posterior pericallosal artery, has a variable origin, frequently from the P3 segment or from the parieto-occipital artery. It runs around the splenium and anastomoses anteriorly with the posterior pericallosal artery branch of the pericallosal artery (Drawing 7.3).

7.5 Anatomical Variations

The P1 segment can be hypoplastic or absent. When the P1 is absent, the PCA takes its origin directly from the ICA via a large PcomA. This aspect, corresponding to the initial embryological pattern, is called the fetal origin of the PCA (Fig. 7.5a, c). It can occur uni- or bilaterally; its frequency is reported as being between 30 and 40% of cases (Zeal and Rhoton 1978; Pedroza et al. 1987). Conversely, when the P1 is well developed, the PcomA is commonly small or can be completely absent (Fig. 7.5d). Among the variants concerning the proximal segment of the PCA, duplication with two separated branches arising from the ICA or one from the ICA and the other from the basilar artery has been reported (Hoyt et al. 1974; Wollschlaeger and Wollschlaeger 1974; Bulsara et al. 2007; Uchino et al. 2016). Other studies have shown that in the majority of these cases, the duplication represents a variant of the AchA, which can be hyperplastic and take over partially or completely the vascular territory of the PCA (see Sect. 2.2.3.3 and Figs. 2.9 and 2.12). Cases of fenestration of P1 have been reported (Wollschlaeger and Wollschlaeger 1974; Uchino et al. 2016). Variants involving the posterior thalamoperforating arteries have already been described.

7.6 Vascular Territories

- The posterior thalamoperforating arteries (P1) supply the medial and anterior parts of the midbrain and medial thalamus. Frequently, a common large trunk supplies both midbrain and thalamus. In other cases several smaller branches supply selectively the midbrain and the thalamus.
- The collicular artery (P1), middle posterior choroidal artery, and small perforators arising directly from the P1–P2 segment supply the lateral and posterior parts of the midbrain.
- The thalamogeniculate artery (P2) supplies the lateral thalamus, optic tract, and posterior limb of the internal capsule. The posterior choroidal arteries (P1–P2) supply the posterior thalamus.
- There are many variants concerning the origin and extension of the vascular territories of all these branches. Furthermore, they form a rich anastomotic network on the surface of the midbrain and posterior thalamus that involves also the superior cerebellar artery and the AchA. This probably explains the absence or the different extension of ischemic lesions in occlusion of the proximal PCA distal to the P1.
- The cortical branches comprise the temporal arteries (P2) supplying the medial inferior surface of the temporal and occipital lobes, the parieto-occipital (P3) supplying the medial

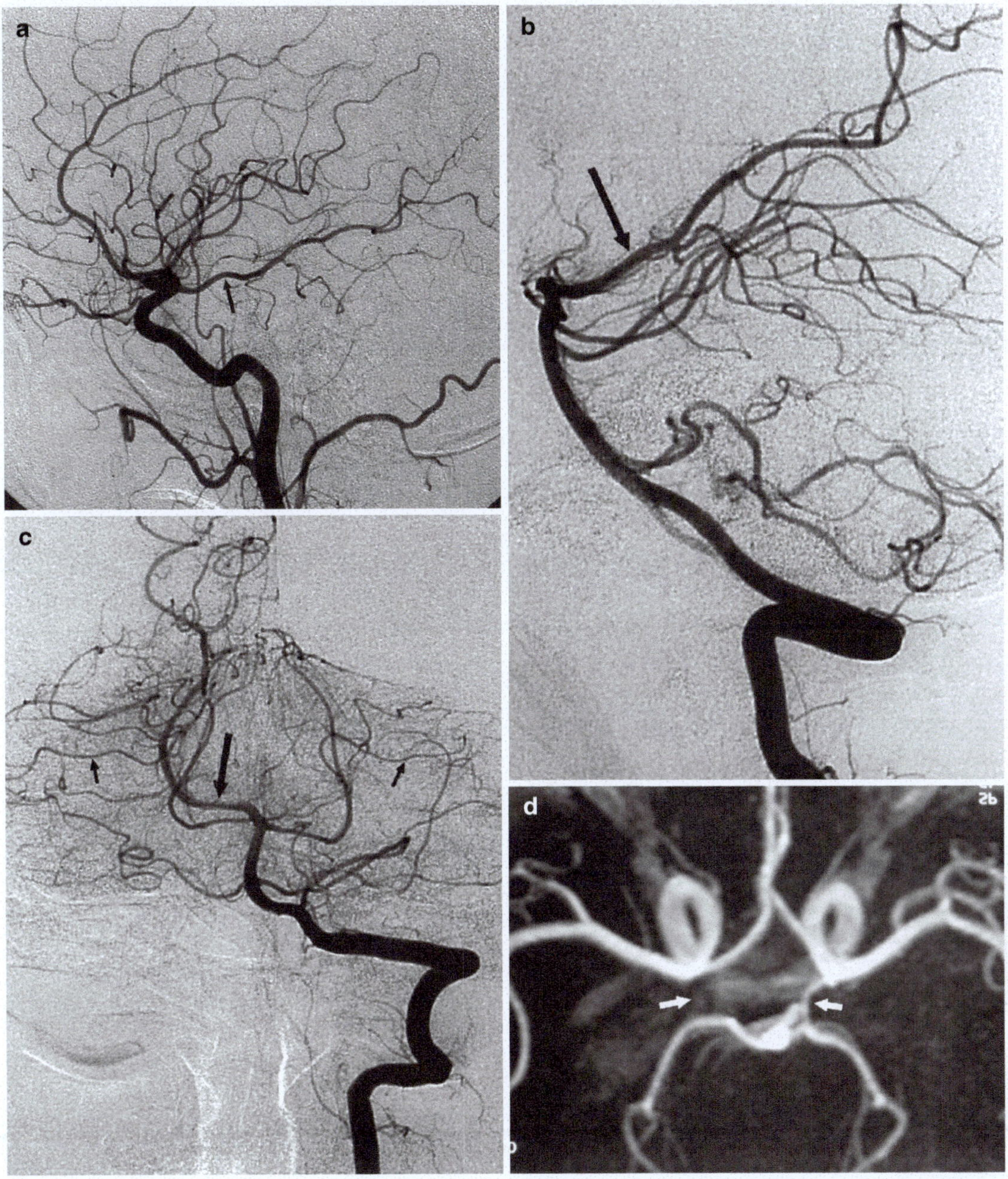

Fig. 7.5 (**a**, **b**, **c**) Angiographic study in the same patient. (**a**) Left PCA arising directly from the ICA (fetal pattern; *arrow*), (**b**) lateral vertebral angiogram showing the right PCA (*arrow*), and (**c**) vertebral angiogram, AP view. Right PCA (*arrow*) arising from the basilar artery. Hypoplastic or absent left P1. The marginal artery of the left superior cerebellar artery (*small arrow*) is easily identifiable. Owing to the minimal overlap, the right marginal artery is also well recognizable (*small arrow*). There are two well-developed AICA. (**d**) MR angiography in another patient. Left PcomA (*arrow*). On the right, absence of the PcomA (*arrow*)

surface of the hemisphere in the precuneus and cuneus areas and partially also the convexity of the posterior parietal and occipital regions, and the calcarine artery (P3) supplying *the primary visual cortex* and adjacent areas. Contributors in the supply of the primary visual cortex arise also from the parieto-occipital and temporo-posterior arteries.

- The splenial artery (P3) supplies the splenium. It can anastomose with the posterior pericallosal artery, branch of the ACA.
- Between the branches of the PCA surrounding the midbrain and running towards the thalamic area, there are many superficial anastomoses. However, the perforating branches which arise from them and penetrate the parenchyma are basically *end arteries* (Duvernoy 1999).
- Between the superficial distal branches of the PCA and those of the ACA and MCA there are potential leptomeningeal anastomoses at the temporo-occipital and parietal convexity. Leptomeningeal connections can develop also in the anterior part of temporo-medial convexity between the temporal branches of the MCA and PCA. From the superficial branches arise small arteries entering the brain parenchyma supplying cortex (cortical branches) and white matter (medullary branches), having the same angio-architecture already described in the vascular territories of ACA and MCA (Sects. 4.4 and 5.4). These arteries are *end arteries* without any possibility of collateral circulation.

7.7 Angiogram

The P1 segment is well demonstrated on the CT or MR angiography and on the AP view of the vertebral angiogram (Figs. 7.2 and 7.6). The perforators appear as small branches directed upward and somewhat posteriorly, respectively,

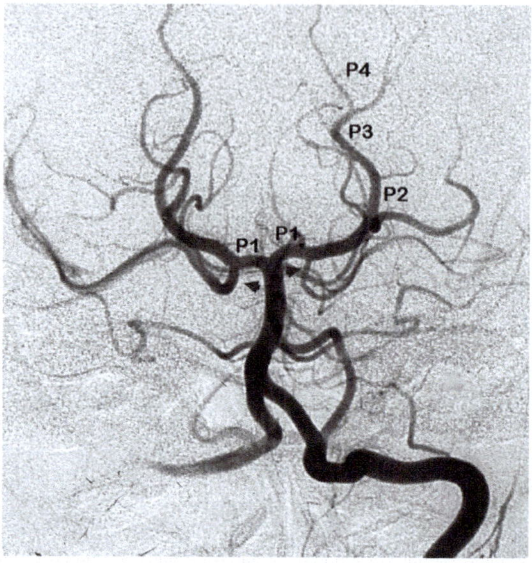

Fig. 7.6 Vertebral angiogram, AP view. The retrograde injection of the PcomA bilaterally (*arrow*) allows the precise definition of the P1 segment of the posterior cerebral artery (PCA). The P2, P3, and P4 are also shown

in the AP (Figs. 6.6, 6.12, and 7.8) and lateral view (Fig. 7.9). The P2 segment is better visualized in the AP view in its course surrounding the midbrain. In the lateral view, the segments of both sides are superimposed. Among the branches arising from the P2, the thalamogeniculate and both medial and lateral choroidal arteries are distinct on the lateral angiogram (Fig. 7.9). The thalamogeniculate arteries consist of one or more small branches running upwards and slightly posteriorly between the posterior thalamo-perforators and the choroidal arteries. The medial choroidal frequently forms a double curve with a posterior convexity, and the lateral choroidal arteries form a simple curve with a posterior convexity located posterior to the medial choroidal arteries (Wackenheim and Braun 1970) (Fig. 7.9). In Figs. 12.5, 12.7, 12.9, 12.26, and 12.27 are presented examples of involvement of the artery in arteriovenous malformation.

The lateral branch of the posterior choroidal artery can lead to a blush in the capillary phase visible in the AP and lateral views corresponding to the plexus of the lateral ventricle.

As far as it concerns the branches supplying the hemispheres, the anterior temporal branches, which arise from P2, are well recognized on the AP and lateral views. On the latter projection they have a course that is anterior to the basilar artery (Fig. 7.9). The more distal branches arising from the P2 and P3 segments can be easily recognized on the AP and lateral views (Figs. 7.6, 7.7, 7.8, and 7.9). A more precise identification, however, is possible on the lateral angiogram, where in particular the calcarine artery can be easily identified owing to its particular course and location between the parieto-occipital superiorly and the temporo-occipital arteries inferiorly (Fig. 7.9). On the AP view, superimposition of these branches makes their precise identification more difficult. Basically, the calcarine artery runs in the paramedian area, medially to the parieto-occipital artery. The splenial artery is easily identified on the lateral angiogram located posteriorly to the choroidal arteries (Fig. 7.9).

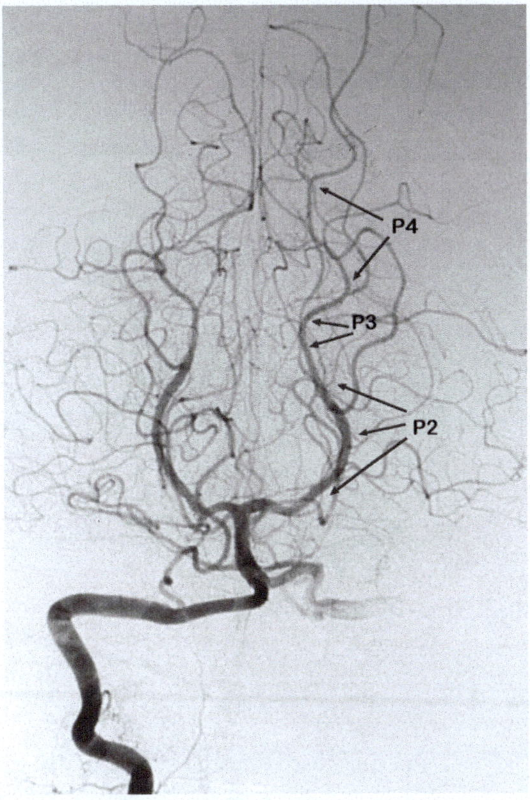

Fig. 7.7 Vertebral angiogram, AP view, showing the typical course of the PCA with its typical P2, P3, and P4 segments

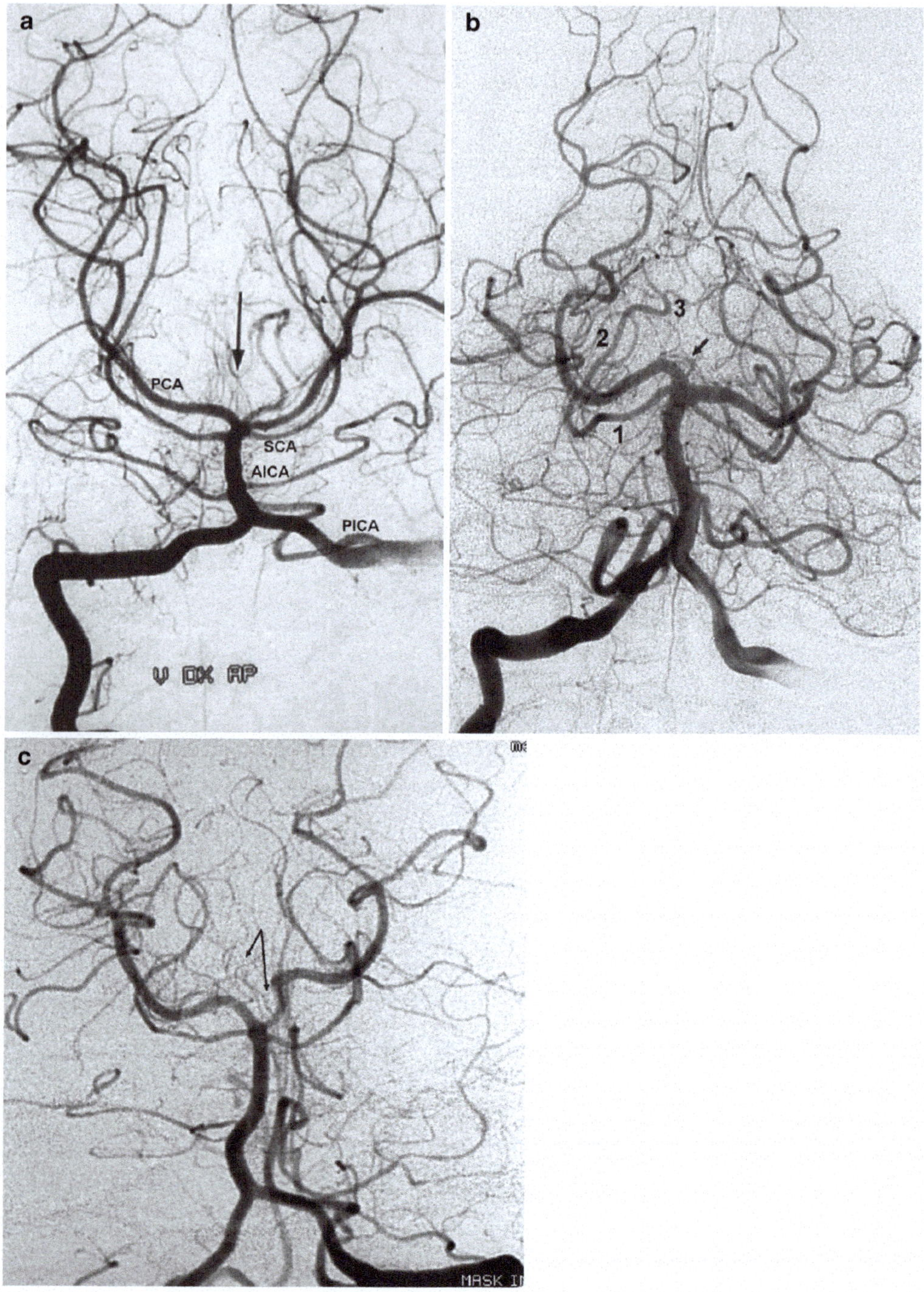

Fig. 7.8 (**a**) Vertebral angiogram, AP view. The posterior thalamoperforating branches (*arrow*) are well visible. They arise prevalently from the right P1. There is a partial overlap with distal branches of the left posterior inferior cerebellar artery (PICA). Anterior inferior cerebellar artery (AICA), superior cerebellar artery (SCA), and posterior cerebral artery (PCA). The PCA and SCA both run around the midbrain. In the AP view, the SCA is medially located. (**b**) Vertebral angiogram, AP view. A faint network of vessels corresponding to the posterior thalamoperforating artery is recognizable, in which a larger branch arising from the right P1 (*arrow*) predominates. The numbers 1, 2, and 3 correspond, respectively, to the pontine, midbrain, and quadrigeminal segments of the SCA. (**c**) Vertebral angiogram, AP view. Ascending course of the P1 owing to a short basilar artery. At least three well-developed posterior thalamoperforating arteries are recognizable: two on the left and one on the right (*double arrow*)

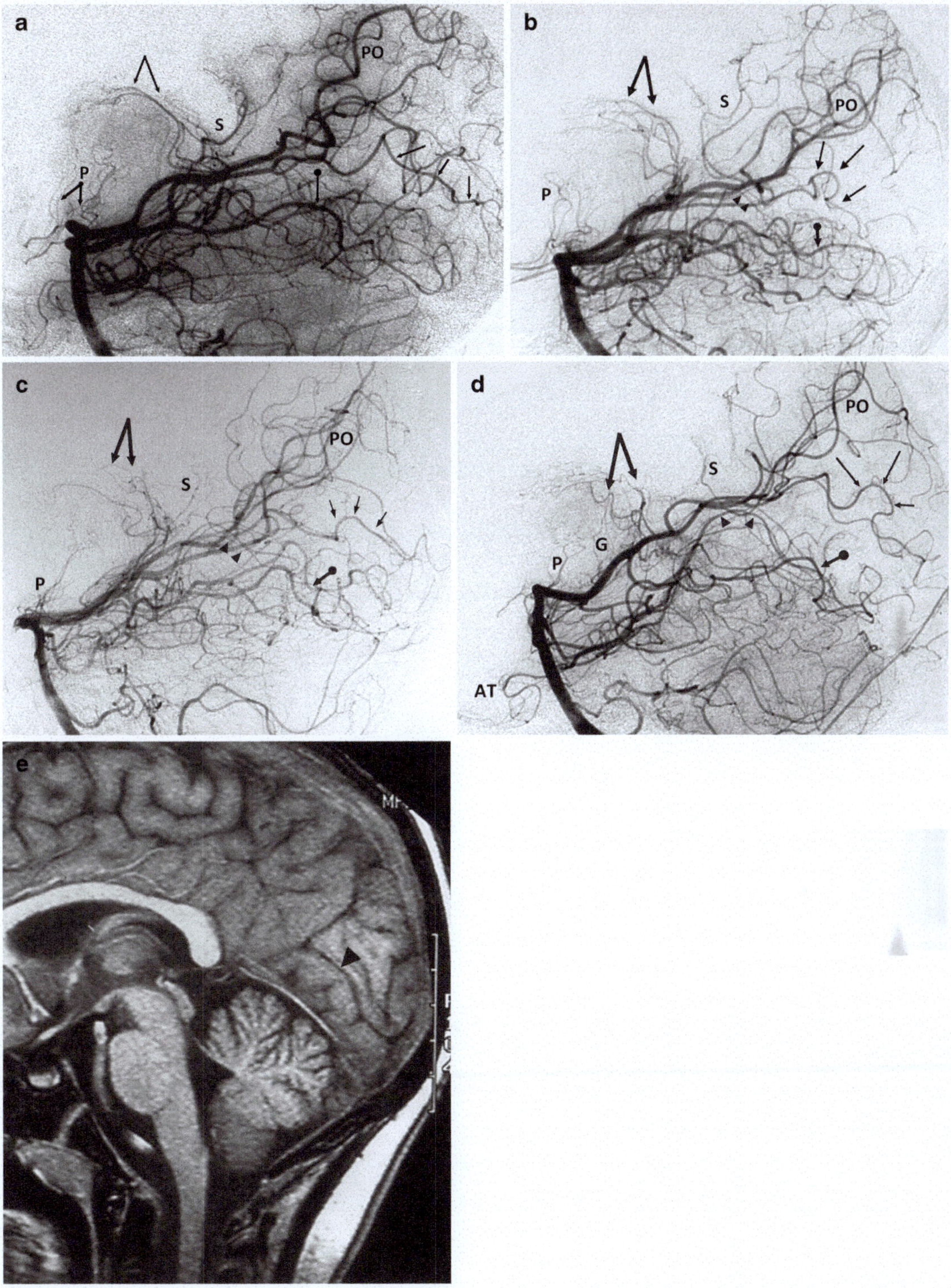

Fig. 7.9 Vertebral angiograms (**a**–**d**), lateral views in four different patients. Thalamoperforating artery arising from P1 and PcomA (P). Medial and lateral posterior choroidal arteries (*arrow with dot*). Splenial artery (S). Thalamogeniculate artery (G) recognizable in (**d**). Parieto-occipital artery (PO), calcarine artery (*arrows*), posterior temporal artery (*arrow with dot*). Anterior temporal branch recognizable in (**d**). There is overlap of the temporal branches of the PCA and the cerebellar branches of the SCA. The latter can be identified owing to their typical superiorly convex curve (*arrowheads*). (**e**) T1-weighted MR showing the typical calcarine fissure (*arrowhead*)

8 Vascular Territories

In this chapter, the vascular territories of the brain as viewed in the axial planes on CT or MRI are presented (Figs. 8.1 and 8.2). The cerebral hemispheres are vascularized by the anterior cerebral (ACA), middle cerebral (MCA), posterior cerebral (PCA), and anterior choroidal (AchA) arteries. From the proximal segments of these arteries (A1, M1, P1) and cisternal segment of the choroidal artery arise small branches (deep perforators) that enter the brain parenchyma at the level of the anterior and posterior perforated substance supplying basal ganglia, thalamus, and internal capsule. Other perforators arise from the anterior communicating (AComA) and posterior communicating (PComA) arteries, which belong to the circle of Willis, and from the distal segment of the internal carotid artery.

The perforators are *end arteries* with no possible collateral circulation. From the more distal branches of ACA, MCA, PCA, and AchA running on the surface of the cerebral hemispheres arise perforating branches supplying cortex (cortical branches) and white matter (medullary branches). Like the deep perforators, these perforators are *end arteries*.

Instead there are potential pial (leptomeningeal) anastomoses between the distal superficial branches of all the cerebral arteries at the border (*border zone*) between their vascular territories (De Reuck 1972). There is a certain degree of variability of the extent of the vascular territory of each artery leading to changes in the location of the border zone. See also Chaps. 4, 5, and 7.

In the posterior fossa, medulla and pons are vascularized by perforators arising from the vertebral basilar, and anterior and posterior spinal arteries. The midbrain is vascularized by perforators of PCA and AchA. Furthermore, perforating branches for medulla, pons, and midbrain arise from the first segments of the three cerebellar arteries (PICA, AICA, and SCA). All these perforating branches are *end arteries*.

Cerebellum is supplied by the distal branches of the three cerebellar arteries running on the surface of the cerebellar parenchyma. From the superficial branches arise perforators supplying the cortex (cortical branches) and the white matter and deep grey nuclei (medullary branches). The perforators are *end arteries*, while potential leptomeningeal anastomoses are present among the distal superficial branches of the three cerebellar arteries (De Reuck 1972). See also Chap. 6.

G.B. Bradac, *Applied Cerebral Angiography*, DOI 10.1007/978-3-319-57228-4_8

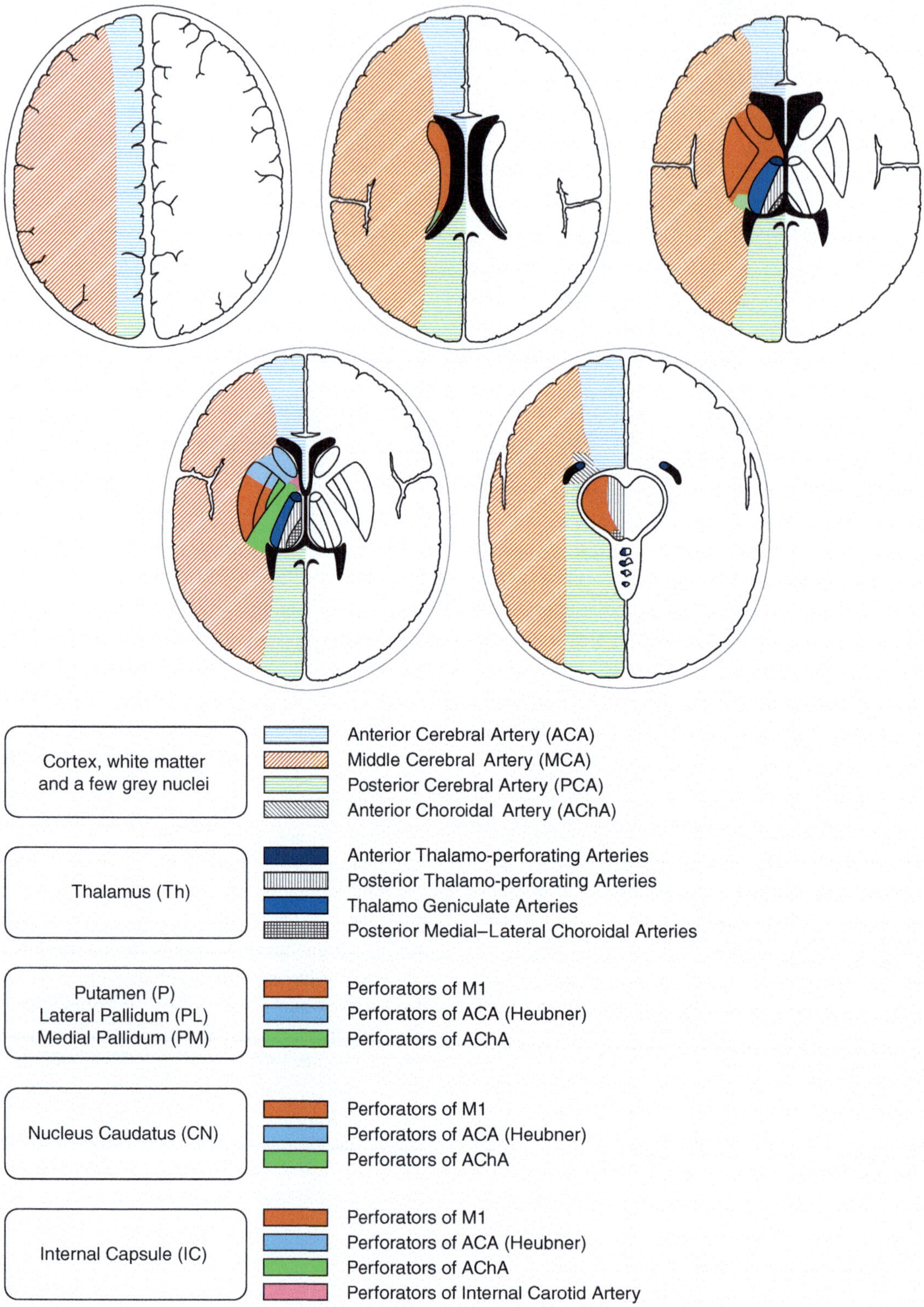

Fig. 8.1 Vascular territories of the cerebral hemispheres

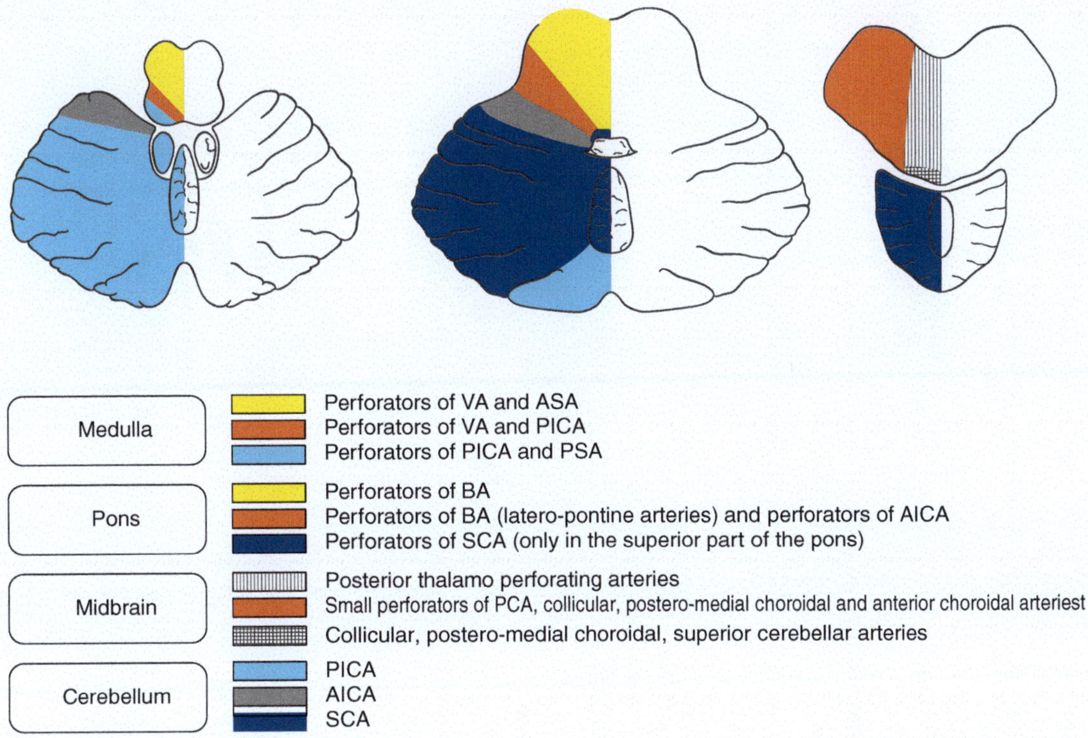

ASA anterior spinal artery
PSA posterior spinal artery
BA basilar artery
VA vertebral artery
PICA postero-inferior cerebellar artery
AICA antero-inferior cerebellar artery
SCA superior cerebellar artery

Fig. 8.2 Vascular territories of the brainstem and cerebellum, with the section at the level of the medulla, pons, and midbrain

Cerebral Veins

9

The very complex embryological evolution of the venous system of the brain has been largely studied by many authors (Evans 1912; Sabin 1917; Streeter 1918; Markowski 1922; Hochstetter 1938; Padget 1956, 1957; Kaplan and Ford 1966; Okudera et al. 1984). We summarize the essential points.

In the early stages, the venous drainage is characterized by *a superficial capillary network draining through three dural plexuses called the anterior, middle, and posterior dural stems in the paired primary head sinus* (*PHS*), *also called vena capitis lateralis, continuing posteriorly in the anterior cardinal veins* (*future internal jugular veins*). The further evolution is characterized by a progressive dwindling and disappearing of the PHS and changes in the three dural stems. From the anterior dural stem arises the *primitive marginal sinus* which extends medially connecting with that of the other side through plexiform anastomoses evolving in the formation of the *superior sagittal sinus* (*SSS*). Through the growth of the brain the SSS extends posteriorly continuing in the *torcular herophili* formed by the distal part of the SSS and the proximal *transverse sinus* derived from the proximal middle dural stem. From the distal middle dural and the posterior dural stems derive the *distal TS and the sigmoid sinus*, respectively, continuing in the internal jugular vein. Frequently, the SSS drains, as occurs in the adult, into the right transverse sinus. This asymmetric drainage has been explained by the more direct vertical drainage of the right cardinal vein (internal jugular vein) towards the heart, while that of the left has a more long course to reach the superior cava vein.

To the transverse sinus converge now two new temporary dural channels (Okudera et al. 1994; Padget 1957). The one is the so-called *pro-otic sinus* directed anteromedially. Its anterior remnant contributes to the formation of the cavernous sinus and for some authors also to the superior petrosal sinus. To the formed cavernous sinus converge anteriorly the developed supraorbital and maxillary veins (see Sect. 10.1) and it is connected posteriorly with the developed superior and inferior petrosal sinuses. The other one is the *primitive tentorial sinus more laterally located* and to which converge the lateral and medial telencephalic veins, precursors of the superficial and deep middle cerebral veins (SMCV, DMCV), respectively. Other tributaries are the diencephalic and mesencephalic veins. Following the growth of the temporal lobe the sinus is displaced medially favoring the connection of the SMCV in the developed CS. The sinus disappears, but in some cases it does not regress or its regression is only partial explaining the various forms of drainage of the SMCV, DMCV, and basal vein (see Sects. 9.1.1.2, 9.1.2.6 and 9.3.10).

To the dural sinuses converges progressively the superficial venous system, including the veins of the white matter (medullary veins) and the cortical veins.

The deep venous system appears later between the 6th and 11th weeks of age. It is

G.B. Bradac, *Applied Cerebral Angiography*, DOI 10.1007/978-3-319-57228-4_9

characterized by several veins draining mainly the blood of the well-developed choroid plexus of both hemispheres. These are the bilateral telencephalic, diencephalic, and mesencephalic veins entering the primitive marginal sinus, and the single median vein known *as the median prosencephalic vein (PV) described by Markowski* (1922) *or as primitive internal cerebral vein* (*Padget* 1956, 1957) *entering the primitive marginal sinus on the right through the primitive straight sinus. Later both marginal sinuses fuse together forming the SSS. The dorsal shift of the SSS leads to a posterior displacement also of the straight sinus.*

The development of the basal ganglia and thalamus and the growth of the brain result in the formation *of the new paired internal cerebral veins* (*ICV*) receiving tributaries of these structures characterized by the medullary, subependymal, and thalamic veins. The drainage of these veins is centripetally directed, differently from the superficial medullary veins which have a centrifugal flow towards the superficial system. The appearance of the ICV leads progressively to a regression of the PV, with exception of its more distal part which together with the distal segments of both ICVs forms *the Galen vein draining into the straight sinus*. In addition to the straight sinus, another primitive dural channel can be present, called the *falcine sinus* running within the falx and connecting the Galen vein with the superior sagittal sinus. It normally disappears. It can persist and be incidentally discovered during normal examinations (see Sect. 9.3.3). It is frequently the main venous drainage often associated with hypo-aplasia of the straight sinus in the Galen vein malformation which is linked to the persistence of the PV (see also Sect. 12.8).

In the meantime important changes involve the telencephalic, diencephalic, and mesencephalic veins which following the growth of the brain and the development of the primitive tentorial sinus are connected with this new structure and no longer enter the marginal sinus. Following the regression of the primitive tentorial sinus all its tributaries anastomose contributing to the formation of the basal vein. Variations in this process explain the great variations of the drainage of the basal vein and as above already mentioned of the SMCV and DMCV (Sects. 9.1.1.2, 9.1.2.6 and 9.3.10).

As far as it concerns the embryogenesis of *the veins of the posterior fossa*, the early-appearing veins are the primitive anterior vein draining the anterior part of the brainstem and the primitive posterior vein draining the cerebellar structure. In the early phases, the anterior is the most voluminous and will be involved also in the formation of the petrosal vein, its tributaries, and the superior petrosal sinus. The posterior vein is first smaller, since in this phase the cerebellum is not yet completely developed. Later, in connection with the growth of the cerebellum, a group of veins develops, draining in the transverse or straight sinuses. A partial part of the drainage of the posterior fossa will be enclosed in the Galen vein.

The pattern of the cerebral veins at the end of the embryological development is not so constant in comparison to that of the arteries. It is characterized by many variations, involving the morphology, flow direction, type of drainage, as well as presence of many connections. In spite of the great variations some typical aspects can always be recognized. The veins are basically grouped in the veins of the cerebral hemispheres (supratentorial) and those of the brainstem and cerebellum (infratentorial or veins of the posterior fossa). The drainage occurs through the sinuses and further extracranially in the jugular veins and vertebral plexuses.

9.1 Supratentorial Cerebral Veins

The supratentorial can be subdivided into the superficial and deep systems (Kaplan and Ford 1966; Salamon and Houang 1976). The first drains the blood towards the superior sagittal, transverse, sigmoid, and cavernous sinuses and the second into the internal cerebral and basal veins continuing in the Galen vein and straight sinus.

9.1.1 The Superficial System

This comprises the superficial medullary veins, which drain the superficial white matter. They run in a centrifugal direction towards the cortex, where they join the pial veins continuing in the cortical veins receiving the blood of the cortex. The latter run within the subarachnoid space along the superficial sulci, frequently crossing the gyri. Close to the sinuses, they perforate the arachnoid, and run in the subdural space (bridging veins) with a less or more long course before entering the sinuses.

9.1.1.1 Superior Group

The veins of this group have an ascending course entering the superior sagittal sinus (SSS). The group includes the veins of the lateral surface of the fronto-parieto-occipital lobes, part of the veins of the medial surface of the hemispheres (see Sect. 9.1.1.4), and those of the anterior part of the frontobasal region. The posterior part of the frontobasal area drains in the system of the basal vein (see Sect. 9.1.2.6). Sometimes, a large vein called the *vein of Trolard* (Fig. 9.1) is present anastomosing the SSS with the veins normally draining into *the superficial*

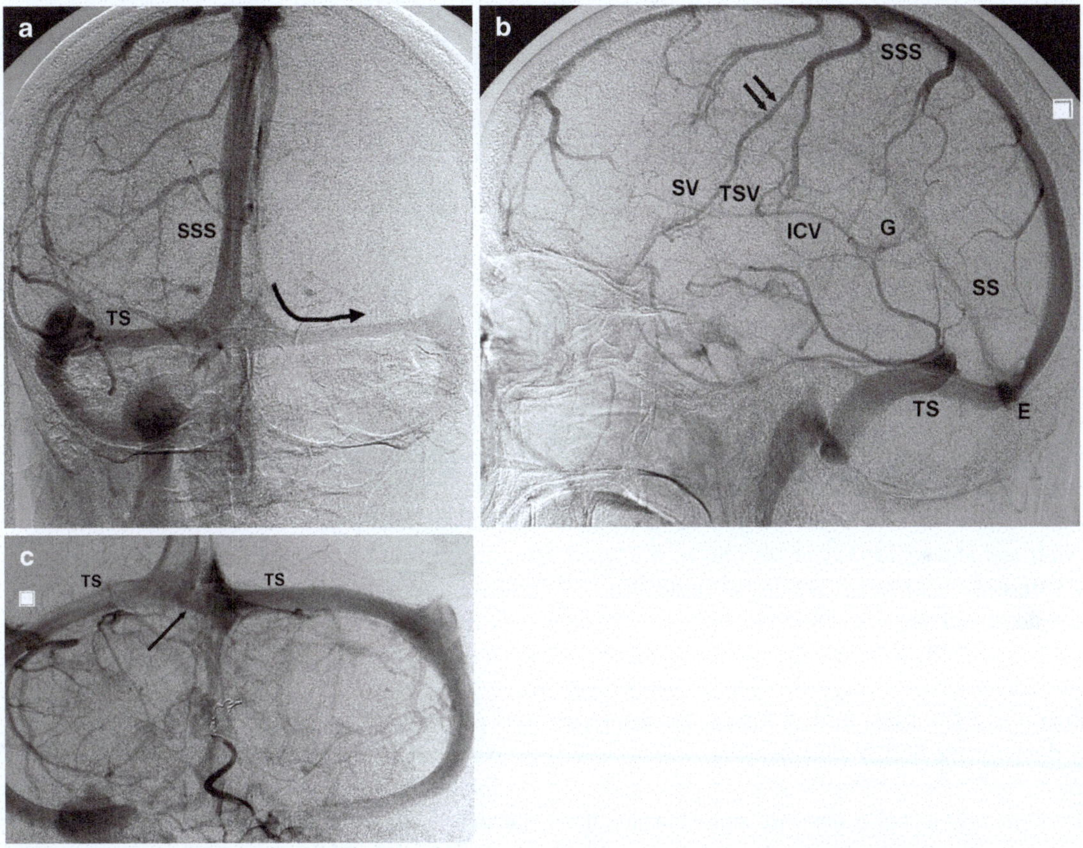

Fig. 9.1 Carotid angiogram, AP (**a**) and lateral (**b**) view, venous phase. The SSS drains predominantly in *the right TS*. The left TS is only poorly filled, since it receives blood predominantly from the SS. The SS continuing in the TS (*curved arrow*) is only faintly filled since to it drain only the right deep veins. Among the cortical veins entering the SSS, there is a large Trolard vein (*double arrow*). The drainage of the temporal region occurs into the TS. Thalamostriate vein (TSV), septal vein (SV), internal cerebral vein (ICV), Galen vein (G), straight sinus (SS), torcular herophili (E). (**c**) Vertebral angiogram in the same patient. On this angiogram is documented the presence of the *left TS*, in which drains predominantly the SS. At the torcular herophili, a complementary dural channel (*arrow*) connects the right and the left TSs (patient treated for distal PICA aneurysm, with occlusion also of the distal PICA with coils)

middle cerebral vein forming a connection between the superior and anterior-inferior group. The flow can predominate towards the SSS and so the superficial middle cerebral vein is small or also absent or the drainage occurs predominantly towards a large superficial middle cerebral vein. In the frontal area the veins enter the SSS at a right angle. The more posterior veins join the sinus, forming a progressively more oblique angle in the direction against the flow of blood into the SSS (Figs. 9.1 and 9.5). This aspect has been interpreted as due to the progressively posterior extension of the brain (Padget 1956, 1957). The veins are well recognizable on the AP and especially on the lateral carotid angiogram.

9.1.1.2 Anterior-Inferior Group

The main vein in this group is the *superficial middle cerebral vein* (*SMCV*) which collects several branches of the lateral inferior surface of the frontal and parietal area and from the lateral superior surface of the temporal lobe. It can be single or represented by a group of veins.

The SMCV runs along the surface of the Sylvian fissure, continuing into the sphenoparietal sinus (see also Sect. 9.3.9). The latter enters the sinuses *of the parasellar region* (*cavernous and lateral cavernous sinuses*) or runs towards the base of the skull, where it divides into small channels that pass through the foramina of the base, reaching the pterygoid plexus. *This variant is called paracavernous drainage* (Wolf and Huang 1963) (see also Sects. 9.3.9 and 9.3.10, Figs. 9.20 and 9.21). It can occur that the sphenoparietal sinus is absent or poorly developed and so then SMCV drains directly into the cavernous, lateral, or paracavernous sinuses (San Millan Ruiz et al. 2004; Tubbs et al. 2007b). In another type of drainage, the SMCV runs *on the floor of the middle cranial fossa reaching the superior petrous sinus or continuing within a tentorial sinus* towards the *transverse sinus*. This is due to the persistence of the primitive tentorial sinus (see also embryogenesis).

The anastomosis of the SMCV with a frontoparietal vein forming *the Trolard vein* has already been described. The SMCV can also anastomose with a temporal vein forming a large channel called *the vein of Labbé*, which runs on the surface of the temporal lobe entering posteriorly the transverse sinus (Hacker 1968, 1974). The vein of Labbé can be the dominant venous drainage of the temporal area when the SMCV is small or absent. It can also occur that the SMCV divides early into two branches, the one continuing in the vein of Labbé and the other in a frontal vein entering the superior sagittal sinus.

Examples of SMCV on AP and lateral angiograms are presented in Figs. 9.4a, b, 9.5, 9.9, and 9.10 and Drawings 9.20, 9.21, and 9.22.

As far as it concerns the *deep middle cerebral vein* and veins of the insula, see Sect. 9.1.2.6.

9.1.1.3 Posterior-Inferior Group

The veins of this group, also called the tentorial group, drain the lower lateral and lateral basal surface of the temporal and occipital lobes. These veins converge to the lateral tentorial sinus which is an intratentorial channel entering the transverse sinus. More rarely the veins drain directly in the transverse sinus or converge in a large vein of Labbé (Figs. 9.4, 9.7, 9.8, and 9.10). The veins of the medial basal surface of the temporal lobe drain into the basal vein, those of the basal occipital area into a lateral tentorial channel entering the transverse sinus.

9.1.1.4 Veins of the Medial Surface of the Hemispheres

The veins of the medial frontoparietal surface run upward, and curve over the superior margin of the hemisphere where they connect with the veins of the lateral surface, and together enter the SSS. The lower part of the medial surface involving the region of the cingulate gyrus and corpus callosum drains into the *anterior pericallosal vein* in the frontal region and into *the posterior pericallosal vein* (*splenial vein*) in the parietal area (Oka et al. 1985). The drainage can occur also partially into the inferior sagittal sinus (ISS).

As far as it concerns the medial surface of the occipital lobe, the venous drainage occurs through the anterior calcarine vein (also called internal occipital vein) which has a course anteriorly directed entering the posterior pericallosal vein, sometimes the basal vein, and through the posterior calcarine vein directed posteriorly draining into the SSS (Oka et al. 1985).

The anterior pericallosal vein is a paired vein which courses around the genu of the corpus callosum and further in the cisterna of the lamina terminalis reaching *the anterior cerebral vein* tributary of the basal vein (Drawing in Fig. 9.2). Sometimes, the anterior pericallosal vein runs upward entering the

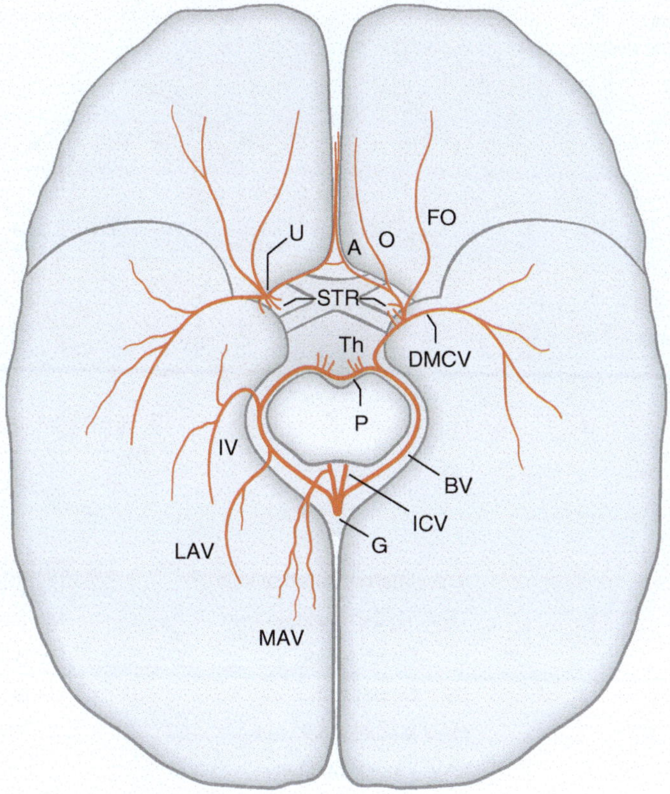

Fig. 9.2 Drawing. The basal vein and its tributaries seen from below

BV	Basal vein
ICV	Internal cerebral vein
G	Galen vein
MAV	Medial atrial vein
LAV	Lateral atrial vein
IV	Inferior ventricular vein
P	Peduncolar vein (both peduncolar vein can be connected by small posterior communicating vein)
Th	Thalamo striate veins (draining in the peduncular veins)
DMCV	Deep middle cerebral vein (draining in the BV on the right and in the uncal vein on the left)
U	Uncal vein
STR	Striate vein (draining in the DMCV)
O	Olfactory vein
Fo	Fronto orbital vein
A	Anterior cerebral vein (both anterior cerebral vein can be connected by a small anterior communicating vein)

ISS. The posterior pericallosal vein is commonly a paired, sometimes a single vein which runs in the pericallosal cistern around the splenium entering the Galen vein, sometimes the basal or internal cerebral veins (Ben-Amor and Billewicz 1970; Duvernoy 1975; Salamon and Huang 1976).

On the angiogram the veins of the medial surface are superimposed with the veins of the lateral surface and so they cannot be commonly identified. Among these veins the splenial vein can sometimes be easily recognized on the lateral carotid and vertebral angiograms (Figs. 9.6, 9.9a, and 9.10).

9.1.2 The Deep System

The deep system drains the blood of the deep structures of the cerebral hemispheres (deep white matter and deep gray nuclei with adjacent structures). The venous drainage occurs through

Fig. 9.3 Drawing. Lateral view of the deep cerebral veins

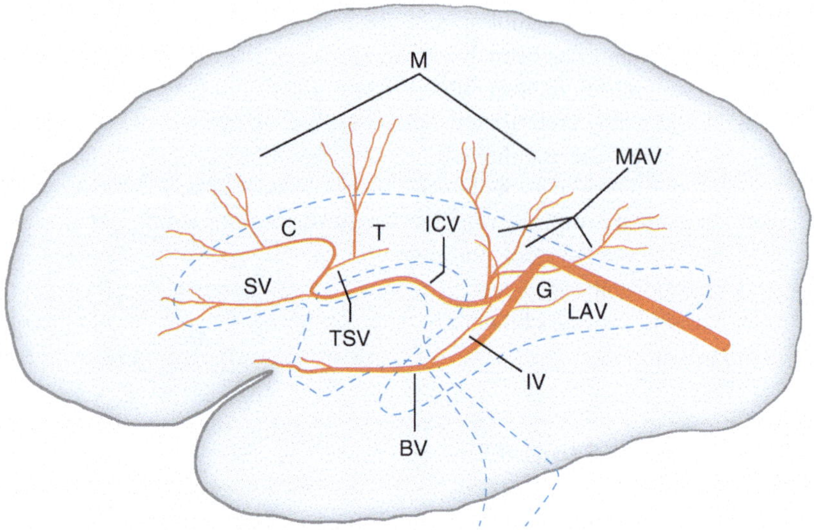

ICV Internal cerebral vein
TSV Thalamo-striate vein
C Caudate vei
SV Septal vein
MAV Medial atrial vein
LAV Lateral atrial vein
IV Inferior ventricular vein
M Deep medullary veins
BV Basal vein
G Galen vein
T Terminal vein

several channels (medullary, subependymal, striate, thalamic, and choroidal veins). The flow is centripetally directed towards main collectors: *the internal cerebral vein and the basal vein both draining into the Galen vein. However, while the internal cerebral vein is a regular tributary of the Galen vein, the drainage of the basal vein and its tributaries in the Galen vein is not constant, since due to several variants, they can take a different way of drainage* (*see Sect. 9.1.2.6*).

9.1.2.1 Medullary Veins

These veins originate in the deep white matter and run towards the lateral ventricles, where they enter the subependymal veins (Huang and Wolf 1964; Wolf and Huang 1964). The deep medullary veins can anastomose with the superficial one, forming a continuous venous channel that extends from the cortex to the subependymal veins (Schlesinger 1939; Kaplan 1959). On the AP, better on the lateral carotid angiogram, the medullary veins especially those in the frontal and parietal regions are frequently well visible as very fine vessels directed centrally with an oblique or a perpendicular course, joining the subependymal veins. The medullary veins of the occipital and temporal areas are commonly not recognizable on the normal angiograms (Drawings 9.3 and Figs. 9.6 and 9.11).

9.1.2.2 Subependymal Veins

The subependymal veins derive their blood from the medullary veins and from the striate veins (Wolf and Huang 1964; Stein and Rosenbaum 1974; Salamon and Huang 1976). Their main function is to relay the blood to the internal cerebral and basal veins. Some of these veins (septal vein, thalamostriate, and caudate veins) in spite of possible variants are relatively constant and easily recognizable on the angiogram. The presence of others is more variable.

The *septal vein* (SV), also called the vein of the septum pellucidum, is a paired vein arising each from several tributaries characterized by deep medullary veins in the frontal region, converging to the anterior lateral corner of the frontal horn. The SV runs further to the medial corner and then turns backward, continuing into the two sleeves of the septum pellucidum, joining the internal cerebral vein at the foramen of Monro where it joins the thalamostriate vein. It can easily be recognizable on the lateral angiogram due to its horizontal course in the frontal region towards the thalamostriate vein. Sometimes it can be duplicated with superior and inferior trunks fusing together in a unique vessel entering the internal cerebral vein. It can have an anomalous course, running in the superior part of the septum pellucidum, turning then downward, reaching the internal cerebral vein behind the foramen of Monro. It can contribute in the formation of the *direct lateral vein* together with caudate vein or be connected posteriorly with the *medial atrial vein*. (Drawing 9.3 and Figs. 9.1, 9.6, 9.11, and 12.23). It can occur in cases of DVA of the frontal region that the majority of the medullary veins or all of these drain into an anomalous common collector and not in the septal vein which in these cases can be minimally filled or completely absent (Fig. 12.25a, c). In other cases the dilated SV can be the main collector of the medullary veins in the frontal region (Fig. 9.25).

The *thalamostriate vein* (TSV) is usually prominent among the subependymal veins. It is formed by the junction of the *terminal vein*, which runs beneath *the stria terminalis* between the thalamus and the body of the nucleus caudatus, and the *caudate veins* (*CV*). The TSV is easily recognizable especially on the lateral ICA angiogram, where it is directed anteriorly and slightly caudally to join the septal vein. The pattern of the CV is more variable. Frequently (long longitudinal caudate vein) it runs on the lateral wall of the frontal horn, curving then medially with a course on the head of the nucleus caudatus joining the TSV. It is well recognizable on the lateral angiogram as a channel running above the septal vein with a course parallel to it forming then a curve towards the TSV. Sometimes the CV extends more posteriorly draining directly in the internal cerebral vein after having made a short course on the floor of the lateral ventricle. This anomalous pattern has been called by some authors *the direct lateral vein* which can receive some tributaries also from the anomalous SV. To the TSV and CV converge the medullary veins of the posterior frontal and anterior parietal regions and the superior striate veins. Despite its name, the TSV does not drain the thalamus (Giudicelli et al. 1970; Salamon and Houang 1976). Examples of TSV and its tributaries are presented in Drawing 9.3 and Figs. 9.1, 9.5, 9.6, 9.7b, 9.11, and 9.25.

Other subependymal veins are the *medial atrial* (MAV) and the *lateral atrial* (LAV) veins and the *inferior ventricular (IVV) and the hippocampal veins*. The MAV and LAV run on the medial and lateral wall of the atrium and occipital horn, respectively. They receive the medullary veins of the posterior parietal, occipital, and posterior temporal regions. The MAV is frequently well recognizable especially on the lateral angiogram as a stem formed by several tributaries entering the internal cerebral vein near the Galen vein. The LAV is a fine vessel difficult to identify on the angiogram. It can reach the MAV forming a common trunk or runs more laterally entering the IVV or the basal vein.

The *inferior ventricular vein (IVV) is the most prominent subependymal vein of the temporal region* to which converge many of the medullary veins of this area. It begins in the body of the lateral ventricle, accompanies the body and tail of the nucleus caudatus, and runs anteriorly in the roof of the temporal horn, forming an anterior concave curve. Short before the uncus it turns medially, and exits at the choroid fissure, entering the basal vein. *The hippocampal veins* are transverse channels running on the surface of the hippocampus, on the floor of the temporal horn. They exit the temporal horn connecting into a longitudinal vein running on the surface of the parahippocampal gyrus entering the IVV or basal vein. The hippocampal veins are very fine vessel, not recognizable on the angiogram. Examples of MAV, LAV, and IVV are presented in Drawings 9.2 and 9.3 and in Figs. 9.6, 9.11, 9.25, and 12.6.

9.1.2.3 Striate Veins

Striate veins drain the blood of the basal ganglia, internal capsule, and claustrum. The superior group runs towards the subependymal veins of the posterior frontal and parietal regions entering mainly the

TSV and the CV. The inferior group drains into the inferior striate veins, which run downward, and pass through the anterior perforated space reaching in the deep portion of the Sylvian fissure, the stem of the deep middle cerebral vein, which is one of the tributaries of the basal vein (Wolf and Huang 1963; Salamon and Houang 1976). These veins are rarely recognizable on the angiogram (Figs. 9.7 and 9.10) (see also Drawing 9.20b–d).

9.1.2.4 Thalamic Veins

These veins can be subdivided *into a superior group draining into the ICV and an inferior group entering the basal vein*. Among the superior group, a few drain the anterior part of the thalamus entering the ICV near the foramen of Monro. The posterior superior part of the thalamus is drained by larger veins which run first parallel to the ICV, entering this near the Galen vein.

A few of the veins of the inferior group emerge at the posterior perforated substance, reaching the peduncular veins, tributaries of the proximal basal vein, or posterior mesencephalic vein. Other drains the posterior part of the thalamus entering the distal basal vein or posterior mesencephalic vein (Giudicelli et al. 1970; Salamon and Huang 1976). They can be sometimes recognizable on the lateral vertebral angiogram (Fig. 9.16b and Drawing 9.13).

9.1.2.5 Choroidal Veins

The choroidal plexus is drained by the superior and inferior choroidal veins which enter, respectively, the ICV and BV (Ben Amor et al. 1971; Salamon and Huang 1976). The choroidal veins belong to the group of the subependymal veins. They receive blood only from the choroidal plexus and are recognizable on the carotid and vertebral angiograms as a faint blush in the capillary phase.

9.1.2.6 Collectors of the Deep System

- *The internal cerebral vein.*

  It is a paired venous channel arising at the level of the foramen of Monro or near to it. It is located near the midline, where it runs in the *tela choroidea* on the roof of the third ventricle. Each vein arises at a junction of the SV with the TSV and receives during its course several subependymal veins. Each ICV ends in the cistern of the quadrigeminal plate, below the splenium, where both veins unite and together with both BV enter the Galen vein. The ICV is always recognizable on the lateral carotid angiogram owing to its typical S-shaped form and on the AP view due to its paramedian position.

  The ICV and some of its tributaries (SV, TSV, and CV) are a constant venous finding. They are better demonstrated on the later lateral angiogram when the cortical veins slowly disappear (Drawing 9.3 and Figs. 9.1, 9.4, 9.5, 9.6, 9.7, 9.10, and 9.11). Differently from the ICV and TSV, the SV appears later on the angiogram. On the AP view commonly only the ICV and the TSV are well recognizable. As far as it concerns the other deep veins, overlapping or variations and different filling make their certain identification in every case impossible. Their identification can be easier when they are enlarged, when involved in the drainage of vascular malformations (see Chap. 12).

- *Basal Vein (BV)*

  First described by Rosenthal in the 1824, the BV is an important, frequently large vein collecting blood from the supratentorial and infratentorial veins. It is formed on the undersurface of the anterior perforated substance (APS) by the union of several veins (Drawings 9.2, 9.20b, and 9.21a). In many cases, the *deep middle cerebral vein* (DMCV) is the most important tributary of the BV, which in these cases can be considered its direct continuation (Duvernoy 1975). The DMCV is formed by several veins running on the surface of the insula (insular veins), directed inferiorly and anteriorly towards the pole of the insula (Wolf and Huang 1963). They can be recognized on the lateral angiogram as small channels directed downward with an oblique course (Figs. 9.8, 9.9b, 9.10, and 9.26c). Among them the most prominent and more frequently recognizable is the posterior insular vein running in the posterior limiting sulcus of the insula. The central insular vein with its course in the central sulcus of the insula and the most anterior insula vein which runs in the anterior limiting sulcus of the insula are more rarely identified with certainty. The DMCV runs further medially in the deep portion of the Sylvian fissure, where it is joined by *the inferior striate veins* running through the APS. These are very fine vessels, sometimes recognizable on the AP view of the angiogram (Figs. 9.7b and 9.10). At the level of the APS, the DMCV is joined by other veins coming from *the frontobasal area* (*olfactory and frontobasal veins*) *and by the anterior cerebral vein*. This latter receives blood from the genu and rostrum of the corpus callosum by the way of the anterior *pericallosal vein* (Sect. 9.1.1.4). This latter continues in the vein of *the lamina terminalis entering the anterior cerebral vein*. The right and left anterior cerebral veins are linked by the anterior communicating vein, running above the optic chiasma (Duvernoy 1975; Ono et al. 1984). The frontobasal veins and the anterior cerebral veins are rarely recognizable on the angiogram (see Figure 9.4b).

By the union of all these venous channels at the level of the APS is formed the BV, which from its point of origin runs first medially towards the anterior surface of the cerebral peduncle continuing then around it directed posteriorly and medially towards the Galen vein. In its course the BV lies laterally to the upper part of the cerebral peduncle and medially and above the uncus. The posterior cerebral artery lies inferiorly, while the anterior choroidal artery crosses the BV superiorly from medial to lateral. In its more distal segment the BV runs slightly upward along the posterior part of the thalamus to join the vein of Galen or the ICV (Drawings in Figs. 9.2, 9.3, 9.20, and 9.21).

The BV receives along its course many tributaries. Near its origin there are *the peduncular veins* located in the interpeduncular fossa, draining the inferior thalamic veins. The peduncular veins of both sides are connected to each other by a small communicating vein. Other tributaries are *the inferior ventricular, the hippocampal, and the lateral atrial veins*, which have already been described (see Sect. 9.1.2.2), and cortical veins of *the inferomedial temporal lobe and of the medial anterior part of the occipital lobe, and sometimes the posterior pericallosal vein, also called splenial vein* (see also Sect. 9.1.1.4). Another vein that can be connected with the BV is the *lateral mesencephalic vein*, which runs in the lateral mesencephalic sulcus and can be connected with the petrosal vein, forming an anastomosis between the supratentorial and infratentorial venous sectors (Wolf et al. 1963; Bradac 1970; Wackenheim and Braun 1970; Salamon and Houang 1976).

Many variants have been described (Johanson 1954; Padget 1956, 1957; Wolf et al. 1963; Bekov 1965; Babin 1971; Duvernoy 1975; Salamon and Huang 1976): one is represented by the *posterior mesencephalic vein* (*PM*). This is a venous channel which can take its origin in the interpeduncular fossa having a long course similar to that of the BV or be present only in its more distal segment as a continuation of the lateral mesencephalic vein. The PM enters the Galen vein. It can replace the BV or be present in addition to it. In this latter case it has been proposed (Duvernoy 1975) to call this vessel the "accessory basal vein."

Not rarely the anterior tributaries mainly the DMCV which normally converge forming the BV do not connect with the BV. Instead they join together forming a common trunk called *uncal vein* due to its relationship with the uncus. This vein, which can be very prominent, drain either in the sinuses of the paraseller area (CS and LCS) or through the paracavernous sinus in the pterygoid sinus (Drawings 9.20 and 9.21 and Figs 9.4a, b, 9.5a–c, 9.7a, 9.8, and 9.10). It can also occur that the uncal vein enters the SMCV short before the latter reaches the sinuses. In these cases the BV can appear on the angiogram as very small due to the partial filling when minimal connections with the anterior tributaries are still present, or its proximal segment is not present and the distal part is replaced by the posterior mesencephalic vein. It can also occur that the distal part of the BV drains through the lateral mesencephalic vein in the petrosal vein. Sometimes the anterior tributaries of the BV do not join together but have each a separate drainage partially in the uncal vein continuing in its ways of drainage as described above and partially into the pontomesencephalic veins (Fig. 9.9a, b).

Connections between these two type of drainage can be present. Rarely the BV can drain posteriorly into the straight or transverse sinuses after a short course within tentorium due to persistence of the primitive tentorial sinus (see also embryogenesis).

On the angiogram, the BV is well recognizable in the lateral view as a venous channel, forming a slight upwardly concave curve in its course towards the Galen vein. In the AP view, the main trunk forms a small curve corresponding to the peduncular vein before showing a straight, medially directed course towards the Galen vein. The presence and recognition of its tributaries are irregular. This typical pattern can change when anatomical variations are present.

In conclusion, to the BV can converge the venous blood of many veins coming from different regions involving basal ganglia, corpus callosum, insular and frontobasal areas, as well as part of the brainstem. The drainage in the BV of these venous channels, however, is not constant, since as described above they can take a different way of drainage leading to changes in the morphological and angiographic pattern of the BV.

As far as it concerns the BV and the posterior mesencephalic vein, considering that one vessel can replace the other, and that both can be present and the numerous variants, it is not possible in the given case to establish which vessel we are dealing with. In the routine daily work, from the practical point of view, it is acceptable to define the vein visible on the carotid angiogram as the BV, since it drains blood from the supratentorial area, while the vessel visible on the vertebral angiogram draining blood predominantly from the infratentorial area can be defined as the posterior mesencephalic vein.

Aspects of the basal vein and its tributaries are presented in Drawings 9.2, 9.3, 9.20, and 9.21 and Figs. 9.4a, 9.5, 9.6, 9.7, 9.8, 9.9, 9.10, 9.25, 9.26, 12.5, and 12.6.

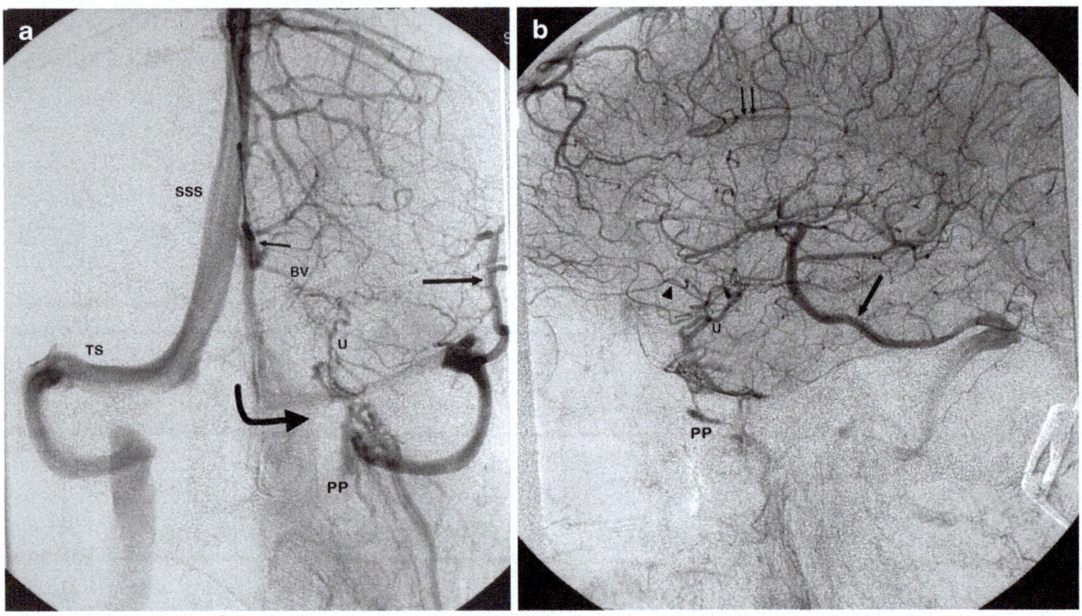

Fig. 9.4 Left carotid and left vertebral angiograms in the same patient. (**a**, **b**) Left carotid angiogram. AP and lateral views of the venous phases. The SSS drains selectively into the right TS. The distal part of the left TS is filled through a large vein of Labbé (*arrow*). Its proximal part is poor filled receiving blood only from the SS (*curved arrow*). Olfactory vein (*arrowhead*). Large uncal vein (U) receiving the anterior tributaries of the basal vein (BV), which is only partially filled. The uncal vein drains into the paracavernous sinus continuing in the pterygoid plexus (PP). Inferior sagittal sinus (*arrows*). (**c**, **d**) Vertebral angiogram. The SS is better filled receiving blood from the thalamic and pontomesencephalic veins. It enters the left TS which is therefore better filled. There is also a small connection (*arrow with dot*) with the right TS. Occipital sinus (OS). Cavernous sinus (CS) continuing in the IPS. This latter enters, on the left, the internal jugular vein more caudally than commonly

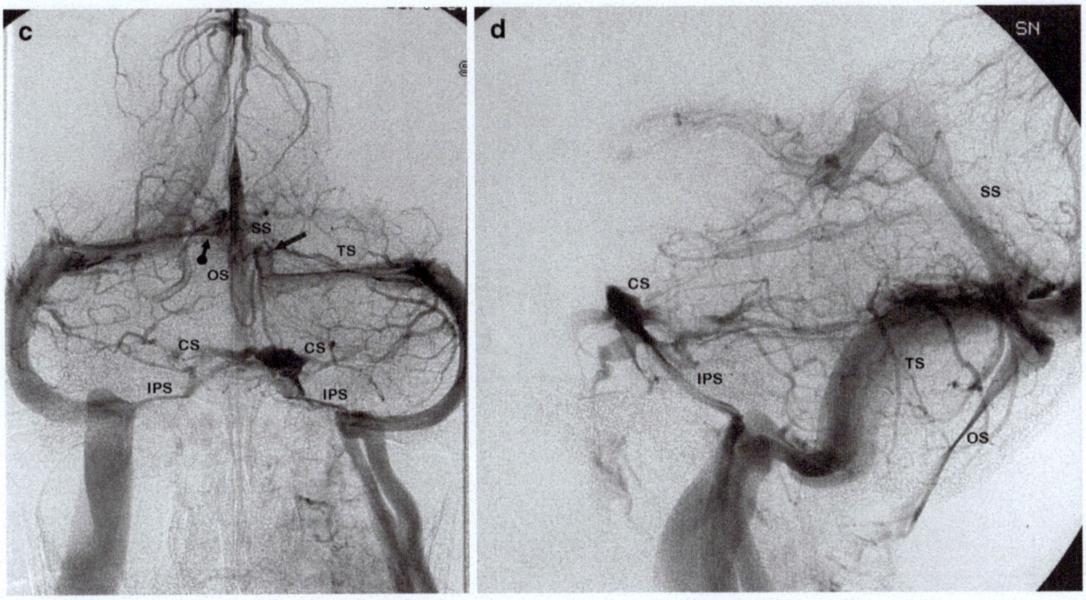

Fig. 9.4 (continued)

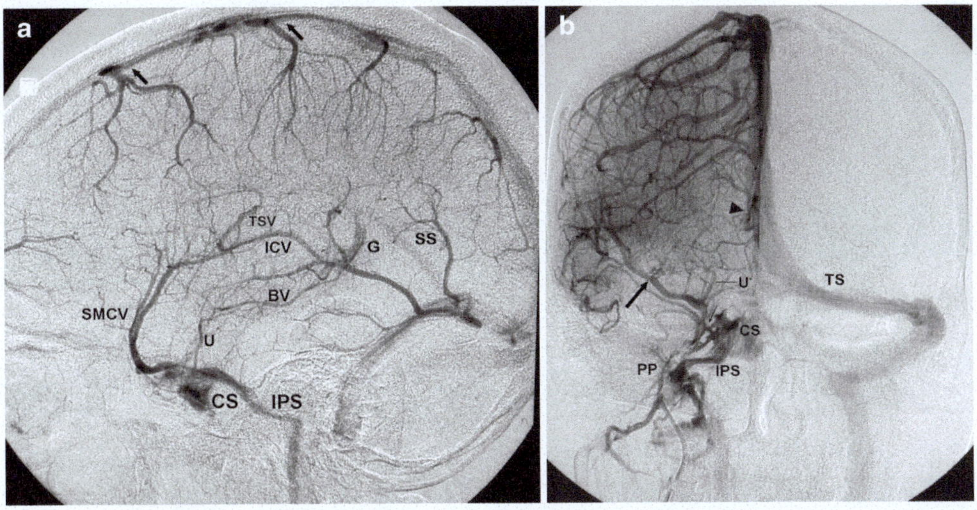

Fig. 9.5 (**a**) Carotid angiogram, venous phase, lateral view. There is a large superficial middle cerebral vein (SMCV) draining into the cavernous sinus (CS), connected with the inferior petrosal sinus (IPS). Small uncal vein (U) receiving partially the anterior tributaries of the basal vein (BV) which is consequently only partially filled. Thalamostriate vein (TSV), internal cerebral vein (ICV), Galen vein (G). Straight sinus (SS). Typical course of the cortical veins (*arrows*) draining into the SSS with frontal to occipital in a progressively reverse direction to the flow of the sinus. (**b**) AP view, right carotid angiogram, venous phase in a different patient. Preferential drainage of the SSS in the left TS. Internal cerebral vein (*arrowhead*) draining in the Galen vein and further in the SS which is, however, not recognizable due to the faint amount of contrast medium. The duplicated SMCV (*large arrow*) drains mainly into a venous channel located laterally to the cavernous sinus (CS), corresponding probably to a lateral cavernous sinus. Anastomoses (*small arrow*) between the sinuses are recognizable. The further drainage occurs into the inferior petrosal sinus (IPS) and into the pterygoid plexus (PP). Uncal vein (U) draining into the SMCV. ICV (*arrowhead*). (**c**, **d**) AP and lateral views, venous phase in another patient. The SSS drains into the right TS. The left TS is poorly injected in its proximal part since it receives blood only from the SS (*curved arrow*). The distal part of the left TS is better filled through posterior temporal veins (*bidirectional arrow*). There is a SMCV composed by several branches (*arrow*) draining into the cavernous sinus and further into of the IPS. Partial drainage also in the pterygoid plexus. On the lateral view the uncal vein (U) draining into the cavernous sinus is well recognizable. Poor injection of the basal vein (BV). Internal cerebral vein (ICV). Galen vein (G)

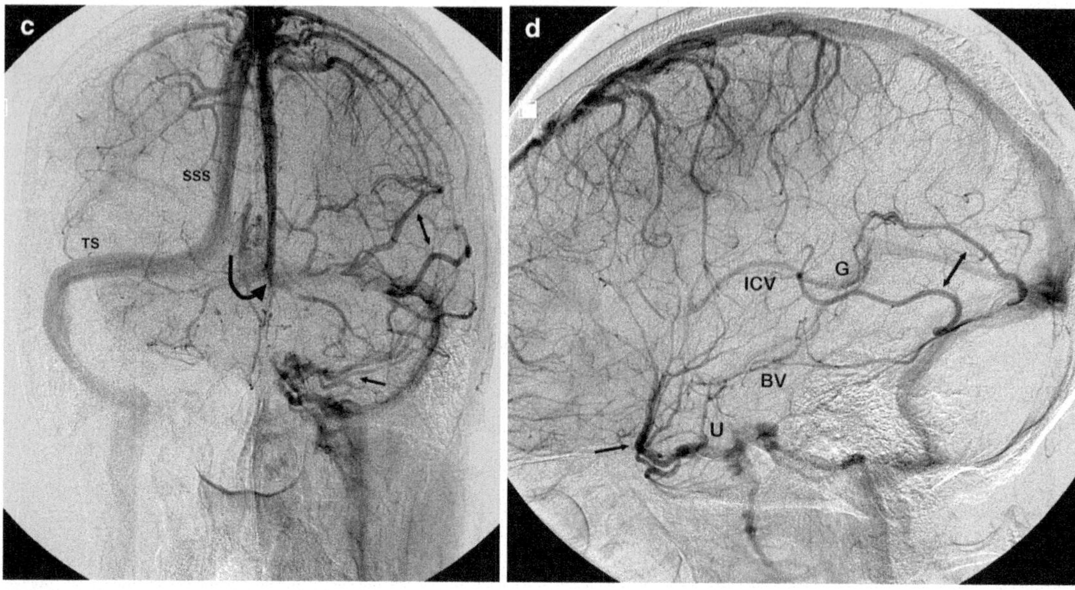

Fig. 9.5 (continued)

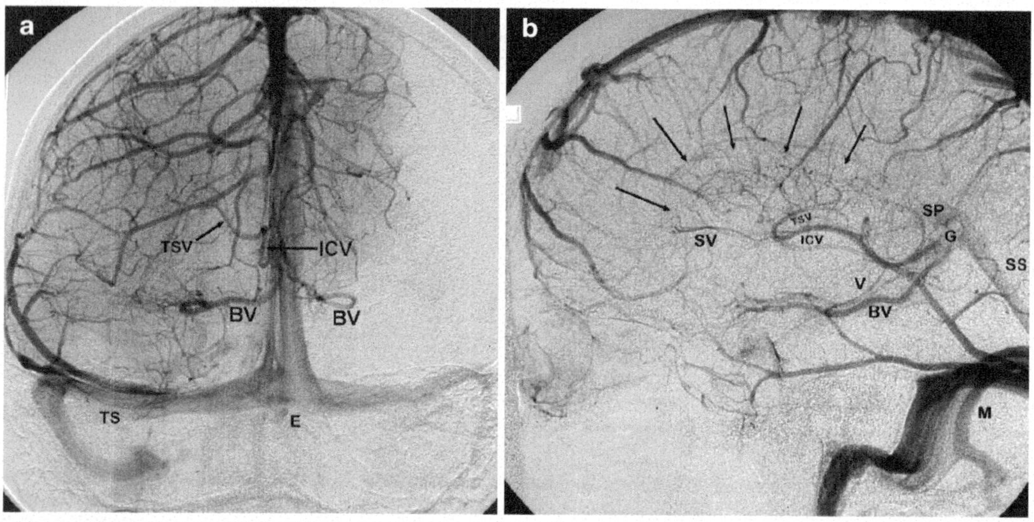

Fig. 9.6 AP (**a**) and lateral (**b**) right carotid angiogram with partial injection also of the left ICA. Venous phase. (**a**) The SSS ends in the torcular herophili (E) from which arise both TSs. The deep system drains into the SS which drains mainly into the right TS. This probably explains the better filling of the right TS. A small connection with the torcular herophili is also present. The tributaries of the deep system (TSV, ICV, and both BVs) are well recognizable. Bilateral filling of the cortical veins in the frontoparietal area draining into the SSS. The drainage of the temporal area occurs through large temporal veins which enter the TS after a short intratentorial course. (**b**) On the lateral angiogram the veins of the deep system are better demonstrated. Faint network corresponding to the medullary veins (*arrows*). Septal vein (SV), internal cerebral vein (ICV), thalamostriate vein (TSV), basal vein (BV), inferior ventricular vein (V), septal vein (SP), Galen vein (G). Straight sinus (SS). There is a large emissary mastoid vein (M)

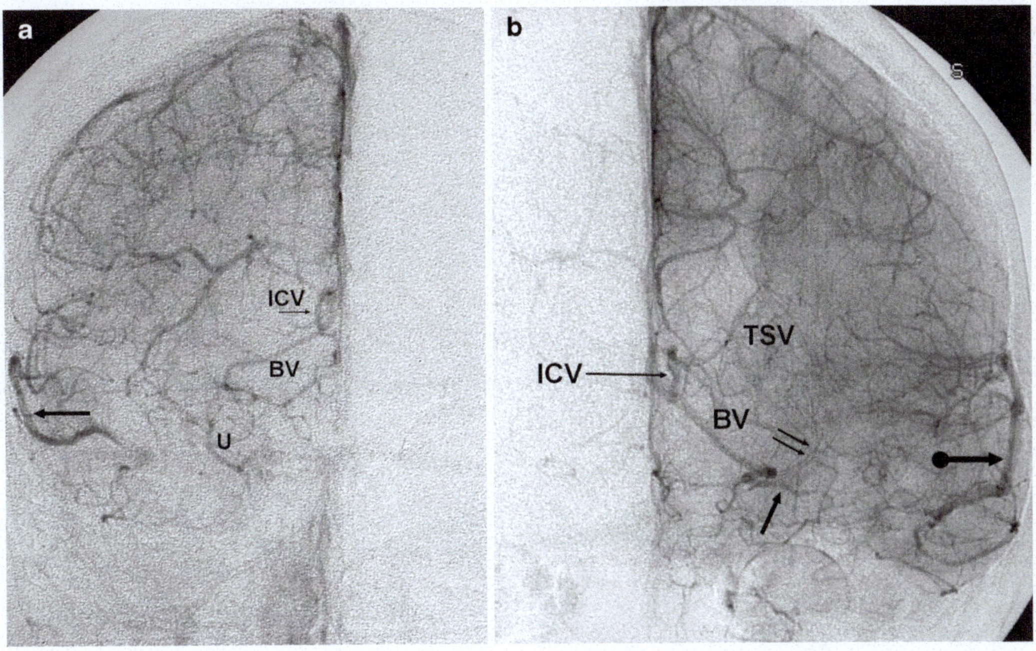

Fig. 9.7 Carotid angiogram AP view in two different patients. (**a**) Large uncal vein (U) draining into the cavernous sinus. Basal vein (BV), internal cerebral vein (ICV). The drainage of the temporal region is directly prevalently into a vein of Labbé (*arrow*). (**b**) To the basal vein (BV) converges the deep middle cerebral vein (*arrow*) collecting the inferior striate veins (*double small arrows*). Internal cerebral vein (ICV) to which converges the thalamostriate vein (TSV). Vein of Labbé (*arrow with dot*)

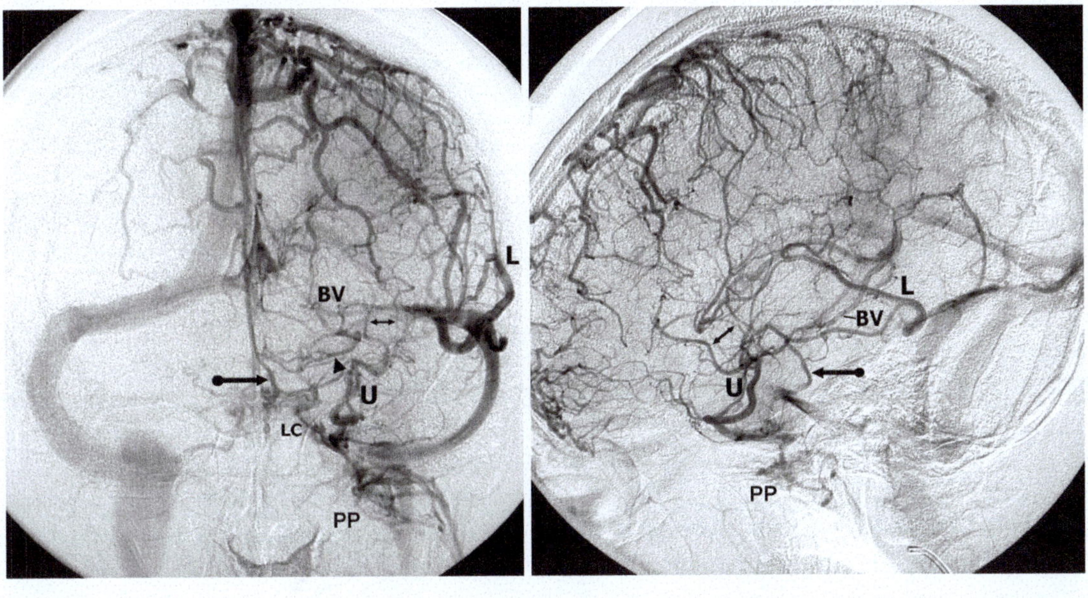

Fig. 9.8 Carotid angiogram. Venous phase AP and lateral views in the same patient (same indications). There is no SMCV; the drainage of the temporal area occurs mainly through a large vein of Labbé (L) draining into the TS. A few tributaries of the BV, among them the insular veins (*bidirectional arrow*), are well recognizable; they converge through the deep middle cerebral vein only partially to the BV. Instead they join together converging into a large uncal vein (U) draining into the paracavernous sinus continuing in the pterygoid plexus. A small component enters the lateral cavernous sinus (LC) draining further into the pterygoid plexus (PP). Another part of the drainage (*arrowhead*) is connected with the peduncular vein continuing in the pontomesencephalic vein (*arrow with dot*) which has a more lateral course than normally. Only the distal part of the BV can be recognized. It could also be the posterior mesencephalic vein

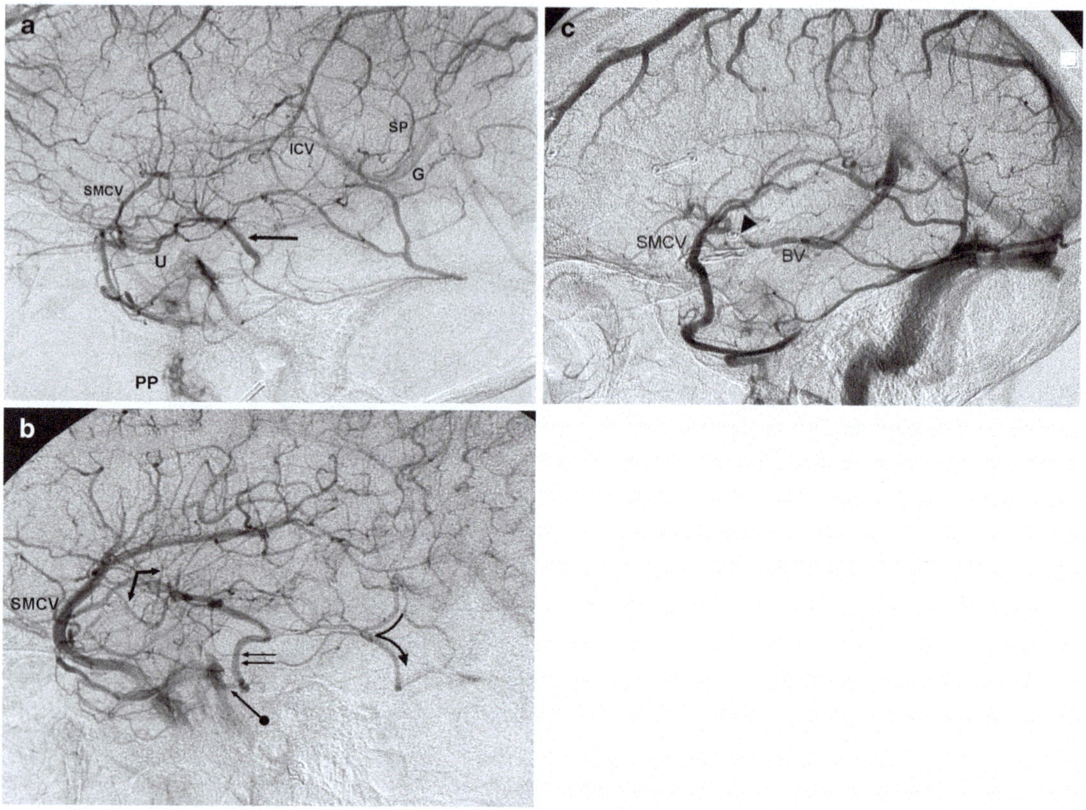

Fig. 9.9 Lateral carotid angiogram, venous phase, in three different patients. (**a**) The anterior tributaries of the basal vein do not join the vein. Instead, they converge partially into the uncal vein (U) and partially into a large lateral pontomesencephalic vein (*arrow*) which have a more lateral course than normally. The uncal vein enters the SMCV which drains in the paracavernous sinus continuing in the pterygoid plexus. BV is not recognizable. Internal cerebral vein (ICV). Galenic vein (G). Splenial vein (SP). (**b**) Large superficial middle cerebral vein (SMCV) entering the cavernous sinus. The anterior tributaries (insular veins, *arrow with angle*), which normally through the deep medial cerebral vein reach the basal vein, converge to a large antero-pontomesencephalic vein (*double arrow*). Bridging veins with the cavernous sinus (*arrow with dot*). Partial filling of the distal BV which drains into a lateral mesencephalic vein (*arrow*). (**c**) The anterior tributaries of the BV unite forming the vein on the undersurface of the anterior perforate substance (*arrowhead*). Large SMCV running through the middle cranial fossa draining partially in the pterygoid plexus continuing further into a tentorial vein draining into the TS

9.1.2.7 Vein of Galen

The vein of Galen is a unique large trunk lying in the cistern called "cistern of Galen vein." It forms a typical curve, running very close to the splenium or at various distances to it (Johanson 1954) entering the straight sinus. It is always well recognizable on the carotid and vertebral angiograms especially on the lateral view. To the Galen vein converges blood deriving from the deep structures of the cerebral hemispheres and partially from the infratentorial area.

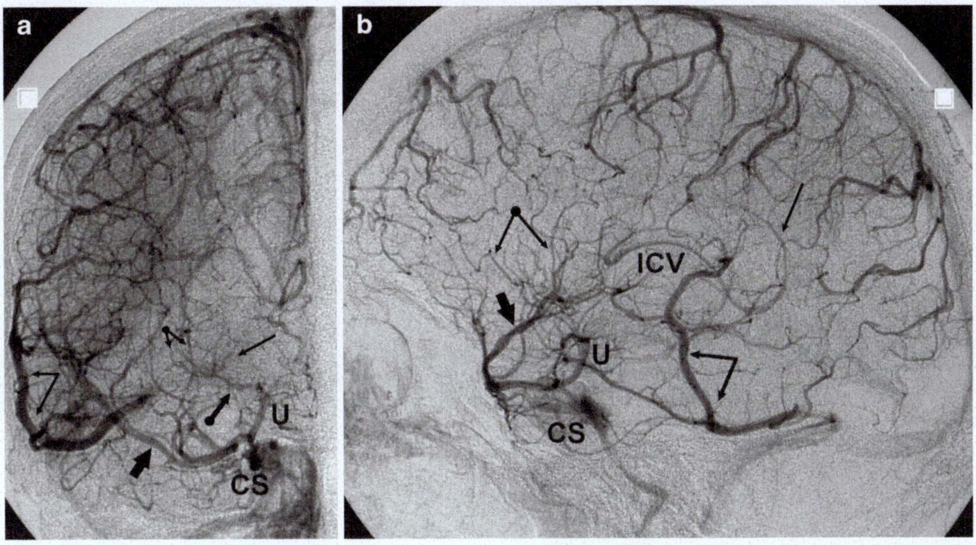

Fig. 9.10 Right carotid angiogram, venous phase AP and lateral views. (**a**) AP view. Insular veins (*arrow angle*), inferior striate veins (*small arrow*) converge to the deep middle cerebral vein (*arrow with dot*) which continues into a large uncal vein (U) which enters the SMCV (*large arrow*) draining into the CS continuing in the IPS. (**b**) Lateral view. SMCV (*large arrow*). Uncal vein (U). Cavernous sinus (CS) continuing in the IPS. The venous drainage of the temporal region occurs also through large superficial temporal veins (*arrow with angle*) which enter the TS after a short course within the tentorium. The typical BV is not recognizable. Internal cerebral vein (ICV). Splenial vein (*small arrow*)

9.2 Infratentorial Cerebral Veins (Veins of the Posterior Fossa)

Owing to the irregularity in the development and course of the veins of the posterior fossa and the frequent overlapping, it is difficult or impossible to identify all these venous channels in a given case. Some patterns, however, permit recognition on a venous angiogram of many of these veins. Three major venous drainage groups can be considered: *the superior or Galen group, draining mainly in the vein of Galen; the anterior or petrosal group, draining into the petrosal vein; and the posterior-tentorial group, draining into the transverse and straight sinuses directly or indirectly after a short course within the tentorium* (Huang and Wolf 1965; Huang et al. 1968, 1969; Bradac et al. 1967, 1971; Bradac 1970; Rosa and Viale 1970; Duvernoy 1975; Salamon and Houang 1976; Wackenheim and Braun 1978; Matsushima et al. 1983) (Drawings 9.12 and 9.13 and Figs. 9.14, 9.15, and 9.16).

9.2.1 Superior Group

This group includes many venous channels draining mainly in the Galen vein. To this group belong *the pontomesencephalic, lateral mesencephalic, medullary, precentral, superior vermian, and brachial veins.*

The anterior pontomesencephalic vein (APM) is a longitudinal venous channel running on the anterior surface of the pons in or adjacent to the midline. On the lateral vertebral and sometimes carotid angiograms the APM defines well the anterior profile of the pons. It can occur, however, that the APM is more laterally located, and so a precise morphological definition of the pons cannot be achieved. Superiorly the APM is often connected through the peduncular vein with the BV or the PM and inferiorly with the anterior medullary (AM) vein, which runs on the midline of the anterior surface of the medulla continuing caudally into the anterior spinal vein. Transverse pontine veins some-

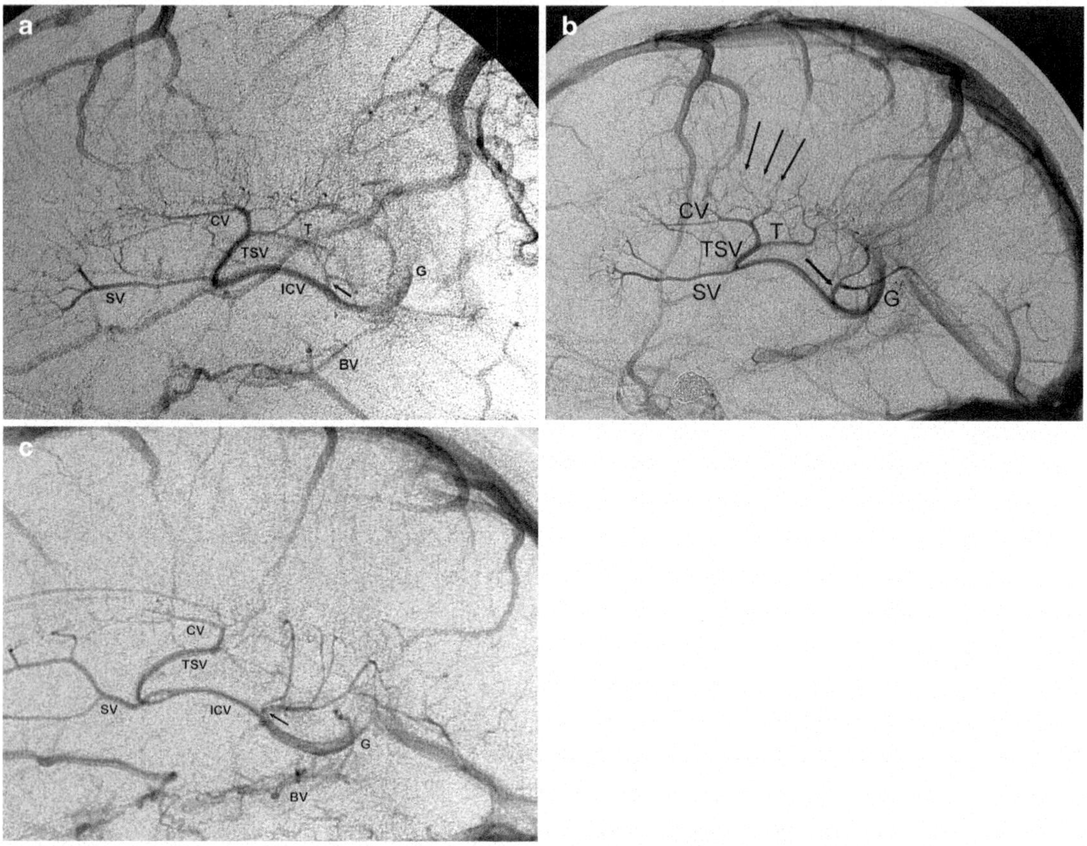

Fig. 9.11 Lateral carotid angiogram in three different patients. Late venous phase, showing detailed deep drainage. (**a**) The diffuse network corresponding to the medullary veins draining in the subependymal veins is well recognizable. Septal vein (SV), thalamostriate vein (TSV), formed by the junction of the terminal vein (T) and caudate vein (CV). Medial atrial vein (*arrow*) to which converge many tributaries. Internal cerebral vein (ICV) entering the Galen vein (G). Small basal vein (BV). (**b**) Subependymal veins. Septal vein (SV), terminal vein (T), caudate vein (CV), thalamostriate vein (TSV), medial atrial vein (*arrow*). Galen vein (G). Small ISS (*arrows*). (**c**) There is an anomaly of the septal vein (SV). Two vessels are present, one superiorly, and the other inferiorly located. Both converge to a unique trunk which enter the ICV behind the foramen of Monro. Caudate vein (CV). Thalamostriate vein (TSV). Medial atrial vein (*arrow*), Galen vein (G). Small basal vein (BV)

times relatively large running on the surface of the pons connect the APM with the petrosal vein. As described also in Sect. 9.3.10, bridging veins connect the APM with the cavernous sinus.

In addition to the anteromedial channel, there is bilaterally another longitudinally running vein or group of small veins laterally located (Tournade 1972; Duvernoy 1975). This venous channel runs on the posterolateral surface of the pons (lateral pontine veins) continuing caudally in the lateral medullary vein which runs on the posterolateral surface of the medulla. This latter leaves then the lateral position, running medially along the floor of the fourth ventricle. This segment, also called marginal vein or vein of the restiform body, fuses then on the midline with the vein of the other side forming the posterior median medullary vein continuing into the posterior spinal vein. On the surface of the medulla there are transverse channels connecting antero-median and lateral medullary veins.

The trunk formed by the lateral medullary and lateral pontine veins joins the anterior cerebellar vein (vein of the horizontal cerebellar fissure), entering the petrosal vein.

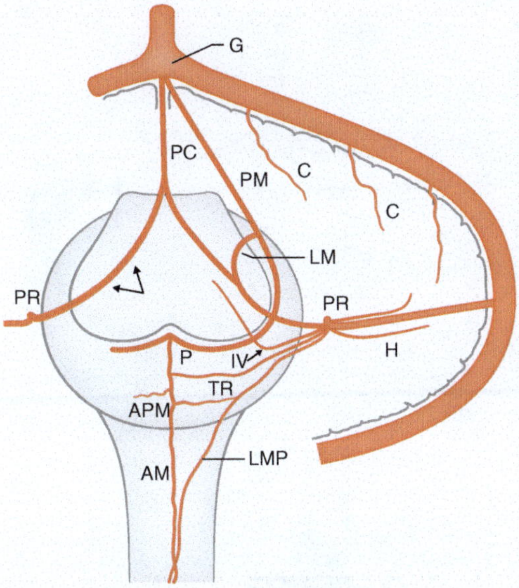

APM	Antero-ponto-mesencephalic vein
LM	Latero-mesencephalic vein
PM	Posterior mesencephalic vein
AM	Anterior medullary vein
LMP	Latero medullo-pontine vein
P	Peduncular vein
TR	Transverse pontine vein
PC	Precentral vein
⊲	Brachial vein
IV→	Vein of lateral recess of the IV ventricle
G	Galen vein
PR	Petrosal vein draining in the superior petrosal sinus
C	Hemispheric cerebellar veins
H	Vein of the horizontal cerebellar fissure

Fig. 9.12 Posterior fossa veins seen in the AP view

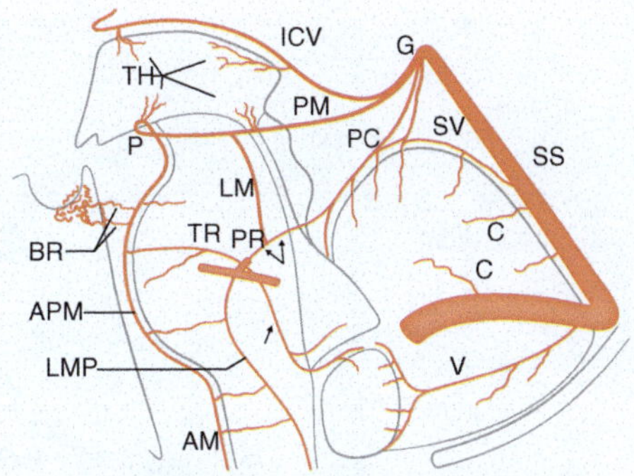

APM	Antero-ponto-mesencephalic vein
LM	Latero-mesencephalic vein
PM	Posterior mesencephalic vein
AM	Anterior medullary vein
LMP	Latero medullo-pontine vein
P	Peduncular vein
TR	Transverse pontine vein
PC	Precentral vein
⊲	Brachial vein
TH	Thalamic veins
ICV	Internal cerebral vein
G	Galen vein
PR	Petrosal vein (→) draining in the superior petrosal sinus
BR	Bridging veins
SV	Superior vermian vein
IV→	Vein of lateral recess of the IV ventricle
C	Hemispheric cerebellar veins
V	Inferior vermian vein
SS	Straight sinus

Fig. 9.13 Lateral view of the posterior fossa veins

The lateral mesencephalic vein (LM) runs in the lateral mesencephalic sulcus, draining superiorly into the BV or posterior mesencephalic vein (PM) or inferiorly into the petrosal vein through the brachial veins. Both connections can be present, and so the LM becomes an anastomotic channel, connecting supra- and infratentorial sectors. It can be a large or frequently a very fine vessel, well recognizable on the lateral angiogram, in particular the segment running in the mesencephalic sulcus. On the AP view, it forms a convex curve medially to the PM running downward towards the petrosal vein.

The precentral vein (PC) is commonly a single midline vessel entering the Galen vein. It originates in the precentral cerebellar fissure between the lingual and central lobule by the junction coming bilaterally of two or three branches called the brachial veins, since they cross the brachium pontis and brachium conjunctivum. Through the brachial veins the PC is connected with the petrosal vein. The brachial veins can be recognized as a channel directed downward and laterally reaching the petrosal vein.

The superior vermian vein (SV) can be a single large trunk or be formed by several small branches, which run median or paramedian over the culmen with a superior convex curve around this structure. They join the precentral vein or the vein of Galen. The precentral vein and the SV define on the lateral angiogram the superior border of the superior vermis.

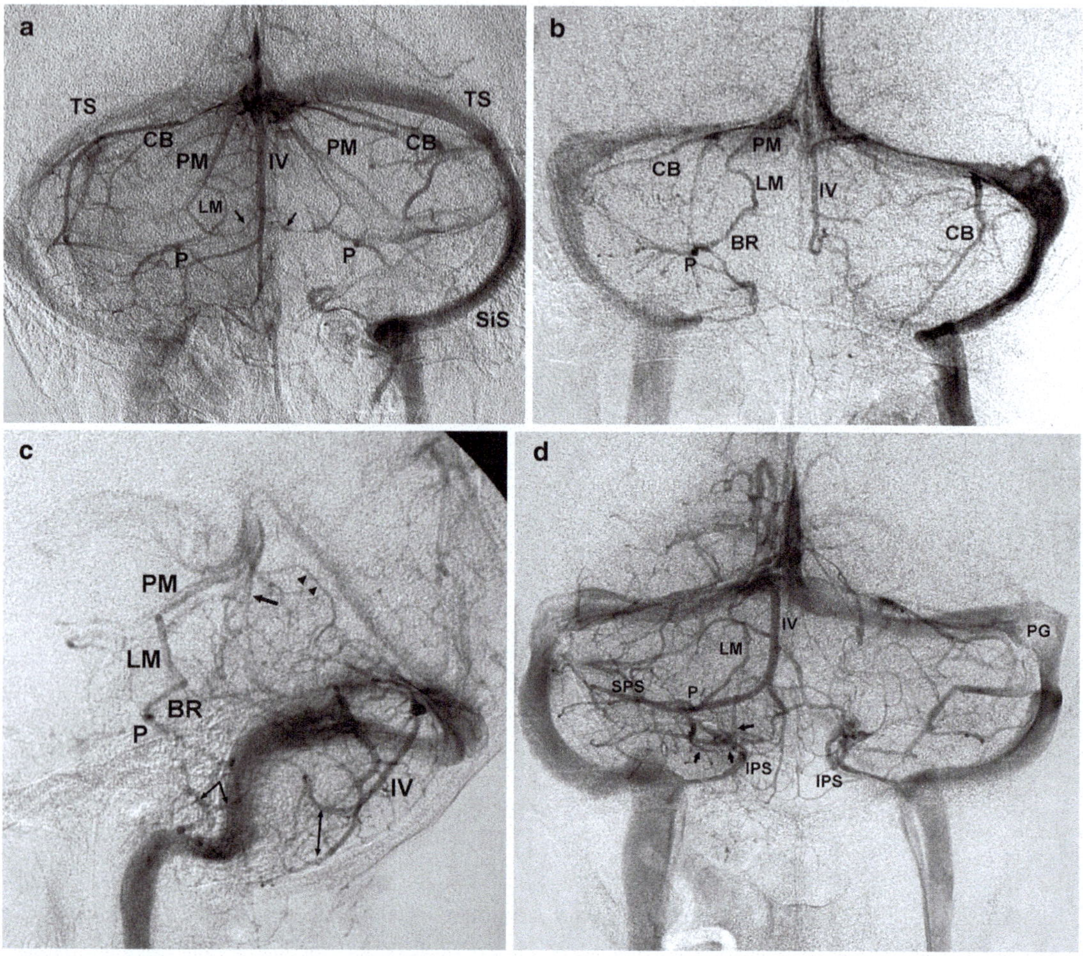

Fig. 9.14 (**a**) Vertebral angiogram, AP view, venous phase. Left and right peduncular veins (*arrow*) continuing in the posterior mesencephalic veins (PM). On the right there is a lateral mesencephalic vein (LM) connected superiorly with the PM and inferiorly through the brachial vein, with the petrosal vein (P). Inferior vermian vein (IV) with its tonsillary tributaries, hemispheric cerebellar veins (CB) draining into the TS after a short transtentorial course. The right TS is less filled since it receives blood predominantly from the SSS leading to a less concentration of contrast medium. The left TS receives blood predominantly from the straight sinus. Sigmoid sinus (SiS) continuing in the internal jugular vein. (**b**, **c**) AP and lateral venous phases in another patient. (**b**) Right lateral mesencephalic vein (LM) connecting PM with the petrosal vein (P) through the brachial vein (BR). Inferior vermian vein (IV). Hemispheric cerebellar veins (CB). (**c**) On the lateral angiogram the courses of PM, LM, BR, and petrosal vein (P) are well recognizable. There is a faint injection of the APM. Precentral vein (*arrow*). Superior vermian vein (*arrowhead*). Inferior vermian vein (IV) with its tonsillar tributaries (*bidirectional arrow*). Vein of the lateral recess of the IV ventricle (*arrow with angle*). Asymmetric filling of the TSs sinuses. (**d**) Vertebral angiogram, AP view, in another patient. The straight sinus enters the torcular herophili leading to symmetric filling of both TSs. Large Pacchionian granulations especially on the left (PG). Large inferior vermian vein (IV). Vein of the lateral recess of the fourth ventricle (*arrows*). Lateral mesencephalic vein (LM) connected caudally with the petrosal vein (P) and distally with the posterior mesencephalic vein. Superior petrosal sinus (SPS) and inferior petrosal sinus (IPS)

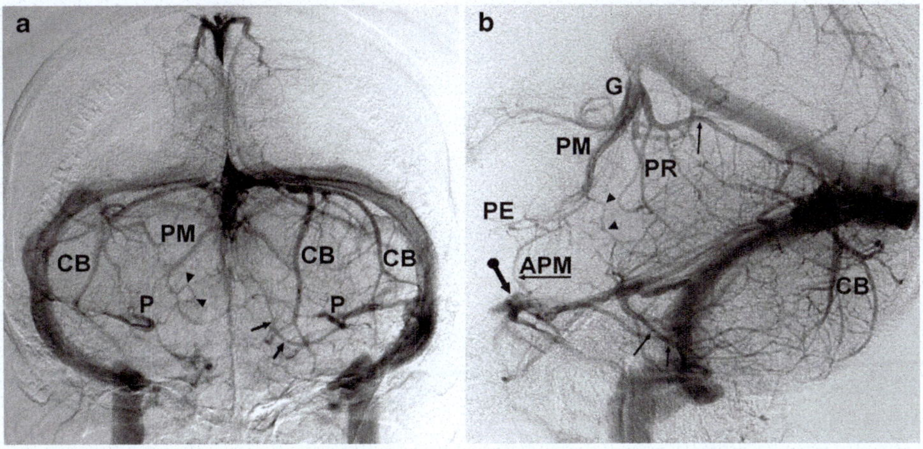

Fig. 9.15 (**a**, **b**) AP and lateral views, venous phases of the vertebral angiogram in the same patient. (**a**) AP view. The left TS is better filled, since in it drains the SS. The right TS receives predominantly the blood from the SSS. A latero-mesencephalic vein (LM) is recognizable. It is connected with the posterior mesencephalic vein (PM) and with the petrosal vein (P) draining in the superior petrosal sinus. Hemispheric cerebellar veins draining in the TS. Vein of the lateral recess of the IV ventricle (*arrows*) converging to the petrosal vein. A typical inferior vermian vein is not recognizable. It is replaced by large hemispheric vein (CB) and by the vein of the lateral recess. (**b**) On the lateral angiogram, the small LM (*arrowheads*) and the large PM are well visible. The LM is connected inferiorly with the petrosal vein to which converges also a large vein of the IV ventricle (*arrows*). Faint injection of the anterior pontomesencephalic veins (APM) connected through the peduncular veins (PE) with the PM. Cerebellar hemispheric veins (CB). Precentral vein (PR). Superior vermian vein (*arrow*). Galen vein (G) draining into the straight sinus. Bridging veins (*arrow with dot*) connecting APM with the cavernous sinus

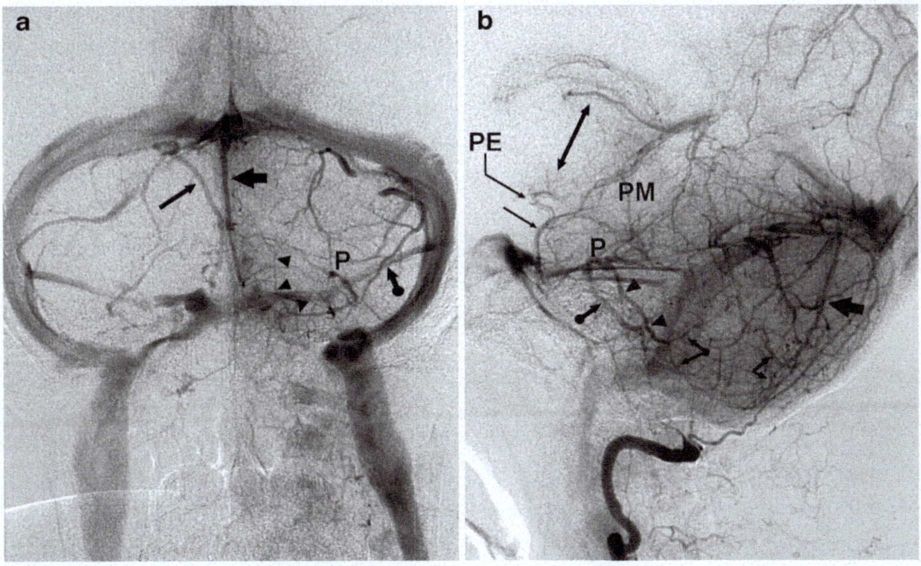

Fig. 9.16 Left vertebral angiogram. AP and lateral views. Venous phase. (**a**) AP view. Well-filled left inferior vermian vein (*large arrow*). The right inferior vermian vein (*small arrow*) is also partially injected. It drains more laterally into the TS. Vein of the lateral recess of the IV ventricle (*arrowhead*) entering the petrosal vein (P). Superior petrosal sinus (*arrow with dot*). Asymmetric filling of both TSs. That of the left is better injected probably because in it drains predominantly the SS. The right TS receives predominantly blood from the SSS. A large venous channel connects both TSs. (**b**) Lateral view. Left inferior vermian vein (*large arrow*), with its retrotonsillary tributaries (*arrow with angle*). Vein of the lateral recess of the IV ventricle (*arrowheads*) with its anterior tonsillar tributaries (*arrow with angle with dot*). Petrosal vein (P). Anterior pontomesencephalic vein (*small arrow*) connected through bridging veins with the cavernous sinus. Peduncular vein (PE). Lateral pontine vein (*arrow with dot*) continuing in the lateral medullary vein. Posterior mesencephalic vein (PM). Thalamic veins (*bidirectional arrow*)

9.2.2 Anterior Petrosal Group

This group consists of several tributaries converging to a single trunk represented by *the petrosal vein* (P), which enters the superior petrosal sinus commonly above the internal auditory meatus (Huang et al. 1968; Bull and Kozlowski 1970; Duvernoy 1975; Naidich et al. 1976; Salamon and Houang 1976). The petrosal vein is a large, but very short venous channel (a few millimeters), located in the cerebellopontine angle, running in close relationship with the trigeminal nerve. It is frequently well recognizable especially on the AP view. Tributaries of the P are the APM via the transverse pontine veins and the lateral mesencephalic and precentral veins through the brachial veins (see also Sect. 9.2.1). Other tributaries of the petrosal vein are those running on the anterior surface of the cerebellum including the superior and inferior branches entering mainly the vein of the great horizontal fissure (anterior cerebellar vein). The last tributary of the petrosal vein to be considered is the vein of the lateral recess of the IV ventricle. To it converge many branches arising especially from the inferior vermis and *tonsilla*. The vein runs first inferiorly and anteriorly, and then turns superiorly to reach the petrosal vein. It is sometimes recognizable on the AP and lateral angiograms owing to its typical course.

9.2.3 Posterior Tentorial Group

The most important vein is the inferior vermian vein. It is formed by the union of the superior and inferior retrotonsillary tributaries. It is a paired vein running posterosuperiorly in the inferior paramedian vermis sulcus, entering the straight or transverse sinuses, sometimes after a short course within the tentorium. A course more lateral is not uncommon. On the lateral angiogram, when the vein has its typical location, it is well recognizable at a variable distance from the occipital bone. On the AP view, it has a median paramedian location. The other veins of this group are the hemispheric cerebellar veins which run on the surface of the hemispheres entering the transverse sinus frequently after a short intratentorial course. The hemispheric veins can replace the inferior vermian vein when this is hypo-aplastic.

9.3 Dural Sinuses

The cranial sinuses are venous channels lying in the dura, enclosed between its external (periosteal) and internal (meningeal) layers. The wall has endothelial lining and no musculature. There is no valve inside, but frequently there are fibrous trabeculae crossing the sinus (Browder et al. 1972). Particular structures extending within the sinuses are *the arachnoid villi* (*arachnoid granulations: AG*) also called Pacchionian granulations or Pacchionian bodies from Antonio Pacchioni who first described this in the 1705. They play an important role in the returning of the cerebrospinal fluid (CSF) towards the vascular system. The AG develop from the arachnoid membrane, and grow under the influence of the pulsation of the cerebrospinal fluid invading the sinuses (Trolard 1892; Le Gros 1920; Cooper 1958). They are commonly found where to the sinus afferent cortical veins join the sinus. Indeed, as it has been demonstrated by ultrastructure microscopic studies (Krisch 1988) and described on CT-MR and angiographic examinations (Leach et al. 1996; Gailloud et al. 2001), the AG are formed by the protrusion into the sinus of the arachnoid layer and arachnoid space surrounding the vein entering the sinus. All sinuses can be involved, but the AG are found mainly in the transverse sinus (Leach et al. 1996; Gailloud et al. 2001). Other more rare locations are in the superior sagittal sinus (SSS), lacunae lateral to the SSS, and occipital and straight sinuses. They can also develop and invade the dura mater far from the sinuses extending towards the calvarium producing bone defects as large as 1 cm in diameter more typically in the frontoparietal region (Grossman and Potts 1974; Tural et al. 2012). They are commonly absent in the early childhood, appearing progressively in the adult age (Le Gros 1920; Leach et al. 1996; Gailloud et al. 2001). Commonly they can be easily differentiated from sinus thrombosis, since they never appear as hyperdensity on CT or hyperintensity on T1-weighted images. Furthermore they are not associated with abnormal flow signs as visible on CT-MR angiography or conventional angiography (Leach et al. 1996).

Another formation which can involve the sinuses *are herniations of brain parenchyma* developing inside of the AG. They have been first described by Wolbach in the 1908 in an autoptic study. Since then, sporadic cases have been reported on angiographic and MR studies by many authors (Kollar et al. 1998; Liang et al. 2002; Chan et al. 2011; Coban et al. 2013; Kocyigit et al. 2015; Asadi et al. 2015). A large number of cases have been reported more recently by Battal et al. (2014, 2015), and by Malekzadehlashkariani et al. (2016). According to these last authors brain herniations occur more frequently in the transverse sinus, straight sinus, and calvarium of the occipital bone. In the majority of the cases, cerebellar or temporal lobe parenchyma is involved. Multiplicity is frequent. The typical density or signal of brain parenchyma on CT or MR, respectively, is identified inside of the AGs (Malekzadehlashkariani et al. 2016). They have been reported as incidental findings or in patients with various pathologies. Their possible clinical relevance has been discussed.

The dural sinuses collect blood from the superficial and deep cerebral veins and from the meninges through the meningeal veins, which accompany the respective meningeal arteries. A connection with the diploic veins and with the extracranial venous system through emissary veins is also present (see Sects. 10.7 and 10.8). The main drainage occurs bilaterally in the internal jugular vein.

9.3.1 Superior Sagittal Sinus (SSS)

The SSS arises anteriorly at the junction of the falx cerebri with the dura lining the inner table of the calvarium. At this level, the SSS communicates with the veins of the nasal cavity and then with the facial vein. It extends on the midline posteriorly, following the calvarium, forming a typical upward convex curve, ending in the torcular herophili where it continues in both transverse sinuses. The SSS is very small, sometimes absent anteriorly, while it increases progressively in caliber in its medial-posterior portion. When the SSS is aplastic in the frontal region the cortical veins, sometime represented by large bilateral paramedian cortical channels, run posteriorly to reach the SSS.

Many variants can occur (Hacker 1974). One of the most frequent is the drainage directed predominantly or entirely in one TS commonly that of the right. In these cases a large part of the blood of the superficial venous system of both hemispheres flows into the involved TS. It can also occur that in the same TS drains predominantly also the straight sinus (SS). In this condition the involved TS receives blood from the superficial and deep venous system of both hemispheres and partially also from the veins of the posterior fossa. *More frequently, however, in the cases of a predominant drainage of the SSS on one TS, the SS drains predominantly in the other TS.*

The SSS can divide into separate segments proximal to the torcular herophili, draining each segment into the corresponding TS or one segment exceptionally into the occipital sinus (Fig. 9.17d). The two segments of the SSS can be connected to each other by a small or large horizontal venous channel. In other cases the distal SSS and the torcular herophili form a complex network of small channels continuing in the TSs.

All these features are well recognizable especially on the AP view of the angiogram (Figs. 9.1, 9.4, 9.5, 9.6, and 9.17a–d).

The SSS is poorly formed in children, characterized by several irregular channels rather than a single sinus. It acquires its final aspect later in the childhood (Harwood-Nash 1974).

There are some other anatomic aspects of the SSS which should be considered, but commonly not recognizable on the angiogram. The sinus can be connected with the ISS through a venous channel running within the falx, which is commonly small, but can occasionally be very large (Oka et al. 1985). Close to the SSS there are enlarged venous pouches called "*lacunae*" connected with sinus. These are venous spaces contained in the dura most frequently in the posterior

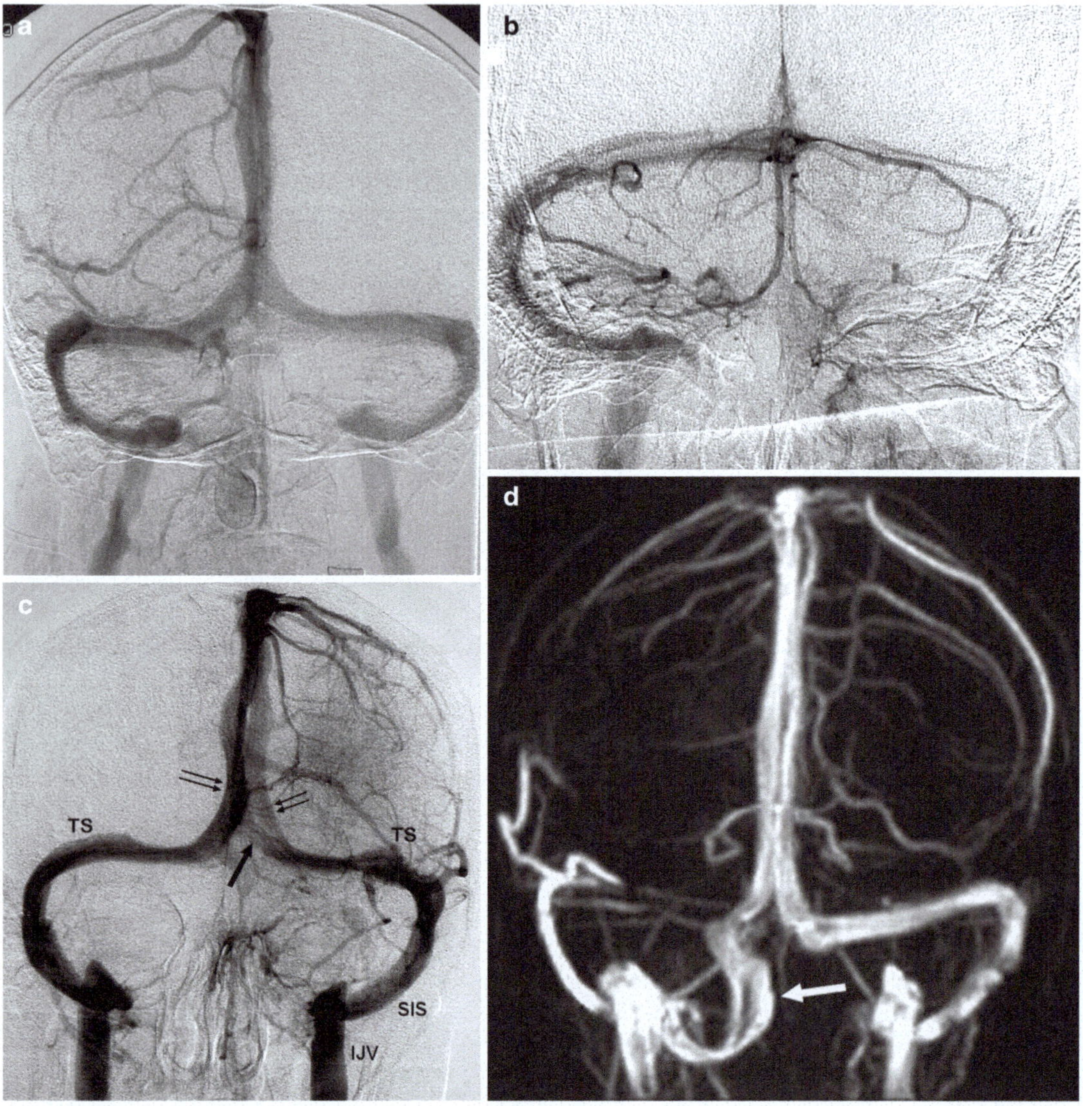

Fig. 9.17 Bilateral carotid (**a**) and vertebral (**b**) angiograms. Venous phases. (**a**) On both carotid angiograms, there is a similar venous pattern with the SSS ending in the torcular herophili. The drainage continues in both TSs, which are equally filled. The superficial venous system is equally distributed in both TSs. Into the torcular herophili drains also the SS receiving the deep venous drainage. (**b**) Vertebral angiogram. The right TS is better filled, probably due to the predominant drainage in it of the SS. This cannot be recognizable on the carotid angiogram owing to the inferior amount of blood of the SS in comparisons to that of the SSS. (**c**) Carotid angiogram. Venous phase. The SSS (*double arrows*) divides at the torcular herophili in two channels, each draining into the corresponding TS. A venous channel (*arrow*) connects both TSs. (**d**) MR angiography. Study of the venous phase. Anomalous drainage of the SSS. One distal segment drains in the left TS, and that on the right into a well-developed occipital sinus (*arrow*). The proximal part of the right TS is aplastic. It is filled by a Labbè vein. (**e**) MR T1-weighted image and angio-MR in a young patient. Incidental discovery of a large falcine sinus in which more than one channel is recognizable (*arrows*). There is a small straight sinus (*arrowhead*) (courtesy of Drs. Gozzoli G. and Boghi A. Neuroradiology. Cuneo)

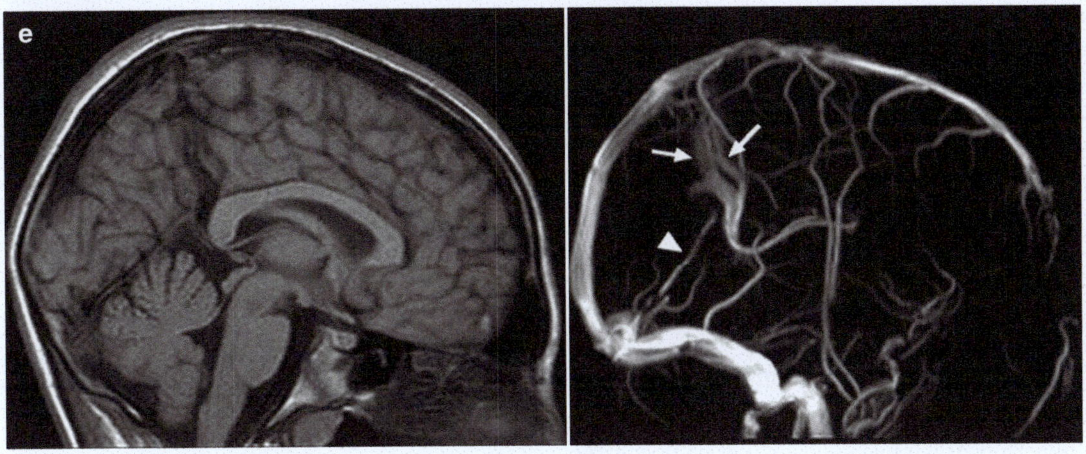

Fig. 9.17 (continued)

frontal and parietal regions. The lacunae drain the blood of the meningeal veins, and there is commonly no connections with the cortical veins which reach the SSS passing beneath the lacunae opening into the sinus separately from the lacunae. In a few cases, however, the cortical vein can empty in the lacunae (Oka et al. 1985).

Finally it should be remembered that the flow in the SSS is laminar, and so on one carotid angiogram, it can appear smaller, since the other part depends on the other carotid sector.

9.3.2 Inferior Sagittal Sinus (ISS)

The inferior sagittal sinus is commonly a small channel running within the inferior free margin of the falx, commonly a few millimeters above the edge of the falx cerebri. It begins in the anterior middle part of the falx above the anterior part of the body of the corpus callosum. It becomes larger in its course posteriorly. It ends in the falco-tentorial apex, joining the vein of Galen. Part of the venous drainage of the corpus callosum and gyrus cingula can occur in the ISS (see Sect. 9.1.1.4). It is not always visible in the angiogram since it can be very small or absent (Hacker 1974; Oka et al. 1985) (Figs. 9.4b and 9.11b). It can occasionally be very large (Figs. 9.26 and 20.5) and dilated when involved as drainage of vascular malformation (Fig. 12.11).

9.3.3 Straight Sinus (SS) and Falcine Sinus (FC)

The SS originates at the confluence of the vein of Galen with the ISS. The SS lies on the midline, at the junction of the falx cerebri with the tentorium. It is commonly one channel but sometimes can be double or triple, running downwards and backwards with an oblique course reaching the torcular herophili. In this case the blood of the SS flows equally in each TS. It can, however, occur that the SS enter selectively one transverse sinus commonly that of the left. In these cases the drainage of the deep venous system of both hemispheres as well as part of the veins of the posterior fossa occurs in one TS.

Since it drains blood from the cerebral hemispheres and also from the posterior fossa, it can be only partially filled and may give a false image of a small sinus when only a carotid or vertebral angiogram is performed (Figs. 9.1, 9.4, 9.6, 9.14, 9.15, and 9.16) (Drawings 9.3 and 9.13).

The straight sinus can rarely be associated in the normal population with *a falcine sinus* (*FC*), an embryonic vessel which runs within the falx connecting the Galen vein or the SS with the SSS. It can also be connected with the ISS. The FC normally disappears, but in some cases can persist together with the SS or replace it (Cai et al. 2009; Ryu 2010). It can be a single vessel or made up by more than one venous channels (Fig. 9.17e) and sometimes can be characterized by an extensive network of veins not necessarily communicating with the SSS (Tubbs et al. 2007a). The FC can be present in cases of acquired occlusion of the straight sinus and typically in the Galen vein malformation (see Sect. 12.8).

9.3.4 Occipital Sinus (OS) and Marginal Sinus (MS)

This is a small channel, exceptionally very large, running on the midline from the torcular herophili, with which it is connected at its posterosuperior end. It extends anteriorly to the posterior margin of the foramen magnum and further towards the sigmoid sinus or jugular vein, draining into these venous channels. It has been reported to be larger and more frequent in neonates and younger children becoming smaller and more rare in older children and adults (Lang 1991; Widjaja and Griffiths 2004). This evolution of the OS can be explained by the upright position progressively taken by the child learning to walk considering that in this position the venous drainage occurs preferentially in the venous vertebral plexuses reducing the flow in the jugular vein and indirectly in the OS (Widjaja and Griffiths 2004).

The OS is frequently well developed in the vein of Galen malformation (Sect. 12.8).

Medially the OS communicates with the marginal sinus (MS), which is a small venous channel that bilaterally surrounds the foramen magnum. The MS is connected anteriorly with the clival plexus and the jugular bulb, and anteroinferiorly with the anterior internal vertebral plexus (McDougall et al. 1997). Examples of OS and MS are presented in Drawing 9.18 and in Figs 9.4 and 9.17d.

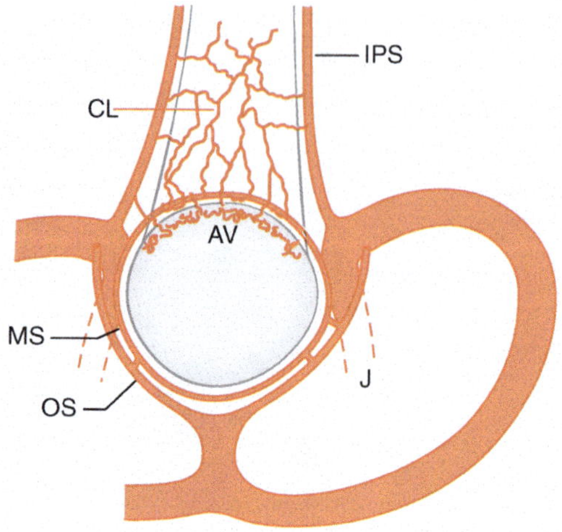

MS	Marginal sinus, with its connections with CL, AV and OS
OS	Occipital sinus, draining in the jugular bulb
IPS	Inferior petrosal sinus
J	Internal jugular vein
CL	Venous plexus of the clivus
AV	Anterior epidural vertebral venous plexus

Fig. 9.18 Drawing showing the marginal and occipital sinuses and tributaries. The marginal sinus (MS) circling the foramen magnum connected anterosuperiorly with the venous plexus of the clivus (CL), anteroinferiorly with the anterior epidural vertebral venous plexus (AV), and posteriorly with the occipital sinus (OS). The latter drains anteriorly into the jugular bulb or sigmoid sinus. The inferior petrosal sinus (IPS) is connected with the clival venous plexus with the anterior epidural vertebral plexus (AV) and jugular vein (J)

9.3.5 Transverse Sinus (TS)

Also called lateral sinus, the TS is contained within two layers of the tentorium where it is connected with the dura of the calvarium. It begins at the torcular herophili and runs laterally and slightly anteriorly along the groove of the squamous portion of the occipital bone. It then turns downward and medially, leaving the tentorium and becoming the sigmoid sinus where the transverse and the superior petrosal sinuses meet.

The two TSs are frequently asymmetric: that of the right is frequently larger. In these cases the larger TS receives the majority of the venous drainage from the SSS which drains the superficial venous system of the hemispheres. The smaller TS, commonly that of the left, receives blood predominantly from the SS to which converge the veins draining the deep structures and those of the posterior fossa. The proximal part of the TS can be hypoplastic-aplastic while its distal part is injected by a large vein of Labbé (Kaplan et al. 1973; Hacker 1974). In some cases, the aplasia of the proximal part of the TS visible on

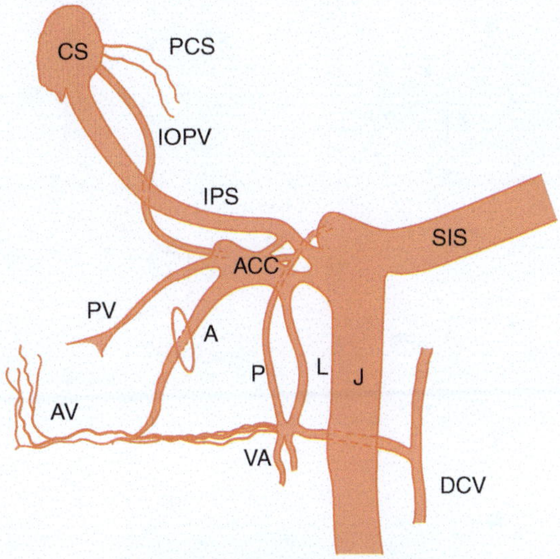

ACC	Anterior condylar confluence
A	Anterior condylar vein running through the hypoglossal canal reaching the AV
L	Lateral condylar vein connecting ACC with the VA
P	Posterior condylar vein connecting jugular bulb with the VA and the DCV
IPS	Inferior petrosal sinus
CS	Cavernous sinus
PCS	Paracavernous sinus
IOPV	Inferior petro occipital vein
J	Jugular vein
SIS	Sigmoid sinus
AV	Anterior epidural vertebral plexus with some connection with the clival venous plexus
VA	Vertebral artery venous plexus
PV	Prevertebral venous plexus
DCV	Deep cervical vein

Fig. 9.19 Drawing showing the venous drainage in the craniocervical region considering especially the anterior condylar confluence and its connections, AP view. Anterior condylar confluence (ACC), anterior condylar vein running through the hypoglossal canal, reaching the AV (A), lateral condylar vein connecting the ACC with the VA (L), posterior condylar vein connecting jugular bulb with the VA and the DCV (P), inferior petrosal sinus (IPS), cavernous sinus (CS), paracavernous sinus (PCS), Inferior Petro Occipital Vein (IOPV), jugular vein (J), sigmoid sinus (SIS), anterior epidural vertebral plexus with some connection with the clival venous plexus (AV), vertebral artery venous plexus (VA), prevertebral venous plexus (PV), deep cervical vein (DCV) (modified from San Millar Ruiz)

the carotid angiogram is only apparent, since it will be filled on the vertebral angiogram through its tributaries represented by the cerebellar veins.

Similarly to that occurring in the SSS and SS, a precise morphological image of the sinus can be obtained in many cases only if both carotid and vertebral angiograms are performed.

Examples are presented in Drawings 9.12, 9.13, 9.18, 9.21, and 9.22 and Figs. 9.1, 9.4, 9.5, 9.6, 9.14, 9.15, 9.16, and 9.17a–d.

9.3.6 Sigmoid Sinus (SiS), and Jugular Vein

The SiS is the direct continuation of the TS. It curves inferiorly and medially behind the inferior part of the petrous bone, reaching the jugular foramen where it ends in the internal jugular vein. The first segment of the jugular vein (jugular bulb) bulges upward in a rounded fossa of the temporal bone, below the internal auditory canal. The bulb is frequently larger on the right since the TS and the SiS are larger on this side (Hacker 1974; Katsuta et al. 1997) (Drawings 9.18, 9.19, 9.21, and 9.22 and Figs. 9.14, 9.15, 9.16, and 9.17a–d).

9.3.7 Superior Petrosal Sinus (SPS)

The superior petrosal sinus is a small channel lying within the attachment of the tentorium to the superior border of the petrous bone. It extends laterally, connecting with the TS. Medially it is connected with the posterior part of the cavernous sinus. The most important tributary of the SPS is the petrosal vein (Drawings 9.12, 9.13, 9.20, 9.21, and 9.22 and Figs. 9.14, 9.15, and 9.16).

9.3.8 Inferior Petrosal Sinus (IPS)

The inferior petrosal sinus lies in the groove between the petrous temporal bone and the clivus. It begins in the posterior part of the cavernous sinus and runs inferiorly and slightly laterally to join the internal jugular vein. Many reports have been devoted to the anatomy around the jugular foramen, considering especially the relationship of the drainage of the IPS into the jugular vein (Shiu et al. 1968; Miller et al. 1993; Gailloud et al. 1997; Benndorf and Campi 2002; Calzolari 2002; Mitsuhashi et al. 2007). These

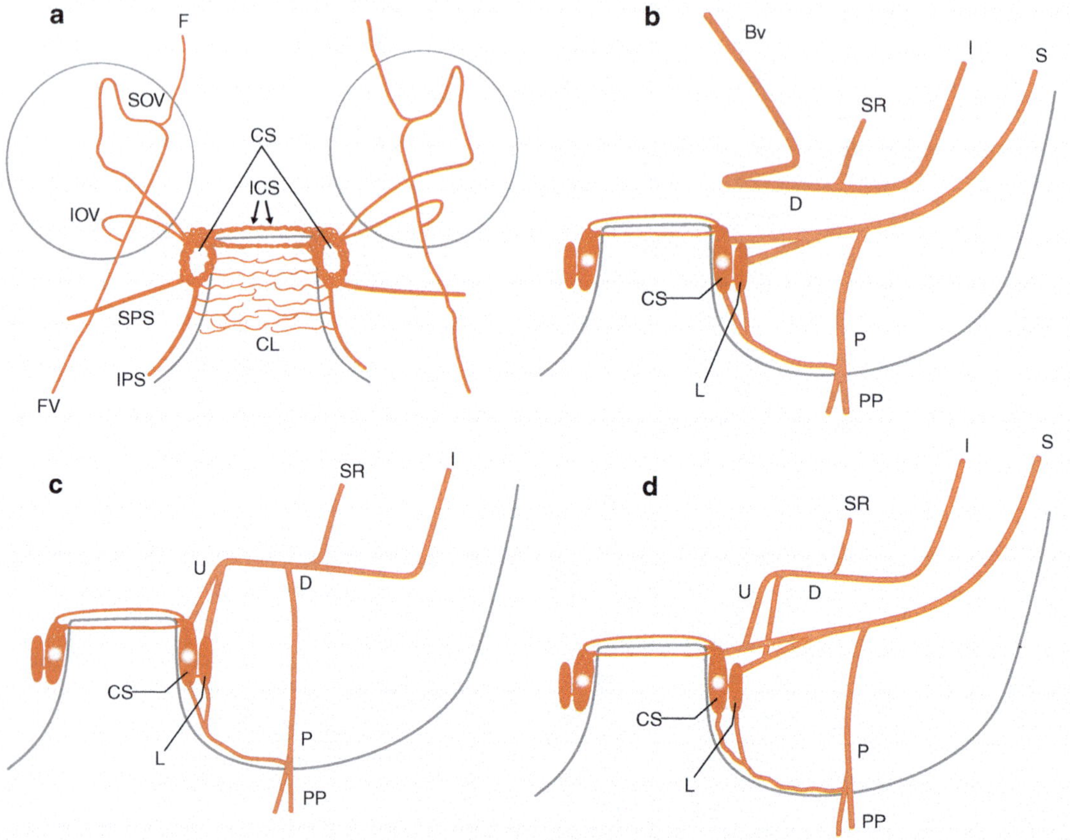

Fig. 9.20 (**a**) AP view. Drawings showing the connections of the *SOV*, *IOV*, and *petrous sinuses* with the *cavernous sinus*. The tributaries of the ophthalmic veins (frontal and facial veins) as well as the intercavernous anastomoses (ICS) and the clival plexus (CL) are also indicated. In the cavernous sinus is indicated the ICA surrounded by the venous plexus. (**b–d**) AP view. Drawings showing the possible ways of drainage of *the deep middle cerebral* (*D*) *and superficial middle cerebral veins* (*S*). (**b**) The deep middle cerebral vein continues in the basal vein. The superficial middle cerebral vein drains either in the cavernous or in the lateral cavernous sinuses. Alternatively, it can drain in the pterygoid plexus through the paracavernous sinus. Possible connections between the CS and LCS and with the paracavernous sinus are also indicated. (**c**) There are no connections of the deep middle cerebral vein with the basal vein. Instead the tributaries of the deep middle cerebral vein drain in the uncal vein, which enter the cavernous or lateral cavernous sinuses. Alternatively the deep middle cerebral vein can drain into the pterygoid plexus through the paracavernous sinus. (**d**) The three possible drainages of the superficial middle cerebral vein in the CS, LCS, and paracavernous sinus are indicated. The uncal vein drains into the SMCV before this terminates in the CS or LCS. Explications of the legend. Superior ophthalmic (SOV) and inferior ophthalmic (IOV) veins. Frontal vein (F). Facial vein (FV). Cavernous sinus (CS). Lateral cavernous sinus (L). Paracavernous sinus (P). Pterygoid plexus (PP). Intercavernous anastomosis (ICS). Superior petrous (SPS) and inferior petrous (IPS) sinuses. Clival venous plexus (CL). Deep middle cerebral vein (D). Superficial middle cerebral vein (S). Basal vein (BV). Insular veins (I). Inferior striate veins (SR)

studies have shown that the IPS is commonly formed by a unique channel, but occasionally it consists of a plexus of veins. It usually enters the jugular bulb, but it can also extend extracranially, joining the jugular vein 3–4 or more centimeters below the base of the skull. It can occasionally drain into the anterior condylar vein, thereby having no connection with the jugular vein or, as we have observed in one of our cases, be connected with the posterior condylar vein. In extremely rare cases, the IPS can be completely absent.

The IPS can be identified on the carotid or vertebral angiograms or by selective catheterization (Figs. 9.4, 9.5, 9.10, 9.23, and 9.24). Drawings are presented in Figs. 9.18, 9.19, 9.20, 9.21, and 9.22.

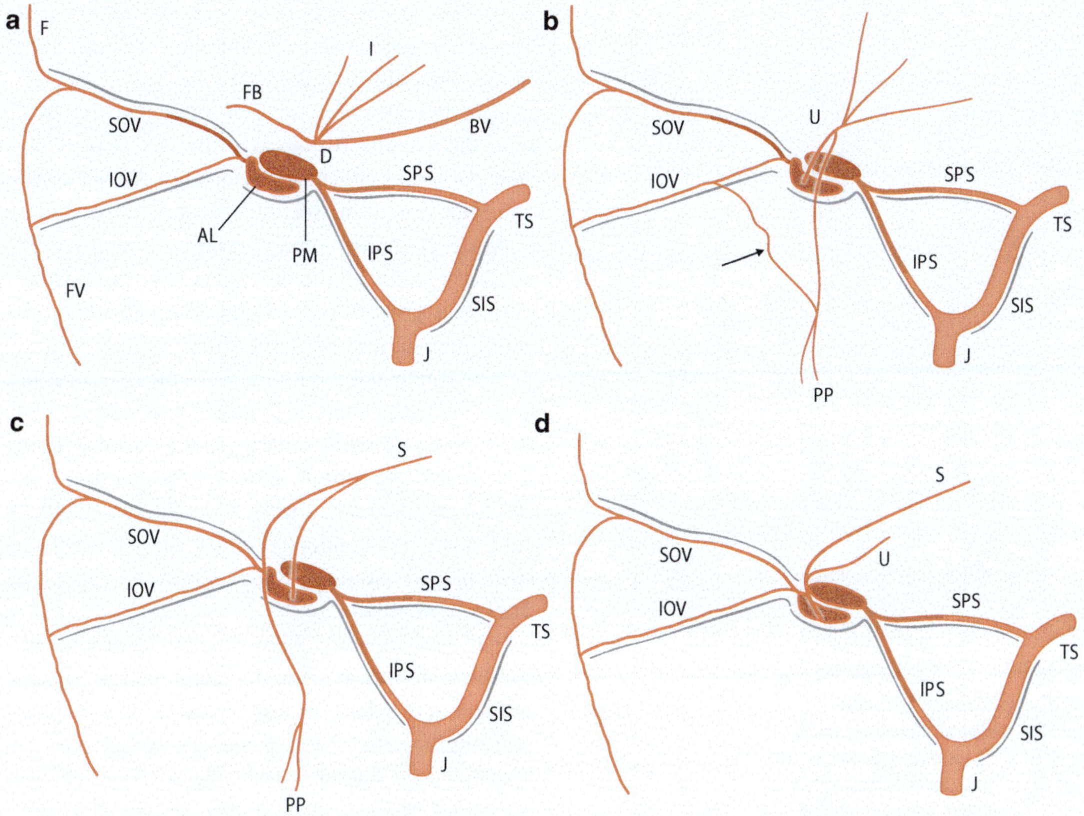

Fig. 9.21 (**a**–**d**) Lateral view. Drawings showing the most frequent ways of drainage *of the deep middle* (*D*) and *superficial middle* (*S*) *cerebral veins. Connections* of *the cavernous sinus with the ophthalmic veins* (*SOV-IOV*) *and petrous sinuses* (*SPS-IPS*) *are also presented. The anterolateral and posteromedial segments of the cavernous sinus between them run the ICA are indicated. The lateral cavernous sinus is not presented in these lateral views.* (**a**) The deep middle cerebral vein with its tributaries (insular and frontobasal veins) drain in the basal vein (BV). The ophthalmic veins drain into the anterolateral part of the cavernous sinus, while the petrous sinuses are connected with its posteromedial part. (**b**) The deep middle cerebral vein receiving the insular veins drains into the uncal vein which can enter the cavernous sinus or alternatively the pterygoid plexus through the paracavernous sinus. Anastomosis of the inferior ophthalmic vein with the pterygoid plexus (*arrow*). Course and drainage of the ophthalmic veins and petrous sinuses. (**c**) Superficial middle cerebral vein draining into the cavernous sinus, or alternatively in the pterygoid plexus through the paracavernous sinus. (**d**) The superficial middle cerebral vein drains into the cavernous sinus. Drainage of the uncal vein into the superficial middle cerebral vein. Explication of the legend. Anterolateral (AL) and posteromedial (PM) segments of the cavernous sinus. Deep middle cerebral vein (D) with a few of its tributaries, insular veins (I), and frontobasal veins (FB). Uncal vein (U). Superficial middle cerebral vein (S). Pterygoid plexus (PP). Superior (SOV) and inferior (IOV) ophthalmic veins. Frontal (F) and facial (FV) veins. Superior (SPS) and inferior (IPS) petrous sinuses. Transverse (TS), and sigmoid (SIS) sinuses. Jugular vein (J)

In examining the venous drainage of the craniocervical region, it should be taken into consideration that this area is a crossing point of several venous channels as described in details by some authors on the basis of anatomical, surgical, and magnetic resonance imaging studies (Katsuta et al. 1997; Arnautovic et al. 1997; San Millan Ruiz et al. 2002). These authors describe the almost constant presence of a venous pouch called the anterior condylar confluence (*ACC*), which has been described already in 1868 by Trolard. It is located extracranially, medially to the jugular vein, in front of the hypoglossal canal, slightly caudal to the junction of the IPS with the jugular bulb. *The ACC is formed by the confluence of several venous channels: the anterior, lateral, and posterior condylar veins and*

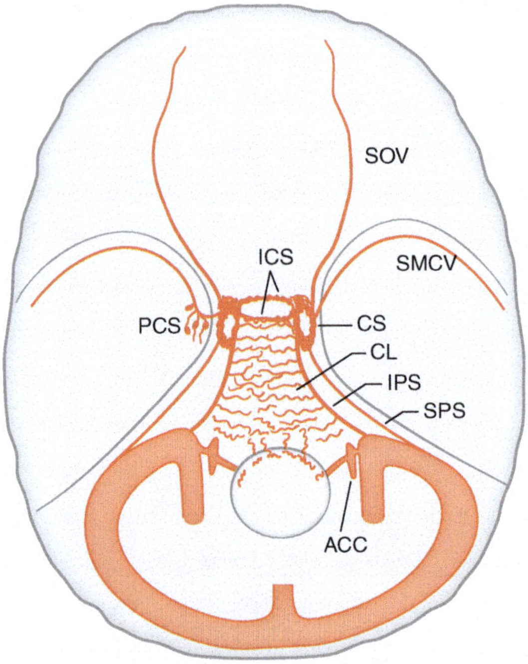

ICS	Intercavernous anastomosis
CS	Cavernous sinus
PCS	Paracavernous sinus
SPS	Superior petrosal sinus
IPS	Inferior petrosal sinus
CL	Clival venous plexus
SMCV	Superficial middle cerebral vein draining in the CS on the right and in the PCS on the left
SOV	Superior ophthalmic vein
ACC	Anterior condylar confluence

Fig. 9.22 Cavernous sinus and its tributaries seen from above. The lateral sinus is not indicated in this drawing

other small anastomoses with the IPS and the jugular bulb. Other connections are the prevertebral venous plexus, the venous plexus surrounding the petrous segment of ICA (Rektorzik plexus) and the inferior occipito-petrosal vein or inferior petro-occipital vein (IOPV). (Drawing 9.19).

The anterior condylar vein passes through the hypoglossal canal, reaching *the anterior internal vertebral plexus (AV)*. This is an epidural venous network connected cranially with the clival venous plexus and marginal sinus, and laterally with *the venous plexus (VA) surrounding the vertebral artery in its horizontal course on the atlas*. This plexus, which has been called by Arnautovic the suboccipital cavernous sinus, has also a connection posterolaterally with the deep cervical vein. It continues downwards in the cervical region with the venous plexus surrounding the vertebral artery in its course through the foramina.

The lateral condylar vein connects the ACC with the VA and further with the AV. *The posterior condylar vein* passes through the posterior condylar foramen, located behind the occipital condyle. It connects the jugular bulb with the VA medially and with the deep cervical vein posteriorly. The posterior condylar vein can also give small connections with the ACC.

The IOPV, named in the literature also inferior petro-occipital sinus or inferior petro-clival vein, is a venous channel running extracranially along the petro occipital suture, connecting the Rektorzik plexus to the ACC or Jugular Bulb. It can, however, connect directly the CS to these structures (Benndorf, 2009; Kurata et al, 2012; Tubbs et al, 2014).

All these venous channels connected with the ACC are not always equally developed or present and so not regularly visible on the carotid or vertebral angiograms. Some connections can only be visualized on selective venograms. However, it is important to be aware of the presence of this venous circuit, since its knowledge is useful in many examinations involving this area. Indeed the IPS is an important way in the cavernous sinus sampling for the diagnosis of pituitary adenoma and can also be used in the endovascular treatment of cavernous sinus fistulae. In this context, it should be mentioned that in the catheterization of the IPS, a frequent occurrence is the entering of the catheter erroneously in one of the condylar veins especially the ACC. Further, the knowledge of the various channels present in the area is essential in the diagnosis and endovascular treatment of dural fistula in the cervico-cranial region (see Sect. 13.7) and of direct fistulae involving the vertebral artery in its course in the area (see Chap. 14).

Finally, as reported in several studies (Epstein et al. 1970; Eckenhoff 1970; Braun and Tournade 1977; Arnautovic et al. 1997; Valdueza et al. 2000; San Millan Ruiz et al. 2002) it should be emphasized that through these connections, the venous drainage of the brain can change with a flow directed prevalently in the internal jugular vein in the supine position of the body or directed in the AV and VA in the upright position, and they can be an important collateral way in case of impaired flow in the internal jugular vein.

9.3.9 Sphenoparietal Sinus (SpS)

The SPS is also known as sinus of Breschet, since it has been described by this author in 1829. It runs within the dura as a continuation of the meningeal veins of the frontoparietal and sometimes also temporal region, accompanied by the anterior branch of the middle meningeal artery (Oka et al. 1985; Tubbs et al. 2007b). Reaching the base of the skull, it separates from the artery and runs in the dura along the undersurface of the lesser wing of the sphenoid bone draining into the cavernous sinus. Alternatively it can turn towards the floor of the middle cranial fossa draining in the paracavernous sinus and further in the pterygoid plexus. More rarely it can run along the middle cranial fossa reaching the superior petrosal or the transverse sinuses. This latter pattern, also called sphenopetrosal sinus, is considered a remnant of the tentorial sinus through which, in the embryonic age, the SMCV drains into the TS (Padget 1956). To the sinus converges commonly the SMCV (see also Sect. 9.1.1.2).

9.3.10 Cavernous Sinus (CS)

The commonly so-called cavernous sinus (*CS*) is a paired structure lying adjacent to either side of the body of the sphenoid bone (Drawings 9.19, 9.20, 9.21, and 9.22). It originates from a fold of the dura which diverges in two layers: one medial (periosteal), adherent to the periosteum of the sphenoid, and one lateral forming the medial wall of the middle cranial fossa. The lateral layer also forms the roof of the sinus, continuing medially into the diaphragma sellae. The CS extends from the superior orbital fissure anteriorly to the petrous apex posteriorly. Actually, as also described in Sect. 2.2.2, it is not a typical venous sinus, but a space containing vascular (ICA, and veins) and nervous elements called by Taptas (1982) *space of the cavernous sinus.*

As far as it concerns the venous component, according to the anatomical studies of Parkinson (1973, 1982), Parkinson and West (1982), and Taptas (1982), it is considered to be a venous plexus formed by a network of several small veins. The plexus of each side communicates via the anterior and posterior small channels running within the dura of the diaphragma sellae forming the so-called coronary sinus which surrounds the pituitary gland. The presence of the ICA leads to the formation in the cavernous space of two venous compartments: one anterolateral and the other posteromedial, which have varying connections to each other and connections with several afferent and efferent veins.

Anteriorly, the venous plexus receives the superior and inferior ophthalmic veins passing through the superior orbital fissure. *Laterally*, it is joined by the SMCV directly or through the SpS and by the uncal vein (see also Sect. 9.1.2.7). The SMCV and the uncal vein can instead drain in the *paracavernous sinus* (*P*) continuing in the *pterygoid plexus* or in the so-called *lateral cavernous sinus* (*LCS*). This latter is a small venous space, described by San Millan Ruiz et al. (1999) and Gailloud et al. (2000), located within the lateral wall of the CS. Many connections can be present between cavernous, lateral cavernous, and paracavernous sinuses in extremely various combinations which can be recognized on the angiogram as has been well demonstrated in the study of Ide et al. (2014). Furthermore the SMCV can drain into the transverse sinus through the persistent primary tentorial sinus.

The posteromedial compartment is connected with the SPS, IPS, and a network of dural veins extending on the clivus down to the foramen magnum (Hanafee et al. 1965; Doyon et al. 1974). The clival plexus is connected caudally with the anterior internal vertebral venous plexus and lateroposteriorly with the marginal sinus.

Bridging veins connect the posteromedial compartment with the pontine veins (Matsushima et al. 1983; Kiyosue et al. 2008) (see also Sect. 9.2.1, Drawing 9.13, and Figs. 9.15 and 9.16).

Aspects of the CS, LCS, and paracavernous sinuses as well as their tributaries are presented in Drawings 9.20, 9.21, and 9.22. Angiographic studies are shown in Figs. 9.4, 9.5, 9.7a, 9.8, 9.9, 9.10, 9.23, and 9.24.

9.3.11 Superior Ophthalmic Vein (SOV)

The superior ophthalmic vein arises near the roof of the nose in the superomedial angle of the orbita, by the junction of the superior tributaries' continuation of the frontotemporal veins and inferior tributaries' continuation of the angular vein, terminal branch of the facial vein (Drawings 9.20, 9.21, and 9.22). It runs backward and laterally, first outside the muscle cone, and then it enters the cone, running further laterally along the undersurface of the superior rectus muscle. In its course the vein crosses superiorly the ophthalmic artery and the optic nerve. The vein takes finally a medial and downward course, leaving again the muscle cone, passing the superior orbital fissure, and entering the anterolateral compartment of the venous plexus of the cavernous sinus (Hanafee et al. 1965, 1968; Lombardi and Passerini 1967; Doyon et al. 1974). *It can occur that the drainage is not in the CS but in the paracavernous sinus. This variant explains probably the filling of the SOV in some type of fistula without involvement of the cavernous sinus as it has been described by some authors* (Theron et al. 1975; Bradac et al. 1981b) (see also Sect. 13.7 and Fig. 13.9). *The same anomaly explains the uncommon venous pattern we have observed in a case of intraorbital angioma* (Chap. 23, Fig. 23.5).

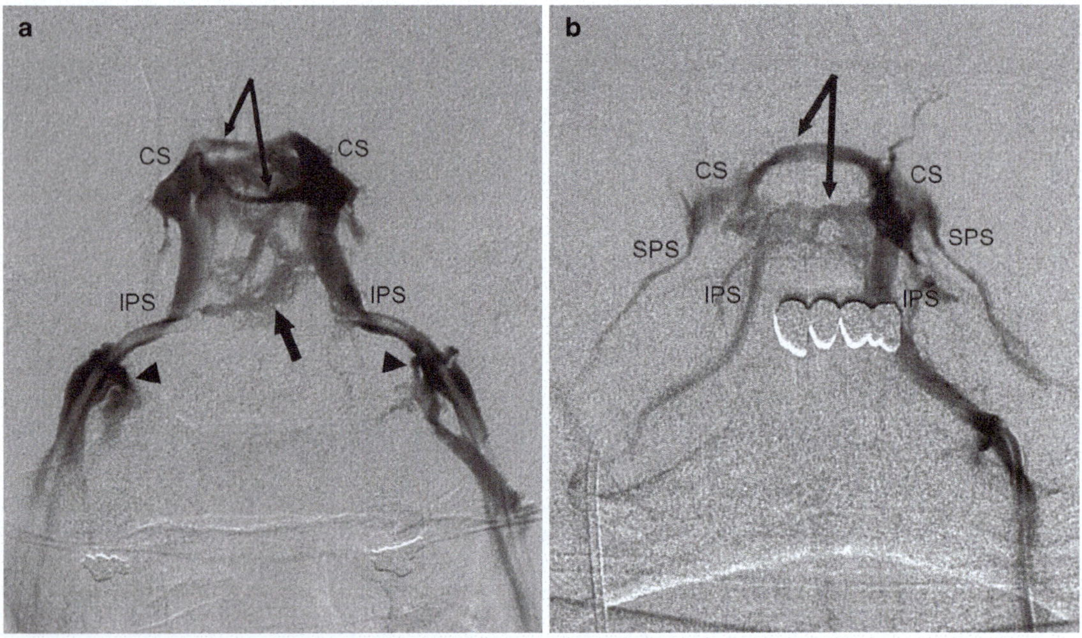

Fig. 9.23 Selective catheterization of the inferior petrosal sinus. (**a**) Catheter in both IPS. Cavernous sinus (CS). Connection between both cavernous sinuses (coronary sinus, *arrow with angle*). Clival venous plexus (*arrow*). Anterior condylar confluence (*arrowhead*). Filling also of the lateral condylar vein, better visible on the left. (**b**) Injection in the left IPS with retrograde injection also of the right IPS. Cavernous sinus (CS), superior petrosal sinus (SPS). Connections between the cavernous sinuses (*arrow with angle*). (**c**) The injection in the jugular bulb shows that the IPS consists of a network of veins. Partial retrograde injection in the contralateral IPS. Anterior condylar confluence (ACC). (**d**) Bilateral selective catheterization of the IPS. The catheters are placed at the distal end of both IPS. Injection of contrast medium through the left catheter. Retrograde filling of the left lateral condylar vein (LCV). Through the ACC there is the injection of the AV connected with the clival venous plexus (*arrow with angle*). Intercavernous sinus connections (*arrows*). Retrograde filling of the right IPS, the right lateral condylar vein (RCV) continuing distally with the vertebral artery venous plexus

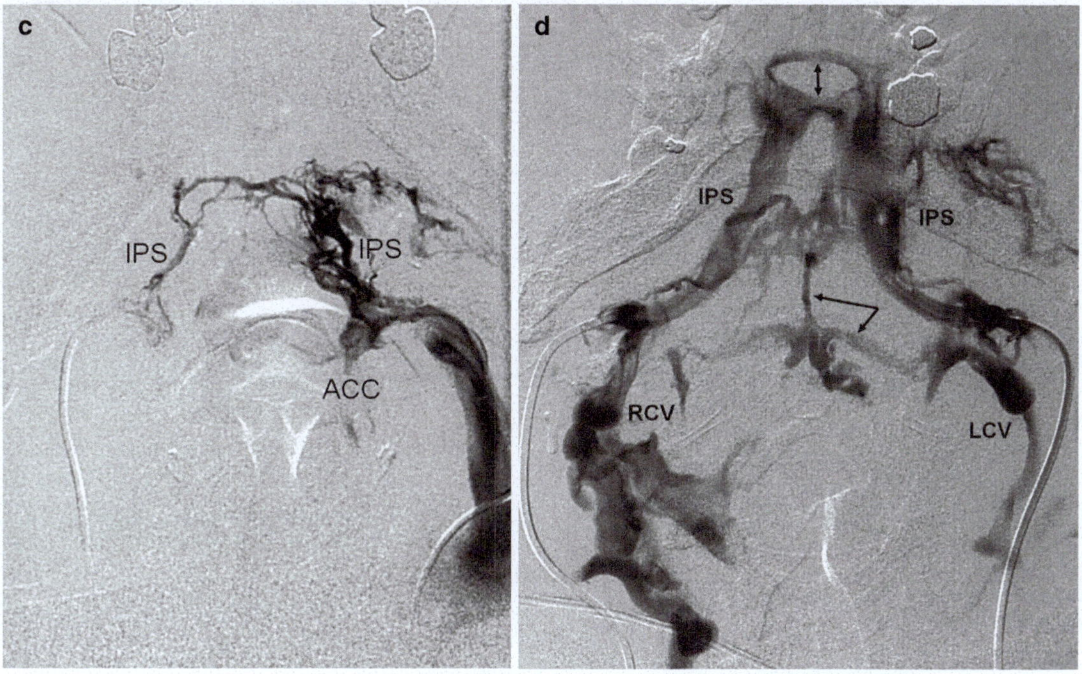

Fig. 9.23 (continued)

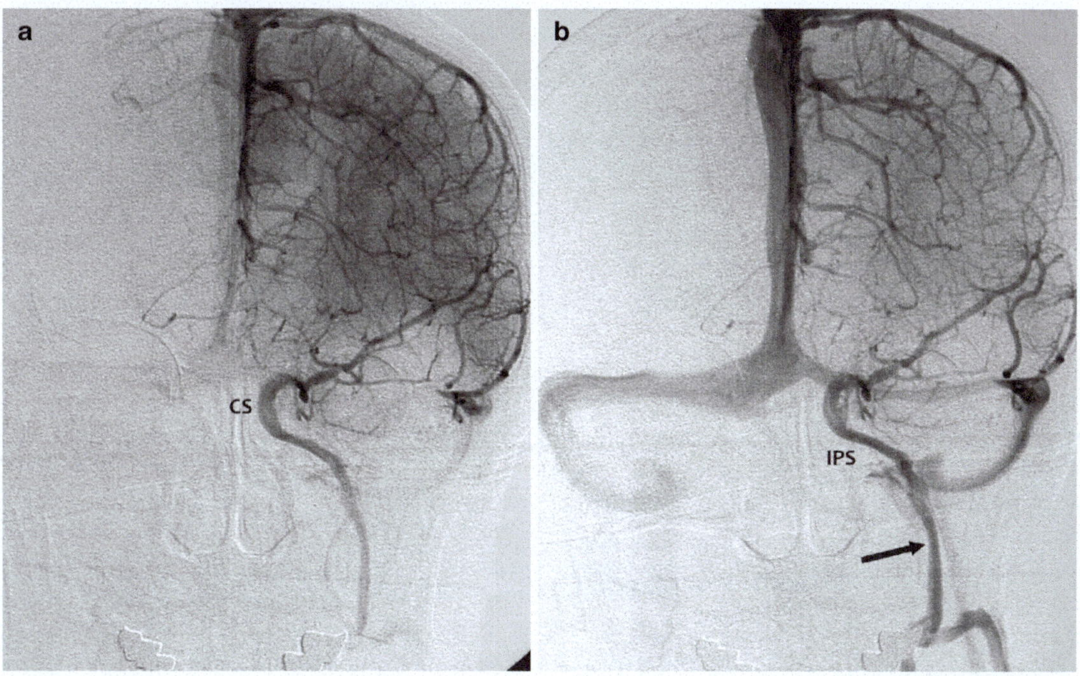

Fig. 9.24 Carotid angiogram: venous phase AP. (**a**, **b**) The cavernous sinus (CS) drains into the inferior petrosal sinus (IPS). This latter seems to have a small connection with the jugular bulb as visible in the late phase (**c**). However, the flow is directed mainly towards a venous channel (*arrow*) not connected with the IJV. Considering its course it seems to be a condylar vein, probably the lateral draining caudally in the vertebral artery venous plexus (VA) which is further connected with the distal portion of the deep cervical vein. (**d**) Lateral view, same indications

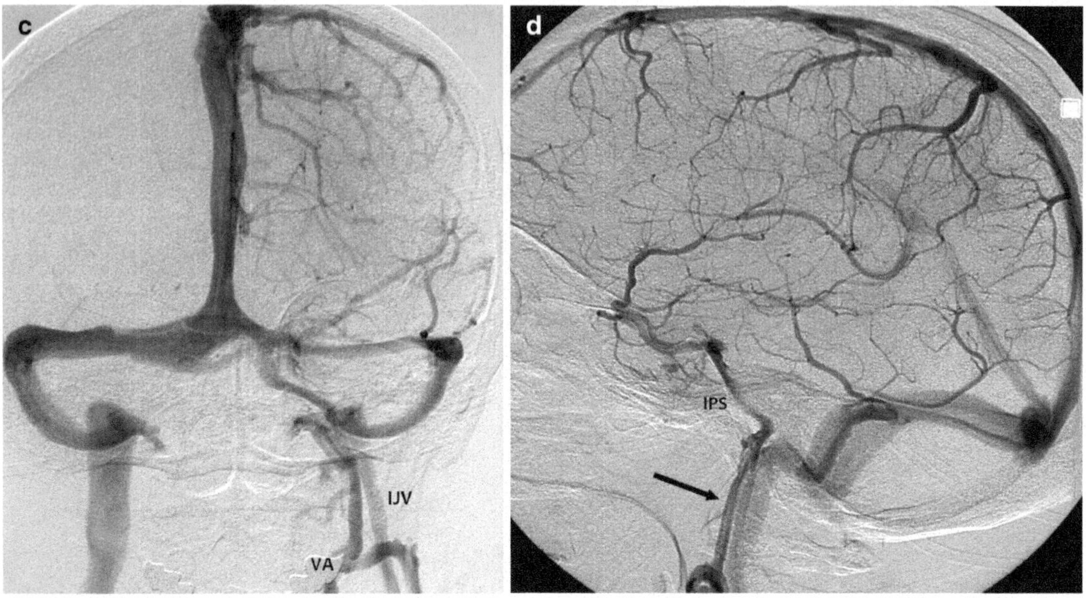

Fig. 9.24 (continued)

9.3.12 Inferior Ophthalmic Vein (IOV)

The inferior ophthalmic is a small vein, sometimes replaced by a network of fine venous channels. It lies within the muscle cone above the inferior rectus muscle. Anteriorly, it communicates with the facial vein and posteriorly it has some connections with the pterygoid plexus through fine anastomoses passing the inferior orbital fissure. The IOV drains into the CS directly or after entering the SOV (Drawings 9.20 and 9.21).

The SOV and IOV drain blood from the intraorbital (globe, muscles, fat, and lacrimal gland) and craniofacial structures through the connections with the facial and frontotemporal veins and pterygoid plexus. This explains also the fact that the ophthalmic veins, especially the SOV, are frequently recognizable in the venous phase of the external carotid angiogram (Tornow and Piscol 1971; Hacker and Porrero 1969; Bradac et al. 1974).

The SOV is an important route in the treatment of fistulae involving the cavernous sinus. This is particularly true when the venous drainage is directed predominantly in the SOV or in the cases in which the SOV is the unique possible route, since the IPS (thrombotic occlusion, aplasia, or anomalous drainage) or the IOPV *cannot be catheterized* (Agid et al. 2004; Kirsch et al. 2006; Kato et al. 2007; Kurata et al, 2012).

9.3.13 Considerations About Vein Variations

- As described in the previous sections, variations in the venous drainage involving the veins of the brain parenchyma and sinuses are frequent and numerous. They appear, however, with a frequent repetitive pattern allowing to consider them as *normal* anomalies.
- In some cases the anomalies can be conspicuous, as it occurs, for example in the DVAs (Chap. 12, Sect. 12.6), which have been interpreted in the past as pathological venous malformations (venous angiomas). Actually they are considered only variations without pathological significance, even if, in rare cases, thrombosis can occur.
- In other very rare cases, the variations are very complex involving a great part of the venous system, including the superficial, deep veins,

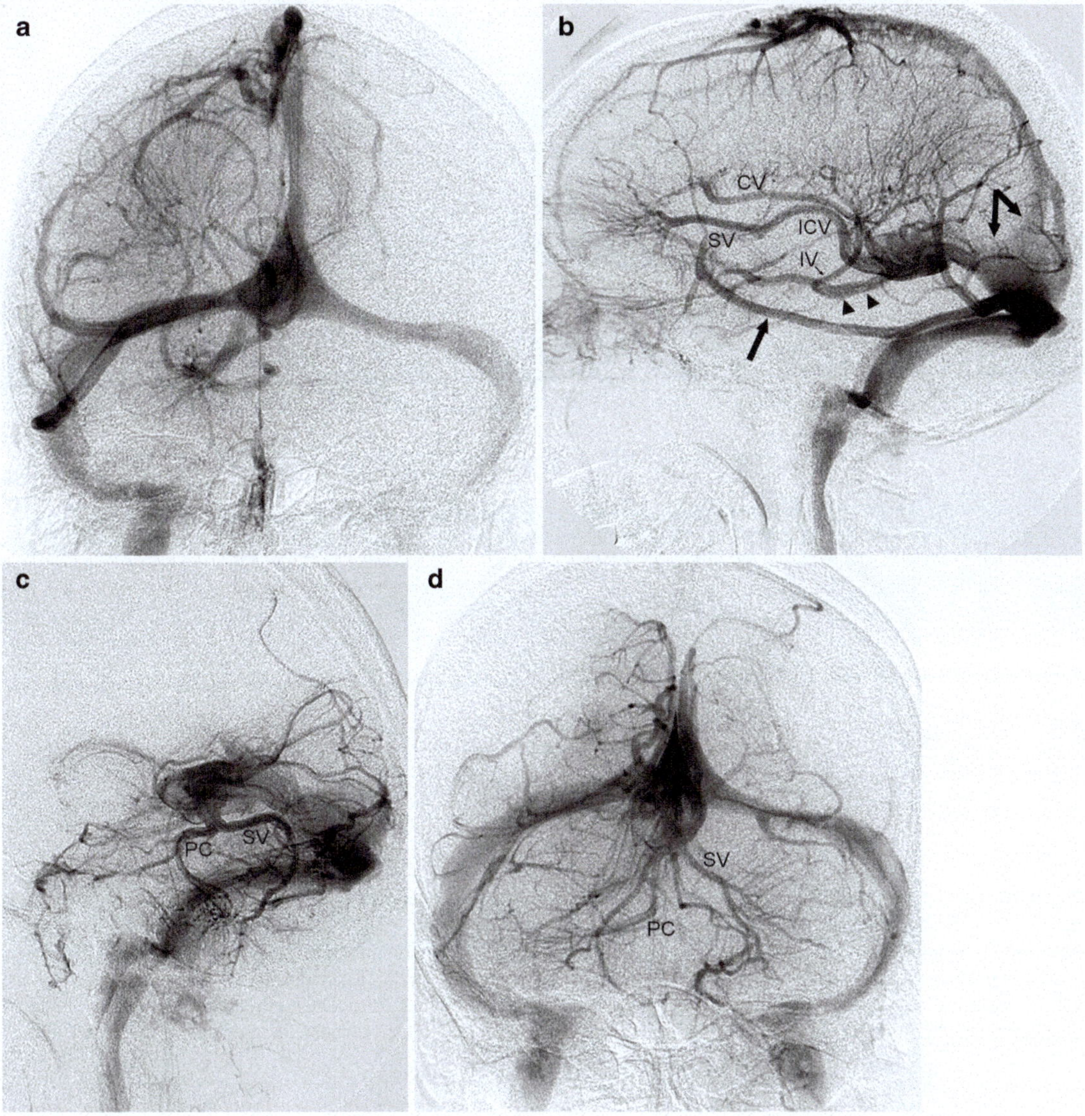

Fig. 9.25 Incidental discovery of grossly anomalies of the venous drainage. (**a**, **b**) On the AP and lateral views of the venous phase of the right carotid angiogram there are a diffuse dilatation of the medullary veins draining into the deep system. The drainage in the superficial system is poor with exception of a large superficial middle cerebral vein which drains posteriorly in the TS after a course in the tentorium (*arrow*). In the frontal and parietal regions the dilated medullary veins converge to dilated collectors represented by the septal vein (SV) continuing in the ICV and the caudate vein (CV) which has an anomalous course entering the ICV. It is possible that also the inferior ventricular vein (IV) and the basal vein (*arrowheads*) act as collectors. This pattern is not much different from that of DVA. There is also a dilatation of the Galen vein and of the distal straight sinus receiving a duplicated SSS (*arrow with angle*). (**c**, **d**) On the AP and lateral views of the venous phase of the vertebral angiogram, dilated medullary veins are also visible. The collectors seem to be a dilated precentral vein (PC) and a superior vermis vein (SV)

and sinuses. They can be an incidental discovery or diagnosed in association with venous thrombosis or other vascular intracranial or intraorbital pathologies (Figs. 9.25, 9.26, 20.5, and 23.4).

- Variations can involve the sinuses in association with other venous anomalies or to be the primary pathology in pediatric dural arteriovenous fistulas (Chap. 13, Sect. 13.8).

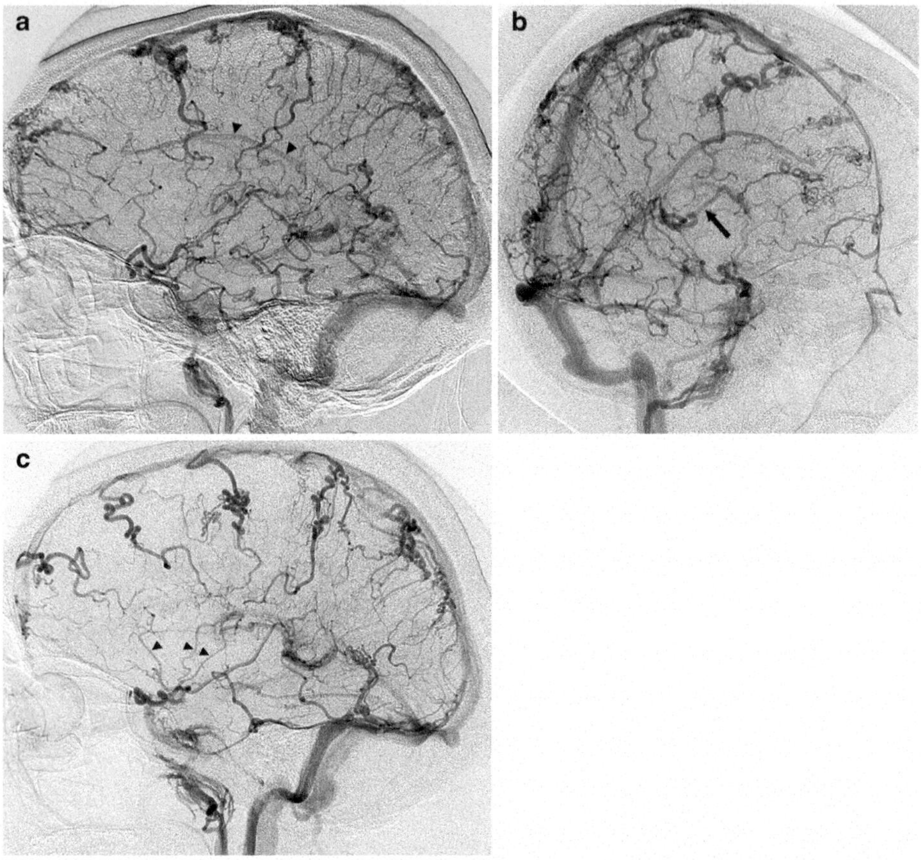

Fig. 9.26 Incidental discovery of grossly venous anomalies involving predominantly the superficial venous system bilaterally. The arterial phases were normal. (**a**) Right ICA lateral angiogram, venous phase. The course of the cortical veins is tortuous forming coiling, sometimes very tight and small network. Minimal tortuosity also of the ICV. Dilated inferior sagittal sinus (*arrowhead*). (**b**) Oblique view showing better the small tortuosity also of the ICV (*arrow*). (**c**) Left ICA lateral angiogram. Venous phase. Similar pattern as on the right. Note the dilated insular veins (*arrowheads*) continuing in a tortuous and dilated venous channel which could be a dilated uncal vein entering the cavernous sinus

10 Extracranial Venous Drainage

10.1 Orbital Veins

The orbital veins develop from the primitive maxillary and supraorbital veins, tributaries of the cavernous sinus developed from the pro-otic sinus (see also embryology, Chap. 9). The final result of the embryological evolution of these venous channels is the superior and inferior ophthalmic veins, draining posteriorly in the cavernous sinus. They are connected anteriorly with the facial and frontal veins. Other connections are those with the pterygoid plexus (see also Sects. “Superior and Inferior Ophthalmic Veins” 9.3.11 and 9.3.12) (Drawing in Fig. 10.1).

10.2 Facial Veins

The anterior facial vein begins in the naso-orbital angle, with the angular vein, which anastomoses with the ophthalmic veins. From its origin, the vein runs inferiorly and posteriorly with an oblique course along the face; it crosses over the mandible, and shortly after it is joined by the retromandibular vein; together they form the common facial vein, which enters the internal jugular vein at the level of the hyoid bone.

The anterior facial vein receives tributaries from the orbit, lips, facial skin, and muscles and from the menton-submental region. Occasionally, the lingual and superior thyroid veins also join the anterior facial vein instead to drain separately into the internal jugular vein. The anterior facial

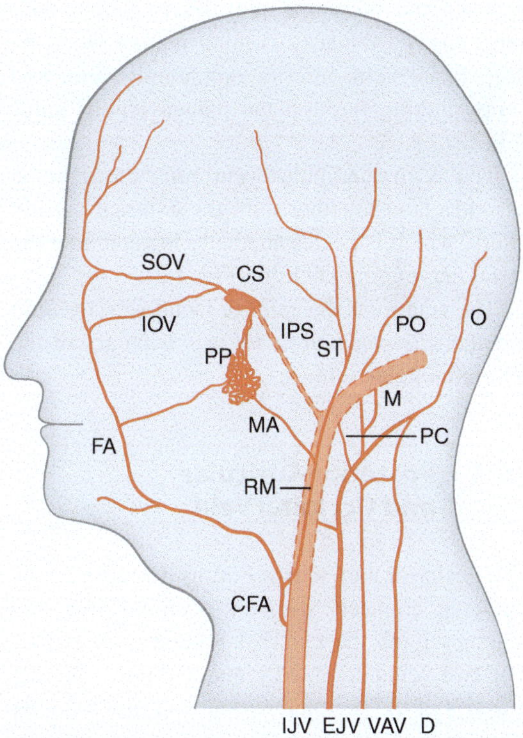

Fig. 10.1 Internal jugular vein (IJV), retromandibular vein (RM), to which converge the superficial temporal vein (ST) and the maxillary vein (MA). Pterygoid plexus (PP), facial vein (FA), common facial vein (CFA), formed by the confluence of the RM and FA entering the IJV. Posterior auricular vein (PO), superficial temporal vein (ST), occipital vein (O), external jugular vein (EJV), to which converge PO and O. Mastoid vein (M), posterior condylar vein (PC), deep cervical vein (D), to which drain partially the occipital vein and venous plexus of the vertebral artery (VAV), with some of its connections. Superior–inferior ophthalmic veins (SOV–IOV), connected anteriorly with the facial and frontal veins and posteriorly with the cavernous sinus (CS). Inferior petrosal sinus (IPS)

G.B. Bradac, *Applied Cerebral Angiography*, DOI 10.1007/978-3-319-57228-4_10

vein is connected posteriorly with the pterygoid plexus through the deep facial vein (Fig. 10.1).

10.3 Retromandibular Vein

Also known as the posterior facial vein, the retromandibular vein arises, below the neck of the mandible at the *junction of the maxillary and the superficial temporal veins*. The superficial temporal vein drains the temporal region and the maxillary vein drains the pterygoid venous plexus. The latter is a venous network lying between the pterygoid muscles to which converge many tributaries coming mainly from the deep facial vein, inferior ophthalmic vein, and veins running through the paracavernous sinus (Fig. 10.1).

The retromandibular vein runs through the parotid gland together with the external carotid artery, which is posteromedially located, and the facial nerve lying laterally (Fig. 2.2).

The vein joins the anterior facial vein forming together *the common facial vein* tributary of the internal jugular vein.

10.4 Posterior Auricular and Occipital Vein

The posterior auricular vein runs posterior to the pinna; it connects with the occipital vein and together joins the external jugular vein (EJV). The two veins drain the pinna and the scalp and muscles of the occipital region, and fuse together entering the EJV. *The EJV, which receives also an anastomotic branch from the retromandibular vein, runs superficially between the platysma and superficial fascia ending in the brachiocephalic vein, near and lateral to the internal jugular vein (IJV)* (Fig. 10.1).

10.5 Deep Cervical Vein

This vein drains the deep muscles of the neck, between which it runs. The deep cervical vein, which can be very large, drains the occipital vein, and is connected with the posterior condylar vein and the venous plexus of the vertebral artery (San Millan Ruiz et al. 2002) (Fig. 9.19). *Distally, it enters the brachiocephalic vein, either as an isolated channel or after joining the venous plexus of the vertebral artery.*

10.6 Venous Plexus of the Vertebral Artery

This is formed by a venous network surrounding the vertebral artery within its course through the transverse foramina. At the level of the horizontal segment of the vertebral artery (see Sect. 9.3.8), the venous plexus is connected medially with the anterior epidural vertebral plexus and posterolaterally with the deep cervical vein. Other connections are with the posterior and lateral condylar veins and indirectly with the ACC, IJV, and SIS (see also Sect. 9.3.8). Distally, it ends together with the deep cervical vein in the brachiocephalic vein (Drawing in Fig. 9.19). *It is together with the IJV, an important extracranial venous drainage of the brain.*

10.7 Emissary Veins

These veins pass through the cranial vault and link the intracranial dural venous sinuses with the extracranial veins. Their presence is not constant. They can be classified with regard to the dural sinus involved:

1. Those related to the superior sagittal sinus (SSS): an emissary vein can connect the SSS with the vein of the nasal cavity passing through the foramen cecum. Another emissary vein can connect the SSS with the scalp vein in the parietal region.
2. Those related to the torcular herophili and/or transverse sinus (TS) and sigmoid sinus (SiS): the most frequent are the mastoid vein, which connects the TS with the posterior auricular or occipital veins and passes through the mastoid foramen, and the posterior condylar vein, which arises from the SiS or jugular bulb and

is connected with the venous plexus surrounding the vertebral artery. In this group, we would include also the anterior condylar vein, which passes through the condylar foramen and connects the ACC with the epidural anterior vertebral plexus (Fig. 9.18). A rare connection can be present between the TS–torcular herophili and occipital vein.

3. Those related with the cavernous sinus (CS): a small anastomosis connects the CS and pterygoid plexus via the paracavernous sinus. An infrequent connection is presented by the petro-occipital sinus, linking the CS with the ACC.

10.8 Diploic and Meningeal Veins

The diploic veins run within the diploe which are connected with the scalp veins, dural sinuses, and meningeal veins. *The meningeal veins* run in the dura accompanying the meningeal arteries. They drain superiorly into the SSS directly or through the venous lacunae, posteriorly in the TS, and inferiorly in the sinuses along the cranial base. The largest meningeal vein is that associated with the anterior branch of the middle meningeal artery continuing in the sphenoparietal sinus (see Sect. 9.3.9).

10.9 Internal Jugular Vein

The IJV is the inferior continuation of the SiS, and it begins in the posterior segment of the jugular foramen, forming at its origin a small dilatation called the jugular bulb, which reaches its final diameter of about 10 mm within the first year of age (Okudera et al. 1994). Extracranially, the IJV runs in the carotid space (Fig. 2.1), lateral to the internal carotid artery, and terminates in the brachiocephalic vein. The latter is very short on the right side. The IJVs are frequently asymmetric, one being larger than the other. The IJV is the main venous extracranial drainage of the brain parenchyma. Among the extracranial tributaries, the most important is the common facial vein. For the connections of the IJV with the inferior petrosal sinus and condylar veins, see Sect. 9.3.8.

The course and the tributaries of the brachiocephalic veins are well known from anatomic and angiographic studies. On the AP view commonly used in the catheterization of the IJV, this arises from the brachiocephalic trunk laterally to the sometimes very large inferior thyroidal veins and medially to the vertebral vein and EJV. More cranially the vertebral vein after receiving frequently as tributary the distal deep cervical vein joins the vertebral artery forming around it, the vertebral venous plexus, and it is then located medially to the IJV.

11 Aneurysms

11.1 Incidence

The precise incidence of cerebral aneurysms is unknown. From autopsy studies, it is estimated at 5% (Stehbens 1972, 1990).

11.2 Type and Location

Cerebral aneurysms are commonly of a saccular type (berry aneurysms) located at the bifurcation of the arteries of the circle of Willis. The most frequent sites are the internal carotid artery (30–35%). Among them, those in the area of the junction of the posterior communicating artery (PcomA) account for more than the half. Also very frequently are aneurysms involving the area of the anterior cerebral artery (ACA: 33–34%) and those of the middle cerebral artery (MCA: 20%).

Aneurysms in the posterior circulation are less common (10%). Of these, half are basilar artery aneurysms (Locksley 1966a, b; Weir 1987; Nakstad et al. 1988).

Multiple aneurysms occur with an incidence that varies from 20 to 40% (Locksley 1966a; Nakstad et al. 1988; Rinne et al. 1994). They are more often in women.

It is thinkable that in those patients in whom there is a tendency to develop multiple aneurysms, these do not appear always simultaneously but in different periods of life. This could explain the appearance, sometimes even years (Wermer et al. 2005) after the diagnosis of the first aneurysm, of the so-called *de novo* aneurysm. The presence of multiple aneurysms suggests that at least in these cases we are dealing not with a focal but with a diffuse progressing vascular disease of the cerebral vessels which needs to be monitored (Bruneau et al. 2011) (Fig. 11.16c–f).

11.3 Anatomopathological Aspects

Unlike aneurysms affecting the aorta and other extracranial arteries, which are commonly fusiform, cerebral aneurysms are in the majority of the cases saccular (berry aneurysm) (Stehbens 1990; Powell 1991). They are characterized by a dilated sac attached to the parent artery by a neck which can be narrow or large, sometimes so large as the base of the sac. The sac can have a regular spherical aspect or appear more elongated or lobulated with two or more lobular components. Commonly the site of rupture is on the top of the sac (dome), where sometimes small extroflexions (blebs) can be recognized, indicating the point of rupture. Rarely the rupture occurs at the neck. They vary in size. The majority have a diameter of 5–7 mm, though smaller and larger aneurysms also occur; those larger than 25 mm are termed *giant aneurysm*.

G.B. Bradac, *Applied Cerebral Angiography*, DOI 10.1007/978-3-319-57228-4_11

Histopathological studies show structural changes in the wall, where there is no internal elastic lamina and the tunica media is absent or very thin. Its smooth muscle cells are replaced by connective tissue.

11.4 Pathogenesis

Some considerations about the cerebral arteries should first be made. They are composed of four layers, represented by the adventitia, media, internal elastic lamina, and intima. *There is no external elastic lamina. The arteries have a thin wall and are located in the subarachnoid space, where they have relatively little external support.*

As far as the pathogenesis of the aneurysms is concerned, this remains a little unclear and controversial. The first hypothesis has been that they were congenital. Indeed in the early 1930s, Forbus showed in histological studies a congenital gap defect of the media in the wall of the cerebral arteries, which was interpreted as a locus *minoris resistentiae*, where aneurysm could develop. Other authors (Carmichael 1950; Crompton 1966a, b; Crawford 1959) explained the formation of aneurysms as due to the association of the inborn defect with an acquired degenerative process linked to hypertension and atherosclerosis. Later, however, it has been demonstrated that these gaps are very frequent in normal individuals without the presence of aneurysms, and it has progressively become accepted that *cerebral aneurysm is an acquired lesion* (Glynn 1940; Stehbens 1959, 1972, 1989, 1990), linked to a degenerative process involving the internal elastic lamina and media weakening the wall. Indeed these changes linked to atherosclerosis can appear already in a young age as demonstrated with microscopical studies by Stehbens (1975). These associated with hemodynamic factors such as hypertension, variations of circle of Willis, sometimes arteriovenous malformations, and perhaps also undetected metabolic disorders would favor the formation of the aneurysms. Other authors (Pope et al. 1984; Pope 1989; Ostergaard and Oxlund 1987; Chyatte et al. 1990; Van den Berg et al. 1999) have demonstrated a reduction of reticular fibers owing to a decrease in type III collagen in patients with aneurysm. In a few of these patients, genetic disorders could also be demonstrated (Chyatte and Mjerzejewski 1991).

There are other more uncommon pathological conditions in which cerebral aneurysm can occur. Among them, there are a few systemic diseases of the connective tissue, such as fibromuscular dysplasia, Ehlers-Danlos syndrome, Marfan syndrome, neurofibromatosis, polycystic kidney, and aortic coarctation. In some of these patients, hypertension is also present as an adjuvant factor (Schwartz and Baronofsky 1960; Handa et al. 1970a; Stehbens 1989; Butler 1996; Osborn 1999; Benyounes et al. 2011).

Aneurysms can occur in different types of arteritis and in cases of cardiac myxoma. Other factors reported to promote aneurysm formation are smoking, consumption of alcohol, and oral contraceptives (Hillbom and Kaste 1982; Lindergärd et al. 1987; Juvela et al. 1993). The incidence of aneurysms is reported to be high in some families (Edelsohn et al. 1972; Hashimoto 1977; Crompton 1979; Norrgard et al. 1987).

Finally intracranial aneurysms can occur following a traumatic lesion of the arterial wall leading to dissection and to formation of a pseudoaneurysm (Birley and Trotter 1928; Benoit and Wortzman 1973; Jacobson et al. 1984; Bozzetto-Ambrosi et al. 2006; De Andrade et al. 2008; Nakstad et al. 2008). In the past, angiography was the method of choice to examine patients with craniocerebral trauma. Aneurysm could be easily detected. Today the diagnosis is performed commonly with CT. With this method, in spite of the improved quality, small aneurysms can escape to the routine CT diagnosis. As reported by others, (Gjertsen et al. 2007; Jussen et al. 2012) we think that in patients with severe trauma especially in whom CT has shown an intracranial hemorrhage a study of the vessels with CT angiography or conventional angiography is mandatory to exclude a vascular lesion especially an aneurysm of the cerebral or meningeal arteries.

As far as aneurysms due to dissection, or developing in AVMs or occurring in other rare angiopathies or cardiac diseases, are concerned see Chaps. 12, 16, 17, and 18.

11.5 Clinical Presentation

Aneurysms are typical in patients of adult age with a predominance among women. Aneurysms in pediatric patients are rare; they are more frequent in boys. In 90% of the cases, the aneurysm presents with subarachnoid hemorrhage (SAH) (10 cases of SAH/year/100.000 people). On CT bleeding in the subarachnoid space and in the ventricular system is commonly easily recognizable. The hemorrhage can involve the adjacent parenchyma especially in aneurysm of the middle and anterior cerebral arteries. Extension to the subdural space can occur in aneurysm near the skull basis.

Morgagni was the first to describe an autopsy case of cerebral aneurysm in the 1761. This pathology, however, has been considered for a long time, as a rarity with little clinical relevance. An exception to this general medical thinking has been Gull, who described in 1858 in an autopsy study a series of aneurysms making the consideration that "*this apparent rarity is doubtful … probably due to careless inquiry*." He also suggested that aneurysm could be the cause of SAH. It has been the merit of Symonds only later in the years 1923 and 1924 to have identified some typical symptoms of the SAH and have correlated these with a rupture of a cerebral aneurysm.

Today this clinical condition with its typical sudden onset of headache, stiff neck vomiting, and frequently disturbances of consciousness (drowsiness, confusion, and coma) is commonly easily recognized and correctly diagnosed. The grade of hemorrhage as well as the involvement of the ventricular system, brain parenchyma, and presence of hydrocephalus can be demonstrated on CT. In some cases, due to the very high increase of the intracranial pressure, death can ensue within minutes and hours. In other patients, the bleeding can be minimal (warning leak), the mild clinical symptoms can be misinterpreted, and the diagnosis of SAH can be eluded. Particular clinical aspects, which occur especially in severe SAH, are pulmonary complications and myocardial dysfunction including arrhythmias, troponin elevation, myocardial infarction, and the syndrome called "neurogenic stunned myocardium" (Kono et al. 1994; Urbaniak et al. 2007; Frontera et al. 2008). All these cardiac complications can lead to cerebral infarct (Mayer et al. 1999; Gaita et al. 2002; Jain et al. 2004; Lee et al. 2006c).

In other cases, the patient presents with neuropathy, particularly oculomotor palsy in PcomA and basilar artery aneurysms. Trigeminal neuralgia can also occur in large aneurysms in this location. Visual symptoms, due to compression of chiasma and/or optic nerve, and sometimes hydrocephalus due to compression of the third ventricle can be present in large aneurysms of the anterior communicating artery and carotid ophthalmic lesions. Symptoms as a result of brain compression can occur in large basilar artery and middle cerebral artery aneurysms. The typical cavernous sinus syndrome occurs in intracavernous aneurysms. In exceptional cases, aneurysms present with ischemia. This occurs especially in nonruptured, partially thrombosed aneurysms (Antunes and Correll 1976; Stewart et al. 1980). Ischemia can be due to distal embolization or involvement of perforators adjacent to the aneurysm.

In addition to aneurysms in patients presenting with related symptoms, an increasing number of asymptomatic aneurysms are today diagnosed with CT and MRI in patients who underwent the examination for other reasons. The clinical problems and therapeutic considerations with such aneurysms are discussed in Sect. 11.10.

11.6 Aneurysm Location

11.6.1 Extracranial ICA Aneurysms

Aneurysms in this location are uncommon. Trauma is a possible etiology. Atherosclerosis can be another cause in the elderly. In younger patients, aneurysms may be associated with collagenopathies such as FMD, Ehlers-Danlos, and Marfan syndromes in which dissecting aneurysms are frequent (see Chap. 16).

11.6.2 Intracranial ICA Aneurysms

These aneurysms have been differently subdivided considering their origin and location. This has been and still is useful for the surgical approach. Today these aneurysms are basically endovascular treated and so this classification is less important, as also emphasized by other authors (Shapiro et al. 2014). For a general nomenclature, however, a subdivision remains of utility.

11.6.2.1 Aneurysms of the ICA Petrous Segment

These are very rare. Spontaneous in some cases traumatic dissection can be assumed to be the most frequent etiological mechanism (Fig. 11.1).

11.6.2.2 Cavernous Aneurysms

These are located within the cavernous sinus and are thus extradural. They are frequently large or giant. Pathogenesis is not completely clear. Spontaneous dissection or degenerative changes due to atherosclerosis are probably the etiology in many cases (see also Sect. 11.8). Association with collagene tissue disease is present in some cases. Traumatic and infectious diseases are other possible causes. They can be asymptomatic and incidentally discovered or especially when large, present with the typical cavernous sinus syndrome,

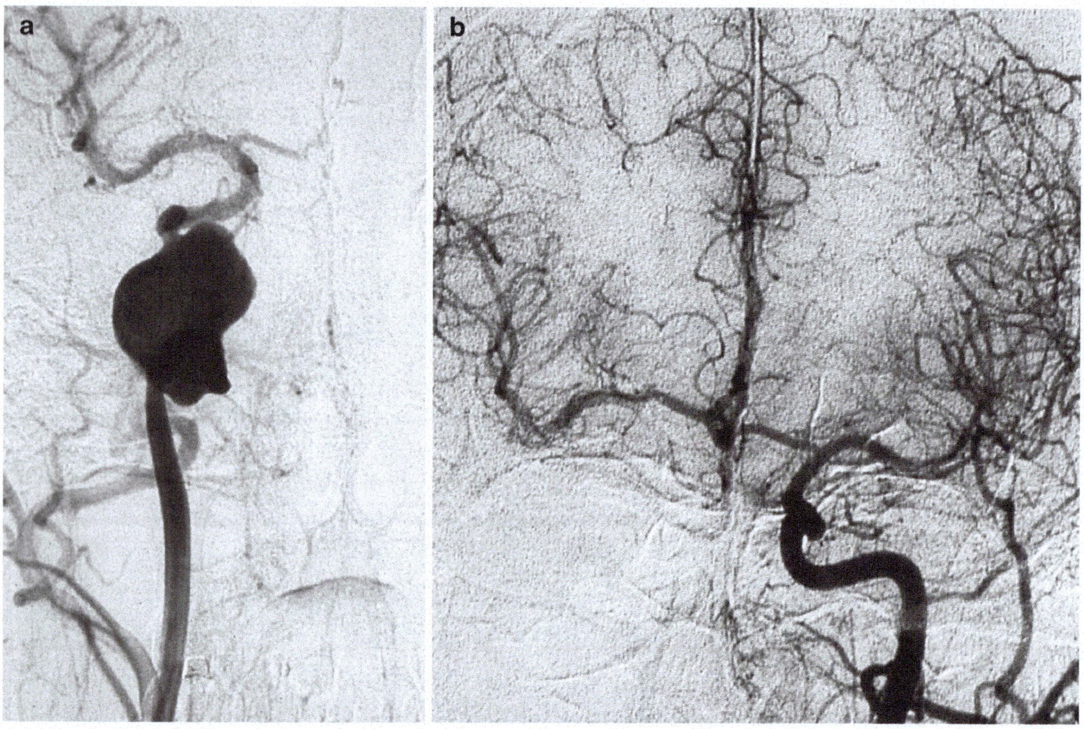

Fig. 11.1 Giant aneurysm of the petrous segment of the right ICA, presenting with palsy of cranial nerve VI. (**a**) ICA angiogram showing the aneurysm which was treated with occlusion of the ICA. (**b**) Left control angiogram, showing the good collateral circulation through the AcomA and exclusion of the aneurysm. The occlusion was preceded by a test as described in Sect. 11.9

first described by Jefferson (1938) and characterized by involvement in various combinations of cranial nerves III, IV, VI, and partially V. Commonly there is no risk of SAH, and in general the clinical prognosis is good, and so in the decision for the treatment of these aneurysms, their size, the symptoms, and the age of the patient should be considered. These aneurysms can rupture in the sphenoid sinus and nasal cavity, leading to a dramatic epistaxis. This is, however, an extremely rare evolution (Linskey et al. 1990).

Occlusion of the aneurysm and ICA or selective occlusion of the aneurysm with stent plus coils and flow-diverter stent are today possible methods (Fig. 11.2a–d).

In some cases, intracavernous aneurysms, even small, can extend upwards, reach the subarachnoid space, and rupture causing SAH. This can occur since the dural ring surrounding the ICA, through which the artery passes to become intradural, is medially not adherent to the artery; a small cavity (cave) is thereby formed. Aneurysms developing here, in spite of their

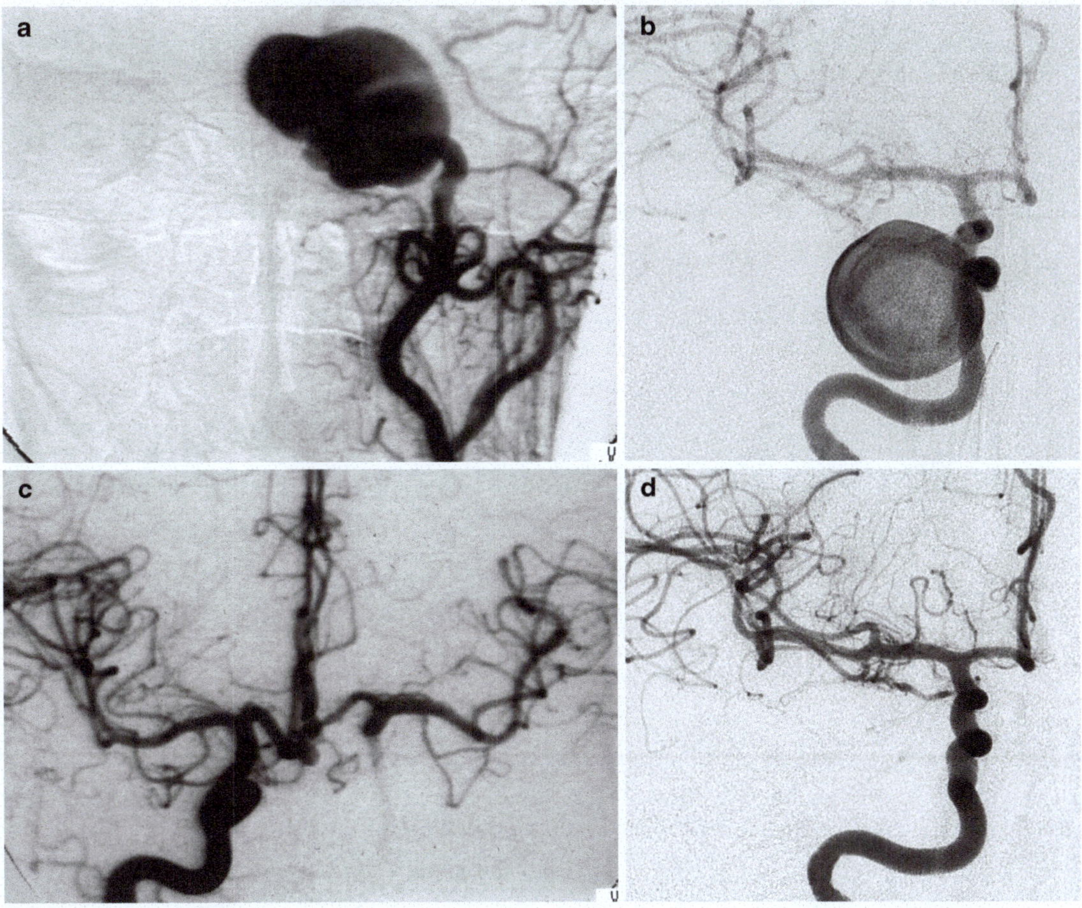

Fig. 11.2 Two examples of giant aneurysms of the cavernous portion of ICA with typical cavernous sinus syndrome treated differently. (**a**) Left ICA angiogram showing the aneurysm. There is no filling of the intracranial branches. (**c**) Right ICA angiogram showing the exclusion of the aneurysm and a good collateral circulation towards the occluded left ICA performed after the test occlusion. Another patient with the same clinical symptoms due to a giant cavernous aneurysm. Right ICA angiogram pretreatment (**b**) and posttreatment (**d**). Occlusion of the aneurysm with flow diverter (silk)

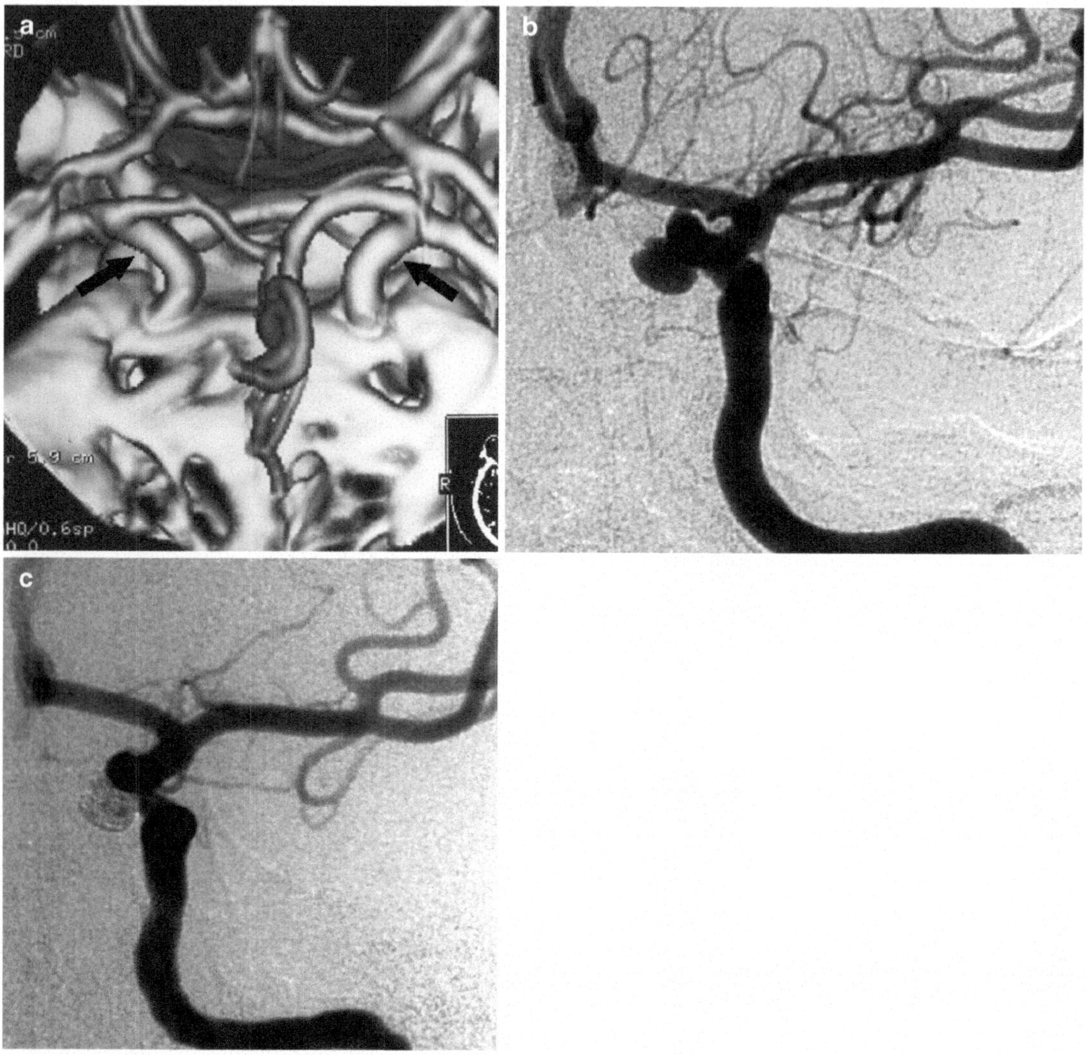

Fig. 11.3 Small left cave aneurysm presenting with SAH. (**a**) On the CT angiography, the aneurysm was not recognizable. The *arrows* show the normal aspect of both ICAs. (**b**) The left angiogram showed the aneurysm which was occluded with coils (**c**)

intracavernous location, can expand intradurally and become responsible of a SHA. These aneurysms are also called *cave aneurysm* (Fig. 11.3).

11.6.2.3 Paraclinoid: Paraophthalmic Aneurysms

In this group, the aneurysms arising in the ophthalmic segment of ICA are described, which are those developing between the origin of the ophthalmic artery and the origin of the PcomA. These comprehend the *carotid-ophthalmic aneurysms*, those *arising from the posterior wall of ICA* among them the superior hypophyseal aneurysms and *those from the superior wall of ICA*.

The carotid-ophthalmic originate at the junction with the ophthalmic artery or close to it. They are directed upward and are frequently large (Figs. 11.4, 11.5, and 11.6).

The aneurysm of the posterior wall arises at the junction of small branches represented by the superior hypophyseal arteries and by other arteries directed to the optic chiasm, optic nerve, and floor of the third ventricle. These aneurysms are very rare; they are directed downward and medially

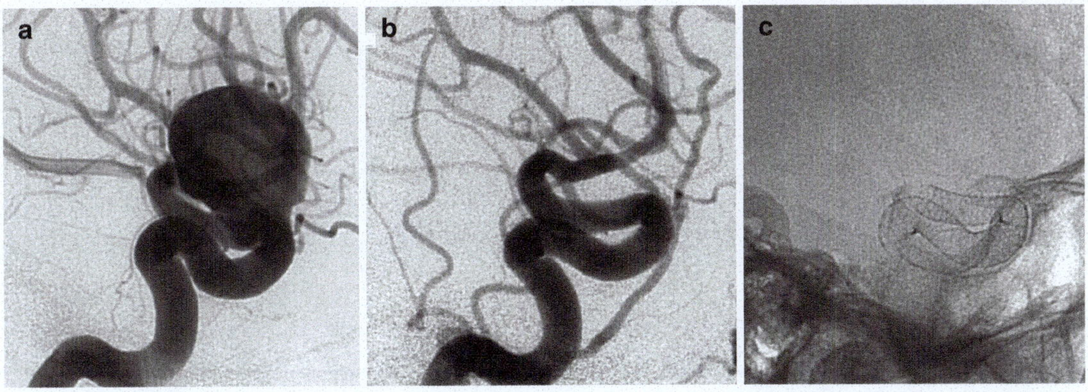

Fig. 11.4 Large carotid ophthalmic aneurysm presenting with progressive worsening of the visual acuity. (**a**) ICA lateral angiogram showing the giant aneurysm. (**b**) ICA angiogram after occlusion of the aneurysm with a flow-diverter stent (pipeline). (**c**) Rx image showing the stent

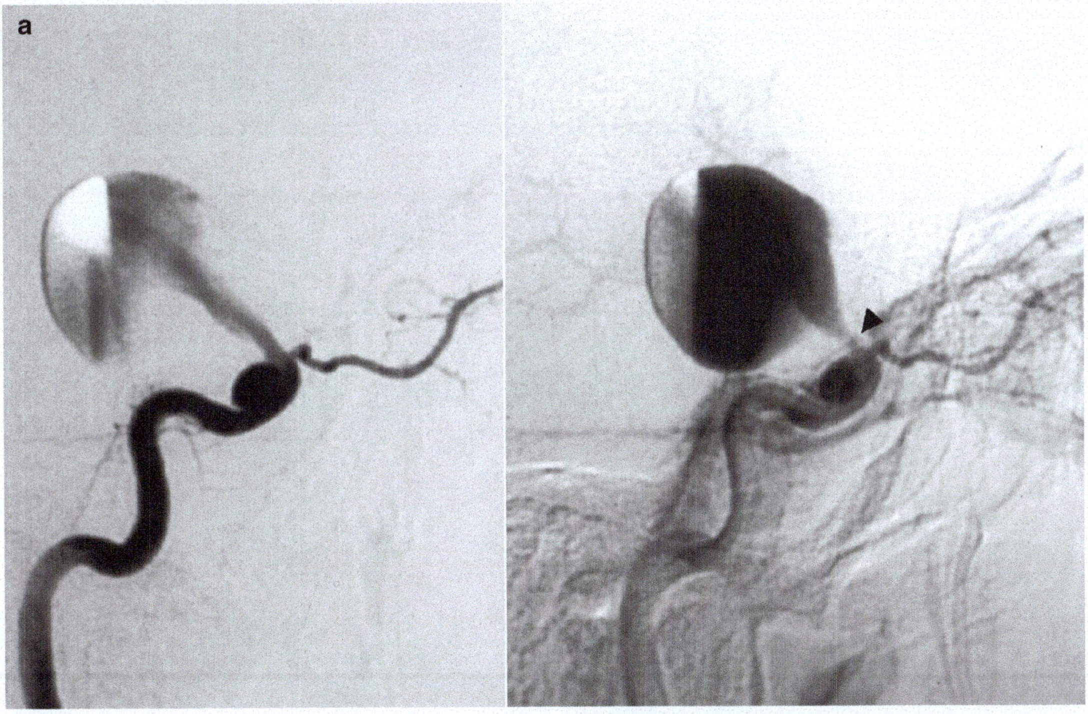

Fig. 11.5 Giant carotid ophthalmic aneurysm presenting acutely with severe disturbances of visual acuity. (**a**) Right lateral carotid angiogram, early and late phases, showing the progressive injection of the aneurysm through a relatively small neck, close to the origin of the ophthalmic artery (*arrowhead*). A second small intracavernous aneurysm is present. (**b**) Left carotid angiogram, upward displacement (*arrowheads*) of both A1 segments and partial filling of the right middle cerebral artery (MCA) through leptomeningeal anastomoses (*arrows*) with the right anterior cerebral artery (ACA). (**c**) Right vertebral angiogram, showing the leptomeningeal collateral circulation towards the right MCA from the posterior cerebral artery (PCA; *arrows*). (**d**) The ICA was occluded with a balloon positioned in the intracavernous segment. The patient recovered completely and a left control angiogram 2 months later showed the normal course of both the A1 and filling of the right MCA. (**e**) Right common carotid angiogram showing the retrograde injection of the distal ICA and MCA through the ophthalmic artery via a collateral circulation involving an anastomosis between the anterior deep temporal artery and the lacrimal branch of the ophthalmic artery (*angled arrow*). The aneurysm is completely excluded

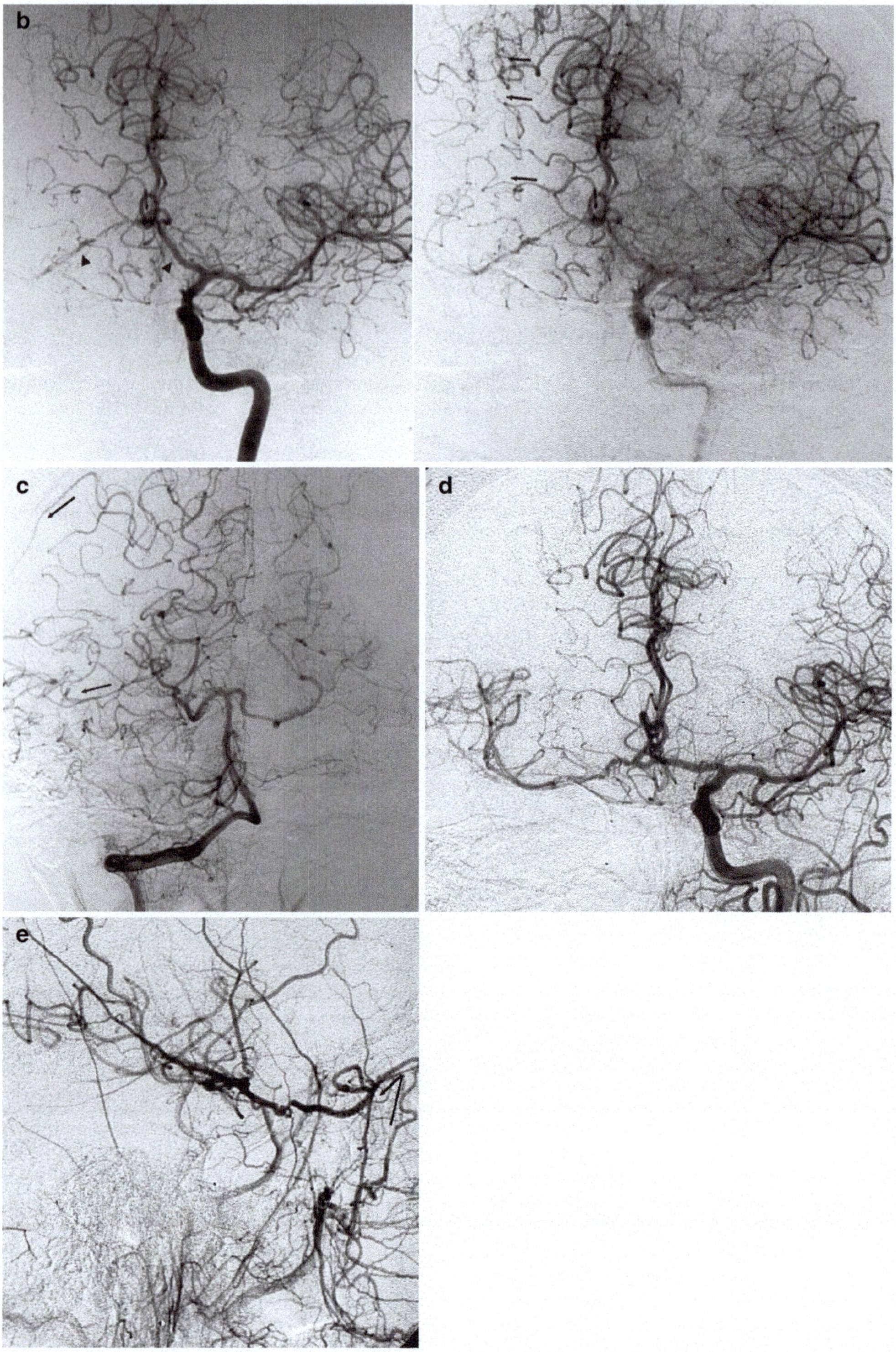

Fig. 11.5 (continued)

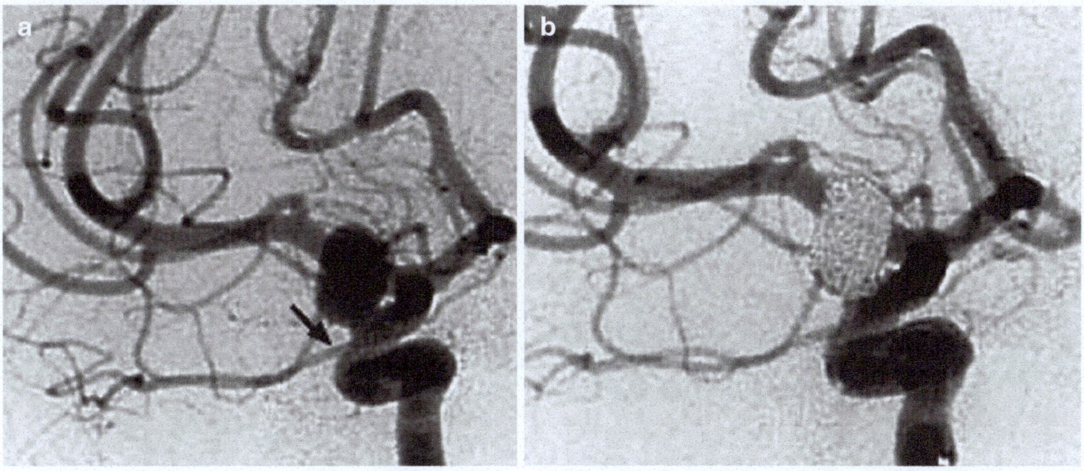

Fig. 11.6 Carotid ophthalmic aneurysm presenting with hemorrhage. (**a**) Carotid angiogram, oblique view showing the aneurysm. Origin of the ophthalmic artery (*arrow*). (**b**) Carotid angiogram showing the occluded aneurysm with coils

toward the chiasma and sella turcica, becoming intrasellar when large (Fig. 11.7).

The aneurysms developing on the anterior wall distally to the carotid-ophthalmic aneurysms are also very rare. They have been called by some authors *nonbranching aneurysms* (Ogawa et al. 2000). These are probably dissecting aneurysms. An example is presented in Fig. 11.8a–b, e–h. In the group of *nonbranching*, a particular form are the *blister-like aneurysms* (Fig. 11.8c, d). These are described more in detail in Chap. 16.

11.6.2.4 Aneurysms of the Communicating and Choroidal Segments

These consist of aneurysms of *the posterior communicating artery (PcomA) and anterior choroidal artery (AchA)*. PcomA aneurysms are *located at the junction of this artery with the ICA*; they are posteriorly directed and can compress cranial nerve III. They can be very large, elongated, and with an irregular wall (Fig. 11.9a, b). More rarely the aneurysm *can arise selectively from the trunk of the PcomA* (true PcomA aneurysm) (Fig. 11.9c). In some of these later cases, the etiopathogenesis is probably a dissection (Mizutani et al. 1998; Nakao et al. 2004; Duncan and Terblanche 2005; He et al. 2011; Kocak et al. 2013). Involvement of the anterior thalamoperforating branches is frequent.

AchA aneurysms arise at the origin of the artery; it can also occur that the aneurysm arises directly from the first segment of the AchA or at the level of perforators of ICA immediately distal to the AchA. They are directed posteriorly and laterally. They are frequently small but large aneurysm can also occur (Figs. 11.10 and 15.25). A variable number of perforators arise from the choroidal segment of the ICA, and can be stretched around the neck of the aneurysm. In endovascular treatment, it is sometimes difficult to identify precisely the neck of the aneurysm despite several projections.

11.6.2.5 Aneurysm of the Carotid Bifurcation

These are typical T-bifurcation aneurysms similar to those located at the basilar-tip (Gibo et al. 1981a; Ingebrigtsen et al. 2004; Van Rooij et al. 2008). They are relatively rare and easily recognizable on the angiogram. These aneurysms can, sometimes, be very large with a large neck which can involve the adjacent A1 and M1 segments. Perforating branches arising from the choroidal segment of ICA and from A1 and M1 are stretched

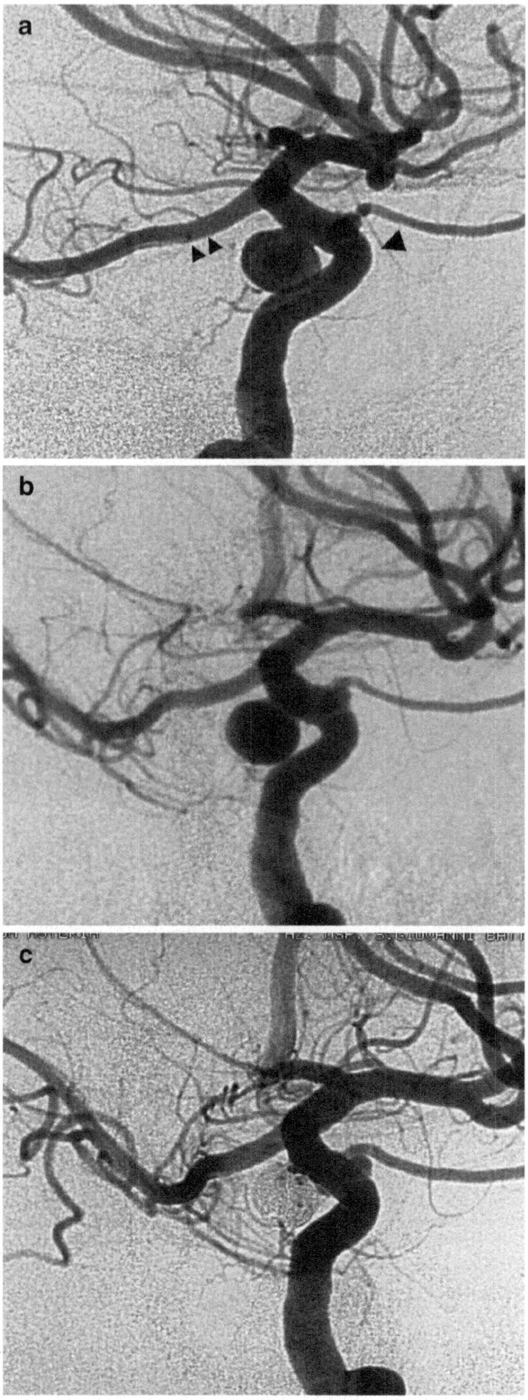

Fig. 11.7 Superior hypophyseal aneurysm presenting with hemorrhage. Two different oblique (**a**, **b**) views of the carotid angiogram to better identify the aneurysm and its neck. PCA (*arrowheads*), ophthalmic artery (*arrowhead*). Control angiogram (**c**) after occlusion of the aneurysm with coils

around the neck and the posterior wall of the aneurysm or in cases of a large neck can be directly involved. Examples are presented in Figs. 11.11 and 5.4b.

11.6.3 Anterior Cerebral Artery Aneurysms

These are very frequent aneurysms arising mainly *in the A1–A2 angle of the ACA*. The neck of the aneurysm can be large and partially involve the adjacent A1–A2 segments or the AcomA. As it has been described in the anatomy, these aneurysms are frequently associated with anomalies of the anterior part of the circle of Willis; furthermore, many perforators arise from the A1 and AcomA. This should be taken into account in the surgical or endovascular treatment of these lesions (Figs. 4.4, 4.9, 4.10, 4.11a, b, 4.13, 4.15, and 11.12a).

The *aneurysm can be found sometimes along the A1 segment*. This is a very rare location (less than 1%). It has been reported that in these cases, the aneurysm takes its origin at the junction with a perforating branch (Handa et al. 1984; Wakabayashi et al. 1985; Suzuki et al. 1992; Wanibuchi et al. 2001). This is certainly true for many of these aneurysms. In other cases, however, a relationship with a perforator could not be demonstrated (Lubicz et al. 2006–2009; Cho et al. 2014; Liu et al. 2016) suggesting a dissection as a possible pathogenesis. They are located more frequently in the proximal part of A1 and can have a saccular or a fusiform shape. Not rarely they are small with a feature similar to the blister-like aneurysm described above. For these aneurysms, surgery has been the therapy of choice (Dashti et al. 2007; Park et al. 2013). With advances in endovascular techniques this treatment has become progressively a good alternative. Flow-diverter stents would be the best treatment Clarençon et al. (2017); however their use in the acute phase is controversial. Depending on the morphological aspect, coils alone or assisted with balloon or stents as well as occlusion of the aneurysm as well as the involved segment of A1 have been used (Lubicz

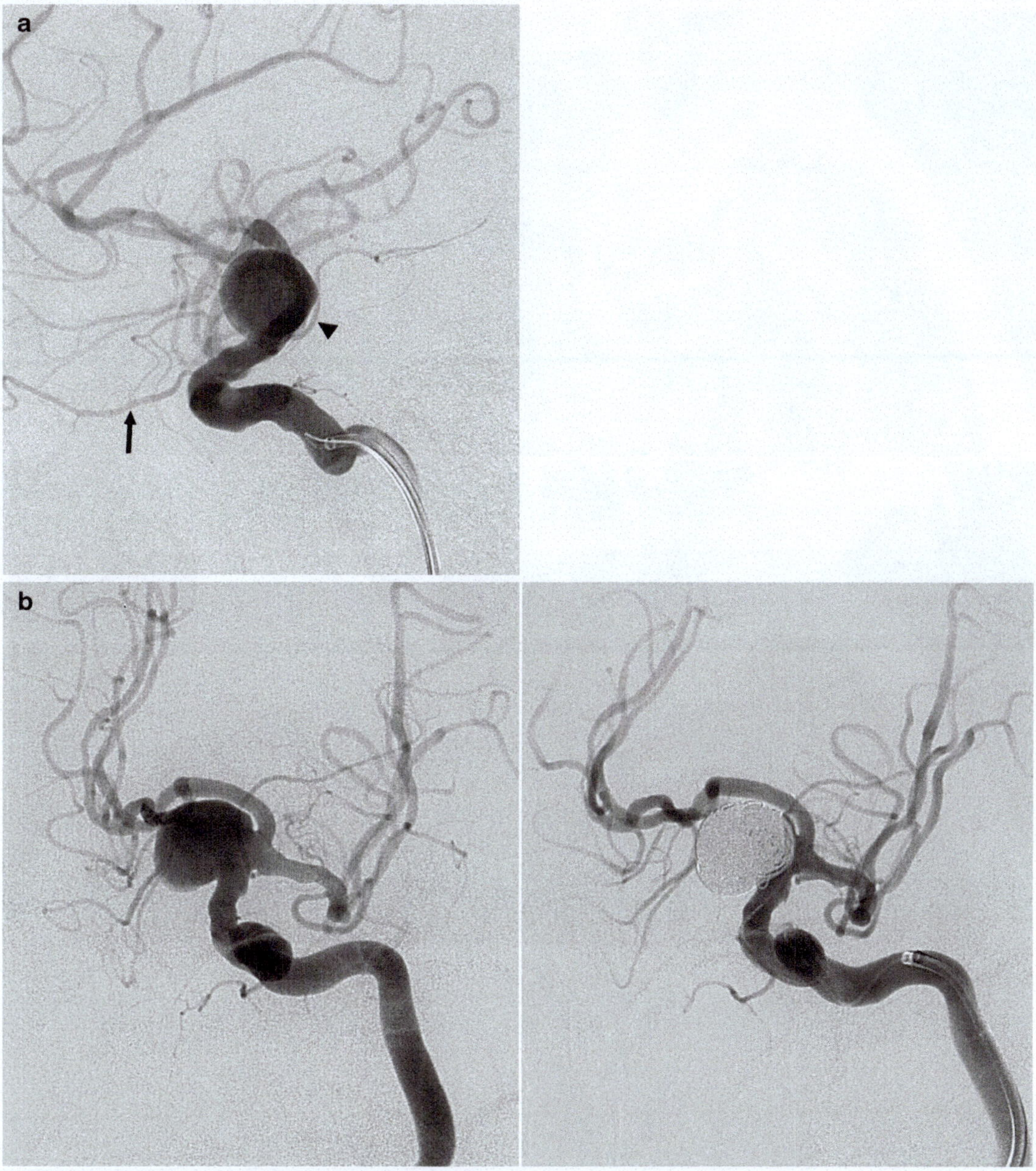

Fig. 11.8 Three different patients with aneurysm arising on the superior wall of ICA. (**a**, **b**) Large aneurysm presenting acutely with disturbances of the visual acuity in a hypertensive 60-year-old man. (**a**) Lateral view. The aneurysm is located on the anterior wall of the ICA near its distal bifurcation. Ophthalmic artery (*arrow*), AchA (*arrowhead*). The image has been performed at the beginning of the treatment. (**b**) Oblique view showing the aneurysm and its large neck. Pretreatment and posttreatment angiograms. Occlusion of the aneurysm with stent (enterprise) and coils. (**c**, **d**) Fifty-year-old woman presenting with SAH. (**c**) Angio-CT, showing a minimal bulge on the superomedial wall of ICA (*arrow*, image above). A week later the bulge is more evident (*arrow*, image below). (**d**) ICA angiogram showing the small bulge (*arrow*, image above), which a week later has become a clear saccular aneurysm (*arrow*, image below) (Courtesy of dott. Venturi F., Neuroradiology of Turin). During the surgical exploration, the aneurysm ruptured. The patient died. (**e**–**h**) Sixty-year-old woman presenting with SAH on the ICA angiogram (oblique view) (**e**), a nonbranching probably dissecting aneurysm was shown on the superior-medial surface of the ICA, which was also a little irregularly fusiform dilated. The aneurysm was acutely treated with coils and stent (enterprise) (**f**). The patient recovered, but a week later she suffered a second severe SAH. The control angiogram (**g**) showed a minimal (*arrow*) recanalization of the aneurysm and irregularity of the ICA wall. A severe spasm was also present. A flow-diverter stent (**h**) was inserted in the already present enterprise. The patient recovered slowly

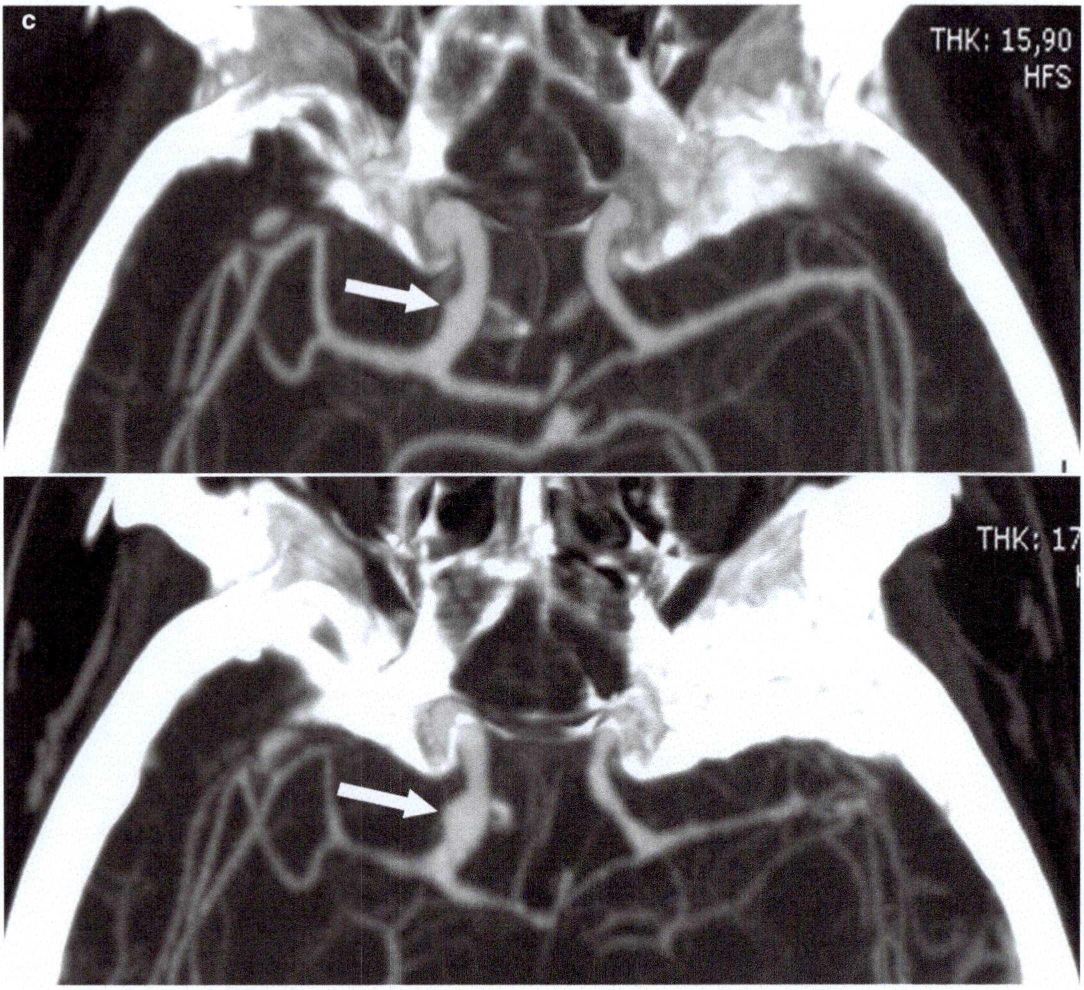

Fig. 11.8 (continued)

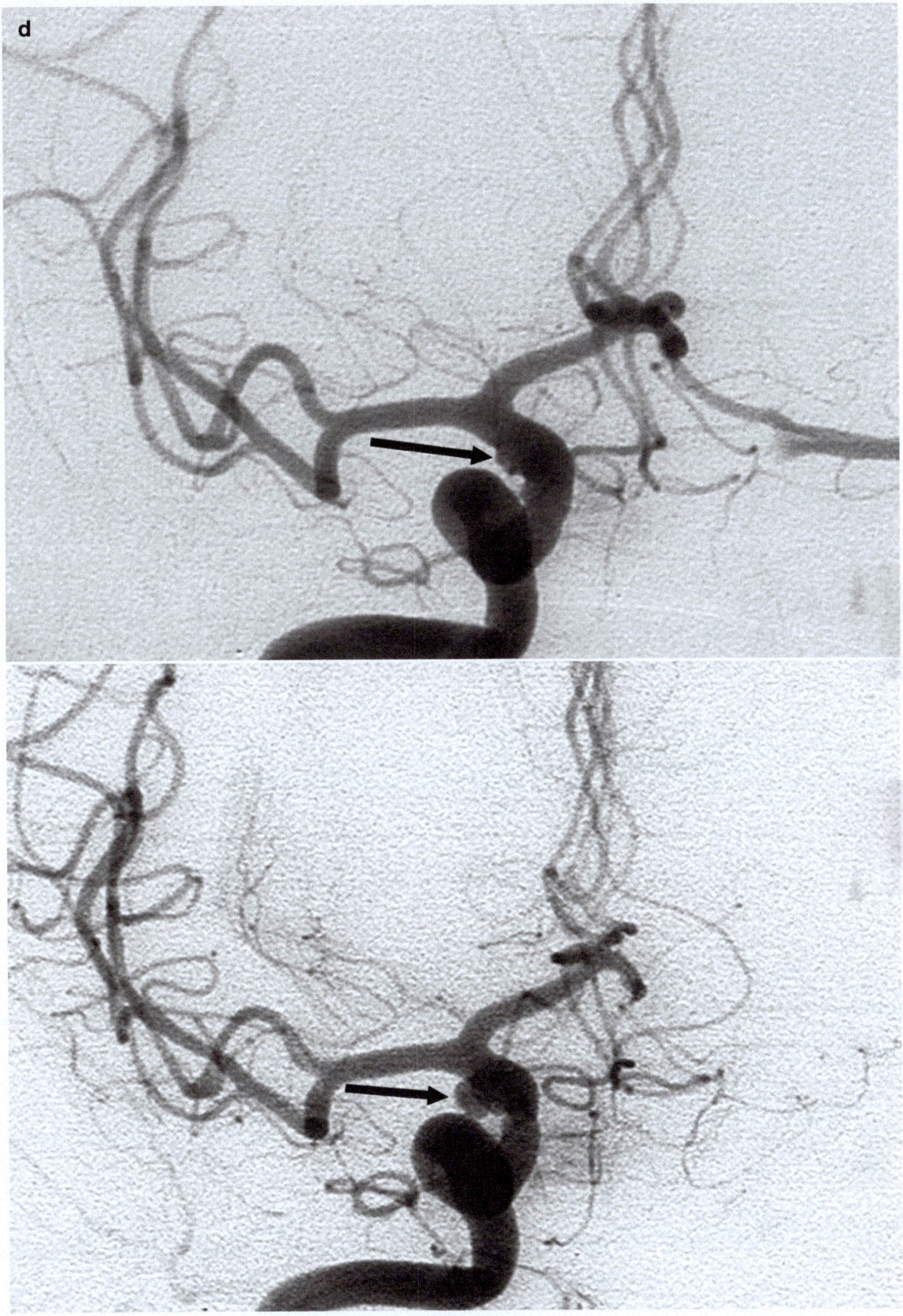

Fig. 11.8 (continued)

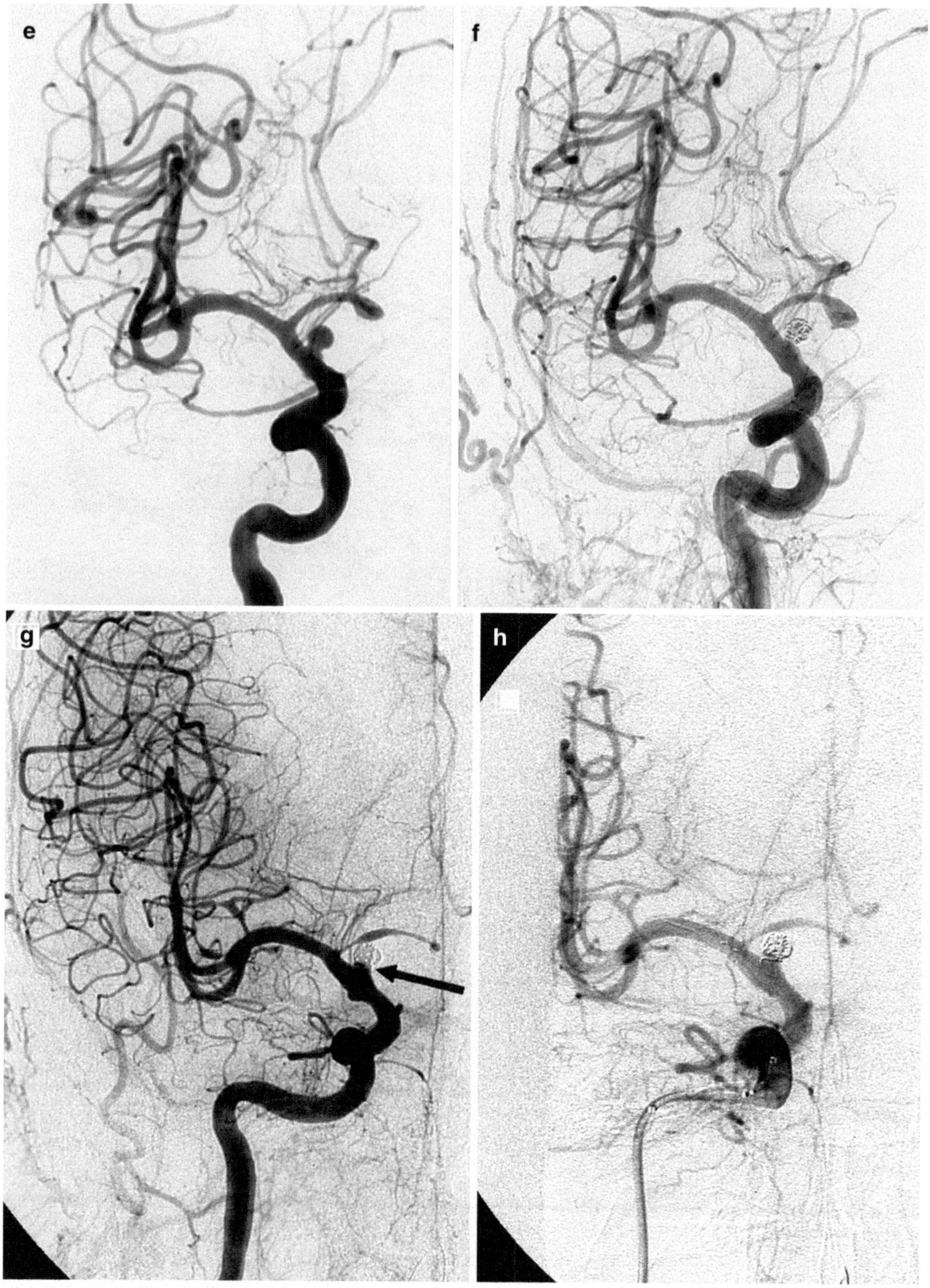

Fig. 11.8 (continued)

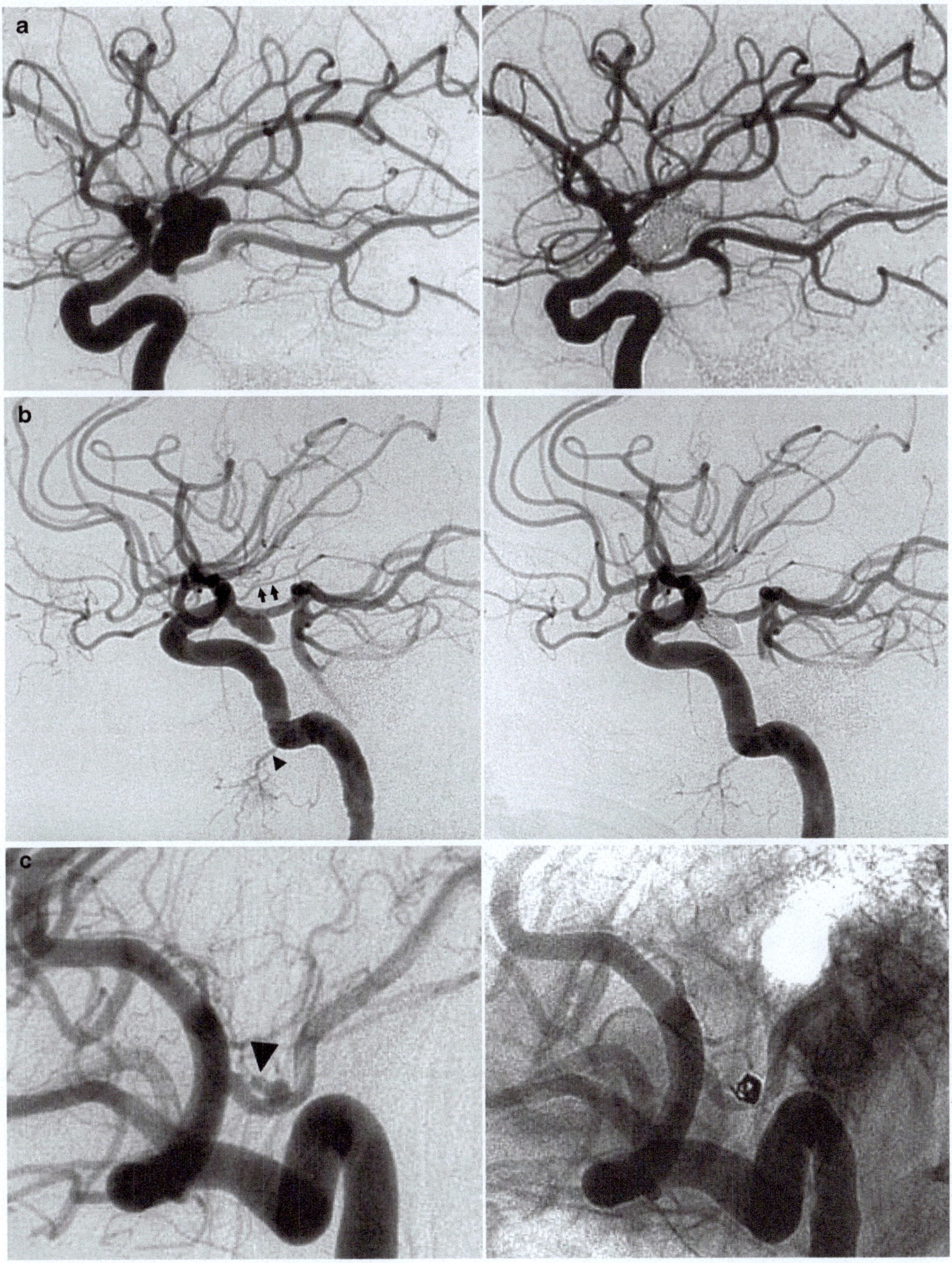

Fig. 11.9 Three examples of aneurysms of the PcomA presenting with SAH. ICA angiogram pre- and posttreatment. (**a**) Lateral ICA angiogram showing a large irregularly shaped aneurysm with blebs. Occlusion with coils. (**b**) Another example of PcomA aneurysm. Note the frequently typical elongated shape of the aneurysm. Through the large PcomA, filling of the Ba and both PCAs. AchA (*arrows*). Persistence of the embryonic mandibular artery (*arrowhead*). Occlusion with coils. (**c**) Example of a small aneurysm arising from the trunk of the PcomA (*arrowhead*). Occlusion of the aneurysm with coils

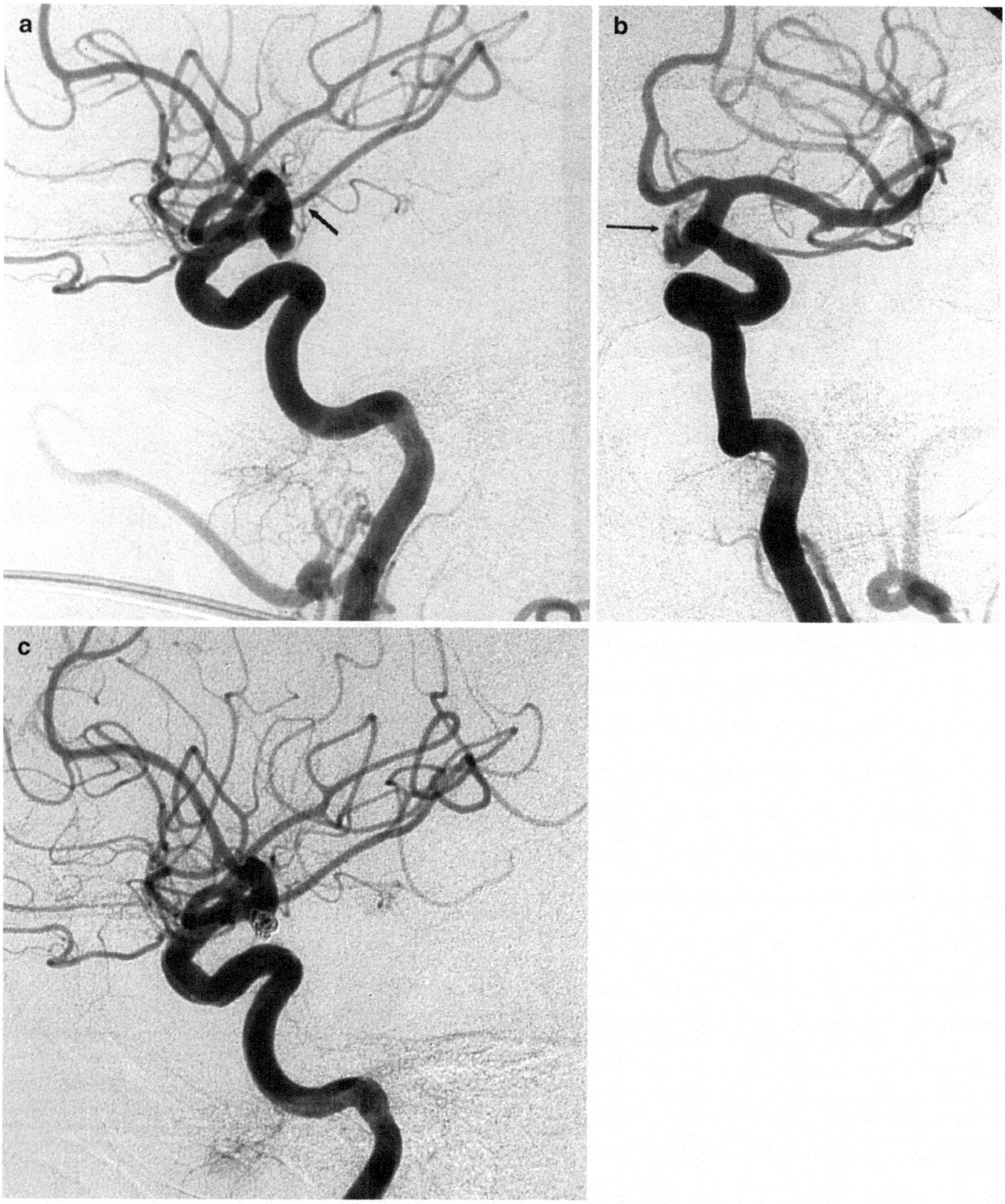

Fig. 11.10 Irregular ruptured aneurysm with neck close to the origin of the anterior choroidal artery (AchA). (**a**) Lateral carotid angiogram, showing the small aneurysm. AchA (*arrow*). (**b**) Oblique view better showing the irregular and elongated part (*arrow*) of the aneurysm. (**c**) Lateral carotid angiogram, posttreatment, showing the occluded aneurysm with coils

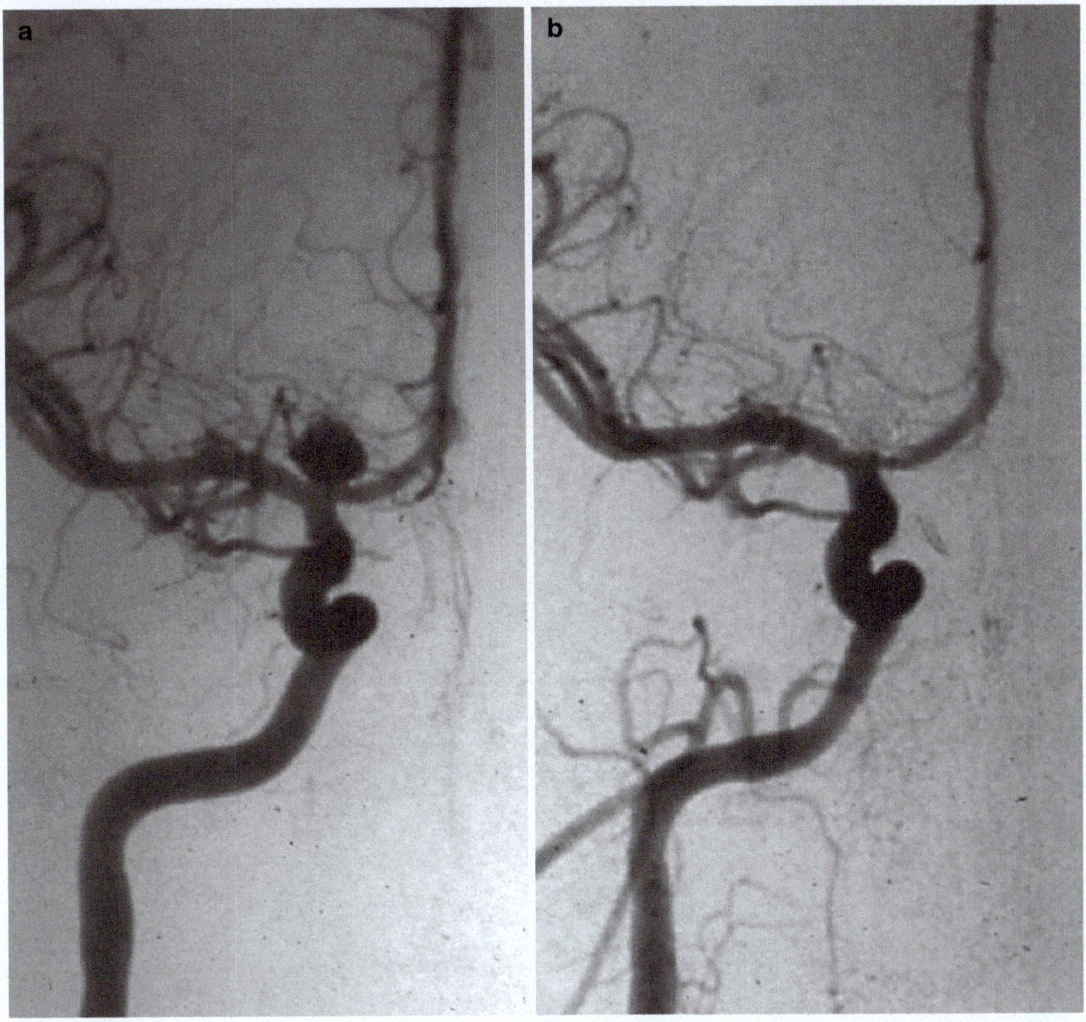

Fig. 11.11 Aneurysm of the distal carotid bifurcation, presenting with hemorrhage. (**a**) Carotid angiogram, AP view, showing the aneurysm. (**b**) Control angiogram showing the occluded aneurysm with coils

et al. 2009; Tollard et al. 2011; Chang et al. 2011; Park et al. 2013; Cho et al. 2014; Kim et al. 2014; Liu et al. 2016). An example is presented in Fig. 11.12b. An unsolved problem is the involvement of the perforators especially in the cases in whom also part of A1 is occluded in the treatment. In this context, it is useful to remember that the perforating branches of M1 can partially arise from A1 and/or from the Heubner artery (see Sect. 5.1). In one of our cases treated with occlusion of the proximal segment of A1, the CT performed routinely a few days later in completely asymptomatic patient showed ischemia in the putamen and pallidum probably due to involvement of the perforators arising in this case from the A1.

Another typical location, though less common, *is the pericallosal-calloso-marginal junction*. Depending on the site of the junction the aneurysm can be located anterior to the genu of the corpus callosum, distal to it or proximal beneath the genu. Examples are presented in Figs. 4.14 and 11.13.

Rarely, other aneurysms involving commonly the distal segment of the ACA, such as dissecting, mycotic, flow dependent, or developing in various arteriopathies as well as following trauma, can occur (Nakstad et al. 1986;

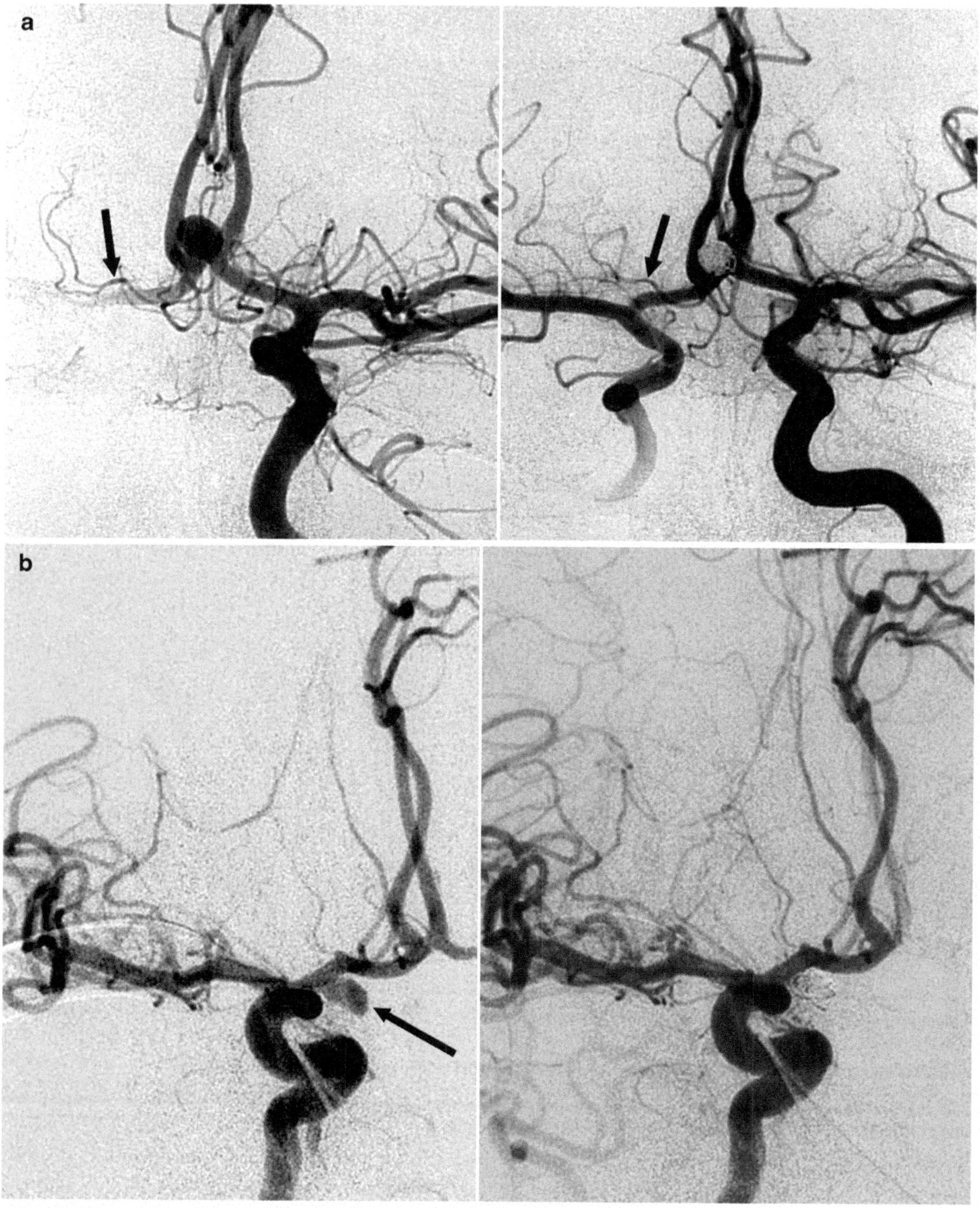

Fig. 11.12 (**a**) ICA angiogram. AP view. Pre- and post-treatment with coils of a ruptured AcomA aneurysm. On the right the artery of Heubner dividing distally into two branches is easily recognizable (*arrow*). On the left, it cannot be identified with certainty due to overlapping with other branches. (**b**) ICA angiogram. AP view in another patient. Angiogram pre- and posttreatment. Aneurysm in the proximal part of A1 directed anteriorly and slightly inferiorly (*arrow*) presenting with SAH. Occlusion with coils

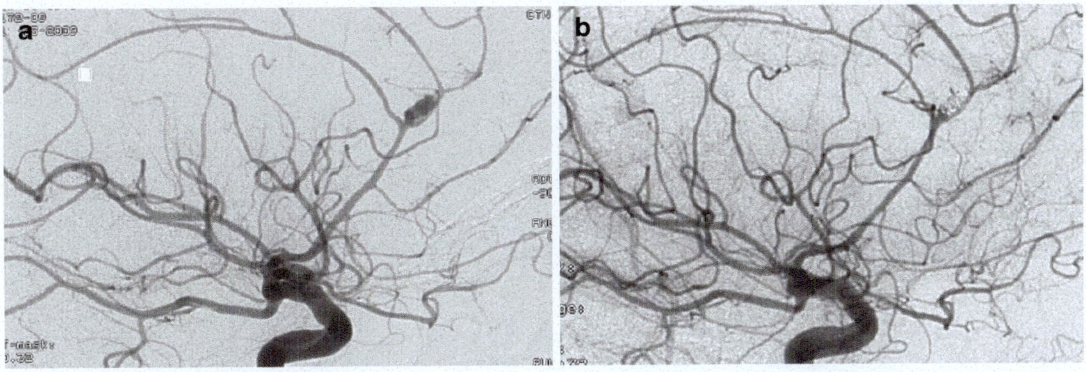

Fig. 11.13 (**a**) ICA lateral angiogram, showing the typical site of a pericallosal-calloso-marginal aneurysm. (**b**) Control angiogram posttreatment with coils

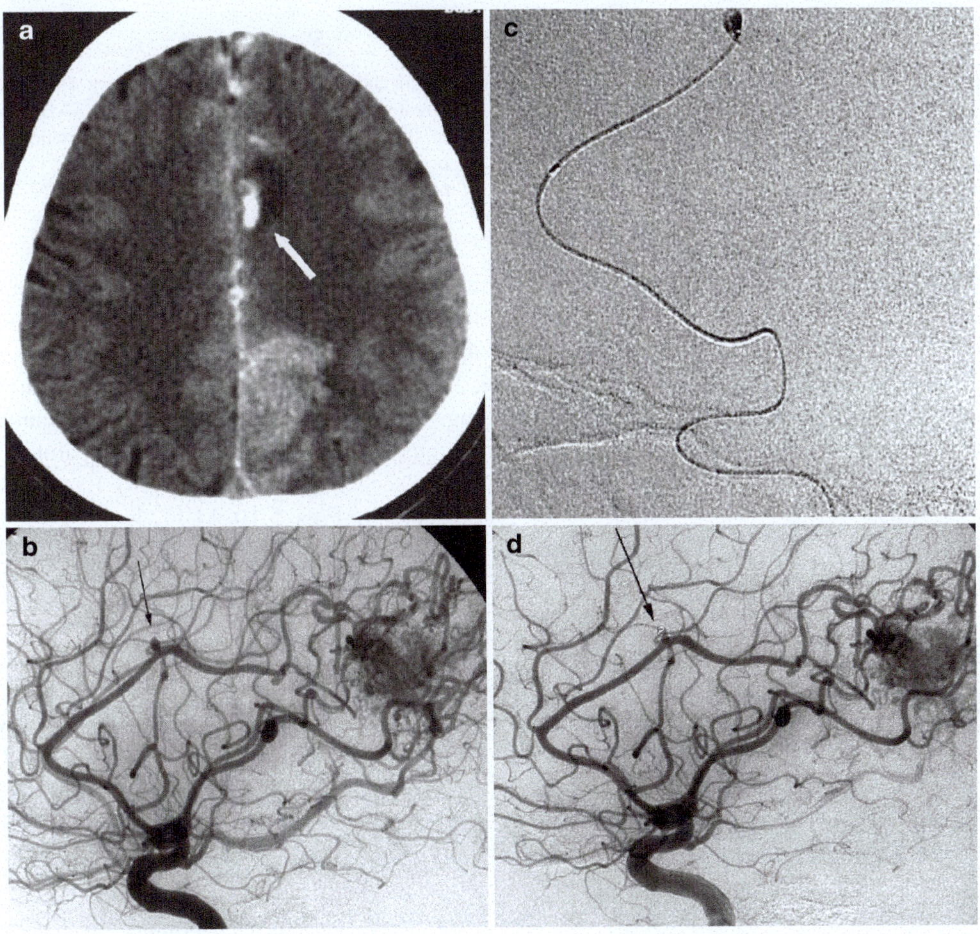

Fig. 11.14 Patient with known parietal AVM presenting with minimal interhemispheric SAH and small hemorrhage involving the parenchyma not related to the AVM. (**a**) CT demonstrating the small hemorrhage with surrounding edema (*arrow*). (**b**) The angiography showed a small pericallosal probably "flow-dependent" aneurysm (*arrow*). (**c**) Selective catheterization of the aneurysm which was occluded with coils. (**d**) Final control angiogram showing the occluded aneurysm

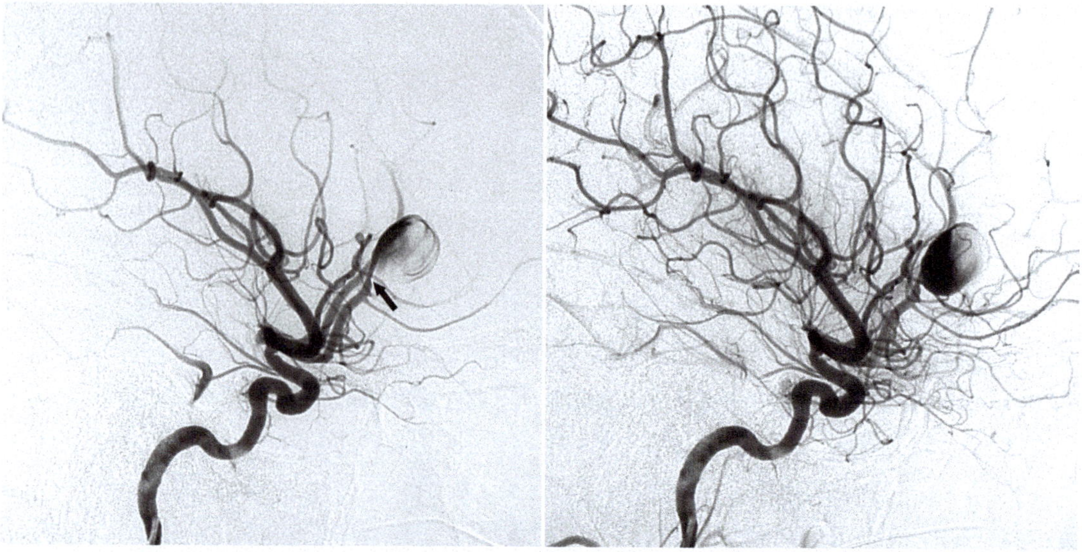

Fig. 11.15 Young patient presenting with severe SAH with a parenchymatous component owing to rupture of an aneurysm of the proximal pericallosal artery. Narrowing of the artery (*arrow*) proximally to the aneurysm suggesting probably dissection. In the later phase progressive filling of the aneurysm and of the distal pericallosal artery. The aneurysm was treated with clipping, but a week later SAH occurred again. The angiography revealed the recanalization of the aneurysm which was now enormously increased. Occlusion of the aneurysm and parent artery with coils was performed

Ohkuma et al. 2003; Cohen et al. 2005; Pandey et al. 2007; Fukuma et al. 2015; Hensler et al. 2016). An example of dissecting aneurysm of the pericallosal artery is presented in Fig. 11.15. A flow-dependent aneurysm is demonstrated in Fig. 11.14 (see also Chaps. 16, 17, and 18).

11.6.4 MCA Aneurysms

Most of these aneurysms arise at the bi-trifurcation of the M1 (Fig. 11.16a–f). It is not always easy to separate the MCA branches from the neck of the aneurysm. Furthermore in cases of early division of the M1 (short M1), the perforators can arise very close to the aneurysm (Figs. 5.4c, d and 5.5a, e). Considering all these morphologic aspects, despite improvement in the endovascular techniques, surgery can be the method of choice in some cases.

Other more rare aneurysms such as dissecting, mycotic, flow dependent, or associated to various arteriopathies and cardiopathies can occur (see Chaps. 16, 17, and 18). Examples are presented in Figs. 5.10, 11.17, 16.17, 17.1, 17.7, 18.1, and 18.2.

11.6.5 Aneurysms of the Posterior Circulation

Basilar aneurysms are the most frequent in the vertebrobasilar sector. They are typically located at the top of the artery. The perforators are commonly shifted laterally and stretched along the sac,

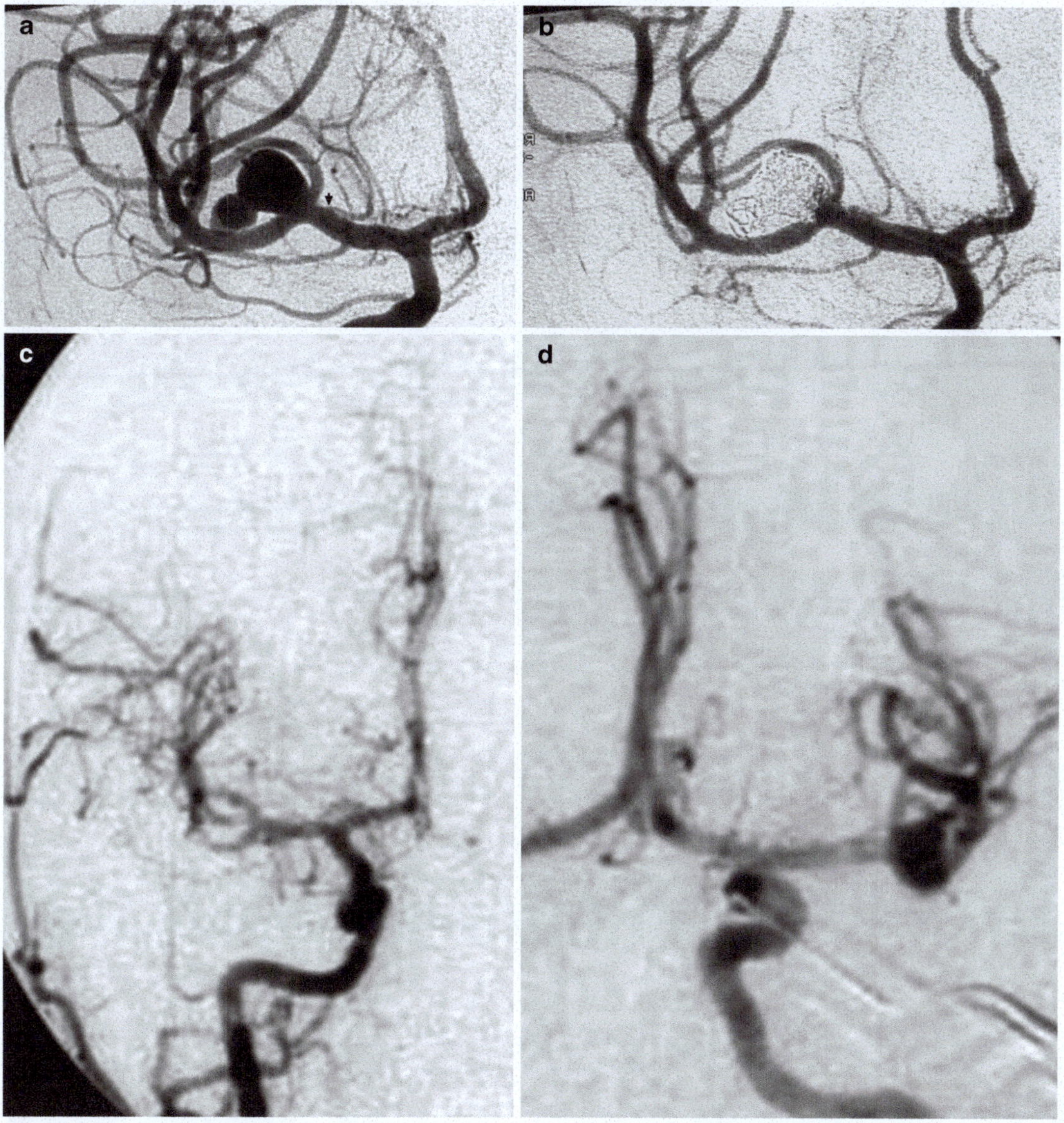

Fig. 11.16 (**a**, **b**) Bilobated large saccular aneurysm at the MCA bifurcation, presenting with hemorrhage. Origin of the distal perforator close to the aneurysm (*arrow*). Pretreatment angiogram (**a**), posttreatment with coils (**b**). Another example of MCA aneurysm in a young patient (**c**–**f**) presenting with SAH. (**c**) Normal right ICA angiogram. (**d**) Left ICA angiogram showing a middle cerebral aneurysm which was treated by surgery. About 8 years later the patient suffered a new SAH. (**e**) A new aneurysm was now recognizable on the right ICA angiogram. (**f**) Left ICA angiogram showing that the treated aneurysm was occluded

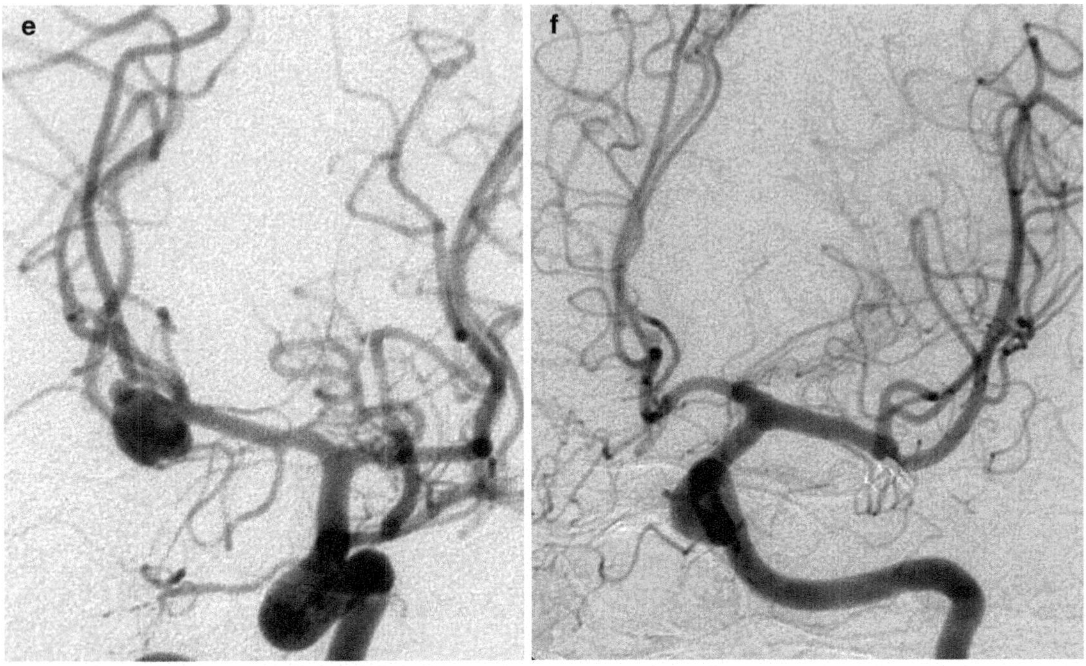

Fig. 11.16 (continued)

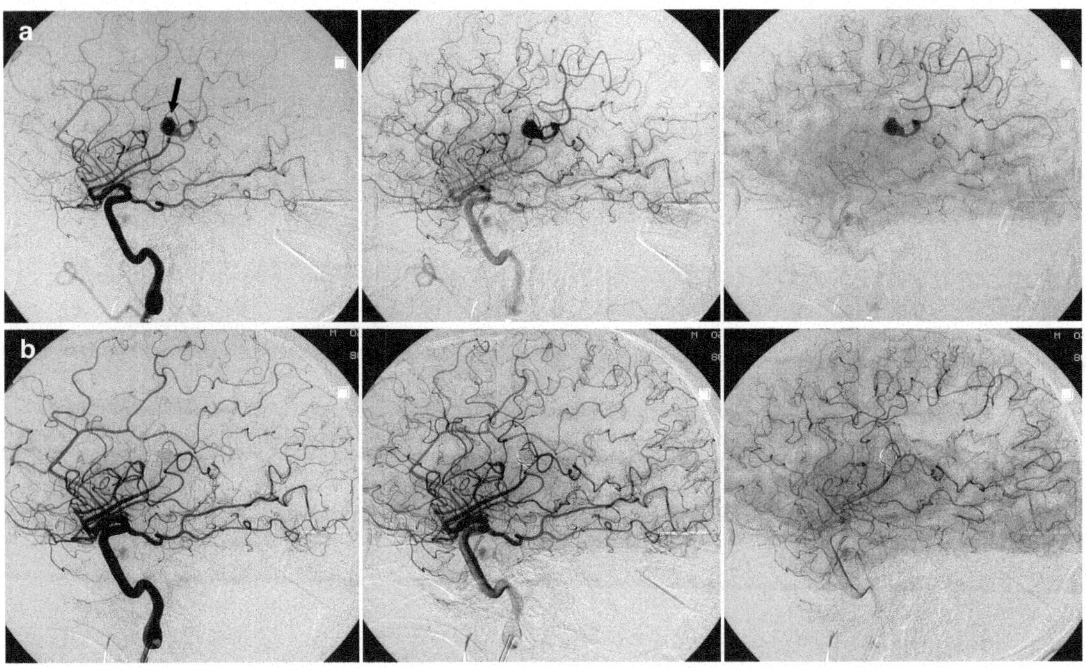

Fig. 11.17 Right carotid angiogram in young patient with transient ischemic episodes. (**a**) Slightly oblique view (early and later phases). There is a large irregular aneurysm (*arrow*) on a distal branch of the MCA. A second small aneurysm is visible more distally. No infection or collagenopathies were present. The aneurysms were thought to be a spontaneous dissection. A decision was taken to occlude the aneurysm and the parent artery which appeared to be a parietal branch. (**b**) Posttreatment angiogram postcoiling. There is a beginning of a retrograde filling of the occluded branch, through leptomeningeal anastomoses. A minimal ischemic lesion developed in the parietal area responsible for a mild motor deficit. Complete recovery in a few weeks

but they are not directly involved since they arise from the P1; this can, however, occur when the neck enlarges involving the proximal part of the P1. Examples are presented in Figs. 11.18, 11.19, and 11.20. Other locations are the distal lateral basilar artery between the origin of the PCA and SCA (Fig. 11.21). Also typical, even if more rare, are the aneurysms of the trunk of basilar artery, more frequently located distally to or close to the origin of the anterior inferior cerebellar artery. These aneurysms are frequently large with a fusiform aspect, often due to a dissection (Figs. 11.29, 11.30, and 16.14). Exceedingly rare (Ding et al. 2013) are aneurysms involving the latero-pontine branches of BA developing at they origin from BA or along their course at the junction with one perforating branch. They are sometimes very small, on the first angiogram, becoming larger on the control. Spontaneous thrombosis is frequent. Flow-related aneurysm associated with AVM can also occur. (Forbrig et al. 2016) Fig. 12.13.

- *Vertebral-PICA Aneurysms.* The vertebral-PICA junction is another typical site of aneurysms. These, however, are sometimes located not exactly at the junction but close to it, in the proximal part of the PICA (medullary segment) (Mukonoweshuro et al. 2003; Bradac and Bergui 2004; Mericle et al. 2006; Cellerini et al. 2008) (Figs. 11.22 and 11.23).
- *Vertebral Artery Aneurysms.* These involve selectively the VA. The pathogenesis is frequently a dissection. They are presented in Chap. 16.
- *Distal Aneurysms of the Cerebellar Arteries.* These are very rare aneurysms, occurring in fewer than 1% of all aneurysms (Lubicz et al. 2003a; Mitsos et al. 2008). The great majority is located in the PICA (Bradac and Bergui 2004; Mitsos et al. 2008), followed by the SCA and AICA aneurysms (Gacs et al. 1983; Chaloupka et al. 1996; Kurosu et al. 2000; Leonardi et al. 2001; Zager et al. 2002; Menovsky et al. 2002; Gonzales et al. 2004; Peluso et al. 2007; Mitsos et al. 2008; Tokimura et al. 2012).
- Unlike the majority of the intracranial aneurysms, which arise at an arterial branching point, distal cerebellar aneurysms arise frequently directly from the artery where it forms a curve, which can be very sharp, especially in the course of the PICA. It has been suggested that this could be the cause of a *hemodynamic stress on the wall* of the artery contributing to the formation of the aneurysm (Lewis et al. 2002; Horiuchi et al. 2003; Mitsos et al. 2008). *Dissection* is another possible pathogenesis, which has been increasingly reported (Lefkowitz et al. 1996; Yamamura et al. 1999; Dinichert et al. 2000; Tawk et al. 2003; Bradac and Bergui 2004; Maimon et al. 2006; Mitsos et al. 2008; Cellerini et al. 2008; Fukushima et al. 2009). *Flow-dependent aneurysms* of the cerebellar arteries related to a more distal AVM, DAVF, and cerebellar hemangioblastoma are relatively frequent in our experience as well as in the reports of several authors (Hudgins et al. 1983; Kaech et al. 1987; Guzman and Grady 1999; Kaptain et al. 1999; Menovsky et al. 2002; Lewis et al. 2002; Peluso et al. 2007). Examples of distal aneurysms of the cerebellar arteries are presented in Figs. 11.24, 11.25, 12.14, 12.15, 12.16, and 16.12. Aneurysms of the distal branches of the cerebellar arteries can rarely also develop in *a great variety of vasculopathies and cardiac diseases* as described in Chaps. 17 and 18.
- In aneurysms involving the distal segment of the cerebellar arteries, the occlusion also of the parent artery can commonly be performed, without or with minimal clinical problems, since the perforators for the medulla arise in the proximal segments (see anatomy). Furthermore, the distal cortical branches are revascularized through a frequently rich collateral circulation via leptomeningeal anastomoses involving the distal branches of the cerebellar arteries.
- *Aneurysms of the Posterior Cerebral Arteries.* These are rare aneurysms with a relatively typical location at the P1–P2 passage. More rare are those located along the course of the P2 and P3. Such aneurysms are frequently large without a well-defined neck, sometimes with a fusiform aspect. Their etiology is thought to be frequently a dissection (Lazinski et al. 2000; Ciceri et al. 2001; Li et al. 2007a;

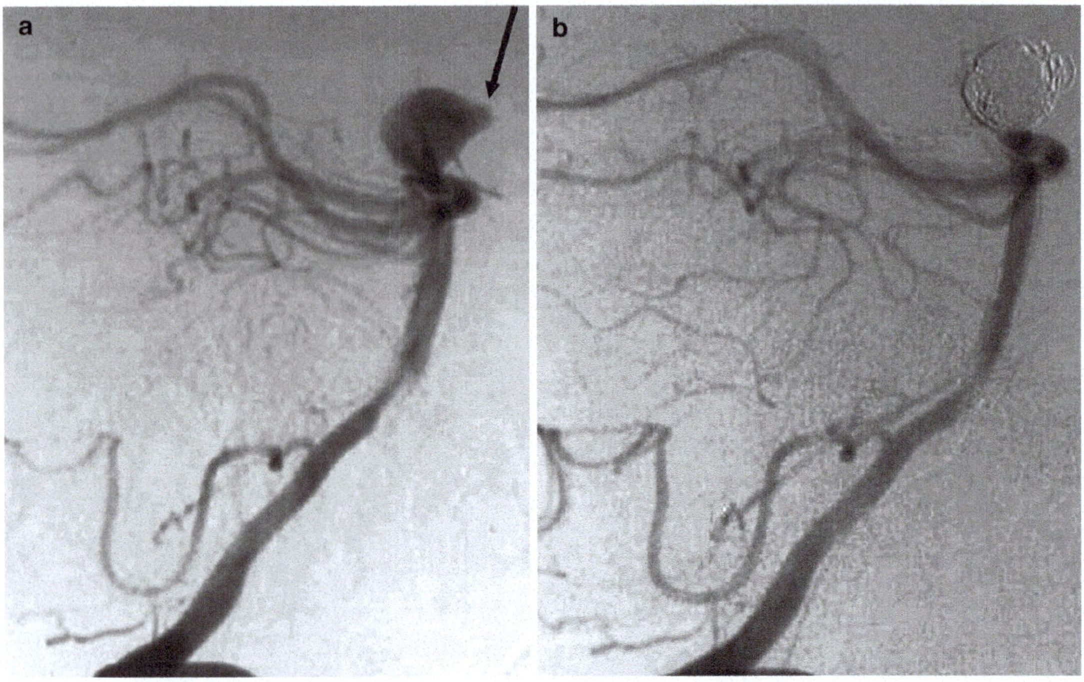

Fig. 11.18 (**a**) Vertebral angiogram in patient presenting with SAH, showing a basilar-tip aneurysm where a bleb (*arrow*) probably indicating the site of rupture is visible. (**b**) Posttreatment angiogram showing the occlusion of the aneurysm with coils

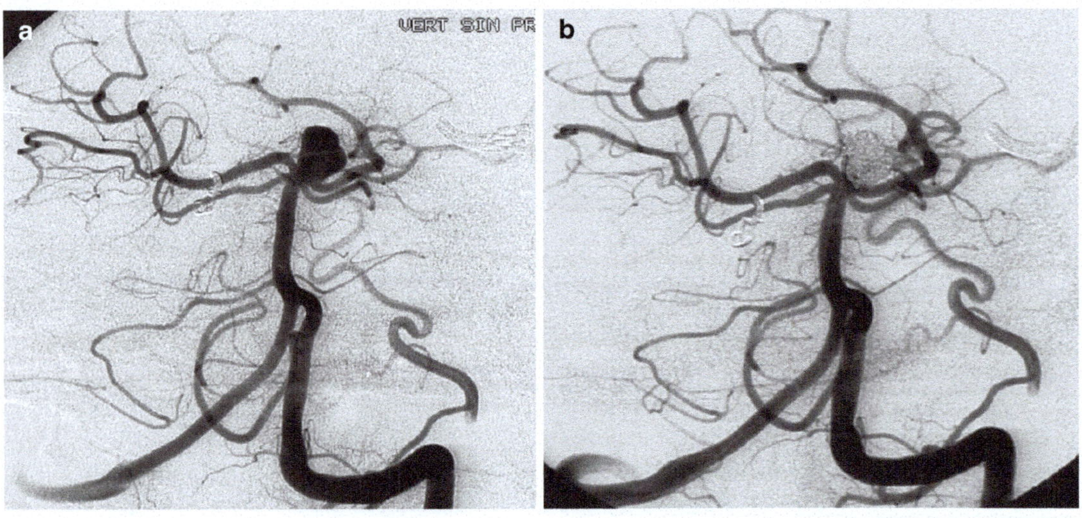

Fig. 11.19 Nonruptured basilar-tip aneurysm with a neck involving partially the left P1. The patient had already been treated for a ruptured aneurysm of the AcomA. (**a**) Angiographic study pretreatment. (**b**) Control angiogram posttreatment. Occlusion of the aneurysm with coils assisted with positioning of a stent (enterprise) in the distal BA and left P1

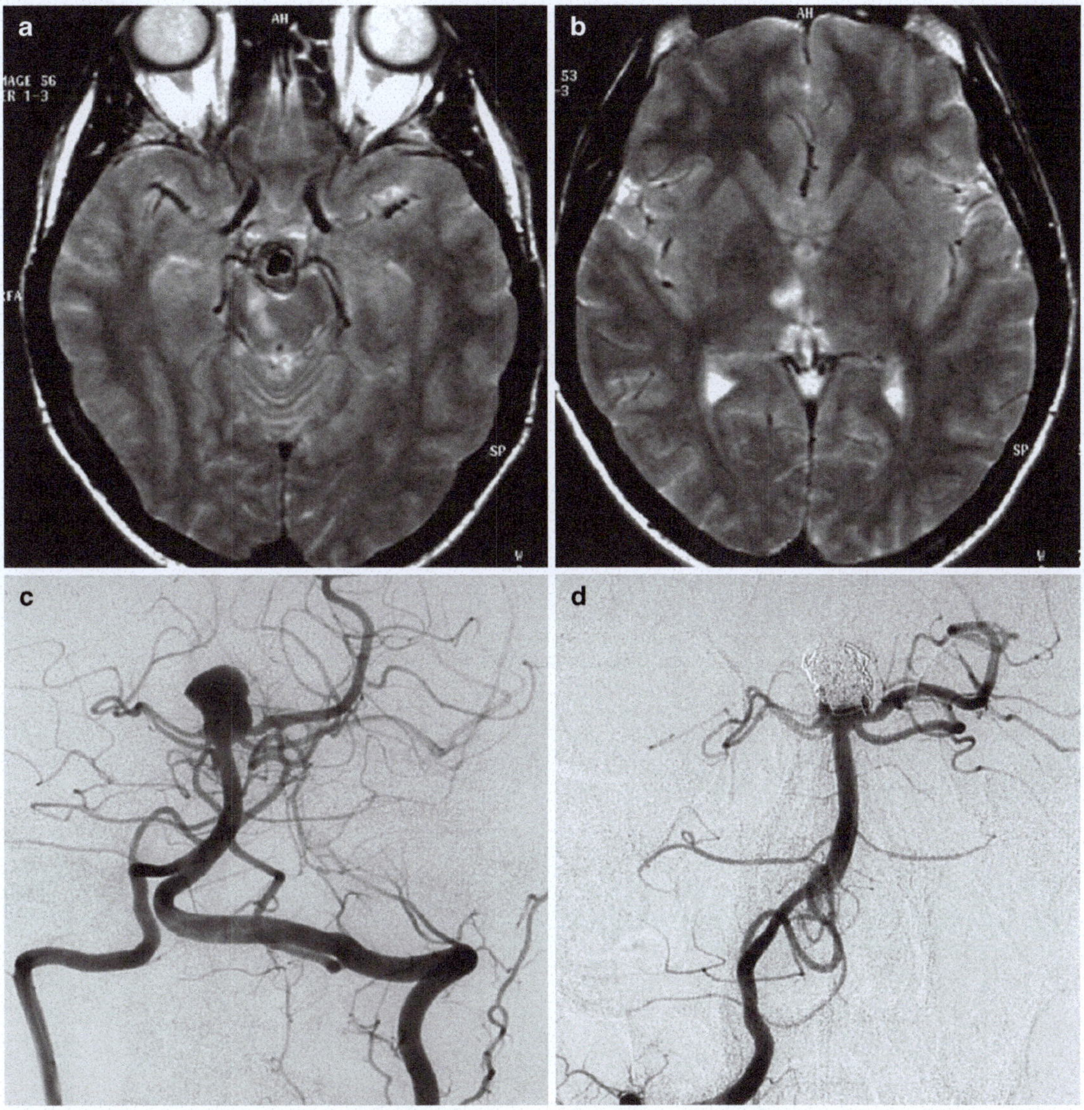

Fig. 11.20 A 40-year-old man presenting with acute midbrain-thalamic syndrome. (**a**, **b**) On MRI small ischemic lesions were recognizable in the right medial part of midbrain and thalamus, corresponding to the vascular territory of perforators arising from P1. A large basilar-tip aneurysm, probably responsible for the thromboembolic occlusion of the right P1, was demonstrated. (**c**) The patient recovered, and the aneurysm was occluded with coils a week later

Roh et al. 2008; Zhao et al. 2013). This is especially true for such aneurysms in pediatric patients (Laughlin et al. 1997; Lasjaunias et al. 2005; Vilela and Goulão 2006; Bradac et al. 2008a).

In many cases, occlusion of the aneurysm and parent artery cannot be avoided by endovascular as well as surgical treatment. The risk of a temporo-occipital infarction is relatively low owing to the rich leptomeningeal collateral circulation. Greater risk is present in more proximal occlusions as a result of involvement of perforators supplying the midbrain and thalamus arising from P1–P2 (Pia and Fontana 1977; Sakata et al. 1993; Ciceri et al. 2001; Arat et al. 2002; Hallacq et al. 2002; Li et al. 2007a; Roh et al. 2008) (see also Chap. 7). Owing to the anatomical variations of the branches from which arise the perforators for midbrain and thalamus, the developing of an ischemic lesion is a little unpredictable. Fortunately, when ischemia occurs, the clinical impairment is commonly mild and resolves over ensuing weeks. Examples of aneurysms of the PCA are presented in Figs. 11.26 and 11.27.

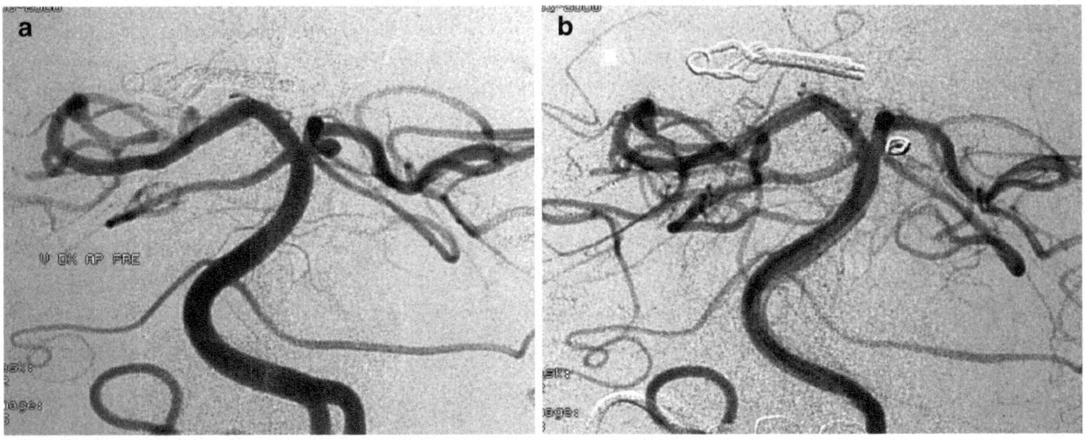

Fig. 11.21 Patient already operated on for an AcomA aneurysm. A second aneurysm was present at the junction of the left PCA and SCA. (**a**) Vertebral angiogram. (**b**) Control angiogram posttreatment showing the occlusion of the aneurysm with coils

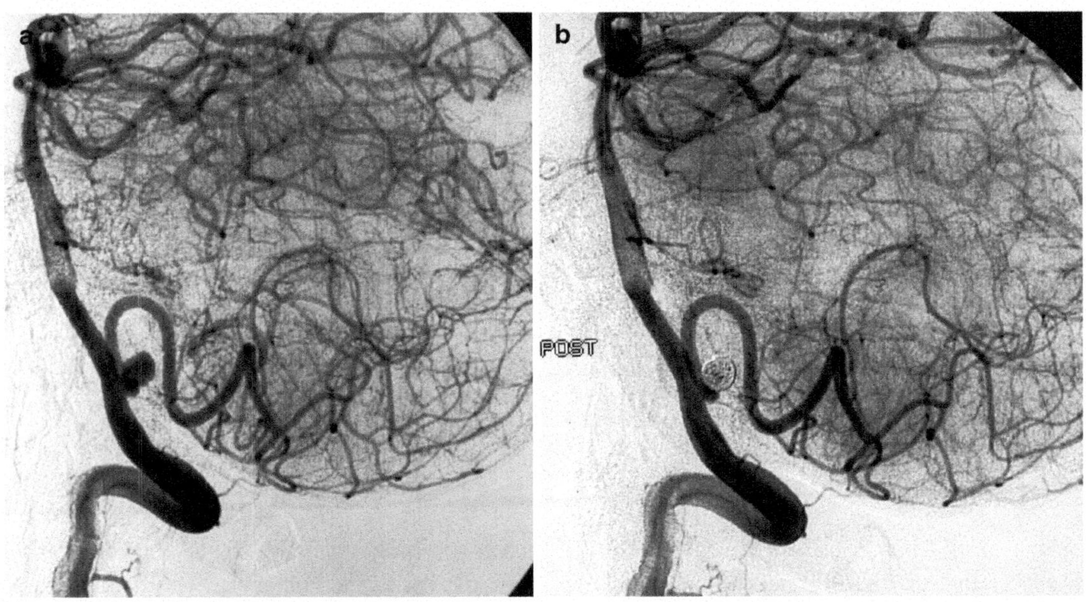

Fig. 11.22 Patient presenting with SAH due to rupture of a typical VA-PICA aneurysm. Pre- and posttreatment vertebral angiogram, with occlusion of the aneurysm with coils

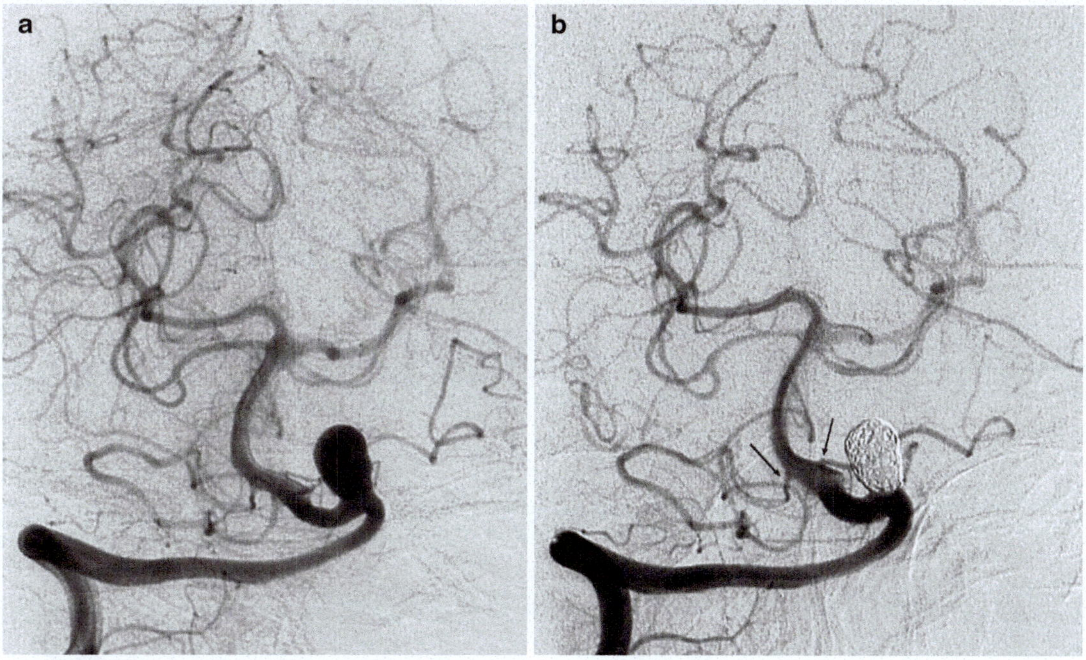

Fig. 11.23 Large unruptured aneurysm in a young patient. There was a well-developed right vertebral artery and a hypoplastic left vertebral artery. The aneurysm was located at the junction of the vertebral artery with a hypoplastic left PICA. There was a bilateral well-developed anterior inferior cerebellar artery (AICA; *arrows*), supplying the vascular territory of both PICAs. Right vertebral angiogram pre- and posttreatment with coils

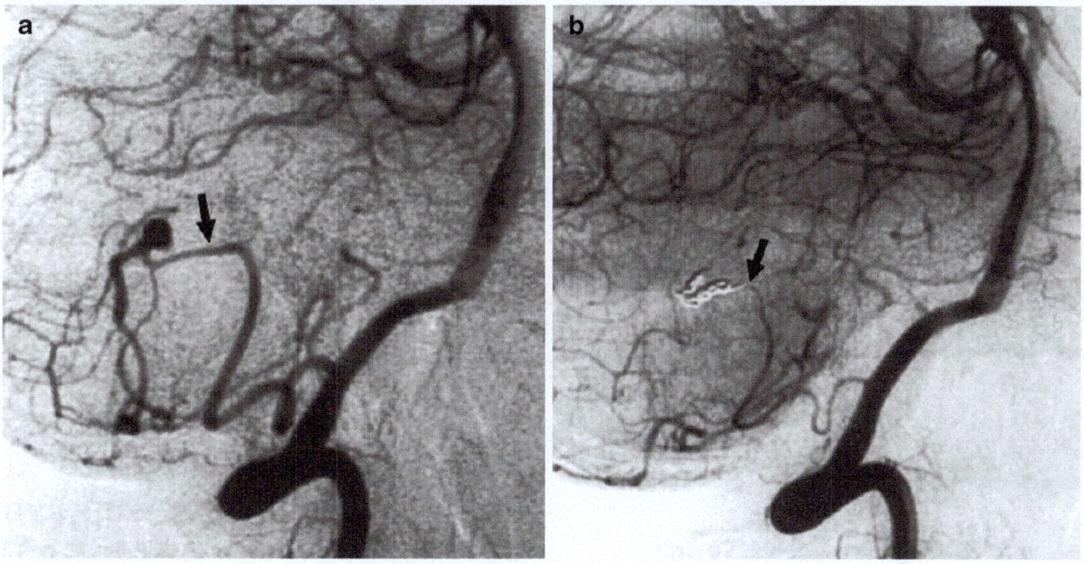

Fig. 11.24 Aneurysm of the supratonsillar segment of the PICA presenting with hemorrhage. (**a**) Vertebral angiogram showing the aneurysm. Supratonsillar segment (*arrow*). There is also a dilatation of the retrotonsillar segment of the PICA

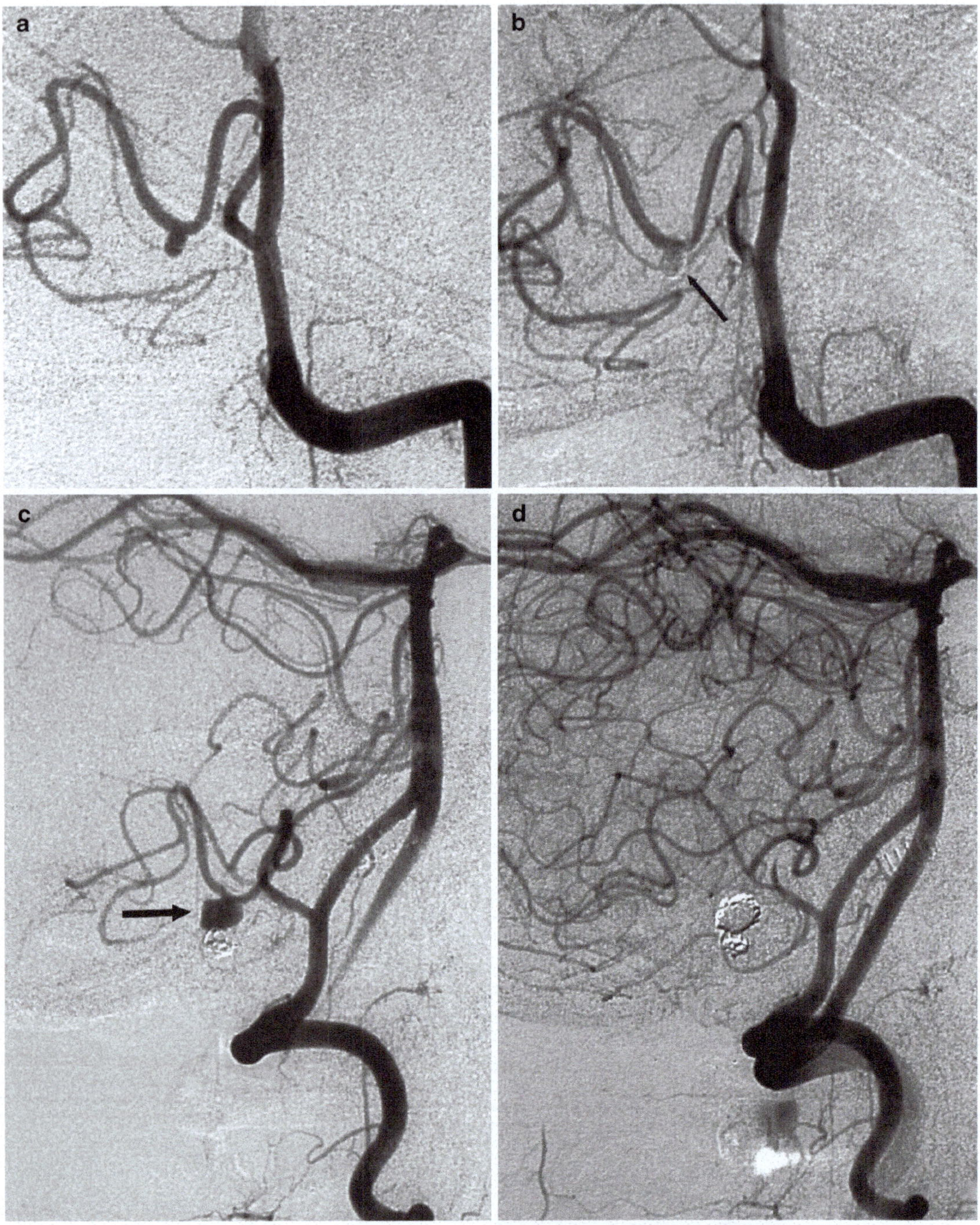

Fig. 11.25 Aneurysm of the retromedullary segment of the PICA presenting with hemorrhage. Pretreatment vertebral angiogram (**a**) presenting the aneurysm and post-treatment angiogram (**b**) showing its occlusion with coils (*arrow*). Two months later, the patient was admitted again owing to a new hemorrhage. The angiogram (**c**) showed regrowth of the aneurysm (*arrow*), which probably was originally a dissecting aneurysm. Occlusion of the aneurysm and of the parent artery was performed (**d**). The patient recovered completely

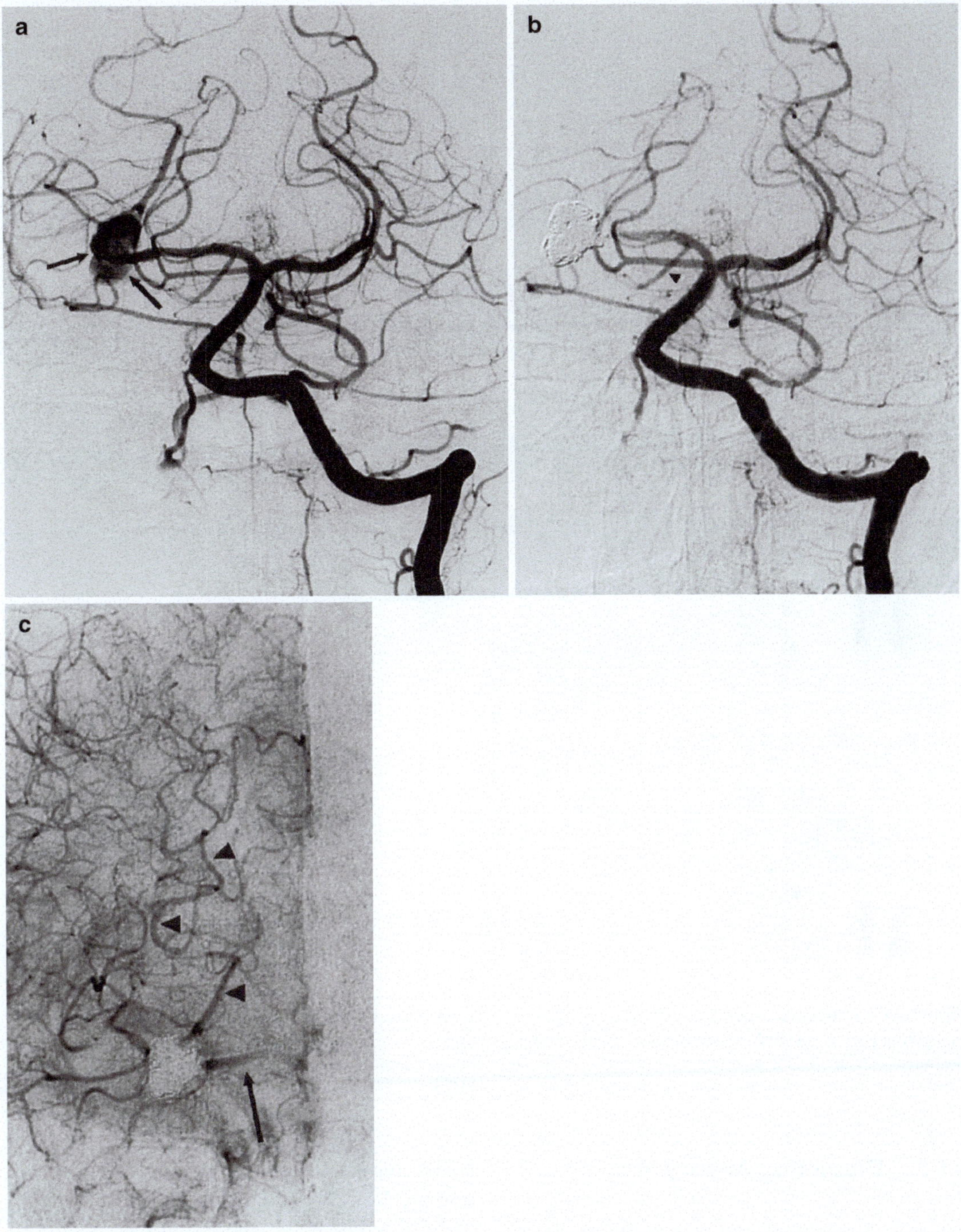

Fig. 11.26 Unruptured irregularly, probably dissecting aneurysm involving the P2 segment of the right PCA in a patient presenting with a transient episode of hemianesthesia. Left vertebral angiogram (**a**) showing the aneurysm (*arrows*). Posttreatment angiogram (**b**) with occlusion of the aneurysm and the P2 segment of the PCA. Note the posterior communicating artery (*arrowhead*). Carotid angiogram (**c**) with retrograde injection of the posterior cerebral artery up to the aneurysm (*arrowheads*), through leptomeningeal anastomosis between the MCA and PCA. There is also injection of the proximal P2 (*arrow*) through the posterior communicating artery. The patient tolerated the treatment well despite a small ischemic lesion in the lateral thalamus owing to involvement of the thalamogeniculate artery, as demonstrated by MRI

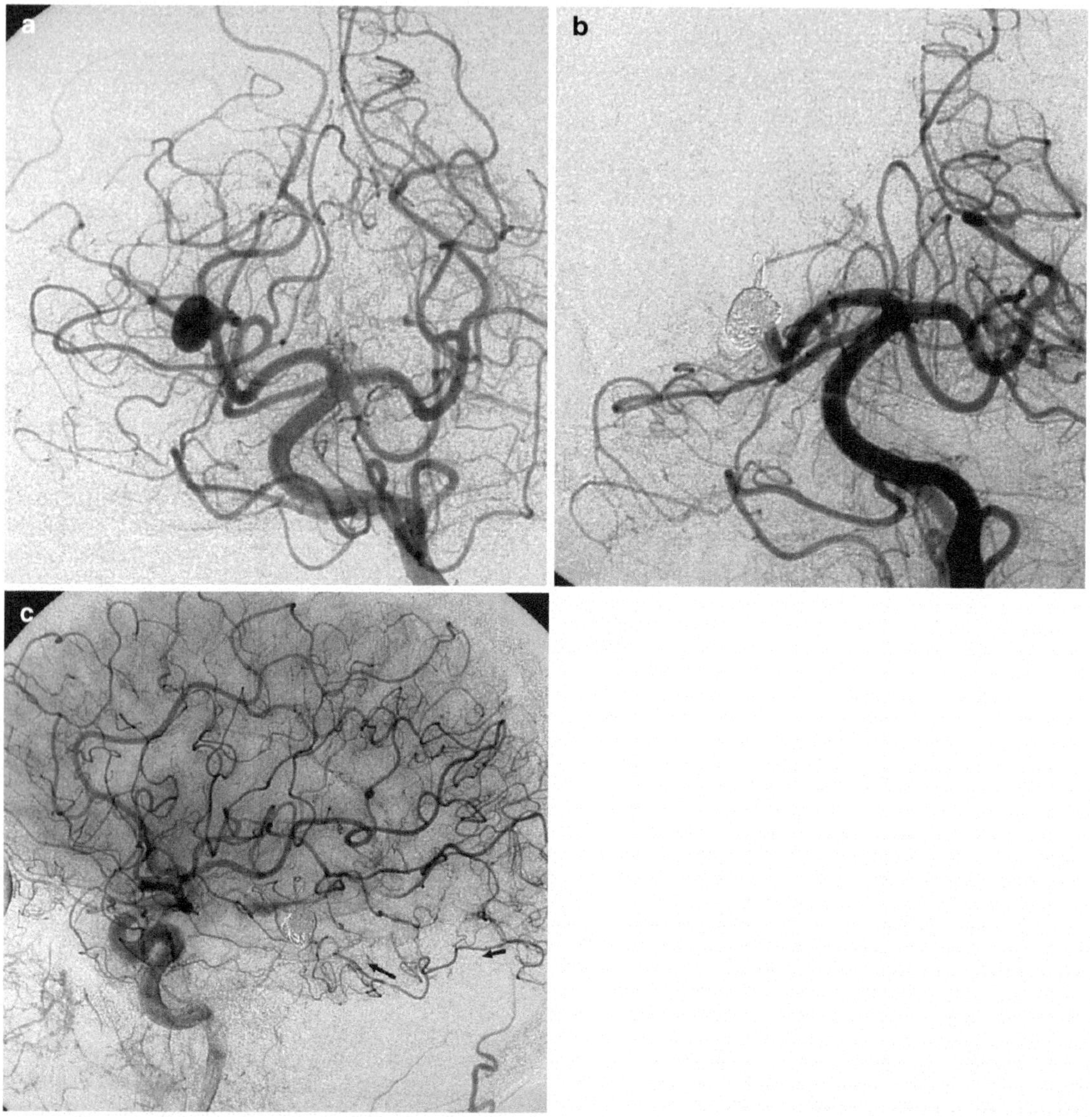

Fig. 11.27 Unruptured aneurysm without a defined neck in a young patient. (**a**) Left vertebral angiogram, showing the aneurysm at the P2 segment of the right PCA. (**b**) Control angiogram after occlusion of the aneurysm and the parent artery. (**c**) Carotid angiogram. Retrograde injection of distal branches of the PCA (*arrows*) through opening of a leptomeningeal anastomosis with branches of the MCA

11.7 Dissecting Aneurysms

They are found most frequent in the posterior circulation, where they typically affect the basilar and vertebral arteries. Aneurysm of distal branches of the cerebellar and posterior cerebral arteries can also be due to dissection. Dissecting aneurysms in the anterior circulation are less common. The distal ICA, proximal MCA and ACA, and PcomA are the most usual locations. Distal branches can occasionally be involved. All aspects of these particular aneurysms are discussed extensively in Chap. 16.

11.8 Large and Giant Aneurysms

Large and giant aneurysms are relatively rare, representing, respectively, about 15 and 5% of all cerebral aneurysms. They can be saccular or fusiform.

Fusiform aneurysms are located predominantly in the vertebrobasilar sector where the most frequent location is the basilar artery followed by the vertebral and the P1–P2 segments of the PCA. Fusiform aneurysm in the anterior circulation is more rare. The most frequent involved artery is the MCA followed by the ICA and ACA (Drake and Peerless 1997). Unlike saccular aneurysm, in which a neck is recognizable, in fusiform a large part or the entire vessel expands. The pathogenesis of these aneurysms cannot be always determined. Considering however their clinical evolution the fusiform aneurysms can be subdivided basically into two groups (Shokunbi et al. 1988; Nakatomi et al. 2000; Leibowitz et al. 2003).

In some cases, the fusiform aneurysm grows gradually. Its pathogenesis is thought to be due to a primary lesion of the internal elastic lamina characterized by its disruption and fragmentation occurring in a particular form of atherosclerosis. Intima hyperplasia, infiltration and fibrosis of the media, and neovascularization causing intramural bleeding follow. All this leads to a progressive dilatation and tortuosity of the artery, promoted also by the frequent presence of hypertension. The increased luminal diameter leads to a slowing down of the circulation and formation of thrombi on the wall which promotes further the pathologic process with progressive rigidity of the wall and dilatation (Hegedus 1985; Echiverri et al. 1989). The prognosis of these patients is commonly very poor. The progressive dilatation leads to compression of the brain parenchyma. Ischemia, due to involving of the perforating branches or due to embolization, occurs. Hemorrhage is uncommon (Little et al. 1981; Nishizaki et al. 1986; Echiverri et al. 1989; Pessin et al. 1989; Leibowitz et al. 2003), even though in some studies it has been reported as being not so rare (Flemming et al. 2004) (Fig. 11.28).

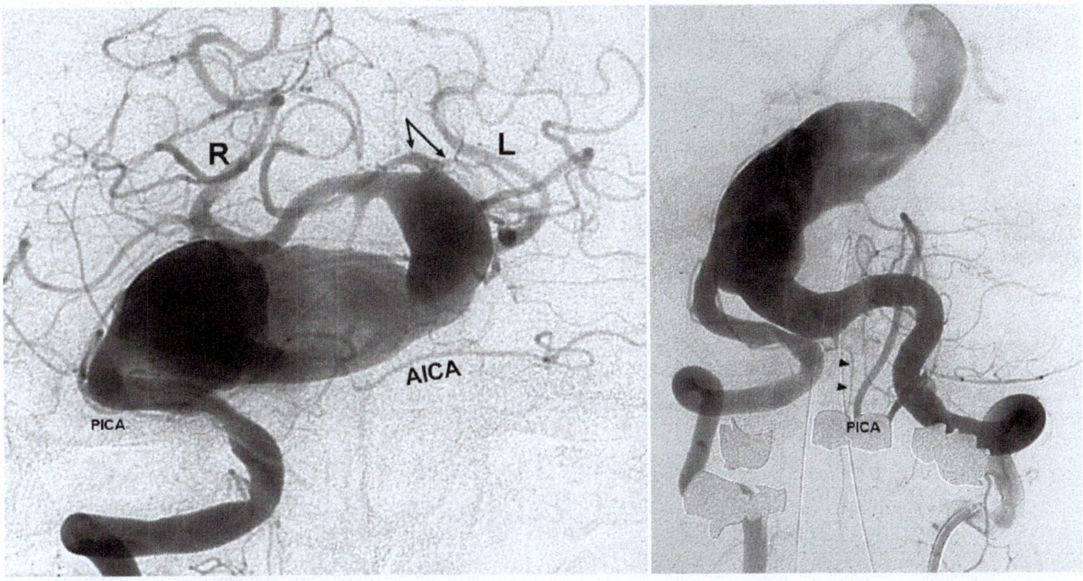

Fig. 11.28 Patient presenting with progressive tetraplegia due to giant fusiform aneurysm of the basilar artery compressing the brainstem. Left and right vertebral angiogram. The left and right PCAs are well recognizable (L and R) as is the right superior cerebellar artery (*arrow with angle*). That on the left is very small. AICA is on the left. Well-developed PICA is on the left and small PICA on the right. Anterior spinal artery (*arrowheads*). Occlusion of the left vertebral artery with a balloon, proximal to the PICA, was performed to reduce the flow in the aneurysm. There was a partial clinical improvement, followed by a severe fatal subarachnoid hemorrhage 1 month later

In other cases, the pathogenesis is probably a dissection due to pathological changes involving the internal elastic lamina and the media occurring in various pathological conditions which are described in detail in Chap. 16. These patients are commonly younger in whom the aneurysm presents frequently with acute symptoms due to compression of brain, ischemia, and frequently with hemorrhage.

Saccular aneurysms can be found in the anterior circulation (cavernous, supraclinoid portion of the ICA proximal ACA and MCA) and in the posterior circulation (VA, BA, and PCA) (Figs. 11.1, 11.2, 11.4, 11.5, 11.29, and 11.30). It is thinkable that they evolve from smaller saccular aneurysm due to the same causes which had led to the formation of the aneurysm, that is, degenerative changes of the internal elastic lamina and media favored by hemodynamic factors (see Sect. 11.4). However, the reason why this occurs in a few aneurysms and while not in the other is not clear. In other cases the pathogenesis is probably dissection. They can present with hemorrhage but frequently the symptoms are due to mass effect. *A particular subgroup of these aneurysms* are those in which the growth appears in spite that the aneurysm is partially or completely thrombosed (Schubiger et al. 1987). It has been suggested that a proliferation of new capillaries within the thrombotic part of the aneurysm occurs. Their rupture causes hemorrhage especially at the periphery of the intra-aneurysmal thrombus leading to a progressive dissection and growth (Katayama et al. 1991; Nagahiro et al. 1995; Yasui et al. 1998; Kaneko et al. 2001). Other authors (Atlas et al. 1987; Zhao et al. 2004; Krings et al. 2007a) have also discussed the possibility that in association of the angiogenesis an

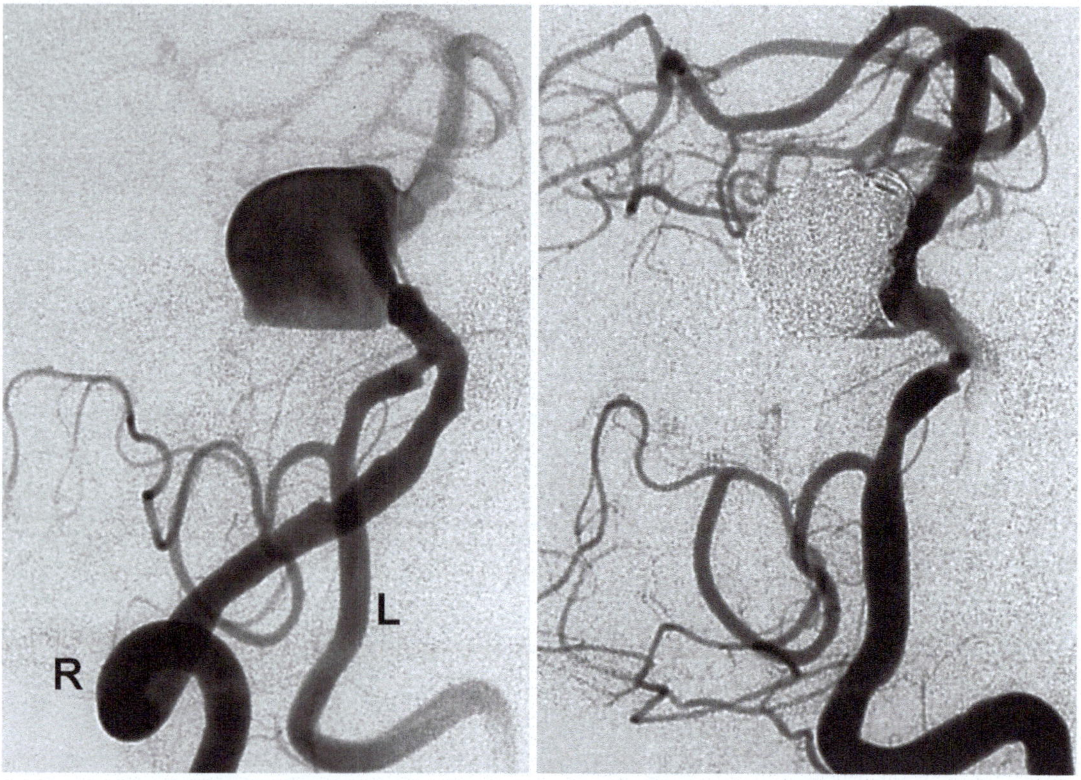

Fig. 11.29 Giant aneurysm of the middle basilar artery presenting with brainstem syndrome in an older patient. A clear neck cannot be recognizable. Atheromatous changes along the course of the VAs and BA are visible. Left (L) and right (R) vertebral arteries. The aneurysm was occluded with coils supported by balloon (remodeling technique). On the right the angiogram posttreatment. There was a clinical improvement of the symptoms

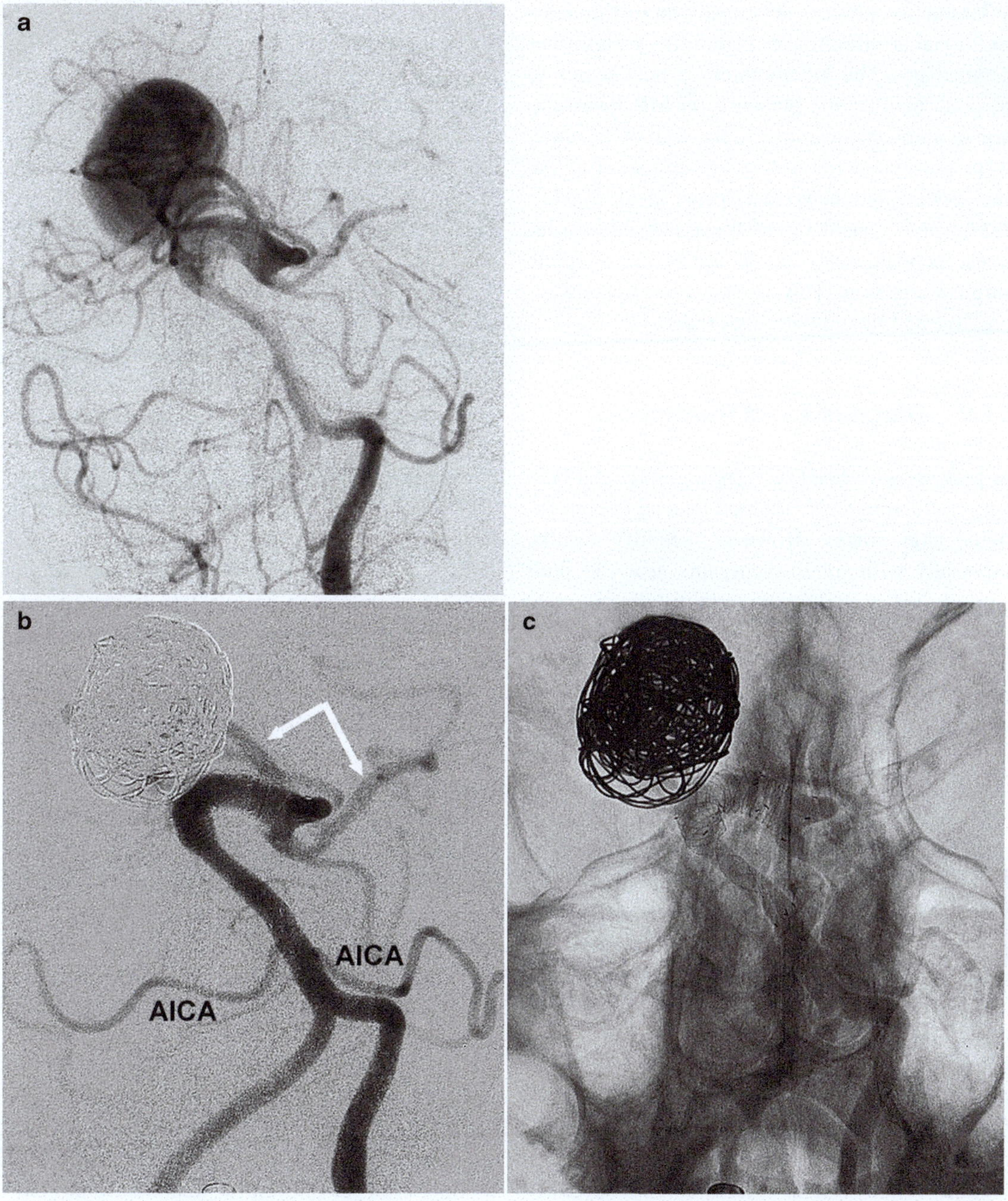

Fig. 11.30 Giant aneurysm of the basilar artery (BA) in middle-aged hypertensive patient presenting with acute cranio-cervical pain. (**a**) Vertebral angiogram showing a giant aneurysm in the middle-superior part of a tortuous and dilated BA. (**b**) Vertebral angiogram posttreatment showing the occlusion of the aneurysm with coils and flow diverter. The latter better visible on the nonsubtracted images. Anterior inferior cerebellar arteries (AICA), PCAs (*arrow with angle*). Due to overlapping SCAs are not clearly identified. Progressive improvement of the clinical condition. Compare this case with that in Fig. 11.29, presenting a similar situation treated years before indicating the evolution in the possibility of treatment

inflammation process develops which stimulates additional proliferation of capillaries with further hemorrhage. The inflammatory aspect seems to be confirmed by the presence on MR investigation of a rim enhancement of the wall of the aneurysm associated frequently by an edema of the surrounding parenchyma (Krings et al. 2007a). Taking into consideration these possible evolutions, the ideal treatment should be the complete surgical excision. This is, however, not always possible and linked with high risks.

11.9 Diagnosis and Treatment

In patients with SAH, CT angiography can be a very useful diagnostic method instead of conventional angiography. However, negative results, especially with small aneurysms near the skull base, do not exclude with certainty the presence of an aneurysm. MR angiography can be employed as a screening method to detect aneurysms in selective group of persons, such as those with polycystic kidney or in family with a rich incidence of aneurysm. MR angiography is not commonly used in the acute phase of SAH. Conventional vessel angiography remains the gold standard that is to be performed every time there is an unclear diagnosis or if an endovascular treatment has been considered.

Cushing (1926), was the first to approach surgically a cerebral aneurysm. Dott (1933), Toennis (1936), and Dandy (1938) were the first to treat a cerebral aneurysms diagnosed with the cerebral angiography introduced by Moniz (1927). Since then many improvements in the surgical technique have been made leading to a progressive morphological and clinical good results.

The endovascular technique especially through the development of the detachable coils (Guglielmi et al. 1991a, b) has opened a new, certainly revolutionary approach for the treatment of cerebral aneurysms. Since then, the increasing quality of coils and microcatheters, the application in selected cases of new techniques (Balloon-coil Moret et al. 1997), and the later introduction of stents in association with coils have progressively expanded the indications for endovascular treatment (Moret et al. 1997; Boccardi et al. 1998; Molineux et al. 2002; Murayama et al. 2003a; Henkes et al. 2004; Bradac et al. 2005, 2007; Gallas et al. 2005; Park et al. 2005; Kurre and Berkefeld 2008; Wanke and Forsting 2008; Loumiotis et al. 2012; Pierot et al. 2012; Fargen et al. 2012; Lubicz et al. 2006, 2017). Further progress has been obtained with the development of flow-diverter stents which allow to treat more difficult aneurysm such as fusiform, those with more complex morphology, or aneurysms presenting with recanalization, as progressively reported (Berge et al. 2012; Martinez et al. 2015; Yavuz et al. 2014; Saleme et al. 2014; Caroff et al. 2015; Clarençon et al. 2017; Gawlitza et al. 2015; Chalouhi et al. 2015; Puri et al. 2016). There is, however, one limitation, which should be considered in the use of flow-diverter stent. Indeed, the need to associate this device to an antiplatelet therapy makes its use commonly not indicated in the acute phase of SAH.

Today the endovascular treatment is the first choice of therapy in the majority of the aneurysms with good morphological and clinical results. Complications mainly characterized by ischemia due to distal embolization, occlusion of perforators or parent artery occurring during or in the next hours after the treatment, as well as hemorrhage due to rupture of the aneurysm have become with increasing experience progressively more rare. An extremely seldom complication has been reported by some authors (Fealey et al. 2008; Ulus et al. 2012; Cruz et al. 2014; Lobotesis et al. 2015). This is characterized by distal parenchymal lesions which present on MR-CT with ischemic-like areas or small enhancing lesions surrounded by edema, appearing weeks or months after an uneventful treatment and commonly reacting positively to corticosteroid therapy. The cause of this strange phenomenon is not clear. Inadvertent foreign body embolization occurring during the treatment has been considered (Shotar et al. 2016). In a few cases where a biopsy has been performed a granulomatous angiitis related to the foreign material has been identified (Fealey et al. 2008; Shapiro et al. 2015; Shotar et al. 2016).

More problematic remains the treatment of giant and fusiform aneurysm. For these, different endovascular strategies can be applied. The most old technique is the occlusion of the parent artery with coils or balloon. This technique has been frequently used in paraclinoid ICA aneurysm. For this treatment the patient has to pass a test of occlusion of the ICA commonly performed with balloon. Once the balloon is inflated, heparin is injected, the clinical condition of the conscious patient is continuously monitored, and an angiographic study of the contralateral carotid and of the vertebral arteries is performed. After 25–30 min if no neurological deficit develops and the angiographic study reveals a good collateral circulation of the circle of Willis including the examination of the venous phase in the hemisphere of the occluded ICA, this can be occluded. The accepted delay of the appearance of the veins in the involved hemisphere indicating tolerance has been differently reported varying from a simultaneous appearance (Lubicz et al. 2003b; Van Rooji et al. 2005) to 1.5–2 s (Abud et al. 2005; Clarençon et al. 2011; Kim et al. 2014; Whisenant et al. 2015). In a recent study (Matouk et al. 2012) 3 s has been accepted. We performed the test of occlusion in a *conscious* patient which allows a precise clinical monitoring. If the clinical condition remains unchanged, the collateral circulation is efficient, and the venous delay is between 1.5 and 2 s, the ICA is occluded definitely. The treatment is characterized, in the great majority of the cases by an excellent result (Van der Schaaf et al. 2002; Abud et al. 2005; Li et al. 2007; Van Rooji et al. 2005; Clarençon et al. 2011; Matouk et al. 2012; Whisenant et al. 2015). Ischemic lesions, however, even if very rare, due to a hemodynamic or embolic mechanism cannot be completely excluded.

The same technique can be useful in the occlusion of one VA in cases of giant or fusiform aneurysms where the posterior circulation is guaranteed by the contralateral VA. Also occlusion of the basilar artery in cases of fusiform or giant aneurysm has been used, provided that the distal part of the BA is revascularized through an efficient circle of Wills (Henkes et al. 2006). Distal aneurysms can be treated with a sacrifice of the parent artery, when a good leptomeningeal collateral circulation is present. If this is insufficient the parent vessel occlusion can be preceded by bypass-surgery (Van Roij and Sluzewski 2009).

The introduction of stent in association with coils and flow-diverter stents has opened new ways in the treatment of large and fusiform aneurysms and progressively those with complex morphology as reported by several authors (Yang et al. 2007; Lubicz et al. 2008; Liebig and Henkes 2008; Gall et al. 2009; Chapot et al. 2009; Fiorella et al. 2009b; Szikora et al. 2010; Leonardi et al. 2011; Briganti et al. (2012); Kalmes et al. (2015); Deutschman et al. 2012) (see also Sect. 16.6).

As far as it concerns the flow-diverter stent, there is no doubt about its utility in the treatment of morphologic complex aneurysms. However, some problems linked with this device remain unsolved at this time. In particular, as demonstrated in the studies of Bing et al. (2013) and Roszelle et al. (2013), improvement should be directed in the attempt to guarantee an adequate porosity which should on the one hand exclude the flow into the aneurysm and on the other hand allow a flow into a small adjacent perforator. Furthermore, this device can be associated with complications responsible for morbidity and mortality with a rate reported to be 3.9% and 6.6%, respectively, in the study of Velioglu et al. (2012), and 3% and 3% in the report of Piano et al. (2013). In the extensive meta-analysis of Brinjikji et al. (2013) morbidity and mortality were, respectively, 5% and 4% and these results remained grossly unchanged in a more recent study of Kalmes et al. (2015) in which the morbidity and mortality were 7.4% and 3.8%, respectively. Complications are due basically to ischemia or hemorrhage.

Ischemia results from thrombus at the site of the stent leading to occlusion of the parent vessel or to distal embolization. Another cause of ischemia is occlusion of the perforating branches, which occurs more frequently in the posterior circulation (Kulcsar et al. 2010; Siddiqui et al. 2012; Meckel et al. 2013b; Brinjikji et al. 2013; Kalmes et al. 2015; Wang et al. 2016a).

As far as it concerns *the hemorrhage*, it can be due to rupture of the aneurysm (Kulksar et al.

2011, 2012; Chow et al. 2012; Kalmes et al. 2015), or occurring as delayed hemorrhage distal to the aneurysm, mainly but not always in the vascular territory of the flow diverter (Velat et al. 2012; Cruz et al. 2012; Leung et al. 2012; Brinjikji et al. 2013). Hemorrhages *occurred predominantly in giant and posterior circulation aneurysms* (Rouchaud et al. 2016; Wang et al. 2016a). The cause of hemorrhage is not completely clear. In cases of *hemorrhage due to aneurysm rupture*, some authors (Cebral et al. 2011) have suggested that the stent leads to flow modifications responsible indirectly for an increased intra-aneurysmal pressure and consequently rupture of the aneurysm. For other authors (Tulamo et al. 2010; Kulksar et al. 2011), the intraluminal thrombus could be the source of proteases with proteolytic activity which could favor the rupture of the wall. The same mechanism could be responsible for the rupture of aneurysms completely occluded with coils and excluded from the circulation.

In cases of distal hemorrhage, this could be due to hemorrhagic transformation of distal ischemic lesions occurring more frequently in giant aneurysms requiring the use of more than one stent and a more longer treatment (Velat et al. 2012; Chiu et al. 2013; Rouchaud et al. 2016). Interesting to note is that the same can occur in the treatment of aneurysm treated with coils in association with stents (flow diverter or not flow diverter), indicating that in some cases, at least, the hemorrhagic transformation of microembolic ischemic lesions favored by the antiplatelet therapy could be the cause of the hemorrhage (Kayan et al. 2016). Other authors (Cruz et al. 2012) examining the distal hemorrhage in aneurysm treated in the anterior circulation (large aneurysms of the distal ICA or proximal MCA) have suggested that the flow diverter could reduce the arterial compliance of these vessels leading to changes of the blood pressure in the form of a higher systolic and lower diastolic values transmitted distally which could favor the formation of the hemorrhage. Also a sudden hyperperfusion distal to the aneurysm as suggested by others (Chiu et al. 2013) could be taken into consideration. Similar mechanism has been reported also in giant aneurysm treated by surgery (Murakami et al. 2002). Another possible complication is a distal SAH which can be severe, but sometimes be minimal, and it is discovered on the routine CT 24 h after the treatment. These patients can have only a mild headache which rapidly disappears or will be completely asymptomatic. All these signs disappear rapidly. As also reported by others (Piano et al. 2013; Gawlitza et al. 2015), we observed this event in two patients. The SAH can be due to minimal damage with partial dissection through the guide wire, or be the result of hyperperfusion in association with a toxicity of the contrast medium due to the frequent angiographic controls damaging the wall of the distal small arteries causing the bleeding.

Another interesting aspect which has become progressively known in patients treated with stents is the narrowing of the lumen of the artery due to *intimal hyperplasia*. In a large study presented by Kim et al. (2016) it has been demonstrated that this is a reversible phenomenon. It was more evident within 8 months after the treatment followed by a progressive regression with almost normalization of the lumen within 24 months. It has been suggested that weeks to months after the treatment, smooth muscle cells migrate from the tunica media into the intima leading to thickening of it. Later this infiltration regresses and the intima returns to its primary structure. In later controls, especially in treated fusiform aneurysm, the stent can be incorporated in the new restructured intima.

In conclusion, the introduction of the endovascular treatment and its positive evolution due to the extremely rapid development of new devices and progressive higher experience have certainly improved the treatment and prognosis of many patients. However, we are far from an ideal endovascular treatment and one should be cautious in the indication of treatment and in the choice of the device, in the attempt to determine the best treatment in the given patient (Van Rooij 2012; Kalmes et al. 2015). *The aneurysm remains an insidious pathology and its treatment is still burdened by a certain degree of morbidity and mortality. This is especially true for large, giant, fusiform dissecting aneurysms and those with a complex morphology which still can have a little unpredictable, sometimes very bad evolution in spite of the apparently technically successful treatment.*

11.10 Unruptured Aneurysms

Modern diagnostic methods have revealed an increase in the number of unruptured aneurysms, raising the question of whether they should be treated or not. The rupture of an aneurysm can have catastrophic clinical consequences for the patient. Technical improvements in surgery and endovascular treatment mean today that unruptured aneurysms can be treated and good results obtained, with a low rate of complications, though they are not completely excluded (Roy et al. 2001; Henkes et al. 2004; Bradac et al. 2007; Naggara et al. 2010).

Some factors that can influence the decision have been reported in an international study (Wiebers et al. 2003) of unruptured aneurysms that appeared in Lancet (2003). According to this study, the risk of hemorrhage is low in aneurysms with a diameter of less than 7 mm, but it increases progressively with greater diameters. This consideration seems to be in contradiction with the experience that the majority of the aneurysms diagnosed in patients with SAH are relatively small. However, it is today commonly accepted that the risk of hemorrhage is particularly high in the acute phase when the aneurysm develops and, at this time, it is commonly relatively small (Wiebers et al. 1987). If the aneurysm does not rupture in this phase, the wall fortifies and the risk of hemorrhage decreases. Later, some of these unruptured aneurysms can grow and this associated with changes in their wall increases progressively the risk of hemorrhage.

Independent of the diameter, the risk of rupture is reported to be greater in aneurysms of the posterior circulation and in patients who have already undergone treatment for another ruptured aneurysm.

In the attempt to clarify this matter, further studies have been performed related to the morphology of the aneurysm (irregular shape, multilobar, presence of blebs, as well as perianeurysmal environment). The latter involves constraints on the shape of the aneurysm that could favor its rupture (Rüfenacht 2005). Another factor, which has been considered, is the transmission of pressure and flow rates within the aneurysm, which has been reported to be higher in bifurcation aneurysms (Sorteberg and Farhoudi 2006). Postprocessing analysis of 3D visualization of the angiogram has shown the possible influence of the flow within the aneurysm being dependent on its location (Cebral et al. 2005) and also on its geometry, particularly when the aneurysm has a main axis parallel to the parent artery (Szikora et al. 2008).

In spite of the Lancet report (2003) and the above-reported interesting studies, which have enhanced some aspects of the aneurysm which can influence the decision to treatment, there is today, in many incidentally discovered aneurysms, no certain criteria to define the risk of rupture and no general consensus for or against the preventive treatment of an unruptured aneurysm (Raymond et al. 2008; Raymond 2009; Nagara et al. 2010). The decision still depends on many factor, among them the experience and attitude of the medical team involved, the situation of the patient, especially the age and his emotional reactions to the informations about the pathology.

11.11 Negative Angiograms in Patients with SAH

In about 15–20% of patients with SAH, the aneurysm is not detected on the angiogram. In some patients, particularly in cases with a perimesencephalic pattern of bleeding, the SAH is thought frequently not due to an aneurysm (Rinkel et al. 1991). In other cases, the aneurysm can be definitely thrombosed at the time of the bleeding and thus no longer recognizable, even in later controls. Vasospasm and large hematoma can, however, temporarily hide the presence of an aneurysm that can be demonstrated later.

In rare cases, no spasm or hematoma is present, but the aneurysm is not visualized on the

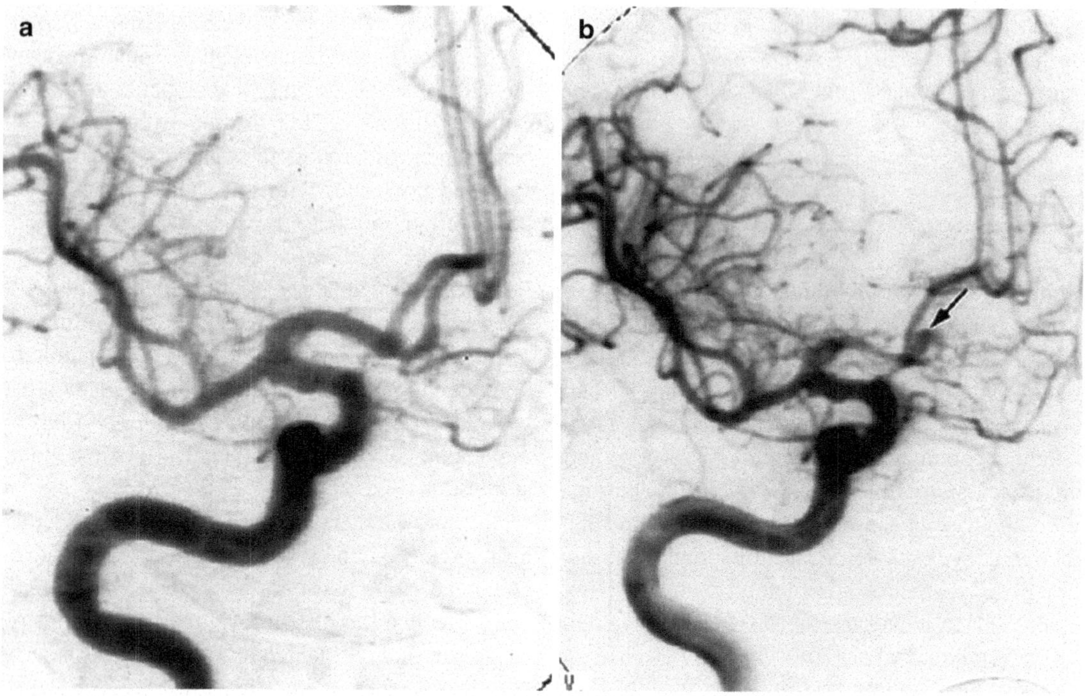

Fig. 11.31 Middle-aged woman with SAH, in whom a complete angiographic study with several projections did not reveal a vascular malformation. A SAH occurred again 2 weeks later. The angiographic study revealed now the aneurysm of the AcomA. (**a**) First right carotid angiogram judged normal. (**b**) The second angiogram showing the aneurysm. Minimal spasm is also present

angiogram performed in the acute phase. It can however be detected 1 or 2 weeks later. This has been reported in about 10–19% of the cases in which the first angiogram was negative (Bradac et al. 1997; Urbach et al. 1998; Alves et al. 2005) (Fig. 11.31). The cause of this phenomenon is not completely known. A temporary thrombosis of the aneurysm can occur. In other cases, the bleeding could be due to a dissection involving the wall of the artery not recognizable in the acute phase. Later this could lead to the formation of the aneurysm. This mechanism probably occurs in the so-called *blister-like aneurysms* (see Sect. 16.5). The extremely rare possibility of SAH due to spinal AVM should also be considered in these cases with a normal angiogram.

11.12 Vasospasm

Vasospasm as a complication of SAH has already been recognized and described in the early 1950s (Ecker and Riemenschneider 1951). Today it is reported to occur in about 70% of the cases. Among them, symptomatic ischemia occurs in about 35% (Wintermark et al. 2006; Komotar et al. 2007; Hanggi et al. 2008). Spasm can occur after every SAH, but younger patients and those with severe hemorrhage, as recognizable on CT, are more at risk of developing vasospasm.

Diagnosis. All patients with SAH should be closely monitored for vasospasm in the days after the acute episode. This can be done using daily transcranial Doppler (TCD), followed on the third or fourth day by CT-perfusion, whenever the TCD shows an increased velocity (more than 120–130 cm/s), especially when this occurs over a short period of time. Independently of and/or in association with these technical controls, every clinical worsening of the patient that is not due to rebleeding or hydrocephalus, excluded by CT examination, can be an indirect sign of spasm. In these cases, angiography should be performed when spasm is confirmed, by endovascular therapy.

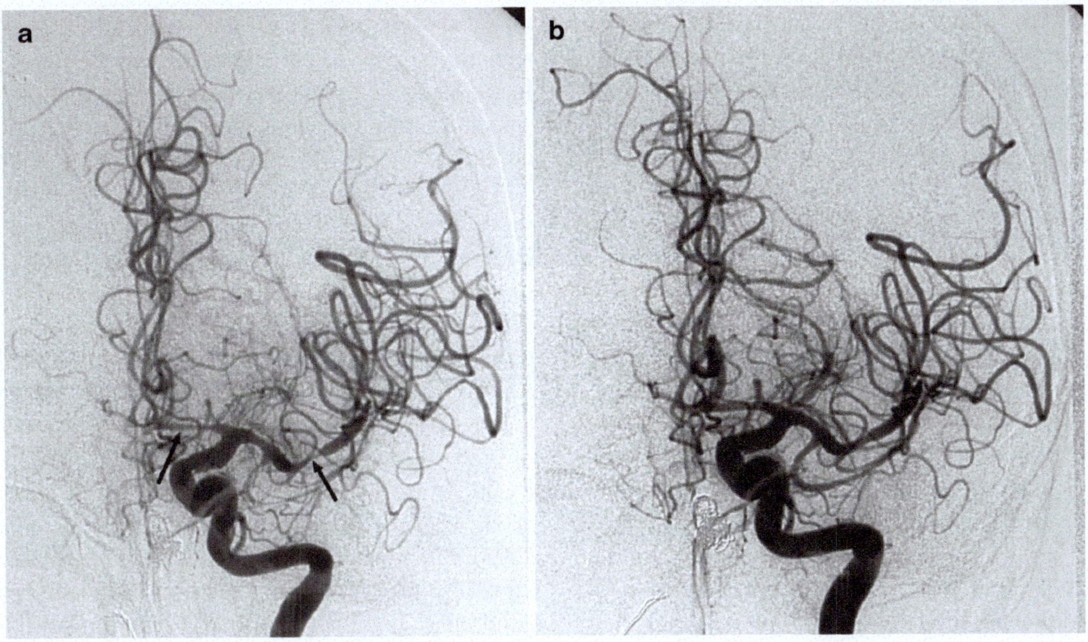

Fig. 11.32 Middle-aged patient with severe SAH due to rupture of a PICA aneurysm that was occluded with coils. The asymptomatic patient deteriorated a few days later owing to severe spasm involving, in particular, the anterior circulation. Left carotid angiogram (**a**) showing the severe spasm involving the A1 and M2 segments (*arrows*). Also, there is minimal spasm of the M1 segment. A similar pattern was seen on the right. The patient was treated with an injection of nimodipine in the ICA, with excellent angiographic results, demonstrated on the left angiogram (**b**). Similar result on the right

Therapy. Medical therapy consists of prophylactic nimodipine, oral or intravenous, depending on the grade of risk for a given patient. Nimodipine is a calcium antagonist which is supposed to act by blocking the effects of many vasoconstrictors of cerebral smooth muscle (Pickard et al. 1987). In many centers, this is associated with monitored hypertensive, hypervolemic, and hemodilution (*triple—H*) therapy. By confirmed spasm on angiography, the most used therapy today is the injection of nimodipine into the ICA uni- or bilaterally at the dose of 1–2 mg per vascular territory (Fig. 11.32). Selective injection in the A1–M1 segments can be useful in certain cases as well as the injection into the vertebrobasilar sector. A partial or complete resolution is obtained in 60–70% of cases (Boker et al. 1985; Bracard et al. 1999; Biondi et al. 2004; Bandeira et al. 2007; Hänggi et al. 2008). The resolution of the spasm may be only temporary, and so a second or more administrations can be necessary in the following days (Biondi et al. 2004). In selected cases of severe spasm not responsive to nimodipine, balloon angioplasty of the distal segment of ICA A1–M1 segments can be performed (Murayama et al. 2003b; Abruzzo et al. 2012). The progressive improvement of the endovascular material makes it possible today to perform this treatment with a low risk.

11.13 Aneurysms in Children

Intracranial aneurysms in children are rare, representing less than 5% of all cerebral aneurysms. They differ from those of the adults in the location, etiology, and clinical presentation. As reported by several authors (Ostergaard and Voldby 1983; Laughlin et al. 1997; Allison et al. 1998; Proust et al. 2001; Lasjaunias et al. 2005; Huang et al. 2005), they are more frequent in boys, whereas in the adults there is a predominance among women. This seems to indicate a gender influence.

The most frequent location of the aneurysms is reported to be the anterior circulation. Among them, ICA (ICA bifurcation, more rarely cavernous) and MCA aneurysms are predominant (Ostergaard and Voldby 1983; Heiskanen 1989; Laughlin et al. 1997; Allison et al. 1998; Proust et al. 2001; Huang et al. 2005; Aryan et al. 2006; Vaid et al. 2008) in the posterior circulation are more rare. These later, however, are more frequent in comparison to the adults. These can involve the BA and VA and their branches. Among these a frequent location is reported in the PCA (Meyer et al. 1989; Laughlin et al. 1997; Huang et al. 2005; Aryan et al. 2006; Vilela and Goulão 2006; Bradac et al. 2008a). The aneurysm can present as in the adult with a SAH (Heiskanen 1989; Proust et al. 2001; Vaid et al. 2008), but differently from aneurysms in the adulthood, in about 50% of the cases the symptoms in children are due to the mass effect as reported by several other authors (Lasjaunias et al. 1997; Massimi et al. 2003; Huang et al. 2005; Aryan et al. 2006; Lv et al. 2009; Gandolfo 2012). This is due to the fact that at the time of diagnosis the aneurysms are frequently large, sometimes giant. Giant aneurysms can be found in the anterior and in the posterior circulation. This later location seems to be more frequent in children.

Unlike in adults, a frequent cause of aneurysm in children is trauma, even minor (Ventureyra and Higgins 1994; Yazbak et al. 1995). Other etiologies are infectious, collagen diseases (FMD, Ehlers–Danlos syndrome, Marfan syndrome, polycystic kidney, pseudoxanthoma elasticum, neurofibromatosis), coarctation of the aorta hemoglobinopathies, as well as a family history of aneurysms (Ostergaard and Voldby 1983; Pasqualin et al. 1986; Roche et al. 1988; Meyer et al. 1989; Allison et al. 1998; Huang et al. 2005). Spontaneous dissection has been increasingly also recognized as a cause of aneurysm in children (Laughlin et al. 1997; Massimi et al. 2003; Lasjaunias et al. 2005; Vilela and Goulão 2006; Bradac et al. 2008a). In some cases, the etiology remains uncertain. Examples of aneurysm in the pediatric age are presented in Figs. 16.16 and 16.17.

Vascular Malformations of the Central Nervous System

12

12.1 Introduction

Rokitansky (1846) is reported to be the first to have described this kind of pathology which he called "vascular brain tumor in pial tissue." It was Virchow (1862–1863) who first clearly differentiated tumors from brain angiomas, which were identified as vascular malformations of congenital derivation. The concept that brain arteriovenous malformations (BAVM) is an anomaly caused by errors during vascular development in the embryo has been suggested by Cushing and Bailey (1928) and Dandy (1928). However, some difficulties in the differential diagnosis between BAVM and tumors remained, as noted by Zülch (1957) and Russell et al. (1959). An accurate description of this pathology as a definite congenital malformation was proposed by McCormick (1966, 1976). The same author made a classification, which, with a few modifications (Challa et al. 1995; Yaşargil 1987, 1999; Chaloupka and Huddle 1998; Valavanis et al. 2004), is still valid today.

12.2 Classification

- Arteriovenous malformation (AVM)
- Vein of Galen malformations
- Cavernous malformations (cavernomas)
- Capillary malformations (telangiectasias)
- Developmental venous anomaly (DVA)
- Transition forms
- Vascular malformations as part of well-defined congenital-hereditary syndromes
- Rendu-Osler syndrome
- Sturge-Weber syndrome
- Wyburn-Mason syndrome
- Klippel-Trenaunay-Weber syndrome

12.3 AVM

12.3.1 Pathogenesis and Pathology

The pathogenesis of AVMs is not completely clear. In this context some aspects of the development of the cerebral vessels should be shortly reconsidered. This occurs in two phases: vasculogenesis and angiogenesis. In the vasculogenesis, angioblasts differentiate into endothelial cells to form the primary vascular plexus. Later, angiogenesis follows, in which the primary plexus undergoes remodeling and organization, leading to the formation of the final cerebral vessels with the typical pattern of arteries, capillaries, and veins (Streeter 1918; Risau and Flamme 1995; Risau 1997). The causes of an aberrant vasculo-angiogenesis leading to AVM are unknown. Many factors are probably involved; among them, some endothelial growth factors (VEGFR1-VEGFR2) and their binding receptors (FLt-1; FLK-1) and a group of cytokines the angiopoietin-1 (Ang1) and angiopoietin-2 (Ang2) and their receptors have been identified as important for the normal development of the cerebral

G.B. Bradac, *Applied Cerebral Angiography*, DOI 10.1007/978-3-319-57228-4_12

vessels (Risau and Flamme 1995; Shalaby et al. 1995; Fong et al. 1995; Sonstein et al. 1996; Davis et al. 1996; Zhang et al. 1996; Maisonpierre et al. 1997; Uranishi et al. 2001; Hashimoto et al. 2001, 2005). Highest levels of these factors, as well as their altered balance, possibly due to gene mutations could lead to aberrant development and formation of AVMs. In this context, it is interesting to note that AVMs can develop in some cases of hereditary hemorrhagic telangiectasia in which a known gene mutation has been identified (Zhang et al. 1996; Matsubara et al. 2000; Letteboer et al. 2006; Saba et al. 2007).

Also not clear is when the AVMs develop. Some authors (Mullan et al. 1996a, b) *have suggested that this occurs in the gestational period.* In some cases the AVMs may be relatively small at birth and grow later. In this context, however, a few consideration should be made. Indeed in spite of the routine use of ultrasound diagnosis during gestation, AVMs are commonly not identified, while the extremely more rare vein of Galen malformation can be easily diagnosed. Furthermore there are reports, describing the appearance of cerebral AVMs later in life in patients, in whom previously performed magnetic resonance imaging (MRI) showed no malformations. In some of these patients, cerebral AVMs develop in the pathologically altered brain as a result of different causes such Moyamoya disease (Schmit et al. 1996), hemangiomatosis (Song et al. 2007), heterotopia (Stevens et al. 2009), and after radiosurgery (Rodriguez-Arias et al. 2000); in others, the brain parenchyma was completely normal (Gonzalez et al. 2005; Bulsara et al. 2002). These observations *raise doubts about the congenital nature* of cerebral AVMs, which, at least in some cases, seem to be acquired lesions caused by different nonspecific insults on the brain, perhaps in patients in whom an altered genetic background preexists (Kim et al. 2011).

The main angioarchitectural characteristic of an AVM is an area called the "*nidus*," in which there is a direct shunting between arteries and veins without interposed capillaries. The elevated intravascular flow leads to changes of the vessels. Histology shows the nidus to be composed basically of dilated arteries and veins. In some vessels, the wall structure is still recognizable, characterized by the presence of a media with smooth muscle cells and an elastic lamina in the arteries and an absence of muscle cells in the veins. In other arteries, prominent changes, characterized by areas of wall thickening caused by the proliferation of fibroblasts, muscle cells, and an increase in connective tissue, are present. Segments with thinning of the wall also occur, which potentially can lead to aneurysm formation. Severe changes take place in the venous sector, forming the so-called arterialized veins, in which a wall thickening is recognizable due to fibroblast proliferation without smooth muscle cells. The interposed parenchyma shows gliosis, hemosiderin pigmentation, and calcification, resulting from ischemia or previous hemorrhages. The surrounding parenchyma may appear normal or show similar changes (Challa et al. 1995; Kalimo et al. 1997; Brocheriou and Capron 2004).

12.3.2 Incidence

The incidence of AVM is reported in studies involving general autopsy to be 0.15–0.8% (McCormick 1984; Jellinger 1986), with an annual detection of about 1–2 AVM per 100,000 persons. Multifocal lesions can occur with a frequency of 1–10% (Perret and Nischioka 1966; Rodesch et al. 1988; Willinski et al. 1990). These latter are reported to be more frequent in *children* in whom multiplicity appears to be twice as common as in adults (Rodesch et al. 1988; Lasjaunais 1997; Berenstein et al. 2010).

12.3.3 Clinical Relevance

Most AVMs are diagnosed in young or middle-aged patients. In about 20% they are discovered in *children*. As far as it concerns the clinical presentation, 5–10% remain asymptomatic and are diagnosed incidentally by CT or MR investigations performed for other reasons. Some 40–50% present with intracranial hemorrhage, 30% with seizures, 10–15% with headaches, and 5–10% with neurological deficits (Perini et al. 1995; Stapf et al.

2002; Hofmeister et al. 2000; Valavanis et al. 2004). The incidence of symptomatic AVM in the adult population is reported to be one-tenth the frequency of intracranial aneurysms (Berenstein and Lasjaunias 1992; Valavanis et al. 2004). The highest risk in AVM is hemorrhage (Graf et al. 1983; Brown et al. 1985; Crawford et al. 1986; Ondra et al. 1990; Mast et al. 1997) which has been calculated to be between 1.7 and 4.1% per year in AVMs presenting without history of hemorrhage. In the reports of the same authors the mortality and morbidity following the first hemorrhage was between 10 and 30% and 10 and 20%, respectively. The risk of a repeated hemorrhage after an initial episode is reported to increase in the first year, later rapidly decreasing (Mast et al. 1997), reaching for some authors (Graf et al. 1983) the level of the initial risk. Hemorrhage is the most frequent initial symptom in *children* (Berenstein and Lasjaunias 1992; Rodesch et al. 1995; Lasjaunais 1997; Di Rocco et al. 2000; Klimo et al. 2007; Berenstein et al. 2010; Soltanolkotabi et al. 2013).

Cases of spontaneous thrombosis of AVMs (Sukoff et al. 1972; Levine et al. 1973; Mabe and Furuse 1977; Pascual-Castroviejo et al. 1977; Sartor 1978; Nehls and Pittman 1982; Omojola et al. 1982; Wakai et al. 1983; Pasqualin et al. 1985; Barker and Anslow 1990; Ezura and Kagawa 1992; Hamada and Yonekawa 1994; Abdulrauf et al. 1999) as well as its possible recanalization occurring even a few years later (Mizutani et al. 1995) have been reported. Also the reappearance of the malformation after its proven complete obliteration after gamma knife radiosurgery can occur (Lindquist et al. 2000). A long follow-up of these patients is mandatory. *All this seems to indicate that AVM, at least in some cases, is a dynamic pathology with an evolution, which is probably influenced by many not precisely known factors.*

12.3.4 Location

The majority of AVMs (85%) are located in the supratentorial area. They can be located in a sulcus (sulcal) in a gyrus (gyral), or involve both (sulcogyral) (Valavanis et al. 2004). They can remain superficial or extend deeply toward the ventricles, basal ganglia and thalamus. AVMs involving primary deep structures are rare (Saatci et al. 2011). These later are more frequent *in pediatric patients* (Berenstein and Lasjaunias 1992).

The infratentorial AVMs are rare with a frequency reported between 5 and 15% (Perret and Nischioka 1966; Yasargil 1999; Batjer and Samson 1986; Valavanis et al. 2004; Saatci et al. 2011). The most frequent are those involving the cerebellum (hemisphere, vermis), located on the superior-inferior convexity or on its anterior surface. The AVM can remain superficial or extend toward the deep structures with extension toward the brainstem. Primary AVM of the brainstem are very rare. Their frequency is in the range of 2–3% of all AVMs (Garcia-Monaco et al. 1990; Liu et al. 2003).

12.3.5 Diagnosis

MRI, including functional studies, provides information about the site and extension of the malformation and its relationship with the adjacent areas. Furthermore, it shows which functional changes have occurred in the affected and unaffected hemisphere (Alkadhi et al. 2000; Bradac et al. 2014). Angiography is essential in defining the angioarchitecture of the malformation. This comprises angiography of the internal and external carotid arteries and vertebral artery, followed, when necessary, by selective examinations aimed to better characterize the supplying arteries, venous drainage, and aspects of the nidus.

12.3.5.1 Supplying Arteries (Feeders)

These are variously dilated and tortuous, unique or multiple, and arise from one or more vascular territories. Cortical branches are involved in superficial AVMs (Figs. 12.1, 12.2, 12.3, 12.4, 12.5, and 12.6). Perforators (deep and medullary branches) and choroidal arteries can be recruited every time deep structures and ventricles are primary or secondary involved by large cortical AVM extending to the depth (Figs. 12.5, 12.7a–d, 12.8a, 12.9, 12.10, 12.11 and 12.20).

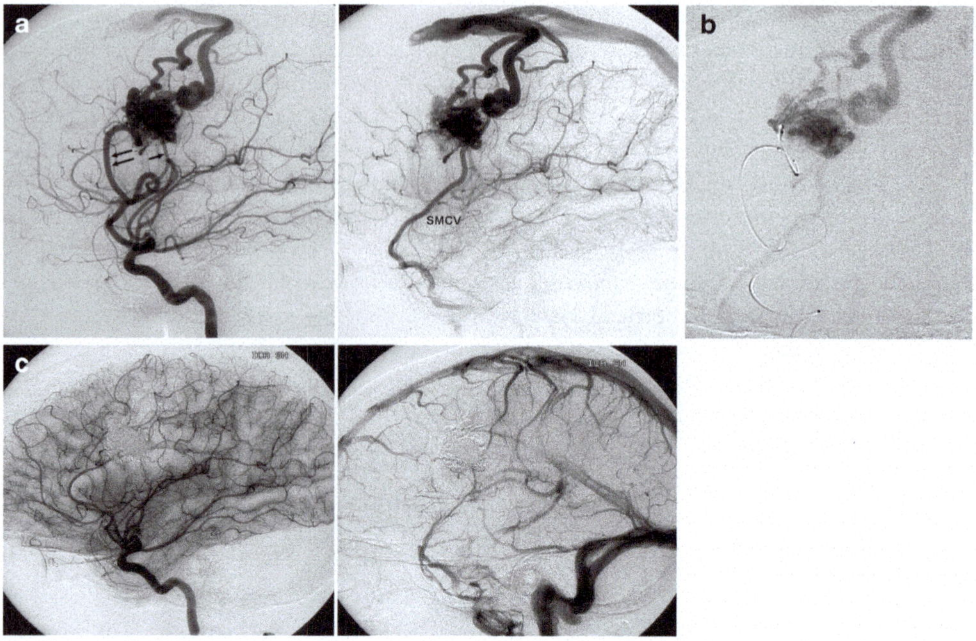

Fig. 12.1 Well-defined nidus of a lateral frontal AVM presenting with epilepsy in a young patient. (**a**) Lateral carotid angiogram, early and late phases. The AVM is supplied by a dilated insular branch (*double arrow*). A second, smaller feeder appears posteriorly (*arrow*). Cortical drainage in the SSS, where a partial retrograde injection of its anterior segment is visible. A second drainage occurs in the SMCV. (**b**) Selective catheterization of the anterior feeder preceding embolization with onix. (**c**) Lateral carotid angiogram, arterial and venous phases performed 2 months later confirming the complete occlusion of the AVM and normalization of the arteries and veins. Clinically: intact patient

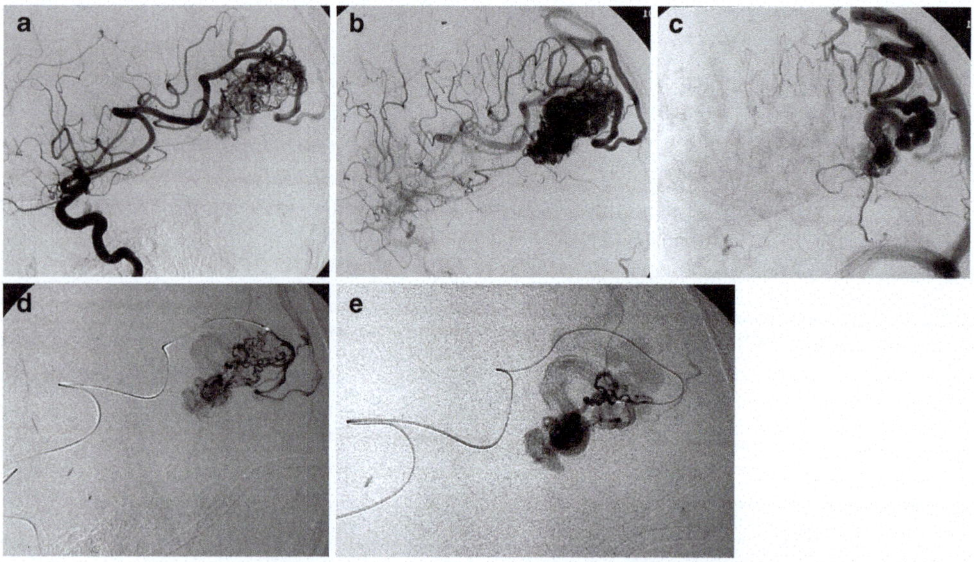

Fig. 12.2 Latero-temporo-occipital AVM, presenting with hemorrhage. Carotid lateral angiogram. (**a**) arterial phase showing the supply through the distal branches of the dilated gyrus angularis artery. (**b**), (**c**) Venous phases, showing the different venous drainage related to the corresponding compartments. These are well demonstrated on super-selective studies (**d**), (**e**). At the periphery of the nidus, an isolated arteriovenous shunt is recognizable (**d**)

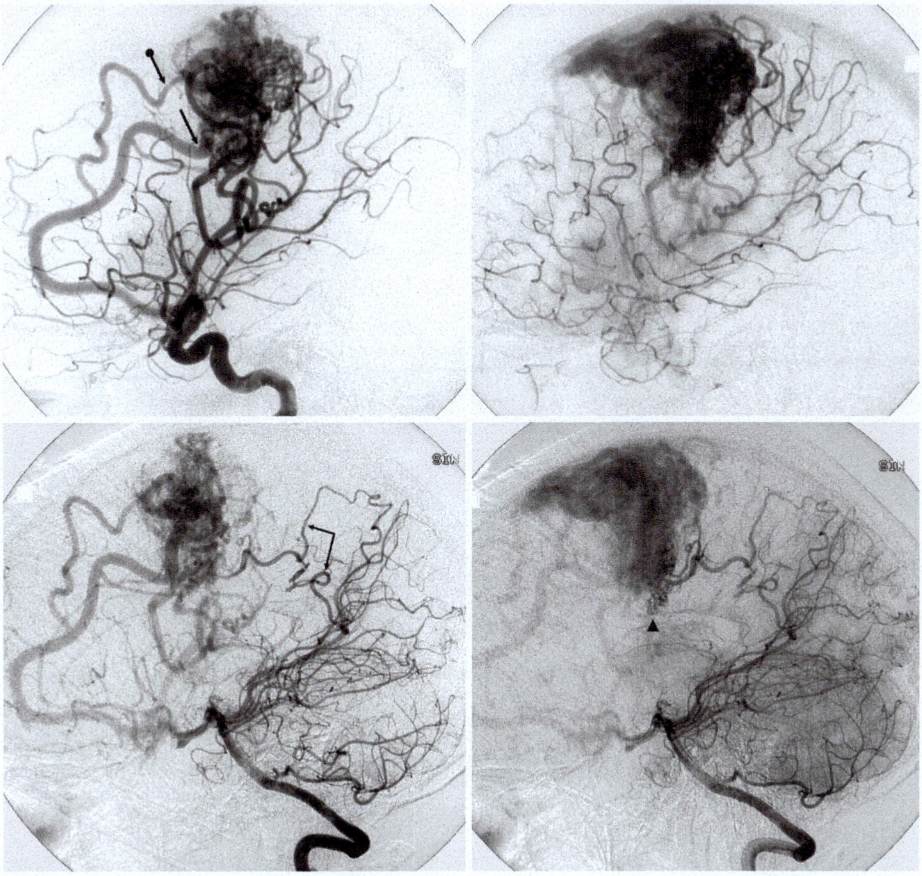

Fig. 12.3 Incidental discovery of a large parietal AVM in patient with headache. (same case as in Fig. 12.8b). Carotid angiogram (*above*) showing the involvement of the enlarged pericallosal artery from which arise the dilated posterior frontal artery (*arrow with dot*) and a large trunk probably the paracentral artery (*arrow*). There is also involvement of branches of the MCA probably the central arteries. On the vertebral angiograms (*below*) an indirect involvement of the PCA is visible. This occurs through opening of leptomeningeal anastomoses between PCA and ACA (*arrow with angle*), with retrograde filling of the distal segment of the pericallosal. In the later phase is recognizable a minimal part (*arrow with dot*) of the AVM, not present on the carotid angiogram

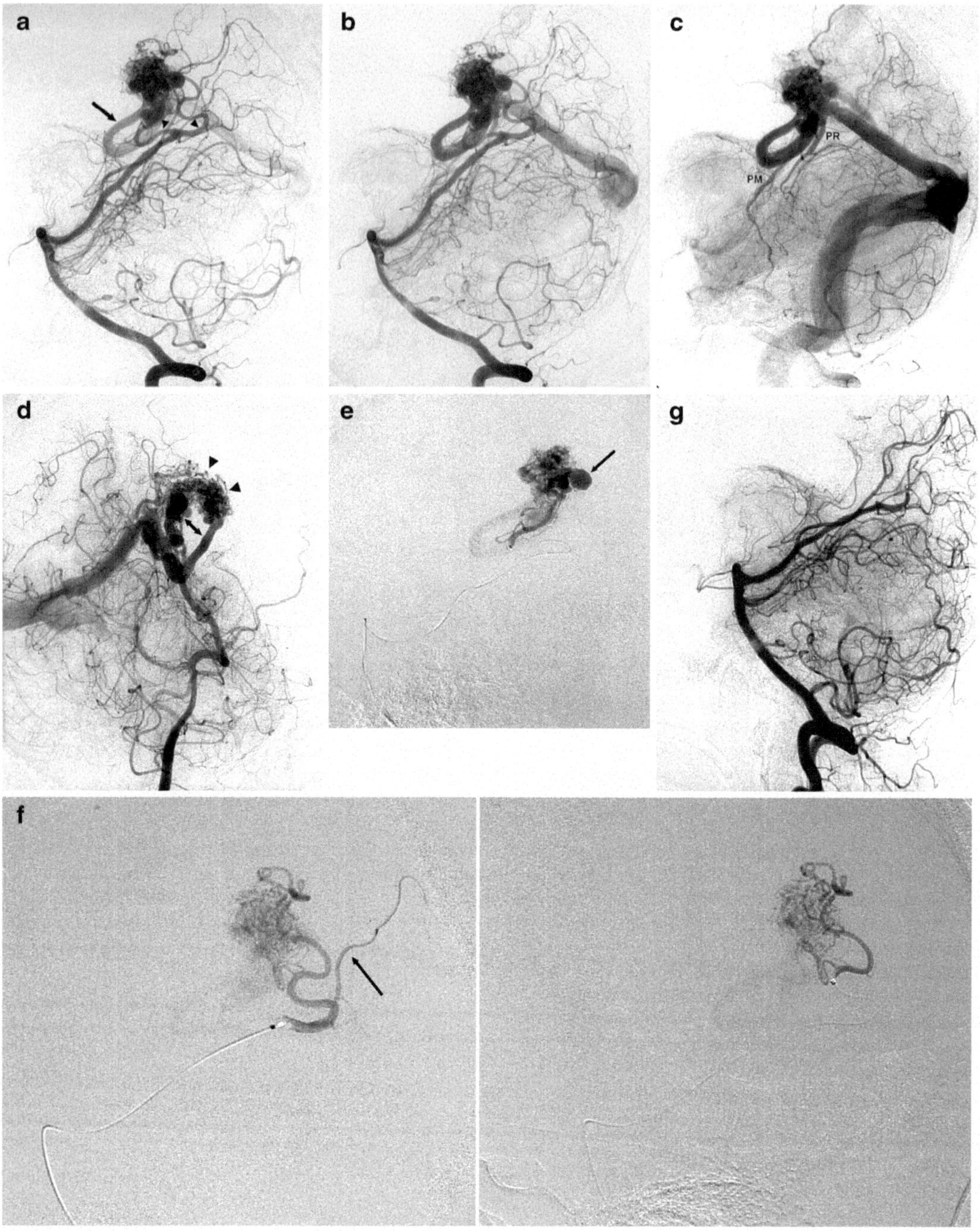

Fig. 12.4 Large hematoma located in the left deep medial occipital retrosplenial area in a young patient, removed in the acute phase. Angiographic study a few days later. (**a**), (**b**) The vertebral angiogram showed an AVM supplied by two feeders (*arrowheads*) arising from the P4 segment of the left PCA. The main drainage occurred into a large venous channel probably a medial atrial vein (*arrow*) continuing in the Galen vein and straight sinus. (**c**) In the later venous phase, a retrograde filling of the precentral vein (PR) and posterior mesencephalic vein (PM is visible). (**d**) Oblique view. Better demonstration of the nidus (*arrowheads*) and of the venous drainage (*arrow*). (**e**) Catheterization of the feeder supplying the compartment with an intranidal aneurysm (*arrow*) occluded with glue. (**f**) Catheterization of the second feeder with a progressive advance of the microcatheter distal to a normal parenchymal branch (*arrow*) before the injection of glue. (**g**) Posttreatment angiogram showing the occlusion of the AVM. The remaining minimal component supplied by the pericallosal artery was treated later by radiosurgery with final good outcome

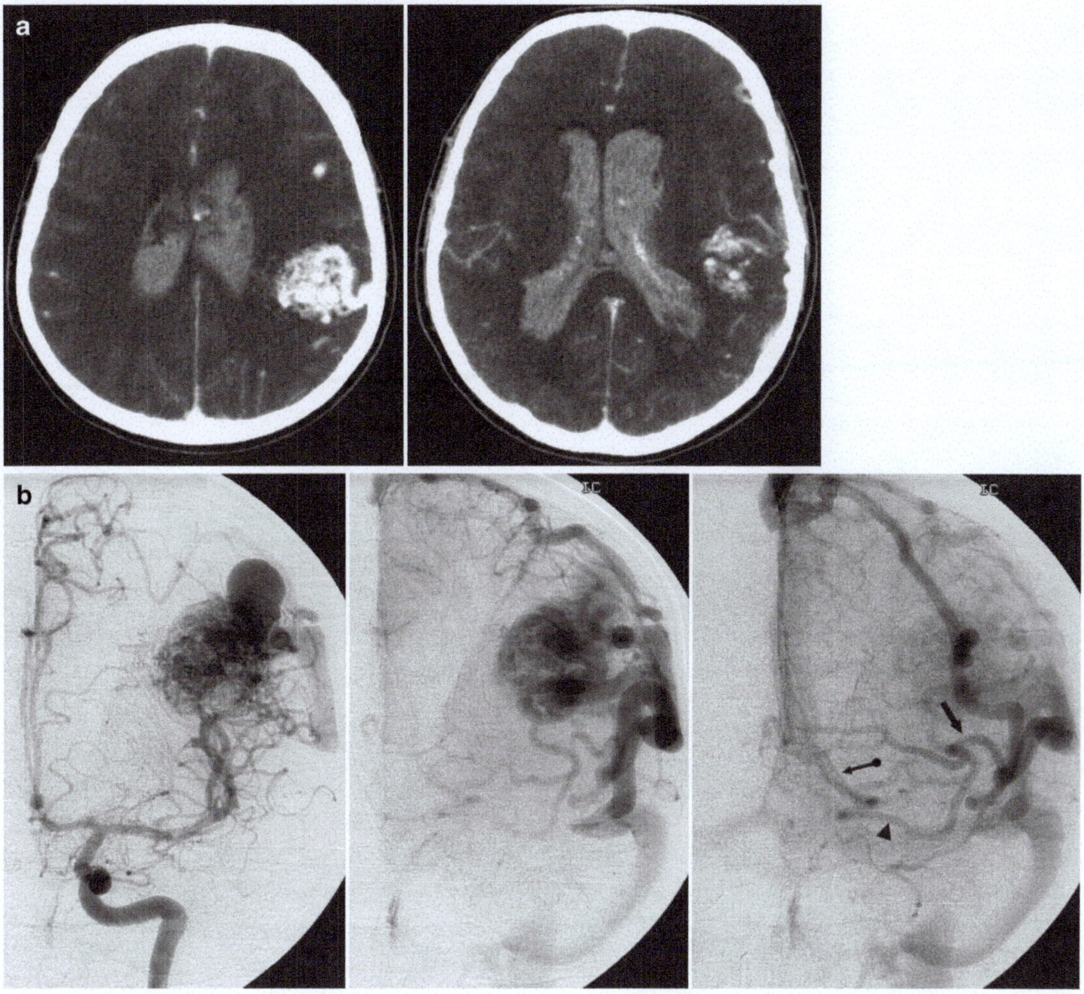

Fig. 12.5 Young patient presenting with severe intraventricular hemorrhage leading to deep coma. (**a**) CT showed a severe intraventricular hemorrhage and a left parietal AVM. (**b**) Left ICA angiogram, AP view. Earlier ad late phases. Parietal AVM supplied by branches of the MCA. On the periphery of the malformation there is a large aneurysm. On the later phases the superficial cortical venous drainage directed toward the SSS and in the SMCV is recognizable. Through connections between SMCV and insular veins (*arrow*) later filling also of the DMCV (*arrow head*) and BV (*arrow with dot*). (**c**) Oblique view, showing better the aneurysm (*arrow*). (**d**) On the vertebral angiogram a small aneurysm (*arrow*) on the course of the left lateral posterior choroidal artery close to the nidus, could be clearly recognized. Thinking that the aneurysm was responsible of the intraventricular hemorrhage, the decision was taken to treat it. Selective angiogram preceding the occlusion with glue. Angiogram posttreatment. (**e**) A few days later ICA angiogram was performed aimed to treat the aneurysm which was spontaneously almost completely thrombosed. This was confirmed on a second angiogram 1 month later in which the aneurysm was no longer visible. The patient did not recover from its primary severe hemorrhage and remained in a vegetative state. No other treatment was performed

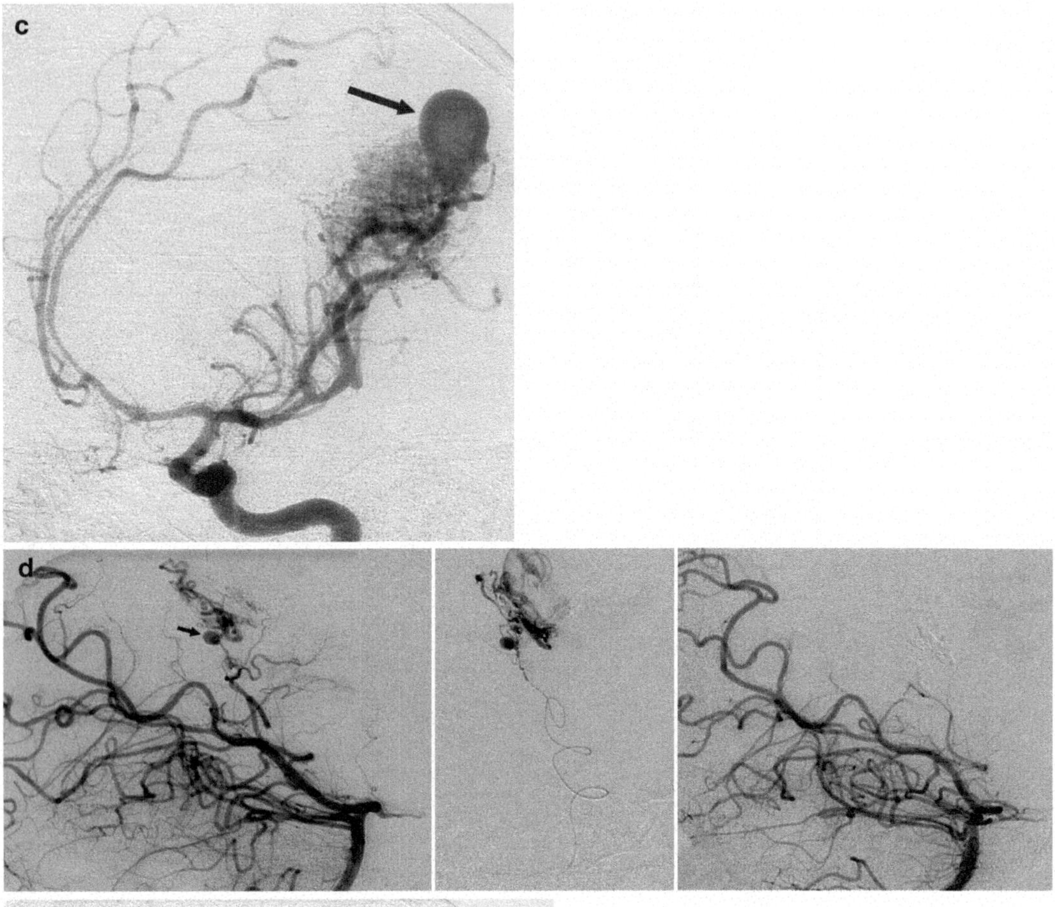

Fig. 12.5 (continued)

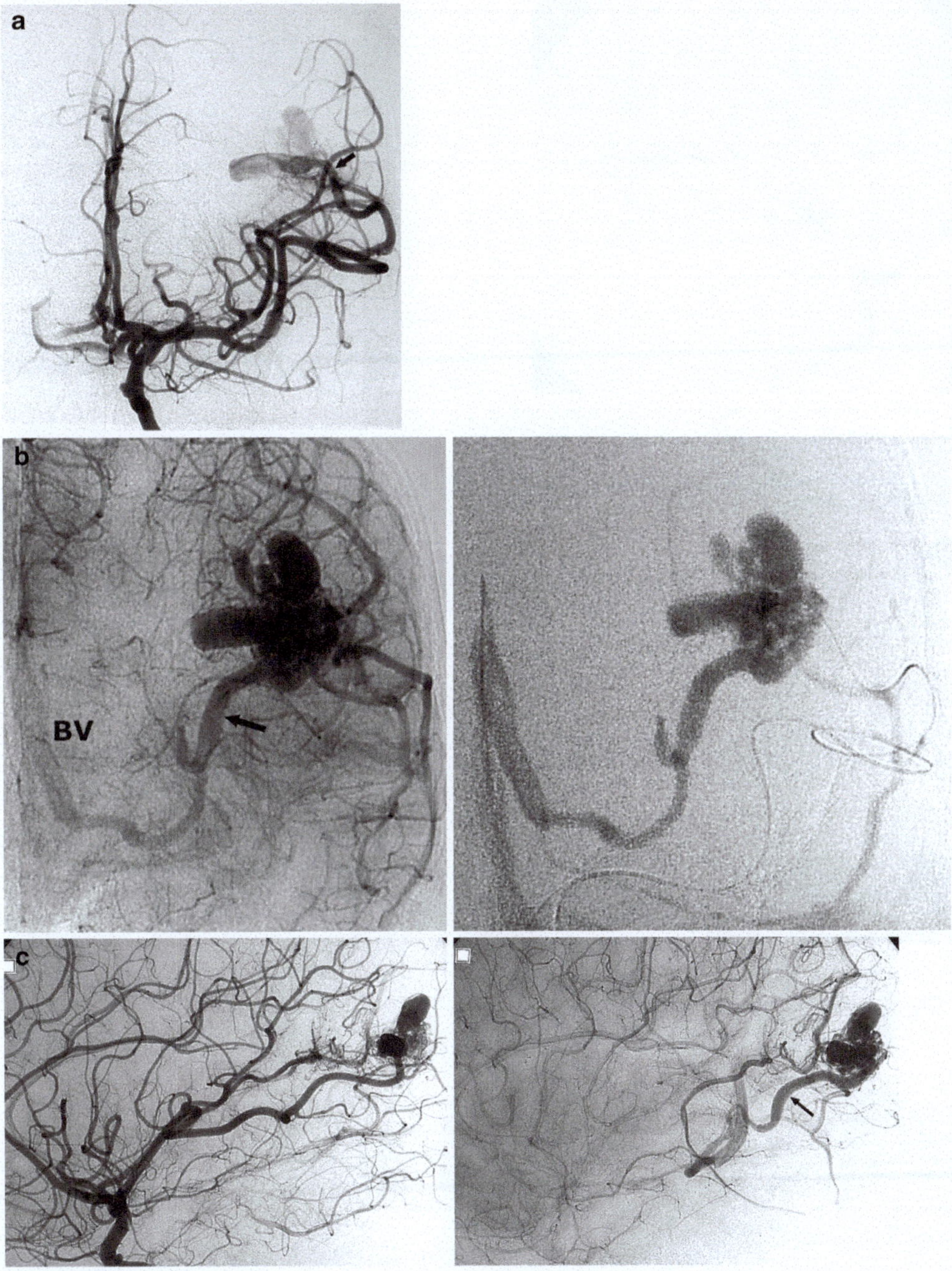

Fig. 12.6 Laterotemporal occipital AVM in a child presenting with epileptic seizures. (**a**) Carotid angiogram AP view. The AVM consists mainly of a direct fistula (*arrow*) between the gyrus angularis artery and the venous sector, which is characterized by a large complex venous pouch. (**b**) Late phase, showing the further cortical and deep drainage. This latter occurs through a dilated medullary branch of the lateral atrial vein (*arrow*) continuing into the distal segment of the basal vein (BV). On the selective study (*on the right*) the deep venous drainage is better demonstrated. (**c**) Lateral ICA angiogram, early and late phases, showing the feeding branches and the venous drainage involving the lateral atrial vein (*arrow*). (**d**) Lateral carotid angiogram, after occlusion of the fistula with glue. A minimal network corresponding to the persistent nidus is still recognizable. This was treated with radiotherapy with a final good clinical outcome

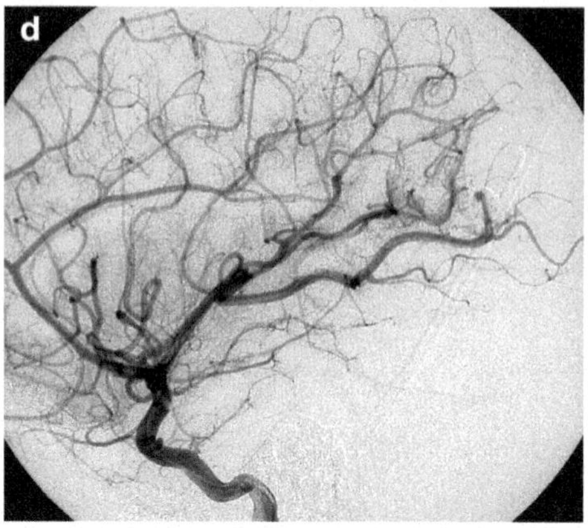

Fig. 12.6 (continued)

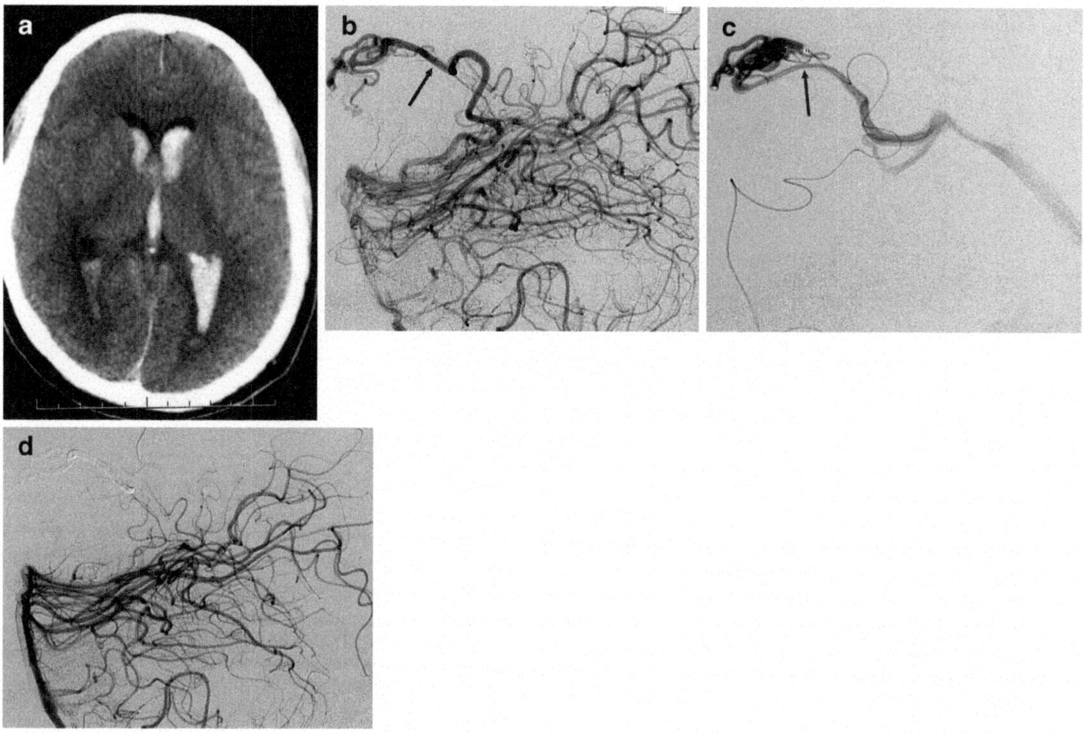

Fig. 12.7 AVM in a young patient presenting with hemorrhage involving the third and lateral ventricles. (**a**) CT showing the hemorrhage. (**b**) Lateral vertebral angiogram shows a dilated posterior medial choroidal artery supplying the small AVM on the roof of the third ventricle. (**c**) On the selective study the nidus and the venous drainage in the internal cerebral vein (*arrow*) continuing in the Galen vein is well visible. (**d**) Control angiogram after occlusion of the AVM with acrylic glue

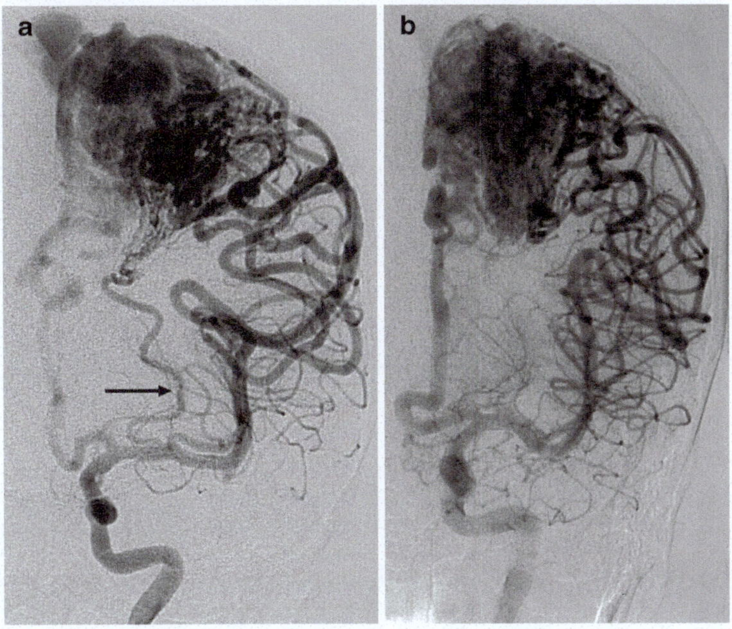

Fig. 12.8 Large superficial AVM extending into the deep white matter in two different patients. (**a**) in addition to the cortical feeders there is the involvement of a large perforator (*arrow*). (**b**) a similar AVM in which, however, the perforators are not involved

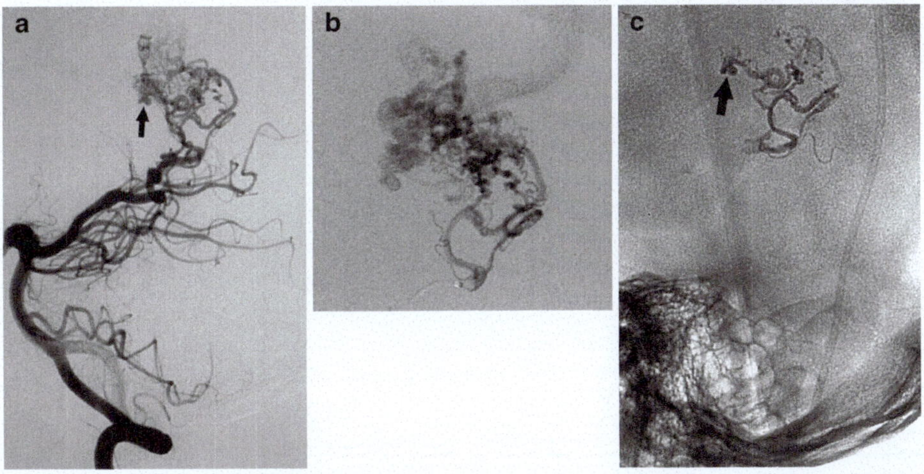

Fig. 12.9 Example of an intranidal aneurysm, probably responsible of repetitive small intraventricular hemorrhage in a young patient with a large right parietal AVM extending deeply toward the lateral ventricle. (**a**) Lateral vertebral angiogram showing the deep part of the AVM where a small aneurysm is visible (*arrow*) supplied by the medial posterior choroidal artery. (**b**) Selective study preceding the injection of acrylic glue. (**c**) Cast of the glue involving part of the nidus and the aneurysm (*arrow*). Despite the only partial treatment, the hemorrhagic episodes arrested completely over a period of many years

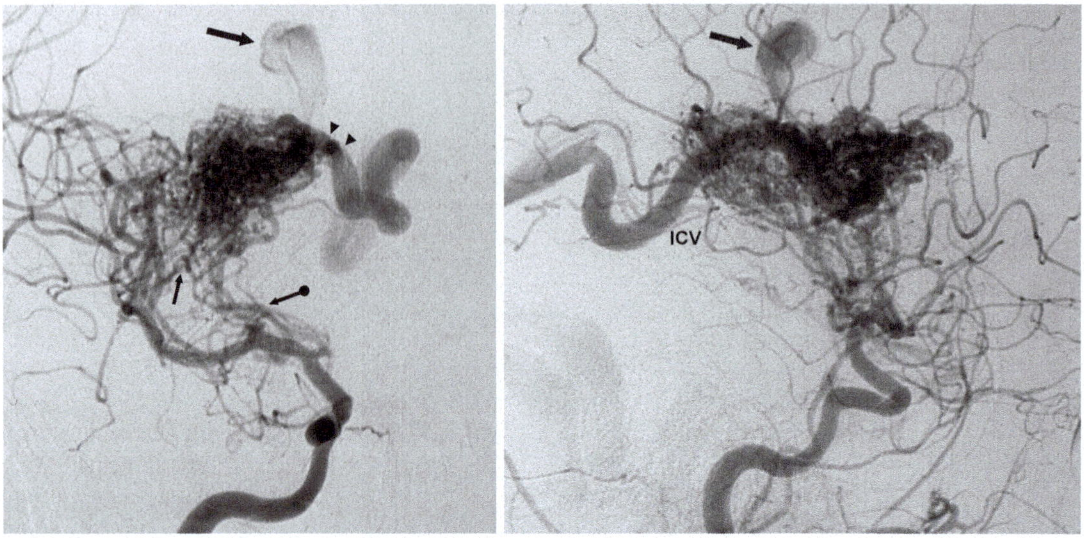

Fig. 12.10 AVM in a child involving the basal ganglia and the adjacent white matter presenting with hemorrhage. Carotid angiogram, AP and lateral views. The AVM is supplied mainly by dilated perforators arising from M1 (*arrow with dot*) and by branches of the M2 segment (*arrow*). The drainage occurs in the thalamostriate vein (*arrowhead*) continuing into the internal cerebral vein (ICV). On the periphery of the AVM, there is a partial injected venous pouch (*large arrow*), which corresponds probably to a venous pseudoaneurysm. Surgery was performed

Fig. 12.11 Young girl presenting with sudden headache followed by coma due to rupture of AVM in the right nucleus caudatus. (**a**) CT showing the hemorrhage in the superior part of right head of the nucleus caudatus and in the ventricular system. (**b**) Right and left AP, ICA angiograms showing the AVM supplied by small branches of the right pericallosal artery and by a dilated perforating branch artery (arrow) arising from the proximal part of the right A1 supplying also a small aneurysm (*arrowhead*). The supplying perforator could be the Heubner artery. Considering, however, that the AVM is located in the superior part of the nucleus caudatus (commonly vascular territory of the MCA), and that the perforators of the MCA can partially arise from the ACA (see also Sect. 5.1), it is thinkable that this perforating branch is not the Heubner artery. (**c**) Right ICA lateral angiogram showing the drainage of the AVM in the SSS and ISS (*bidirectional arrows*) and partially also in the deep system through the by the hemorrhage deformed caudate vein (*arrowheads*). (**d**) Acute treatment aimed to occlude the aneurysm thought responsible of the hemorrhage. AP view showing the origin of the perforator from the proximal A1 (*arrow*). Detail demonstrating the microcatheter (*arrowheads*) which was advanced first in the left A1, then through the AcomA in the right A1 and in the right perforating artery and further close to the aneurysm where glue was injected. (**e**) Control angiogram showing the occlusion of the distal segment of the artery (*arrow*) with exclusion of the aneurysm. The patient recovered slowly and the small remaining part of the AVM was later treated with surgery

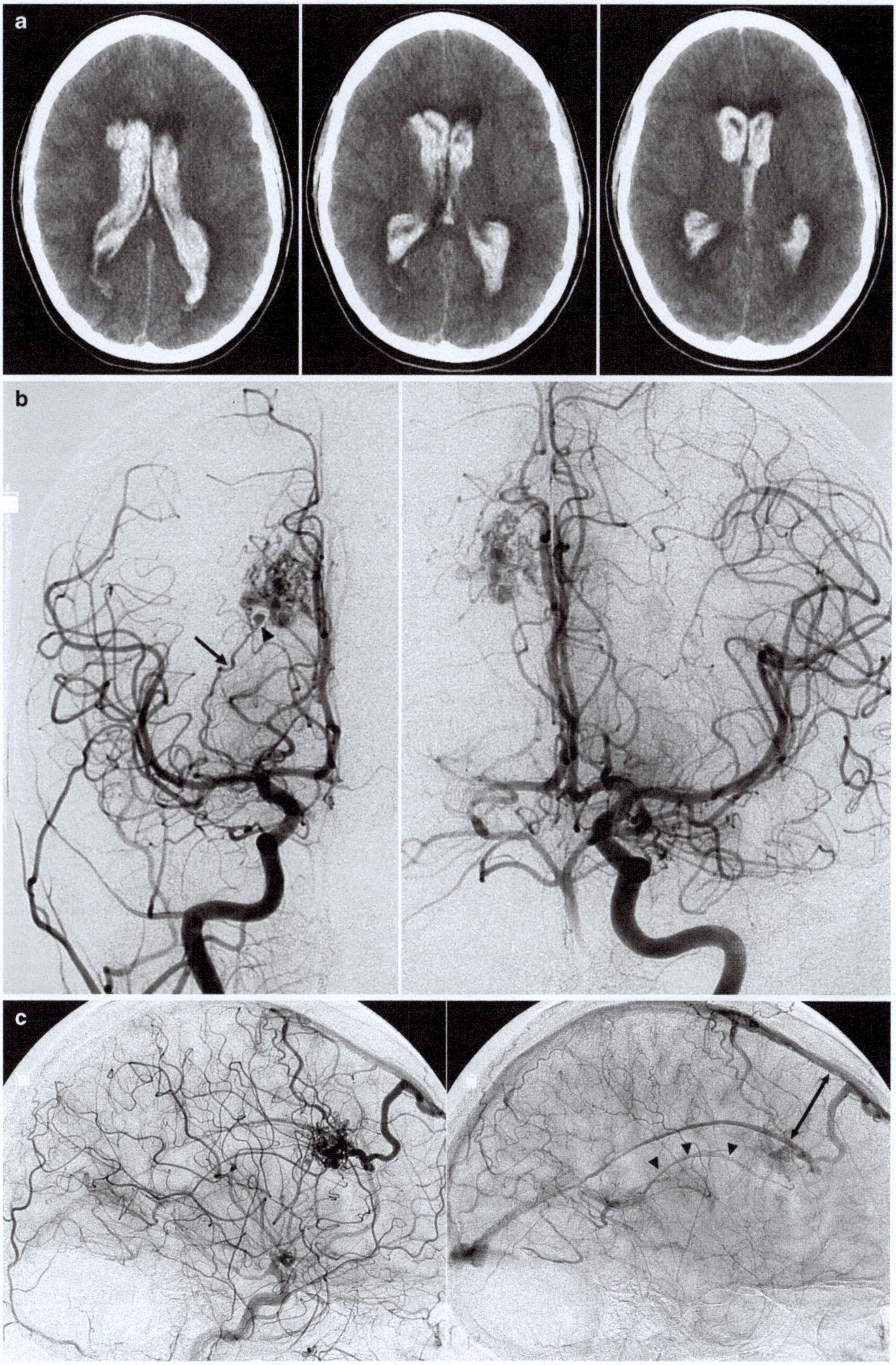
a
b
c

Fig. 12.11 (continued)

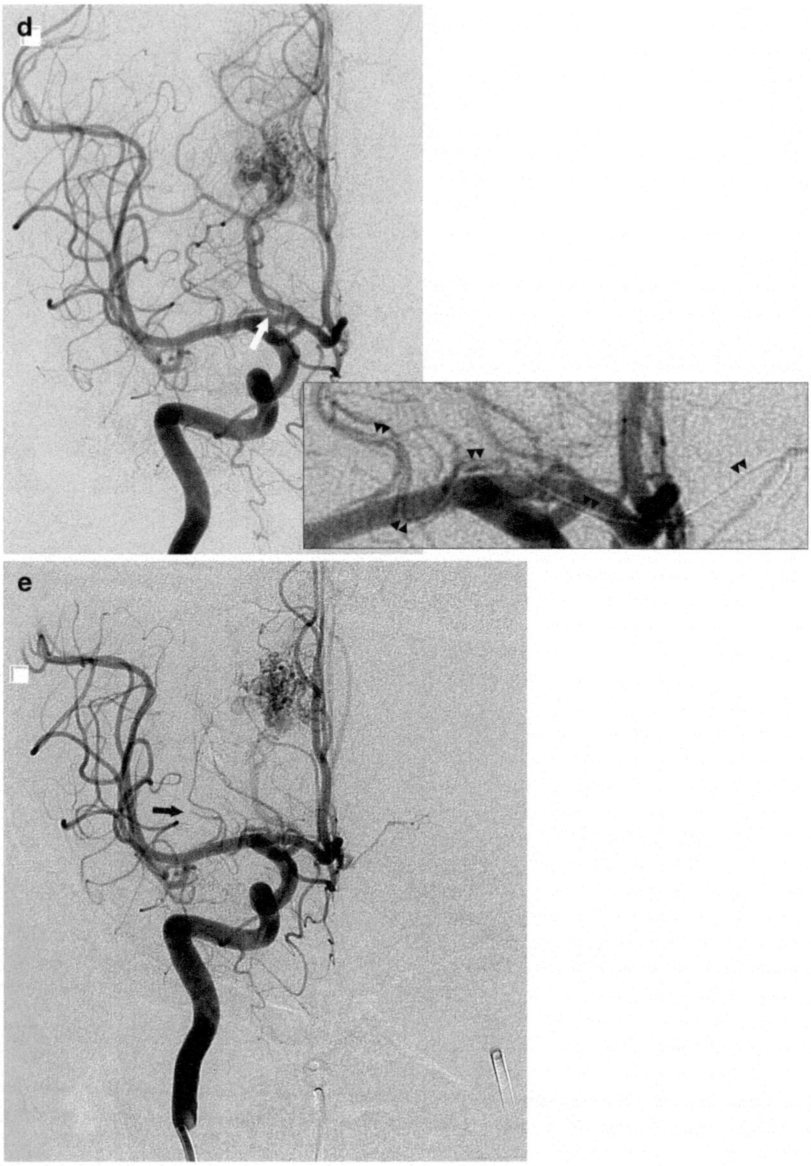

Each feeder can end in the nidus, connected through one or more small branches with one or more venous channels, in various combinations (Houdart et al. 1993), forming what is termed the plexiform aspect of the nidus (Figs. 12.1 and 12.2). Otherwise, after giving branches to the AVM, the feeders continue distally to supply the normal parenchyma (Fig. 12.4). It can occur that on the angiogram, they appear to end in the nidus, though they do in fact run further distally. The distal part, however, is not recognizable, owing to the steal phenomenon present in the nidus. In other cases, a large artery "*en passage feeder*," running adjacent to the nidus, can give some

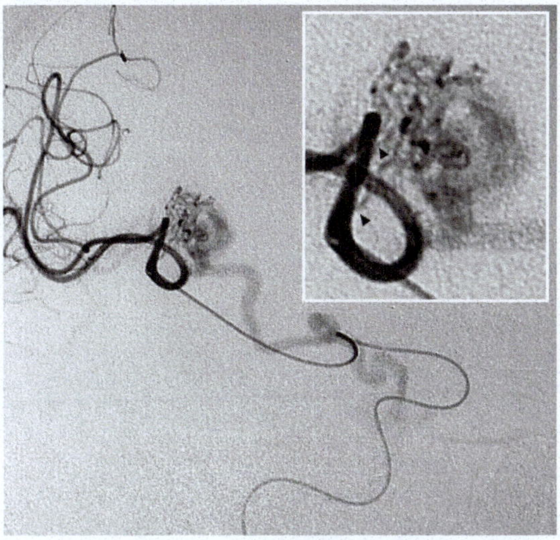

Fig. 12.12 *(Fig. 12.17a of the Cerebral Angiography 2nd Edition). Example of "en passage feeder." From the proximal part of a branch of MCA which distally supplies the normal parenchyma arise small branches (*arrowhead*) supplying a temporo-insular AVM

small branches to the nidus, coursing further to the normal parenchyma (Fig. 12.12). All these aspects should be carefully studied with selective injections since embolization of these feeders carries the risk of ischemia of the normal parenchyma (Berenstein and Lasjaunias 1992; Valavanis 1996; Chaloupka and Huddle 1998; Pierot et al. 2004; Valavanis et al. 2004).

Sometimes, indirect feeders can reach the nidus through the opening of leptomeningeal (pial) anastomoses (Fig. 12.3). This occurs when an important branch supplying the AVM ends completely in the nidus and no branches reach the distal normal parenchyma, which is supplied indirectly by the developed collateral circulation. The latter can extend to the AVM and supply indirectly its distal part (Berenstein and Lasjaunias 1992; Chaloupka and Huddle 1998; Valavanis et al. 2004).

Involvement of meningeal branches is reported in about 30% of the cases (Newton and Cronquist 1969; Rodesch and Terbrugge 1993). This occurs through anastomoses between the meningeal arteries and the pial branches involved in the vascularization of the AVM. In this context, it should be remembered that dilated dural branches can be a cause of headache. In selected cases, these branches can be catheterized and used to reach the nidus of the AVM where embolic material can be injected.

Finally an interesting aspect, occurring in the cerebral arteries as well as in the branches of ECA, when involved, and in the draining veins, is that their dilatation due to the increased in-out flow disappears with return to normalization, when the vascular malformation is eliminated. This is due to the specific characteristic of the cerebral vessels to adapt to the different vascular conditions (Fig. 12.1).

12.3.5.2 Aneurysms

Many type of aneurysms can be identified in the examination of AVMs. There are aneurysms which are located far from the nidus on one or more supplying arteries. They are thought to be due to the increased flow (*flow-related or "stress" aneurysm*) and frequently, though not always, disappear when the AVM is excluded (Berenstein and Lasjaunias 1992; Valavanis and Yasargil 1998). They can be an incidental discovery in the study of an AVM, but can be also the cause of a subarachnoid or parenchymatous hemorrhage (Stapf et al. 2006). The frequency of these aneurysms is reported to increase with the age of the AVM. This probably means that the development of these aneurysms is due to the high flow associated with the AVM and especially to the long-lasting AV shunt (Valavanis 1996). In our experience the majority of these aneurysms occur in old patients especially in the vertebrobasilar sector. Examples are presented in Figs. 11.14, 12.13, 12.14, and 12.16. Similar aneurysms can develop in other conditions. (Fig. 12.15) (See Sect. 11.6.5).

Other small aneurysms are located close or within the nidus (*intranidal aneurysm)*. These can be better identified by selective studies. They are very frequent and are thought to be responsible for the hemorrhage of the AVM in many cases (Willinsky et al. 1988; Marks et al. 1992; Turjman

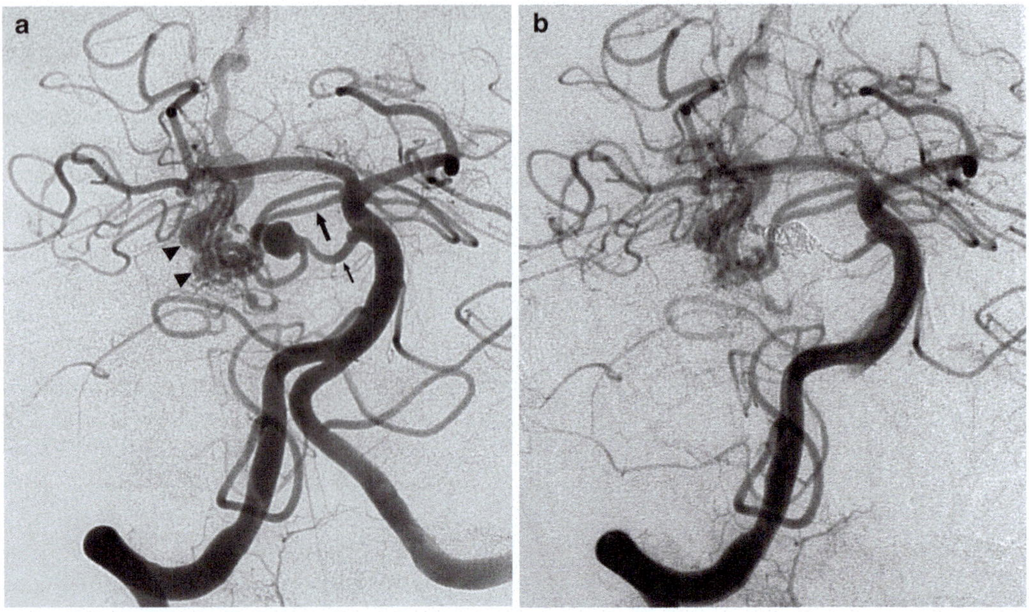

Fig. 12.13 Old patient presenting with severe SAH involving predominantly the right cerebellopontine angle. (**a**) Vertebral angiogram showing a small AVM in the cerebellopontine angle (*arrowheads*) supplied by a double superior cerebellar artery (*large arrow*). A dilated lateropontine artery (*small artery*). On the course of this latter a flow-dependent aneurysm is recognizable. This was thought to be responsible for the SAH and acutely occluded with coils together with the parent artery. (**b**) Posttreatment angiogram. The comatose patient recovered also from his small pontine infarct. The AVM was treated by surgery

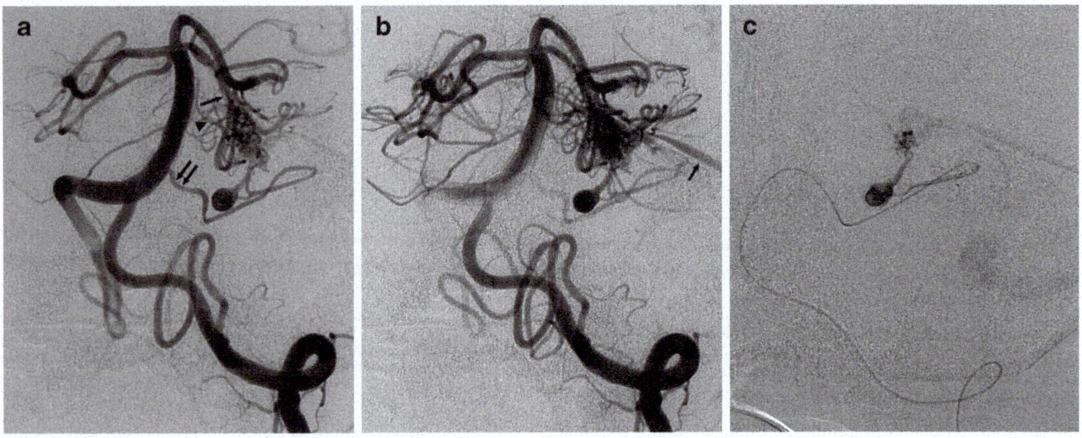

Fig. 12.14 Old patient presenting with severe SAH involving predominantly the left cerebellopontine angle. (**a**) Vertebral angiogram (*partial oblique view*), showing the AVM supplied by branches of the superior cerebellar artery (*arrow*). A dilated lateral pontine artery (*arrow head*) is also involved. There is also an important supply from the AICA (*two arrows*). On its course a large flow-dependent aneurysm is recognizable. (**b**) In the later phase the drainage in the superior petrosal sinus (*arrow*) is visible. (**c**) Selective study of AICA, preceding the occlusion of the aneurysm and of the distal AICA with coils. The patient recovered, and the AVM was treated later with surgery

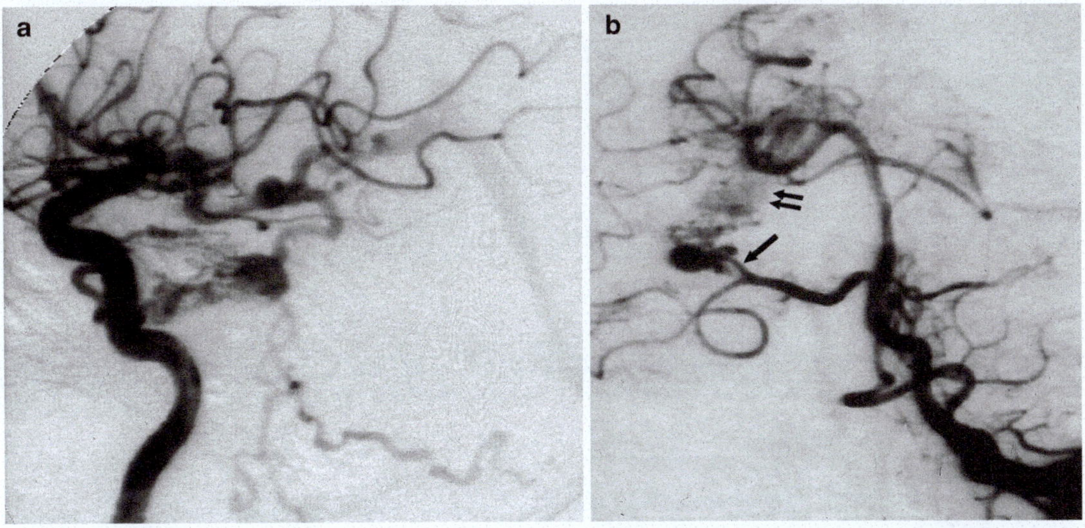

Fig. 12.15 Old patient presenting with SAH involving predominantly the posterior fossa. An angiographic study showed the pattern of a typical petro-tentorial DAVF (see Chap. 13, Sect. 13.7). (**a**) Lateral ICA angiogram showing the fistula at the level of the superior petrous sinus and the drainage in the lateral mesencephalic vein. (**b**) Vertebral angiogram, AP view, showing the supplying of the fistula (*arrows*) through the dilated AICA. On its rostro-lateral branch a flow-dependent aneurysm (*arrow*) is recognizable

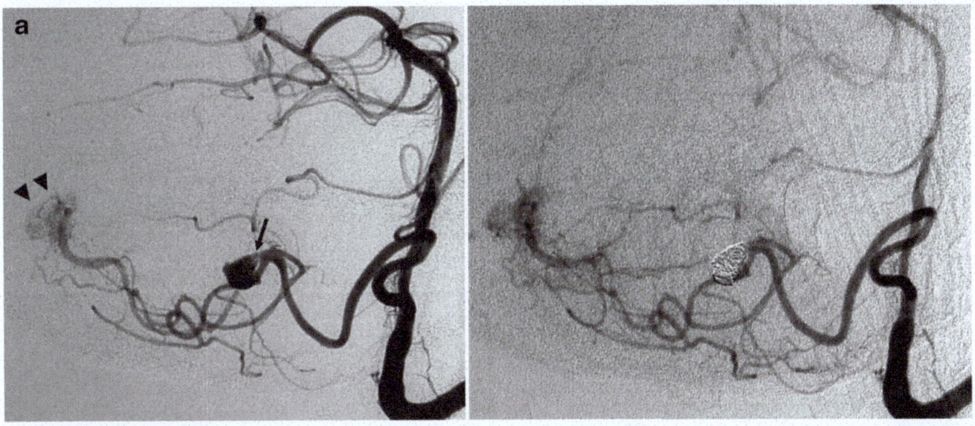

Fig. 12.16 Examples of flow-dependent aneurysms in two different patients. (**a**) Severe SAH in an old patient predominantly in the posterior fossa with involvement of the fourth ventricle. On the vertebral angiogram a small AVM (*arrowheads*) supplied by distal branches of the PICA was demonstrated. A flow-dependent aneurysm (*arrow*) is recognizable on the supratonsillar segment of the PICA. This was occluded with coils. No further hemorrhage occurred, but the patient did not recovered from the SAH. He remained comatose and died 3 months later. (**b**)–(**d**) Another old patient presenting with SAH, less severe than the above described case. The angiographic study revealed a small AVM involving the inferior cerebellar vermis and partially the right cerebellar hemisphere. (**b**) Right VA angiogram. The PICA is replaced by a well-developed AICA supplying the AVM. On its distal cerebellar branch a flow-dependent aneurysm (*arrow head*), better visible on the selective study is recognizable. (**c**) Control angiogram after occlusion of the aneurysm with onyx. (**d**), left VA angiogram. Flow-dependent aneurysms are visible on the supratonsillar segment (*arrow head*) and on the vermis branch (*arrows*) with dot of the PICA. Note the normal posterior meningeal artery (*arrow*). Selective angiogram, preceding the distal occlusion of the PICA and aneurysms with Onix. The patient recovered. The AVM was treated by surgery

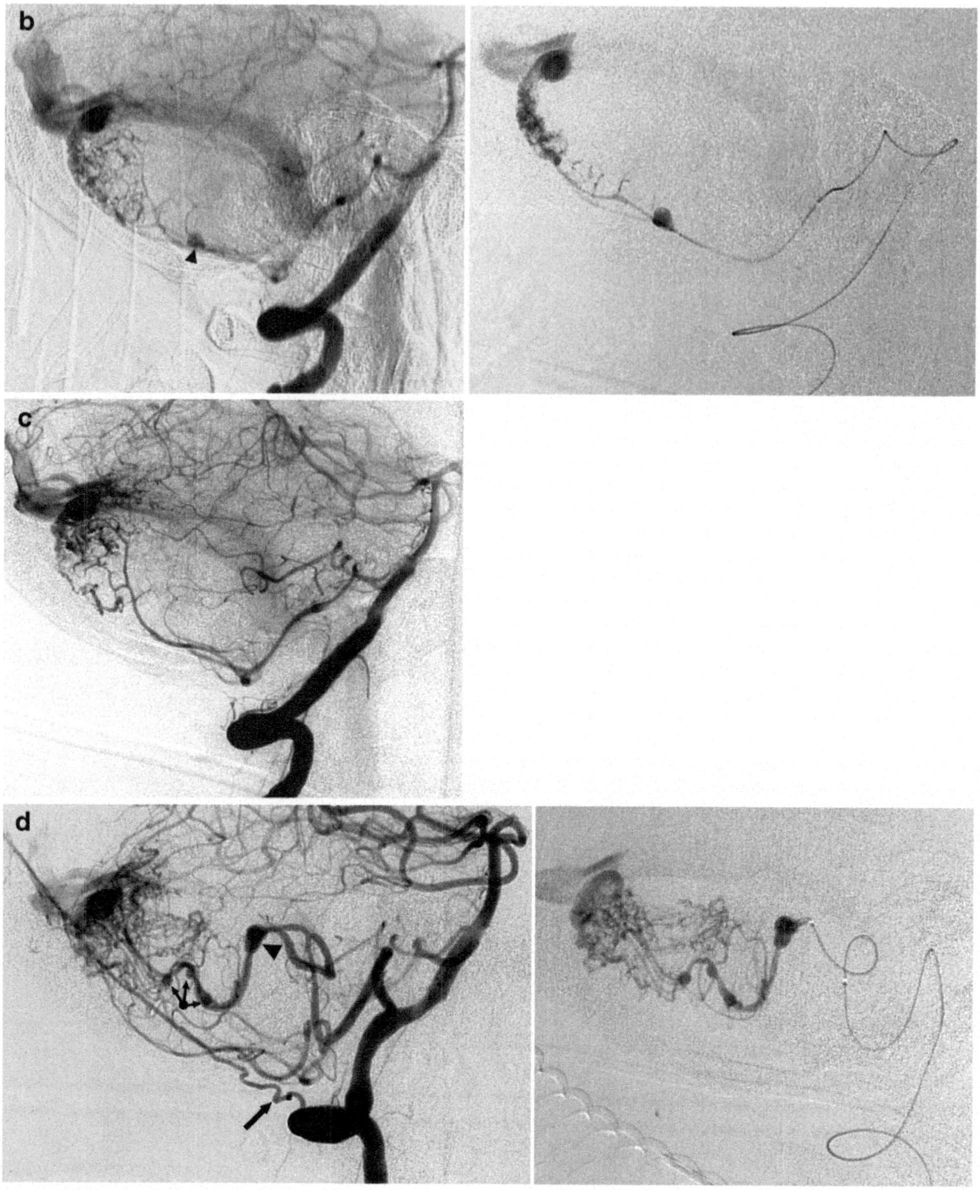

Fig. 12.16 (continued)

et al. 1994; Pollock et al. 1996; Redekop et al. 1998; Bradac et al. 2001; Pierot et al. 2004; Valavanis et al. 2004, Signorelli et al. 2014), (Figs. 12.4, 12.5, 12.9, 12.11, 12.17, and 12.18).

Rarely aneurysms can be found on an arterial branch independent of the AVM. The pathogenesis of these is probably the same of other aneurysms, as described in Sect. 11.4.

One notable type is the *pseudoaneurysm*, which develops at the site of rupture of the AVM. This type is detected in patients presenting clinically with a recent AVM rupture. Pseudoaneurysms lack a true vessel wall and consist of a pouch arising from a partially reabsorbed hematoma which still communicates with the lumen of the ruptured vessel. They can be

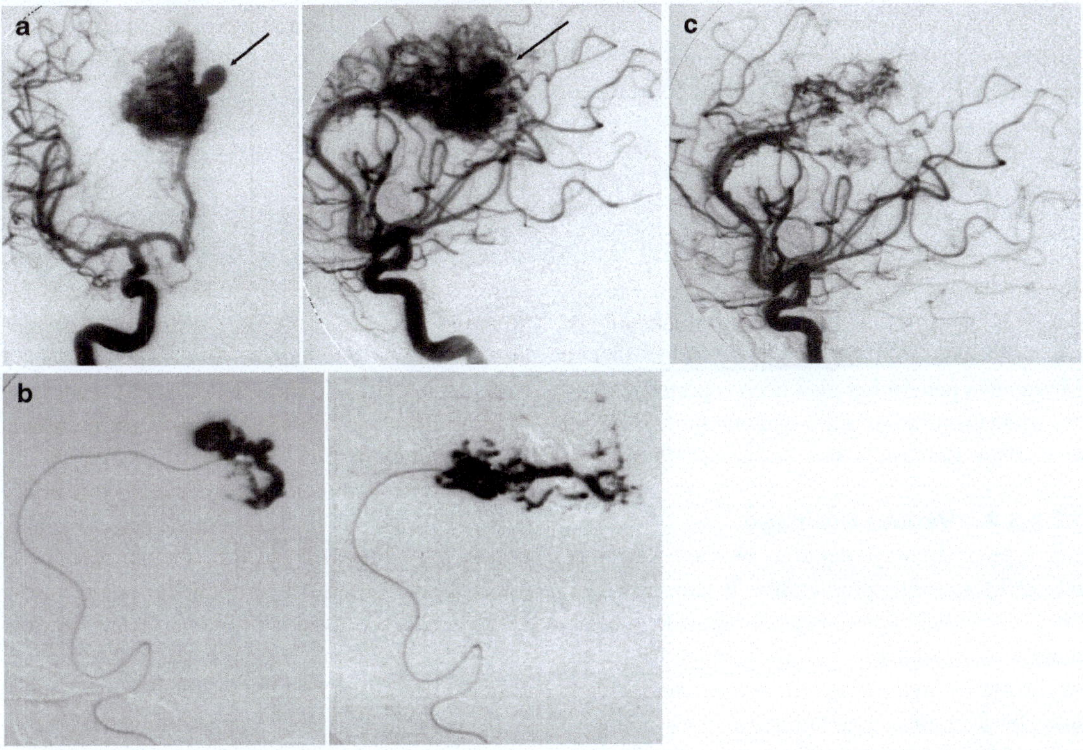

Fig. 12.17 AVM involving the corpus callosum and adjacent gyrus cingula in a young patient presenting with hemorrhage. (**a**) Right carotid angiogram, AP and Lateral views. A dilated pericallosal artery supplies the well-defined nidus. In its posterior part a large aneurysmal dilatation is visible. (**b**) Two selective studies of distal branches of the pericallosal artery preceding the injection of acrylic glue aimed to occlude partially the nidus and especially the aneurysm. (**c**) Control angiogram, post-treatment, well tolerated by the patient, who was operated 1 month later with a final clinical good result

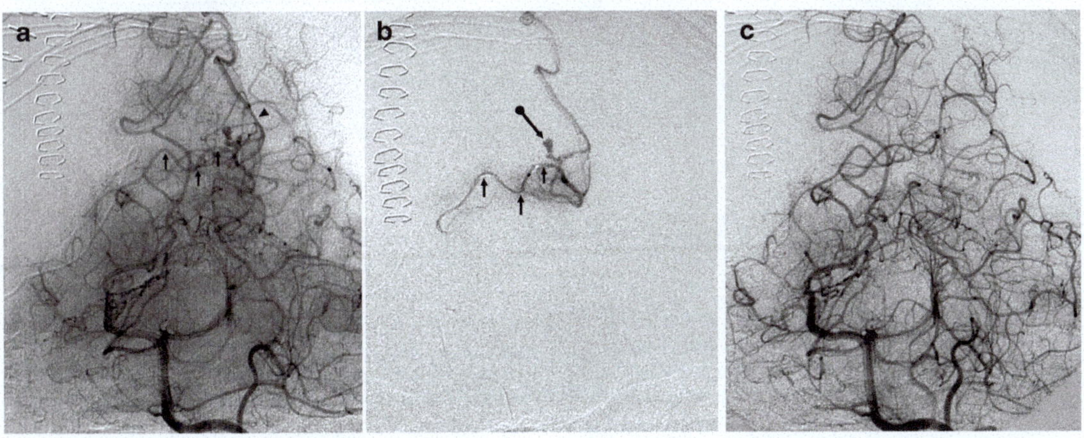

Fig. 12.18 Very small medial occipital AVM presenting with a large hemorrhage in a young man. The parenchymatous hematoma was acutely operated on. Two days later the vertebral angiography showed the AVM. (**a**) Oblique view. A small supplying branch (*arrow*) arising from the P4 segment of the PCA is recognizable. This seems to continue directly in the venous drainage (*arrowhead*). (**b**) On the selective study the fistulous pattern is better identified. Feeding artery (*arrows*), continuing in the single draining vein. A small aneurysmatic dilatation (*arrow with dot*) along the course of the feeding artery is also visible. The AVM was occluded with glue. (**c**) VA angiogram posttreatment

angiographically identified by their irregular shape and location at the margin of the recent hematoma diagnosed with CT or MR (Berenstein and Lasjaunias 1992; Garcia-Monaco et al. 1993; Valavanis 1996, 2004) (Fig. 12.10).

12.3.5.3 Other Changes of the Arteries

Among other changes of the arteries there are stenosis (Willinsky et al. 1998) due to thickening of the wall (see also Sect. 12.3.1) (Fig. 12.19). Moyamoya pattern has also been reported, probably being the result of hemodynamic stress (Mawad et al. 1984; Berenstein and Lasjaunias 1992).

12.3.5.4 Venous Drainage

The type of drainage depends on the location of the AVM and thus predictable. It can, however, be aberrant due to preexistent variants or the formation of a collateral venous drainage following occlusion or stenosis in the venous sector or it may be a venous adaptation in the attempt to reduce the high intranidal pressure. The venous drainage can be superficial, deep, or both, and can consist of one single draining vein or several venous channels. It can also occur that a unique large drainage vein can divide early giving the erroneous appearance of a multiple venous drainage. Examples are presented in Figs. 12.1, 12.2, 12.4, 12.5, 12.6, 12.7a–d, 12.10, 12.11, 12.18, and 12.20.

The veins draining the AVM are always dilated. The dilatation is sometimes enormous, forming large pouches (Fig. 12.6) which can be due to the presence of *a large fistula which occur more frequently in children* (Rodesch et al. 1995; Lasjaunais 1997). In these latter cases, according to some authors (Krings et al. 2005), the presence of a hereditary hemorrhagic telangiectasia should be suspected. Another cause of venous dilatation is a distal stenosis or thrombotic occlusion. The cause of stenosis may differ. It can be due to hyperplasia of the wall components as a reaction to the increased flow and pressure. Otherwise, the stenosis occurs when the vein enters the dura or may the result of a kinking or bone compression.

Some aspects of the venous drainage (unique vein, deep venous drainage, stenosis, large pouches) are considered potential risks of hemorrhage (Vinuela et al. 1985, 1987; Berenstein and

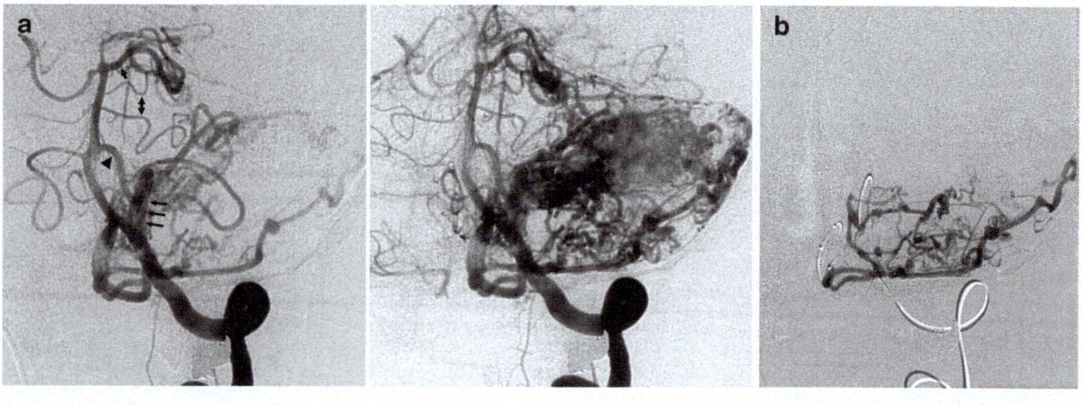

Fig. 12.19 AVM with a diffuse character, involving largely the left cerebellar hemisphere, presenting with SHA in a young woman. (**a**) Left AP vertebral angiogram (*early and later phase*), showing the AVM supplied by branches of the PICA (*three arrows*), AICA (*arrowhead*) and SCA (*arrow*). The dilated lateral pontine arteries (*bidirectional arrow*) are also involved. Wall irregularities and several aneurysms are recognizable along the course of the branches of the PICA supplying the AVM. These are better visible on the selective study (**b**) preceding the treatment with glue, followed by complete surgical eradication of the AVM

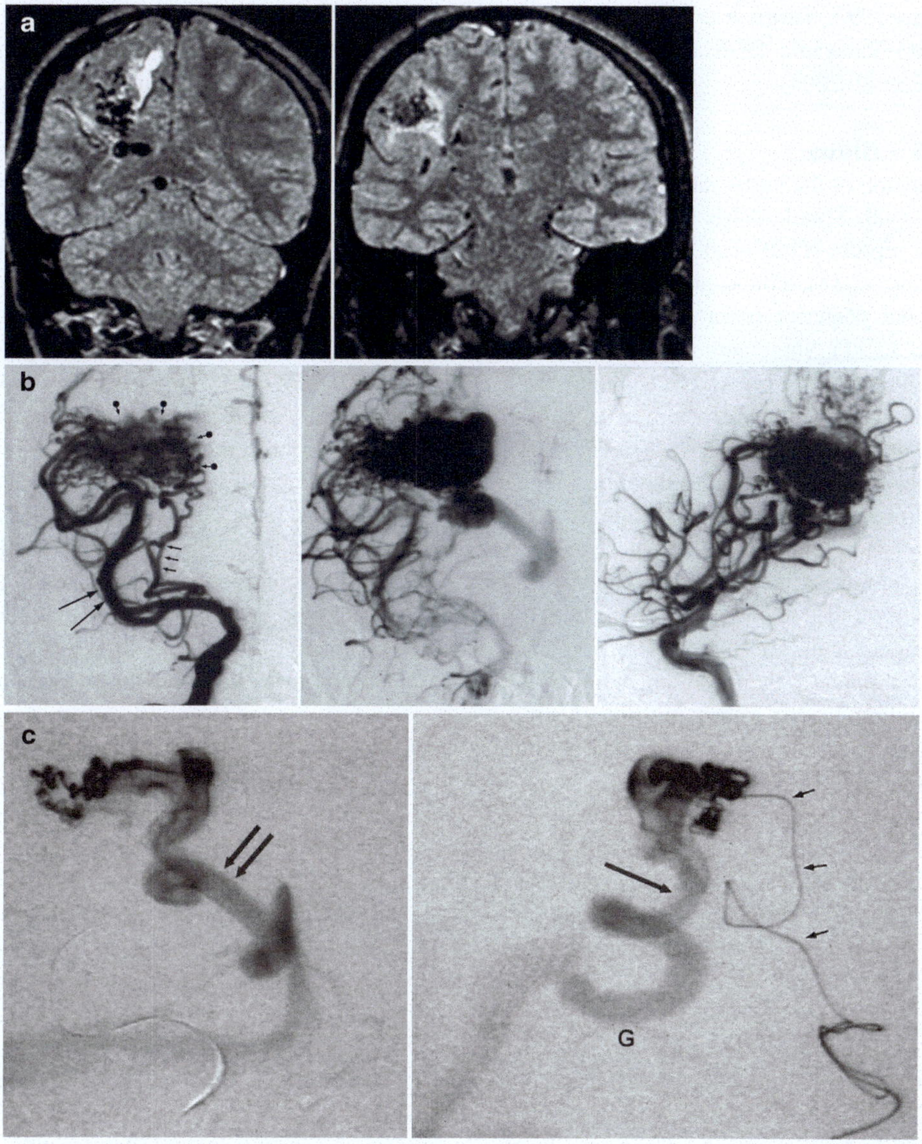

Fig. 12.20 Parietal AVM with a history of a previous bleeding in a young patient. (**a**) Coronal MRI, T2-weighted image showing the extension of the AVM toward the ventricle and an area of gliosis due to the previous hemorrhage. (**b**) Carotid angiogram, AP view, early and late phases, and lateral angiogram showing the nidus (*arrows with dot*) and the supplying branches of the MCA (*cortical, large arrows* and *perforators, small arrows*). There is a predominant large deep venous drainage (**c**) AP and Lateral views showing part of the selective study. Microcatheter (*small arrows*). Portion of the nidus draining into the dilated medial atrial vein (*arrow*) entering the Galen vein (G). Surgery was performed after partial embolization with glue

Lasjaunias 1992; Turjman et al. 1995; Mueller-Forell and Valavanis 1996; Pierot et al. 2004; Valavanis et al. 2004).

12.3.5.5 Nidus

The extension of the nidus varies from very large to very small. Small AVMs have a greater tendency to rupture (Graf et al. 1983; Pierot et al. 2004) (Fig. 12.18). The same is true for deeply located and posterior fossa AVMs (Figs. 12.5, 12.9, and 12.11). Some authors (Garcia-Monaco et al. 1990; Berenstein and Lasjaunias 1992) reported a higher tendency to hemorrhage also in temporo-insular and callosal AVMs (Fig. 12.17).

On the angiographic studies the nidus appears as formed by one or more compartments. AVMs with only one compartment are those in which the same unique vein of drainage appears after injection of all feeders involved. Those with multiple compartments are characterized by a specific and separate venous drainage linked to a corresponding feeder. When multiple compartment are present, one can be larger and predominant to the other (Yasargil 1987; Berenstein and Lasjaunias 1992; Valavanis et al. 2004). Small connections between the compartments are frequently present as recognizable in the endovascular treatment. Indeed, the very slow injection of embolic material can pass from one compartment to the adjacent one.

There are other aspects of the nidus which should be considered. One are the intranidal aneurysms which have already been described in Sect. 12.3.5.2. The other is the presence of very large arteriovenous shunts, leading to the formation of large fistulas. The fistula can be the unique feature of the AVM or be associated to the plexiform component (Fig. 12.6).

The nidus of the AVM is commonly characterized by well-defined borders, but in some cases its boundaries cannot be precisely recognizable. In these cases, many feeders can be present which are, however, not particularly dilated and can present with dysplastic aspects characterized by narrowings and small aneurysms. The venous drainage is slower than that of the typical AVMs (Fig. 12.19). In some rare cases this aspect is prominent with some clinical and structural features very different from the typical AVM. Some authors (Lasjaunias et al. 2008) have proposed to call this kind of malformation "*cerebral proliferative angiopathy*." Histopathological studies have demonstrated that it consists of a diffuse network of capillaries and venous ectasias in which areas of normal brain parenchyma are present. The malformation is usually very large, involving frequently more than one lobe sometimes the whole hemisphere. Children and young female are predominantly involved. The clinical presentation is commonly characterized by epileptic seizure, headache and stroke-like symptoms. Hemorrhages are rare. On MR, the typical aspect of the AVM with a nidus with dilated supplying arteries and draining veins is not recognizable. Instead, a diffuse network of vascular structures intermingled by areas of normal parenchyma and frequently a diffuse irregular enhancement on T1-weighted images indicating probably alterations of the blood-brain barrier is recognizable. Furthermore on perfusion studies (Lasjaunias et al. 2008) an increase of the blood volume and a longer mean transit time differently from AVM, can be demonstrated, probably due to the richness of capillaries and to the absence of the typical AV shunts. On the angiogram there is a diffuse vascular network supplied by arteries which are not particularly dilated. Some of the draining vein can be dilated, but their appearance is not so rapid as in AVMs. A large transdural supply is frequently present. Treatment (surgery, radiotherapy, and endovascular) considering the extent of the malformation, the absence of certain supplying arteries, and as demonstrated in histopathological studies the presence within the abnormal vascular network of normal brain tissue is commonly not indicated. In cases of hemorrhage, the endovascular treatment can be considered and directed to possible source of the bleeding represented by intranidal aneurysms. An example is presented in Fig. 12.21.

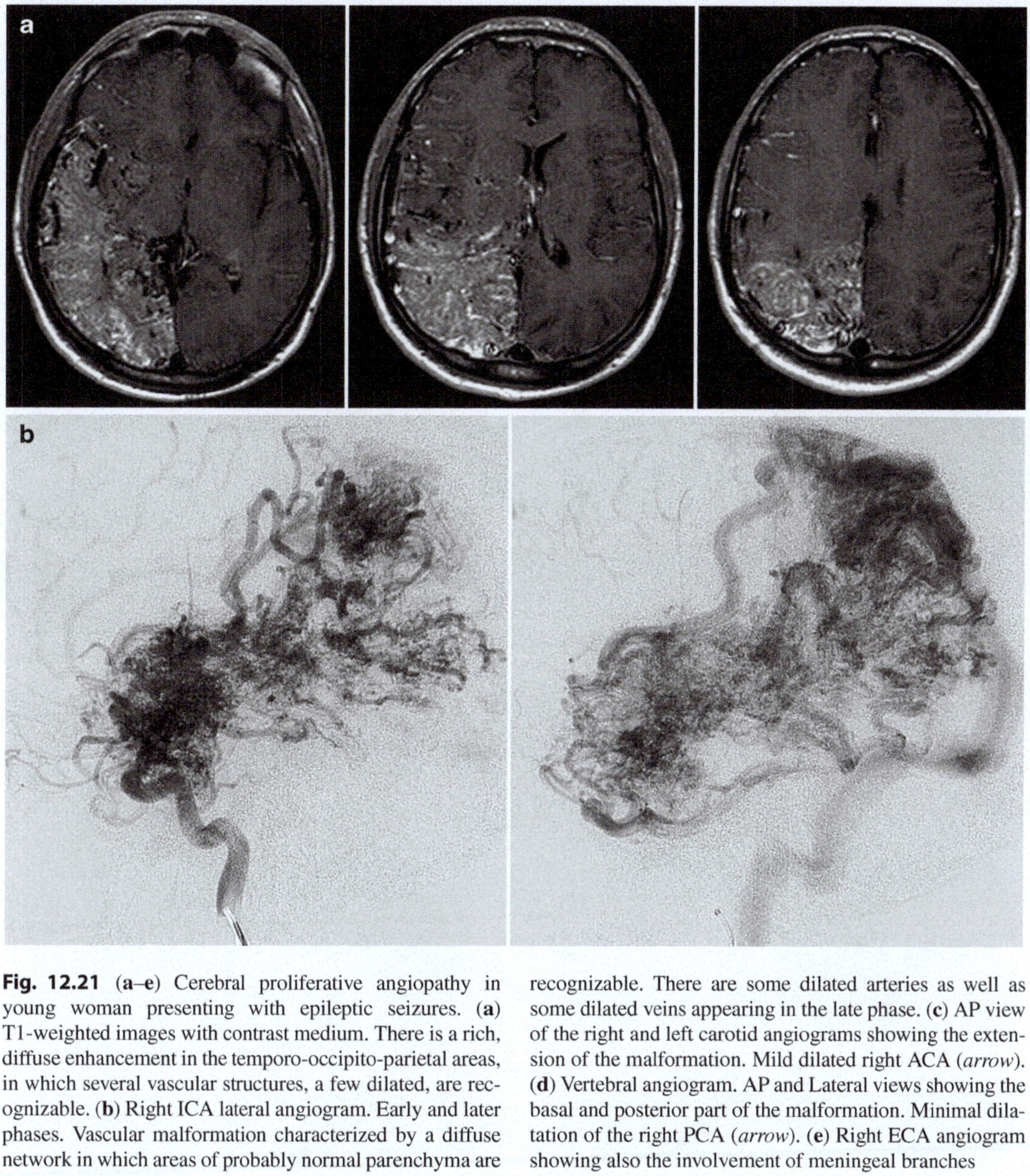

Fig. 12.21 (**a–e**) Cerebral proliferative angiopathy in young woman presenting with epileptic seizures. (**a**) T1-weighted images with contrast medium. There is a rich, diffuse enhancement in the temporo-occipito-parietal areas, in which several vascular structures, a few dilated, are recognizable. (**b**) Right ICA lateral angiogram. Early and later phases. Vascular malformation characterized by a diffuse network in which areas of probably normal parenchyma are recognizable. There are some dilated arteries as well as some dilated veins appearing in the late phase. (**c**) AP view of the right and left carotid angiograms showing the extension of the malformation. Mild dilated right ACA (*arrow*). (**d**) Vertebral angiogram. AP and Lateral views showing the basal and posterior part of the malformation. Minimal dilatation of the right PCA (*arrow*). (**e**) Right ECA angiogram showing also the involvement of meningeal branches

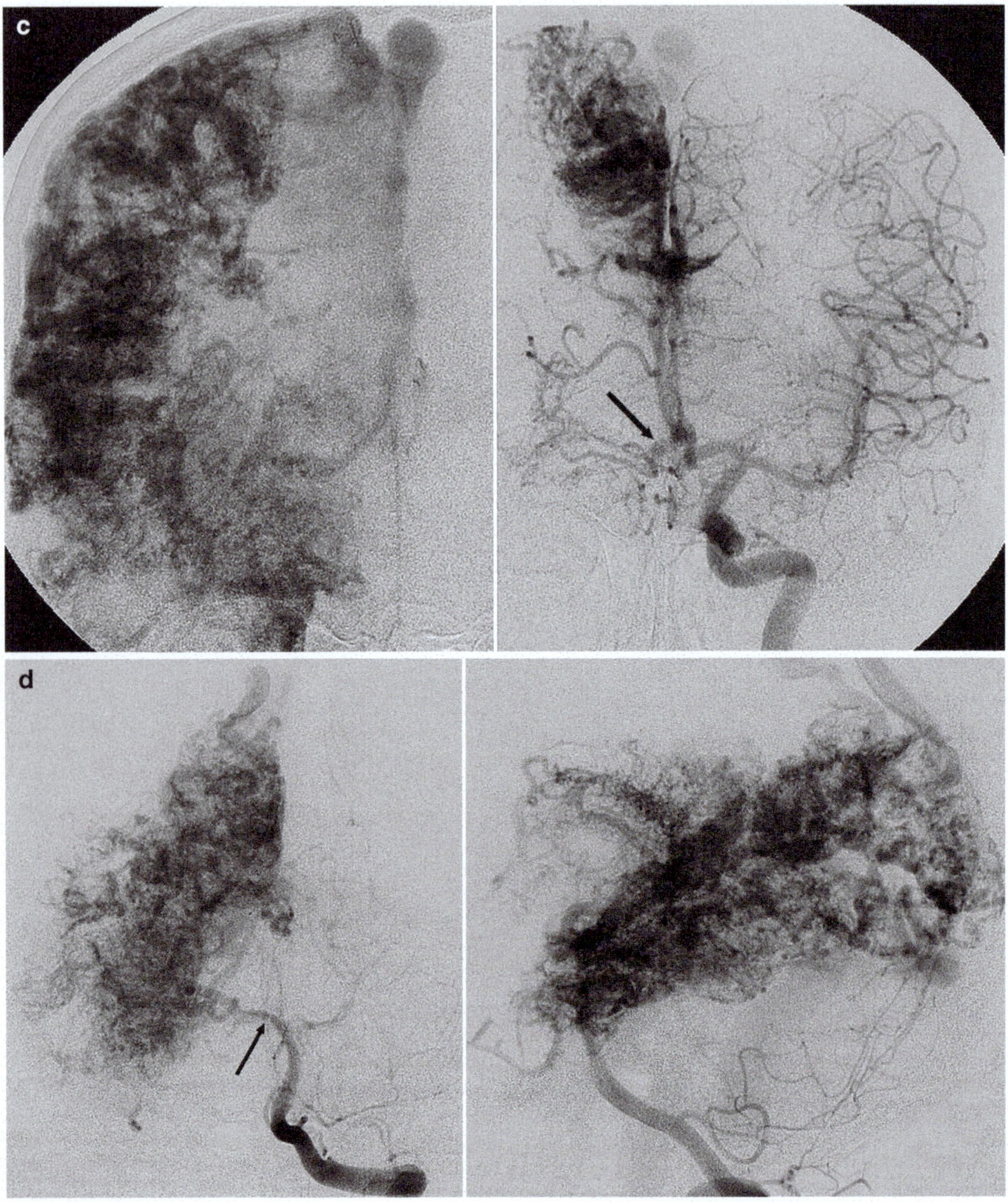

Fig. 12.21 (continued)

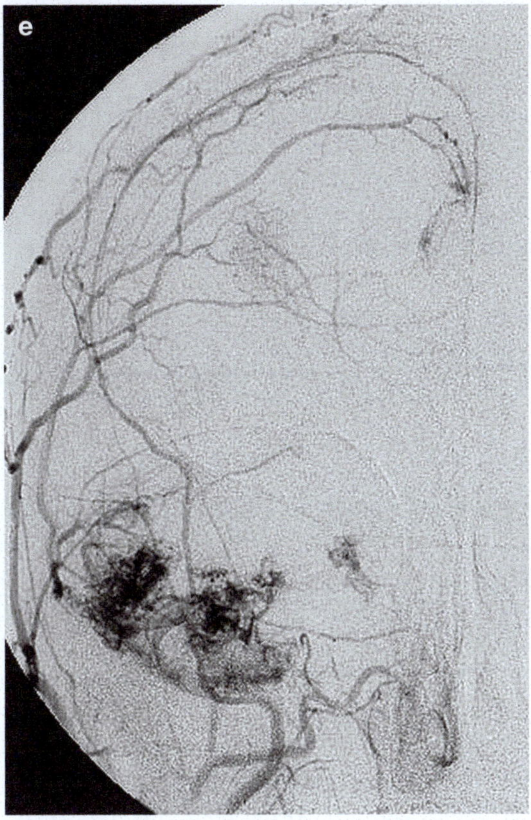

Fig. 12.21 (continued)

12.3.5.6 Perinidal Changes

In a number of cases, one can observe around the nidus the presence of a rich vascular network, consisting of tiny vessels; this is usually due to dilatation of collateral following the demand of blood flow from the AVM. Some authors have described the development of new vessels as expression of angiogenesis, in response to chronic ischemia of the perinidal parenchyma (Berenstein and Lasjaunias 1992; Valavanis et al. 2004).

12.3.6 Treatment

In spite of the introduction of cerebral angiography which made the diagnosis of AVMs easier, this pathology was considered for many years beyond of any useful surgical therapy. This attitude changed in the 1950s following the experiences published by some authors (Olivecrona and Riives 1948; Olivecrona 1950; Toennis and Lange-Cosack 1953) who showed that a complete surgical extirpation was possible with an acceptable risk of morbidity and mortality. Since then, better definition of the site, size, morphology, and hemodynamic aspects of the AVM, combined with an improved knowledge of its pathophysiology and the technical improvement in surgical and endovascular treatments as well as in radiosurgery, applied in various combinations has progressively improved the treatment allowing to achieve in many patients a complete cure with relatively low rates of morbidity and mortality as reported by many authors (Spetzler and Martin 1986; Berenstein and Lasjaunias 1992; Colombo et al. 1994; Valavanis 1996; Debrun et al. 1997; Valavanis and Yasargil 1998; Valavanis et al. 2004; Beltramello et al. 2005; Picard et al. 2005; Vinuela et al. 2005; Raymond et al. 2005; Nagaraja et al. 2006; Mounayer et al. 2007; Panagiotopoulos et al. 2009).

Considering especially the endovascular treatment, the rate of morbidity (0-8%) and mortality (0-4%) reported in the literature varies greatly. (Katsaridis et al. 2008; Berenstein et al. 2010; Saatci et al. 2011; Pierot et al. 2009, 2013; Van Rooji et al. 2012a, b; Soltanolkotabi et al. 2013; Baharvahdat et al. 2014; Crowley et al. 2015; de Castro-Alfonso et al. 2016). These different results can be explained by the different complexity and localization of the treated AVM, the different strategy of treatment aimed to try to occlude completely the AVM or only partial of it, and the experience of the operators. The most feared complication is the hemorrhage which can be due to mechanical lesion of a feeding artery or to premature occlusion of a principal draining vein. Also an extensive but not complete occlusion of the AVM can lead to flow slowing and thrombosis of the draining veins responsible of impairment of the drainage of the remaining part of the AVM responsible for hemorrhage.

As far as it concerns the endovascular treatment, considering that *in the nidus no active brain parenchyma* is present, the injection of the embolic material, strictly confined into the nidus, avoids the damage of normal parenchyma and so reduces the risk of complications. This approach should be considered especially in the treatment of AVM in so-called *eloquent areas*. In this context, however, it should be taken into account that all regions of the brain are eloquent, even if the clinical symptoms due to damage of the brain parenchyma are in some areas more evident than in other.

Among the factors playing a role in the positive results are the technical improvements characterized by the introduction of more suitable microcatheters and guide wire and the use of better liquid agents such the acrylic glue (Glubran) and more recently the onyx. This latter permits a slower injection with a larger penetration of the embolic agent in the nidus. In cases of small- or medium-sized AVMs, the endovascular treatment could be performed trying to occlude completely the malformation. In cases of large or deeply located AVMs, the embolization can be directed to eliminate those components of the AVMs which are considered a risk of hemorrhage such aneurysms (flow related, intranidal, or pseudoaneurysm), or to reduce partially the volume of the malformation to facilitate surgery or radiosurgery.

It is still open to question whether an invasive therapy or a noninvasive management should be chosen in cases of asymptomatic AVM or those with minimal symptoms. The age of the patient, location and extension of the AVM, and anticipated difficulty of treatment will play a role in the decision. Also, aspects of the angioarchitecture thought to increase the risk of hemorrhage should be considered. In this context studies have been made trying to identify gene factors which could favour hemorrhage in AVMs. Some authors (Pawlikowska et al. 2004; Chen et al. 2008) have found polymorphism in the inflammatory cytokine IL6 gene (IL6-174 GG) in AVMs presenting with hemorrhage. Furthermore association of genotypes of tumor necrosis factor-alpha. 238 G > A and apolipoprotein E2 polymorphisms with hemorrhage in AVMs has been demonstrated (Achrol et al. 2006, 2007).

More information is certainly needed about the evolution of this very complicated pathology. Such data could emerge from a multicenter, randomized trial still ongoing (Fiehler and Stapf 2008; Mohr et al. 2014), assessing possible invasive treatment and noninvasive management of patients with AVM. The first results of this study shows a superiority of the medical management concerning death and stroke in comparison to the interventional therapy (Mohr et al. 2014). The study is continuing and enlarged to additional 5 years to see if these preliminary results will be confirmed.

12.4 Cavernous Malformations (Cavernomas)

12.4.1 Pathology

These appear as well-circumscribed masses formed by dilated vascular channels without intervening brain parenchyma. The wall of the channels is lined by a single layer of vascular endothelium, surrounded by fibrous tissue. Some of these channels show thrombosis. Evidence of hemosiderin due to previous hemorrhage is present within and around the malformation. There may be calcification. Cavernous malformations can grow following hemorrhage or because of their intrinsic activity.

12.4.2 Incidence

Based on MRI and anatomo-pathologic studies, the incidence is reported to be 0.4–0.9% of the general population (McCormick 1984; Otten et al. 1989; Robinson et al. 1991; Maraire and Awad 1995). Cavernous malformations can be single or, frequently multiple. Familial cases have been increasingly recognized, especially in patients of Hispanic-American origin (Rigamonti and Brown 1994; Zambranski et al. 1994; Gunnel et al. 1996). In this group of patients an autosomal dominant pattern of inheritance linked to a gene mutation known CCM1 in the chromosome 7 has been identified (Gunnel et al. 1996). Other more rare mutations are the CCM2 in chromosome 7 and CCM3 in chromosome 3. In other families these mutations have not been found, suggesting that probably other not known mutations occur. Cavernomas have been considered to be congenital. However, "de novo" lesions appearing during life can develop in families in whom the cavernomas are known, but also as sporadic cases (Pozzati et al. 1996; Tekkök and Ventureyra 1996; Porter et al. 1997; Brunereau et al. 2000; Massa-Micon et al. 2000). Furthermore, the possibility that, at least in some cases, venous hemodynamic changes induce the development of cavernoma have been considered. This could explain the frequent association of cavernoma with DVA (Dillon 1995; Hong et al. 2010b). Cases of "de novo" cavernoma have been described in pathological conditions leading to changes of the venous circulation. Some authors (Desal et al. 2005) reported a case of cavernoma probably induced by many triggers factors acting in different phases including the presence of a DVA, a "de novo" formation of a DAV fistula and a thrombophlebitis. Other authors (Janz et al. 1998; Ha et al. 2013) have described the appearance of cavernomas in patients with DAVF especially in those cases associated with venous reflux. In conclusion, it is thinkable that the formation of cavernomas is linked to a genetically predisposition, to which different factors can act favoring their development.

12.4.3 Location

Cavernomas can be found throughout the brain and spinal cord. They are more frequent in the subcortical white matter and pons. Extraparenchymal lesions can occur (Meyer et al. 1990; Sepehrnia et al. 1990) and are particularly frequent in the cavernous sinus, especially in women.

12.4.4 Diagnosis and Clinical Relevance

In CT, cavernomas appear as rounded, hyperdense masses, sometimes with calcifications. In MR, they are hypointense on T1- and hyperintense on T2-weighted images. Not rarely the signal is inhomogeneous. In large cavernomas, the pattern is frequently characterized by a mass made of several rounded cavities. Very useful for the diagnosis are the T2*-weighted gradient echo and the susceptibility-weighted images (SWI) (Haacke et al. 2009 Bradac et al. 2014). With this technique the cavernoma appears as a lesion characterized by a signal loss due to the deoxyhemoglobin in the venous channels and hemosiderin linked to previous hemorrhages. Enhancement in CT and MR is typical (Rigamonti et al. 1987). Frequently the presence of an associated DVA can be demonstrated. Examples are presented in Figs. 12.22 and 12.24. Angiograms are commonly negative. In cases of cavernoma within the cavernous sinus, the differential diagnosis with cavernous sinus meningioma can be very difficult (Bradac et al. 1987).

Cavernomas are frequently asymptomatic and discovered incidentally on CT or MR. They can present clinically with seizures, hemorrhage, or impairment of brain parenchyma due to its compression. With regard to the association with DVA, see also Sect. 12.6. The alternative management between microsurgical resection or conservative therapy remains controversial: the clinical presentation, location, aspect of the lesion, and experience of the team involved are the essential factors influencing the decision (Mouchtouris et al. 2015).

12.5 Capillary Malformations (Telangiectasias)

The telangiectasias are similar to the cavernous angiomas. They are formed by a collection of dilated capillary, between those, however, unlike cavernomas there is brain parenchyma. The incidence on autopsy is reported to be 0.1–0.15% (McCormick 1984; Jellinger 1986). They are frequently associated with cavernomas, and some authors have suggested that these lesions represent the phenotypic spectrum within a single pathological entity (Rigamonti et al. 1991; Chaloupka and Huddle 1998).

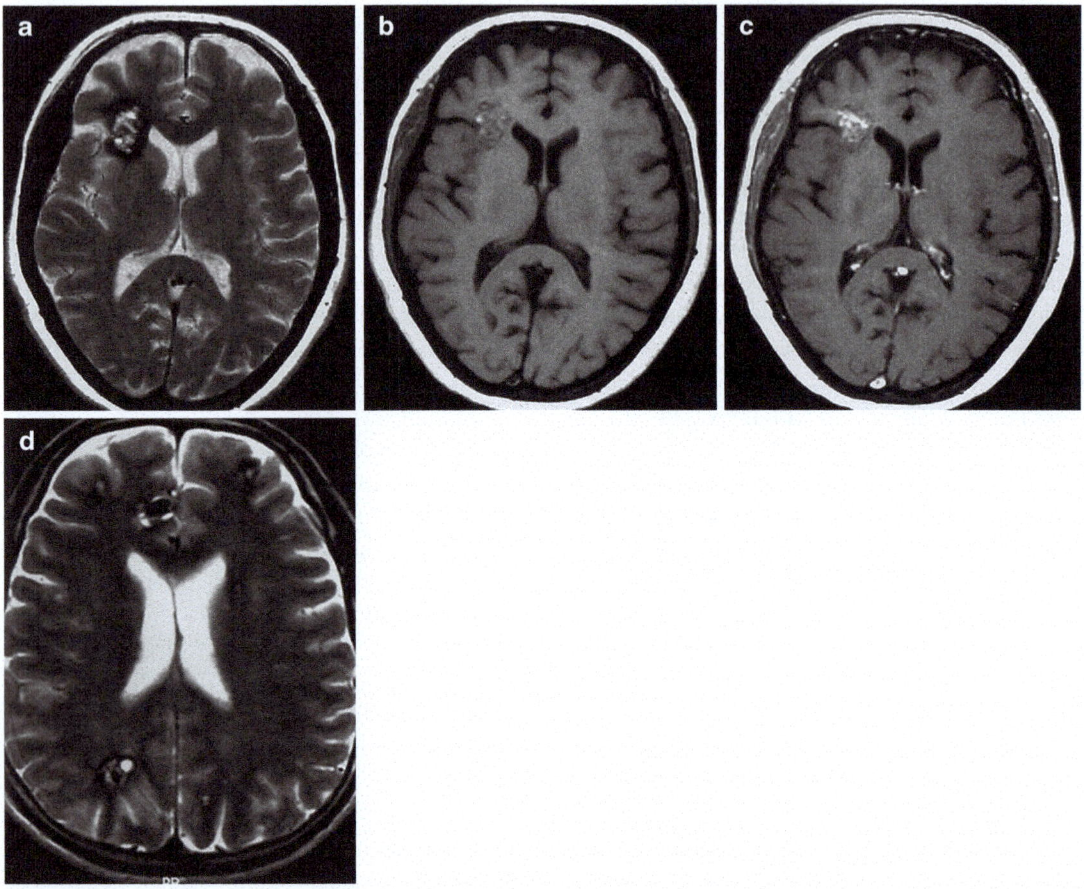

Fig. 12.22 Incidental discovery of cavernoma in two different patients. (**a**)–(**c**) MRI study of the lesion located in the frontal region. T2-weighted image (**a**) showing the irregular foci of hyperintensity surrounded by hypointense rim of hemosiderin. T1-weighted images without (**b**) and with (**c**) contrast medium showing the rich enhancement of the cavernoma. (**d**) Another patient in whom multiple cavernomas have been diagnosed. T2-weighted image showing the inhomogeneous pattern of the lesions

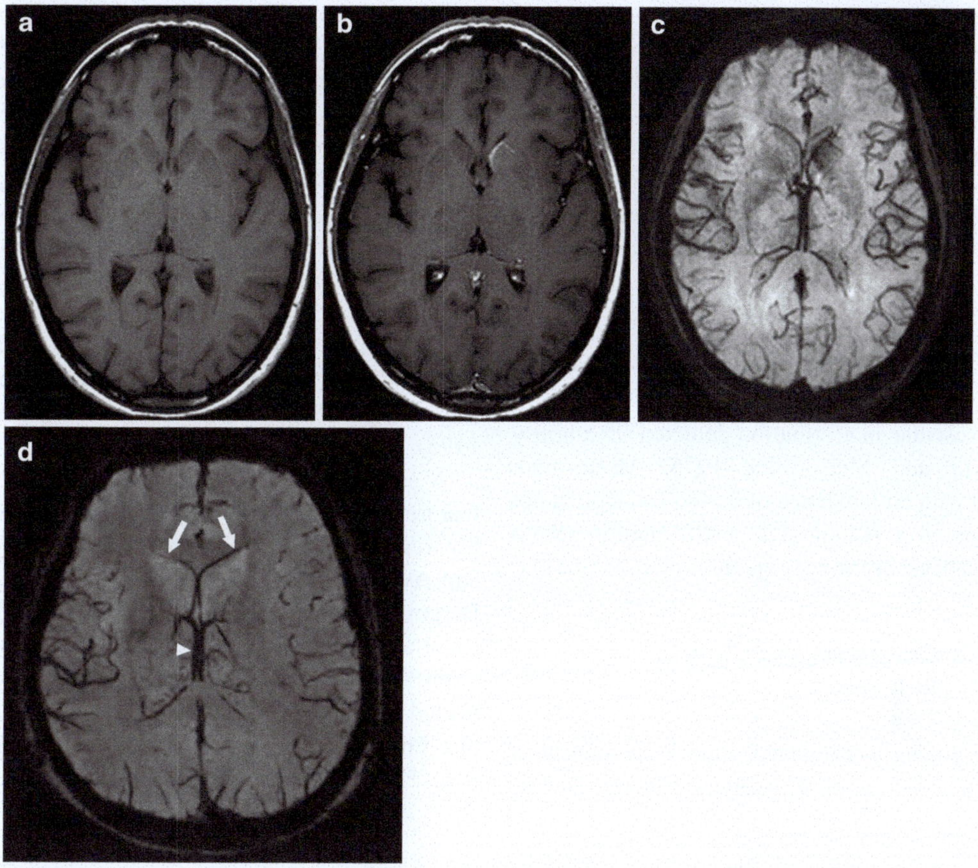

Fig. 12.23 Suspected telangiectasia studied with T1-weighted images without (**a**) and with (**b**) contrast medium. There is a small enhancement in the region of the caudate nucleus. On the SWI (**c**), vascular structures (*small arrow*) are recognizable connected with the septal vein. (**d**) Study of the normal anatomy. Same section as in (**c**) showing the course of both septal veins (*arrows*) running from the lateral to the medial corner of the frontal horn. The veins turn back along the septum pellucidum joining the ICV (*arrowhead*)

Telangiectasias can be found everywhere in the brain parenchyma and spinal cord, with a predominance in the pons and basal ganglia. The neuroradiological diagnosis is similar to that for cavernous angiomas. Also in these cases the use of SWI can be very useful in detecting small lesions not recognizable on T1- and T2-weighted images (El-Koussy et al. 2012) (Fig. 12.23).

12.6 Developmental Venous Anomaly (DVA)

12.6.1 Pathology

Also called venous angioma, DVA is prevalently located in the white matter of the cerebral hemisphere, whereby several medullary veins converge to a unique collector draining further superficially in one of the sinuses or in one of the subependymal vein continuing in the ICV or BV. Another typical location is the white matter of the cerebellar hemisphere, where medullary veins converge commonly to the vein of Galen through the precentral vein or into the petrosal vein. DVA is considered to be the result of a focal abnormal development of the medullary veins (Saito and Kobayashi 1981).

12.6.2 Incidence

DVA has been reported as being the most common vascular malformation detected on autopsy (Sarwar and McCormick 1978; McCormick 1984) with an incidence of 2.6%. Today, it is not considered a pathological malformation (Saito and Kobayashi 1981; Lasjaunias et al. 1986a).

12.6.3 Diagnosis and Clinical Relevance

DVAs appear as enhanced venous channels after contrast medium on T1-weighted images and as hypointense channels on T2* gradient echo and SWI images (Fig. 12.24). On the angiogram, DVAs are typically recognizable in the capillary-venous phases, characterized by several medullary veins converging on a large collector (Figs. 20.5, 12.25 and 9.25).

Most DVAs are asymptomatic. When hemorrhages occur, these are considered commonly due to the associated cavernous angiomas (Ostertun and Solymosi 1993; Forsting and Wanke 2006). Rarely thrombosis of the main collector can lead to hemorrhagic ischemia (Ostertun and Solymosi 1993; Field and Russell 1995).

A few DVAs do not completely fit the typical features described above. These lesions probably represent a transition-form between DVA and AVM (Awad 1993; Mullan et al. 1996a, b; Bergui and Bradac 1997; Komiyama et al. 1999; Im et al. 2008; Oran et al. 2009). On the angiogram, several not dilated arterial feeders are connected with veins that have the feature of DVA, but unlike these appear on an early phase (Fig. 12.25f).

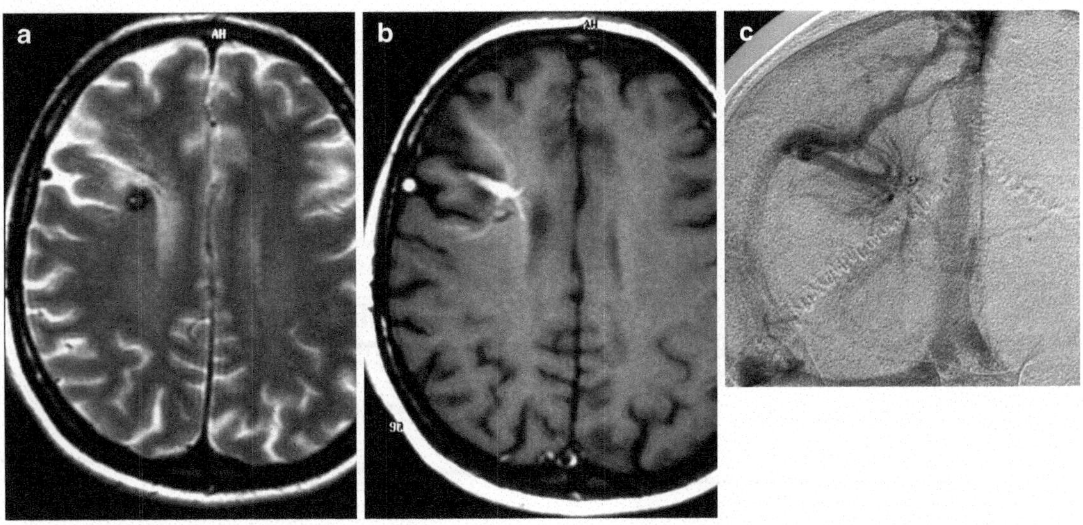

Fig. 12.24 Incidental discovery of a cavernoma associated to a DVA. (**a**) T2* gradient echo MR image showing a small cavernoma surrounded by a nonspecific area of hyperintensity corresponding to a gliosis or edema. (**b**) T1-weighted MR images after contrast medium. Partial enhancement of the cavernoma and structures having the feature of vessels. (**c**) Venous phase of the carotid angiogram confirming the presence of a large DVA

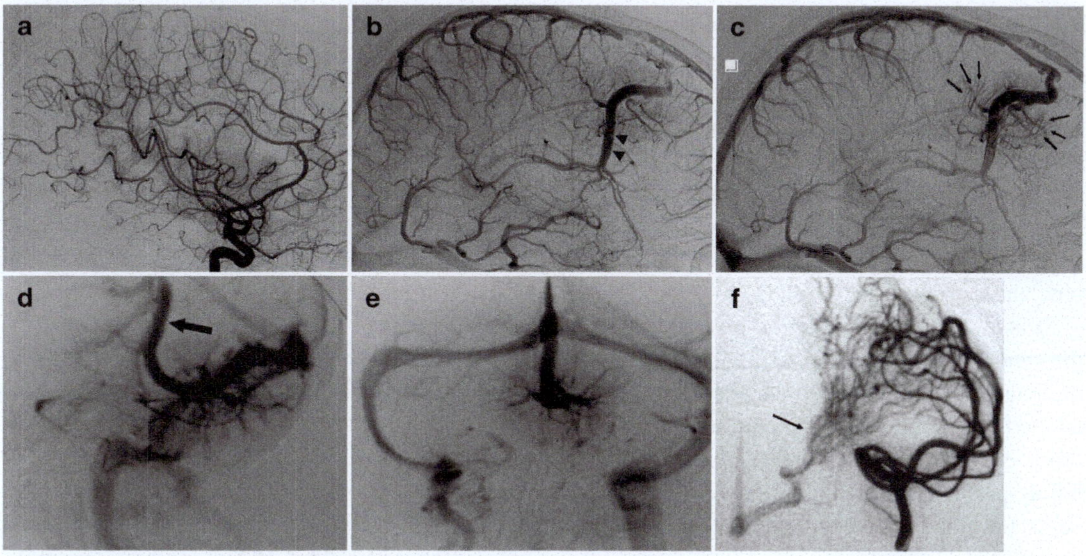

Fig. 12.25 (**a**–**e**) Two example of DVAs. Carotid angiogram, (**a**) normal arterial phase; (**b**), (**c**) in the late venous phase a large frontal cortical vein, draining also partially the temporal region is recognizable (*arrowheads*). To this converge all the medullary veins of the area (*arrows*). The septal vein is not recognizable, probably absent. Further drainage occurs in the superior sagittal sinus. (**d**), (**e**) Venous phase of a vertebral angiogram in another patient. Lateral (**d**) and (**e**) AP view. There is an enlarged precentral vein (*arrow*) to which converge the majority of the medullary veins of both cerebellar hemispheres. (**f**) Example of a mixed angioma. Carotid angiogram. AP view. Selective study of the middle cerebral artery. There is a pathological network (*arrow*), consisting of a medullary veins injected early through connections with medullary branches of the middle cerebral artery. All the veins converge on a dilated atrial vein continuing into the Galen vein

12.7 Central Nervous System Vascular Malformation (Part of Well-Defined Congenital or Hereditary Syndrome)

12.7.1 Rendu-Osler Syndrome (Hereditary Hemorrhagic Telangiectasias)

Rendu-Osler is an autosomal dominant neurocutaneous disease caused by mutations of two genes, the endoglin gene (ENG) and the active like kinase gene (ALK). It is characterized by telangiectasias of the skin and mucosa of the oral-nasal cavities and gastrointestinal tract which can be responsible of severe hemorrhages (Fig. 3.21). Arteriovenous angiomas and fistulae are also frequently present in the lung and liver. In addition, different types of vascular malformations can involve the central nervous system. The most frequent are AVMs, which are often relatively small and multiple (Putman et al. 1996; Chaloupka and Huddle 1998; Berenstein and Lasjaunias 1992; Garcia-Monaco et al. 1995). According to some authors (Krings et al. 2005), the presence of AVMs characterized mainly by a large fistula occurring in children should arise the suspicion of an underlying hereditary hemorrhagic telangiectasia. See also Sect. 12.3.5.4.

12.7.2 Sturge-Weber Syndrome (Encephalotrigeminal Angiomatosis)

Sturge-Weber syndrome is a familial neurocutaneous disease, characterized by a facial vascular nevus, in the trigeminal distribution, mainly in the first branch, a retinal angioma, and leptomeningeal angiomatosis.

12.7.3 Pathology and Diagnosis

The pathology consists of a rich network of thin-walled capillaries and venules, lying between the pial and subarachnoid membrane. There is also typically a paucity of cortical veins which is probably responsible for the stasis and progressive hypoxia of the cortex, which becomes atrophic and partially calcified.

It is assumed today that the primary cause of the syndrome is a maldevelopment of the venous drainage, involving the cortical veins and the distal part of the superficial medullary veins. The drainage is redirected through a collateral circulation, particularly involving the deep medullary veins toward the subependymal veins and choroid plexus.

The dystrophic, calcified cortex can be well studied with CT and MRI. The contrast medium shows the enhancement of the angiomatous plexus and the redirected venous drainage (Smirniotopoulos and Murphy 1992; Chaloupka and Huddle 1998). On the angiogram the absence or paucity of the cortical veins as well as the redirected venous drainage toward the deep venous system can be demonstrated (Berenstein and Lasjaunias 1992).

12.7.4 Wyburn-Mason Syndrome

Wyburn-Mason syndrome is an exceptionally rare neurocutaneous disease, characterized by a cutaneous facial nevi in the distribution of the trigeminal nerve and an extensive AVM typically involving the visual pathways, including the retina, the optic nerve, optic tract and sometimes the diencephalon an the occipital lobe (Patel and Gupta 1990; Chaloupka and Huddle 1998). Some authors have proposed for this syndrome the term unilateral retinocephalic vascular malformation (Theron et al. 1974). Unusual variants, include bilateral intracranial vascular anomalies (Patel and Gupta 1990). Other variants of this disease have been described characterized by involvement of hypothalamus, corpus callosum, hypophysis or in other cases of pons, cerebellum associated with facial bone malformations. Considering the multiple aspects of the disease and the probable metameric development, the term "*cerebrofacial AV metameric syndrome*" (*CAMS*) has been proposed by some authors (Krings et al. 2007c). CT and MRI are useful diagnostic tools, but angiography is the essential diagnostic step to decide if there is a possibility of partial endovascular treatment.

12.7.5 Klippel-Trenaunay-Weber Syndrome

This is a very rare syndrome characterized by unilateral soft tissue or osseous limb hypertrophy, cutaneous telangiectasia, and venous malformations and/or occasional high-flow AVM of the affected limb. Commonly the lower limb is affected but involvement of the upper limb can also occur (Chaloupka and Huddle 1998). The typical syndrome can be associated by many other pathologies. Vascular malformation and hemangiomas can be present in other organs including head, neck, abdominal, pelvic organ, and thorax (Samuel and Spitz 1995). Intracranial hemorrhage due to aneurysms and AVMs have been described (Makiyama et al. 1994). Cases of cerebral infarct due to carotid dissection and spinal cord ischemia due to occlusion of the anterior spinal artery (Brunaud et al. 1994) as well as the association with Kasabach-Merritt syndrome (Samuel and Spitz 1995) and with Sturge-Weber syndrome have been reported (Kanterman et al. 1996). CT and MRI are useful in detecting all these different lesions; angiography is essential in the evaluation of the vascular pathology especially when endovascular treatment has been taken into consideration.

12.8 Arteriovenous Shunts Involving the Vein of Galen

The real incidence of this pathology is unknown. It can be estimated to be less than 1 per 1,000,000. The first description of "aneurysms of the vein of

Galen" was made in 1937 by Jaeger et al. Today we know that these malformations constitute a group of lesions that have been more precisely classified only in the recent years.

- **Vein of Galen aneurysmal malformations**. Nowadays, there is a common agreement that the pathogenesis of this malformation, which can be termed "true vein of Galen malformation," is a malfunction in the embryogenesis, involving the median prosencephalic vein (Raybaud et al. 1989; Mickle and Quisling 1994; Burrows et al. 1996; Brunelle 1997; Lasjaunais 1997; Chaloupka and Huddle 1998; Rao and Mathuryia 2011). As it has already been described in Chap. 9, *the prosencephalic vein (PV)* is an embryological vessel which receives the venous drainage of the deep cerebral structures and choroid plexus, continuing into the falcine sinus. The vein disappears in a period between the sixth and eleventh weeks, and it is replaced by the vein of Galen, arising from the unification of the caudal remnant of the Prosencephalic vein with the in the meantime developed internal cerebral veins. In the malformation of the Galen vein, the PV does not regresses and appears as a dilated vein draining in the straight sinus. Frequently, however the drainage occurs in the persistent falcine sinus, while the straight sinus is hypo or aplastic (Raybaud et al. 1989; Mickle and Quisling 1994; Burrows et al. 1996). The pathogenesis of the arteriovenous shunt remains unknown. Raybaud et al. (1989) suggested that it may be linked to an embryogenetic error involving the choroidal arteries, which, in the same embryonic period when the PV is prominent, are the most active arterial vessels; thus they are the most vulnerable to maldevelopment.

12.8.1 Diagnosis and Treatment

CT and MR allow easy identification of this kind of malformation. Selective and superselective angiograms are essential for a precise study. Angiography in patients with this pathology present a series of technical problems, since in the majority of the cases they are neonatal or infants aged 1 to 2 years. The femoral approach is more difficult and there is the necessity to limit the quantity of contrast medium; thus, it is prudent and rational to postpone the angiography until the patient is at least 5–6 months old. The angiographic study and the endovascular treatment should be performed earlier if rapid clinical deterioration, particularly as a result of heart failure, occurs, which can rapidly improve after embolization with complete occlusion or only partial reduction of the shunt (Lylyk et al. 1993; Lasjaunais 1997; Bhattacharaya and Thammaroj 2003; McSweeney et al. 2010; Khullar et al. 2010; Boccardi et al. 2013). In this context, it should, however, be taken into consideration, as reported by some authors (Brevis-Nunez et al. 2013), that the normalization of the heart activity in the days postembolization can be followed by a severe myocardial dysfunction similar to that occurring in patients with SAH (see Sect. 11.5). The cause of this condition is not completely clear; temporary increase of the hydrocephalus observed in some of these cases after the embolization has been suggested (Brevis-Nunez et al. 2013).

On the angiogram, two type of shunts can be recognized (Lasjaunais 1997): *the choroidal type*, characterized by the presence of many feeders shunting with the PV, commonly on its anterior surface, and the *mural type*, in which one or two feeders are connected with the inferolateral part of the PV. The feeders can be uni- or bilateral. The most common vessels involved are the posterior choroidal and the distal segment of the pericallosal arteries. In the embryo, the distal segment of the pericallosal artery runs around the splenium and extends anteriorly into the *tela choroidea*, anastomosing with the posterior medial choroidal artery. This connection normally disappears, but it can persist forming in the vein of Galen malformation the so-called limbic arch (Raybaud et al. 1989). The collicular and posterior thalamic branches, the anterior choroidal as well as the superior cerebellar arteries can also be involved; sometimes, secondary feeders can arise from peripheral branches of the middle cerebral artery and its perforators and from meningeal branches.

The venous drainage occurs into a dilated PV, which appears as a rounded or elongated channel, with the greatest dimension along the sagittal axis. In at least 50% of the cases the further drainage occurs into the Falcine sinus (Raybaud et al. 1989) which replaces the hypo-aplastic straight sinus. In other cases, this latter can be the main drainage appearing frequently dilated, and sometimes duplicated. Other anomalies of the dural sinuses are frequently present. Among them there is a large occipital sinus and the presence of two segments of falcine sinus, one directed toward the SSS and the other back to the distal straight sinus forming a kind of loop (Raybaud et al. 1989). Aplasia and thrombosis of the transverse and sigmoid sinuses uni- or bilaterally as well as aplasia of the internal jugular vein can occur leading in these cases to a rerouting of the venous drainage anteriorly toward the cavernous sinus and ophthalmic vein and the pontomesencephalic veins.

Endovascular treatment is today the therapy of choice and has certainly improved the prognosis of these patients. The transarterial route alone or combined with the venous approach allows today a complete or partial occlusion of the malformation. The venous approach occurs through catheterization of the jugular vein via the femoral vein and a further catheterization of the transverse sinus and straight or falcine sinus followed by positioning of coils in the dilated prosencephalic vein . In cases of occlusion of the TS sinus the route through the occipital sinus can be used. In some cases a direct approach of the straight or falcine sinus after craniectomy is necessary. Depending on the morphological aspect of the shunt, and on the difficulty of the treatment, this can be performed in one or more sessions. Anatomical and clinical cure can be obtained in about 50–70% of the cases (Lylyk et al. 1993; Lasjaunais 1997; McSweeney et al. 2010; Khullar et al. 2010; Boccardi et al. 2013). In other cases, in spite of the technical good result, a progression of the disease occurs. In this context, in some cases, occlusion of the dural sinuses with progressively thrombosis of the cortical and medullary veins occurs in spite of the occlusion of the shunt. In general the worst results occur in neonatal patients in comparison with infants aged 1–2 years or with older children who have a clearly better outcome (Lylyk et al. 1993; Lajaunais 1997; Khullar et al. 2010; McSweeney et al. 2010; Boccardi et al. 2013). Examples are presented in Figs. 12.26 and 12.27.

- **Vein of Galen dilatation.** In this group of lesions, which occur in older patients, the ectatic vein is the normal developed great vein of Galen, into which drains an AVM. The dilatation can be associated with an obstruction of the venous drainage, involving the straight sinus or torcular herophili leading to a rerouting of the drainage in the superficial and deep veins. The filling of these latter does not occur in the true vein of Galen malformation, since they are not developed replaced by the prosencephalic vein (Lasjaunais 1997).

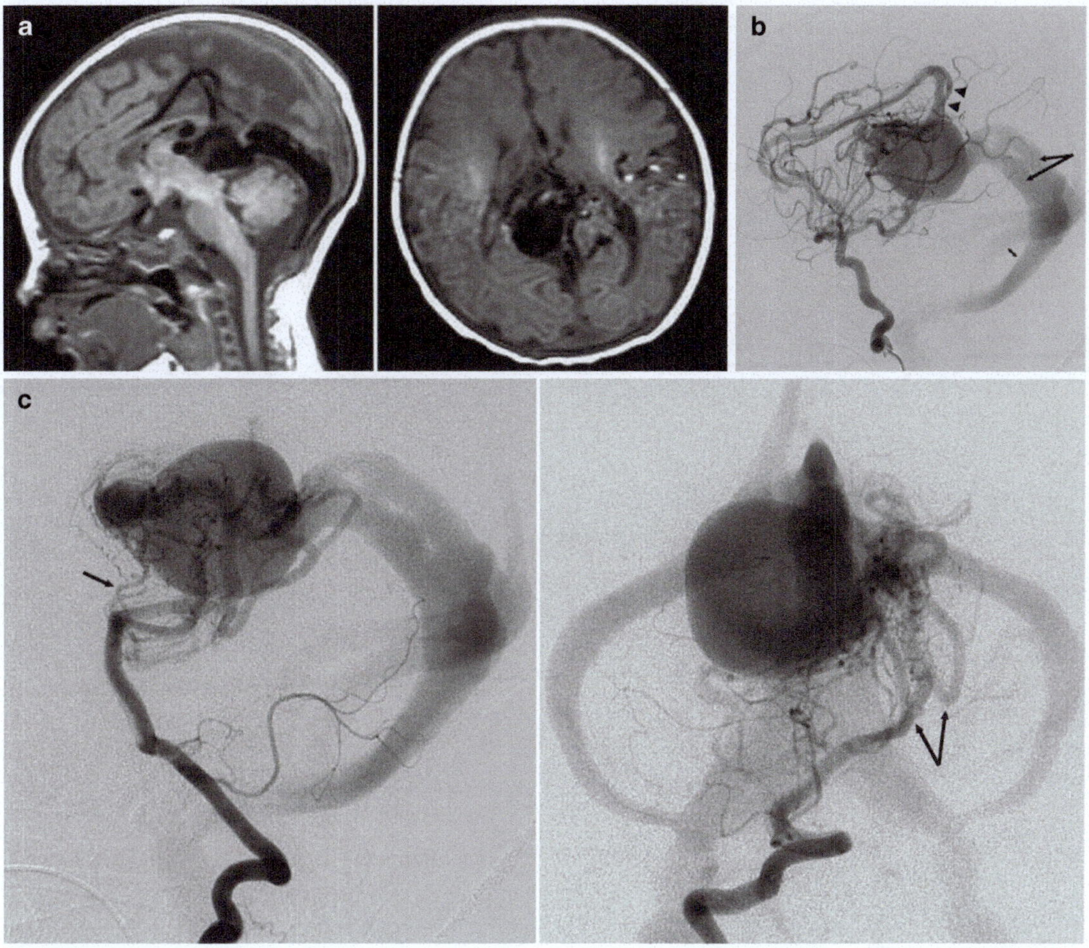

Fig. 12.26 Vein of Galen malformation in a 3-month-old child. Treatment owing to heart failure. (**a**) MRI T1-weighted images showing the malformation. (**b**) Right ICA angiogram, lateral view, showing the dilated prosencephalic vein draining into the dilated and fenestrated (*arrow with angle*) straight sinus. The malformation is supplied by the distal pericallosal artery (*arrowheads*) and by the posterior choroidal arteries of the PCA. Anterior and posterior systems converge forming the so-called limbic arch. The bidirectional arrow shows the further drainage into the transverse sinus and dilated occipital sinus. (**c**) Right VA angiograms, AP and Lateral views, showing the supplying arteries arising from the left PCA. Posterior thalamoperforating branches (*arrow*). Medial and Lateral choroidal arteries (*arrow with angle*). Following selective catheterization, coils were placed in the choroidal arteries, leading to a great decrease of the flow with significant clinical improvement. (**d**) Two years later, a control vertebral angiogram showed the partial persistence shunt. (**e**), after selective catheterization of the remaining branches, the shunt was completely occluded with glue. On the non-subtracted image the cast of the coils and glue is demonstrated. (**f**) Final control of the left ICA and Left VA. Normal clinical development of the child

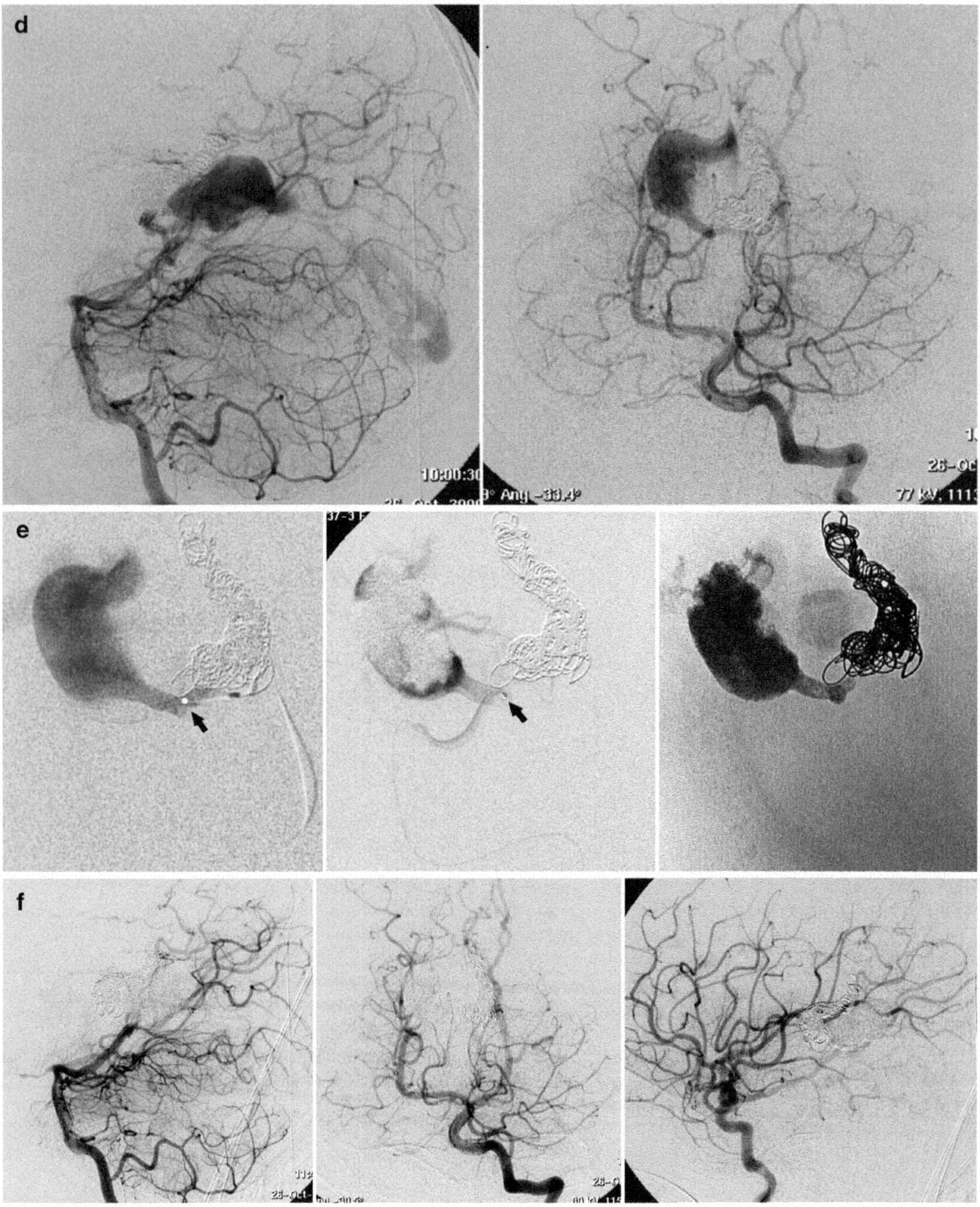

Fig. 12.26 (continued)

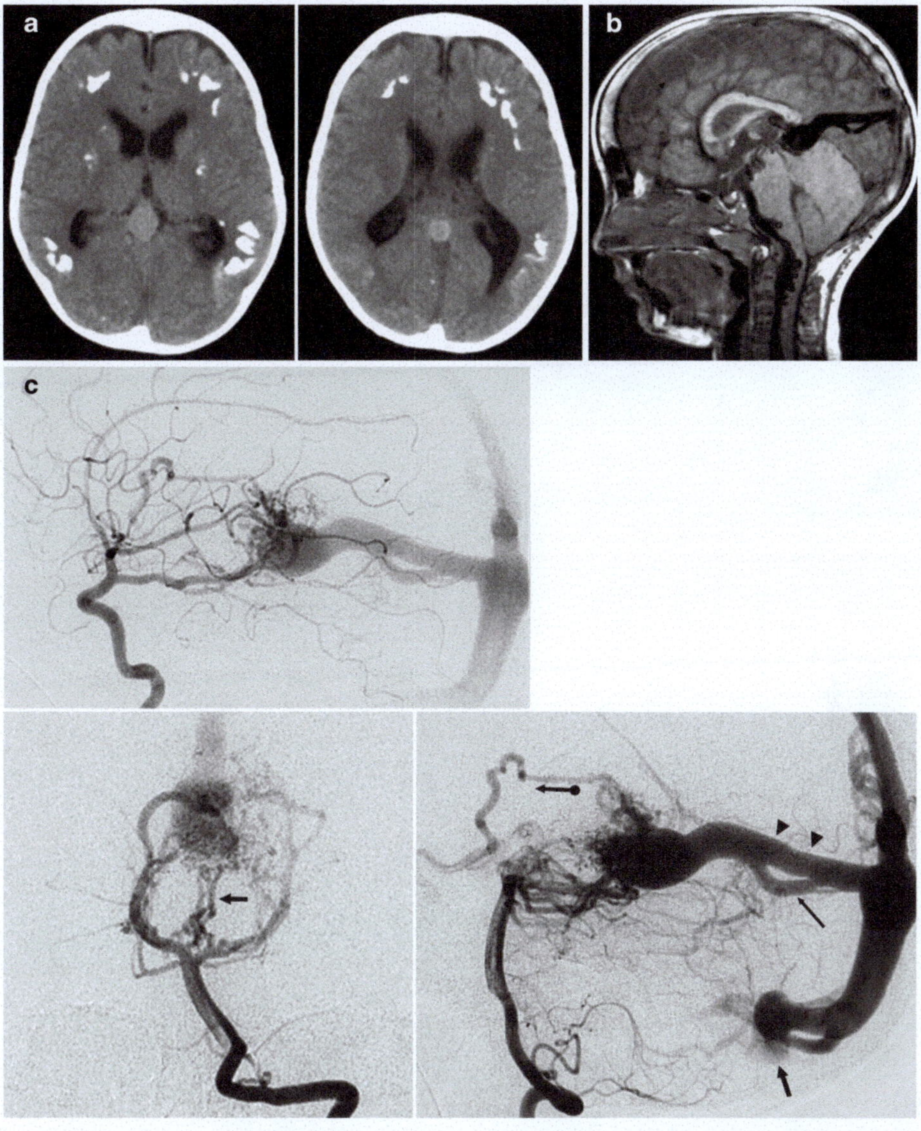

Fig. 12.27 Vein of Galen malformation in a 5-year-old child discovered incidentally after examination for craniocerebral trauma. On the clinical examination the child was normal. (**a**) On CT a large rounded probably vascular structure in the cistern of the Galen vein was recognizable. In addition there were several small calcifications involving the subcortical white matter and basal ganglia. These are thought to be due to a chronic ischemia linked to the venous flow impairment involving the medullary and subependymal veins. There was no hydrocephalus. (**b**) MRI T1-weighted image showing the dilated prosencephalic vein draining into the falcine sinus. There is a partial anterior rerouting of the venous drainage in the dilated pontomesencephalic veins. (**c**) Left lateral ICA angiogram, and vertebral angiogram (*lateral and AP view*). Supply of the malformation through the posterior choroidal arteries and posterior thalamoperforating branches (*arrow*) better demonstrated on the AP view. The prosencephalic vein and the falcine sinus (*arrow head*) as well as a duplication of the sinus (*arrow*) are clearly visible. Bilateral thrombosis of transverse sinuses. Anteriorly rerouting of the venous drainage (*arrow with dot*). The left ICA angiogram showed no involvement of the pericallosal artery. Similar pattern on the right ICA angiogram. (**d**) VA angiogram, showing the partial occlusion of the malformation after injection of glue in the selectively catheterized posterior choroidal arteries. (**e**) One year later, a control angiogram showed no changes of the partial treated malformation and unchanged normal clinical condition. A decision was taken to complete the treatment using the venous route. A catheterization of the falcine sinus (*arrowhead*) was performed after a surgical craniectomy. Coils were positioned in the prosencephalic vein (*arrow*) with complete occlusion of the malformation. (**f**) Final normal ICA and VA angiograms

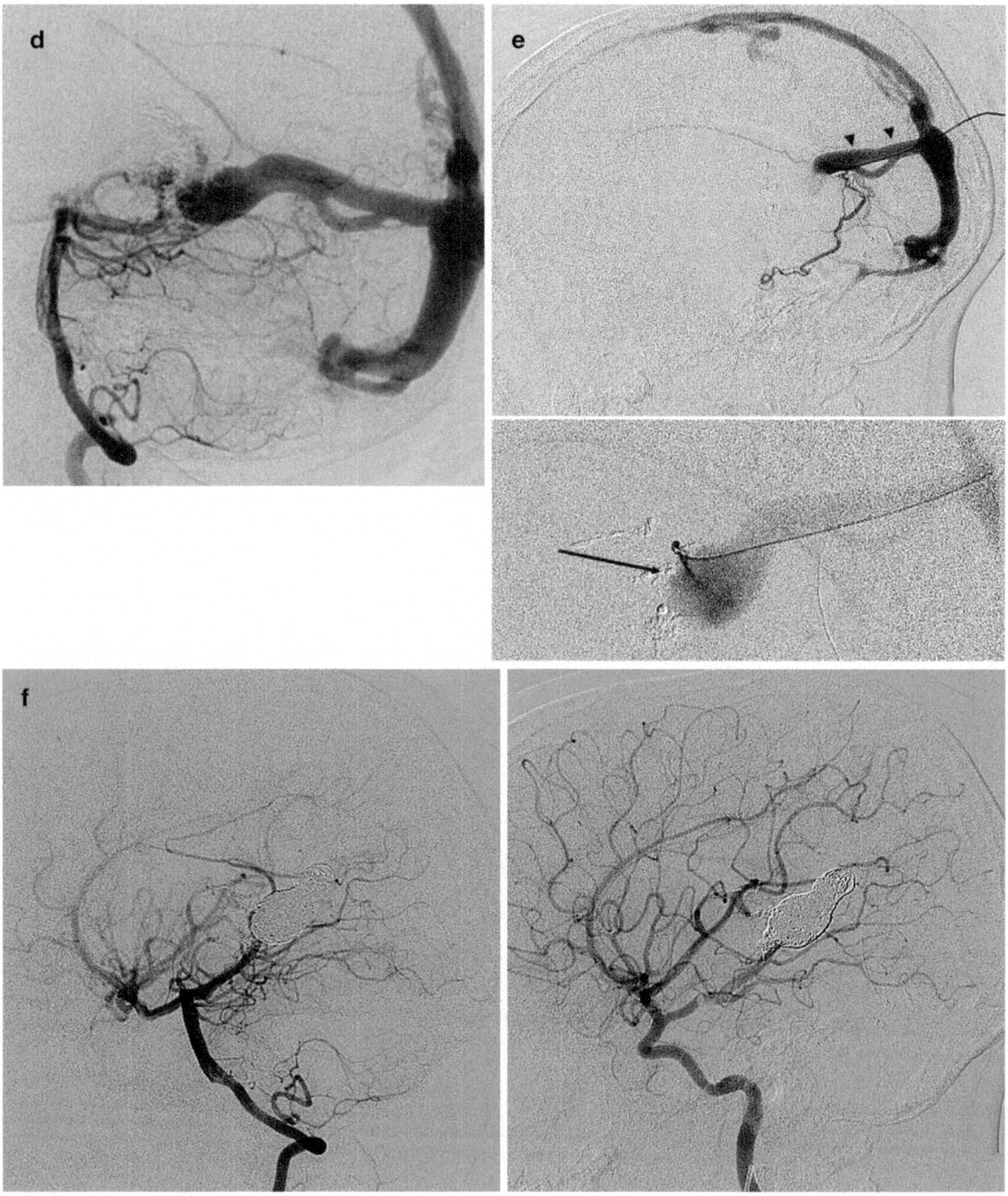

Fig. 12.27 (continued)

13 Dural Arteriovenous Fistulas

Isolated cases of dural arteriovenous fistula (DAVF) have been reported by some authors in the 1950s (Verbiest 1951; Obrador and Urquiza 1952). Progressive identification and precise description of this pathology came at the end of the 1960s, especially through selective and superselective angiographic studies, allowing to differentiate it as a special pathological entity (Hayes 1963; Laine et al. 1963; Newton et al. 1968; Newton and Hoyt 1970; Djindjian et al. 1968, 1973).

13.1 Incidence

DAVFs account for 10–15% of all intracranial vascular malformations. Both sexes are affected, but a sexual predominance occurs in some type of DAVF. The fistulas are more frequent in middle-aged and older patients, though younger patients and children can be affected.

13.2 Pathology and Pathogenesis

DAVFs consist of a shunt between dural (meningeal) arteries and sinuses, either directly or mediated by pial veins. DAVF is considered to be an acquired pathology (Houser et al. 1979; Chaudary et al. 1982). The specific pathogenetic process, however, leading to a formation of the fistula is still debated. Some authors believe (Houser et al. 1979; Mironov 1995), DAVFs are preceded by thrombosis of a sinus. When the lumen recanalizes, microscopic AV shunts, normally present within the wall of the sinus (Kerber and Newton 1973) may enlarge and open into the sinus. Considering, however, that sinus thrombosis is not necessarily associated with DAVF and that DAVFs are not always associated with sinus thrombosis, other authors (Chaloupka and Huddle 1998; Chaloupka et al. 1999) have suggested that the primary cause is angiogenesis within the sinus wall. This leads to the formation of abnormal arteriovenous connections and finally to DAVF. Indeed, abnormal artery-vein connections have been demonstrated in histological studies of resected specimens of sinuses of patients with DAVF (Nishijima et al. 1992; Hamada et al. 1997). The cause of these microfistulas in the sinus wall is not clear. Some authors have shown with studies in animal models that venous hypertension can lead to DAVF formation more frequently in animals with induced hypertension than in those without (Terada et al. 1994; Lawton et al. 1997). In this context, flow changes in the venous sector due to partial venous thrombosis and venous hypertension has been suggested to be at least one of the cause of the "de novo" developing of dural arteriovenous fistula occurring after endovascular occlusion of brain AVM (Paramasivan et al. 2013). Other studies (Uranishi et al. 1999; Shin et al. 2007a) have demonstrated in resected dural specimens of DAVF of patients and in rat models the presence of an angiogenetic growth factor that can participate in the fistula formation. Other authors

G.B. Bradac, *Applied Cerebral Angiography*, DOI 10.1007/978-3-319-57228-4_13

(Klisch et al. 2005) examined, in blood of patients with DAVFs, the concentration of the endothelial growth factor (VEGF) and the angiopoietic receptor (TIE-2). These factors were increased in the pretreatment phase and decreased after the endovascular treatment.

Another difficulty is that to explain the developing of DAVFs located in the dura close to the sinus, but not draining directly into it. The outflow is characterized by pial veins, which, after a course more or less long, enter a near or distant sinus. In these cases a nonspecific thrombophlebitic process involving the pial veins has been suggested (Djindjian and Merland 1978).

13.3 Clinical Relevance

Many DAVFs remain asymptomatic or have a benign course. In other cases, DAVFs can have a more aggressive course, characterized by cranial nerve palsy, ischemia, hemorrhage, and cognitive disorders. In this context, it has become increasingly clear that the main factor responsible for the symptoms and evolution of DAVFs is the pattern of venous drainage (Lasjaunias et al. 1986b; Halbach et al. 1987, 1988c, 1989, 1990; Awad et al. 1990; Awad 1993; Barnwell et al. 1991; Lasjaunias and Rodesch 1993; Cognard et al. 1995; Davies et al. 1996).

13.4 Location

DAVfs can occur everywhere. The most frequent are those involving the transverse-sigmoid sinuses followed by the cavernous sinus. Less common are DAVFs of the superior sagittal sinus, those in the tentorial area, and in the anterior cranial fossa. Very rare are DAVFs located in the region of the foramen magnum.

13.5 Diagnosis

DAVVs involve the preexisting vascular structures (meningeal branches, dural sinuses cortical/pial veins) of the area in which they develop, and so a typical repetitive pattern can be expected, corresponding to the site and type of fistula. However, it should be noted that *the pathogenetic mechanism leading to the fistula can vary and there are many variants concerning the arteries supplying the dura and the venous drainage of the area involved. Furthermore the pattern of the fistula can be altered by thrombosis involving the sinuses, and so it is not surprising that fistulas at the same site can have some different angiographic patterns.*

A complete angiographic study with examination of the ECAs, ICAs, VAs, and in particular cases also cervical arteries is essential for the diagnosis. This provides information about all the meningeal sometimes leptomeningeal branches involved and the venous drainage. The site, the extension of the DAVF, and in particular the type of the venous drainage explain the symptoms and offer information about the risk and prognosis of the fistula. Furthermore, a decision can be made about whether the fistula should be treated and, if so, whether surgical or endovascular therapy should be used; in the case of endovascular treatment which route—arterial or venous—is the most appropriate.

13.6 Classification

Castaigne et al. (1976) were the first to distinguish DAVFs draining directly into the sinus from those in which the drainage into the sinus was mediated by cortical/pial veins. Taking into consideration mainly the type of venous drainage, Djindjian and Merland (1978) made the first classification. This was revised by Cognard et al. (1995) and Borden et al. (1995). Five types of DAVFs can be identified:

1. Drainage into a main sinus with an anterograde flow.
2. (a) Drainage into a main sinus with possible retrograde reflux within the sinus, but not into the pial veins; the latter drain normally into the affected sinus.
   (b) Drainage into the sinus with reflux into the pial veins with or without association of sinus reflux.
3. Drainage into the sinus is mediated by pial veins.

4. Drainage into the sinus is mediated by pial veins, which present large ectasia.
5. Drainage involves the spinal perimedullary veins.

The type 1 has commonly a benign course. In types 2 and 3, the impaired normal drainage can lead progressively to venous congestion, ischemia, and intracranial hypertension. In types 3 and 4, hemorrhage is frequent owing to rupture of the pial veins draining the shunt. Cranial nerve palsy can occur, particularly in DAVFs of the cavernous and DAVFs located close to the brainstem. Drainage involving the veins of the spinal cord can lead to a myelopathy.

The fistulae are not static lesions, but they can change passing from one type to another with morphological and clinical improvement or worsening. In the study of some authors (Kim et al. 2010), this occurred in 16% of the cases, more frequent in the DAVS of the transverse-sigmoid sinuses and in those involving the cavernous sinus. All these aspects should be considered in the management of this pathology.

13.7 Detailed Considerations in the Most Frequent Fistulas

- *DAVFs involving the transverse sinus (TS) and the sigmoid sinus (SiS)* are the most frequent, occurring in about 63% of all fistulas (Halbach et al. 1987; Awad 1993). In many cases, these belong to the type 1 and so can be characterized by a benign course. The only symptom is bruit, since the fistula is close to the petrous bone, containing the auditory apparatus. In some cases, the bruit can become very loud and intolerable for the patient, necessitating treatment. Some of these fistulas can change (Piton et al. 1984) and become of the type 2, with a large reflux in the SSS and in the veins of the brain parenchyma especially in the temporal veins, leading to a progressively venous congestion. This occurs, especially, when a distal or proximal thrombosis of the sinus occurs, or in the very rare cases in which the sinus is thrombosed proximally and distally (*isolated sinus*). The symptoms in such patients can be very severe, characterized by cognitive disorders, epileptic seizures, or other neurological deficit (Naito et al. 2001; Bradac et al. 2002, Kiura et al. 2007). Examples are presented in Figs. 13.1, 13.2, and 13.3.

In some cases, the fistula can be localized in the proximal part of the TS *near the torcular herophili or in the torcular itself* (Figs. 13.4 and 13.5).

Several arteries can be involved in the supply of these fistulas. These are commonly branches of the ECA (occipital, ascending pharyngeal, middle meningeal arteries). Meningeal branches of the cavernous portion of the ICA and meningeal branches of the VA are also frequently involved. A supply from pial branches of the cerebellar and posterior cerebral arteries, especially in fistula of the torcular herophili can also occur.

In fistulas of the torcular herophili thrombosis of the proximal TS uni- or bilaterally, and of the distal SSS is not rare. This leads to a rerouting of the venous drainage in the straight sinus and then in the basal and deep cerebral veins. The impaired venous drainage can lead to congestion in the cerebral and cerebellar hemispheres (Fig. 13.4). In other cases the venous drainage occurs in one TS (Fig. 13.5). The other TS is not involved, due to duplication of the distal SSS, each segment entering separately one TS. It can occur in these cases that the SS enters the not involved TS avoiding impairment of the depp venous drainage. This latter, however, can be severely impaired in cases of thrombosis extending towards the SS and Galen vein.

A particular aspect, especially recognizable in some cases of fistula of the TS or SiS, is the changes of the sinuses, which *appear divided in two or more compartments* (Figs. 13.2 and 13.3). Why this occur is not known. It has been suggested that this could be related to a partial thrombosis of the sinus followed by its recanalization and formation of two or more separated venous channels (Piske et al. 2005). An accessory dural sinus separated but communicating with the main sinus can also develop and become the main venous drainage.

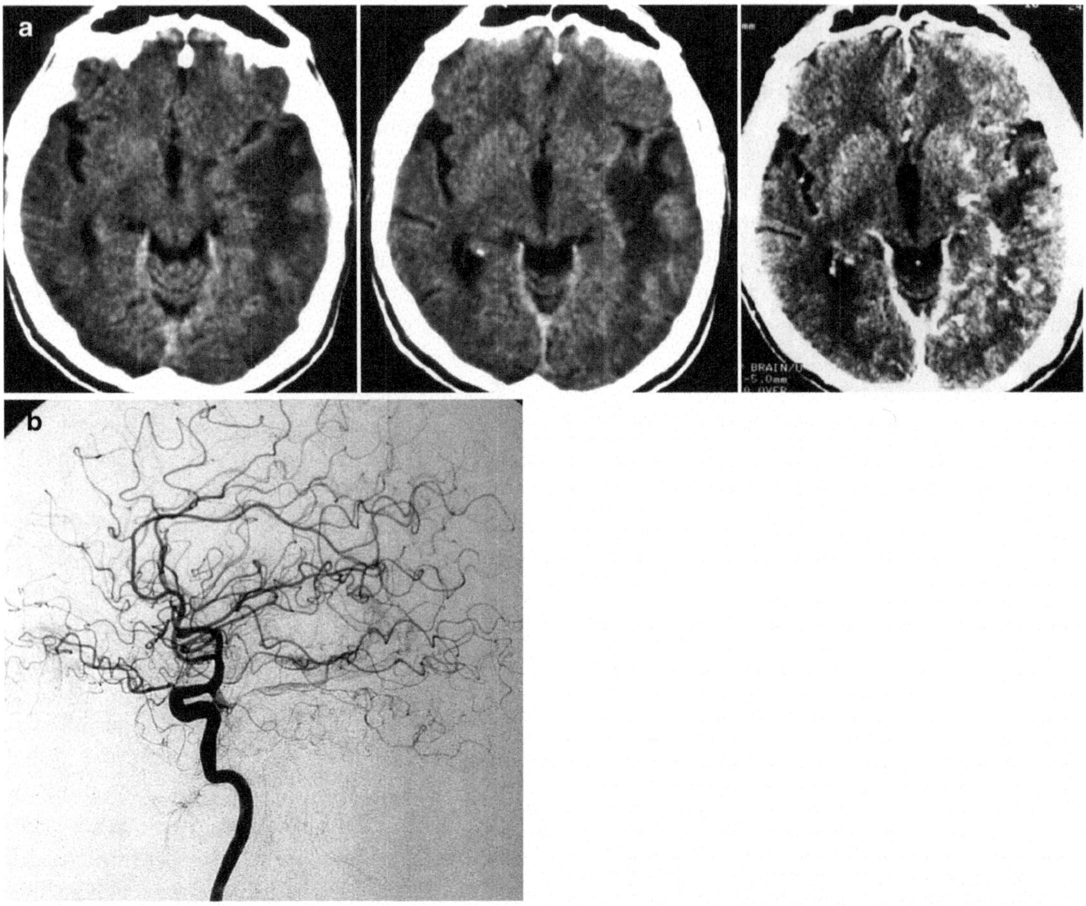

Fig. 13.1 Patient with mild aphasia and cognitive disorders. (**a**) CT showed changes involving predominantly the white matter in the left temporo-insular area with rich contrast enhancement suggesting vascular malformation. (**b**) Normal ICA angiogram. (**c**) On the ECA angiogram a DAVF involving the distal transverse sinus was demonstrated. The sinus was proximally and distally occluded. The main supplying artery was the occipital artery with its stylomastoid (S) and mastoid (M) branches. (**d**) A partial supply came from the middle meningeal artery (MMA). (**e**) On the VA angiogram the involvement of the posterior meningeal artery (*arrow*) and small C1 branches (*small arrows*) was shown. (**f**) selective catheterization of the stylomastoid and MMA arteries preceding the injection of glue. The Branches arising from the VA were occluded with polyvinyl alcohol (PVA). (**g**) The occlusion of the fistula was confirmed on a control angiogram of the ECA and VA 3 months later. (**h**) Normalization of the CT. Complete recovery of the patient

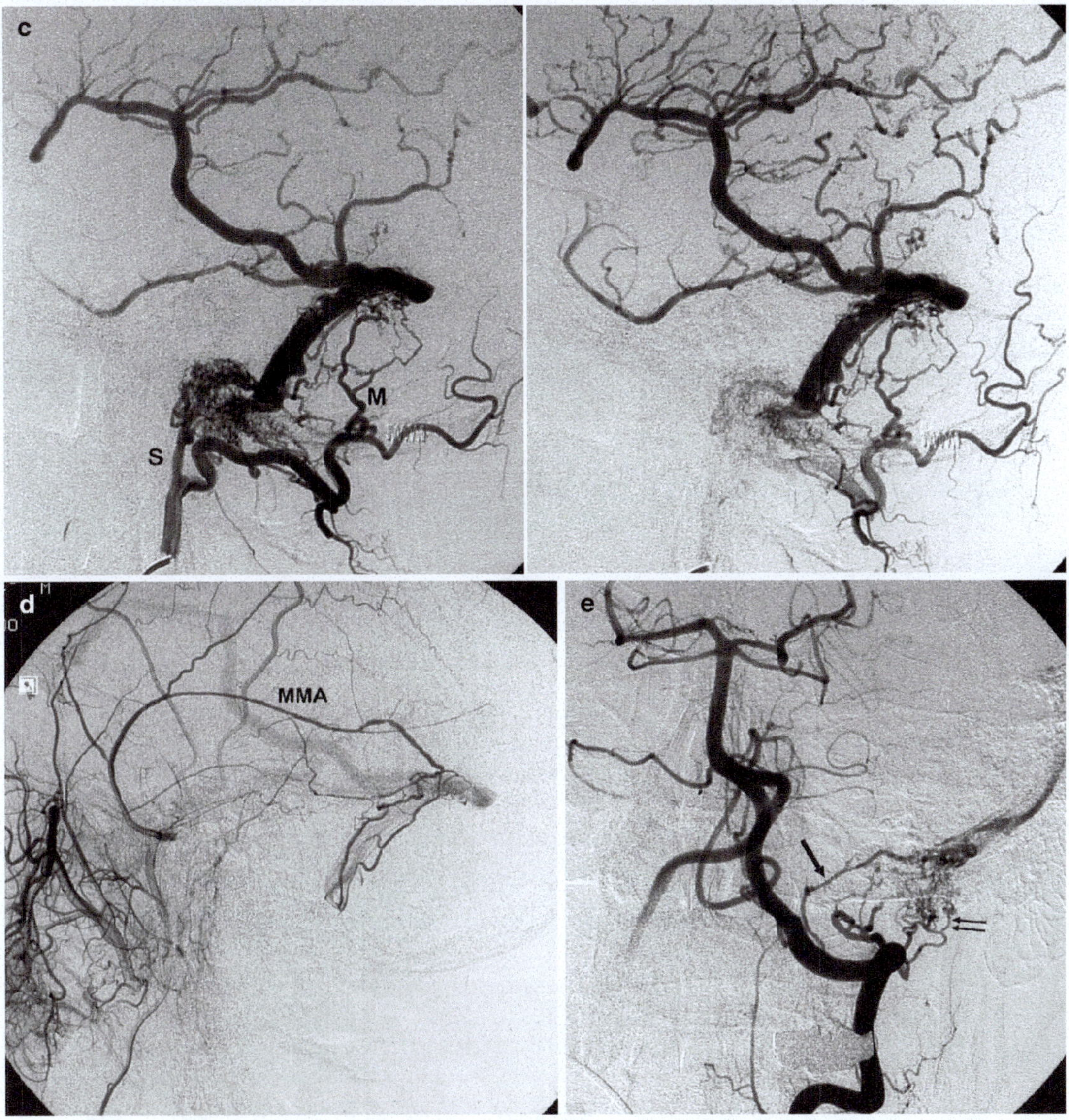

Fig. 13.1 (continued)

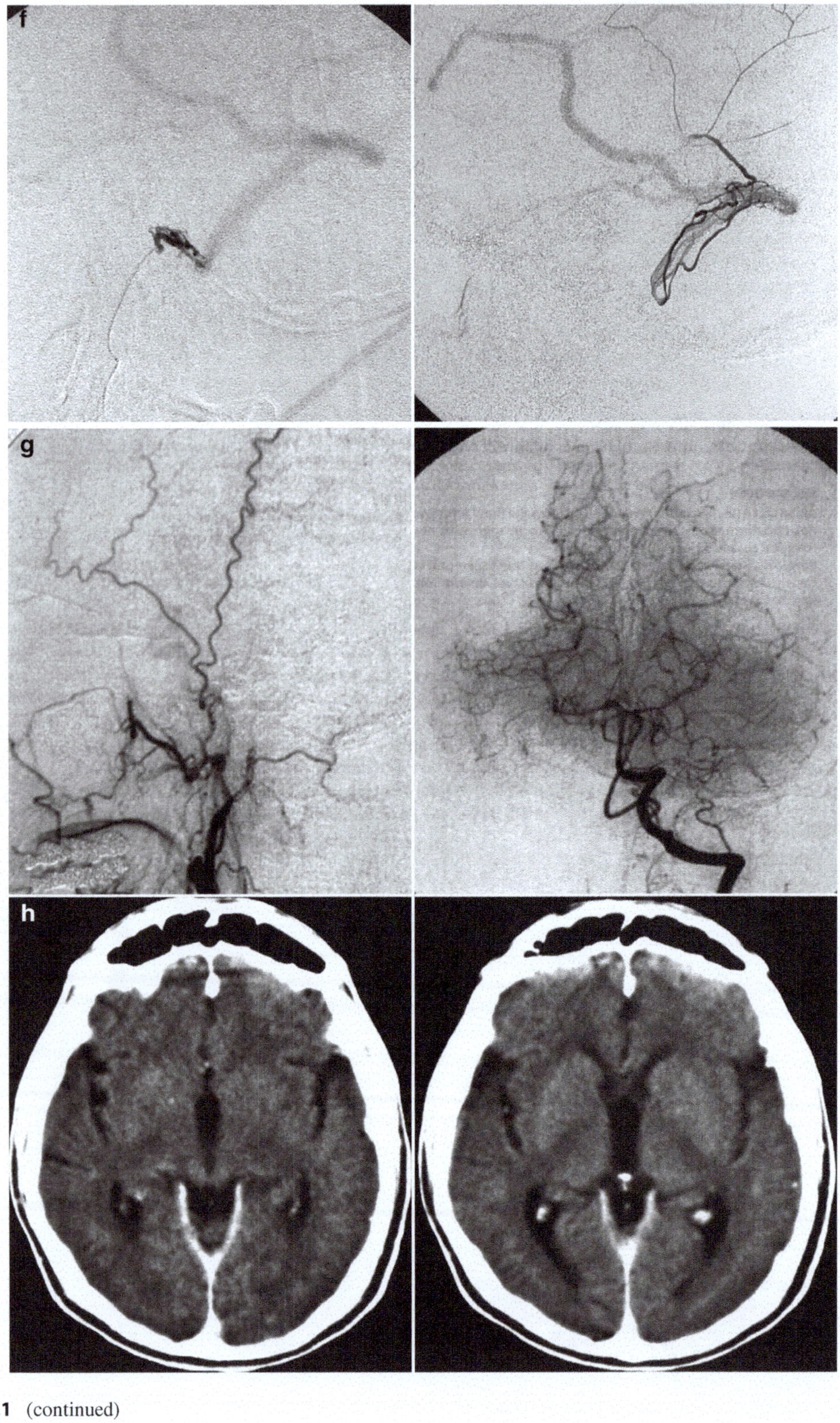

Fig. 13.1 (continued)

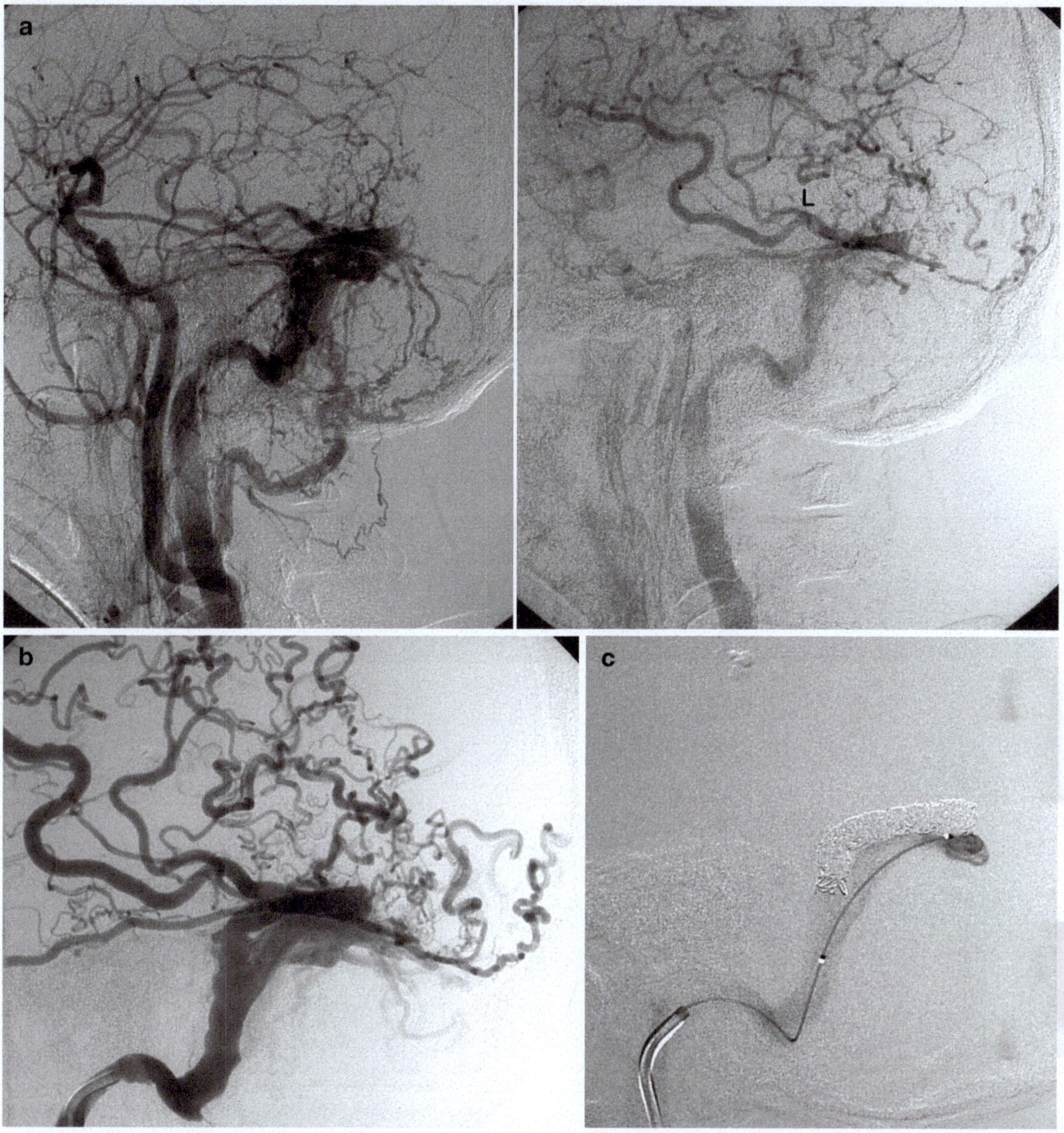

Fig. 13.2 Old patient with known DAVF involving the transverse sinus already treated with occlusion of several branches of ECA with PVA. The patient returned 6 months later suffering from the same symptoms, characterized by high bruit and headache. (**a**) The angiogram of the common carotid artery showed a complete recanalization of the fistula. The transverse sinus was proximally occluded. In the later phase a retrograde injection of the temporal veins including a large vein of Labbé (L) was visible. (**b**) Venous angiogram after catheterization of the sigmoid and further of the proximal transverse sinus. The retrograde filling of the cortical veins is better visible as well as the duplication of transverse sinus. (**c**), (**d**) A microcatheter was advanced first in the superior segment of the duplicated sinus were coils were placed (**c**) followed by catheterization and occlusion of the inferior segment with coils (**d**). (**e**) Control angiogram showing complete occlusion of the fistula

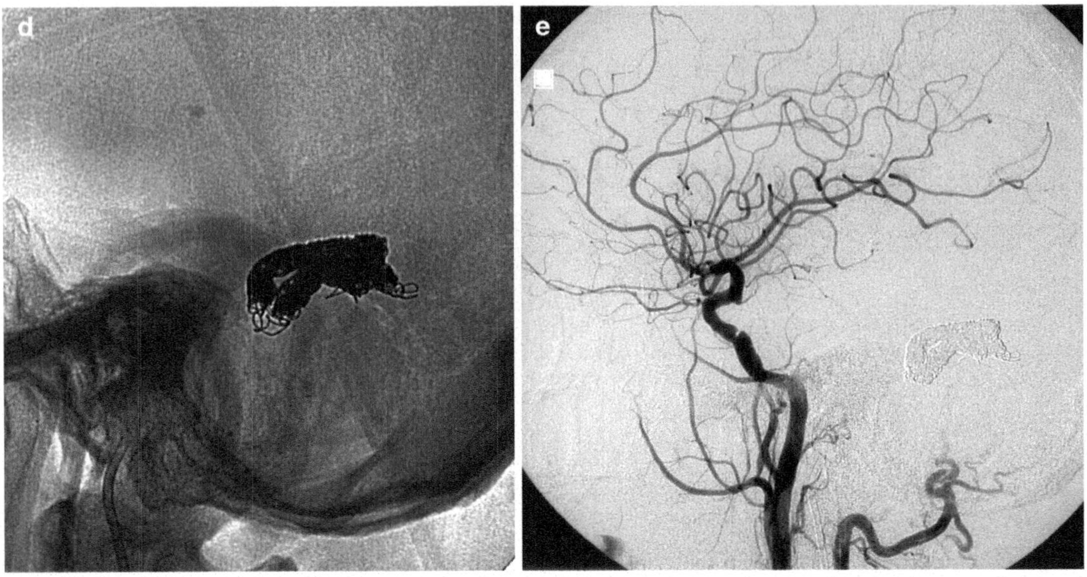

Fig. 13.2 (continued)

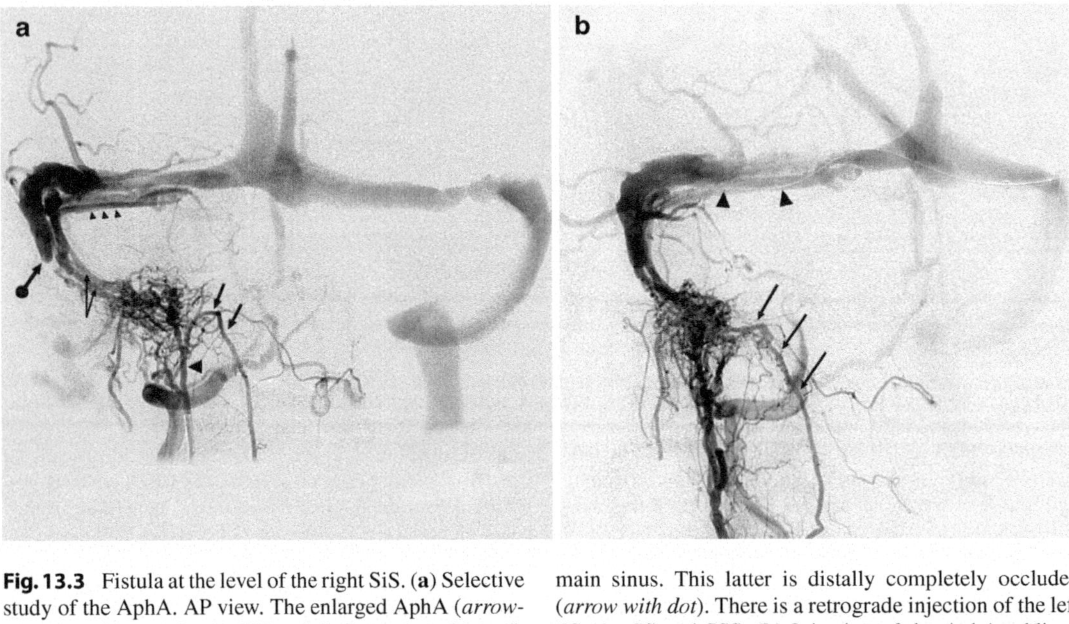

Fig. 13.3 Fistula at the level of the right SiS. (**a**) Selective study of the AphA. AP view. The enlarged AphA (*arrowhead*) supplies the fistula. Through its hypoglossal branch (*arrows*) there is a large connection with the radiculomeningeal branch of the VA. The venous drainage occurs through two separated dural venous channels (*arrow with angle*) converging to a dural sinus (accessory sinus, arrow heads) running parallel to the main transverse sinus (TS). There is a connection between the accessory sinus and the main sinus. This latter is distally completely occluded (*arrow with dot*). There is a retrograde injection of the left TS, the SS and SSS. (**b**) Injection of the AphA, oblique view. The connection with the VA is better recognizable. Through the left TS, a microcatheter has been advanced in the accessory right dural sinus (*arrowheads*) and further in the two distal dural channels were coils were placed occluding the fistula

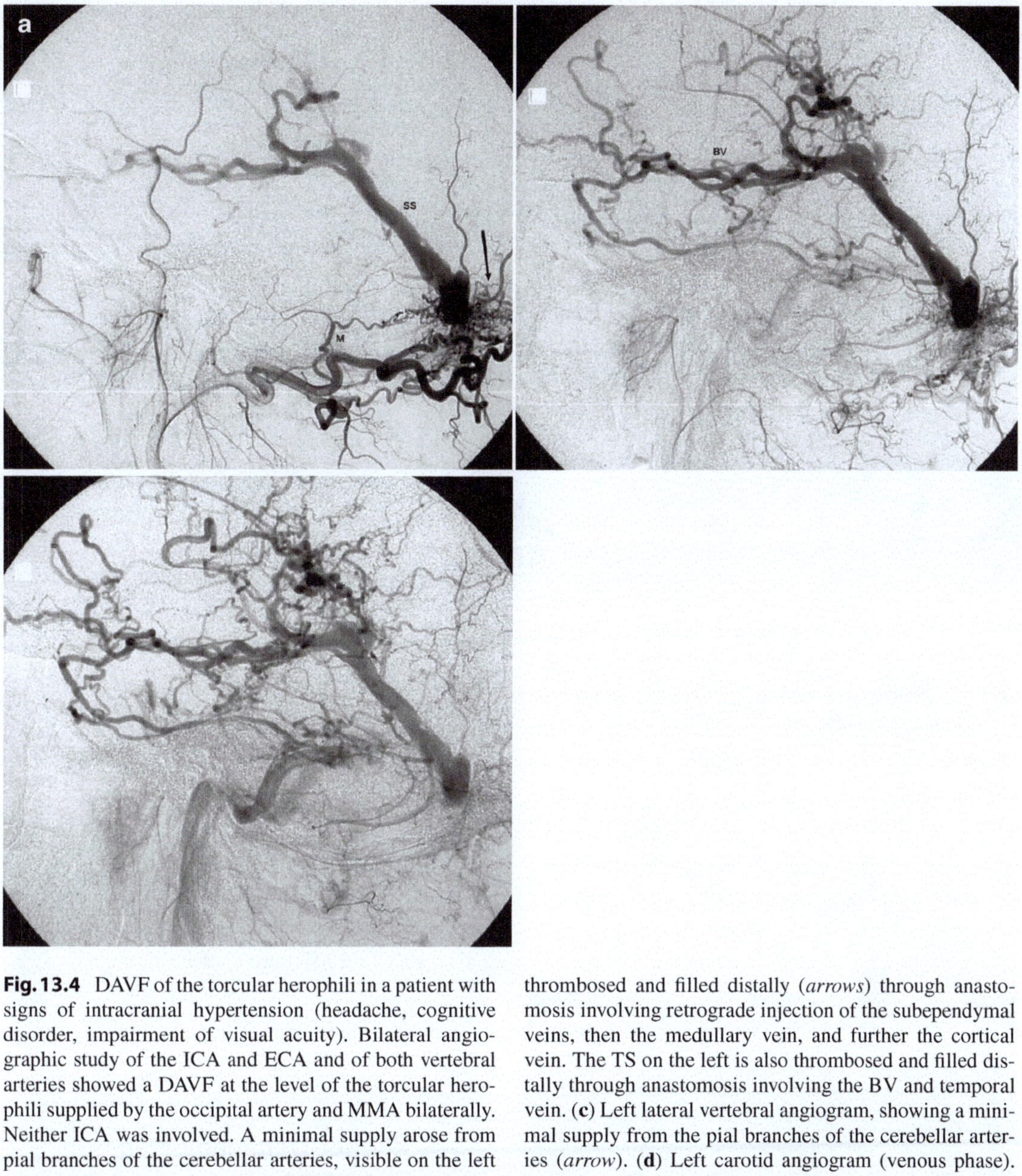

Fig. 13.4 DAVF of the torcular herophili in a patient with signs of intracranial hypertension (headache, cognitive disorder, impairment of visual acuity). Bilateral angiographic study of the ICA and ECA and of both vertebral arteries showed a DAVF at the level of the torcular herophili supplied by the occipital artery and MMA bilaterally. Neither ICA was involved. A minimal supply arose from pial branches of the cerebellar arteries, visible on the left vertebral angiogram. Occlusion of the fistula was obtained by selective injection of acrylic glue in both occipital arteries and the MMA. Complete recovery of the patient. (**a**) Lateral left occipital angiogram, showing a rich network of branches arising from the mastoid (M) and distal transosseous (*arrow*) branches of the occipital artery, converging on the torcular herophili, probably thrombosed. There is retrograde injection of the straight sinus (SS) and both basilar veins (BV), continuing to the anterior tributary and then to the temporal vein. There is also retrograde injection of the internal cerebral vein, dilated medial atrial vein, and inferior sagittal sinus. (**b**) Right occipital angiogram (AP views) showing the shunt and retrograde injection of the SS. The TS on the right is proximally thrombosed and filled distally (*arrows*) through anastomosis involving retrograde injection of the subependymal veins, then the medullary vein, and further the cortical vein. The TS on the left is also thrombosed and filled distally through anastomosis involving the BV and temporal vein. (**c**) Left lateral vertebral angiogram, showing a minimal supply from the pial branches of the cerebellar arteries (*arrow*). (**d**) Left carotid angiogram (venous phase). There is a diffuse venous congestion owing to impairment of the deep venous drainage and partial thrombosis also of the distal superior sagittal sinus (SSS). (**e**) Left vertebral angiogram, AP view, venous phase. There is diffuse congestion involving the drainage of the cerebellar vein as a result of thrombosis of the proximal transverse sinus bilaterally. Normal filling of the distal transverse sinus (TS). (**f**) Left common carotid angiogram, lateral view, arterial and venous phases, showing occlusion of the fistula and normalization of the venous drainage after treatment. Similar results on the right common carotid angiogram, not shown. (**g**) Vertebral angiogram, presenting occlusion of the fistula after treatment

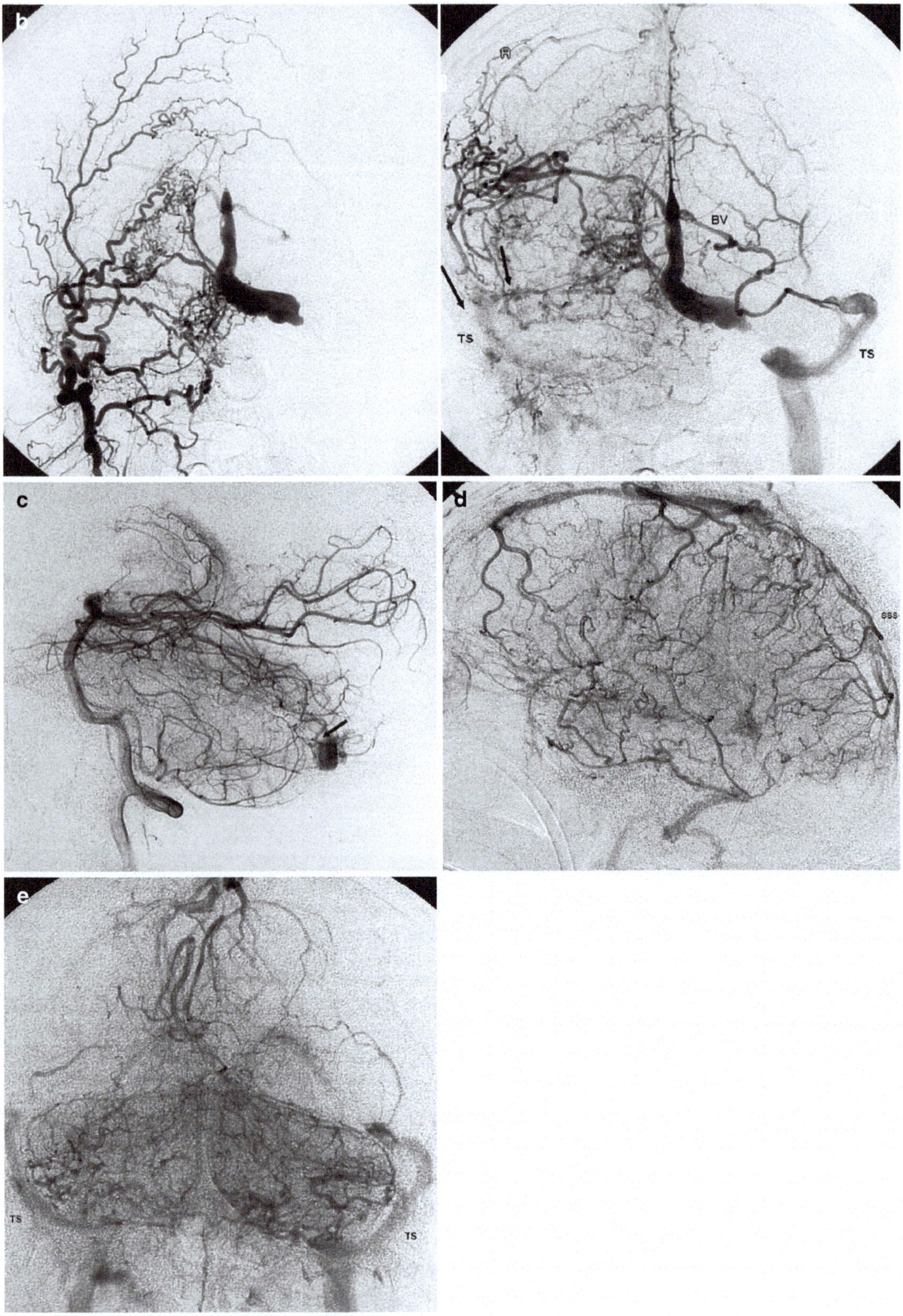

Fig. 13.4 (continued)

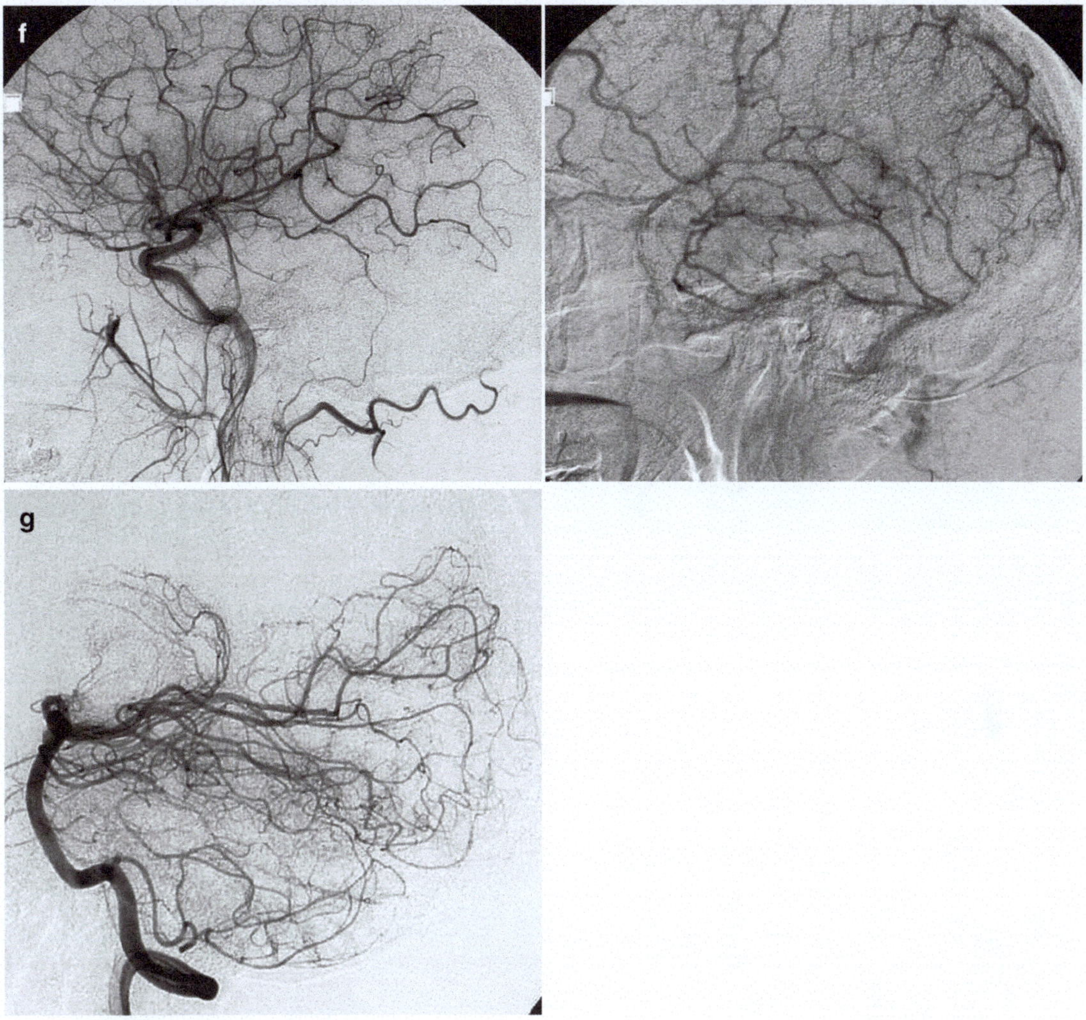

Fig. 13.4 (continued)

These possible patterns should be taken into account since are important in the endovascular treatment of these fistulae.

Occlusion of these fistulae can be today achieved by endovascular treatment either through selective catheterization of the feeders, followed by injection of acrylic glue, onyx, or similar embolic material or by a venous approach through the sinuses following by the placing of coils (Kirsch et al. 2009; Macdonald et al. 2010). Surgery can be reserved for those cases in which the endovascular treatment has failed in leading to a complete occlusion or in those cases in which it appears not appropriate due to an unfavorable anatomical/morphological condition.

- *DAVFs involving the cavernous sinus* (CS) are the second-most frequent fistulas, with a frequency reported being 12% of all fistulae (Awad 1993; Cognard et al. 1995). Women are predominantly affected. The fistula can be uni- or bilateral. Clinical symptoms, can vary, being minimal or severe due to involvement of

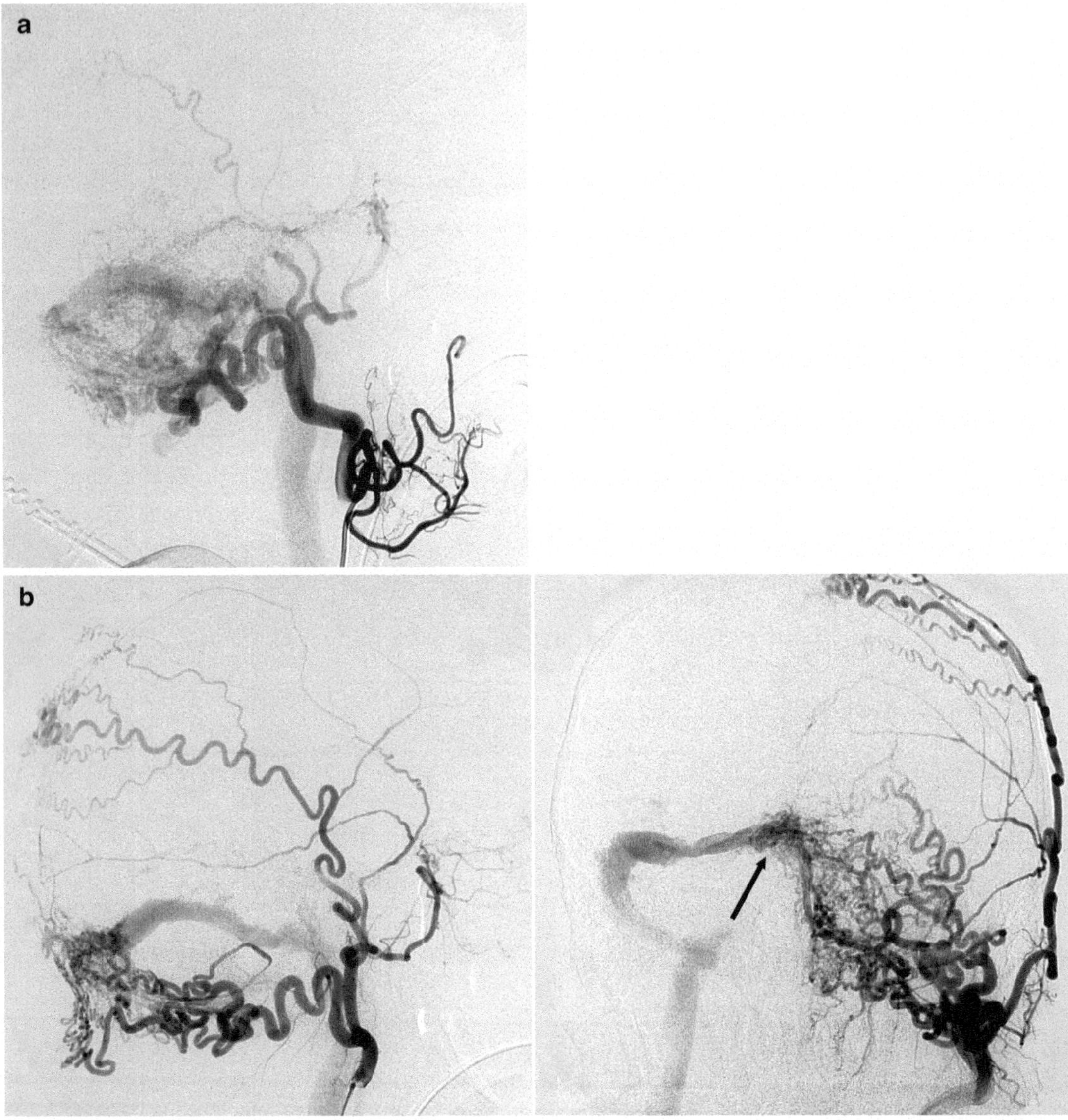

Fig. 13.5 DAVF involving the torcular herophili. Many feeders arising from ECA, ICA, and VA bilaterally were present; not all are shown in the figures. The main supplying branches were the right and left occipital arteries. There was also involvement of the deep cervical artery. (**a**) Lateral angiogram of the large right occipital artery with many branches converging to the torcular. Filling of the MMA which was also involved. (**b**) Similar pattern visible on the lateral and AP angiograms of the left occipital artery. The site of the fistula (*arrow*) is well visible on the AP angiogram. (**c**) Angiogram of the deep left cervical artery with branches converging to the torcular. In all studies the drainage occur in the right TS continuing in the SiS. (**d**) AP of the venous phase of the right and left ICA angiograms showing that the main venous drainage of both hemispheres occur through the SSS continuing in the left TS (*arrow*). It is thinkable that there were two separate distal segments of then SSS draining each in the corresponding TS. (**e**) Right ICA angiogram. Venous phase. There is a normal drainage of the deep venous system in the SS draining to the left TS. There were no connections with the involved right TS. (**f**) Vertebral angiogram. Venous phase. The main venous drainage converge to the left TS. After partial devascularization with PVA injected in the distal occipital arteries. Complete occlusion was obtained with coils put in the distal right TS

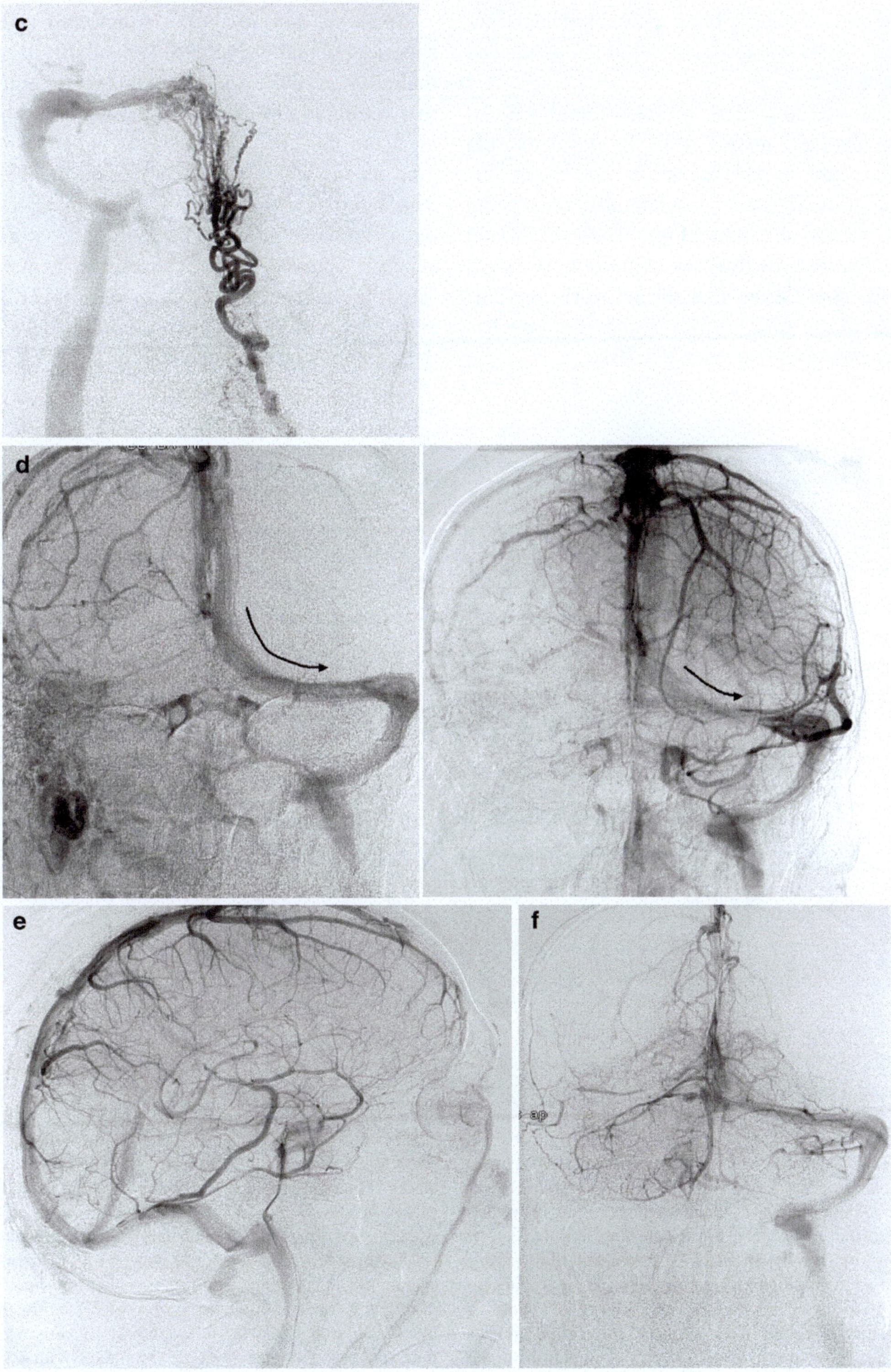

Fig. 13.5 (continued)

the intraorbital or brain parenchymal structures. Spontaneous occlusion of the fistula can occur.

Many arterial feeders in various combinations can be involved, arising from the distal internal maxillary artery, MMA, AphA, and cavernous portion of the ICA uni- or bilaterally, converging to the cavernous sinus (Figs. 13.6, 13.7, and 13.8). The venous drainage can show different patterns, considering that the network of veins running in the *space of the cavernous sinus* has vary connections (see Sect. 9.3.10).

Since the anterior part is connected with the superior ophthalmic vein (SOV) and inferior ophthalmic vein (IOV), a fistula in this sector will lead to a retrograde filling of these vessels. It should be noted that the IOV connects with the pterygoid plexus which can be indirectly involved. *A fistula posteriorly located* will be characterized by a drainage into the inferior petrosal (IPS), superior petrosal (SPS) sinuses and sometimes into the inferior petro-occipital vein (IOPV), since this posterior location communicates with these venous channels (Cheng et al. 1999; Agid

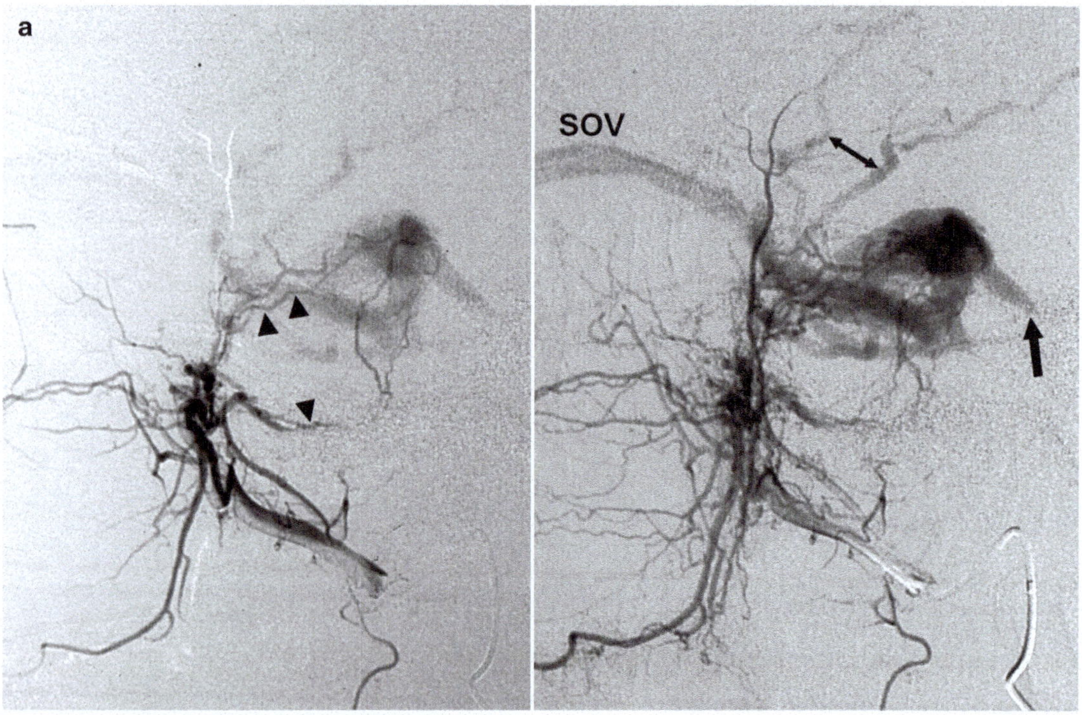

Fig. 13.6 Old woman presenting with bilateral exophthalmos, chemosis, and diplopia. The angiographic study of the bilateral ICA, ECA and vertebral artery revealed DAVF at the level of the left cavernous sinus (CS) supplied by branches of the distal internal maxillary artery (IMA) and ascending pharyngeal artery. (**a**) Lateral angiogram of the distal left IMA showing the involvement of the foramen rotundum (*arrowheads*) and vidian arteries (*arrowhead*). In the later phase, partial injection of the inferior petrosal sinus, which appears to be thrombosed (*arrow*). There is also a partial retrograde injection of the superficial middle cerebral vein an uncal vein (*bidirectional arrow*). (**b**) Lateral angiogram of the ascending pharyngeal artery (APhA), showing the supply through its clival branches (*arrows*). (**c**) Lateral angiogram of the MMA involved with several branches (*arrows*). (**d**) Left ECA angiogram, AP view, showing involvement of the left cavernous sinus. Through intercavernous anastomoses (*arrowhead*), injection of the right cavernous sinus (CS) of the right superior ophthalmic vein (SOV) and further of the facial vein (F). The left ophthalmic vein was not injected, probably thrombosed. Occlusion with PVA of the supplying arteries was performed, followed by treatment with low doses of heparin. Clinical improvement

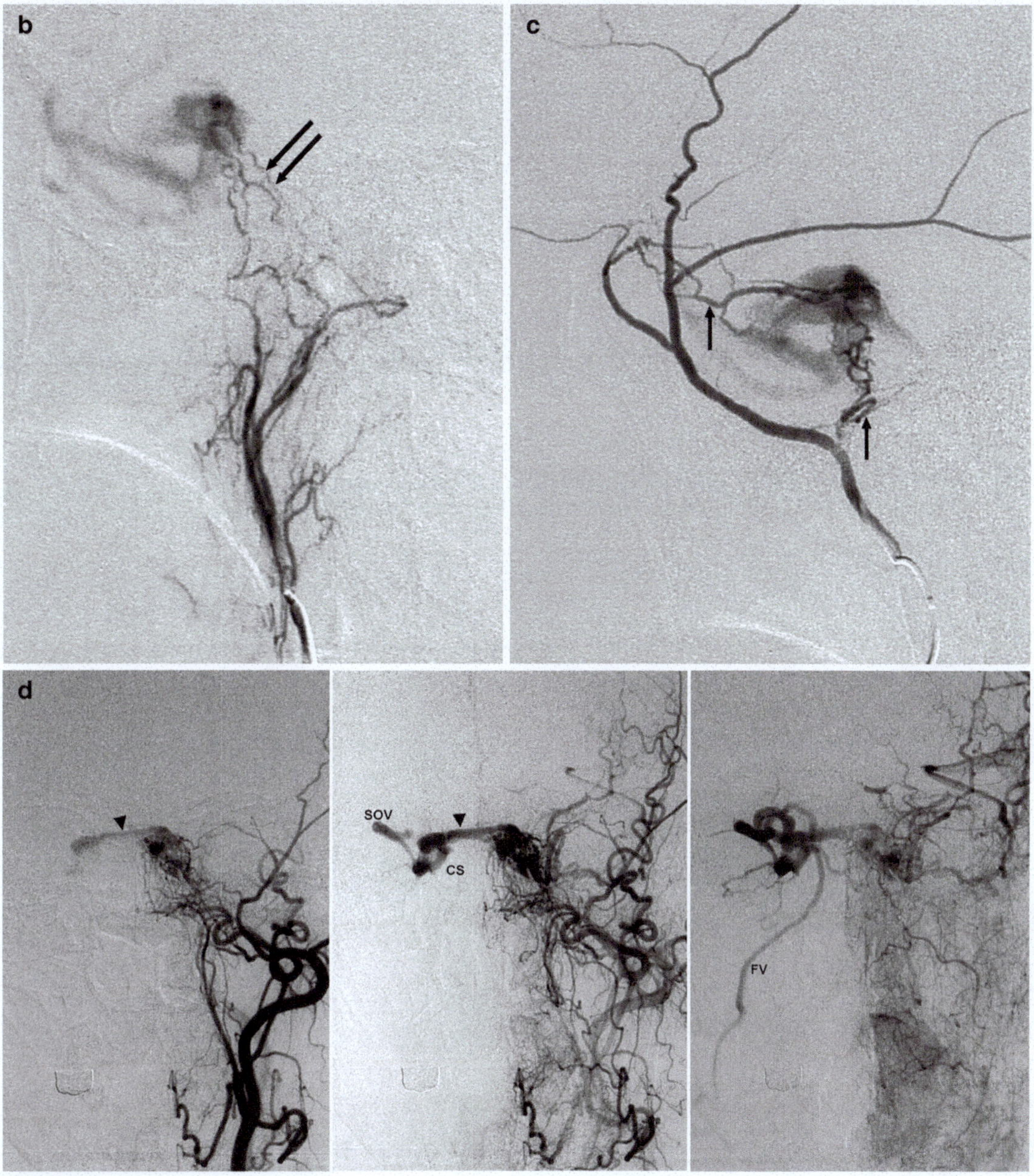

Fig. 13.6 (continued)

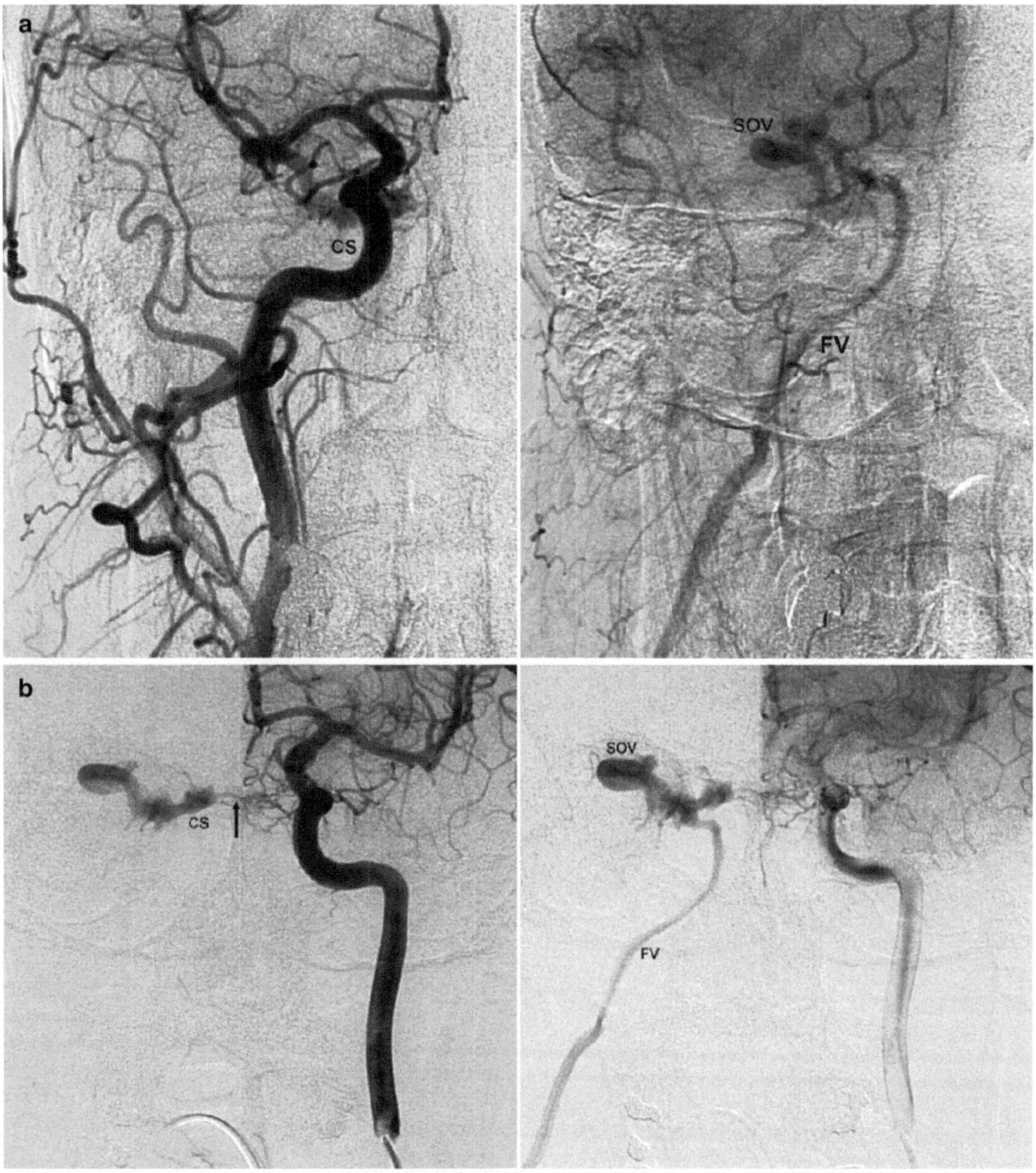

Fig. 13.7 Middle-aged man with sudden onset of exophthalmos and diplopia due to a cavernous fistula located in the right sinus, supplied mainly by cavernous branches of both ICAs. There was a minimal involvement of the ECA. The drainage occurred anteriorly in the superior ophthalmic vein and further in the facial vein. This venous route was used to occlude the fistula with coils. (**a**) Right carotid angiogram, AP view. Cavernous sinus (CS), superior ophthalmic vein (SOV), facial vein (FV). (**b**) Left carotid angiogram, AP view. Left cavernous branches of the ICA supplying the fistula (*arrow*). Right cavernous sinus (CS), right superior ophthalmic vein (SOV), facial vein (FV). (**c**) Injection of contrast medium into the ophthalmic vein, near the cavernous sinus, during selective catheterization with microcatheter (*arrows*). The small arrow shows the distal microcatheter during its progress. (**d**) Right and left carotid angiogram posttreatment

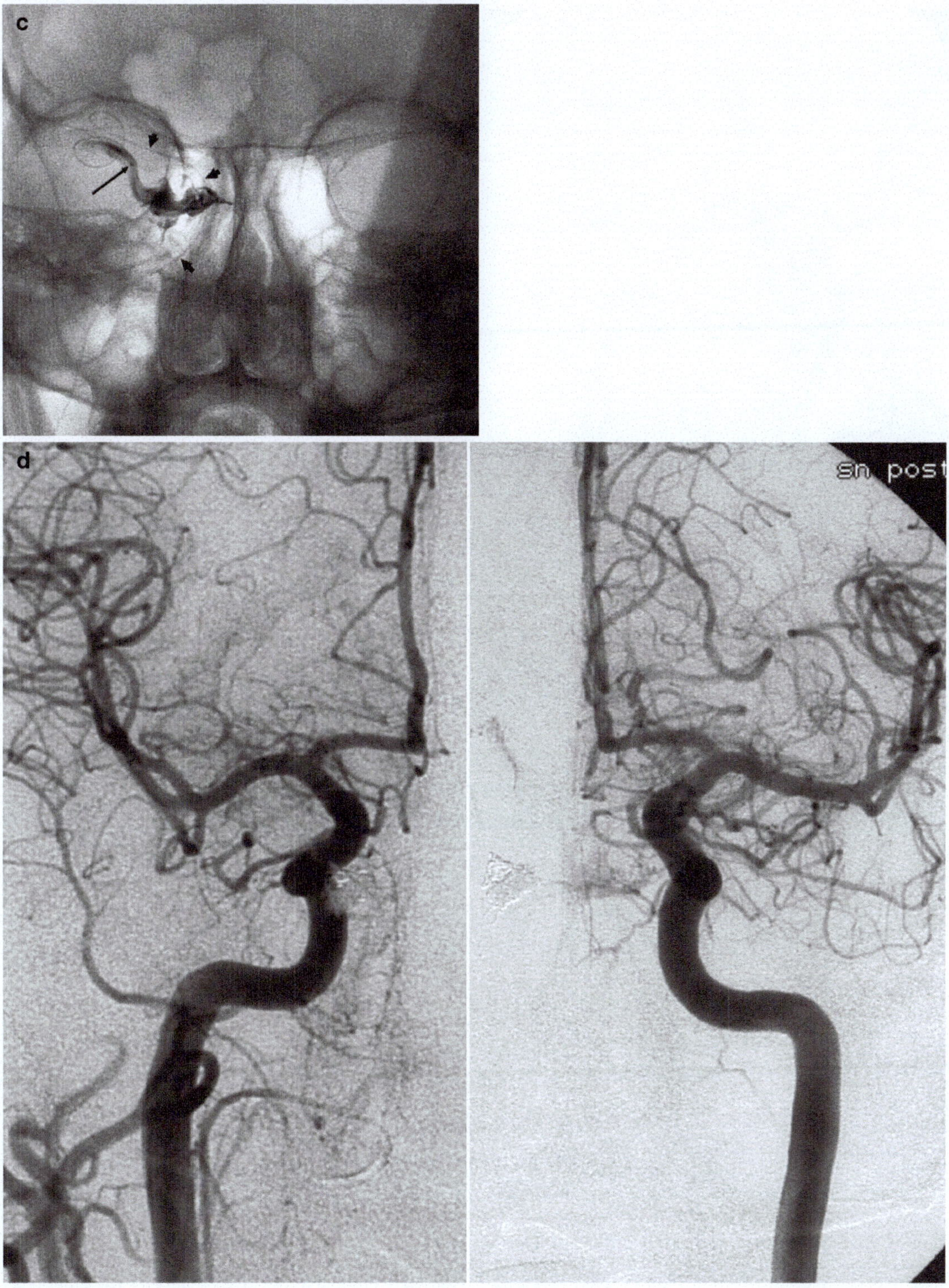

Fig. 13.7 (continued)

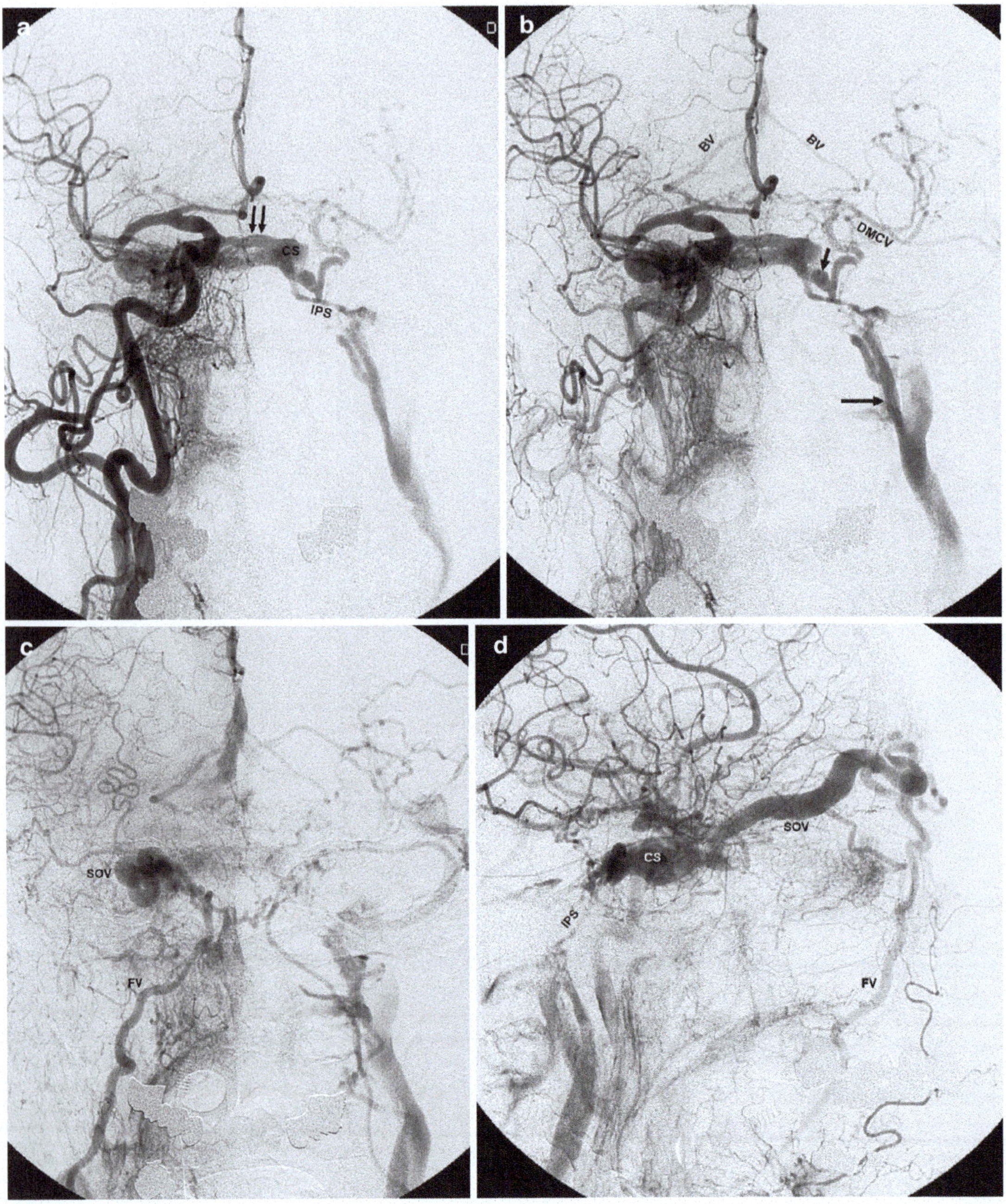

Fig. 13.8 Large DAVF involving the right CS in an old woman presenting with bilateral exophthalmos and diplopia. Several branches of IMA, AphA, and intracavernous branches of ICA bilaterally are involved. The patient has already been treated with occlusion of ECA branches with PVA. Reappearance of symptoms after a short period of clinical improvement. (**a**), (**b**), (**c**) Right ICA angiogram, AP view. There is immediate filling of the right CS and of the superior ophthalmic vein. Through intercavernous anastomoses (*double arrows*), there is a rapid filling of the left CS, then of the inferior petrosal sinus (IPS). This latter does not enter the jugular bulb but reaches the IJV more distally (*arrow*). Through a large connections between the left CS and the paracavernous sinus (*arrow*) there is a retrograde filling of the dilated uncal vein and further of the deep middle cerebral vein (DMCV) with its insular tributaries and of the basal vein (BV) bilaterally. On the right through the retrograde filling of the superior ophthalmic vein (SOV), filling of the facial vein (FV). (**d**) Right ICA angiogram. Lateral view, showing the drainage of the CS in the right ophthalmic vein (SOV) and facial vein (FV) and in the left IPS. (**e**) Selective catheterization of the right CS, advancing the microcatheter first in the right IJV, then in the facial vein and finally in the SOW. Occlusion of the fistula by placing coils in the right CS

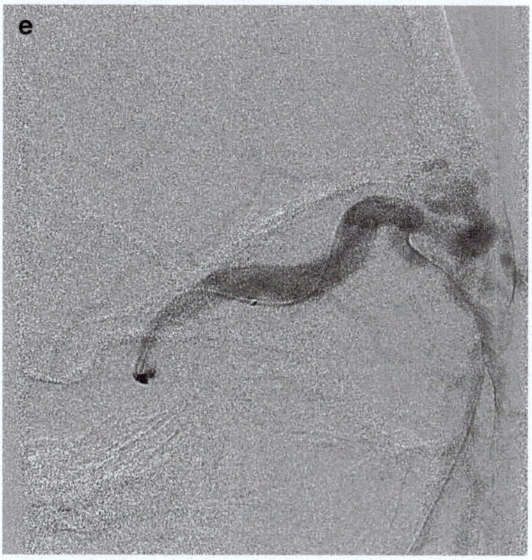

Fig. 13.8 (continued)

et al. 2004; Kurata et al. 2012). Such a situation occurs when the two compartments do not communicate to each other or the links are minimal. A dominant drainage can also occur when one of the routes (ophthalmic veins or IPS, Fig. 13.6) is occluded by thrombosis, which leads to a rerouting of the drainage. In other cases the two compartments, probably, are largely in communication and so both draining patterns are present.

Since the CSs are connected by a large intercavernous anastomotic channel, the contralateral sinus is also frequently involved.

Furthermore, as it has already been described in Sect. 9.3.10, the CS is connected with several veins of the brain parenchyma which can be indirectly involved in the cases of CS fistula (Figs. 13.6 and 13.8). As described below, various patterns can develop.

The superficial and deep middle cerebral as well as the uncal veins can have different types of drainage, but frequently drain into the CS, and so they can be retrograde filled in the case of CS fistula.

Involvement of the deep middle cerebral vein, which is the most important tributary of the basal vein (BV), can lead to a retrograde injection of the basal vein.

The BV can also be involved, considering that the CS is linked through bridging veins with the pontomesencephalic veins connected with the peduncular veins tributaries of the BV.

The drainage in the SPS can lead to a retrograde injection of the petrosal vein, which is connected with the lateral mesencephalic vein and further with posterior mesencephalic or BV.

To the petrosal vein converge also many cerebellar veins which can be indirectly involved.

All these possible pathological venous drainages explain the various clinical symptoms, which are commonly proptosis, chemosis, nerve palsy, and pain (Vinuela et al. 1984; Cheng et al. 1999). Glaucoma and involvement of visual acuity can also occur probably due to impairment of the venous drainage of the central retinal vein. Involvement of the veins of the brain parenchyma can lead to a venous congestion with ischemia or hemorrhage in the temporal area, basal ganglia, brainstem, and cerebellum (Takahashi et al. 2001; Suh et al. 2005; Kim et al. 2006; Kiyosue et al. 2008; Miyamoto et al. 2009). It should furthermore be mentioned that all these clinical symptoms can be present in all DAVFs involving the CS, primarily or secondarily.

There is a rich literature dedicated to the endovascular treatment of these fistulae, in which the arterial or venous route can be chosen (Halbach et al. 1989; Quinones et al. 1997; Cheng et al. 1999; Benndorf et al. 1999, 2001a; Agid et al. 2004; Satomi et al. 2005; Kirsch et al. 2006; Kato et al. 2007; Yung et al. 2011; Yu et al. 2011). Examples of endovascular treatment are presented in Figs. 13.6, 13.7, and 13.8. The treatment is indicated in patients with severe clinical presentation and in cases with involvement of veins of cerebral parenchyma. In the cases in which the symptoms are mild and there is not a large shunt, a conservative therapy can be chosen, considering also that a possible spontaneous occlusion and/or improvement of the symptoms can occur (Satomi et al. 2005; Bujak et al. 2010; Kim et al. 2010; Choi et al. 2015). However, also a clinical worsening and changes in the venous drainage especially with involvement of brain parenchyma veins can occur (Choi et al. 2015). A close control for these untreated patients is mandatory.

- *DAVFs in the middle cranial fossa.* This is a very rare location of fistula, involving typically the venous channels forming the paracavernous sinus (PCS) to which can converge the SMCV (directly or through the SpS), the DMCV, the uncal vein, and occasionally also the ophthalmic veins (see Sects. 9.1.1.2, 9.3.9, 9.3.10, and 9.3.11). Therefore it is not surprising that the venous drainage can present with various patterns. The fistula is supplied in various combinations by branches arising from ECA (AphA, MMA, AMA, foramen rotundum artery) and from ICA (MHT, ILT). They present frequently with hemorrhage.

Examples are presented in Figs. 13.9 and 13.10. In the case presented in Fig. 13.9, the involvement of the paracavernous sinus led to a retrograde injection of the SpS and of the SOV, with clinical symptoms mimicking a cavernous sinus fistula. It is thinkable that this type of fistula is due to the preexistent anomalous drainage of the SOV in the paracavernous sinus, instead of draining in the cavernous sinus. A few cases of this kind of fistula frequently traumatic are reported in the literature (Pakarinen 1965; Theron et al. 1975; Bradac et al. 1981b; Freckmann et al. 1981; Unterhofer et al. 2009; Shi et al. 2013).

Another example of fistula in the middle cranial fossa presenting with hemorrhage is shown in Fig. 13.10. the fistula involved the SpS in its course on the floor of the middle cranial fossa towards the TS. The SpS was partially thrombosed with reflux in the SMCV.

- *DAVFs of the SSS* occur with a frequency of about 7% of all DAVFs (Awad 1993). They can be of type 1, but it is not seldom to be of types 2 and 3. In the type 2, the venous reflux can be extensive and involve a large part of the cortical veins of the hemisphere, leading to an extensive venous congestion. In type 3 the drainage in the sinus is mediated by dilated cortical veins, which enter the sinus after a fairly long, tortuous course. Branches of the ECA, mainly the MMA, are commonly involved, frequently bilaterally. Transosseous branches of the superficial temporal and occipital arteries can be present. The anterior meningeal artery, arising from the ophthalmic artery, and the meningeal branches of the VA may sometimes be involved. Epileptic seizures,

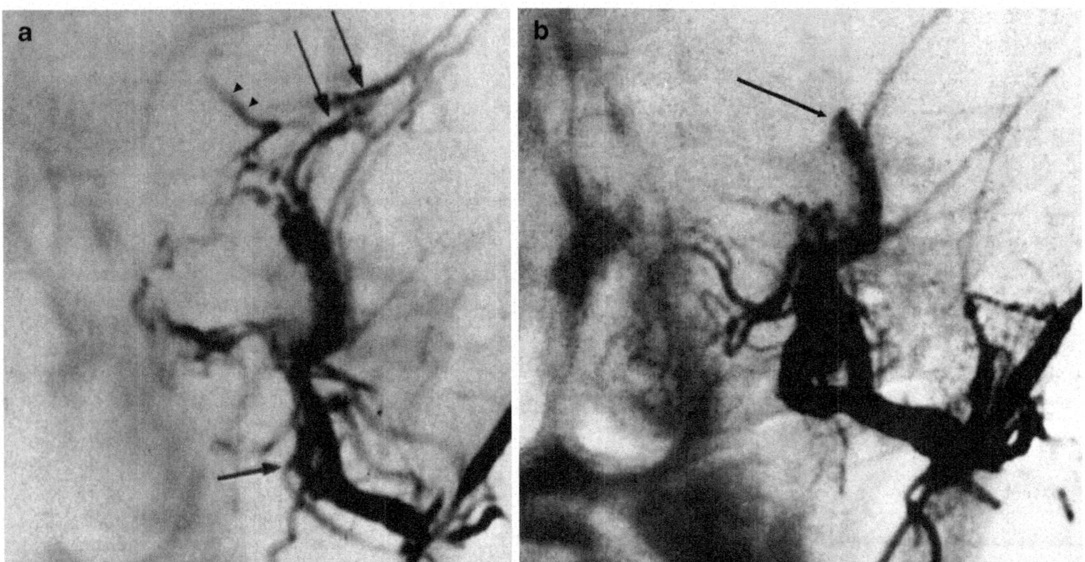

Fig. 13.9 Fistula in the middle cranial fossa developing a few days after a blunt head trauma, presenting clinically with symptoms suggesting a cavernous sinus fistula. ICA angiogram was normal. (**a**) The selective angiogram of the middle meningeal artery showed a fistula involving a network of veins (*arrow*) in the middle cranial fossa, corresponding to dilated veins of the paracavernous sinus. The drainage continued in the sphenoparietal sinus (*arrows*) and in the superior ophthalmic vein (*arrowheads*). The CS was not involved. (**b**) Control angiogram after occlusion (*arrow*) of the fistula with PVA

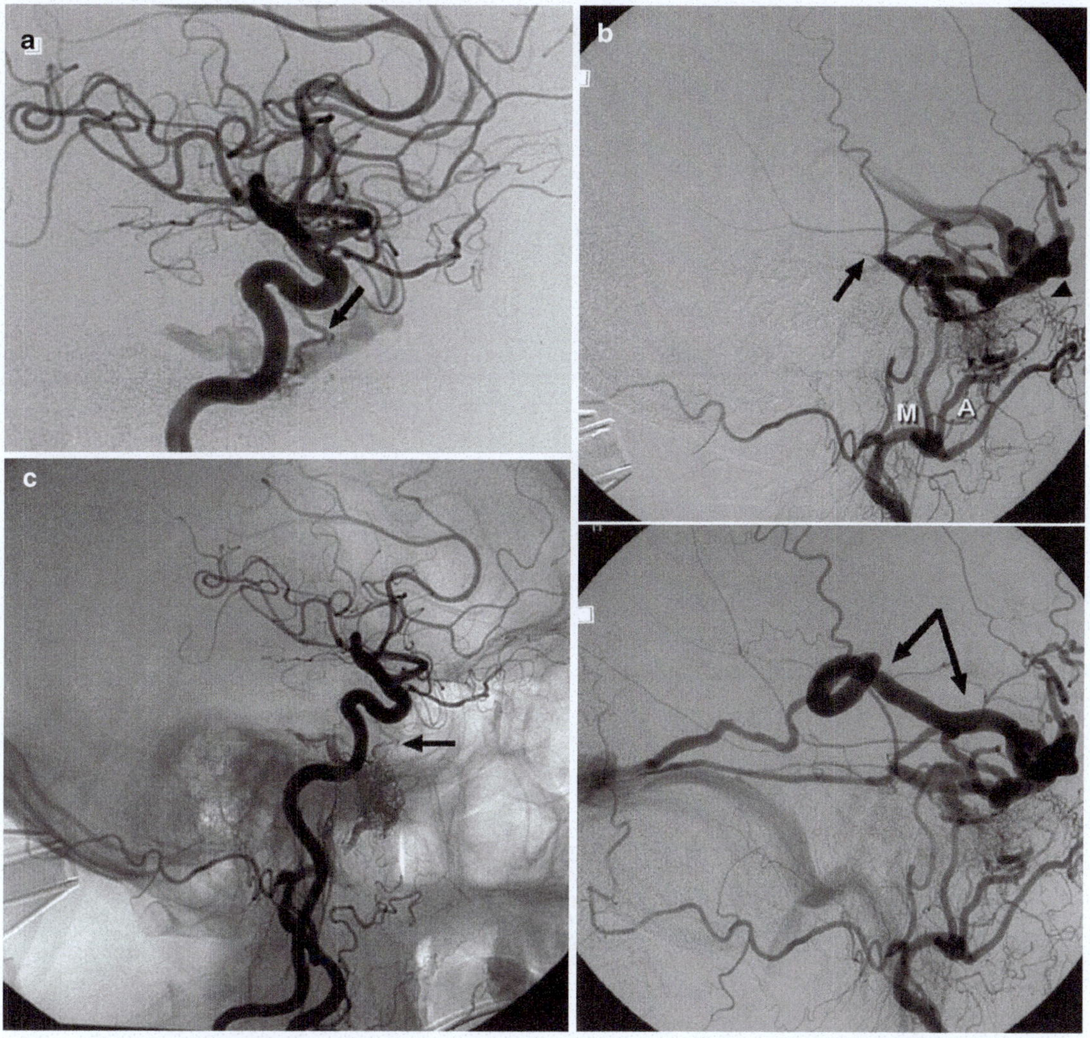

Fig. 13.10 Fistula in the middle cranial fossa presenting with hemorrhage. (**a**) Lateral ICA angiogram showing the dilated ILT (*arrow*). (**b**) ECA angiogram, early and later phases. The middle meningeal (M), the accessory meningeal (A), and the forum rotundum (*arrowhead*) arteries supply the fistula. The shunt seems to involve predominantly the SpS in its course on the floor of the middle cranial fossa draining in this case in the TS through tentorial channels. One of these is thrombosed (*arrow*). There is a retrograde filling of the SMCA (*arrow with angle*). A treatment with onyx injected in the middle meningeal artery was performed. *Considering the risk of a retrograde injection of the embolic material in the ILT and in the ICA, several angiographic controls of the ICA were performed during the treatment.* (**c**) Common carotid angiogram post treatment showing the occlusion of the fistula. Note the small ILT (*arrow*)

hemorrhage, and/or clinical signs of intracranial hypertension, such as headache and cognitive disorders, are present (Halbach et al. 1988c; Riva et al. 1991). Surgical disconnection of the shunt or selective embolization with acrylic glue or onyx may be effective treatments (Collice et al. 2000; Van Dijk et al. 2004; Rodesch et al. 2009). Examples are presented in Figs. 13.11 and 13.12.

- *Tentorial DAVFs* occur with a frequency of about 8%. They are commonly of type 3, characterized by an high hemorrhagic risk (Picard et al. 1990; Awad et al. 1990; Awad 1993; King and Martin 1992; Cognard et al. 1995; Deasy et al. 1999; Tomak et al. 2003; Seong et al. 2006; Jiang et al. 2009). Tentorial DAVFs can be schematically divided into an anterior and posterior group. Those of the *anterior*

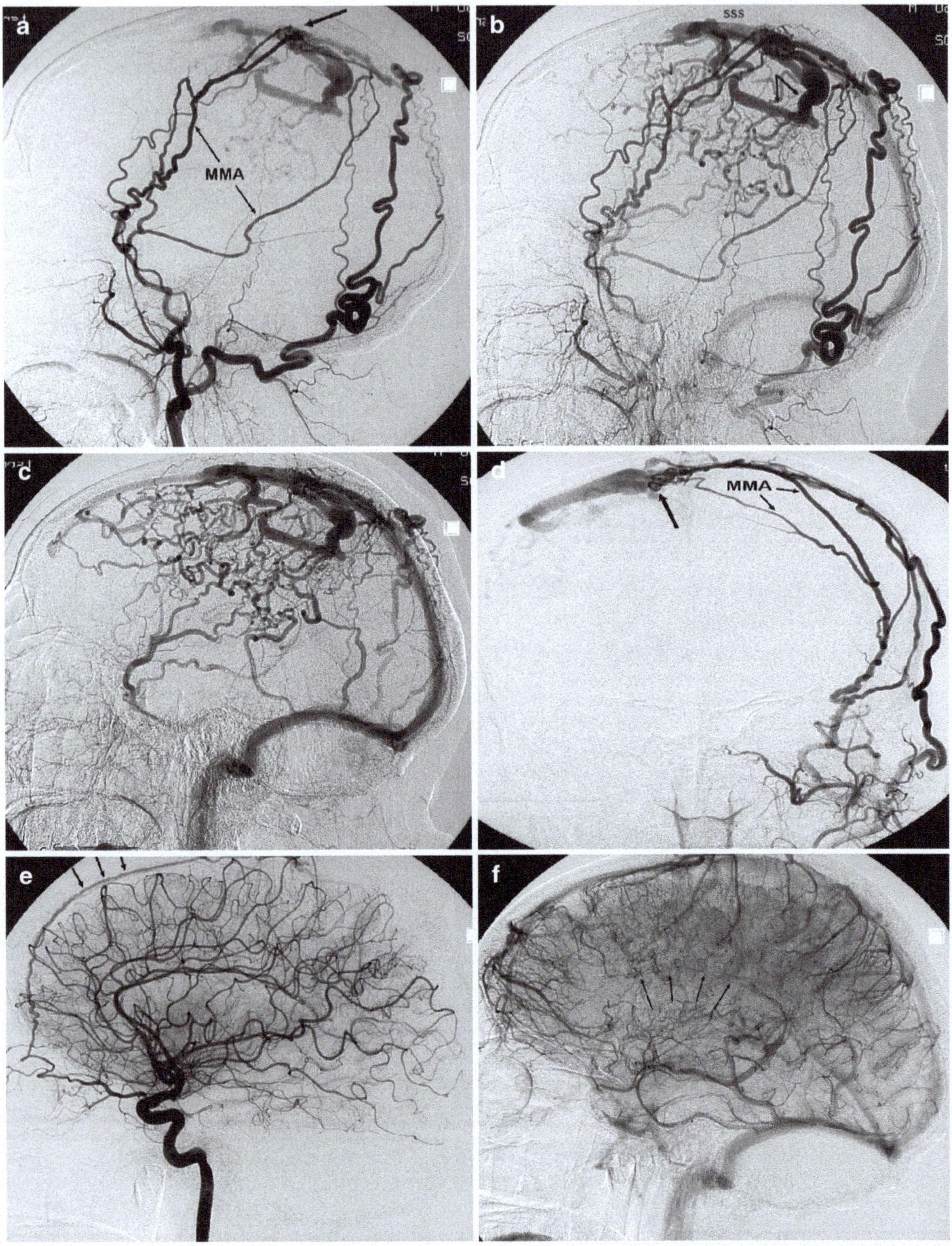

Fig. 13.11 Middle-aged patient presenting with increased headache and suspected epileptic seizures. The angiographic study shows DAVF located in the right parietal area, close to the SSS. (**a**), (**b**), (**c**) Right ECA angiogram showing the DAVF supplied by branches of both middle meningeal artery (MMA) and by transosseous branches of the occipital artery. Site of the shunt (*arrow*). The drainage occurs in the enlarged cortical veins (*arrow with angle*), which after a tortuous course enter the superior sagittal sinus (SSS). There is an extensive retrograde injection of the cortical veins of the right hemisphere. (**d**) Left ECA angiogram, oblique view. Involvement of the MMA (*arrows*). Site of the shunt (*arrow*). (**e**) On the ICA angiogram, a minimal contribute through the anterior meningeal artery (*arrows*) is visible. (**f**) In the venous phase the venous congestion demonstrated in (**a**), (**b**), (**c**) explains the poor filling of the frontoparietal veins (*arrows*). The patient underwent surgical treatment

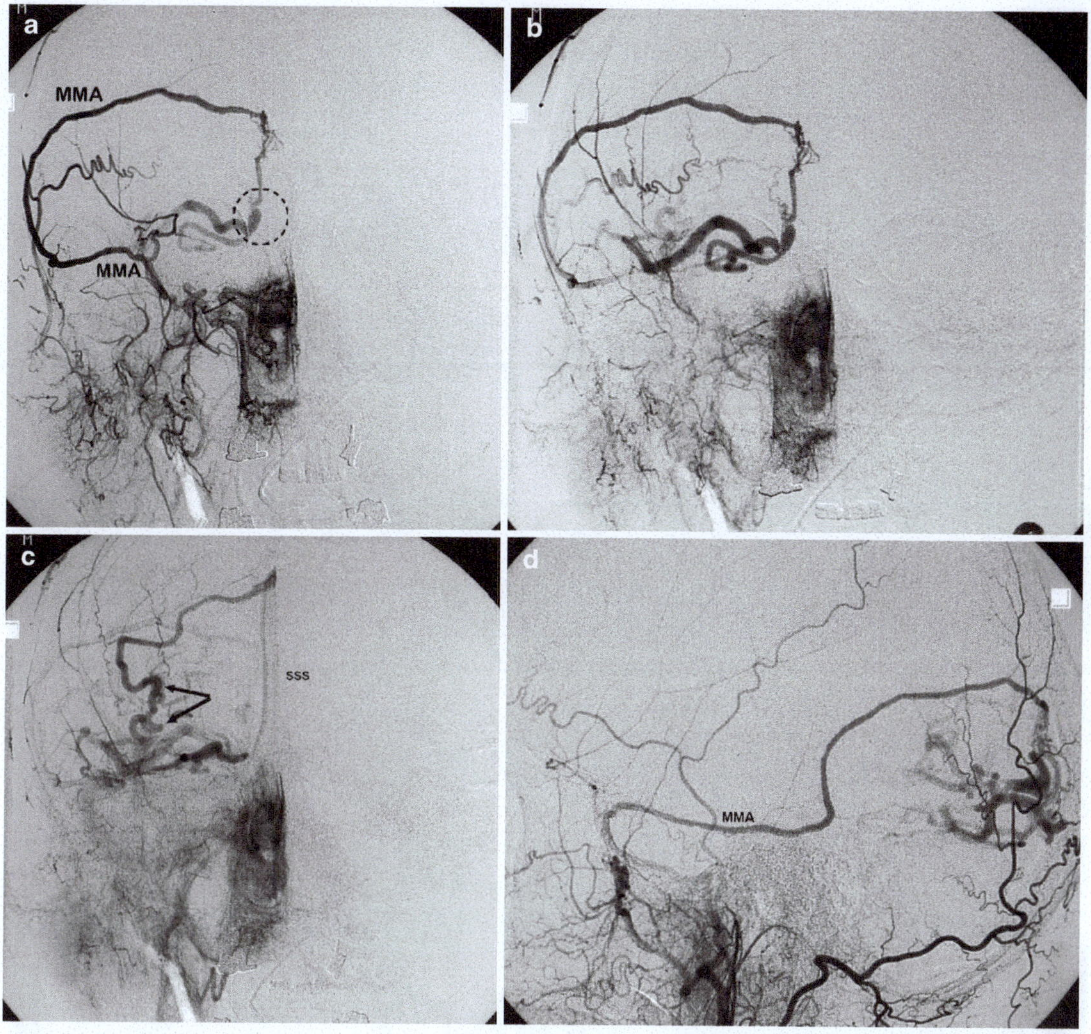

Fig. 13.12 Old patient suffering from acute headache. CT showed a suspected vascular malformation in the right occipital area. ICA angiogram was normal. (**a**), (**b**), (**c**) Angiogram of right ECA. AP view. There is a large middle meningeal artery (MMA) supplying the fistula in the paramedian right occipital region. Site of the shunt (*circle*). The venous drainage occur through cortical veins (*arrow with angle*) which, after a long, tortuous course enter the superior sagittal sinus (SSS). (**d**), (**e**) ECA angiogram, followed by selective MMA study. Site of the shunt (*arrows*) and venous drainage. (**f**) Vertebral angiogram showing the involvement of the falx cerebellar artery (FCA). Endovascular treatment with injection of onyx in the MMA allowed complete occlusion of the fistula. Carotid (**g**) and vertebral (**h**) angiograms posttreatment

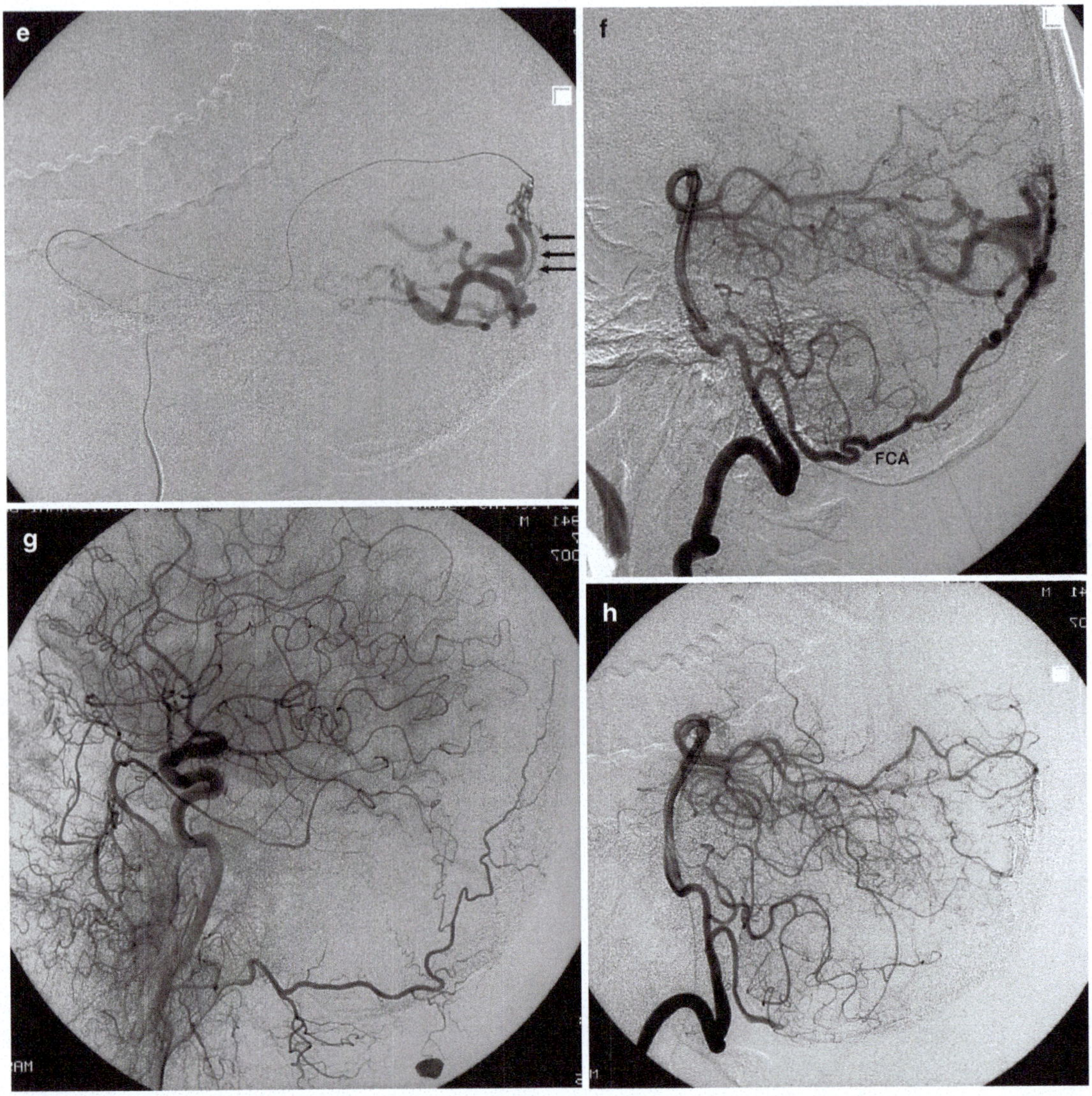

Fig. 13.12 (continued)

group, also called petrotentorial fistulas, are commonly located in the area corresponding to the attachment of the tentorium to the superior ridge of the petrous bone, involving the SPS which is commonly partially or completely thrombosed. The pattern of these fistulas is frequently (Figs. 13.13 and 13.14) characterized by many arterial tributaries represented in various combinations, by branches of ECA (petrosquamosal branch of the MMA, the AMA, the artery of the foramen rotundum, and branches of the occipital artery and AphA), cavernous branches of the ICA, leptomeningeal branches of the PCA, and cerebellar arteries. A rich network of vessels can be recognizable at the level of the superior ridge of the petrous bone converging to the thrombosed SPS and so the venous drainage occurs in the petrosal vein and continues retrogradely through the brachial veins to the dilated lateral mesencephalic vein, then to the basal vein or posterior mesencephalic vein, and further to the Galen vein and straight sinus. In some cases, the venous drainage can be directed more medially towards the pontomesencephalic veins and further through the peduncular vein in the basal vein (Fig. 13.15).

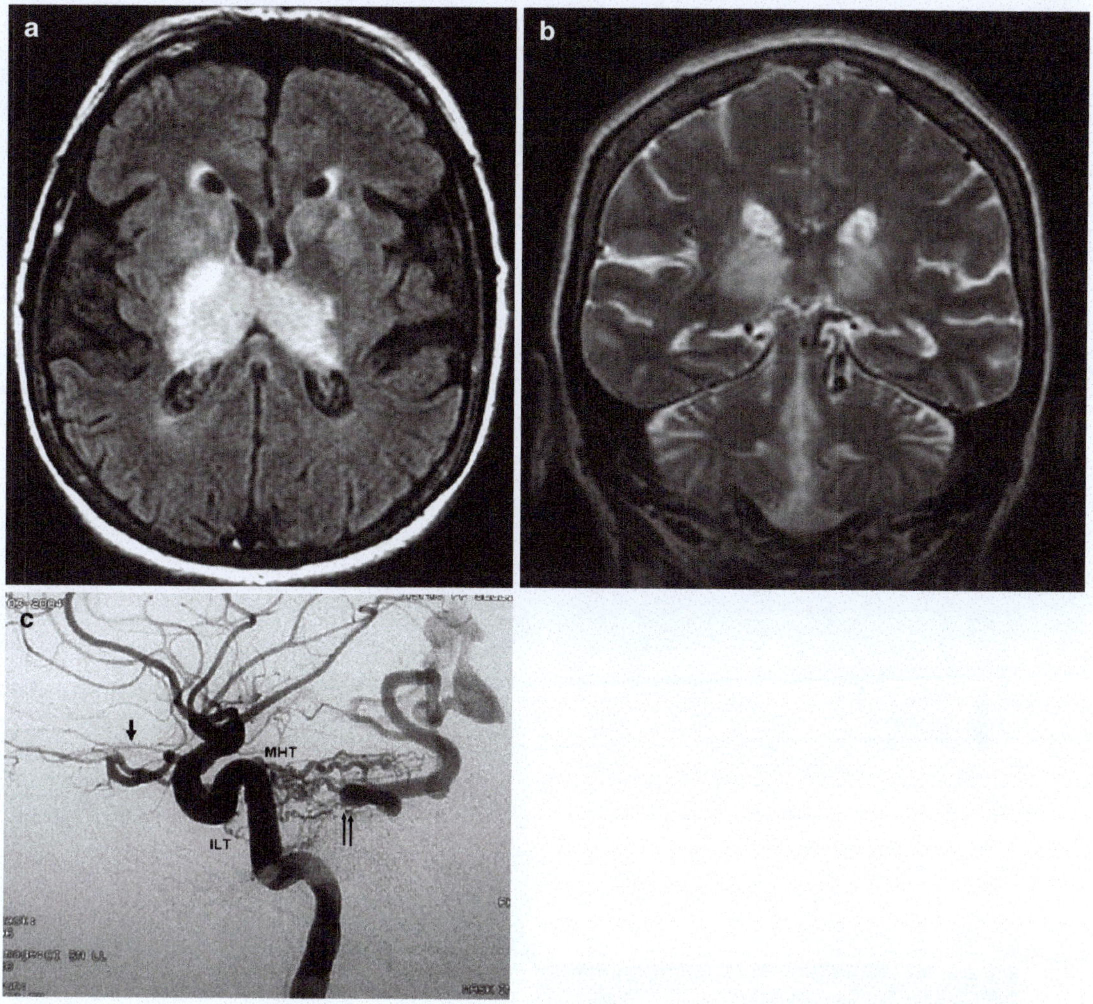

Fig. 13.13 Petrotentorial DAVF in a patient with slowly progressive cognitive disorders. (**a**), (**b**) Axial MR Flair image and coronal T2-weighted image showed nonspecific hyperintensity involving both thalami and partially both basal ganglia. On the coronal image anomalous vascular structures lateral to the midbrain suggesting vascular malformation are recognizable. (**c**) Angiogram of ICA showing a DAVF with a shunt (*arrows*) in the area where the tentorium attaches to the superior ridge of the petrous bone, supplied by dilated branches of the MHT and ILT. A recurrent branch of the ophthalmic artery is also involved (*large arrow*). The venous drainage occurs through the petrosal vein in the lateral mesencephalic vein. (**d**) ECA angiogram. Lateral and AP views. The petrosquamous branch of the MMA (*arrowhead*), the accessory middle meningeal artery (*arrow*) and the mastoid branch of the occipital artery supply the DAVF. Projecting on the superior petrous sinus which is probably completely or partially thrombosed (*arrows*) there is a large petrosal vein continuing in the lateral mesencephalic vein (LMV) and further into the posterior mesencephalic vein entering the Galen vein. The straight sinus (SS) is thrombosed and so the venous drainage is rerouted anteriorly in the contralateral basal vein (BV) and further in the deep (*arrow with dot*) and in the superficial (*arrows with dot*) middle cerebral vein. (**e**) Selective study (AP and lateral views) of the MMA better showing the shunt (*arrow*). The fistula was occluded with acrylic glue. (**f**) MRI Flair image performed a few weeks later showing the progressive reduction of the hyperintensity of the thalami. Gradually, clinical improvement of the patient

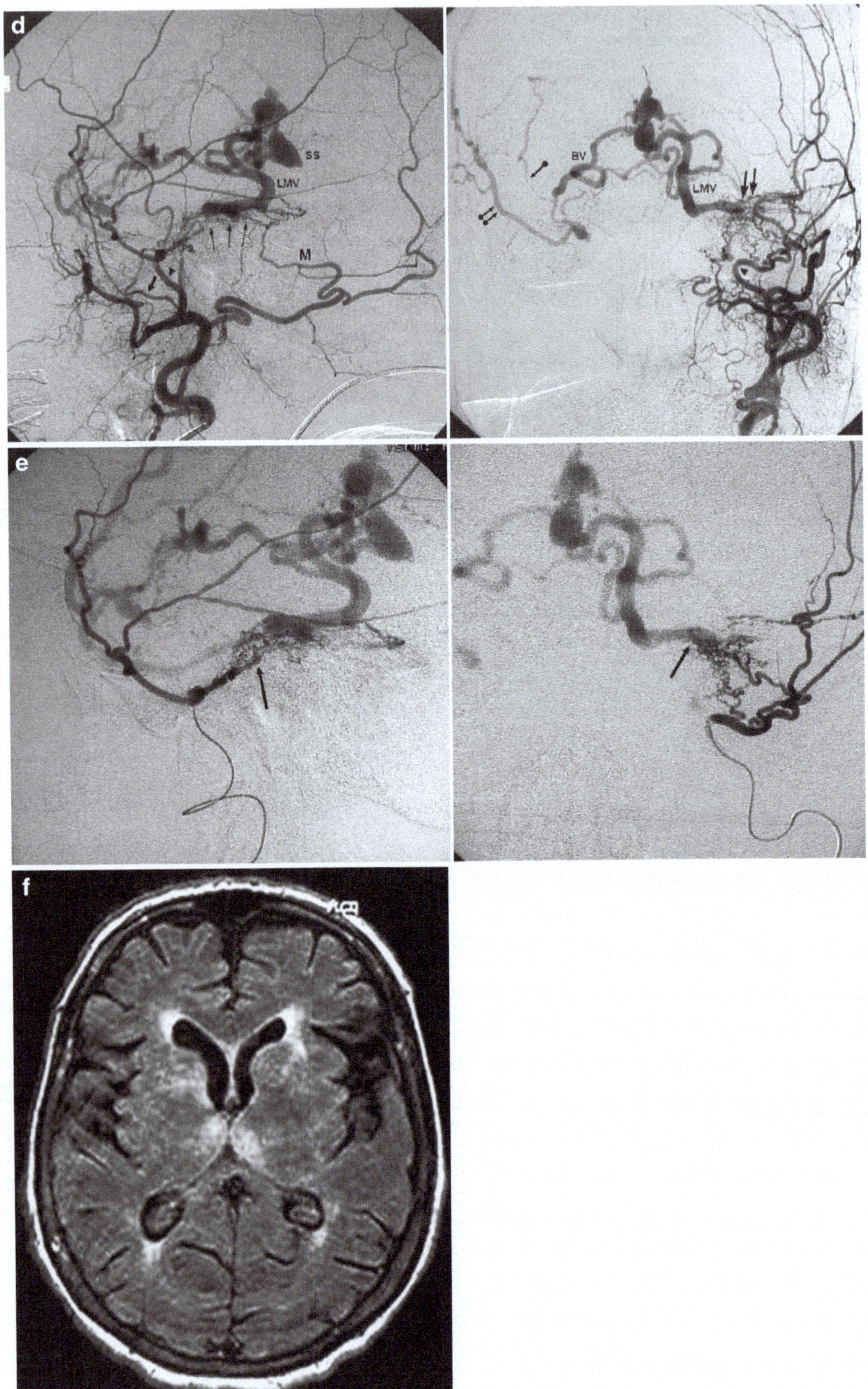

Fig. 13.13 (continued)

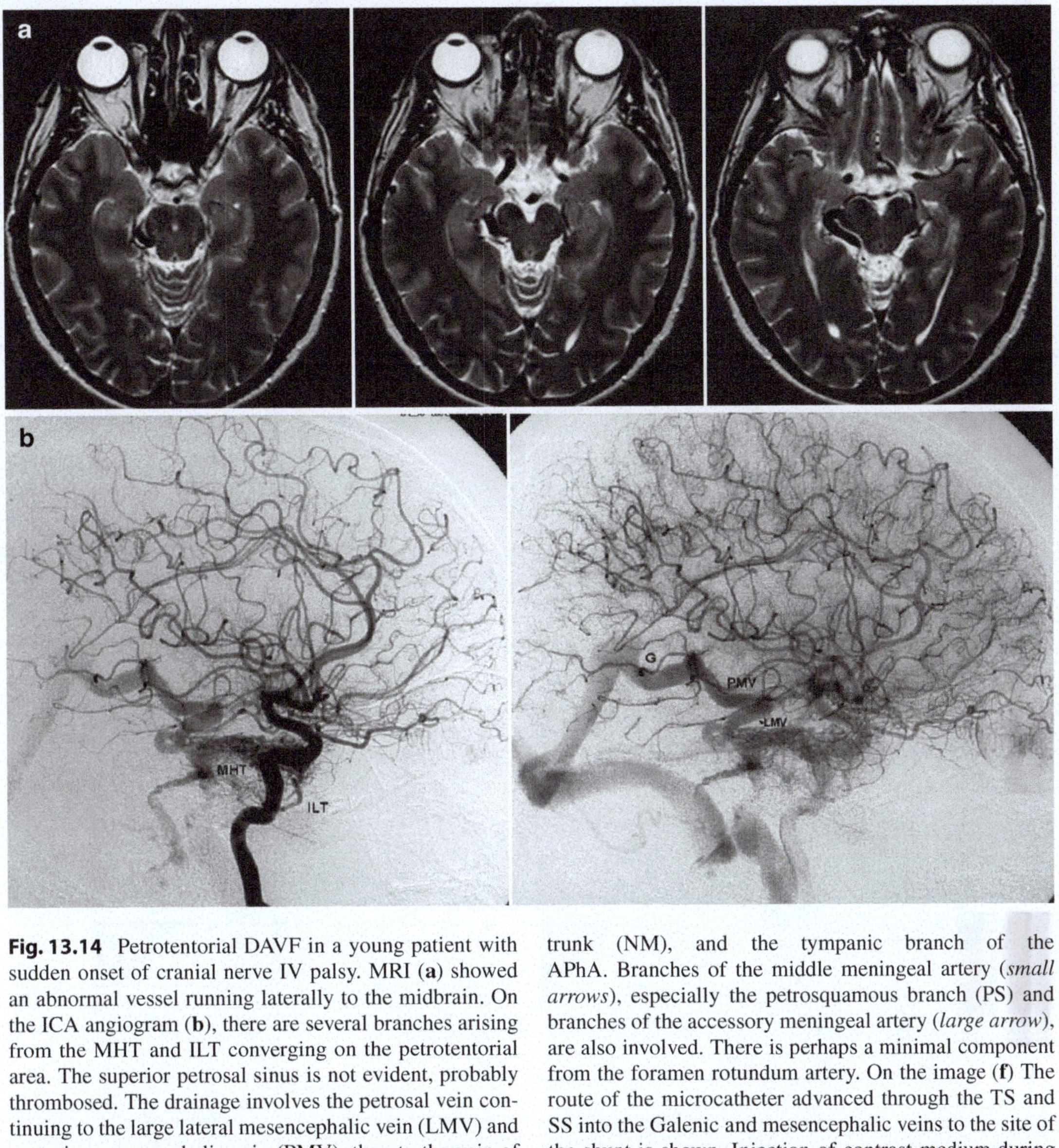

Fig. 13.14 Petrotentorial DAVF in a young patient with sudden onset of cranial nerve IV palsy. MRI (**a**) showed an abnormal vessel running laterally to the midbrain. On the ICA angiogram (**b**), there are several branches arising from the MHT and ILT converging on the petrotentorial area. The superior petrosal sinus is not evident, probably thrombosed. The drainage involves the petrosal vein continuing to the large lateral mesencephalic vein (LMV) and posterior mesencephalic vein (PMV), then to the vein of Galen (G) and straight sinus. A minimal drainage is directed towards the transverse sinus. Other feeders are (**c**), (**d**), and (**e**), the mastoid branch (M) of the occipital artery, the clival branches (*arrow*) of the neuromeningeal trunk (NM), and the tympanic branch of the APhA. Branches of the middle meningeal artery (*small arrows*), especially the petrosquamous branch (PS) and branches of the accessory meningeal artery (*large arrow*), are also involved. There is perhaps a minimal component from the foramen rotundum artery. On the image (**f**) The route of the microcatheter advanced through the TS and SS into the Galenic and mesencephalic veins to the site of the shunt is shown. Injection of contrast medium during treatment characterized by placement of coils. Angiogram of the ICA and ECA posttreatment (**g**) showing occlusion of the fistula

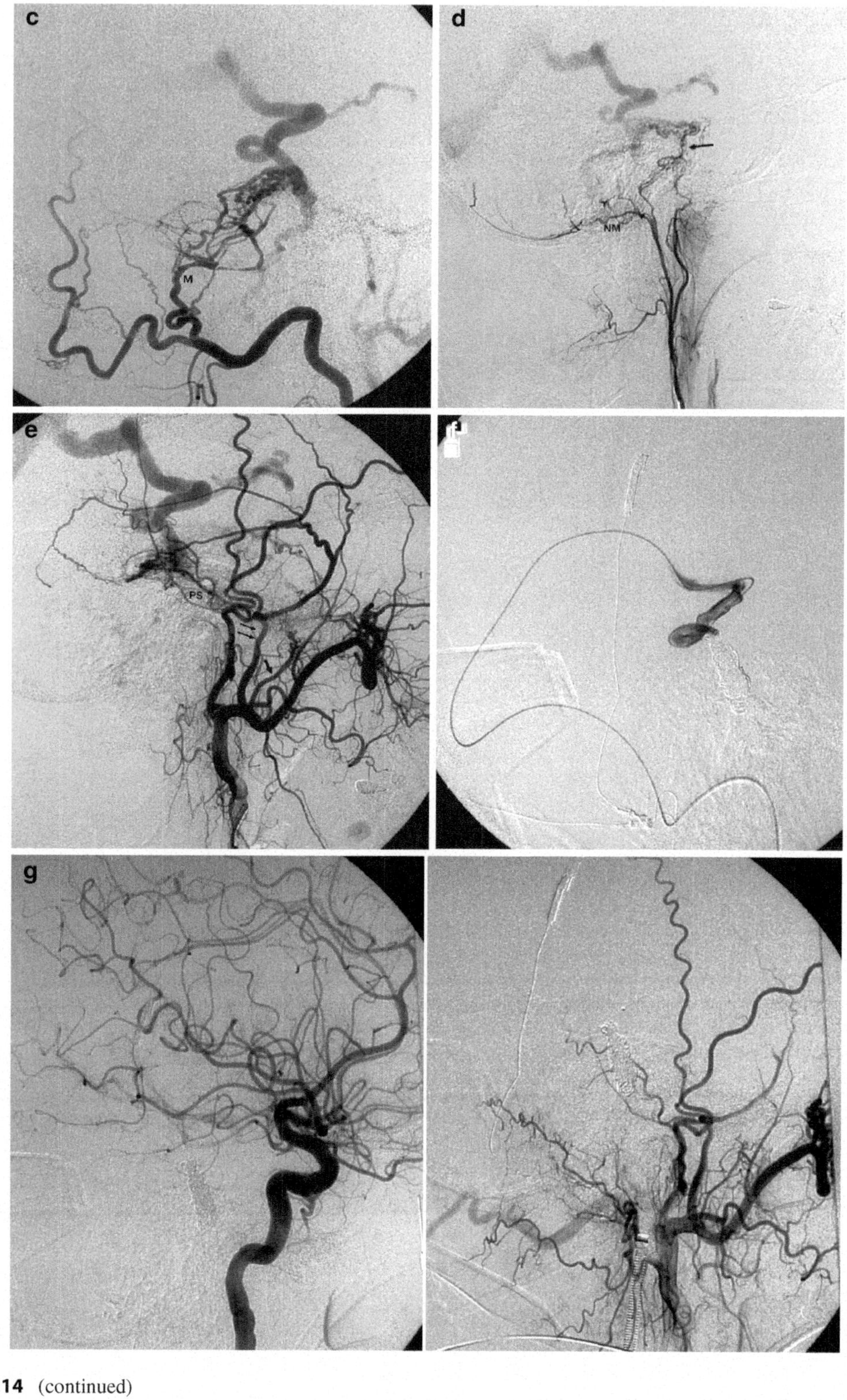

Fig. 13.14 (continued)

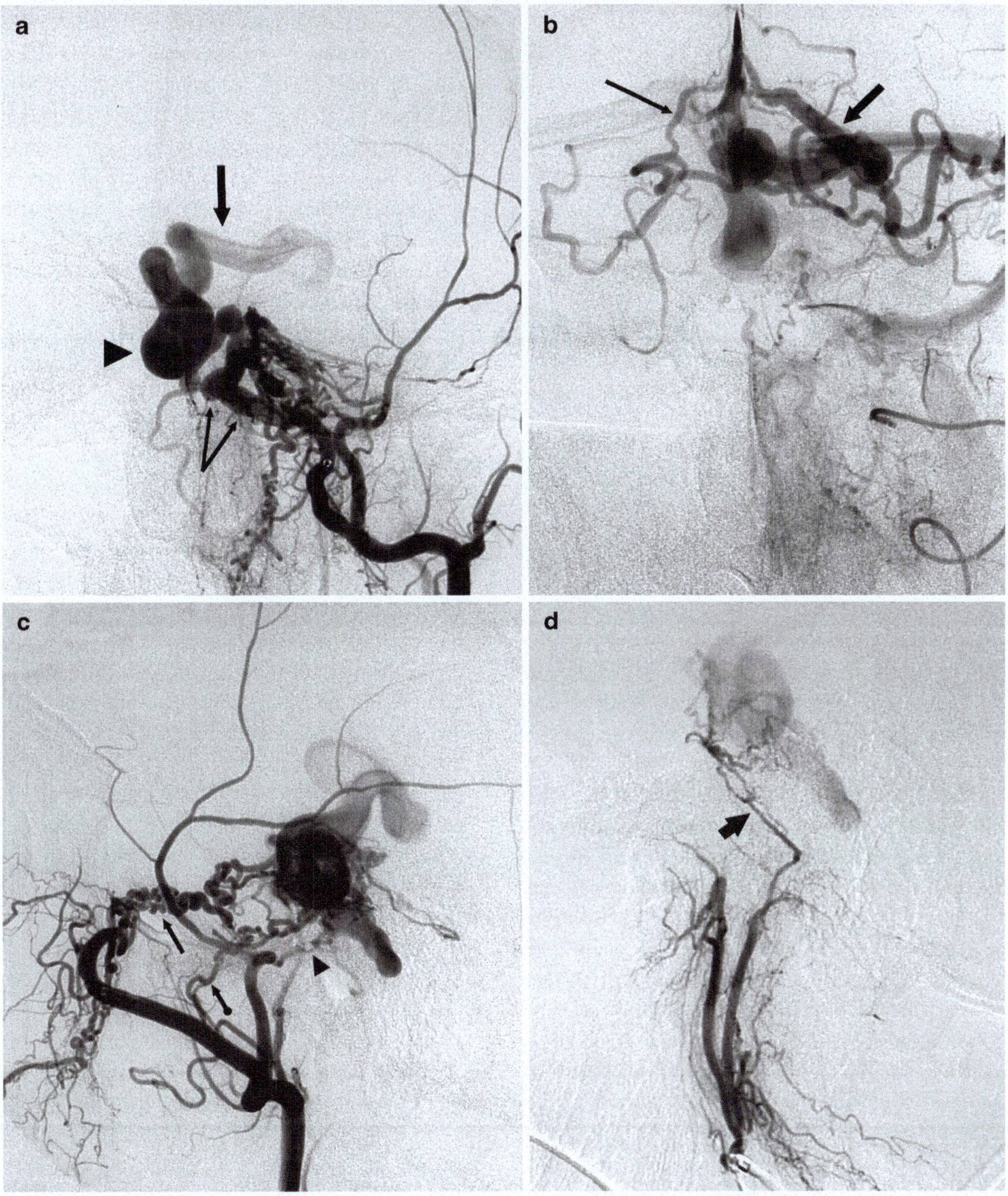

Fig. 13.15 Another example of petrotentorial fistula presenting with trigeminal neuralgia. (**a**), (**b**) Angiogram of the IMA. Earlier and later phases. AP view. Several dilated distal branches of IMA supply a pathological network projecting in the area of the medial petrous ridge draining in a dilated venous channel (*arrow with angle*) directed medially continuing in enormously dilated and tortuous probably pontomesencephalic veins (*arrowhead*). The drainage continues in the left peduncular vein (*arrow*). In the later phase the drainage in both basal veins, left (*arrow*) and right (*small arrow*) and further in the SS and left TS is recognizable. (**c**) IMA angiogram. Lateral view showing better the supplying branches: foramen rotundum artery (*arrow*), accessory meningeal artery (*arrow with dot*), and petrosquamous branch of the MMA (*arrowhead*). (**d**) Angiogram of the AphA which contributes with its clival branches (*arrow*). (**e**) ICA angiogram showing the involvement of the cavernous and petrous branches of ICA and also of the recurrent meningeal branch of the ophthalmic artery. Endovascular devascularization with PVAs followed occlusion of the fistula with coils after reaching the shunt through the venous route as shown in Fig. 13.14

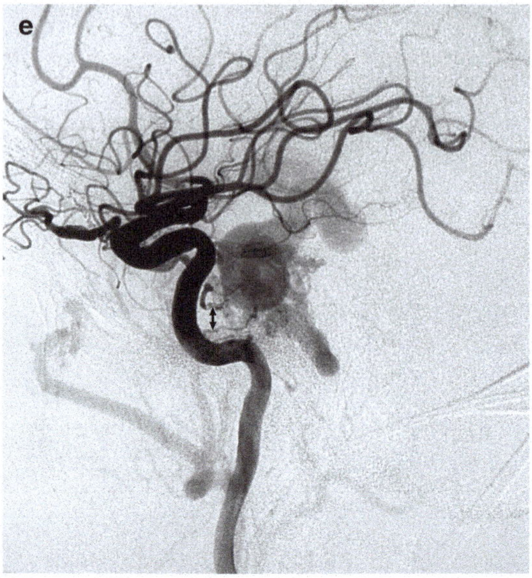

Fig. 13.15 (continued)

Tentorial DAVFs can be located more posteriorly along the tentorial incisura or in the area of the Galen vein. The supplying arteries are similar to those already described, but there is a greater involvement of the meningeal branches of the VA and more frequently of pial branches of the superior cerebellar and posterior cerebral arteries. Furthermore there is a frequent involvement of *a meningeal branch* arising from the posterior cerebral artery (see also Sect. 7.1). The precise point of the shunt is not always possible to be identified. The fistula can involve the Galen vein with further drainage in the SS. This channels, however, can be thrombosed and so the venous drainage is rerouted anteriorly with retrograde injection of the deep venous system ICVs and BVs. In other cases the shunt involves selectively the distal segment of the basal vein which is retrograde injected, while the Galen vein and the SS are patent.

Examples are presented in Figs. 13.16, 13.17, and 13.18.

Both surgical and endovascular treatment are difficult. With endovascular treatment good results can be obtained by injecting acrylic glue in the shunt reached through the feeding arteries. The venous route when possible can be an excellent alternative. The association of both technique can be useful in some cases.

- DAVFs of the *anterior fossa*, also called ethmoidal amount to around 6% of all DAVFs (Picard et al. 1987; Kobayashi et al. 1988; Cognard et al. 1995; Awad 1993): They are reported more frequently in males (Martin et al. 1990; Van Dijk et al. 2004). They are of types 3 and 4 and present commonly with hemorrhage (Ito et al. 1983; Martin et al. 1990; Awad 1993). The supplying arteries are the posterior and anterior ethmoidal arteries, branches of the ophthalmic artery. Among the anterior ethmoidal arteries, the anterior meningeal branch is often largely involved. This artery supplies normally the dura of the falx and adjacent dura of the frontal convexity. Since it anastomoses with the corresponding falx branch of the MMA, this latter may also be involved. The ethmoidal arteries anastomose with the ethmoidal arteries arising from the sphenopalatine branches of the IMA, which consequently become involved. Drainage is via the pial veins of the frontal convexity, which are frequently dilated, forming large venous pouches and entering then SSS. More rarely the drainage is directed posteriorly towards the cavernous sinus (Martin et al. 1990). The involvement of the ophthalmic artery and the drainage in the cavernous sinus, explain the impairment of the visual acuity and the ocular motility dysfunction. An example of this type of fistula is presented in Fig. 13.19.

  Surgical excision has been the therapy of choice. Due to the improvement of the endovascular technique with selective catheterization of the ophthalmic artery, followed by injection of glue or Onyx™, this approach has become a possible good alternative treatment as reported also by some authors (Lv et al. 2008; Agid et al. 2009).
- Another complex group of DAVFs is that *located at or close to the foramen magnum* (Barnwell et al. 1990, 1991; McDougall et al. 1997; Ernst et al. 1999; Miyachi et al. 2008; Abiko et al. 2008; Manabe et al. 2008; Choi et al. 2012). Pulse-synchronous tinnitus is a common clinical symptom. Palsy of the cranial nerves especially of the hypoglossal nerve can occur. Retrograde filling of the IPS

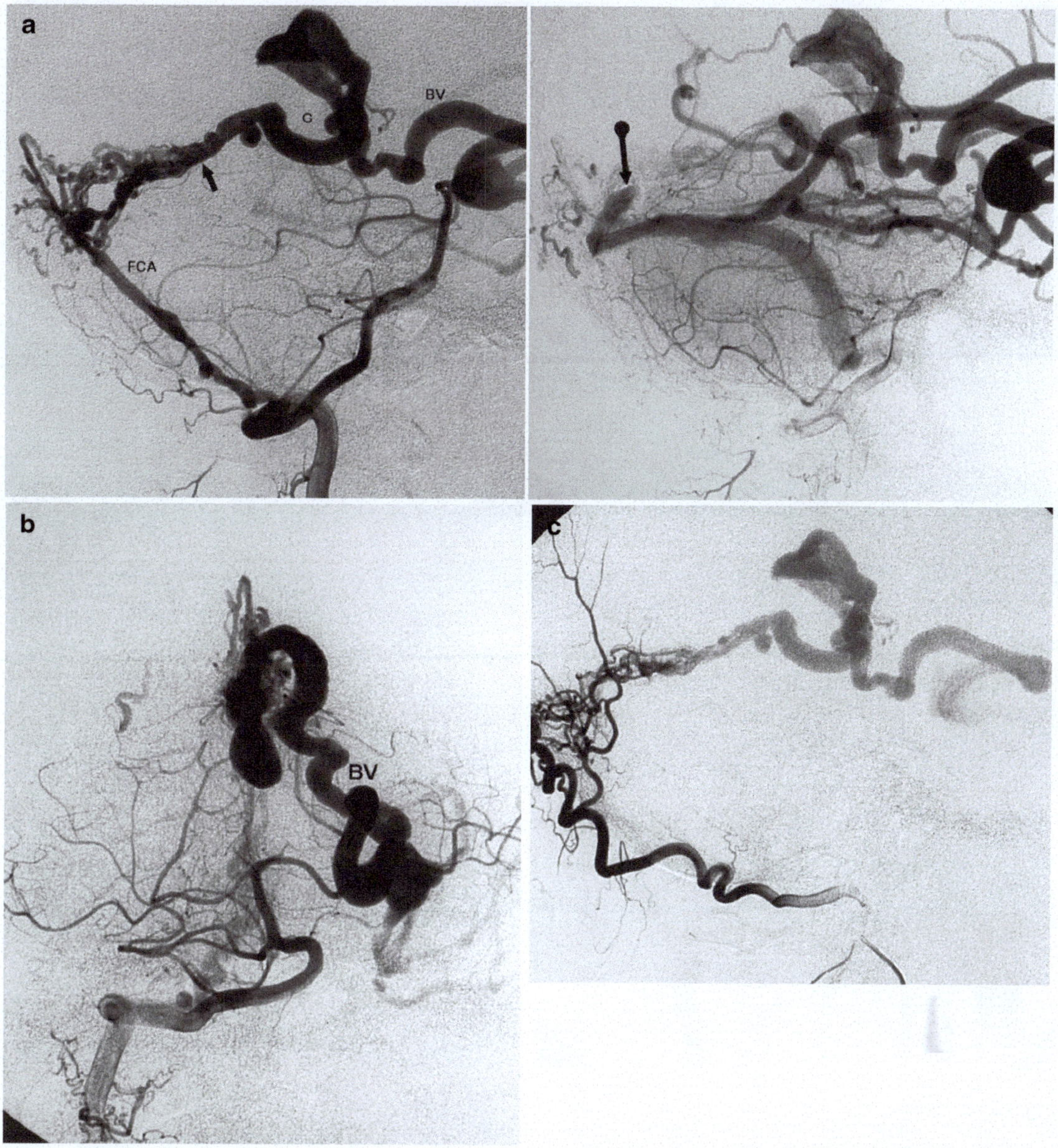

Fig. 13.16 Posterior tentorial DAVF in a patient suffering from headaches and cognitive disorder. The malformation was supplied mainly by the falx cerebelli artery (FCA) arising from the right VA. A small meningeal branch arose also from the left VA. Other feeders were the bilateral occipital and MMA arteries. The ICA was not involved. (**a**) Lateral right vertebral angiogram showing a dilated falx cerebelli artery (FCA), continuing into branches (*arrow*), running in the dura along the straight sinus, connected with the vein of Galen (G). Retrograde filling of the dilated tortuous basal vein (BV). This drains further into the temporal veins. In the later venous phase, the stump of the occluded straight sinus (*arrow with dot*) is recognizable. (**b**) Right vertebral angiogram (AP view) showing the drainage to the retrograde injected BV. (**c**) Left occipital artery and (**d**) selective left MMA supplying the DAVF. (**e**) ICA angiogram venous phase, showing the diffuse venous congestion due to impairment of the deep venous drainage. (**f**) Right vertebral angiogram after endovascular treatment. Acrylic glue was injected close to the shunt after selective catheterization of the right FCA and left MMA. The shunt is not completely occluded. The remaining supply (*arrow*) may be due to either pial branches of the SCA or a meningeal branch. A selective study was not performed. The patient improved and refused further treatment

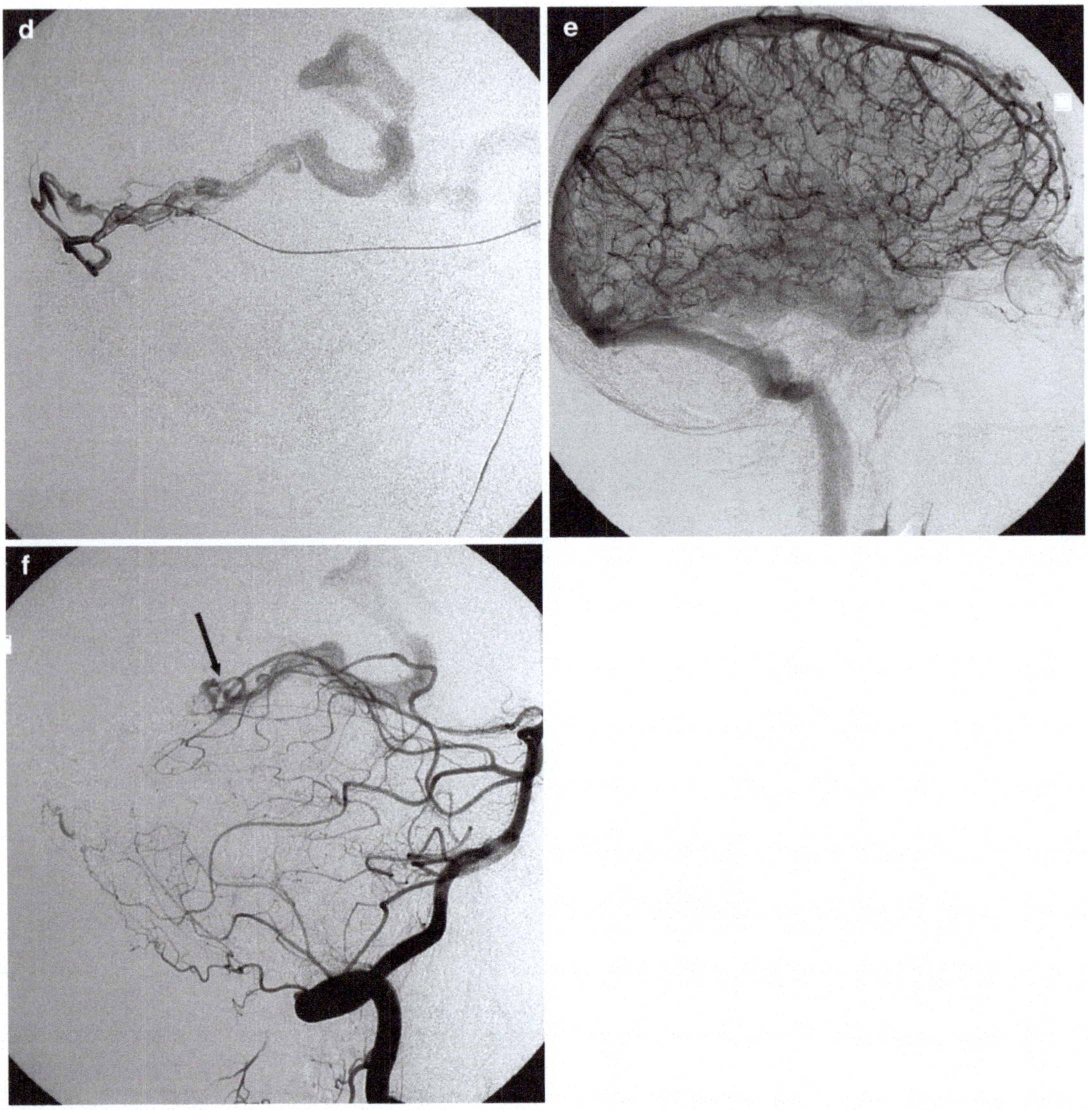

Fig. 13.16 (continued)

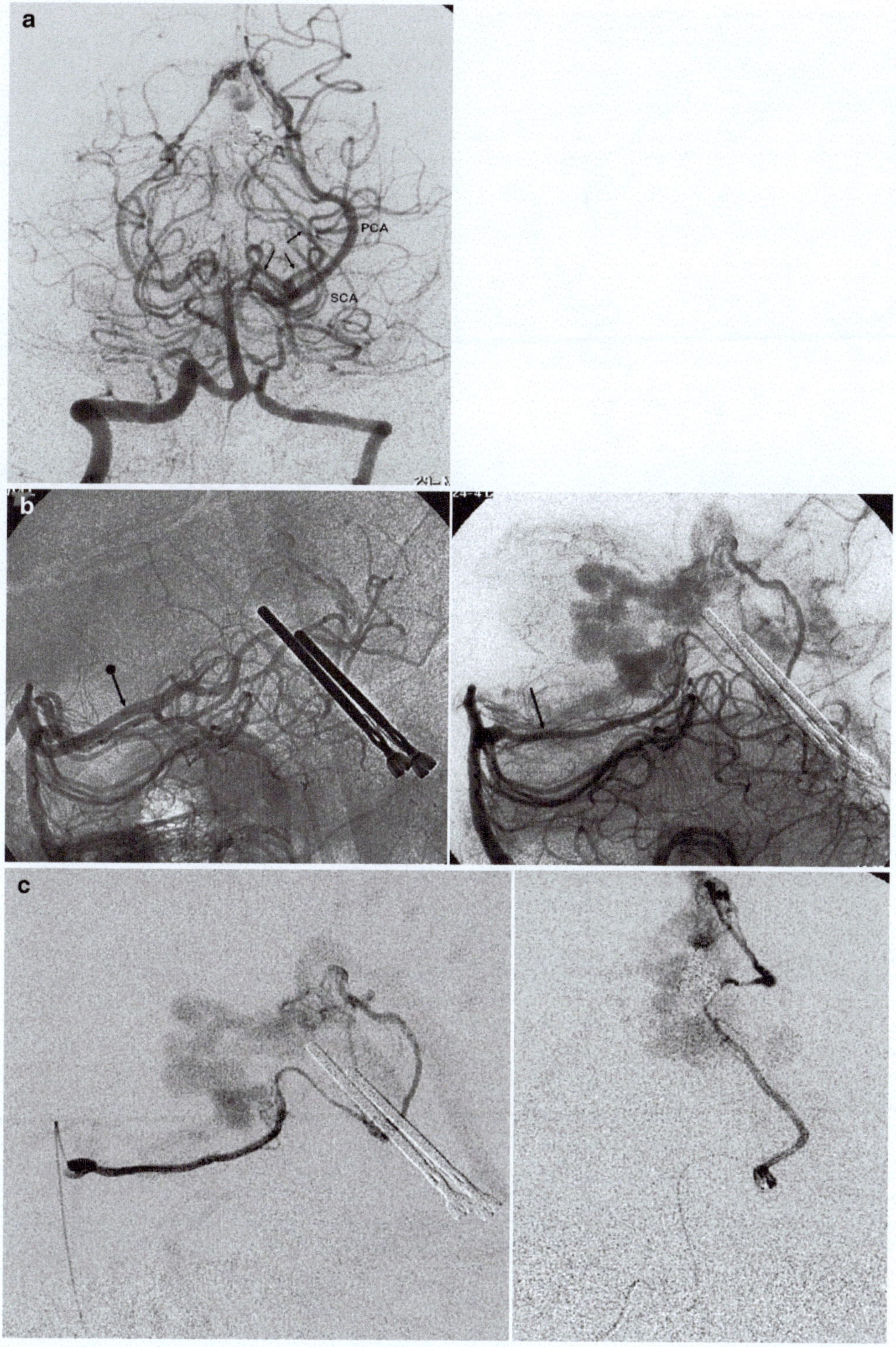

Fig. 13.17 Posterior tentorial DAVF. (**a**) Vertebral angiogram, AP view. Left posterior cerebral artery (PCA), superior cerebellar artery (SCA), meningeal branch (*arrows*) arising from the PCA close to its origin from the basilar artery. (**b**) Lateral vertebral angiogram. Owing to overlap with the PCA (*arrow with dot*), the meningeal branch (*arrow*) connected with the vein of Galen is better evident only later, when the contrast medium has left the PCA. Clips of previous surgery are present. (**c**) Selective catheterization of the meningeal branch, lateral and AP view, showing its supply of the DAVF. Injection of onyx followed. The patient recovered

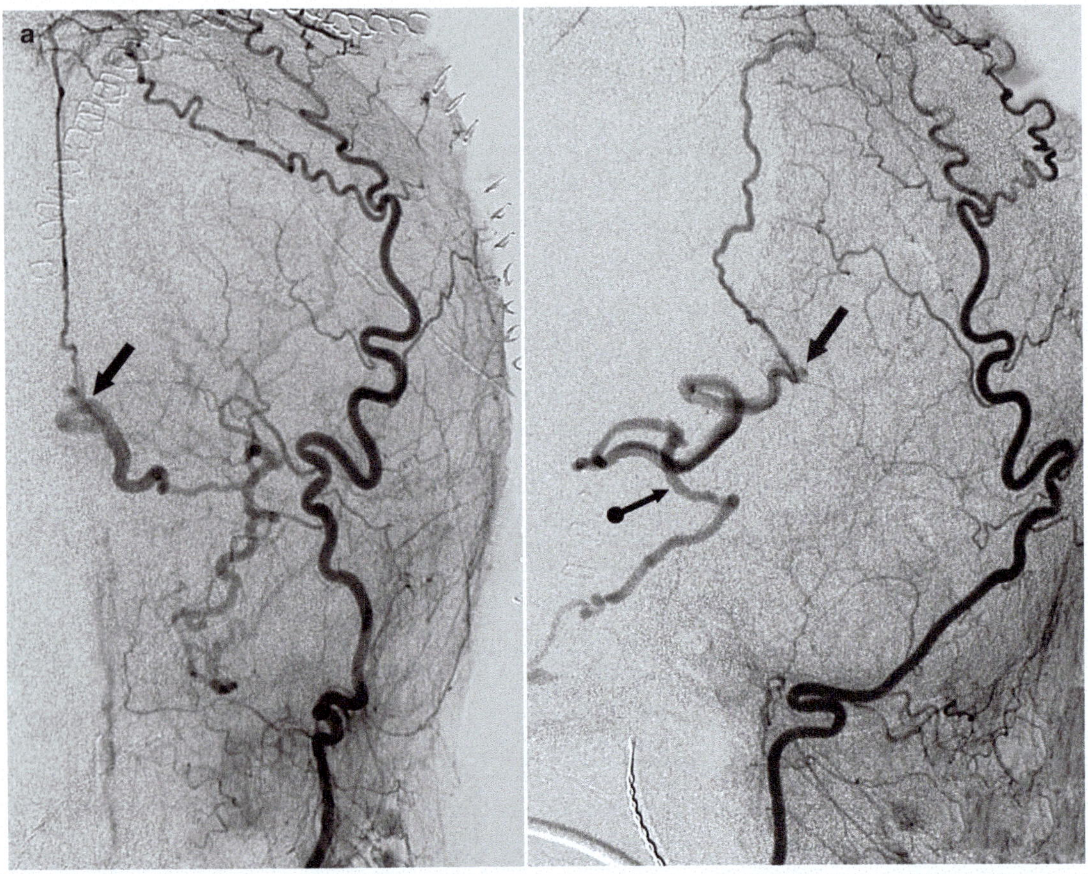

Fig. 13.18 Middle-aged man presenting with temporo-badsal hemorrhage due to a posterior tentorial DAVF supplied by meningeal branches of the ECA and VA. There are no involvement of the ICA. (**a**) Angiogram of the left occipital artery (AP and Lateral view). Through transosseous branches, injection of a meningeal branch of the falx, connected (*arrow*) with the distal portion of the BV or posterior mesencephalic vein, which with a tortuous retrograde course continues into the lateral mesencephalic vein (*arrow with dot*). Galen vein is not involved. (**b**) Left VA angiogram. AP view (early and late phases). The fistula is supplied by a branch, probably a meningeal artery (*arrow with angle and dot*) arising from the left P1 near its origin from the basilar artery. There is a possible involvement of the FCA. Venous drainage along the midbrain (*arrow*). The petrosal vein and superior petrosal sinus are probably thrombosed and so the venous drainage (*arrow with angle*) is diverted towards the superficial middle cerebral vein (SMCV). (**c**) Left VA angiogram. AP view with a different projection (early and late phases) showing the diverted venous drainage (*arrow*) towards the SMCV. There is no filling of the left TS and SPS. There was a minimal filling of the left internal jugular vein (JV). The extracranial drainage occurred through the right internal jugular vein (JV) and the right vertebral artery venous plexus (*arrow with dot*). (**d**) Left VA angiogram, lateral view, showing the drainage involving BV and lateral mesencephalic vein (LM) and its diversion towards the SMCV (*arrowheads*). In a later phase normal filling of the Galen vein (G) and straight sinus (SS). (**e**) Selective study (*arrow*) of the meningeal branch arising from P1, preceding the injection of glue leading to a closure of the fistula, demonstrated on the control angiogram. Complete occlusion of the fistula confirmed also through study of the ECA and ICA. Complete recovery of the patient

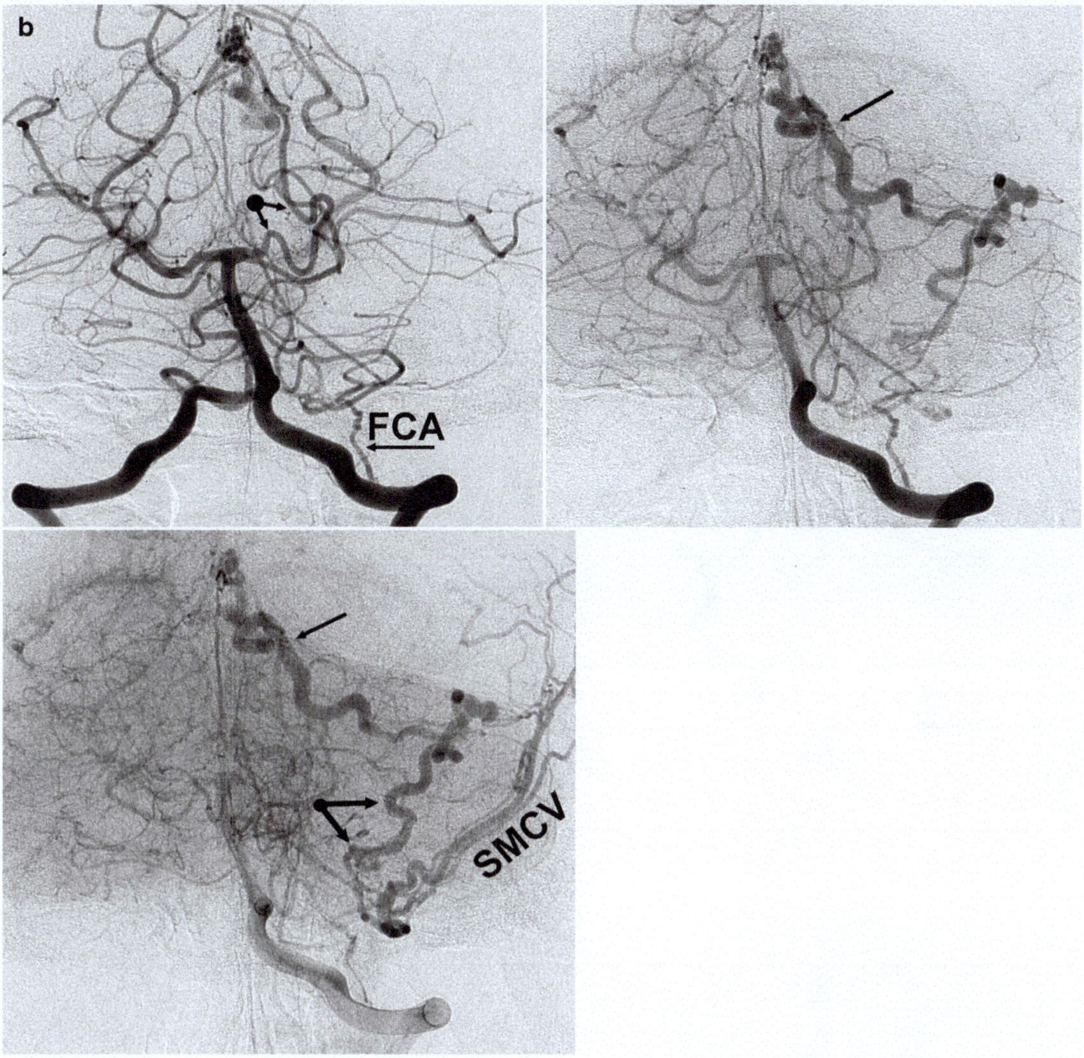

Fig. 13.18 (continued)

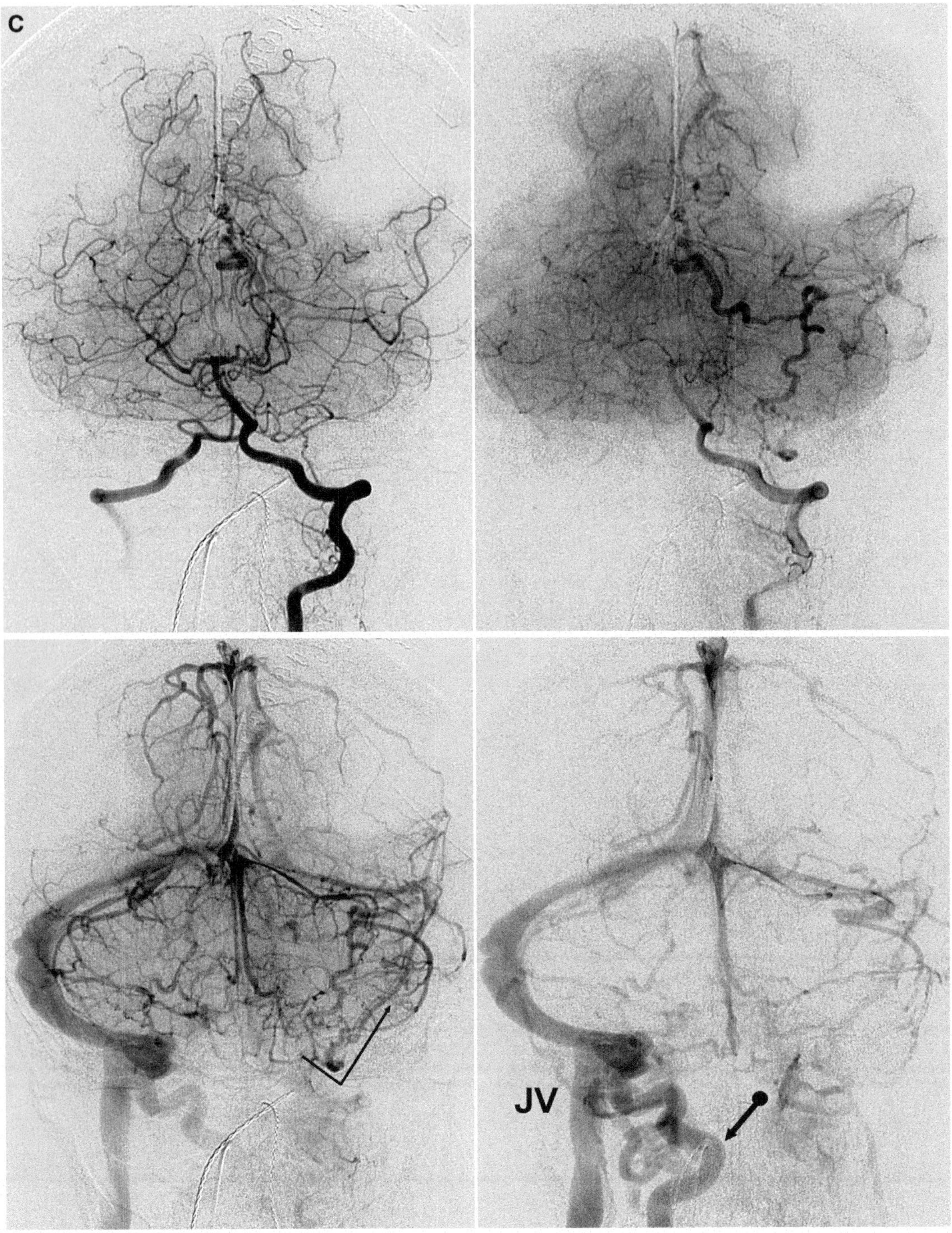

Fig. 13.18 (continued)

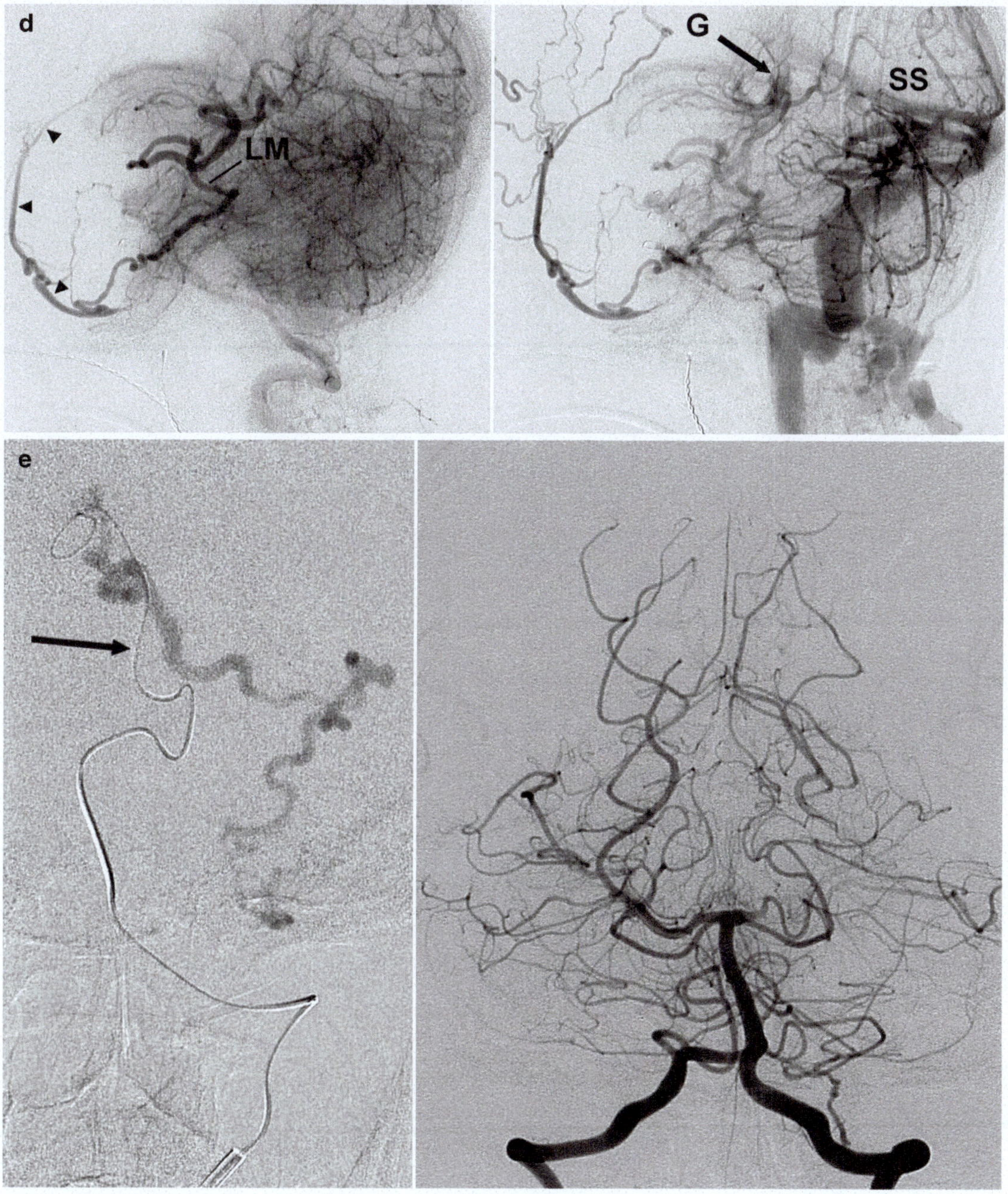

Fig. 13.18 (continued)

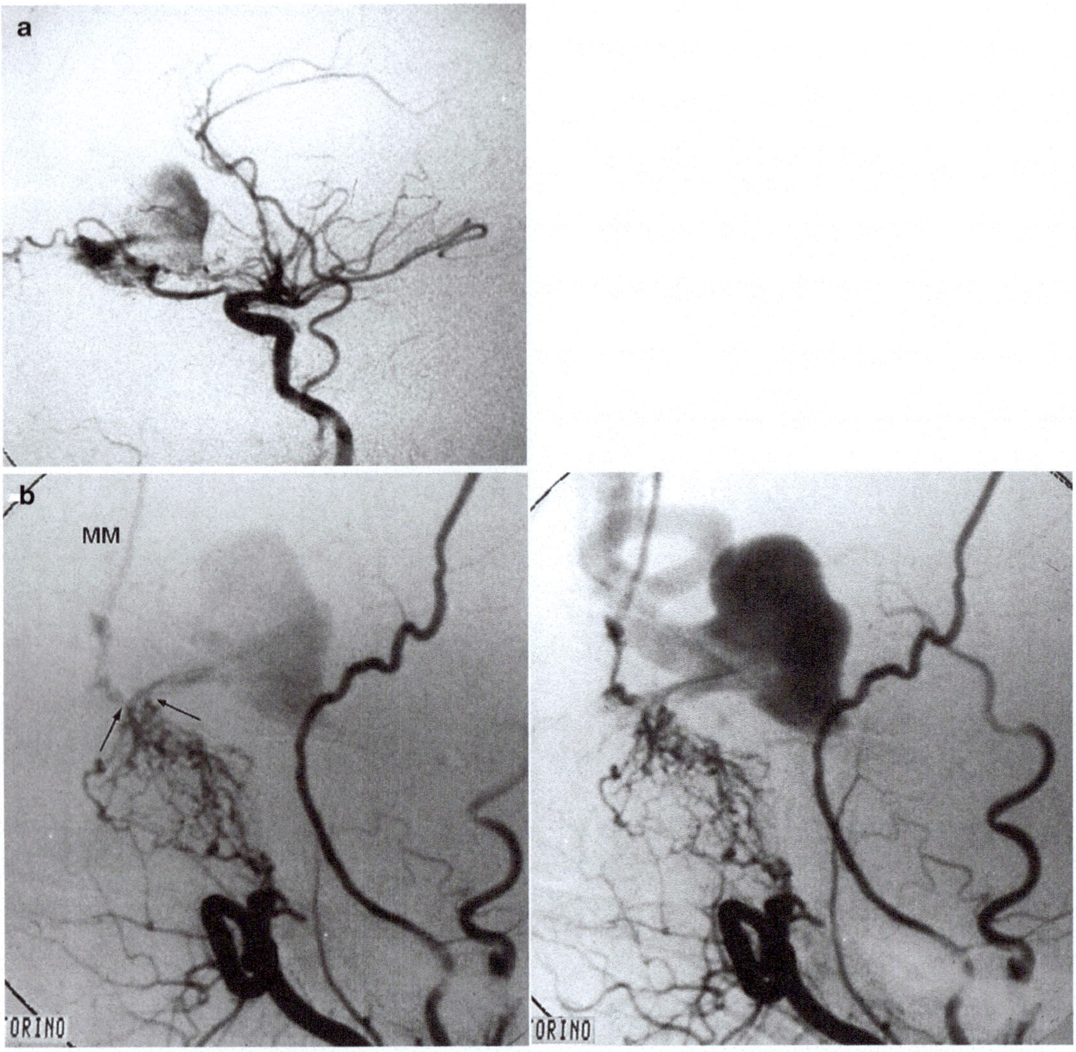

Fig. 13.19 DAVF of the anterior cranial fossa presenting with hemorrhage, supplied by anterior ethmoidal branches of the ophthalmic artery as visible on the lateral ICA angiogram. (**a**) On the lateral angiogram of the internal maxillary artery (**b**) branches of the sphenopalatine artery anastomosing with ethmoidal branches of the ophthalmic artery are also involved as well as the anterior falx branch of the middle meningeal artery. (MM). *Arrows* show the site of the shunt. The drainage occurred in the pial veins, forming a large pouch entering the SSS. Selective angiographic study (**c**) in the lateral and AP views of the MMA better showing its supply to the fistula. Treatment by surgery

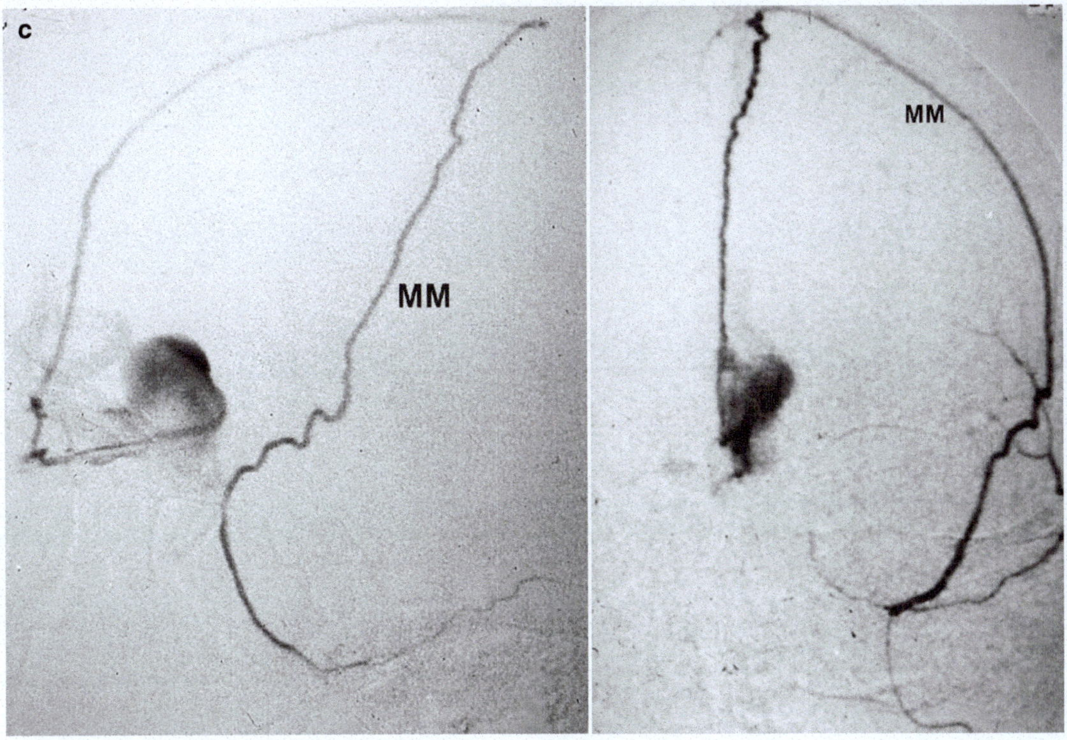

Fig. 13.19 (continued)

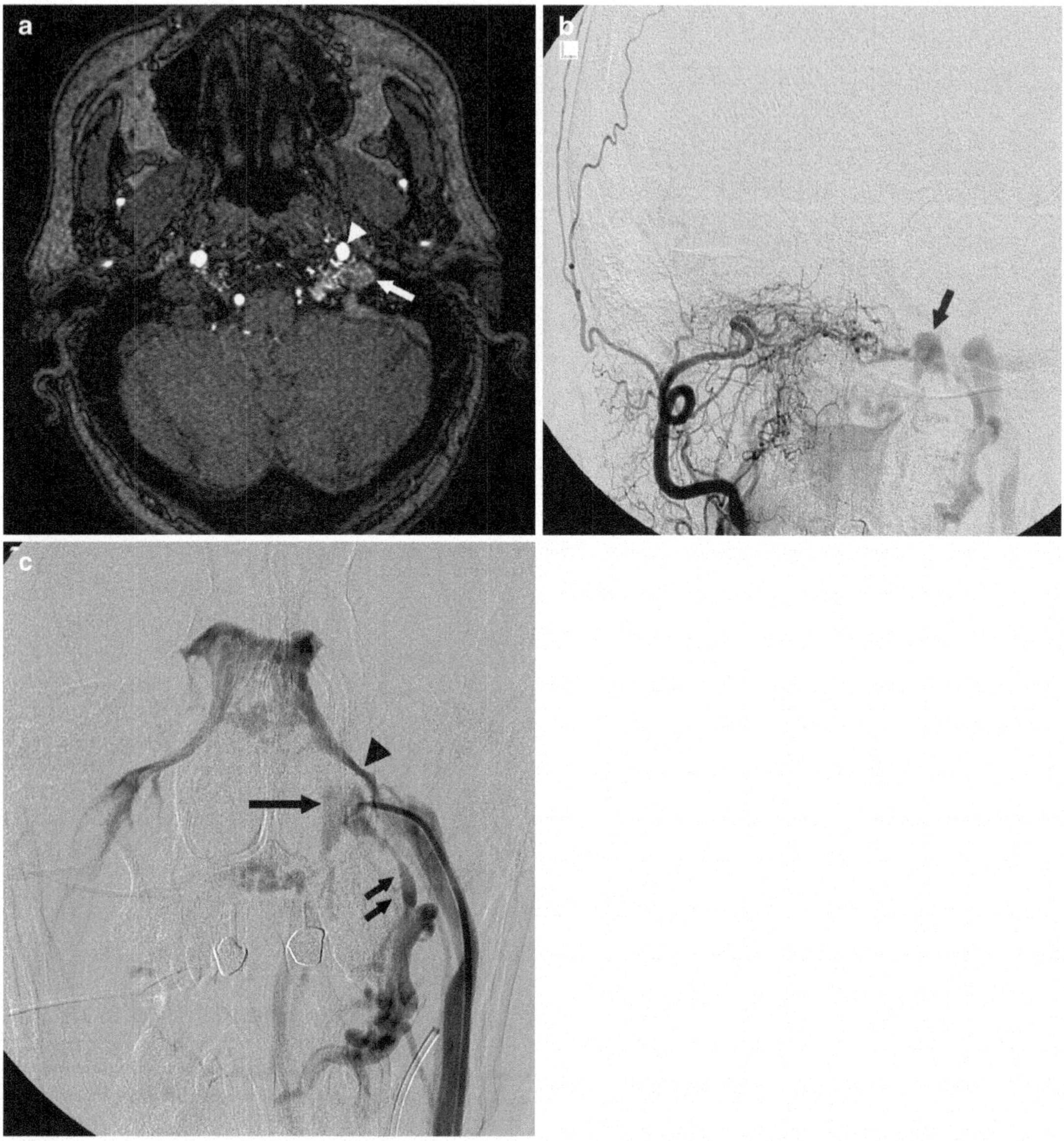

Fig. 13.20 Middle-aged woman presenting with headache and left pulse-synchronous tinnitus. (**a**) MR study (fast sequence with contrast) showed on the left a pathological early enhancement (*arrow*) involving the IJV, ACC, and hypoglossal foramen, suggesting DAVF. A retrograde injection of the sigmoid sinus is also present. ICA (*arrowhead*). On bilateral ICA, ECA, and vertebral angiograms (*not shown*), the fistula at the level of the ACC was demonstrated. There was a bilateral involvement of branches of the APhA, IMA, and of the radiculomeningeal branch of the left VA. Particularly useful to localize the shunt were the angiograms of the contralateral ECA with the selective studies of APhA. (**b**) Angiogram of the right distal ECA showing a rich network arising mainly from the AphA running along the inferior clivus and connected with the ACC (*arrow*). (**c**) Left venous angiogram. Catheterization of the IJV with the tip of the catheter in the ACC (*arrow*). Filling of the lateral condylar vein (arrows). Retrograde injection of the IPS (*arrowhead*). Coils were placed in the ACC leading to occlusion of the fistula. (**d**) Right AP common carotid and left lateral common carotid angiograms showing the occlusion of the fistula. Coils in the ACC (*arrow*) "Courtesy of dott. Gozzoli and dott. Boghi, Neuroradiology-Cuneo"

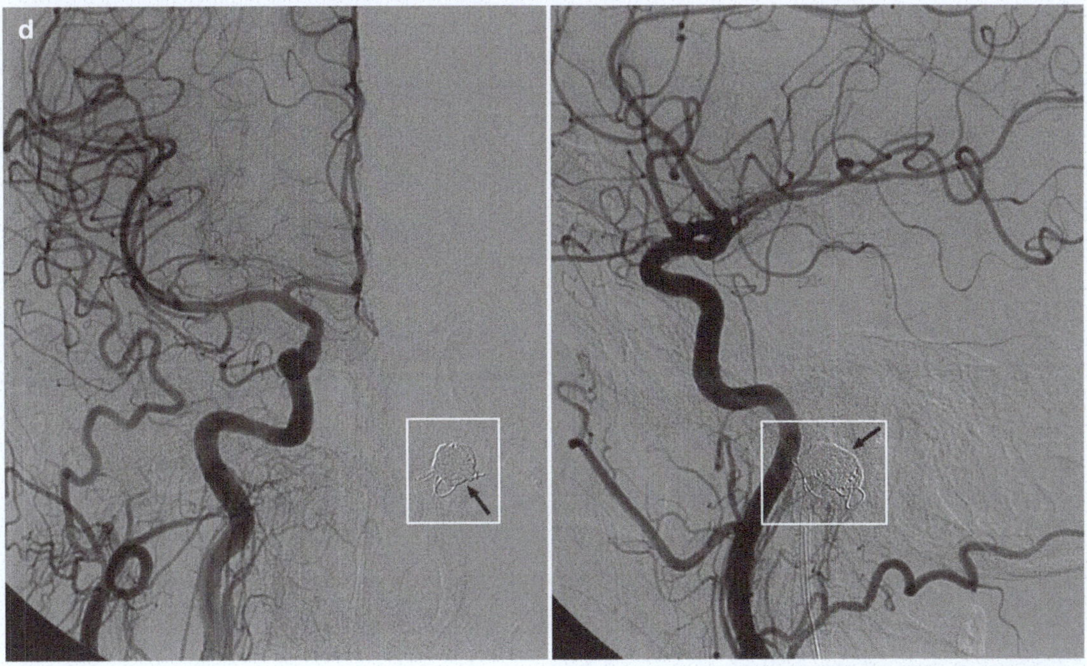

Fig. 13.20 (continued)

and CS can cause symptoms similar to that occurring in CS fistula. These DAVFs involve several uni-or bilateral feeders that arise from the AphA especially from its hypoglossal branch, occipital artery, VA and, more rarely, from IMA and ICA. The venous drainage can vary, since this area is a crossing point for several venous channels represented by the IPS, jugular vein, sigmoid sinus, anterior condylar confluence with its connections, and marginal sinus with its connections (for anatomical aspects of the venous channels see also Sect. 9.3.8). MR can be a useful complementary study in identifying the precise site of these fistulas. An example is shown in Fig. 13.20.

In some of these fistulas, as reported also by other authors (Rodesch et al. 1991b; Pierot et al. 1992) the shunt involves directly the pial veins surrounding the medulla, continuing further caudally in the perimedullary veins of the cervical spinal cord and cranially in the ponto-mesencephalic veins. Examples are presented in Figs. 13.21, 13.22, and 13.23.

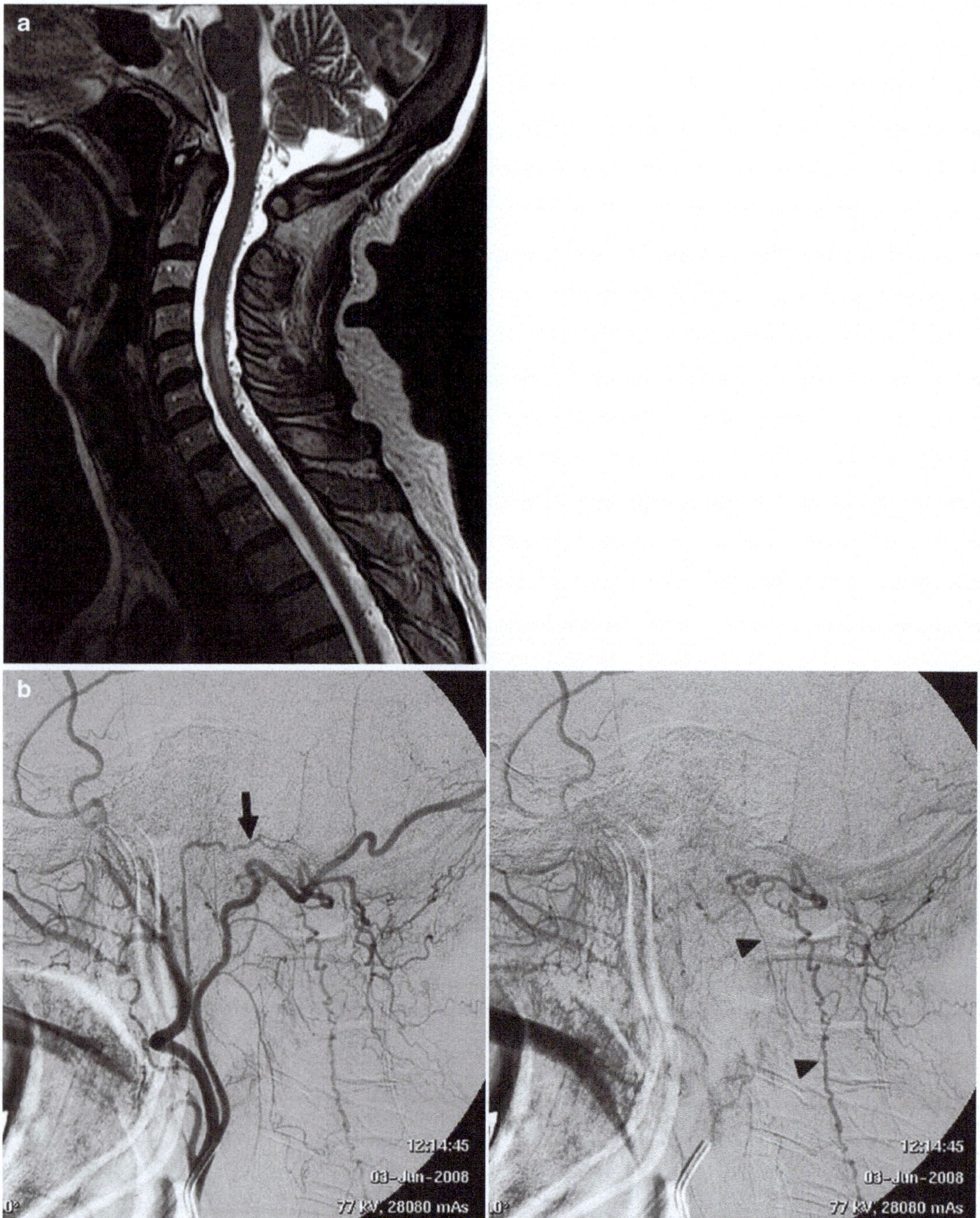

Fig. 13.21 Middle-aged patient presenting with paraparesis. (**a**) MRI T2-weighted image, showing hyperintensity lesion in the cervical spinal cord. (**b**) Lateral external carotid angiogram. There is a fistula involving the hypoglossal branch (*arrow*) of the ascending pharyngeal artery (APhA) connected with dilated medullary veins continuing caudally in the perimedullary veins of the cervical cord (*arrowheads*). (**c**) Oblique view angiogram of the occipital and APhA (*arrow*) arising as a common trunk. The hypoglossal branch of the APhA supplies the fistula (*arrowhead*). (**d**) Selective study (lateral oblique view) preceding the injection of Onyx. Distal tip of the microcatheter in the hypoglossal branch at the site of the shunt (*arrow*). (**e**) MRI T2-weighted image showing the disappearance of the lesion with complete recovery of the patient

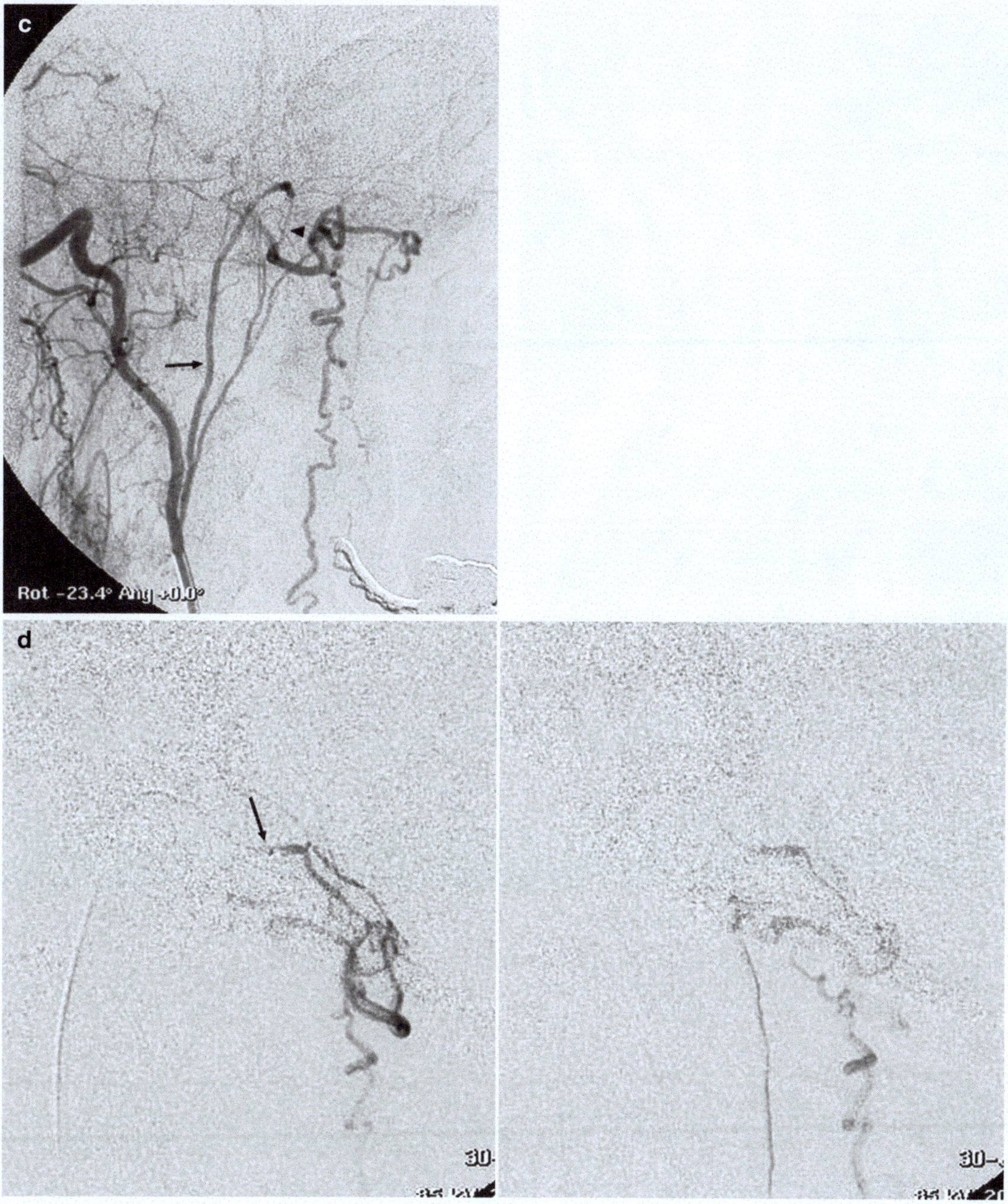

Fig. 13.21 (continued)

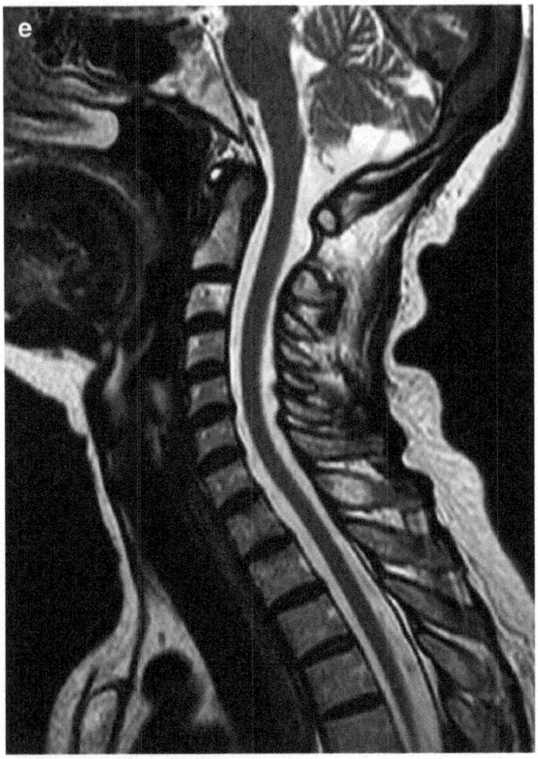

Fig. 13.21 (continued)

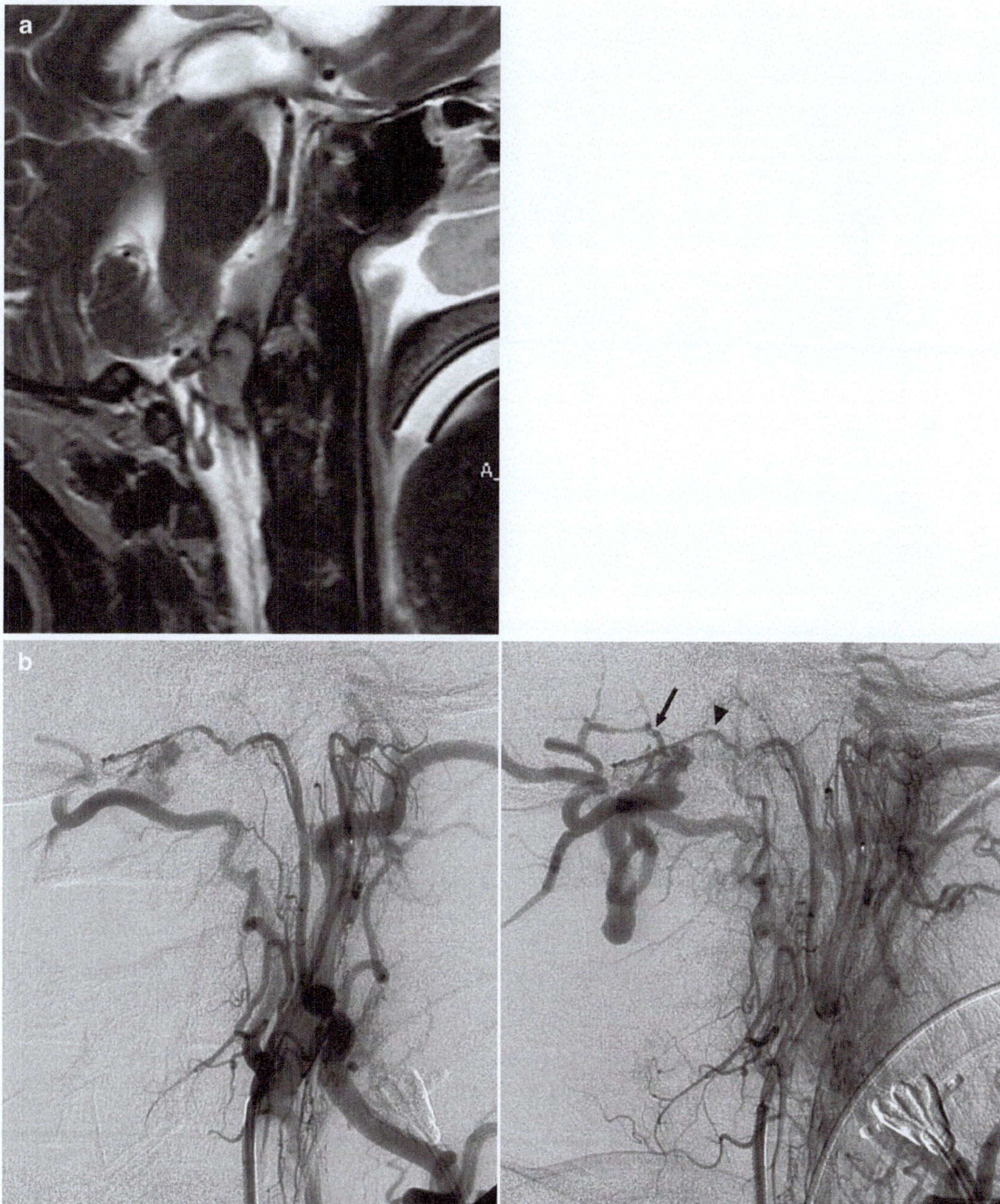

Fig. 13.22 DAVF of the foramen magnum presenting with SAH and subdural hematoma localized in front of the medulla in a young patient. (**a**) MRI T2-weighted sagittal image showing the hematoma and dilated perimedullary veins. (**b**) External carotid angiogram, lateral views, early and later phases, showing the fistula supplied by the hypoglossal branch of the APhA (*arrowhead*) and mastoid branch of the occipital artery (*arrow*). (**c**) Selective study of the APhA showing the site of the shunt (*circle*) and venous drainage involving dilated veins lateral to the medulla, continuing caudally to the perimedullary veins (*arrows*) of the spinal cord. The drainage is also directed intracranially, involving, among others, the pontine veins (*arrowheads*). (**d**) The same site of the shunt visible on the selective occipital artery angiogram. (**e**) Right vertebral angiogram showing a small meningeal component (*arrowhead*). In the later phases, the drainage (*inclined arrows*) in the enormous dilated pontine vein is visible. Endovascular treatment with almost complete occlusion of the fistula with acrylic glue followed by surgical evacuation of the hematoma was performed. The deeply comatose patient recovered

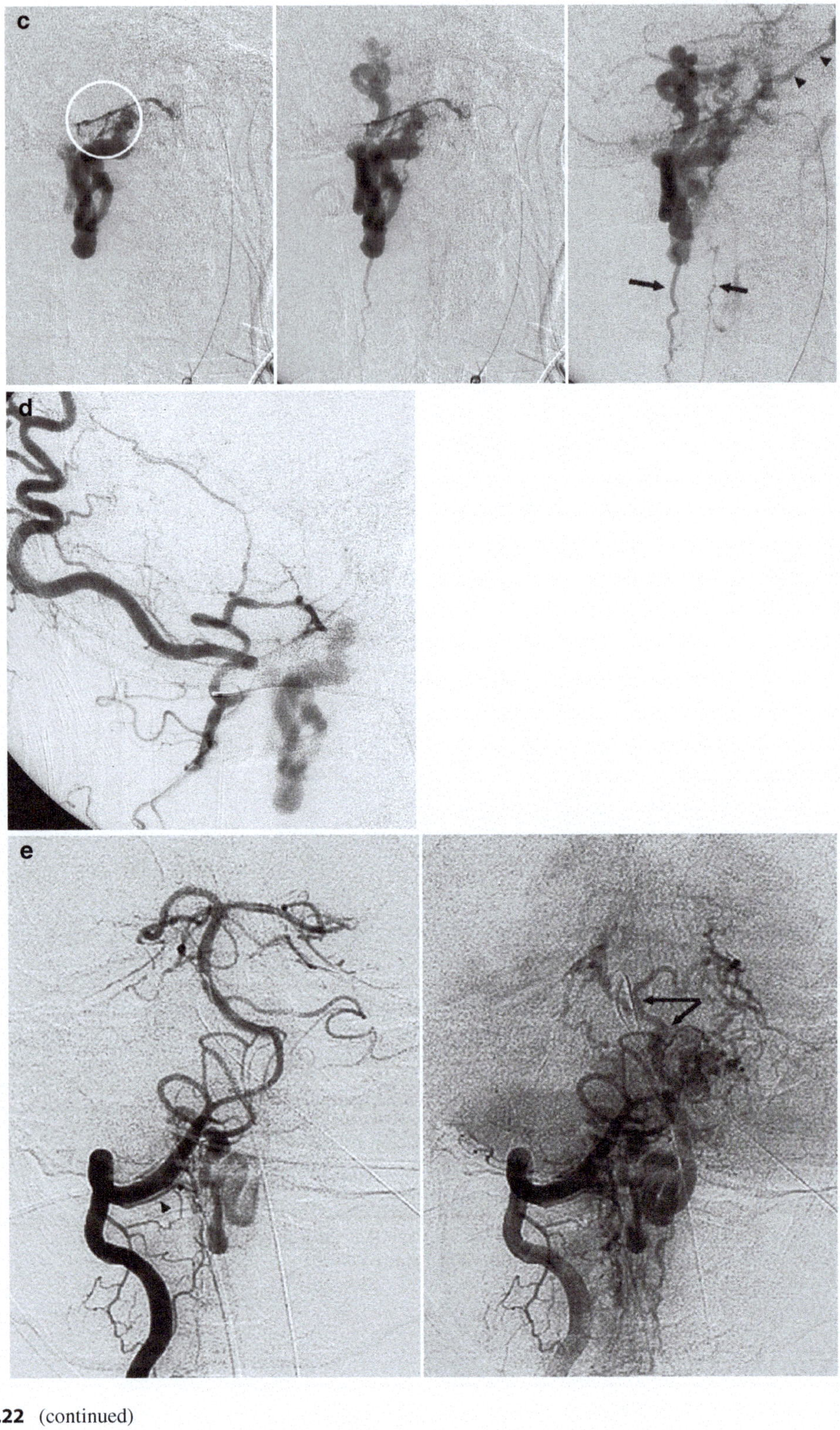

Fig. 13.22 (continued)

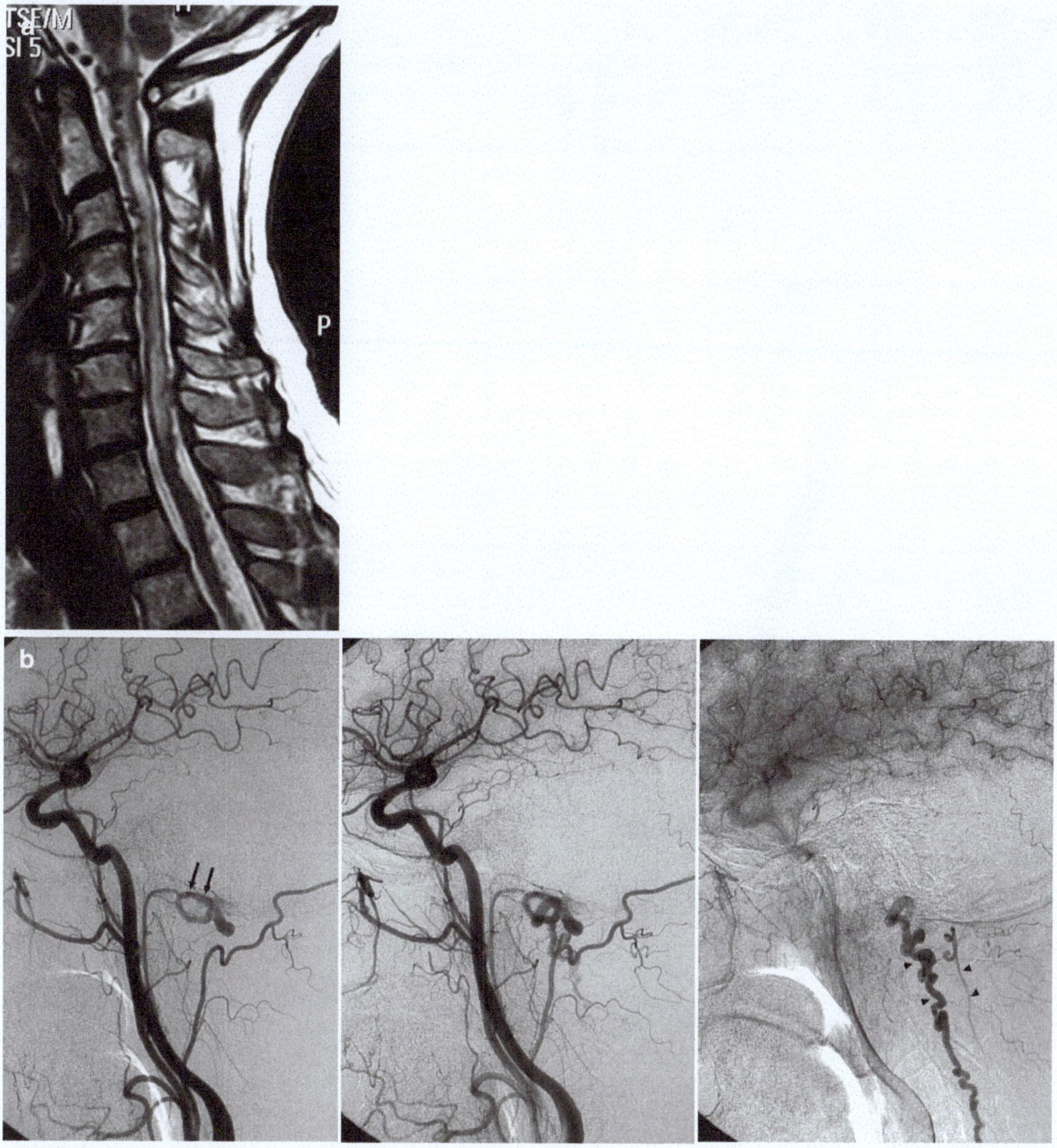

Fig. 13.23 Old woman presenting with rapid progressive tetraplegia. (**a**) MRI T2-weighted image. Extensive hyperintensity of the cervical spinal cord associated with dilated vascular structures is shown, suggesting vascular malformation. (**b**) Angiogram of the common carotid artery. Lateral view. A DAVF supplied by the hypoglossal branch of the AphA is identified. Site of the shunt (*arrows*). In the later phases the drainage in the dilated perimedullary veins is recognizable (*arrowheads*). (**c**) Vertebral angiogram. Lateral view. Supply from a small meningeal branch (*arrow*). Perimedullary drainage (*arrowheads*). Endovascular treatment with catheterization of the hypoglossal branch and injection of glue was soon performed. The patient, however, did not recover and died

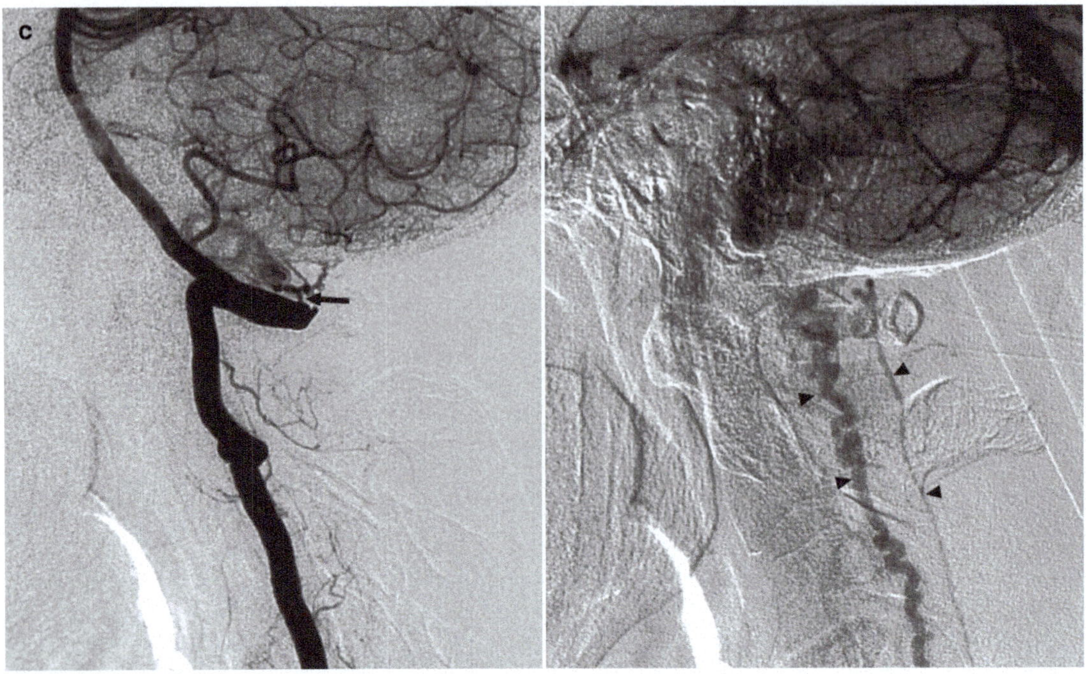

Fig. 13.23 (continued)

13.8 DAFVs in Pediatric Patients

These are rare malformations. According to Lasjaunais (1997), they can be divided into the following two groups:

- DAVFs in *neonates and younger children.* These very rare fistulas, and are characterized by huge dural lakes which are thought to be the primary pathology, commonly involving the SSS, TS, and SiS, associated with thrombotic occlusion or hypo-agenesis of the jugular vein. There is a slow-flow communication between the sinuses. Secondary to the sinus malformation, many shunts develop within the wall of the sinus. Pathogenesis is unclear. They are considered as congenital malformation developing in the prenatal period. An example is presented in Fig. 13.24.
- DAVFs *in older children.* They can share the same features as those in adults. In one third of the cases, however, some particular aspects are present (Garcia-Monaco et al. 1991; Lasjaunais 1997). The DAVFs are multifocal and show in addition the development of shunts involving the pial arteries. The sinuses to which the high-flow dural shunts converge can be dilated without, however, presenting the huge lakes of the previous group. Thrombosis of the sinuses is frequently present. This can be due to many factors; among them trauma and a systemic hypercoagulable state have been considered (Kraus et al. 2000; Walcott et al. 2013; Gross et al. 2013). Other authors (Morales et al. 2010) have reported a case of DAVF in a child developed subsequent to a sinus thrombosis. This raises the question if in children, similarly to the adults, theses fistulae at least in some cases are acquired lesions. An Example is shown in Fig. 13.25.

Progress has been made in the endovascular treatment of this complex pathology. This treatment alone or associated with surgery has improved the prognosis in many of these patients.

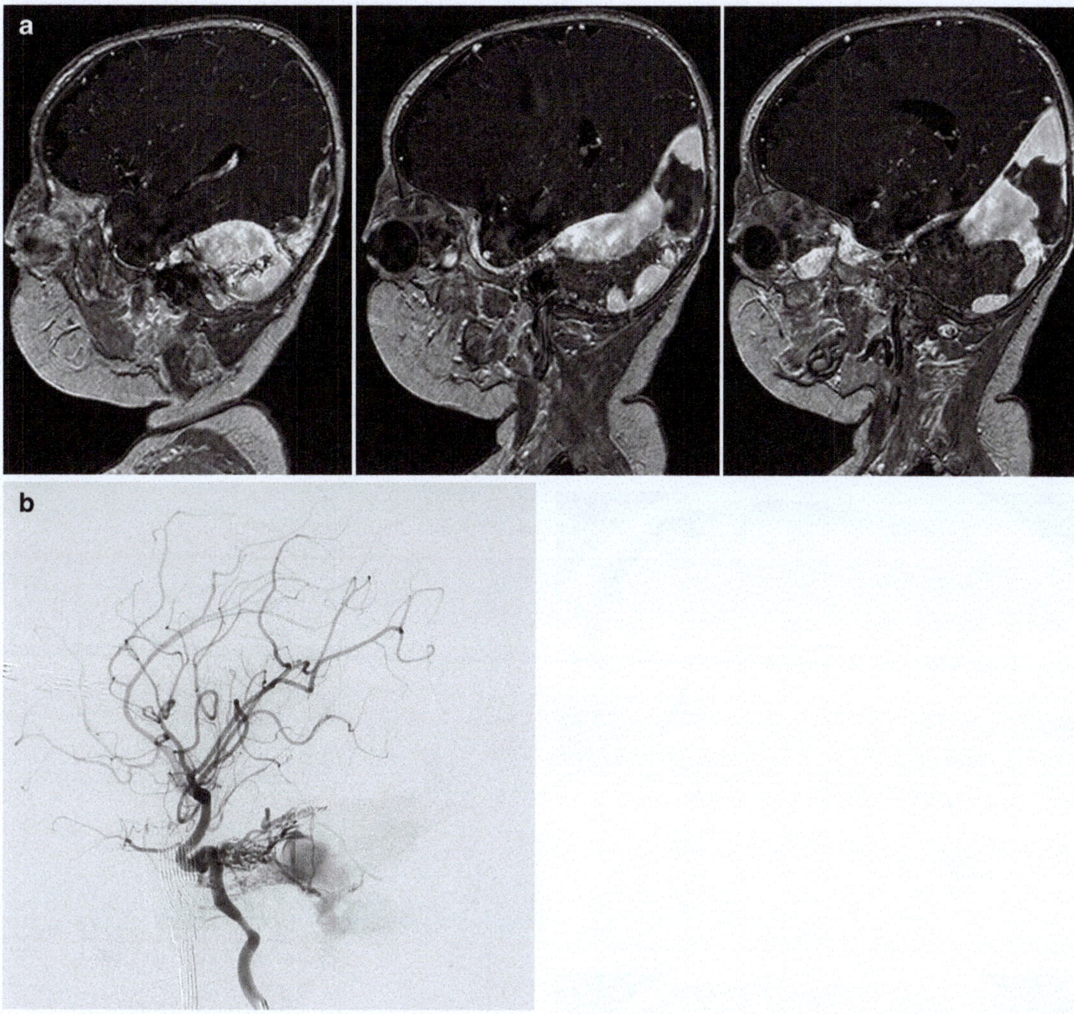

Fig. 13.24 One-year-old child presenting with epileptic seizures. (**a**) MR study T1-weighted images with contrast medium. Lateral view. A venous lake due to an huge dilatation of the TS and SiS extending anteriorly to the SPS and posteriorly to the torcular herophili and posterior part of the SSS is recognizable. An intraorbital pathology (lymphangioma) (*arrow*) was also present. (**b**) On the ICA angiogram there are several branches arising from the cavernous segment of the artery converging to the SPS and TS. The several components of the dilated sinuses fill slowly and only partially (**c**). On the later phases of the ICA angiogram, the normal antero-middle portion of the SSS is filled. The distal dilated segment communicates slowly (*arrow*) with the huge lake. There is a partial rerouting of the venous drainage towards the SMCV (*arrow with dot*) draining into the CS further continuing into the retrograde filled SOV. The jugular vein is not recognizable. (**d**) ECA angiogram showing a second rich supply from the petrosquamous branch of the MMA. Occlusion of the supplying vessels, followed by surgical eradication of the huge lake was performed

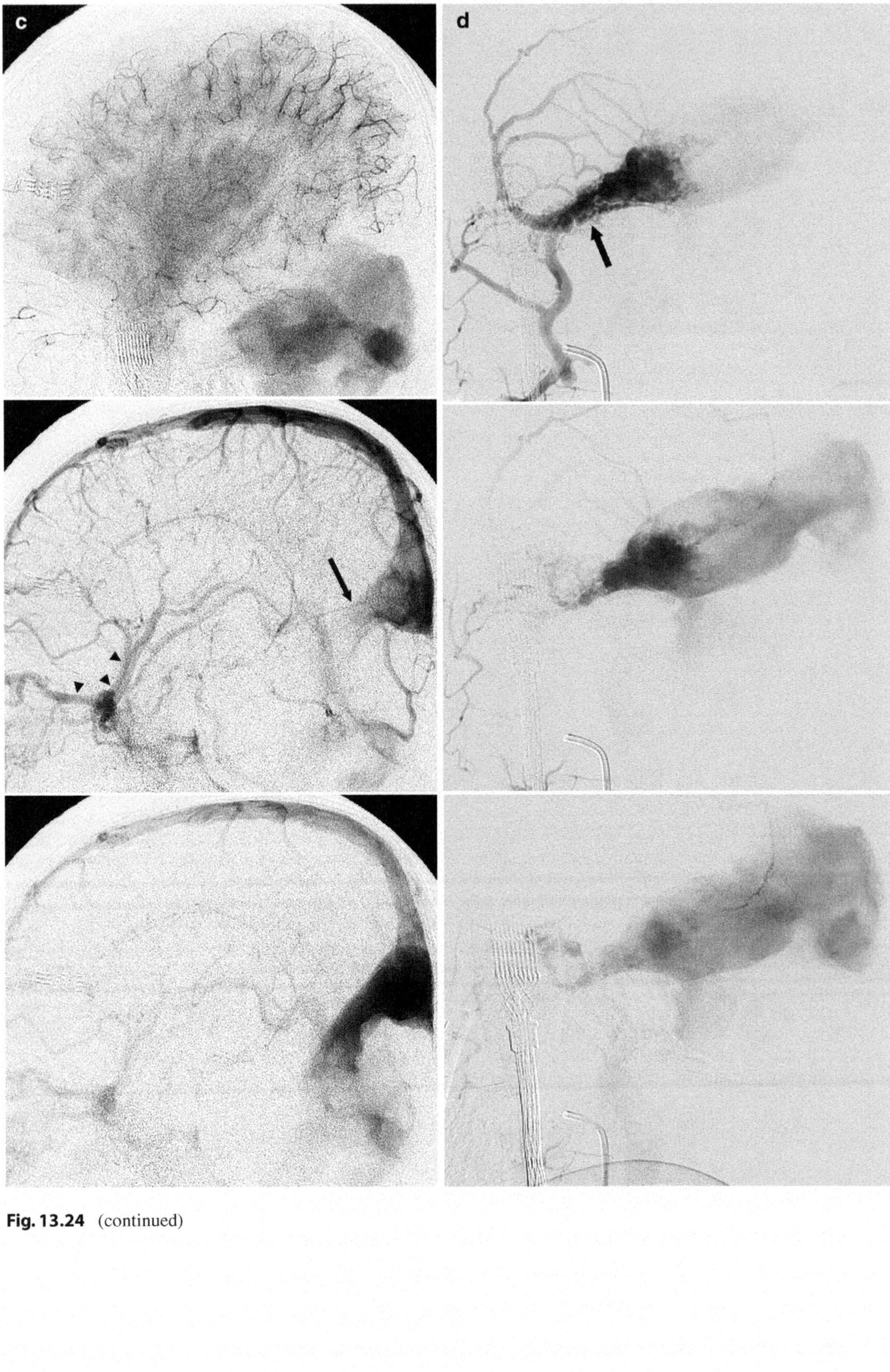

Fig. 13.24 (continued)

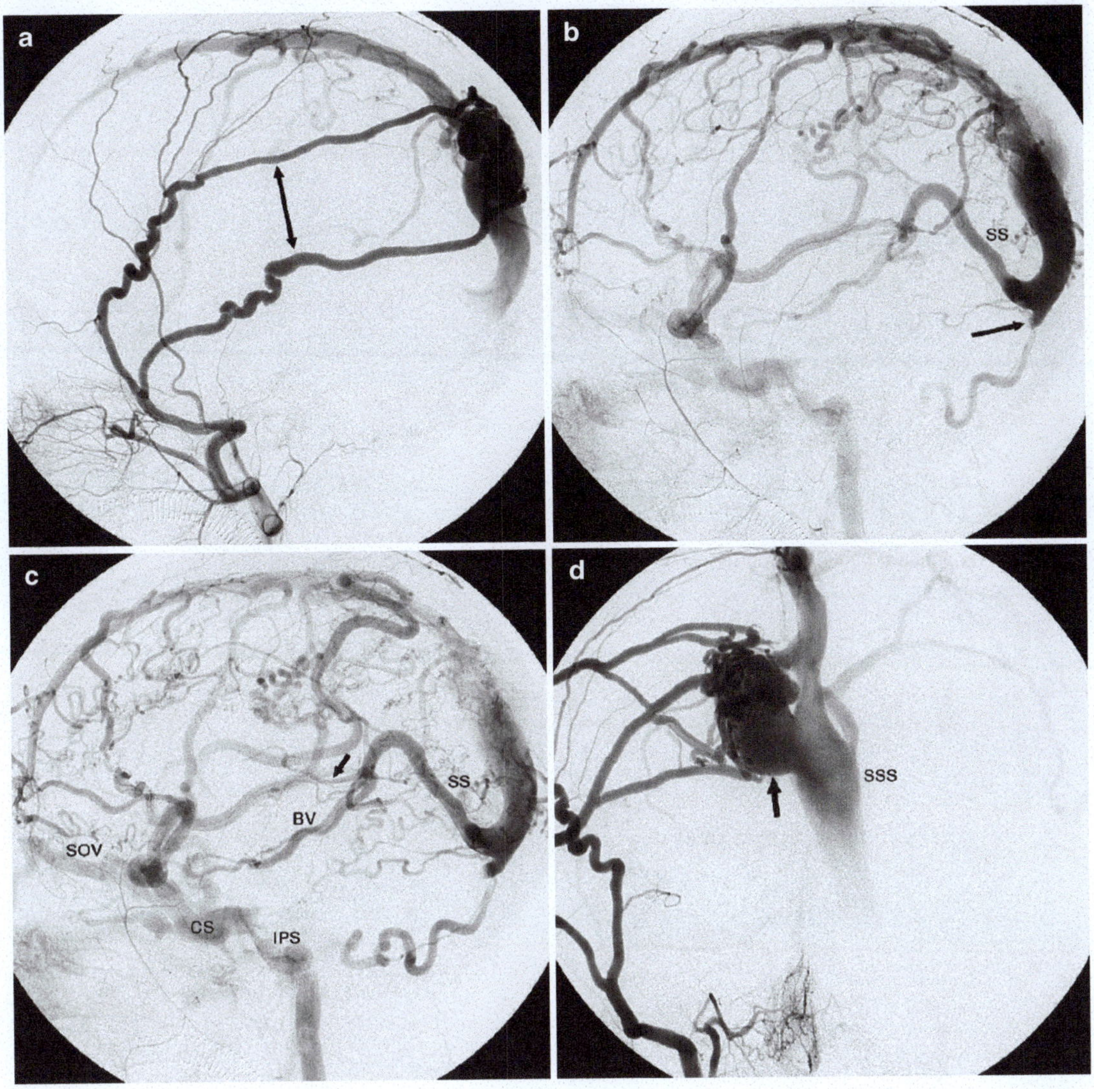

Fig. 13.25 Dural arteriovenous fistula at the level of the SSS in a child. Angiogram (**a**) of the right external carotid artery (lateral view). Dilated branches of the middle meningeal artery (MMA) (*arrow*), supplying the fistula in the middle portion of the enlarged SSS. A meningeal supply also came from the left MMA. Early (**b**) and late (**c**) venous phase. The SSS is distally occluded at the level of the torcular Herophili (*arrow*). There is a venous congestion, with rerouting of the venous drainage partially in the straight sinus (SS) and further in the internal cerebral vein (*arrow*), especially in the basal vein (BV) bilaterally, continuing to its anterior tributaries. The superficial venous system drains mainly into the cavernous sinus (CS) and further to the inferior petrosal sinus (IPS), jugular vein, and superior ophthalmic vein (SOV). There is also a retrograde injection of the dilated cerebellar vein. Right external carotid angiogram, AP view (**d**) showing that the connection to the SSS occurs through a large venous pouch (*arrow*), to which converge several branches of the MMA. In the late phase (**e**), the duplicated SSS, dilated on the left, is visible. Both TSs are proximally occluded (*arrows*). Retrograde injection of the basal veins (BV)

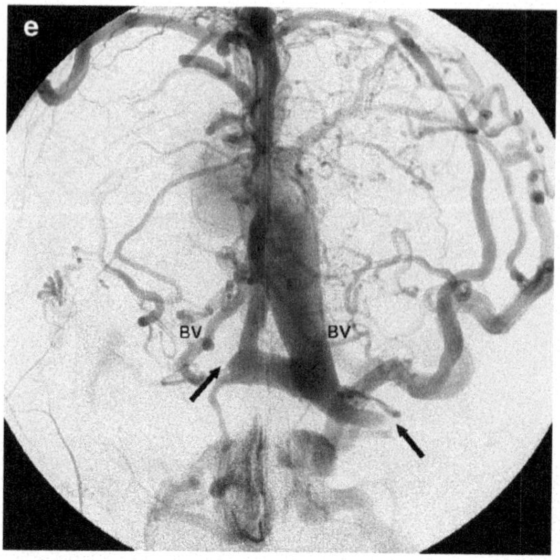

Fig. 13.25 (continued)

14 Arteriovenous Fistulas

Arteriovenous fistulas are direct abnormal communication between arteries and veins. They can occur in any locations where both vessels run very close to each other.

14.1 Carotid-Cavernous Fistulas

These are the most frequent type of arteriovenous fistulas, characterized by a direct shunt between the intracavernous segment of the internal carotid artery (ICA) and the surrounding venous plexus of the cavernous sinus. The pathogenesis is commonly a rupture of the artery and veins, after a penetrating or blunt trauma. Spontaneous fistulas can also occur: these are commonly due to rupture of an intracavernous aneurysm of ICA, frequently linked to an associated angiodysplasia such as FMD, Ehlers-Danlos syndrome and neurofibromatosis (Kanner et al. 2000). Fistulas resulting from spontaneous dissection of ICA have been described (Bradac et al. 1985), and cases developing from laceration of the artery during transsphenoidal surgery have also been reported.

14.1.1 Clinical Presentation

Carotid-cavernous fistulas are characterized by a typical cavernous sinus syndrome with ophthalmoplegia, visus involvement, pulsating exophthalmos, chemosis, and bruit. The symptoms can appear acutely or slowly, progressively days or weeks after the trauma, or have a spontaneous onset. Commonly adults, but sometimes also children can be involved. Like in cases of DAVFs involving the cavernous sinus also in direct carotid-cavernous fistulas, ischemic or hemorrhagic cerebral complications can develop as a result of the several connections of the cavernous sinus as described in Sects. 9.3.10 and 13.7.

14.1.2 Diagnosis and Treatment

The dilated cavernous sinus and superior ophthalmic vein can be easily recognized on CT or MRI. Angiography is essential for a precise diagnosis and in planning treatment. Carotid-cavernous fistulas are high-flow shunts with a rapid injection on the angiogram of the cavernous sinus. Depending on the location of the shunt and anatomical, further drainage can prevalently be directed as follows: anteriorly in the superior-inferior ophthalmic veins, in which the flow is reversed extracranially, draining into the facial vein system; posteriorly into the superior-inferior petrosal sinuses. Through intercavernous anastomoses, there can be involvement of the contralateral cavernous sinus. The pontomesencephalic, the basal, and the superficial and deep middle cerebral veins can also be involved (see also Sects. 9.3.10 and 13.7).

G.B. Bradac, *Applied Cerebral Angiography*, DOI 10.1007/978-3-319-57228-4_14

In cases of a large laceration of the ICA, the blood flow passes completely in the venous sector, and on the angiogram no cerebral arteries are recognizable. The angiographic study consists of a bilateral carotid examination to exclude a possible bilateral lesion and to check for a collateral circulation. The ECA should be also investigated since in some cases its branches, especially the internal maxillary and ascending pharyngeal arteries can be involved. To precisely identify the site of the shunt, a vertebral angiogram with compression of the involved ICA is very useful.

Endovascular treatment with a detachable balloon to occlude the shunt proposed and developed by Serbinenko (1974) and Debrun et al. (1975a, b) has progressively become the therapy of choice. The treatment is performed today with a balloon or coils with good clinical and anatomical results and limited morbidity and mortality (Berenstein and Kricheff 1979; Debrun et al. 1981; Scialfa et al. 1983; Kendall 1983; Lewis et al. 1995; Gupta et al. 2006). An alternative route, introduced by Mullan (1979), Manelfe and Berenstein (1980), and Halbach et al. (1988a) and increasingly used today, is the endovascular venous approach. With this technique coils are directed positioned in the cavernous sinus reached with a microcatheter pushed from the IJV in the facial and further in the superior ophthalmic veins. Another way is that through the inferior petrosal sinus.

Examples are presented in Figs. 14.1, 14.2, and 14.3.

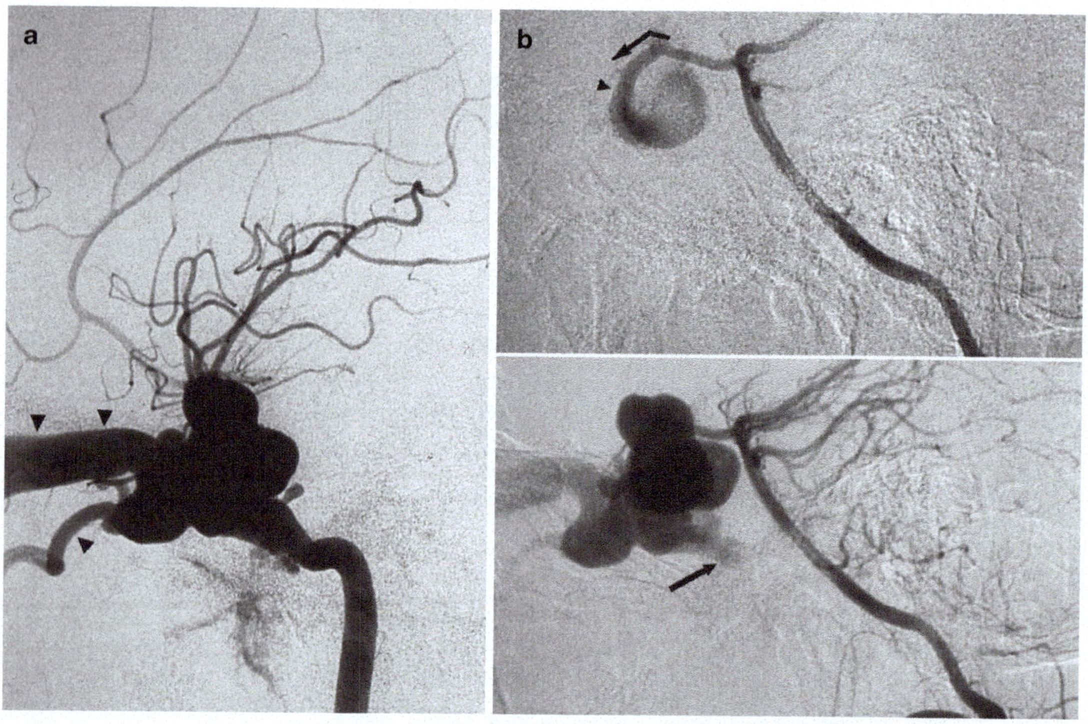

Fig. 14.1 Traumatic carotid-cavernous fistula. (**a**) Internal lateral carotid angiogram. There is an immediate injection of the dilated and deformed cavernous sinus and filling of the dilated superior (*arrowheads*) and inferior ophthalmic veins. (**b**) Lateral vertebral angiogram with compression of the involved internal carotid artery. Through the posterior communicating artery, there is injection of the internal carotid artery and fistula. Site of the shunt (*arrowhead*). In a later phase, there is a partial injection of the inferior petrosal sinus, which could be thrombosed (*arrow*)

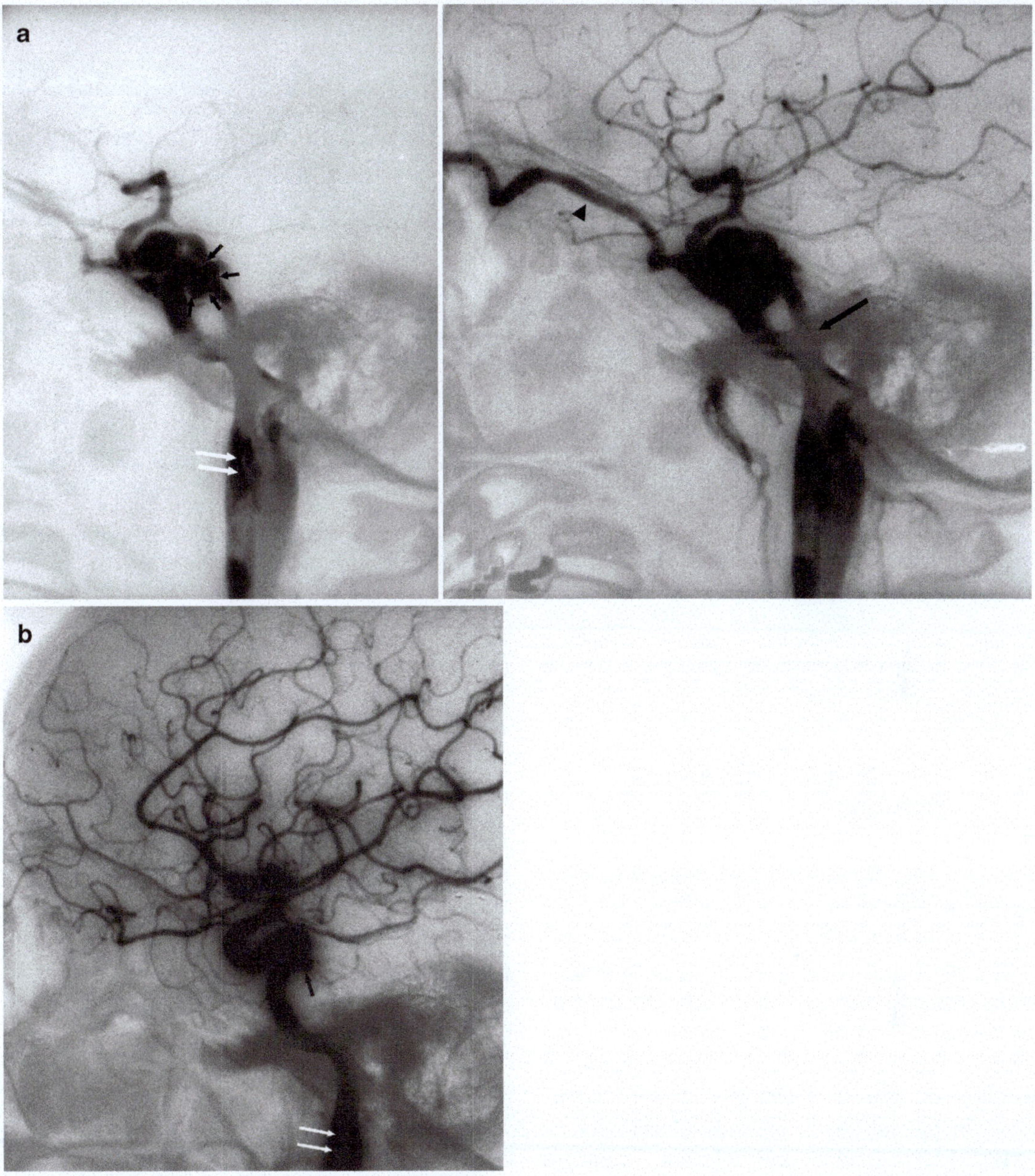

Fig. 14.2 Spontaneous carotid-cavernous fistula in a young patient, which appeared after heavy vomiting caused by alcohol intoxication. (**a**) Lateral carotid angiogram. The small arrows show the site of the shunt. There is drainage in the superior ophthalmic vein (*arrowhead*) and inferior petrosal sinus (*arrow*). There is also a minimal injection of the pterygoid plexus. The extracranial internal carotid artery is slightly enlarged, and there is an intraluminal defect (*white arrows*), suggesting dissection. It is possible that the dissection has extended intracranially, leading to formation of the fistula. The shunt was occluded with a balloon. (**b**) Control angiogram 2 months later. The balloon is now deflated. A small pseudoaneurysm (*arrow*) at the site of the fistula is recognizable. The dissection is unchanged (*white arrow*)

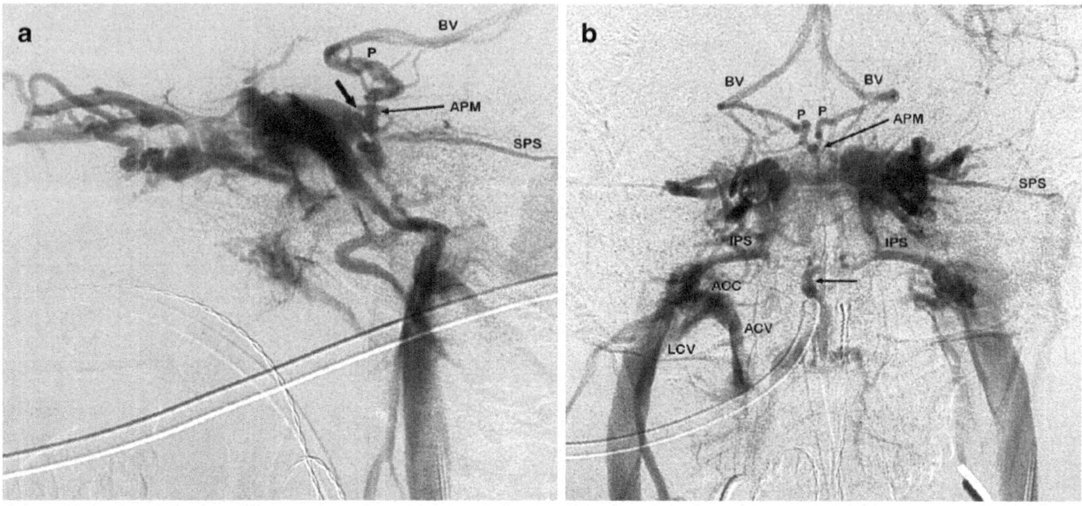

Fig. 14.3 Large traumatic carotid-cavernous fistula. Left carotid angiogram, lateral (**a**) and AP (**b**) 2 s after injection into the internal carotid artery. Extensive anterior (ophthalmic veins) and posterior (petrosal sinuses) drainage. There is also an injection of the contralateral sinus. Bridging veins visible in the lateral view (*arrow*) connect the cavernous sinus with the anterior pontomesencephalic veins (APM), which via the peduncular veins (P) continue into the basal vein (BV) bilaterally. Inferior petrosal sinus (IPS), superior petrosal sinus (SPS), and anterior condylar confluence (ACC) continuing into the anterior condylar vein (ACV) and lateral condylar vein (LCV). Through injection of the clival venous plexus, filling also of the anterior epidural vertebral plexus (*arrow*)

14.2 Vertebral Arteriovenous Fistulas

This is a very uncommon pathology, characterized by a shunt between the vertebral artery and the surrounding vertebral venous plexus. The pathogenesis is traumatic due to a penetrating or blunt cervico-vertebral trauma (Goodman et al. 1975; Halbach et al. 1988b), associated sometimes with bony fractures. Cases of spontaneous fistulas can occur (Halbach et al. 1988b), sometimes in patients with angio-dysplasia such as FMD, Ehlers-Danlos and Marfan syndrome (Bahar et al. 1984), and neurofibromatosis (Deans et al. 1982; Kahara et al. 2002). It can develop at any level of the vertebral artery, but C1, C2, and C3 are the most frequent locations.

14.2.1 Clinical Presentation

Progressive vertebrobasilar insufficiency leading to ischemia in the brainstem and/or spinal cord can occur. The dilated venous sector can compress the nerve root. Paraspinal bruit and cervical pain are present.

14.2.2 Diagnosis and Treatment

Angiography is essential to define the lesion precisely and to plan the treatment, which will be basically endovascular (Moret et al. 1979; Miller et al. 1984). Examples are presented in Figs. 14.4 and 14.5.

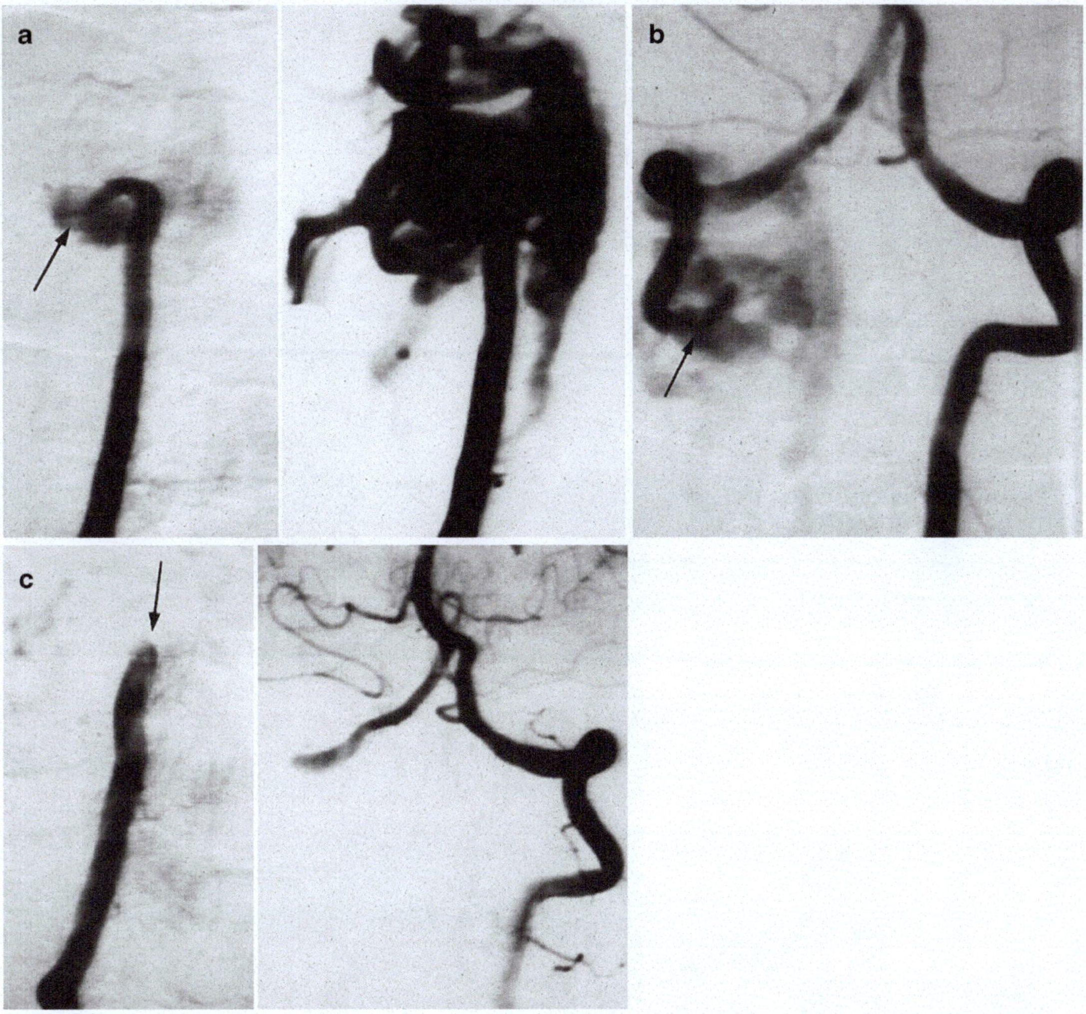

Fig. 14.4 Spontaneous vertebral arteriovenous fistula. (**a**) Right vertebral angiogram, showing the site of the shunt (*arrow*) at the C2 level. There is immediate extensive venous drainage. (**b**) Left vertebral angiogram with retrograde injection of the right vertebral artery better displaying the site of the shunt (*arrow*). Occlusion of the fistula and right vertebral artery with a balloon. (**c**) Control angiograms of the right occluded vertebral artery (*arrow*) and left vertebral artery showing the retrograde filling of the occluded right VA down to the site of occlusion

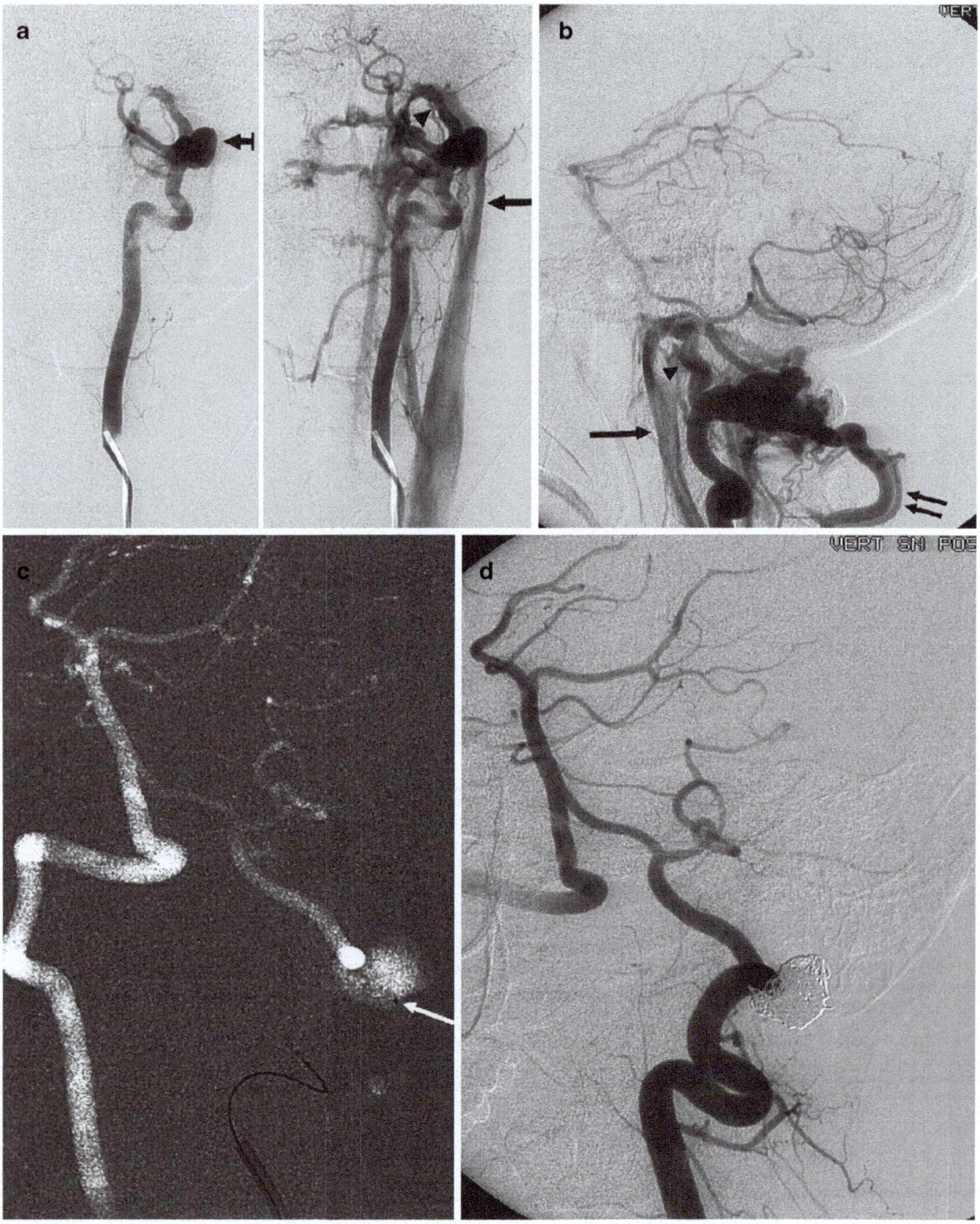

Fig. 14.5 Spontaneous fistula between the left vertebral artery and surrounding venous plexus at the level of C1. Left vertebral angiogram, AP view (**a**), lateral view (**b**). There is a shunt between the dilated vertebral artery and venous plexus, which is dilated and forming a large pouch (*arrow with line*). The venous drainage continues into the retrogradely injected and dilated lateral or posterior condylar vein (*arrowhead*) and further into the internal jugular vein (*large arrow*). There is also drainage into the deep cervical vein (*double arrows*). Right vertebral angiogram (**c**) prior to endovascular treatment. There is retrograde injection of the left vertebral artery and venous pouch site of the shunt. A microcatheter has been advanced into the left vertebral artery and placed in the venous pouch. The white arrow shows the distal marker of the microcatheter. Left vertebral angiogram (**d**) post treatment. Occlusion of the fistula with coils

15 Ischemic Stroke

Stroke is the third cause of death and disabling in the general population after heart disease and cancer. It can be basically divided into one group due to occlusion of the vessels (arteries or veins) leading to ischemia and another in which rupture occurs resulting in hemorrhage. The latter is discussed in Chaps. 11, 12, 13 16, 17, 21, and 22. In this chapter the ischemic stroke due to a pathology involving the arteries is described. Thrombosis in the venous sector is discussed in Chap. 20.

It is slightly more frequent in men than in women. It is more rare in young patients. Some racial differences occur. In general it is less frequent in Caucasian than in non-Caucasian and there is an higher incidence in blacks in comparison to the whites (Sacco et al. 1991; Sacco 1993). The main pathological process responsible for stroke is atherosclerosis affecting the extra-intracranial arteries. Less commonly, stroke is due to other pathologies, such as spontaneous dissection and other nonatherosclerotic arteriopathies as well as heart diseases. These latter will be described in Chaps. 16–18, and 19.

Atherosclerosis is a generalized condition involving predominantly certain vascular territories: the aorta, the iliac, the coronary, and the extra-intracranial arteries. It is a typical pathology of the adulthood, and the likelihood of occurrence increases with the age. There is a slight predominance among men.

15.1 Pathology

It is not the aim of this monography to treat the pathology of the atherosclerosis, for which we refer to specific works (Miller and Sobey 2011). We describe here only the basic point. The pathologic changes begin with a dysfunction of the endothelial cells, due to various stress factors such as hypertension, diabetes, elevated cholesterol level especially of the low-density lipoproteins, obesity, hyperhomocysteinemia, and cigarette smoking, which promote the passage and deposit of cholesterol in the subendothelial space. Further injury of the endothelial cells promotes the passage into the subendothelial space of monocytes, which progressively engulf lipid and become foam cells. Dysfunction and proliferation of smooth muscle cells, migration of lymphocytes, cell necrosis, and collagen production follows (Witznum and Steinberg 1991; Ohara et al. 1993; Libby and Clinton 1993; Garcia et al. 1998). All these processes lead to the formation of the atheromatous plaque covered by a well-formed fibrous cap, which can extend progressively into the lumen of the artery, resulting in stenosis. The evolution of the plaque can vary from a stable course to a slow or rapid growth leading to a progressive narrowing and occlusion. Hemodynamically, stenosis, leading to a major reduction of flow and thereby, increasing the risk of ischemia, has to be rather severe,

G.B. Bradac, *Applied Cerebral Angiography*, DOI 10.1007/978-3-319-57228-4_15

as previously described (Brice et al. 1964; Archie and Feldman 1981). It is important to note, however, that in a certain number of cases, the plaque can slowly progress up to complete occlusion of the vessel, but remains asymptomatic. *This occurs when the collateral circulation is efficient and there has been no distal embolization during the progression of the plaque.*

Erosion of the endothelial cells and breakdown of the fibrous plaque leads to exposure and delivery of the material inside the plaque, favoring the formation of thromboembolic material. Further reduction of the lumen up to occlusion can occur due to formation of an intraplaque hemorrhage. This latter can be due to the entry of blood into the plaque through erosion of the fibrous cap or disruption of the vasa vasorum which are abundant in advanced atherosclerotic lesions (Fisher and Ojeman 1986; Bo et al. 1989; Beach et al. 1993). The improvement of the diagnostic tools allows today to study the structure of the plaque, giving useful information about its possible evolution (see also Sect. 15.4.1).

15.2 Location

Atherosclerosis affects mainly the extracranial segments of the arteries. Especially involved are *the carotid* bifurcation and the adjacent segments of the internal and common carotid arteries, and *the proximal vertebral* and the subclavian arteries. Other less frequent sites, however, not so rare as previously thought are the *aortic arch and the proximal common carotid artery*. Multiple vessels are frequently involved. Atherosclerosis can also involve *the intracranial arteries*. According to some reports, the incidence of significant intracranial atherosclerosis in patients with transient ischemic attacks (TIAs) or stroke has been given as 4% (Thijs and Albers 2000). The incidence of these lesions is reported to be higher among Blacks and Asians (Caplan et al. 1986; Caplan 1989; Feldmann et al. 1990; Wong et al. 1998; Min et al. 2000; Thijs and Albers 2000). The most common site are the distal ICA and the basilar and intracranial vertebral arteries. Less frequent involved are the first segment of the middle and anterior cerebral arteries, the pericallosal artery around the genu of then corpus callosum and the first segments (P1–P2) of the posterior cerebral artery (Fisher et al. 1965; Fisher and Caplan 1971; Castaigne et al. 1973, 1981; Kavase et al. 1979; Hinton et al. 1979; Bradac and Oberson 1983; Caplan et al. 1986; Caplan 1989). Today, the intracranial lesions are the object of greater attention.

15.3 Pathogenesis of Ischemia and Basic Diagnostic and Therapeutic Approach

The mechanism leading to ischemia can be grouped into three categories:

- Thromboembolic occlusion or severe stenosis linked to atheromasic changes of the arteries or occlusion due to cardiac embolization leads to ischemia strictly related to the supplied territory. These infarcts are also called by some authors territorial infarcts (Ringelstein et al. 1983, 1985; Weiller et al. 1991).
- The occlusion or severe stenosis can involve the arteries in their proximal segments and the suffering territory is more distal and not in the vascular territory of a defined artery. Instead, the infarct is located at the borders of two different vascular territories (*border-zone infarcts*), where the collateral circulation due to the hypoperfusion is insufficient. This occurs typically in occlusion or severe stenosis of the ICA with cortical infarcts in the border zones between ACA, MCA and PCA. Similar infarcts can occur in the cerebellum in the distal border-zone areas between the PICA, AICA, and SCA in severe stenosis or occlusion of the vertebral arteries. Distal infarcts can occur also in the vascular territories of the deep and superficial perforators since these arteries are end arteries where no collateral circulation is present. The same type of infarct can also occur in hypoperfusion due to cardiac failure.

- A particular form of infarct is *the lacunar infarct* (from Latin, *Lacuna-ae*, small cavity). The term was first used by Dechambre in the 1838 and later by Durand-Fardel in the 1843 to express pathological findings due to small infarcts, hemorrhages or other causes. More precise reports were presented by Pierre Marie (1901) and Ferrand (1902), who used the term *lacunae* to describe small infarcts in the basal ganglia, capsula interna, and pons which presented some typical clinical aspects. However, the clinical relevance and the peculiar aspects of this pathology, termed *lacunar syndrome*, were emphasized and comprehensively discussed by Fisher (1965, 1968, 1979). CT and MRI have later played an important role in the diagnosis and identification of these lesions.
  Today lacunar infarcts are reported to account for about 25% of all cerebral infarcts (Bamford et al. 1987; Bougosslavsky et al. 1988b; Arboix et al. 1990; Boiten and Lodder 1991). Lacunar infarcts are due to involvement of the perforators (deep and superficial) due to various pathogenetic mechanisms. The vessels can be occluded directly by lipohyalinosis or indirectly through a microatheroma involving the parent artery from which they arise. Their vascular territory can also be involved due to hypoperfusion in cases of occlusion or severe stenosis located more proximal or in cases of cardiac failure. Microemboli from atheromasic plaques or in cardiac diseases can also lead to occlusion of the perforators.
- All these mechanisms can act in isolation or be variously associated with one another. Taking this into consideration, the neuroradiological diagnosis of these patients should include CT and MRI to examine the brain parenchyma, followed by a complete study of the cerebral vessels including the extra and intracranial sectors. The latter can be performed today with highly accurate CT and MR angiography and for the extracranial arteries also with ultrasound examinations. However, catheter angiography should be instituted every time the diagnosis is insufficiently clear or more information about the vessels are needed or in cases where endovascular treatment is considered. In addition to the study of the brain parenchyma and cerebral vessels, the aortic arch and a cardiac examination should be routinely performed.
  Considering, in particular the patients with an acute stroke, the type, site, size of the ischemia, and the identification of the artery and vascular territory involved can be today studied rapidly with CT. This should include CT-perfusion indicating which part of the parenchyma can be considered as lost and which can be savaged, and CT angiography. These give the information which are essential for the therapy especially in the cases of endovascular treatment.

15.4 Mechanism of Ischemia in the Anterior Circulation

The atheromasic process leading to ischemia can be various and complex due to the different location of the atheromasic plaque, which can involve typically the ICA bifurcation, but also, even if more rarely, all segments of the carotid from its origin from the aortic arch to its end. Plaques can develop in the aortic arch and in the intracranial branches of the ICA. The size and the evolution of the atheroma can be extremely various. Occlusion can also occur due to cardiac embolism or other vasculopathies (see Chaps. 16, 17, and 18). All these aspects can make sometimes difficult a precise diagnosis of the pathogenetic process responsible for the stroke in a given case. The angiographic studies can add some more information useful in the comprehension of the mechanism responsible for the ischemia and which are essential in planning the therapy.

15.4.1 Carotid Artery

15.4.1.1 Locations and Aspects of the Atheromasic Plaques

- *The most frequent and typical atheroma develops at the extracranial ICA bifurcation*, which can be responsible for narrowing of the lumen. Four degrees of stenosis have been distinguished: *mild* when the lumen is reduced by less than 30%; *moderate* with 30–70% reduction of the lumen; *severe* if the reduction is greater than 70%; and *preocclusion state*, when the narrowing is more than 90%. In a few of these cases with a very severe stenosis, only a filiform column of contrast medium is visible in the late phase of the angiogram, demonstrating that the artery is patent but only minimally filled by contrast medium layering in the dorsal part of the lumen. This finding has been called in the past pseudo-occlusion (Sekhar et al. 1980; Bradac and Oberson 1983) (Fig. 15.4).

Today CT-MR angiography are valid diagnostic tools to study and measure the stenosis (Anzalone et al. 2005, Weber et al. 2015) being frequently a good alternative to the angiography. A few methods have been proposed: among them, the European Carotid Surgery Trial (ECST 1998) and North American Symptomatic Carotid Endarterectomy Trial (NASCET 1991) are the most used (Fig. 15.1) Differences in the applied criteria lead to differences in the results with this methods.

There is broad agreement that a stenosis of 70% with NASCET (corresponding to a stenosis of about 80% with ECST) is a clear indication for endarterectomy, thereby reducing significantly the risk of stroke. Alternatively, endovascular treatment has increasingly been successfully performed (Wholey et al. 2000; Bonaldi 2002; Ringleb et al. 2006; Wittkugel et al. 2008; Stingele et al. 2008; Clark 2010) (Figs. 15.2 and 15.3).

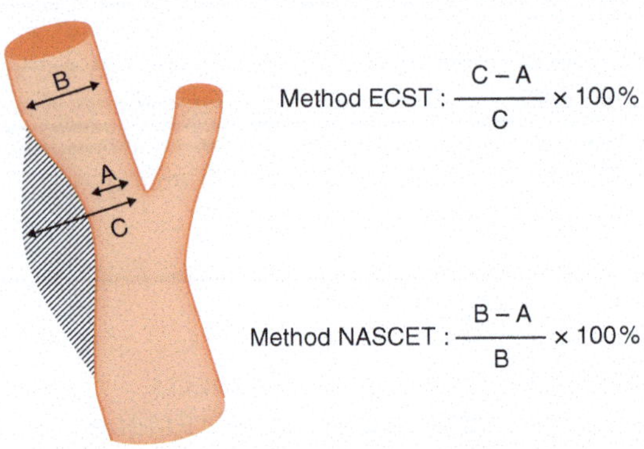

Fig. 15.1 Methods to measure ICA stenosis on the angiogram: European Carotid Surgery Trial (ECST) and North American Symptomatic Carotid Endarterectomy Trial (NASCET)

Another important aspect, as it has already been described in Sect. 15.1, is the evolution of the plaque which can be extremely different (Arbeille et al. 1996). There are *stable plaques* which remain unchanged for years leading to a minimal narrowing of the lumen whose surface remain regular and smooth. When growth occurs, this is slow allowing a good collateral circulation to develop. Stable plaques can remain asymptomatic and discovered incidentally during vessels studies performed for other reasons. *That is true also for severe stenosis or occlusion provided that no embolization occurred and there is a good collateral circulation*. Other plaques are *unstable* characterized by a progressively sometimes even acute growth or ulcerations on which a superimposed thrombus develops. Distal embolization or acute hypoperfusion or both lead to ischemia presenting in various form of stroke (TIAs, mild, severe) (Figs. 15.2 and 15.4). An anterograde extension of the thrombus can also rarely develop (Fisher 1954; Castaigne et al. 1970). Furthermore it should be considered that changes of the plaque can occur which modify its aspect and so a stable plaque can become unstable or vice versa.

Taking this into account, attention has been progressively given to the components of the atheroma. These information can be obtained today using many methods (ultrasound, CT, and especially MR) in various combinations to each other. Progressively experience has shown that some components of the plaque such the fibrous cap, the amount of lipids, the presence of hemorrhage can be better studies with MRI (Ouhlous et al. 2005; Cappendijk et al. 2005; Cai et al. 2005; Wasserman et al. 2008; Gao et al. 2009; Watanabe and Nagayama 2010; Kurosaki et al. 2011; Liu et al. 2012; Yoon et al. 2012; Ture et al. 2012), while CT is more useful in demonstrating calcifications (Kwee 2010; Korn et al. 2015). The characterization of the plaque will give information about its evolution, considering that the presence of calcifications is reported to be a sign of a more stable plaque (Kwee 2010; Korn et al. 2015), while thin or ruptured fibrous cap a rich lipid core and intraplaque hemorrhage occur more frequently in unstable plaques and so increasing the risk of stroke (Watanabe and Nakayama 2010; Kurosaki et al. 2011; Liu et al. 2012; Treiman et al. 2015). The morphologic aspect of the plaque, together with other aspects of the cerebral circulation especially the presence of absence of a collateral circulation as well the age, gender, and general clinical condition of the patient will influence the kind of therapy (medical or more aggressive approach), and also which kind of approach either the endovascular or the endarterectomy should be chosen. All these aspects should be considered in the attempt to offer the patient the more appropriate treatment (Rosencranz and Gerloff 2010).

- *Atheroma can also occur in the intracranial segments of the ICA*, typically in the distal petrous and cavernous segment. These changes have already been described in the past by some authors on autoptic studies (Yates and Hutchinson 1961; Castaigne et al. 1970). Recently, some authors have shown on autoptic (Mazighi et al. 2008) and on MR studies (Lee et al. 2015) that even subtle atheromasic changes without clear stenotic character are frequently present in the distal

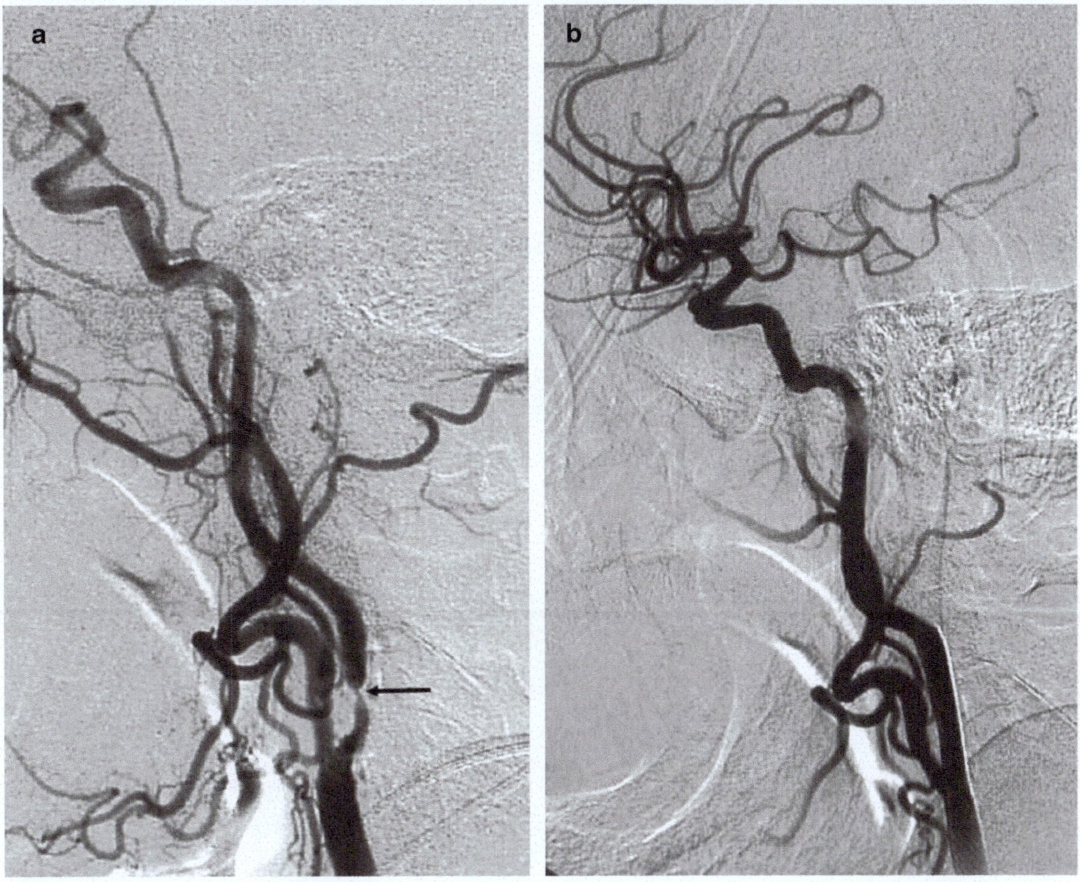

Fig. 15.2 Two examples of atherosclerotic plaque leading to severe stenosis of the ICA in patients with TIAs. (**a**) Carotid angiogram showing the stenosis (*arrow*) with slowing of the flow. (**b**) Angiogram after angioplasty and stent showing normalization of the lumen and flow. (**c**) Carotid angiogram showing the stenosis (*arrow*). (**d**) Angiogram after angioplasty and stent with normalization of the lumen and improvement in flow

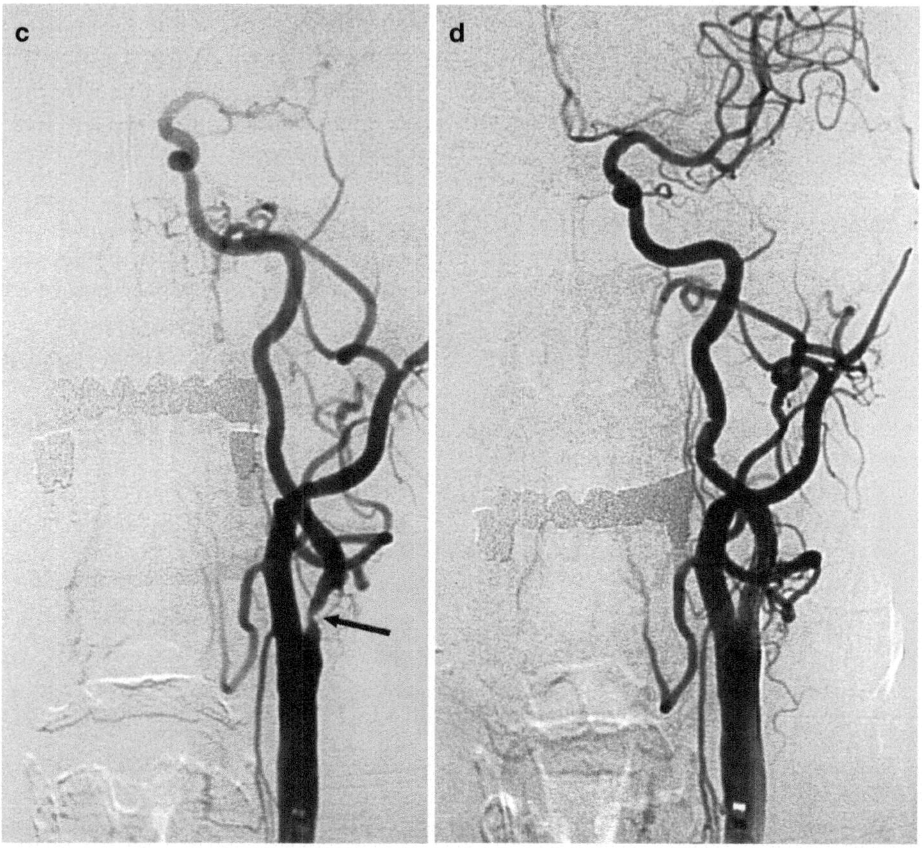

Fig. 15.2 (continued)

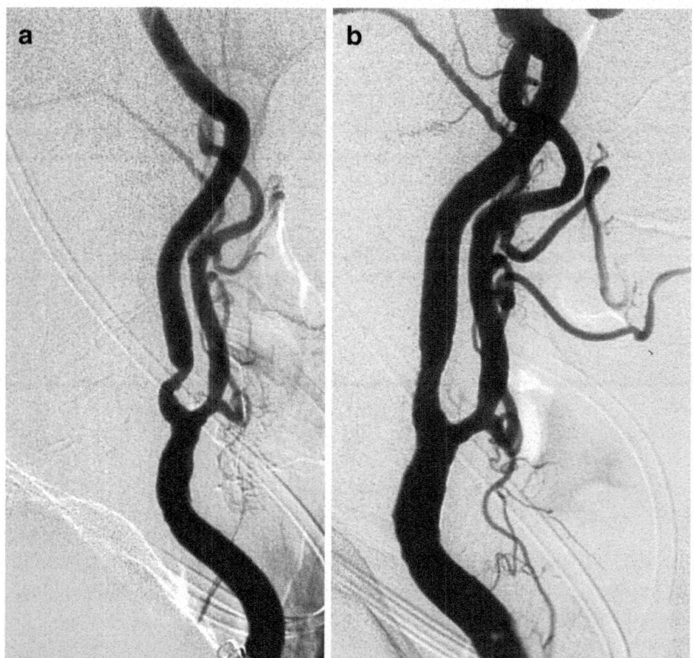

Fig. 15.3 Very old female patient with asymptomatic moderate to severe stenosis of the left ICA (**a**). The ICA was an important collateral way through the circle of Willis for the right carotid sector since the right ICA was occluded. A preventive endovascular treatment with stent and angioplasty (**b**) was performed without any clinical problems

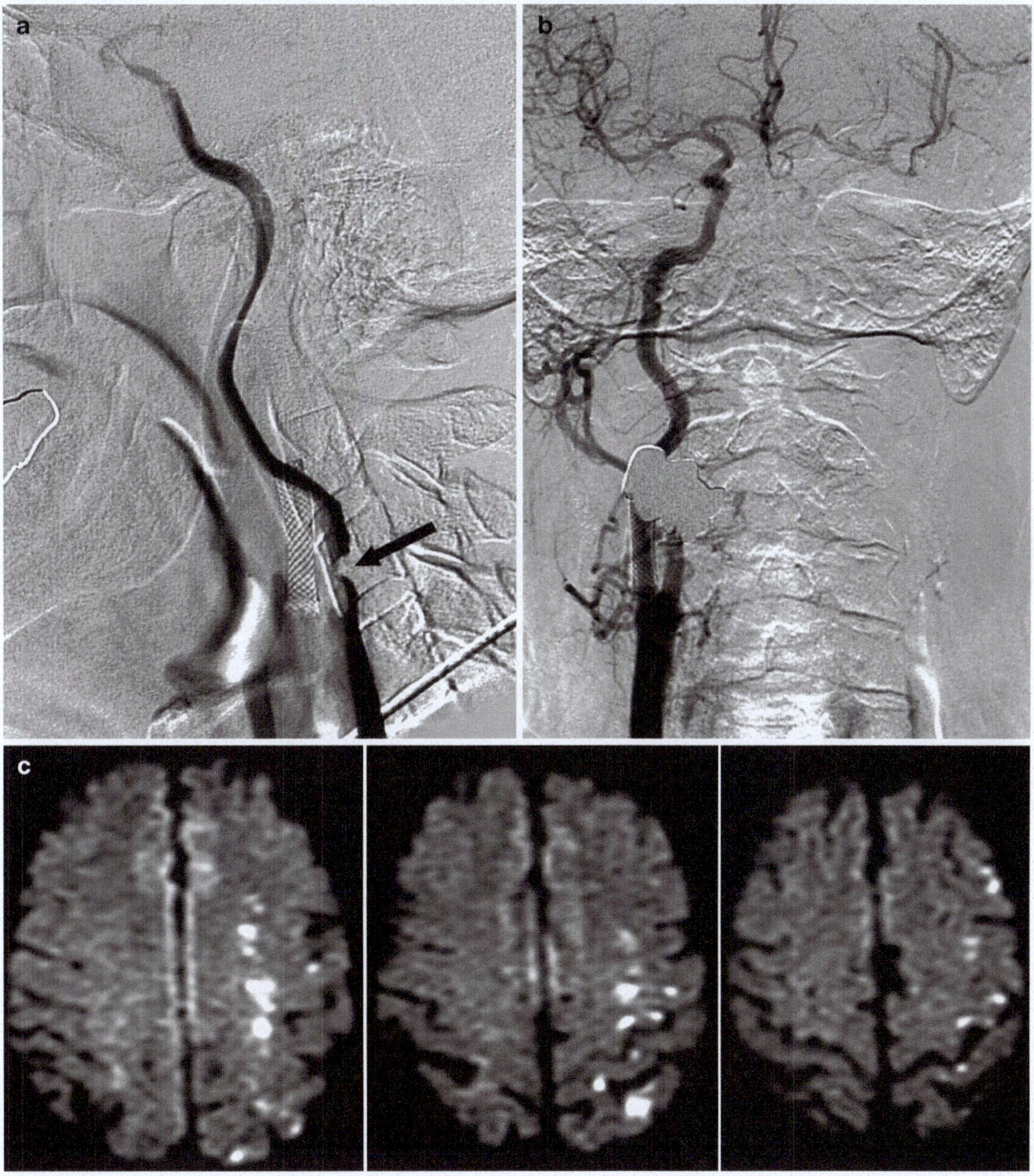

Fig. 15.4 Patient already treated with stent for stenosis of the right ICA. Acute mild neurological symptoms owing to impairment of the circulation in the left cerebral hemisphere. (**a**) left ICA angiogram showing subocclusion of the ICA due to a large atheroma (*arrow*) responsible for severe flow impairment. (**b**) Right ICA angiogram showing a good collateral circulation to the left vascular territory. (**c**) CT was normal. MRI, diffusion-weighted images showed several ischemic lesions in the vascular territory of the medullary arteries, probably due to hypoperfusion. The small cortical lesions are probably embolic. Surgical treatment of the stenosis was successfully performed a week later, with good clinical results

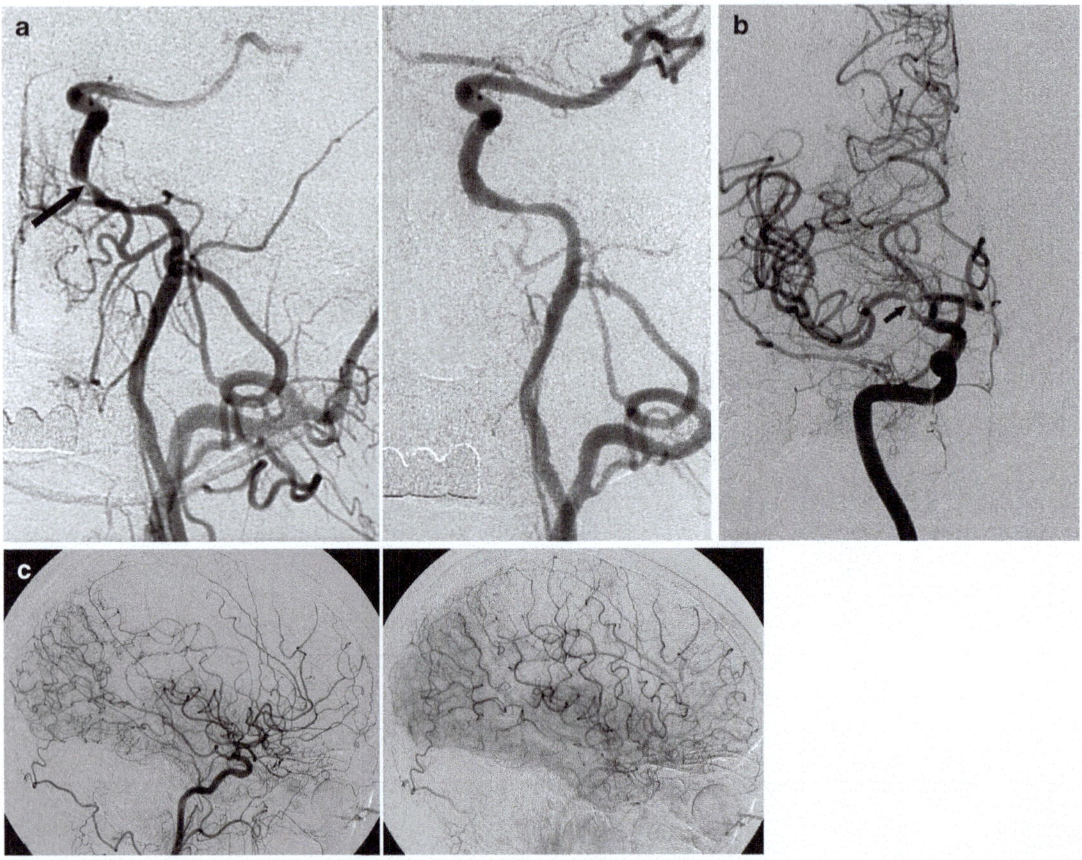

Fig. 15.5 (**a**) Middle-aged patient with TIAs not responsive to medical therapy. Carotid angiogram. AP view. On the left, angiogram showing severe stenosis of the proximal intracavernous portion of the ICA (*arrow*). There is impairment of the intracranial flow. On the right, angiogram postendovascular treatment with angioplasty and stent with restoration of the intracranial circulation. (**b**), (**c**) Middle-aged hypertensive black woman presenting with stroke characterized by left hemiplegia. (**b**) On the AP view of the right carotid angiogram a severe stenosis of the distal ICA (*arrow*) involving also partially the M1 segment is recognizable. (**c**) Lateral view. Early and late phases. There is a slowing down in the filling of distal branches of the MCA and ACA. Retrograde injection of some branches of the MCA (*arrows*) through opening of leptomeningeal anastomoses

ICA in patient who suffered an ischemic stroke. Other authors (Hong et al. 2011) showed a correlation with the degree of the carotid siphon calcification and the occurrence of lacunar infarct and concluded that this finding can be used to predict the risk of lacunar infarct.

The lesion can be associated with an extra cranial plaque *(tandem lesions)*. This condition has been reported to occur with a frequency of about 10% (Craig et al. 1982; Kappelle et al. 1999). In a more recent study (Marquering et al. 2013) performed in patients with recent TIAs or stroke, examined with CT, the association of extra-intracranial stenosis was diagnosed in 38% of the cases. The intracranial stenosis was between 30 and 50% of the lumen. Its frequency and severity increased in patients with an extracranial stenosis of or more than 70%.

All these studies indicate that attention should be paid to the presence of atheromasic changes of intracranial ICA. The clinical relevance of these lesions is, however, not definitely clear. Certainly, in some cases, these can grow to severe stenosis responsible for stroke (Fig. 15.5a–c). Furthermore it is possible that even subtle changes can favor the distal occlusion of ICA through emboli arising from plaques of the extracranial segment or of cardiac origin.

- *Another site of atheromasic plaques which has progressively gained more attention and can be the source of emboli is the aortic arch.* These atheromas are more frequent in old patients and their frequency increases in patients older than 60 years of age (Amarenco et al. 1992, 1994a, b). Other studies have shown that the risk of stroke increases when the plaque is larger (thickness more than 4 mm) and is ulcerated or there is a superimposed thrombus (Uemo et al. 2007; Di Tullio et al. 2008). In the diagnosis of these lesions, the transesophageal echocardiography and MRI as a screening method and in the acute phase CT angiography are very accurate diagnostic tools.

15.4.1.2 Findings in Angiographic Studies in Patients with Stroke

- *The study can show stenosis of different degrees at the ICA bifurcation,* frequently not associated with pathological findings intracranially. These patients present frequently with mild stroke or transient ischemic attacks (TIAs). The cause of the transient ischemia can be acute hypoperfusion or a temporary occlusion of cerebral arteries followed by a spontaneous rapid thrombolysis. Distal embolization and hypoperfusion can occur together (Fig. 15.4). Among the involved branches, there is the ophthalmic artery. Its occlusion leads to "amaurosis," which frequently is transient (*fugax*) due to the spontaneous recanalization; unfortunately sometimes the lesion is irreversible.
- In the acute phase of the stroke a medical therapy is commonly the treatment of choice. In the following days or weeks with the clinical improvement, the type of therapy will change depending on the grade of stenosis, the morphology of the plaque, and the type and extension of the ischemia. Taking all these aspects into consideration in patients with moderate–severe stenosis and in those in whom the plaque appears unstable a more aggressive therapy (angioplasty with stent, or endarterectomy) will be considered. Whether an endovascular or a surgical approach is chosen varies in the different centers depending mainly on the experience and obtained results of the teams involved. Some criterions, however, are commonly accepted. Endarterectomy is favored in older patients and in those in whom diffuse atherosclerotic changes with great tortuously especially of the carotid artery will make difficult and more risky an endovascular approach. This latter is performed in patients with high carotid bifurcation as well as in those in whom the contralateral carotid is occluded or other severe comorbidities are present. The type of plaque will also influence the type of approach considering that as also reported by other authors (Mansur et al. 2011; Yoon et al. 2012) the risk of endovascular treatment can be higher in the cases of unstable plaques.
- Endovascular or surgical therapy can be indicated in asymptomatic patients with severe stenosis. This can be performed also in stenosis of lower grade in those patients in whom the involved carotid, considering also the circle of Willis, is the main perfusion way of the brain since the contralateral carotid is occluded. Examples of ICA stenosis and treatment are presented in Figs. 15.2, 15.3, and 15.4.

- *In other cases, a major stroke occurs. The angiographic pattern can vary. Frequently, there is occlusion of the extracranial ICA, due to a thrombus superimposed on an atherosclerotic plaque. More rarely, the occlusion is due to a big embolus originating from the heart or from aortic arch, arrested often at the level of a tortuosity of the ICA. In both cases, the thrombus can extends intracranially or, more often, be the source of a distal embolization (tandem lesions). In these cases, the MCA and its branches are commonly* the vessels involved (see Sect. 15.4.2). Not so rare, however, there is occlusion of the distal ICA (Moehlenbruch et al. 2012; Mpotsaris et al. 2013; Maurer et al. 2015; Cohen et al. 2015) (see below). Improvement of the endovascular technique allows today, in many of these cases, a rapid reopening of the extra cranial (stenting with or without angioplasty) and intracranial a mechanical thrombectomy) occlusion. The occlusion at the ICA bifurcation can frequently be easily overcome by the devices used today, indicating that the occlusion is due to a fresh thrombus, formed on a preexisting atheromasic plaque. An interesting aspect, as described also by other authors, (Malik et al. 2011; Maurer et al. 2015) is that in a certain number of these patients, the opening of the extracranial ICA leads to a spontaneous resolution also of the intracranial lesion, probably due to the re-stablished flow favouring the thrombolysis of the distal fresh embolus.
- More rarely *the primary site of the occlusion is located in the intracranial segment of ICA, or more frequently the occlusion involves the MCA (see Sect.* 15.4.2). *The pathogenesis is thought to be in the majority of the cases a cardiac embolization, since no pathological changes are recognizable at the ICA bifurcation or along its course, while a cardiac disease can be identified in many of these patients* (Zaidat et al. 2002; Flint et al. 2007; Mpotsaris et al. 2013). *The source of emboli from atheromasic plaque in the aortic arch should also be considered.* The different location of the occlusion, either in the ICA or in the MCA depends probably on the size of the embolus which is larger in ICA occlusion. As described above, the frequent presence in the distal ICA, of atheromasic changes not necessarily leading to stenosis can be a factor favoring the occlusion. It can also occur that a retrograde extension of the thrombus down to the ICA bifurcation develops, making it difficult, in these cases to establish where the primary occlusive process has started (Castaigne et al. 1970).

In the *occlusions of the intracranial segment of ICA due to a primary occlusion or in tandem lesions, two different condition can basically be recognized*:

In one, the occlusion is located at the level of the ophthalmic artery commonly proximal to the origin of the artery. More rarely the occlusion is more distal involving the origin of the ophthalmic artery and the supraclinoid segment of ICA, but not its terminal branches (A1 and M 1) (Drawings 15.6a, c). *In these cases, the collateral circulation can be guaranteed through the circle of Willis provided that no anomalies are present and also through anastomoses between ECA and ophthalmic artery when the occlusion is proximal to the ophthalmic artery*. In many cases the collateral circulation is very efficient to protect the brain parenchyma avoiding ischemia. In cases of stroke, this can be due to the acute hypoperfusion linked to an insufficient collateral flow or frequently due to occlusion of intracranial branches by emboli arising from the occluding thrombus. These are sometimes indirectly recognizable on the performed contralateral angiograms of the ICA, VA, or ECA. *In some of these cases, especially when the patient is clinically stable, it is probably better to perform a more conservative therapy, instead to try to reopen the ICA, with the risk of further embolization*. This approach can be reconsidered when the collateral circulation is impaired due to anomalies of the circle of Willis or the presence of other lesions involving especially the other ICA, or there is a progressive clinical worsening.

In other cases the occlusion is more distally located, involving not only the supraclinoid segments of ICA (communicating and choroidal), but also the first segment of the MCA (M1) and the first segment of the ACA (A1) (Drawings 15.6b, d). *In some cases the occlusion does not extend to the more distal segments of the ACA which can be filled through the AcomA and so contribute to the collateral circulation through pial anastomoses with the MCA. This way of collateral circulation can be impaired when the occlusion extends also towards the A2.* The angiographic pattern in which both M1 and A1 are involved is called "*T occlusion.*" This is a rare condition, since in the majority of the cases the ICA occlusion extend only to the M1 forming a pattern called *"L occlusion"* or *partial "T occlusion"* (Liebeskind et al. 2015; Bradac et al. 2017). It is thinkable that in some of these cases a partial involvement of the proximal A1 is present, but this finding is difficult to be demonstrated.

The only possible collateral flow is through leptomeningeal anastomoses between ACA, PCA, and MCA. It can occur, in a few of these patients, that the collateral circulation is already rich, immediately after the occlusion explaining the mild clinical symptoms and the good outcome (Arnold et al. 2003; Flint et al. 2007; Bradac et al. 2017; Liebeskind et al. 2015) (Fig. 15.8). In the majority of the cases, however, the collateral circulation is poor and it is associated with a bad prognosis. The endovascular treatment leading to a reopening of the artery with complete or partial revascularization of the hemisphere had improved the prognosis in many of these patients (Castano et al. 2010; Moehlenbruch et al. 2012; Machi et al. 2012; Berkhemer et al. 2015; Bradac et al. 2017).

There are, however, some "unfavorable conditions" which can impair the final result. Among them there are anomalies of the circle of Willis: both ACAs arise from the occluded ICA and the A1 segment of the normal contralateral ICA is hypo-aplastic or the A1 segment is completely thrombosed up to the A1–A2 angle and so no collateral flow is possible from the normal A1 towards A2 of the affected side through the AcomA. The PCA (fetal origin) arises from the occluded ICA and the P1 is hypo-aplastic. The collateral flow through leptomeningeal anastomoses from PCA towards ACA and MCA is impaired. This latter condition can also be unfavorable for a collateral circulation towards the AchA since in this case this artery cannot be retrograde injected by the way of anastomosis with branches of PCA, especially through the posterior choroidal arteries.

Another possible condition influencing negatively the result is the involvement of the collateral circulation through a distal embolization leading to occlusion of the arteries involved in the leptomeningeal collaterals. Emboli can arise from the occluding thrombus or occurring during the endovascular treatment. This involvement of the collateral circulation leads to a suffering of the brain parenchyma and this can explain, in these cases, the failure of its recovery, in spite of the rapid proximal revascularization obtained with the endovascular treatment, as also reported by other authors (Bang et al. 2011).

Examples of distal ICA involvement in cases of tandem lesions or primary occlusion are presented in Figs. 15.7, 15.8, 15.9, 15.10, 15.11, 15.12, and 15.13. In a few of these cases, endovascular treatment makes it possible to reopen the distal ICA with restoration of the collateral circulation through the circle of Willis and ECA, while the proximal ICA remain occluded. In accordance with other authors (Jakubowska et al. 2008) we think that this can be an acceptable result, which can be sufficient to avoid a bad clinical course (Fig. 15.14a– d).

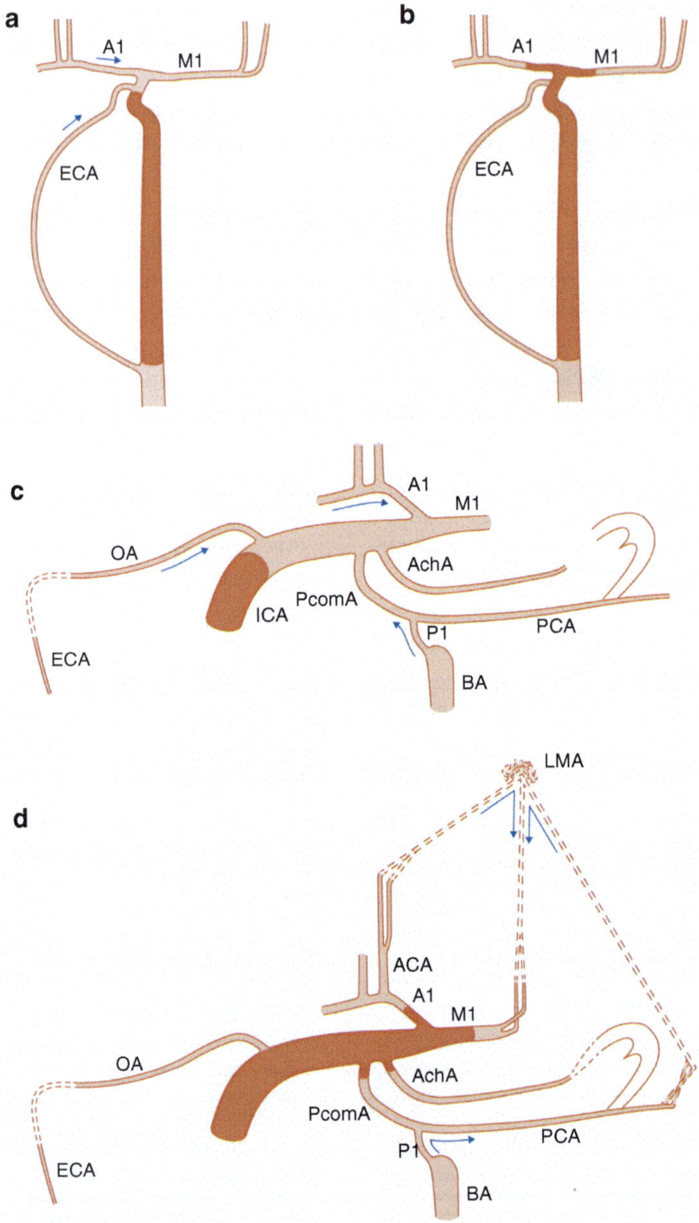

Fig. 15.6 Drawing, AP view. Different findings in occlusion of the ICA proximal to the ophthalmic artery (**a**) and distal to it with involvement of communicating, choroidal, A1 and M1 segments (T occlusion) (**b**). In (**a**) a collateral circulation occurs through the ECA-ophthalmic artery anastomosis and through the circle of Willis provided that this is not impaired by hypo/aplasia of the A1. In (**b**), the above-described collateral ways are involved in the thrombotic process. The only possible collateral circulation is leptomeningeal anastomoses between distal branches of anterior cerebral (ACA) and posterior cerebral arteries (PCA) with middle cerebral artery (MCA). (**c**) Lateral view. Occlusion of the internal carotid artery (ICA) proximal to the ophthalmic artery (OA). Possible collateral circulation through OA and circle of Willis. External carotid artery (ECA); first segment of anterior cerebral artery (A1); first segment of middle cerebral artery (M1); posterior cerebral artery (PCA); posterior communicating artery (PcomA); basilar artery (BA) and P1 segment; and anterior choroidal artery (AchA). (**d**) Distal occlusion of ICA involving the communicating, choroidal, M1, and A1 segments. Possible collateral circulation only through leptomeningeal anastomoses (LMA). Possible anastomoses between posterior and anterior choroidal arteries. For the role played by anomalies of Circle of Willis see text.

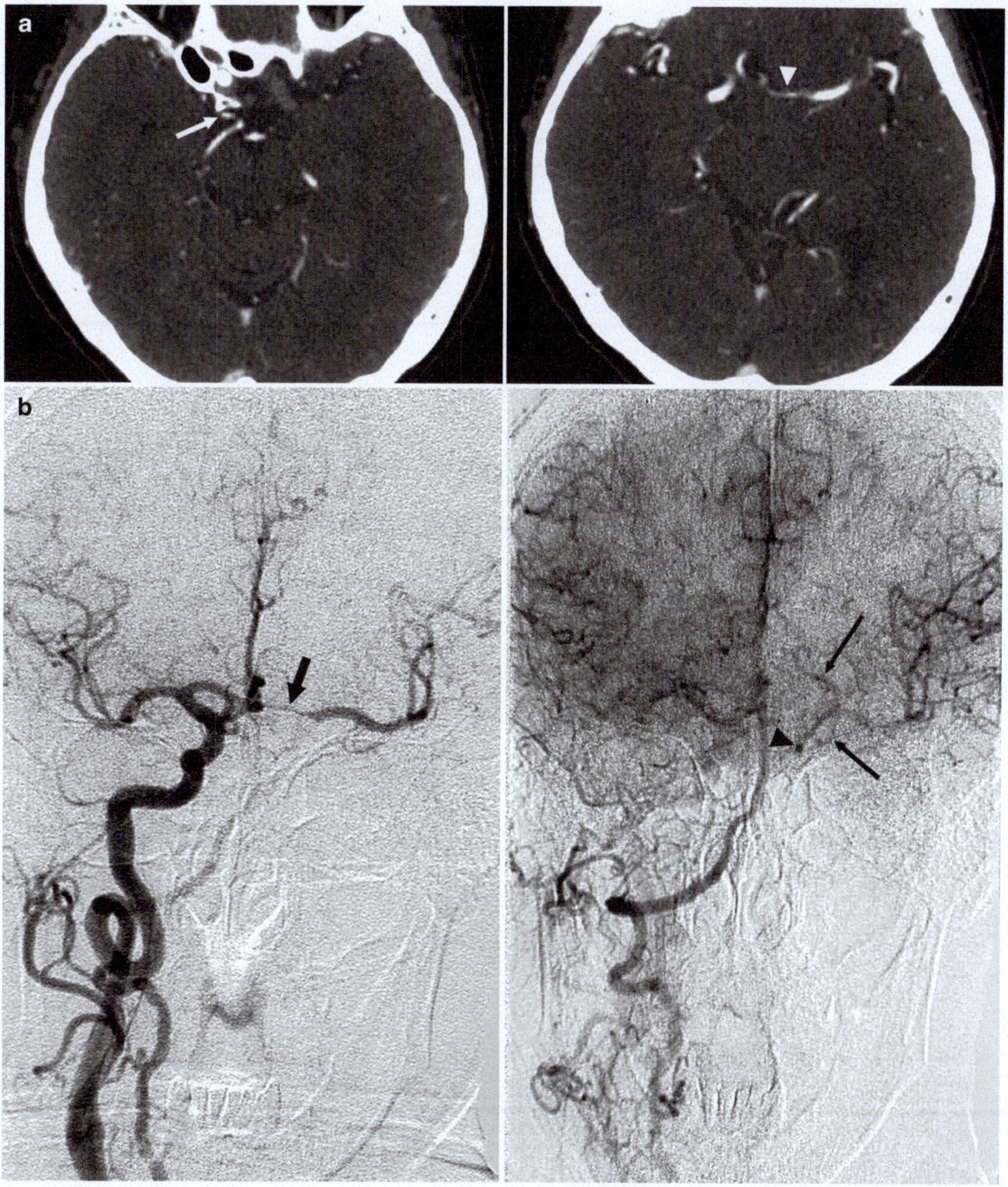

Fig. 15.7 Acute occlusion of the left extracranial ICA, probably embolic in a cardiopathic middle aged woman, at the level of tight coiling. Further distal embolization and occlusion of the distal intracranial ICA (tandem lesion). (**a**) Angio-CT. The right PCA (fetal type) is filled by a large PcomA (*arrow*). Distal left ICA occlusion. Good filling of the left MCA through A1 (*arrowhead*) in spite of its small size. The left PCA is of a fetal type, arising mainly from the occluded ICA. (**b**) Right carotid angiogram. Tight coiling of the ICA. Hypoplastic left A1 (*arrow*). There is, however, a good filling of the left MCA and a faint retrograde injection of the left PCA and PcomA (*arrows*) to the point of the occlusion of ICA (*arrowhead*). (**c**) Angiogram of the left ICA, showing the occlusion at the very tight coiling (*arrow*) associate with a distal occlusion (*arrow with dot*). Progressively proximal and distal reopening with aspiration device. (**d**) Left ICA angiogram posttreatment. Now the anomaly of the circle of Willis with the fetal origin of the PCA is better demonstrated. On the posttreatment CT, ischemic lesions were recognizable on the basal ganglia. The patient had a mild motor deficit which improved in the next months with a final good clinical outcome (Courtesy of Prof. Bergui M and Dr Ventilii G, Neuroradiology of Turin)

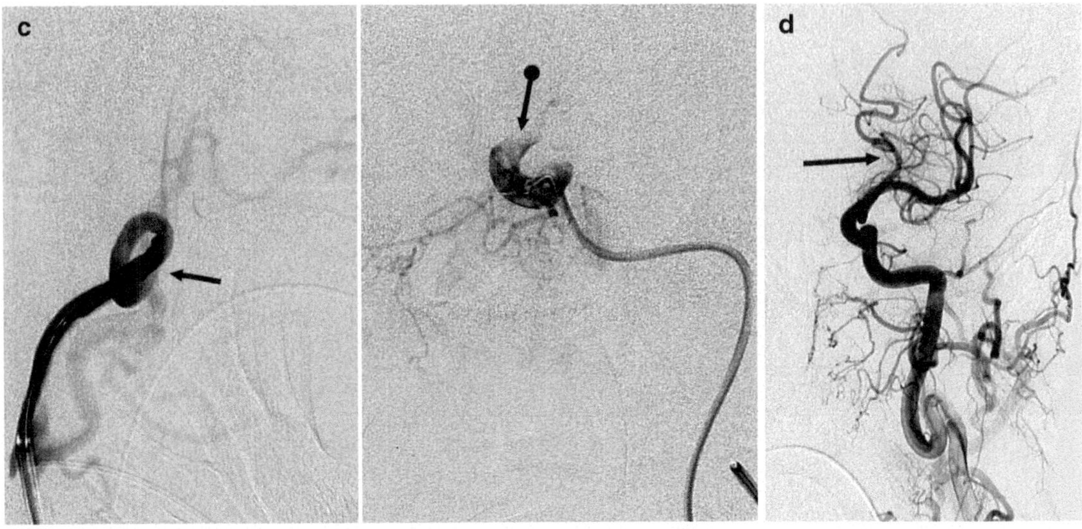

Fig. 15.7 (continued)

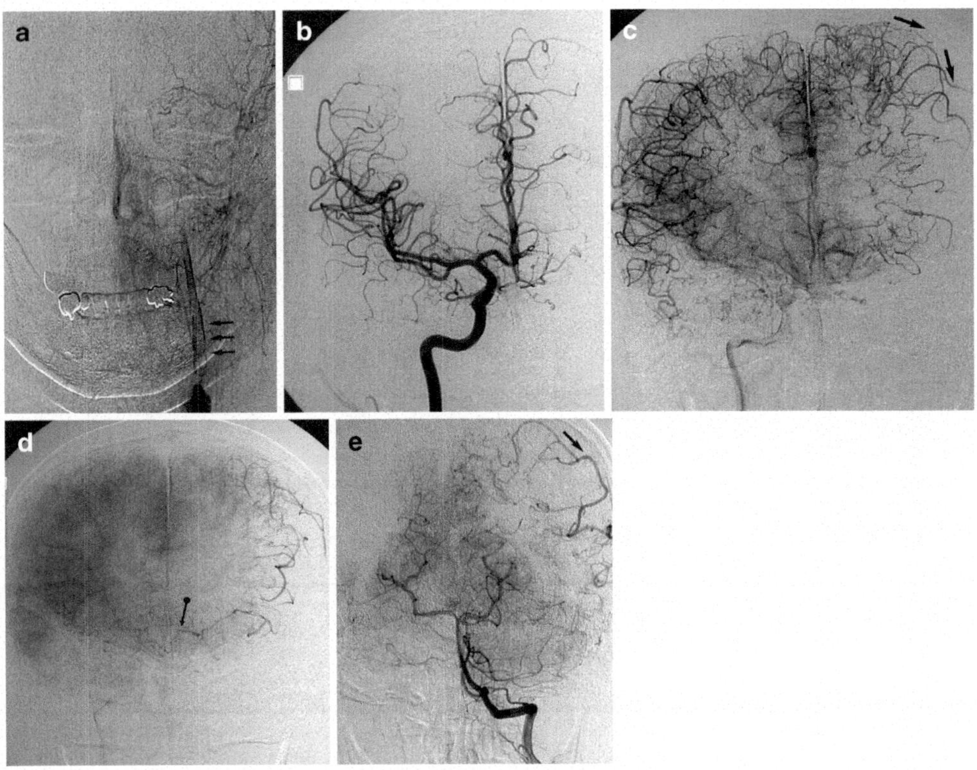

Fig. 15.8 Middle-aged patient with acute onset of right hemiplegia and minimal aphasic disturbances. (**a**) Severe stenosis of left ICA bifurcation with superimposed thrombus extending intracranially. (**b**)–(**d**) Right ICA angiogram (early and late phases). Through the AcomA partial filling of the left A1 (*arrow with dot*). There is a rich leptomeningeal collateral circulation (*arrow*) through distal branches of left A.C.A towards the left MCA. (**e**) On the vertebral angiogram the origin of the PCAs from the BA is demonstrated. A collateral circulation from the left PCA towards the MCA is recognizable (*arrow*). The patient presented with a stable mild clinical symptomatic. Medical therapy was chosen and the patient recovered with minimal neurological deficit

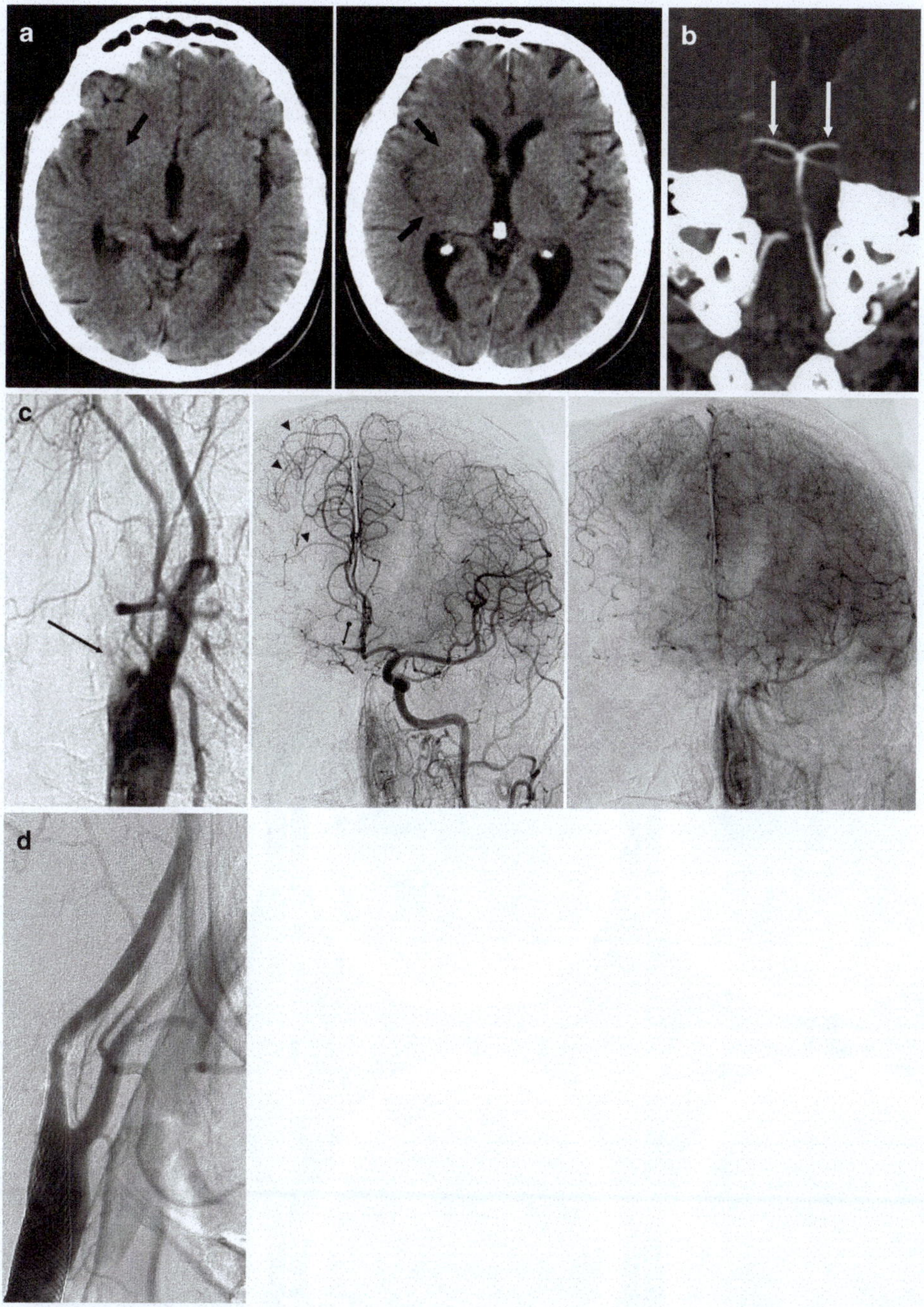

Fig. 15.9 70-year-old patient examined about 6 h after the stroke presenting with a mild agitation and left hemiplegia. (**a**) CT showing a probable ischemia involving the insula and right basal ganglia (*arrows*). (**b**) On the CT angiography, occlusion of the right ICA was visible. There was a partial collateral circulation through the anterior and posterior part of the circle of Willis. Both PCA arise from the BA (*arrow*). (**c**) Right and left carotid angiogram showed, occlusion of the proximal right ICA (*arrow*) and partial filling of the right A1 (*arrow with dot*) through the AcomA. Partial collateral leptomeningeal circulation (*arrowheads*) from distal right ACA towards the right MCA. (**d**) Reopening of the right ICA with angioplasty and stent. (**e**) Injection of the distal occluded ICA (*arrow*) and progressively reopening with a mechanical thrombectomy (solitaire) of the distal ICA and further of the MCA with complete recanalization. (**f**) Posttreatment CT showing presence of contrast in the infarcted areas (*above*) involving the basal ganglia and the insular and parietal areas. Progressively disappearing of the contrast in the following days (*below*). Slowly clinical improvement of the left hemiplegia in the following months

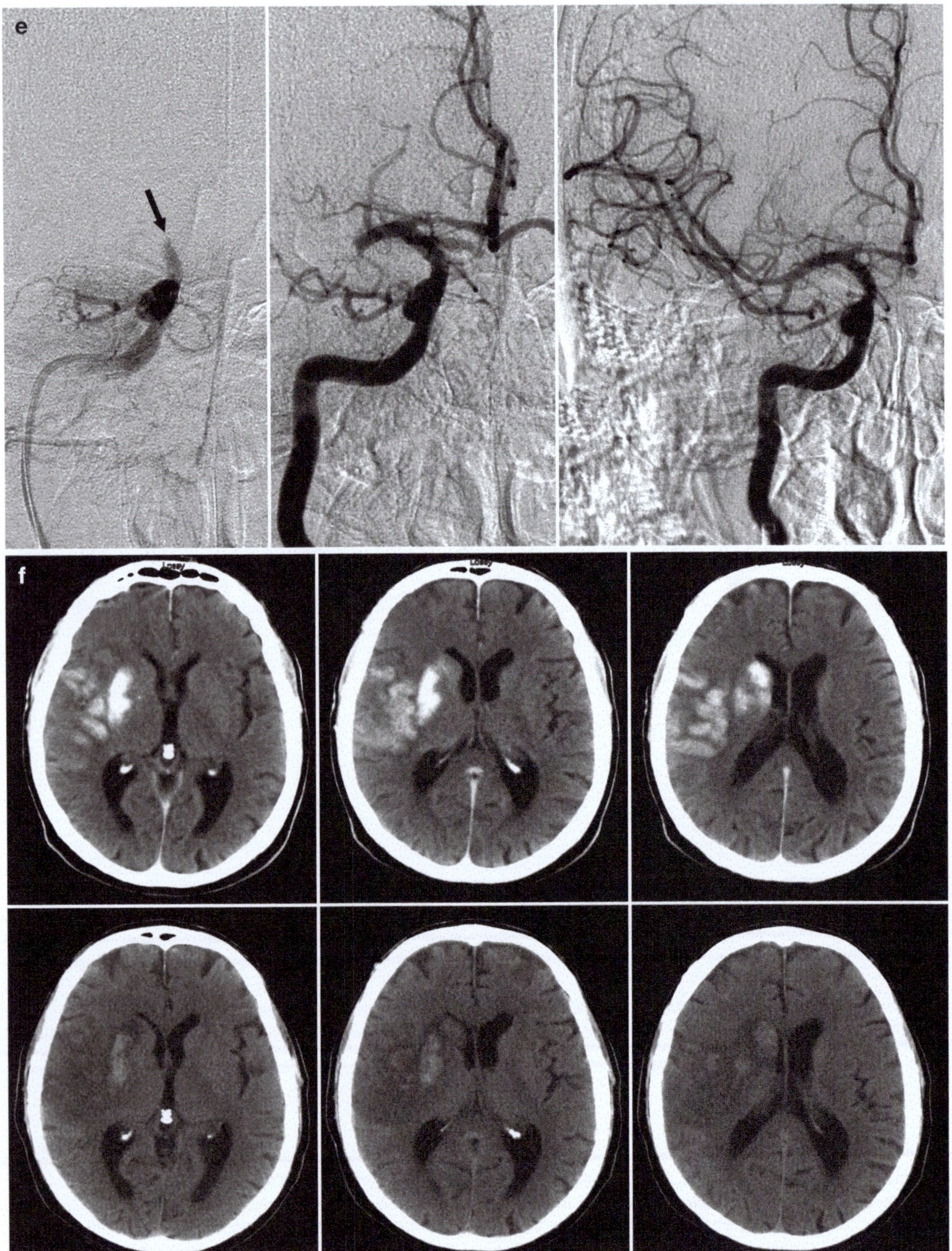

Fig. 15.9 (continued)

Furthermore, it should be remembered that in ICA "*distal occlusion*" the perforators of ACA, MCA, AchA, and distal ICA and depending on the preexisting anomalies of the circle of Willis, also of PcomA, are always involved. This explains the frequent presence of ischemic lesions in the basal ganglia, even in the most favorable cases with a rapid and complete recanalization.

- A particular rare condition is that occurring in patients, who have a known occluded ICA, in whom ischemia occurs months or years later. In such cases, the etiopathogenesis is frequently linked to emboli arisen from atheroma in the stump of the occluded ICA, in the distal common carotid artery or ECA, which can reach the intracranial vessels though anastomoses between the branches of ECA and ophthalmic artery (Barnett et al. 1978; Countee and Vijayanathan 1979; Countee et al. 1981; Bradac and Oberson 1983).

15.4.2 Middle Cerebral Artery

- The MCA is the most frequently vessel involved in stroke of the anterior circulation. Similarly to the distal ICA occlusion, also the occlusion of the MCA, in the majority of the cases, is embolic (Zaidat et al. 2002; Mpotsaris et al. 2013). The source of the emboli is thought to be most commonly cardiac occurring in patients with cardiac diseases. Another possible mechanism are emboli arising from thrombi formed at the ICA bifurcation typically in the cases of "tandem lesions." In a certain number of cases, however, the source of the emboli remains unclear, since neither a cardiac nor a carotid pathology can be identified. It could be that in these patients, the thromboembolic material has been formed on minimal atheromasic changes no longer visible on the angiogram or the source was more proximal, located in the aortic arch.

The occlusion involves the M1 segment (Figs. 15.15, 15.16, 15.17, 15.18, 15.19, 15.20, and 15.21) or more distal branches Figs. 18.2, 18.3, and 18.4.

In some cases, after successful reopening of the vessel, it appears clear that the embolus has been arrested at the level of a stenosis of M1 due to a microatheroma (Fig. 15.16a, b). In occlusion of the M1, the perforators are always involved. The degree of involvement depends on many factors (location in the proximal or more distal M1 of the embolus, morphological aspects of M1, short or long and on the site of origin of the perforators (see Sect. 5.1). *Perforators are end arteries, and so their occlusion, even for a relatively short time, can lead to ischemia in their territories.* This is confirmed in those cases where despite a rapid reopening of the occluded M1 through the endovascular treatment, ischemic lesions in the basal ganglia are frequently demonstrated in CT or MR, which should always performed 24–48 h later (Bradac et al. 2008b) (Figs. 15.15, 15.18, 15.19, 15.20, and 15.21). Frequently, these patients remain asymptomatic or only with minimal neurological deficit. Certainly the lesions are sometimes very limited. However, the absence or the presence only of minimal clinical symptoms in these patients is a little surprising, since it is well known that the basal ganglia are involved in brain activity concerning movements, cognition and behavior. Indeed many connections between associations cortical areas and deep structures in form of cortico-striato-pallido-nigral-thalamic-cortico circuit are well known (Yeterian and Van Hoesen 1978; Alexander et al. 1986; DeLong 2000). Complementary functional MR study could be useful in the follow-up of these patients.

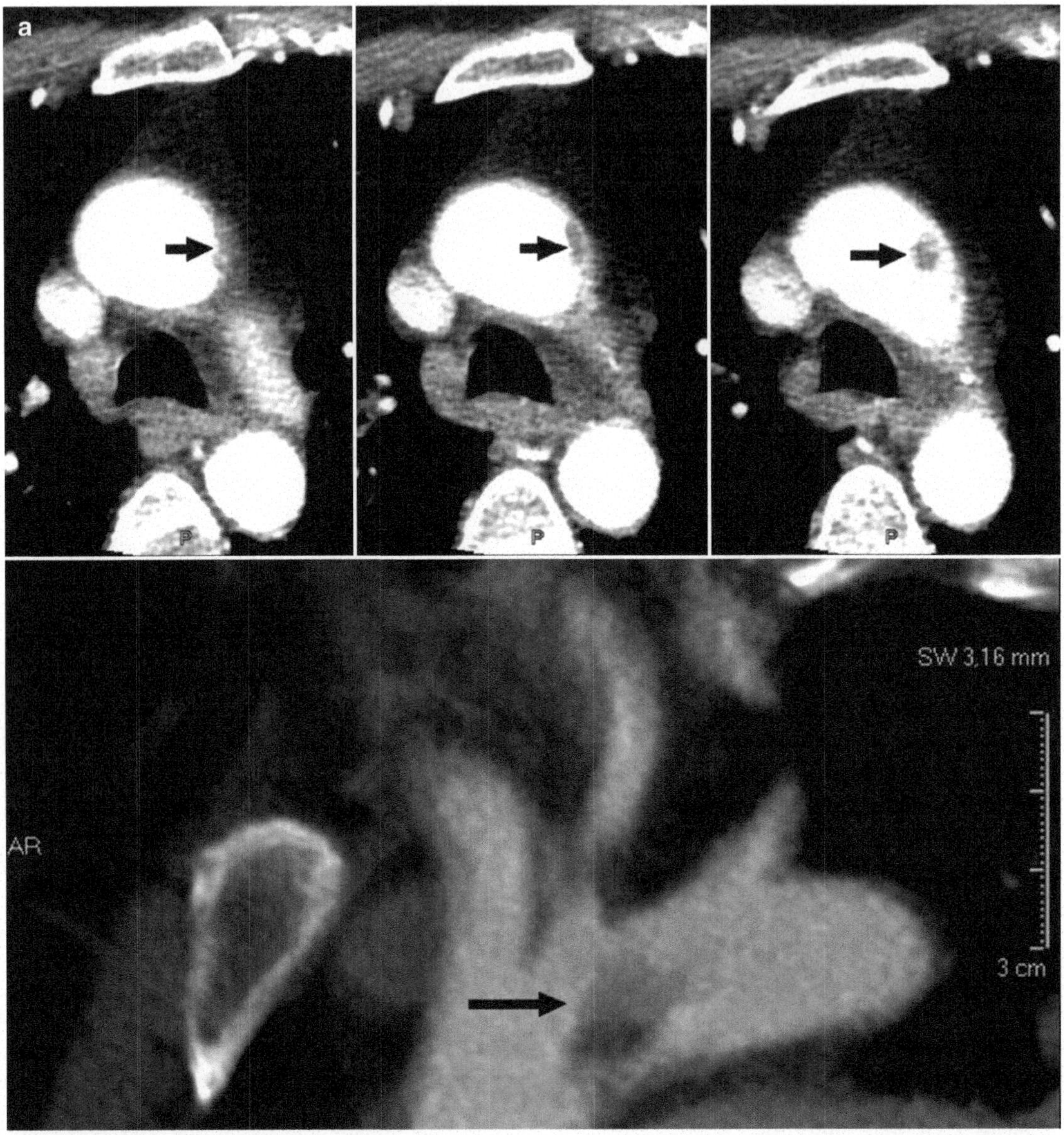

Fig. 15.10 50-year-old woman admitted in a comatose state with right hemiplegia about 3 h after the stroke. (**a**) On CT no certain ischemic lesions were visible. The CT angiography showed a large plaque with a superimposed thrombus (*arrows*) in the aortic arch. (**b**) The intracranial study (volume rendering) showed a distal occlusion probably embolic of the left ICA (*arrow*) (sides inverted). The occlusion was located distally to the origin of the small PcomA (*arrowhead*) connected with the PCA arising mainly from the BA. There was a good filling of the right ACA but no filling of the left A1 and M1. The leptomeningeal collateral circulation towards the occluded ACA and MCA probably only possible through the left PCA was very poor. (**c**) The right and left angiograms confirmed the "T Occlusion" of the left ICA (*arrow*). The occlusion was distal to the origin of the small PcomA (*arrowhead*) continuing in the PCA. Retrograde filling of the large P1 (*arrow with dot*). (**d**) Rapid reopening of the occlusion using a solitaire stent. Note also the filling of the perforators. The distal branches of ACA and MCA, however, remained occluded (*arrows*) in spite to a few attempts to reopen them, probably occluded by emboli arising from the proximal thrombus and/or due to severe impairment of the microvasculature (see Sect. 15.8). (**e**) CT performed 24 h later showed a large infarct involving the ACA, MCA, and AchA vascular territories, sparing the territory of the PCA. Minimal hemorrhagic component in the basal ganglia. The patient died

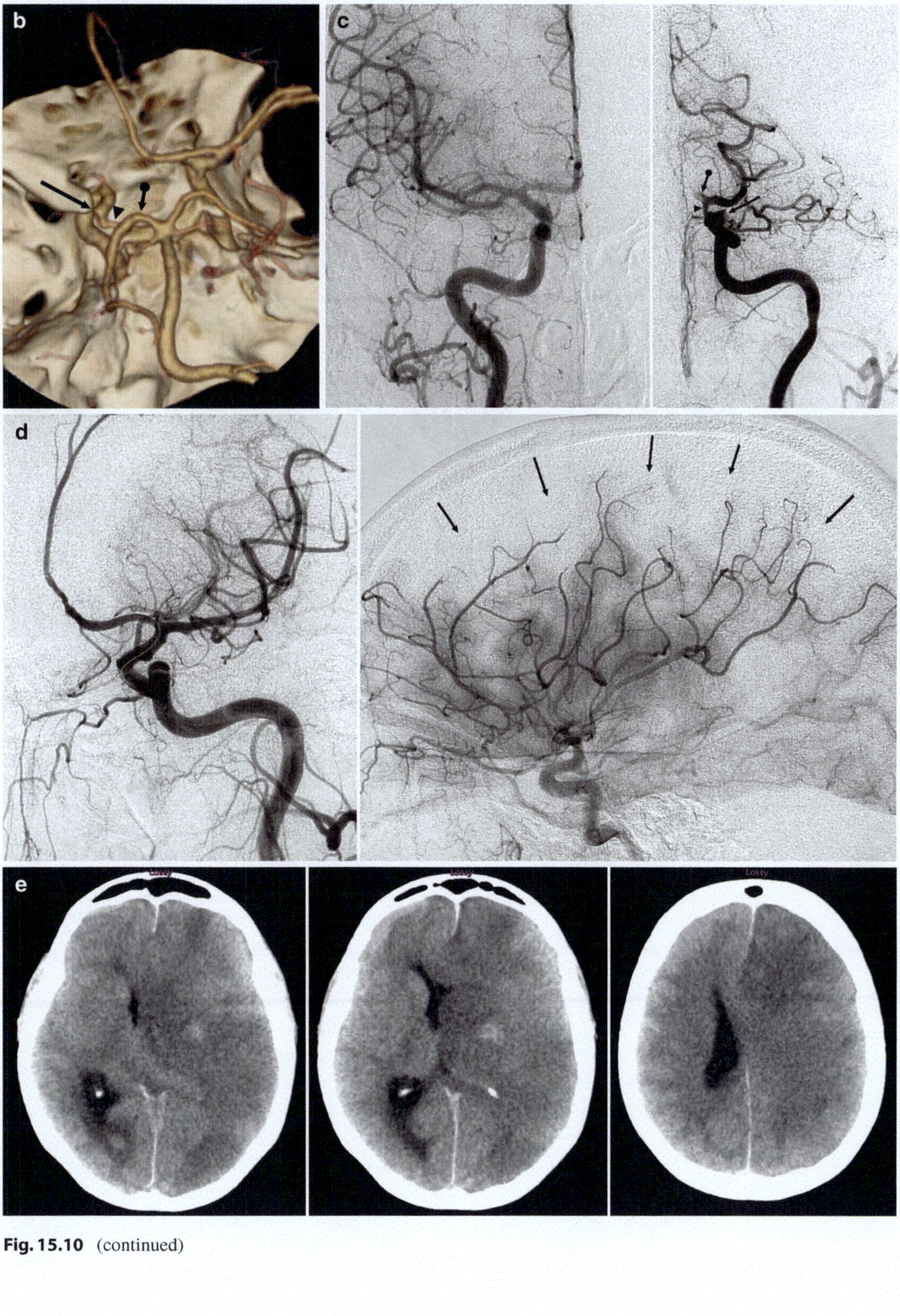

Fig. 15.10 (continued)

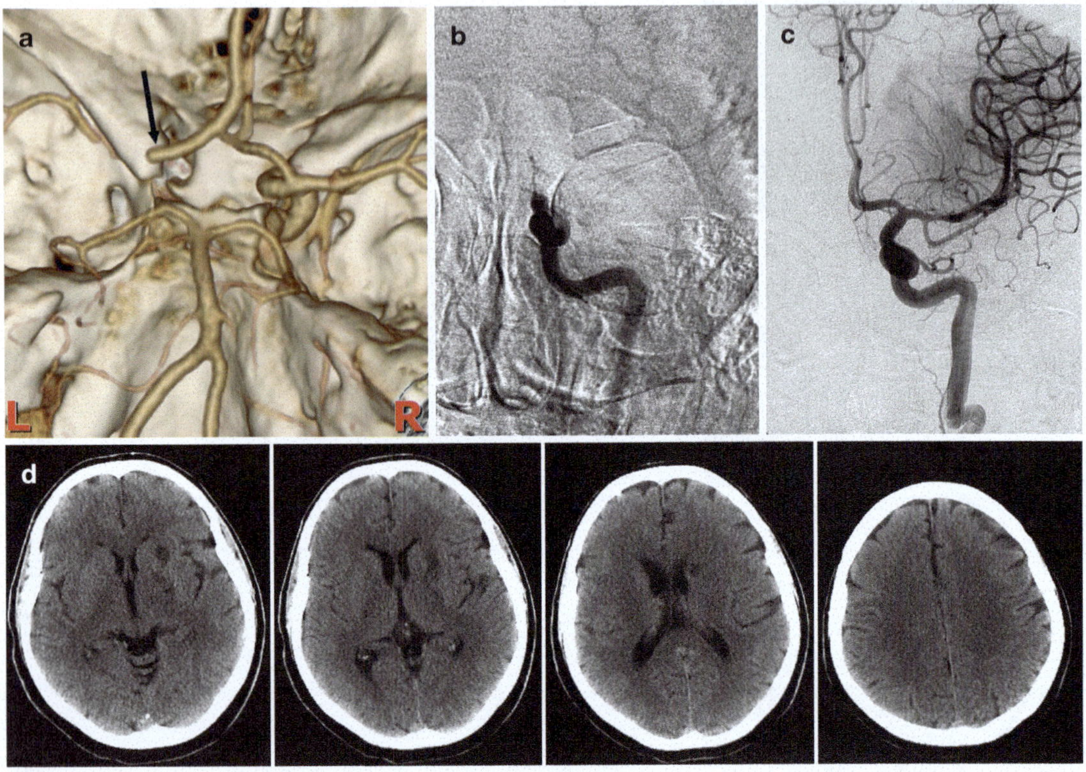

Fig. 15.11 Distal intracranial ICA occlusion in a 58-year-old woman. (**a**) Angio-CT represented by volume rendering technique (the sides are inverted). Occlusion of the left ICA. Through the AcomA filling of a well-developed left ACA and left A1 (*arrow*). There is a distal filling of M1, indicating the presence of a rich leptomeningeal collateral circulation towards the left MCA from the left ACA and probably also left PCA, considering that both PCAs arise from the basilar artery. (**b**) Left carotid angiogram showing the distal ICA occlusion. (**c**) Left posttreatment angiogram showing the complete revascularization obtained with a solitaire stent. Blush in the vascular territory of the perforators. (**d**) Posttreatment CT. Small ischemia at the level of the basal ganglia. Rapid clinical complete recovery

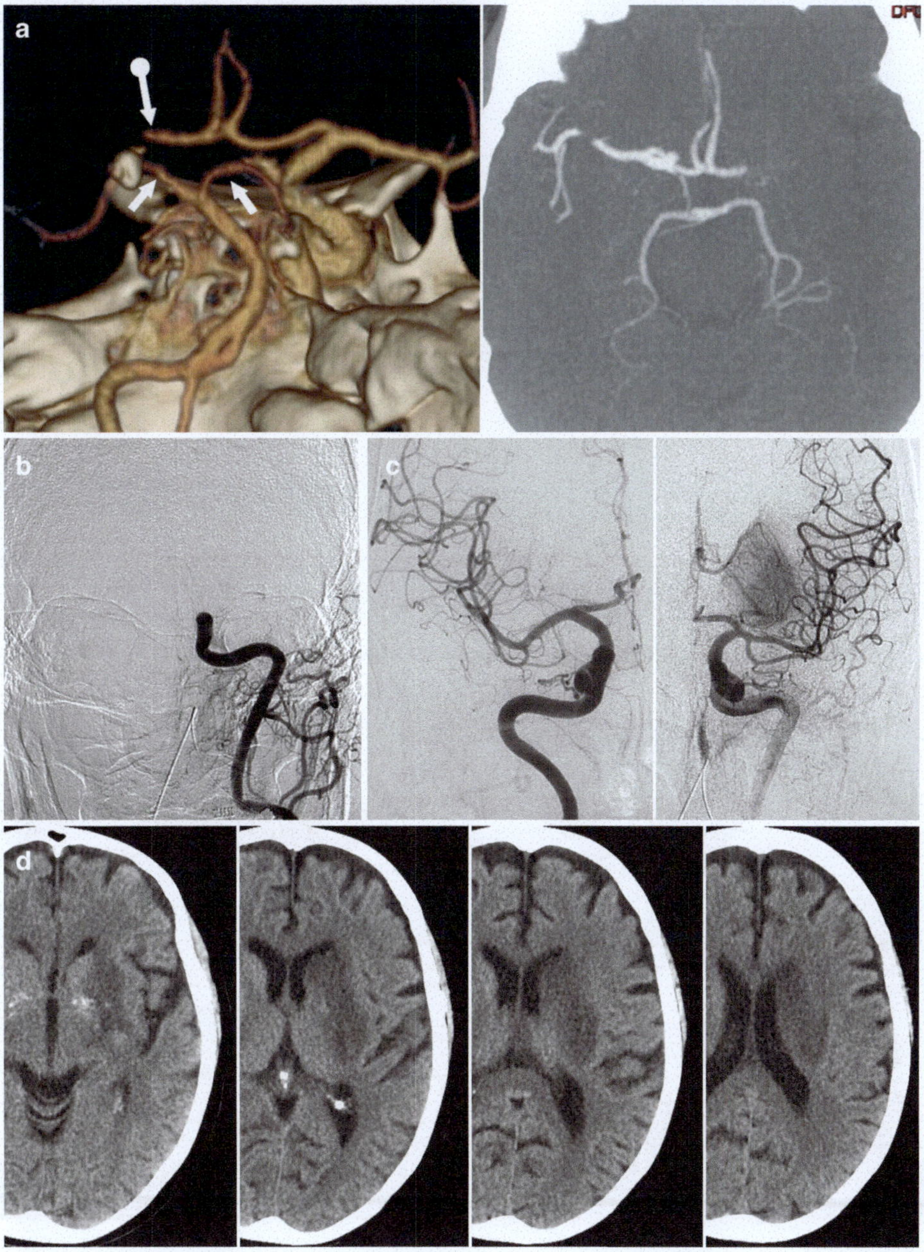

Fig. 15.12 Distal intracranial ICA occlusion in a 68-year-old woman. (**a**) Angio-CT represented by volume rendering (*left*) and multiplanar reconstruction (*right*). Occlusion of the left ICA. Through an efficient AcomA filling of the left ACA and left A1 (*arrow with dot*). Fenestration of the BA from which arise Both PCAs (*arrows*). There is also a right PcomA on the right. There was a collateral circulation towards the MCA. (**b**) Left carotid angiogram showing the ICA occlusion. (**c**) Post-treatment left ICA angiogram showing the complete revascularization after thrombectomy with solitaire stent. Note the rich blush in the vascular territory of the perforators and early filling of the inferior striate veins draining in the BV and superior striate veins draining in the thalamo-striate vein. Right ICA angiogram (*right on the figure*). (**d**) Posttreatment CT. Ischemia involving pallidum, putamen, and superior part of the head of the nucleus caudatus. Improvement of the motor deficit and mild aphasia with final good recovery 3 months later

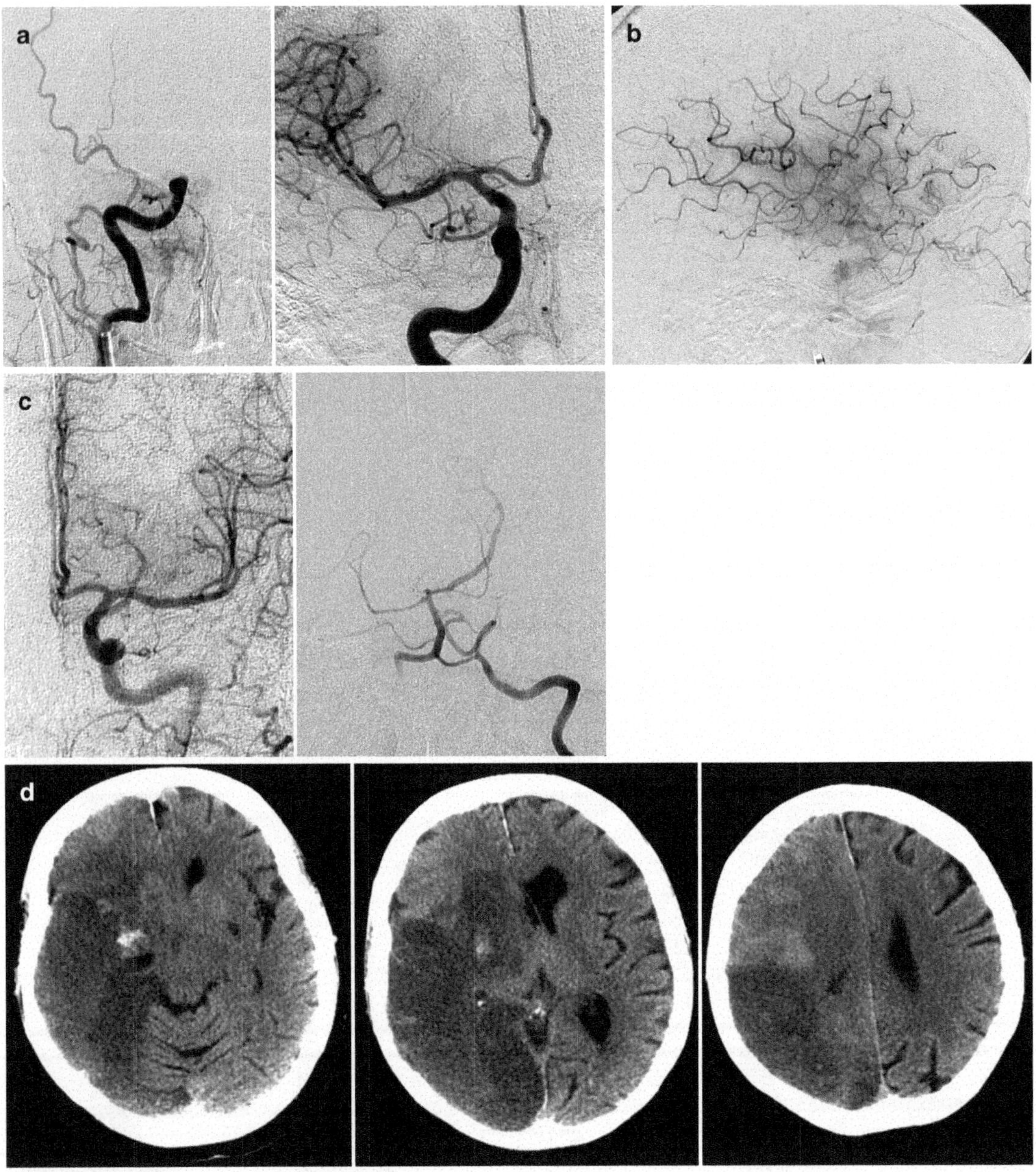

Fig. 15.13 Distal intracranial ICA occlusion in a 70-year-old woman. After exclusion of ischemia on CT, the patient was soon transferred in the angiographic room. (**a**) Right carotid angiogram showing the ICA occlusion (*on the left of the figure*). Carotid angiogram posttreatment (*on the right*), showing the revascularization achieved rapidly with a solitaire stent. (**b**) On the lateral right angiogram, however, occlusion of several distal branches of MCA was visible. (**c**) The posttreatment left carotid and vertebral angiograms showed the left PCA filled on both the ICA and VA angiograms. There was no right PCA on the vertebral angiogram, indicating that this artery arose from the right occluded ICA and remained occluded. (**d**) Posttreatment CT. There is a large ischemia in the vascular territories of MCA and PCA. Minimal hemorrhagic component in the basal ganglia. The patient survived with severe motor deficit. It is thinkable that in this case the impairment of the leptomeningeal collateral circulation due to diffuse occlusion of the distal leptomeningeal anastomoses was probably due to many factors acting together such distal embolization, the unfavorable condition in the circle of Willis and damage of the microvasculature due to the ischemia (see also Sect. 15.8)

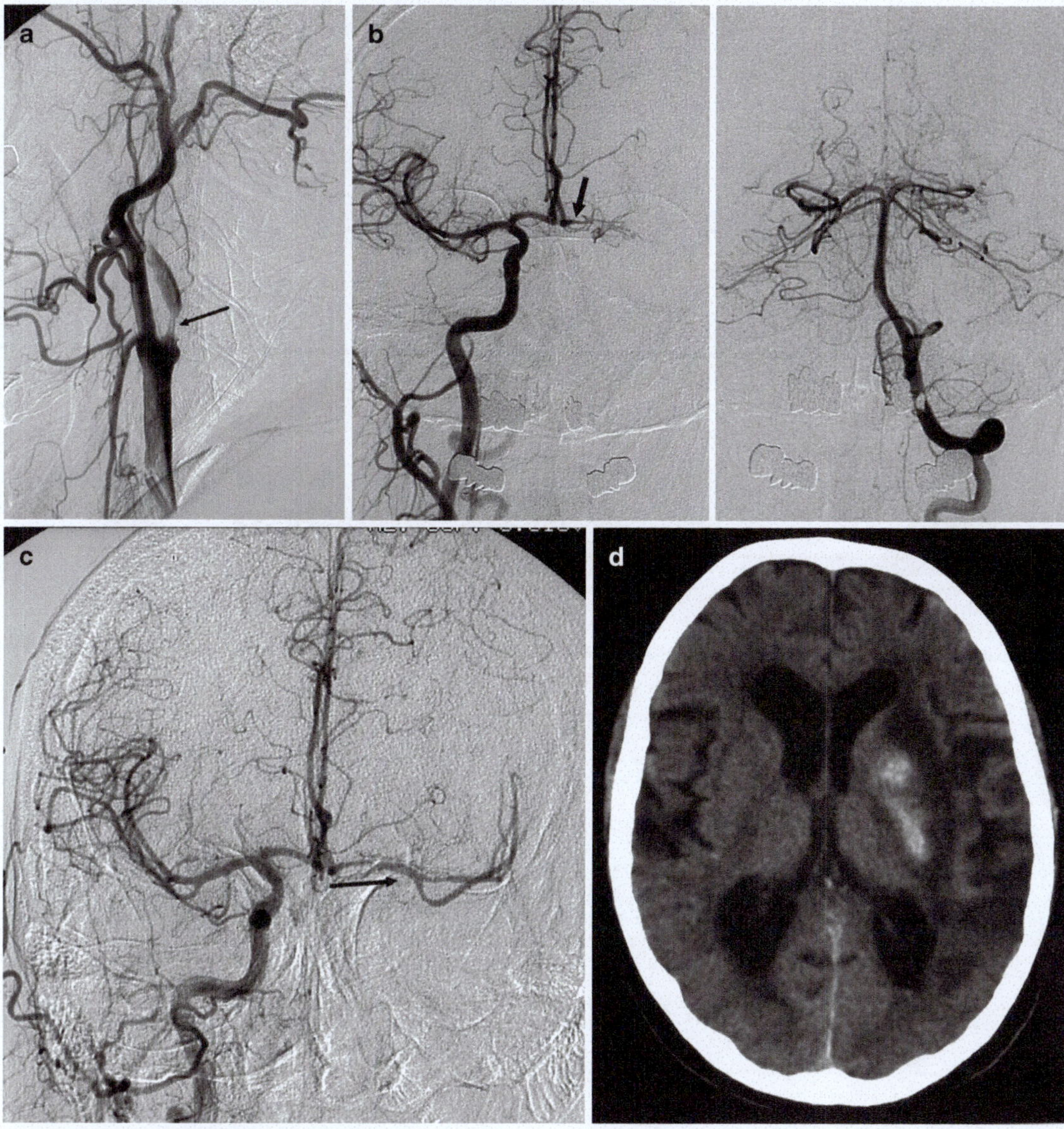

Fig. 15.14 75-year-old man presenting with aphasia right hemiplegia and progressive cognitive disturbances. Intravenous thrombolysis was first performed without positive response. (**a**) On the left carotid angiogram a severe stenosis of the ICA (*arrow*) was demonstrated. There was no filling of the distal ICA and no collateral circulation through the ECA. (**b**) Right carotid and left vertebral angiogram. Through the AcomA filling of the left ACA and filling of the distal part of a small left A1 (*arrow*). On the vertebral angiogram both PCAs arose from The basilar artery. There was a poor leptomeningeal collateral circulation. (**c**) After a stent application at the left ICA bifurcation a microcatheter was advanced in the distal ICA, followed by a mechanical thrombectomy. This led to a reopening of the distal ICA, A1, and MCA and improvement of the collateral circulation through the circle of Willis. There was a reocclusion of the ICA bifurcation. (**d**) Posttreatment CT showed an ischemic lesion with an hemorrhagic component in the basal ganglia. Ischemic foci in the insular an parietal areas. There was a partial slow recovery

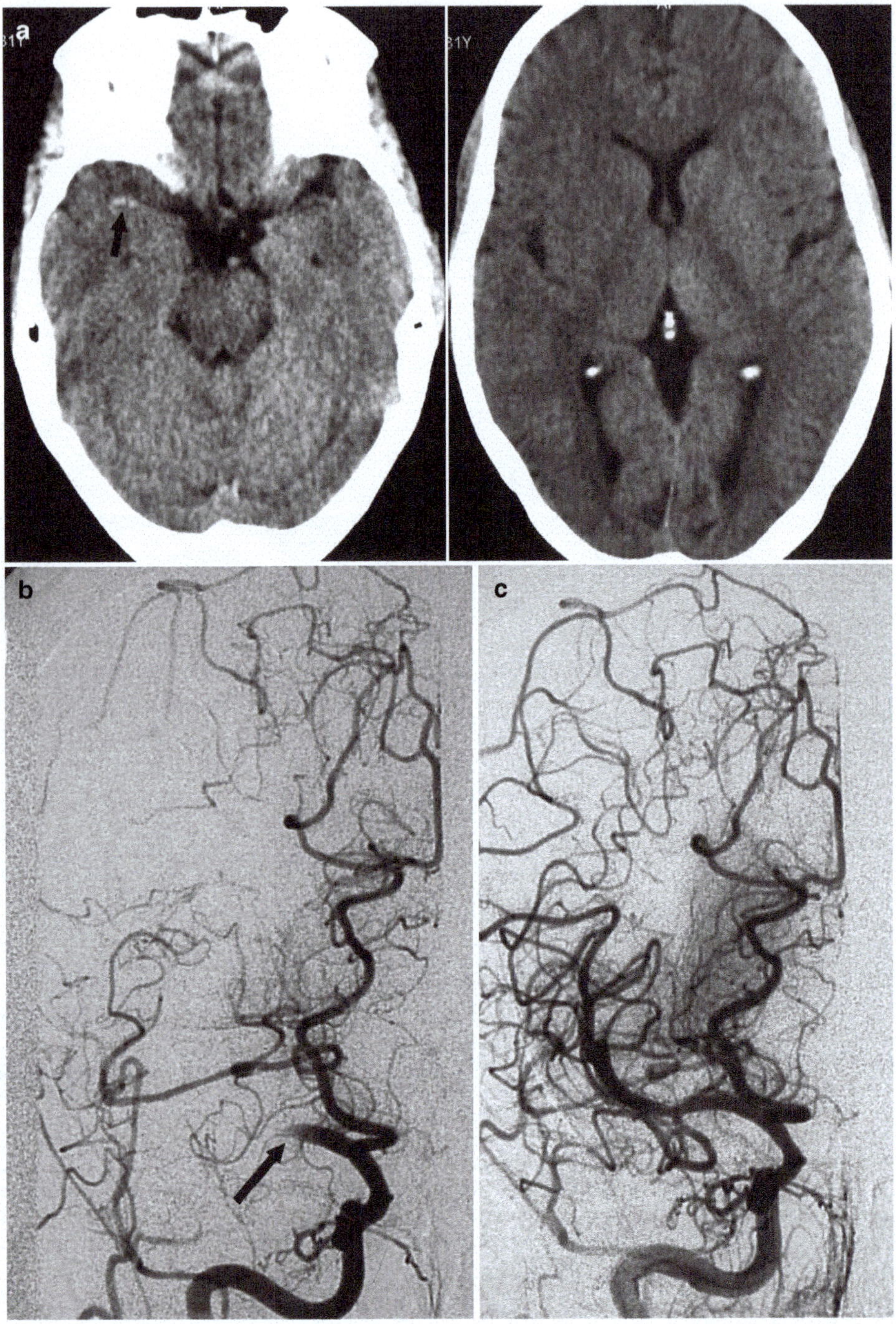

Fig. 15.15 Young cardiopathic patient with sudden onset of left hemiplegia. (**a**) On CT, hyperdensity of the MCA (*arrow*) and ill-defined borders of the right basal ganglia are recognizable. (**b**) On the right ICA angiogram, on occlusion of the proximal right M1 is visible (*arrow*). There is a fetal origin of the PCA, not involved in the occlusion. The left ACA arises from the left ICA. There is a rich leptomeningeal collateral circulation from the PCA towards the MCA. (**c**) Control angiogram after selective pharmacological thrombolysis showing the reopening of the occlusion where a minimal irregularity of the wall remains. (**d**) On CT performed 3 days later, there is a small ischemic lesion in the putamen. Rapid complete recovery of the patient

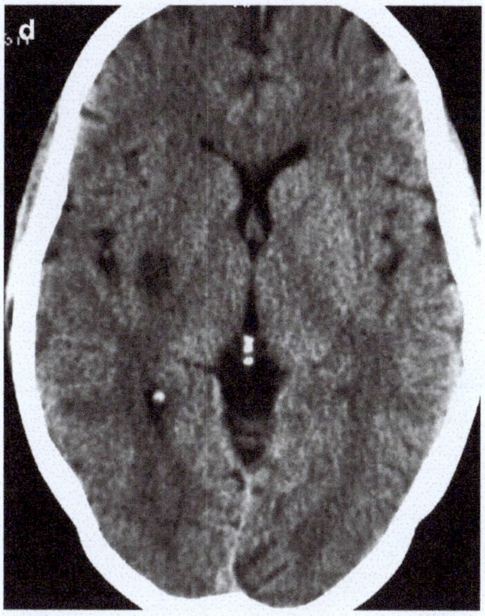

Fig. 15.15 (continued)

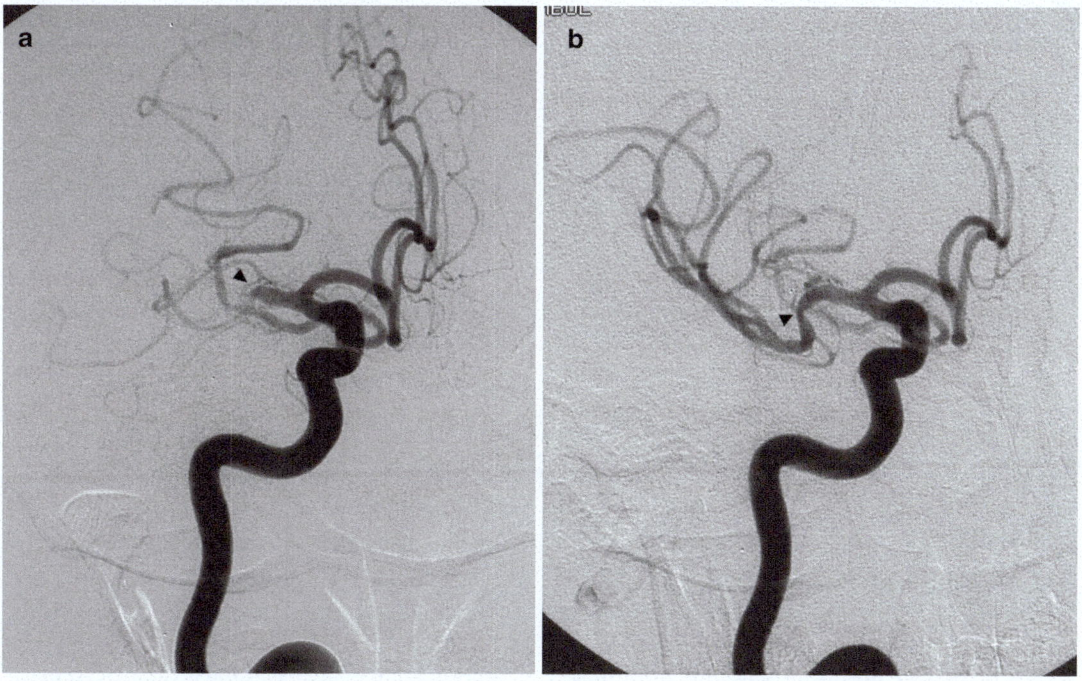

Fig. 15.16 Older woman with heart dysrhythmia presenting with sudden onset of hemiplegia and cognitive disorders. (**a**) The right ICA angiogram disclosed occlusion of the proximal M1 (*arrow head*). Note the presence of a fetal PCA not involved in the occlusion (compare this case with that in Fig. 18.3). Rapid reopening of the occlusion with a solitaire stent. (**b**) The posttreatment angiogram showed a microatheroma (*arrowhead*) associated with a tortuous distal M1, which probably favored the arrest of the embolus. A small infarct in the basal ganglia was visible on the posttreatment CT. The patient recovered rapidly

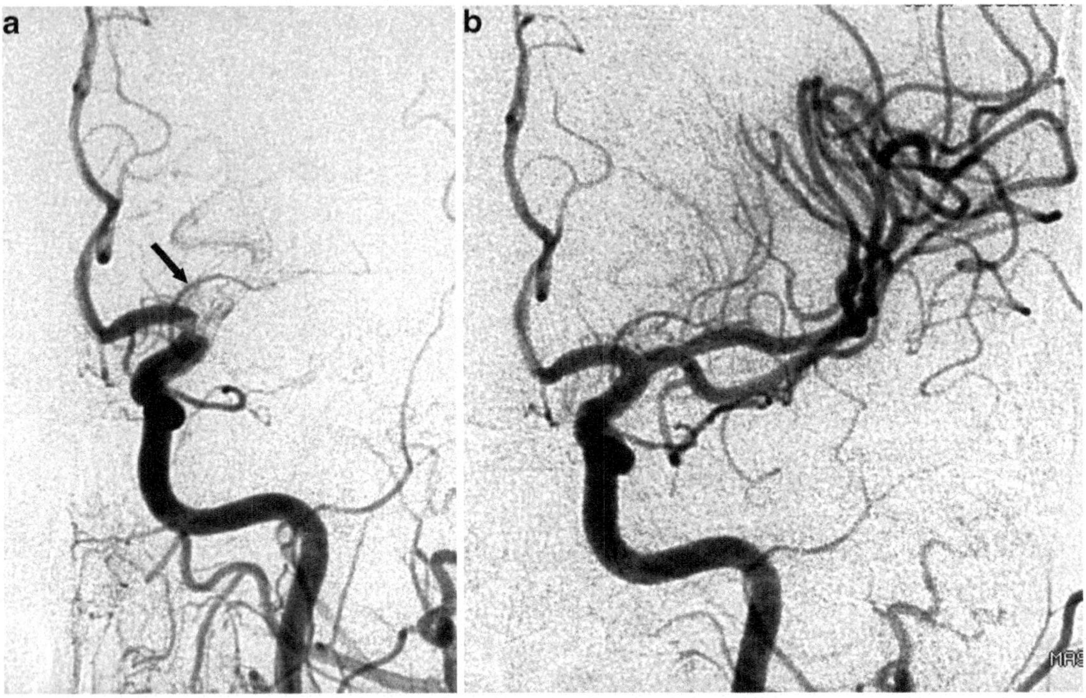

Fig. 15.17 Young patient with acute occlusion of proximal M1, due to a cardiac embolism occurring during a cardiac vascular intervention. (**a**) Left carotid angiogram showing the occlusion. Note the filling of the AchA (*arrow*). (**b**) Complete recanalization achieved with pharmacological selective thrombolysis. Note the early bifurcation due to the short M1, where the embolus was lodged. The perforators arise from the superior branch. On the posttreatment CT small ischemic lesions were visible in the basal ganglia. The patient recovered completely

In other cases, in CT performed after successful endovascular reopening of the M1, small hemorrhages are recognizable in the basal ganglia following the hemorrhagic transformation of the ischemic area. Contrast medium present in the infarcted area can have a similar pattern which, however, rapidly disappears in the CT controls.

In proximal occlusion of MCA the efficiency of the leptomeningeal collateral circulation between branches of ACA, PCA, and distal branches of MCA plays an important role to protect the distal vascular territories. These, however, can be impaired by embolization involving the distal branches, which can be due to emboli arising from the thrombus lodged in the proximal MCA. Distal embolization can also occur during endovascular treatment with disruption or asportation of the thrombus lodged in the proximal MCA. Furthermore distal branches can be selectively involved by small emboli, arising from ICA or heart and passing through the proximal ICA involving its distal branches (Figs. 18.3, 18.4, 18.5).

The therapeutic endovenous or selectively intra-arterial thrombolysis-thrombectomy has improved in many cases the prognosis of these patients.

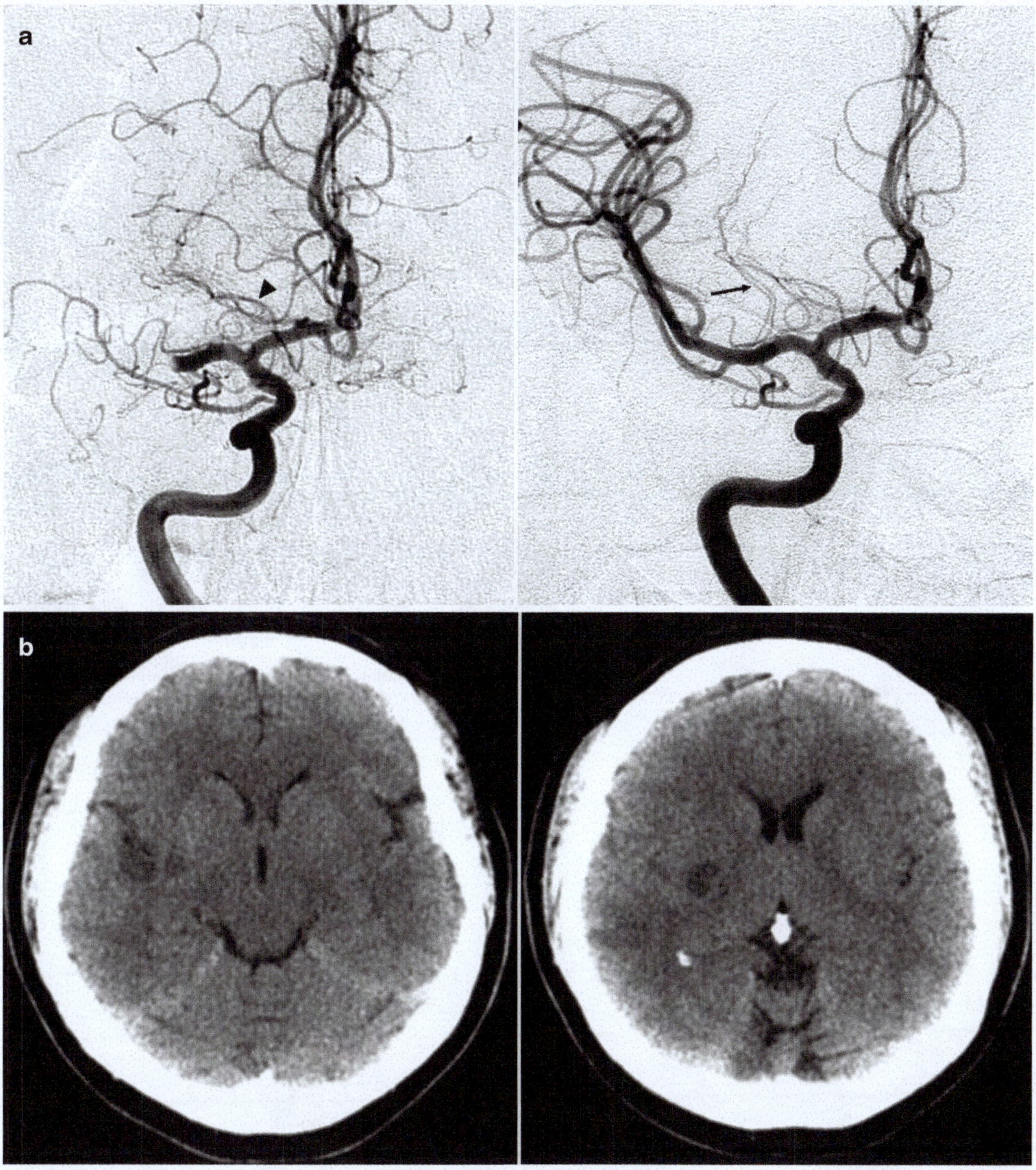

Fig. 15.18 (**a**) Probably embolic occlusion in middle-aged cardiopathic patient. Angiogram pretreatment (*on the left*). Occlusion of the proximal M1. A temporal branch and a small artery directed towards the insula are not involved. The AchA (*arrowhead*) as well as the perforators arising from the ACA are well recognizable. On the angiogram on the right is demonstrated the complete recanalization achieved rapidly with a solitaire stent. The perforators are now well filled (*arrow*). (**b**) The CT post-treatment showed small ischemic foci in the basal ganglia and in the insula region. The patient recovered completely

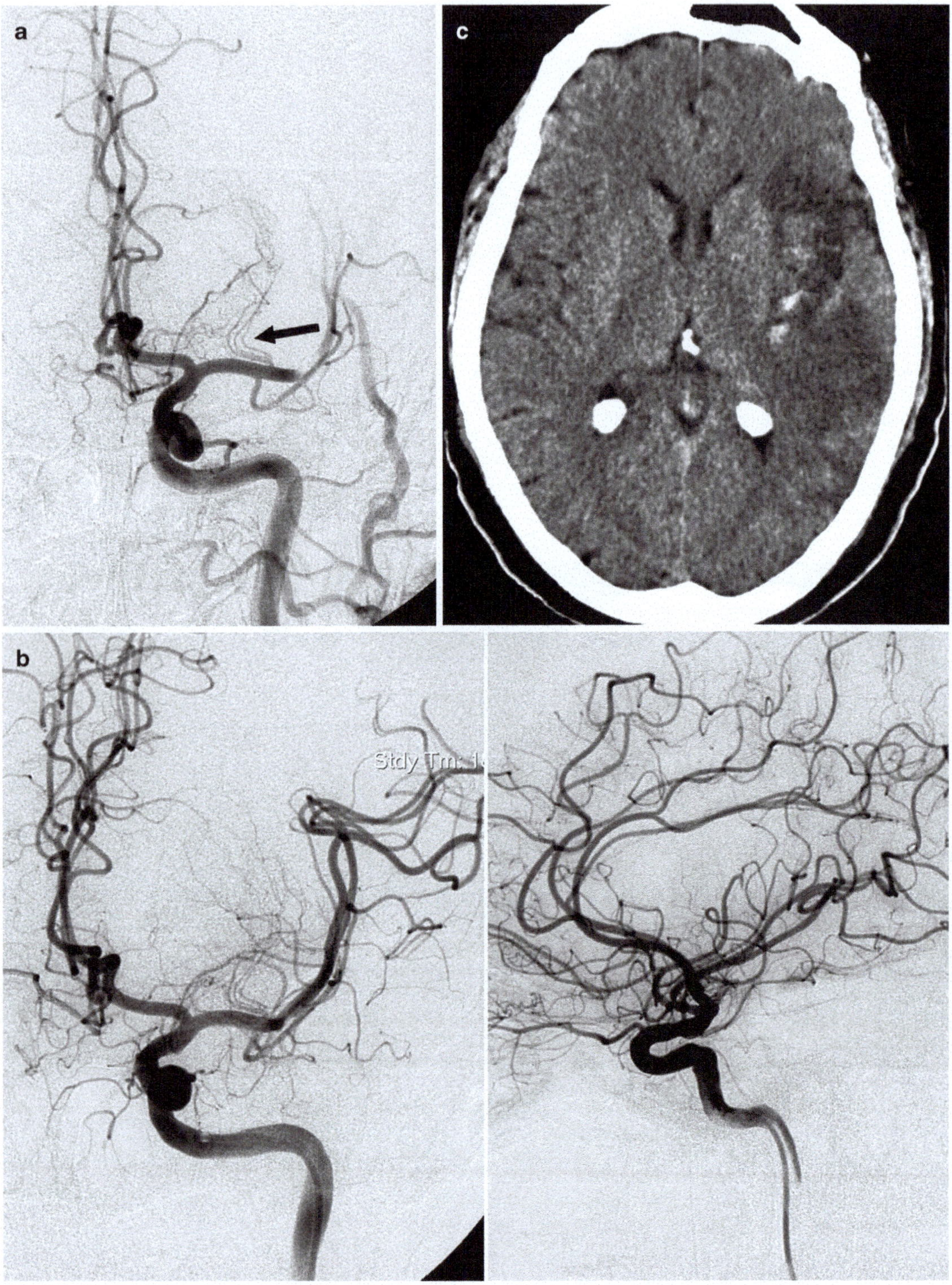

Fig. 15.19 Probably embolic occlusion (middle-aged diabetic and cardiopathic patient), presenting with aphasia and hemiplegia. Intravenous thrombolysis was performed without clinical improvement. The patient was transferred in the angiographic room for treatment. (**a**) The left carotid angiogram showed the occlusion of the MCA at the passage M1–M2. The perforators (*arrow*) were not involved. (**b**) AP and lateral views, after mechanical thrombolysis (solitaire stent) with reopening of the occlusion. As visible on the lateral angiogram, some distal branches remained occluded, probably due to embolization from thrombus in the M1 or following endovascular treatment. (**c**) Posttreatment CT showed an insular ischemia with partial involvement of the temporal and parietal areas. Minimal hemorrhagic component. The basal ganglia were not involved. Slowly, progressively recovery in the following months

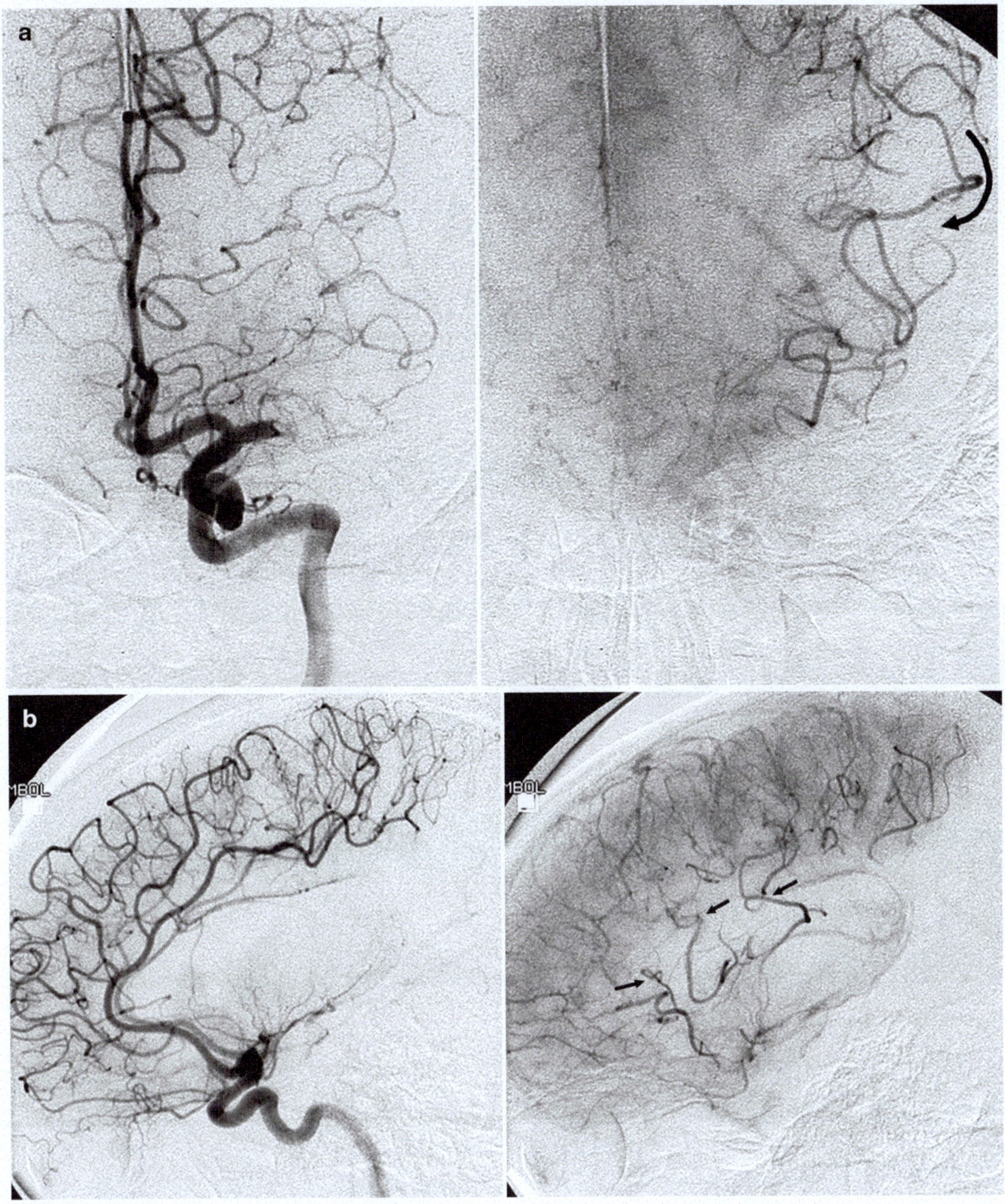

Fig. 15.20 Young woman presenting with right hemiplegia and aphasia. (**a**) Left carotid angiogram. Occlusion of the proximal M1. Rich collateral circulation (*arrow*). through opening of leptomeningeal anastomoses between ACA and MCA. (**b**) Lateral angiogram showing the collateral circulation (*arrows*). Faint filling of the deep venous system. (**c**) Reopening of the occluded M1 with aspiration device. There is a poor filling of the distal perforators (*arrowheads*). (**d**) CT performed 24 h later showed ischemia involving lateral putamen (*arrow*). Good and rapid recovery. Only a minimal motor deficit was present a month later

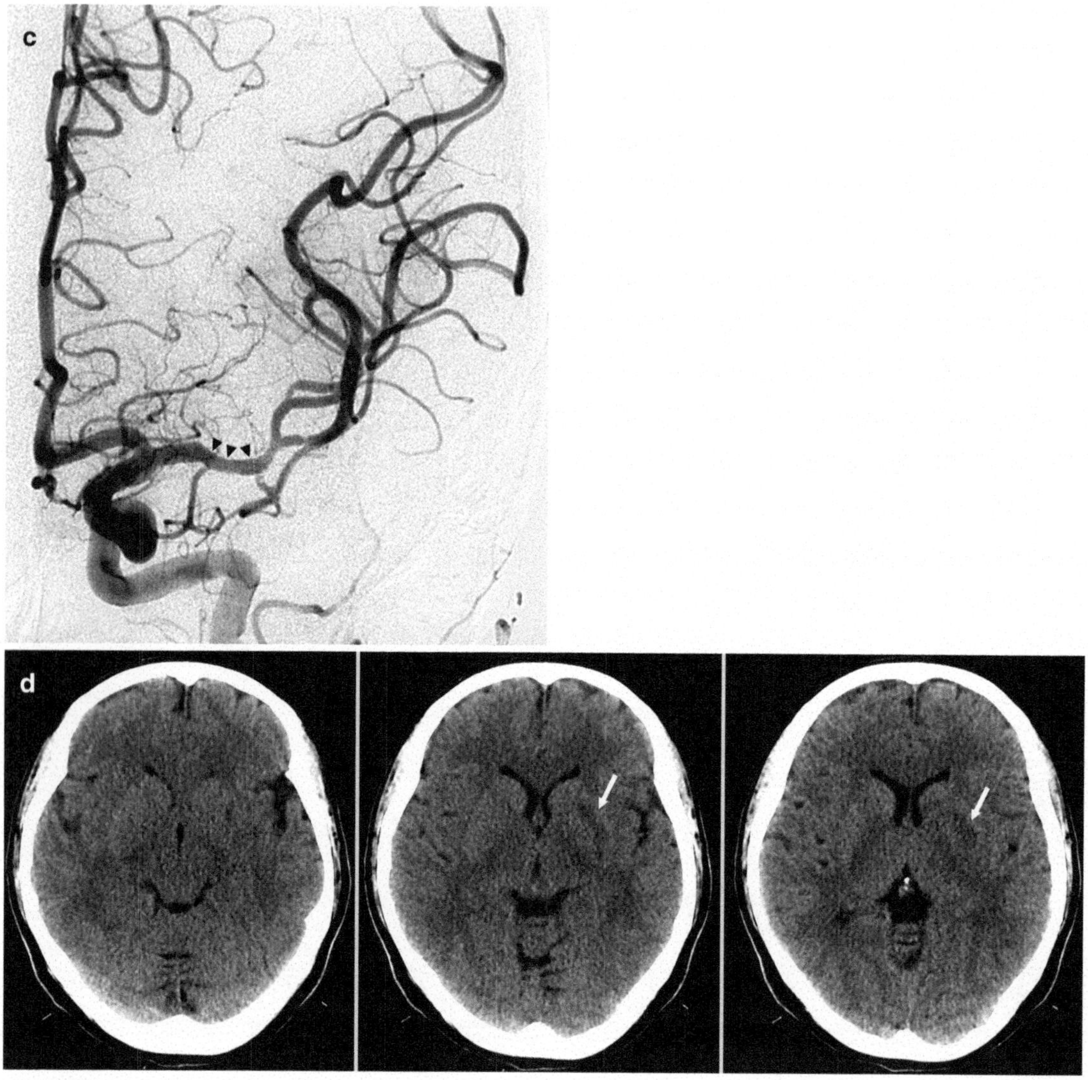

Fig. 15.20 (continued)

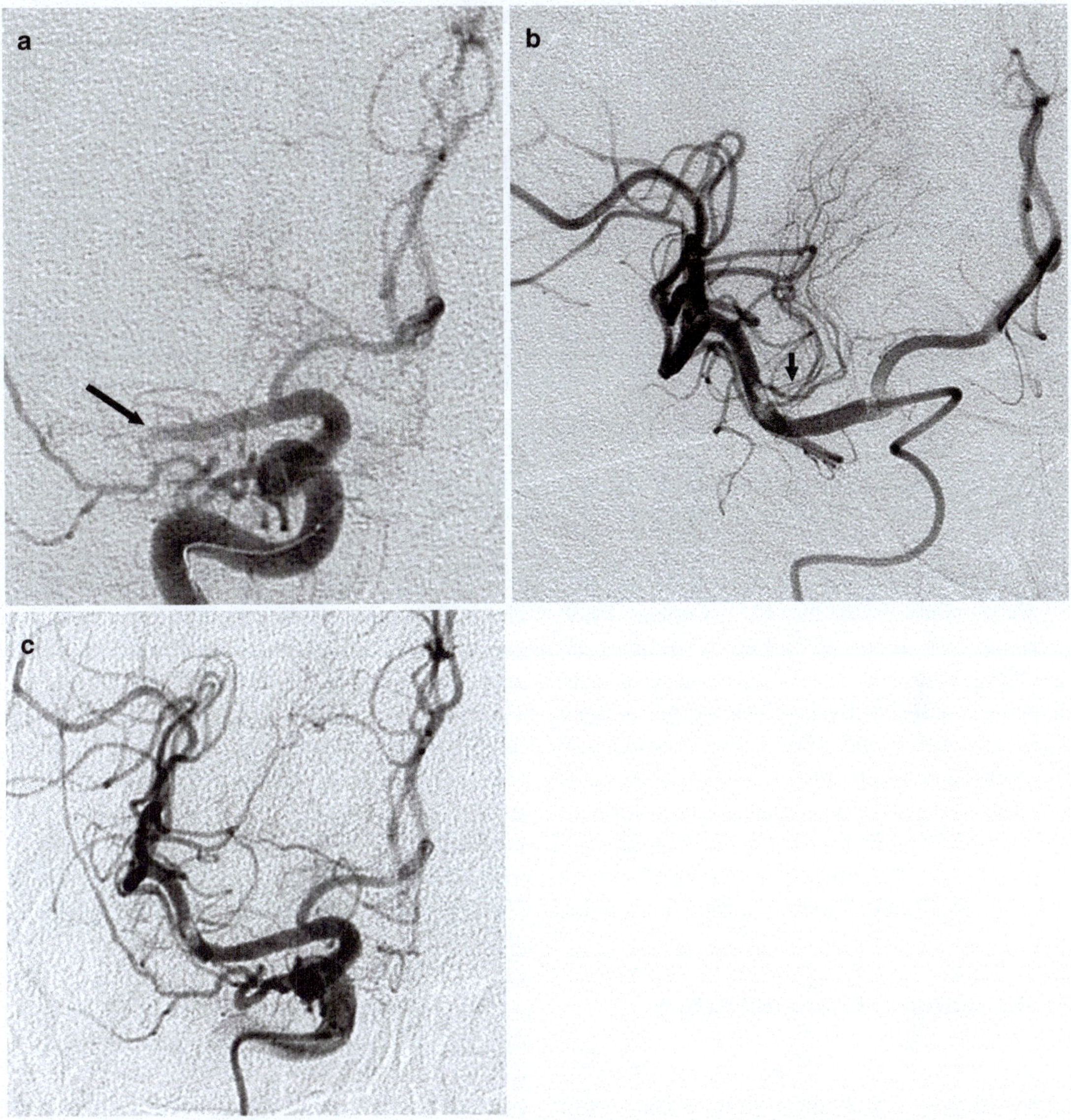

Fig. 15.21 Old cardiopathic woman presenting with hemiplegia due to a probably embolic occlusion of the right M1. (**a**) Right carotid angiogram showing the occlusion (*arrow*). (**b**) Angiogram after mechanical reopening (solitaire stent) of the M1. In the distal segment, a small thrombus is still recognizable. Note that at this level originate the perforators (*arrow*). (**c**) After a further passage with the solitaire, the thrombus is smaller and the distal flow is better. The perforators, however, are now only poorly injected. (**d**) CT performed 2 days later showed ischemia in the basal ganglia with a partially hemorrhagic transformation. Small edema is present. The ischemia involved also the upper part of the nucleus caudatus (MCA territory). Small ischemic foci in the parietal area. There is an old ischemia on the left. The patient recovered slowly. She presented with a mild left hemiplegia 4 months later

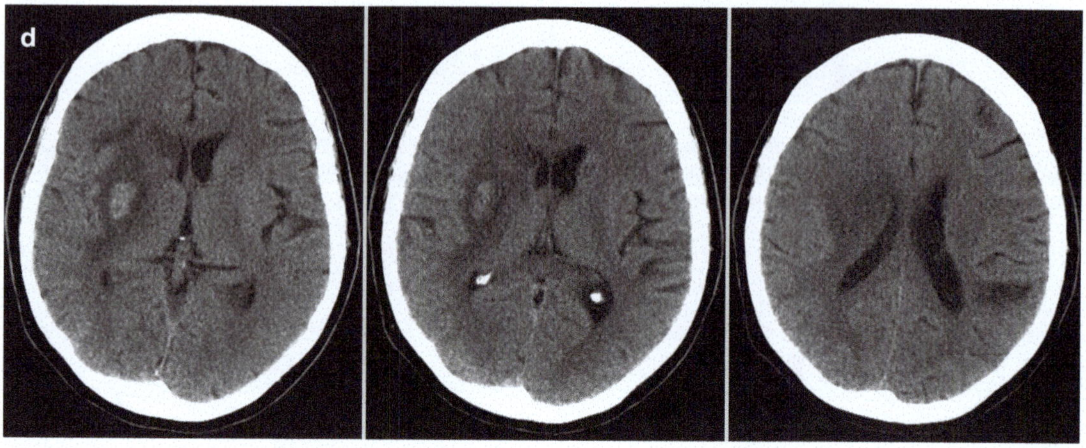

Fig. 15.21 (continued)

- *Microatheroma* of the M1 (Caplan 1989; Bougosslavski et al. 1991; Mazighi et al. 2008) is another less common cause of ischemia which directly involves the perforators. Distal cortical and white matter vascular territories can also be involved in these cases by either hypoperfusion or distal embolization. In selected cases, endovascular treatment (angioplasty, stent) has opened new ways of therapy (Figs. 15.22 and 15.23) with promising results.

15.4.3 Anterior Choroidal Artery

Infarcts in the territory of the anterior choroidal artery (AchA) are frequently associated with ischemia in adjacent vascular territories, especially in the MCA and PCA. The pathogenesis is carotid or cardiac embolism, or atheroma in the supraclinoid segment of ICA (Helgason et al. 1986; Boiten and Lodder 1991; Hupperts et al. 1994; Levy et al. 1995). Involvement of the AchA in *distal ICA occlusion* has already been described (Sect. 15.4.1). Ischemia due to an isolated embolization (Fig. 15.24) as well as that due to an inadvertent occlusion of the artery or its perforators (Friedman et al. 2001; Küker and Schulz 2014) in surgical or endovascular treatment of aneurysm of AchA or located in adjacent area can occur (Fig. 15.25).

The vascular territories of the perforators of the AchA can be partially replaced by other arteries. Furthermore there are many anastomoses connecting AchA and PCA. This explains the irregular presence of ischemic lesions in cases of occlusion of AchA (see Sects. 2.2.3.3, 15.7.1).

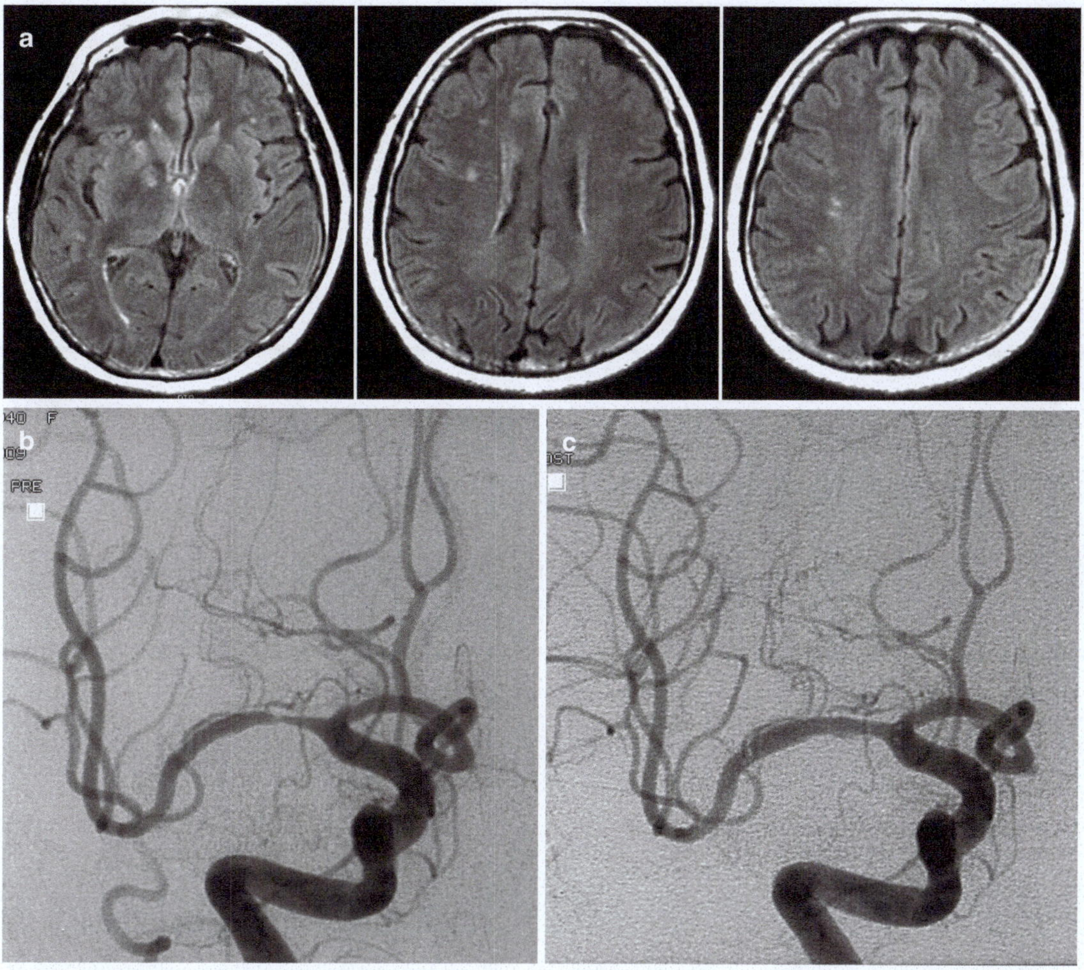

Fig. 15.22 Middle-aged woman presenting with mild neurological deficit, indicating impairment of the right ICA vascular territory. (**a**) MRI showed small ischemic lesions in the right basal ganglia and white matter. Minimal lesions also on the left. (**b**) On the right ICA angiogram severe stenosis of the proximal M1 was recognizable. There is a normal filling of the distal perforators. Endovascular treatment (angioplasty and stent) was performed a week later. (**c**) Angiogram posttreatment. It is conceivable that the ischemic lesions were due partially to involvement of the proximal perforators occluded by the microatheroma. Hypoperfusion in the distal perforators and medullary arteries was responsible of the more distal lesions

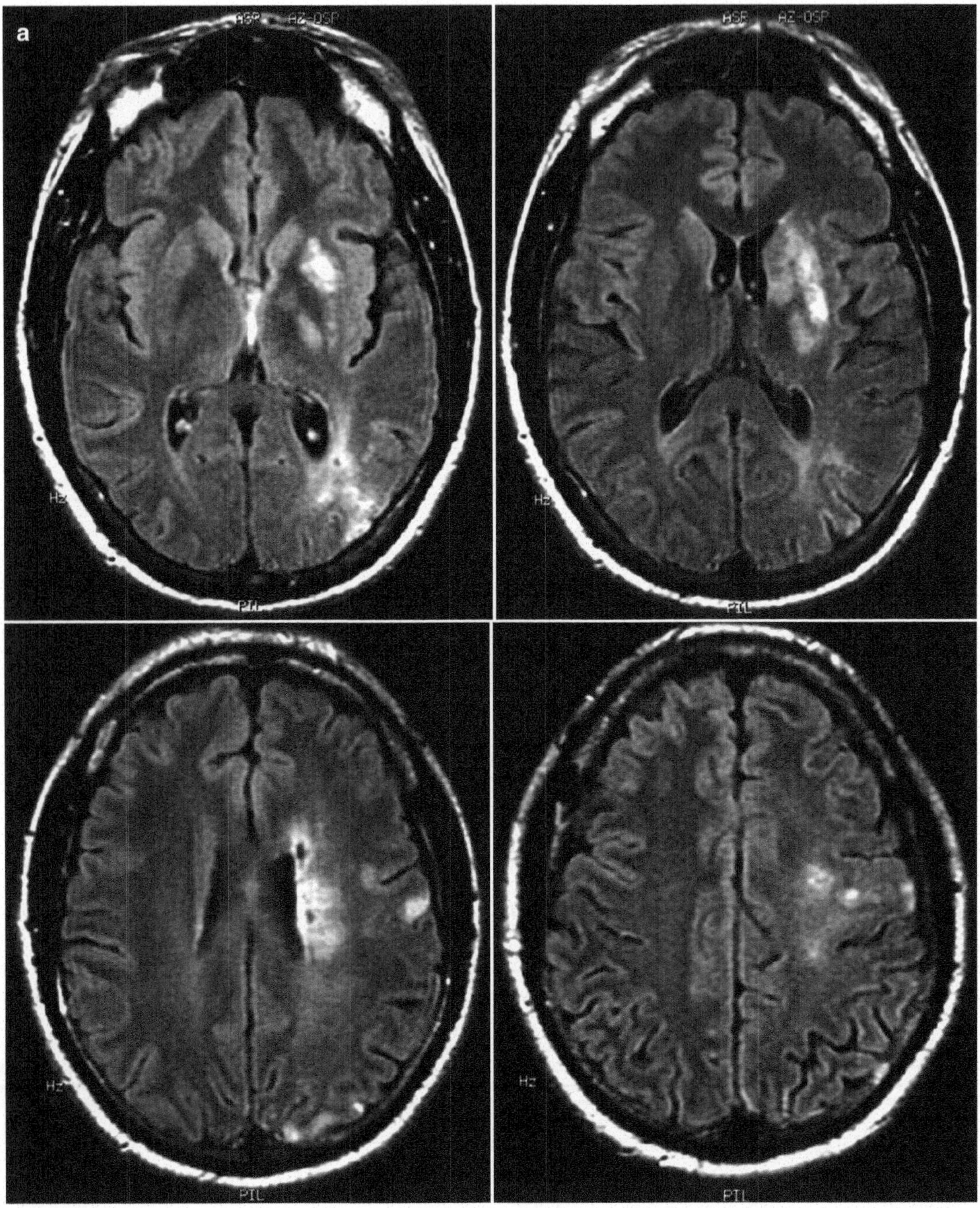

Fig. 15.23 Acute ischemia in a middle-aged woman involving the left basal ganglia, white matter, and partially the cortex. (**a**) MRI showing the ischemic lesions. (**b**) On the angiogram of the left ICA, a severe stenosis of the proximal M1 (*arrow*) was visible. The proximal perforators are probably occluded by the microatheroma. The more distal are well filled. Ischemia was probably due to direct occlusion of some perforators and hypoperfusion involving the distal perforators and medullary arteries. The cortical lesions could be embolic. Three weeks later endovascular treatment (angioplasty and stent) was performed. (**c**) Angiogram posttreatment. Recovery of the patient

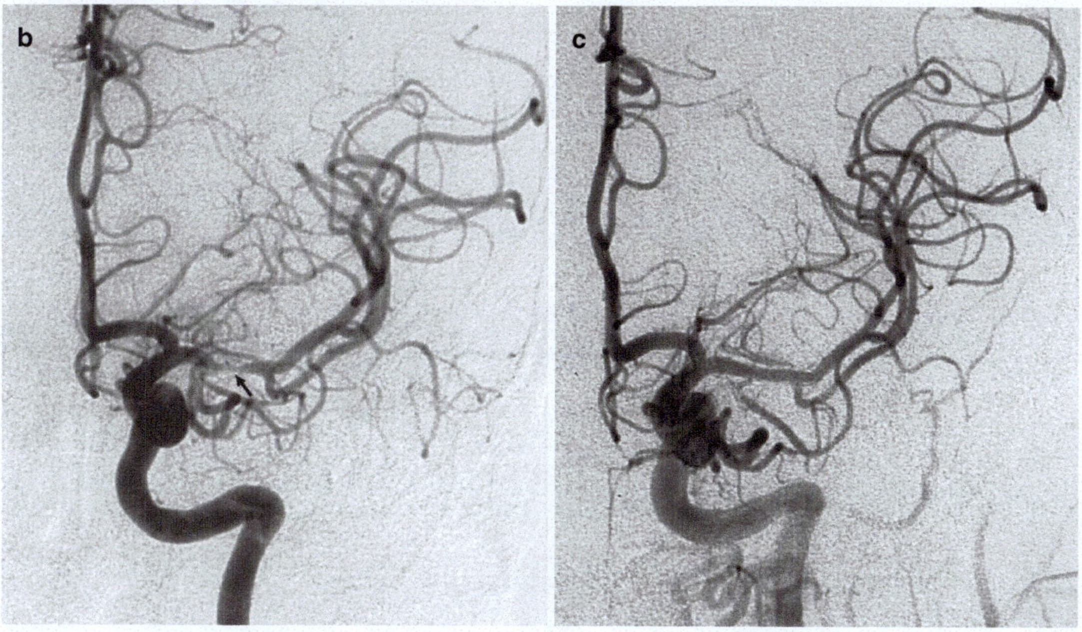

Fig. 15.23 (continued)

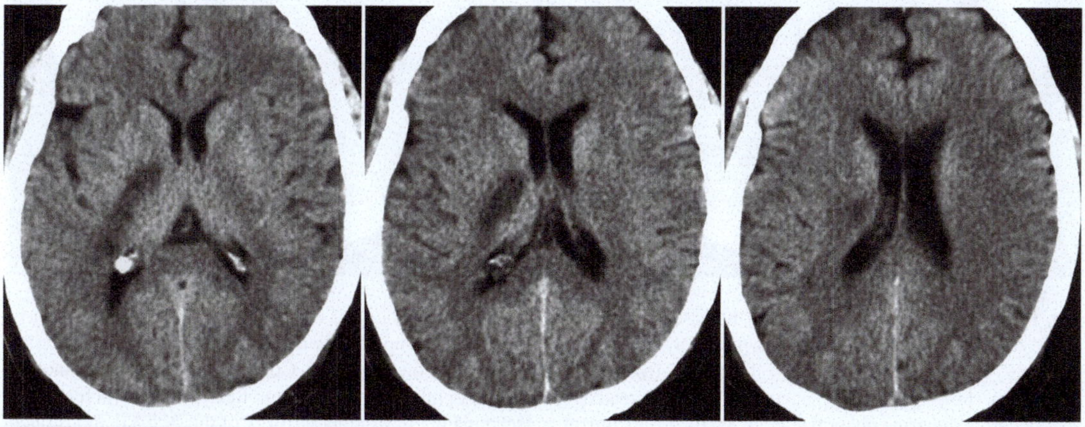

Fig. 15.24 Middle-aged woman with cardiac dysrhythmia presenting with stroke due to occlusion, probably embolic of the AchA. CT showing the ischemia in the corresponding vascular territory

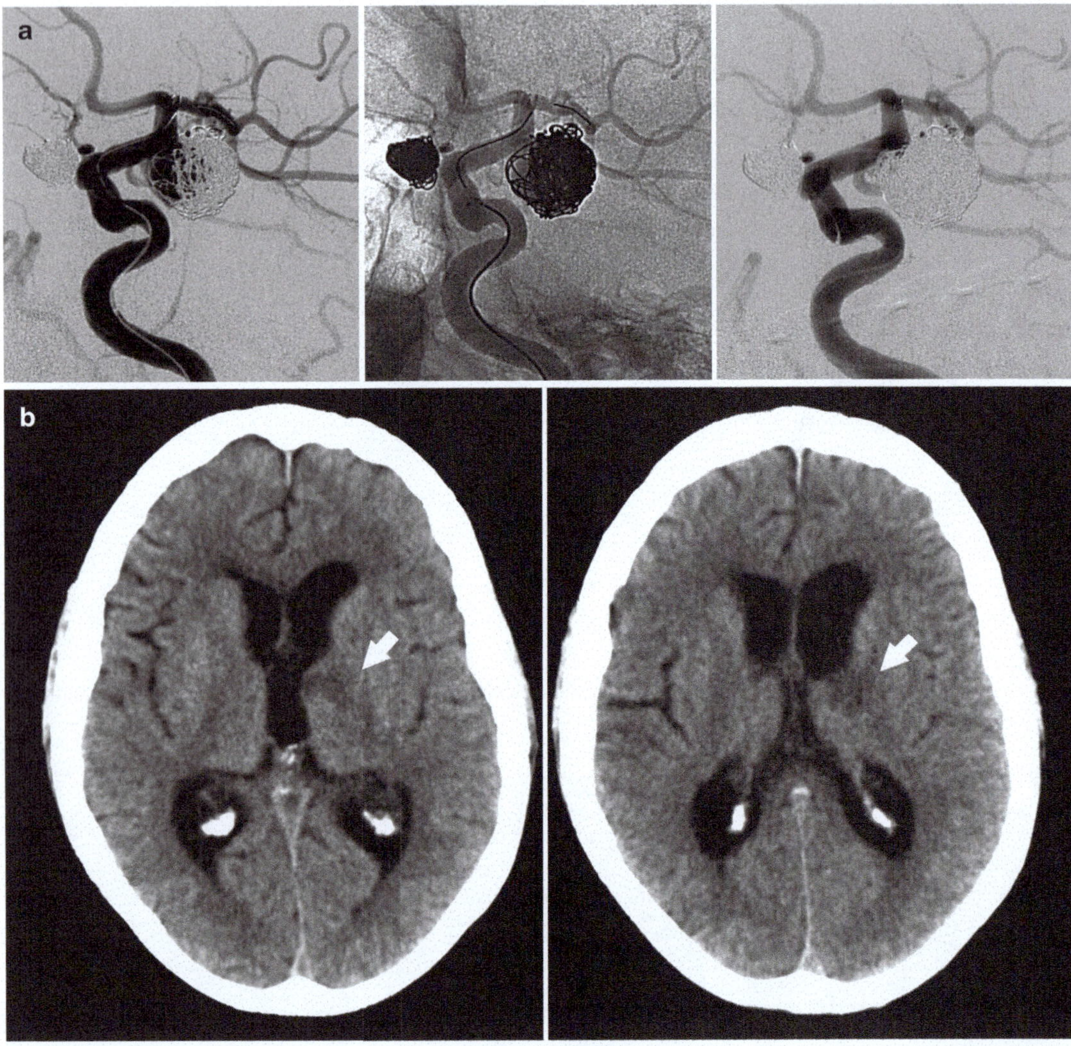

Fig. 15.25 Symptomatic ischemia in the vascular territory of the AchA after endovascular treatment of large aneurysm of the ICA involving probably also the AchA. (**a**) Carotid angiogram showing the large aneurysm already treated 1 year before in the acute phase due to an SAH. A retreatment was performed due to the partial recanalization. In spite of several projections the origin of the AchA was not clearly identified. Complete occlusion of the Aneurysm with coils (angiogram on the right). Coils are also visible in another carotid-ophthalmic aneurysm treated years before. (**b**) Ischemia in the vascular territory of the AchA demonstrated on CT performed 24 h after the treatment

15.4.4 Anterior Cerebral Artery

Infarcts in the territory of the ACA can be due to selectively occlusions of the artery (Fig. 15.26) but more frequently are associated with ischemia involving other vascular territories after carotid or cardiac embolism (Figs. 18.3, 18.4 and 18.5). Embolization is frequently associated with variants in the anterior portion of the circle of Willis, which favor the passage of emboli in the ACA (Kazui et al. 1993). Also variations of the distal branches can explain the type and extension of the infarct (Boccardi et al. 2002) (see also Sect. 4.3). The vascular territories of perforators and/or peripheral branches in varying combinations can be involved depending on the proximal or more distal location of the occlusion.

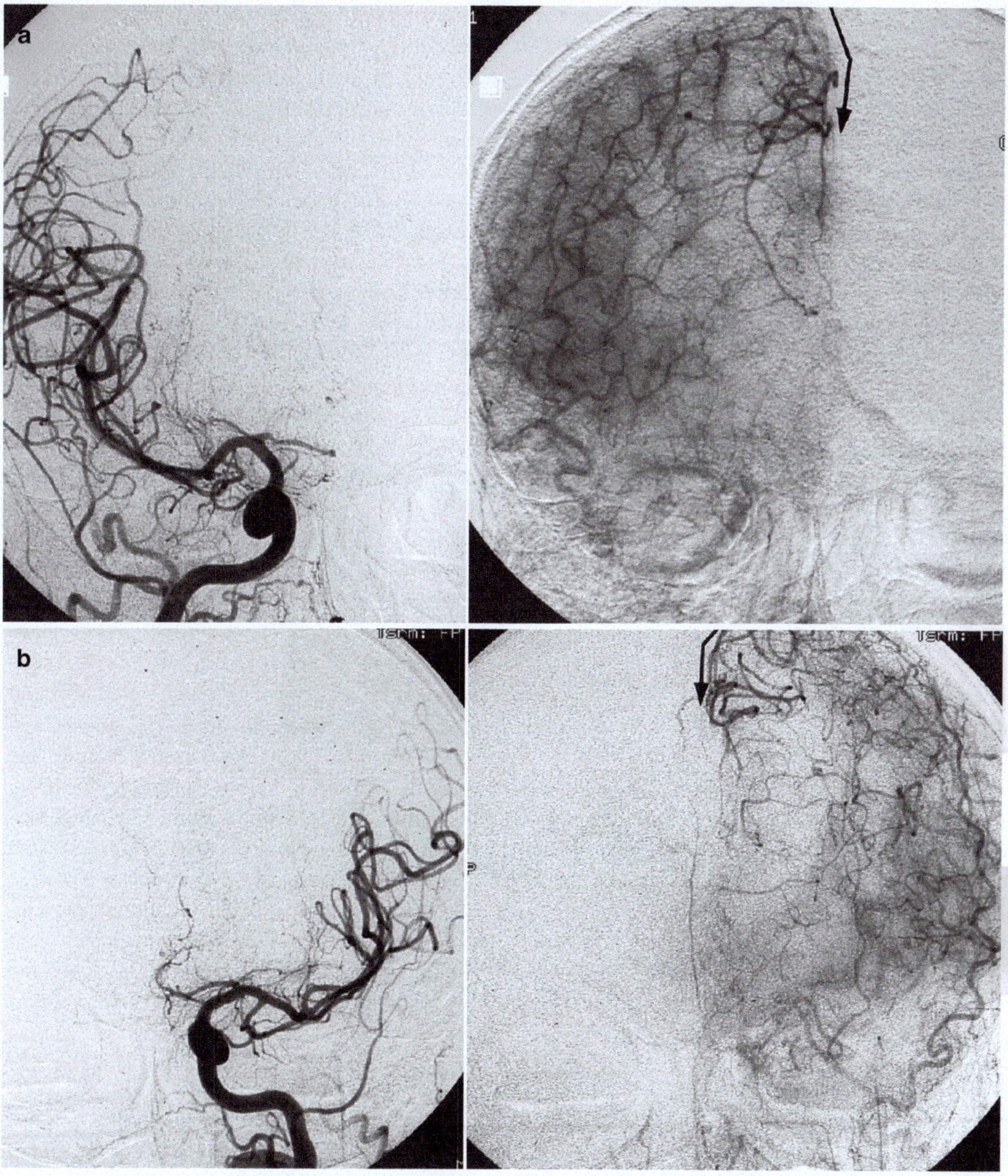

Fig. 15.26 Old female patient with stroke due to bilateral occlusion of the ACA. (**a**) right ICA angiogram showing the occlusion at the passage of the A1–A2 segments. In the late phase, beginning collateral circulation (*arrow*) through opening of anastomoses with distal branches of the MCA. (**b**) Left ICA angiogram showing a similar finding

Considering especially the artery of Heubner, its involvement also in proximal occlusion of the ACA can vary due to its possible different site of origin (see also Chap. 4). Also the inadvertent occlusion of the artery of Heubner as well as of the other perforating branches arising from A1 and/or AcomA especially the subcallosal artery in treatments of aneurysms of AcomA, can lead to ischemia involving, respectively, the head of the nucleus caudatus and adjacent areas and the lamina terminalis, preoptic part of the hypothalamus, anterior commissure, anterior pillars of fornix, and genu of corpus callosum (see Chap. 4, Sects. 4.1 and 4.4).

Microatheroma typically located in the pericallosal segment can be another rare cause of ischemia (Fig. 15.27).

15.4.5 Lacunar and Other Deep Infarcts in the Anterior Circulation

These infarcts are due to involvement of the deep perforators supplying the basal ganglia and capsula interna and perforators (medullary arteries) supplying the superficial and deep white matter. As already described, these arteries do not anastomose to each other, and neither there are connections between the deep perforators and medullary arteries (De Reuck 1972; Moody et al. 1988, 1990, 1991). Both vascular territories or only one can be involved. The typical lacunar infarct is due to the occlusion of a single perforator and it is small, reaching the maximal diameter of 15 mm. This pathological condition, also called "*small vessel disease*," presents with clinical symptoms which depend on the site and extension of the damaged parenchyma. Isolated lesions can be asymptomatic. Stroke can occur followed by improvement. Multiple lesions can develop presenting with or without stroke leading to a progressive decline of different neurological function with cognitive impairment and finally to dementia.

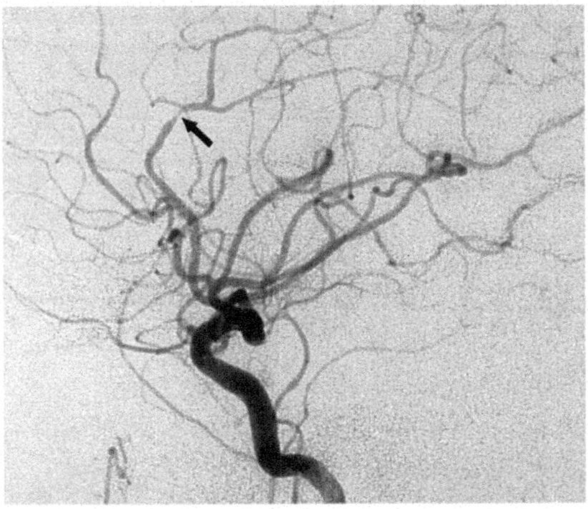

Fig. 15.27 Typical location of a microatheroma in the pericallosal artery in an asymptomatic patient

A frequent pathogenesis is a special type of atherosclerosis, occurring particularly in patients with hypertension and diabetes called by Fisher (1965, 1968) *lipohyalinosis*, which is characterized by a large presence in the wall of the vessels of hyaline material. In other cases the histological examination has shown a finding characterized by eosinophilic material representing accumulation of necrotic smooth muscles cells and extravasated plasma proteins, called fibrinoid necrosis. Both processes lead to stenosis and occlusion of the artery causing ischemia and sometimes rupture of the wall and hemorrhage. The lesions can be very numerous and disseminated in both hemispheres. MR demonstrates on T2-weighted images the hyperintense ischemic lesions not rarely associated with small hypointense foci due to old microhemorrhages well defined with T2*-weighted gradient-echo and SW-weighted images. Pathological examinations disclose infarcts commonly small (less than 15 mm in diameter) in the basal ganglia and white matter. In this latter, also areas of rarefaction due to destruction of fibers and area of gliosis have been demonstrated (Brun and Englund 1986; Pantoni et al. 1996). No pathological changes of the arteries can be detected on the angiogram in many of these patients.

Another pathogenesis of deep infarcts involving predominantly the white matter, and which are commonly larger than the typical lacunar infarct is thought to be occlusion or severe stenosis of the ICA leading to hypoperfusion or embolization or both, as reported by several authors (Levine et al. 1988; Waterston et al. 1990; Angeloni et al. 1990;

Weiller et al. 1990, 1991; Bougosslawski et al. 1991; Bougosslavsky and Regli 1992; Bladin and Chambers 1993; Horowitz and Turim 1997; Boiten et al. 1997; Read et al. 1998; Bradac et al. 2008b). Romero et al. 2009; Enzinger et al. 2010). In this context, the type of ICA plaque as source of emboli plays probably an important role (Altaf et al. 2006; Baradaran et al. 2017)(See also Sects. 15.4.1). Furthermore, as suggested by some authors (Patankar et al 2006), the presence of ICA atheromasic lesions could be an indirect sign of diffuse atherosclerosis involving not only the extracrtanial ICA but also its intracranial perforators branches. Cardiac emboli or emboli arising from aortich arch atheroma are also a possible pathogenesis.

Another type of ischemic lesion is that termed *striatocapsular infarct*, also called by Fisher (1965, 1979) "giant" or "super" lacunas. These differently from the small lacunas involve the vascular territory of more than one deep perforators arising from ICA, AchA, MCA, and ACA. (Ghika et al. 1989; Boiten and Lodder 1991; Nicolai et al. 1996; Horowitz and Turhim 1997; Bradac et al. 2008b). The pathogenesis of these lesions varies. They can be due to occlusion of the M1 segment of the MCA due to emboli arising from atheroma of the ICA or cardiac in origin. Another cause can be a severe stenosis due to microatheroma located at the ostium of the perforators arising from M1 or distal ICA. The striatocapsular infarct can be associated with lesions in the white matter for involvement of the medullary arteries due to hypoperfusion or distal embolization (Figs. 15.22 and 15.23).

Association of ischemic lesions and microhemorrhages can occur in patients with amyloid angiopathy and in other nonatherosclerotic vasculopathies (see also Chaps. 17, 18, 19 and 22).

In conclusion, the pathogenetic process leading to ischemia in the vascular territory of the deep or superficial perforators cannot always be assessed with certainty since several mechanisms can be involved and these can also occur together (Jackson and Sudlow 2005). *In this context, also morphological variants should be considered, as already described (see anatomy). In particular, it should be noted that the lenticulostriate arteries can arise as a common trunk* (Fisher 1979; Umansky et al. 1985) *responsible for the vascularization of a large territory. In case of its occlusion due to lipohyalinosis leading to a large lacunas, it will be difficult to make a differential diagnosis with a striatocapsular infarct in which more than one perforators are involved in which the pathogenesis is commonly different (see above). We mention here that lacunar infarcts in the anterior circulation are frequently associated with similar lesions in the posterior circulation (see Sects.* 15.5.3, 15.5.4, *and* 15.5.5). *Finally it should be taken into account that similar lesions can occur in other vascular pathologies not linked to arteriosclerosis and in other diseases.*

Two other particular conditions deserve a short description. The one is a very rare type of infarct located lateral to the basal ganglia involving the *white matter of the external capsule* in the border area between the deep perforators arising from M1 and the medullary arteries arising from the insular branches. These lesions have been termed subinsular infarcts by Pullicino et al. (1992). The other is the so-called *Binswanger's disease,* described by the author in 1894 on the basis of anatomo-pathologic studies, characterized mainly by diffuse ischemic lesion involving the white matter. Further histological and in vivo MR examinations have shown that the parenchymal lesions and the clinical symptoms are basically not different from those due to involvement of deep and superficial (medullary) perforators occurring in arteriosclerotic patients with small vessels disease. Binswanger's disease is today considered by many authors as an aspect of the disease and not as a specific entity.

15.5 Posterior Circulation

The most frequent sites of atherosclerotic plaques in the vertebrobasilar sector are at the origin of the VA from the subclavian artery and at the intracranial VA shortly after the artery perforates the dura (Castaigne et al. 1973; Caplan 1996). The third frequent location is the basilar artery, particularly in its middle segment (Castaigne et al. 1973; Pessin et al. 1987; Caplan 1996). Atheroma can be found also at the P1–P2 segment of the PCA (Bradac and Oberson 1983; Fisher 1986; Pessin et al. 1987). The composition of the atheroma, basically, does not differ from that in the carotid sector (Fisher and Ojeman 1986; Amarenco et al. 1990). Commonly, the plaques are less ulcerated than that in the carotid sector, but ulceration is not rare in the plaques of the subclavian arteries near the origin of the VA (Amarenco et al. 1998).

15.5.1 Subclavian and Innominate Arteries

Atheroma of the subclavian and innominate arteries are frequently and commonly asymptomatic. The atheroma can grow, leading to severe stenosis or occlusion of the artery. When this occurs proximal to the origin of the VA, blood flow from the normal VA intracranially and then down into the contralateral VA, further filling the distal subclavian artery. This situation, described first by Reivich et al. (1961), is termed the subclavian steal syndrome. This can present with TIAs involving the brainstem or pain in the ischemic arm, but frequently it remains completely asymptomatic. Symptoms occur commonly in patients with diffuse atherosclerotic lesions that involve also the carotid sector (Hennerici et al. 1998). Since it has a benign course, this syndrome only rarely needs aggressive treatment. Examples are presented in Fig. 15.28.

15.5.2 Vertebral Artery

Severe stenosis or occlusion of one VA at its origin remains frequently asymptomatic provided that no embolization occurs and the other VA as well as the circle of Willis sufficiently supply the vertebrobasilar sector. In contrast to thrombus of the ICA bifurcation, which can extend intracranially, thrombus of the extracranial VA does not extend intracranially, probably because the rich collateral circulation that arises from the other VA, the deeply and ascending cervical arteries, and from ECA, maintains sufficient flow.

Flow impairment can occur when stenosis of one VA at its origin is associated with occlusion or hypoplasia of the other or due to bilateral severe stenosis and in cases in which the vertebrobasilar sector is an important collateral way towards the severe affected carotid sector. In these cases endovascular treatment (stenting with or without angioplasty) has progressively become an accepted therapeutic approach (Taylor et al. 2008; Karameshev et al. 2010; Stayman et al. 2011; Edgell et al. 2013). Examples are presented in Fig. 15.29.

The atherosclerotic plaque can also be the source of emboli which are not so rare as previously thought (Wityk et al. 1998). In this context see also Sect. 15.5.4.

Another very rare cause of stroke in the vertebrobasilar sector is the involvement of the extracranial VA due to degenerative processes (hyperplasia of the facets, lateral disc herniation, fibrous band) leading to stenosis or temporary occlusion during neck rotation (Mapstone and Spetzler 1982; Vilela et al. 2005; Andereggen et al. 2012). This condition can be the cause of stroke due distal embolism or hemodynamic impairment of flow. Surgical treatment is considered the most appropriate therapy.

Occlusion of the intracranial VA due to artery-to-artery or cardiac embolism or thrombus developed on a preexisting microatheroma is a more serious condition since it involves the perforators of the VA that supply the anterolateral medulla. Depending on the distal extension of the superimposed thrombus, the PICA, supplying prevalently the dorsal medulla and cerebellum, and the anterior spinal artery supplying the anterior medulla can be involved. However, probably owing to a rapid spontaneous lysis of the embolus or thrombus in the VA, the ischemia is frequently only temporary or limited only to the PICA territory, which has commonly a benign course.

The thrombus can extend to the basilar artery. This is a very critical condition which needs a rapid endovascular treatment leading to a reopening of the artery. Examples of occlusion involving primary the VA with different distal extension of the thrombus are presented in Figs. 15.30, 15.31, and 15.32).

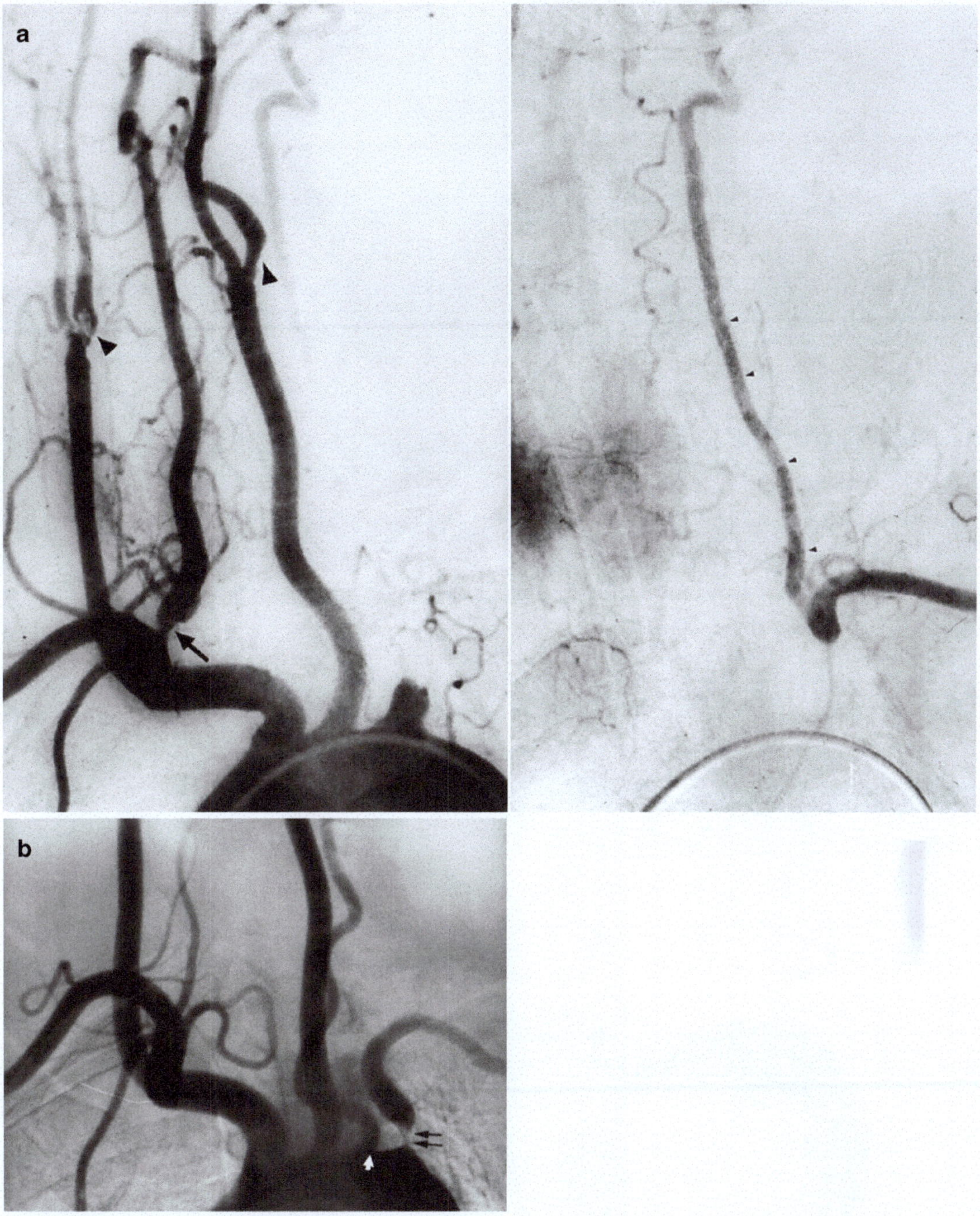

Fig. 15.28 (**a**) Subclavian steal syndrome, aortic arch angiogram. Occlusion of the proximal left subclavian artery. There is a well-developed right VA, stenotic at its origin (*arrow*). The blood flows in the right VA intracranially, and then down into the left VA (*small arrowheads*) and further into the distal left subclavian artery. Atheromasic changes at the carotid bifurcations bilaterally (*arrowheads*). (**b**) Aortic arch angiogram in a different patient. Severe stenosis of the proximal left subclavian artery (*arrows*). A subclavian steal syndrome is not present, since the left VA originates directly from the aortic arch

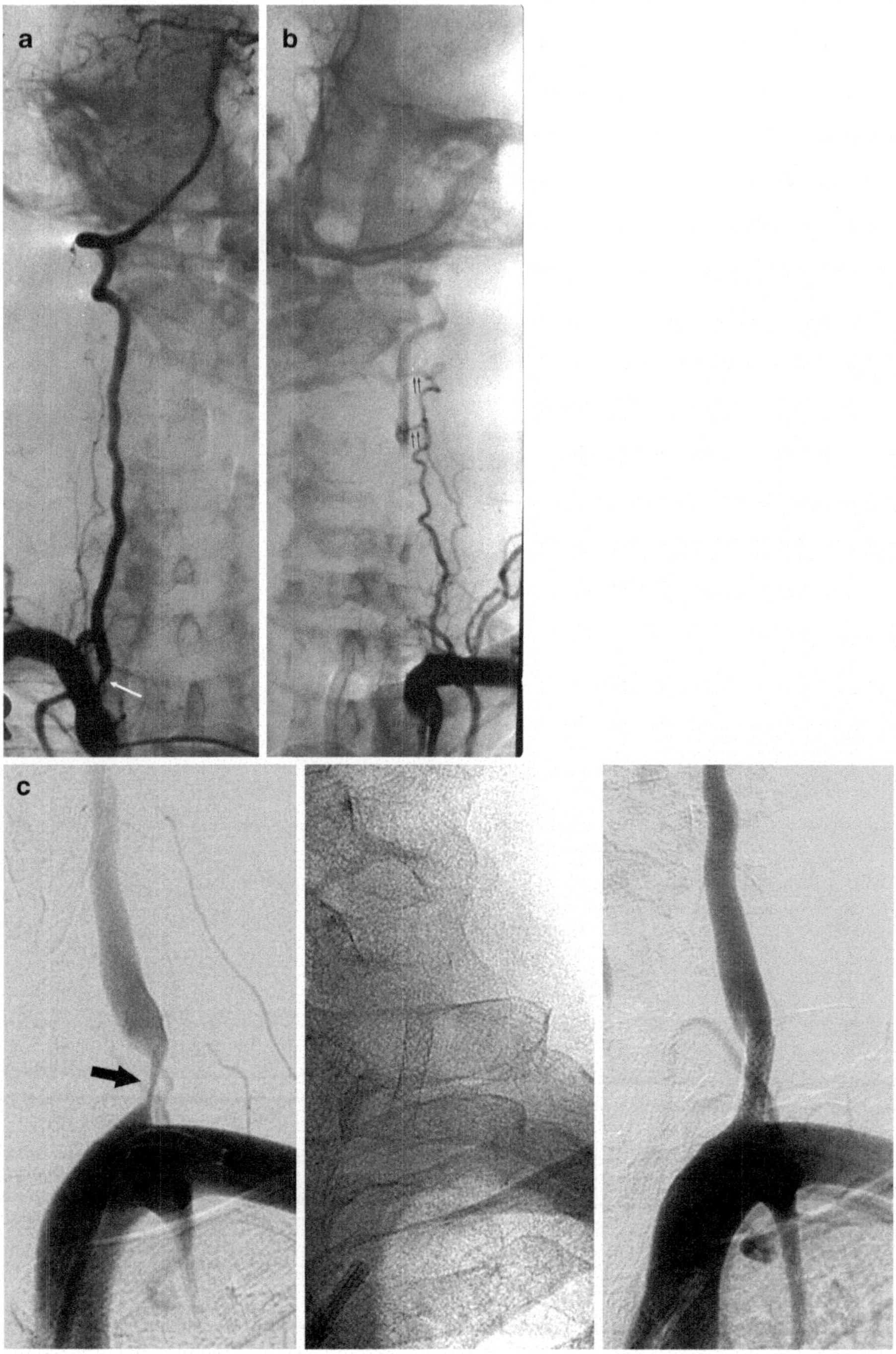

Fig. 15.29 (**a**), (**b**) Middle-aged patient presenting with vertigo, ataxia, and nystagmus. (**a**) On the right VA angiogram a moderate stenosis of the VA at its origin was visible (*arrow*). There was a normal flow in the intracranial vertebrobasilar sector. The PICA was not recognizable, but a well-developed AICA was present. This was visible on a more selective study. (**b**) The left VA was occluded. There was a partial distal revascularization through anastomosis with the ascending cervical artery. The patient recovered and the stenosis was later treated with angioplasty and stent. (**c**) Severe impairment of flow in the vertebrobasilar sector in another patient with severe stenosis of the dominant left VA (*arrow*) and hypoplastic right VA. Angiograms pre- and poststenting

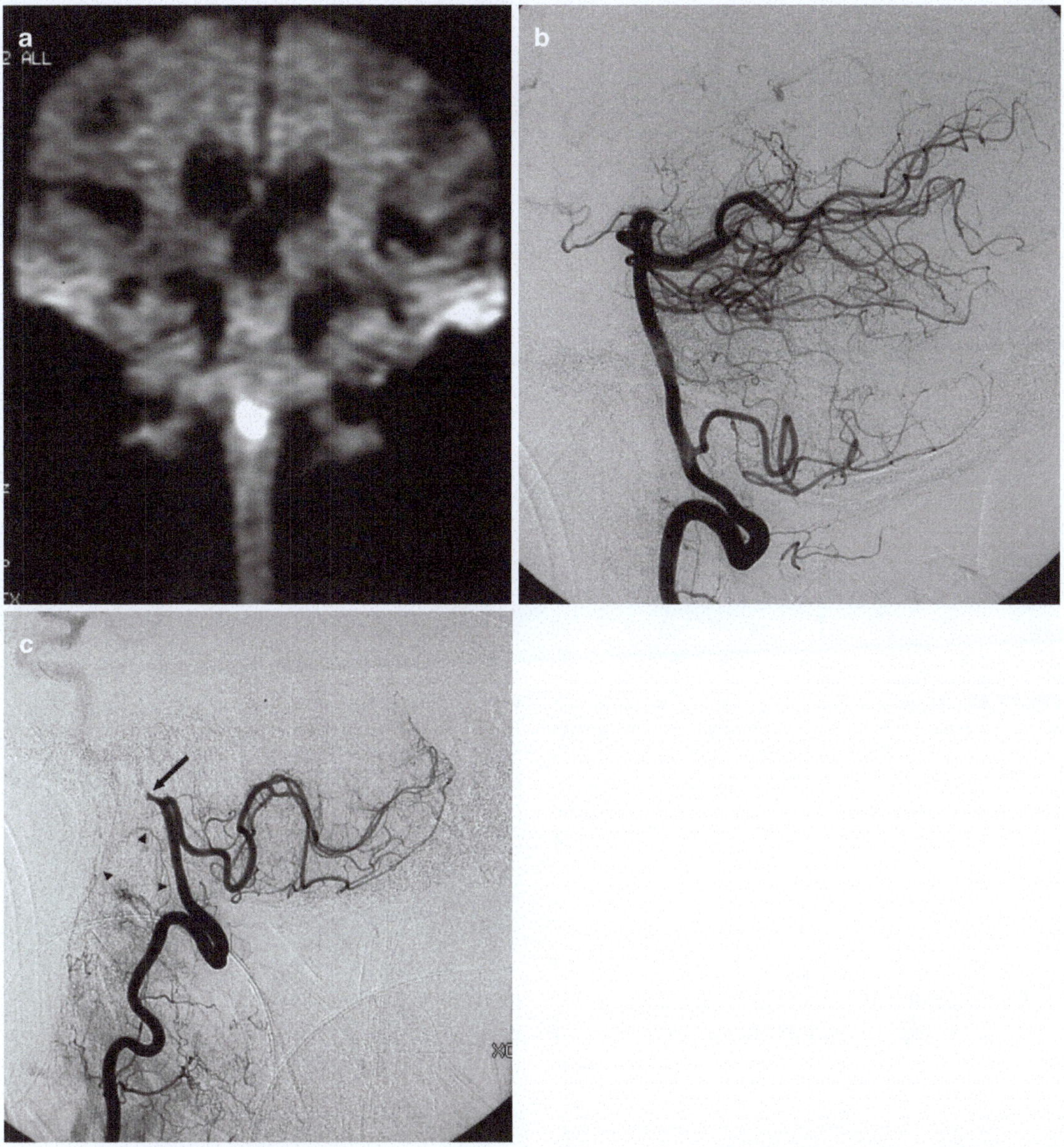

Fig. 15.30 (**a**) Acute ischemia involving the anterior medulla, corresponding to the vascular territory of the anterior spinal artery, demonstrated on the MRI with diffusion-weighted image. (**b**) Angiography shows a normal left angiogram and an occlusion of the right VA distal to the PICA (**c**). On both angiograms the anterior spinal artery was not identified, probably occluded. Note the presence of the odontoid arch (*arrowhead*)

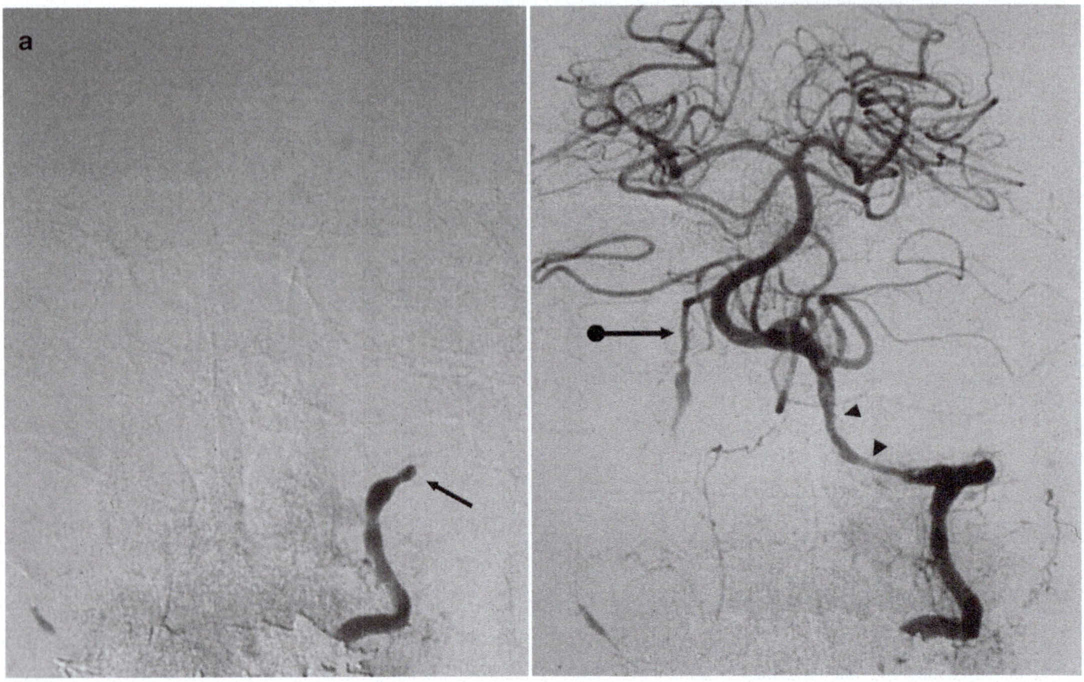

Fig. 15.31 Acute thromboembolic occlusion of the left VA at the extra-intracranial passage in an old patient with an old occlusion of the contralateral VA. Left vertebral angiogram showing the occlusion of the VA (*arrow*). Control angiogram (*right on the figure*) after selective pharmacological thrombolysis. Reopening of the artery with complete revascularization of the vertebrobasilar sector. Severe atheromatous changes of the VA (*arrowhead*) on which the thrombus developed are now recognizable. There is a partial retrograde filling in the occluded right VA (*arrow with dot*)

Endovascular treatment can also be considered in cases of symptomatic *intracranial microatheroma* of one VA especially if associated with lesions or hypoplasia of the contralateral VA, not responsive to medical therapy (Fig. 15.33).

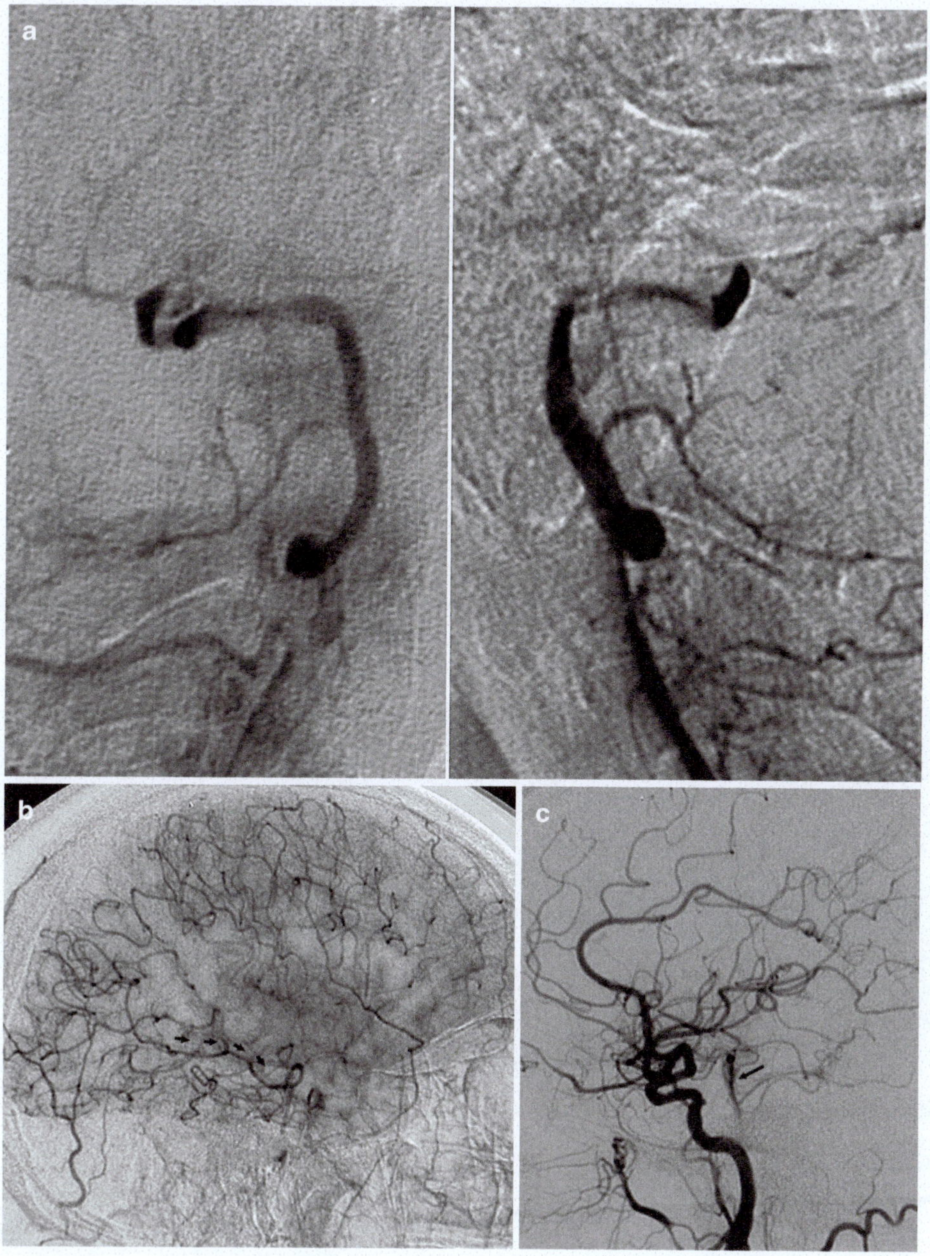

Fig. 15.32 60-year-old hypertensive, diabetic patient presenting with symptoms indicating involvement of posterior circulation, progressively worsening. Consciousness was not affected. CT and CT angiography showed small ischemic area in the cerebellum. Brainstem appeared normal. Basilar artery was not recognizable. (**a**) Left and right vertebral angiograms showed occlusion of both VAs at the extra-intracranial passage. (**b**) right carotid angiogram, late phase. Through leptomeningeal anastomoses (*arrow*) there is a retrograde filling of the right PCA and distal BA. (**c**) left carotid angiogram. Filling of the distal BA through a small PcomA (*arrow*). (**d**) Mechanical thrombectomy (aspiration device and solitaire stent) with reopening of the right VA and restoring of the intracranial circulation. Partial retrograde injection of the occluded left VA. (**e**) DWI-MRI performed 3 days later showed several ischemic lesions due to involvement of median, paramedian perforators of BA, as well as its lateropontine branches. Small infarcts are also recognizable in the cerebellum. Severe tetraparesis and partial involvement of the last cranial nerves in a conscious patient were present. Slowly improvement occurred in the following months

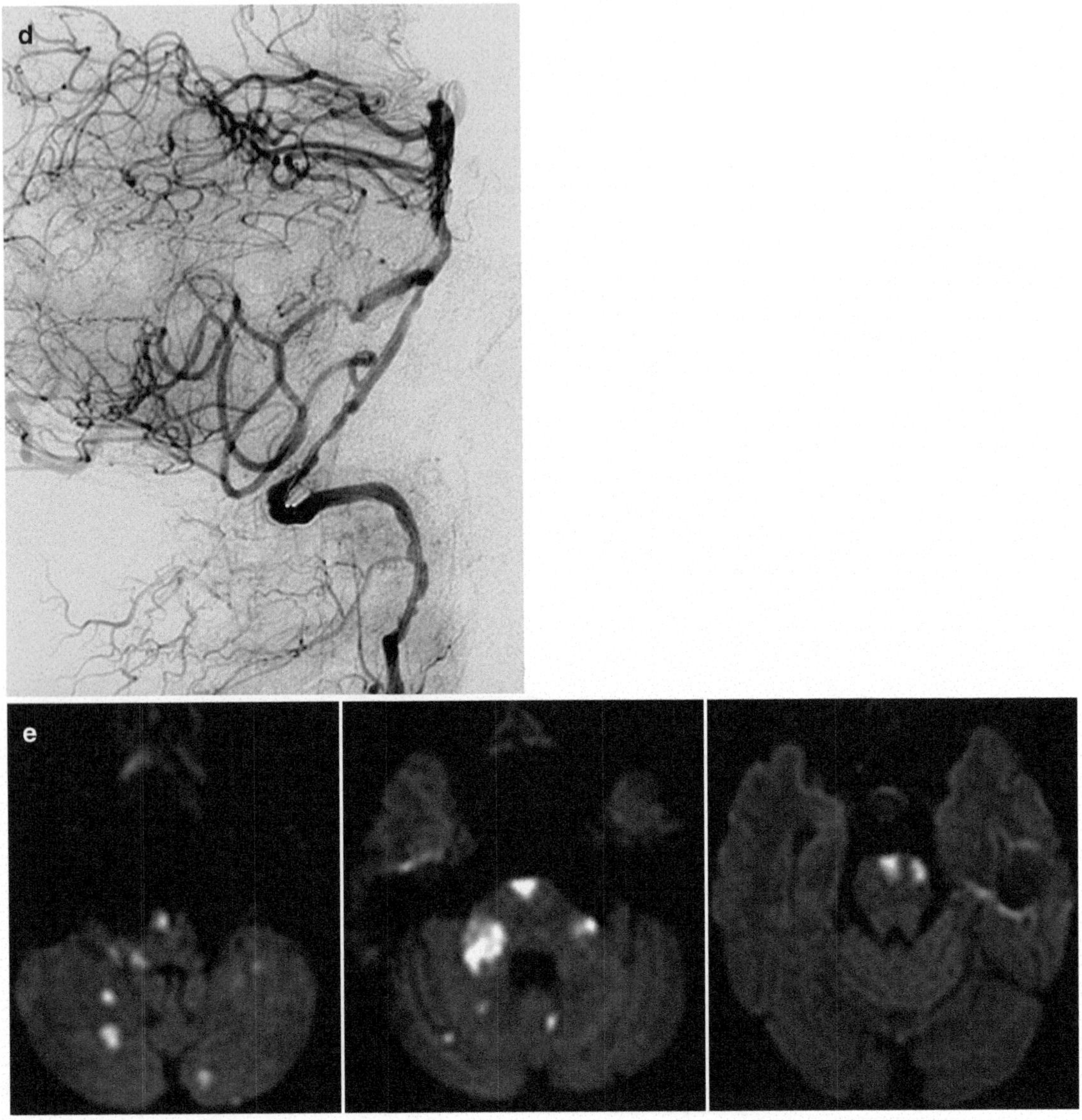

Fig. 15.32 (continued)

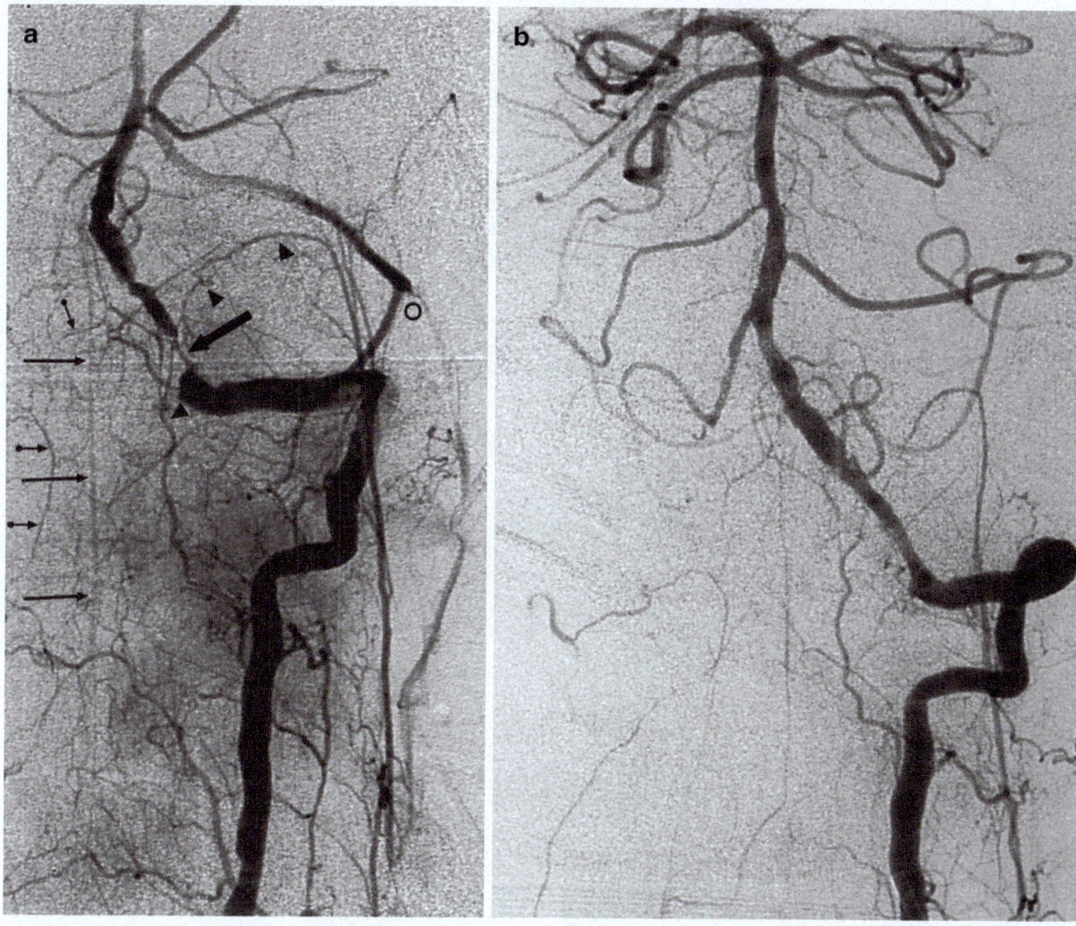

Fig. 15.33 TIAs in the vertebrobasilar sector in middle-aged hypertensive patient, in whom MRI disclosed an ischemia in the inferior pons. (**a**) Left vertebral angiogram showed a severe stenosis (*arrow*) of the VA at its entrance in the cranial cavity. The PICA is not recognizable, though the ASA is well injected (*small arrows*). There is a slowing down of the distal flow. The right VA was occluded. Probably due to the impairment of the flow in the VA, there is a good filling of the occipital artery (o) through opening of the C1 anastomosis. It seems that the proximal pharyngo-occipital trunk is retrograde injected. There is a clear connection (*arrowhead*) between the hypoglossal branch of the ascending pharyngeal artery and the radiculomeningeal branch of the VA. Connection of the left radiculomeningeal branch of the VA with the contralateral one (*arrow with dot*). (**b**) Endovascular treatment of the stenosis with angioplasty and stent with improvement of the flow in the vertebrobasilar sector

15.5.3 Basilar Artery

Involvement of the basilar artery (BA) is another very serious condition. It can occur as an extension of the thrombus in the case of occlusion of the intracranial VA (Castaigne et al. 1973). Another very frequent cause is embolism cardiac in origin or arising from plaques located in the extra or intra cranial segment of the VA. Emboli flow frequently into the distal part of the BA (Caplan 1980) provided that no stenotic plaque is located along the artery arresting their course, or the emboli are large occluding the middle portion of the BA. Some examples are presented in Figs. 15.34, 15.35, 15.36, 15.37, 15.38, 15.39, and 15.40.

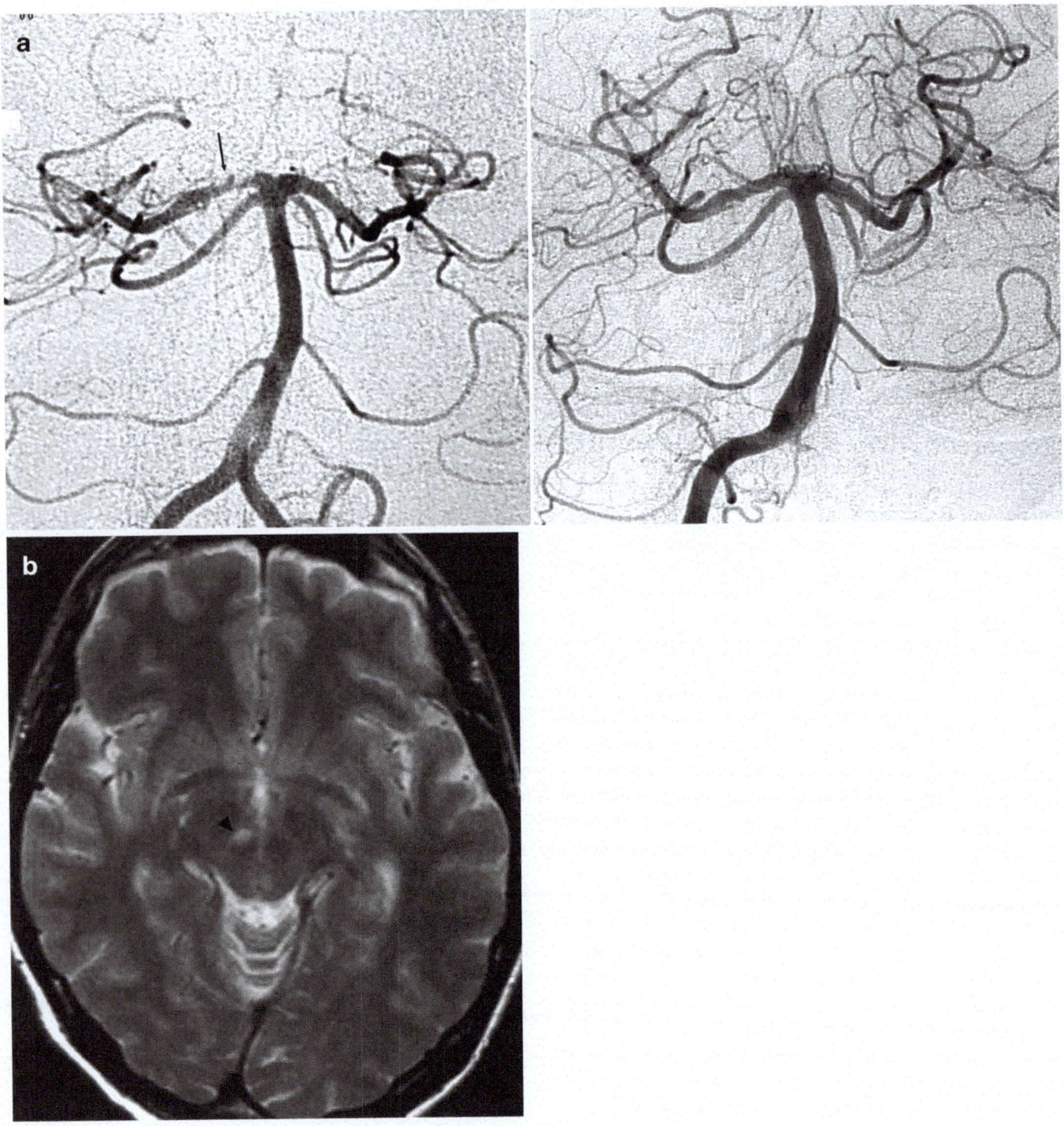

Fig. 15.34 Acute "top of the basilar" syndrome in a middle-aged patient. (**a**) On the right vertebral angiogram there is a normal filling of the basilar artery. Fenestration in its proximal segment. Small irregularities are visible on the right P1 segment (*arrow*), and there is no filling of the perforators. Selective pharmacological thrombolysis was performed with normalization of the P1 segment and reappearance of the perforators (*figure on the right*). The comatose patient recovered completely. (**b**) On the MRI performed 3 days later in an asymptomatic patient a small ischemic lesion was recognizable in the medial right midbrain (*arrowhead*)

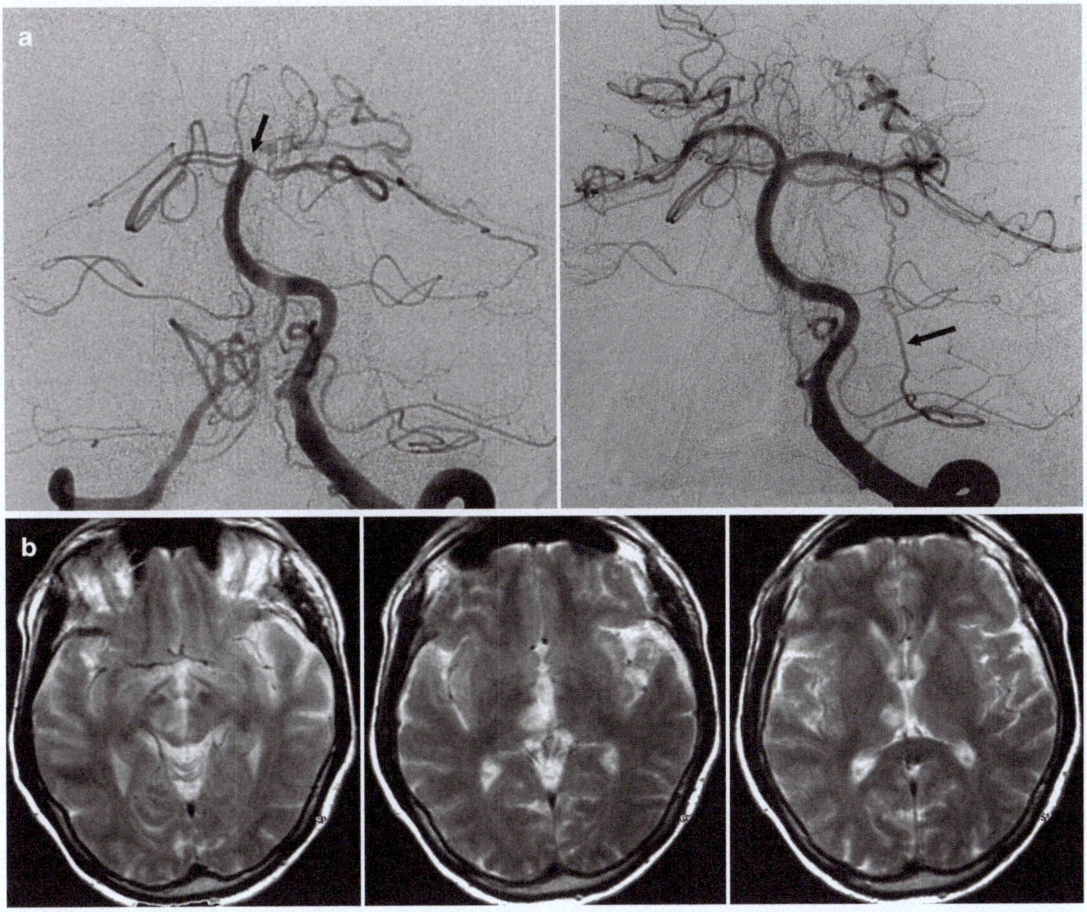

Fig. 15.35 Acute "top of the basilar" syndrome. (**a**) Angiogram of the left well-developed VA, with retrograde injection of the hypoplastic right VA. The distal basilar artery is occluded (*arrow*). There is no injection of either PCA. Normal right SCA, and partial occlusion of the left SCA. (**b**) After selective pharmacological thrombolysis, reopening of the distal basilar artery and its branches. Note the injection of the perforators arising from both P1. There is a good filling of the posterior meningeal artery (*arrow*). (**c**) On the MRI performed a few days later, ischemic lesions involving the right medial midbrain and medial thalamus were recognizable. The comatose patient recovered slowly with a favorable outcome in the following months

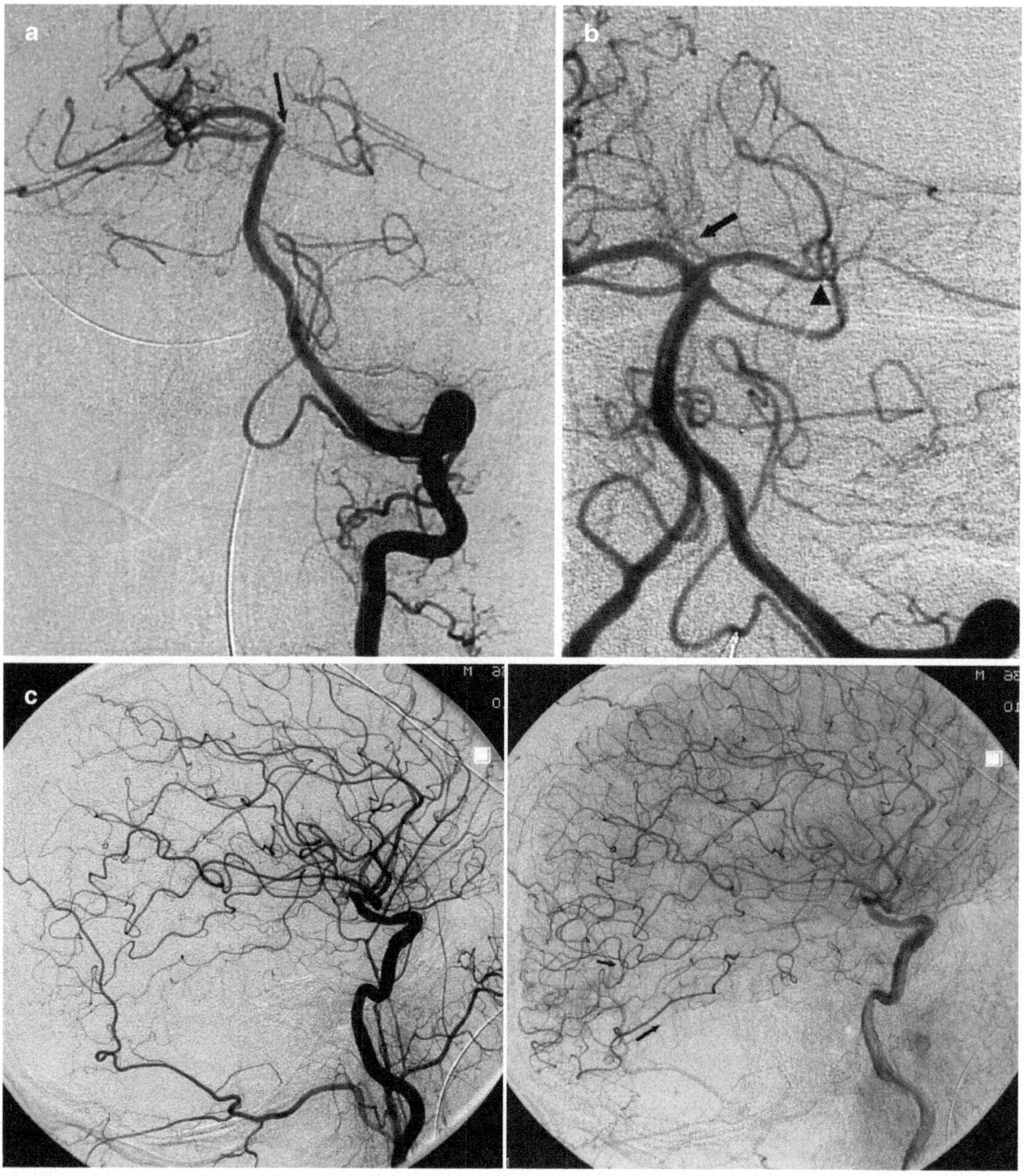

Fig. 15.36 Cardioembolic occlusion of the top of the basilar artery in an old patient presenting with acute symptoms indicating progressively vertebrobasilar insufficiency, followed by coma. (**a**) Occlusion of the BA (*arrow*) with involvement also of the left PCA and SCA was shown on the left VA angiogram. (**b**) VA angiogram post selective pharmacological fibrinolysis, showing the reopening of the BA and of the perforating branches (*arrow*) which seem to originate predominantly from the left P1. There is also revascularization of the left SCA and PCA. This latter, however, remains occluded at its P2 segment (*arrowhead*). (**c**) On the left carotid angiogram, a partial retrograde filling of the left PCA through leptomeningeal anastomoses with the MCA was recognizable (*arrows*). (**d**) MRI 24 h later showed some ischemic lesions bilaterally in the medial thalamus owing to the temporary occlusion of the posterior thalamoperforating arteries arising mainly from the occluded left PCA. Small ischemia also in the left posterior thalamus. Old right frontal ischemia. Progressively recovery of the patient

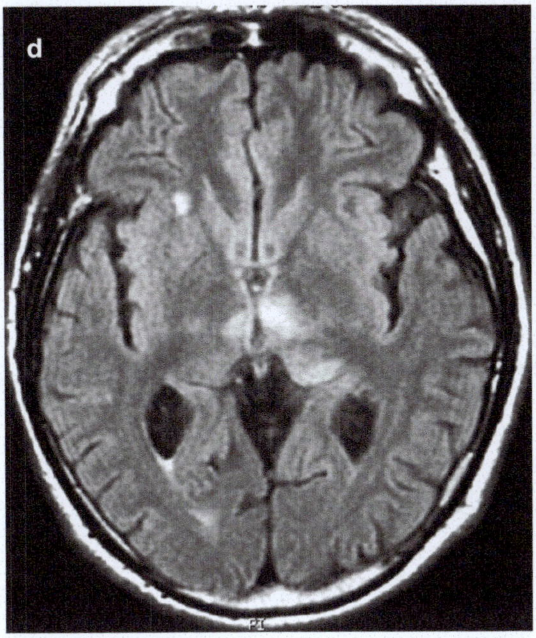

Fig. 15.36 (continued)

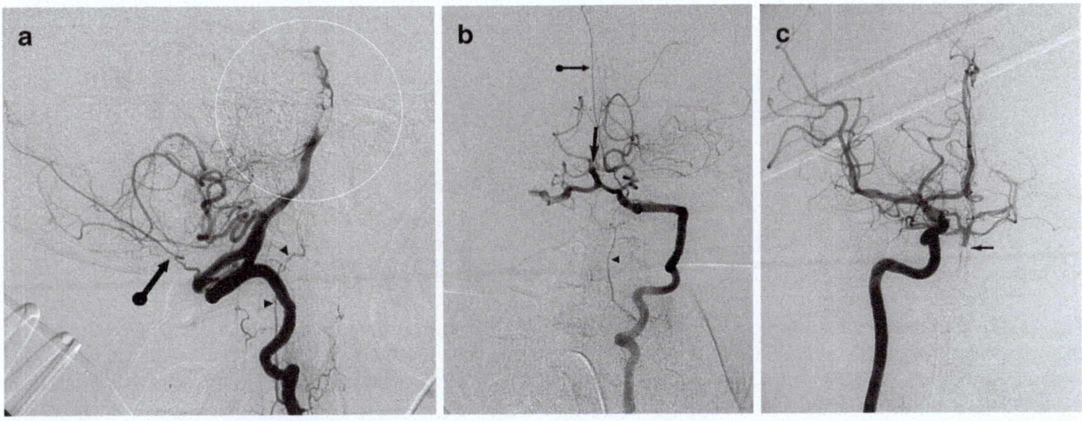

Fig. 15.37 Acute brainstem syndrome. (**a**), (**b**) Vertebral angiogram. AP and lateral views, showing the occlusion due to a large thrombus in the middle segment of the basilar artery (*arrow*) distal to the origin of the AICA. Radiculomeningeal branch (*arrowhead*), falx cerebelli meningeal artery (*arrow with dot*). (**c**) The right carotid angiogram showed through the PcomA a patent distal segment of the basilar artery (*arrow*) and also injection of the PCAs and SCAs. The thrombus in the BA is also well defined. (**d**) Detail obtained during selective study with a microcatheter, showing better the thromboembolic material in the BA. (**e**) Selective pharmacological thrombolysis allowed a complete recanalization of the BA. Small pontine ischemic lesions were visible on CT a few days later. There was a good recovery

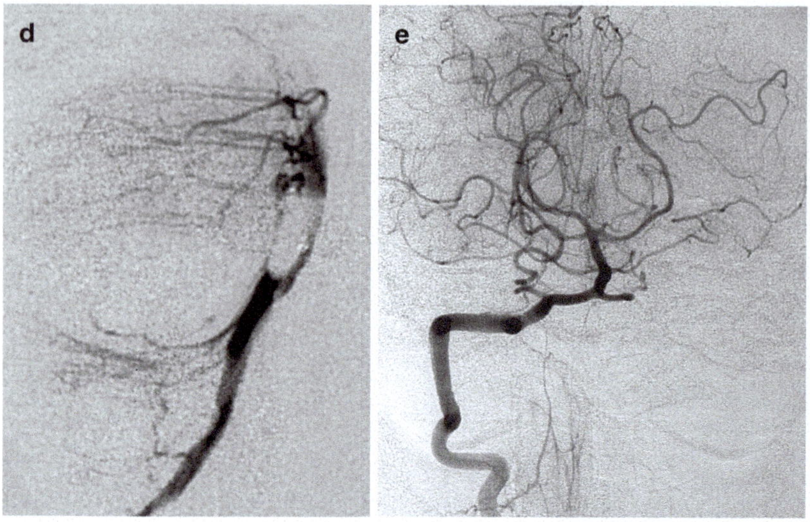

Fig. 15.37 (continued)

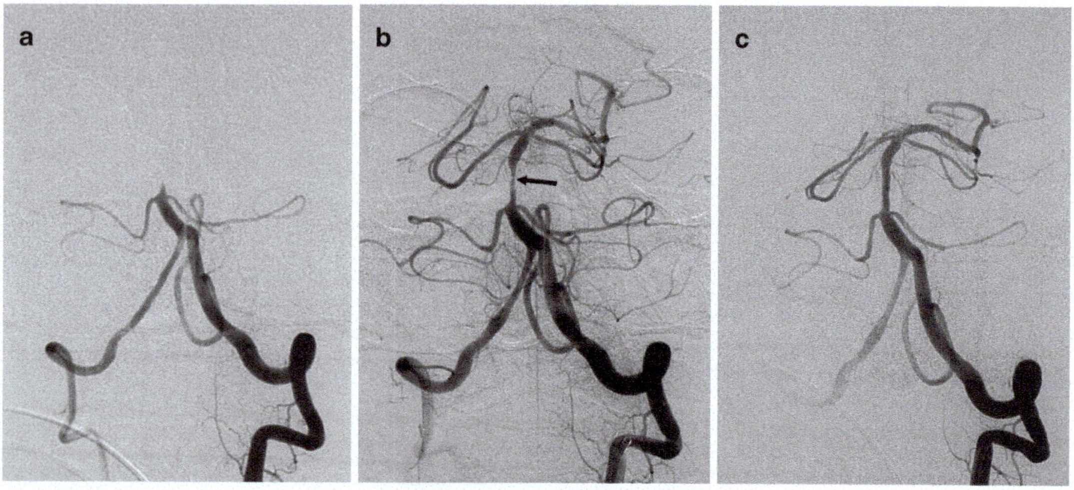

Fig. 15.38 Old patient with acute occlusion of the middle basilar artery owing to a thrombus superimposed on a long microatheroma. (**a**) Left VA angiogram. Occlusion of the BA distal to the origin of the AICAs. Retrograde injection of the hypoplastic right VA. (**b**) Reopening of the basilar artery with a selective pharmacological thrombolysis, showing the presence of the long stenosis (*arrow*), which was treated with angioplasty. (**c**) Angiogram after treatment

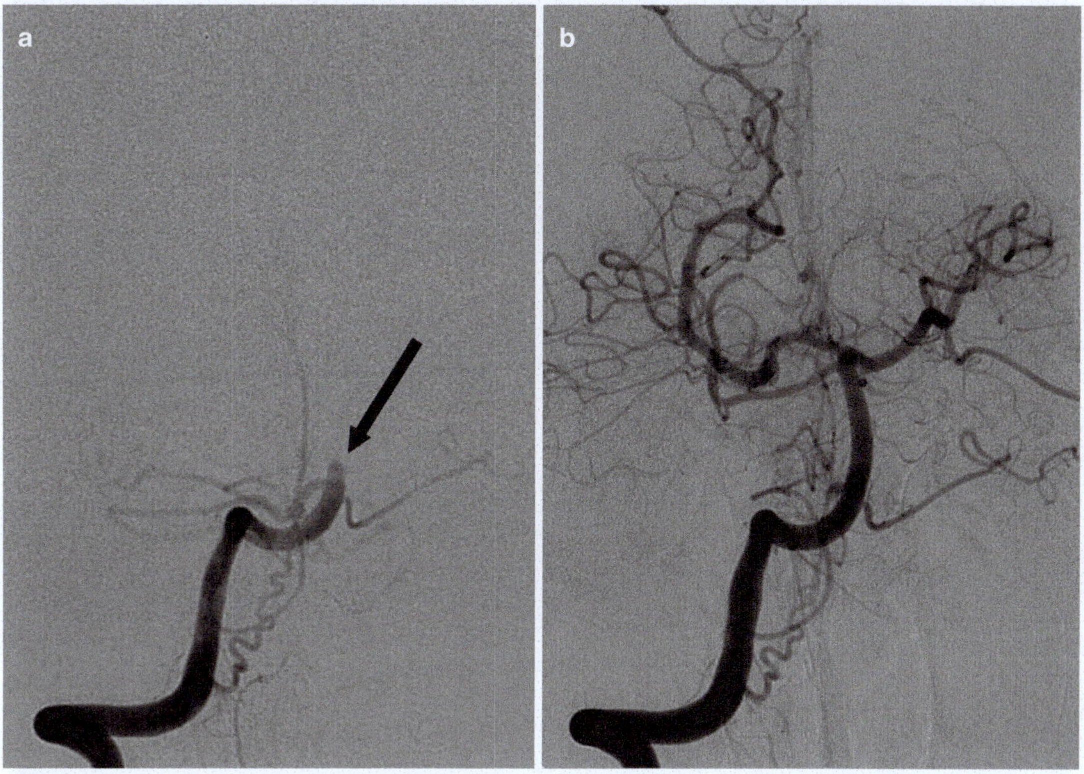

Fig. 15.39 Young woman admitted in a peripheral hospital with unclear symptoms suggesting, brainstem syndrome progressively leading to coma. On the admission in our department ca 7 h later, CT showed no certain ischemic lesions. (**a**) On the vertebral angiogram the occlusion of the BA distal to the origin of the AICAs was recognizable (*arrow*). Certainly the time passed from the start of the symptoms was relatively long. However, considering the young age of the patient and the experience that in occlusion of the BA positive results can be obtained also by relatively delayed treatment, endovascular treatment was decided. (**b**) Control angiogram performed after the rapid reopening of the BA with a solitaire stent. The left PCA was better filled on the left carotid angiogram. The left SCA remained partially occluded. Also the right AICA was suffering, probably occluded by embolus. The patient regained rapidly the consciousness, but was tetraplegic with palsy of cranial nerves with exception of eye movements. (**c**) MRI (DWI) performed 48 h later showed a large anterior and paramedian infarct in the pons. There was no lesion in the medulla, in the upper pons and mesencephalon sparing the reticular formation responsible of the consciousness. Several small cerebellar and a small left occipital lesions were also present. A few months later, after intensive physiotherapy, the patient was severely dysarthric and able to perform some minimal movements with the hands and finger. One year later however, she was independent, presenting with moderate right motor deficit and mild dysarthria

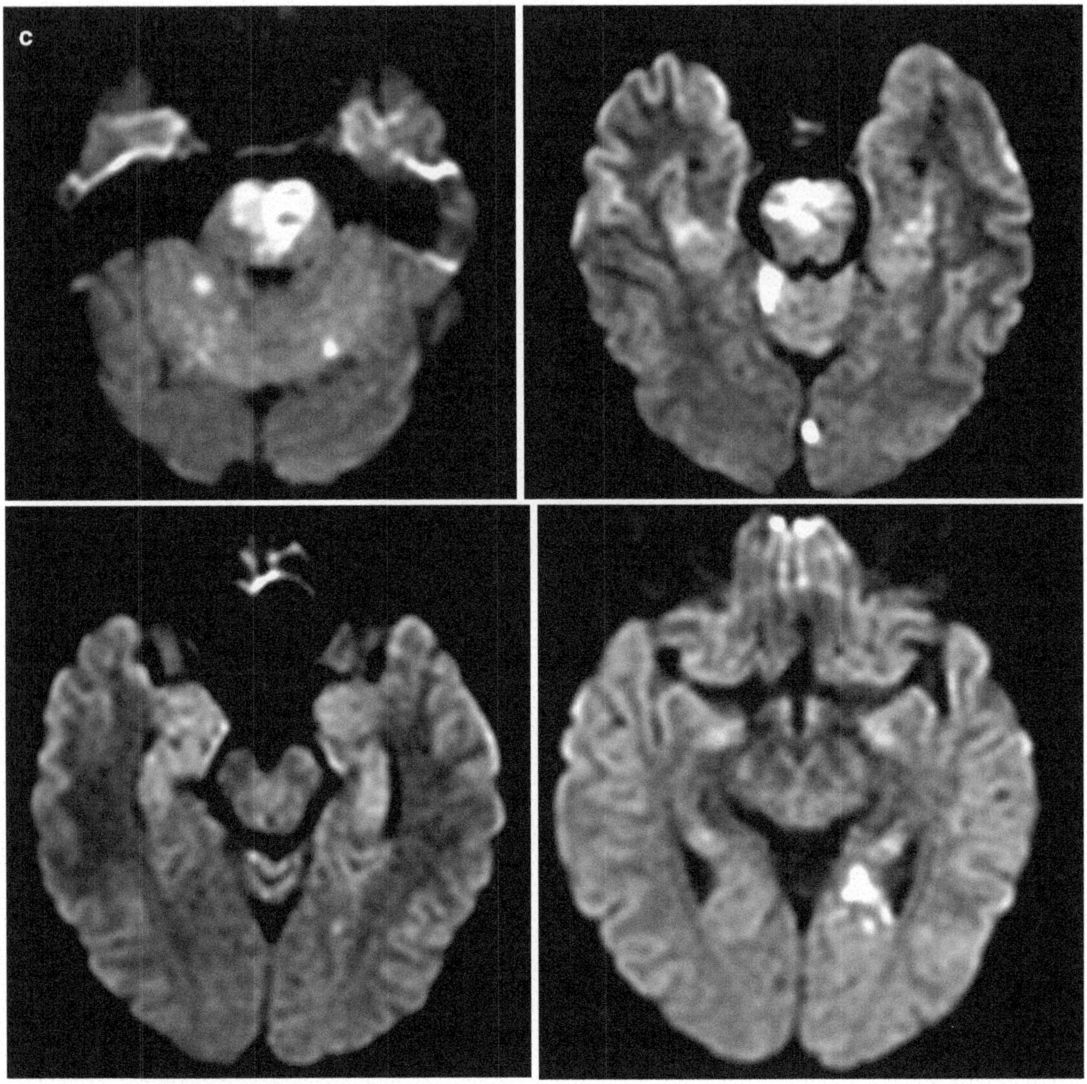

Fig. 15.39 (continued)

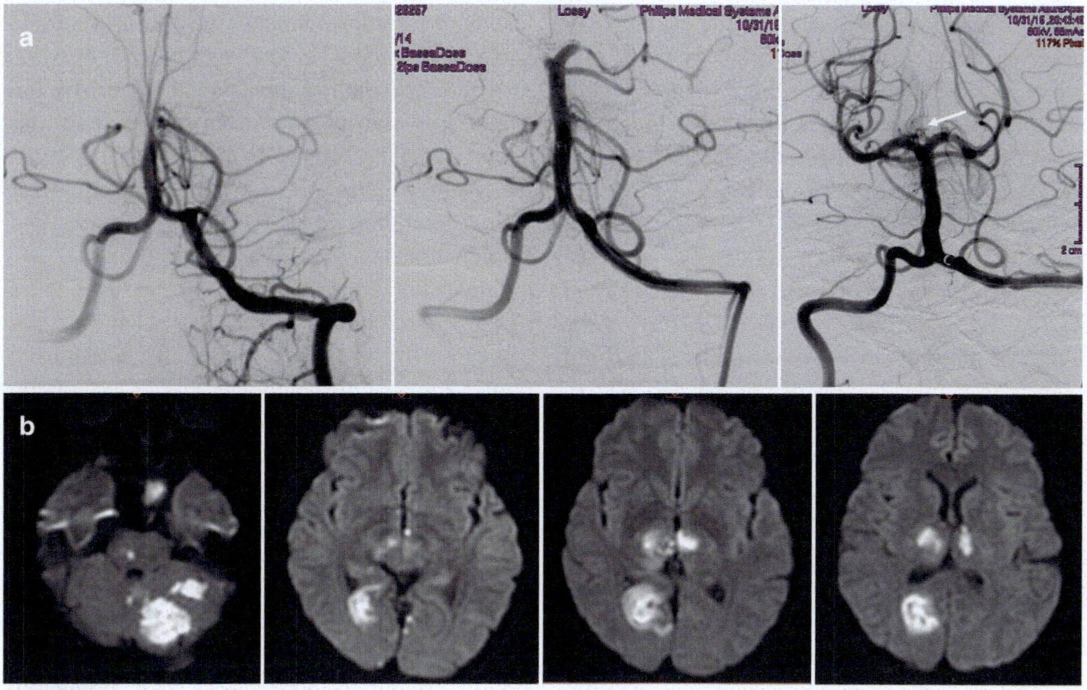

Fig. 15.40 Probably embolic occlusion of the BA in its middle segment due to embolization following a small dissection of the right VA occurring in a young woman. (**a**) Vertebral angiogram pre- and posttreatment with progressively reopening of the BA and PCAs with a solitaire stent. Note the reopening of the perforators (*arrow*). (**b**) Ischemic lesions were visible on CT and better on MRI (diffusion-weighted images) performed a few days later. These involved the cerebellum and the right occipital area. A small ischemia was recognizable in the pons. Severe lesions were recognizable on the medial hypothalamus and thalamus bilaterally. The patient regained very slowly the consciousness. Two months later, she was able to stand up and walk with minimal help. The striking symptoms were disturbances of the wake-sleep cycle, characterized by frequent episodes of sleep during the day probably a consequence of the bilateral thalamic lesions

The occlusion of the BA is always a very serious condition, since it involves, depending on the site and extension of the occlusion, the perforators for upper medulla and pons. The frequent involvement of the PCA leads to further ischemia in its vascular territory (see Sect. 15.5.5). Collateral circulation can occur only through leptomeningeal anastomoses between the cerebellar arteries. Another important collateral way can be represented by the PCA, if the artery (fetal type) arises from the ICA, and so it is not involved in the occlusion. In these cases a retrograde filling of the distal BA can develop through the PCA connected with the P1 provided that this is not absent (Figs. 15.32, 15.36, and 15.37). This is a relatively favorable condition. However, this can be impaired by distal embolization through emboli arising from the BA thrombus or during the endovascular treatment.

Zeumer et al. (1982) performed the first selective endovascular treatment of BA occlusion. Since then the advantages of this type of therapy has progressively become evident, and it is today the treatment of choice, improving in many cases the prognosis of the patients. Unfortunately, in some cases, in spite of the obtained reopening of the BA, severe clinical symptoms develop, due to impairment of the perforators. Nevertheless, in some of these patients a remarkable clinical improvement due to intensive care and rehabilitation can be achieved (Fig. 15.39 and 15.40).

Microatheroma of the BA can involve perforators at their origin and lead to pontine lacunar infarct (Fisher and Caplan 1971; Caplan 1989).

The occlusion usually involves the paramedian branches; less commonly it involves those that are more lateral. Microatheroma can be very thin and thus escape neuroradiological diagnosis. In other cases, it can be large enough to lead to stenosis of various degree of the BA, which is recognizable on MRI and CT angiography; in selected cases, microatheroma can be subjected to endovascular treatment (Figs. 15.41 and 15.42). The same risks we have described in the treatment of atheroma of M1 and distal ICA are present in similar lesions of the VA and BA. In particular, there is a risk of occlusion of the perforating branches. Also intrastent restenosis or thrombosis can occur; however, this seems to be rarer in comparison to that occurring in the anterior circulation (see also Sect. 15.8).

As in the white matter and basal ganglia also in brainstem especially in the pons lacunar infarcts can occur due to *lipohyalinosis*. The association of these lesions with similar microinfarcts in the anterior circulation is frequent. They often occur in hypertensive and diabetic patients (Fisher and Caplan 1971; Caplan 1996; Bradac et al. 2008b). The prognosis of these lacunar lesions are "quoad vitam" better than those due to occlusion of the branches by microatheroma. An example is presented in Fig. 15.43.

15.5.4 Cerebellar Arteries

There are many pathological conditions which can lead to impairment of the cerebellar arteries.

PICA, AICA, and SCA can be involved in acute thromboembolic occlusion of the intracranial VA and/or BA. The site and extent of the ischemia can vary considering that the cerebellar arteries supply the cerebellum but partially also the brainstem. The cerebellar parenchyma can be partially protected by the leptomeningeal collateral circulation between the distal branches of the cerebellar arteries. This can be insufficient if more than one cerebellar arteries is involved. This probably occur in the cases of very acute large cerebellar infarct with important swelling leading to compression of the fourth ventricle and hydrocephalus.

Microatheroma of intracranial VA or BA can occlude or narrow the ostium of the cerebellar arteries causing ischemia in their vascular territories due to hypoperfusion or distal embolization (Amarenco et al. 1990; Amarenco and Caplan 1993).

Stenosing or occluding atheromasic plaques of the extracranial VA especially when bilateral can cause hypoperfusion and infarct in the distal border zones of the cerebellar arteries (Savoiardo et al. 1987; Amarenco et al. 1994a, b, 1998). Emboli can arise from the plaques or from Heart, reaching the VA or BA and enter selectively the cerebellar arteries. Considering that the emboli can be very small and distal cerebellar parenchyma can be protected due the presence of leptomeningeal anastomoses between the PICA, AICA, and SCA, the infarcts are commonly small located in the border-zone areas, frequently asymptomatic and discovered incidentally. It is, however, important to recognize these microlesions, since, as emphasized by some authors (Park et al. 2009a; De Cocker et al. 2015, 2016), they are a sign of an already present atheromasia of the VAs or cardiac pathology.

Lacunar infarcts in the deep cerebellar structures (white matter and deep grey nuclei) can also occur even if they are more rare in comparisons to that in the cerebral hemispheres (Fisher 1965, 1968).Hypoperfusion and microembolization can be considered but as pathological studies have shown the lesion are frequently also due to intrinsic atherosclerotic changes (lipohyalinosis). Some authors (Park et al. 2009a; De Cocker et al. 2016) have described the frequent association of deep cerebellar with deep cerebral ischemic lesions. Lipohyalinosis is a well-known pathology involving the perforators of the cerebral hemispheres responsible for small ischemic lesions and for microhemorrhages (see also Sect. 15.4.5). The same findings can be found in cerebellum. It is therefore conceivable that we are dealing with the same basic pathology.

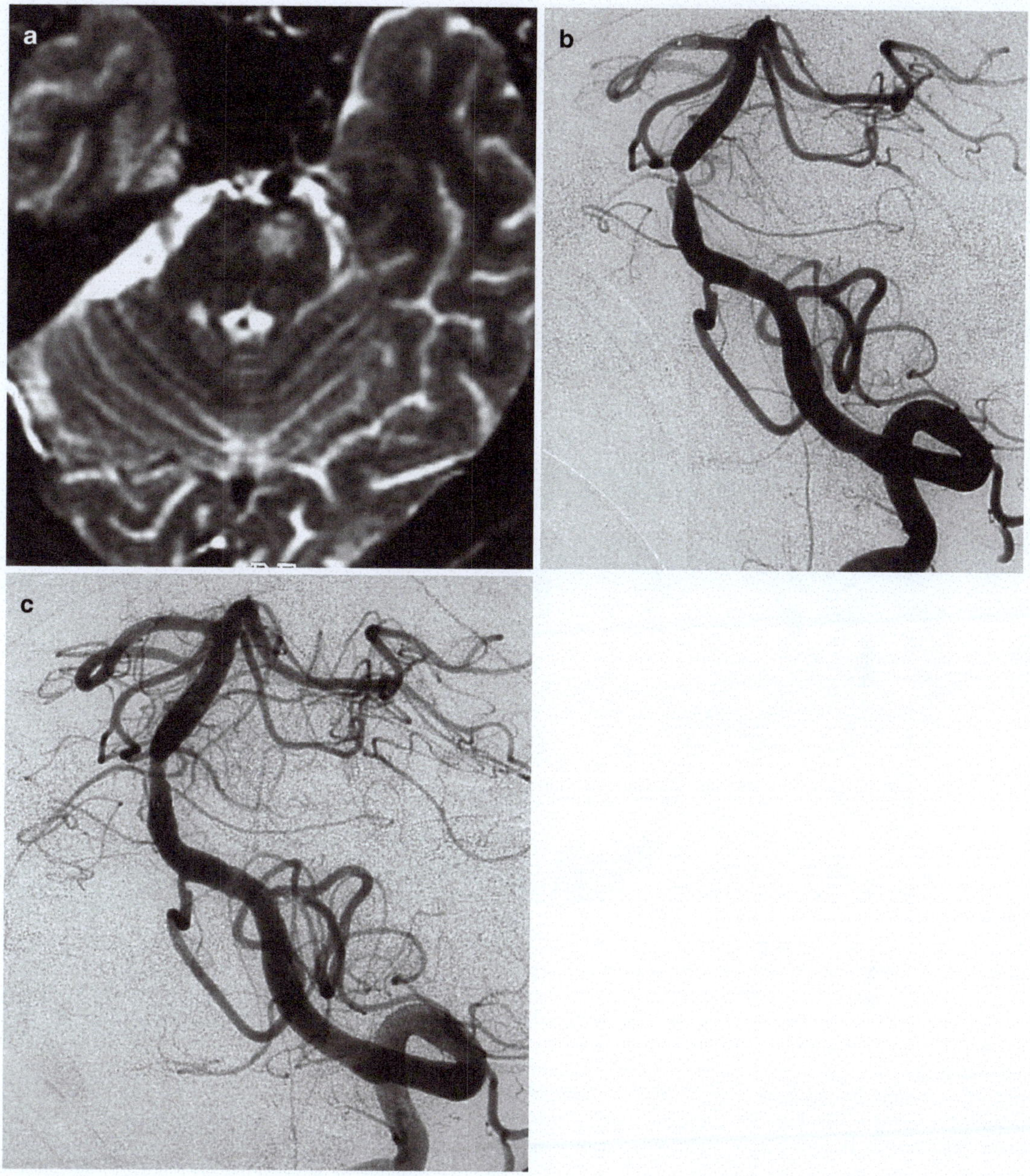

Fig. 15.41 Middle-aged patient with repeated mild stroke episodes involving the posterior circulation, not responsive to medical therapy. (**a**) MRI disclosed a small paramedian pontine infarct. (**b**) The vertebral angiogram showed a severe stenosis of the BA. (**c**) Posttreatment angiogram after angioplasty and application of a stent. The patient tolerated well the treatment

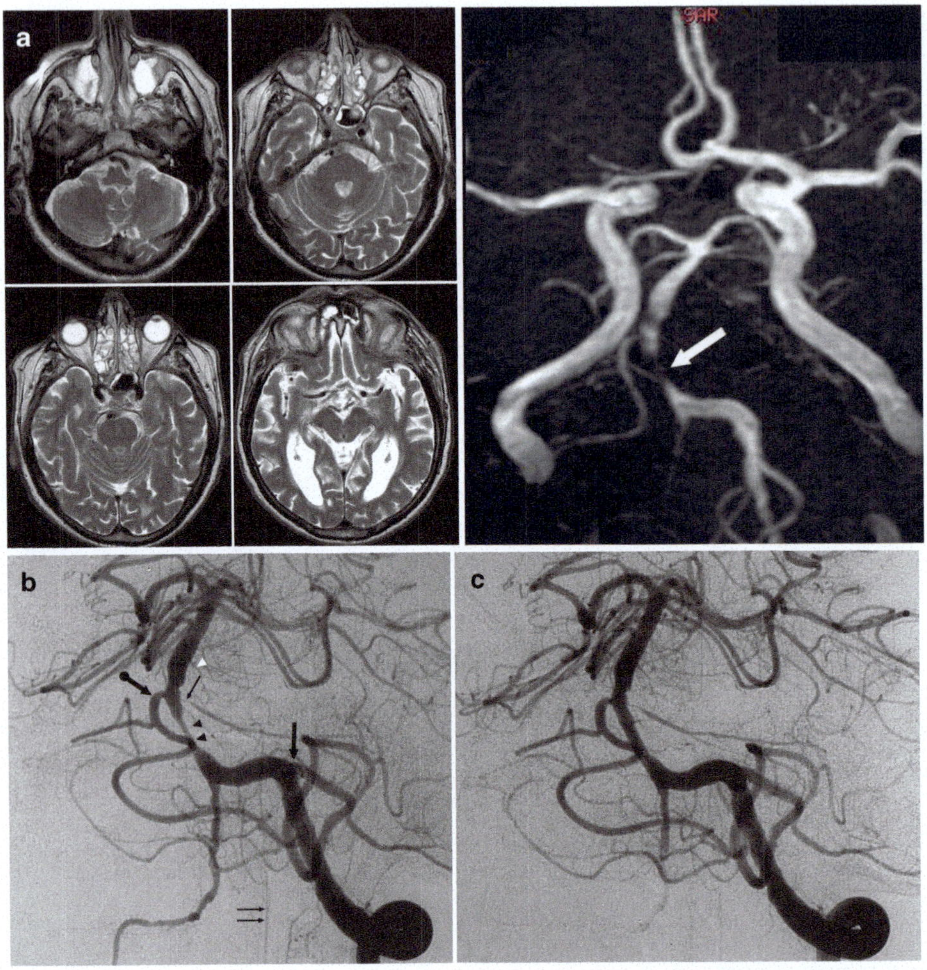

Fig. 15.42 Old hypertensive patient with repeated TIAs suggesting involvement of the posterior circulation. (**a**) MRI showed extensive micro-ischemic lesions in the white matter of both cerebral hemispheres. Brainstem and cerebellum were normal. On the MR angiography a severe stenosis of the BA was visible (*arrow*). Considering the risk of occlusion, the decision to perform endovascular treatment was taken. (**b**) Left VA angiogram with retrograde injection of an hypoplastic right VA. There is a severe stenosis of the BA (arrow heads) involving also partially the left AICA (*small arrow*). The distal BA is normal. Lateropontine branches (*white arrowhead*). Unilateral origin of the ASA (*small arrows*). Well-developed left PICA (*large arrow*). Well-developed right AICA (*arrow with dot*) supplying also the territory of the absent or occluded right PICA. (**c**) Angiogram post treatment (angioplasty and stent). The patient tolerated well the procedure

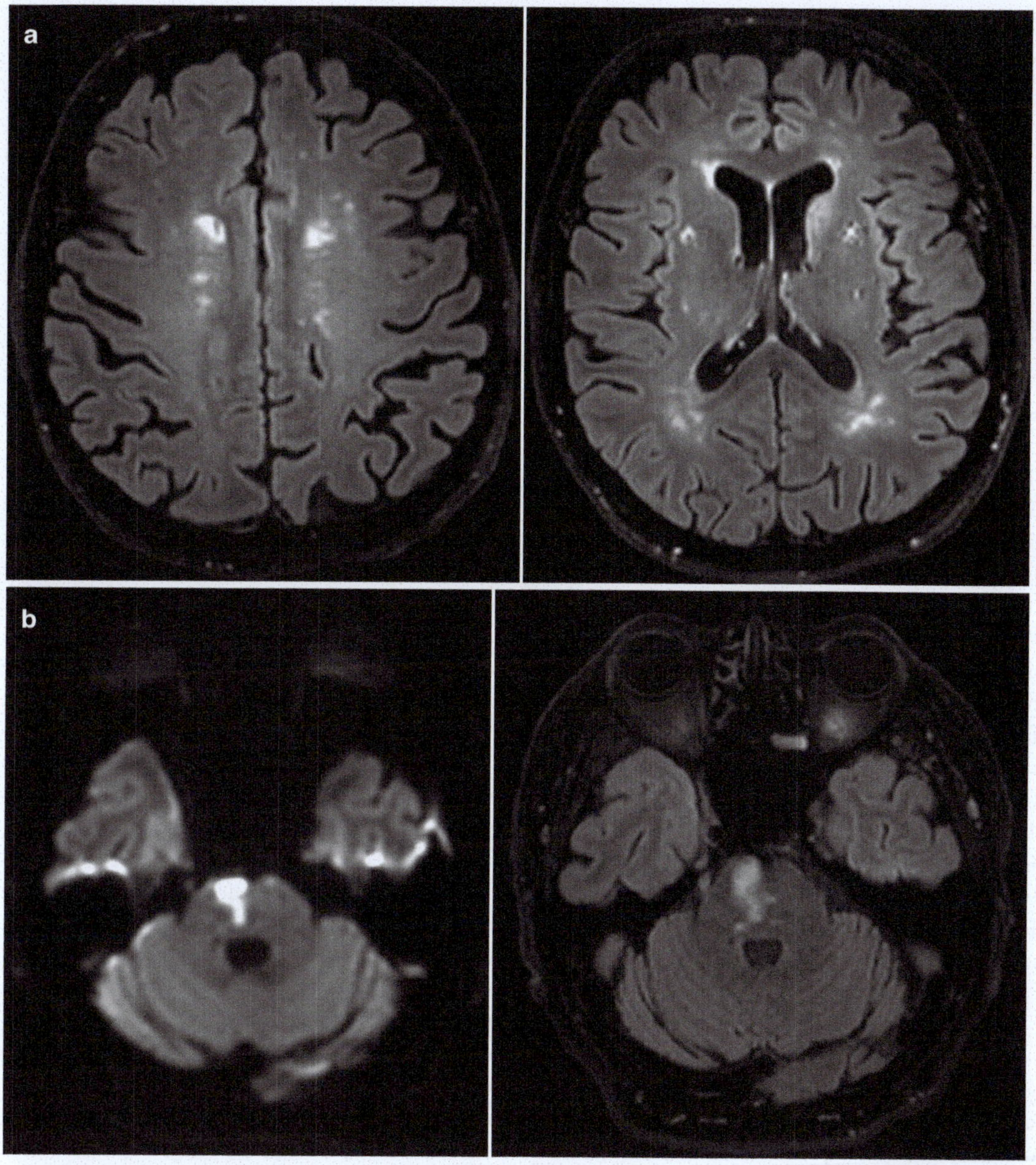

Fig. 15.43 80 year old hypertensive patient presenting with acute left hemiplegia. (**a**) MRI. T2 weighted images, showing several old ischemic lesions in the white matter and basal ganglia of both hemispheres. Small hypointensity lesions corresponding to old microhemorrhages are also present. (**b**) MRI T2-weighted and Diffusion Weighted images showing acute ischemia in the right pons. (**c**) MRI-Angiography showing no alterations of the vertebral and basilar arteries. The ischemia was probably due to occlusion of the right paramedian perforating branches in a patient with diffuse small vessels disease due to lipohyalinosis. The patient recovered slowly

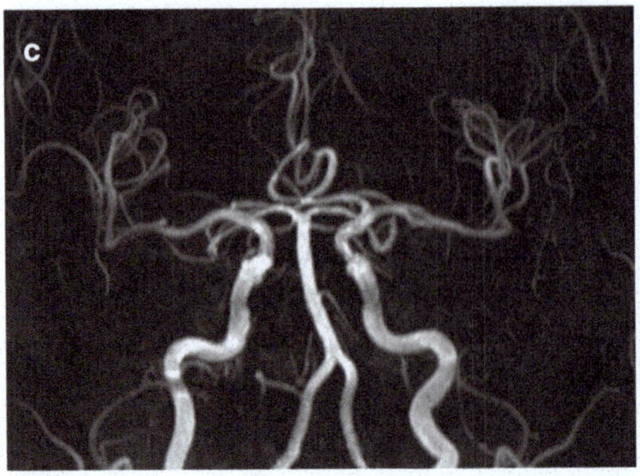

Fig. 15.43 (continued)

15.5.5 Posterior Cerebral Artery

Ischemia in the vascular territory of the PCA uni- or bilaterally occurs commonly in association with the occlusion of the BA and the extension of the thrombosis to the PCA through the P1. The most frequent cause of isolated occlusion of the PCA is embolism from plaques of the extra-or intracranial VA or it can be cardiac in origin (Castaigne et al. 1973, 1981; Milandre et al. 1994; Caplan 1996; Yamamoto et al. 1999; Brandt et al. 2000; Kumral et al. 2004). Embolism to the PCA can arise from the ICA in cases of fetal origin of the PCA and in cases of persistent embryonic carotid-basilar and carotid vertebral anastomoses.

Another cause of ischemia is a primary micro-atheroma that is commonly located at the P1–P2 (Bradac and Oberson 1983; Fisher 1986; Pessin et al. 1987; Caplan 1996), which can be complicated by a superimposed thrombosis. Ischemia can also occur in cases of aneurysm of the BA or VA due to direct involvement of perforators or to embolization.

The ischemic lesion can differ according to the site and extent of the stenosis/occlusion; it can involve the deep or cortical territories or both (Bougosslavsky et al. 1988a; Caplan et al. 1988b; Caplan 1996; Brandt et al. 2000). Involvement of the posterior thalamoperforating branches (P1) leads to ischemia of the medial midbrain and medial thalamus. The ischemia may be bilateral when both P1s are involved or when the perforators arise unilaterally just from the affected P1. Occlusion of the thalamogeniculate artery (P2) is responsible for ischemia in the lateral thalamus. Perforators arising from the P2 segment supplying the midbrain and the branches for the posterior thalamus (posterior choroidal arteries arising from P1–P2) can also be involved. These areas, however, may be more protected since they have a complementary supply from the AchA and SCA.

The distal territories have greater protection owing to the presence of leptomeningeal anastomoses between PCA, MCA, and ACA: however, ischemia can occur due to further embolism occluding the distal branches. Among the distal vascular territories, the primary visual cortex is frequently involved, with an incidence of up to around 90% of the cases (Kumral et al. 2004).

Lipohyalinosis involving perforators is another cause of ischemia which can be responsible for small lacunar infarcts in the midbrain and thalamus.

Examples of ischemia due to involvement of the PCA are presented in Figs. 15.34, 15.35, 15.36, 15.37, 15.40, 15.44, 15.45, and 11.20.

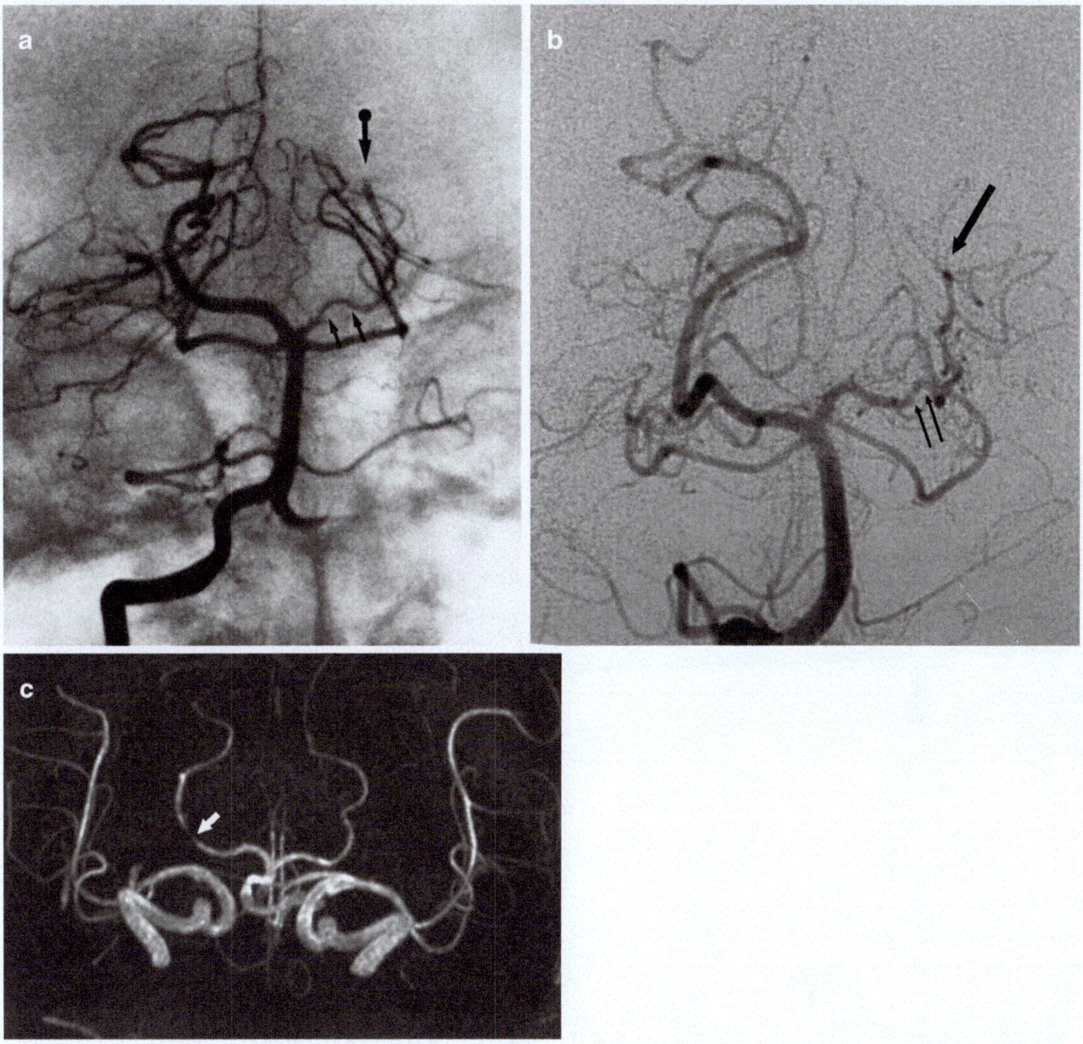

Fig. 15.44 Atheromatous changes involving the P1-P2 segments of the left PCA (*arrows*) demonstrated on the angiograms in (a and b). The distal PCA (*arrow*) is probably occluded by emboli. (**c**) Atheromasic plaque in the P2 segment in old patient demonstrated on MRI angiography

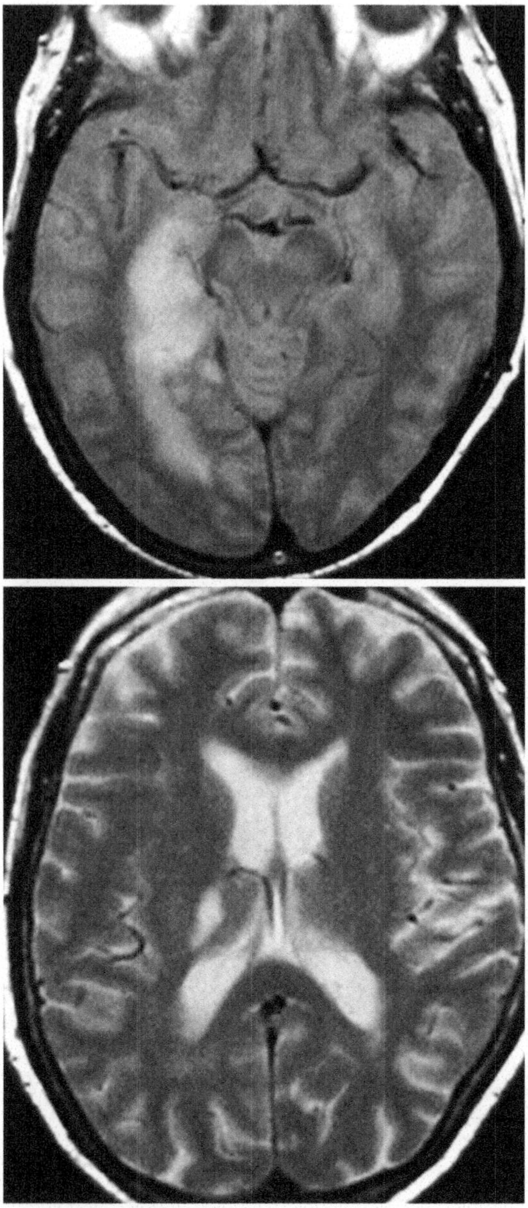

Fig. 15.45 Old cardiopathic patient with embolic occlusion of the right PCA in its P2 segment. MRI study showing ischemia in the medial temporal and occipital regions and in the lateral thalamus (thalamogeniculate artery). The medial midbrain and medial thalamus corresponding to the vascular territories of the perforators arising from P1 are spared

15.6 Changes in the Venous Sector

On the normal angiogram, the filled veins strictly correlated with the injected arterial sector appearing typically in the venous phase after the arterial and capillary phases. On the angiogram of patients with focal cerebral ischemia, *"early filling of veins"* in the ischemic and peri-ischemic areas associated with a capillary blush of the involved parenchyma are often recognizable in the acute and sometimes subacute phase of the ischemia; later this phenomenon disappears. This angiographic sign which has been correlated to a local increased perfusion called "*luxury perfusion syndrome*" (Lassen 1966) has been in the past the object of several and intensive studies, when the cerebral angiography was the main diagnostic method in the diagnosis of ischemia (Cronquist and Laroche 1967, 1969; Taveras et al. 1969; Bradac et al. 1975; Huber 1979; Bradac and Oberson 1983) and redescribed more recently (Dorn et al. 2012b). It has been interpreted as due to metabolic changes occurring in the ischemic area, characterized by lowered oxygen tension and acidosis, leading to vasodilatation and vasoparalysis of the small arteries, facilitating the rapid passage of contrast medium in the venous sector.

15.7 Collateral Circulation

One important factor which can avoid or reduce the ischemia in cases of occlusion of the extra-intracranial segments of the carotid and vertebral arteries, is the presence of an efficient collateral circulation. Many collateral ways can be involved.

15.7.1 Collateral Circulation Between Intracranial Arteries

- The most important in this group is the circle of Willis, involving A1, AcomA, P1, and PcomA (Fig. 1.2), connecting both carotid sectors and these with the vertebrobasilar system. However, as it has been described in this and in previous chapters, variants are frequent which can impair, in the given case, partially or severely the collateral way.

- A second important collateral circulation are the leptomeningeal (pial) anastomoses between ACA, MCA, PCA, and AchA developing on the cortex of the cerebral hemispheres and between PICA, AICA and SCA on the cerebellar cortex.
- Perforators supplying the deep structures (grey nuclei and capsula interna) of the cerebral hemispheres, those responsible for the vascularization of the brainstem, and the medullary arteries supplying the white matter of the cerebral hemispheres and the white matter and grey nuclei of the cerebellum, are considered *end arteries* without any possibility of collateral circulation. Their involvement can occur in occlusion or stenosis of the parent artery from which they arise, or selectively due to emboli, atherosclerotic plaques at their origin or lipohyalinosis.

  In a few reports (Kodama and Suzuki 1974; Umansky et al. 1985), a few connections between single deep perforators and between these and medullary arteries have been identified. Occasionally, on the angiogram, connections between perforating branches can be demonstrated or at least suspected (Fig. 5.15). These anastomoses have been postulated to explain the network of vessels present in Moyamoya disease (see also Sect. 17.3.1). In a similar pathological condition, these connections could be clearly demonstrated on the angiogram (Fig. 17.7).

  Minimal anastomoses have also been identified between perforators of the brainstem (Kalimo et al. 1997). These connections and the corresponding flow, however, are minimal and do not play any role in developing a collateral circulation in case of an acute occlusion, but they could increase the flow and act as a collateral way when the occlusion is a long-lasting process.

  In this context, considering the patients in whom a flow-diverter stent has been used, the perforators can be saved by the flow passing through the strut of the stent, or in cases of occlusion of some perforators, these could be revascularized by the developing of a collateral circulation favored by the long-lasting occlusion process.
- Other more rare possibilities of collateral circulation are those between the choroidal arteries involving the AchA and the posteromedial and posterolateral choroidal arteries of the PCA and the connections between ICA and the vertebrobasilar sector through embryonic anastomoses (Sect. 2.3).

15.7.2 Collateral Circulation Between Extracranial and Intracranial Arteries

- Several branches of ECA are connected with the petrous-cavernous segments of ICA and can develop and enlarge in case of ICA occlusion. An important way is the connection of ECA with ICA through the ophthalmic artery.
- Potential large connections can develop between ECA, especially through branches of the APhA and occipital arteries and the VA.
- A particular very rare condition is the occlusion of the common carotid artery. In this case, a complex collateral circulation frequently develops, involving several arteries showing frequently the patency of the ICA (Mishkin and Schreiber 1974; Bradac and Oberson 1983). One way is that in which anastomoses develop between the VA or the cervical arteries and the ECA. This latter is retrograde injected down to the bifurcation, followed by an anterograde filling of the ICA. We also observed a case of anterograde filling of the ICA through anastomoses between VA and APhA, the latter arising as a variant from the ICA (Bradac and Oberson 1983; Pelz et al. 1987). Another collateral way in case of occlusion of the common carotid artery is that in which there is involvement of the circle of Willis. In this condition the ICA is retrograde filled down to the bifurcation through the contralateral ICA by the way of the AcomA or through the vertebrobasilar sector by the way of the PcomA.

– Connections between meningeal and transosseous temporal and occipital branches of ECA and meningeal branches of ICA and vertebrobasilar sector can develop. This commonly occurs in DAVFs. Meningeal branches of ECA can also be connected with pial arteries in some pathological conditions such as AVM and in Moyamoya disease.

15.7.3 The Vertebrobasilar Sector Deserves a Few More Considerations

An occlusion of one VA can be frequently well tolerated provided that the other VA is well developed. In this context, it is essential that the vascular territory of the PICA of the involved VA is sufficiently vascularized either by a retrograde injection of the occluded VA or by a large AICA or by a well-developed contralateral PICA extending also to the involved vascular territory. Also important is the sparing of ASA due to its bilateral origin or if unilateral arising from the nonoccluded VA. An important collateral circulation to the ASA can develop from the radiculomedullary branch arising from C3/C4 (Fig. 16.9). Finally, vertebral to vertebral artery anastomoses by the way of radicular branches and anastomoses with branches of the deep and ascending cervical arteries can also play a role in the revascularization of the occluded VA (Sect. 3.9). As far as it concerns the subclavian steal syndrome see Sect. 15.5.1.

In conclusion: the type of collateral circulation is strictly connected with the vascular territory involved. Its efficiency, however, depends on many factors, mainly on the rapidity of the occlusive process, distal embolization, anatomical variations, as well as the presence of comorbidity (cardiac and vascular). Furthermore, it should be take into account that for some territories (perforators), there is no collateral circulation.

15.8 General Considerations About Endovascular Treatment

- The introduction of thrombolysis has marked a big step forward in the treatment of this pathology. Positive results have been obtained using pharmacological drugs either intravenously (IV) or selectively injected in the occluded artery (Zeumer et al. 1982, 1993; Jansen et al. 1995; Von Kummer et al. 1995; Urbach et al. 1997; Qureshi et al. 2001, 2002; Arnold et al. 2002, 2003, 2004; Noser et al. 2005; Zaidat et al. 2005; Nogueira et al. 2008; Tauntopoulou et al. 2008; Ciccone et al. 2013). Progressively the selective endovascular treatment has changed passing from pharmacological to the mechanical thrombectomy (stent retriever and/or aspiration devices), sometimes preceded by IV application of thrombolytic agents, association called "bridging therapy."
- *Today, mechanical thrombectomy is considered superior to the IV and to the selective pharmacological thrombolysis when large vessels are involved (proximal-distal ICA; trunk of MCA; VA-BA). The devices used have become progressively more easily and safe to handle allowing in many cases a rapid recanalization with an evident improvement of the clinical results as reported* by many authors (Ringer et al. 2001; Song et al. 2002; Bergui et al. 2006; Eckert 2009; Choi et al. 2009; Kashiwagi et al. 2010, Roth et al. 2010; Castano et al. 2010; Moehlenbruch et al. 2012; Machi et al. 2012; Nogueira et al. 2012; Andersson et al. 2013; Davalos et al. 2012; Dorn et al. 2012c; Pereira et al. 2013; Kuntze-Soederquist et al. 2014; Kurre et al. 2014; Schwaiger et al. 2016) and confirmed in a large randomized study concerning the treatment of patients with acute stroke in the anterior circulation (Goyal et al. 2015; Berkhemer et al. 2015; Jovin et al. 2015).

Considering this group of patients, those with a favourable outcome i.e. without or only with a minimal disability after thrombectomy is now estimated between 50 and 60%.

Also improved positive results, can today be obtained in more critical conditions such distal ICA occlusion and in tandem occlusions with stenting of the extracranial ICA and thrombectomy of its intracranial segments. The rate of good outcome reported in this group of patients is less homogeneous varying from about 25 to 40% with improving of the positive results in the last years (Flint et al. 2007; Malik et al. 2011; Fesl et al. 2011; Mpotsaris et al. 2013; Cohen et al. 2013, 2015; Behme et al. 2015; Liebeskind et al. 2015; Maurer et al. 2015; Kappelhof et al. 2015; Bradac et al. 2017; Frahm et al. 2016).

Considering the cases involving the posterior circulation positive results can be obtained as reported by the majority of the authors in about 30% of the cases (Machi et al. 2012; Mordasini et al. 2012; Dorn et al. 2012c; Nagel et al. 2013; Gerber et al. 2017), with exceptionally good results reported by Roth et al. (2010) and Andersson et al. (2013) with almost 60% of the cases.

- *Many factors can influence negatively the final results*: among them there are the *severe clinical condition* from the onset, *the extent of the brain parenchyma definitely lost* demonstrated with CT perfusion and the *long interval* between the appearing of symptoms and the *achieved complete or partial recanalization. This basically should not be longer than 6 hours in the anterior circulation. A more longer interval is accepted in the posterior circulation.* The reopening of the vessel is an important step, but it is not a guarantee for a good clinical result. Indeed, there is a discrepancy between the rate of recanalization obtained today in about 70–80% of all the cases and the favourable clinical results as described above.
- An important role is played by the presence of an *efficient collateral circulation* which can partially protect the brain parenchyma in the interval between stroke and revascularization. In this context, an important factor is certainly the morphologic aspect of the *circle of Willis*. As described in Sects. 15.4.1.2 and 15.5.3, the same variations of the circle of Willis can act negatively or positively on the collateral circulation depending whether the occlusion is in anterior or posterior circulation. Another important form of collateral circulation is represented by the opening of *leptomeningeal anastomoses*. Its efficiency depends on many factors such the above described aspects of circle of Willis. Hypotension and anastomoses developing in a border zone located far from the occluded artery could also play a role in the given case. Furthermore this type of collateral way can be impaired by distal embolization which can reduce or annul the expected positive results obtained with the reopening of the occluded proximal artery (Bradac et al. 2014, 2017). Distal embolization can occur through emboli developing during the endovascular treatment or arising from the occluding thrombus immediately after the occlusion. More recently some authors (Pham and Bendszus 2016) have emphasized that also other factors can be involved in the failure of this kind of collateral circulation. Indeed, already in the 1960s some authors (Ames et al. 1968) have suggested that the failure of the collateral circulation could be due to the narrowing and occlusion of the microvasculature following the swelling of the endothelium and of the glial cell in the ischemic area. They called this phenomenon *no-reflow*. Other studies have shown that obstruction of the microcirculation can be due to the increased adhesion of white blood cells and platelets favoring the formation of thrombus (Okada et al. 1994; Choudri et al. 1998; Kleinschnitz et al. 2009). Other authors have suggested that

the pathological changes involving the microvasculature could be due to the association of thrombosis and inflammatory mechanisms, which have been called "thromboinflammation" (Nieswandt et al. 2011). In another study the role of hyperglycemia leading to pathological changes of the microvasculature has been emphasized (Van Seeters et al. 2016).

- This could explain the bad clinical result occurring in some patients in spite of the technically obtained good revascularization (Bang et al. 2011; Gratz et al. 2015). In this context as also reported by some authors (Nelles et al. 2014) a more accurate selection of the patients could be useful performing CT perfusion and CT angiography before treatment, aimed to identify large irreversible damaged area and the presence and extent of the collateral circulation.

Complications are another factor which can worse the prognosis. Among them there are dissections, occlusion, spasm of the treated artery, as well as distal embolization into new territories and hemorrhages occurring with a frequency which varies in the reports of different authors. It can be valued occurring in about 10% of the cases. Fortunately, the majority of these events are asymptomatic. In a certain number of patients, however, complications have a clinical relevance and so influence negatively the prognosis.

Symptomatic new ischemic lesions occur in a rate of 1.1–5.6% (Kurre et al. 2014, Behme et al. 2014; Berkhemer et al. 2015), and *symptomatic parenchymal hemorrhage* is reported in a rate of 1.5–10% of the cases (Castano et al. 2010; Pereira et al. 2013; Behme et al. 2014; Soize et al. 2014; Goyal et al. 2015; Behme et al. 2015; Berkhemer et al. 2015; Jovin et al. 2015; Maurer et al. 2015). Some authors reported an higher rate of hemorrhage in tandem and "carotid-T" occlusion (Fesl et al. 2011, Behme et al. 2014, 2015). Hemorrhage can be due to vessel injury or reperfusion of the infarcted area. The risk increases in relation to the extent of the area of ischemic brain and also to the time passing between the occlusion and recanalization (Kuntze-Soderquist et al. 2014; Nelles et al. 2014). The role of the associated I.V. fibrynolitic therapy has been discussed. The rate of hemorrhage is lower in patients with vertebrobasilar occlusions, probably due to the increased density of the white matter tracts.

More recently also the *presence of subarachnoid hemorrhage (SAH)*, associated to the parenchymal but frequently also isolated, has been reported by several authors (Shi et al. 2010; Parrilla et al. 2012, Yoon et al. 2013; Yilmaz et al. 2014; Nikoubashman et al. 2014; Schwaiger et al. 2016). The pathogenesis of SAH is not completely clear. A vessel injury is a probable mechanism in many cases. Indeed it occurs more frequently where a mechanical thrombolysis is performed. Another pathogenesis could be the high concentration of contrast medium acting as a toxic element in the area already suffering of hypoperfusion. Both factors could favor the disruption of the blood brain barrier with leakage of blood sometimes associated with contrast medium (Nikoubashman et al. 2014). In the cases in which the finding is due only or predominantly to contrast medium, it rapidly disappears on CT controls made 24–48 h after treatment. It is worthwhile noticing that similar cases due to the toxicity of contrast medium in the absence of ischemia have been reported in patients who received an high amount of contrast during cardiovascular procedures (Sharp et al. 1999; Khan et al. 2014).

A particular aspect, as already described in Sects. 15.4.1, 15.4.2, 15.4.3, 15.4.4, 15.4.5, 15.5.2, 15.5.3, 15.5.4, and 15.5.5 *is the involvement in* the *ischemic stroke of the perforators*. These are *end arteries* and so their impairment in acute occlusion also for a short time, can lead to ischemia in their vascular territories. In addition should be note that gray matter tolerates hypoperfusion worse in comparison to the white matter.

While the infarcts in the vertebrobasilar sector are always more or less symptomatic, those involving the basal ganglia are often "asymptomatic" or only linked with mild neurological deficit.

- Another cause of ischemia, frequently presenting with TIAs or minimal stroke is that due to *microatheroma* involving the intracranial arteries as described in Sects. 15.4.1, 15.4.2, 15.4.4, 15.5.2, 15.5.3, and 15.5.4. These lesions which can be symptomatic but also incidentally discovered are reported to have a bad prognosis (Thijs and Albers 2000; WASID 2003). Some of these lesions, such those located in the petrous-cavernous-supraclinoid segments of ICA, in the M1 segment and in the intracranial VA and BA, can today be treated, by endovascular approach, with a good morphological and clinical results (Phatouros et al. 2000; Berkefeld et al. 2003; Gupta et al. 2003; Bose et al. 2007; Turk et al. 2008; Bradac et al. 2008b; Qureschi et al. 2008; Berkefeld and Zanella 2009; Wang et al. 2009; Miao et al. 2011; Dorn et al. 2012a; Derdeyn et al. 2013, Wang et al. 2016b). The rate of complications, however, is still relatively high: hemorrhage due to vessel rupture or reperfusion of the acute ischemic lesion, and new ischemia can occur due to distal embolization or occlusion of the perforating branches from the struts crossing their origin or displacement of microatheroma material. Another unsolved problem is the *in-stent restenosis or thrombosis* (Kurre et al. 2010). This negative evolution involves all treated intracranial stenosis, but it is reported to be more frequent in the anterior circulation, especially in those lesions located in the intradural ICA and in young patients (Levy et al. 2007; Turk et al. 2008). There is not a clear explication for this. It has been suggested that the intradural ICA stenosis, in this group of patients represents a distinct pathology not linked to atherosclerosis, but to inflammatory diseases (Levy et al. 2007; Turk et al. 2008), which react differently to the endovascular treatment. *Taking these considerations into account, in the attempt to reduce complications and improve the clinical results, it seems to be appropriated as also recommended by several authors* (Chaturvedi and Caplan 2003; Kurre et al. 2010; Fiehler 2014; Reith et al. 2015) *to perform an accurate selection of patients, considering for endovascular treatment only symptomatic patients with severe stenosis not responsive to medical therapy, and not undertake the treatment in the acute phase of the stroke. Furthermore, it is thinkable that improvement of this treatment can be achieved by increasing experience of the operators and by the introduction of more appropriate devices.*

16 Spontaneous Dissection of Carotid and Vertebral Arteries

16.1 Introduction

Jentzer (1954) is the first author to have reported a case of spontaneous dissection of the carotid artery. Since then, more detailed studies have described this pathology (Bostrom and Liliequist 1967; Ehrenfeld and Wylie 1976; Fisher et al. 1978; Mokri et al. 1979; Anderson et al. 1980; Friedmann et al. 1980; Bradac et al. 1981a) which have helped to widen the understanding of the clinical and angiographic aspects of this disease. Spontaneous dissection is today a well-recognized pathology that is responsible for stroke in many cases. Its incidence is reported to be 2.5–3 cases of carotid dissection and 1–1.5 cases of vertebral dissection per 100,000 persons yearly (Schievink 2001; Menon and Norris 2008; Redekop 2008). Young and middle-aged patients are predominantly affected. Multiple lesions occur about 20% of the cases (Pelkonen et al. 2003; Bejot et al. 2014).

16.2 Pathology and Pathogenesis

The cause of dissection can be *traumatic* occurring in blunt trauma of the neck due to forced flexion-extension or rotation movements, sometimes associated with fractures of the cervical spine or the skull basis (Redekop 2008). In the majority of the cases the dissection is *spontaneous* whose pathogenesis is not completely clear. Underlying structural changes of the arterial wall associated with mechanical factors, are probably involved. Indeed, it is well known that dissection occurs frequently in patients with connective-tissue disorders such fibromuscular dysplasia (FMD), Ehlers-Danlos syndrome, Marfan syndrome, alpha-glucosidase deficiency, polycystic kidney disease, osteogenesis imperfecta, and lupus erythematous (Makos et al. 1977; Anderson et al. 1980; Mitsias and Levine 1994; Schievink et al. 1994a; Schievink 2001; North et al. 1995). Association with cystic medial necrosis has also been demonstrated in some cases (Schievink et al. 1994a). Dissections are reported to be more frequent in patients with migraine and in women using oral contraceptives (Mokri et al. 1986; D'Anglejan-Chatillon et al. 1989). A family history has also been reported (Shievink and Mokri 1995). Trauma, even minor as occurring in neck manipulation (Hufnagel et al. 1999; Nadgir et al. 2003) or associated with sport, sexual activity, lifting of heavy objects, or in forced vomiting and coughing, is present in the clinical history of some patients (Dittrich et al. 2007). In this context, the association of trauma, even minor, and raised blood alcohol as a cause of dissection, particularly in the vertebrobasilar sector, have been emphasized (Hiraiwa et al. 2005). Also the association with recent infection, especially of the upper respiratory tract has been reported (Guillon et al. 2003). Some authors (Konrad et al. 2003; Vila et al. 2003) have reported in cases of dissection a deficiency of alpha 1-antitrypsin. This is a proteinase that has a protective action on collagen

G.B. Bradac, *Applied Cerebral Angiography*, DOI 10.1007/978-3-319-57228-4_16

and elastin, which are important components of the connective tissue of the arterial wall. The same deficiency has been identified also in cases of FMD, which has an high tendency for dissection (Schievink et al. 1998). In many cases a specific etiologic cause is not found.

From the pathological point of view, the lesion is characterized by a hemorrhage within the layer of the media from which it can extend towards the intima or towards the adventitia. In the first case one speaks of an *subintimal* dissection leading commonly to a stenosis or occlusion of the artery, in the second case of a *subadventitial* dissection in which frequently a pseudoaneurysm develops. The hemorrhage is due to a tears of the intima allowing the blood to pass into the wall of the artery. Commonly the dissection can present only with a point of entrance, but sometimes there is an entry and an exit point leading to the formation of a pseudo-lumen. Both the true lumen and the pseudo-lumen can be patent. This latter can sometimes be predominant. Dissection can be due to a primary hemorrhage within the wall of the artery not associated to a rupture of the wall. In these cases the cause is thought to be due to rupture of the vasa vasorum.

In this context, there are interesting autoptic studies in patients with intracranial dissecting aneurysms involving the vertebrobasilar sector (Sasaki et al. 1991a; Mizutani et al. 2001, Endo et al. 1993). An anatomopathological study of dissecting aneurysm of the anterior cerebral artery has also been reported (Mizutani et al. 2001). The main remarkable finding was a disruption of the internal elastic lamina (IEL) associated with a different grade of disruption of the media. If the disruption of the media was minimal the hemorrhage was between the media and IEL. When the disruption involved largely the media the hemorrhage extended towards the adventitia. In a few cases, however, there was no connections between the lumen and the intramural hemorrhage indicating that the hemorrhage develop primary in the wall of the artery probably due to rupture of the vasa vasorum.

Vasa vasorum deserves some specific considerations. They are present in extracerebral arteries, but their presence in the intracranial sector has not been well-defined and has been controversial. In this context very useful have been the anatomical and immunohistochemical studies of some authors (Aydin 1998) who have demonstrated the presence of vasa vasorum in the first segments of intracranial ICA and VA, more rarely in the more distal branches. According to the studies of these authors, the frequency of vasa vasorum increases in atherosclerosis and other cerebral vasculopathies. It is then thinkable that vasa vasorum can develop in pathological changes involving the wall of the artery leading to intramural hemorrhage and dissection. The presence of new formed vessel in and around the wall have been demonstrated in pathological specimen in dissecting intracranial vertebral artery where the intramural hemorrhage had no connections with the lumen of the artery (Endo et al. 1993).

As far as it concerns the types of dissection, it should be considered that an important factor is the structure of the wall of the arteries which is different in the extracranial or intracranial sector. The wall of the extracranial carotid and vertebral arteries is composed of five layers (intima, internal elastic lamina, media, external elastic lamina, and adventitia). In the intracranial arteries, as it has already been described in Chap. 11 there is no external elastic lamina. This explain the more frequent extent of the hemorrhage towards the adventitia with the formation of pseudoaneurysm in dissection of intracranial arteries.

16.3 Location

- *Extracranial internal carotid artery* (*ICA*), 2–3 cm distal to the bifurcation is the most frequent site of dissection. This ends abruptly with rare exception, where the artery enters the base of the skull. *The extracranial vertebral artery* (VA) is the second frequent location. Any part of the VA can be involved, but the lesions involve more often the distal segment of the artery, at the C1–C2, and at the atlas loop level, probably because this segment is more exposed to mechanical trauma (Chiras et al. 1985; Mokri et al. 1988).

- *Intracranial dissection is more rare. It involves typically the vertebrobasilar sector* usually affecting the first intracranial segment of the VA, sometimes as an extension of an extracranial dissection. The VA dissection differs from that of the extracranial ICA, which usually do not extend intracranially. Dissection of the basilar artery is less common; it can be primary or a secondary extension of a dissected VA. Progressively more attention has been given to this pathology which explains the increase of studies appeared in the literature in the last years (Alexander et al. 1979; Shimoji et al. 1984; Berger and Wilson 1984; Friedman and Drake 1984; Caplan et al. 1988a; Yamamura et al. 1990, 1999; Pozzati et al. 1995; Shin et al. 2000; Lacour et al. 2000; Manabe et al. 2000; Kurata et al. 2001; Anxionnat et al. 2003; Sugiu et al. 2005; Ramgren et al. 2005; Lee et al. 2006a; Zhao et al. 2007; Bhogal et al. 2015, 2016). Depending on the site of the dissection the cerebellar arteries can be indirectly involved. Selective dissection of these arteries can also occur (see Sect. 11.6.8). For dissecting aneurysms of the posterior cerebral artery and PcomA, see Sects. 11.6.5 and 11.6.2.4, respectively.
- *Intracranial dissection in the anterior circulation* occurs less frequently in comparison to the vertebrobasilar sector, but it is not so rare as commonly thought. The terminal ICA, M1, or A1 is the typical location. Dissections involving also the M2, A2, or more distal branches are increasingly reported (Ramsey and Mosquera 1948; Kunze and Schiefer 1971; Hochberg et al. 1975; Fisher et al. 1978; Sasaki et al. 1991b; Massoud et al. 1992; Bassetti et al. 1994; Nakatomi et al. 1997; Mizutani 1998; Mizutani et al. 2001; Kurino et al. 2002; Ohkuma et al. 2003; Lee JS et al. 2006a, b; Thines et al. 2006; Lv et al. 2009; Küker et al. 2012; Fukuma et al. 2015; Gao et al. 2016; Hensler et al. 2016).

16.4 Morphological Diagnostic Appearance

- *In the extracranial sector*, occlusion with a typical tapered appearance is frequent as well as an irregular stenosis, which extends for various lengths. Minor changes in the form of slight irregularities of the wall can occur. Demonstration of a double lumen or an intimal flap can be pathognomonic, but it is rare. Pseudosaccular aneurysms can also develop. Examples of dissection of the ICA are presented in Figs. 11.1, 16.1, 16.2, 16.3, 16.4, 16.5, 16.6, 16.7, and 17.5. Dissections of the VA are shown in Figs. 16.2, 16.8, 16.9, and 16.10.
- *Intracranial dissection* appears as occlusion, irregular stenosis (Figs. 16.9 and 16.15) or aneurysms (pseudoaneurysms) of different size which can be fusiform involving part or entire the lumeno of the artery, which can in some parts present saccular dilatation. Saccular aneurysm can also develop. They have frequently an ill-defined neck and changes in form of narrowing or dilatation are visible on the artery in the area adjacent to the aneurysm. They can grow and become very large (see also Sect. 11.8). Examples of dissecting aneurysms in the anterior circulation are presented in Figs. 5.10, 11.8a, b, 11.12b, 11.15, 11.17, and 16.17. Those in the posterior circulation are demonstrated in Figs. 11.9c, 11.24, 11.25, 11.26, 11.27, 11.29, 11.30, 16.11, 16.13, 16.14, and 16.16. *Not infrequently the arterial changes are minimal. Repeated examinations days or weeks later can show more conspicuous alterations* (Fig. 16.12a, b).

CT and MRI demonstrate the ischemic lesions or the subarachnoid bleeding in case of ruptured dissecting aneurysm. Furthermore MR can show hyperintensity or dilatation of the involved artery. Eccentric signal void (corresponding to the residual lumen) surrounded by a semilunar hyperintensity (corresponding to the mural hematoma) can be evident (Fig. 16.5). This, however, may be visible only 1 or 2 days after the acute episode first on

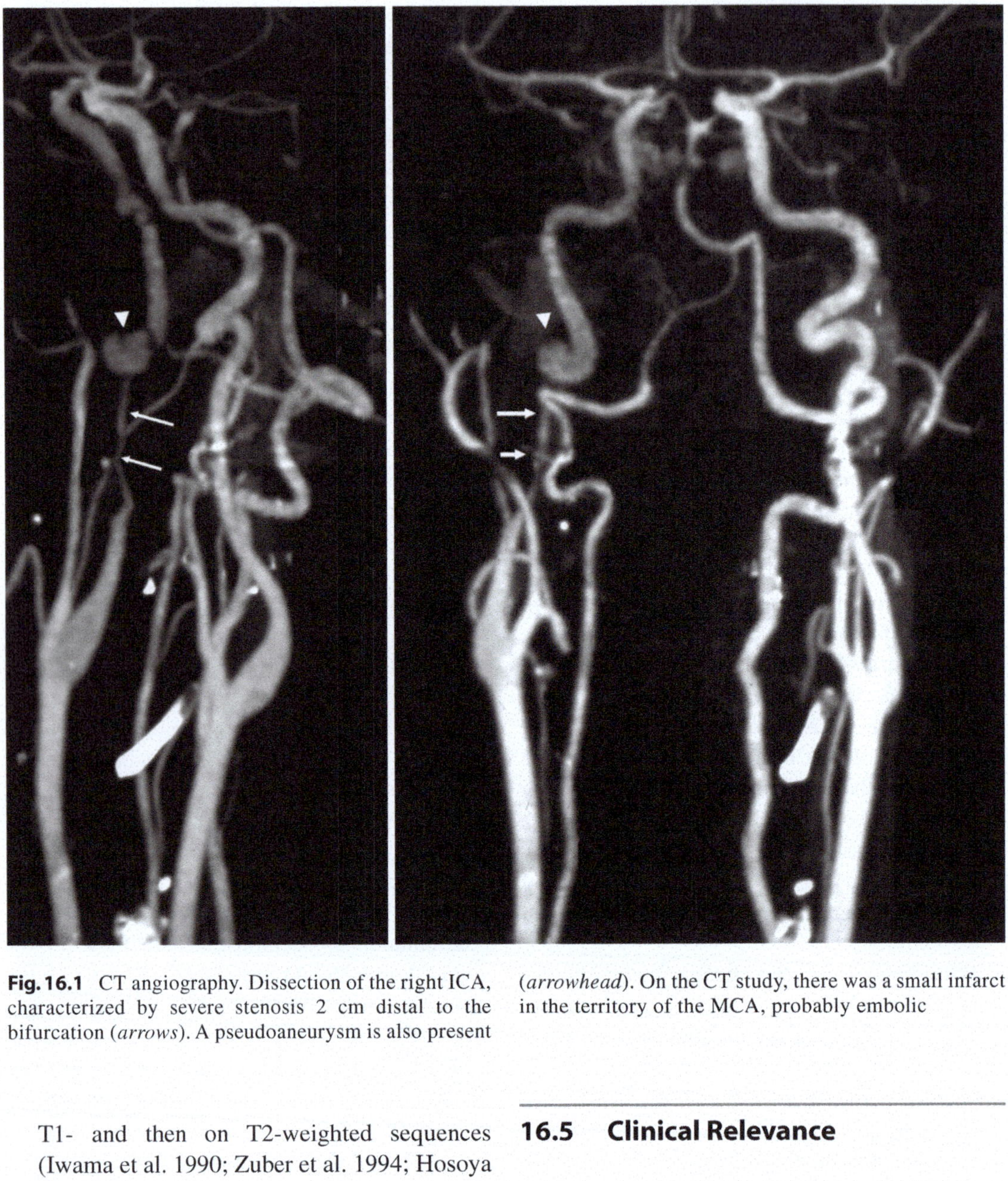

Fig. 16.1 CT angiography. Dissection of the right ICA, characterized by severe stenosis 2 cm distal to the bifurcation (*arrows*). A pseudoaneurysm is also present (*arrowhead*). On the CT study, there was a small infarct in the territory of the MCA, probably embolic

T1- and then on T2-weighted sequences (Iwama et al. 1990; Zuber et al. 1994; Hosoya et al. 1999; Vertinsky et al. 2008; Provenzale 2009).

MRI and CT angiography can replace conventional angiography in many cases. But the latter, owing to its precise morphological demonstration of the arteries, still remains the gold standard and it is to be used every time the diagnosis is not sufficiently clear or when a treatment (surgical or endovascular) is planned.

16.5 Clinical Relevance

- *Dissection most commonly occurs extracranially*, and it is responsible for 2% of all ischemic strokes and for about 20% of all strokes in patients under 45 years of age (Bogousslavsky et al. 1987; Schievink 2001; Redekop 2008). Some dissections may be accompanied only by cervical and facial pain (Buyle et al. 2001), which are always present and by Horner's syndrome as a result of a palsy of the sympathetic fibers sur-

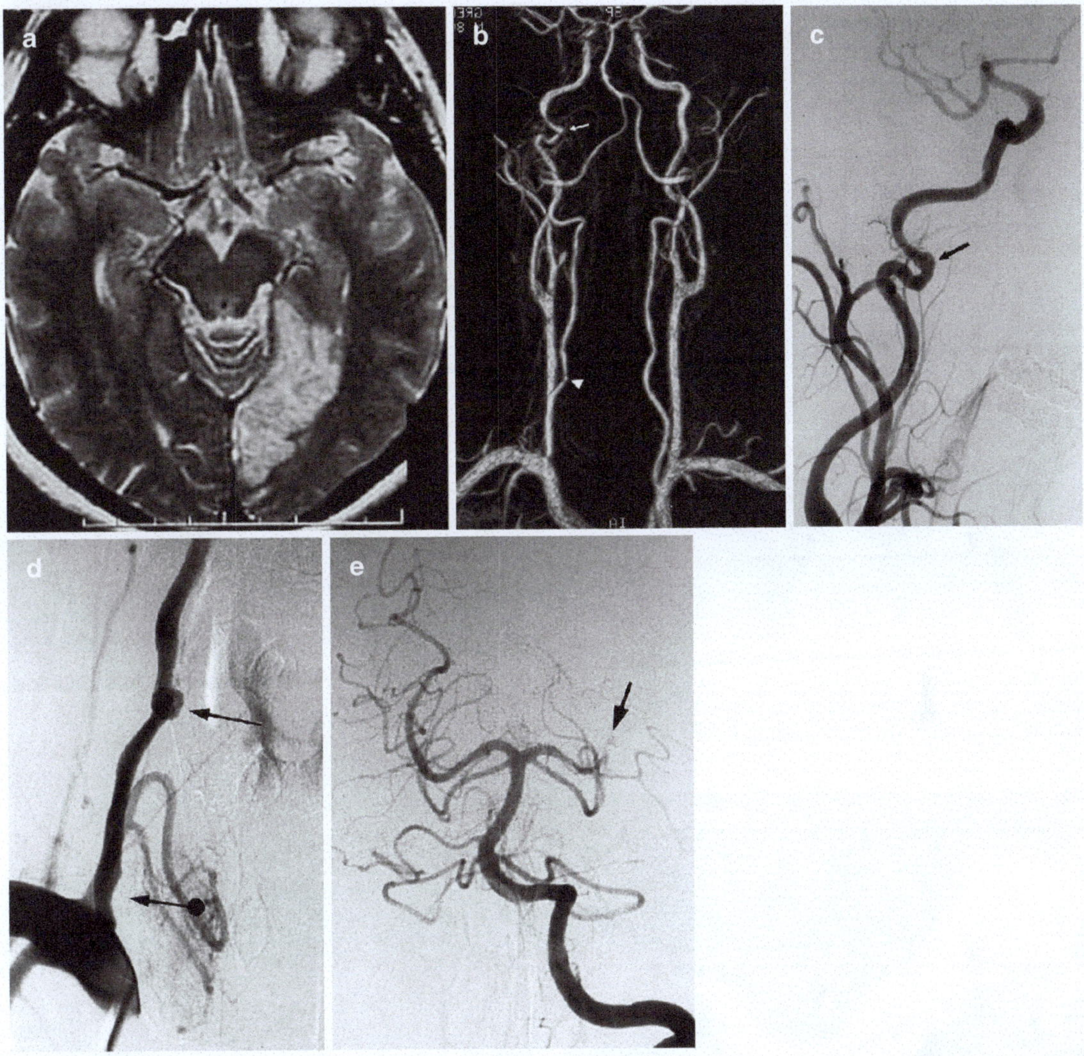

Fig. 16.2 Patient admitted with acute ischemia in the vascular territory of the left PCA. (**a**) MRI showing infarct in the left occipital region. (**b**) MR angiography. Irregularity of the right ICA (*arrow*). Doubtful finding on the right vertebral artery (*arrowhead*). (**c**) The angiogram of the right ICA confirmed the changes of the artery, suggesting dissection (*arrow*). (**d**) The right VA angiogram showed the stenosis of the artery at its origin (*arrow with dot*) and the rounded dissecting pseudo-aneurysm at the level of C6 (*arrow*). (**e**) Probably embolic occlusion of the left PCA (*arrow*)

rounding the dissected ICA. These minimal signs can elude an imprecise neurological examination.

Dissection can leads to ischemia, varying from TIAs to major strokes, due to hypoperfusion, intracranial embolization, or both, involving the anterior or posterior circulation, depending on the artery involved. Ischemia of the spinal cord can also occur in dissection of the VA (Weidauer et al. 1999; Crum et al. 2000) (Fig. 16.8).

An interesting aspect is palsy of the last four cranial nerves in various combination in dissection of the ICA (Bradac et al. 1981a, 1989, 2000; Maitland et al. 1983; Hommel et al. 1984; Lieschke et al. 1988; Nusbaum et al. 1998; Waespe et al. 1988; Vargas et al. 1992; Schievink et al. 1993; Mokri et al. 1996;

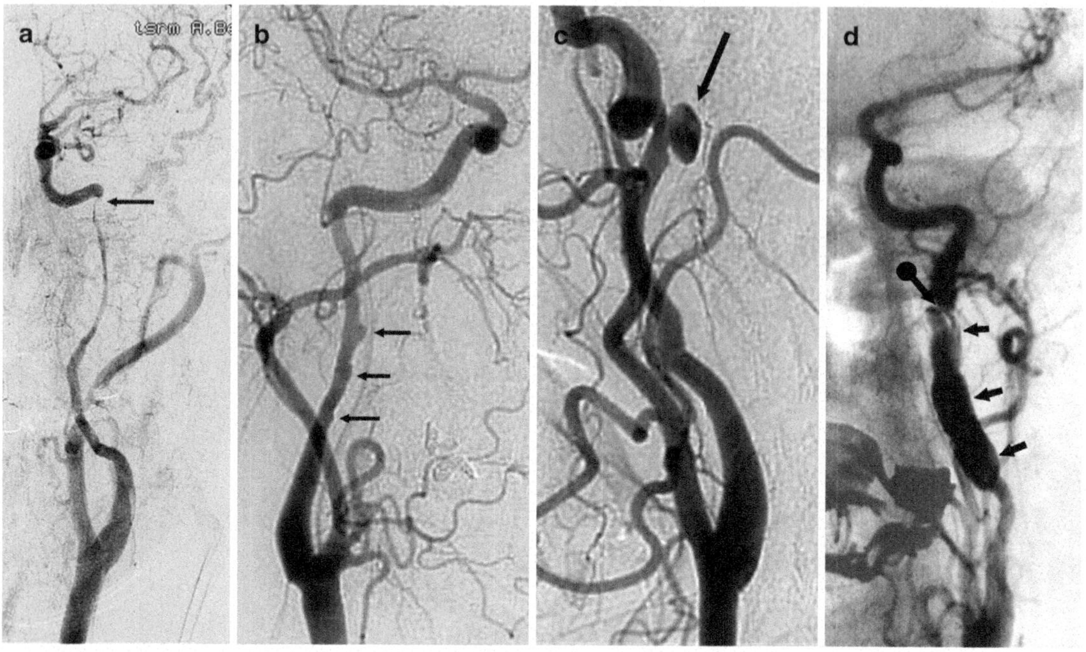

Fig. 16.3 Examples of ICA dissections in four different patients. (**a**) Severe stenosis which abruptly ends at the skull basis (*arrow*). (**b**) Wall irregularity (*arrows*). Note the adjacent course of the AphA. (**c**) Pseudoaneurysm (*arrow*). (**d**) Irregular fusiform dilatation (*arrows*). A flap (*arrow with dot*) is recognizable

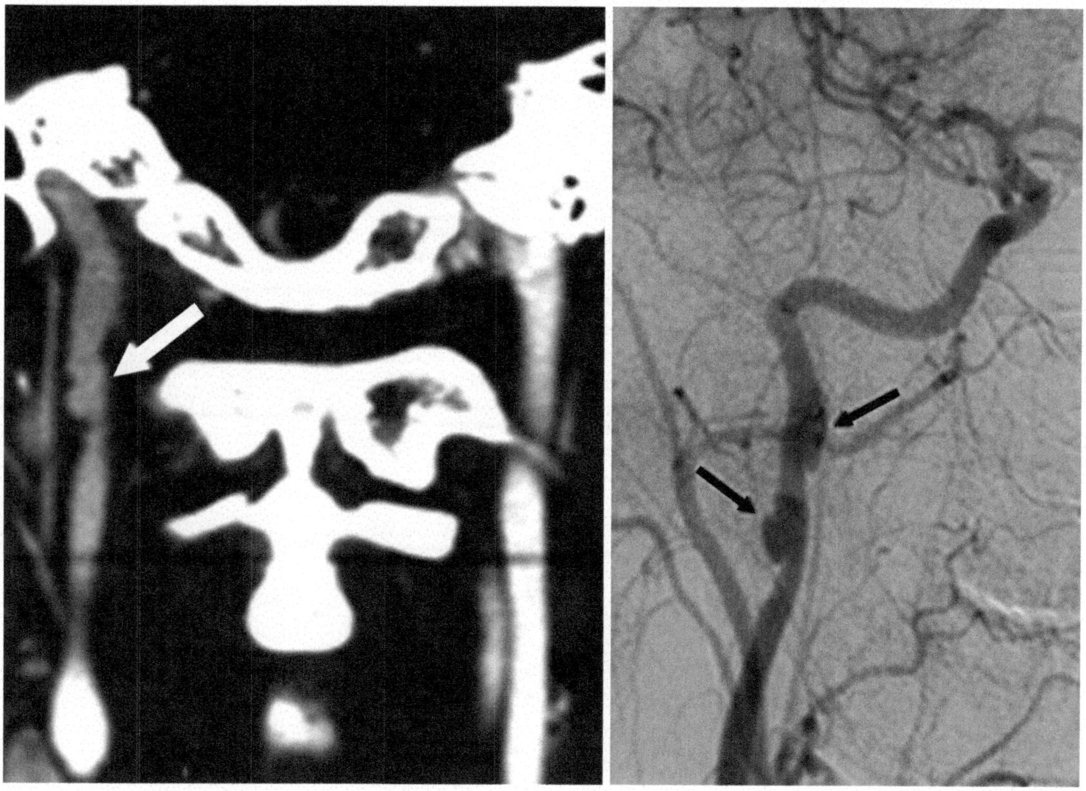

Fig. 16.4 ICA dissection. CT angiography and conventional angiography. The irregularity and the pseudoaneurysms (*arrows*) are better demonstrated on the angiogram

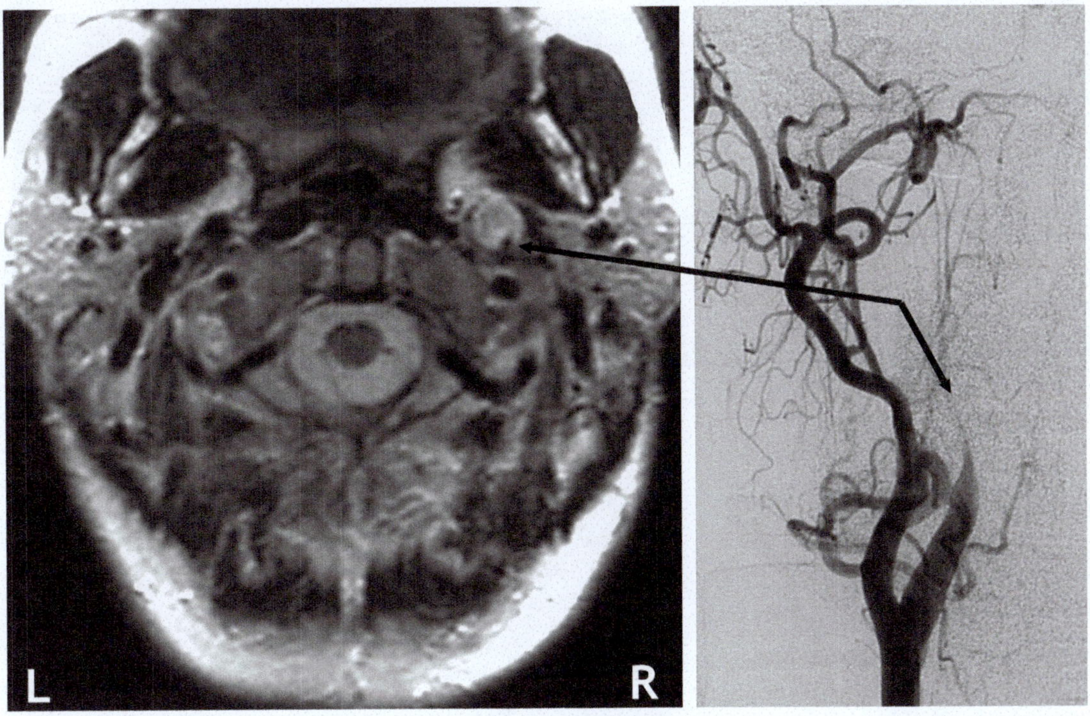

Fig. 16.5 Typical tapered occlusion in ICA dissection shown on the angiogram. On the MR examination the eccentric signal void and the surrounding hyper-intensity corresponding to the residual lumen and the mural hemorrhage respectively are well visible

Sturzenegger et al. 1995). Increased volume of the artery due to the intramural hematoma or the presence of pseudoaneurysm can lead to compression of the nerves. Another possibility which could be considered is the involvement of the ascending pharyngeal artery which runs parallel and adjacent to the ICA (Bradac et al. 1989, 2000). It is known that this artery supplies these nerves (Lasjaunias and Doyon 1978) and so its compression could lead to impairment in the supply of the nerves. The rare occurring palsy of the cranial nerves III, IV, and VI supplied by branches of the inferolateral trunk could be explained by involvement of these fine arteries due to minimal extension of the dissection to the cavernous portion of the ICA or due to hypoperfusion or distal embolization. Indeed, involvement of these nerves has been described in cases of ICA occlusion (Wilson et al. 1989; Kapoor et al. 1991).

- *Intracranially, dissection involves predominantly the vertebrobasilar sector*. The lesion can lead to *ischemia* in the cerebellum and/or brainstem with varying clinical symptoms frequently associated with abrupt cervico-occipital pain (Yoshimoto and Wakai 1997; Hosoya et al. 1999; Lee et al. 2006a). In many other cases, as increasingly reported in the literature dissection leads to a rupture of the artery and *to subarachnoid bleeding*. Some authors (Ro et al. 2009) in an histopathological study of dissection of intracranial VA resulted in fatal bleeding have demonstrated along with the ruptured dissection a few other partially or completely repaired dissections involving also the contralateral VA. According to these authors, this finding seems to demonstrate the existence of a specific vulnerability and tendency to dissections in certain patients.

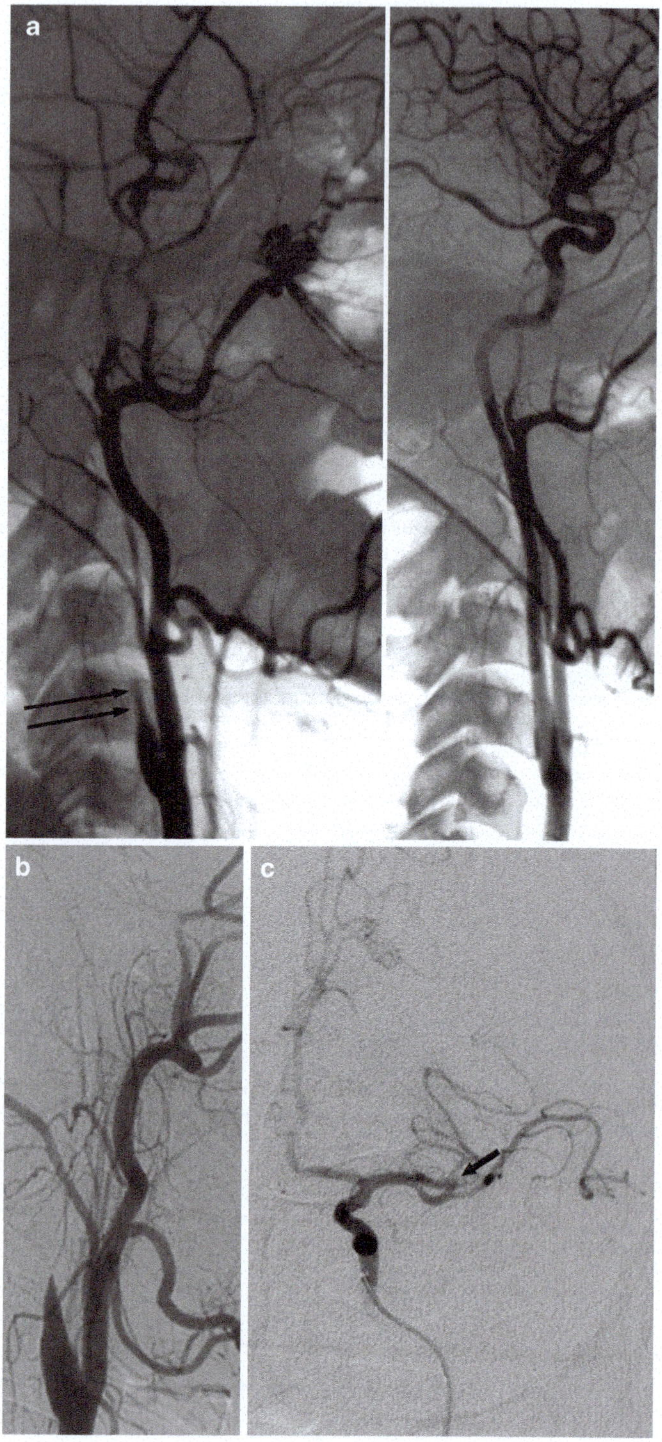

Fig. 16.6 (**a**) Occlusion of the Left ICA due to traumatic dissection (*arrows*), without involvement of the intracranial circulation. Complete recanalization 4 weeks later. (**b–d**) Occlusion of the left ICA due to spontaneous dissection in a young patient presenting with stroke. (**b**) ICA angiogram. (**c**) It was not difficult to overcome the dissection with a microcatheter showing the patent distal ICA and occlusion of M2 branches of the MCA (*arrow*) which were reopened by aspiration. (**d**) The treatment was completed with stenting of the ICA. Good clinical outcome

Fig. 16.6 (continued)

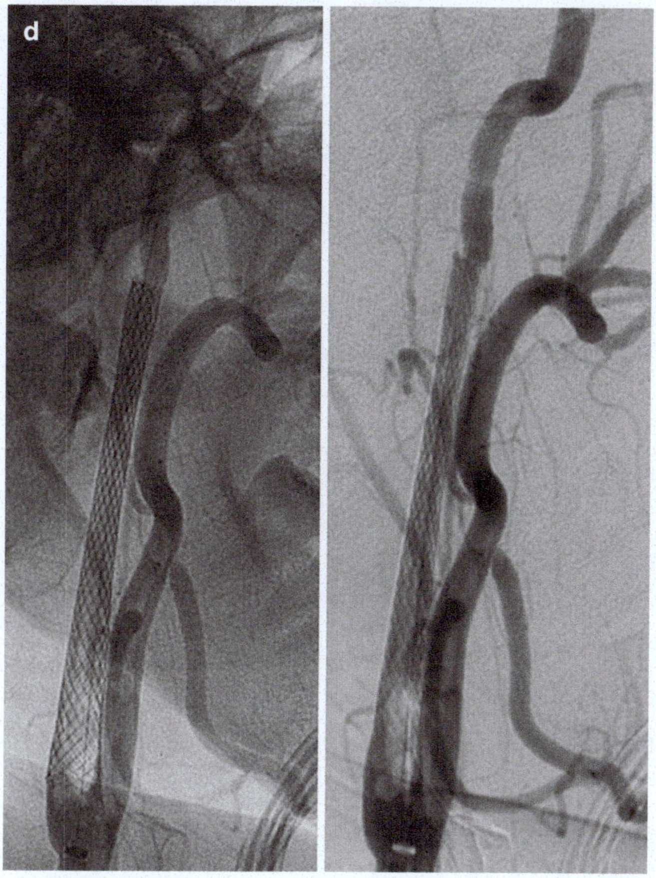

- *Dissection even if more rare can occur also in the anterior circulation where it is progressively recognized as a cause of ischemia* (*Sato* et al. *2010; Nagamine* et al. *2014; Fukuma* et al. *2015*). *Subarachnoid bleeding can also develop* (*see also Chap. 11*). A little curious aspect in patients with dissecting aneurysm in the anterior circulation *presenting with ischemia is its association with a subarachnoid bleeding* (Ohkuma et al. 2003, Fukuma et al. 2015; Hensler et al. 2016). In these cases, the hemorrhage is localized commonly in the frontoparietal cortical area distally to the diagnosed dissection, in a territory different of the ischemia, suggesting that probably in these cases multiple dissections occur (Fukuma et al. 2015). This is indeed conceivable, considering, as described above, that multiple dissections have been demonstrated on pathological studies in some patients. These patients are reported to be younger in comparison to those with dissection in the vertebrobasilar sector and seem to be more frequent in Asiatic population (Yamaura et al. 1999; Kurino et al. 2002). Among other pathologies presenting with cortical hemispheric hemorrhage, the RCVS should be especially taken into consideration in the differential diagnosis in these cases (see also Sect. 17.6).

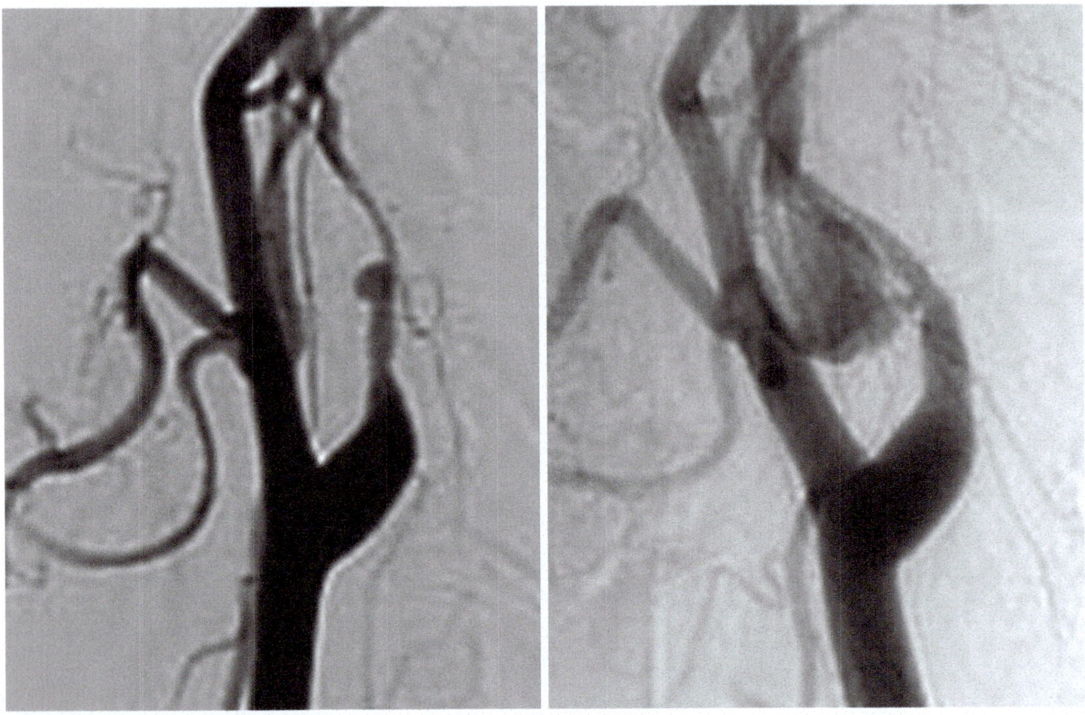

Fig. 16.7 Severe stenosis and pseudoaneurysm (*arrow*) in a case of ICA dissection. Enlargement of the aneurysm on the control angiogram 3 months later

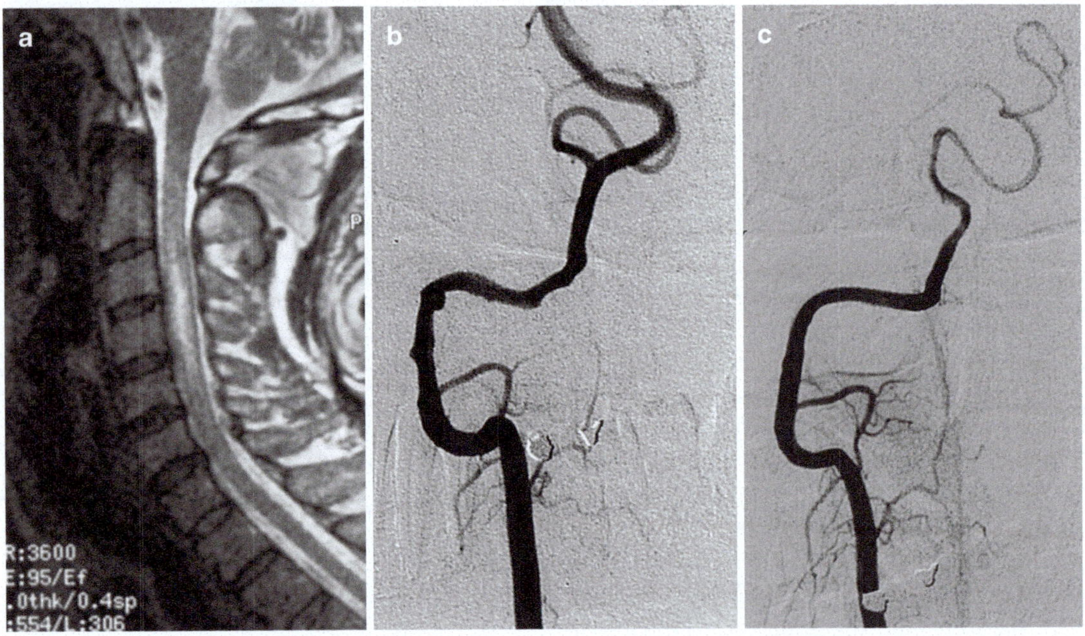

Fig. 16.8 Spinal cord ischemia in a young patient due to involvement of the anterior spinal artery following dissection of the right VA. (**a**) MRI, T2-weighted images showing the hyperintense lesion of the spinal cord. (**b**) Right VA angiogram showing the irregularity with small pseudoaneurysms of the distal extracranial VA with partial involving also of the intracranial segment. The left VA was normal. (**c**) Normalization of the right VA 4 months later. Clinical improvement

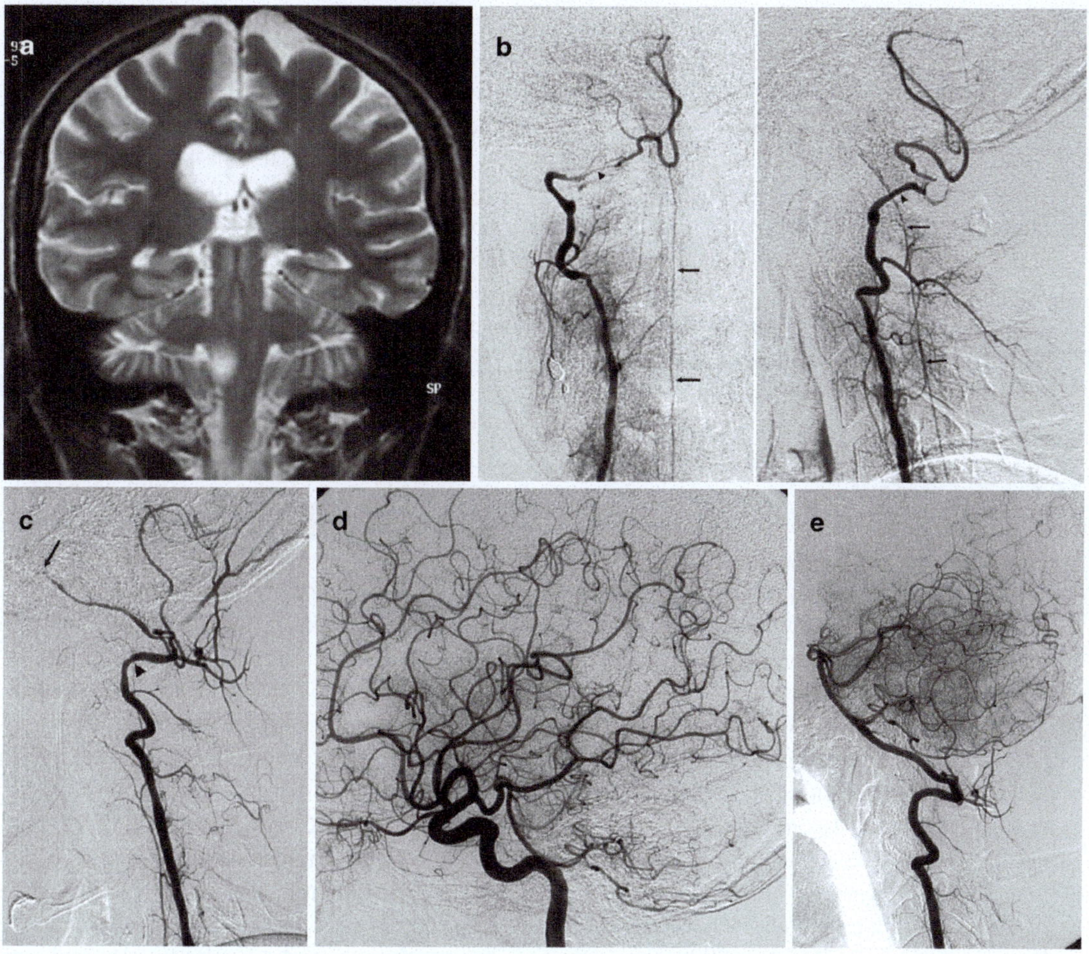

Fig. 16.9 Young patient with brainstem stroke due to dissection of both vertebral arteries (VAs) with involvement also of the basilar artery (BA). (**a**) MRI T2-weighted image showing a small lateral bulbar ischemia. (**b**) Angiogram of the right VA (AP and lateral view). Irregularity and narrowing of the distal VA (*arrowhead*). The VA ends in the PICA. Injection of the anterior spinal artery (*arrows*) through a radiculo-medullary artery. (**c**) Left VA angiogram. Narrowing of the distal VA (*arrowhead*) and BA which is distally occluded. (**d**) Carotid angiogram. Good collateral circulation through the circle of Willis, with retrograde filling of the BA. Medical therapy was performed. Clinical improvement of the patient who was symptom free 2 month later. (**e**) Normal control vertebral angiogram

- A few specific considerations should be made for the so-called *blood blister-like aneurysm* which has been reported to occur typically on the superior-anterior wall of the supraclinoid ICA, predominantly in women. With acquired experience similar aneurysm has been identified also in the ACA and MCA as well as in the vertebrobasilar sector (Park et al. 2009b; Peschillo et al. 2015). The pathogenesis of these lesions is not completely clear. They are probably a particular form of dissection (Ishikawa et al. 1997; Okuchi et al. 1999; Ogawa et al. 2000). Some other authors (Abe et al. 1998) have suggested that an important role is played by the atherosclerosis, where ulcerated plaques lead progressively to a focal defect in the artery wall which eventually is replaced by hematoma covered by fibrous tissue. In the acute phase the angiogram can be normal or only show an irregularity or a small bulge on the artery wall. Over subsequent days, the pattern commonly changes, with increasing of the bulge or forming a saccular aneurysm (Fig. 11.8c, d).

- The cause of dissection can also be *a traumatic lesion* of the artery. In patients with severe craniocerebral trauma, as well as in those with blunt neck injuries, or cervico-vertebral or basal skull fracture the neuroradiological investigation should be directed not only to exclude intracranial parenchymal and/intracranial vessels lesions but also to examine the extracranial cervical arteries showing sometimes unexpected dissections, as also reported by others (Biffl et al. 2000; Miller et al. 2001; Schneidereit et al. 2006; Redekop 2008).

16.6 Treatment

- *The treatment of extracranial dissection* is controversial, since an important aspect is normalization of the vessel lumen in the majority of the cases (Kasner et al. 1997; Schievink 2000, 2001; Djouhri et al. 2000; Touzé et al. 2001; Pelkonen et al. 2003). This is particularly true when stenosis is present (80–90% of the cases of Pelkonen et al. (2003), Wessels et al. (2008), and Rederkop (2008) showed normalization). Recanalization can also occur in a certain number of occlusion. In the report of the same authors (Pelkonen et al. 2003; Wessels et al. 2008; Redekop 2008) this latter occurred in 30–50% of the cases (Fig. 16.6). Pseudoaneurysms can thrombose, but commonly remain unchanged or enlarge (Fig. 16.7). The risk, however, that remaining irregularities of the wall as well as the persistence of pseudoaneurysm are the cause of stroke due to distal embolization is minimal.
  Taking this into consideration, in the acute phase, in asymptomatic patients or in those with mild neurological symptoms, conservative medical therapy, to avoid embolization is commonly preferred. Endovascular treatment may be performed later, if stenosis or pseudoaneurysm remains. There is, however, a group of patients in whom dissection leads to a major stroke, due to hypoperfusion or due to intracranial embolization (tandem lesions) or both. In these cases the efficiency of the medical therapy is limited (Engelter et al. 2012) and there is a need of a more aggressive approach. While in patients presenting with these clinical conditions due to atherosclerotic pathology, endovascular treatment is the undiscussed therapy of choice (see also Sect. 15.4.1.2) the indication of this treatment in patient presenting similar clinical conditions due to dissection, has not been clearly defined. Progressively, positive results of endovascular approach have appeared in many reports, describing this treatment in ICA dissections, showing the utility of this therapy (Ohta et al. 2011; Juszkat et al. 2015; Haussen et al. 2016; Kurre et al. 2016; Murias Quintana et al. 2016; Hoving et al. 2017). The same authors, reported however the not seldom re-occlusion of the artery, due to the formation of an intra-stent thrombus. This occurs more frequently than in similar patients treated for atherosclerotic pathology, indicating that the damage of the wall of the artery reacts differently in case of dissection.
  Depending on the pathological findings and on the presenting difficulties, the endovascular treatment can involve the extracranial dissection or the intracranial occlusion or both (Fig. 16.6b–d).
  The same endovascular approach can be applied in tandem lesions in the vertebrobasilar sector (Cohen et al. 2003). In these cases, frequently one of the VAs is normal and can be used to reach and to treat the intracranial occlusion. This avoids the risk to pass through the dissected VA, with the risk of further embolization. Examples are represented in Figs. 15.40, 16.2, and 16.10.
- *In cases of intracranial dissection with an angiographic pattern of stenosis or occlusion with ischemia,* considering the frequent spontaneous healing of the artery, as reported by some authors (Yoshimoto and Wakai 1997; Kurino et al. 2002) and also observed in our cases, the best treatment is probably a medical therapy (Figs. 16.8 and 16.9). However, it should be considered that the aspect of these lesion can change with formation of aneurysm which can modify the therapeutic approach.
- *In cases in which the intracranial dissection presents with SAH with clearly diagnosed aneurysm, the lesion should be treated rapidly,* since the risk of rebleeding is high especially

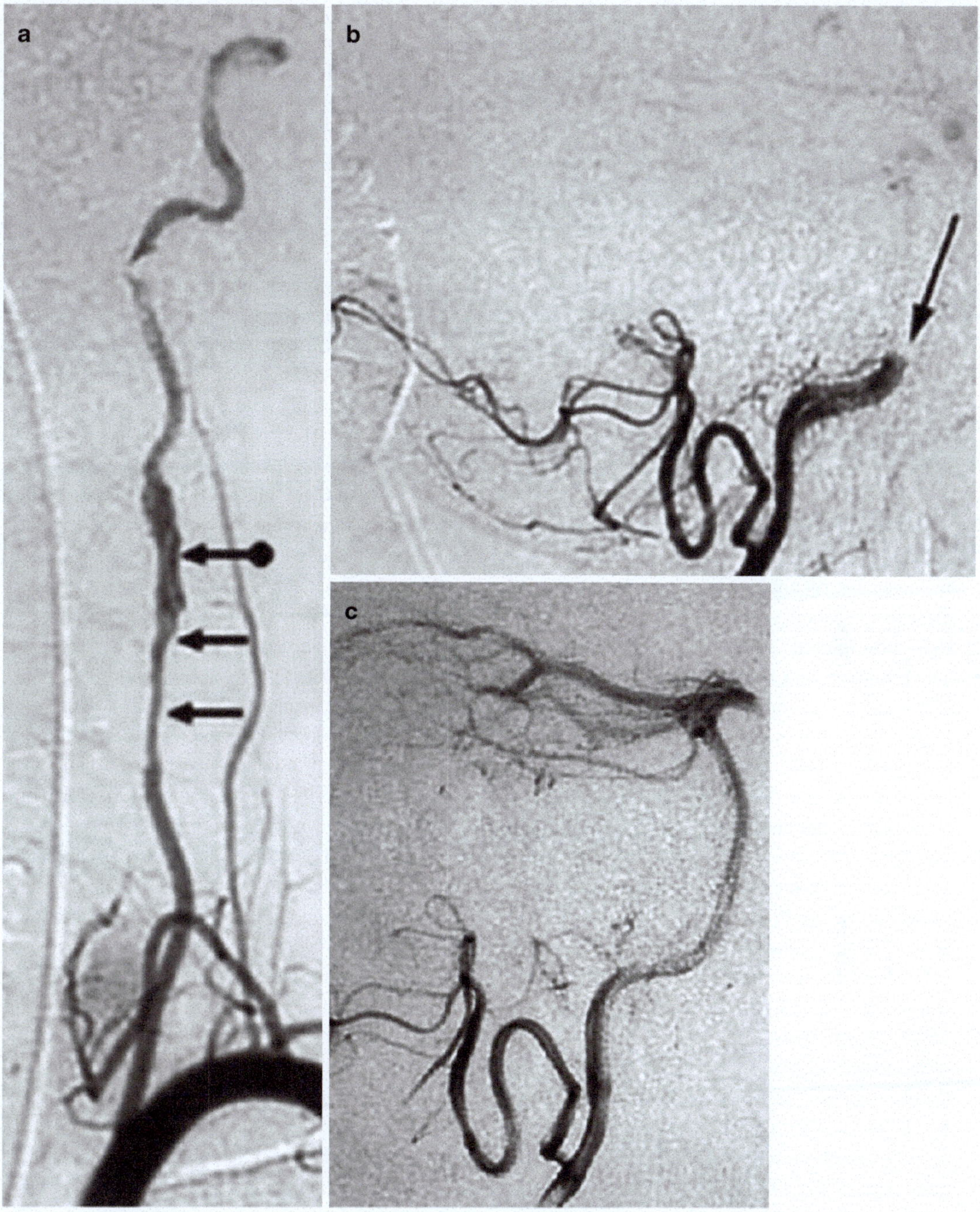

Fig. 16.10 Spontaneous dissection of the left vertebral artery in a very young patient presenting with clinical symptoms of brainstem involvement. (**a**) Left VA angiogram showing irregularity of the artery with stenosis (*arrow*) and dilatation. A double lumen is also recognizable (*arrow with dot*). (**b**) Occlusion probably embolic of the basilar artery (*arrow*). (**c**) A micro-catheter was advanced in the BA through the right VA, followed by pharmacological thrombolysis. Rapid reopening of the BA. Complete recovery of the patient within 2 months

in the first days. The risk of bleeding decreases progressively in the next weeks (Mizutani et al. 2001). Sometimes, however, the vascular changes are minimal on the first angiogram making a certain diagnosis very difficult. The dissection appears more evident in the following days, showing, sometimes very surprising change (Fig. 16.12). This evolution has been described typically in the so-called *blister-like aneurysms* commonly observed in the supraclinoid ICA. It is, however, conceivable as also reported by some authors that they can be develop also elsewhere. It can also occur, as reported by some authors (Mizutani 1998) that the angiographic study in patients with SAH does not show a pseudo-aneurysm, but stenosis or occlusion. The latter can evolve later forming an aneurysm (Hensler et al. 2016). Also spontaneous occlusion of the dissected artery followed by recanalization demonstrating the dissecting aneurysm has been reported (Matsumoto et al. 2008; Akiyama et al. 2012).

- The treatment can be surgical or endovascular. This latter is today commonly preferred. Depending on the site of the dissection, its morphological aspect, the pattern of the vascular anatomy, and the presenting clinical situation, occlusion of the aneurysm along with the parent artery or selective repair with stent and coils, flow-diverter stent, or stent-in-stent technique can be used (Sugiu et al. 2005; Ramgren et al. 2005; Henkes et al. 2006; Yang et al. 2007; Fiorella et al. 2009a; Fischer et al. 2012; Bhogal et al. 2015, 2016) (Figs. 11.12b, 16.11, 16.12, 16.13, 16.14, and 16.17). For examples of treatment of dissecting aneurysms involving distal branches see Figs. 5.10, 11.9c, 11.15, 11.17, 11.24, 11.25, 11.26, 11.27, and 16.16).

  A few extra observations should be made for the treatment of the *blister-like* aneurysms, which remain very critical lesions due to their morphological and structural aspects. Surgery (clipping, trapping, or arterial suturing) has been commonly used as treatment. Endovascular approach has been mainly characterized by occlusion of the aneurysm together with the ICA. This technique remains a good alternative approach as confirmed in a recent work (Kim et al. 2014). In the study, of these authors, in which 11 patients were considered, a good outcome was reported in about 70% of the cases. Improvement of the endovascular technique with the introduction of stent with coils and later of the flow-diverter stent has added new possibility of treatment. A few cases of treatment with flow-diverter has been reported (Kulcsar et al. 2010; Rasskazoff et al. 2010). In the last years, the number of cases has progressively increased. Cinar et al. (2013) reported their experience in the treatment of seven patients presenting with SAH due to "blister-like" aneurysms of distal ICA. One patient was treated with occlusion of ICA. In the other six a flow-diverter was used. Treatment was performed in the subacute phase (3, 4 weeks after the bleeding). The final result was a good clinical outcome with occlusion of the aneurysm in 70% of the cases. Chalouhi et al. (2014) reported the successful treatment of height "blister-like" aneurysms. Seven were located in the ICA and one in the BA. Among them five presented with SAH. In the patients with SAH the treatment was performed acutely (within 2 days) in two cases and in the subacute phase (2 weeks) in the other three patients, using a flow diverter with a final good outcome in all these patients. The treatment (surgical or endovascular) of these aneurysms has been described by Gonzales et al. (2014) in an extensive review of the literature to which the authors added their own positive experience with stent with coils and flow-diverter stents, showing the advantage of the endovascular treatment in comparison to the surgery. The necessity to associate the flow diverter to an antiplatelet therapy applied shortly before or during the treatment appears not to have negative consequences also in the acute phase of treatment. In conclusion, all these reports seem to show the great utility of the flow diverter, which appears to be the best promising therapy to be used at least as the first choice in the treatment of this very critical pathology.

In conclusion: In the diagnosis and treatment of dissections it should be taken into consideration that these are in the acute-subacute phase unstable lesions which can modify their morphological feature, showing not rarely very surprising vascular changes. In cases of an uncertain diagnosis, a close monitoring is mandatory. This can today be performed with angio-CT or angio-MR, followed by conventional angiography when necessary.

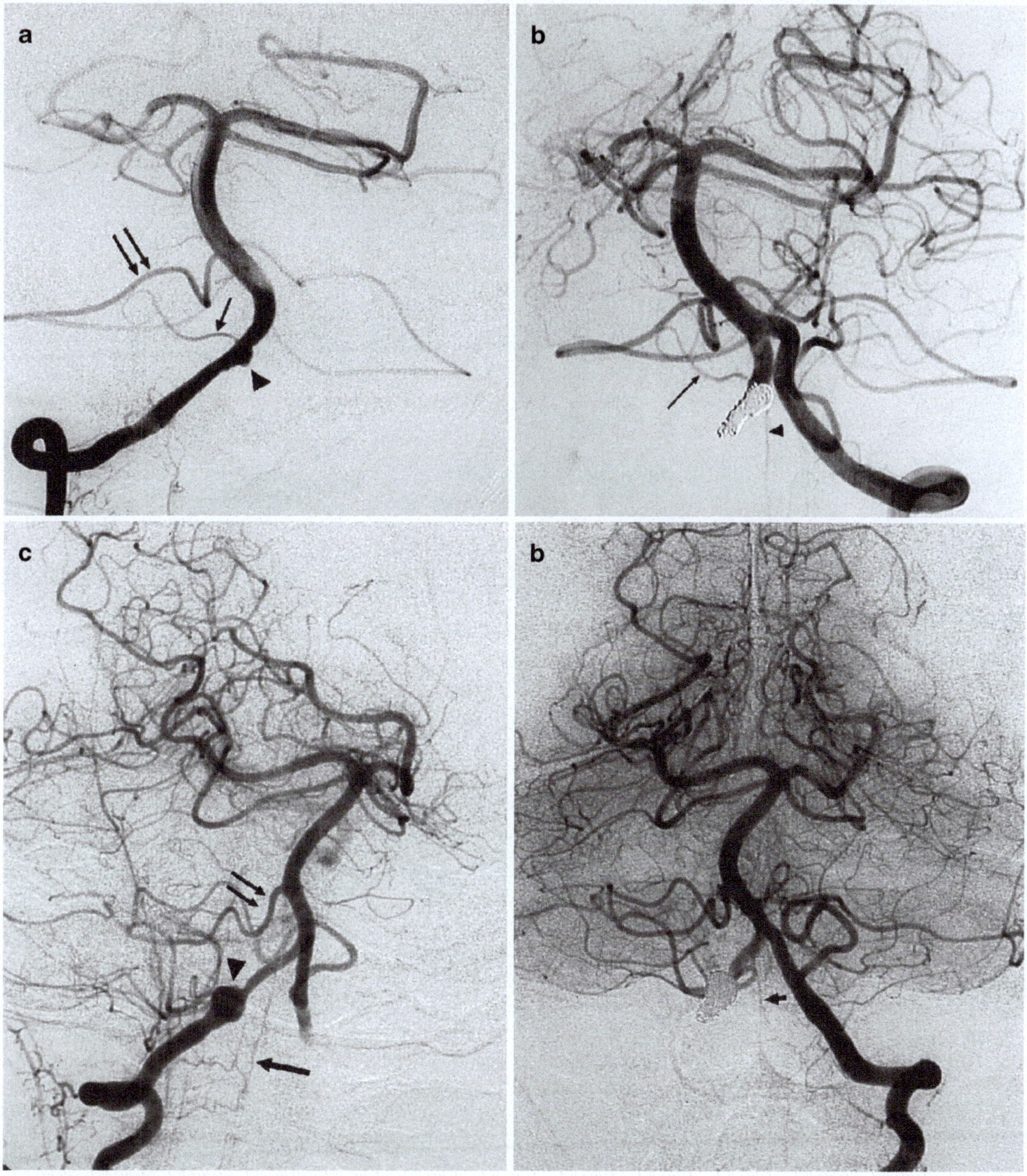

Fig. 16.11 Two examples (**a**, **b** and **c**, **d**) of dissecting aneurysms of the right VA presenting with SHA. (**a**) Right VA angiogram showing the irregular small fusiform aneurysm (*arrow head*) proximal to the origin of the small PICA (*small arrow*). There is a large AICA (*arrows*) supplying also the territory of the PICA. (**b**) Left VA angiogram after occlusion of with coils of the aneurysm and the VA. The right PICA was preserved (*arrow*) as well as the anterior spinal artery (*arrowhead*). The patient recovered well. (**c**) Another with a similar lesion and clinical presentation. On the right VA angiogram (oblique view) round dilatation of the VA (*arrowhead*) suggesting dissection. There is a large AICA (*double arrows*) supplying completely the vascular territory of the PICA. Anterior spinal artery (*arrow*). Well-developed PICA on the left. (**d**) Left VA angiogram after occlusion of the aneurysm and right VA with coils. Anterior spinal artery (*arrow*). Good recovery of the patient

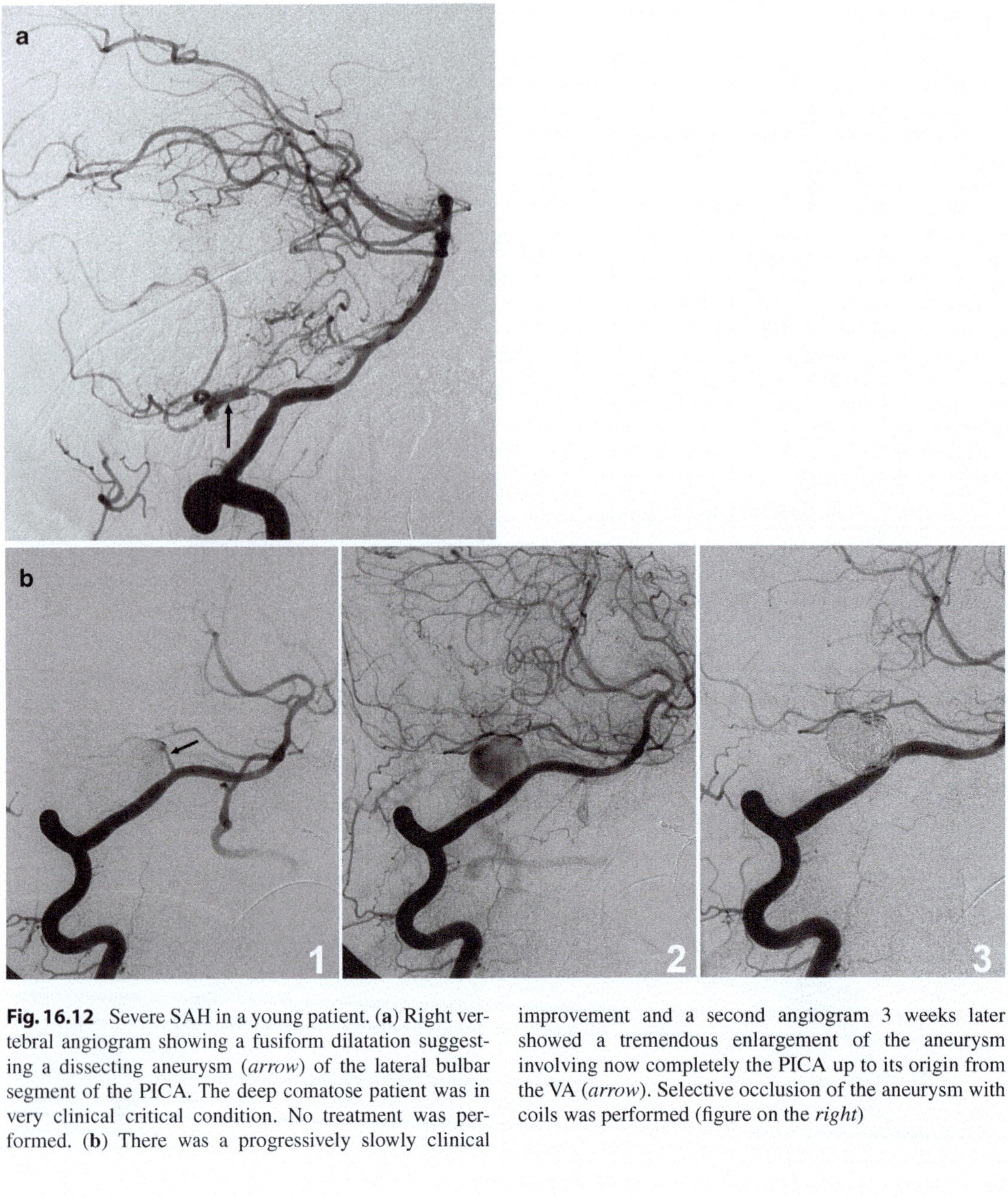

Fig. 16.12 Severe SAH in a young patient. (**a**) Right vertebral angiogram showing a fusiform dilatation suggesting a dissecting aneurysm (*arrow*) of the lateral bulbar segment of the PICA. The deep comatose patient was in very clinical critical condition. No treatment was performed. (**b**) There was a progressively slowly clinical improvement and a second angiogram 3 weeks later showed a tremendous enlargement of the aneurysm involving now completely the PICA up to its origin from the VA (*arrow*). Selective occlusion of the aneurysm with coils was performed (figure on the *right*)

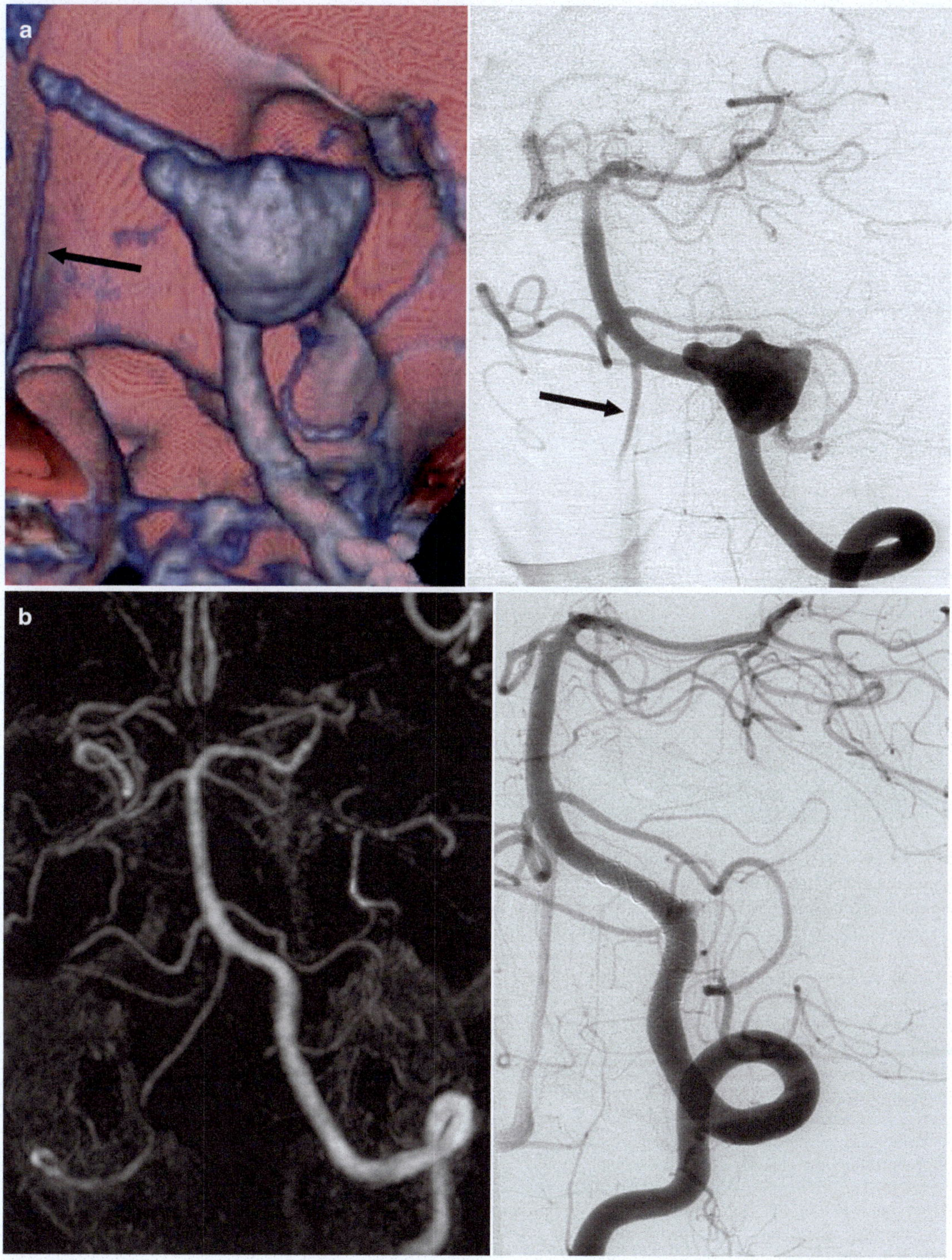

Fig. 16.13 Large dissecting aneurysm of the left VA incidentally discovered in a young patient with a Marfan syndrome. (**a**) CT angiography showing the aneurysm. There is no filling of the PICA. The right VA is hypoplastic (*arrow*). The conventional angiogram (*right* in the figure) shows that the absence of the PICA is compensated by a bilateral well-developed AICA. The patient was treated with a flow-diverter stent (Sylk). (**b**) MR angiography and left vertebral angiogram performed 1 year later. The aneurysm is no longer visible. Normalization of the vertebral artery. On the angiogram the stent is well visible

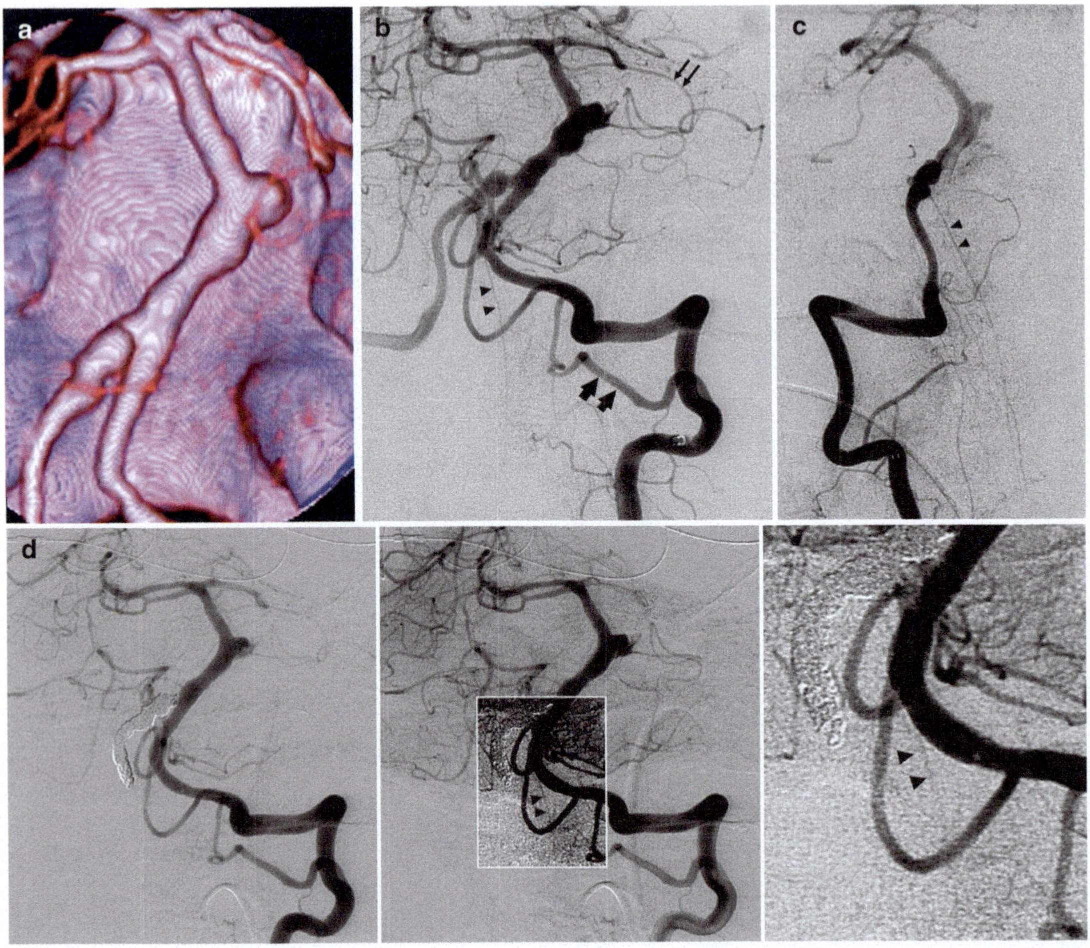

Fig. 16.14 Middle-aged black patient presenting with hemorrhage due dissecting aneurysm involving the distal right VA and proximal and middle BA. (**a**) CT angiography, showing the extensive dissection of the right distal VA and proximal and middle BA, with associated rounded saccular dilatation. (**b**) Left vertebral angiogram with a reflux in the smaller right VA. There is a very large left PICA (*arrows*) arising extracranially at the C1–C2 level supplying also the contralateral territory of the right PICA and probably also of the right AICA. The left SCA (*small arrows*) through its marginal branch supplies probably partially the territory of the left AICA also involved in the dissection. The origin of the ASA is not recognizable. It appears to be unilateral arising from the right VA (*arrowhead*). (**c**) The origin of the ASA from the right VA (*arrowhead*) seems to be confirmed on the angiogram of the right VA. On this angiogram two flow-diverter stents have already be positioned in the BA. *The unilateral right origin of the ASA was an unfavorable condition. However, considering the presence of a large left VA and Left PICA and the possibilities of anastomoses on the surface of the medulla (see also Sects. 6.2.1.1 and 15.7.3), the decision to occlude the right VA with coils was taken.* (**d**) On the left VA angiogram (early and late phases with magnification) performed immediately posttreatment shows improvement of the lumen of the BA and presence of the ASA. The patient tolerated well the treatment. He had deglutition troubles which improved gradually. (**e**) The MRI performed 10 days later showed a small ischemic lesion in the right lateral medulla and in the right splenium, the latter probably embolic. Small old lesion in the white matter. (**f**) MR angiography a month later showed a progressive morphological improvement of the BA. (**g**) Control angiogram 1 year later showed the normalization of the BA and of the left AICA. On the non-subtracted image the position of stents and coils is recognizable. Note the slight intima hyperplasia (*arrows*) which has incorporated the stent. The patient was clinically free of symptoms

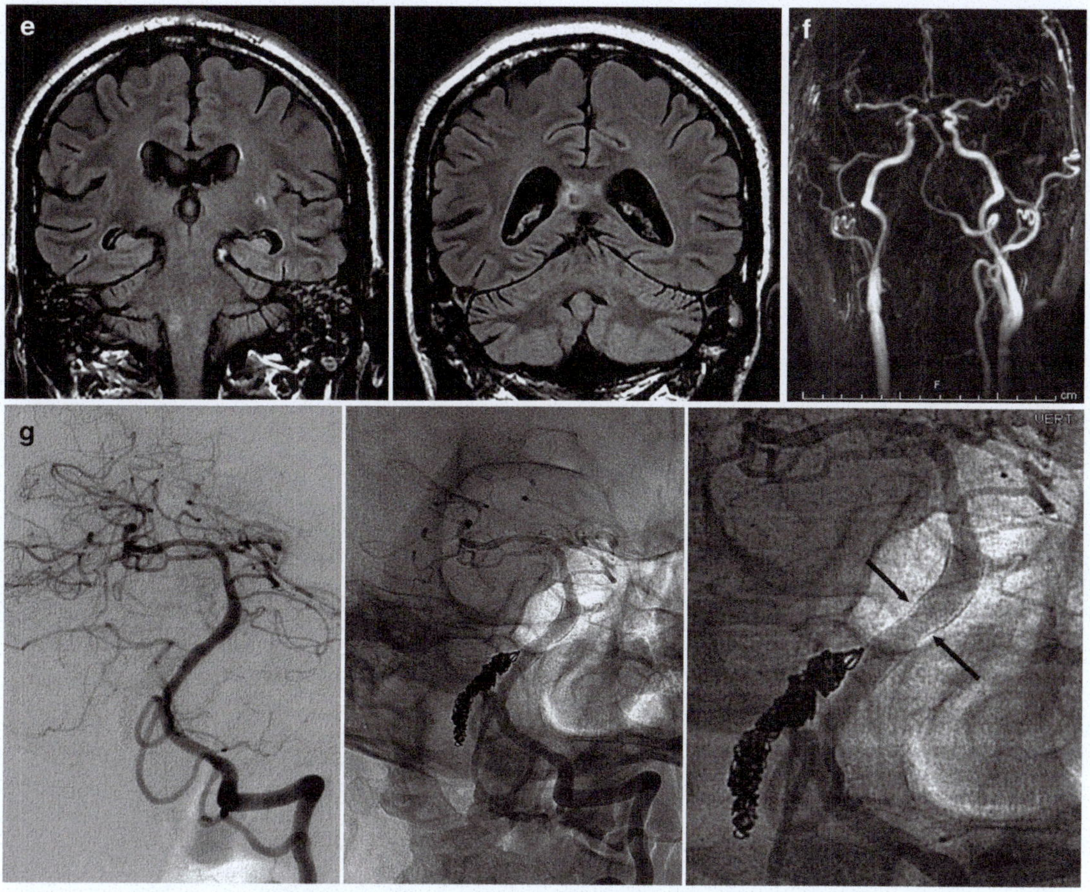

Fig. 16.14 (continued)

16.7 Dissection and Dissecting Aneurysms in Children

Dissection in childhood is relatively rare, accounting for about 7% of all dissections (Schievink et al. 1994b). As in adults, dissection in children can be *traumatic* or *spontaneous*. In children with extracranial dissection the pathogenesis is prevalently traumatic involving ICA or VA with a pattern similar to that described in adults. Instead, the intracranial dissection is predominantly spontaneous and is reported to be relatively more common than in adults (Schievink et al. 1994b; Fullerton et al. 2001). It typically involves the distal ICA and MCA appearing as stenosis or occlusion with a pattern not different to that occurring in other intracranial vasculopathies (Fig. 16.15) (see also Chap. 19), leading to ischemia. It is noteworthy that in cases of extracranial dissection the typical symptoms characterized by head and neck pain associated with the Horner syndrome is frequently absent and the clinical symptoms are due to the cerebral ischemia due to hypoperfusion or embolization (Fullerton et al. 2001; Rafay et al. 2006). The pathogenesis of the spontaneous dissections are pathological changes involving the wall of the arteries as described in Sect. 16.2.

Dissection can lead to the formation of pseudoaneurysms. Unlike in adults, in whom this kind of aneurysms are prevalently in the posterior circulation, in children they are also frequent in the anterior circulation (Lasjaunias et al. 2005).

Among the arteries typically involved in the anterior circulation, the most frequent are the terminal ICA and MCA, and in the posterior circulation the VA, BA, and PCA (Hochberg et al. 1975; Nass et al. 1982; Graber et al. 1992; Laughlin et al. 1997; Massimi et al. 2003; Vilela and Goulao 2006; Bradac et al. 2008a; Gandolfo 2012). Intracranially large pseudoaneurysms have frequently a mass effect on the surrounding parenchyma. They can cause distal embolization and depending on the site can involve the perforators (Figs. 16.16 and 16.17). Intracranial hemorrhage due to rupture of the aneurysm can occur, but it is less common than in the adults.

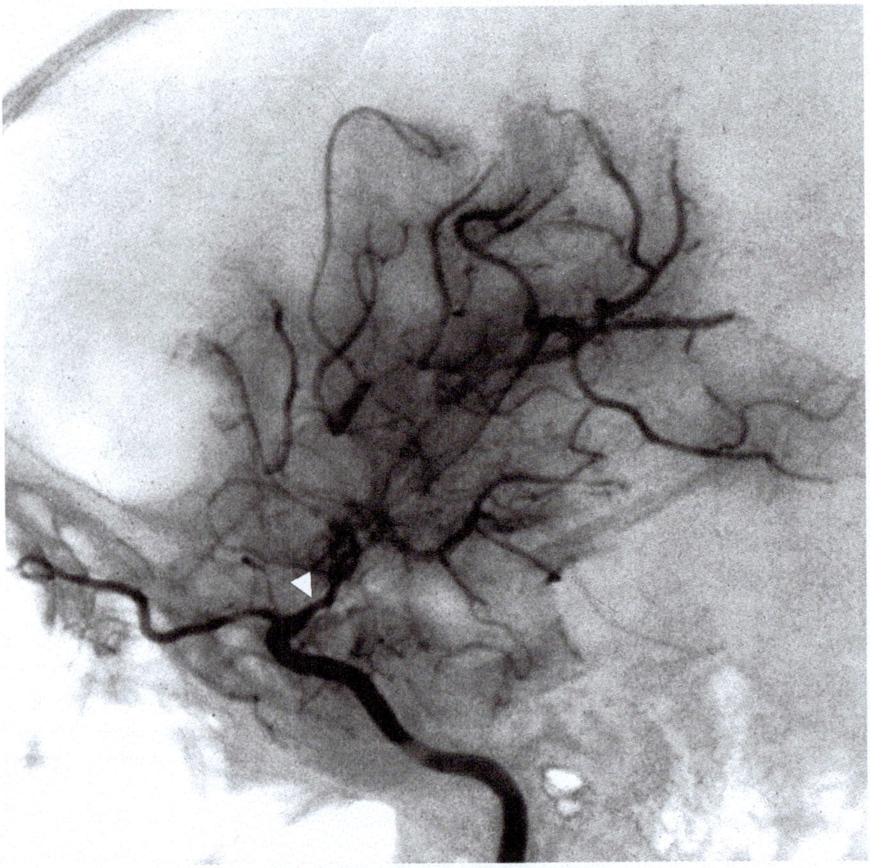

Fig. 16.15 Lateral carotid angiogram in a 5-year-old child with a sudden onset of hemiplegia and aphasia. Left lateral carotid angiogram. Narrowing of the terminal segment of the ICA (*arrowhead*). Irregularity and occlusion of many branches of the middle cerebral artery. A spontaneous dissection with distal embolization was suspected

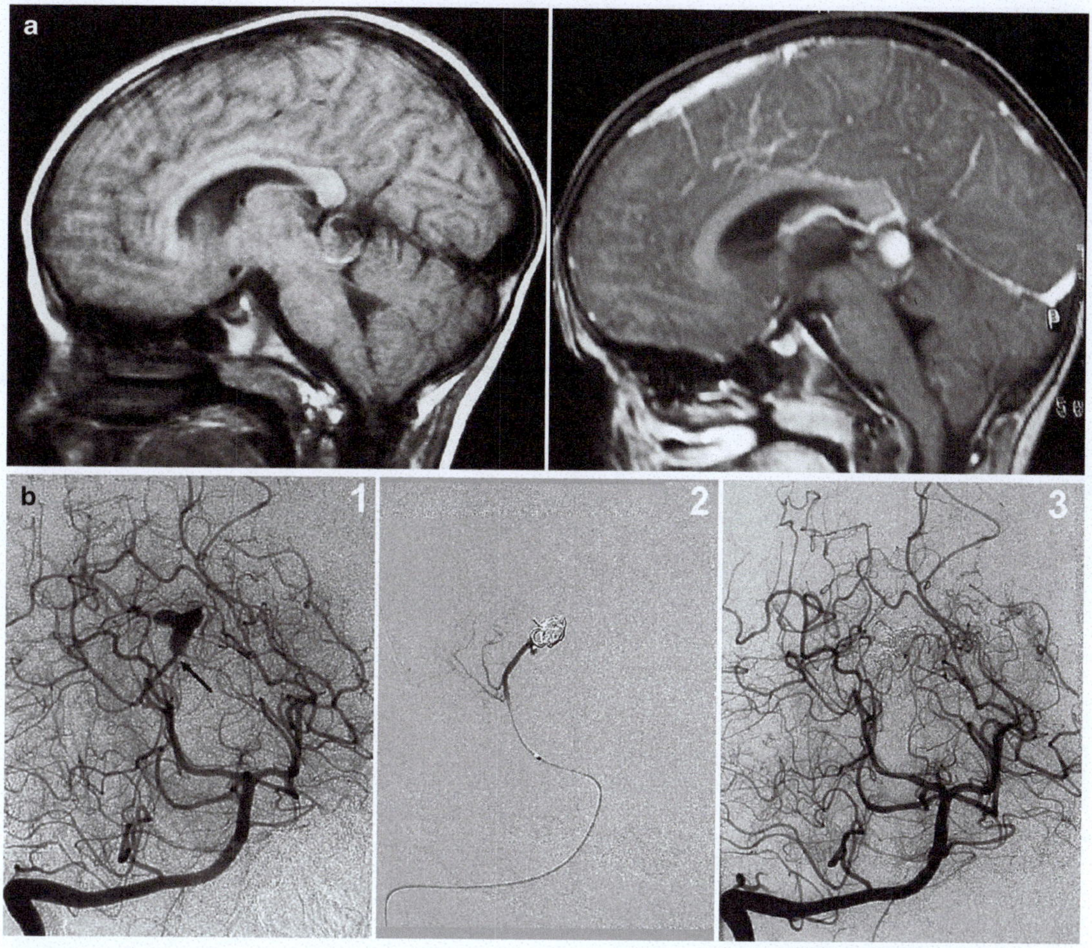

Fig. 16.16 Seven-year-old boy with a short clinical history of severe headache. (**a**) MRI T1-weighted sagittal images. A mass lesion, characterized by an inhomogeneous signal and an hyper-intense peripheral rim, corresponding probably to a large aneurysm with an intramural thrombus of different ages, is visible. After contrast medium enhancement of the patent part of the aneurysm is demonstrated. Compression of the aqueduct and midbrain. (**b**) The vertebral angiogram (1) shows an irregularly shaped aneurysm arising from a small branch of the P3–P4 segment of the right PCA. Focal stenotic segment proximal to the aneurysm (*arrow*). Selective catheterization of the aneurysm and its occlusion with coils (2). Final control angiogram (3). Full recovery of the patient

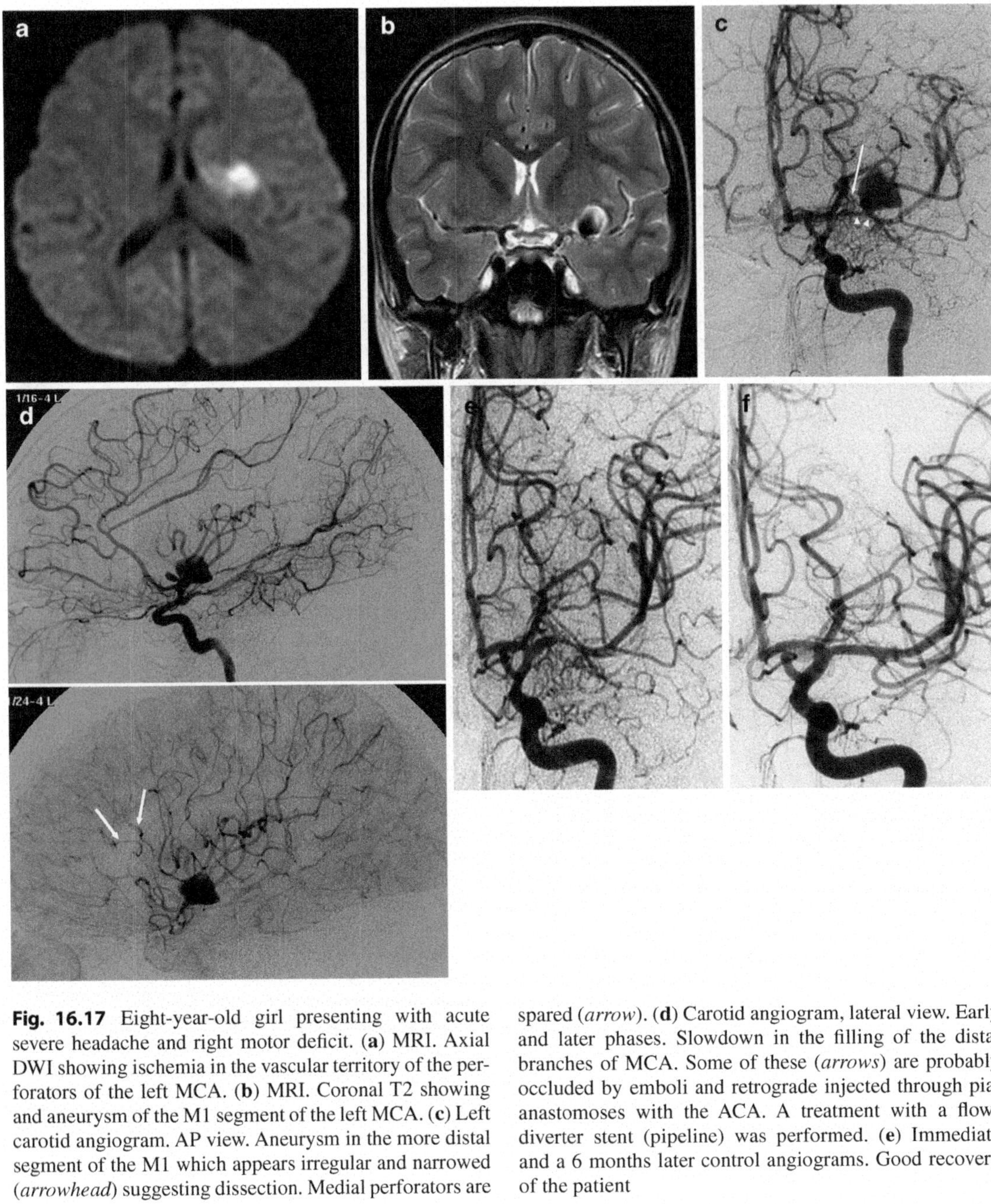

Fig. 16.17 Eight-year-old girl presenting with acute severe headache and right motor deficit. (**a**) MRI. Axial DWI showing ischemia in the vascular territory of the perforators of the left MCA. (**b**) MRI. Coronal T2 showing and aneurysm of the M1 segment of the left MCA. (**c**) Left carotid angiogram. AP view. Aneurysm in the more distal segment of the M1 which appears irregular and narrowed (*arrowhead*) suggesting dissection. Medial perforators are spared (*arrow*). (**d**) Carotid angiogram, lateral view. Early and later phases. Slowdown in the filling of the distal branches of MCA. Some of these (*arrows*) are probably occluded by emboli and retrograde injected through pial anastomoses with the ACA. A treatment with a flow-diverter stent (pipeline) was performed. (**e**) Immediate and a 6 months later control angiograms. Good recovery of the patient

Other Nonatherosclerotic Vasculopathies

17

17.1 Introduction

Nonatherosclerotic vasculopathies comprise a great variety of diseases, as listed in Table 17.1, that result from different etiologies; they include collagenopathies, immunological, hematological, and infection diseases, and other rarer pathologies that have in common the possibility of affecting the cerebral vessels, leading to ischemia and sometimes hemorrhage. Indeed cerebral

Table 17.1 Non-atherosclerotic vasculopathies

Primary angiitis of the CNS
Granulomatous vasculitis, affecting predominantly the precapillary arterioles, occasionally large vessels (such as ICA, PCA, MCA, ACA)
Small aneurysms have also been reported. Angiography frequently normal
Systemic vasculitis that can affect the CNS
Giant cell arteritis: granulomatous vasculitis involving predominantly superficial temporal ophthalmic arteries, occasionally extracranial ICA and VA
Takayasu arteritis: granulomatous vasculitis
Polyarteritis nodosa: necrotizing vasculitis; large intracranial vessels can be involved; aneurysm may be present
Wegener's granulomatosis: necrotizing vasculitis; can involve intracranial arteries with ischemia due to occlusion and hemorrhage due to rupture of arteries
Systemic diseases that can affect the CNS
Systemic lupus erythematosus
Behçet's disease: no signs of necrotizing or granulomatous vasculitis are found; frequently perivenular infiltrations are present; angiography often
Normal
Rheumatic disease
Sjogren's syndrome
Crohn's disease
Sarcoidosis
Hematological conditions
Sickle cell disease, polycythemia vera, Moschcowitz disease, platelet hyperaggregability, disseminated intravascular coagulation, antiphospholipid syndrome, deficiency of factors involved in coagulation (deficiency of proteins C and S, antithrombin III, factor V Leyden), homocystinuria
Vasculopathies not linked to inflammatory processes
Neurofibromatosis, intracranial vessel occlusion, possible intracranial aneurysm
Ehlers Danlos, Marfan syndrome (aneurysm, frequently dissecting, of the extra- and intracranial arteries; dissecting aneurysm of the intracavernous ICA with spontaneous cavernous fistulas)
Fibromuscular dysplasia
Scleroderma
Livedo reticularis (Sneddon's syndrome)
Moyamoya disease
Infection diseases
Bacterial, fungal, parasitic, viral (varicella zoster; herpes simplex; cytomegalovirus; HIV)
Others
Vasculopathies related to drugs, radiotherapy, intracranial tumors, trauma
Migraine stroke

G.B. Bradac, *Applied Cerebral Angiography*, DOI 10.1007/978-3-319-57228-4_17

arteries can be involved in many systemic angiitis or other generalized vasculopathies and can be selectively affected in some diseases such as the primary angiitis of the central nervous system (PACS), the Moyamoya diseases, and the reversible cerebral vasoconstriction syndrome (RCVS). In the majority of these diseases, the angiogram shows findings such as occlusion, irregular narrowing, alternating with dilatation, and sometimes aneurysms. These changes are not typical of one specific disease, but are common to all, independent of their basic pathology. Furthermore, in many cases where the involved vessel is in the precapillary-capillary sector, the angiogram can be normal, and so the diagnosis is made on clinical, laboratory findings and neuroradiological examinations of the brain parenchyma including CT and MR. In some cases brain biopsy is the essential final step.

It is beyond the scope of this book to treat each of these pathologies, since there is already a very rich literature about it. A few specific studies are here reported (Hayreh 1981; Tan et al. 1982; Collado et al. 1989; Beattie et al. 1995; Singleton et al. 1995; Govoni et al. 2001; Hoffman et al. 2002; Weygand and Goronzy 2003; Lucivero et al. 2004). Some angiographic examples are presented in Figs. 17.1, 17.2, and 17.3.

A few of these diseases will be described more in detail considering their clinical relevance and particular clinic-pathological aspects.

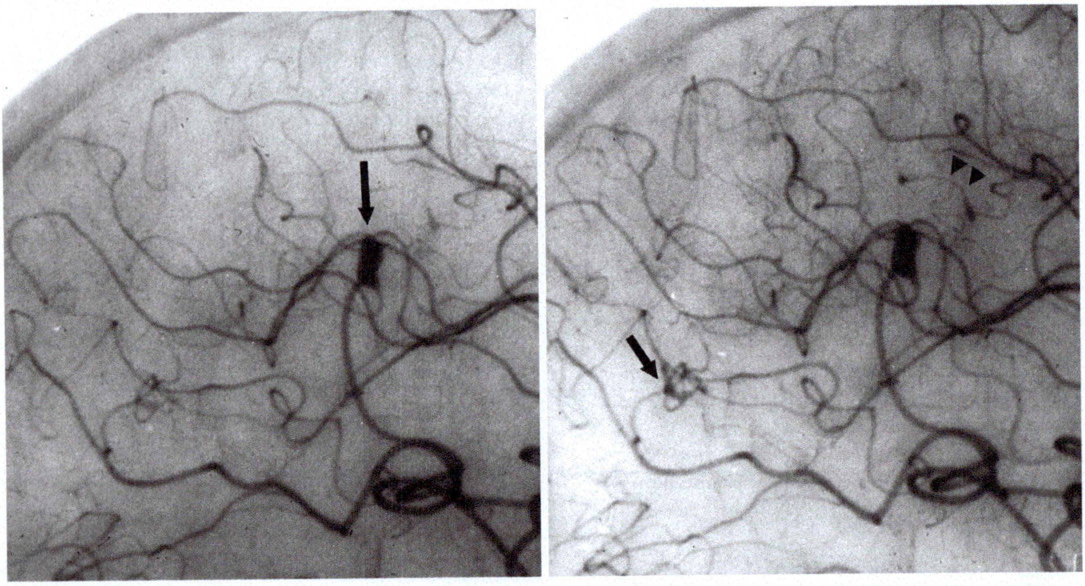

Fig. 17.1 Lateral carotid angiogram (detail with magnification) in a patient with polyarteritis nodosa. Early and later phases. Aneurysm dilatations (*arrow*) and stenotic changes (*arrowheads*) are present

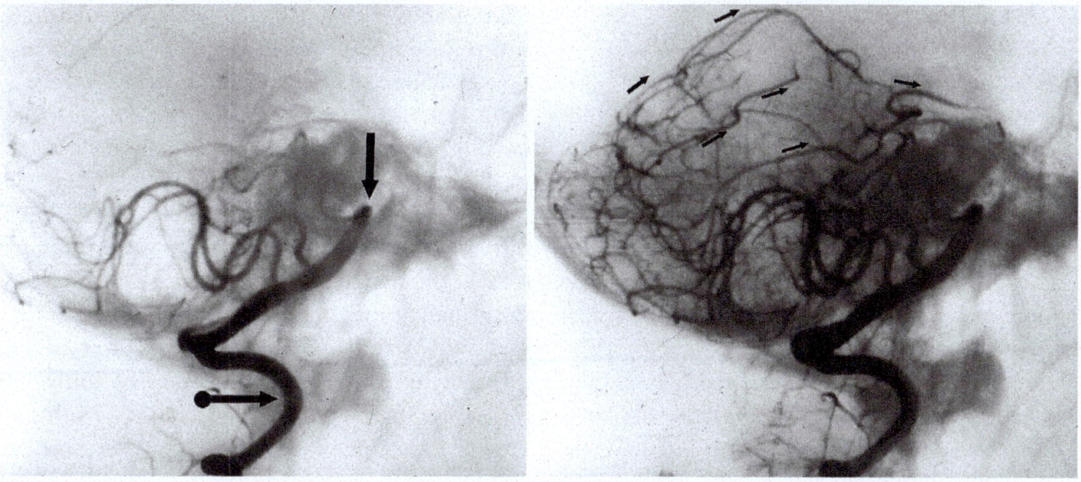

Fig. 17.2 A young boy with neurofibromatosis presenting with acute occlusion of the basilar artery. VA angiogram showing the occlusion (*arrow*). The rounded feature of the occlusion let think that this was probably preceded by a narrowing of the lumen due to pathological changing in the wall linked to the primary pathology. In the later phase, there is a retrograde injection of the SCA and AICA through opening of anastomoses with the PICA (*arrows*). There is a slight displacement of the vertebral artery at the C1–C2 owing to dysplasia of the cervical spine (*arrow with dot*)

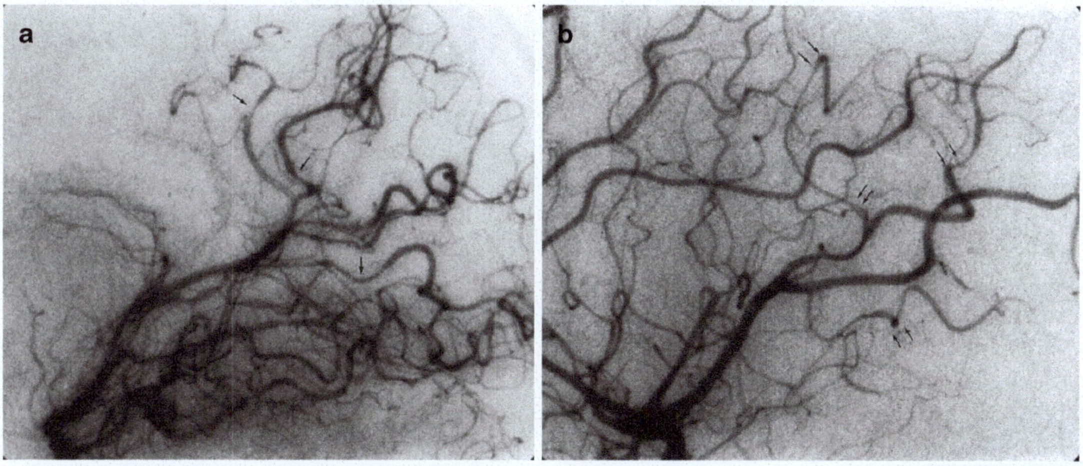

Fig. 17.3 Suspected primary angiitis in a young boy presenting with sudden onset of epileptic seizures and psychic impairment. Lateral VA (**a**) and (**b**) ICA angiograms. Irregularity and focal narrowings of several distal branches in the anterior and posterior circulations are visible (*arrows*)

17.2 Cerebrovascular Fibromuscular Dysplasia

Fibromuscular dysplasia (FMD) is a systemic vascular disease, characterized by nonatherosclerotic changes involving the wall of medium-small sized. All territories can be affected; among them, FMD is seen more frequently in the renal arteries, responsible very often for hypertension. Palubinskas and Rypley (1964) and Palubinskas and Newton (1965) were the first to report the angiographic aspects of this disease in the cephalic arteries. Numerous reports followed (Palubinskas et al. 1966; Andersen 1970; Wylie et al. 1966; Houser and Baker 1968; Bradac and Heymat 1970; Stanley et al. 1974; Manelfe et al. 1974; Osborn and Anderson 1977; Bradac and Oberson 1983), clearly distinguishing FMD from atherosclerotic and other nonatherosclerotic lesions and its possible relationship with stroke.

The true incidence of FMD is unknown. In some reports, it is cited up to 10% of angiographic studies (Chiras et al. 1985). It is more frequent in women, and it is rare in children (Andersen 1970).

17.2.1 Pathology and Etiopathogenesis

FMD is due to structural changes involving the medial layer, characterized by a fibrous proliferation and hyperplasia of the smooth muscle cells. More rarely also involvement of the intima and adventitia occur. These pathological changes lead to thickening and fibrosis of the layers involved. The precise etiopathogenesis has not identified yet. A deficiency of alpha1-antitrypsin has been reported (Schievink et al. 1998).

17.2.2 Diagnosis

The aspect of the arteries involved is typical, characterized on CT-MR angiography and conventional angiography by multifocal stenosis, sometimes severe, alternating with areas of mural dilatation, leading to the so-called *string-of-beads* appearance. Sometimes, it can present with tortuosity, pseudoaneurysm, and tubular stenosis. In the latter situation, the differential diagnosis with spontaneous dissection can be sometimes difficult. On control angiograms, the findings remain unchanged in FMD, while in dissection a normalization frequently occurs. The extracranial ICA in its midportion is the artery most often involved, frequently bilaterally, followed by the extracranial vertebral artery. Multiple vessels are commonly affected. A more rare atypical form, as already been described in the past, and which has gained a new attention more recently is that characterized by the presence at the ICA bifurcation of a small defect frequently thin called differently (septal fibromuscular dysplasia, septa, diaphragm, web and pseudovalvular fold or simply atypical FMD) (Ehrenfeld et al. 1967; Rainer et al. 1968; Houser et al. 1971; Gee et al. 1974; Lipchik et al. 1974; Osborn and Anderson 1977; So et al. 1979; Wirth et al. 1981; Kliever and Carroll 1991; Morgenlander and Goldstein 1991, Kubis et al. 1999; Joux et al. 2014; Boesen et al. 2017). Similar findings involving the distal extracranial VA (Lenk et al. 2014) and the common carotid artery (So et al. 1979) have been reported. It is important to think of FMD in this atypical form and location especially when this is present in young patients presenting with stroke. The defect can be very small and escape to the diagnosis with CT-MR angiography, requiring in these patients a conventional angiography. Differently from the majority of the authors Lenk et al. (2014) consider these changes as an independent entity and not a subtype of FMD.

Intracranial vessels may also be affected, even if it is relatively rare (Elias 1971; Frens et al. 1974; Tomasello et al. 1976; Saygi et al. 1990).

Examples are presented in Figs. 17.4a, b, 17.5, and 17.6.

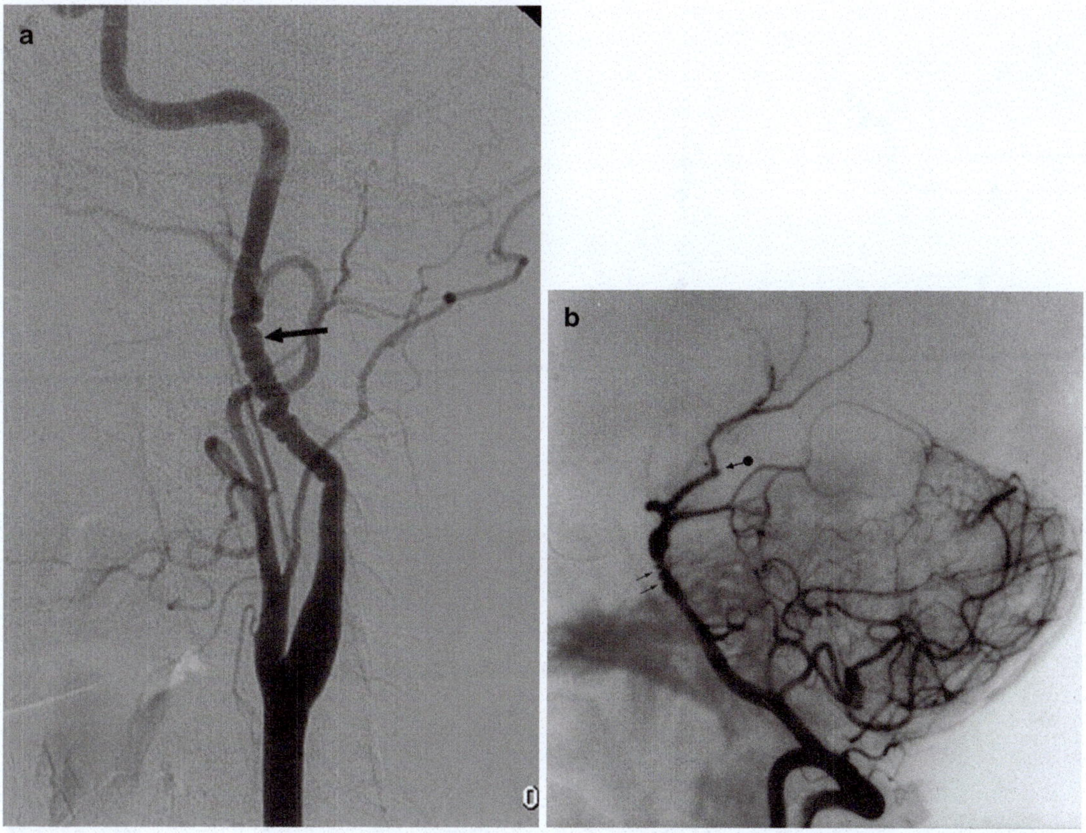

Fig. 17.4 (**a**) Typical "string of beads" appearance of the internal carotid artery (*arrow*). (**b**) "String-of-beads" appearance at the level of distal basilar artery (*arrows*) associated to a similar finding involving PCO. Occlusion of branches of the PCO (*arrow with dot*) and of the SCA probably embolic

FMD is frequently associated with cerebral aneurysms (Handa et al. 1970a; Hirsch and Roessmann 1975; Osborn and Anderson 1977; Paulson et al. 1978). Carotid-cavernous sinus and vertebral artery with peri-vertebral veins fistulas have also been reported (Geraud et al. 1973; Bradac et al. 1985; Hieshima et al. 1986).

17.2.3 Clinical Relevance

The extracranial changes of FMD can be an incidental discovery in patients examined for other pathologies. It can, however, be the cause of ischemic stroke owing to the formation of thromboembolic material at the site of the extracranial-intracranial lesions. The close relationship of FMD with dissection is well known. This can be in the extra or intracranial sector and be responsible of ischemia. Intracranial dissection can be the cause of hemorrhage (Küker et al. 2011). Association with intracranial berry aneurysms is frequent.

The treatment (medical, surgical endovascular) depends on the clinical symptoms and type of lesion.

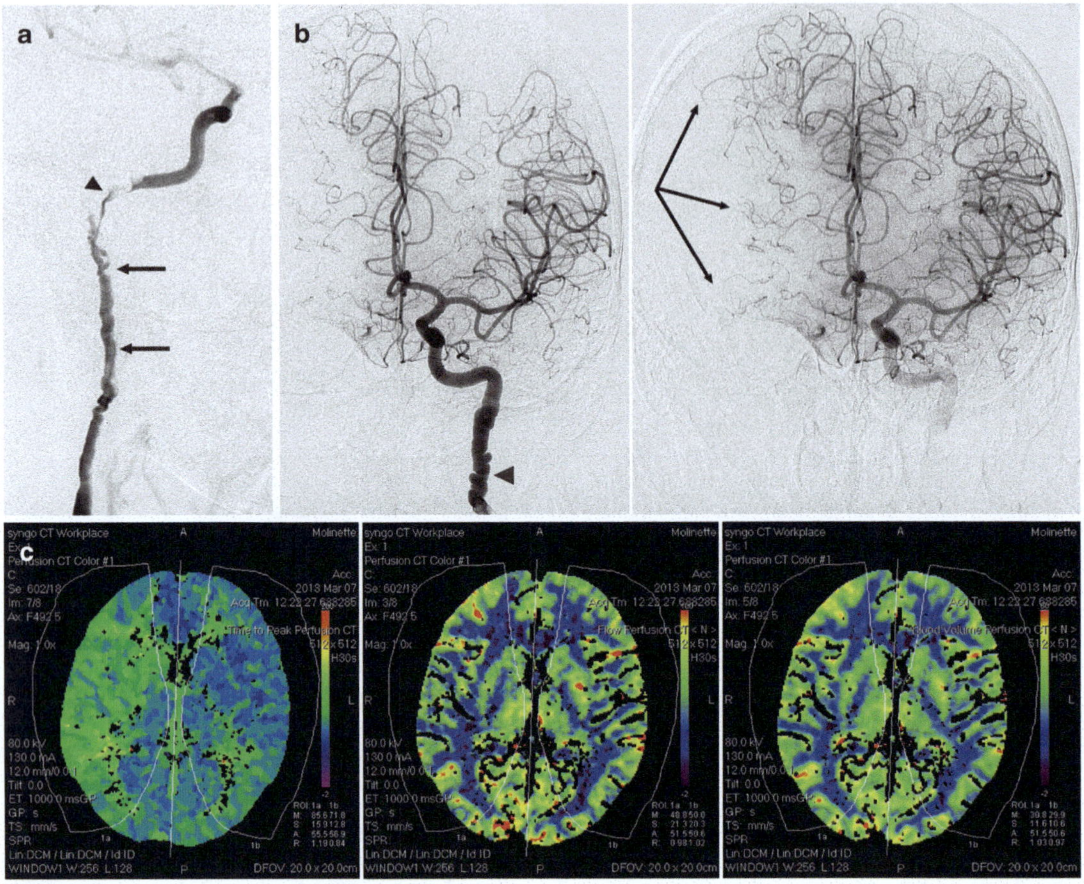

Fig. 17.5 (**a**–**c**) Middle-aged woman presenting with one episode of right amaurosis fugax. On the admission in the hospital, she was clinically normal. The CT angiography followed by a complete conventional angiographic study revealed a severe fibromuscular dysplasia with dissection of the right extracranial ICA. (**a**) Right ICA angiogram. Typical string-of-beads appearance (*arrows*). Dissection close to the skull base where probably an intraluminal thrombus is present (*arrowhead*). There is a slowdown of the intracranial circulation. The A1 is not recognizable. (**b**) Left ICA angiogram (early and later phases) showing similar changing of the extracranial ICA (*arrowhead*). There is aplasia of the right A1. Pial anastomoses from right ACA toward MCA (*arrows*). A rich collateral circulation through pial anastomoses between right PCA and right MCA was visible on the vertebral angiogram. There was aplasia of P1. A rich collateral circulation occurred also through the right ECA (see Fig. 3.13). (**c**) CT perfusion study showed, on the image on the *left*, a slight delay of the *time to peak* on the right hemisphere, but no changes of the *CBF* and *CBV* were demonstrated on the two images on the *right*. The patient underwent medical therapy

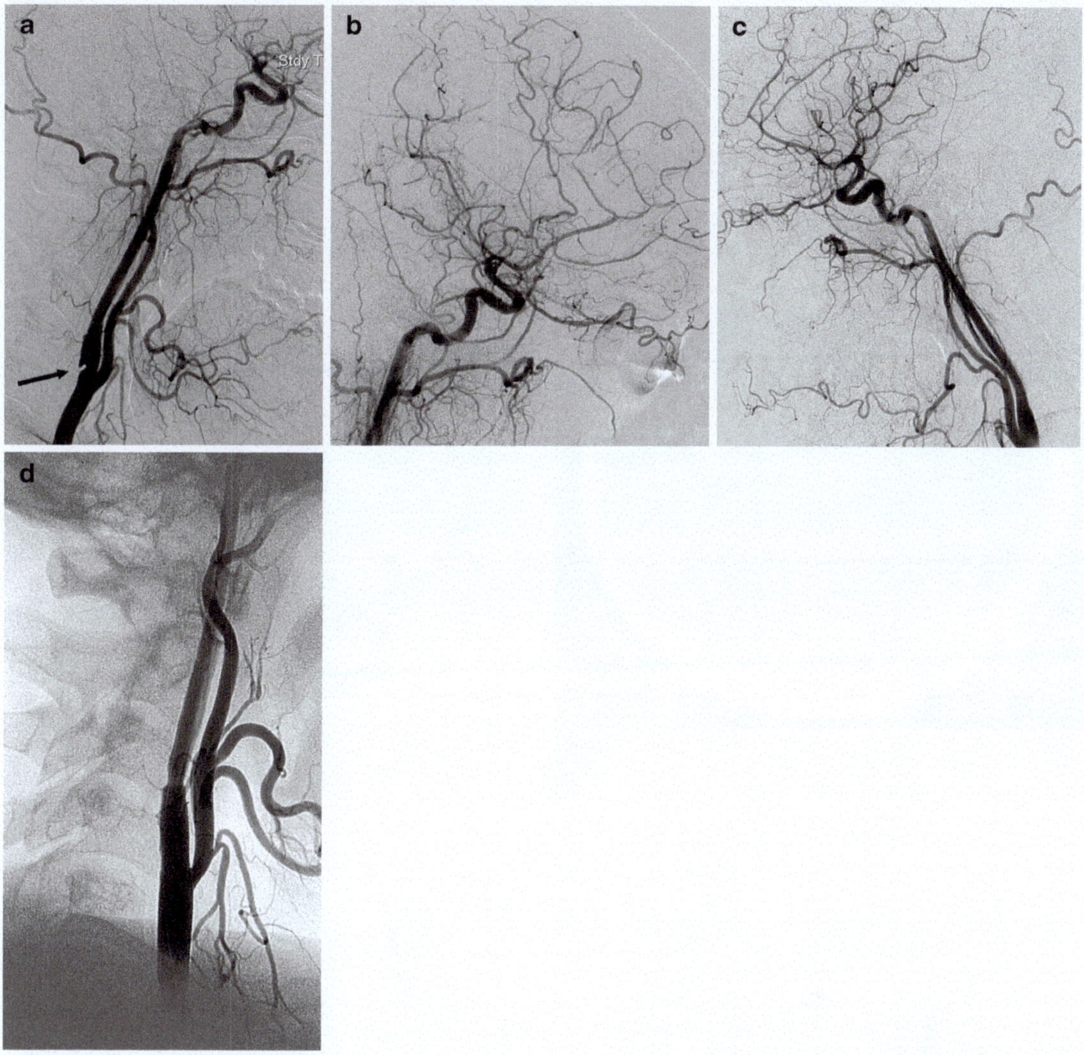

Fig. 17.6 Young man presenting with hypertension and acute left motor deficit. On CT small infarct in the vascular territory of the MCA was shown. A carotid and vertebral angiograms were performed. (**a**) Right carotid angiogram showing atypical defect on the extracranial right ICA, associated (**b**) with occlusion probably embolic of a few branches of the MCA. (**c**) The left carotid angiogram was normal as well as the VA angiogram. The diagnosis of atypical FMD was made. (**d**) Two weeks later after improvement of the deficit the right ICA lesion was treated with stent (courtesy of Prof. Bergui M and Dr. Stura G, Neuroradiology of Turin)

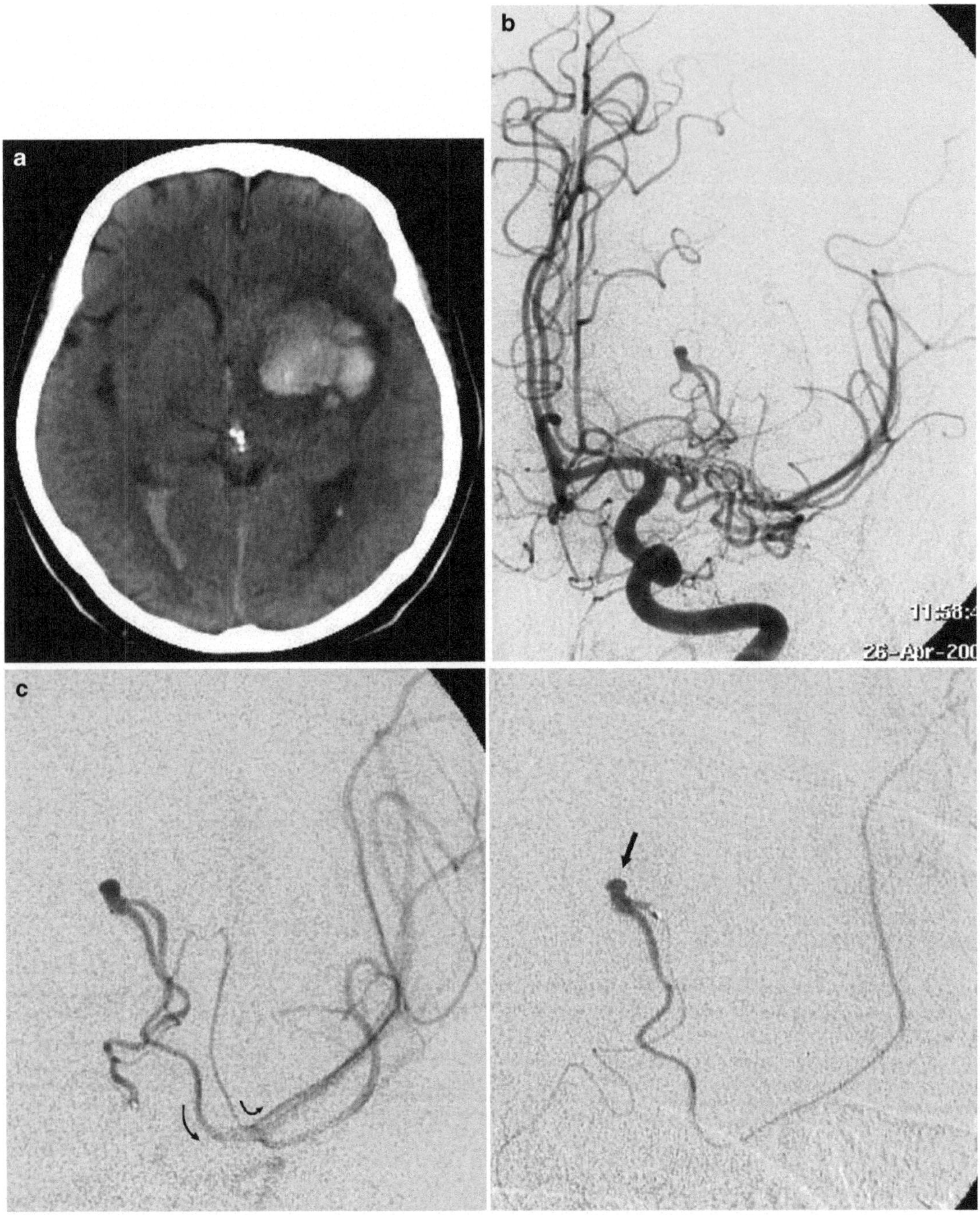

Fig. 17.7 Sixty-year-old woman presenting with sudden hemorrhage in the left basal ganglia. CT (**a**). Left carotid angiogram, AP view (**b**). Several changes are recognizable in the first segment of the middle cerebral artery (MCA). The M1 appears tortuous and very small, continuing to the normal M2 segment. There is also a well-developed network involving perforators with a pattern resembling that in Moyamoya disease. As demonstrated in the selective study (**c**), at least two perforators (*curved arrows*) anastomose to each other, contributing to revascularization of the M2 segments. On one of these perforators, a small aneurysm is recognizable (*arrow*), which was probably responsible for the hemorrhage. This was occluded with glue. Final control angiogram (**d**)

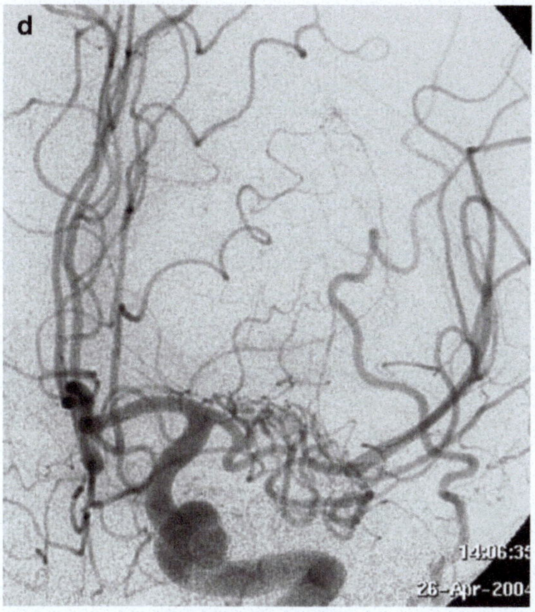

Fig. 17.7 (continued)

17.3 Moyamoya Disease

This is a chronic, progressive cerebrovascular disease, first described in Japanese patients by Takeuchi and Shimizu in 1957. Other Japanese reports followed (Kudo 1968; Nishimoto and Takeuchi 1968). The term Moyamoya, proposed by Suzuki and Takaku (1969), means *vague puff of smoke*. Despite the fact that the disease is more frequent in Japanese patients, later studies showed that it can occur also in non-Asians (Taveras 1969; Galligioni et al. 1971; Picard et al. 1974a, b; Huber 1979; Bradac and Oberson 1983; Masuda et al. 1993). Children and young patients are prevalently involved, but the disease may also be seen in older patients.

17.3.1 Pathology and Etiopathogenesis

Intimal thickening due to proliferation and migration of smooth muscle cells leads to progressive narrowing and final occlusion of the distal intracranial ICA with extension also to the M1 and A1. A rich network of vessels develops, involving perforators of the middle and anterior cerebral arteries as well as of the posterior communicating and anterior and posterior choroidal arteries. The significance of this network is not completely clear. Some authors (Taveras 1969; Crouzet et al. 1974; Picard et al. 1974a, b) have suggested that this is a kind of collateral circulation formed through anastomoses between the deep and superficial vascular territories. This interpretation is little intriguing since deep perforators are commonly thought to be end arteries with no connections neither to one another nor to the medullary arteries. Microangiographic studies, however (Kodama and Suzuki 1974; Umansky et al. 1985) (see also Sect. 15.7.1), have shown that a few anastomoses can be demonstrated between the single deep perforators and between these and the medullary arteries. It is therefore conceivable that in slowly occurring occlusive disease like Moyamoya, anastomoses develop. In Fig. 17.7 is shown a patient with an angiographic study with a pattern resembling a Moyamoya in which these anastomoses could be identified.

Basically, two phases of the disease can be recognized. The first is characterized by a progressive occlusion of the distal ICA and its branches at the circle of Willis with formation of the typical basal network involving the deep perforators. In second phase, the occlusion of the ICA is complete, and a leptomeningeal collateral circulation develops between the distal branches of the cerebral arteries and meningeal and transosseous branches of the external carotid artery and before complete occlusion of the ICA with meningeal branches arising from the ophthalmic artery (Fig. 17.8).

The posterior circulation, at least in the first phases of the disease, is not affected. It is, however, involved in the collateral circulation through the perforators arising from PcomA and P1 and through leptomenigeal anastomoses between PCA and ACA and MCA. Later, the P1 also is occluded, so that only branches of the external carotid artery are responsible for the intracranial circulation in the supratentorial sector.

The etiology of the disease is not known. Moyamoya disease can occur in different pathological conditions: vasculitis with or without an autoimmune mechanism, a postirradiation state, neurofibromatosis, hemoglobinophaties, and atherosclerosis. A genetic predisposition plays probably also an important role.

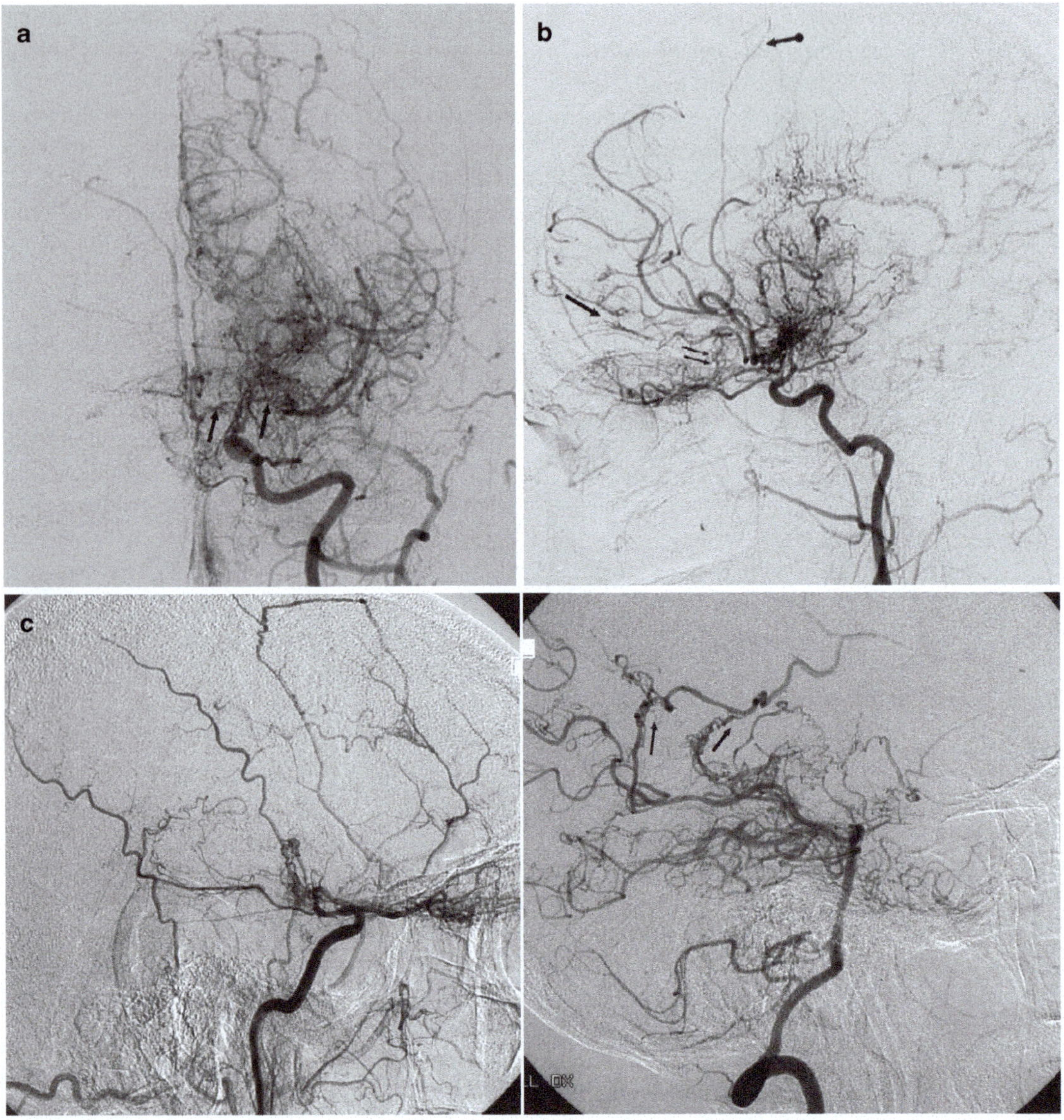

Fig. 17.8 Moyamoya disease in a young patient. (**a**) Common carotid angiogram, AP view. There is an irregular severe stenosis of the proximal M1 and A1 (*arrow*). Rich basal network involving perforators, acting probably as a collateral circulation toward the distal segments of the MCA and anterior cerebral artery (ACA). (**b**) On the lateral angiogram, there is the beginning of collateral flow toward the ACA through opening of an anastomosis from the anterior falx artery (*arrow*) and posterior ethmoidal arteries (*arrows*). There is also an initial collateral flow from the middle meningeal artery (MMA; *arrow with dot*). A partial filling of the distal pericallosal artery through anastomosis with the PCA is present. (**c**) Bilateral Moyamoya disease in another patient. Lateral carotid angiogram. Complete occlusion of the ACA and MCA. Persistence of a minimal basal network. Beginning of a collateral circulation toward the anterior circulation from the left PCA visible on the left carotid angiogram and from the right PCA (*arrow*) visible on the vertebral angiogram. Partial involvement also of the MMA, recognizable on the left carotid angiogram

17.3.2 Diagnosis and Clinical Relevance

Angiography is the essential method for identifying the disease. It has a progressive character, leading to ischemic stroke and hemorrhage. The latter is frequently due to small aneurysms present in the basal network (Figs. 17.7 and 17.8).

17.4 Takayasu's Arteritis

Takayasu's arteritis is a systemic granulomatous vasculitis that affects the aorta and its main branches. The pathological changes are characterized by inflammatory infiltration of the artery wall by lymphocytes, plasma cells, and presence of granulomas composed by giant cells. The pathological changes are similar to those found in giant cells arteritis, but the vessels involved are different and the affected patients in Takayasu's arteritis are younger. Children can also be involved (Al Abrawi et al. 2008).

Retinal ischemia and cerebral infarction are the typical symptoms. Hemorrhage due to a ruptured intracranial aneurysm can occurs. Hypertension due to involvement of the renal arteries and symptoms linked to aorta aneurysm are present. On the angiogram, occlusion or stenosis of the common carotid artery and vertebral artery at its origin are typical findings. A rich collateral circulation involving especially the ECA develop. An example is presented in Fig. 17.9.

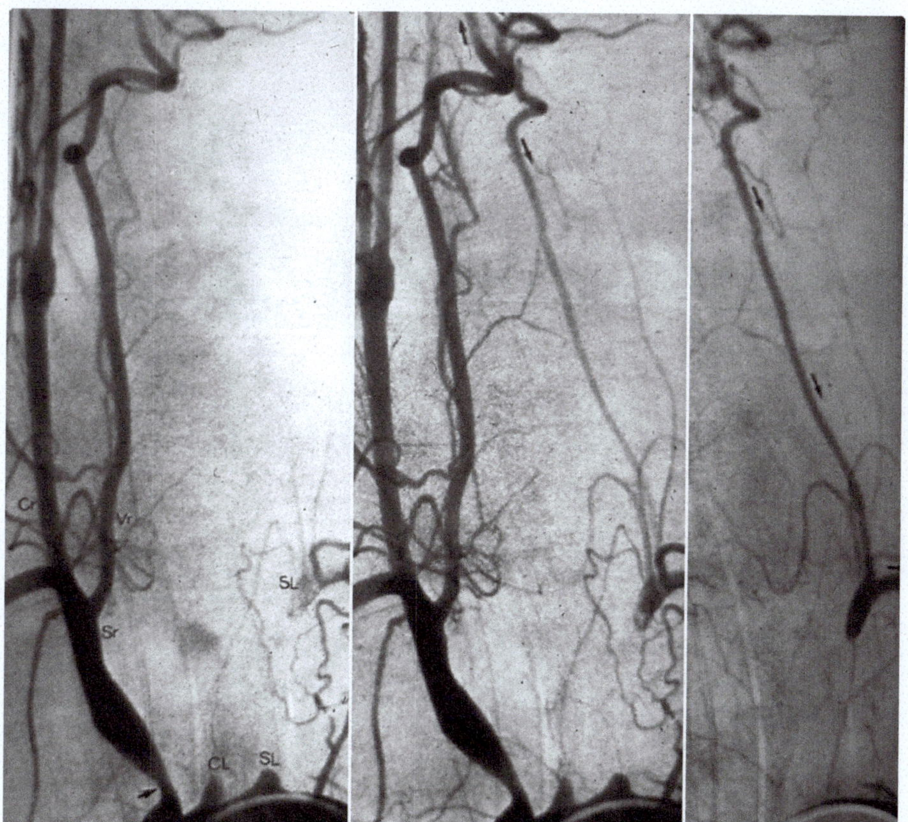

Fig. 17.9 Aortic arch angiogram in a patient with Takayasu's arteritis. Occlusion of the left subclavian (SL) and left common carotid artery (CL). Severe stenosis of the brachiocephalic trunk. In the late phases, retrograde injection of the left vertebral and left subclavian artery (*arrows*) from the right vertebrobasilar artery (subclavian steal syndrome)

17.5 Sneddon's Syndrome

This syndrome is characterized by pathological changes involving the cerebral vessels in patients by *Livedo reticularis* (Sneddon 1965). This is a cutaneous pathology characterized by a bluish reticulated aspect of the skin as a result of impaired superficial venous drainage. The pathogenesis is unknown. In some of these patients antiphospholipid antibodies have been demonstrated. The cerebral angiogram of these patients shows occlusions of the distal branches of the cerebral vessels, with a formation close to the occlusions of a typical network of capillaries and venules (Fig. 17.10). The clinical symptoms are characterized by several stroke which can lead in some patients to a progressive dementia.

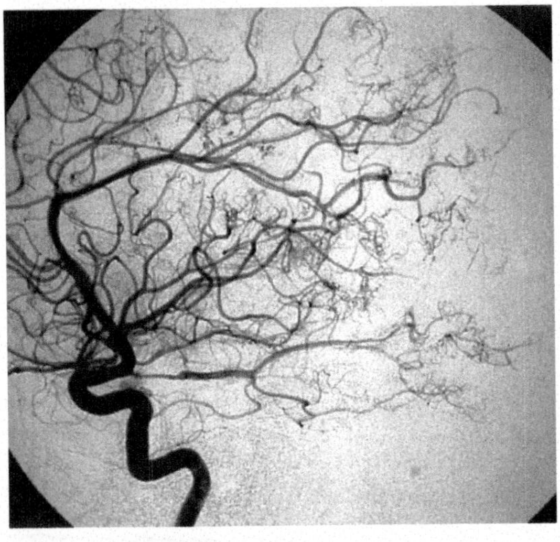

Fig. 17.10 Lateral carotid angiogram in a patient with Sneddon's syndrome. Several areas with the typical capillary network associated with vessel occlusion are evident

17.6 Reversible Cerebral Vasoconstriction Syndrome (RCVS)

This is a very rare pathological condition characterized by a sudden onset of extreme intense headache, sometimes with agitation and confusion. CT can show a minimal SAH on the convexity of the cerebral hemispheres. More rarely infarct or intracerebral hemorrhage occur. On the angiogram, segmental narrowings of medium-sized arteries are visible. This angiographic finding is not different from that presented in vasculitis and spasm in cases of SAH. The syndrome is commonly benign, and all symptoms as well as the morphologic changes of the vessels disappear within weeks. It was described first in women during pregnancy and in the postpartum (Fisher 1971; Rascol et al. 1979). Later, it became evident that it can occur in several other conditions, among them in association with a sympathomimetic therapy using drugs such as ergot derivate or following use of drugs such as cocaine and amphetamine or also triggered by factors such as sexual and physical activity (Calabrese et al. 2007; Singhal et al. 2009). The diagnosis can present some difficulties. It is important to exclude a SAH due to a ruptured aneurysm. In this context it is useful to remember that some dissecting aneurysm in the anterior circulation can present with a pattern of minimal SHA involving the cerebral convexity similar to that visible in cases of RCVS (Kumar et al. 2010; Obusez et al. 2014; Fukuma et al. 2015) (see also intracranial dissection in Chap. 16). The absence of inflammation signs in labor examinations will help in the differential diagnosis with angiitis. Cases of amyloid angiopathy with bleeding localized on the cortical region have been also reported (Apoil et al. 2013). These patients are older in comparison to those with RCVS. In the differential diagnosis also cerebral venous thrombosis should be considered. Indeed, as reported by some authors (Boukobza et al. 2016) thrombosis of cortical veins, alone or associated with sinus thrombosis, can present with cortical SAH. An example of RCVS is presented in Fig. 17.11.

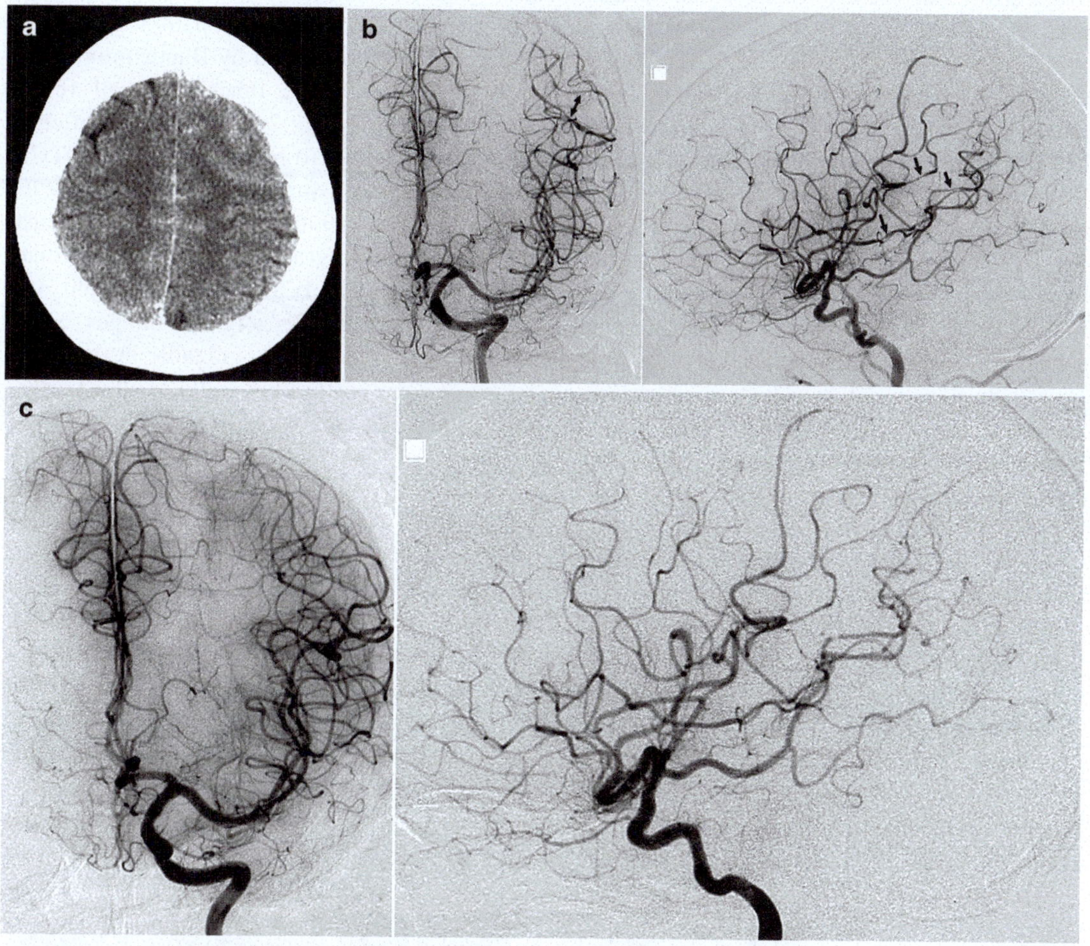

Fig. 17.11 (RCVS) Middle-aged woman with acute onset of severe headache followed by intense physical activity. (**a**) CT showed diffuse mild SAH. (**b**) left ICA angiogram AP and Lateral view. Irregularity and narrowing of several MCA branches are visible (*bidirectional* and *simple arrows*). Similar changes were visible on the contralateral ICA angiogram. The VA angiogram was normal. No vascular malformations were detected. The patient recovery completely and the control angiograms (**c**) two months later showed clear improvement of the morphological changes of the arteries

17.7 Primary Angiitis of the CNS (PACNS)

This is a rare angiitis involving selectively the cerebral arteries, which is increasingly recognized and diagnosed (Calabrese and Mallek 1988; Salvarani et al. 2007; Birnbaum and Hellmann 2009; Hajj-Ali and Calabrese 2009; Magnus et al. 2012; Taschner et al. 2013). No other arteries outside of the CNS are affected. All ages are involved including children and old patients. The disease can present with acute headache disorientation, and coma followed by death, or can have a slower indolent course with episodes of stroke, seizures, and multifocal neurological symptoms. The pathologic aspects are commonly that of a granulomatous infiltration of small, sometimes medium-sized arteries. Labor examinations show signs of inflammation. Small ischemic lesions, sometimes hemorrhages, in the vascular territories of deep perforators and medullary arteries are recognizable on MR images. Large infarcts are rare.

Useful for the diagnosis are T1-weighted images with contrast medium showing enhancement in the wall of the arteries suggesting inflammation (Küker 2007; Küker et al. 2008). As reported by some authors (Mineyko et al. 2013) the latter finding should be considered with caution since it can be observed also in normal conditions. Angiography is frequently normal. Pathological changes of the vessels can be identified when medium-sized arteries are involved (Fig. 17.3).

17.8 Autosomal Dominant Arteriopathy with Subcortical Infarct and Leukoencephalopathy (CADASIL)

The term CADASIL for this pathological condition has been proposed by Baudrimont et al. (1993). This is a rare inherited small vessel arteriopathy affecting 20/30-year-old patients. The disease is due to mutation of the NOTCH 3 gene on chromosome 19 (Joutel et al. 1996; Chabriat et al. 1996). Clinically the disease presents with migraine attacks and strokes leading progressively to cognitive deficit and dementia. On MR T2-weighted images, areas of hyperintensity are recognizable in the territories of deep perforators and medullary arteries, indicating ischemia. This finding is not different of that occurring in patients with severe small vessel disease due to arteriosclerosis (Sect. 15.4.5). However, differently from these, in CADASIL the lesions extend frequently and typically to the external capsule, to the anterior white matter of the temporal lobe, and sometimes to the corpus callosum (Chabriat et al. 2000; Van Den Boom et al. 2002) (Fig. 17.12). Angiography is normal.

Studies with the electron microscopy have shown the presence of granular osmiophilic material in the media of the arteries, where the muscles cells appear swollen and degenerated.

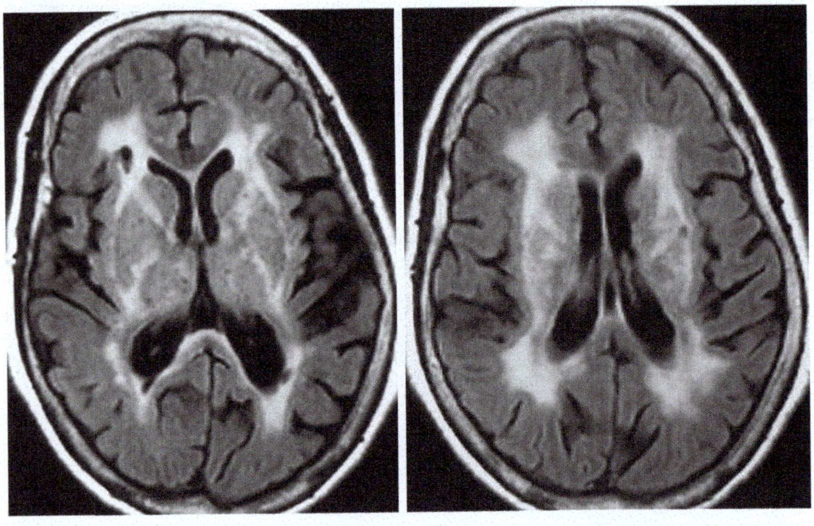

Fig. 17.12 40-year-old patient with episodes of stroke and progressive cognitive deficit. MR T2-Flair sequences. Diffuse changes involving the white matter and basal ganglia. The external capsule is also affected. The diagnosis of CADASIL was confirmed by biopsy and genetic studies

These changes can be associated with hyalinosis and fibrosis (Gutierrez et al. 1994; Zhang et al. 1994; Dong et al. 2012), leading progressively to occlusion of the arteries.

An interesting observation has been made by Ruchoux et al. (1994, 1995). These authors demonstrated that the changes present in the cerebral vessels could also be demonstrated in other organs, including spleen, liver, kidneys, muscles, and skin. Other authors (Schroder et al. 1995) identified the same pathological changes in the sural nerve. These studies showed the systemic character of the disease, allowing a more easy diagnosis with biopsy.

Similar diseases with, however, some different genetic and clinical features have also been described by other authors (Fukutake and Hirayama 1995).

17.9 Migraine and Stroke

Stroke occurring in migraine has been described in the past. MR studies in patients with migraine have shown the presence of microlesion in the white matter. However, it is not clear whether these are due to a previous infarcts or are simple micro-areas of gliosis due to other causes. Furthermore, various vascular disorders are present being responsible for the stroke in many migrainous patients: dissection, dysfunction of the venous drainage, Cadasil and the coexistence of other factors such as the use of oral contraceptives, cigarettes, and various other drugs (Diener and Kurth 2011). Today, there are many doubts whether migraine can lead to a stroke. Which are the possible intervening factors or possible mechanisms remains open to question and should be clarified with further studies.

18 Cardiac Diseases

There are many cardiac diseases responsible for cerebral ischemia, which is frequently the result of emboli in patients with cardiac dysrhythmia, myocardial infarction, atrial septal aneurysm, cardiac valves diseases, prosthetic valves, endocarditis of various origin, and congenital heart diseases. Paradoxical embolism through a *patent foramen ovale (PFO)* has become also a frequent explanation for embolism. It should, however, be remarked that the role played by the PFO in stroke remains a little controversial. Indeed, it should be considered that it can be detected in the general population with a frequency up to 25% (Lechat et al. 1988) and some authors have found no clear relationship between PFO and stroke (Kutty et al. 2009, 2012). Others have linked the possibility of stroke in cases of large PFO (Schuchlenz et al. 2000).

Cardiac abnormalities responsible for embolization have been identified in systemic lupus erythematosus and in primary antiphospholipid antibody syndrome. Another rare pathology is *myxoma*. This is a tumor that commonly originates in the left atrium in young to middle-aged patients. The emboli consist of myxomatous tissue that leads to parenchymal tumoral localization and vascular changes characterized by occlusion and aneurysmal dilatation (New et al. 1970; Roeltgen et al. 1981) (Fig. 18.1).

The commonest site of cardiac emboli is the anterior circulation, especially in the distal internal carotid artery and in the middle cerebral artery territory. The posterior circulation is less involved with a typical location being that of the top of the basilar artery (see Chap. 15). In this context it should be noticed, that the distribution of the emboli can be influenced by the presence of anomalies in the origin of the arteries from aortic arch, anomalies of the circle of Willis and persistence of embryonic arteries.

The incidence of cardio-embolism is reported to be 15–20% of all strokes (see also Chap. 15). It is frequent in young adults (Barnett 1974; Barnett et al. 1980; Bogousslavsky et al. 1988a; Kittner et al. 1991; Broderick et al. 1992). In case of small emboli, spontaneous thrombolysis and rapid recanalization can occur, so the angiography performed later appears normal. In other cases, occlusion is visible and a cardiogenic brain embolism can be clearly suspected when there is a *specific cardiac pathology, no atherosclerotic changes or other pathologies of carotid or vertebral arteries are present, more than one vascular territory,* is involved, or the stroke has occurred in *patients younger than 45 years of age*, where atherosclerotic changes of the extra-intracranial arteries are less frequent than in the elderly. In some cases the cardiac embolism can be only suspected, but not

G.B. Bradac, *Applied Cerebral Angiography*, DOI 10.1007/978-3-319-57228-4_18

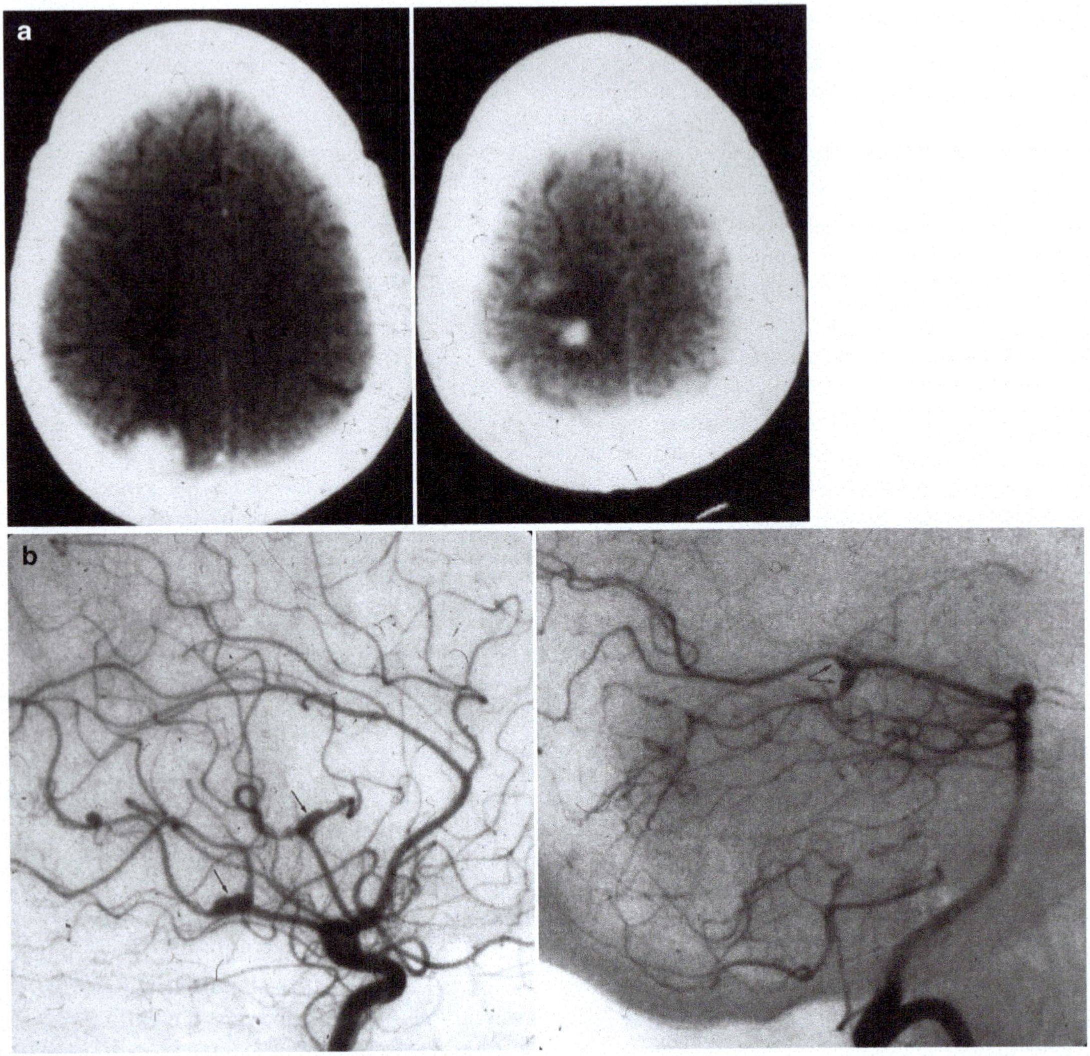

Fig. 18.1 Cerebral embolization in a patient with cardiac myxoma. (**a**) CT showing parenchymal localization of tumor emboli. (**b**) Carotid and vertebral angiogram disclosing aneurysmal dilatation due to tumoral infiltration of the vessel wall (*arrows*)

confirmed. In other cases, it can be difficult to establish whether the stroke has a cardiogenic origin or it is due to atherosclerotic changes of the extra-intracranial arteries when both pathologies are present in the given patient. (In addition to Figs.18.1, 18.2, 18.3, 18.4, and 18.5 see also Sects. 15.4.1, 15.4.2, 15.5.2, 15.5.3, 15.5.4, and 15.5.5).

Cerebral infarcts due to cardiac embolism are commonly superficial. However, even if they are more rare, small lacunar infarcts in the deep structures can occur. These are thought to be due to microemboli especially in patients with atrial fibrillation and they are diagnosed progressively more often in the elderly considering that this cardiac pathology increases with the age (Kempster et al. 1988; Arboix et al. 2000) (See also Sects. 15.4.5 and 15.5.4).

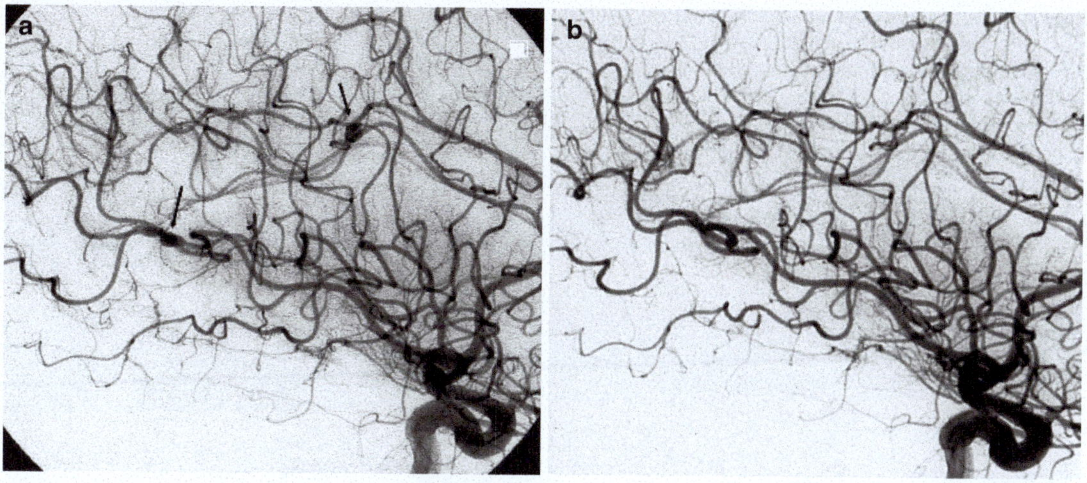

Fig. 18.2 Mycotic aneurysm in a patient with bacterial endocarditis. (**a**) Lateral carotid angiogram showing the aneurysm involving branches of the MCA and ACA (*arrows*). (**b**) Control angiogram 2 months later after intensive antibiotic therapy, showing the disappearance of the aneurysms

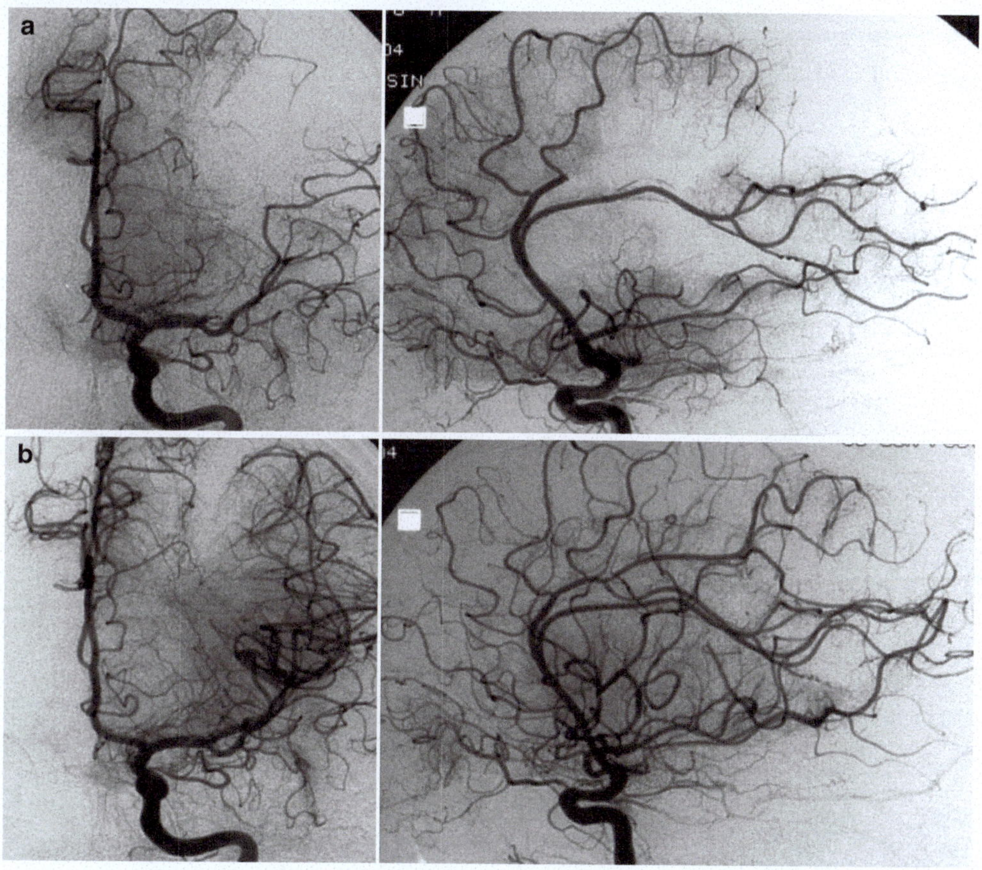

Fig. 18.3 Acute multiple occlusions involving branches of the MCA and ACA in a middle-aged woman presenting with right hemiplegia and partial aphasia. (**a**) Left carotid angiogram. AP and lateral views showing the occlusions. (**b**) Selective pharmacological thrombolysis, with a quick reopening of the occlusions. There was a rapid clinical improvement. A cardiac pathology was suspected. However, no cardiac disease was found

Fig. 18.4 Cardiac embolization in an old patient involving several branches of the MCA and fetal PCA (*arrow*). Partial retrograde filling of the occluded branches (*arrows*) through opening of anastomoses with the ACA. Compare this figure with Figs. 15.15 and 15.16

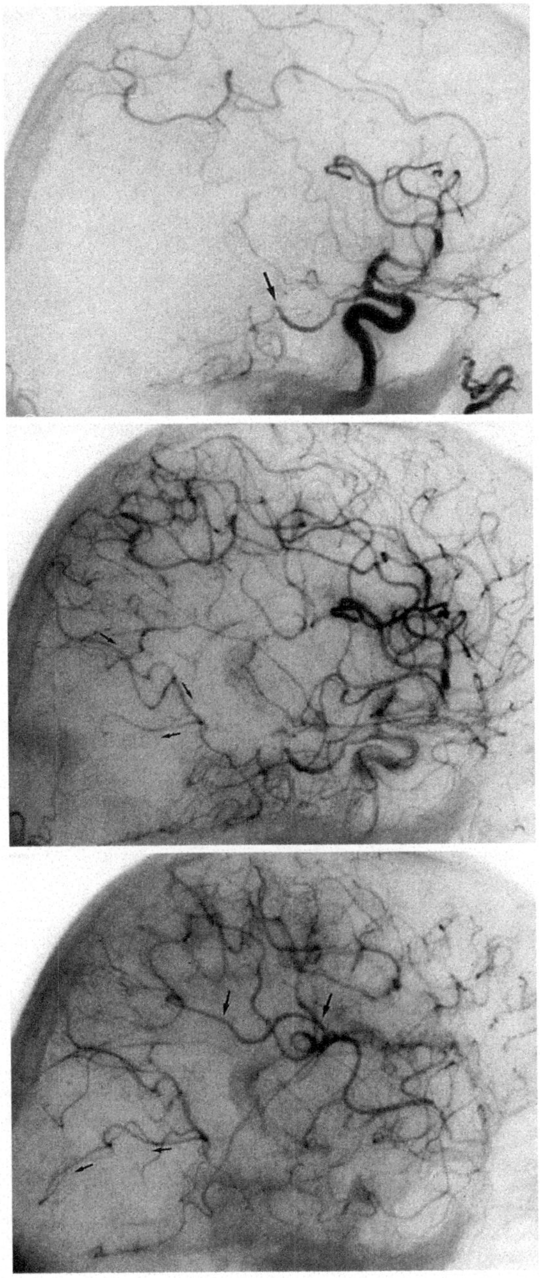

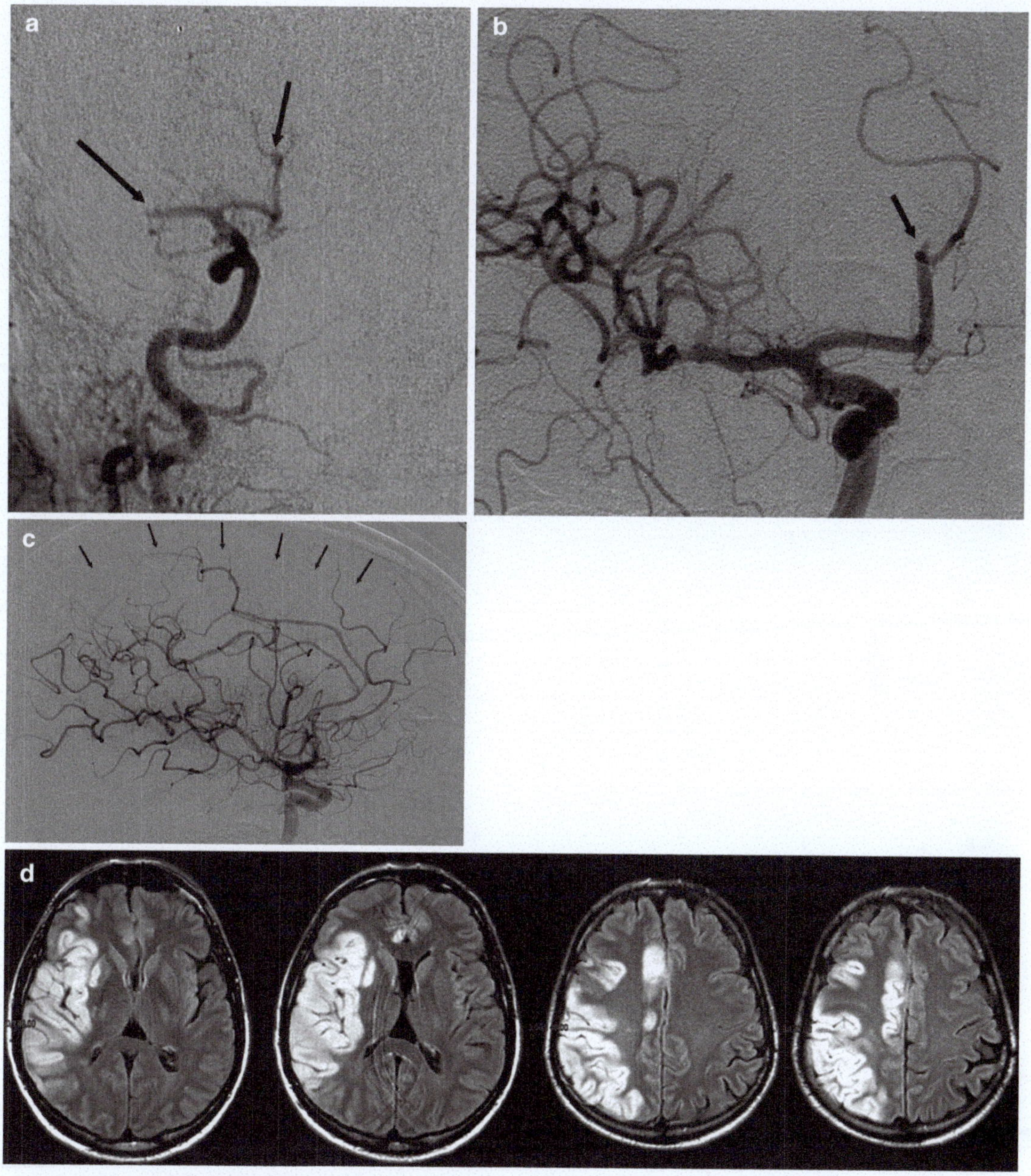

Fig. 18.5 Acute occlusion, probably cardioembolic, of proximal right MCA and more distal ACA in a middle-aged patient. (**a**) Angiogram of the right carotid artery showing the occlusions (*arrows*). (**b**) Rapid recanalization with mechanical thrombolysis (solitaire) of the M1 with apparently good filling of the distal branches of MCA. An attempt to open the occlusion of the distal pericallosal artery (*arrow*) was also performed. (**c**) final lateral carotid angiogram showing persistent occlusion of a few distal branches of ACA and MCA (*arrowheads*). (**d**) CT performed 24 h later showed ischemic lesions, which were better defined with MR 4 days later. The ischemia involved largely the distal vascular territory of the MCA, with partial involvement of the basal ganglia (pallidum). There was ischemia in the distal territory of the ACA. The extensive ischemia in spite of the rapid proximal reopening of the occluded arteries can be explained by impairment of the leptomeningeal collateral circulation due to many factors, among them a distal embolization and the unfavorable contemporaneous occlusion of the MCA and ACA (see also text 15.8). The left hemiplegic patient recovered slowly. After intensive physiotherapy he was able to walk with some difficulties 6 months later

Arterial Occlusive Diseases in Children

19

Ischemia in children deserves some specific considerations. It is not as rare as it is commonly thought (Barnes et al. 2004; Amlie-Lefond et al. 2008, 2009) with a frequency reported to be 2–6/100,000 children per year. Unlike in adults, it is rarely linked to factors related to atherosclerosis. Instead, it is associated with a variety of pathological processes, including cardiac diseases and other non-atherosclerotic arteriopathies which are listed in Table 17.1 (Chap. 17). In some cases, a precise etiology can be identified; in many others, however, it may only be suspected or remain unknown (Barnes et al. 2004; Jones et al. 2010).

Among the most frequent causes are infections, especially of the upper respiratory tract. Arteritis due to a previous varicella zoster infection can occur (Sebire et al. 1999; Barnes et al. 2004; Lauthier et al. 2005; Miravet et al. 2007; Jones et al. 2010).

Primary cerebral angiitis (PACNS) has been emphasized by some authors (Elbers and Benseler 2008; Hajj-Ali and Calabrese 2009) (see also Sect. 17.7). An example is presented in Fig. 17.3.

Stroke occurs frequently in sickle cell disease (Jones et al. 2010). This is due to the altered red blood cells causing an endothelia damage leading to an intimal hyperplasia and a progressive narrowing of the lumen of the artery. Among the intracranial vessels, the supraclinoid internal carotid artery is the most frequently involved. Furthermore, MR studies (Pegelow et al. 2002; Stem et al. 2003) in children with sickle cell disease have shown with a high frequency asymptomatic microischemic lesions in the white matter of the cerebral hemispheres indicating an involvement also of the superficial perforators.

Many pathological conditions altering the blood coagulation (see Table 17.1) can be the cause of stroke in adults. The same can occur, even if more rarely, in childhood. In the study of Barnes and Deveber (2006), prothrombotic abnormalities (elevated lipoprotein, protein C deficiency, factor V Leiden mutation, antiphospholipid antibodies) were identified in 25–50% of children presenting with stroke.

Another pathology which can be associated with occlusion of the cerebral vessels is the neurofibromatosis type 1. The pathogenesis is unclear. It has been suggested that the loss of the neurofibromin, a special protein, would favor the proliferation of the smooth muscle cells leading to occlusion of the artery (Rea et al. 2009). Since it can be asymptomatic, at least in the first phases of the pathological process, it can escape the diagnosis. Rea et al. (2009) recommend in all these patients to perform MR angiography to study the cerebral vessels. An example is presented in Fig. 17.2.

Moyamoya disease is another pathology of the cerebral vessels involving frequently children (see Sect. 17.3).

Among other pathologies, spontaneous dissection is another increasingly recognized cause of stroke in children (Schievink et al. 1994a;

G.B. Bradac, *Applied Cerebral Angiography*, DOI 10.1007/978-3-319-57228-4_19

Fullerton et al. 2001; Rafay et al. 2006) (see Sect. 16.7 for dissection in children).

As far as it concerns the diagnosis, in addition to the CT/MR examination for the identification of the ischemia, CT/MR angiography can be frequently sufficient to show the lesions of the vessels in the form of occlusion, stenosis, or in some case of dissection also the presence of pseudoaneurysm. For a more detailed diagnosis, angiography remains the examination of choice.

20 Cerebral Venous Thrombosis

A venous thrombosis involving the SSS linked to an infectious disease has been described, in an autoptic study, about 200 years ago in 1825 by Ribes. Since then, a progressive improvement in diagnostic methods, in particular angiography (Huhn 1957, 1962; Krayenbühl 1961, 1967; Vines and Davis 1971) and later CT, MRI, and MR angiography, has made it possible to better identify this pathology. The true incidence of cerebral venous thrombosis (CVT), however, remains unknown, even if there is no doubt that it is more frequent than had previously been thought (Krayenbühl 1967; Bousser et al. 1985; Ameri and Bousser 1992; Einhäupl and Masuhr 1994; Bousser and Russell 1997; Linn and Brueckmann 2010).

20.1 Etiopathogenesis

A frequent cause is infection, either due to intracranial extension of infectious diseases involving the skin or bone cavities of the craniofacial area or in cases of general bacterial septicemia or viral infection, especially due to HIV and cytomegalovirus (CMV). In young women, CVT can occur during pregnancy, in puerperium, and due to use of oral contraceptives. Other causes are pathologies of the red blood cells, such as thrombophilia, polycythemia, sickle cell disease, leukemia, and lymphoma and many coagulation disorders, such as protein C and S deficiency and disseminated intravascular coagulation. CVT is also frequent in patients with Behçet's disease, systemic lupus erythematosus, and in patients with severe dehydration and cardiac failure. Another rare cause, described by a few authors (Savoiardo et al. 2006; Katoaka et al. 2007; Richard et al. 2007; Bousser and Ferro 2007; Schievink and Maya 2008; Yoon et al. 2011; Qin et al. 2012), is that occurring in patients with spontaneous intracranial hypotension. It has been suggested that, in these patients, the fall of volume and pressure of cerebrospinal fluid (CSF) is compensated by a dilatation of the sinuses commonly the SSS or the TS which reduces the flow velocity of the blood in the sinuses. This, associated with a decreased absorption of CSF in the sinuses due to the loss of CSF, results in an increased blood viscosity, favoring the thrombosis. Other causes of CVT are cranial trauma and neurosurgical intervention. Finally, intracranial tumors, specially meningiomas, can involve the adjacent sinus and cause thrombosis. As far as it concerns sinus thrombosis in patients with DAVfs, see Chap. 13.

It should be taken into consideration that in many cases—up to 35%—the cause remains unknown (Milandre et al. 1988; Ameri and Bousser 1992; Cantu and Barinagarrementeria 1993; Einhäupl and Masuhr 1994).

20.2 Location

The superior sagittal sinus (SSS) is the venous channel most commonly involved, followed by the transverse sinus. Both are frequently affected together (Ameri and Bousser 1992; Cantu and

G.B. Bradac, *Applied Cerebral Angiography*, DOI 10.1007/978-3-319-57228-4_20

Barinagarrementeria 1993; Tsai et al. 1995; Linn and Brükmann 2010). The thrombosis can be limited to the sinus and the clinical presentation may frequently be characterized by symptoms owing to the intracranial hypertension, such as headache and visual disturbances. Cortical vein tributaries of the thrombosed sinus can be secondarily involved as a result of retrograde propagation of the thrombus, leading to ischemia. Many anastomoses are present in the venous system which can act as a venous collateral circulation, in case of isolated sinus thrombosis. This collateral, however, can be impaired in cases of thrombosis of the sinuses involving also of the cerebral veins (Bergui et al. 1999; Bradac and Bergui 2001).

Isolated cortical veins thrombosis may also occur. This has been thought to be very rare (Ameri and Bousser 1992; Einhäupl and Masuhr 1994). Through improvement in MRI technique and progressive awareness of this pathology, cases have been increasingly reported (Sagduyu et al. 2006; Boukobza et al. 2009; Albayram et al. 2009; Linn and Brükmann 2010).

Thrombosis of the deep venous system, involving the straight sinus and the vein of Galen, with retrograde extension to the internal cerebral and basal veins, is another usually severe condition that is, fortunately, less common. In these situations, transcerebral anastomoses via connections between the superficial and deep medullary veins as described in Sect. 9.1.2.1 can act as an important collateral circulation (Bergui et al. 1999; Bradac and Bergui 2001). Another important aspect which can play a role in the location of the ischemia are the variations of the drainage of the SSS and SS as described in Sects. 9.3.1, 9.3.3, and 9.3.5.

Cavernous sinus thrombosis is another frequent localization leading to a typical cavernous sinus syndrome. The lesion can extend to the superior-inferior petrosal sinuses. Involvement of the intracavernous internal carotid artery due to arteritis, especially in the infectious form of thrombosis can lead to stenosis/occlusion of the artery (Segall et al. 1982). Indeed the cavernous sinus thrombosis is commonly due to infectious processes involving the skin of the facial area, nose, paranasal sinuses, orbita, teeth, and middle ear.

20.3 Diagnosis

CT and MRI allow the detection of changes to the brain parenchyma in the form of hemorrhagic and/or nonhemorrhagic infarcts, uni- or bilateral, single or multiple, with various locations depending on the site and extension of the thrombosis. Hemorrhagic ischemia can be easily diagnosed on CT studies, presenting as hyperdense area. White matter hypodensity on CT and hyperintensity on T2-weighted images on MR, which indicate edema preceding the venous infarct, is also a finding which can suggest venous thrombosis.

On CT, an abnormal hyperdensity can be recognized at the level of the torcular herophili, SSS and TS. This is difficult to differentiate in the majority of cases from the normally slight hyperdensity of these structures. Though it is not always present, a more reliable indicator is the so-called empty sign, which is visible after contrast injection, and it is due to opacification of the sinus wall contrasting with the low density of the thrombus in the lumen (Buonanno et al. 1978; Kingsley et al. 1978). MRI can be a very useful alternative method (Bousser and Ferro 2007). In the acute phase, it shows the hypointensity and isointensity of the thrombus in the involved sinus and occasionally in the cortical or deep veins, respectively, on T2- and T1-weighted images. After 3–7 days, the clot becomes progressively hyperintense on both T1 and T2 sequences. In this context, the high diagnostic value of the T2*-weighted gradient echo in identifying also isolated cortical vein thrombosis has been emphasized by many authors. On these sequences the thrombotic vessel is characterized by hypointensity which is already present in the acute phase and persist for several months differently from the changes of signal of T1 and T2 images (Fellner et al. 2005; Urban and Müller-Forrell 2005; Boukobza et al. 2009; Linn and Brükmann 2010).

Finally, in cases of suspected venous thrombosis, MRI angiography should be performed. In doubtful cases or whenever a more specific diagnosis is required, conventional angiography remains the essential diagnostic method. A rapid and precise diagnosis is important since anticoagulant therapy can improve the prognosis in many of these patients (Villringer et al. 1994; Einhäupl et al. 2006).

As far as the differential diagnosis with arachnoid granulations and brain herniations protruding in the sinuses see Sect. 9.3.

Examples of venous thrombosis are shown in Figs. 20.1, 20.2, 20.3, 20.4, and 20.5.

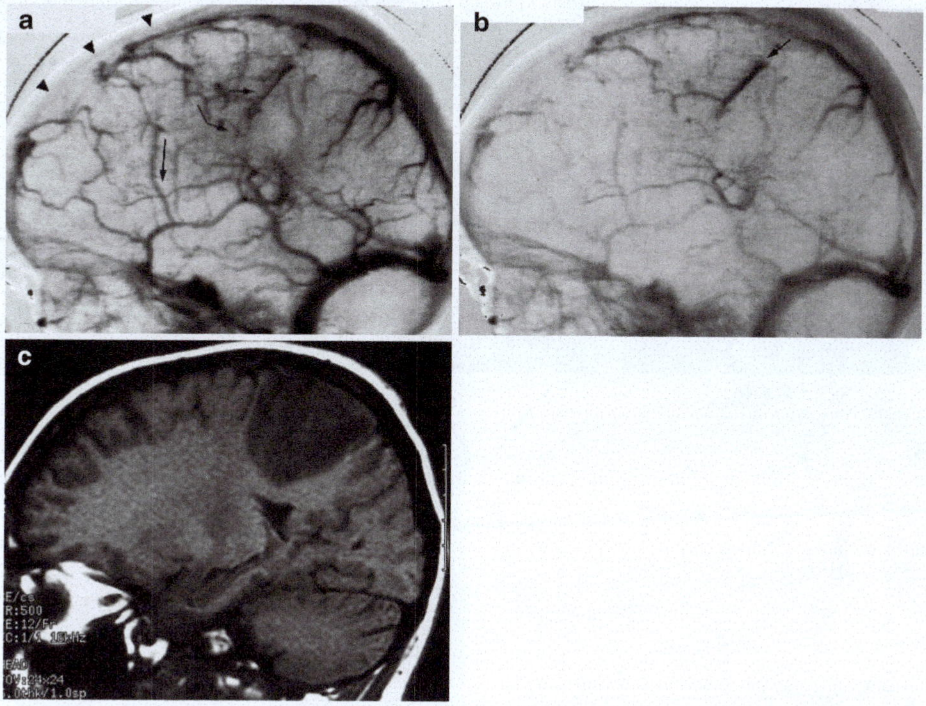

Fig. 20.1 Thrombosis of the anterior and middle segments of the superior sagittal sinus (SSS). (**a**) Carotid angiogram, lateral view, venous phase. Occlusion of the SSS (*arrowheads*). Rerouting of the venous drainage through cortical veins toward the parietal and temporal area (*arrows*). Later phase (**b**). Distal occlusion of the cortical veins (*arrow*) probably explaining the parietal ischemia, visible on MRI (**c**)

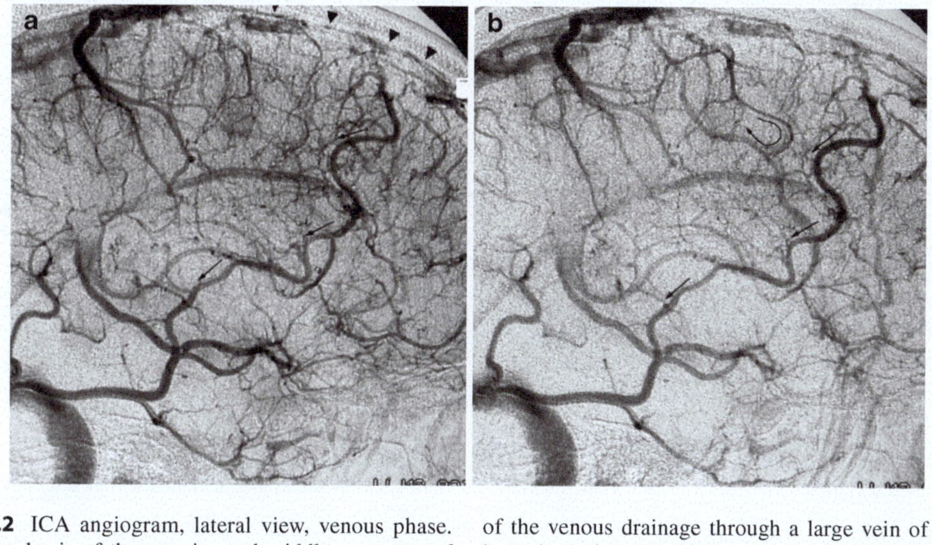

Fig. 20.2 ICA angiogram, lateral view, venous phase. (**a**) Thrombosis of the anterior and middle segments of the SSS (*arrowheads*) involving also the distal segment of the draining cortical veins. (**b**) Later phase. Rerouting of the venous drainage through a large vein of Trolard into the vein of Labbé (*arrows*), partial reorganized venous drainage in the parietal area (*curved arrow*) toward the patent segment of the SSS

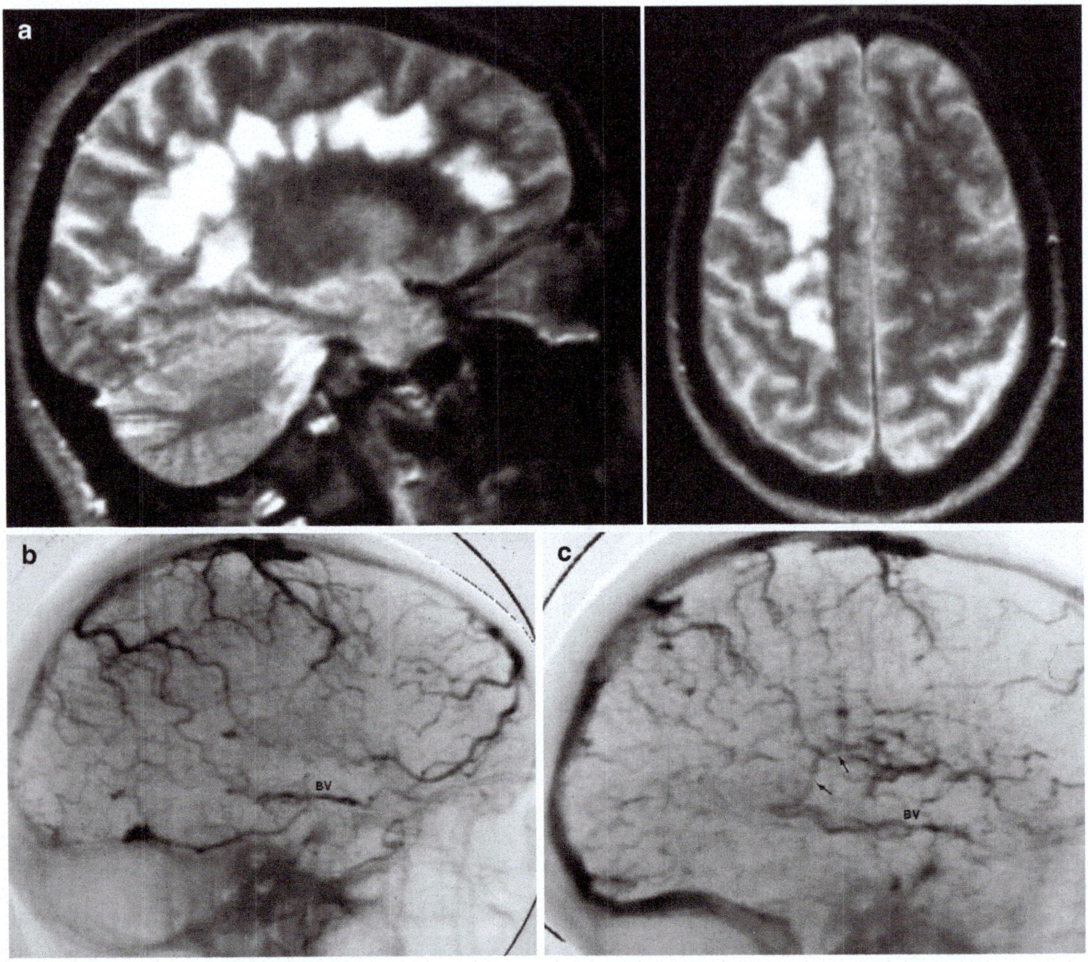

Fig. 20.3 Venous thrombosis involving the deep venous system. (**a**) MRI, sagittal and axial views, T2-weighted images showing changes corresponding to venous ischemia involving predominantly the white matter of the right hemisphere. (**b**) Right and (**c**) left carotid angiograms, venous phase. The vein of Galen and straight sinus are not injected. On the right, only the proximal part of the basal vein (BV) is recognizable, filled through its anterior tributaries. On the left, in addition to the basal vein reaching the proximal part of the vein of Galen, there is filling of the internal cerebral vein (*arrows*). It is conceivable that a collateral circulation involving the subependymal and medullary veins toward the cortical veins has developed, explaining the absence of ischemia in the left hemisphere

Catheterization of TS and/or SSS followed by injection of thrombolytic agents has been proposed in those cases in which the anticoagulant therapy fails to improve the clinical condition of the patient (Hocker et al. 2008; Sujith et al. 2008; Gala et al. 2013).

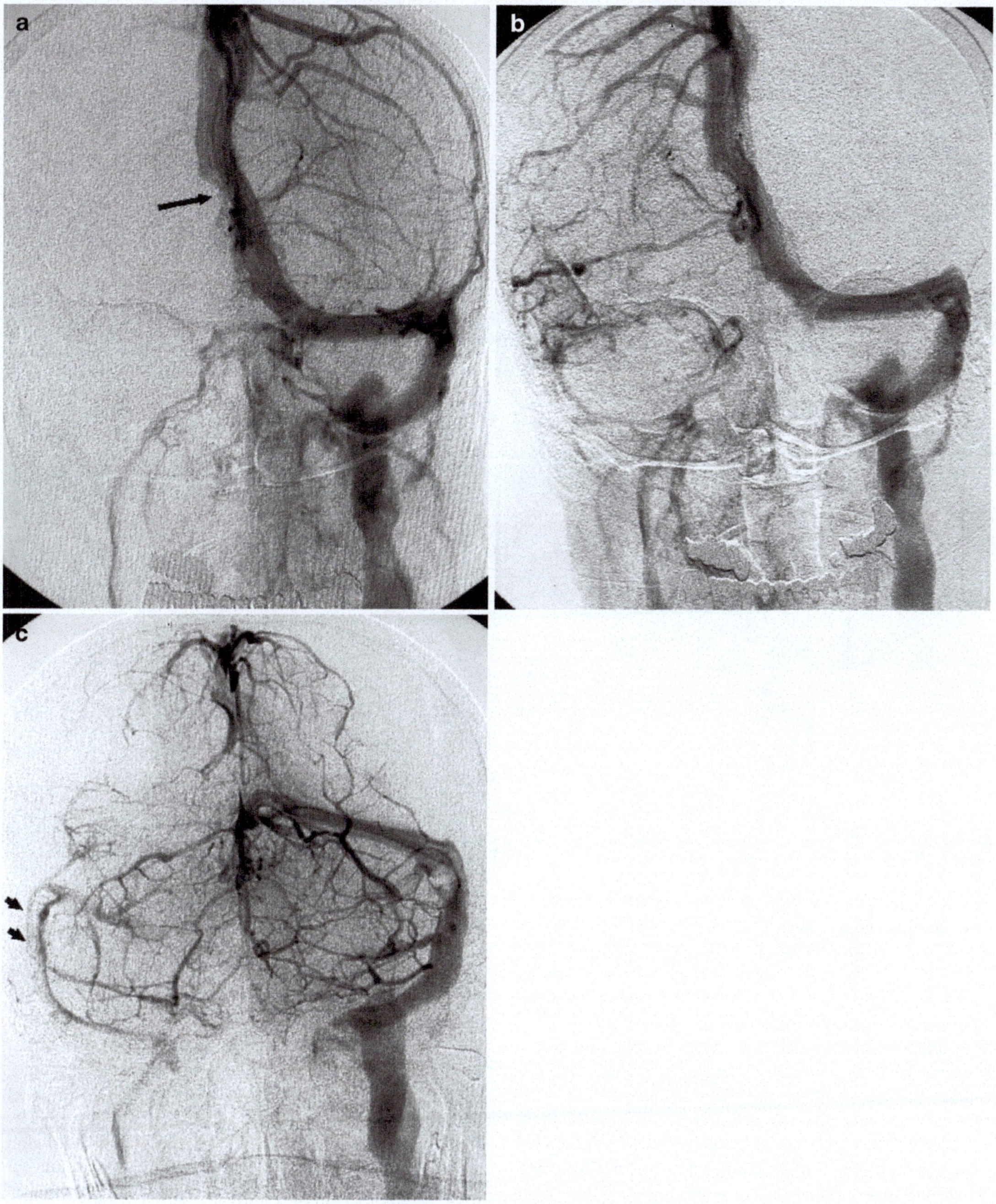

Fig. 20.4 Venous thrombosis involving the distal portion of the right segment of the duplicated SSS extending to the transverse sinus (TS). Old female patient with headache. The MRI study disclosed hyperintensity on T1-weighted images of the right TS. Angiography confirmed the suspected diagnosis of sinus thrombosis. (**a**, **b**) Left and right carotid angiogram, venous phase. There is no opacification of the distal right segment of the SSS (*arrow*) and right TS. Both internal cerebral veins are recognizable. They drain into the vein of Galen, continuing to the straight sinus well visible on the lateral angiogram (not demonstrated) entering the left TS. (**c**) Vertebral angiogram. The minimal laminar injection of the right TS (*arrows*) allows identification of the thrombus

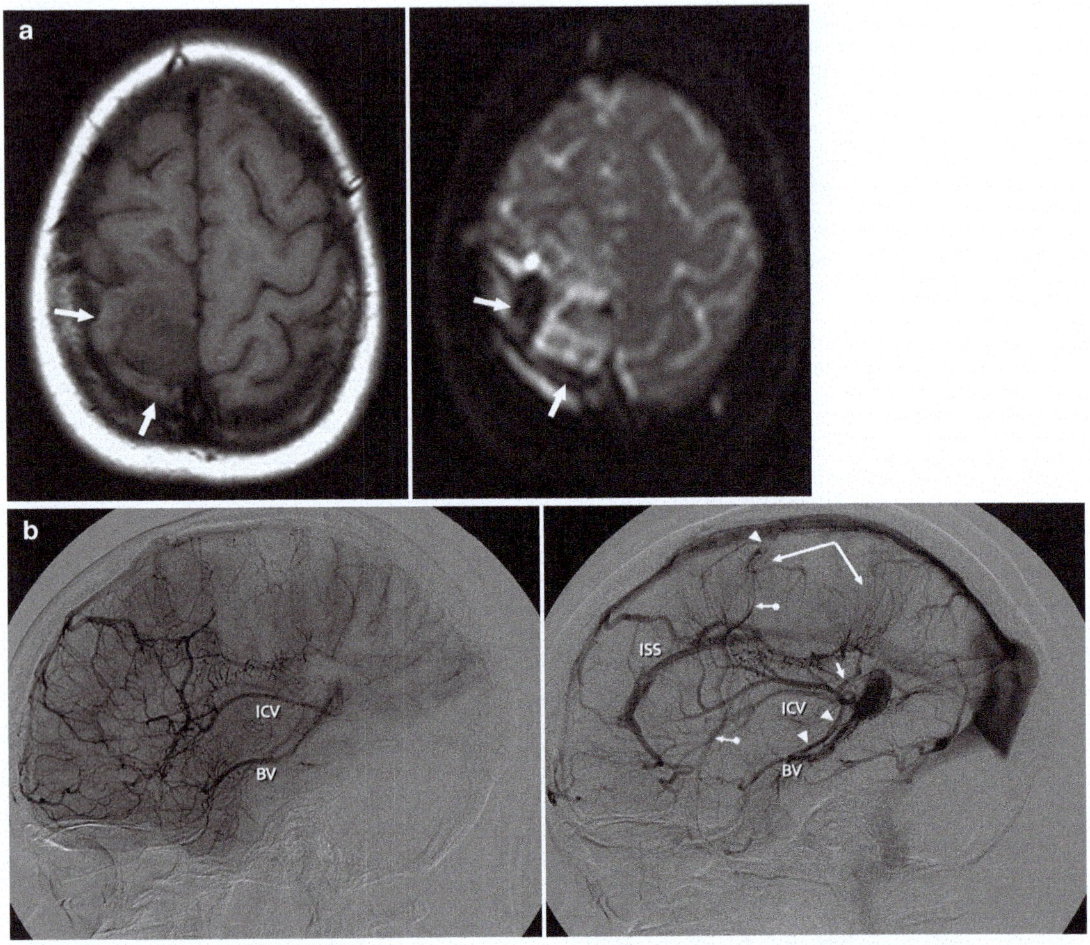

Fig. 20.5 Ischemia due to venous thrombosis occurring in a middle-aged patient with a congenital complex anomaly of the venous system. (**a**) MR performed a few days later. A large parietal vein is occluded by a thrombus appearing slightly hyperintense on the T1 image (*arrow*). The same is partially hypointense (*arrow*) on the T2* gradient-echo image. Ischemia in the surrounding area is present. (**b**) Lateral right ICA angiogram. Early and late phases. In the early phase, the venous drainage in the frontal area is recognizable occurring in the SSS, in the internal cerebral vein (ICV), and basal vein (BV). In the later phase appears the venous drainage in the parietal area, where the superficial veins are poorly filled. There is an anterior parietal vein occluded near the SSS (*arrowhead*) corresponding to that visible on the MR. Its drainage is diverted toward the SMCV (*arrow with dot*). The dilated medullary veins (*arrow with angle*) converge to dilated subependymal veins (*medial atrial*: *arrow*; *inferior ventricular*: *coupled arrowheads*) ending in the ICV and BV, respectively. The distal BV close to the Galen is dilated. The drainage continues in the straight sinus to which converges a dilated ISS. (**c**) AP view of the right ICA angiogram. Early and late phases. Aplasia of the right TS. Occlusion of the parietal vein (*arrowhead*). The dilated medullary veins converge to the medial atrial vein (*arrow*) and the basal vein (BV). (**d**) AP view of the left ICA confirming the aplasia of the right TS, absent also on the VA angiogram. Greatly dilated left TS

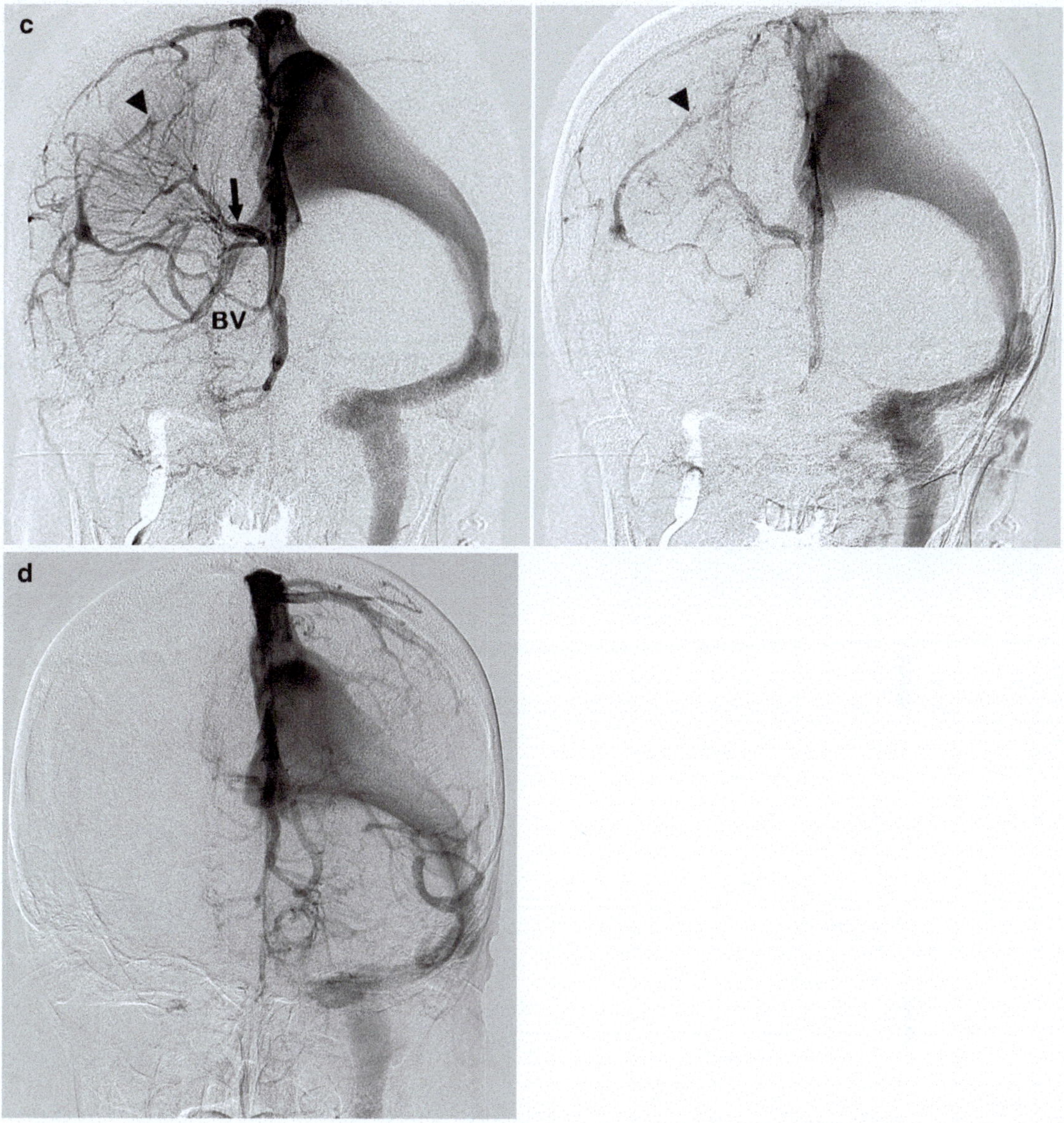

Fig. 20.5 (continued)

21 Association of Venous Sinus and IJV Stenosis and Some Clinical Pathological Conditions

The relationship between Idiopathic Intracranial Hypertension (IIH) and stenosis of the transverse sinus has already been described in the past (Johnston et al. 2002; Farb et al. 2003). Venous sinus stenting, as a possible method of therapy, has also been reported (Higgins et al. 2002; Tsumoto et al. 2003) and proposed again, recently, by Aguilar - Perez et al. 2017. This last author have treated with stenting 51 patients with IIH reporting in improvement or resolution of headache or papilledema in about 80% of the cases. This is certainly an interesting study which needs, however, further confirmation. More recently, the association of bilateral stenosis of transverse sinus and migraine has been emphasized by some authors (Bono et al. 2006; De Simone et al. 2012; Fofi et al. 2012). Other authors (Donnet et al. 2013) have reported a study in patients with a special type of headache provoked by several triggering factors such as cough, physical exertion, and sexual activity, well known as *a primary cough, exertional, and sexual activity headache* (Headache Classification 2004). In many of these patients examined with MR and MR venography, stenosis of TS, and/or IJV was detected. The authors discussed the possibility that headache could be due to a preexistent asymptomatic idiopathic intracranial hypertension which could be increased by cough or physical or sexual activity favored by the preexistent venous abnormalities.

The recognition of these venous abnormalities in the above-described pathological conditions is certainly an interesting aspect. Their causative effect, however, is still controversial and needs further studies for confirmation. In this context, the efficiency of a collateral outflow involving the condylar veins, the anterior epidural venous plexus, and the vertebral artery venous plexus should be examined (see also Sect. 9.3.8).

In another recent study some authors (Malekzadehlashkariani et al. 2016) in patients with idiopathic intracranial hypertension have demonstrated the presence of brain herniations within arachnoid granulations, and have discussed the possibility that these could lead to flow impairment in the sinus responsible of the clinical syndrome. However, it is not completely clear whether the brain herniations are the cause or the result of the intracranial hypertension or a simple incidental discovery. Indeed, it should be considered that brain herniations can be found in patients examined for different clinical reasons. Among them there are cases with secondary intracranial hypertension due to intracranial tumors, but also patients with nonspecific symptoms such headache, vertigo, cognitive disorders, epileptic seizures, and vascular pathologies, which raises doubts, at least in the majority of the cases, about the real clinical relevance of the herniations (see also Sect. 9.3).

We only mention here that similar venous abnormalities including stenosis or brain herniations have been described in patients with multiple sclerosis. Today, however, there is no scientific confirmation that these are responsible in the development of the disease.

G.B. Bradac, *Applied Cerebral Angiography*, DOI 10.1007/978-3-319-57228-4_21

22 Cerebral Hemorrhage

The cerebral hemorrhage can have different patterns involving the cerebral parenchyma, the ventricular system, or the subarachnoid space in varying combinations and at different locations. The presence of bleeding can be diagnosed using CT or MR. More difficult is the diagnosis of its pathogenesis. In many cases, the angiography can clearly show the cause of the hemorrhage, but in other cases it does not. As we have described previously (Chaps. 11, 12, 13, and 16) in cases of bleeding due to aneurysm, AVM, and DAVF, the malformation can easily be diagnosed with angiography. Angiography can also show the venous thrombosis responsible of the hemorrhagic ischemia. Hemorrhages occur in many cases of arteriopathies and arteritis described in Chap. 17. To this group belong also hemorrhages due to drug abuse such as sympathomimetic drugs, heroin, cocaine, LSD, as well as those linked to alcohol abuse. In all these cases the angiography shows nonspecific changes of the vessels characterized by narrowing, dilatations, and sometimes aneurysms involving commonly medium-sized arteries. The angiography, however, fails to demonstrate some vascular malformations such as the cavernous angioma and the telangiectasia. In spite of this the angiogram can be useful, excluding the presence of other vascular malformations such as an AVM. Also in some cases of SAH, the suspected aneurysm cannot be demonstrated. This uncommon condition has been discussed in Sect. 11.11.

Furthermore, there are other pathological conditions in which the cause of bleeding cannot be demonstrated on the angiogram. The most frequent in this group is the hemorrhage occurring in hypertensive commonly old patients reported to be more frequent in the Asian population. It involves typically the putamen and thalamus; more rarely the hemorrhage is lobar or located in the pons or cerebellum. In the latter the hemorrhage is commonly in the hemisphere and more rarely is located in the vermis. Depending on the site, the ventricular system or the subarachnoid space can be secondarily involved. Typically it is a monophase bleeding as opposed to rebleeding in case of aneurysm, AVM, or amyloid angiopathy. The etiopathogenesis has been thought to be the rupture of microaneurysms of deep or superficial perforators. Some authors, however (Challa et al. 1992) failed to demonstrate aneurysms on histological studies, interpreting the bleeding as due to the extensive atherosclerotic changes (lipohyalinosis) in the wall of the arteries, similar to that responsible of lacunar infarct (Sect. 15.4.5). Indeed, in these patients, the association of lacunar infarcts and old microhemorrhage is not rare, indicating probably that this specific pathology of the arteries can be the cause either of ischemia or of hemorrhage (see also Sects. 15.4.5 and 15.5.4). In elderly, hypertensive patients with typical location of the hemorrhage, the angiography is commonly not indicated.

G.B. Bradac, *Applied Cerebral Angiography*, DOI 10.1007/978-3-319-57228-4_22

A second less frequent cause is the cerebral amyloid angiopathy responsible of lobar hemorrhages in the elderly. The hemorrhages are frequently multiple, occurring typically in the posterior regions (parieto-occipital and temporo-occipital) (Massaro et al. 1991). The superficial location explains the frequent extension of the bleeding to the subarachnoid space. A few cases presenting only with a minimal subarachnoid bleeding involving the cerebral convexity have been reported (Apoil et al. 2013). The angiopathy is characterized by the deposition of amyloid material in the wall of the arteries, followed by destruction of the smooth muscle, dilatation of the artery, and formation of microaneurysms. The lobar location indicates an involvement of the superficial medullary arteries. In these patients, but also in some healthy old persons, small foci suggesting the presence of old and new microhemorrhages in the superficial white matter can be demonstrated on MRI studies using T2 gradient-echo or SWI sequences (Roobs et al. 1999; Greenberg et al. 1999, 2004). In these patients angiography is commonly not performed.

The risk of hemorrhage due to anticoagulant therapy usually in old cardiovascular patients is known. In these patients also subdural hematomas following minimal craniocerebral trauma can develop.

Finally, hemorrhage can be the first clinical manifestation of intracranial tumors linked to the rich vascularization and the presence of pathological altered vessels. Commonly this occurs in malignant primary or secondary lesions such as glioblastoma or metastasis of choriocarcinoma, melanoma, thyroid, renal, and lung carcinomas. Sometimes, also benign tumors such as meningiomas can present with bleeding (Bradac et al. 1990).

23 Vascular Pathology Involving the Intraorbital Vessels

The anatomical situation of the ophthalmic artery and ophthalmic veins and their connections with arterial and venous channels of the intracranial and craniofacial areas explain the frequent involvement of the ophthalmic artery and ophthalmic veins in vascular malformations located outside of the orbit.

In this chapter we summarize all these pathological conditions already presented in detail in the corresponding chapters. In addition we describe some specific pure intraorbital vascular malformations.

As far as it concerns the ophthalmic artery and its branches, they can be involved in DAVFs especially of the frontobasal area (Sect. 13.7) and in brain AVM such as in the Wyburn-Mason syndrome (Sect. 12.7.4), and can contribute to the supply of AVM of the craniofacial region (Sect. 3.9.1 and Fig. 3.19). Pure intraorbital AVMs are rare (Smoker et al. 2008). An example is presented in Fig. 23.5. Angiography is essential in the diagnosis of this pathology, and it is also useful in the differential diagnosis of other richly vascularized orbital tumors such as hemangiopericytoma, hemangioblastoma, and meningioma.

The most frequent and typical aneurysm involving the ophthalmic artery is the carotid-ophthalmic arising at the origin of the ophthalmic artery from ICA or close to it (Sect. 11.6.2.3). More distally located aneurysms are very rare. They can originate from the intracranial portion of the artery close to the optic canal (Yanaka et al. 2002) or be more distally intraorbitally located (Meyerson and Lazar 1971; Day 1990; Ogawa et al. 1992; Kikuchi and Kowada 1994; Ernemann et al. 2002; Dehdashi et al. 2002). Cases of fusiform, probably dissecting aneurysms extending through the optic canal, have also been described (Piche et al. 2005; Choi et al. 2008). Visual disturbances and SAH in cases of rupture of the intracranial segment are the typical symptoms.

The ophthalmic veins can be involved as a venous drainage in DAVFs involving directly or indirectly the cavernous and paracavernous sinuses (Sects. 13.7 and 13.8) and in direct carotid-cavernous sinus fistulae (Chap. 14). Craniofacial AVMs can drain in the ophthalmic veins.

A specific intraorbital malformation is the so-called orbital varix. This is characterized by a single dilated venous channel or by a network of small veins all draining commonly, but not always in the SOV (Smoker et al. 2008). It can intermittently fill and collapse leading to an acute temporary exophthalmos. The latter occurs in particular conditions such as lowering the head or following a cough or exertion activities. Hemorrhage and thrombosis can occur (Mafee 2003). The malformation can be associated with other intracranial venous anomalies. CT and/or MR without and with contrast medium in supine and prone positions or associated with Valsalva maneuver or jugular vein compression can clarify the diagnosis (Figs. 23.1 and 23.2). Commonly, the carotid angiogram fails to demonstrate the

G.B. Bradac, *Applied Cerebral Angiography*, DOI 10.1007/978-3-319-57228-4_23

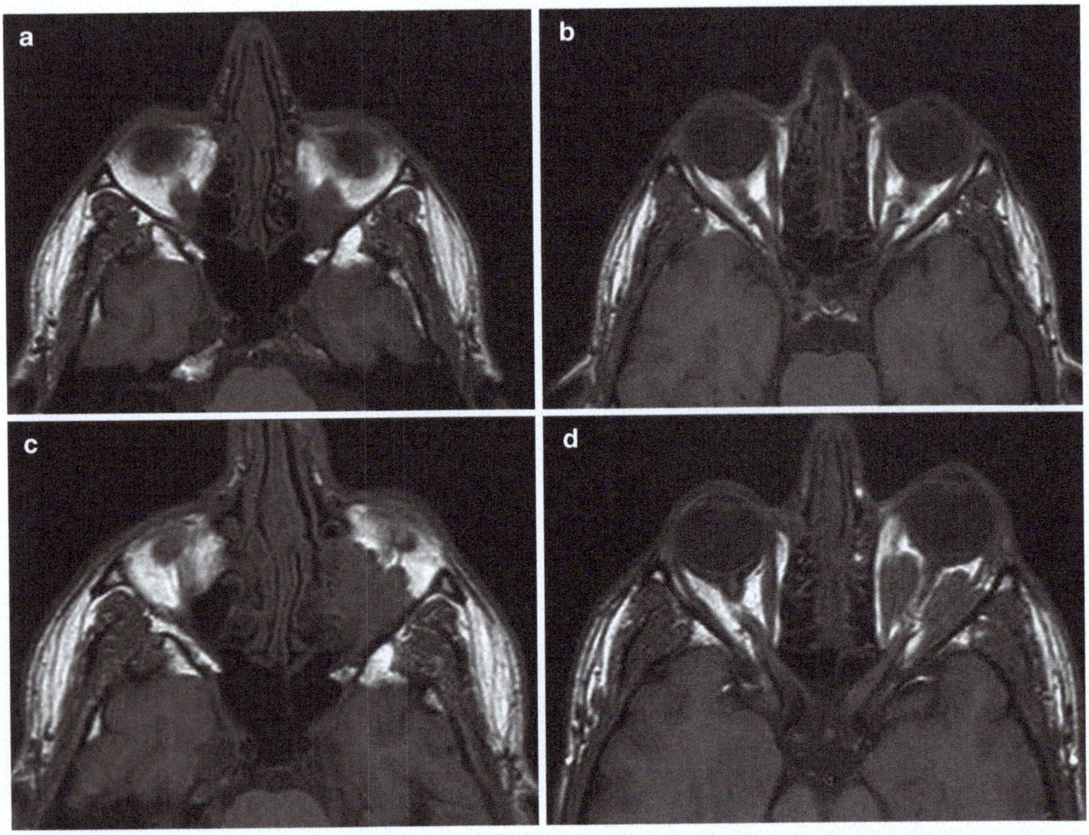

Fig. 23.1 Orbital varix. MR T1-weighted images. (**a**, **b**) Supine position. (**a**) Image near the floor of the orbita. (**b**) Image more cranial, at the level of the optic nerve. (**c**, **d**) Prone position. (**c**) Image near the floor of the orbita. (**d**) Image at the level of the optic nerve. The varix is already partially recognizable in the supine position. It becomes very large in the prone position

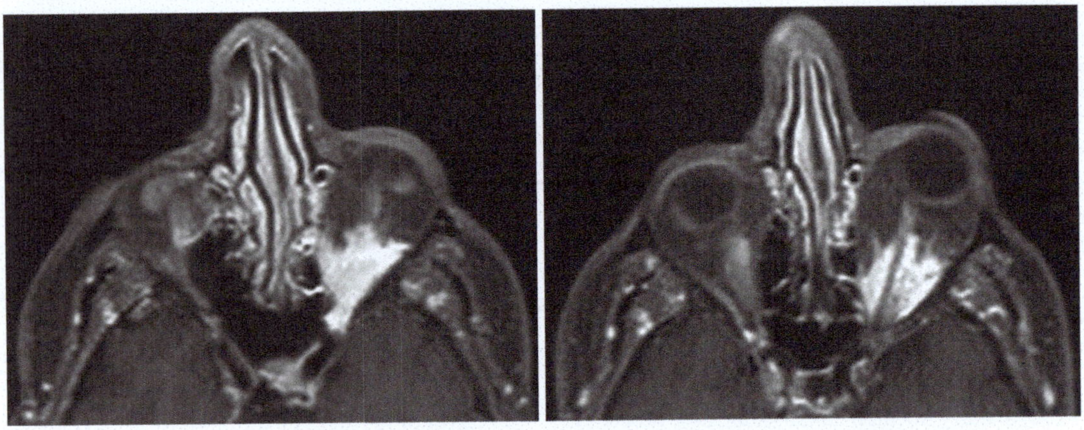

Fig. 23.2 Same patient as in Fig. 23.2. Enhancement of the varix after contrast medium in the prone position

malformation, which is sometimes visible when large connections with the intracranial venous system are present. Furthermore, the angiogram can show the presence of other associated intracranial venous malformations (Figs. 23.3 and 23.4).

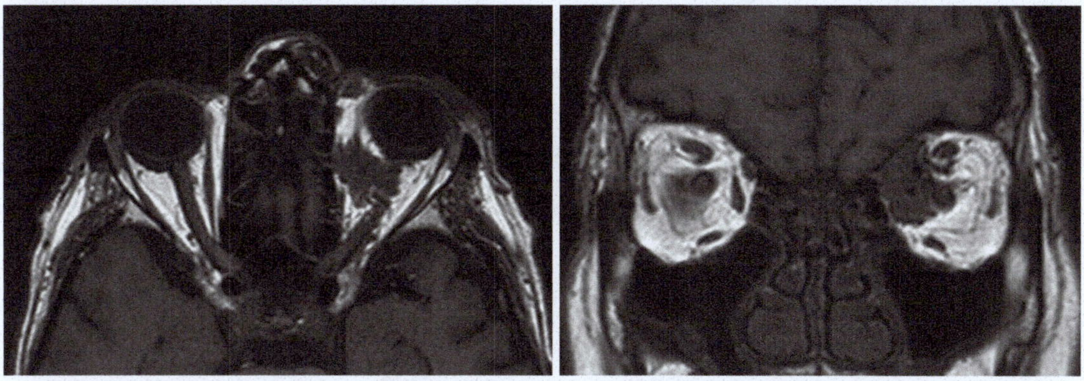

Fig. 23.3 Another case of orbital varix recognizable on MR T1-weighted images in the supine position. Axial and coronal studies. The medially located, predominantly intraconal lesion compresses the optic nerve

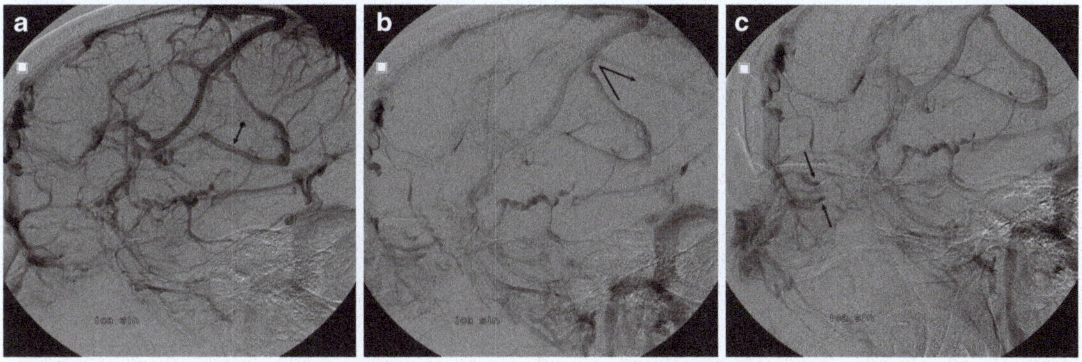

Fig. 23.4 Same patient of Fig. 23.3. Carotid angiogram, early and later phases (**a**, **b**). Complex intracranial venous anomaly. Several veins are dilated and a small varix is present. The straight sinus is absent. The enlarged internal cerebral vein (*arrow with dot*) drains into the Galen vein and further in an anomalous sinus, probably the falcine sinus (*arrow with angle*). Repeated angiogram to better study the orbita (**c**). The intraorbital dilated veins, partially filled, are recognizable (*arrows*). There are connections with the superior sagittal sinus. The varix drains in a large extraorbital pouch continuing in the facial vein (courtesy of Dr. Gozzoli and Dr. Boghi, Neuroradiology, Cuneo)

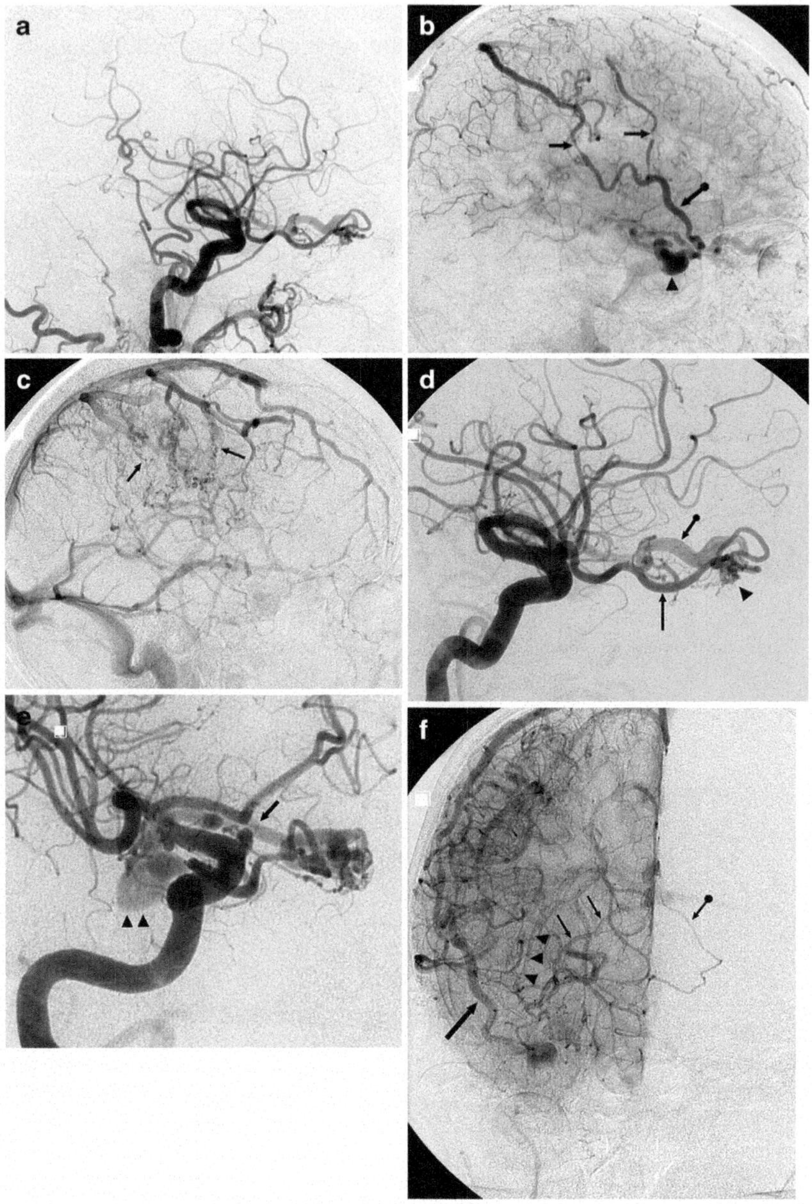

Fig. 23.5 Middle-aged man examined for acute transient left hemiparesis. The angiographic study showed an intra-orbital arteriovenous angioma having an anomalous intracranial venous drainage. (**a**) Carotid angiogram. There is a dilated ophthalmic artery supplying a small AVM located in the anterior part of the orbita, draining in a dilated superior ophthalmic vein. (**b**, **c**) ICA angiogram, late phases. (**b**) The ophthalmic vein having passed the superior orbital fissure, is not connected with the cavernous sinus, instead it continues in a tortuous intracranial venous channel, ending in a temporo-basal venous pouch (*arrowhead*). This continues in the superficial middle cerebral vein (SMCV, *arrow with dot*) which drains further in parietal cortical veins ending in the SSS. Where the distal segment of the SMCV is connected with the parietal veins a narrowing is present (*arrows*). Partial thrombosis? (**c**) Appearing of a venous network indicating the venous congestion in the parietal area (*arrows*). (**d**) Lateral carotid angiogram. Magnification. Ophthalmic artery (*arrow*), small Nidus (*arrowhead*). Ophthalmic vein (*arrow with dot*). (**e**) Oblique view showing the dilated ophthalmic vein (*arrow*), reaching the temporo-basal venous pouch (*arrowhead*). (**f**) AP view, later phase, showing that in addition to the superficial middle cerebral vein (*arrow*), there is involvement of the deep middle cerebral veins (*arrowheads*) which are connected with the right basal vein (*arrows*). There is also a minimal injection of the left basal vein (*arrow with dot*). The patient recovered completely and 2 months later the AVM was occluded with glue after selective endovascular catheterization of the ophthalmic artery

Bibliography

Abbie AA (1933) The clinical significance of the anterior choroidal artery. Brain 56:233

Abbie AA (1934) The morphology of the forebrain arteries with especial reference to the evolution of the basal ganglia. J Anat 68:28

Abdulrauf SI, Malik GM, Awad IP (1999) Spontaneous angiographic obliteration of cerebral arteriovenous malformations. Neurosurgery 44:280

Abe M, Tabuchi K, Yokoyama H et al (1998) Blood blister like aneurysms of the internal carotid artery. J Neurosurg 89:419

Abiko M, Ikawa F, Ohbayashi N et al (2008) Endovascular treatment for dural arteriovenous fistula of the anterior condylar confluence involving the anterior condylar vein. A report of two cases. Interv Neuroradiol 14:313

Abruzzo T, Moran CH, Blackham K et al (2012) Invasive interventional management of post-hemorrhagic cerebral vasospasm in patients with aneurismal subarachnoid hemorrhage. J Neurointerv Surg 4:169

Abud DG, Spelle L, Piotin M et al (2005) Venous phase timing during balloon test occlusion as a criterion for permanent internal carotid artery sacrifice. AJNR Am J Neuroradiol 26:2602

Achrol AS, Kim H, Pawlikowska L et al (2007) Association of tumor necrosis factor-alpha-238>A and apolipopretein E2 polymorphisms with intracranial hemorrhage after brain arteriovenous malformation treatment. Neurosurgery 61:731

Achrol AS, Pawlikowska L, Mc Culloch CE et al (2006) Tumor necrosis factor-alpha-238 G>A promoter polymorphism is associated with increased risk of new haemorrhage in the natural course of patients with brain arteriovenous malformations. Stroke 37:231

Adachi B (1928) Das Arteriensystem der Japaner. Kaiserlich – Japanischen Universitaet zu Kyoto, Kyoto

Agid R, Willinsky RA, How C et al (2004) Targeted compartmental embolization of cavernous sinus dural arteriovenous fistulae using transfemoral medial and lateral facial vein approaches. Neuroradiology 46:156

Agid R, TerBrugge KG, Rodesch G et al (2009) Management strategies for anterior cranial fossa (ethmoidal) dural arteriovenous fistulae with emphasis on endovascular treatment. J Neurosurg 110:79

Aguillar-Pérez M, Martinez-Moreno R, Kurre W et al (2017) Endovascular treatment of idiopathic intracranial hypertension: retrospective analysis of immediate and long-term results in 51 patients. Neuroradiology. 59:277

Ahmad I, Tominaga T, Suzuki M et al (1994) Primitive trigeminal artery associated with cavernous aneurysm. Case report. Surg Neurol 41:75

Akar ZC, Dujovny M, Gomez-Tortosa E et al (1995) Microvascular anatomy of the anterior surface of the medulla oblongata and olive. J Neurosurg 82:97

Akers DL, Fowl RJ, Plettner J et al (1991) Complications of anomalous origin of the right subclavian artery: case report and review of literature. Ann Vasc Surg 5:385

Akiyama T, Onozuka S, Horiguchi T et al (2012) Recanalisation of spontaneously occluded vertebral artery dissection after subarachnoid haemorrhage. Neurosurg Rev 35:615

Al Abrawi S, Fouillet-Desjouqueres M, David L et al (2008) Takayasu arteritis in children. Pediatr Rheumatol 6:17

Albayram S, Gailloud P, Wasserman BA (2002) Bilateral arch origin of the vertebral arteries. AJNR Am J Neuroradiol 23:455

Albayram S, Kara B, Ipek H et al (2009) Isolated cortical venous thrombosis associated with intracranial hypotension syndrome. Headache 49:916

Alexander CB, Burger PC, Goree JA (1979) Dissecting aneurysms of the basilar artery in two patients. Stroke 10:294

Alexander GE, DeLong MR, Strick PL (1986) Parallel organization of functionally segregated circuits linking basal ganglia and cortex. Ann Rev Neurosci 9:357

Alexander MW, Chui MC, Gray T (1984) Transsellar intercavernous carotid anastomosis with agenesis of the internal carotid artery and right-sided descending aorta: case report. J Can Assoc Radiol 35:92

Alkadhi H, Kollias SS, Crelier GR et al (2000) Plasticity of the human motor cortex in patient with arterio—venous malformations: a functional MR imaging study. AJNR Am J Neuroradiol 21:1423

Allison JW, Davis PC, Sato Y et al (1998) Intracranial aneurysms in infants and children. Pediatr Radiol 28:223

Alsuwaidan SA (2012) PHACE syndrome in association with airway hemangioma: first report from Saudi Arabia and literature review. Ann Thorac Med 7:44

G.B. Bradac, *Applied Cerebral Angiography*, DOI 10.1007/978-3-319-57228-4

Altaf N, Daniels L, Morgan PS et al (2006) Cerebral white matter hyperintense lesions are associated with unstable carotid plaque. Eur J Vasc Endovasc Surg 31:8

Alves JV, Andersson T, Edner G et al (2005) Subarachnoid haemorrhage from a large cerebral aneurysm visible only on repeated angiography. Interv Neuroradiol 11:59

Amagasaki K, Yagishita T, Kawataki T et al (1998) Middle cerebral artery aplasia associated with an aneurysm of the proximal anterior cerebral artery. Acta Neurochir 140:1313

Amarenco P, Hauw JJ, Gautier JC (1990) Arterial pathology in cerebellar infarction. Stroke 21:1299

Amarenco P, Duyckaerts C, Tzourio C et al (1992) The prevalence of ulcerated plaques in the aortic arch in patients with stroke. N Engl J Med 326:221

Amarenco P, Caplan L (1993) Vertebrobasilar occlusive disease: review of selected aspects. 3. Mechanisms of cerebellar infarctions. Cerebrovasc Dis 3:66

Amarenco P, Cohen A, Tzourio C et al (1994a) Atherosclerotic disease of the aortic arch and the risk of ischemic stroke. N Engl J Med 331:1474

Amarenco P, Levy C, Cohen A et al (1994b) Causes and mechanisms of territorial and non territorial cerebellar infarcts in 115 consecutive cases. Stroke 25:105

Amarenco P, Caplan L, Pessin MS (1998) Vertebrobasilar occlusive disease. In: Barnett HJM, Mohr JP, Stein BM, Yatsu FM (eds) Stroke. Churchill Livingstone, New York, pp 551–570

Amat EF, Armengol JP, Solivellas IM et al (2015) Gingival bleeding of high-flow mandibular arteriovenous malformation in a child with 8-year follow-up. Case Rep Pediatr 2015:745718

Ameri A, Bousser MG (1992) Cerebral venous thrombosis. Neurol Clin 10:87

Ames A III, Wright RL, Kowada M et al (1968) Cerebral ischemia. II The no-reflow phenomenon. Am J Pathol 52:437

Amlie-Lefond C, Sebire G, Fullerton HJ (2008) Recent developments in childhood arterial ischemic stroke. Lancet Neurol 7:425

Amlie-Lefond C, Bernard TJ, Sebire G et al (2009) International pediatric study subgroup. Predictors of cerebral arteriopathy in children with arterial ischemic stroke: results of the international pediatric stroke study. Circulation 119:1417

Anderegen L, Arnold M, Andres RH et al (2012) Bow hunter's stroke due to prominent degenerative spinal disorder. Clin Neuroradiol 22:355

Andersen PE (1970) Fibromuscular hyperplasia of the carotid arteries. Acta Radiol 10:90

Anderson CA, Collins GJ Jr, Rich NM (1980) Spontaneous dissection of the internal carotid artery associated with fibromuscular dysplasia. Am Surg 4:263

Andersson T, Kuntze-Söderquist A, Söderman M et al (2013) Mechanical thrombectomy as the primary treatment for acute basilar artery occlusion: experience from 5 years of practice. J Neurointerv Surg 5:221

Andoh K, Tanohata K, Moriya N et al (2001) The posterior inferior cerebellar artery arising from the extracranial segment of the internal carotid artery via the hypoglossal canal without an interposed segment of the basilar artery a persistent primitive hypoglossal artery variant. Clin Imaging 25:86

Angeloni V, Bozzao L, Fantozzi LM et al (1990) Internal border-zone infarction following acute middle cerebral artery occlusion. Neurology 40:1196

Antunes JL, Correll JW (1976) Cerebral emboli from intracranial aneurysms. Surg Neurol 6:7

Anxionnat R, de Melo F, Neto J et al (2003) Treatment of hemorrhagic intracranial dissections. Neurosurgery 53:289

Anzalone N, Scomazzoni F, Castellano R et al (2005) Carotid artery stenosis: intraindividual correlationof 3D time-of-flight MRA, contrast-enhanced MRA, conventional DSA, and rotational angiography for detection and grading. Radiology 236:204

Apley J (1949) Congenital anomalies of the aortic arch and its branches. Proc R Soc Med 42:918

Apoil M, Cogez j DL et al (2013) Focal cortical subarachnoid hemorrhage revealed by recurrent paresthesias: a clinic-radiological syndrome strongly associated with cerebral amyloid angiopathy. Cerebrovasc Dis 36:138

Arat A, Jslak C, Saatci I (2002) Endovascular parent artery occlusion in large-giant or fusiform distal posterior cerebral artery aneurysms. Neuroradiology 44:700

Arbeille P, Bounin-Pineau MH, Philippot M et al (1996) Suivi des paramètres morphologiques des plaques d'atherome sur 24 mois. J Echograph Med Ultrason 17:337

Arboix A, Marti-Vilalta JL, Garcia JH (1990) Clinical study of 227 patients with lacunar infarcts. Stroke 21:842

Arboix A, Garcia-Eroles L, Masson J et al (2000) Lacunar infarcts in patients aged 85 years and older. Acta Neurol Scand 101:95

Archie JP, Feldtman RW (1981) Critical stenosis of the internal carotid artery. Surgery 89:67

Arey LB (1965) Developmental anatomy. A Textbook and laboratory manual of embryololgy, 7th edn. Saunders, Philadelphia

Arnautovic KI, Al Mefty O, Glenn Pait T et al (1997) The suboccipital cavernous sinus. J Neurosurg 86:252

Arnold M, Schroth G, Nedeltchev K et al (2002) Intraarterial thrombolysis in 100 patients with acute stroke due to middle cerebral artery occlusion. Stroke 33:1828

Arnold M, Nedeltchev K, Mattle HP et al (2003) Intraarterial thrombolysis in 24 consecutive patients with internal carotid artery T occlusions. J Neurol Neurosurg Psychiatry 74:739

Arnold M, Nedeltchev K, Schroth G et al (2004) Clinical and radiological predictors of recanalisation and outcome of 40 patients with acute basilar artery occlusion treated with intraarterial thrombolysis. J Neurol Neurosurg Psychiatry 75:857

Aryan HE, Giannotta S, Fukushima T et al (2006) Aneurysms in children: review of 15 years' experience. J Clin Neurosci 13:188

Asadi H, Morokoff A, Gaillard F (2015) Occult temporal lobe encephalocele in the transverse sinus. J Clin Neurosci 22:1202

Asai K, Hasno K, Hara T et al (2010) Traumatic persistent trigeminal artery cavernous sinus fistula treated by trans-catheter arterial embolization. A case report. Int Neuroradiol 16:93

Atlas SW, Grossman RI et al (1987) Partially thrombosed giant intracranial aneurysms. Correlation of Mr and pathologic findings. Radiology 162:111

Awad IA (1993) Dural arteriovenous malformations with aggressive clinical course. In: Awad IA, Barrow DL (eds) Dural arteriovenous malformations. AANS, Park Ridge

Awad IA, Little JR, Akravi WP et al (1990) Intracranial dural arteriovenous malformation: factors predisposing to an aggressive neurological course. J Neurosurg 72:839

Aydin F (1998) Do human intracranial arteries lack vasa vasorum? A comparative immunohistochemical study of intracranial and systemic arteries. Acta Neuropathol 96:22

Babin E (1971) Contribution a l'etude radioanatomique de la veine basilaire normale et des ses affluents. Diplome d' Electroradiologie, Strasbourg

Bahar F, Chiras JP, Carpena JP et al (1984) Spontaneous vertebral arteriovenous fistula associated with fibromuscular dysplasia. Neuroradiology 26:45

Baharvahdat H, Blanc R, Termechi R et al (2014) Hemorrhagic complications after endovscular treatment of cerebral arteriovenous malformations. AJNR Am J Neuroradiol 35:978

Bamford J, Sandercock P, Jones L et al (1987) The natural history of lacunar infarction. The Oxfordshire Community Stroke Project. Stroke 18:5

Banach MJ, Flamm ES (1993) Supraclinoid internal carotid artery fenestration with associated aneurysm. J Neurosurg 79:438

Bandeira A, Ribeiro C, Reis J (2007) Treatment of vasospasm secondary to subarachnoid haemorrhage using intraarterial nimodipine in low dosage. Interv Neuroradiol 13:403

Bang OY, Saver JL, Kim SS et al (2011) Collateral flow predicts response to endovascular therapy for acute ischemic stroke. Stroke 42:693

Baptista AG (1963) Studies of the arteries of the brain. II. The anterior cerebral artery: some anatomic features and their clinical implications. Neurology 23:825

Baradaran H, Mtui EE, Richardson JE et al (2017) Hemispheric differences in Leukoaraiosis in patients with carotid artery stenosis: a systemic review. Clin Neuroradiol 27:7

Barker CS, Anslow PL (1990) Case report: spontaneous thrombosis of a cerebral arteriovenous malformation. Clin Radiol 42:130

Barnes C, Newall F, Furmedge J et al (2004) Arterial ischemic stroke in children. J Paediatr Child Health 40:384

Barnes C, Deveber G (2006) Prothrombotic abnormalities in childhood ischemic stroke. Thromb Res 118:67

Barnett HJM (1974) Transient cerebral ischemia: pathogenesis, prognosis and management. Ann R Coll Physicians Surg Can 7:153

Barnett HJM, Perless SJ, Kaufmann JC et al (1978) "Stump" of internal carotid artery—a source for further cerebral embolic ischemia. Stroke 9:448

Barnett HJM, Boughner DR, Taylor WD et al (1980) Further evidence relating mitral-valve prolapse to cerebral ischemic events. N Engl J Med 302:149

Barnwell SL, Halbach VV, Dowd CF et al (1990) Dural arteriovenous fistulas the inferior petrosal sinus: angiographic finding in 6 patients. AJNR Am J Neuroradiol 11:511

Barnwell SL, Halbach VV, Dowd CF et al (1991) A variant of arteriovenous fistulas within the wall of dural sinuses. J Neurosurg 74:199

Barrow DL, Spector RH, Braun IF et al (1985) Classification and treatment of spontaneous carotid-cavernous sinus fistulas. J Neurosurg 62:248

Barry A (1951) The aortic arch derivatives in the human adult. Anat Res 111:221

Bassetti C, Bougouslavsky J, Eskenasy-Cottier AL et al (1994) Spontaneous intracranial dissection in the anterior circulation. Cerebrovasc Dis 4:170

Batjer H, Samson D (1986) Arteriovenous malformations of the brain: natural history study. J Neurosurg 64:849

Battal B, Castillo M (2014) Brain herniations into the dural venous sinuses or calvarium. MRI of a recently recognized entity. J Neuroradiol 27:55

Battal B, Hamcan S, Akgun V et al (2015) Brain herniations into the dural venous sinus or calvarium: MRI findings, possible causes and clinical significance. Eur Radiol. doi:10.1007/s00330-015-3959-x

Baudrimont M, Dubas F, Joutel A et al (1993) Autosomal dominant leukoencephalopathy and subcortical ischemic stroke. A clinicipathological study. Stroke 24:122

Bayford D (1789) An account of a singular case of obstructed deglutition. Mem Med Soc London 2:271

Beach KW, Hatsukami T, Detmer PR et al (1993) Carotid artery intraplaque haemorrhage and stenotic velocity. Stroke 23:314

Beattie DK, Hellier WP, Powell MP (1995) Stroke-induced cardiovascular changes: a rare cause of death from polyarteritis nodosa. Br J Neurosurg 9:223

Behme D, Gondecki L, Fiethen S et al (2014) Complications of mechanical thrombectomy for acute ischemic stroke. A retrospective single-center study of 176 consecutive cases. Neuroradiology 56:467

Behme D, Mpotsaris P, Zeyen P et al (2015) Emergency stenting of the extracranial in ternal carotid artery in combination with anterior circulation thrombectomy in acute ischemic stroke: a retrospective multicenter study. AJNR Am J Neuroradiol 36:2340

Beigelman C, Mowrey-Gerosa I, Gamsu G et al (1995) New morphologic approach to the classification of anomalies of the aortic arch. Eur Radiol 5:435

Béjot Y, Aboa-Eboulé C, Debette S et al (2014) Characteristics and outcomes of patients with multiple cervical artery dissection. Stroke 45:37

Bekov DB (1965) Some structural differences among posterolateral tributaries of the great cerebral vein. Fed Proc Transl Suppl 24:166

Beltramello A, Zampieri P, Ricciardi GK et al (2005) Combined treatment of brain AVMs: analysis of 5 years (2000–2004) in the Verona experience. Interv Neuroradiol 11(Suppl 1):63

Ben Amor M, Maion CH, Heldt N (1971) Normal and pathological radioanatomy of the superior choroid vein. Neuroradiology 3:16

Ben-Amor M, Billewicz O (1970) The posterior cerebral vein. Neuroradiology 1:179

Benndorf G, Lehman TN, Molsen HP et al (1999) Puncture of the superficial sylvian vein for embolization of cavernous dural arteriovenous fistula. Interv Neuroradiol 5:167

Benndorf G, Bender A, Campi A et al (2001a) Treatment of cavernous sinus dural arteriovenous fistula by deep puncture of the superior ophthalmic vein. Neuroradiology 43:499

Benndorf G, Campi A, Hell B et al (2001b) Endovascular management of a bleeding mandibular arteriovenous malformation by transfemoral venous embolization with NBCA. AJNR Am J Neuroradiol 22:359

Benndorf G, Campi A (2002) Aberrant inferior petrosal sinus: unusual transvernous approach to the cavernous sinus. Neuroradiology 44:158

Benndorf G (2009) Dural cavernous sinus fistulas. Springer, Berlin

Benoit B, Wortzman G (1973) Traumatic cerebral aneurysms. Clinical features and natural history. J Neurol Neurosurg Psychiatry 36:127

Benyounes N, Blanc R, Boissonnet H et al (2011) Subarachnoid hemorrhage revealing aortic coarctation in a young man. Neuroradiology 53:931

Berenstein A, Kricheff II (1979) Balloon catheter of the investigating of carotid cavernous fistulas. Radiology 132:762

Berenstein A, Lasjaunias P (1992) Endovascular treatment of cerebral lesions, vol 4. Springer, Berlin

Berenstein A, Ortiz R, Niimi Y et al (2010) Endovascular management of arteriovenous malformations and other intracranial arteriovenous shunts in neonates, infants and children. Childs Nerv Syst 26:1345

Berge J, Biondi A, Machi P et al (2012) Flow-diverter silk stent for the treatment of intracranial aneurysms: 1-year follow-up in a multicentric study. AJNR Am J Neuroradiol 33:1150

Berger MS, Wilson CB (1984) Intracranial dissecting aneurysms of the posterior circulation. Report of six cases and review of the literature. J Neurosurg 61:882

Bergeron M, Cortes M, Dolen Y et al (2013) Extensive bilateral arteriovenous malformation of the mandibule successfully controlled by combined endovascular and transcutaneous angio-embolization: case report and review of the literature. Int J Pediatr Otorhinolaryngol 77:130

Bergquist E (1975) Anomalous posterior cerebral artery. Neuroradiology 8:213

Bergui M, Bradac GB (1997) Uncommon symptomatic cerebrovascular malformations. AJNR Am J Neuroradiol 18:779

Bergui M, Bradac GB, Daniele D (1999) Brain lesions due to cerebral venous thrombosis do not correlate with sinus involvement. Neuroradiology 41:419

Bergui M, Stura G, Daniele D et al (2006) Mechanical thrombolysis in ischemic stroke attributable to basilar artery occlusion as first-line treatment. Stroke 37:145

Berkefeld J, Du Mesnil de Rochemont R, Rohde S et al (2003) Intracranial stenting of high grade atherosclerotic stenoses. Follow up results in elective cases. Riv Neuroradiol 16:1330

Berkefeld J, Zanella FE (2009) Intracranial stenting of atherosclerotic stenosis, current status and perspectives. Clin Neuroradiol 19:38

Berkhemer OA, Fransen PSS, Beumer D et al (2015) A randomized trial of intra-arterial treatment for acute ischemic stroke. N Engl J Med 372:11

Bernasconi V, Cassinari V (1957) Caratteristiche angiografiche dei meningiomi del tentorio. Radiol Med 43:1015

Bernini FP, Cioffi FA, Muras I et al (1982) Carotid syphon pericallosal arterial anastomosis. Case report. Acta Neurochir 66:61

Bhattacharaya JJ, Luo CB, Alvarez H et al (2004) PHACES syndrome: a review of eight previously un reported cases with late arterial occlusions. Neuroradiology 46:227

Bhattacharaya JJ, Thammaroj J (2003) Vein of Galen malformations. J Neurol Neurosurg Psychiatry 74:42

Bhogal P, Brouwer PA, Kuntze-Söderquist A et al (2015) Patients with subarachnoid haemorrhage from vertebrobasilar dissection: treatment with stent—in stent technique. Neuroradiology 57:605

Bhogal P, Aguilar-Pérez M, Ganslandt O et al (2016) Treatment of posterioir circulation non-saccular aneurysm with flolw-diverter. A single-center experience and review of 56 patients. J Neurointerv Surg 9(5):471–481

Biffl WL, Moore EE, Elliott J et al (2000) The devastating potential of blunt vertebral artery injuries. Ann Surg 231:672

Bing F, Darsaut TE, Salakin I et al (2013) Stents and flow diverters in the treatment of aneurysms: device deformation in vivo may alter porosity and impact efficacy. Neuroradiology 55:85

Binswanger OL (1894) Die Abngrenzung der allgemeinen progressive paralyse. Berl Klin Wochenschr 31. 1102, 1137, 1180

Biondi A, Ricciardi GK, Puybasset L et al (2004) Intraarterial nimodipine for the treatment of symptomatic cerebral vasospasm after aneurysmal sub-

arachnoid haemorrhage: preliminary results. AJNR Am J Neuroradiol 25:1067
Birley J, Trotter W (1928) Traumatic aneurysm of the intracranial portion of the internal carotid artery. Brain 51:184
Birnbaum J, Hellmann DB (2009) Primary angiitis of the central nervous system. Arch Neurol 66:704
Bisaria KK (1984) Anomalies of the posterior communicating artery and their potential clinical significance. J Neurosurg 60:572
Blackburn IW (1907) Anomalies of the encephalic arteries among the insane. J Comp Neurol Psychiat 17:493
Bladin GF, Chambers BR (1993) Clinical features, pathogenesis and computed tomographic characteristic of internal watershed infarction. Stroke 24:1925
Bloch S, Danziger J (1974) Proatlantal segmental artery. Neuroradiology 7:5
Bo WJ, McKinney WM, Bowden RL (1989) The origin and distribution of vasa vasorum at the bifurcation of the common carotid artery with atherosclerosis. Stroke 1484:1989
Boccardi E, Branca V, Valvassori L et al (1998) Endovascular treatment with GDCs: results in 100 patients. J Neurosurg Sci 42(Suppl 1):127–129
Boccardi E, Della Sala S, Motto C et al (2002) Utilisation behaviour consequent to bilateral SMA softening. Cortex 38:289
Boccardi E, Valvassori L, Piano MG (2013) Galen vein malformation. Endovascular treatment. Technique and results. Report about 21 cases. Personal communication. Ospedale Niguarda
Boesen ME, Eswaradass PV, Singh D (2017) MR Imaging of carotid webs. Neuroradiology 59:361
Bogousslavsky J, Deruaz JP, Regli F (1985) Bilateral obstruction of internal carotid artery from giant-cell arteritis and massive infarction limited to the vertebrabasilar area. Eur Neurol 24:57
Bogousslavsky J, Regli F (1986) Borderzone infarctions distal to internal carotid artery occlusion: prognostic implications. Ann Neurol 20:346
Bogousslavsky J, Despland PA, Regli F (1987) Spontaneous carotid dissection with acute stroke. Arch Neurol 44:137
Bogousslavsky J, Regli F, Huske A (1988a) Thalamic infarcts: clinical syndromes etiology and prognosis. Neurology 38:837
Bogousslavsky J, van Melle G, Regli F (1988b) The lacunar stroke registry: analysis of 1000 consecutive patients with first stroke. Stroke 19:1083
Bogousslavsky J, Regli F, Maeder PH (1991) Intracranial large artery disease and lacunar infarction. Cerebrovasc Dis 1:154
Bogousslavsky J, Regli F (1992) Centrum ovale infarcts: subcortical infarction in the superficial territory of the middle cerebral artery. Neurology 42:1992
Boiten J, Lodder J (1991) Lacunar infarcts. Pathogenesis and validity of the clinical syndromes. Stroke 22:1374
Boiten J, Rotrhwell PM, Slattery J et al (1997) Frequency and degree of carotid stenosis in small centrum Ovale infarcts as compared to lacunar infarct. Cerebrovasc Dis 7:138
Bojanowski MW, Laroche M, Leveque M et al (2009) Surgical treatment of blister-like aneurysms. Interv Neuroradiol 15(Suppl 1):126
Boker DK, Solymosi L, Wassmann H (1985) Immediate post angiographic intraarterial treatment of cerebral vasospasm after subarachnoid haemorrhage with nimodipine, report of three cases. Neurochirurgia (Stuttg) 28:118
Bonaldi G (2002) Angioplasty and stenting of the cervical carotid bifurcation: report of a 4-year series. Neuroradiology 44:164
Bono F et al (2006) Bilateral transverse sinus stenosis predict IIH without papilledema in patients with migraine. Neurology 8:67
Borden JA, Wu JK, Schucart WA (1995) A proposed classification for spinal and cranial dural arteriovenous fistulous malformations and implications for treatment. J Neurosurg 82:166
Bose A, Hartmann M, Henkes H et al (2007) A novel self expanding nitinol stent in medically refractory intracranial atherosclerotic stenosis: the Wingspan study. Stroke 38:1531
Bosniak MA (1964) An analysis of some anatomicroentgenol. Aspect of the brachiocephalic vessels. Am J Roentgenol 91:1222
Bostrom K, Liliequist B (1967) Primary dissecting aneurysm of the extracranial part of the internal carotid and vertebral artery. Neurology 17:179
Boukobza M, Crassard I, Bousser MG et al (2009) MR images features of isolated cortical vein thrombosis: diagnosis and follow up. AJNR Am J Neuroradiol 30:344
Boukobza M, Crassard I, Bousser MG et al (2016) Radiological findings in cerebral venous thrombosis presenting as subarachnoid hemorrhage. A series of 29 cases. Neuroradiology 58:11
Bousser MG, Chiras J, Sauron B et al (1985) Cerebral venous thrombosis. A review of 38 cases. Stroke 16:199
Bousser MG, Ferro JM (2007) Cerebral venous thrombosis: an update. Lancet Neurol 6:162
Bousser MG, Russell RR (1997) Cerebral venous thrombosis. Saunders, London
Bouthillier A, van Loveren HR, Keller JT (1996) Segments of the internal carotid artery: a new classification. Neurosurgery 38:425
Bozzetto-Ambrosi P, Andrade G, Azevedo-Filho H (2006) Traumatic pseudoaneurysm of the middle meningeal artery and cerebral intraparenchymal hematoma. Case report. Surg Neurol 66(Suppl 3):29
Bracard S, Arrue P, Banal FC et al (1999) Management of vasospasm from subarachnoid haemorrhage: attitude of French Center-French Society of Neuroradiology. J Neuroradiol 26(Suppl 1):44
Bradac GB, Wackenheim A, Braun JP (1967) Contribution a l'etude des phlebogramme de l'angiographie vertebrale. Neurochirurgia (Stuttg) 12:1
Bradac GB (1970) The pontomesencephalic veins (radioanatomical study). Neuroradiology 1:52

Bradac GB, Heymat F (1970) Considerations concerning a case of fibromuscular hyperplasia of the carotid arteries. Neuroradiology 1:217

Bradac GB, Holdorff B, Simon RS (1971) Aspects of the venous drainage of the pons and mesencephalon. Neuroradiology 3:102

Bradac GB, Simon RS, Leonhardt W (1974) The ophthalmic vein in the carotid angiogram. Neuroradiology 8:39

Bradac GB, Simon RS, Fiegler W (1975) The early filling of a vein in the carotid angiogram. Neuroradiology 9:13

Bradac GB, Kaernbach A, Bolk-Weiscedel D et al (1981a) Spontaneous dissecting aneurysm of cervical cerebral arteries. Report of six cases and review of the literature. Neuroradiology 21:149

Bradac GB, Schramm J, Kaernbach A (1981b) Traumatic exophthalmos due to an uncommon arteriovenous fistula without involvement of the cavernous sinus. J Neurosurg 55:993

Bradac GB, Oberson R (1983) Angiography and computed tomography in cerebro-arterial occlusive diseases. Springer, Berlin

Bradac GB, Bender A, Curio G et al (1985) Report of two cases of spontaneous direct carotido-cavernous fistula. Diagnostic and therapeutic considerations. Neuroradiology 27:436

Bradac GB, Riva A, Schörner W et al (1987) Cavernous sinus meningiomas: an MRI study. Neuroradiology 29:578

Bradac GB, Riva A, Stura G et al (1989) Dissection spontanée de la carotide interne aves paralyse du XII. J Neuroradiol (Masson) 16:197

Bradac GB, Ferszt R, Kendall BE (1990) Cranial meningiomas diagnosis, biology, therapy. Springer, Berlin

Bradac GB, Bergui M, Ferrio MF et al (1997) False-negative angiograms in subarachnoid haemorrhage due to intracranial aneurysms. Neuroradiology 39:772

Bradac GB, Bergui M, Ferrio MF et al (2000) Cranial nerve palsy in spontaneous dissection of the internal carotid artery. Klin Neuroradiol 10:1

Bradac GB, Bergui M (2001) Role of the different venous structures involved in cerebral venous thrombosis. Klin Neuroradiol 11:20

Bradac GB, Bergui M, Stura G (2001) Endovascular treatment of brain arteriovenous malformations. Riv Neuroradiol 14:373

Bradac GB, Bergui M, Genovese E (2002) Considerations about ischemia in intracranial dural arteriovenous fistula. Klin Neuroradiol 12:40

Bradac GB, Bergui M (2004) Endovascular treatment of the posterior inferior cerebellar artery aneurysms. Neuroradiology 46:1006

Bradac GB, Bergui M, Fontanella M (2005) Endovascular treatment of cerebral aneurysms in elderly patients. Neuroradiology 47:938

Bradac GB, Bergui M, Stura G et al (2007) Periprocedural morbidity and mortality by endovascular treatment of cerebral aneurysms with GDG: a retrospective 12-year experience of a single center. Neurosurg Rev 30:117

Bradac GB, Peretta P, Stura G et al (2008a) Paediatric dissecting aneurysms of the posterior cerebral artery, case report and review of the literature. Interv Neuroradiol 14:325

Bradac GB, Daniele D, Bergui M et al (2008b) Lacunes and other holes: diagnosis, pathogenesis, therapy. Neuroradiol J 21:35

Bradac GB, Boghi A, Corsico M et al (2014) Applicazioni cliniche della RM in neuroradiologia. In: Coriasco M, Rampado O, Bradac GB (eds) Elementi di risonanza magnetica. Springer, New York

Bradac GB, Venturi F, Stura G et al (2017) Considerations about occlusion of the intracranial distal internal carotid artery. Clin Neuroradiol. 27:169

Brandt T, Sternke W, Thie A et al (2000) Posterior cerebral artery territory infarcts: clinical features, infarct topography, causes and outcome. Cerebrovasc Dis 10:170

Brassier G, Morandi X, Fournier D et al (1998) Origin of the perforating arteries of interpeduncular fossa in relation to the termination of the basilar artery. Interv Neuroradiol 4:109

Braun JP, Tournade A (1977) Venous drainage in the craniocervical region. Neuroradiology 13:155

Breschet G (1829) Research of the anatomy, physiology, and pathology of the venous system with special emphasis on the veins of bones (in French). Villeret et rouen, Paris

Brevis-nunez F, Tschiedel E, Felderhoff-Mueser U et al (2013) Neurogenic stunned myocardium after Embolization in two children with vein of Galen aneurysmal malformation. Neuroradiology 55:213

Brice JG, Dowsett DJ, Lowe RD (1964) Haemodynamic effect of carotid artery stenosis. Br Med J 2:1363

Briganti F, Napoli M, Tortora F et al (2012) Italian multi-center experience with flow-diverter devices for intracranial unruptured aneurysms treatment with periprocedural complications—a retrospective data analisys. Neuroradiology 54:1145

Brinjikji W, Murad MH, Lanzino G et al (2013) Endovascular treatment of intracranial aneurysms with flow diverters: a meta analysis. Stroke 44:442

Brismar J (1976) Persistent hypoglossal artery, diagnostic criteria. Acta Radiol Diagn 17:160

Brocheriou I, Capron F (2004) Malformations arteriovenouses intracraniennes. Aspects anatomopathologiques. J Neuroradiol (Masson) 31:359

Broderick JP, Phillips SJ, O'Fallon M et al (1992) Relationship of cardiac disease to stroke occurrence, recurrence and mortality. Stroke 23:1250

Browder J, Browder A, Kaplan HA (1972) The venous sinuses of the cerebral dura mater. 1 Anatomical structures within the superior sagittal sinus. Arch Neurol 26:175

Brown RD, Wiebers DO, Forbes G et al (1985) The natural history of unruptured intracranial arteriovenous malformations. J Neurosurg 16:421

Brun A, Englund E (1986) A white matter disorder in dementia of the Alzheimer type: a patho-anathomical study. Ann Neurol 19:253

Brunaud V, Delerue O, Muller JP et al (1994) Klippel-Trenaunay syndrome and ischemic neurologic complications. Rev Neurol (Paris) 54:150

Bruneau M, Rynkowski M, Smida-Rynkowska K et al (2011) Long-term follow-up survey reveals a high yield, up to 30% of patients presenting newly detected aneurysms more than 10 years after ruptured intracranial aneurysms clipping. Neurosurgery 34:485

Brunelle F (1997) Arteriovenous malformation of the vein of Galen in children. Pediatr Radiol 27:501

Brunereau L, Levy C, Labrerge S et al (2000) De novo lesions in familial form of cerebral cavernous malformations: clinical and MR features in 29 non-hispanic families. Surg Neurol 53:475

Brunsteins DB, Ferreri AJ (1990) Microsurgical anatomy of VII and VIII cranial nerves and related arteries in the cerebellopontine angle. Surg Radiol Anat 12:259

Bujak M, Margolin E, Thompson A et al (2010) Spontaneous resolution of two dural carotid-cavernous bfistulas presenting with optic neuropathy and marked congestive ophthalmopathy. J Neuroophthalmol 30:222

Bull J, Kozlowski P (1970) The angiographic pattern of the petrosal veins in the normal and pathological. Neuroradiology 1:20

Bulsara KR, Alesander MJ, Villavicencio AJ et al (2002) De novo cerebral arteriovenous malformations: case report. Neurosurgery 50:1137

Bulsara KR, Zomorodi A, Provenzale JM (2007) Anatomic variants of the posterior cerebral artery. AJR Am J Roentgenol 188:w 395

Buonanno F, Moody DM, Ball MR et al (1978) Computed cranial tomographic findings in cerebral sinuses-venous occlusion. J Comput Assist Tomogr 2:281

Burger IM, Siclari F, Gregg L et al (2007) Bilateral segmental agenesis of the vertebrobasilar junction: developmental and angiographic anatomy. AJNR Am J Neuroradiol 28:2017

Burrows PE, Robertson RL, Barnes PD (1996) Angiography and the evaluation of cerebrovascular disease in childhood. Neuroimaging Clin N Am 6:561

Butler WE, Barker FG, Crowell RM (1996) Patients with polycystic kidney disease would benefit from routine magnetic resonance angiography screening for intracerebral aneurysms: a decision analysis. Neurosurgery 38:506

Buyle M, Engelborghs S, Kuneren J et al (2001) Headache as only symptom in multiple cervical artery dissection. Headache 41:509

Cai CQ, Zhang QJ, Yang WD et al (2009) Neuroimages of persistent falcine sinus in children. World J Pediatr 5:63

Cai J, Hatsukami TS, Ferguson MS et al (2005) In vivo quantitative measurement of intact fibrous cap and lipid—rich necrotic core size in atherosclerotic carotid plaque: comparison of high-resolution contrast enhanced magnetic resonance imaging and histology. Circulation 112:3437

Calabrese LH, Mallek JA (1988) Primary angiitis of the central nervous system. Report of 8 new cases; review of the lilterature and proposal for diagnostic criteria. Medicine 67:20

Calabrese LH, Dodick DW, Schwedt TJ et al (2007) Narrative review: reversible cerebral vasoconstriction syndromes. Ann Intern Med 146:34

Cali RL, Berg R, Rama K et al (1993) Bilateral internal carotid agenesis: a case study and review of the literature. Surgery 113:227

Calzolari F (2002) Unusual termination of the inferior petrosal sinus. Neuroradiology 44:796

Campos J, Fox AJ, Vinuela F et al (1987) Saccular aneurysms in basilar artery fenestration. AJNR Am J Neuroradiol 8:233

Cantu C, Barrinagarrementeria F (1993) Cerebral venous thrombosis associated with pregnancy and puerperium. Review of 67 cases. Stroke 24:1880

Caplan LR (1980) "Top of the basilar" syndrome: selected clinical aspects. Neurology (NY) 30:72

Caplan LR, Gorelick PB, Hier DB (1986) Race, sex, and occlusive cerebrovascular disease: a review. Stroke 17:648

Caplan LR, Baquis GD, Pessin MS et al (1988a) Dissection of intracranial vertebral artery. Neurology 38:868

Caplan LR, Dewitt LD, Pessin MS et al (1988b) Lateral thalamic infarcts. Arch Neurol 45:959

Caplan LR (1989) Intracranial branch atheromatous disease: a neglected, understudied and underused concept. Neurology (NY) 39:1246

Caplan LR, Schmahmann JD, Kase CS et al (1990) Caudate infarcts. Arch Neurol 47:133

Caplan LR (1996) Posterior circulation disease. Blackwell, Cambridge

Cappendijk VC, Clentjens KB, Kessels AG et al (2005) Assessment of human atherosclerotic carotid plaque components with multisequence MR imaging. Initial experience. Radiology 234:487

Caranci F, Napoli M, Cirillo M et al (2012) Basilar artery hypoplasia. Neuroradiol J 25:739

Carmichael R (1950) The pathogenesis of non inflammatory cerebral aneurysms. J Pathol Bacterial 62:1

Caroff J, Neki H, Mihalea C et al (2015) Flow-diverter stents for the treatment of saccular middle cerebral artery bifurcation aneurysms. AJNR Am J Neuroradiol 37:279

Carpenter MB, Noback CR, Moss ML (1954) The anterior choroidal artery. The origin, course, distribution and variations. AMA Arch Neurol Psychiatry 71:714

Castaigne P, Lhermitte F, Gauthier JC et al (1970) Internal carotid artery occlusion. Brain 93:231

Castaigne P, Lhermitte F, Gautier JC et al (1973) Arterial occlusions in the vertebral basilar system. Brain 96:133

Castaigne P, Bories J, Brunet P et al (1976) Les fistulas arterioveneuses meningées a drainage venoux cortical. Rev Neurol 132:169

Castaigne P, Lhermitte F, Buge A et al (1981) Paramedian thalamic and midbrain infarcts: clinical and neuropathological study. Ann Neurol 10:127

Castano C, Dorado L, Guerrero C et al (2010) Mechanical thrombectomy with the solitaire AB device in large artery occlusions of the anterior circulation: a pilot study. Stroke 41:1836

Cebral JR, Castro MA, Burgess JE et al (2005) Characterization of cerebral aneurysms for assessing risk of rupture by using patient-specific computational hemodynamics models. AJNR Am J Neuroradiol 26:2550

Cebral JR, Mut F, Rashi M et al (2011) Aneurysm rupture following treatement with flow-diverting stents: computational hemodynamic analysis of treatment. AJNR Am J Neuroradiol 32:27

Cellerini M, Mangiafico S, Ammannati F et al (2008) Ruptured dissecting posterior inferior cerebella artery aneurysms: endovascular treatment without parent vessel occlusion. Neuroradiology 50:315

Chabriat H, Joutel A, Vahedi K et al (1996) CADASIL (cerebral autosomal dominant arteriopathy with subcortical infarcts and leukoencephalopathy). J Mal Vasc 21:277

Chabriat H, Joutel A, Vahedi K et al (2000) CADASIL (cerebral autosomal dominant arteriopathy with subcortical infarcts and leukoencephalopathy): clinical features and neuroimaging. Bull Acad Natl Med 184:1523

Chakraborty S, Fanning NF, Lee SK et al (2006) Bilateral infraoptic origin of the anterior cerebral arteries: a rare anomaly and its embryological and clinical significance. Interv Neuroradiol 12:155

Challa VR, Moody DM, Bell MA (1992) The Charcot-Bouchard aneurysm controversy: impact of a new histologic technique. J Neuropathol Exp Neurol 51:264

Challa VR, Moody DM, Brown WR (1995) Vascular malformations of the central nervous system. J Neuropathol Exp Neurol 54:609

Chalouhi N, Zanaty M, Tjoumakaris S et al (2014) Treatment of Blister-like Aneurysms with the pipeline embolization device. Neurosurgery 74:527

Chalouhi N, Zanaty M, Whiting A (2015) Safety and efficacy of the Pipeline Embolization Device in 100 small intracranial aneurysms. J Neurosurg 122:1498

Chaloupka J, Putman C, Awad I (1996) Endovascular therapeutic approach to peripheral aneurysm of the superior cerebellar artery. AJNR Am J Neuroradiol 17:1338

Chaloupka JC, Huddle DC (1998) Classification of vascular malformations of the central nervous system. Neuroimaging Clin N Am 8:295

Chaloupka JC, Huddle DC, Aldermann J (1999) Local induction of angiogenesis within the walls of a dural sinus results in the creation of dural arteriovenous fistula. Demonstration by polymeric release implantation model in swine. J Neurosurg 90:194

Chan WC, Lai V, Wong YC et al (2011) Focal brain herniations into giant arachnoid granulations: a rare occurrence. Eur Radiol Radiol Extra 78:111

Chang HW, Youn SW, Jung CH et al (2011) Technical strategy in endovascular treatment of proximal anterior cerebral artery aneurysms. Acta Neurochir 153:279

Chapot R, Stracke P, Meyn H et al (2009) Experience with Silk in 25 aneurysms. Interv Neuroradiol 15(Suppl 1):112

Chase NE, Taveras JM (1963) Temporal tumors studied by serial angiography. A review of 150 cases. Acta Radiol Diagn 1:225

Chaturvedi S, Caplan LR (2003) Angioplasty for intracranial atherosclerosis. Is the treatment worse than the disease? Neurology 61:1647

Chaudary MY, Sachdev VP, Cho CH et al (1982) Dural arteriovenous malformation of the major venous sinuses. An acquired lesion. AJNR Am J Neuroradiol 3:13

Chen CJ, Chen ST, Hsieh FY et al (1998) Hypoplasia of the internal carotid artery with intercavernous anastomosis. Neuroradiology 40:252

Chen Y, Zhu W, Bollen AW et al (2008) Evidence of inflammatory cell involvement in brain arteriovenous malformations. Neurosurgery 62:1340

Cheng KM, Chan CM, Cheung JL et al (1999) Transvenous embolization of spontaneous carotid-cavernous fistulas by sequential occlusion of the cavernous sinus. Interv Neuroradiol 5:225

Chiras J, Launay M, Gaston A et al (1982) Thoracic vertebral artery. Neuroradiology 24:67

Chiras J, Marciano S, Vega Molina J et al (1985) Spontaneous dissecting aneurysms of the extracranial vertebral artery (20 cases). Neuroradiology 27:327

Chiu AH, Wenderoth J (2013) Cerebral hyperperfusion after flow diversion of large intracranial aneurysm. J Neurointerv Surg. doi:10.1136/bcr-2012-010479

Cho YD, Ahn JH, Jung SC et al (2014) Coil embolization in precommunicating (A1) segmenht aneurysms of anterior cerebral artery. Neuroradiology 56:219

Choi BK, Lee TH, Choi CH et al (2008) Fusiform intracanalicular ophthalmic artery aneurysm; case report and review of literature. J Korean Neurosurg Soc 44:43

Choi HS, Kim DI, Kim BM et al (2012) Endovascular treatment of dural arteriovenous fistula involving marginal sinus with emphasis on the routes of transvenous embolization. Neuroradiology 54:163

Choi JHO, Jo IK, Kim KH et al (2015) Spontaneous angiographic changes in venous drainage patterns related to symptom changes in patients with untreated cavernous sinus dural arteriovenous fistula. Neuroradiology 57:1153

Choi JW, Kim JK, Choi BS et al (2009) Adjuvant revascularization of intracranial artery occlusion with angioplasty and/or stenting. Neuroradiology 51:33

Choudhri TF, Hoh BL, Zerwes HF et al (1998) Reduced microvascular thrombosis and improved outcome in acute murine stroke by inhibiting GB IIb/IIIa receptor-mediated platelet aggregation. J Clin Invest 102:1301

Chow M, McDougall CO, Kelly C et al (2012) Delayed spontaneous rupture of a posterior inferior cerebellar artery aneurysm following treatment with flow diversion: a clinicopathologic study. AJNR Am J Neuroradiol 33:46

Christoforidis GA, Kollias SS, Valavanis A (2002) Identification of the ascending middle cerebral artery branches on digital subtraction angiography. Neuroradiology 44:803

Church DG, Lowe LH (2006) Case 103: PHACE syndrome. Radiology 241:939

Churojana A, Khumtong R, Songsaeng D et al (2012) Life-threatening arteriovenous malformation of the maxillomandibular region and treatment outcomes. Interv Neuroradiol 18:49

Chyatte D, Brophy C, Reilly J et al (1990) Morphometric analysis of reticular elastic and Type III collagen fibers in the cerebral arteries of patients with intracranial aneurysms. Stroke 21:161

Chyatte D, Mjerzejewski C (1991) Novel restriction fragment length polymorphism associated with gene for type III collagen (COL3A1) in patients with cerebral aneurysms. Congress of Neurological Surgeons, Orlando, p 163

Ciccone A, Valvassori L, Nichelatti M et al (2013) Endovascular treatment for acute ischemic stroke. N Engl J Med 368:904

Ciceri EF, Klucznik RP, Grossman RG et al (2001) Aneurysms of the posterior cerebral artery: classification and endovascular treatment. AJNR Am J Neuroradiol 22:27

Çinar C, Oran I, Bozkaya H et al (2013) Endovascular treatment of ruptured blister-like aneurysms with special reference to the flow-diverting strategy. Neuroradiology 55:441

Clarençon F, Bonnevile F, Boch AL et al (2011) Parent artery occlusion is not obsolete in giant aneurysms of the ICA. Experience with very-long term follow-up. Neuroradiology 53:973

Clarençon F, Di Maria F, Gabrieli J et al (2017) Flow Diverter Stents for the treatment of anterior cerebral artery aneurysms: safety and effectiveness. Clin Neuroradiol 27:51–56

Clark WM, On behalf of the CREST Investigators (2010) Carotid revascularization endarterectomy versus stenting trial. Paper presented at the international stroke conference, San Antonio, 2010

Coban G, Yildirim E, Horasanil B et al (2013) Unusual case of dizziness: occult temporal lobe encephalocele into transverse sinus. Clin Neurol Neurosurg 115:1911

Cognard C, Gobin YP, Pierot L et al (1995) Cerebral dural arteriovenous fistulas: clinical and angiographic correlations with revised classification of venous drainage. Radiology 194:671

Cohen JE, Gomori JM, Umansky F (2003) Endovascular management of spontaneous bilateral symptomatic vertebral artery dissections. AJNR Am J Neuroradiol 24:2052

Cohen JE, Raiz G, Itshayek E et al (2005) Endovascular management of traumatic and iatrogenous aneurysms of the pericallosal artery. Report of two cases. J Neurosurg 102:555

Cohen JE, Gomori JM, Rajz G et al (2013) Emergent stenassisted angioplasty of extracranial internal carotid artery and intracranial stent-based thrombectomy in acute tandem occlusive disease: technical considerations. J Neurointerv Surg 5:440

Cohen JE, Gomori JM, Rajz G et al (2015) Extracranial carotid artery stenting followed by intracranial stent-based thrombectomy for acute tandem occlusive disease. J Neurointerv Surg 7:412

Collado A, Santamaria J, Ribalta T et al (1989) Giant-cell arteritis presenting with ipsilateral hemiplegia and lateral medullary syndrome. Eur Neurol 29:266

Collice M, D'Aliberti G, Arena O et al (2000) Surgical treatment of the intracranial dural arteriovenous fistulae: role of the venous drainage. Neurosurgery 47:56

Colombo F, Pozza F, Chierego G et al (1994) Linear accelerator radiosurgery of cerebral arteriovenous malformations: an update. Neurosurgery 34:14

Congdon ED (1922) Transformation of the aortic-arch system during the development of the human embryo. Contrib Embryol 14:47

Cooper ER (1958) Arachnoid granulations in man. Acta Anat (Basel) 34:187

Countee RW, Vijayanathan T (1979) External carotid in internal carotid artery occlusion. Angiographic, therapeutic and prognostic considerations. Stroke 10:450

Countee RW, Vijayanathan T, Chavis P (1981) Recurrent retinal ischemia beyond cervical carotid occlusion. Clinical-angiographic correlation and therapeutic implications. J Neurosurg 55:532

Cowan JA, Barkhoudarian G, Yang LJ et al (2004) Progression of a posterior communicating artery infundibulum into an aneurysm in a patient with ALAGILE syndrome. Case report. J Neurosurg 101:694

Craig DR, Meguro K, Watridge C et al (1982) Intracranial internal carotid artery stenosis. Stroke 13:825

Crawford T (1959) Some observations on the pathogenesis and natural history if intracranial aneurysms. J Neurol Neurosurg Psychiatry 22:259

Crawford PM, West CR, Chadwick DW et al (1986) Arteriovenous malformations of the brain: the natural history in unoperated patients. J Neurol Neurosurg Psychiatry 49:1

Critchley M (1930) The anterior cerebral artery, and its syndrome. Brain 53:120

Crompton MR (1962) The pathology of ruptured middle cerebral artery aneurysms with special reference to the differences between the sexes. Lancet 2:421

Crompton MR (1966a) Mechanisms of growth and rupture in cerebral berry aneurysms. Br Med J 1:1138

Crompton MR (1966b) The pathogenesis of cerebral aneurysms. Brain 89:797

Crompton MR (1979) Familial incidence of cerebral aneurysms. Br Med J 1:292

Cronquist M (2009) Endovascular stenting combined with modified antiplatelet regime in acute blister like aneurysms. A multi-center presentation. Interv Neuroradiol 15(Suppl 1):125

Cronquist S, Laroche F (1967) Transitory hyperaemia in focal cerebral vascular lesions studied by angiography and regional blood flow measurements. Br J Radiol 40:270

Cronquist S, Laroche F (1969) Venous abnormalities in cerebrovascular disorders. Acta Radiol Diagn (Stockh) 9:251

Crouzet G, Agnettaz G, Pellat J et al (1974) Les voies de supplence en cours de Moya-Moya. J Neuroradiol (Masson) 1:87

Crowley RW, Ducruet AF, Kalani MY et al (2015) Neurological morbidity and mortality associated with the endovascular treatment of cerebral arteriovenous malformations before and during the Onyx era. J Neurosurg 122:1492

Crum B, Mokri B, Fulgham J (2000) Spinal manifestation of vertebral artery dissection. Neurology 55:304

Cruz JP, Chow M, O'Kelly C et al (2012) Delayed ipsilateral parenchymal hemorrhage following flow diversion for the treatment of anterior circulation aneurysms. AJNR Am J Neuroradiol 33:603

Cruz JP, Marotta T, O'Kelly C et al (2014) Enhancing brain lesions after endovasculare treatment of aneurysms. AJNR Am J Neuroradiol 35:1954

Cushing H (1926) Cited by Toennis W and Walter W: Die Behandlung der Sackfoermigen intrakraniellen Aneurysmen. In: Olivecrona H, Toennis W (eds) Handbuch der Neurochirurgie. Springer, Berlin

Cushing H, Bailey P (1928) Tumours arising from the blood vessels of the brain, angiomatous malformations and haemangioblastomas. Balliere, Tindall and Cox, London

D'Anglejan-Chatillon J, Ribeiro V, Mas JL et al (1989) Migraine—a risk factor for dissection of cervical arteries. Headache 29:560

Dabus G, Walker MT (2010) Right vertebral artery arising from the aortic arch distal to the left subclavian artery diagnosed with magnetic resonance angiography. Arch Neurol 67:508

Dandy WS (1928) Arterio-venous aneurysms of the brain. Arch Surg 17:190

Dandy WE (1938) Intracranial aneurysm of internal carotid artery cured by operation. Ann Surg 107:654

Danziger J, Bloch S, Hefer AG (1972) Bilateral rete carotis in man. S Afr Med J 46:1487

Dashti R, Hernesniemi J, Lehto H et al (2007) Microneurosurgical management of proximal anterior cerebral artery aneurysm. Surg Neurol 68:366

Davalos A, Mendes Pereira V, Chapot R et al (2012) Retrospective multicenter study of solitaire FR for revascularization in treatment of acute ischemic stroke. Stroke 43:2699

Davies MA, Terbrugge K, Willinsky R et al (1996) The validity of classification for the clinical presentation of intracranial dural arteriovenous fistulas. J Neurosurg 85:830

Davis S, Aldrich TH, Jones PF et al (1996) Isolation of angiopoietin-1, a ligand for the TIE2 receptor, by secretion—trap expression cloning. Cell 87:1161

Day AL (1990) Aneurysms of the ophthalmic segment: a clinical and anatomical analysis. J Neurosurg 72:677

De Almeida F (1931) Note sur les collaterales de l'artère communicante cerebrale anterieure. ASrch Anat Anthrop (Lisboa) 13:551

De Andrade AF, Figueiredo EG, Caldas JG et al (2008) Intracranial vascular lesions associated with small epidural hematomas. Neurosurgery 62:416

De Blasi R, Medicamento N, Chiumarullo L et al (2009) A case of aneurysm on a persistent hypoglossal artery treated by endovascular coiling. Interv Neuroradiol 15:175

De Castro-Alfonso LH, Nakiri GS, Oliveira RS et al (2016) Curative embolization of pediatric intracraniual arteriovenous malformations using Onyx: the role of new embolization technique on patient's outcomes. Neuroradiology 58:585

De Cocker LJL, Kloppenborg RP, van dr Graaf Y et al (2015) Cerebellar cortical infarct cavities. Correlation with risk factors and MRI Markers of cerebrovascular disease. Stroke 45:3154

De Cocker LJL, Compter A, Kappelle L et al (2016) Cerebellar cortical infarct cavities and vertebral artery disease. Neuroradiology 58:853

De La Torre E, Netzky MG (1960) Study of persistent primitive maxillary artery in human fetus: some homologies of cranial arteries in man and dog. Am J Anat 106:185

DeLong MR (2000) The basal ganglia. In: Kandel ER, Schwartz JH, Jessel TM (eds) Principle of neural science. Mc Graw-Hill, New York

De Reuck J (1972) The cortico-subcortical arterial angioarchitecture in the human brain. Acta Neurol Belg 72:323

De Simone R, Ranier A, Montella S et al (2012) Sinus venous stenosis-associated idiopathic intracranial hypertension without papilledema as a powerful risk factor for progression and refractoriness of headache. Curr Pain Headache Rep 16:261

De Vriese B (1905) Sur la signification morphologique des artères cerebrales. Arch Biol 21:357

Deans WR, Block S, Leibrock L et al (1982) Arteriovenous fistula in patients with neurofibromatosis. Radiology 144:103

Deasy NP, Gholkar AR, Cox TCS et al (1999) Tentorial dural arteriovenous fistulae: endovascular treatment with transvenous coil embolization. Neuroradiology 41:308

Debrun G, Lacour P, Caron JP et al (1975a) Inflatable and released balloon technique. Experimental in dog-application in men. Neuroradiology 9:267

Debrun G, Lacour P, Caron JP et al (1975b) Traitment des fistules arterioveneiseuses et des aneurysms par balloon gonflable et largable. Nouv Presse Med 8:2315

Debrun G, Lacour P, Vinuela F et al (1981) Treatment of 54 traumatic carotid-cavernous fistulas. J Neurosurg 55:678

Debrun G, Aletich V, Ausman J et al (1997) Embolization of the nidus of brain AV malformations with butylcyanoacrylate. Neurosurgery 40:112

Dechambre A (1939) Memoire sur la curabilitè du rammollissement cerebral. Gaz Med Fr 6:305

Dehdashi AR, Safran AB, Martin JB et al (2002) Intraorbital ophthalmic artery aneurysm associated with basilar tip saccular aneurysm. Neuroradiology 44:600

Del Zotto E, Ritelli M, Pezzini A et al (2012) Clinical neuroradiological and molecular feature of a patient affected by pseudoxhantoma elasticum associated to carotid rete mirabile: case report. Clin Neurol Neurosurg 114:758

Derdeyn CP, Fiorella D, Lynn MJ et al (2013) Stenting and aggressive medical management for preventing recurrent stroke in intracranial stenosis. Trial Investigators. Mechanism of stroke after intracranial angioplasty and stenting in the Sammpris trial. Neurosurgery 72:777

Desal HA, Lee SK, Kim BS et al (2005) Multiple de novo vascular malformations in relation to diffuse venous occlusive disease: a case report. Neuroradiology 47:38

Deutschmann HA, Wehrschuetz M, Augustin M et al (2012) Long-term follow-up after treatment of intracranial aneurysms with pipeline embolization device: results from a single center. AJNR Am J Neuroradiol 33:481

Di Rocco C, Tamburini G, Rollo M (2000) Cerebral arteriovenous malformations in children. Acta Neurochir 142:145

Di Tullio MR, Homma S, Jin Z et al (2008) Aortic atherosclerosis hypercoagulabily and stroke: the aortic plaques and risk of ischemic stroke (APRIS) study. J Am Coll Cardiol 52:855

Diener HC, Kurth T (2011) Migraine and stroke. In: Mohr JP, Wolf PA, Grotta JC, Moskowitz MA, Mayberg MR, Von Kummer R (eds) Stroke pathophysiology, diagnosis and management. Elsevier, Philadelphia

Dillon WP (1995) Criptic vascular malformations: controversies in terminology, diagnosis, pathophysiology and treatment. AJNR Am J Neuroradiol 18:1839

Dimmick SJ, Faulderf KC (2009) Normal variants of the cerebral circulation at multidetector CT angiography. Radiographics 29:1027

Ding D, Starke RM, Jensen ME et al (2013) Perforator aneurysms of the posterior circulation: case series and review of the literature. J Neurointerv Surg 5:546

Dinichert A, Rufenacht DA, Tribolet N (2000) Dissecting aneurysms of the posterior inferior cerebellar artery: report of four cases and review of the literature. J Clin Neurosci 7:515

Dittrich R, Rohsbach D, Heidbreder A et al (2007) Mild mechanical traumas are possible risk factors for cervival artery dissection. Cerebrovasc Dis 23:275

Djindjian R, Cophignon J, Comoy J et al (1968) Polimorphisme neuroradiologique des fistules carotido-cavernouse. Neurochirurgie 14:881

Djindjian R, Manelfe CF, Picard L (1973) Fistules arterioveneuses carotide externe-sinus caverneux; etude angiographique à propos de 6 observations et revue de la literature. Neurochirurgie 19:91

Djindjian R, Merland JJ (1978) Superselective arteriography of the external carotid artery. Springer, Berlin

Djouhri H, Guillon B, Brunereau L et al (2000) MR Angiographhy for the long-term follow-up of dissecting Aneurysms of the extracranial internal carotid artery. AJR Am J Roentgenol 174:1137

Dmytriw AA, Terbrugge KG, Krings T et al (2014) Endovascular treatment of head and neck arteriovenous malformations. Neuroradiology 56:227

Dong H, Blaivas M, Wang MM (2012) Bidirectional encroachment of collagen into the tunica media in cerebral Autosomal Dominant Arteriopathy with Subcortical Infarcts and Leucoencephalopathy. Brain Res 1456:64

Donnet A, Valade D, Houdart E et al (2013) Primary cough headache, primary exertional headache, and primary headache associated with sexual activity; a clinical and radiological study. Neuroradiology 55:297

Dorn F, Prothmann S, Wunderlich S et al (2012a) Stent angioplasty of intracranial stenosis. Single center experience of 54 cases. Clin Neuroradiol 22:149

Dorn F, Kuntze-Söderquist A, Popp S et al (2012b) Early venous drainage of the successful endo-vascular recanalization in ischemic stroke-a predict for final infarct volume? Neuroradiology 54:745

Dorn F, Stehle S, Lockau H et al (2012c) Endovascular treatment of acute intracerebral artery occlusion with the solitaire stent: single-center experience with 108 recanalization procedures. Cerebrovasc Dis 34:70

Dott NM (1933) Intracranial aneurysms: cerebral arterioradiography: surgical treatment. Tr Med Chir Soc Edinburgh 47:219

Doyon DL, Aron Rosa DS, Ramèe A (1974) Orbital veins and cavernous sinus. In: Newton T, Potts DG (eds) Radiology of skull and brain, Book 3. Mosby, St. Louis

Drake CG, Peerless SJ (1997) Giant fusiform in tracranially aneurysms: review of 120 patients treated surgically from 1965 to 1992. J Neurosurg 87:141

Duncan IC, Terblanche JM (2005) Spontaneous isolated posterior communicating artery dissection in a young adult with hyperhomocysteinemia. AJNR Am J Neuroradiol 26:2030

Dunker RO, Harris AB (1976) Surgical anatomy of the proximal anterior cerebral artery. J Neurosurg 44:359

Durand-Fardel M (1843) Traitè du rammollissement du cerveau. Baillière J. B, Paris

Duret H (1874) Recherches anatomiques sur la circulation de l'encephale. Arch Physiol Norm Pathol 1:60–91

Duvernoy HM (1975) The superficial veins of the human brain. Springer, Berlin

Duvernoy HM (1999) The human brain stem vessels. Springer, Berlin

Echiverri HC, Rubino FA, Gupta SR et al (1989) Fusiform aneurysms of the vertebrobasilar arterial system. Stroke 20:1741

Eckenhoff J (1970) The physiologic significance of the vertebral venous plexus. Surg Gynecol Obstet 131:72

Ecker A, Riemenschneider PA (1951) Arteriographic demonstration of spasm of intracranial arteries: with special reference to saccular arterial aneurysms. J Neurosurg 8:660

Eckert B (2009) Acute stroke therapy 1981–2009. Clin Neuroradiol 19:8

Edelsohn L, Caplan L, Rosenbaum AE (1972) Familial aneurysms and infundibular widening. Neurology 22:1056

Edgell RC, Zaidat OO, Gupta R et al (2013) Multicenter study of safety in stenting for symptomatic vertebral artery origin stenosis. Results from the Society of Vascular and Interventional Neurology Research Consortium. J Neuroimaging 23:170

Edwards JE (1948) Anomalies of the derivatives of the aortic arch system. Med Clin North Am 32:925

Ehrenfeld WK, Stoney RJ, Wylie WJ et al (1967) Fibromuscular hyperplasia of the in ternal carotid artery. Arch Surg 95:284

Ehrenfeld WK, Wylie EJ (1976) Spontaneous dissection of the internal carotid artery. Arch Surg 111:1294

Einhäupl K, Bousser MG, de Brujin SF et al (2006) EFNS guideline on the treatment of cerebral venous and sinus thrombosis. Eur J Neurol 13:553

Einhäupl KM, Masuhr F (1994) Cerebral venous and sinus thrombosis. An update. Eur J Neurol 1:109

Eisenberg RA, Vines FS, Taylor SB (1986) Bifid origin of left vertebral artery. Radiology 159:429

Elbers J, Benseler SM (2008) Central nervous system vasculitis in children. Curr Opin Rheumatol 20:47

Elefante R, Fucci G, Granata F et al (1983) Agenesis of the right internal carotid artery with an unusual transsellar intracavernous intercarotid connection. AJNR Am J Neuroradiol 4:88

Elias WS (1971) Intracranial fibromuscular hyperplasia. JAMA 218:254

El-koussy M, Schrott G, Gralla J et al (2012) Susceptibility-weighted MR imaging for diagnosis of capillary telangiectasia of the brain. AJNR Am J Neuroradiol 33:715

Endo S, Nishijima M, Nomura H et al (1993) A pathological study of intracranial posterior circulation dissecting aneurysms with subarachnoid hemorrhage: report of three cases and review of the literature. Neurosurgery 33:732

Engelter S, Dallongeville J, Kloss M et al (2012) Thrombolysis in cervical artery dissection—data from the cervical artery dissection and ischaemic stroke patients (CADISP) database. Eur J Neurol 19:1199

Enomoto T, Sato A, Maki Y (1977) Carotid-cavernous sinus fistula caused by rupture of a primitive trigeminal artery aneurysm. Case report. J Neurosurg 46:373

Enzinger C, Ropele S, Gattringer T et al (2010) High-grade internal carotid artery stenosis and chronic brain damage; a volumetric magnetic resonance imaging study. Cerebrovasc Dis 30:540

Epstein F, Ronsohoff J, Budzilovich G (1970a) The clinical significance of junctional dilatation of the posterior communicating artery. J Neurosurg 33:529

Epstein HM, Linde HW, Crampton AR et al (1970b) The vertebral venous plexus as a major cerebral venous outflow tract. Anesthesiology 32:332

Ernemann U, Freudenstein D, Pitz S et al (2002) Intraorbital aneurysm of the ophthalmic artery: a rare cause of apex orbitae compression syndrome. Graefes Arch Clin Exp Ophthalmol 240:575

Ernst R, Bulas R, Tomsick T et al (1999) Three cases of dural arteriovenous fistula of the anterior condylar vein within the hypoglossal canal. AJNR Am J Neuroradiol 20:2016

European Carotid Surgery Trialists' Collaborative Group (1998) Randomized trial of endarterectomy for recently symptomatic carotid stenosis: final results of the MRC European Carotid Surgery Trial (ECST). Lancet 351:1379

Evans HM (1912) The development of the vascular system. Keibel–Mall Manual Hum Embryol 2:570–709

Ezura M, Kagawa S (1992) Spontaneous disappearance of a huge cerebral arteriovenous malformation. Case report. Neurosurgery 30:595

Fan X, Zhang Z, Zhang C et al (2002) Direct-puncture embolization of intraosseous arteriovenous malformation of jaws. J Oral Maxillofac Surg 60:890

Farb RI, Vanek I, Scott JN et al (2003) Idiopathic intracranial hypertension: the prevalence and morphology of sinovenous stenosis. Neurology 60:1418

Fargen KM, Hoh BL, Welch BG et al (2012) Long-term results of enterprise stent-assisted coiling of cerebral aneurysms. Neurosurgery 71:239

Fealey ME, Edwards WD, Giannini C et al (2008) Complications of endovascular polymers associated with vascular introducer sheats and metallic coils in 3 patients, with literature review. Am Surg Pathol 32:1310

Feldmann E, Daneault N, Kwan E et al (1990) Chinese-white differences in the distribution of occlusive cerebrovascular disease. Neurology 40:1541

Fellner FA, Fellner C, Aichner FT et al (2005) Importance of T2*-weighted gradient echo MRI in diagnosis of cortical vein thrombosis. Eur J Radiol 56:235

Ferrand J (1902) Essai sur l'hémiplegie des vieillards, les lacunes de désintegration cérébrale. Thesis Rousset, Paris

Fesl G, Wiesmann M, Patzig M et al (2011) Endovascular mechanical recanalization of acute carotid T-occlusions: a single—center retrospective analysis. Cardiovasc Interv Radiol 34:280

Fiehler J (2014) Studies in intracranial stenting: we must not retire from the germ of science. Clin Neuroradiol 24:309

Fiehler J, Stapf C (2008) ARUBA-beating natural history in unruptured brain AVMs by intervention. Neuroradiology 50:465

Field LR, Russell EJ (1995) Spontaneous hemorrhage from a cerebral venous malformation related to thrombosis of the central draining vein: demonstration with angiography and serial MR. AJNR Am J Neuroradiol 16:1885

Fine AD, Cardoso A, Rhoton AL Jr (1999) Microsurgical anatomy of the extracranial-extradural origin of the posterior inferior cerebellar artery. J Neurosurg 91:645

Fiorella D, Kelly ME, Albuquerque FC et al (2009a) Curative reconstruction of a giant midbasilar trunk aneurysm with the Pipeline embolization device. Neurosurgery 64:212

Fiorella D, Lylyk P, Szikora I et al (2009b) Curative cerebrovascular reconstruction with the pipeline embolization device: the emergence of definitive endovascular therapy for intracranial aneurysms. J Neurointerv Surg 1:56

Fischer S, Hopf N, Henkes H (2011) Evolution from an Infundibulum of the posterior communicating artery to a saccular aneurysm. Clin Neuroradiol 21:87

Fischer S, Vajda Z, Aguilar-Pérez M et al (2012) Pipeline embolization device (PED) for neurovascular reconstruction: initial experience in the treatment of 101 aneurysms and dissections. Neuroradiology 54:369
Fisher CM (1954) Occlusion of the carotid arteries: further experiences. AMA Arch Neurol Psychiatry 72:187
Fisher CM (1965) Lacunes: small deep cerebral infarcts. Neurology 13:30
Fisher CM, Gore I, Okabe N et al (1965) Calcification of the carotid siphon. Circulation 32:538
Fisher CM (1968) The arterial lesion underlying lacunes. Acta Neuropathol 12:1
Fisher CM (1971) Cerebral ischemia-less familiar types (review). Clin Neurosurg 18:267
Fisher CM, Caplan L (1971) Basilar artery branch occlusion: a cause of pontine infarction. Neurology 21:900
Fisher CM, Ojeman RG, Roberson GH (1978) Spontaneous dissection of cervico-cerebral arteries. Can J Neurol Sci 5:9
Fisher CM (1979) Capsular infarcts. Arch Neurol 36:65
Fisher CM (1986) The posterior cerebral artery syndrome. Can J Neurol Sci 13:232
Fisher CM, Ojeman RG (1986) A clinico-pathologic study of carotid endarterectomy plaques. Rev Neurol (Paris) 142:153
Flandroy P, Lacour P, Marsault C et al (1987) The intravascular treatment of a cavernous fistula caused by rupture of a traumatic carotid trigeminal aneurysm. Neuroradiology 29:308
Flemming KD, Wieber DO, Brown RD Jr et al (2004) Perspective risk of hemorrhage in patients with vertebrobasilar non-saccular intracranial aneurysms. J Neurosurg 101:82
Flint AC, Duckwiller GR, Budzik RF et al (2007) Mechanical thrombectomy of intracranial internal carotid occlusion: pooled results of the Merci and multi Merci part I trials. Stroke 38:1274
Fofi L, Giugni E, Vadalà R et al (2012) Cerebral transverse sinus morphology as detected by MR venography in patients with chronic migraine. Headache 52:1254
Foix C, Hillemand P (1925a) Les artères de l'axe encefalique jusq'au diencephale inclusivement. Rev Neurol 41:705
Foix C, Hillemand P (1925b) Les syndromes de la region thalamique. Presse Med 1:113
Fong GH, Rossant J, Gertsenstein M et al (1995) Role of the FLt-1 receptor tyrosine kinase in regulating the assembly of vascular endothelium. Nature 376:66
Forbus WD (1930) On the origin of military aneurysms of the superficial cerebral arteries. Bull Johns Hopkins Hosp 47:239
Forbrig R, Eckert B, Ertl L et al (2016) Ruptured basilar artery perforator aneurysms--treatment regimen and long-term follow-up in eight cases. Neuroradiology. 58:285
Forsting M, Wanke I (2006) Development venous anomalies. In: Forsting M (ed) Intracranial vascular malformations and aneurysms. Springer, Berlin
Frahm D, Wunderlich S, Schubert MI et al (2016) Mechanical trombectomy in acute occlusion of the Carotid-T: a retrospective single centre study in 51 patients. Clin Neuroradiol 26:23
Freckmann N, Sartor K, Hermann HD (1981) Traumatic arteriovenous fistulae of the middle meningeal artery and neighbouring veins or dural sinuses. Acta Neurochir 55:273
Freed K, Low VHS (1997) The aberrant subclavian artery. AJR Am J Roentgenol 168:481
Frens DB, Petajan JH, Anderson R et al (1974) Fibromuscular dysplasia of the posterior cerebral artery: report of a case and review of the literature. Stroke 5:161
Frieden IJ, Reese V, Cohen D (1996) Phace syndrome. The association of posterior fossa brain malformation, hemangiomas, arterial anomalies, coarctation of the aorta and cardiac defects, and eye abnormalities. Arch Dermatol 132:307
Friedlander RM, Ogilvy CS (1996) Aneurysmal subarachnoid haemorrhage in patients with bilateral A1 fenestrations associated with an Azygos anterior cerebral artery. Case report and literature review. J Neurosurg 84:681
Friedman AH, Drake CG (1984) Subarachnoid hemorrhage from intracranial dissecting aneurysm. J Neurosurg 60:325
Friedman JA, Pichelmann MA, Piepgras DG et al (2001) Ischemic complications of surgery for anterior choroidal artery aneurysms. J Neurosurg 94(4):565–572
Friedmann WA, Day AL, Quisling RG Jr et al (1980) Cervical carotid dissecting aneurysms. Neurosurgery 7:207
Frontera JA, Parra A, Shimbo D et al (2008) Cardiac arrhythmias after subarachnoid hemorrhage: risk factors and impact on outcome. Cerebrovasc Dis 26:71
Fujii K, Lenkey C, Rhoton A Jr (1980) Microsurgical anatomy of the choroidal arteries: lateral and 3rd ventricles. J Neurosurg 52:165
Fukuma K, Ihara M, Tanaka T et al (2015) Intracranial cerebral artery dissection of anterior circulation as a cause of convexity subarachnoid hemorrhage. Cerebrovasc Dis 40:45
Fukushima S, Hirohata M, Okamoto Y et al (2009) Anterior inferior cerebellar artery dissecting aneurysms in a juvenile—case report. Neurol Med Chir (Tokyo) 49:81
Fukutake T, Hirayama K (1995) Familial young-onset arteriosclerotic leukoencephalopathy with alopecia and lumbago without arterial hypertension. Eur Neurol 35:69
Fullerton HJ, Johnston S, Smith ZS (2001) Arterial dissection and stroke in children. Neurology 57:1155
Gacs G, Vinuela F, Fox AL et al (1983) Peripheral aneurysms of the cerebellar arteries. J Neurosurg 58:63
Gade A (1982) Amnesia after operations of aneurysms of the anterior communicating artery. Surg Neurol 18:46
Gailloud P, Fasel JH, Muster M et al (1997) Termination of the inferior petrosal sinus. An anatomical variant. Clin Anat 10:92
Gailloud P, San Millan Ruiz D, Muster M et al (2000) Angiographic anatomy of the latero-cavernous sinus. AJNR Am J Neuroradiol 21:1923
Gailloud P, Muster M, Khan N et al (2001) Anatomic relationship between arachnoid granulations in the transverse sinus and the termination of the vein of Labbé: an angiographic study. Neuroradiology 43:139

Gaita F, Formato L, Bianchi F et al (2002) Role of the cardiologist in management of subarachnoid hemorrhage. Rivista di Neuroradiologia 15:625

Gala NB, Agarwal N, Barrese J et al (2013) Current endovascular treatment options of dural venous sinus thrombosis: a review of the literature. J Neurointerv Surg 5:28

Gall G, Nepper Rasmussen J (2009) Initial experience with the silk stent. Interv Neuroradiol 15(Suppl 1):115

Gallas S, Pasco A, Cottier JP et al (2005) A multicenter study of 705 ruptured intracranial aneurisms treated with Guglielmi Detachable Coils. AJNR Am J Neuroradiol 26:1723

Galligioni F, Andrioli GC, Marin G et al (1971) Hypoplasia of the internal carotid artery associated with cerebral pseudoangiomatosis. Report of 4 cases. Am J Roentgenol Radium Ther Nucl Med 112:251

Galloway JR, Greitz T (1960) The medial and lateral choroid arteries; an anatomic and roentgenographic study. Acta Radiol 53:353

Gandolfo C (2012) Giant vertebrobasilar aneurysm in a child: a challenging management. Neuroradiology 54:505

Gao T, Zhang Z, Yu W et al (2009) Atherosclerotic carotid vulnerable plaque and subsequent stroke: a high-resolution MRI study. Cerebrovasc Dis 7:345

Gao M, Gemmette JJ, Chaudhairy N et al (2013) A comparison of particulate and onyx embolization in pre-operative devascularization of juvenile nasopharyngeal angiofibromas. Neuroradiology 55:1089

Gao PH, Yang L, Wang G et al (2016) Symptomatic unruptured isolated middle cerebral artery dissection: clinical and magnetic resonance imaging features. Clin Neuroradiol 26:81

Garcia JH, Ho KL, Pantoni L (1998) Pathology. In: Barnett HJM, Mohr JP, Stein BM, Yatzu FM (eds) Stroke. Churchill Livingstone, New York, pp 139–140

Garcia-Monaco R, Alvarez H, Goulau Z et al (1990) Posterior fossa arteriovenous malformations. Angioarchitecture in relation to their hemorrhagic episodes. Neuroradiology 31:471

Garcia-Monaco R, Rodesch G, Terbrugge K et al (1991) Multifocal dural arteriovenous shunts in children. Childs Nerv Syst 7:425

Garcia-Monaco R, Rodesch G, Alvarez H et al (1993) Pseudoaneurysm within ruptured intracranial arteriovenous malformations: diagnosis and early endovascular management. AJNR Am J Neuroradiol 14:345

Garcia-Monaco R, Taylor W, Rodesch G et al (1995) Pial arteriovenous fistula in children as presenting of Rendu-Osler-Weber disease. Neuroradiology 37:60

Gawlitza M, Januel AC, Tall PH et al (2015) Flow diversion treatment of complex bifurcation aneurysm beyond the circle of Willis: a single-center series with special emphasis on covered cortical branches and perforating arteries. J Neurointerv Surg 8:481–487

Gee W, Burton R, Stoney RJ (1974) Atypical fibromuscular hyperplasia involving the carotid artery. Ann Surg 180:136

Geibprasert S, Jiarakongmun P, Krings T et al (2008) Trigeminal fistula treated by combined transvenous and transarterial embolisation. Acta Neurochir 150:583

Geibprasert S, Pongpech S, Armstrong D et al (2009) Multifocal Dangerous extra-intracranial anastomoses and supply to the cranial nerves: vessels the neurointerventionalist needs to know. AJNR Am J Neuroradiol 30:1459

Gemmette JJ, Patel S, Pandey AS et al (2012) Preliminary experience with the percutaneous embolization of juvenile angiofibromas using only ethylene vinyl alcohol co-polymer (Onyx) for pre-operative devascularization prior to surgical resection. AJNR Am J Neuroradiol 33:1669

Geraud J, Manelfe C, Caussaud JP et al (1973) Fistule arteroveneuse spontanèe de l'artère vertebrale: role eventuelle de la displasie fibromuscolaire dans ça patologie. Rev Neurol (Paris) 128:206

Gerber JC, Daubner D, Kaiser D et al (2017) Efficacy and safety of direct aspiration first pass technique versus stent-retriever thrombectomy in acute basilar artery occlusion-a retrospective single center experience. Neuroradiology. 59:297–304

Ghika J, Bougosslavsky J, Regli F (1989) Infarcts in the territory of the deep perforators from the carotid system. Neuroradiology 39:507

Gibo H, Lenkey C, Rhoton AL Jr (1981a) Microsurgical anatomy of the supraclinoid portion of the internal carotid artery. J Neurosurg 55:560

Gibo H, Carver CC, Rhoton A et al (1981b) Microsurgical anatomy of the middle cerebral artery. J Neurosurg 54:151

Giudicelli G, Faure I, Salamon G (1970) The veins of the thalamus. Neuroradiology 1:92

Given CA, Huang-hellinger F, Baker MD et al (2001) Congenital Absence of the internal Carotid artery: case reports and review of the collateral circulation. AJNR Am J Neuroradiol 22:1953

Gjertsen O, Nakstad PHJ, Pedersen HKR et al (2007) Carotid pseudo-valvular fold: a probable cause of ischemic stroke. J Neurol 242:351–353

Glastonbury CM, Harnsberger HR, Hudgins PA et al (2012) Lateralized petrous internal carotid artery: imaging features and distinction from the aberrant internal carotid artery. Neuroradiology 54:1007

Glynn LE (1940) Medial defects in the circle of Willis and their relation to aneurysm formation. J Pathol Bacterial 51:213

Goddard AJP, Annesley-Williams D, Guthrie JA et al (2001) Duplication of the vertebral artery: report of two cases and review of the literature. Neuroradiology 43:477

Goldberg HL (1974) The anterior choroidal artery. In: Newton T, Potts DG (eds) Radiology of the skull and brain. Mosby, St. Louis

Goldstein JH, Woodcock R, Doa HM et al (1999) Complete duplication or extreme fenestration of the basilar artery. AJNR Am J Neuroradiol 20:149

Gomes F, Dujovny M, Umansky F (1984) Microsurgical anatomy of the recurrent artery of Heubner. J Neurosurg 60:130

Gonzales LF, Alexander MJ, McDougall CG et al (2004) Anteroinferior cerebellar artery aneurysms. Surgical approaches and outcomes-a review of 34 cases. Neurosurgery 55:1025

Gonzales AM, Narata AP, Ylma H et al (2014) Blod blister-like aneurysms: single center experience and systematic literature review. Eur J Radiol 83:197

Gonzalez LF, Bristol RE, Porter RW et al (2005) De novo presentation of an arteriovenous malformation. Case report. J Neurosurg 102:726

Goodman SJ, Hasso A, Kirkpatrick D (1975) Treatment of vertebro-jugular fistula by balloon occlusion. J Neurosurg 43:362

Govoni M, Padovan M, Rizzo N (2001) CNS involvement in primary Sjogren's syndrome: prevalence, clinical aspects, diagnostic assessment and therapeutic approach. CNS Drugs 15:597

Goyal M, Demchuk AM, Bk M et al (2015) Randomized assessment of rapid endovascular treatment in ischemic stroke. N Engl J Med 372:1019

Gozzoli L, Crasto S, Leombruni S et al (1998) Magnetic resonance angiography in a case of agenesis the left carotid artery associated with transellar intercarotid anastomosis. Riv Neuroradiol 11:883

Graber D, Flurin-Chollet V, Chaix Y et al (1992) Dissection de l'artere sylvienne avec anevrysme sequellaire chez un enfant de 6 ans. Arch Fr Pediatr 49:445

Graf CJ, Perret GE, Torner JC (1983) Bleeding from cerebral arteriovenous malformations as part of their natural history. J Neurosurg 58(3):331

Gratz PP, Schroth G, Gralla J et al (2015) Whole-Brain susceptibility—weighted thrombusi maging in stroke: fragmented thrombi predict worse outcome. AJNR Am J Neuroradiol 36:1277

Gray H (1859) Anatomy, descriptive and surgical. Blanchard and Lea, Philadelphia

Greenberg SM, Eng JA, Ning M et al (2004) Hemorrhage burden predict recurrent intracerebral hemorrhage after lobar hemorrhage. Stroke 35:1415

Greenberg SM, O'Donnell HC, Schaefer PW et al (1999) MRI detection of new hemorrhages: potential marker of progression in cerebral amyloid angiopathy. Neurology 53:1135

Greitz T, Lauren T (1968) Anterior meningeal branch of the vertebral artery. Acta Radiol 7:2

Gross BA, Akgoz A, Orbach DB (2013) Dural arteriovenous shunts in children. J Pediatr Neuroradiol 2:263

Grossman CB, Potts DG (1974) Arachnoid granulations: radiology and anatomy. Radiology 113:95

Grzyska U, Fieler J (2009) Pathophysiology and treatment of brain AVMs. Clin Neuroradiol 19:82

Guglielmi G, Vinuela F, Dion J et al (1991a) Electrothrombosis of saccular aneurysms via endovascular approach. II: preliminary clinical experience. J Neurosurg 75:8

Guglielmi G, Vinuela F, Sepetka I et al (1991b) Electrothrombosis of saccular aneurysms via endovascular approach. I: electrochemical basis, technique and experimental results. J Neurosurg 75:1

Guillon B, Berthet K, Benslamia L et al (2003) Infection and the risk of spontaneous cervical artery dissection. A case-control study. Stroke 34:e34–e81

Gull W (1859) Cases of aneurysm of the cerebral vessels. Guys Hosp Rep 5:281

Gunnel M, Awad IA, Finberg K et al (1996) A gender mutation as a cause of cerebral cavernous malformation in Hispanic Americans. N Engl J Med 334:946

Gupta R, Schumacher HG, Mangla S et al (2003) Urgent endovascular revascularization for symptomatic intracranial atherosclerotic stenosis. Neurology 61:1729

Gupta A, Purkayastha S, Krishnamoorthy T et al (2006) Endovascular treatment of direct carotid cavernous fistulae: a pictorial review. Neuroradiology 48:831

Gupta A, Shah AD, Zhang Z et al (2013) Variability in the position of the retropharyngeal Internal Carotid Artery: a potential surgical hazard. Laryngoscope 123:401

Gutierrez-Molina M, Caminero Rodriguez A, Martinez Garcia C et al (1994) Small arterial granular degeneration in familial Binswanger's syndrome. Acta Neuropathol 87:98

Guzman R, Grady MS (1999) An intracranial aneurysm on the feeding artery of cerebellar hemangioblastoma. J Neurosurg 91:136

Ha SY, Kim DI, Kim BM et al (2013) Cavernous malformations associated with dural arteriovenous shunts in the central nervous system. Neuroradiology 55:187

Haacke EM, Mittal S, Wu Z et al (2009) Susceptibility-weighted imaging: technical aspects and clinical application, part 1. AJNR Am J Neuroradiol 30:19

Hacker H (1968) Abflusswege der sylvischen Venengruppe. Radiologie 8:383

Hacker H (1974) Superficial supratentorial veins and dural sinuses. In: Newton T, Potts AG (eds) Angiography, Book 3. Mosby, St. Louis

Hacker H, Porrero M (1969) Darstellung und Bedeutung der vena ophthalmica im Carotids Angiogram. Fortschr Roentgen 110(5):656

Hajj-Ali RA, Calabrese LH (2009) Central nervous system vasculitis. Curr Opin Rheumatol 21:10

Halbach VV, Higashida RT, Hieshima GB et al (1987) Dural fistulas involving the transverse and sigmoid sinus: results of treatment in 28 patients. Radiology 163:443

Halbach VV, Higashida RT, Hieshima GB et al (1988a) Transvenous embolization of direct carotid cavernous fistula. AJNR Am J Neuroradiol 9:741

Halbach VV, Higashida RT, Hieshima GB (1988b) Treatment of vertebral arteriovenous fistulas. AJR Am J Roentgenol 150:405

Halbach VV, Higashida RT, Hieshima GB et al (1988c) Treatment of dural arteriovenous malformations involving the superior sagittal sinus. AJNR Am J Neuroradiol 9:337

Halbach VV, Higashida RT, Hieshima GB et al (1989) Transvenous embolization of dural fistulas involving the cavernous sinus. AJNR Am J Neuroradiol 10:337

Halbach VV, Higashida RT, Hieshima GB et al (1990) Dural arteriovenous fistulas supplied by ethmoidal arteries. Neurosurgery 26:816

Hallacq P, Piotin M, Moret J (2002) Endovascular occlusion of the posterior cerebral artery for the treatment of p2 segment aneurysms: retrospective review of a 10-year series. AJNR Am J Neuroradiol 23:1128

Hamada J, Yonekawa Y (1994) Spontaneous disappearance of a cerebral arteriovenous malformation: case report. Neurosurgery 34:171

Hamada Y, Goto K, Inoue T et al (1997) Histopathological aspects of dural arteriovenous fistulas in the transverse-sigmoid sinus region in nine patients. Neurosurgery 40:452

Han DH, Gwak HS, Chung CK (1993) Aneurysm at the origin of accessory middle cerebral artery associated with middle cerebral artery aplasia. Surg Neurol 42:388

Hanafee W, Rosen LM, Weidner GH et al (1965) Venography of the cavernous sinus orbital veins and basal venous plexus. Radiology 84:751

Hanafee W, Shiu P, Dayton GO (1968) Orbital venography. AJR Am J Roentgenol 104:29

Handa J, Kamijyo Y, Handa H (1970a) Intracranial aneurysm associated with fibro-muscular hyperplasia of renal and internal carotid arteries. Br J Radiol 43(511):483

Handa J, Shimizu Y, Matsuda M et al (1970b) The accessory middle cerebral artery. Report of further two cases. Clin Radiol 21:415

Handa J, Nakasu Y, Matsuda M et al (1984) Aneurysms of the proximal anterior cerebral artery. Surg Neurol 22:486

Hanggi D, Turowski B, Beseoglu K et al (2008) Intra-arterial nimodipine for severe cerebral vasospasm after aneurysmal subarachnoid hemorrhage: influence on clinical course and cerebral perfusion. AJNR Am J Neuroradiol 29(6):1053

Hardy DG, Rhoton AL Jr (1978) Microsurgical relationships of the superior cerebellar artery and the trigeminal nerve. J Neurosurg 49:669

Harnsberger HR (1995) Handbook of head and neck imaging (carotid space). Mosby, St Louis

Harwood-Nash DC (1974) Cerebral vessels in childhood. In: Newton TH, Potts DG (eds) Angiography, Book 1. Mosby Company, St. Louis

Hashimoto I (1977) Familial intracranial aneurysms and cerebral vascular anomalies. J Neurosurg 46:419

Hashimoto H, Ohnishi H, Yuasa T et al (1987) Duplicate origin of the vertebral artery: report of two cases. Neuroradiology 29:301

Hashimoto T, Lam T, Boudreau NJ et al (2001) Abnormal balance in the Angiopoietin-Tie 2 system in human brain arteriovenous malformations. Circ Res 89:111

Hashimoto T, Wu Y, Lawton MT et al (2005) Co-expression of angiogenetic factors in brain arteriovenous malformation. Neurosurgery 56:1058

Hassler O, Salzmann GF (1967) Angiographic and histological changes in infundibular widening of the posterior communicating artery. Acta Radiol 1:321

Hassler W, Zentner J, Voigt K (1989) Abnormal origin of the ophthalmic artery from the anterior cerebral artery: neuroradiological and intraoperative findings. Neuroradiology 21:85

Haughton VM, Rosenbaum EE (1974) The normal and anomalous aortic arc and brachio-cephalic arteries. In: Newton T, Potts DG (eds) Radiology of skull and brain. Mosby, St. Louis

Haughton VM, Rosenbaum EE, Pearce J (1978) Internal carotid artery origins of the inferior cerebellar arteries. Am J Roentgenol 130:1191

Haussen D, Jadhav A, Jovin T et al (2016) Endovascular management vs intravenous thrombolysis for acute stroke secondary to carotid artery dissection: local experience and systemic review. Neurosurgery 78:709

Hayes GJ (1963) External carotid cavernous sinus fistulas. J Neurosurg 20:692

Hayreh SS (1962) The ophthalmic artery. III-Branches. Br J Ophthalmol 46:212

Hayreh SS (1981) Anterior ischemic optic neuropathy. Arch Neurol 38:675

Hayreh SS, Dass R (1962a) The ophthalmic artery II—Intraorbital course. Br J Ophthalmol 46:165

Hayreh SS, Dass R (1962b) The ophthalmic artery. I—Origin, intracranial and intracanalicular course. Br J Ophthalmol 46:65

He W, Gandhi CD, Quinn J et al (2011) True aneurysms of the posterior communicating artery: a systematic review and meta-analysis of individual patient data. World Neurosurg 75:64

Headache Classification Committee of the International Headache Society (2004) Classification and diagnostic criteria for headache disorders, cranial neuralgias and facial pain, second edition. Cephalalgia 24(Suppl 1):1

Hegedus K (1985) Ectasia of the basilar artery with special reference to possible pathogenesis. Surg Neurol 24:463

Heiskanen O (1989) Ruptured intracranial arterial aneurysm of children and adolescents. Surgical and total management results. Childs Nerv Syst 5:66

Helgason C, Caplan LR, Goodman J et al (1986) Anterior choroidal artery territory infarction: case reports and review. Arch Neurol 43:681

Henkes H, Fisher S, Weber W et al (2004) Endovascular coil occlusion of 1811 intracranial aneurysms: early angiographic and clinical results. Neurosurgery 54:268

Henkes H, Liebig T, Reinartz J et al (2006) Endovaskulaerer Verschluss der a. basilaris zur Behandlung dissektionsbedingter und dysplastischer fusiformer Aneurysmen. Nervenarzt 77:192

Hennerici M, Klemm C, Rautemberg W (1988) The subclavian steal phenomenon: a common vascular disorder with rare neurologic deficits. Neurology 38:669

Hensler J, Jensen-Kondering U, Ulmer ST et al (2016) Spontaneous dissections of bthe anterior cerebral artery: a meta-analysis of the literature and three recent cases. Neuroradiology 58:997

Herman LH, Fernando OU, Gurdjian ES (1966) The anterior choroidal artery: an anatomical study of its area of distribution. Anat Rec 154:95

Heubner O (1872) Zur topographie der Ernährungsgebiet der einzelnen Hirnarterien. Cent Med Wissen 52:816

Hieshima GB, Cahan LD, Mehringer CM et al (1986) Spontaneous arteriovenous fistulae of cerebral vessels in association with fibromuscular dysplasia. Neurosurgery 18(4):454

Higgins JN, Owler BK, Cousins C et al (2002) Venous sinus stenting for refractory benign intracranial hypertension. Lancet 359:228

Hillbom M, Kaste M (1982) Alcohol intoxication: a risk factor for primary subarachnoid haemorrhage. Neurology 32:706

Hinton RG, Mohr JP, Hackerman RH et al (1979) Symptomatic middle cerebral artery stenosis. Ann Neurol 5:152

Hiraiwa K, Sato T, Sasaki T et al (2005) Medico-legal aspects of traumatic injury of the vertebro basilar artery. Neurol Med Chir (Tokyo) 45:549

Hirsch CS, Roessmann U (1975) Arterial dysplasia with ruptured basilar artery aneurysm: report of a case. Hum Pathol 6:749

Hochberg FH, Bean C, Fisher M et al (1975) Stroke in a 15-year old girl secondary to terminal carotid dissection. Neurology 25:725

Hochstetter F (1938) Ueber eine Varietaet der Vena cerebrali basialis des Menschen nebst Bemerkungen ueber der Entwickung bestimmter Hirnvenen. Z Anat Entwicklungsgesch 108:311

Hocker SE, Dafer RM, Hacein-Bey L (2008) Successful delayed thrombolysis for cerebral venous and dural sinus thrombosis: a case report and review of the literature. J Stroke Cerebrovasc Dis 17:429

Hoffman G, Cid M, Hellmann D et al (2002) A multicentric randomized, double-blind placebo-controlled study of adjuvant methotrexate treatment for giant cell arteritis. Arthritis Rheum 46:1309

Hofmeister C, Stapf C, Hartmann A et al (2000) Demographic morphological and clinical characteristics of 1289 patients with brain arteriovenous malformations. Stroke 31:1307

Hoh BL, Rabinov JD, Pryor JC et al (2004) Persistent non—fused segments of the basilar artery: longitudinal versus axial non—fusion. AJNR Am J Neuroradiol 25:1194

Hommel M, Pollak P, Gaio JM et al (1984) Paralyses du nerf grand hypoglosse par deux aneurysms et un aneurysme dissequant de l'artère carotide interne. Rev Neurol 140:415

Hong JM, Lee JH, Yong SW (2010a) Neurological picture. Asymptomatic rete mirabile in carotid and vertebral circulation systems. J Neurol Neurosurg Psychiatry 81:1022

Hong YJ, Chung TS, Suh SH et al (2010b) The angioarchitectural factors of the cerebral developmental venous anomaly: can they be the cause of concurrent sporadic cavernous malformations? Neuroradiology 52:883

Hong NR, Seo HS, Lee YH et al (2011) The correlation beween carotid siphon calcification and lacunar infarct. Neuroradiology 53:643

Horiuchi T, Tanaka Y, Hongo K et al (2003) Characteristics of distal posteroinferior cerebellar artery aneurysms. Neurosurgery 53:589

Horowitz DR, Tuhrim S (1997) Stroke mechanisms and clinical presentation in large subcortical infarction. Neurology 49:153

Hosoya T, Adachi M, Yamaguchi K et al (1999) Clinical and neuroradiological features of intracranial vertebrobasilar artery dissection. Stroke 30:1083

Houdart E, Gobin YP, Casasco A et al (1993) A proposal angiographic classification of intracranial arteriovenous fistulae and malformations. Neuroradiology 35:381

Houser OW, Baker HL Jr, Sandok BA et al (1971) Cephalic arterial fibromuscular dysplasia. Radiology 101:605

Houser OW, Baker HL Jr (1968) Fibromuscular dysplasia and other uncommon disease of the cervical carotid artery: angiographic aspects. Am J Roentgenol 104:201

Houser OW, Campbell JK, Campbell RJ et al (1979) Arteriovenous malformations affecting the transverse sinus. An acquired lesion. Mayo Clin Proc 54:651

Hoving JW, Marquering HA, Majoie Ch BL 2017. Endovascular treatment in patients with carotid artery dissection and intracranial occlusion: a systematic review. Neuroradiology 59:641

Hoyt F, Newton TH, Margolis TM (1974) The posterior cerebral artery. In: Newton TH, Potts DG (eds) Radiology of skull and brain, Angiography, 2nd edn. Mosby, St. Louis

Huang J, McGirt MJ, Gailloud P (2005) Intracranial aneurysms in the pediatric population: case series and literature review. Surg Neurol 63(5):424

Huang YP, Wolf BS (1964) Veins of the white matter of the cerebral hemispheres (the medullary veins): diagnostic importance in carotid angiography. Am J Roentgenol 92:739

Huang YP, Wolf BS (1965) The veins of the posterior fossa; superior or galenic draining group. Am J Roentgenol 95:808

Huang YP, Wolf BS (1967) The veins of the lateral recess of the IV ventricle and its tributaries; roentgen appearance and anatomic relationships. Am J Roentgenol 101:1

Huang YP, Wolf BS, Autin BS et al (1968) The veins of the posterior fossa, anterior or petrosal draining group. Am J Roentgenol 104:36

Huang YP, Wolf BS, Okudera T (1969) Angiographic anatomy of the inferior vermian vein of the cerebellum. Acta Radiol Diagn 9:327

Huber P (1979) Zerebrale Angiographie für Klinik und Praxis. Thieme, Stuttgart

Hudgins RJ, Day AL, Quisling RG et al (1983) Aneurysms of the posterior inferior cerebellar artery. A clinical and anatomical analysis. J Neurosurg 58:381

Hufnagel A, Hammers A, Schönle PW et al (1999) Stroke following chiropractic manipulation of the cervical spine. J Neurol 246(8):683

Huhn A (1957) Die Bedeutung der Serienangiographie fuer die Diagnose der Hirnvenen und sinus thrombose. Dtsch Z Nervenheilkd 17:7

Huhn A (1962) Ueber den diagnostischen Wert der Serienangiographie bei den intrakraniellen venoesen Thrombose. Fortschr Neurol Psychiatr 30:65

Hupperts RMM, Lodder J, Hents van Raak EPM et al (1994) Infarcts with anterior choroidal artery territory. Anatomical distribution, clinical syndromes, presumed pathogenesis and early outcome. Brain 117:825

Hutchings M, Weller RO (1986) Anatomical relationships of the pia mater to cerebral blood vessels in man. J Neurosurg 65:316

Hyogo T, Nakagawara J, Nakamura J et al (1996) Multiple segmental agenesis of the cerebral arteries: case report. Neuroradiology 38:433

Ide S, Kiyosue H, Tanoue S et al (2014) Anatomical variations in termination of the uncal vein and its clinical implications in cavernous sinus dural arteriovenous fistulas. Neuroradiology 56:661

Im SH, Han MH, Kwon BJ et al (2008) Venous-predominant parenchymal arteriovenous malformation: a rare subtype with a venous drainage pattern mimicking developmental venous anomaly. J Neurosurg 108:1142

Ingebrigtsen T, Morgan MK, Faulder K et al (2004) Bifurcation geometry and the presence of cerebral artery aneurisms. J Neurosurg 101:108

Ionete C, Omojola MF (2006) MR angiography demonstration of bilateral duplication of the extracranial vertebral artery: unusual course and review of the literature. AJNR Am J Neuroradiol 27:1304

Ishikawa T, Nakamura N, Houkin K et al (1997) Pathological consideration of a "blister-like" aneurysm at the superior wall of the internal carotid artery. Neurosurgery 40:403

Ito J, Imamura H, Kobayashi K et al (1983) Dural arteriovenous malformations of the anterior cranial fossa. Neuroradiology 24:149

Iwama T, Andoh T, Sakai N (1990) Dissecting and fusiform aneurysms of vertebro-basilar systems. Neuroradiology 32:272

Jackson C, Sudlow C (2005) Are lacunar strokes really different? A systematic review of differences in risk factor profiles between lacunar and non-lacunar infarcts. Stroke 36:891

Jackson RS, Wheeler AH, Darden BV (2000) Vertebral artery anomaly with atraumatic dissection causing thromboembolic ischemia: a case report. Spine 25:1989

Jacobson KE, Carlsson C, Elverson J et al (1984) Traumatic aneurysms of cerebral arteries. A study of five cases. Acta Neurochir 71:91

Jaeger JR, Forbes RP, Dandy WE (1937) Bilateral congenital cerebral arteriovenous communication aneurysm. Trans Am Neurol Assoc 63:173

Jain R, Deveikis J, Thompson BG (2004) Management of patients with stunned myocardium associated with subarachnoid hemorrhage. AJNR Am J Neuroradiol 25:126

Jakubowska MM, Michels P, Müller-Jensen A et al (2008) Endovascular treatment in proximal and intracranial carotid occlusion 9 hours after symptom onset. Neuroradiology 50:599

Jansen O, von Kummer R, Forsting M et al (1995) Thrombolytic therapy in acute occlusion of the intracranial internal carotid artery bifurcation. AJNR Am J Neuroradiol 10:1977

Janz C, Meisel J, Brock M et al (1998) De novo appearance of three cavernomas in an infant with dural AV shunts. Case report. Interv Neuroradiol 4:85

Jefferson G (1938) On the saccular aneurisms of the internal carotid artery in the cavernous sinus. Br J Surg 26(102):267–302

Jellinger K (1986) Vascular malformations of the central nervous system: a morphological overview. Neurosurg Rev 9:177

Jentzer A (1954) Dissecting aneurysm of the left internal carotid artery. Angiology 5:232

Jiang CH, Lv X, Li Y et al (2009) Endovascular treatment of high-risk tentorial dural arteriovenous fistulas: clinical out-comes. Neuroradiology 51:103

Johanson C (1954) The central veins and deep dural sinuses of the brain. Acta Radiol Suppl 107:1

Johnston I, Kollar C, Dunkley S et al (2002) Cranial venous outflow obstruction in the pseudotumour syndrome: incidence, nature and relevance. J Clin Neurosci 9:273

Jones BP, Ganesan V, Saunders DE et al (2010) Imaging in childhood arterial ischemic stroke. Neuroradiology 52:577

Joutel A, Corpechot C, Ducros A et al (1996) Notch3 mutations in CADASIL, a hereditary adult-onset condition causing stroke and dementia. Nature 383:707

Joux J, Chausson N, Jeannin S et al (2014) Carotid-bulb atypical fibromuscular dysplasia in young afro-caribbean patients with stroke. Stroke 45:3711

Jovin TG, Chamorro A, Cobo E et al (2015) Throbectomy within 8 hours after symptom onset in ischemic stroke. N Engl J Med 11:2296

Judd CD, Chapman PR, Koch B et al (2007) Intracranial infantile hemangiomas associated with PHACE syndrome. AJNR Am J Neuroradiol 28:25

Jussen D, Wiener E, Vajkoczy P et al (2012) Traumatic middle meningeal artery pseudoaneurysms. Diagnosis and endovascular treatment of two cases and review of the literature. Neuroradiology 54:1133

Juszkat R, Liebert W, Stanislawska K et al (2015) Extracranial internal carotid artery dissection treated with self—expandable stents: a single-centre experience. Cardiovasc Interv Radiol 38:1451

Juvela S, Hillbom M, Numminen H et al (1993) Cigarette smoke and alcohol consumption as risk factors for aneurismal subarachnoid haemorrhage. Stroke 24:639

Kaech D, De Tribolet B, Lasjaunias P (1987) Anterior inferior cerebellar arteries aneurysm, carotid bifurcation aneurysm and dural-venous malformation of the tentorium in the same patient. Neurosurgery 21:575

Kahara V, Letho U, Ryymin P et al (2002) Vertebral epidural arteriovenous fistula and radicular pain in neurofibromatosis type 1. Acta Neurochir 144:433

Kakou M, Destrieux CH, Velut ST (2000) Microanatomy of the pericallosal arterial complex. J Neurosurg 93:667

Kalimo H, Kaste M, Haltia M (1997) Vascular disease. In: Graham DI, Lantos PI (eds) Greenfield neuropathology, vol 1. Arnold, London. pp 317–318, 345–347

Kalmes DF, Hanel R, Lopes D et al (2015) International retrospective study of the pipeline embolization device: a multicenter aneurysm treatment study. AJNR Am J Neuroradiol 36:108

Kanai H, Nagai H, Wakabanashi S et al (1992) A large aneurysm of the persistent primitive hypoglossal artery. Neurosurgery 30:794

Kaneko T, Nomura M et al (2001) Serial neuroimaging of a growing thrombosed giant aneurysm of the distal anterior cerebral artery—case report. Neurol Med Chir 41:33

Kanner AA, Maimon S, Rappaport ZH (2000) Treatment of spontaneous carotid-cavernous fistula in Ehlers-Danlos syndrome by transvenous occlusion with Guglielmi detachable coils. Case report and review of the literature. J Neurosurg 93:689

Kanterman RY, Witt PD, Hseieh PS et al (1996) Klippel-Trenaunay syndrome: imaging findings and percutaneous intervention. AJR Am J Roentgenol 167:989

Kaplan HA (1959) The transcerebral venous system. An anatomical study. Arch Neurol Psychiatry 1:148

Kaplan HA (1965) The lateral perforating branches of the anterior and middle cerebral arteries. J Neurosurg 23:305

Kaplan HA, Browder A, Browder J (1973) Narrow and atretic transverse subdural sinuses: clinical significance. Ann Otol Rhinol Laryngol 82:351

Kaplan HA, Ford DH (1966) The brain vascular system. Elsevier, Amsterdam

Kapoor R, Kendall BE, Harrison MJG (1991) Permanent oculomotor palsy with occlusion of the internal carotid artery (letter). J Neurol Neurosurg Psychiatry 54:745

Kappelhof M, Marquering HA, Berkhemer OA et al (2015) Intra-arterial treatment of patients with acute ischemic stroke and internal carotid artery occlusion: a literature review. J Neurointerv Surg 7:8

Kappelle LS, Eliasziw M, Fox AJ et al (1999) Importance of intracranial arteriosclerotic disease in patients with intracranial stenosis of the internal carotid artery. Stroke 30:282

Kaptain GJ, Lanzino G, Do HM et al (1999) Posterior inferior cerebellar artery aneurysms associated with posterior fossa arteriovenous malformation: report of five cases and literature review. Surg Neurol 51:146

Karameshev A, Schroth G, Mordasini P et al (2010) Long term outcome of symptomatic severe ostial vertebral artery stenosis. Neuroradiology 52:371

Karcaaltincaba M, Haliloglu M, Ozkan E et al (2009) Non-invasive imaging of aberrant right subclavian artery pathologies and aberrant right vertebral artery. Br J Radiol 82:73

Kashiwagi J, Kiyosue H, Hori Y et al (2010) Endovascular recanalization of acute intracranial vertebrobasilar artery occlusion using local fibrinolysis and additional balloon angioplasty. Neuroradiology 52:361

Kasner SE, Hankins LL, Bratina P et al (1997) Magnetic resonance angiography demonstrates vascular healing of carotid and vertebral artery dissections. Stroke 28:1993

Katayama Y, Tsubokawa T, Miyazaki S et al (1991) Growth of totally thrombosed giant aneurysms within the posterior cranial fossa. Diagnostic and therapeutic considerations. Neuroradiology 33:168

Kato S, Ishihara H, Nakayama H et al (2007) Transvenous embolization of dural arteriovenous shunts of the cavernous sinus. Interv Neuroradiol 13:353

Katoaka H, Tanizawa E, Ueno S et al (2007) Spontaneous intracranial hypotension is associated with a risk of venous thrombosis and subdural hematoma. Cerebrovasc Dis 23:315

Katsaridis V, Papagiannaki C, Aimar E (2008) Curative embolization of cerebral arteriovenous malformations (AVMs) with Onyx in 101 patients. Neuroradiology 50:589

Katsuta T, Rhoton AL Jr, Matsushima T (1997) The jugular foramen: microsurgical anatomy and operative approaches. Neurosurgery 41:149

Kawase T, Mizukami M, Tazawa T (1979) The significance of lenticulostriate arteries in transient ischemic attack—neuroradiological and regional cerebral blood flow studies. Brain Nerve 31:1033

Kayan Y, Delagado Almandoz JE, Fease JL et al (2016) Incidence of delayed ipsilateral intraparenchymal hemorrhage after stent-assisted coilinjg of intracranialm aneurysms in high-volume single center. Neuroradiology 58:261

Kazui S, Saweda T, Naritomi H et al (1993) Angiographic evaluation of brain infarction limited to the anterior cerebral artery territory. Stroke 24:549

Kemmetmueller H (1911) Ueber eine seltene Varietaet der Arteria vertebralis. Anat Heft Wiesb 44:305

Kempster PA, Genaty RP, Gates PC (1988) Asymptomatic cerebral infarction in patients with chronic atrial fibrillation. Stroke 19:955

Kendall B (1983) Results of treatment of arteriovenous fistulae with the Debrun technique. AJNR Am J Neuroradiol 4:405

Kerber CW, Newton TH (1973) The macro- and microvascularization of the dura mater. Neuroradiology 6:175

Khan SM, Ho DW, Lazar JM et al (2014) Cerebral contrast retention after difficult cardiac catheterization: case report. SAGE Open Med Case Rep. 2. doi:10.1177/2050313X14530283

Khullar D, Andeejani AM, Bulsara KR (2010) Evolution of treatment options for vein of Galen malformations. J Neurosurg Pediatr 6:444

Kier EL (1974) Fetal cerebral arteries: a phylogenetic and ontogenetic study. In: Newton T, Potts DG (eds) Radiology of skull and brain. Mosby, St. Louis

Kikuchi K, Kowada M (1994) Case report: saccular aneurysm of the intraorbital ophthalmic artery. Br J Radiol 67:1134

Kilic K, Orakodogen M, Bakirci A et al (2005) Bilateral internal carotid to anterior cerebral artery anastomosis with anterior communicating artery aneurysm: technical case report. Neurosurgery 57(suppl 4):E400

Kim BC, Known OK, Oh CW et al (2014) Endovascular ICA trapping for ruptured blood blister like aneurysms: long-term results from a single center. Neuroradiology 56:211

Kim DJ, Kim DI, Shu SH et al (2006) Results of transvenous embolization of cavernous dural arteriovenous fistula: a single centre experience with emphasis on complications and management. AJNR Am J Neuroradiol 27:2078

Kim DJ, Terbrugge K, Krings T et al (2010) Spontaneous angiographic conversion of intracranial dural arteriovenous shunt: long-term follow-up in nontreated patients. Stroke 41:1488

Kim H, Su H, Weinsheimer S et al (2011) Brain arteriovenous malformation pathogenesis: a response-to-injury paradigm. Acta Neurochir Suppl 111:83

Kim JT, Heo SH, Lee SH et al (2009) An uncommon anastomosis of the posterior inferior cerebellar artery and the external carotid artery with the patent vertebrobasilar system. Br J Radiol 82:e171

King WA, Martin NA (1992) Intracerebral hemorrhage due to dural arteriovenous malformations and fistulae. Neurosurg Clin N Am 3:577

Kingsley DPE, Kendall BE, Moseley IF (1978) Superior sagittal sinus thrombosis, an evaluation of the changes demonstrated on computed tomography. J Neurol Neurosurg Psychiatry 41:1065

Kirsch M, Henkes H, Liebig T et al (2006) Endovascular management of dural carotid-cavernous sinus fistulas in 141 patients. Neuroradiology 48:486

Kirsch M, Liebig T, Kuehne D et al (2009) Endovascular management of dural arteriovenous fistulas of the transverse and sigmoid sinus in 150 patients. Neuroradiology 51:477

Kishore PRS, Kaufman AB, Melichar FA (1979) Intrasellar carotid anastomosis simulating pituitary adenoma. Radiology 132:381

Kittner SJ, Sharkness CM, Price TR et al (1991) Infarcts with a cardiac source of embolism in the NINCDS Stroke Data Bank: historical features. Neurology 40:281

Kiura Y, Ohba S, Shibukawa M et al (2007) Transfemoral transvenous embolization of dural arteriovenous fistulas involving the isolated transverse-sigmoid sinus. Interv Neuroradiol 13(Suppl 1):109

Kiyosue H, Tanoue S, Sagara Y et al (2008) The anterior medullary-anterior pontomesencephalic venous system and its bridging veins communicating to the dural sinuses: normal anatomy and drainage routes from dural arteriovenous fistulas. Neuroradiology 50:1013

Kleinschnitz C, De Mayer SF, Schwarz T et al (2009) Deficiency of von Willebrand factor protects mice from ischemic stroke. Blood 113:3600

Kliever MA, Carroll BA (1991) Ultrasound case of the day, internal carotid artery web (atypical fibromuscular dysplasia). Radiographics 11:504

Klimo P Jr, Rao G, Brockmeyer D (2007) Pediatric arteriovenous malformations: a 15-year experience with an emphasis on residual and recurrent lesions. Childs Nerv Sist 23:31

Klinkhamer AC (1969) Esophagography in anomalies of the aortic arch system. Lippincott Williams & Wilkins, Baltimore

Klisch J, Kubalek R, Scheufler KM et al (2005) Plasma vascular endothelial growth factor and serum soluble angiopoietin receptor sTIE-2 in patients with dural arteriovenous fistulas: a pilot study. Neuroradiology 47:10

Kobayashi H, Hasayashi M, Noguchi Y et al (1988) Dural arteriovenous malformations in the anterior cranial fossa. Surg Neurol 30:396

Kobayashi N, Miyachi S, Oi S et al (2011) Traumatic carotid-cavernous fistula associated with persistent primitive trigeminal artery treated by transarterial coil embolization—case report. Neurol Med Chir 51:37

Kobayashi S, Koike G, Orz Y et al (1995) Juxta-dural ring aneurysms of the internal carotid artery. J Clin Neurosci 2:345

Kocak B, Tureci E, Kizilkilic O et al (2013) Dissecting aneurysms of the posterior communicating artery itself: anatomical, diagnostic, clinical and therapeutical considerations. Neuroradiology 55:1103

Kocyigit A, Herek D, Balci YI (2015) Focal herniation of cerebral parenchyma into transverse sinus. J Neuroradiol 42:126

Kodama N, Suzuki J (1974) Cerebrovascular Moya-Moya disease. Third report-the study of the aging of the perforating branches and the possibility of collateral pathway. Neurol Med Chir 14:55

Kollar C, Johnston I, Parkweer G et al (1998) Dural arteriovenous fistula in association with heterotopic brain nodule in the transverse sinus. AJNR Am J Neuroradiol 19:1126

Komiyama M, Yamanaka K, Iwai Y et al (1999) Venous angiomas with arteriovenous shunts: report of three cases and review of the literature. Neurosurgery 44:1328

Kommerell B (1936) Verlagerung des Ösophagus durch eine abnorm verlaufende Arteria subclavia dextra (Arteria lusoria). Fortschr Geb Roentgenstrahlen 54:590–595

Komotar RJ, Zacharia BE, Valhora R et al (2007) Advances in vasospasm treatment and prevention. J Neurol Sci 261:134

Kono T, Morita H, Kuroiwa T et al (1994) Left ventricular wall motion abnormalities in patients with subarachnoid hemorrhage: neurogenic stunned myocardium. J Am Coll Cardiol 24:636

Konrad C, Nabavi DG, Junker R et al (2003) Spontaneous internal carotid artery dissection and alpha-1-antitrypsin deficiency. Acta Neurol Scand 107:233

Korn A, Bender B, Brodoefel H et al (2015) Grading of carotid artery stenosis in the presence of extensive calcifications: Dual-Energy CT Angiography in comparison with contrast-enhanced MR Angiography. Clin Neuroradiol 25:33

Kornieieva MA, Hadidy AM, Hinno SH (2017) Median pericallosal artery with double proximal origin: case report and clinical consideration. Surg Radiol Anat. doi: 10.1007/s00276-017-1852-9

Kraus JA, Stuper BK, Nahser HC et al (2000) Significantly increased prevalence of factor V Leiden in patients with dural arteriovenous fistulas. J Neurol 247:521

Krayenbühl H (1961) Die Thrombose der Hirnvenen und sinus. Acta Neurochir Suppl 7:248

Krayenbühl HA (1967) Cerebral venous and sinus thrombosis. Clin Neurosurg 14:1

Krayenbühl HA, Yaşargil MG, Flamm ES et al (1972) Microsurgical treatment of intracranial saccular aneurysms. J Neurosurg 37:678

Krings T, Ozanne A, Chug SM et al (2005) Neurovascular phenotypes in hereditary hemorrhagic telangiectasia patients according to age. Review of 50 consecutive patients aged 1 day-60 years. Neuroradiology 47:711

Krings T, Alvarez H, Reinacher P et al (2007a) Growth and rupture mechanisms of partially thrombosed aneurysms. Interv Neuroradiol 13:117

Krings T, Baccin CE, Alvarez H et al (2007b) Segment unfused basilar artery with kissing aneurysms: report of three cases and literature review. Acta Neurochir 149:567

Krings T, Geibprasert S, Luo CB et al (2007c) Segmental neurovascular syndrome in children. Neuroimaging Clin N Am 17:245

Krisch B (1988) Ultrastructure of the meninges at the site of penetration of veins through the dura mater with particular reference to Pacchionian granulations. Cell Tissue Res 251:621

Kubis N, Von Langsdorff D, Petitjean C et al (1999) Thrombotic carotid megabulb: fibromuscular dysplasia, septae and ischemic stroke. Neurology 52:883

Kudo T (1968) Spontaneous occlusion of the circle of Willis: a disease apparently confined to Japanese. Neurology 18:485

Küker W, Schulz U (2014) Choroid artery occlusion after treatment of ICA termination aneurysm: was it cause by stent induced spasm? Neuroradiology 56:259

Küker W (2007) Cerebral vasculitis: imaging signs revisited. Neuroradiology 49:471

Küker W, Gaertner S, Nagele T (2008) Vessel wall contrast enhancement: a diagnostic sign of cerebral vasculitis. Cerebrovascular Dis 26:23

Küker W, Downer J, Cellerini M (2012) Fatal haemorrhage after MCA dissection: how aggressive should we treat? Neuroradiology 54:65

Küker W, Downer J, Cellerini M et al (2011) Dissecting aneurysm of a dominant intracranial vertebral artery in fibromuscular dysplasia: flow diversion using multiple conventional stents. Neuroradiology 53:93

Kulcsar Z, Wetzel SG, Augsburger L et al (2010) Effect of flow diversion treatment on very small ruptured aneurysms. Neurosurgery 67:789

Kulcsar Z, Houdart E, Bonafé A et al (2011) Intra-aneurysmal thrombosis as a possible cause of delayed aneurysm rupture after flow-diversion treatment. AJNR Am J Neuroradiol 32:20

Kumar S, Goddeau RP Jr, Selim MH et al (2010) Atraumatic convexal subarachnoid hemorrhage: clinical presentation, imaging patterns and etiologies. Neurology 74:893

Kumar S, Justin EM, Mishra NK (2012) Fenestrated posterior inferior cerebellar artery with concomitant vertebro-basilar junction fenestration and vertebral artery aneurysm. Clin Neuroradiol 22:235

Kumar S, Mishra NK (2012) Middle meningeal artery arising from the basilar artery: report of a case and its probable embryological mechanism. J Neurointerv Surg 4:43

Kumral E, Bayulchem G, Ataç C et al (2004) Spectrum of superficial posterior cerebral artery infarcts. Clinical and outcome correlates. Eur J Neurol 11:237

Kuntze-Söderquist A, Kaijser M, Soederman M et al (2014) Mechanical thrombectomy in acute ischemic stroke. Experience from 6 years of practice. Neuroradiology 56:477

Kunze ST, Schiefer W (1971) Angiographic demonstration of a dissecting aneurysm of the middle cerebral artery. Neuroradiology 2:201

Kurata A, Ohmomo T, Miyasaka Y et al (2001) Coil embolization of the treatment of ruptured dissecting vertebral aneurysms. AJNR Am J Neuroradiol 22:11

Kurata A, Suzuki S, Iwamoto K et al (2012) A new transvenous approach to the carotid-cavernous sinus via the inferior petrooccipital vein. J Neurosurg 116:587

Kurino M, Yoshioka S, Ushio Y (2002) Spontaneous dissecting aneutrysms of anterior and middle cerebral artery associated with brain infarction. A case report and review of the literature. Surg Neurol 57:428

Kurosaki Y, Yoshida K, Endo H et al (2011) Association between carotid atherosclerotic plaque with high signal intensity on T1 weighted imaging and subsequent ipsilateral ischemic stroke. Neurosurgery 68:62

Kurosu A1, Fujii T, Ono G (2000) Distal superior cerebellar artery aneurysm. Br J Neurosurg 14:244

Kurre W, Berkefeld J (2008) Material and technique for coiling of cerebral aneurisms: how much scientifical evidence do we have. Neuroradiology 50:909

Kurre W, Chapot R, Mesnil de Rochemont R et al (2010) Intracranial stenting in atherosclerotic disease-recent results and challenges to face. Neuroradiology 52:633

Kurre W, Aguilar-Peréz M, Schmid E et al (2014) Clinical experience with the Reset stent retriever for the treatment of acute ischemic stroke—a review of 271 consecutive cases. Neuroradiology 56:397

Kurre W, Bansemir K, Aguilar-Peréz M et al (2016) Endovascular treatment of acute internal carotid artery dissections: technical considerations, clinical and angiographic outcome. Neuroradiology 58:1167

Kutty S, Brown K, Qureshi AM et al (2009) Maximal potential foramen diameter does not correlate with the type or frequency of neurological event prior to closure. Cardiology 113:111

Kutty S, Sengupta PP, Khandheira BK (2012) Patent Foramen Ovale. The known and the to be known. J Am Coll Cardiol 59:1665

Kwee RM (2010) Systemic review on the association between calcification in carotid plaques and clinical ischemic symptoms. J Vasc Surg 51:1015

Lacour JC, Ducrocq X, Anxionnat R et al (2000) Isolated dissection of the basilar artery. Rev Neurol 156:654

Lacout A, Khalil A, Figl A et al (2012) Vertebral arteria lusoria: a life-threatening condition for oesophageal surgery. Surg Radiol Anat 34:381

Laine E, Galibert P, Lopez C et al (1963) Aneurysmes arterioveineux intraduraux (developpèes dans l'epeisseur de la dure mère) de la fosse posterieure. Neurochirurgie 9:147

Lang J, Schaffrath H, Fisher G (1987) Weitere Befunde zu den rami diencephalici. Neurochirurgia 30:103

Lang J (1991) The floor of the posterior cranial fossa. In: Clinical anatomy of the posterior cranial fossa and its foramina. Thieme Medical, New York

Lantos PL, Vandenberg SR, Kleihues P (1997) Tumours of the nervous system. In: Graham DI, Lantos PL (eds) Greenfield's neuropathology. Arnold, London

Lasjaunias P, Moret J, Manelfe C et al (1977) Arterial anomalies at the base of the skull. Neuroradiology 13:267

Lasjaunias P, Doyon D (1978) The ascending pharyngeal arteries and the blood supply of the lower cranial nerves. J Neuroradiol 5:287

Lasjaunias P, Moret J, Theron J (1978a) The so-called anterior meningeal artery of the cervical vertebral artery. Neuroradiology 17:51

Lasjaunias P, Theron J, Moret J (1978b) The occipital artery. Neuroradiology 15:31

Lasjaunias P, Manelfe C, Roche A et al (1979) Segmental aplasia of the basilar artery in man. J Neuroradiol 6:127

Lasjaunias P (1981) Craniofacial and upper cervical arteries. Lippincott Williams & Wilkins, Baltimore

Lasjaunias P, Guibert-Tranier F, Braun JP (1981) The pharyngo-cerebellar or ascending pharyngeal artery origin of the posterior inferior cerebellar artery. J Neuroradiol 8:317

Lasjaunias P, Burrows P, Planet C (1986a) Developmental venous anomalies (DVA). The so called venous angioma. Neurosurg Rev 9:233

Lasjaunias P, Chiu M, Terbrugge K et al (1986b) Neurological manifestations of intracranial dural arteriovenous malformations. J Neurosurg 64:724

Lasjaunias P, Berenstein A (1990) Surgical neuroangiography, vol 3. Springer, Berlin

Lasjaunias P, Rodesch G (1993) Lesion types, hemodynamics and clinical spectrum. In: Awad IA, Barrow DL (eds) Dural arteriovenous malformations. AANS, Park Ridge

Lasjaunias P (1997) Vascular disease in neonates, infants and children. Springer, Berlin

Lasjaunias P, Berenstein A, Ter Brugge KG (2001) Surgical neuroangiography. Clinical, vascular anatomy, and variations, 2nd edn. Springer, Berlin

Lasjaunias P, Wuppalapati S, Alvarez H et al (2005) Intracranial aneurysms in children aged under 15 years: review of 59 consecutive children with 75 aneurysms. Childs Nerv Syst 21:437

Lasjaunias PL, Landrieu P, Rodesch G et al (2008) Cerebral proliferative angiopathy. Clinical and angiographic description of an entity different from cerebral AVMs. Stroke 39:878

Lassen NA (1966) The luxury perfusion syndrome and its possible relation to acute metabolic acidosis localized within the brain. Lancet 2:1113

Laughlin S, Terbrugge KG, Willinsky RA et al (1997) Endovascular management of pediatric intracranial aneurysms. Interv Neuroradiol 3:205

Lauthier S, Armstrong D, Domi T et al (2005) Post-varicella arteriopathy of childhood: natural history of vascular stenosis. Neurology 64:660

Lawton MT, Jacobowitz R, Spetzler RF (1997) Redefined role of angiogenesis in the pathogenesis of dural arteriovenous malformations. J Neurosurg 87:267

Lazinski D, Willinsky RA, Terbrugge K et al (2000) Dissecting aneurysms of the posterior cerebral artery: angioarchitecture and a review of the literature. Neuroradiology 42:128

Lazorthes G (1961) Vascularisation et circulation cerebrales. Masson, Paris

Lazorthes G, Gouaze A, Salamon G (1976) Vascularisation et circulation de l'encephale. Masson, Paris

Lazorthes G, Salamon G (1971) The arteries of the thalamus: an anatomical and radiological study. J Neurosurg 34:23

Lazzaro NA, Wright B, Castillo M et al (2010) Artery of percheron infarction: imaging patterns and clinical spectrum. AJNR Am J Neuroradiol 31(7):1283–1289. Published 18/3/2010

Le Gros Clark WE (1920) On the pacchionian bodies. J Anat 55:40

Leach JL, Jones BV, Tomsick TA et al (1996) Normal appearance of arachnoid granulations on contrast-enhanced CT and MR of the brain: differentiation from dural sinuses disease. AJNR Am J Neuroradiol 17:1523

Lechat P, Mas JL, Lescault G et al (1988) Prevalence of patent foramen ovale in patients with stroke. N Engl J Med 318:1148

Lee JH, CW O, Lee SH et al (2003) Aplasia of the internal carotid artery. Acta Neurochir 145:117

Lee JS, Yong SW, Bang OY et al (2006a) Comparison of spontaneous intracranial vertebral artery dissection with large artery disease. Arch Neurol 63:1738

Lee JS, Bang OY, Lee PH et al (2006b) Two cases of spontaneous middle cerebral artery dissection causing ischemic stroke. J Neurol Sci 250:162

Lee KS, Kelly DL Jr (1989) Intrasellar persistent trigeminal artery associated with a pituitary adenoma. Case report. J Neurosurg 70:271

Lee VH, JK O, Mulvagh SL et al (2006c) Mechanism in neurogenic stress cardiomyopathy after aneurysmal subarachnoidal hemorrhage. Neurocrit Care 5:243

Lee WJ, Choi HS, Jang J et al (2015) Non-stenotic intracranial arteries have atherosclerotic changes in acute stroke patients: a 3T MRI Study. Neuroradiology 57:1007

Lefkowitz MA, Teitelbaum GP, Giannotta SL (1996) Endovascular treatment of a dissecting posteroinferior cerebellar artery aneurysm. Case report. Neurosurgery 39:1036

Leibowitz R, Do HM, Marcellus ML et al (2003) Parent vessel occlusion for vertebrobasilar fusiform and dissecting aneurysms. AJNR Am J Neuroradiol 24:902

Lemke AJ, Benndorf G, Liebig T et al (1999) Anomalous origin of the right vertebral artery: review of the literature and case report of right vertebral origin distal to the left subclavian artery. AJNR Am J Neuroradiol 20:1318

Lenk S, Labeyrie MA, Saint-Maurice JP et al (2014) Diaphragms of the carotid and vertebral arteries: an underdiagnosed cause of ischemic stroke. Euro J Neurol 21:586

Leonardi M, Simonetti L, Andreoli A (2001) Endovascular treatment of a distal aneurysm of the superior cerebellar artery by intra-aneurysmal injection of glue. Interv Neuroradiol 7:343

Leonardi M, Cirillo l TF et al (2011) Treatment of intracranial aneurysms using flow-diverting silk stents (Balt): a single centre experience. Interv Neduroradiol 17:306

Lesley WS (2008) Fenestration of the posterior inferior cerebellar artery. Cerebellum 240:1

Letteboer TG, Mayer JJ, Snjjder RJ et al (2006) Genotype-phenotype relationship in hereditary hemorrhagic telangiectasia. J Med Genet 43:371

Leung GK, Tsang AC, Lui WM (2012) Pipeline embolization device for intracranial aneurysm: a systematic review. Clin Neuroradiol 22:295

Levine J, Misko JC, Seres JL et al (1973) Spontaneous angiographic disappearance of a cerebral arteriovenous malformation. Arch Neurol 28:195

Levine RL, Lagreze HL, Dobkin JA et al (1988) Large subcortical hemispheric infarctions: presentation and prognosis. Arch Neurol 45:1074

Levy C, Laissy JP, Raveau V et al (1994) Carotid and vertebral artery dissections: three dimensional time of flight MR angiography and MR imaging versus conventional angiography. Radiology 190:97

Levy R, Diuyckaerts C, Hauw JJ (1995) Massive infarcts involving the territory of the anterior choroidal artery and cardioembolism. Stroke 26:609

Levy EI, Turk AS, Albuquerque FC et al (2007) Wingspan in-stent restenosis and thrombosis: incidence. Clinical presentation and management. Neurosurgery 61:644

Lewis AI, Tomsick TA, Tew JM (1995) Management of 100 consecutive direct carotid—cavernous fistulas: results of treatment with detachable balloon. Neurosurgery 36:239

Lewis SB, Chang DJ, Peace DA et al (2002) Distal posterior inferior cerebellar aneurysms: clinical feature and management. J Neurosurg 97:756

Li MH, Li YD, Fang C et al (2007b) Endovascular treatment of giant or very large intracranial aneurysms with different modalities: an analysis of 20 cases. Neuroradiology 49:819

Li TL, Fang B, He XY et al (2005) Complication analysis of 469 brain AV malformations treated with N-butyl-cyanoacrylate. Interv Neuroradiol 11:141

Li Y, Lv X, Liu A et al (2007a) Endovascular treatment strategies for posterior cerebral artery aneurysms. Interv Neuroradiol 13(Suppl 2):208

Liang L, Korogi Y, Sugahara T et al (2002) Normal strurctures in the intracranial dural sinuses: delineation with 3D contrast-enhanced magnetization prepared rapid acquisition gradient-echo imaging sequence. AJNR Am J Neuroradiol 23:1739

Libby P, Clinton SW (1993) The role of macrophages in atherogenesis. Curr Opin Lipidol 4:355

Lie TA (1968) Congenital anomalies of the carotid arteries. Excerpta Medica Foundation, Amsterdam

Liebeskind DS, Flint AC, Budzik RF et al (2015) Carotid I's, L's and T's: collaterals shape the outcome of intracranial carotid occlusion in acute ischemic stroke. J Neurointerv Surg 7:402

Liebig T, Henkes H (2008) Re- and deconstruction: staged treatment of large fusiform aneurysm of the vertebrobasilar junction by stent deployment and subsequent coil occlusion of one vertebral artery. Clin Neuroradiol 18:255

Lieschke GJ, Davis S, Tress BM et al (1988) Spontaneous intracranial carotid artery dissection presenting as hypoglossal nerve palsy. Stroke 19:1151

Lin E, Linfante I, Dabus G (2013) Unilateral Rete Mirabile as a result of segmental agenesis of the ascending petrous segment of the internal carotid artery: embryology, differential diagnosis and clinical implications. Interv Neuroradiol 19:73

Lindergärd B, Hillbom M, Brody S (1987) High dose estrogen-progestogen oral contraceptives: a risk factor for aneurysmal subarachnoid haemorrhage? Acta Neurol Scand 76:37

Lindquist M, Karlsson B, Guo WY et al (2000) Angiographic long-term follow-up data for arteriovenous malformations. Neurosurgery 46:803

Linn JP, Kriceff I (1974) Normal anterior cerebral artery complex. In: Newton TH, Potts DG (eds) Radiology of skull and brain. Mosby, St. Louis

Linn J, Brückmann H (2010) Cerebral venous and dural sinus thrombosis. Clin Radiol 20:25

Linskey ME, Sekhar LN, Hirsch WJ et al (1990) Aneurysms of the intracavernous carotid artery: clinical presentation, radiographic features and pathogenesis. Neurosurgery 26:71

Lipchik EO, De Weese JA, Schenk EA et al (1974) Diaphragm-like obstructions of the human arterial tree. Radiology 113:43

Little JR, St LP, Weinstein M et al (1981) Giant fusiform aneurysms of the cerebral arteries. Stroke 12:183

Liu D, Ma X, Zhao F et al (2009) Intraosseous embolitherapy of central arteriovenous malformation in a Jaw. Long term experience with 8 cases. J Oral Maxillofac Surg 67:2380

Liu JP, Kricheff II (1974) The anterior cerebral artery complex. In: Newton T, Potts DG (eds) Radiology of the skull and brain, vol 2. Mosby, St. Louis

Liu HM, Shih HG, Huang YC et al (2001) Posterior cranial fossa arteriovenous fistula with presenting as carotico cavernous fistula. Neuroradiology 43:405

Liu HM, Wang YH, Chen YF et al (2003) Endovascular treatment of brainstem arteriovenous malformations. Safety and efficacy. Interv Neuroradiol 45:644

Liu XS, Zhao HL, Cao Y et al (2012) Comparison of atherosclerotic plaque characteristics by high-resolution black-blood MR imaging between patients with first-time and recurrent acute ischemic stroke. AJNR Am J Neuroradiol 33:1257

Liu P, Lv X, Li Y et al (2016) Endovascular treatment of A1 aneurysms of the anterior cerebral artery. Neurol India 64:694

Lo WWM, Solti-Bohmen LG, McElveen JT Jr (1985) Aberrant carotid artery: radiological diagnosis with emphasis on high resolution computed tomography. Radiographics 5:985

Lobotesis K, Mahady K, Ganesalingam J et al (2015) Coiling associated delayed cerebral hypersensitivity: is nickel the link? Neurology 84:97

Locksley HB (1966a) Natural history of subarachnoid hemorrhage, intracranial aneurysms and arteriovenous malformations. J Neurosurg 25:321

Locksley HB (1966b) Report on cooperative study of intracranial aneurysms and subarachnoid haemorrhage. Section V. J Neurosurg 25:219

Lombardi G, Passerini SA (1967) The orbital veins. Am J Ophthalmol 64:440

Loumiotis I, D'Urso PI, Tawk R et al (2012) Endovascular treatment of ruptured paraclinoid aneurysms: results, complications and follow-up. AJNR Am J Neuroradiol 33:632

Lu X, Jang C, Li Y et al (2009) Endovascular treatment for pediatric intracranial aneurysms. Neuroradiology 51:749

Lubicz B, Leclerc X, Gauvrit IY et al (2003a) Endovascular treatment of peripheral cerebellar artery aneurysms. AJNR Am J Neuroradiol 24:1209

Lubicz B, Gauvrit JY, Leclerc X et al (2003b) Giant aneurysms of the internal carotid artery: endovascular treatment and long-term follow-up. Neuroradiology 45:650

Lubicz B, Lefranc F, Levivier M et al (2006) Endovascular treatment of intracranial aneurysms with a branch arising from the sac. AJNR Am J Neuroradiol 27:142

Lubicz B, Collignon L, Lefranc F et al (2008) Circumferential and fusiform intracranial aneurysms: reconstructive endovascular treatment with self-expandable stents. Neuroradiology 50:499

Lubicz B, Bruneau M, Dewindt H et al (2009) Endovascular treatment of proximal anterior cerebral artery aneurysms. Neuroradiology 51:99

Lubicz B, Kadou A, Morais R et al (2017) Leo Stent for endovascular treatment of intracranial aneurysms; very long – term results in fifty patients with 52 aneurysms and literature review. Neuroradiology. 59:271

Lucivero V, Mezzapesa DM, Petruzzellis M et al (2004) Ischaemic stroke in progressive systemic sclerosis. Neurol Sci 25:230

Lv M, Lv X, Li Y et al (2009) Dissecting aneurysm at the proximal anterior cerebral artery treated by parent artery occlusion. Interv Neuroradiol 15:123

Lv MM, Fan XD, LX S et al (2013) Pre-operative direct puncture embolization of advanced juvenile nasopharyngeal angiofibroma in combination with transarterial embolization: an analysis of 22 consecutive patients. Cardiovasc Intervent Radiol 36:111

Lv X, Li Y, Liu A et al (2007) Endovascular treatment of intracranial Giant serpentine aneurysms. Neuroradiol J 20:237

Lv X, Li Y, Wu Z (2008) Endovascular treatment of anterior cerebral fossa dural arteriovenous fistula. Neuroradiology 50:433

Lylyk P, Vinuela F, Dion JE et al (1993) Therapeutic alternatives of vein pf Galen vascular malformations. J Neurosurg 78:438

Ma D, Min L, Yang H et al (2010) Diagnosis and surgical treatment of carotid body tumor: a report of 18 cases. J Cardiovasc Dis Res 1:122

Mabe H, Furuse M (1977) Spontaneous disappearance of a cerebral arteriovenous malformation in infancy: case report. J Neurosurg 46:811

Macdonald JHM, Millar JS, Barker CS (2010) Endovascular treatment of cranial dural arteriovenous fistulae: a single center, 14-years experience and the impact of onyx on local practice. Neuroradiology 52:387

Machi P, Constalat V, Lobotesis K et al (2012) Solitarie FR thrombectomy system: immediate results in 56 consecutive acute ischemic stroke patients. J Neurointerv Surg 4:62

Mafee MF (2003) Orbit: embryology, anatomy, and pathology. In: Som PM, Curtin HD (eds) Head and neck imaging. Mosby, St. Louis

Magnus T, Mellms A, Kotter I et al (2012) Neurological manifestations of vasculitis an primary central nervous system vasculitis. Z Rheumatol 71:551

Mahadevan J, Batista L, Alvarez H et al (2004) Bilateral segmental regression of the carotid and vertebral arteries with rete compensation in a western patient. Neuroradiology 46:444

Maimon S, Saraf-Lavi E, Rappaport ZH et al (2006) Endovascular treatment of isolated dissecting aneurysm of the posterior inferior cerebellar artery. AJNR Am J Neuroradiol 27:527

Maisonpierre PC, Suri C, Jones PF et al (1997) Angiopoietin-2, a natural antagonist for TIE-2 that disrupts in vivo angiogenesis. Science 277:55

Maitland CG, Black JL, Smith WA (1983) Abducens nerve palsy due to spontaneous dissection of internal carotid artery. Arch Neurol 40:448

Makiyama Y, Nishimoto H, Fukaya C et al (1994) Massive intracerebral hematoma in a child with Klippel-Trenaunay syndrome. Surg Neurol 42:392

Makos MM, McComb RD, Hart MN et al (1977) Alphaglucosidase deficiency and basilar artery aneurysm. Report of a sibship. Ann Neurol 22:629

Malekzadehlashkariani S, Wanke I, Ruefenacht DA et al (2016) Brain herniations into arachnoid granulations: about 68 cases in 38 pqatients and review of the literature. Neuroradiology 58:443

Malik AM, Vora NA, Lin R et al (2011) Endovascular treatment of tandem ecxtracranial—intracranial anterior circulation occlusions: preliminary single—center experience. Stroke 42:1653

Manabe H, Hatayama T, Hasegawa S et al (2000) Coil embolisation for ruptured vertebral artery dissection distal to the origin of the posterior inferior cerebellar artery. Neuroradiology 42:384

Manabe S, Satoh K, Matsubara S et al (2008) Characteristics, diagnosis and treatment of hypoglossal canal dural arteriovenous fistula: report of nine cases. Neuroradiology 50:715

Manelfe C, Berenstein A (1980) Treatment of carotid cavernous fistulas by venous approach. J Neuroradiol 7:13

Manelfe C, Clausse J, Fredy D et al (1974) Dysplasie fibromuscolaire des artères cervicocephaliques. A propos de 70 cas. J Neuroradiol (Masson, Paris) 1:149

Mangiafico S, Cellerini M, Nencini P et al (2005) Intravenous glycoprotein IIb/IIIa inhibitor (Tirofiban) followed by intraarterial urokinase and mechanical thrombolysis in stroke. AJNR Am J Neuroradiol 26:2515

Mani RL, Newton TH, Glickmann MG (1968) The superior cerebellar artery: an anatomical roentgenographic correlation. Radiology 91:1102

Mansur OY, Weber J, Niesen W et al (2011) Carotid angioplasty and stenting without protections devices. Clin Neuroradiol 21:65

Mapstone T, Spetzler RF (1982) Vertebrobasilar insufficiency secondary to vertebral artery occlusion from a fibrous band. Case report. J Neurosurg 56:581

Maraire JN, Awad IA (1995) Intracranial cavernous malformations: lesion behavior and management strategies. Neurosurgery 37:591

Marcucci C, Thomas F, Sewell DA (2009) Retropharyngeal carotid artery: an important anatomic variation for the anesthesiologist. Anesthesiology 111:454

Margolis MT, Newton TH (1974) The posterior inferior cerebellar artery. In: Newton TH, Potts DG (eds) Radiology of the skull and brain. Mosby, St. Louis

Margolis MT, Newton TH, Hoyt WT (1974) The posterior cerebral artery. In: Newton TH, Potts DG (eds) Radiology of the skull and brain. Mosby, St. Louis

Marie P (1901) Des foyers lacunaire de dèsintegration et de differents autres ètats cavitaire du cerveau. Rev Med 21:281

Marinković S, Milisavljevic M, Marinković Z (1990) Branches of the anterior communicating artery. Acta Neurochir 106:78

Markowski J (1922) Bull. Inte. Acad. Pol. Scien. et Lettres. Classe des sciences mathematiques et naturelles. Cracovie

Marks MP, Lane B, Steinberg GK et al (1992) Intranidal aneurysms in cerebral arteriovenous malformations: evaluation and endovascular treatment. Radiology 183:355

Marquering HA, Nederkoorn PJ, Bloeker L et al (2013) Intracranial carotid artery disease in patients with recent neurological symptoms High prevalence on CTA. Neuroradiology 55:179

Marshman L, Ward P, Walter P et al (1998) The progression of an infundibulum to aneurysm formation and rupture: case report and literature review. Neurosurgery 43:1445

Martin NA, King WA, Wilson CB et al (1990) Management of dural arteriovenous malformations of the anterior cranial fossa. J Neurosurg 72:692

Martinez-Galdamez M, Romance A, Vega P et al (2015) Pipeline endovascular device for the treatment of intracranial aneurysms at the level, of the circle of Willis and beyond: multicenter experience. J Neurointerv Surg 7:816

Massa-Micon B, Luparello V, Bergui M et al (2000) De novo cavernoma: case report and review of literature. Surg Neurol 53:484

Massaro AR, Sacco RL, Mohr JP et al (1991) Clinical discriminators of labor and deep hemorrhages. The stroke Data Bank. Neurology 41:1881

Massimi L, Moret J, Tamburini G (2003) Dissecting giant vertebro-basilar aneurysm. Childs Nerv Syst 19:204

Massoud TF, Anslow P, Molyneux AJ (1992) Subarachnoid haemorrhage following spontaneous intracranial carotid artery dissection. Neuroradiology 34:33

Mast H, Young WL, Koennecke HC et al (1997) Risk of spontaneous haemorrhage after diagnosis of cerebral arteriovenous malformation. Lancet 350:1065

Masuda J, Ogata J, Yutani C (1993) Smooth muscle cell proliferation and localization of macrophages and T-cell in the occlusive intracranial major arteries in moya-moya disease. Stroke 24:1960

Matouk CC, Kaderali Z, Ter Brugge KG et al (2012) Long term clinical and imaging follow up of complex intracranial aneurysms treated by endovascular parent vessel occlusion. AJNR Am J Neuroradiol 33:1991

Matsubara S, Mandzia JI, Ter Brugge K et al (2000) Angiographic and clinical characteristic of patients with cerebral arteriovenous malformation associated with hereditary hemorrhagic telangiectasia. AJNR Am J Neuroradiol 21:1016

Matsumoto Y, Nagashima H, Toriyama T et al (2008) Uncommon course for a vertebral artery dissection: rupture, occlusion and recanalisation. J Clin Neurosci 15:700

Matsushima T, Rhoton AL Jr, de Oliveira E et al (1983) Microsurgical anatomy of the veins of the posterior fossa. J Neurosurg 59:63

Matula C, Trattnig S, Tschabitscher M et al (1997) The course of the prevertebral segment of the vertebral artery: anatomy and clinical significance. Surg Neurol 48:125

Maurer CJ, Joachimski F, Berlis A (2015) Two in one: endovascular treatment of Acute Tandem Occlusions in the Anterior Circulation. Clin Neuroradiol 25:397

Mawad ME, Hilal SK, Michelson WJ et al (1984) Occlusive disease with cerebral arteriovenous malformations. Radiology 153:401

Mayer SA, Lin J, Homma S et al (1999) Myocardial injury and left ventricular performance after subarachnoid hemorrhage. Stroke 30:780

Mazighi M, Labreuche J, Gongora-Rivera F et al (2008) Autopsy prevalence of intracranial atherosclerosis in patients with fatal stroke. Stroke 39:1142

McCormick WF (1966) The pathology of vascular arteriovenous malformations. J Neurosurg 24:807

McCormick WF, Schochet SS Jr (1976) Atlas of cerebral disease. Saunders, Philadelphia

McCormick WF (1984) Pathology of vascular malformations of the brain. In: Wilson CB, Stein BM (eds) Intracranial arteriovenous malformations. Lippincott Williams & Wilkins, Baltimore, pp 44–63

McSweeney N, Brew S, Bhate S et al (2010) Management and outcome of vein of Galen malformation. Arch Dis Child 95:903

McDougall CG, Halbach VV, Christopher F et al (1997) Dural arteriovenous fistulas of the marginal sinus. AJNR Am J Neuroradiol 18:1565

McLennan JE, Rosenbaum AE, Haughton VM (1974) Internal carotid origin of the middle meningeal artery. The ophthalmic middle meningeal and stapedial–middle meningeal arteries. Neuroradiology 7:265

Meckel S, Spittau B, McAuliffe W (2013a) The persistent trigeminal artery: development, imaging anatomy, variants, and associated vascular pathologies. Neuroradiology 55:5

Meckel S, McAuliffe W, Fiorella D et al (2013b) Endovascular treatment of complex aneurysms at the vertebrobasilar Junction with flow-diverting stents: initial experience. Neurosurgery 73:386

Meila D, Saliou G, Krings T (2015) Subcallosal artery stroke: infarction of the fornix and the genu of then corpus callosum. The importance of the anterior communicating complex. Case series and review of the literature. Neuroradiology 57:4

Meila D, Tysiac M, Petersen M et al (2012) Origin and course of the extracranial vertebral artery: CT findings and embryologic considerations. Clin Neuroradiol 22:327

Mellado JM, Merino X, Ramos A et al (2001) Agenesis of the internal carotid artery with a trans-sellar anastomosis: CT and MRI findings in late onset congenital hypopituitarism. Neuroradiology 43:237

Mendez MF, Adams NL, Lewandowski KS (1989) Neurobehavioral changes associated with caudate lesions. Neurology 39:349

Menon RK, Norris JW (2008) Cervical arterial dissection. Current concepts. Ann N Y Acad Sci 1142:200

Menovsky T, Grotenhuis JA, Bartels RHMA (2002) Aneurysm of the anterior inferior cerebellar artery (AICA) associated with high-flow lesion: report of two cases and review of literature. J Clin Neurosci 9:207

Mercier P, Brassier G, Cronier D et al (1993) Microsurgical anatomy of the perforating arteries of the anterior part of the circle of Willis. Rivista di Neuroradiologia 6(Sup 3):31

Mericle RA, Reig AS, Burry MV et al (2006) Endovascular surgery for proximal posterior inferior cerebellar artery aneurysms: an analysis of Glasgow outcome score by Hunt-hess grades. Neurosurgery 58:619

Metry DW, Dowd CF, Barkovich AJ et al (2001) The many faces of PHACE syndrome. J Pediatr 139:117

Meyer FB, Sundt TM, Fode NC et al (1989) Cerebral aneurysms in childhood and adolescence. J Neurosurg 70:420

Meyer FB, Lombardi D, Scheithauer B et al (1990) Extraaxial cavernous hemangiomas involving the dural sinuses. J Neurosurg 73:187

Meyerson L, Lazar SJ (1971) Intraorbital aneurysm of the ophthalmic artery. Br J Ophthalmol 55:199

Miao Z, Wang B, Feng L et al (2011) Primary angioplasty for a subtype of symptomatic middle cerebral artery stenosis. Neuroradiology 53:651

Michotey P, Grisoli F, Raybaud C et al (1974) Etude anatomique et radiologique de l'artère cèrèbrale moyenne. Procèdè de rerpèrage. Ann Radiol 17:721

Mickle JP, Quisling RG (1994) Vein of Galen fistulas. Neurosurg Clin N Am 5:529

Midkiff RB, Boykin MW, McFarland DR et al (1995) Agenesis of the in ternal carotid artery with intercavernous anastomosis. AJNR Am J Neuroradiol 16:1356

Milandre L, Gueriot C, Giraud N et al (1988) Les thromboses veineuses cèrèbrales de l'adulte. Ann Med Interne 139:544

Milandre L, Brosset C, Botti G et al (1994) Etude de 82 infarctus du territoire des artères cerebrales posterieures. Rev Neurol 150:133

Milenkovic Z (1985) Anastomosis between internal carotid artery and anterior cerebral artery with other anomalies of the circle of Willis in a fetal brain. J Neurosurg 55:701

Miller AA, Sobey Ch G (2011) Vascular biology and atherosclerosis of cerebral arteries. In: Mohr JP, Wolf Ph A, Grotta JC, Moskowitz MA, Mayberg MR, Von Kummer R (eds) Stroke, Pathphysiology, diagnosis and management. Elsevier Saunders, Philadelphia

Miller DL, Doppman JL, Chang R (1993) Anatomy of the junction of the inferior petrosal sinus and the internal jugular vein. AJNR Am J Neuroradiol 14:1075

Miller P, Fabian T, Bee T et al (2001) Blunt cerebrovascular injuries: diagnosis and treatment. J Trauma 51:279

Miller RE, Hieshima GB, Giannotta SL et al (1984) Acute traumatic vertebral arteriovenous fistula: balloon occlusion with use of a contralateral approach. Neurosurgery 14:225

Min WK, Park KK, Kim JS et al (2000) Atherosclerotic middle cerebral artery territory infarction. Topographic diversity with common occurrence of concomitant small cortical and subcortical infarcts. Stroke 31:2055

Minagi H, Newton TH (1966) Carotid rete mirabile in man. A case report. Radiology 86:100

Mineyko A, Kirston A, Ng D et al (2013) Normal intracranial periarterial enhancement on pediatric brain MR imaging. Neuroradiology 55:1161

Miravet E, Dauchaivijitz N, Basu H et al (2007) Clinical and radiological features of childhood cerebral infarction following varicella-zoster virus infection. Dev Med Child Neurol 49:417

Mironov A (1995) Classification of spontaneous dural arteriovenous fistulas with regard to their pathogenesis. Acta Radiol 36:582

Mishkin MM, Schreiber MN (1974) Collateral circulation. In: Newton TH, Potts AG (eds) Angiography, Book 4. Mosby, St. Louis

Mitsias P, Levine SR (1994) Large cerebral vessel occlusive disease in Systemic Lupus Erythematosus. Neurology 44:385

Mitsos AP, Corkhill RA, Lalloo S et al (2008) Idiopathic aneurysms of distal cerebellar arteries: endovascular treatment after rupture. Neuroradiology 50:261

Mitsuhashi Y, Nishio A, Kawahara S et al (2007) Morphologic evaluation of the caudal end of the inferior petrosal sinus using 3D rotational venography. AJNR Am J Neuroradiol 28:1179

Miyachi S, Ohshima T, Isumi T et al (2008) Dural arteriovenous fistula at the anterior confluence. Interv Neuroradiol 14:303

Miyamoto N, Naito I, Takatama S et al (2009) Clinical and angiographic characteristics of cavernous sinus dural arteriovenous fistulas manifesting as venous infarction and/or intracranial hemorrhage. Neuroradiology 51:53

Mizutani T, Tanaka H, Aruga T (1995) Total recanalization of a spontaneously thrombosed arteriovenous malformation. J Neurosurg 82:506

Mizutani T (1998) Subarachnoid haemorrhage associated with angiographic "stenotic" or "occlusive" lesions in the carotid circulation. Surg Neurol 49:495

Mizutani T, Kojima H, Asamoto S et al (2001) Pathological mechanism and three-dimensional structure of cerebral dissecting aneurysms. J Neurosurg 94:712

Moehlenbruch M, Seifert M, Okulla T et al (2012) Mechanical thrombectomy compared to local intraarterial thrombolysis in carotid T and middle cerebral artery occlusions. Clin Neuroradiol 22:141

Moffat AB (1959) Developmental changes in the aortic arch system of the rats. Am J Anat 105:1

Mohr JP, Moskovitz AF, Stapf CH et al (2010) The ARUBA trial: current status, future hopes. Stroke 41:537

Mohr JP, Parides MK, Stapf C et al (2014) Medical management with or without interventional therapy for unruptured brain arteriovenous malformations (ARUBA). A multicenter, non—blinded, randomized trial. Lancet 383:614

Mokri B, Sundt TM Jr, Houser OW et al (1979) Spontaneous internal carotid dissection, hemicrania and Horner's syndrome. Arch Neurol 36:677

Mokri B, Sundt TM Jr, Hauser OW et al (1986) Spontaneous dissection of the cervical internal carotid artery. Ann Neurol 19:126

Mokri B, Houser OW, Sandok BA et al (1988) Spontaneous dissection of the vertebral arteries. Neurology 38:880

Mokri B, Siebert PL, Schievink WI et al (1996) Cranial nerve palsy in spontaneous dissection of the extracranial internal carotid artery. Neurology 46:356

Molineux AJ, Kerr SC, LM Y et al (2002) International subarachnoid aneurism trial (ISAT) of neurosurgical clipping versus endovascular coiling in 2143 patients with ruptured intracranial aneurisms; a randomized comparison of effect on survival dependency seizures rebleeding subgroup and aneurism occlusion. Lancet 366:809

Moniz E (1940) Die Cerebrale Arteriographie and Phlebographie. Springer, Berlin

Moody DM, Bell MA, Challa VR (1988) The corpus callosum, a unique white matter tracyt anatomic features that may explain sparing in Binswanger disease and resistance to flow on fluid masses. Am J Neuroradiol 9:1051

Moody DM, Bell MA, Challa VR (1990) Features of the cerebral vascular pattern that predict vulnerability to perfusion in oxygenation deficiency: an antomic study. Am J Neuroradiol 11:431

Moody DM, Santamore VP, Bell MA (1991) Does tortuosity in cerebral arterioles inpair down autoregulation in hypertensive and elderly normotensive? A hypothesis and computer model. Clin Neurosurg 37:372

Moon WJ, Porto L, Laufermann H et al (2002) Agenesis of internal carotid artery associated with congenital anterior hypopituitarism. Neuroradiology 44:138

Morales H, Jones BV, Leach JL et al (2010) Documented development of a dural arteriovenous fistula in an infant subsequent to sinus thrombosis: case report and review of the literature. Neuroradiology 52:225

Mordasini P, Brekenfeld C, Byrne JV et al (2013) Technical feasibility and application of mechanical thrombectomy with the solitaire FR revascularization device in acute basilar artery occlusion. AJNR Am J Neuroradiol 34:159

Morello A, Cooper IS (1955) Angiographic anatomy of the anterior choroidal artery. Am J Roentgenol 73:748

Moret J, Lasjaunias P, Theron J et al (1977) The middle meningeal artery; its contribution to the vascularization of the orbit. J Neuroradiol 4:225

Moret J, Lasjaunias P, Doyon D (1979) Occipital approach for treatment of arteriovenous malformations of the vertebral artery by balloon occlusion. Neuroradiology 17:269

Moret J, Lasjaunias P, Theron J (1980) Vascular compartments and territory of timpano-giugular tumors. J Belge Radiol 63:321

Moret J, Cognard C, Weill A et al (1997) The "remodelling technique" in the treatment of wide neck intracranial aneurisms. Interv Neuroradiol 3:21

Morgagni GB (1761) De sedibus, et causis morborum, per anatomen indagatis. Typographia Remondiniana, Venice

Morgenlander JC, Goldstein LB (1991) Recurrent transient ischemic attacks and stroke in association with an internal carotid artery. Stroke 22:94

Morimoto T, Nitta K, Kazekawa K et al (1990) The anomaly of a non bifurcating cervical carotid artery. Case report. J Neurosurg 72:130

Morris P (1997) Practical neuroangiography. Lippincott Williams & Wilkins, Baltimore

Moscow NP, Michotey P, Salamon G (1974) Anatomy of the cortical branches of the anterior cerebral artery. Section II. In: Newton TH, Potts AG (eds) Radiology of the skull and brain. Mosby, St. Louis

Mosimann PJ, Saint-Maurice JP, Leuck S et al (2012) Fornix infarction and Korsakoff dementia after coiling of a large anterior communicating artery aneurysm. Neurol Clin Pract 2:260

Mouchtouris N, Chalouhi N, Chitale A et al (2015) Management of cerebral cavernous malformations: from diagnosis to treatment. Scientific World Journal

Mounayer C, Hammami N, Piotin M et al (2007) Nidal embolization of brain arteriovenous malformations using Onyx in 94 patients. AJNR Am J Neuroradiol 28:518

Moyer DJ, Flamm ES (1992) Anomalous arrangement of the origins of the anterior choroidal and posterior communicating arteries. J Neurosurg 76:1017

Mpotsaris A, Bussmeyer M, Bucher H et al (2013) Clinical outcome of Neurointerventional Emergency Mechanic Treatment of Extra-or Intracranial Tandem occlusions in acute Major Stroke: anterograde approach with Wallstent and Solitaire Stent Retriever. Clin Neuroradiol 23:207

Mueller M, Schmitz BL, Pauls S et al (2011) Variations of the aortic arch-a study on the most common branching patterns. Acta Radiol 52:738

Mukherji SK (2003) Pharynx in head and neck imaging. In: Som PM, Curtin HD (eds) . Mosby, St. Louis

Mukonoweshuro W, Laitt RD, Hughes DG (2003) Endovascular treatment of PICA aneurysms. Neuroradiology 45:188

Mullan S (1979) Treatment of carotid-cavernous fistulas by cavernous sinus occlusion. J Neurosurg 50:131

Mullan S, Mojtahedii S, Johnson DL et al (1996a) Cerebral venous malformations-arteriovenous malformations transition forms. J Neurosurg 85:9

Mullan S, Mojtahedii S, Johnson DL et al (1996b) Embryological basis of some aspects of cerebral vascular fistulas and malformations. J Neurosurg 85:1

Müller DP, Sato Y, WTC Y (1991) Accessory middle cerebral artery as a source of collateral blood flow. AJNR Am J Neuroradiol 12:1223

Muller-Forell W, Valavanis A (1996) Bestimmung des Blutungsrisiko bei zerebralen arteriovenösen Malformationen (AVM). Klin Neuroradiol 6:71

Mulliken JB, Glowacki J (1982) Hemangiomas and vascular malformations in infants and children. A classification based on endothelial characteristics. Plast Reconstr Surg 69:348

Munoz A, De Vergas J, Crespo J et al (2010) Imaging and clinical findings in patients with aberrant course of the cervical internal carotid artery. Open Neuroimag J 4:174

Murakami H, Inaba M, Nakamura A et al (2002) Ipsilateral hyperperfusion after neck clipping of a giant internal carotid artery aneurysm. Case report. J Neurosurg 97:1233

Murayama Y, Nien YI, Duckwiller G et al (2003a) Guglielmi detachable coils embolization of cerebral aneurysms: 11 years' experience. J Neurosurg 98:959

Murayama Y, Song TK, Uda K et al (2003b) Combined endovascular treatment for both intracranial aneurysm and symptomatic vasospasm. AJNR Am J Neuroradiol 24:133

Murias Quintana E, Vega Valdés P, Morales Deza E et al (2016) Endovascular reconstruction of internal carotid artery dissection in patients with acute ischemic stroke using the Wingspan stent. Int Neuroradiol 22:700

Nadgir RN, Loevner LA, Ahmed T et al (2003) Simultaneous bilateral internal carotid and vertebral artery dissection following chiropractic manipulation: case report and review of the literature. Neuroradiology 45:311

Nagahiro S, Takada A, Goto S et al (1995) Thrombosed growing giant aneurysms of the vertebral artery: growth mechanisms and management. J Neurosurg 82:796

Nagamine Y, Fukuoka T, Hayashi T et al (2014) Research article. Clinical characteristics of isolated anterior cerebral artery territory infarction due to arterial dissection. J Stroke Cerebrovasc Dis 23:2907

Nagara ON, White PHM, Guilbert F et al (2010) Endovascular treatment of intracranial unruptured aneurysms: systematic review and meta-analysis of the literature on safety and efficacy. Radiology 256:887

Nagaraja S, Lee KJ, Coley SC et al (2006) Stereotactic radiosurgery for brain arterio—venous malformations: quantitative MR assessment of nidal response at 1 year and angiographic factors predicting early obliteration. Neuroradiology 48:821

Nagel S, Kellert L, Moehlenbruch M et al (2013) Improved clinical outcome after acute basilar artery occlusion since the introduction of endovascular thrombectomy devices. Cerebrovasc Dis 36:394

Naidich T, Kricheff II, George AE et al (1976) The normal anterior inferior cerebellar artery. Radiology 119:355

Naito I, Iwai T et al (2001) Percutaneous transvenous embolization through the occluded sinus for transverse-sigmoid dural arteriovenous fistulas with sinus occlusion. Neuroradiology 43:672

Nakai K, Kaji T, Uchino A et al (2012) Congenital external carotid-internal artery anastomosis associated with contralateral non bifurcating cervical carotid artery. Neuroradiology 54:21

Nakao Y, Watanabe M, Yamamoto T et al (2004) Dissecting aneurysm of the posterior communicating artery. Acta Neurochir 146:1365

Nakatomi H, Nagata K, Kawamoto S et al (1997) Ruptured dissecting aneurysm as a cause of intracranial haemorrhage of verified etiology. Stroke 28:1278

Nakatomi H, Segawa H, Kurata A et al (2000) Clinicopathological study of intracranial fusiform and dolichoectatic Aneurysms. Insight on the mechanism of growth. Stroke 31:896

Nakazawa T, Suzuki F, Miyamoto Y et al (1985) Aplasia of the proximal middle cerebral artery associate with an aneurysm of the proximal anterior cerebral artery. Case report. Nippon Geka Hokan 54:521

Nakstad P, Nornes H, Hauge HN (1986) Traumatic aneurysm of the pericallosal arteries. Neuroradiology 28:335

Nakstad P, Nornes H, Hauge HN et al (1988) Cerebral panangiography in spontaneous subarachnoid haemorrhage from intracranial aneurysms. Acta Radiol 29:633

Nakstadt PH, Gjertsen O, Pedersen HK (2008) Correlation of head trauma and traumatic aneurysms. Interv Neuroradiol 14:33

Nass R, Hays A, Chutorian A (1982) Intracranial dissecting aneurysms in childhood. Stroke 13:204

Nehls DG, Pittman HW (1982) Spontaneous regression of arteriovenous malformations. Neurosurgery 11(6):776–780

Nelles M, Greschus S, Moehlenbruch M et al (2014) Patients selection for mechanical thrombectomy. Clin Neuroradiol 24:239

Nelson MD, Kendall BE (1987) Intracranial catecholamine secreting paragangliomas. Neuroradiology 29:277

Neumeier-Probst E (2009) Dural arteriovenous fistulas. Clin Neuroradiol 19:91

New PFS, Price DL, Carter B (1970) Cerebral angiography in cardiac myxoma; correlation of angiographic and histopathological findings. Radiology 96:335

Newton TH, Cronquist S (1969) Involvement of the dural arteries in intracranial arteriovenous malformations. Radiology 1071:1093

Newton TH, Hoyt WF (1970) Dural arteriovenous shunts in the region of the cavernous sinus. Neuroradiology 1:71

Newton TH, Weidner W, Greitz T (1968) Dural arteriovenous malformations in posterior fossa. Radiology 90:27

Newton TH, Young DA (1968) Anomalous origin of the occipital artery from the internal carotid artery. Radiology 90:550

Nicolai A, Lazzarino LG, E Biasutti E (1996) Large striatocapsular infarcts: clinical features and risk factors. J Neurol 243:44

Nieswandt B, Kleinschnitz C, Stoll G (2011) Ischemic stroke: a thrombo-inflammatory disease? J Physiol 589:4115

Nikoubashman O, Reich A, Pjontek R et al (2014) Postinterventional subarachnoid haemorrhage after endovascular stroke treatment with stent retrievers. Neuroradiology 56:1087

Nishijima M, Takaku A, Endo S et al (1992) Etiological evaluation of dural arteriovenous malformations of the lateral and sigmoid sinus based on histopathological examination. J Neurosurg 76:600

Nishimoto A, Takeuchi S (1968) Abnormal cerebrovascular network related to the internal carotid arteries. J Neurosurg 29:255

Nishio A, Yoshimura M, Yamauchi S et al (2009) Anomalous origin of the anterior choroidal artery. Interv Neuroradiol 15:73

Nishizaki T, Tanaki N, Takeda N et al (1986) Dolichoectatic basilar artery. A review of 23 cases. Stroke 17:1277

Nogueira RG, Schwamm LH, Buonanno FS et al (2008) Low pressure balloon angioplasty with adjuvant pharmacological therapy in patients with acute ischemic stroke caused by intracranial arterial occlusion. Neuroradiology 50:331

Noreau G, Landry PP, Morais D (2001) Arteriovenous malformation of the mandible: review of literature and case history. J Can Dent Assoc 67:646

Norrgard Ö, Angquist KA, Fodstad H et al (1987) Intracranial aneurysms and heredity. Neurosurgery 20:236

North American Symptomatic Carotid Endarterectomy Trial (NASCET) Collaborators (1991) Beneficial effect of carotid endarterectomy in symptomatic patients with high grade stenosis. N Engl J Med 325:445

North KN, Whiteman DAH, Pepin MG et al (1995) Cerebrovascular complications in Ehlers-Danloss syndrome type IV. Ann Neurol 38:960

Noser EA, Shaltoni HM, Hall CE et al (2005) Aggressive mechanical clot disruption. A safe adjunct to thrombolytic therapy in acute stroke? Stroke 36:292

Nusbaum AO, Som PM, Dubois P et al (1998) AJNR Am J Neuroradiol 1845:19

Nutik ST, Dilenge D (1976) Carotid—anterior cerebral artery anastomoses. J Neurosurg 44:378

Obholzer RJ, Hornigold R, Connor S et al (2011) Classification and management of cervical paragangliomas. Ann R Coll Surg Engl 93:596

Obrador A, Urquiza P (1952) Angioma arteriovenoso de la tienda del cerebelo. Rev Esp Oto Neuro Oftal 10:387

Obusez EC, Hui F, Hajj-Ali RA et al (2014) High-resolution MRI vessel wall imaging spatial and temporal patterns of reversible cerebral vasoconstriction syndrome and central nervous system vasculitis. AJNR Am J Neuroradiol 35:1527

Ogawa A, Tominaga T, Yoshimoto T et al (1992) Intraorbital ophthalmic artery aneurysm: case report. Neurosurgery 31:1102

Ogawa A, Suzuki M, Ogasawara K (2000) Aneurysms at non-branching sites in the supraclinoid portion of the internal carotid artery: internal carotid trunk aneurysms. Neurosurgery 47:578

Ohara Y, Petersen TE, Harrison DG (1993) Hypercholesterolemia increases endothelial superoxide anion production. J Clin Invest 91:2546

Ohkuma H, Suzuki S, Shimamura N et al (2003) Dissecting aneurysms of middle cerebral artery: neuroradiological and clinical features. Neuroradiology 45:143

Ohshiro S, Inoue T, Hamoda Y et al (1993) Branches of the persistent primitive trigeminal artery. An autopsy case. Neurosurgery 32:144

Ohta H, Natarajan SK, Hauck E et al (2011) Endovscular stent therapy for extracranial and intracranial carotid artery dissection: single-center experience. J Neurosurg 115:91

Oka J, Sodmoto K, Tadanowa M et al (2000) Transvenous embolization of carotid—cavernous sinus fistula associated with a primitive trigeminal artery. Case report. Neurologia Medico Chirurgica 40:61

Oka K, Rhoton AL, Barry M et al (1985) Microsurgical anatomy of the superficial veins of the cerebrum. Neurosurgery 17:711

Okada Y, Brian RC, Fitridge R et al (1994) Fibrin contributes to microvasculaturte obstruction and oarenchymal damage during early focal cerebral ischemia. Stroke 25:1847

Okuchi K, Nagata K, Fujioka M et al (1999) Rapid growth of the internal carotid anterior wall aneurysm. Riv Neuroradiol 12(Suppl 2):9

Okudera T, Huang YP, Ohta T et al (1994) Development of posterior fossa dural sinuses, emissary veins and jugular bulb: Morphological and Radiologic study. AJNR Am J Neuroradiol 15:1871

Okudera T, Ohta T, Huang UP (1984) Embryology of the cranial venous system. In: Kapp JP, Schmidek HH (eds) The cerebral venous system and its disorders. Grune and Stratton, Orlando

Olivecrona H, Riives J (1948) Arteriovenous aneurysms of the brain: their diagnosis and treatment. Arch Neurol Psychiatry 59:567

Olivecrona H (1950) Die arteriovenoesen Aneurysmen des Gehirns. Dtsch med Wschr 75:1169

Omojola MF, Fox AJ, Vinuela FV (1982) Spontaneous regression of intracranial arteriovenous malformations. J Neurosurg 57:818

Ondra SL, Traupp H, George ED et al (1990) The natural history of symptomatic arteriovenous malformations of the brain: a 24-year follow-up assessment. J Neurosurg 73:387

Ono M, Ono M, Rhoton AL Jr et al (1984) Microsurgical anatomy of the region of the tentorial incisura. J Neurosurg 60:365

Oran J, Kiroglu Y, Yurt A et al (2009) Developmental venous anomaly (DVA) with arterial component. A rare cause of intracranial hemorrhage. Neuroradiology 51:25

Osborn DA (1959) So called juvenile angiofibroma of nasopharynx. J Laryngol Otol 73:295

Osborn AG, Anderson RE (1977) Angiographic spectrum of cervical and intracranial fibromuscular dysplasia. Stroke 8:617

Osborn AG (1980) The vidian artery: normal and pathologic anatomy. Radiology 136:373

Osborn AG (1999) Diagnostic cerebral angiography. Lippincott Williams & Wilkins, Philadelphia

Ostergaard JR, Voldby B (1983) Intracranial arterial aneurysms in children and adolescence. J Neurosurg 58:832
Ostergaard JR, Oxlund A (1987) Collagen type III deficiency in patients with rupture of intracranial saccular aneurysms. J Neurosurg 67:690
Ostertun B, Solymosi L (1993) Magnetic resonance angiography of cerebral developmental venous anomalies. Its role in differential diagnosis. Neuroradiology 35:97
Otomo E (1965) The anterior choroidal artery. Arch Neurol 13:656
Otten P, Pizzolato GP, Rilliet B et al (1989) A propos de 131 cas d'angiomes caavernoux (cavernoma) du SNC, repérés par l'analyes retrospective de 24535. Neurochirurgie 35:82
Ouhlous M, Flach HZ, de Weert TT et al (2005) Carotid plaque composition and cerebral infarction. MR imaging study. AJNR Am J Neuroradiol 26:1044
Pacchioni A (1705) Dissertatio epistolaris de glandulis conglobatis durae meningis humanae
Padget DH (1944) The circle of Willis, its embryology and anatomy. Comstock, Ithaca
Padget DH (1948) The development of the cranial arteries in the human embryo. Contrib Embryol 32:205
Padget DH (1954) Designation of the embryonic intersegmental arteries in reference to the vertebral artery and subclavian stem. Anat Rec 119:349
Padget DH (1956) The cranial venous system in men in reference to developmental, adult configuration, and relation to the arteries. Am J Anat 98:307
Padget DH (1957) The development of the cranial venous system in man from the viewpoint of comparative anatomy. Contr Embriol Carnegie Inst 36:79
Pakarinen S (1965) Arteriovenous fistula between the middle meningeal artery and the sphenoidal sinus. A case report. J Neurosurg 23:438
Palubinskas AJ, Newton TH (1965) Fibromuscular hyperplasia of the internal carotid arteries. Radiol Clin Biol 34:365
Palubinskas AJ, Perloff P, Newton TH (1966) Fibromuscular hyperplasia: an arterial dysplasia of increasing clinical importance. Am J Roentgenol 98:907
Palubinskas AJ, Ripley AR (1964) Fibromuscular hyperplasia in extra-renal arteries. Radiology 82:451
Panagiotopoulos V, Gizewski E, Asgari S et al (2009) Embolization of intracranial arterio-venous malformations with ethylene—vinyl alcohol copolymer (ONYX). AJNR Am J Neuroradiol 30:99
Pandey A, Rosenwasser RH, Veznedaroglu E (2007) Management of distal anterior cerebral artery aneurysms: a single institution retrospective analysis (1997–2005). Neurosurgery 61:909
Pantoni L, Garcia JH, Brown GG (1996) Vascular pathology in three cases of progressive cognitive deterioration. J Neurol Sci 135:131
Paramasivan S, Yoma N, Niimi Y et al (2013) De novo development of dural arteriovenous fistula after endovascular embolization of pial arteriovenous fistula. J Neurointerv Surg 5:321
Park HK, Horowitz M, Jungreis C et al (2005) Periprocedural morbidity and mortality associated with endovascular treatment of intracranial aneurysms. AJNR Am J Neuroradiol 26:506
Park KY, Chung PW, Kim YB et al (2009a) Association between small deep cerebellar ischemic lesion and small-vessel disease. Cerebrovasc Dis 28:314
Park SI, Kim BM, Kim DI et al (2009b) Clinical and angiographic follow-up of stent only therapy for acute intracranial vertebrobasilar dissecting aneurysms. AJNR Am J Neuroradiol 30:1351
Park HS, Choi JH, Kang M et al (2013) Management of aneurysms of the proximal (A1) segment of the anterior cerebral artery. J Cerebrovasc Endovasc Neurosurg 15:13
Parkin AJ, Leng NR, Stanhope N et al (1988) Memory impairment following ruptured aneurysm of the anterior communicating artery. Brain Cogn 7:231
Parkinson D (1964) Collateral circulation of cavernous carotid artery: anatomy. Can J Surg 7:251
Parkinson D (1973) Carotid-cavernous fistula: direct repair with preservation of the carotid artery. J Neurosurg 38:99
Parkinson D, Shields CB (1974) Persistent trigeminal artery: its relationship to the normal branches of the cavernous carotid. J Neurosurg 40:244
Parkinson D (1982) Comments on the so-called cavernous sinus. Neurosurgery 11:716
Parkinson D, West M (1982) Lesions of the cavernous plexus region. In: Youman JR (ed) Neurological surgery. Saunders WB, Philadelphia
Parlato C, di Nuzzo G, Luongo M et al (2011) Anatomical variant of origin of ophthalmic artery: case report. Surg Radiol Anat 33:275
Parrilla G, Garcia-Villalba B, Espinosa de Rueda M et al (2012) Hemorrhage-contrast staining areas after mechanical intra-arterial thrombectomy in acute ischemic stroke: imaging findings and clinical significance. AJNR Am J Neuroradiol 33:1791
Pascual-Castroviejo I, Pascual-Pascual I, Blazqueq MG et al (1977) Spontaneous occlusion of an intracranial arteriovenous malformation. Childs Brain 3:169
Pascual-Castroviejo I (1978) Vascular and non-vascular intracranial malformation associated with external capillary hemangiomas. Neuroradiology 16:58
Pasqualin A, Vivenza L, Rosta R et al (1985) Spontaneous disappearance of intracranial arteriovenous malformations. Acta Neurochir 76:50
Pasqualin A, Mazza C, Cavazzani P et al (1986) Intracranial aneurysms and subarachnoid haemorrhage in children and adolescence. Childs Nerv Syst 2:185
Patankar T, Widjaia E, Chant H et al (2006) Relationships of deep white matter hyperintensities and cerebral blood flow in severe carotid artery stenosis. Eur J Neurol 13:10
Patel U, Gupta SC (1990) Wyburn-Mason syndrome. A case report and review of the literature. Neuroradiology 31:544

Paulson G, Boesel C, Evans W (1978) Fibromuscular dysplasia. Arch Neurol 35:287

Paulus WS, Glenn Pait T, Rhoton AL (1977) Microsurgical exposure of the petrous portion of the carotid artery. J Neurosurg 47:713

Pawlikowska L, Tran MN, Achrol S et al (2004) Polymorphism in genes involved in inflammatory and angiogenic pathways and the risk of hemorrhagic presentation of brain arteriovenous malformations. Stroke 35:2294

Pedroza A, Dujovny M, Artero JC et al (1987) Microanatomy of the posterior communicating artery. Neurosurgery 20:228

Pegelow CH, Macklin EA, Moser FG et al (2002) Longitudinal changes in brain magnetic resonance imaging findings in children with sickle cell disease. Blood 99:14

Pelkonen O, Tikkakoski T, Leinonen S et al (2003) Extracranial internal carotid and vertebral artery dissection: angiographic spectrum, course and prognosis. Neuroradiology 45:71

Peluso JPP, van Rooij WJ, Sluzewski M et al (2007) Distal aneurysms of cerebellar arteries: incidence, clinical presentation, and outcome of endovascular parent vessel occlusion. AJNR Am J Neuroradiol 28:1573

Pelz DM, Fox AJ, Vinuela F et al (1987) The ascending pharyngeal artery: a collateral pathway in complete occlusion of the internal carotid artery. AJNR Am J Neuroradiol 8:177

Percheron G (1976a) Les artère du thalamus humain II. Artères et territoires thalamiques paramedians de l'artère basilaire communiquante. Rev Neurol 132:309

Percheron G (1976b) Les artères du thalamus humain. I. Artères et territoires thalamiques polaires de l'artères communiquante posterieure. Rev Neurol 132:297

Pereira VM, Gralla J, Davalos A et al (2013) Prospective, multicenter, single-arm study of mechanical thrombectomy using solitaire flow restoration in acute ischemic stroke. Stroke 44:2802

Perini S, Talamini G, Pasqualin A et al (1995) Arteriovenous malformations of the brain. Risk of first bleeding, rebleeding and related risk factors in 168 untreated patients. Neuroradiology 37(Suppl 1):120

Perlmutter D, Rhoton AL Jr (1976) Microsurgical anatomy of the anterior cerebral-anterior communicating-recurrent artery complex. J Neurosurg 45:259

Perlmutter D, Rhoton AL Jr (1978) Microsurgical anatomy of the distal anterior cerebral artery. J Neurosurg 49:204

Perret G, Nischioka H (1966) Report on the cooperative study of intracranial aneurysms and subarachnoid haemorrhage. Section VI Arteriovenous malformations. J Neurosurg 25:467

Perrini P, Cardia A, Fraser K et al (2007) A micro-surgical study of the anatomy and course of the ophthalmic artery and its possibly dangerous anastomoses. J Neurosurg 106:142

Peschillo S, Miscusi M, Capolingua A et al (2015) Blister-Like aneurysms in atypical location: a single center experience and comprehensive literature review. World Neurosurg 84:1070

Pessin MS, Kwan ES, Dewm LD et al (1987) Posterior cerebral artery stenosis. Ann Neurol 21:85

Pessin MS, Chimowitz MI, Levine SR et al (1989) Stroke in patients with fusiform vertebrobasilar aneurysm. Neurology 39:16

Pham M, Bendszus M (2016) facing time in ischemic stroke: an anternative hypothesis for collateral failure. Clin Neuroradiol 26:141

Phatouros CC, Lefler JE, Higashida RT (2000) Primary stenting for high-grade basilar artery stenosis. AJNR Am J Neuroradiol 21:1744

Pia HW, Fontana H (1977) Aneurysms of the posterior cerebral artery. Acta Neurochir 38:13

Piano M, Valvassori L, Quilici L et al (2013) Midterm and long-term follow-up of cerebral aneurysms treated with flow diverter devices: a single-center experience. J Neurosurg 118:40

Picard L, Floquet J, Andrè JM et al (1974a) Syndrome Moya-Moya. Etude anatomopatologique. J Neuroradiol (Masson) 1:113

Picard L, Levesque M, Crouzet G et al (1974b) The Moya Moya syndrome. J Neuroradiol (Masson) 1:47

Picard L, Roy D et al (1993) Aneurysm associated with a fenestrated basilar artery: report of two cases treated by endovascular detachable balloon embolization. AJNR Am J Neuroradiol 14:591

Picard L, Bracard S, Moret J et al (1987) Spontaneous dural arteriovenous fistula. Sem Interv Radiol 4:219

Picard L, Bracard S, Islak C et al (1990) Dural fistulae of the tentorium cerebelli. J Neuroradiol (Masson) 17:161

Picard L, Bracard S, Anxionnat R et al (2005) Brain AVM embolization. Retrospective study concerning 728 patients, followed between 1984–2004. Interv Neuroradiol 11:45

Piche SL, Haw CS, Redekop GJ et al (2005) Rare intracanalicular ophthalmic aneurysm: endovascular treatment and review of the literature. AJNR Am J Neuroradiol 26:1929

Pickard JD, Walker V, Vile J et al (1987) Oral nimodipine reduces prostaglandin and thromboxane production by arteries chronically exposed to a periarterial haematoma and the antifibrinolytic agent tranexamic acid. J Neurol Neurosurg Psychiatry 50:727

Pierot L, Chiras J, Meder JF et al (1992) Dural arteriovenous fistulas of the posterior fossa draining into subarachnoid veins. AJNR Am J Neuroradiol 13:315

Pierot L, Cognard C, Spelle L (2004) Malformations arterioveneuses cérébrales: evolution du risque hémorragique et de sa morbiditè. J Neuroradiol (Masson) 31:369

Pierot L, Januel AC, Herbretau D et al (2009) Endovascular treatment of brain arteriovenous malformations using onyx: results of a prospective multicentric study. Neuroradiology 36:147

Pierot L, Rajpal G, Kadziolka K et al (2012) The place for remodelling technique and stenting in the endo-vascular management of intracranial aneurysms: a single-center analysis from 2008 to 2010. Neuroradiology 54:973

Pierot L, Cognard C, Herbreteau D et al (2013) Endovascular treatment of brain arteriovenous malformations using a liquid embolic agent: results of porospective, multicenter study (BRAVO). Eur Radiol 23:2838

Piotin M, Miralbes S, Cattin F et al (1996) MR and MR angiography of persistent trigeminal artery. Neuroradiology 38:730

Piske RL, Monteiro Siqueira Campos C, Chaves JBL et al (2005) Dural sinus compartment in dural arteriovenous shunts: a new angioarchitectural feature allowing superselective transvenous dural sinus occlusion treatment. AJNR Am J Neuroradiol 26:1715

Pitanguy I, Caldeira AML, Calixto CA et al (1984) Clinical evaluation and surgical treatment of hemangiomata. Head Neck Surg 7:47

Piton J, Guilleux H, Guibert-Trainier F et al (1984) Fistules du sinus lateral. J Neuroradiol (Masson) 11:143

Poetke M, Frommeld T, Berlien HP (2002) PHACE syndrome: new views on diagnostic criteria. Eur J Pediatr Surg 12:366

Pollock BE, Flickinger JC, Lunsford LD et al (1996) Factors that predict the bleeding risk of cerebral arteriovenous malformations. Stroke 27:1

Pope FM, Nicholls AC, Narisi P et al (1984) Some patients with cerebral aneurysms are deficient in type III collagen. Lancet 1:973

Pope FM (1989) Type III collagen mutations and cerebral aneurysms. Stroke 20:1432

Porter PJ, Willinsky RA, Harper W et al (1997) Cerebral cavernous malformations: natural history and prognosis after clinical deterioration with or without hemorrhage. J Neurosurg 2:190

Powell J (1991) Model of arterial aneurysms: for the investigation of pathogenesis and pharmacotherapy: a review. Atherosclerosis 87:93

Pozzati E, Andreoli A, Padovani R (1995) Dissecting aneurysms of the basilar artery. Neurosurgery 36:254

Pozzati E, Acciarri N, Tognetti F et al (1996) Growth, subsequent bleeding and de novo appearance of cerebral cavernous angiomas. Neurosurgery 38:662

Proust F, Tousseint P et al (2001) Pediatric cerebral aneurysms. J Neurosurg 94:733

Provenzale JM (2009) MRI and MRA for evaluation of dissection of craniocerebral arteries. Lessons from the medical literature. Emerg Radiol 16(3):185

Pullicino P, Miller LL, Munschauer EE et al (1992) Linear subinsular MR hyperintensity. Ann Neurol 32:267

Puri AS, Massari F, Asai T et al (2016) Safety, efficacy and short-term follow-up of the use of Pipeline™ embolization device in small (<2.5 mm) cerebral vessels for aneurysm treatment: single institution experience. Neuroradiology 58:267

Putman CM, Chaloupka JC, Fulbright RK et al (1996) Exceptitional multiplicity of cerebral arteriovenous malformations associated with hereditary hemorrhagic telangiectasia. Osler-Weber-Rendu syndrome. AJNR Am J Neuroradiol 17:1733

Qin X, Zhang X, Zhang Q (2012) Cerebral venous sinus thrombosis caused by spontaneous intracranial hypotension. Neurol Asia 17:227

Quain R (1844) The anatomy of the arteries of the human body. Taylor and Walton, London

Quinones D, Duckwiller G, Gobin PY et al (1997) Embolization of dural cavernous fistulas via superior ophthalmic vein approach. AJNR Am J Neuroradiol 18:921

Qureschi AI, Hussein HM, El-Gengaihy A et al (2008) Concurrent comparison of outcomes of primary angioplasty and of stent placement in high-risk patients with symptomatic intracranial stenosis. Neurosurgery 62:1053

Qureshi AI, Ali Z, Suri MF et al (2001) Intra-arterial third-generation recombinant tissue plasminogen activator (reteplase) for acute ischemic stroke. Neurosurgery 49:41

Qureshi AI, Siddiqui AM, Suri MF et al (2002) Aggressive mechanical clot disruption and low-dose intraarterial third-generation thrombolytic agent for ischemic stroke: a prospective study. Neurosurgery 51:1319

Radulovic D, Nestoric B, Rakic M et al (2006) Enlargement to a saccular aneurysm and subsequent rupture of infundibulum widening of posterior communicating artery posterior artery. Neurochirurgie 52:525

Rafay MF, Armstrong D, De Veber J et al (2006) Craniocervical arterial dissection in children: clinical and radiographic presentation and outcome. J Child Neurol 21:8

Rainer WG, Cramer GG, Newby JP et al (1968) Fibromuscular hyperplasia of the carotid artery causing positional cerebral ischemia. Ann Surg 167:444

Ramgren B, Cronquist M, Rommer B et al (2005) Vertebrobasilar dissection with subarachnoid haemorrhage: a retrospective study of 29 patients. Neuroradiology 47:97

Ramsey TL, Mosquera VT (1948) Dissecting aneurysm of the middle cerebral artery. Ohio State Med J 44:168

Rao VRK, Mathuryia SN (2011) Pediatric aneurysms of vein of Galen malformation. J Pediatr Neurosci 6:109

Rascol A, Guiraud B, Manelfe C et al (1979) Accidents vadsculaires cerebraux de la grossesse e du post partum. In: II conference de la Salpetriere sur les maladies vasculaires cerebrales 1979. Hospital de la Salpetriere. JB Balliere, Paris

Rasskazoff S, Silvaggio J, Brouwer PA et al (2010) Endovascular treatment of a ruptured blood blister—like aneurysm with a flow diverting stent. Interv Neuroradiol 16:255

Raybaud CA, Strother CM, Hald JK (1989) Aneurysms of the vein of Galen: embryonic considerations and anatomical features relating to the pathogenesis of the malformation. Neuroradiology 31:109

Raymond J, Iancu D, Weill A et al (2005) Embolization as one modality in a combined strategy for the management of cerebral AV malformation. Interv Neuroradiol 11:57

Raymond J, Guillemin F, Proust F et al (2008) Unruptured intracranial aneurysms. A critical review of the International Study of Unruptured Intracranial Aneurysms (ISUIA) and of appropriate methods to address the clinical problem. Interv Neuroradiol 14:85

Raymond J (2009) Incidental intracranial aneurysms: rationale for treatment. Curr Opin Neurol 22:96

Rea D, Brandsema JF, Armstrong D et al (2009) Cerebral arteriopathy in children with neurofibromatosis type1. Pediatrics 124:e476

Read SJ, Pettigrew L, Schimmel MV et al (1998) White matter medullary infarct: acute subcortical infarction in the centrum ovale. Cerebrovasc Dis 8:289

Redekop G, Terbrugge K, Montanera W et al (1998) Arterial aneurysms associated with cerebral arteriovenous malformations: classification, incidence and risk of haemorrhage. J Neurosurg 89:539

Redekop GJ (2008) Extracranial carotid and vertebral artery dissection: a review. Can J Neurol Sci 35:146

Reith W, Berkefeld J, Dietrich P et al (2015) Diagnosis and treatment of intracranial stenosis. Clin Neuroradiol (Suppl) 25:307

Reivich M, Holling E, Roberts B et al (1961) Reversal of blood flow through the vertebral artery and its effects on cerebral circulation. N Engl J Med 265:88

Rennert J, Ullrich O, Schuierer G (2013) A rare case of supraclinoid Internal Carotid Artery (ICA) fenestration in combination with duplication of the middle cerebral artery (MCA) originating from the ICA fenestration and an associated aneurysm. Clin Neuroradiol 23:133

Reynolds AF Jr, Stovring J, Turner PT (1980) Persistent otic artery. Surg Neurol 13:115

Rhoton AL Jr (2002) Aneurysms. Neurosurgery 51(Suppl 1):121

Rhoton AL Jr, Fujiki K, Frad B (1979) Microsurgical anatomy of the anterior choroidal artery. Surg Neurol 12:171

Ribes M (1825) Des recherches faites sur la phlebite. Revue medicale francaise et etrangere et journal de clinique de l'Hotel de dieu et de la Charite de Paris 3:5

Richard S, Kremer S, Lacour JC et al (2007) Cerebral venous thrombosis caused by spontaneous intracranial hypotension: two cases. Eur J Neurol 14:1296

Richardson DN, Elster AD, Ball MR (1989) Intrasellar trigeminal artery. AJNR Am J Neuroradiol 10:205

Ricolfi F, Decq P, Brugieres P et al (1996) Ruptured fusiform aneurysm of the superior third of the basilar artery associated with the absence of the midbasilar artery. Case report. J Neurosurg 85:961

Rieger P, Huber G (1983) Fenestration and duplicate origin of the left vertebral artery in angiography. Neuroradiology 25:45

Rigamonti D, Drayer BP, Johnson PC et al (1987) The MR imaging appearance of cavernous malformations (angiomas). J Neurosurg 67:518

Rigamonti D, Johnson PC, Spetzler RF et al (1991) Cavernous malformation and capillary telangiectasias. A spectrum within a single pathological entity. Neurosurgery 28:60

Rigamonti D, Brown GB (1994) The natural history of familial cavernous malformations. Results of an ongoing study. J Neurosurg 80:422

Rinaldi I, Harris WO, Kopp JE et al (1976) Intracranial fibromuscular dysplasia: report of 2 cases; one with autopsy verification. Stroke 7:511

Ring BA (1974) The middle cerebral artery. In: Newton TH, Potts DG (eds) Radiology of skull and brain. Mosby, St. Louis

Ring BA, Waddington MM (1967) Ascending frontal branch of the middle cerebral artery. Acta Radiol Diagn 6:209

Ring BA, Waddington MM (1968) Roentgenographic anatomy of the pericallosal artery. Am J Roentgenol 104:109

Ringelstein EB, Zeumer H, Angelow D (1983) The pathogenesis of strokes from internal carotid artery occlusion: diagnostic and therapeutic implications. Stroke 14:867

Ringelstein EB, Zeumer H, Schneider R (1985) Der Beitrag der zerebralen Computertomographie zur Differentialdiagnose und Differntialtherapie des ischemischen Grosshirninfarktes. Fortschr Neurol Psychiatry 53:315

Ringer AJ, Qureschi AI, Fessler RD et al (2001) Angioplasty of intracranial occlusion resistant to thrombolysis in acute ischemic stroke. Neurosurgery 48(6):1282

Ringleb PA, Allenberg J, Bruckmann H et al (2006) 30 day results from the space trial of stent protected angioplasty versus carotid endarterectomy in symptomatic patients: a randomized non-inferiority trial. Lancet 368:1239

Rinkel GJE, Wijdicks EFM, Vermeulen M et al (1991) Nonaneurysmal perimesencephalic subarachnoid hemorrhage: CT and MR patterns that differ from aneurismal rupture. AJNR Am J Neuroradiol 12:829

Rinne J, Hernesniemi J, Puranen M et al (1994) Multiple intracranial aneurysms in a defined population: prospective angiographic and clinical study. Neurosurgery 35:803

Risau W (1997) Mechanisms of angiogenesis. Nature 386:671

Risau W, Flamme I (1995) Vasculogenesis. Annu Rev Cell Dev Biol 11:73

Riva A, Bradac GB, Riccio A et al (1991) Demenza in un paziente con fistula durale. Rivista di Neuroradiologia 4:369

Rizek P, Pasternak S, Leung A et al (2013) Acute onset anterograde amnesia caused by isolated bilateral fornix infarct. Can J Neurol Sci 40:738

Ro A, Kageyama N, Abe N et al (2009) Intracranial vertebral artery dissection resulting in fatal subarachnoid haemorrhage: clinical and histopathological investigations from a medicolegal perspective. J Neurosurg 110:948

Robinson JR Jr, Awad IA, Little JR (1991) Natural history of the cavernous angioma. J Neurosurg 75:709

Robinson LR (1959) An unusual human anterior cerebral artery. J Anat 93:131

Roche JL, Choux M et al (1988) Intracranial arterial aneurysms in children. A cooperative study. A propos of 43 cases. Neurochirurgie 34:243

Roche PH, Paris J, Regis J et al (2007) Management of invasive juvenile naso-pharyngeal angiofibromas: the role of multimodality approach. Neurosurgery 61:768

Rodesch G, Lasjaunias P, Terbrugge K (1988) Lesions vasculaires arteroveineuses intracraniennes de l'enfant. Place des techniques endovsasculaires. A propos de 44 cases. Neurochirurgie 34:293

Rodesch G, Comoy J, Hurth M et al (1991a) Jugular foramen arteriovenous shunt with subarachnoid haemorrhage. Skull Base Surg 1:132

Rodesch G, Choi IS, Lasjaunias P et al (1991b) Complete persistence of the joido-stapedial artery in man. Case report. Intrapetrous origin of the maxillary artery from ICA. Surg Radiol Anat 13:63

Rodesch G, Terbrugge K (1993) Arterial and venous angioarchitecture of CAVMs in adults. Rivista di Neuroradiologia 6:37

Rodesch G, Malherbe V, Alvarez H et al (1995) Non-galenic cerebral arteriovenous malformations in neonates and infants. Childs Nerv Syst 11:231

Rodesch G, Guedin P, Gaillard S et al (2009) Therapeutic management of intracranial dural arteriovenous shunts with leptomeningeal venous drainage: report of 53 consecutive patients with emphasis on transarterial embolization with acrylic glue. Interv Neuroradiol 5(Suppl 1):106

Rodriguez-Arias C, Martinez R, Rey G et al (2000) Recurrence in different location of a cerebral arteriovenous malformation in a child after radiosurgery. Childs Nerv Syst 16:363

Roeltgen DP, Wemer GR, Patterson LF (1981) Delayed neurologica complications of left atrial myxoma. Neurology 31:8

Roh HG, Kim SS, Han H et al (2008) Endovascular treatment of posterior cerebral artery aneurysms using detachable coils. Neuroradiology 50:237

Rokitansky C (1846) Handbuch der allgemeinen pathologischen Anatomie. Braumueller and Seidel, Vienna

Romero JR, Beiser A, Seshdri S et al (2009) Carotid artery atherosclerosis, MRI indices of brain ischemia, aging and cognitive impairment: The Framingtan study. Stroke 40:1590

Roobs G, Schmidt R, Kapelle P et al (1999) MRI evidence of post cerebral microbleeds in a healthy elderly population. Neurology 52:991

Rosa M, Viale GL (1970) Diagnostic value of vertebral phlebogram. Neuroradiology 1:147

Rosenkranz M, Gerloff CH (2010) Diagnostic workup in carotid stenosis. A neurologist's perspective. Neuroradiology 52:619

Rosenorn J, Ahlgren P, Ronde F (1985) Pre-optic origin of the anterior cerebral artery. Neuroradiology 27:275

Rosenthal F (1824) De intimis cerebri venis seu de venae magnae Galeni ramis. Nova Acta Phisiocomedica Academiae Caesareae Leopoldino-Carolinae Naturae Curiosorum 12:302

Rosner SS, Rhoton AL Jr, Ono M et al (1984) Microsurgical anatomy of the anterior perforating arteries. J Neurosurg 61:468

Roszelle BN, Babiker MH, Hafner W et al (2013) In vitro and silico study of intracranial stent treatments for cerebral aneurysms: effects on perforating vessels flows. J Neurointerv Surg 5:354

Roth C, Papanagiotou P, Behnke S et al (2010) Stent-assisted mechanical recanalization for treatment of acute intracerebral artery occlusions. Stroke 41:2559

Rouchaud A, Brinjikji W, Lanzino G et al (2016) Delayed hemorrhagic complications after flow diversion for intractranial aneurysms: a literatrure overview. Neuroradiology 58:171

Roy D, Milot G, Raymond J (2001) Endovascular treatment of unruptured aneurysms. Stroke 32:1998

Ruchoux MM, Chabriat H, Bousser MG et al (1994) Presence of ultrastructural arterial lesions in muscle and skin vessels of patients with CADASIL. Stroke 25:2291

Ruchoux MM, Guerouaou D, Vandenhaute B et al (1995) Systemic vascular smooth muscle cell impairment in cerebral autosomal dominant arteriopathy with subcortical infarcts and leukoencephalopathy. Acta Neuropathol 89:500

Rüfenacht DA (2005) Aneurysms: the role of the perianeurysmal environment. Interv Neuroradiol 11(Suppl 2):50

Russell DS, Rubinstein LJ, Lumsden CF (1959) Tumours and hamartomas of blood vessels. In: Russel DS, Rubinstein LJ (eds) Pathology of tumours of the nervous system. Arnold, London, pp 72–92

Ryu CW (2010) Persistent falcine sinus: it is really rare? AJNR Am J Neuroradiol 31:367

Saatci I, Geyik S, Yavuz K et al (2011) Endovascular treatment of brain arteriovenous malformations with prolonged intranidal onyx injection technique: long term results in 350 consecutive patients with completed endovascular treatment course. J Neurosurg 115:78

Saba C, Pasculli G, Lenato GM et al (2007) Hereditary hemorrhagic telangiectasia: clinical feature in ENG and ALK1 mutations carriers. J Thromb Haemost 5:1149

Sabin FR (1917) Origin and development of the primitive vessels of the chick and of the pig. Contribution to Embryology. Carnegie Inst. Wash. Pub. No. 226

Sacco RL, Hauser WA, Mohr JP (1991) Hospitalized stroke incidence in blacks and Hispanics in northern Manhattan. Stroke 22:1491

Sacco RL (1993) Current epidemiology of stroke. In: Fisher M, Bogousslavsky J (eds) Current review of cerebrovascular disease. Current Medicine, Philadelphia, PA, pp 3–14

Sade B, Tampieri D, Mohr G (2004) Ophthalmic artery origin from basilar artery, rare variant. AJNR Am J Neuroradiol 25:1730

Saeki N, Rhoton AL Jr (1977) Microsurgical anatomy of the upper basilar artery and the posterior circle of Willis. J Neurosurg 46:53

Sagduyu A, Sirin H, Mulayim S et al (2006) Cerebral cortical and deep venous thrombosis without sinus

thrombosis: clinical MRI correlates. Acta Neurol Scand 114:254
Saini J, Kesavadas C, Thomas B et al (2008) Aberrant petrous internal carotid artery with cochlear. Anomaly-an unusual association. Surg Radiol Anat 30:435–437
Saito Y, Kobayashi N (1981) Cerebral venous angiomas. Clinical evaluation and possible etiology. Radiology 139:87
Sakata S, Fujii K, Matsushima T et al (1993) Aneurysms of the posterior cerebral artery: report of eleven cases-surgical approaches and procedures. Neurosurgery 32:163
Salamon G, Faure J, Raybaud C et al (1974) The external carotid artery. In: Newton TH, Potts DH (eds) Radiology of skull and brain. Mosby, St. Louis
Salamon G, Huang YP (1976) Radiologic anatomy of the brain. Springer, Berlin
Salas E, Ziyal IM, Sekhar LN et al (1998) Persistent trigeminal artery: an anatomic study. Neurosurgery 43:557
Saleme S, Losif CH, Ponomarjova S et al (2014) Flow—diverter stents for intracranial bifurcation aneurysm treatment. Neurosurgery 75:623
Salmon M (1936) Artères des muscles de la tete et du cou. Masson, Paris
Saltzman GF (1959) Patent primitive trigeminal artery studied by angiography. Acta Radiol 51:329
Salvarani C, Brown RD Jr, Calamia KT et al (2007) Primary central nervous system vasculitis: analysis of 101 patients. Ann Neurol 62:442
Samuel M, Spitz L (1995) Klippel-Trenaunay syndrome: clinical features, complications and management in children. Br J Surg 82:757
San Millan Ruiz D, Gailloud P, De Miquel Miquel MA (1999) The latero-cavernous sinus: an anatomical study. Anat Rec 254:7
San Millan Ruiz D, Gailloud P, Rüfenacht DA et al (2002) The craniocervical venous system in relation to the cerebral venous drainage. AJNR Am J Neuroradiol 23:1500
San Millan Ruiz D, Fasel JH, Ruefenacht DA et al (2004) The sphenoparietal sinus of Breschet: does it exist? An anatomic study. AJNR Am J Neuroradiol 25:112
Sartor K (1978) Spontaneous closure of cerebral arteriovenous malformation demonstrated by angiography and computed tomography. Neuroradiology 15:95
Sarwar M, McCormick WF (1978) Intracerebral venous angioma: case report and review. Arch Neurol 35:323
Sasaki O, Ogawa H, Koike T et al (1991a) A clinicopathological study of dissecting aneurysms of the intracranial vertebral artery. J Neurosurg 75:874
Sasaki O, Koike T, Tanaka R et al (1991b) Subarachnoid haemorrhage from a dissecting aneurysm of the middle cerebral artery. J Neurosurg 74:504
Sato S, Toyoda K, Matsuoka H et al (2010) Isolated anterior cerebral artery territory infarction: dissection as an etiological mechanism. Cerebrovasc Dis 29:170
Satomi J, Satoh K, Matsubara S et al (2005) Angiographic changes in venous drainage of cavernous sinus dural arteriovenous fistulae after palliative transarterial embolization or observational management: a proposed stage classification. Neurosurgery 56:494
Satti SR, Cerniglia CA, Koenisberg RA (2007) Cervical vertebral artey variations: an anatomic study. AJNR Am J Neuroradiol 28:976
Sauvaget E, Paris J, Kici S et al (2006) Aberrant internal carotid artery in the temporal bone: imaging findings and management. Arch Otolaryngol Head Neck Surg 132:86
Savoiardo M, Bracchi M, Passerini A et al (1987) The vascular territory in the cerebellum and brainstem. CT and MR study. AJNR Am J Neuroradiol 8:199
Savoiardo M, Armenise S, Spagnolo P et al (2006) Dural sinus thrombosis in spontaneous intracranial hypotension. J Neurol 253:1197
Saygi S, Bolay H, Tekkok IH et al (1990) Fibromuscular dysplasia of the basilar artery: a case with brainstem stroke. Angiology 41:658
Schievink WI, Mokri B, Ganity JA et al (1993) Ocular motor nerve palsy in spontaneous dissections of the cervical internal carotid artery. Neurology 43:1938
Schievink WI, Björnsson J, Piepgras DG (1994a) Coexistence of fibromuscular dysplasia and cystic medial necrosis in a patient with Marfan's syndrome and bilateral carotid artery dissection. Stroke 25:2492
Schievink WI, Mokri B, Pipgras D (1994b) Spontaneous dissections of cervico-cephalic arteries in childhood and adolescence. Neurology 44:1607
Schievink WI, Mokri B (1995) Familial aortocervical cephalic arterial dissection and congenital bicuspid aortic valve. Stroke 26:1935
Schievink WI, Meyer FB, Parisi JE et al (1998) Fibromuscular dysplasia of the internal carotid artery associated with alpha- 1-antitripsin deficiency. Neurosurgery 43:229
Schievink WI (2000) The treatment of spontaneous carotid and vertebral artery dissections. Curr Opin Cardiol 15:316
Schievink WI (2001) Spontaneous dissection of the carotid and vertebral arteries. N Engl J Med 344:898
Schievink WI, Maya MM (2008) Cerebral venous thrombosis in spontaneous intracranial hypotension. Headache 48:1511
Schiffer D (1993) Brain tumor. Pathology and its biological correlates. Springer, Berlin
Schlesinger B (1939) The venous drainage of the brain with special reference to the galenic system. Brain 62:274
Schmeidel G (1932) Die Entwickung der arteria vertebralis des Menschen. Morph Jahrb 71:315
Schmit BP, Burrows PE, Kuban K et al (1996) Acquired cerebral arteriovenous malformation in a child with Moya-Moya disease. Case report. J Neurosurg 84:677
Schneidereit N, Simons R, Nicolau S et al (2006) Utility of screening for blunt vascular neck injuries with computed tomographic angiography. J Trauma 60:209
Schroder JM, Sellhaus B, Jorg J (1995) Idendification of the characteristic vascular changes in a sural nerve biopsy of a case with cerebral autosomal dominant

arteriopathy with subcortical infarcts and leukoencephalopathy (CADASIL). Acta Neuropathol 89:116
Schubiger O, Valavanis A, Wichmann W (1987) Growth mechanism of giant intracranial aneurysm; demonstration by CT and MR imaging. Neuroradiology 29:266
Schuchlenz HW, Weihs W, Horner S et al (2000) The association of between the diameter of a patent foramen ovale and risk of embolic cerebrovascular events. Am J Med 109:456
Schumacher M, Wakhloo AK (1994) An orbital arteriovenous malformation in a patient with origin of the ophthalmic artery from the basilar artery. AJNR Am J Neuroradiol 15:550
Schwaiger BJ, Kober F, Gersing AS et al (2016) The pREset stent retriever for endovascular treatment of stroke caused by MCA occlusion: safety and clinical outcome. Clin Neuroradiol 26(1):47
Schwartz MJ, Baronofsky ID (1960) Ruptured intracranial aneurysm associated with coarctation of the aorta: a report of a patient treated by hypothermia and surgical repair of the coarctation. Am J Cardiol 6:982
Scialfa G, Bank W, Megret M et al (1976) The posterior fossa arteries: the morphology and variations of the anterior inferior cerebellar artery. The arteriographic localization of the fourth ventricle. In: Salamon G (ed) Advances in cerebral angiography. Springer, Berlin
Scialfa G, Valsecchi F, Scotti G (1983) Treatment of vascular lesions with balloon catheter. AJNR Am J Neuroradiol 4:395
Scotti G (1975) Anterior inferior cerebellar artery originating from the /cavernous portion of the internal carotid artery. Radiology 116:93
Sebire G, Meyer L, Chabrier S (1999) Varicella as a risk factor for cerebral infarction in childhood: a case-control study. Ann Neurol 45:679
Seeger JF, Hemmer JF (1976) Persistent basilar/middle meningeal artery anastomosis. Radiology 118:367
Seehra J, Horner K, Coulthard P (2006) Arteriovenous malformation of the mandible—a case report. Br Dent J 201:25
Segall HD, Ahmodi J, McComb JC et al (1982) Computed tomographic observations pertinent to intracranial venous thrombotic and occlusive disease in childhood. Radiology 143:441
Sekhar LN, Heros RC, Lotz PR et al (1980) Atheromatous pseudoocclusion of the internal carotid artery. J Neurosurg 52:782
Seong SO, David C, Choi IS (2006) Bilateral petrous ridge dural arteriovenous malformations treated by a combination of endovascular embolization and surgical excision. A case report. Interv Neuroradiol 12:269
Sepehrnia A, Tatagiba M, Brandis A et al (1990) Cavernous angioma of the cavernous sinus. Case report. Neurosurgery 27:151
Serbinenko FA (1974) Balloon catheterization and occlusion of major cerebral vessels. J Neurosurg 41:125
Shah QA, Hurst RW (2007) Anomalous origin of the middle meningeal artery from the basilar artery: a case report. J Neuroimaging 17:261
Shalaby F, Rossant J, Yamaguchi TP et al (1995) Failure of blood-island formation and vasculogenesis in FLk-1-deficient mice. Nature 376:62
Shapiro M, Becske T, Riina HA et al (2014) Toward an endovascular internal carotid artery classification system. AJNR Am J Neuroradiol 35:230
Shapiro M, Ollenschleger MD, Baccin C et al (2015) Foreign body emboli following cerebrovascular interventions: clinical, radiographic and histopathologic features. AJNR Am J Neuroradiol 36(11):2121–2126. doi:10.3174/ajnr.A4415
Sharp S, Stone J, Beach R (1999) Contrast agent neurotoxicity presenting as subarachnoid hemorrhage. Neurology 52:1503
Shi ZS, Liebeskind DS, Loh Y et al (2010) Endovascular stroke therapy investigators poor predictors of subarachnoid hemorrhage in acute ischemic stroke with endovascular therapy. Stroke 56:1087
Shi ZS, Ziegler J, Feng L et al (2013) Middle cranial fossa sphenoidal dural arteriovenous fistulas: anatomic and treatment considerations. AJNR Am J Neuroradiol 34:373
Shimoji T, Bando K, Nakajima K et al (1984) Dissecting aneurysm of the vertebral artery: report of seven cases and angiographic findings. J Neurosurg 61:1038
Shin JH, Suh DC, Choi CG et al (2000) Vertebral artery dissection: spectrum of imaging findings with emphasis on angiography and correlation with clinical presentation. Radiographics 20:1687
Shin YS, Kim SY, Kim BM et al (2005) Ruptured aneurysm of the anomalous cerebella artery originating from internal carotid artery presenting with carotid cavernous fistula: a case report. AJNR Am J Neuroradiol 26:1849
Shin Y, Nakase H, Nakamura M et al (2007a) Expression of angiogenic growth factor in the rat DAVF model. Neurol Res 29:727
Shin YS, Kim AS, Kim SY (2007b) Stenting for vertebrobasilar dissection: a possible treatment option for nonhemorrhagic vertebrobasilar dissection. Neuroradiology 49:149
Shiu PC, Hanafee WN, Wilson GH et al (1968) Cavernous sinus venography. Am J Roentgenol 104:57
Shokunbi MT, Vinters KV, Kaufmann JC (1988) Fusiform intracranial aneurysms: clininicopathologic features. Surg Neurol 29:263
Shotar E, Law-Ye B, Baronnet-Chauvet F et al (2016) Non-ischemic cerebral enhancing lesions secondary to endovascular aneurysm therapy: nickel allergy or foreign body reactions? Case series and review of the literature. Neuroradiology 58:877
Siddiqui AH, Abla AA, Kan P et al (2012) panacea or problem: flow diverters in the treatment of symptomatic large or giant fusiform vertebrobasilar aneurysms. J Neurosurg 116:1258
Signorelli F, Gory B, Pelissou-Guyottot I et al (2014) Ruptured brain arteriovenous malformations associated with aneurysms: safety and efficacy of selective embolization in the acute phase of hemorrhage. Neuroradiology 56:763

Singh NG, Sarkar C, Sharma MC et al (2005) Paraganglioma of cauda equine: report of seven cases. Brain Tumor Pathol 22:15

Singhal AB, Hajj Ali R, Calabrese LH (2009) Reversible cerebral vasoconstriction syndrome: two center experience of 139 cases. J Neurosci 285(Suppl 1):S 101

Singleton JD, West SG, Reddy VV et al (1995) Cerebral vasculitis complicating rheumatoid arthritis. South Med J 88:470

Siniscalchi EN, Minutoli F, Catalfamo L et al (2009) Intraosseous mandibular artero-venous malformations: case report. J Craniomaxillofac Surg 37:106

Siqueira M, Piske R, Ono M et al (1993) Cerebellar arteries originating from the internal carotid artery. AJNR Am J Neuroradiol 14:1229

Sitzer M, Müller W, Siebler M et al (1995) Plaque ulceration and lumen thrombus are the main sources of cerebral microemboli in high-grade internal carotid stenosis. Stroke 26:1231

Sjoegren SE (1956) The anterior choroidal artery. Acta Radiol 46:143

Smirniotopoulos JG, Murphy FM (1992) The phakomatoses. AJNR Am J Neuroradiol 13:725

Smoker WRK, Gentry LR, Yee NK et al (2008) Vascular lesions of the orbit: more than meets the eye. Radiographics 28:185

Sneddon JB (1965) Cerebro-vascular lesions and livedo reticularis. Br J Dermatol 77:180

So EL, Toole JF, Moody DM et al (1979) Cerebral embolism from septal fibromuscular of the common carotid artery. Ann Neurol 6:75

Soize S, Kadziolka K, Estrade L et al (2014) Outcome after mechanical thrombectomy using a stent retriever under conscious sedation; comparison between tandem and single occlusion of the anterior circulation. J Neuroradiol 41:136

Soltanolkotabi M, Schoeneman S, Alden TD et al (2013) Onyx embolization of intracranial arteriovenous malformation in pediatric patients. J Neurosurg Pediatr 11:437

Som PM, Curtin HD (2003) Fascia and spaces of the neck. In: Som PM, Curtin HD (eds) Head and neck imaging. Mosby, St. Louis

Som PM, Smoker WRK, Balboni A et al (2003a) Embryology and anatomy of the neck. In: Som PM, Curtin HD (eds) Head and neck imaging. Mosby, St. Louis

Som PM, Smoker WRK, Curtin HD et al (2003b) Congenital lesions. In: Som PM, Curtin HD (eds) Head and neck imaging. Mosby, St. Louis

Song JK, Cacayorui ED, Campbell MS et al (2002) Intracranial balloon angioplasty of acute terminal internal carotid artery occlusion. AJNR Am J Neuroradiol 23:1308

Song JK, Niimi Y, Kupersmith MJ (2007) Postnatal growth and development of a cerebral arteriovenous malformation on serial magnetic resonance imaging in a child with haemangiomatosis. Case report. J Neurosurg 106:384

Sonneland PRL, Scheithauer BW, Le Chago J et al (1986) Paraganglioma of the cauda equine region. Clinicopathoilogical study of 31 cases with special reference to immunocytology and iltrastructure. Cancer 58:1720

Sonstein WJ, Kader A, Michelsen WJ et al (1996) Expression of vascular endothelial growth factor in pediatric and adult cerebral arteriovenous malformations: an immunocytochemical study. J Neurosurg 85:838

Sorteberg A, Farhoudi D (2006) The influence of aneurysm configuration on intra-aneurysmal pressure and flow. Interv Neuroradiol 12:203

Spetzler RF, Martin NA (1986) A proposed grading system for arterio—venous malformations. J Neurosurg 65:4–76

Stanley JC, Fry WJ, Seeger JF et al (1974) Extracranial internal carotid and vertebral artery fibrodysplasia. Arch Surg 109:215

Stapf C, Labovitz DL, Sciacca RR et al (2002) Incidence of adult brain arteriovenous malformation haemorrhage in a prospective population-based stroke survey. Cerebrovasc Dis 13:43

Stapf C, Mast H, Sciacca R et al (2006) Predictors of haemorrhage in patients with untreated brain arteriovenous malformations. Neurology 66:1350

Staples GS (1979) Transsellar intracavernous intercarotid collateral artery associated with agenesis of the internal carotid artery. Case report. J Neurosurg 50:393

Stark MM, Skeik N, Almandoz JED et al (2013) Concurrent basilar artery double fvenestration with aneurysm and vertebral artery dissection: case report and literature review of rare cerebrovascular abnormalities. Ann Vasc Surg 27:497

Stayman AN, Nogueira RG, Gupta R et al (2011) A systemic review of stenting and angioplasty of symptomatic extracranial vertebral artery stenosis. Stroke 42:2212

Stehbens WE (1959) Medial defects of cerebral arteries of man. J Pathol Bacteriol 78:179

Stehbens WE (1972) Pathology of cerebral blood vessels. Mosby, St. Louis

Stehbens WE (1975) Cerebral atherosclerosis. Intimal proliferation and atherosclerosis in the cerebral arteries. Arch Pathol 99:582

Stehbens WE (1989) Etiology of intracranial berry aneurysms. J Neurosurg 70:823

Stehbens WE (1990) Pathology and pathogenesis of intracranial berry aneurysms. Neurol Res 12:29

Stein RL, Rosenbaum AE (1974) Normal deep cerebral venous system. In: Newton TH, Potts DG (eds) Angiography of the skull and brain. Mosby, St. Louis

Stem RG, Xiong X, Langston JW et al (2003) Brain injury in children with sickle cell disease: prevalence and etiology. Ann Neurol 54:564

Stephen RB, Stilwell DL (1969) Arteries and veins of the human brain. Charles C. Thomas Publisher, Springfield, IL

Stevens J, Leach JL, Abruzzo T et al (2009) De novo cerebral arteriovenous malformations: case report and literature review. AJNR Am J Neuroradiol 30:111

Stewart RM, Samson D, Diehl J et al (1980) Unruptured cerebral aneurysms presenting as recurrent transient neurologic deficit. Neurology 30:47

Stingele R, Berger J, Alfke K et al (2008) Clinical and angiographic risk factors for stroke and death within 30 days after carotid endarterectomy and stent-protected angioplasty: a subanalysis of the SPACE study. Lancet Neurol 7:216

Stoesslein F, Porstmann W, Schueler F et al (1982) Aberrant vertebral artery originating from the descending aorta: a new congenital steal syndrome in coarctation. Eur J Radiol 2:157

Streeter GL (1918) The developmental alterations in the vascular system of the brain of the human embryo. Contrib Embryol 8:5

Sturzenegger M (1995) Spontaneous internal carotid artery dissection: early diagnosis and management in 44 patients. J Neurol 242:231

Sugiu K, Tokunaga K, Watanabe K et al (2005) Emergent endovascular treatment of ruptured vertebral artery dissecting aneurysms. Neuroradiology 47:158

Suh DC, Lee JC, Kim SJ et al (2005) New concept in cavernous sinus dural arteriovenous fistula: correlation with presenting symptom and venous drainage pattern. Stroke 36:1134

Sujith OK, Krishnan R, Asaraf V et al (2008) Local thrombolysis in patients with dural venous thrombosis unresponsive to heparin. J Stroke Cerebrovasc Dis 17:95

Sukoff MH, Barth B, Moran T (1972) Spontaneous occlusion of a massive arteriovenous malformation. Case report. Neuroradiology 4:121

Suttner N, Mura J, Tedeschi H et al (2000) Persistent trigeminal artery: a unique anatomic specimen-analysis and therapeutic implications. Neurosurgery 47:428

Sutton D (1950) Anomalous carotid-basilar anastomosis. Br J Radiol 23:617

Suzuki J, Takaku A (1969) Cerebrovascular "Moya-moya disease": a disease showing abnormal net-like vessels in base of brain. Arch Neurol 20:288

Suzuki S, Kuwabara Y, Hatano R et al (1978) Duplicate origin of the left vertebral artery. Neuroradiology 15:27

Suzuki S, Nobechi T, Itoh I et al (1979) Persistent proatlantal intersegmental artery and occipital artery originating from internal carotid artery. Case reports. Neuroradiology 17:105

Suzuki M, Onuma T, Sakurai Y et al (1992) Aneurysms arising from the proximal (A1) segment of the anterior cerebral artery. J Neurosurg 76:455

Symonds CP (1923) Contribution to the clinical study of intracranial aneurysms. Guys Hosp Rep 63:139

Symonds CP (1924) Spontaneous subarachnoid hemorrhage. Q J Med 18:93

Szikora I, Paal G, Ugron A et al (2008) Impact of aneurismal geometry on aneurysmal flow: a computerized flow simulation study. Neuroradiology 50:411

Szikora I, Berentei Z, Kulcsar Z et al (2010) Treatment of intracranial aneurysms by functional reconstruction of the parent artery: the Budapest experience with the pipeline embolization device. AJNR Am J Neuroradiol 31:1139–1147

Tacconi L, Johnston FG, Symon L (1995) Accessory middle cerebral artery. J Neurosurg 83:916

Takahashi M, Wilson G, Hanafee W (1968) The anterior inferior cerebellar artery. Its radiographic anatomy and significance in the diagnosis of extraaxial tumors of posterior fossa. Radiology 90:281

Takahashi M, Tamakawa Y et al (1973) Fenestration of the basilar artery. Report of three cases and review of the literature. Radiology 109:79

Takahashi M, Arii H, Tamakawa Y (1980) Anomalous arterial supply of temporal and occipital lobes by anterior choroidal artery. Angiographic study. AJNR Am J Neuroradiol 1:537

Takahashi S, Hoshino F, Vemura K et al (1989) Accessory middle cerebral artery: is it a variant form of the recurrent artery of Heubner? AJNR Am J Neuroradiol 10:563

Takahashi S, Suga T, Kawata Y et al (1990) Anterior choroidal artery: angiographic analysis of variations and anomalies. AJNR Am J Neuroradiol 11:719

Takahashi S, Kato K, Tomura N et al (2001) Dural arteriovenous fistula of the cavernous sinus with cortical venous reflux of the posterior fossa via a bridging vein. Radiat Med 19:219

Takahashi T, Tominaga T, Hassan T et al (2003) Cervical cord compression with myelopathy caused by bilateral persistence of the first intersegmental arteries. Neurosurgery 53:234

Takasato Y, Hayashi H, Kobayashi T et al (1992) Duplicate origin of right vertebral artery with rudimentary and accessory left vertebral arteries. Neuroradiology 34(4):287–289

Takeuchi K, Shimizu K (1957) Hypoplasia of the bilateral internal carotid arteries. No To Shinkei 9:37

Tan EM, Cohen AS, Fries JF et al (1982) The 1882 revised criteria for the classification of systemic lupus erythematosus. Arthritis Rheum 25:1271

Tanaka M, Kikuchi Y, Ouchi T (2006) Neuroradiological analysis of 23 cases of basilar artery fenestration based on 2280 cases of MR angiographies. Interv Neuroradiol 12(1):39

Tao X, XJ Y, Bhattarai B et al (2006) Microsurgical anatomy of the anterior communicating artery complex in adult chinese heads. Surg Neurol 65:155

Taptas JN (1982) The so called cavernous sinus: a review of the controversy and its implications for neurosurgeons. Neurosurgery 11:712

Taschner CA, Brendecke S, Freiman T et al (2012) An intradural Mass lesion of the lumbar spine. Clin Neuradiol 22:175

Taschner CA, Brendecke S, Trippel M et al (2013) Freiburg Neuropathology case conference: contrast enhancing white matter lesions adjiacent to the lateral ventricle. Clin Neuroradiol 23:165

Tasker AD, Byrne JV (1997) Basilar artery fenestration in association with aneurysms of the posterior cerebral circulation. Neuroradiology 39:185

Tatu L, Moulin T, Bougosslavsky J et al (2001) Arterial vascular territory of human brain. In: Bougosslawsky J, Caplan L (eds) Stroke syndromes. Cambridge University Press, Cambridge

Tauntopoulou A, Ahl B, Weissenborn K et al (2008) Intra arterial thrombolysis using rtPA in patients with acute stroke due to vessel occlusion of anterior and—or posterior cerebral circulation. Neuroradiology 50:75

Taveras JM, Gilson J, Davis D et al (1969) Angiography in cerebral infarction. Radiology 93:549

Taveras JM (1969) Multiple progressive intracranial arterial occlusion: a syndrome of children and young adults. Am J Roentgenol Radium Ther Nucl Med 106:235

Tawk RG, Bendok BR, Qureshi AI (2003) Isolated dissection and dissecting aneurysms of the posterior inferior cerebellar artery: topic and literature review. Neurosurg Rev 26:180

Taylor RA, Siddig F, Memon MZ et al (2008) Vertebral artery ostial stent placement for atherosclerotic stenosis in 72 consecutive patients: clinical outcomes and follow-up results. Neuroradiology 51:531

Teal JS, Rumbaugh CL, Segall HD et al (1973) Anomalous branches of the internal carotid artery. Radiology 106:567

Tekkök IH, Ventureyra ECV (1996) De novo familial cavernous malformation presenting with hemorrhage 12.5 years after the initial hemorrhagic Ictus: natural history of an infantile form. Pediatr Neurosurg 25:151

Terada T, Higashida RT, Halbach VV et al (1994) Development of acquired arteriovenous fistulas in rats due to venous hypertension. J Neurosurg 80:884

Theodosopoulos PV, Lawton MT (2000) Fenestration of the posterior inferior cerebellar artery: case report. J Neurosurg 47:463

Theron J, Newton TH, Hoyt WF (1974) Unilateral retinocephalic vascular malformations. Neuroradiology 7:185

Theron J, Clay C, Djindjian R (1975) Etude angiographique et embolisation d'une fistule arterio-veineuse durale paracaverneuse d'une type anatomique particulier. Ann Radiol 18:729

Theron J, Newton TH (1976) Anterior choroidal artery. 1. Anatomic and radiographic study. J Neuroradiol 3:5

Thiex R, Wu I, Mulliken JB et al (2011) Safety and clinical efficacy of Onix for embolization of extracranial head and Neck vascular anomalies. AJNR Am J Neuroradiol 32:1082

Thijs VN, Albers GW (2000) Symptomatic intracranial atherosclerosis, outcome of patients who fails antithrombotic therapy. Neurology 55:490

Thines l ZF, Taschner C et al (2006) Subarachnoid hemorrhage from spontaneous dissection of the anterior cerebral artery. Cerebrovasc Dis 22:452

Thron AK (1988) Vascular anatomy of the spinal cord: neuroradiolohgical investigations and clinical syndromes. Springer, Heidelberg

Tiedermann F (1822) Tabulae arteriarum corporis humani. Karlsruhe CF Mueller

Tode JC (1787) Medizinische Chirurgische Bibliothek. Kopenhagen 10:408

Toennis W (1936) Erfolgreich Behandlung eines Aneurysma der Art. communicans anterior. Zbl Neurochir 1:39

Toennnis W, Lange-Cosack H (1953) Klinik operative Behandlung und Prognoses der arterio-venoesen Angiome des gehirns und seiner Haeute. Dtsch Z Nervenklinik 170:460

Tokimura H, Ishigami T, Yamahata H et al (2012) Clinical presentation and treatment of distal anterior inferior cerebellar artery aneurysms. Neurosurg Rev 35:497

Tokuda K, Miyasaka K, Abe H et al (1985) Anomalous atlantoaxial portion of vertebral and posterior inferior cerebellar arteries. Neuroradiology 27:410

Tokunaga K, Sugiu K, Kameda M et al (2004) Persistent primitive trigeminal artery-cavernous sinus fistula with intracerebral hemorrhage: endovascular treatment using detachable coils in a transarterial double-catheter technique. Case report and review of the literature. J Neurosurg 101:697

Tollard E, Niemtschik L, Darsaut TE et al (2011) Endovascular parent artery occlusion for the treatment of wide neck A1 segment aneurysms: a single center experience. AJNR Am J Neuroradiol 32:174

Tomak PR, Cloft HJ, Kaga A et al (2003) Evolution of the management of tentorial dural arteriovenous malformations. Neurosurgery 52:750

Tomasello F, Cioff FA, Albanese V (1976) Fibromuscular dysplasia of the basilar artery. Neurochirurgie 19:29

Tornow K, Piscol K (1971) The evaluation of the superior ophthalmic vein on the carotid angiogram. Neuroradiology 2:30

Tortori-Donati P, Fondelli MP, Rossi A et al (1999) Intracranial contrast-enhancing masses in infants with capillary haemangioma of the head and neck: intracranial capillary haemangiomas? Neuroradiology 41:369

Tournade A (1972) Les veines superficielles du bulbe et du pont chez l'homme. Thesis, Strasbourg

Touzé E, Randoux B, Meary E et al (2001) Aneurysmal forms of cervical artery dissection. Associated factors and outcome. Stroke 32:418

Tracy PT (1987) Unusual intercarotid anastomosis associated with anterior communicating artery aneurysm. J Neurosurg 67:765

Treiman GS, Scott Mcnally J, Kim SE et al (2015) Correlation of carotid intraplaque hemorrhage and stroke using 1.5T and 3T MRI. Magn Reson Insighgts 8:1

Trivelato FP, Abud DG, Nakiri GS et al (2016) Basilar artery fenestration aneurysms: endovascular treatment strategies based on 3D morphology. Clin Neuroradiol 26:73

Trolard P (1868) Anatomie du système veneux de l'encephale et du crane. These de la facultè de medecine de Paris, Paris

Trolard P (1892) Les granulations de Pacchioni. Les lacunes veineuses de la dure – mère. J Anat Physiol 28:28–57

Tsai FY, Wang AM, Matovich VB et al (1995) MR staging of acute dural sinus thrombosis: correlation with venous pressure measurements and implications for treatment and prognosis. AJNR Am J Neuroradiol 16:1021

Tsuji T, Abe M, Tabuchi K (1995) Aneurysm of a persistent primitive olfactory artery. J Neurosurg 83:138

Tsumoto T, Miyamoto T, Shimizu M et al (2003) Restenosis of the sigmoid sinus after stenting for treatment of intracranial venous hypertension: case report. Neuroradiology 45:911

Tubbs RS, Loukas M, Louis RG Jr et al (2007a) Anatomy of the falcine venous plexus. J Neurosurg 107:155

Tubbs RS, Salter EG, Wellons JC (2007b) The sphenoparietal sinus. Neurosurgery 60:ONS9

Tubbs RS, Watanabe K, Loukas M et al (2014) Anatomy of the inferior petro-occipital vein and its relation to the base of the skull: application to surgical and endovascular procedures of the skull base. Clin Anat 27:698

Tulamo R, Froesen J, Hernesniemi J et al (2010) Inflammatory changes in the aneurysm wall: a review. J Neurointerv Surg 2:120

Tural Emon S, Orakdogen M, Akpinar E et al (2012) Arachnoid granulations: a rare cause of lytic occipital bone lesion. Neurol Neurochir Pol 46:603

Ture G, Oppenheim C, Naggara O et al (2012) Relationship between recent intraplaque hemorrhage and stroke risk factors in patients with carotid stenosis: the HIRISC study. Arterioscler Thromb Vasc Biol 32:492

Turjman F, Massoud TF, Vinuela F et al (1994) Aneurysms related to cerebral arteriovenous malformations: superselective angiographic assessment in 58 patients. AJNR Am J Neuroradiol 15:1601

Turjman F, Massoud TF, Vinuela F et al (1995) Correlation of the angioarchitectural features of cerebral arteriovenous malformations with clinical presentation of haemorrhage. Neurosurgery 37:856

Turk AS, Levy EI, Albuquerque FC et al (2008) Influence of patient age and stenosis location on Wingspan in-stent restenosis. AJNR Am J Neuroradiol 29:23

Turnbull I (1962) Agenesis of the internal carotid artery. Neurology 12:588

Uchino A, Kato A, Takase Y et al (2000) Persistent trigeminal artery variants detected by MR angiography. Eur Radiol 10:1801

Uchino A, Saito N, Okada Y et al (2012a) Carotid –anterior cerebral artery anastomosis on MR angiography: a university hospital-based study. Neuroradiology 54:13

Uchino A, Saito N, Okada Y et al (2012c) Fenestration of the intracranial vertebrobasilar system diagnosed by MR angiography. Neuroradiology 54:445

Uchino A, Saito N, Okada Y et al (2013a) Persistent hypoglossal artery and its variants diagnosed by CT and MR angiography. Neuroradiology 55:17

Uchino A, Saito N, Takahashi M et al (2013b) Variations in the origin of the vertebral artery and its level of entry into the transverse foramen diagnosed by CT angiography. Neuroradiology 55:585

Uchino A, Saito N, Watadani T et al (2011) Non-bifurcating cervical carotid artery diagnosed by MR angiography. AJNR Am J Neuroradiol 32:1119

Uchino A, Saito N, Watadani T et al (2012b) Vertebral artery variations at the C1-2 level diagnosed by magnetic resonance angiography. Neuroradiology 54:19

Uchino A, Saito N, Takahashi M et al (2016) Variations of the posterior cerebral artery diagnosed by MR Angiography at 3 Tesla. Neuroradiology 58:141

Uemo Y, Kimuna K, Iguchi Y et al (2007) Mobile aortic plaques are a cause of multiple brains infarcts seen on diffusion-weighted imaging. Stroke 38:2470

Ulus S, Yakupoglu A, Karaslan E et al (2012) Reversible intracranial parenchymal changes in MRI after MCA aneurysm treatment with stent—assisted coiling technique; possible nickel allergy. Neuroradiology 54:897

Umansky F, Nathan H (1982) The lateral wall of the cavernous sinus. J Neurosurg 56:228

Umansky F, Gomes FB, Dujovny M et al (1985) The perforating branches of the middle cerebral artery: a microanatomical study. J Neurosurg 62:261

Umansky F, Dujovny M, Ausman J et al (1988) Anomalies and variations of the middle cerebral artery; a microanatomical study. Neurosurgery 22:1023

Unterhofer C, Chemelli A, Waldenberger P et al (2009) Traumatic fistula between the middle meningeal artery and the sphenoparietal sinus. Acta Neurochir 151:1301

Uranishi R, Nakase H, Sakaki T (1999) Expression of angiogenic growth factors in dural arteriovenous fistulas. J Neurosurg 91:781

Uranishi R, Baew NI, Py NG et al (2001) Expression of endothelial cell angiogenesis receptors in human cerebrovascular malformations. Neurosurgery 48:359

Urbach H, Ries F, Ostertun B et al (1997) Local intra-arterial fibrinolysis in thromboembolic "T" occlusions of the internal carotid artery. Neuroradiology 39:105

Urbach H, Zentner J, Solymosi L (1998) The need for repeat angiography in subarachnoid haemorrhage. Neuroradiology 40(1):6–10

Urban PP, Müller-Forrell W (2005) Clinical and neuroradiological spectrum of isolated cortical vein thrombosis. J Neurol 252:1476

Urbaniak K, Merchant AI, Amin-Hanjani S et al (2007) Cardiac complications after aneurysmal subarachnoid hemorrhage. Surg Neurol 67:21

Vaid VK, Kumar R, Kalra SK et al (2008) Pediatric intracranial aneurysms: an institutional experience. Pediatr Neurosurg 44:296

Valavanis A (1996) The role of angiography in the evaluation of cerebral vascular malformations. Neuroimaging Clin N Am 6:679

Valavanis A, Christoforidis G (2000) Application of interventional neuroradiology in the head and neck. Semin Roentgenol 35:72

Valavanis A, Pangalu A, Tanaka M (2004) Endovascular treatment of cerebral arteriovenous malformations

with emphasis on the curative role of embolization. Swiss Arch Neurol Psychiatry 7:341
Valavanis A, Yaşargil MG (1998) The endovascular treatment of brain arteriovenous malformations. Adv Tech Stand Neurosurg 24:131
Valdueza JM, Von Munster T, Hoffman O et al (2000) Postural dependency of the cerebral venous outflow. Lancet 355:200
Van den Berg JS, Pals G, Arwert F et al (1999) Type III collagen deficiency in saccular intracranial aneurysms: defect in gene regulation? Stroke 30:1628
Van Den Boom R, Lesnik Oberstein SA, Van Duinen SG et al (2002) Subcortical lacunar lesions: an MR imaging finding in patients with cerebral autosomal dominant arteriopathy with subcortical infarcts and leucoencefalopathy. Radiology 224:791
Van der Eecken H (1954) Signification morphologique des anastomoses leptomeningées aux confins du territoire des artères cérébrales. Acta Neurol Belg 54:525
Van der Schaaf IC, Brilstra EH, Buskens E et al (2002) Endovascular treatment of aneurysms in the cavernous sinus: a systematic review on balloon occlusion of the parent vessel and embolization with coils. Stroke 33:313
Van Dijk JMC, Terbrugge KG, Willinsky RA et al (2004) Selective disconnection of cortical venous reflux as treatment for cranial dural arteriovenous fistulas. J Neurosurg 101:31
Van Rooji WS, Sluzewski M, Slob MJ et al (2005) Predictive value of angiographic testing for tolerance to therapeutic occlusion of the carotid artery. AJNR Am J Neuroradiol 26:175
Van Rooij WJ, Sluzewski M, Beute GH (2008) Internal carotid bifurcation aneurysms: frequency, angiographic anatomy and results of coiling in 50 aneurysms. Neuroradiology 50:583
Van Rooij WJ, Sluzewski M (2009) Endovascular treatment of large and giant aneurysms. AJNR Am J Neuroradiol 30:12
Van Rooij WJ (2012) Flow diverters for unruptured aneurysms: are they safe enough? Neuroradiology 54:1179
Van Rooij WJ, Jacobs S, Sluzewski M et al (2012a) Endovascular treatment of ruptured brain AVMs in the acute phase. AJNR Am J Neuroradiol 33:1162
Van Rooij WJ, Jacobs S, Sluzewski M et al (2012b) Curative embolization of brain arteriovenous malformations with onyx: patient selection, embolization technique, results. AJNR Am J Neuroradiol 33:1299
Van Seeters T, Biessels GJ, Kappelle LJ et al (2016) Determinants of leptomeningeal collateral flow in stroke patients with a middle cerebral artery occlusion. Neuroradiology 58:969
Vargas ME, Desrouleaux JR, Kupersmith MJ (1992) Ophthalmoplegia as a presenting manifestation of internal carotid artery dissection. J Clin Neuroophthalmol 12:268
Velat GJ, Fargen KM, Lawson MF et al (2012) Delayed intraparenchymal hemorrhage following pipeline embolization device treatment for a giant recanalized ophthalmic aneurysm. J Neurointerv Surg 4(5):e24
Velioglu M, Kizilkilic O, Selcuk H et al (2012) Early and midterm results of complex cerebral aneurysms treated with silk stent. Neuroradiology 54:1355
Ventureyra ECG, Higgins MJ (1994) Traumatic intracranial aneurysms in childhood adolescence. Case report and review of the literature. Childs Nerv Syst 10:361
Verbiest MH (1951) Intradural arteriovenous aneurysm. Rev Neurol 85:189
Vertinsky AT, Schwartz NE, Fischbein NJ et al (2008) Comparison of multidetector CT angiography and MR imaging of cervical artery dissection. AJNR Am J Neuroradiol 29:1753
Vignaud J, Clay C, Aubin M et al (1972) Orbital arteriography. Radiol Clin North Am 10:39
Vila N, Millan M, Ferrer X et al (2003) Levels of alpha-1-antitripsin in plasma and risk of spontaneous cervical artery dissections. Stroke 34:168
Vilela MD, Goodkin R, Lundin DA et al (2005) Rotational vertebrobasilar ischemia: hemodynamic assessment and surgical treatment. Neurosurgery 56:36
Vilela P, Goulao A (2006) Pediatric dissecting posterior cerebral aneurysms: report of two cases and review of the literature. Neuroradiology 48:541
Villringer A, Mehraein S, Einhäupl KM (1994) Treatment of sinus venous thrombosis. Beyond the recommendation of anticoagulation. J Neuroradiol 21:72
Vines FS, Davis DO (1971) Clinical-radiological correlation in cerebral venous occlusive disease. Radiology 98:9
Vinuela F, Fox AJ, Debrun GM et al (1984) Spontaneous carotid cavernous fistulas: clinical, radiological and therapeutic considerations. J Neurosurg 60:976
Vinuela F, Nombela L, Roach MR et al (1985) Stenotic and occlusive disease of the venous drainage system of deep brain AVMs. J Neurosurg 63:180
Vinuela F, Drake CG, Fox AJ et al (1987) Giant intracranial varices secondary to high flow arteriovenous fistulae. J Neurosurg 66:198
Vinuela F, Duckwiller GR, Johan R et al (2005) Therapeutic management of cerebral arterio—venous malformations. Present role of interventional neuroradiology. Interv Neuroradiol 11:13
Virchow R (1864) Die Krankhaften Geschwulste. Dreissig Vorlesungen gehalten wahrend des Wintersemester 1862–1863 an der Universitat zu Berlin. Verlag von August Hirschwald, Berlin
Vitek JJ, Reaves P (1973) Thoracic bifurcation of the common carotid artery. Neuroradiology 5:133
Von Kummer R, Holle R, Rosin L et al (1995) Does arterial recanalization improve outcome in carotid territory stroke? Stroke 26:581
Wackenheim A, Braun JP (1970) Angiography of the mesencephalon. Normal and pathologic findings. Springer, Berlin
Wackenheim A, Braun JP (1978) The veins of the posterior fossa. Springer, Berlin
Waespe W, Niesper J, Imhoff HG et al (1988) Lower cranial nerve palsies due to internal carotid artery dissection. Stroke 19:4561

Wakabayashi T, Tamaki N, Yamashita H et al (1985) Angiographic classification of aneurysms of the horizontal segment of the anterior cerebral artery. Surg Neurol 24:31

Wakai S, Chin-Hsiung C, KY W et al (1983) Spontaneous regression of a cerebral arteriovenous malformation. Report of a case and review of the literature. Arch Neurol 40:377

Walcott BP, Smith ER, Scott RM et al (2013) Dural arteriovenous fistulae in pediatric patients: associated conditions and treatment outcomes. J Neurointerv Surg 5:6

Wang B, Miao ZR, Li G et al (2009) Treatment of symptomatic complex posterior circulation cerebral artery stenosis with balloon–mounted stents: technique feasibility and outcome. Neuroradiology 51:319

Wang C-B, Wen-Wen S, Zhang GX et al (2016a) Flow diverter treatment of posterior circulation aneurysms. A meta-analysis. Neuroradiology 58:391

Wang Z-L, Gao BL, Li TX et al (2016b) Outcomes of middle cerebral artery angioplasty and stenting with wingspan at a hich-volume center. Neuroradiology 58:161

Wanibuchi M, Kurokawa Y, Ishiguro M et al (2001) Characteristics af aneurysms arising from the horizontal portion of the anterior cerebral artery. Surg Neurol 55:148

Wanke I, Forsting M (2008) Stents for intracranial wide-necked aneurysms: more than mechanical protection. Neuroradiology 50:991

Warfarin-Aspirin Symptomatic Intracranial Disease (WASID) Trial Investigators (2003) Design, progress and challenges of double-blind trial of warfarin versus aspirin for symptomatic intracranial arterial stenosis. Neuroepidemiology 22:106

Wasserman BA, Sharrett AR, Shenghan L et al (2008) Risk factor associations with the presence of a lipid core in carotid plaque of asymptomatic individuals using high-resolution MRI. The multi-ethnic study of atherosclerosis (MESA). Stroke 39:329

Watanabe Y, Nagayama M (2010) MR plaque imaging of the carotid artery. Neuroradiology 52:253

Waterston JA, Brown MM, Butler P et al (1990) Small deep cerebral infarcts associated with occlusive internal carotid artery disease. Arch Neurol 47:953

Weber J, Veith P, Jung B et al (2015) MR- Angiography at 3 Tesla to assess proximal internal carotid artery stenosis: contrast enhanced or 3D Time of flight MR angiography? Clin Neuroradiol 25:41

Weidauer S, Claus D, Gartenschlager M (1999) Spinal sulcal artery syndrome due to spontaneous bilateral vertebral dissection. J Neurol Neurosurg Psychiatry 67:550

Weiller C, Ringelstein EB, Reiche W et al (1990) The large striatocapsular infarct. A clinical and pathophysiological entity. Arch Neurol 47:1085

Weiller C, Ringelstein EB, Reiche W et al (1991) Clinical and hemodynamic aspects of low-flow infarcts. Stroke 22:1117

Weinstein M, Stein R, Pollock J et al (1974) Meningeal branch of the posterior cerebral artery. Neuroradiology 7:129

Weir B (1987) Aneurysms affecting nervous system. Lippincott Williams & Wilkins, Baltimore

Wermer MJ, Van der Schaaf IC, Velthuis BK et al (2005) Follow-up screening after subarachnoid hemorrhage: frequency and determinants of new aneurysms and enlargement of existing aneurysm. Brain 128:2421

Wessels T, Mosso M, Krings T et al (2008) Extracranial and intracranial vertebral artery dissection: long-term clinical and duplex sonographic follow-up. J Clin Ultrasound 36:472

Westberg G (1963) The recurrent artery of Heubner and the arteries of the central ganglia. Acta Radiol Diagn 1:949

Westberg G (1966) Arteries of the basal ganglia. Acta Radiol Diagn 5:581

Weygand CM, Goronzy JJ (2003) Medium and large-vessel vasculitis. N Engl J Med 349:160

Whisenant JT, Kadkhodayan Y, Crfoss DT et al (2015) Incidence and mechanism of stroke after permanent carotid artery occlusion following temporary occlusion testing. J Neurointerv Surg 7:395

Wholey MH, Wholey M, Mathias K et al (2000) Global experience in cervical carotid artery stent placement. Catheter Cardiovasc Interv 50:160

Widjaja E, Griffiths PD (2004) Intracranial MR Venography in children: normal anatomy and variations. AJNR Am J Neuroradiol 25:1557

Wiebers DO, Whisnant JP, Huston J III, International Study of Unruptured Intracranial Aneurysms Investigators et al (2003) Unruptured intracranial aneurysms: natural history, clinical outcome and risk of surgical and endovascular treatment. Lancet 362:103

Wiebers DO, Whisnant JP, Sundt TM Jr et al (1987) The significance of unruptured intracranial saccular aneurysms. J Neurosurg 66:23

Willinsky R, Lasjaunias P, Terbrugge K et al (1988) Brain arteriovenous malformations. An analysis of the angioarchitecture in relationship to haemorrhage. J Neuroradiol 15:225

Willinsky R, Lasjaunias P, Terbrugge K et al (1990) Multiple cerebral arteriovenous malformations. Review of our experience from 203 patients with cerebrovascular lesions. Neuroradiology 32:207

Willis Th (1684) Cerebri anatomiae nervorumque descritpio et usus. Willis Th (1964) (trans: Prodage S). J. Fletcher, London

Wilson WB, Laevengood JM, Ringel SP (1989) Transient ocular motor paresis associated with acute internal carotid artery occlusion. Ann Neurol 25:286

Wintermark M, Ko NU, Smith WS et al (2006) Vasospasm after subarachnoid haemorrhage: utility of perfusion CT and CT angiography on diagnosis and management. AJNR Am J Neuroradiol 27:26

Wirth FP, Miller WA, Russel AP (1981) Atypical fibromuscular hyperplasia. Report of two cases. J Neurosurg 54:685

Wittkugel O, Gbadamosi J, Rosenkranz M et al (2008) Long-term outcome after angioplasty of symptomatic internal carotid artery stenosis with and without stent. Neuroradiology 50:243

Wityk RJ, Chang HM, Rosengart A et al (1998) Proximal extracranial vertebral artery disease in the New England Medical Center Posterior Circulation Registry. Arch Neurol 55:470

Witznum JL, Steinberg D (1991) Role of oxidized low density lipoprotein in atherogenesis. J Clin Invest 88:1785

Wolbach SB (1908) Multiple hernias of the cerebrum and cerebellum due to intracranial pressure. J Med Res 19:153

Wolf BS, Huang YP (1963) The insula and deep middle cerebral venous drainage system: normal anatomy and angiography. Am J Roentgenol 90:472

Wolf BS, Huang YP (1964) The subependymal veins of the lateral ventricle. Am J Roentgenol 91:406

Wolf BS, Huang YP, Newman CM (1963) The lateral anastomotic mesencephalic vein and other variations in drainage of the basal cerebral vein. Am J Roentgenol 89:411

Wollschlaeger G, Wollschlaeger PB (1964) The primitive trigeminal artery as seen angiographically and at postmortem examination. Am J Roentgenol Radium Ther Nucl Med 92:761

Wollschlaeger G, Wollschlaeger PB (1974) The circle of Willis. In: Newton TH, Potts DG (eds) Radiology of skull and brain, Angiography, vol 2. Mosby, St. Louis

Wollschlaeger PB, Wollschlaeger G (1965) Eine infratentorielle meningeale Arterie. Radiologe 5:451

Wong KS, Huang YN, Gao S et al (1998) Intracranial stenosis in Chinese patients with acute stroke. Neurology 50:812

Ka-Tak W, Lam WWM, Simon CHY (2007) MDCT of an aberrant right subclavian artery and of bilateral vertebral arteries with anomalous origin. AJR Am J Roentgenol 188:W274–W275

Wong ST, Yuen SC, Fok KF et al (2008) Infraoptic anterior cerebral artery: review, report of two cases and an anatomical classification. Acta Neurochir 150:10807

Wylie EG, Binkley FM, Palubinkas AJ (1966) Extracranial fibromuscular hyperplasia. Am J Surg 112:149

Yamamoto Y, Georgiades A, Chang HM et al (1999) Posterior cerebral artery territory infarcts in the New England Medical Center (NEMC). Posterior circulation registry. Arch Neurol 56:824

Yamamura A, Watanabe Y, Saeki N (1990) Dissecting aneurysms of the intracranial vertebral artery. J Neurosurg 72:183

Yamamura I, Tani E, Yokota M et al (1999) Endovascular treatment of ruptured dissecting aneurysms aimed at occlusion of the dissected site by using Guglielmi detachable coils. J Neurosurg 90:853

Yamaura I, Tani E, Yokota M (1999) Endovascular treatment of ruptured dissecting aneurysms aimed at occlusion of the dissected site by using Guglielmi detachable coils. J Neurosurg 90:853

Yanaka K, Matsumaru Y, Kamezaki T et al (2002) Ruptured aneurysm of the ophthalmic artery trunk demonstrated by three-dimensional rotational angiography: case report. Neurosurgery 51:1066

Yang X, Mu S, Lu M et al (2007) Endovascular treatment of huge dissecting aneurysms involving the basilar artery. Experience and lessons from two cases. Interv Neuroradiol 13:369

Yaşargil MG, Carter LP (1974) Saccular aneurysms of the distal anterior cerebral artery. J Neurosurg 40:218

Yaşargil MG (1984a) Microneurosurgery I: microsurgical surgical anatomy of the basal cisterns and vessels of the brain. Thieme, New York

Yaşargil MG (1984b) Microneurosurgery II: clinical considerations, surgery of the intracranial aneurysms and results. Thieme, New York

Yaşargil MG (1987) Microsurgery. AVM of the brain. History, embryology, pathological considerations, hemodynamics, diagnostic studies, microsurgical anatomy. Thieme Medical Publisher, New York

Yaşargil MG (1999) A legacy of microsurgery: memoirs, lessons and axioms. Neurosurgery 45:1025

Yasui T, Sakamoto H et al (1998) Rupture machanism of a thrombosed slow—growing giant aneurysm of the vertebral artery. Case report. Neurol Med Chir (Tokyo) 38:860

Yates PO, Hutchinson EC (1961) Cerebral infarction: the role of stenosis of the extracranial cerebral arteries. Memo Med Res Counc 300:1–95

Yavuz K, Geyik S, Saatci I et al (2014) Endovascular treatment of middle cerebral artery aneurysm with flow modification with the use of the pipeline embolizxation device. AJNR Am J Neurorradiol 35:529

Yazbak PA, Mc Comb JC et al (1995) Paediatric traumatic intracranial aneurysm. Pediatr Neurosurg 223:15

Yeterian EH, Van Hoesen GW (1978) Corticostriate projections in the Rhesus monkey: the organization of certain cortico—caudate connections. Brain Res 139:43

Yilmaz U, Walter S, Koerner H et al (2014) Peri-interventional subarachnoid hemorrhage during mechanical thrombectomy with stent retrievers in acute stroke: a retrospective case control study. Clin Neuroradiol 25:173

Yock AH Jr (1984) Fenestration of the supraclinoid internal carotid artery with rupture of associated aneurysm. AJNR Am J Neuroradiol 5:634

Yoon KW, Cho MK, Kim JI et al (2011) Sinus thrombosis in a patient with intracranial hypotension: a suggested hypothesis of venous stasis. A case report. Intervent Neuroradiol 17:248

Yoon W, Kim SK, Park MS et al (2012) Safety of protected carotid artery stenting in patients with severe carotid artery stenosis and carotid intraplaque hemorrhage. AJNR Am J Neuroradiol 33:1027

Yoon W, Jung MY, Jung SH et al (2013) Subarachnoid hemorrhage in a multimodal approach heavily weighted toward mechanical thrombectomy with solitaire stent in acute stroke. Stroke 44:414

Yoshimoto Y, Wakai S (1997) Unruptured intracranial vertebral artery dissection. Clinical course and serial radiographic imagings. Stroke 28:370

Yu SCH, Cheng KM, Tam PHT (2011) A venographic operational classification for transvenous

Embolization of dural carotid-cavernous fistula. Neuroradiology 53:993

Yung KH, Kwon BJ, Chu K et al (2011) Clinical and angiographic factors related to the prognosis of cavernous sinus dural arteriovenous fistula. Neuroradiology 53:983

Zager EL, Shaver EG, Hurst RV et al (2002) Distal anterior inferior cerebellar artery aneurisms: report of 4 cases. J Neurosurg 97:692

Zaidat OO, Suarez JI, Santillan C et al (2002) Response to intra-arterial and combined intravenous and intra-arterial thrombolytic therapy in patients with distal internal carotid artery occlusion. Stroke 33:1821

Zaidat OO, Suarez JI, Sunshine JL et al (2005) Thrombolytic therapy of acute ischemic stroke: correlation of angiographic recanalisation with clinical outcome. AJNR Am J Neuroradiol 26:880

Zak F, Lawson W (1982) Paraganglionic chemoreceptor system: physiology pathology and clinical medicine. Springer, New York

Zambranski JM, Wascher TM, Spetzler RF et al (1994) The natural history of familial cavernous malformations. Results of an ongoing study. J Neurosurg 80:422

Zeal AA, Rhoton AJ Jr (1978) Microsurgical anatomy of the posterior cerebral artery. J Neurosurg 48:534

Zeumer H, Hacke W, Kolmann HL et al (1982) Lokale Fibrinolyse Therapie bei Basilaris Thrombose. Dtsch Med Wochenschr 107:728

Zeumer H, Freitag HJ, Zanella F et al (1993) Local intra-arterial fibrinolytic therapy in patients with stroke: urokinase versus recombinant tissue plasminogen activator (rTPA). Neuroradiology 35(2):159

Zhang H, Shaw AR, Mak A et al (1996) Endoglin is a component of the transforming growth factor (TGF)—beta receptor complex of human pre-B leukemic cells. J Immunol 156:564

Zhang WW, Ma KC, Andersen O et al (1994) The microvascular changes in cases of hereditary multi-infarct disease of the brain. Acta Neuropathol 87:317

Zhao L, Moos MP et al (2004) The 5-lipoxygenase pathway promotes pathogenesis of hyperlipidemia—dependent aortic aneurysm. Nat Med 10:966

Zhao WY, Krings T, Alvarez H et al (2007) Management of spontaneous haemorrhagic intracranial vertebrobasilar dissection: a review of 21 consecutive cases. Acta Neurochir 149:585

Zhao R, Li W, Hong B et al (2013) The safety and efficacy of stenting in the treatment of complex posterior cerebral artery aneurysms: a seven-case report and literature review. Clin Neuroradiol 23:175

Zuber M, Meary E, Meder JF et al (1994) Magnetic resonance imaging and dynamic CT scan in cervical artery dissections. Stroke 25:576

Zülch KJ (1957) Brain tumors: their biology and pathology. Springer, Berlin

Zyial IM, Salas E, Wright DL et al (1998) The petrolingual ligament: the anatomy and surgical exposure of the posterolateral landmark of the cavernous sinus. Acta Neurochir 140:201

Index

A

Amyloid angiopathy, 341, 404, 427–428
Aneurysms
    aneurysms in children, 204–205, 389–392
    in arteritis, 394
    in azygos, 76–79
    basilar artery aneurysms, 186, 190–192, 198–199
    distal aneurysms of cerebellar arteries, 189
    flow related aneurysms
        in AVM, 189, 221
        in DAVF, 189, 223
    fusiform and giant aneurysms, 197, 200
    ICA aneurysms
        cavernous, 170
        extracranial, 169
        of ICA bifurcation, 175
        paraclinoid, 172–175
        petrous segment, 170
        superior hypophyseal artery aneurysm, 172
    and ischemic stroke, 169, 338
    MCA aneurysms, 186
    multiple cerebral aneurysms, 167
    mycotic aneurysms, 168, 409, 411
    negative angiography in SAH, 203
    PCA aneurysm, 189, 195, 196
    pericallosal aneurysm, 183, 185
    unruptured aneurysms, 169, 203
    vasospasm in SAH, 204
    vertebral–PICA aneurysms, 189
Anterior cerebral artery (ACA)
    callosomarginal artery, 70–72
    dissecting aneurysm, 183–184, 186, 373, 379
    duplication of A1, 76
    Heubner's artery, 68, 76–77, 80, 89–91, 183, 218–219, 338, 340
    hypoplasia of A1, 72, 313
    paracentral artery, 70–73, 77, 211
    perforators of A1, 67, 77
    pericallosal artery, 70–71
        aneurysm of, 183, 185–186
    subcallosal artery, 68, 73, 76, 340
    triplex percallosal A., 68, 72, 76, 80
Anterior choroidal artery (AchA)
    aneurysms, 175, 182, 338
    cisternal segment, 23–25, 29
    occlusion, 334, 337, 338
    perforators, 23, 25
    variants, 23
Anterior communicating artery (AcomA)
    aneurysms, 72, 176
    hypoaplasia, 72
    multiple, 72
    perforators of, 67, 77
Anterior inferior cerebellar artery (AICA)
    in AVM, 222, 224
    in fenestration of basilar artery (BA), 106
    flow-dependent aneurysm in AVM, 189, 222, 224
    with flow dependent aneurysm in DAVF, 189, 223
    labyrinth artery, 103, 104
    occlusion, 358
    perforators for pons, 104
    relationship with PICA, 104, 366
    rostro-lateral and caudomedial branches, 103
Aortic arch (embryogenesis), 1–4
Arterial occlusive diseases in children, 415–416
Arteriovenous fistula
    carotid–cavernous fistulas, 297–300
    vertebral arteriovenous fistulas, 300–302
Atherosclerosis
    in black and Asian people, 304
    location, 304
    pathology, 304–305
Autosomal dominant arteriopathy with subcortical infarct and leucoencefalopathy (CADASIL), 406–407

B

Basilar artery (BA)
    aneurysms, 186, 190–192, 197–199
    aplasia, 112–113
    dissection of BA, 373, 388
    duplication, 112
    embolic occlusion of BA in dissection of VA, 383
    embryogenesis, 4

G.B. Bradac, *Applied Cerebral Angiography*, DOI 10.1007/978-3-319-57228-4

Basilar artery (BA) (*cont.*)
fenestration of BA, 110
lateral pontine arteries, 101, 222
microatheroma of basilar artery, 357
perforators of BA, 100–101
thromboembolic occlusion of BA, 350
tortuous BA, 98
Blister-like aneurysms, 175, 381, 384
Brainstem, vascular territories of, 126, 127

C
CADASIL, 406–407
Cardiac diseases
cardiac emboli, 409–410
foramen ovale, 409
myxoma, 409–410
Cerebral hemorrhage, 427–428
Collateral circulation, 364–366
Common carotid artery (CCA)
bifurcation, 11
embryogenesis, 1–3, 11

D
Dissection
blood blister-like aneurysms, 381–384
in children, 389–416
of extracranial arteries, 372
and haemorrhage, 377
of intracranial arteries, 373
and ischemic stroke, 374–377
morphological diagnostic appearance, 373
pathology and pathogenesis, 371–372
treatment of dissection, 382–384
Ductus caroticus, 1–3, 11, 25, 31
Dural arteriovenous fistulas (DAVF)
classification, 246–247
involving anterior fossa, 274
involving CS, 255
involving foramen magnum, 274
involving middle cranial fossa, 263–264
involving SSS, 264
involving tentorium, 265, 268–274
involving TS and SiS, 247–255
in pediatric patients, 292–296
perimedullary drainage in DAVF, 285

E
External carotid artery (ECA)
ascending pharyngeal artery (AphA)
in angiofibroma, 54–55
and cranial nerves, 39–40
in DAVF, 252, 259, 272, 273, 286–291
in Rendu-Osler, 53
hypoglossal branch of, 39
jugular branch of, 40
musculospinal arteries of, 38
neuromeningeal trunk of, 38
in paragangliomas, 57–60
pharyngeal branches, 38
dangerous anastomosis of, 63–64
facial artery, 37
angular artery in AVM, 51
angular artery in ICA occlusion, 37
internal maxillary artery (IMA)
in AVM, 42, 51–52
collateral circulation in ICA occlusion, 45
deep temporal branches, 43
descending palatine A., 41, 42, 44, 47
foramen rotundum A., 15, 45, 63, 258, 265, 273
inferior alveolar artery, 41, 42
infraorbital A., 41, 47, 51
posterior superior alveolar A., 41, 42, 44, 47, 51, 52
pterygoid, masseter, buccal A., 43–44
pterygovaginal A., 45, 46, 63
in Rendu Osler, 53
sphenopalatine A., 46, 282
vidian (pterygoid A.), 45, 46
lingual artery, 36
occipital artery
anomalous origin, 40
branches, 40, 41
in DAVF, 249–257, 269–272, 275, 278, 289
in paraganglioma, 58
transosseous branches in DAVF, 253, 256
posterior auricular A., 41
superficial temporal A.
collateral circulation in ICA occlusion, 46
transverse facial A., 41, 47, 62
superior thyroid artery, 35, 36

F
Fibromuscular dysplasia (FMD)
and aneurysm, 168, 397
in dissection, 371, 398
and fistula, 297, 397
Flow-dependent (related) aneurysms, 221

H
Herophili (torcular), 129, 149, 151
Hyoid A., 19
Hypophyseal arteries, 15, 20
and aneurysms, 172

I
Infarcts
hemodynamic, territorial, lacunar, 304, 305
Inferolateral trunk (ILT), 15
in angiofibroma, 55
in DAVF, 265, 269, 271
Internal carotid artery (ICA)
aberrant course, 33

aplasia of ICA, 26
cavernous segment, 15
cervical segment, 11–13
clinoid segment, 13, 15
embryonic connection between ICA/vertebrobasilar, 28–33
lacerum segment, 13
petrous segment, 13–14
supraclinoid segment, 15
Intraorbital vessels, vascular pathology involving, 429–432

L

Leptomeningeal (pial) anastomosis, 365
Limbic arch in Galen vein malformation, 239
Lipohyalinosis, 305, 340, 358
Luxury perfusion syndrome, 364

M

Mandibular A., 13, 14
in angiofibroma, 55
connections with pterigo-vaginal A. and Vidian A., 13, 14, 63
Maxillary artery (primitive), 26
Medullary arteries, 77, 82, 91, 120, 125, 358, 365
Meningeal arteries
of ACA, 71
accessory meningeal A., 43
anterior artery of falx, 18
falx cerebelli A., 96
middle meningeal artery (MMA), 42
origin from ICA, 20
of PCA, 116
petrosquamous A., 42
posterior meningeal A., 97
recurrent meningeal A., 18
of SCA, 108
Meningohypophyseal trunk (MHT), 15–16
in angiofibroma, 55
in DAVF, 269, 271, 274
Middle cerebral artery (MCA)
accessory middle cerebral A., 89–91
aneurysms of perforators, 85, 400
dissecting aneurysm, 392
distal aneurysms, 186, 188
fenestration/duplication of M1, 89
microatheroma, 334
perforators of M1, 82
Migraine and stroke, 407
Moyamoya disease, 401
haemorrhage in Moyamoya, 403
pathology, pathogenesis, 401–402

N

Neurofibromatosis, 393, 395, 415

O

Occlusive diseases in children, 415–16
Odontoid arch, 39, 64, 96, 97
Ophthalmic A. (OA)
anomalous origin of OA, 16, 20, 22, 75
in DAVF, 282
embryogenesis, 19–20
ethmoidal arteries, 18
in meningioma, 61
Ophthalmic vein (superior-inferior), 158
carotid–cavernous fistulas, 297
in DAVF, 258–262, 293–296
embryogenesis, 163
Orbital varix, 429–431

P

PHACE, 52
Pituitary gland vascular supply, 20
Polyarteritis nodosa, 393, 394
Posterior cerebral artery (PCA)
aneurysms of PCA, 175
calcarine, parieto-occipital temporal A., 118
collicular A., 116
fetal, 1, 21, 115
microatheroma, 362
posterior medial/lateral choroidal A., 116–117
posterior thalamoperforating A., 116
relationship with AchA, 23–25
splenial A., 118
thalamogeniculate A., 117
Posterior communicating A. (PcomA)
aneurysms of, 175
anterior thalamoperforating A., 115
origin from ophthalmic A., 21
Posterior inferior cerebellar A. (PICA)
aneurysms, 189
dissecting aneurysms, 189
in occlusive disease, 358
perforators for medulla, 99–100
relationship with AICA segments, 100, 104
Primary angiitis of the CNS (PACNS), 406

R

Rektorzik venous plexus, 156
Rendu-Osler syndrome, 53, 237
Rete mirabile, 26, 28
Reversible cerebral vasoconstriction syndrome (RCVS), 404–5

S

Sinuses (venous)
cavernous sinus (CS), 157
and bridging veins with pontine veins, 157

Sinuses (venous) (*cont.*)
in carotid–cavernous fistula, 297
in DAVF, 255–263
falcine sinus, 130–151
in Galen vein malformation, 240
inferior petrosal sinus (IPS), 153–157
inferior petro-occipital sinus (inferior occipito-petrosal sinus) IOPV, 153–157
inferior sagittal sinus, 151
occipital sinus, marginal sinus, 152
in Galen vein malformation, 240
paracavernous sinus (PCS), 132, 154, 157
primary head sinus, 129
primary marginal sinus, 129
primitive tentorial sinus, 129
pro-otic sinus, 129
sigmoid sinus (SiSs), 153
spheno-parietal sinus, 157
straight sinus (SS), 129, 151
in Galen vein malformation, 240
superior ophthalmic vein (SOV), 158
superior petrosal sinus (SpS), 153
superior sagittal sinus (SSS), 149
transverse sinus (TS), 152
isolated transverse sinus in DAVF, 247
Sneddon's syndrome, 404
Stapedial artery (StA), 19
Stylomastoid A., 40
Subclavian steal syndrome, 342–343
Superior cerebellar A. (SCA), 105
duplicated, 107–108
marginal A., 106
meningeal A., 108 (*see also* Meningeal arteries)
origin from P1, 101, 105
perforators for midbrain, 108

T
Takayasu's arteritis, 403
Tympanic arteries
anterior tympanic (IMA), 41, 58
caroticotympanic artery (ICA), 13–14, 20, 33, 64
inferior tympanic (AphA), 38, 43, 60
posterior tympanic (stylomastoid A.), 40
superior tympanic (MMA), 42

V
Vascular malformations
arteriovenous malformations
aneurysms and AVM, 221–225
AV fistula, 226
AVM in children, 208, 209, 226
AVM treatment, 231, 232
cerebral proliferative angiopathy, 228–229
eloquent areas, 232
feeders, 209
flow-related aneurysms, 221
intranidal aneurysms, 221
nidus, 228–230
pathogenesis, 207–208
perinidal changes, 231
pseudoaneurysms, 218, 224
venous drainage of AVM, 226–228
venous pouch, 226
capillary malformations (telangiectasias), 233, 235
cavernous malformations (cavernomas), 232–233
in cavernous sinus, 233
familiarity, 232–233
pathology, 232
developmental venous anomaly (DVA), 235–237
Klippel–Trenaunay–Weber syndrome, 238
Rendu–Osler syndrome, 237
Sturge–Weber syndrome, 237
transition form of AVM (mixed AVM), 236–237
vein of Galen dilatation, 240, 244
vein of Galen malformation239–244
Vascular pathology, involving intraorbital vessels, 429–432
Veins
anterior cardinal vein, 129
anterior condylar confluence (ACC), 153–155
anterior, lateral, posterior condylar veins, 155–156
atrial veins, 135
basal vein, 136–138
basal vein and tributaries, 136–137
basal vein and variants, 137
basal vein in carotido-cavernous fistula, 297
basal vein in DAVF of tentorium and torcular, 269–281
basal vein in DAVS of CS, 263
choroidal veins, 136
deep cervical vein, 164, 302
deep middle cerebral vein (DMCV), 136–137, 154–155
diploic vein, 165
embryogenesis (veins), 129–130
emissary veins, 164–165
external jugular vein (EJV), 164
facial vein, 163
in fistula of VA, 302
Galen vein, 142
infratentorial cerebral veins
anterior medullary (AM) vein, 143
anterior pontomesencephalic vein (APM), 143
brachial veins, 145
cerebellar veins, 148
lateral medullo-pontine vein, 144
lateral mesencephalic vein, 145
peduncular vein, 143
petrosal vein, 148
posterior mesencephalic vein, 137–138
precentral vein (PC), 143
Vermian veins (superior-inferior), 143–145
internal cerebral vein (ICA), 136
internal jugular vein (IJV), 165
medullary veins, 134
occipital vein, 164
posterior auricular vein, 164
primitive maxillary vein, 163

prosencephalic vein, 129, 240
pterygoid venous plexus, 132, 154, 155, 157
retromandibular vein, 164
striate veins, 135–136
subependymal veins, 134–135
superficial middle cerebral vein (SMCV), 129, 132, 154, 157
superficial temporal vein, 164
thalamic veins, 136
uncal vein, 137, 154, 155
vein of corpus callosum, 132, 133
vein of Labbè, 132
vein of Trolard, 131
Venous sinus stenosis and headache, 425
Venous thrombosis
etiopathogenesis, 417
thrombosis of cortical vein, 418
thrombosis of deep venous system, 418
Vertebral artery (VA)
aneurysms of VA, 189
dissection of VA
extracranial, 372, 373
intracranial, 373
and spinal cord ischemia, 380
embryogenesis, 2–5
extracranial branches, 95, 96
falx cerebelli and posterior meningeal arteries, 96, 97
intracranial branches, 98–100
occlusive disease
intracranial microatheroma of VA, 346
occlusion, 342
perforators of VA, 98
posterior spinal artery (PSA), 100
unilateral origin of ASA, 100
variants of VA
anomalous origin, 110
duplication, 108
fenestration, 110
Vertebral artery (venous plexus), 153–156
Vertebral venous (anterior internal plexus), 156
Virchow–Robin spaces, 77

W

Willis (circle of), 4, 5, 364
variants, 365
Wyburn–Mason syndrome, 238

The manufacturer's authorised representative in the EU is Springer Nature Customer Service Centre GmbH, Europaplatz 3, 69115 Heidelberg, Germany. If you have any concerns regarding our products, please contact ProductSafety@springernature.com

Printed and bound by CPI Group (UK) Ltd, Croydon, CR0 4YY
25/09/2026
02222880-0004